ICE Manual of Geotechnical Engineering

emerald PUBLISHING
ice Publishing

ICE Manual of Geotechnical Engineering

Second edition

Volume I

Geotechnical engineering principles, problematic soils and site investigation

Edited by

Michael Brown
John Burland
Tim Chapman
Kelvin Higgins
Hilary Skinner
David Toll

ice | **manuals**

Published by Emerald Publishing Limited, Floor 5, Northspring, 21-23 Wellington Street, Leeds LS1 4DL.

ICE Publishing is an imprint of Emerald Publishing Limited

Other ICE Publishing titles:

UK Specification for Ground Investigation
The Association of Geotechnical and Geoenvironmental Specialists. ISBN 978-0-7277-6523-9

Environmental Geotechnics in Practice
Robert W Sarsby. ISBN 978-0-7277-6363-1

Disturbed Soil Properties and Geotechnical Design, Second edition
Andrew Schofield and Stuart Haigh. ISBN 978-0-7277-6155-2

A catalogue record for this book is available from the British Library

ISBN 978-0-7277-6681-6 (volume I)
ISBN 978-0-7277-6683-0 (volume II)
ISBN 978-0-7277-6685-4 (two-volume set)

Cover photo: Shutterstock/FWStudio

Commissioning Editor: Viktoria Hartl-Vida
Assistant Editor: Cathy Sellars
Production Editor: Sirli Manitski

Typeset by Nova Techset
Index created by Thomas Michael Allerton
Printed and bound in Great Britain by Henry Ling, Dorset

Contents

Foreword and endorsement

Geotechnical engineering quite literally underpins the construction industry.

Using a combination of geophysics, structural engineering and materials science, geotechnical engineers explore the interactions between soil, rock, water and structures, and develop solutions that enable buildings, tunnels, bridges, roads and other forms of infrastructure to carry the loads they are designed for, under all the conditions they may be expected to experience, over their planned lifetime and beyond.

Throughout my career, I have worked on numerous projects which have relied heavily on the skills of geotechnical engineering, so I am personally aware of the need to ensure the next generation of engineers continues to develop the knowledge and skills necessary to ensure our infrastructure continues to be built on safe, sustainable and long-lasting foundations.

I therefore commend this impressive and comprehensive work to all those who wish to develop and improve their knowledge of geotechnical engineering.

Keith Howells BSc, MBA, CEng, FICE, FCIWEM, FREng
ICE President, 2022–2023

The *ICE Manual of Geotechnical Engineering* is an essential reference document for experienced geo-professionals and non-specialists seeking to improve their knowledge of the subject. It covers historical aspects of how the subject evolved, the principles of soil mechanics, their application to designing and building in and on the ground, and how to manage and report geotechnical aspects of projects. The British Geotechnical Association is proud to remain affiliated with this updated edition of the document.

Dr Andrew Ridley
Chair of the British Geotechnical Association

Preface

We began to formulate the initial ideas for the first edition of this Manual as early as 2006. It had become apparent to us that civil and structural engineers not specialising in geotechnics face a daunting knowledge gap when they come up against a geotechnical problem. Most civil engineers leave university with very little grounding in geotechnical engineering. They will have a fair grasp of applied mechanics (mainly aimed at structural engineering). They will have had a basic introduction to geology and they will have studied the elements of soil mechanics and rock mechanics. But a recent graduate usually lacks a coherent understanding of the approach to, and methods of, geotechnical engineering and how these differ from other more widely practised branches of engineering. A survey carried out by ICE Publishing showed that information tends to be obtained from a wide range of sources through word of mouth, the internet and various publications. For the young practitioner this leads to a fragmented approach. Much of the geotechnical material is written by specialists for specialists, and its ad hoc application by a general practitioner is often inappropriate and can be extremely dangerous. We felt that it would be of great benefit to our profession to provide a single first-port-of-call authoritative reference source aimed at informing the less experienced engineer. To our delight this concept was endorsed by the ICE Best Practice Panel and the British Geotechnical Association and has offered a unique opportunity to provide authoritative guidance within a coherent framework of good geotechnical engineering.

It is most gratifying to note that the first edition of this Manual proved to be the most downloaded e-book of ICE Publishing's portfolio in 2022. Most of the chapters in this second edition have been extensively revised and updated to include the latest guidance documents and references. A major advance is the inclusion of carbon as a key metric in the choice of foundations and basement systems. In this regard, great emphasis is placed on avoiding unnecessary carbon emissions by avoiding over-conservatism in new designs and the application of codes sensitively in the remediation of old structures. Another important emerging challenge is the impact of climate change which will affect earthworks and slope engineering in particular.

As with the first edition, this second edition of the *ICE Manual of Geotechnical Engineering* has been a labour of love! The contribution of 99 contributors and ten section editors has made it possible to distil a great deal of experience from the profession into the chapters you see here. Don't imagine this will cover everything that a geotechnical engineer will face in their career – but it provides a 'starting point' from which to build experience while remaining grounded in robust fundamentals.

As mentioned previously, the Manual is aimed at people in the early stage of their careers who need a readily accessible source of information when working in new aspects of geotechnical engineering. However, it is expected that it should also prove valuable to all geotechnical engineering professionals. The aim has been to produce a manual that addresses the practice of geotechnical engineering in the twenty-first century including contemporary procurement, process and design standards and procedures. The grouping of chapters has been carefully chosen to facilitate a multidisciplinary and holistic approach to the solution of construction challenges. A key message is the importance of drawing on 'well-winnowed experience' for the smooth and reliable execution of projects. Such experience is best gained by working closely with a suitably experienced design or construction team.

It is hoped that this Manual will help in the training and development of the next generation of geotechnical engineers and will act as a useful source of reference to those with more experience.

The editors are grateful to all those contributors and section editors who have generously given so much of their time and knowledge in producing such a comprehensive book.

Michael Brown, John Burland, Tim Chapman, Kelvin Higgins, Hilary Skinner and David Toll

List of contributors

GENERAL EDITORS:
M Brown, University of Dundee, UK
JB Burland, Imperial College London, London, UK
T Chapman, Arup, London, UK
K Higgins, Geotechnical Consulting Group LLP, London, UK
HD Skinner, COWI UK, London, UK
DG Toll, Durham University, UK

SECTION EDITORS:
M Brown, University of Dundee, UK
JB Burland, Imperial College London, London, UK
JA Davis, Geotechnical Consulting Group, London, UK
M Devriendt, Arup, London, UK
S Hardy, Laing O'Rourke, Dartford, UK
I Jefferson, University of Birmingham, UK
T O'Brien, Mott MacDonald, UK
W Powrie, University of Southampton, UK

CONTRIBUTORS:
P Bailie, Arup, London, UK
J Ball, Roger Bullivant Piling, Swadlincote, UK
D Baxter, ArcelorMittal, UK
A Bell, Cementation Skanska Ltd, Doncaster, UK
A Berry, Arup, London, UK
D Bhuva, GSS Piling Ltd, Welwyn Garden City, UK
J Binns, Byland Engineering, York, UK
C Brewis, Arup, London, UK
M Brown, University of Dundee, UK
S Bullivant, Taranto Ltd, Craigavon, UK
JB Burland, Imperial College London, London, UK
F Chamley, Arup, London, UK
T Chapman, Arup, London, UK
J Chick, Expanded Ltd (Geotechnical), Dartford, UK
B Clarke, University of Leeds, UK
P Coney, Jacobs, UK
J Cowburn, Roger Bullivant Piling, Swadlincote, UK
JC Cripps, University of Sheffield, UK
MA Czerewko, AECOM Ltd, Chesterfield, UK
JA Davis, Geotechnical Consulting Group, London, UK
AM Daykin, Atkins, Leeds, UK
MH de Freitas, Imperial College London and First Steps Ltd, London, UK
M Devriendt, Arup, London, UK
C Doulala-Rigby, Tensar International Ltd, Blackburn, UK
A Duley, Jacobs, UK
PG Dumelow, Balfour Beatty plc, Thame, UK
C Edmonds, Stantec, Reading, UK
RD Essler, RD Geotech, Skipton, UK
I Farooq, Mott MacDonald, UK
ER Farrell, AGL Consulting, Dublin, Republic of Ireland
M Free, Arup, London, UK
S French, BES Group Ltd, Warrington, UK
R Garland, BAM Nuttall Ltd, UK
B Ghosh, Mott MacDonald, UK
P Gilbert, Jacobs, UK
S Glover, Arup, London, UK
S Hardy, Laing O'Rourke, Dartford, UK
A Harwood, Arup, London, UK
T Hayward, Laing O'Rourke, Dartford, UK
A Heathcote, Keller, Coventry, UK
G Herrmann, Keller, Coventry, UK
G Horgan, Huesker, Warrington, UK
C Hughes, Arup, London, UK
P Ingram, Arup, London, UK
I Jefferson, University of Birmingham, UK
LD Jones, British Geological Survey, Nottingham, UK
M Kemp, Atkins, Epsom, UK
V Kumar, Mott MacDonald, UK
M Large, Keller, Coventry, UK
R Lindsay, Atkins, Epsom, UK
S Markvukaj, Mott MacDonald, UK
WA Marr, Geocomp Corporation, Massachusetts, USA
P Marsden, Keller, Coventry, UK
BT McGinnity, retired formerly London Underground, London, UK
D Milo, Balfour Beatty, UK
P Morrison, Arup, London, UK
S Mortimore, Atkins, London, UK
D Mothersille, Geoserve Global Limited, London, UK
J Murray Reid, retired formerly TRL Limited, Wokingham, UK
R Nicholson, AECOM Ltd, Chesterfield, UK
PA Nowak, Atkins Ltd, Epsom, UK
T O'Brien, Mott MacDonald, UK
T Orr, Trinity College, Dublin, Republic of Ireland
H Pantelidou, Arup, London, UK
D Patel, Arup, London, UK
S Pennington, Arup, London, UK
G Pittaro, Mott MacDonald, Singapore
D Potts, Imperial College London, London, UK
JJM Powell, Geolabs Ltd, Watford, UK

W Powrie, University of Southampton, UK
D Preece, Laing O'Rourke, Manchester, UK
M Preene, Richter Associates, Huddersfield, UK
JA Priest, University of Calgary, Canada
D Puller, Geotechnical Experts, Sevenoaks, UK
N Pye, Mott MacDonald, UK
JM Reynolds, Reynolds Geo-Solutions Ltd, Mold, UK
C Robinson, Roger Bullivant Limited, Swadlincote, UK
CDF Rogers, University of Birmingham, UK
A Royal, University of Birmingham, UK
CS Russell, Geotechnical Consulting Group, London, UK
N Saffari, London Underground, London, UK
H Shields, Arup, London, UK
HD Skinner, COWI UK, London, UK
JA Skipper, Geotechnical Consulting Group, London, UK
P Sloan, COWI, London, UK
P Smith, Geotechnical Consulting Group, London, UK
L Sodano, Mott MacDonald, UK
J Standing, Imperial College London, London, UK
M Stanley, Bachy Soletanche Ltd, Camberley, UK
J Strange, Card Geotechnics Ltd, Godalming, UK
J Taggart, Arup, London, UK
R Talby, Mott MacDonald, UK
B Thompson, Keller, Coventry, UK
DG Toll, Durham University, UK
M Walsh, Arup, London, UK
MJ Whitbread, Wales, UK
A Willoner, Mott MacDonald, UK
D Woods, Huesker, Warrington, UK
L Zdravkovic, Imperial College London, London, UK

About the editors

Hilary Skinner (Editor)

Hilary Skinner graduated in Engineering from the University of Cambridge in 1990. After a period of research at BRE, she joined Whitby Bird (later Ramboll) as Associate Director in Geotechnics. Now a Director, she has been at COWI (previously Donaldson Associates) since 2010. She has authored or co-authored a number of papers and books for which she has been awarded the BGA Research Medal and the Halcrow Prize from the ICE. All have involved collaboration with UK and European consultants, contractors and researchers. She continues to collaborate on research into working platforms, piling and tunnelling with universities and industry bodies. Hilary is a Chartered Engineer, a Fellow of the ICE and a past Chair of the British Geotechnical Association (BGA). The importance of measurement and case history data has been critical to the development of our science, which marries theory and judgement, and this has been central to a lifelong pursuit of technical excellence in her work and our industry.

David Toll (Editor)

David Toll is Chair of the British Geotechnical Association (BGA) and Professor of Engineering and Co-Director of the Institute of Hazard, Risk and Resilience at Durham University. He is a member of ISSMGE's Technical Committee TC 107 on Tropical Soils, has been Chair of the ISSMGE's Technical Committee TC 106 on Unsaturated Soil and was the inaugural Chair of the Joint Technical Committee JTC2 of FedIGS on Geo-Engineering Data. David is a Fellow of the ICE and has been Chair of both ICE North East and the Northern Geotechnical Group. He is a Fellow of the Learned Society of Wales. He has been carrying out research into climate resilience of infrastructure and information technology applied to civil engineering for over 30 years and has published over 200 papers and 12 books. He has held Visiting Professor and Research Fellow posts at Tongji University, China, National University of Singapore, Nanyang Technological University, University of Sydney, University of Western Australia and University of Newcastle, Australia. He was the founding editor of the journal *Geotechnical and Geological Engineering* and has been a member of the editorial boards for *Géotechnique*, *Transportation Geotechnics* and the *Quarterly Journal of Engineering Geology and Hydrogeology.*

Kelvin Higgins BSc, MSc, DIC, CEng, FREng, FICE, FCIHT (Editor)

Kelvin Higgins is a Senior Partner at Geotechnical Consulting Group LLP (GCG) and a Visiting Professor at Imperial College London. In 2019 Kelvin became a Fellow of the Royal Academy of Engineering and since 2012 he has been a Visiting Professor at Imperial College London. Until recently he was Chair of the British Geotechnical Association and Chair of CIRIA's Geotechnical Advisory Panel. Since joining GCG in 1986 Kelvin has worked on the application of numerical methods to engineering problems, including the design of tunnels, retaining structures, highways, dams, embankments, cuttings and foundations (offshore and onshore). During his career he has gained broad practical and design experience of foundations, highway construction earthworks (cuttings, embankments and drainage schemes), deep basements and the interaction between structures. He has acted as an expert witness in legal disputes. In addition, he has directed and facilitated research in different aspects of geotechnical engineering. Kelvin has extensive experience of the assessment of the effects of underground construction, (excavations, shafts and tunnels) on adjacent structures. This includes developing alternative designs, construction techniques and sequencing of works to mitigate the effects of construction on adjacent structures and services by predicting the effects of construction on adjacent structures, services, operational tunnels and installations within these structures (e.g. escalators, permanent way, etc.). In many cases he has been able to demonstrate economic benefits to alternative sequences and designs. Kelvin has investigated the causes of the failures of foundations and slopes, designed remedial schemes and monitored construction. His research and publications cover earth-retaining structures, slopes, sheet piling, new monitoring technologies, soil behaviour, the performance and analysis of foundations, tunnels, embankment dams, slope stability, thermal piles, monitoring and the use of advanced numerical methods. Prior to joining GCG, he worked for Sir Alexander Gibb & Partners in the UK and internationally, where he gained experience in foundation design and the design of embankment dams. He oversaw the construction of highway schemes and other facilities in remote and challenging environments, having to adapt designs to accommodate local conditions.

Michael Brown BEng (Hons), PhD, GMICE (Editor and Section Editor)

Michael Brown is Professor of Geotechnical Engineering in the School of Science and Engineering at the University of Dundee, UK. Mike is a researcher, former practitioner and an educator at both undergraduate and post-graduate level. As Professor of Geotechnical Engineering his focus is on translating the complexities and specialist testing approaches of academia into practical outputs that can be easily deployed by practising engineers. Currently his research interests lie in offshore renewable energy, with the development of novel deep foundation systems and anchoring solutions for future floating wind. Onshore he is looking at efficiencies in deep foundation excavation and support and the development of efficient ground energy systems for deployment in diaphragm walls for heating and cooling in the UK and South America. Mike undertook his PhD at the University of Sheffield, UK, looking at the rapid load testing of piles in clay and soil rate effects in general. Based upon this work, the rapid load testing technique was made more accessible in the UK with inclusion in the *ICE Specification for Piling and Embedded Retaining Walls* and Federation of Piling Specialist (FPS) load testing guidance. Internationally he has contributed to national guidance on rapid load testing in the Netherlands and chaired the CEN working group that was responsible for the creation of ISO/CEN standards on rapid load pile testing and dynamic piles testing. These were subsequently adopted as British Standards. Based upon his expertise in deep foundation testing and developing novel foundation deployment, he is currently a UK representative on the ISSMGE Technical Committee 212 Deep Foundations, as well as a member of the International Press-in Association (IPA): IPA-TC4 concerned with press-in and rotary pile installation. Since joining the University of Dundee he has undertaken a significant amount of contract testing and research direct for industry based around specialist physical modelling and centrifuge testing. This type of work keeps Mike industrially relevant and allows him to appreciate the needs of both industry and academia. These specialist interests and skills have recently been recognised, with him being appointed as secretary of ISSMGE TC 104: Physical Modelling in Geotechnics.

Professor John Burland (Editor and Section Editor)

Professor John Burland is Emeritus Professor of Soil Mechanics and Senior Research Investigator at Imperial College London. Born in the UK, Professor Burland was educated in South Africa and studied Civil Engineering at the University of the Witwatersrand. He returned to England in 1961 and worked with Ove Arup and Partners for a few years. After studying for his PhD at Cambridge University, John Burland joined the UK Building Research Station in 1966, became Head of the Geotechnics Division in 1972 and Assistant Director in 1979. In 1980 he was appointed to the Chair of Soil Mechanics at Imperial College London, before becoming Emeritus Professor and Senior Research Investigator there. In addition to being very active in teaching and research, John Burland has been responsible for advising on the design of many large ground engineering projects worldwide, including the underground car park at the Palace of Westminster and the foundations of the Queen Elizabeth II Conference Centre in London. He specialises in problems relating to the interaction between the ground and masonry buildings. He was London Underground's expert witness for the Parliamentary Select Committees on the Jubilee Line Extension underground railway and has advised on many geotechnical aspects of that project, including ensuring the stability of the Big Ben Clock Tower. He was a member of the international board of consultants advising on the stabilisation of the Metropolitan Cathedral of Mexico City and was a member of the Italian Prime Minister's Commission for stabilising the Leaning Tower of Pisa. He has received many awards and medals, including the Gold Medal for Engineering Excellence of the World Federation of Engineering Organisations and the Gold Medals of the UK Institution of Structural Engineers and of the UK Institution of Civil Engineers. In 1994 he was awarded the Kevin Nash Gold Medal of the International Society of Soil Mechanics and Geotechnical Engineering 'In recognition of outstanding services to ISSMGE, to

International Goodwill and to International Geotechnical Practice and Education'. In 1996 he was awarded the Harry Seed Memorial Medal of the American Society of Civil Engineers 'for distinguished contributions as an engineer, scientist and teacher in soil mechanics'. He is a Fellow of both the UK Royal Academy of Engineering and of the Royal Society of London, a Member of the US National Academy of Engineering and was appointed Commander of the Most Excellent Order of the British Empire in 2005.

Tim Chapman BE, MSc, DIC, CEng, FICE, FIEI, FREng (Editor)

Tim Chapman is a Director at Arup, London, with deep expertise in all aspects of geotechnical engineering – his particular areas of expertise relate to ground risk minimisation for large projects and for the design of deep underground structures with and around tunnels. He has had extensive experience in the design of all types of infrastructure and devising how future infrastructure can best serve and shape future society. A current goal is helping clients to choose the right projects to meet their desired outcomes and setting up those projects to succeed far more fundamentally from the outset. He advocates that our infrastructure systems should increase national happiness by promoting prosperity and well-being as well as being fit for the future in terms of decarbonisation, social value, biodiversity and circular economy. Among other Arup roles, he led its London-based Infrastructure Design group for over a decade until April 2020 and was Office Leader for its London office, Arup's largest globally, through the COVID-19 pandemic and was also client for the creation of its new flagship head office, the smartest office space in the world and lowest carbon office building in London, designed for collaboration and inclusion, to attract staff back into a conducive space – delivered on time and under budget. Tim is a Fellow of the ICE and of the Royal Academy of Engineering. In 2011 he was awarded the ICE President's Medal for producing the institution's first significant thought piece on how to decarbonise the infrastructure sector. He is currently Co-Chair of the ICE Community Advisory Board for Fundamentals and Behaviours and Chair of the Association of Consulting and Engineering's (ACE) Risk Task Force, charged with reducing project risks for the benefit of all. For the RAEng he chairs the Sainsbury Management Fellowship helping to equip engineers with more business-facing skills and represents the RAEng on EuroCASE, the federation of European engineering academies, which seeks to influence the European Commission on key topics, one of which is 'sustainable materials for sustainability'.

Stuart Hardy MEng, ACGI, PhD, DIC, CEng, FICE (Section Editor)

Stuart Hardy joined Laing O'Rourke's central technical group as their Geotechnical Lead in 2021, with an overview of all geotechnical works for Laing O'Rourke, from tender through to completion and beyond. After completing an undergraduate degree in Civil Engineering and a PhD in geotechnical engineering at Imperial College London, Stuart joined Mott MacDonald in Croydon and worked on the design of the new Wembley Stadium and Heathrow T5. After leaving Mott MacDonald, Stuart joined the geotechnics department of Arup's office in London and stayed for nearly 16 years, working on a number of challenging basement and foundation projects located mainly in the Middle East and London. These included the Post Building, the Exhibition Road Quarter for the Victoria and Albert Museum, Elizabeth House, the ArcelorMittal Orbit sculpture in the Olympic Park and the redevelopment of the Shell Centre building in Waterloo. For the last five years at Arup Stuart led the geotechnical design of the S1 and S2 contracts on HS2, working with the Main Civils Contractor SCS JV. Stuart is one of the UK's representatives to CEN on the redrafting of the Eurocodes and chairs the parallel BSI committee. Stuart was one of the leading authors on the CIRIA C760 publication on the design of embedded retaining walls and is a RoGEP registered Advisor and a Fellow of the ICE.

Tony O'Brien FREng (Section Editor)

Tony O'Brien is a Fellow of the Royal Academy of Engineering and a Fellow of the ICE. He is the Global Practice Leader for Geotechnics at Mott MacDonald and a Visiting Professor at the University of Southampton. He has provided technical leadership on several major projects across Europe, North America and the Asia-Pacific region and worked as an expert during litigation/dispute resolution. Tony has contributed to best practice guidance, including CIRIA Guide C791 on advanced numerical modelling, and recently published a book, *The Observational Method in Civil Engineering*, and contributed to several international ISSMGE committees.

Mike Devriendt MSc, DIC, CEng, MICE (Section Editor)

Mike Devriendt is Associate Director at Arup, London, and is a geotechnical and tunnelling engineer with over 25 years of experience. He is responsible for leading large multidisciplinary design teams on urban infrastructure projects such as Crossrail 2, HS2, Tideway and Crossrail, and a range of commercial and residential building projects. Mike specialises in working collaboratively with clients and contracting organisations to develop cost-effective and buildable designs. Mike is author of over 25 journal or conference papers. As a recognised technical expert, Mike regularly leads and contributes to national and international steering committees. He has been appointed to a number of industry-wide ICE and CIRIA roles and has been an expert panel member or author of geotechnical and tunnelling standards for BSI, CIRIA and HS2 Ltd. Mike has particular research interests in the practical application of new technologies in the field of monitoring and asset management. He has led a number of consulting commission and collaborative research projects with industrial partners such as UK Power Networks, HS1, National Highways and with academic partners such as the University of Cambridge, University of Oxford and University College London. As part of these projects Arup have provided consultancy advice and developed new techniques for application of photogrammetry, machine learning, laser and thermal scanning to achieve health and safety benefits and operational cost and programme savings associated with infrastructure monitoring and maintenance.

William Powrie (Section Editor)

William Powrie is Professor of Geotechnical Engineering at the University of Southampton, where he served as Dean of the Faculty of Engineering and the Environment 2010–2018. His main technical expertise is in transportation geotechnics and waste/resource management, for which he was elected Fellow of the Royal Academy of Engineering in 2009. William's research on geotechnical transport infrastructure encompasses groundwater control, in-ground construction, understanding and mitigating vegetation and climate change effects, and fundamental soil behaviour. Major projects on which he has worked include the A55 Conwy Crossing, the Jubilee Line extension and HS1. He chairs HS2's Independent Geotechnical Expert Panel. He is Convenor of the 14-university UK Collaboratorium for Research on Infrastructure and Cities (UKCRIC), and leads the UK Rail Research and Innovation Network (UKRRIN) Centre of Excellence in Infrastructure and the Infrastructure for Port and Coastal Cities and Towns Network. William's research in waste/resource management focuses on landfill engineering and science-based policy and practice. He worked on the design and engineering risk assessment of low-level radioactive waste repositories at Drigg, and chaired the Technologies Advisory Committee for Defra's £30M research and demonstrator programme for new technologies for treating biodegradable waste. He is the author of the widely respected textbook *Soil Mechanics: Concepts and Applications.*

John A Davis BSc, MSc, DIC, CGeol, EurGeol (Section Editor)

John A Davis is a Senior Partner at Geotechnical Consulting Group LLP. John has worked for contractor, consultant, client and government organisations. He has had two extended periods of secondment, one into London Underground to work on the potential effects of rising groundwater on below-ground assets, and another to the Crossrail client organisation where he was responsible for geotechnical matters in the eastern half of the project. He has interests in commercial management of ground risk and quaternary engineering geology. John is currently Chair of the Engineering Group of the Geological Society.

Ian Jefferson BEng, DIS, PhD, FGS (Section Editor)

Ian Jefferson is Deputy Head at the College of Engineering and Physical Sciences, University of Birmingham, UK and Professor of Geotechnical Engineering. He has research interests that span geotechnical engineering and engineering geology, addressing issues associated with geohazards and problematic soils, from assessments through modelling to mitigation, using both traditional and novel areolation techniques. Past Vice President (Northern Europe) of the International Association of Engineering Geology (2011–2015), he has some 30 years' experience working on several problematic soils, most notably collapsible and expansive soils. More recently he has focused on sustainable geotechnics and urban near-subsurface, serving on the Technical Committee TC 307 on Sustainable Geotechnics of the International Society of Soil Mechanics and Geotechnical Engineering. Recently he has led European-funded work, ARLI (Alternative Resource of Low Impact) supporting small and medium enterprises to develop new materials from waste streams, supporting the net-zero agenda. Since 2017 he has been Co-Director of the National Buried Infrastructure Facility, where he leads the work on geostructures and works with several industrial partners both in the UK and internationally. He has served on regional, national and international technical committees across both geotechnical engineering and engineering geology, including as a member of the ASCE Geo-institute's Engineering Geology and Site Characterization committee.

Section 1: Context

Section editors: **John B Burland and William Powrie**

doi: 10.1680/icemge.66816.0003

Chapter 1
Introduction to Section 1

John B Burland Imperial College London, London, UK
William Powrie University of Southampton, Southampton, UK

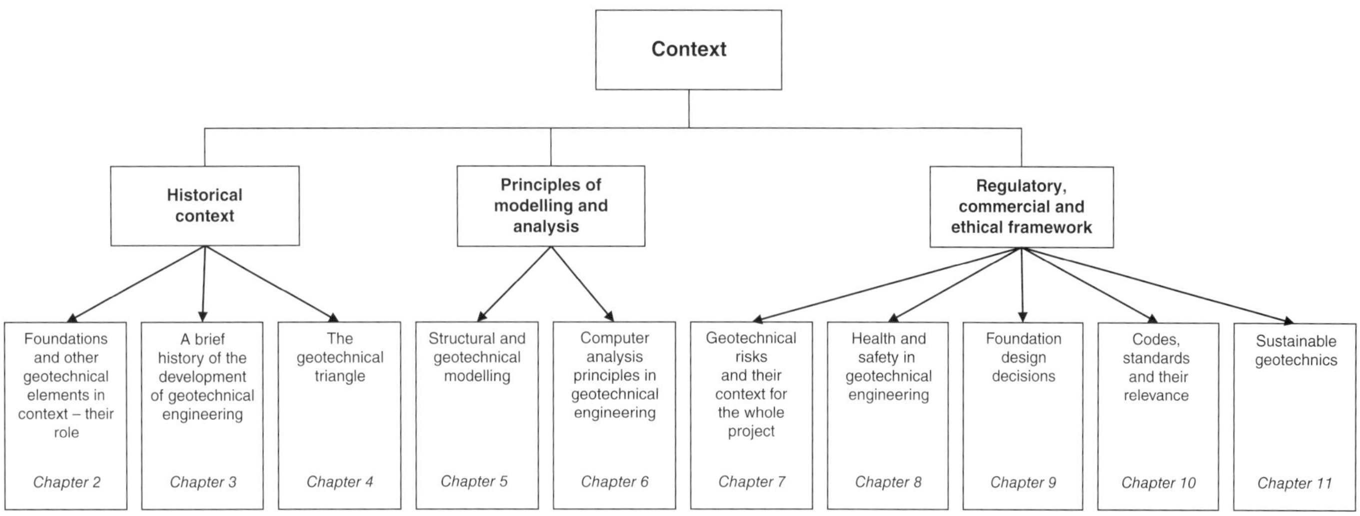

Figure 1.1 Layout of chapters in Section 1

Figure 1.1 outlines the layout and contents of Section 1 *Context*.

The first three parts of the section address the development of soil mechanics and geotechnical engineering as a distinct discipline over the past 80 years or so. Chapter 2 *Foundations and other geotechnical elements in context – their role* explains the importance of foundations and structures built in or of the ground within civil engineering and construction, and the need for a formal and holistic geotechnical engineering design process. Chapter 3 *A brief history of the development of geotechnical engineering* gives a history of the development of geotechnical engineering at the borderline between science and art, with the latter defined by Terzaghi in 1957 as 'mental processes leading to satisfactory results without the assistance of step-for-step logical reasoning'. This is reflected in the 'geotechnical triangle', described in Chapter 4 *The geotechnical triangle*, which emphasises the essential elements of successful geotechnical engineering as understanding the ground, material properties and relevant precedents (well-winnowed experience), connected by an appropriate model for analysis.

Chapters 5 *Structural and geotechnical modelling* and 6 *Computer analysis principles in geotechnical engineering* address some important principles of modelling and analysis in geotechnical engineering. Particular attention is paid in Chapter 6 *Computer analysis principles in geotechnical engineering* to computer methods, for which it is essential that an engineer has a sound understanding of the basis of the method of analysis, the influence of the material properties used and the shortcomings and limitations of the approach.

Finally, Chapters 7: *Geotechnical risks and their context for the whole project* to 11 *Sustainable geotechnics* discuss key aspects of the regulatory, commercial and ethical framework with which geotechnical engineering practice must comply, now and into the future. Chapter 9 *Foundation design decisions* emphasises that foundation engineering is a 'process' involving a number of interlinked operations, and design decisions need to take account of this. The other chapters address understanding and apportioning of geotechnical risk within a whole-project context (Chapter 7 *Geotechnical risks and their context for the whole project*); health and safety (Chapter 8 *Health and safety in geotechnical engineering*); current standards and codes of practice (Chapter 10 *Codes and standards and their relevance*); and sustainability (Chapter 11 *Sustainable geotechnics*). These issues are especially important, given that the place occupied by geotechnical engineering at the borderline between science and art means that risks are often more difficult to foresee and quantify than other areas of civil engineering, and the discipline does not lend itself to highly detailed codes of practice.

doi: 10.1680/icemge.66816.0005

Chapter 2
Foundations and other geotechnical elements in context – their role

John B Burland Imperial College London, London, UK
Tim Chapman Arup, London, UK
Marianne Walsh Arup, London, UK

CONTENTS

The purpose of this chapter is to describe the basic principles of geotechnical design and construction in the context of the whole project. At a given site the ground conditions have resulted from millions of years of natural geological processes (which are seldom simple) and have sometimes been modified by humans (e.g. by mining or many other processes). As a consequence there are always inherent uncertainties and risks, and the art of geotechnical engineering is to make informed allowance for these, in both design and construction. The key requirements for all geotechnical elements of a project are described. Emphasis is placed on the importance of constructive and positive interaction with professionals engaged in the many other contributing disciplines. The design life of the geotechnical elements is considered and the important concepts of geotechnical design and the construction cycle are introduced. Various managerial approaches to identifying the key elements of the complete design and construction process are described. The chapter concludes with a summary of the factors common to most geotechnical design and construction projects that are necessary for a successful outcome.

2.1. Geotechnical elements in the context of the rest of the whole structure

All built structures touch the ground in some way and hence all need some form of foundation. Other geotechnical elements include retaining walls and ground anchors. Sometimes they can be shallow (e.g. pad footings or gravity-retaining walls), other times they are deep (e.g. piles or embedded retaining walls). Often they rely on geotechnical processes such as ground improvement to produce a geotechnical element.

All foundations and other geotechnical elements have a number of characteristics that distinguish them from other parts of the structures that they support.

- They tend to be among the most heavily loaded elements in any structure.
- Their installation process is less amenable to factory-style production.
- Their capacity is very dependent on the ground of the site, which is always characterised by few observations and tests, and is normally very heterogeneous and may contain hazards that are difficult to foresee.
- Their capacity is strongly influenced by the method of construction and how well it is controlled.

Hence, the risk of failure tends to be significantly higher than that for other parts of the structure. The management of ground uncertainty is an important part of the design and construction process in order to produce elements that have the required degree of reliability.

A vital consideration in geotechnical design is the interaction of the structural element that is inserted into the ground with the ground itself – so-called 'soil–structure interaction'. Structural loads are applied to the element and the ground resists – generally either by friction along the element, or by bearing of the element against the ground. Both of these resistances can occur vertically or horizontally, as shown in **Figure 2.1**. Normally ground stresses are maintained within failure limits, so the resulting displacements depend on the stiffness of both the element and the ground.

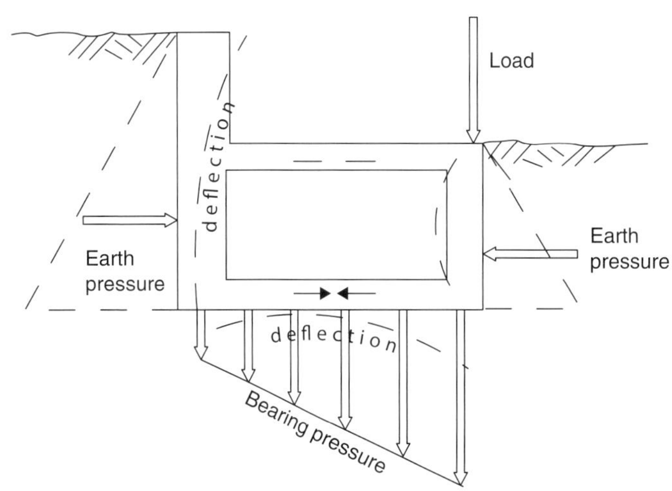

Figure 2.1 Soil–structure interaction

Loads can be imparted into foundations directly, as vertical or horizontal loads, or as imposed bending moments. They can also be imposed indirectly as displacements, which induce loads into the structural parts of the foundation. Eurocode 7 (BS EN 1997 – BSI, 2004, 2007b) treats these in a similar way by introducing the Newtonian concept of an 'action'.

Geotechnical elements are characterised by much higher degrees of uncertainty than other structural elements. The uncertainty is derived from

- the inevitable significant assumptions and idealisations that underlie the overall geotechnical design
- natural variability in the ground being modelled, often manifested by vast scatter in data points on graphs
- inherent variability in the construction processes, and their control, that are used to install the geotechnical elements and foundations, which may also have a profound effect on their eventual performance in service
- usual variability in the applied loads transmitted through a normally highly structurally indeterminate construction meaning that the actual loads on each foundation may be very uncertain
- the possibility of the construction and performance being heavily influenced by an unexpected feature of a site – for example a major geological discontinuity such as a sinkhole or a fault. It is not unusual for major construction problems to arise owing to adverse groundwater conditions.

All of these uncertainties mean that engineers who presume that their calculations are precise and predictable are deluded and may expose themselves to much more onerous load combinations than they could imagine. The wise geotechnical engineer should be humble and make prudent allowances for uncertainty. This is explained in more detail in Chapter 7 *Geotechnical risks and their context for the whole project.*

2.2. Key requirements for all geotechnical elements

2.2.1 General

All foundations or other geotechnical elements must fulfil a number of essential criteria.

- They must not fail, or else the structure they support will also fail. In terms of limit state design, failure by any mode is termed reaching or exceeding an 'ultimate' limit state, and may involve failure of a structural element or rupture along a soil–structure or a soil–soil interface.
- They must not move excessively or else the structure they support may become impaired or fail to operate as intended. In terms of limit state design, excessive deflection involves breaching a 'serviceability' limit state.
- They must last for as long as intended. Unlike many other building elements, foundations are hugely difficult to upgrade or repair, so their longevity will often dictate the life of the structure that they support.

2.2.2 Ultimate limit state modes of failure

There is a range of ways in which geotechnical structures can fail an ultimate limit state, and these are shown in **Figure 2.2**.

2.2.3 Serviceability limit state and displacements

Failure of a serviceability limit state is usually less serious than failure of an ultimate limit state, and is often repairable. It usually occurs when excessive displacements have taken place that impair the function of the structure. In addition to excessive movement, it also includes other forms of unacceptable tolerances, such as moisture penetration into basements. Examples are illustrated in **Figure 2.2**.

2.2.4 Design life and modes of deterioration

Design lives are covered in more detail in Section 2.4. Other than failures of a limit state, the life of a structure can be reached when

- it has become affected by material deterioration processes, such as corrosion of steel, carbonation of concrete or rotting or insect infestation of timber
- it has been subjected to physical processes, such as repeated loading cycles causing fatigue, or excessive damage from impacts
- it fails to meet new design or material standards and so offers a less than acceptable level of resistance against load, corrosion and so on.

2.3. Interaction with other professionals

2.3.1 General

The geotechnical engineer is seldom the professional solely in charge of a complete project – almost invariably the function sought of a new structure or facility goes beyond merely geotechnical considerations. Therefore, the role of a geotechnical engineer should be to support the wider design and construction process. They are likely to have most influence if the input complements the work of other professionals rather than seeking to dominate or failing to engage.

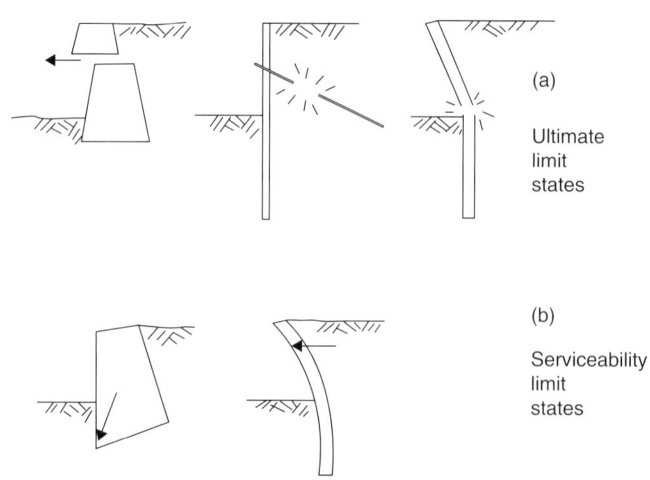

Figure 2.2 Limit state failures (Adapted from BS EN 1997)

The geotechnical engineer should be alive to the broader design process and any failures or gaps in those wider processes. The ICE (2018) report *In Plain Sight: Assuring the Whole-Life Safety of Infrastructure* highlighted a series of broad lessons that the construction industry and civil engineers should learn from the Grenfell Disaster – many projects have similar failings, so-called 'soft hazards' – and it is the responsibility of the professional to be alert to wider problems and ensure that the whole team react positively to resolve them.

2.3.2 General construction process

All project programmes involve controlled phases of input so that the optimum solution is devised with mutually consistent input from all members of the project team, through design procurement, implementation and commissioning. Once commissioned there is a need for maintenance, upgrades and repairs through the working life of the structure and finally decommissioning and sometimes reuse of the original elements in a fresh structure.

Client Best Practice Guide (ICE, 2009) provides a generalised approach for such inputs, based on the systems used for buildings (Royal Institute of British Architects, 2007 and updated in 2020) and the rail industry (Network Rail, 2007; previously known as the *Governance for Railway Investment Projects*, replaced in 2021 with *Project Acceleration in a Controlled Environment*), as well as guidance from the UK's former Office of Government Commerce (OGC) (now part of the Cabinet Office Efficiency and Reform Group). More detail on these processes is given in Section 2.5.

In the early stages there will be little certainty about the optimum solution, and many competing options will be considered and dismissed, often for reasons not obvious to individual members of the technical team. The design team can contribute best to this process if they understand the required function and the client's business drivers for the new structure or facility.

The early stages are also the period when it is easiest for the project to take on board geotechnical constraints and design and construction requirements. Hence, the geotechnical engineer is likely to have most success in integrating these requirements into the rest of the structure if they are identified early on, although this will often be in advance of the acquisition of data and completion of analyses. Therefore, experienced geotechnical input is valuable in these early stages and will often lead to a better solution, simpler design and construction processes and less costly subsequent geotechnical involvement.

The work of the geotechnical engineer is seldom completed in isolation. Normally there is much interaction with structural engineers. Sometimes the work of the geotechnical engineer is subsumed beneath that of the structural engineer. For simple structures this may be appropriate and sufficient, but for more complex geotechnical structures, or where the ground conditions are complex, it is likely that the geotechnical engineer will be better able to explain the particular issues directly to other members of the design team.

2.4. Design lives for geotechnical elements

Relevant design lives for structures are defined in Eurocode 'Basis of structural design' (BSI, 2010) and are generally distinguished according to whether the structure is a building or a piece of infrastructure. In its Table 2.1 it defines design life categories as follows.

- Category 1: Temporary structures, not including structures or parts of structures that can be dismantled with a view to being reused – 10 years.
- Category 2: Replaceable structural parts, such as gantry girders, bearings – 10–25 years (the UK National Annex to BS EN 1990:2002 modifies this to 10–30 years).
- Category 3: Agricultural and similar buildings – 15–30 years (modified in the UK to 15–25 years).
- Category 4: Building structures and other common structures – 50 years.
- Category 5: Monumental building structures, bridges and other civil engineering structures – 100 years (currently modified in the UK to 120 years to bring it into line with traditional National Highways bridge design life; it is possible that the UK may revert to 100 years if National Highways changes their guidance).

As foundations are difficult to repair or upgrade and as most structures last for longer than the period for which they were designed, allowing for a longer life is prudent.

Foundations are difficult and expensive to remove and so some consideration should be given to what will occur after the life expiry of the structure they support. This was addressed by the RuFUS (Reuse of Foundations on Urban Sites) project (see Butcher *et al*., 2006; Chapman *et al*., 2007) and in the paper *A Short Guide to Reusing Foundations* (Tayler, 2020). Where it is likely that a new structure with new foundations could become a major obstruction, the potential for future development of the site may be inhibited.

To prevent abandoned foundations from becoming an insidious form of ground contamination, the RuFUS project advocated that all foundations should be designed with the intention of allowing subsequent reuse. This mainly relates to the recording and saving of records so that the old foundations can be assessed by the future design team.

While most foundation and geotechnical elements are provided for 'permanent' elements that are required to persist for a normal structural life, sometimes foundations are required for shorter periods. These include the following.

- Interim structures, required for a significant life, perhaps 10 years.
- Contractor's 'temporary' works – structures that fulfil a temporary function during construction of a more significant structure, such as thrust blocks and crane bases. Sometimes a life of 1–2

years is specified. For these, Eurocode 7 Part 1 (BS EN 1997-1:2004; BSI, 2004) Clause 2.4.7.1(5) states 'Less severe values than those recommended in Annex A [for partial factors] may be used for temporary structures or transient design situations, where the likely consequences justify it'. This guidance is current at the time of writing and the next update of Eurocode 7 is due for publication in 2024.

- Demountable structures, such as scaffolding or temporary stands as may be required at sport or music venues, where a life of only perhaps weeks is required. The Institution of Structural Engineers (2017) guide, *Temporary Demountable Structures: Guidance on Procurement, Guidance and Use*, defines typical foundation concerns for such structures.

Design life can influence choice of factor of safety. However, it needs to be considered with the following points.

- The frequency with which the most onerous design load combination occurs – if it is very infrequent compared with the design life, some reduction in margin against failure may be possible. This may occur with 1-in-100-year wind or flood events or with 1-in-475-year seismic events.
- The consequences of failure – where the consequences are mild and do not threaten safety, then a lower factor may be permissible provided the economic consequences are judged and agreed as acceptable.

The design life also to some extent governs the measures required to limit deterioration of the foundation materials. Conventional structural design codes such as Eurocodes 2 (BS EN 1992; BSI, 2006) and 3 (BS EN 1993; BSI, 2007a) contain implicit requirements to ensure longevity of foundations; principally, crack width criteria for reinforced concrete, intended to limit the intrusion of air and water which then can come into contact with the reinforcing steel, and whose effect can be exacerbated by the presence of chloride ions from salt. Where potential corrosion processes will be slow compared with the foundation design life, it may be possible to relax these structural requirements for structures intended to have short lives, provided that code non-compliance is acceptable to the owner and any approver.

Temporary structures may sometimes have less onerous movement limits. Their nature may mean that larger movements are easy to accommodate; for instance, in temporary stands where shims and jacks can be used to compensate for differential foundation movements or where there are no rigid finishes that make obvious the effects of differential settlements.

2.5. Design in a time of climate and biodiversity emergency

It is now very clear that the planet is going through a period of unprecedented climate change, resource depletion and huge impacts on biodiversity – all caused by human activity, and the construction industry is responsible for a significant part of those effects. It is important that geotechnical designers (operating as part of a wider team) take into account

- a need for much greater circular economy so resources are used more sustainably
- as part of that, a much stronger emphasis on refurbishment of structures rather than new build, with reuse of foundations a major enabler for that
- a need to integrate better with nature
- a need to always reduce carbon, both capital and whole life, following the principles of PAS 2080 (BSI, 2023)
- much stronger emphasis on resilience – and potentially adaptation – to new climate conditions, for instance much higher sustained temperatures or much greater rainfall.

These are discussed in more detail in Chapter 11 *Sustainable geotechnics*.

2.6. The geotechnical design and construction cycle

Geotechnical design should be carried out in conjunction with the design of the whole structure, as explained in Section 2.1. Geotechnical design must always be carried out considering the source of the data and how the design will be implemented, with all three parts of this process being inextricably linked, as illustrated in **Figure 2.3**.

2.6.1 Project phases

All construction projects go through a number of distinct phases

- planning
- development
- implementation
- operation
- decommissioning.

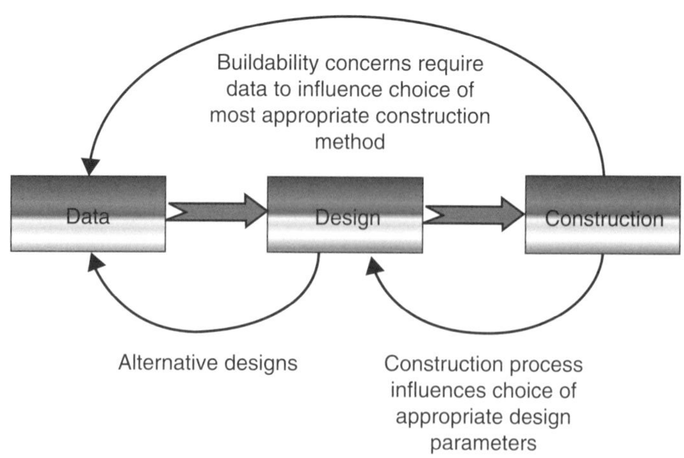

Figure 2.3 The geotechnical design and construction cycle

A number of organisations have devised 'stages' (e.g. RIBA, 2020) or 'gateways' (OGC, UK) and have further subdivided projects to suit these. Those work stages identified by RIBA (0–7, see **Figure 2.4**) as most commonly used in general design and construction are used here. Various stages/divisions are shown in **Figure 2.4** framed in the context of PAS 2080. The framework described in PAS 2080 is similar to that set out in the Value Toolkit (Construction Innovation Hub, 2022) which provides broader descriptors for the various stages of decision-making and activity within a construction project. This framing is helpful as it was designed to fit in with the climate emergency: most other process systems are predicated on something being built – the Value Toolkit is about achieving outcomes for the client, and so its initial 'need' phase is about getting those outcomes agreed in the context of being agnostic about the eventual form of the solution. Far too often in our industry, the solution or output is the first notion, and then justifications are fitted around it – which always leads to a heavy investment in Capital Carbon emissions. This altered approach has two useful by-products. First, it is much more fundamental – and also easier – to achieve a consensus with stakeholders about the outcomes sought rather than the shape of a specific solution with all players having different aims in mind. Second, achieving that consensus at the outset is much more likely to lead to smooth delivery of the whole project, successfully achieving those outcomes and also on time and to budget, with less project change to contend with. The types of input needed at each stage are shown in **Figure 2.5**.

The modes of input needed at each stage are different. During the *preparation and briefing* phase, input tends to be strategic and predominantly consultative – usually senior experienced input is needed to guide the project away from the more common pitfalls or significant hazards. By the end of this phase, the design should have progressed close to the stage of a single preferred option having been identified.

In the *design* phase, the design issues are resolved and all disciplines should produce a single coherent and coordinated output. In the latter stages, tender documents are produced which clearly describe the works to be undertaken on site. A design freeze is needed relatively early in this phase (end of RIBA stage 2 or 3, depending on the procurement strategy) so that in subsequent stages coherent documents

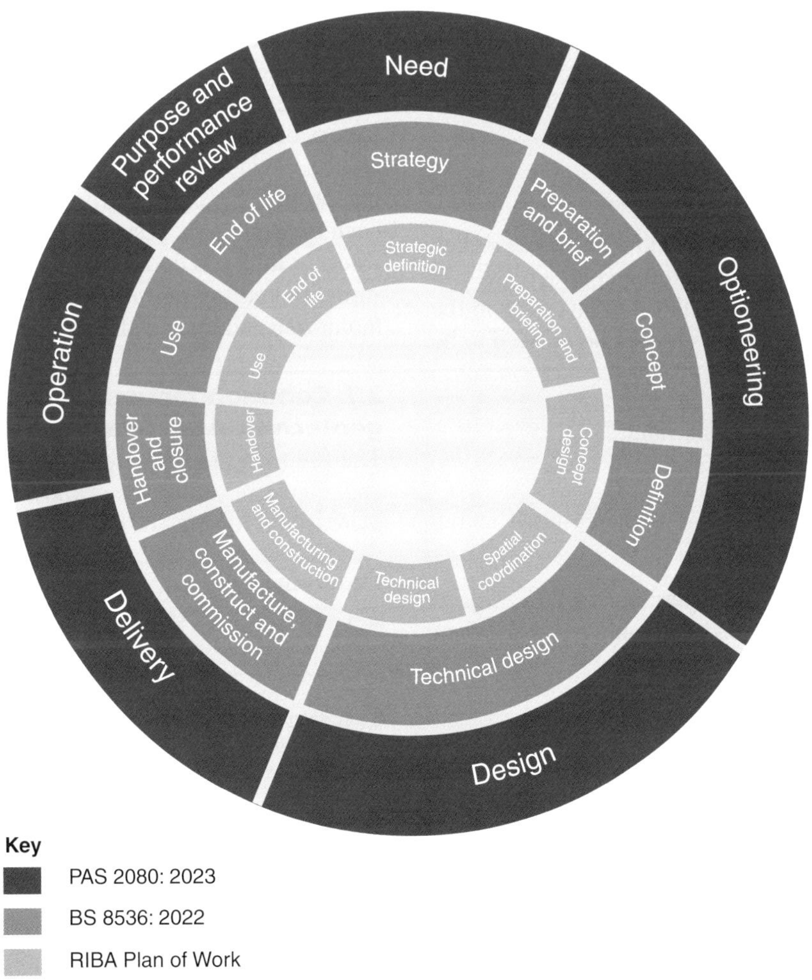

Figure 2.4 The work stages from RIBA, BS 8536 and PAS 2080 (Reproduced from PAS 2080)

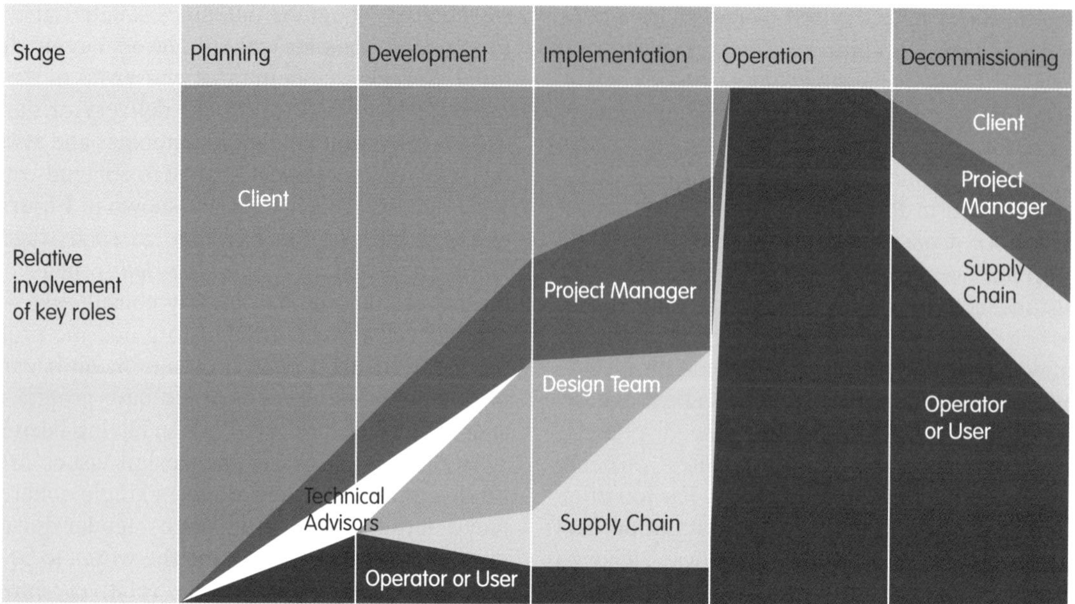

Figure 2.5 Input required at each stage of a project's life cycle (Reproduced from ICE, 2009, after Martin Barnes)

for tender and construction can be produced. The importance of a proper design freeze cannot be underestimated, as continued changes to the design jeopardise the production of an integrated and mutually consistent set of contract documents.

2.6.2 Importance of clarity in the tender process

In RIBA stages 3 *design development* and 4 *technical design* the tender documents are produced and the contract is let. In these stages the client becomes committed to investing in the whole of the project; therefore, at this stage clients need to be confident that

- the investment decision criteria remain valid
- they have funding
- they are content that the project has been organised to run smoothly.

This is a key time for geotechnical input; many well-designed projects rush production of the tender documents and do not adequately communicate

- the design intent and constraints
- the inherent risks and means for allocating responsibility for those (including those under Construction (Design and Management) Regulations 2015 (HMG, 2015)).

2.6.3 Construction control

Unusual geotechnical construction processes need knowledgeable control, both for the contractor and for the client, the latter sometimes provided by a resident engineer or individual with similar technical understanding.

The observational method (see Nicholson *et al.*, 1999 (CIRIA R185) and Chapter 100 *Observational method*) can be an excellent way to build efficiently. The observational method can be used to create something that conventional factors of safety may rule out. However, there must be well-founded evidence and experience that those factors would be inappropriately conservative. The observational method has been used most successfully for deep excavation in stiff clays where stability is controlled by slow dissipation of pore pressures, something that is difficult to rely on, but which can be monitored and managed by a well-trained and dedicated team.

2.7. Common factors associated with geotechnical success

The following factors are associated with successful projects and, conversely, absence of these factors is correlated with an increased frequency of failures.

Planning

- Use a good desk study to identify key hazards appropriate for the planned development – no skimping (e.g. historical map coverage extends back to first significant site occupation).
- Gather constraint/hazard data early and present them clearly to the wider client and design team.
- Address the likely construction issues in the ground investigation.

Development

- Beware trimming of important parts of the design service to save money, particularly the omission of knowledgeable input.
- Choose foundation type to explicitly reduce risks (e.g. avoid bored piles in silty soil) – this applies to safety risks as well as physical/design risks.

- Consider buildability as a critical element of the design process.
- Ensure continuity of process, ideally by the same capable individuals.
- Be clear on responsibilities, especially at gaps where responsibilities are divided or can be confused.
- Beware of designs based on optimistic assessment of parameters based on inappropriate extrapolation of test results.
- Bring experience to bear early to identify the likely key design issues that affect other members of the design team – for instance space-proofing with the architect and structural engineer.
- Watch waterproofing measures for basements, a frequent source of client disappointment.
- Ensure designers understand key assumptions from the analysis process and that contractors understand key assumptions from the design process.

Implementation

- Produce clear specifications highlighting important issues and communicating the design intent.

In general, beware overconfidence and adopt humility – consider the occurrence of the unexpected and do not rely on everything happening as planned. There is the well-known story told of the famous tunnelling engineer Sir Harold Harding who, on being informed of an overconfident young graduate in the design office, growled 'Bring him to me and I will put him down a big hole in the ground and teach him humility'.

References

BSI (British Standards Institution) (2004) BS EN 1997-1:2004: Eurocode 7: Geotechnical design – General rules. BSI, London, UK.

BSI (2006) BS EN 1992: Eurocode 2: Design of concrete structures (Parts 1–3). BSI, London, UK.

BSI (2007a) BS EN 1993: Eurocode 3: Design of steel structures (Parts 1–6). BSI, London, UK.

BSI (2007b) BS EN 1997-2:2007: Eurocode 7: Geotechnical design: Ground investigation and testing. BSI, London, UK.

BSI (2010) BS EN 1990:2001+A1:2005: Eurocode: Basis of structural design. BSI, London, UK.

BSI (2023) PAS 2080: Carbon management in infrastructure verification. BSI, London, UK.

Butcher AP, Powell JJM and Skinner HD (eds) (2006) *Reuse of Foundations for Urban Sites: Proceedings of the International Conference*. BRE Press, Watford, UK.

Chapman T, Anderson S and Windle J (2007) *Reuse of Foundations*. CIRIA Report C653. CIRIA, London, UK.

Construction Innovation Hub (2022) Value Toolkit Overview V2.0.

HMG (Her Majesty's Government) (2015) Construction (Design and Management) Regulations 2015. The Stationery Office, London, UK.

ICE (Institution of Civil Engineers) (2009) *Client Best Practice Guide*. Thomas Telford, London, UK.

ICE (2018) *In Plain Sight: Assuring the Whole-Life Safety of Infrastructure. Final Report*. ICE, London, UK. See https://www.ice.org.uk/news-insight/policy-and-advocacy/policy-engagement/in-plain-sight-assuring-the-whole-life-safety-of-infrastructure-final-report (accessed 19/04/2023)

Institution of Structural Engineers (2017) *Temporary Demountable Structures: Guidance on Procurement, Design and Use*, 4th edn. Institution of Structural Engineers, London, UK.

Network Rail (2007 and 2012) *GRIP Standard (Governance for Railway Investment Projects)*. Network Rail, London, UK.

Network Rail (2021) *Project Acceleration in a Controlled Environment (PACE)*. Network Rail, London, UK.

Nicholson D, Tse C-M and Penny C (1999) *The Observational Method in Ground Engineering: Principles and Applications*. CIRIA Report R185. CIRIA, London, UK.

Royal Institute of British Architects (2020) *RIBA Outline Plan of Work 2020*. RIBA Publishing, London, UK.

Tayler H (2020) A Short Guide to Reusing Foundations. www.istructe.org/journal/volumes/volume-98-(2020)/issue-11/a-short-guide-to-reusing-foundations (accessed 19/04/2023).

Useful websites

Efficiency and Reform Group, Cabinet Office, UK: http://www.cabinetoffice.gov.uk/government-efficiency

Royal Institute of British Architects (RIBA): http://www.architecture.com

All chapters within Sections 1 *Context* and 2 *Fundamental principles* together provide a complete introduction to the Manual and no individual chapter should be read in isolation from the rest.

doi: 10.1680/icemge.66816.0013

Chapter 3
A brief history of the development of geotechnical engineering

John B Burland Imperial College London, London, UK

This chapter traces the development of the craft and science of geotechnical engineering from early history to the recent past. The story is told of how Terzaghi struggled to couple engineering geology with the science of soil mechanics so as to provide the necessary rigour for modern geotechnical modelling and analysis. These lessons from the past are very important and a major aim of this manual is to provide the framework and knowledge necessary for sound geotechnical design and construction.

CONTENTS

3.1. Introduction

Foundation engineering is as old as the art of building and, like building, it developed largely on the basis of accumulated experience and empirical procedures. Because ground conditions vary so much from one locality to another, foundation practice varied widely. Moreover, the extrapolation of experience from one locality to another was fraught with uncertainty.

Parry (2004) describes how the ancient Egyptians learned from the foundation failure of the South Dahshur Pyramid. This was built during the reign of Pharaoh Snofru (2575–2551 BC) using a central core supported by a series of inclined buttress walls, as had become the tradition. The pyramid was founded on a clay layer. Large differential foundation movements resulted in significant structural distress in the tomb chambers and their access passages. As a result the pyramid was finished off so that the upper portions have a considerably reduced slope, giving rise to the structure being named the 'Bent Pyramid'. Abandoning the concept of buttress walls, future pyramids were built tier by tier, each tier consisting of a single layer of blocks of uniform thickness across the full width of the structure. Greater care was also taken in dressing and placing the blocks.

Kerisel (1987) describes how in Mesopotamia, over a period of three millennia, the art of building ziggurats was developed, in most cases on very weak alluvial soils. Because of the scarcity of stone, these high, massive structures were made of sun-baked bricks laid out in successive courses. As construction proceeded the underlying alluvium yielded under the weight, causing the base to spread laterally. Work progressed very slowly, with long pauses, so that little by little the rate of settlement and spreading diminished. Eventually it was possible to build a small temple at the top (**Figure 3.1**). Around 2100 BC the Sumerians began to place thick layers of woven reeds between every six to eight courses of sun-dried brickwork. In this way the horizontal tensions caused by the tendency of the foundations to spread were resisted. As a result ziggurats could be built with nearly sheer sides and massive temples on top. This innovation is often cited as the earliest example of reinforced earth.

Two early examples of successful foundation engineering in China are given by Kerisel (1987). The elegant early seventh-century Zhaozhou (otherwise known as Anji) Bridge (**Figures 3.2** and **3.3**) is founded on clay, which was treated by digging it out beneath the abutments and recompacting it in layers interspersed with compacted layers of broken bricks. The late-tenth-century 44 m high Pagoda of Longhua is founded on a thick layer of soft clay extending to a depth of about 30 m. The foundations are of brick laid on a wooden raft, which in turn rests on wooden piles driven at very close spacing – perhaps one of the earliest examples of a piled raft. The foundations remain unchanged since they were constructed over 1000 years ago.

3.2. Geotechnical engineering in the early twentieth century

It is not widely appreciated what a parlous state ground engineering was in prior to Terzaghi's contributions. Recently, as part of its centenary celebrations, the author was given the interesting task of tracing the development of foundation engineering over the last 100 years through the papers published in *The Structural Engineer* (Burland, 2008). Many of the early papers describe various techniques of foundation construction such as piles, sheet pile wall sections, cofferdams and caissons. But these papers make little reference to the mechanical properties of the ground and how its response can be assessed. For example Brooke-Bradley (1932–1934) states that:

> If the bearing power of sub-soil should prove to be inadequate to carry the proposed loads, it must be artificially strengthened.

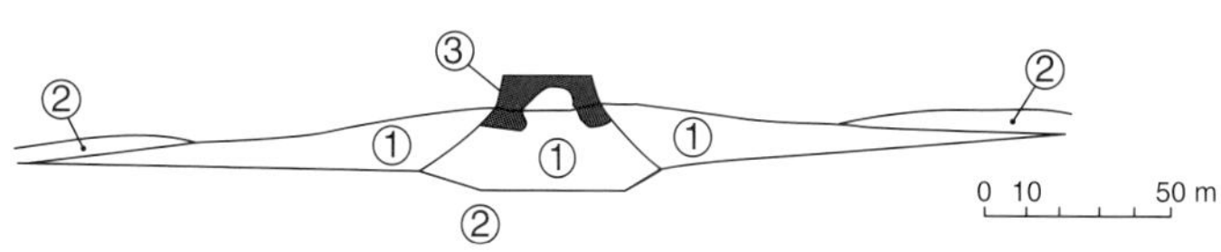

Figure 3.1 Construction of an early ziggurat. (1) Fill, (2) soft alluvium, (3) temenos – sacred enclosure (Reproduced from Kerisel (1987); Taylor & Francis Group)

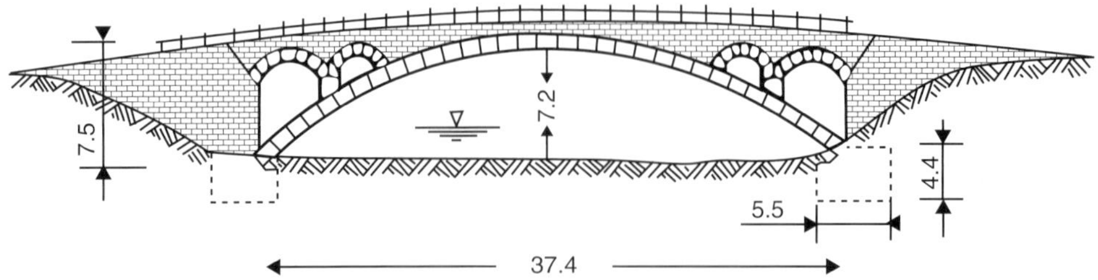

Figure 3.2 Zhaozhou Bridge – early seventh century (Reproduced from Kerisel (1987); Taylor & Francis Group)

Figure 3.3 Zhaozhou Bridge

Methods of doing this are then described, together with the various types of piles available for this purpose. Nowhere does one find how the 'bearing power' of the ground can be assessed in the first place. It is also stated that 'all settlement should be avoided if possible'; examples are given of damaging settlement but no guidance is given on how it could be estimated.

In the early issues of *The Structural Engineer* some space is given to the design and construction of retaining walls. In 1915 Wentworth-Shields wrote a paper on 'The stability of quay walls on earth foundations' (Wentworth-Shields, 1915).

He opens with the following memorable statement:

> In spite of the large amount of experience which has been gained in the construction of quay walls, it is still one of the most difficult problems in engineering to design a wall on an earth foundation with confidence that it will be stable when completed. … Even if the designer of such a wall is assured that it will stand, he cannot with any confidence tell you what factor of safety it possesses.

In 1928 Moncrieff published a major paper in *The Structural Engineer* (Moncrieff, 1928) on earth pressure theories in relation to engineering practice. He summarises the various approaches to calculating earth pressures from Coulomb (1773) through to Bell (1915). At that time the angle of friction was generally equated with the angle of repose, and Moncrieff refers to the difficulty of determining this angle for clayey soils. He cites a cutting in clay in which the side slopes varied from vertical to 1 vertical in 1½ horizontal, while in places the clay was 'running down like porridge'.

It is all too clear from these early papers that, in spite of significant, even heroic, engineering achievements in the construction of major foundations, retaining structures, tunnels and dams, there was little understanding of the factors that control the mechanical behaviour of soil in terms of its strength and stiffness. Moreover, there is almost no reference to the influence of groundwater on strength, stability, or earth pressures. It is hardly surprising that there were frequent failures, particularly of slopes and retaining walls. This was the muddle that Terzaghi found when he first began to practise as a civil engineer.

3.3. Terzaghi, father of geotechnical engineering

Because of his work in developing the scientific and theoretical framework of soil mechanics and foundation engineering, Terzaghi (**Figure 3.4**) is often regarded as essentially a theoretician. Nothing could be further from the truth. It is, therefore, worth reflecting on Terzaghi's struggles to develop the craft and the science of ground engineering for they have relevance in both the teaching and the practice of the discipline.

Goodman (1999) has written a most illuminating and thoroughly researched narrative of Terzaghi's life, *Engineer as Artist*. Terzaghi was born in Prague in 1883. He showed an early interest in geography, especially field exploration, and later astronomy, which evolved into a passion for mathematics. Later at school he was inspired by the natural sciences and performed brilliantly.

3.3.1 Terzaghi's education

Terzaghi went on to read mechanical engineering at the Technical University of Graz. For a time he lost his way, engaging in drinking and duelling. He found the lectures were simply a set of prescriptions, which he claimed he could read for himself. Ferdinand Wittenbauer, a wise teacher, challenged Terzaghi to do better and go back to the original sources – in particular Lagrange's *Analytical Mechanics* (Lagrange, 2001).

Figure 3.4 Karl von Terzaghi (By kind permission of the Norwegian Geotechnical Institute)

Wittenbauer led Terzaghi gently on, guiding him not only into the excitement of scientific creativity but also in the very real social and cultural issues of the day. It was Wittenbauer who saved Terzaghi from being expelled after an over-exuberant student prank. Wittenbauer pointed out to the authorities that in the history of the university there had been only three expulsions: Tesla, who went on to revolutionise electrical technology; Riegler, who created the steam turbine; and a third who developed into a leading church architect. He pointed out that the university was not good at choosing candidates for expulsion. Terzaghi was reprieved!

Although reading mechanical engineering, Terzaghi attended courses in geology. He was keen on climbing and it is related that he made every climbing expedition into a joyous adventure in field geology. During his compulsory year of military service he translated the *Outlines of Field-Geology* by Archibald Geikie (1896; director of the British Geological Survey) into German. In a second edition, he actually extended it to a fuller coverage of karst features and the geomorphology of glaciated country, replacing the English examples with Austrian ones.

3.3.2 The switch to civil engineering

Terzaghi's interest in geology persuaded him that mechanical engineering was not for him. Switching to civil engineering, he returned to Graz for an extra year. He went to work for a firm specialising in hydroelectric power generation. Although his main activity was in the design of reinforced concrete, the planning of the structures was, of course, intimately involved with geology. But frequently he found the guidance of expert geologists unhelpful. He encountered many cases of failure. Significantly, these were mainly due to the lack of ability to predict and control groundwater – piping failures were abundant. He also encountered many slope failures, bearing capacity failures and structures undergoing excessive settlement.

3.3.3 Geology on its own

Recognising the difficulties that civil engineers experienced in dealing with the ground and also the obvious influence of geological factors, he concluded that it was necessary to collect as many case records as possible so as to correlate failures with geological conditions. It is well known that he then spent two intense years (1912–1914) in the western USA observing and recording: two years that ended in disillusion and depression. The following quote from his presidential address to the Fourth International Conference on Soil Mechanics and Foundation Engineering sums up his mood at that time (Terzaghi, 1957):

> At the end of the two years I took my bulky collection of data back to Europe, but when I started separating the wheat from the chaff I realised with dismay that there was practically no wheat. The net result of two years of hard labour was so disappointing that it was not even worth publishing it.

So much for geology on its own! So much for precedent and case histories on their own!

To quote Goodman (1999), the problem lay in the fact that:

> … the names geologists give to different rocks and sediments have developed mainly from a scientific curiosity about the geologic origin of these materials, whereas Terzaghi was aiming towards discerning the differences in their engineering properties.

This is still true today – the engineer needs to understand the key geotechnical properties that affect the response of the ground.

3.3.4 The birth of the science of soil mechanics

Shortly after his appointment to the Royal Ottoman Engineering University in Constantinople in 1916, Terzaghi began to search the literature for insights into the mechanical behaviour of the ground. He became increasingly frustrated. What he witnessed was a steady decline from 1880 in recorded observations and descriptions of behaviour. This was replaced by myriads of theories postulated and published without adequate supporting evidence. This experience must have been uppermost in his mind when, in his presidential address to the First International Conference on Soil Mechanics and Foundation Engineering, he stated the following (Terzaghi, 1936):

> In pure science a very sharp distinction is made between hypothesis, theories, and laws. The difference between these three categories resides exclusively in the weight of sustaining evidence. On the other hand, in foundation and earthwork engineering, everything is called a theory after it appears in print, and if the theory finds its way into a text book, many readers are inclined to consider it a law.

Thus, Terzaghi was emphasising the enormous importance of assembling and examining factual evidence to support empirical procedures. He also brings out the importance of instilling rigour. This is often equated with mathematics, but there is at least as much rigour in observing and recording physical phenomena, developing logical argument and setting these out on paper clearly and precisely.

In 1918 Terzaghi began to carry out experiments on forces against retaining walls. He then moved on to piping phenomena and the flow beneath embankment dams. He used Forchheimer's flow net construction to analyse his observations and apply them in practice – methods that were themselves adapted from the flow of electricity. Here is the interplay between experiment and analytical modelling.

Over this period Terzaghi came to realise that geology could not become a reliable and helpful tool for engineers unless and until the mechanical behaviour of the ground could be quantified – this required systematic experimentation. On a day in March 1919, and on a single sheet of paper, he wrote a list of experiments that would have to be performed.

Terzaghi then entered an intense period of experimental work in which he carried out oedometer (confined compression) tests and shear tests on clays and sands, thereby developing his physical understanding of the principle of effective stress (the cornerstone of soil mechanics), excess pore water pressures and the time-rate of consolidation – this was the birth of soil mechanics. To make headway with modelling the consolidation phenomenon analytically he turned to the mathematics of heat conduction. Again, here is the interplay between experiment and analytical modelling. This intense period of experimental work and theoretical modelling culminated in the publication of his seminal book *Erdbaumechanik* (Terzaghi, 1925).

3.4. The impact of soil mechanics on structural and civil engineering

In 1933, a soil physics section was established at the Building Research Station (BRS) in the UK, and Dr Leonard Cooling was put in charge of it. He set up the first proper soil mechanics laboratory in Britain, equipped with the apparatus necessary to classify soils, measure their basic mechanical properties and carry out sampling. By 1935 the first investigations of civil engineering problems had begun and the group moved to the engineering division of BRS and was renamed as the soil mechanics section. It was in August 1937 that the well-known Chingford embankment dam failure occurred and the team from BRS carried out the investigation. Terzaghi was called in to redesign the embankment, and the necessary testing and analysis was carried out at BRS. This gave great impetus to the acceptance of soil mechanics as a key discipline in civil engineering in the UK.

On 6 December 1934, Terzaghi delivered a lecture before the Institution of Structural Engineers in London, with the title 'The actual factor of safety in foundations' (Terzaghi, 1935). He illustrated his lecture with a large number of case histories of measured distributions of settlement across buildings and their variation with time. He was able to explain the broad features of behaviour using the basic principles of soil mechanics and foundation analysis, demonstrating how vital it is to establish the soil profile with depth and across the plan area of the building. Even so, he showed that local variations in soil properties and stratification make it impossible to predict settlement patterns with any precision. Without actually using the term, he drew attention to the important concept of ground–structure interaction, pointing out that the structure of a building should not be treated in isolation from its foundations. He even drew attention to the fact that reinforced-concrete beams can yield plastically without impairing the stability or appearance of a frame building, provided the cracking is not excessive. It is of interest to note that, in their seminal paper on the allowable settlement of buildings, Skempton and MacDonald (1956) drew extensively on the case histories provided by Terzaghi in this lecture.

Towards the end of his lecture he made the following important assertion:

> Experience alone leads to a mass of incoherent facts. But theory alone is equally worthless in the field of foundation engineering, because there are too many factors whose relative importance can be learned only from experience.

On 2 May 1939, Terzaghi delivered the Forty-fifth James Forrest Lecture at the Institution of Civil Engineers, London with the title 'Soil mechanics – a new chapter in engineering science' (Terzaghi, 1939). The lecture summarised in simple terms the basic elements of the discipline of soil mechanics and its application to a number of engineering problems, ranging from earth pressure against retaining walls and the failure of earth dams due to piping, through to the phenomenon of consolidation and the settlement of foundations. Early in the lecture Terzaghi made the memorable statement that:

> in engineering practice difficulties with soils are almost exclusively due, not to the soils themselves, but to the water contained in their voids. On a planet without any water there would be no need for soil mechanics.

He was a forceful and charismatic figure, and this lecture had a very profound impact on the structural and civil engineers in the UK. The late Peter Dunican, past president of the Institution of Structural Engineers, attended as a young man and told the author of how Terzaghi had electrified the audience. Many leading geotechnical engineers, including the late Sir Alec Skempton, stress what a pivotal role this lecture played in the development of soil mechanics in the UK. As with his earlier lecture to the Institution of Structural Engineers, Terzaghi emphasised very strongly the importance of retaining a balance between theory and practice in soil mechanics. He stressed most strongly that precision of prediction was not possible due to the inherent variability of the ground and construction processes.

It is clear that Terzaghi is very much more than the father of the science of soil mechanics. His contribution was to place ground engineering on a rational basis, with geology as a key supporting discipline and soil mechanics providing the scientific framework for understanding the mechanical response of the ground. He is indeed the father of geotechnical engineering, which embraces engineering geology and soil mechanics and arguably rock mechanics as well.

3.5. Conclusions

Terzaghi's development of the science and art of geotechnical engineering grew out of his experiences as a civil engineer and his gradual realisation that the underlying principles governing the mechanical properties of soil were not understood. Although his contributions are often regarded as primarily theoretical, in reality this is anything but the case. A close study of his work reveals a brilliant and passionate engineer who at all times tried to maintain a balance between underlying theoretical principles, practical experience and the handling of the uncertainties that are always present when dealing with the ground in its natural state.

It is hoped that this chapter will provide a helpful summary that puts into context Terzaghi's struggles to provide a scientific and rigorous basis for geotechnical engineering. It demonstrates his grounding in geology; the importance of gaining an understanding of the mechanical behaviour of the ground and groundwater by means of experiment and testing; the need to develop an analytical framework for predictive purposes; and, very importantly, the key role that experience plays and the importance of case histories. Time and time again he insisted that soil mechanics is not a precise science because of the inherent variability of the ground and the uncertainty of many factors associated with construction.

References

Bell AL (1915) The lateral pressure and resistance of clay and the supporting power of clay foundations. *Minutes of the Proceedings of the Institution of Civil Engineers* **199**: 233–272.

Brooke-Bradley HE (1932–1934) Bridge foundations: parts I and II. *The Structural Engineer* **10(10)**: 417–426; **11(12)**: 508–521; **12(1)**: 18–26; **12(2)**: 96–105; **12(3)**: 130–140.

Burland JB (2008) Ground–structure interaction: designing for robustness. *Proceedings of the Institute of Structural Engineers Centenary Conference, Hong Kong*, pp. 211–234.

Coulomb CA (1773) Essai sur une application des règles des maximis et minimis à quelques problèmes de statique relatifs à l'architecture. *Mémoires de l'Académie royale des sciences* **7**: 342–382. See translation by J. Heyman (1997).

Geikie A (1896) *Outlines of Field-Geology*, 5th edn. Macmillan, London, UK.

Goodman RE (1999) *Karl Terzaghi: The Engineer as Artist*. ASCE Press, Reston, VA.

Heyman J (1997) *Coulomb's Memoir on Statics: An Essay in the History of Civil Engineering*. Imperial College Press, London, UK.

Kerisel J (1987) *Down to Earth: Foundations Past and Present – The Invisible Art of the Builder*. Balkema, Rotterdam, the Netherlands.

Lagrange JL (2001) *Analytical Mechanics*. Springer, Heidelberg, Germany.

Moncrieff JM (1928) Some earth pressure theories in relation to engineering practice. *The Structural Engineer* **6(3)**: 59–84.

Parry RH (2004) *Engineering the Pyramids*. Sutton, Stroud, UK.

Skempton AW and MacDonald DH (1956) Allowable settlement of buildings. *Proceedings of the Institution of Civil Engineers – Part 1* **5**: 727–784.

Terzaghi K (1925) *Erdbaumechanik auf bodenphysikalischer Grundlage*. Franz Deuticke, Leipzig, Germany.

Terzaghi K (1935) The actual factor of safety in foundations. *The Structural Engineer* **13(3)**: 126–160.

Terzaghi K (1936) Presidential address. *Proceedings of the 1st International Conference on Soil Mechanics and Foundation Engineering, Cambridge, MA*, vol. 3, pp. 13–18.

Terzaghi K (1939) Soil mechanics: a new chapter in engineering science. *Proceedings of the Institution of Civil Engineers* **12**: 106–141.

Terzaghi K (1957) Presidential address. *Proceedings of the 4th International Conference on Soil Mechanics and Foundation Engineering, London, UK*, vol. 3, pp. 55–58.

Wentworth-Shields FE (1915) The stability of quay walls on earth foundations. *Proceedings of the Concrete Institute* **20(2)**: 173–222.

All chapters within Sections 1 *Context* and 2 *Fundamental principles* together provide a complete introduction to the Manual and no individual chapter should be read in isolation from the rest.

Chapter 4
The geotechnical triangle

John B Burland Imperial College London, London, UK

doi: 10.1680/icemge.66816.0019

CONTENTS

This chapter outlines a coherent approach to the key aspects of ground engineering by making use of a simple aide-memoire termed the geotechnical triangle. The distinct and rigorous activities undertaken to obtain (*a*) the ground profile, (*b*) the measured behaviour of the ground and (*c*) an appropriate model are represented as the apexes of an equilateral triangle with empirical procedures and 'well-winnowed experience' linked to all three at the centre of the triangle. The ground profile is placed at the top apex of the triangle because of its crucial importance, and experience is located at the centre of the triangle because it is an essential aspect of all ground engineering and relates to the other three activities. Each of the above activities has a distinct methodology, each has its own rigour and each is interlinked with the other. Successful ground engineering requires that each activity is properly considered so that a coherent approach is adopted and the triangle is 'kept in balance'. The application of the geotechnical triangle is illustrated by revisiting the well-known case history of the underground car park at the Palace of Westminster in London.

4.1. Introduction

Geotechnics is a difficult subject and is regarded by many engineers as a kind of black art. The author used to think that this was due to the nature of the ground, which is a two- or even three-phase material. It is true that it is much more complex than the more classical structural materials of steel, concrete and even timber with which most engineers are familiar. As explained in Chapter 14 *Soils as particulate materials*, soil is a particulate material with little or no bonding between the particles. As a consequence

- The stiffness and strength of a given soil is not fixed but depends on the confining pressure.
- When it is deformed, soil will tend to contract or dilate depending on how dense it is, and this can profoundly influence its properties.
- During shearing, the particles tend to change their orientation, which, in turn, has a big influence on the shearing resistance of the material.
- Most important of all – water pressures acting within the pores of the soil are just as important as the applied boundary stresses.

This complex particulate material has to be modelled as if it were a continuum, but it is essential never to forget that in reality it is a particulate material.

As difficult as soil is as an engineering material, the problems confronted by the engineer in tackling a ground engineering problem are more subtle. Chapter 3 *A brief history of the development of geotechnical engineering* describes the struggles experienced by Terzaghi in establishing geotechnical engineering as a valid engineering discipline. After a careful study of the opinions expressed by Terzaghi and others, and from his own experience, the author came to the view that the main problem confronting the engineer is not so much the complex nature of soil as a material (difficult as that is), but a lack of appreciation of the number of aspects that have to be considered in tackling a ground engineering problem (Burland, 1987).

A close study of Terzaghi's struggles towards establishing the subject reveals that there are three distinct but interlinked aspects that have to be considered in tackling any ground engineering problem

- the ground profile including groundwater conditions
- the measured behaviour of the ground
- the appropriate model for assessing and predicting performance.

All three of these aspects are supported by empirical procedures, judgement based on case histories, and what the author has termed 'well-winnowed experience'.

The boundaries between these aspects often become confused. For example, it is frequently not clear whether a particular design approach is based on analysis or is mainly empirical. Moreover, one or more of the aspects is frequently completely neglected. The first three aspects may be depicted as forming the apexes of an equilateral triangle, with empiricism and experience occupying the centre and linked to the apexes as shown in **Figure 4.1**. This representation was originally developed as a teaching aid, but it has since become apparent that it also provides a valuable aide-memoire for geotechnical practice.

Associated with each of the above aspects are distinct and rigorous activities and procedures, which are shown in **Figure 4.1** by the broken circles. Each activity is a major discipline in its own right. Successful ground engineering requires that all aspects of the geotechnical triangle should be considered. Excessive reliance on one aspect (e.g. computer modelling or empirical procedures) can be disastrous. The four aspects – the

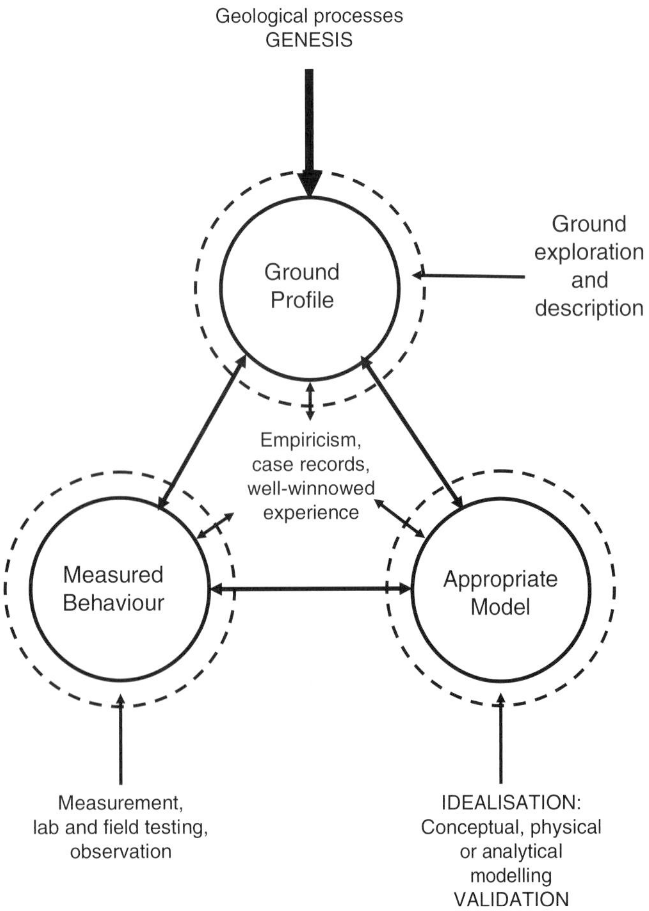

Figure 4.1 The geotechnical triangle. Each activity has its own distinct methodology and its own rigour

three apexes and the centre – depicted in the geotechnical triangle and their associated activities will now be briefly described. The order in which the activities are carried out will depend on the problem and may involve an iterative approach. For most problems the establishment of the ground profile is an early priority. It is also very important to establish at an early stage what experience is available for the problem being tackled and the ground conditions encountered.

4.2. The ground profile

Establishing the ground profile is a key outcome of the ground investigation as described in Chapter 13 *The ground profile and its genesis*. The ground profile is a description of the successive strata in simple engineering terms together with the groundwater conditions and the variations across the site. The importance of handling and describing the soil so as to establish the ground profile cannot be overemphasised. Also it is vital to understand the geological processes and man-made activities that formed the ground profile (i.e. its genesis). The author believes that nine times out of ten, the key conceptual design decisions for a project can be made on the basis of a good description of the ground profile. Similarly, nine failures out of ten result from a lack of knowledge about the ground profile – often the groundwater conditions. The importance of the ground profile is emphasised by placing it at the top apex of the triangle.

Peck (1962) argued that the methodology of the engineering geologist in establishing the geological model for a site consists in making observations, organising and assembling these, formulating a hypothesis and then critically testing the hypothesis. However, the civil engineer, and in particular the structural engineer, is not usually trained in this methodology, which is at the heart of much geotechnical engineering.

4.3. The measured or observed behaviour of the ground

The activity associated with this aspect involves making measurements of the mechanical behaviour and interpreting the measurements. It includes laboratory and field testing, carrying out field observations of the ground and structural movements (e.g. back-analysis of a landslip or of settlement observations) and measurements of groundwater pressures and flow. Rigorous methodologies and advanced instrumentation are often required for this work. The measurements require interpretation, which needs an appropriate analytical framework. A good basic understanding of the mechanical behaviour of soils and rocks is essential to these activities. Section 2 *Fundamental principles* of this manual provides a practical introduction to the fundamentals of soil and rock behaviour.

4.4. Appropriate model

When the author first put forward the original *soil mechanics triangle* in 1987 (now referred to as the geotechnical triangle), the term 'applied mechanics' was used for the bottom right-hand corner of the triangle. However, the term 'appropriate model' is a much better description of this aspect and its associated activities. The procedures involved in geotechnical and structural modelling are described in Chapter 5 *Structural and geotechnical modelling*. Modelling is the process of

- identifying what is to be modelled – for example, the part of the structure, the particular type of behaviour or limit state
- idealising or simplifying the geometry, material properties and loading
- assembling these idealisations appropriately into a model, which is then amenable to analysis and, hence, can be used to predict responses.

The modelling process is not complete until the responses have been validated and assessed. The procedure may involve a number of iterations.

It is important to note that the process of modelling is very much more than simply carrying out an analysis. A model can be a very simple conceptual one, it can be a physical 1*g* model or a centrifuge model, or it can be a very sophisticated numerical model. By using the term 'model' the emphasis is on the idealisation process and demystifying the analytical process. The geotechnical triangle helps in this. It is also important to appreciate that a variety of models and idealisations will usually be needed for different aspects of a given project or structure.

4.5. Empirical procedures and experience

With materials as complex and varied as the ground, empiricism is inevitable and it is (and will always remain) an essential aspect of geotechnical engineering. That is why this aspect is placed at the centre of the triangle and is linked to the other three aspects. Many of our design and construction procedures are the product of what may be termed 'well-winnowed experience' – that is, experience which results from a rigorous sifting of all the facts that relate to a particular empirical procedure or case history. The author chose this term having read Terzaghi's description of his attempts to 'separate the wheat from the chaff' following his two years in the USA collecting case records (see Chapter 3 *A brief history of the development of geotechnical engineering*).

Empirical procedures are often regarded as somehow inferior to analytical procedures. The fact is that both have their essential place in geotechnical engineering and both need to be carried out with care and with understanding of their limitations.

4.6. Summary of the geotechnical triangle

In summary, depicted within the *geotechnical triangle* illustrated in **Figure 4.1** are four key aspects of geotechnical engineering, each associated with distinct types of activity with different outputs. Each activity has a distinct methodology, each has its own rigour, and each is interlinked with the others. Terzaghi's approach to ground engineering reveals a coherence and integration, which is reflected in a 'balanced' geotechnical triangle. The geotechnical triangle provides a useful aide-memoire when tackling any ground engineering problem, whether it is in feasibility, design, construction, forensic investigation or teaching. It is possible to adapt the geotechnical triangle for particular applications. For example, in Chapter 9 *Foundation design decisions* an adaptation is described for the vital task of verifying design assumptions when construction work has started and the influence of construction processes on the ground needs to be assessed.

4.7. Revisiting the underground car park at the Palace of Westminster

Many of the above aspects can be illustrated by means of the well-known case history of the underground car park at the Palace of Westminster, London.

4.7.1 Background

This case history is described in detail by Burland and Hancock (1977). The site finally chosen for the 18.5 m deep car park for the Members of Parliament was in New Palace Yard, presenting major engineering challenges due to the proximity of the Big Ben Clock Tower, the Palace of Westminster and the fourteenth-century Westminster Hall, as shown in **Figure 4.2**. The sensitive, load-bearing masonry construction of the buildings and their priceless historic value demanded a structural form and construction procedure for the car park which would result in minimal ground movements, both during construction and in the long term. The structural form chosen consisted of a reinforced-concrete diaphragm wall, strutted at all stages of excavation and construction by the permanent reinforced-concrete floors, as shown in **Figure 4.3**. The columns are founded on under-reamed bored piles.

Figure 4.2 The site of the underground car park in New Palace Yard, Westminster

4.7.2 The importance of the ground profile

A preliminary site investigation had been carried out prior to the author's involvement in the project. The results showed that the top 10 m consisted of fill and soft alluvium, overlying medium-dense sand and gravel. London Clay extended from 10 to 44 m, at which depth the Woolwich and Reading beds were encountered (now known as the Lambeth Group). A new site investigation was planned, aimed at ascertaining the detailed soil profile, its variation across the site and the groundwater conditions. Altogether 14 boreholes were sunk on the site. The author personally carried out a detailed visual examination on open drive samples from a number of the boreholes by splitting the samples longitudinally with a knife so as to overcome the smearing on the outside of the samples. Visual and tactile descriptions were carried out using a simple scheme described by Burland (1987), which is consistent with BS 5930:1999, 'Code of practice for site investigations'. See

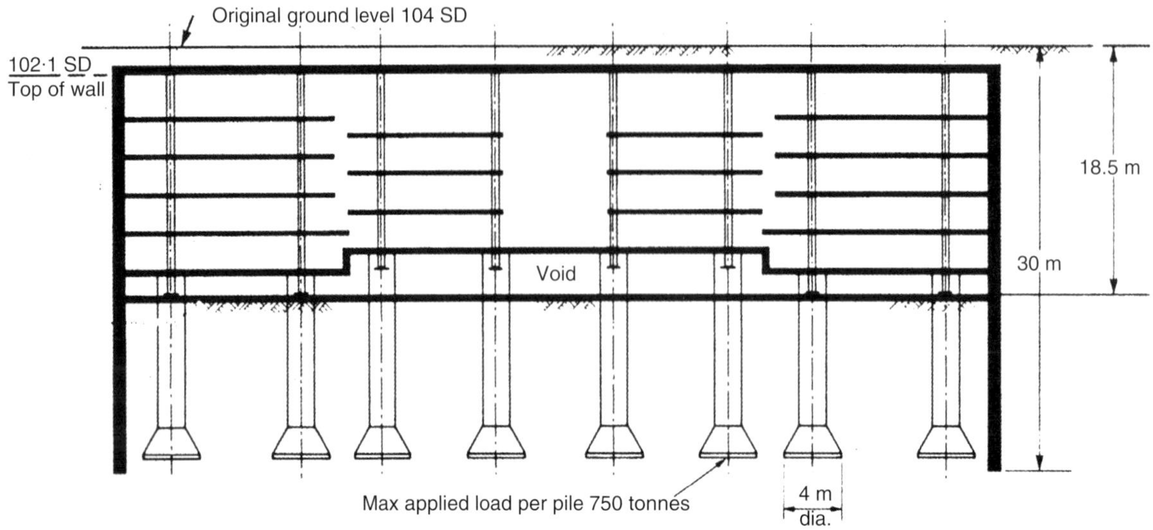

Figure 4.3 Section through the underground car park in New Palace Yard

Chapter 13 *The ground profile and its genesis* for more details of the information required in a description of the soil profile.

Good correlation of the various strata was found between the boreholes and the ground profile, as shown in **Figure 4.4**. A visual inspection revealed that immediately beneath the lowest level of the proposed car park, and extending from 19 to 30 m, there exists a layer of London Clay containing partings of fine sand and silt up to 10 mm thick and at 50 mm spacing near the top of the layer. The frequency of the partings reduces with depth. This layer is immediately underlain by a 4 m thick layer of very stiff intact clay. Casagrande standpipes installed in most of the soil exploration boreholes showed that the groundwater pressures down to a depth of about 35 m were hydrostatic and in equilibrium with the water table in the overlying alluvium and gravel.

At the time, the finding of the layer containing silt and sand partings came as a surprise. Subsequently it has come to light that it had been encountered during tunnelling elsewhere in London and had caused problems of face instability. Recently, Standing and Burland (2006) encountered the same layer in nearby St James's Park in connection with the Jubilee Line Extension underground railway, where excessive volume losses were encountered during tunnelling. The discovery of this layer containing silt and sand partings had very important implications for the conceptual design of the retaining walls and the foundations of the New Palace Yard Car Park. The relatively high horizontal permeability of this layer, coupled with hydrostatic water pressures in the surrounding ground, meant that high water pressures could develop beneath the excavation level, leading to hydraulic uplift and possible base failure. Moreover, seepage from the sand layers could have caused difficulties in the formation of the under-reams of the bored piles. The development of high water pressures beneath the bottom of excavations due to the presence of water-bearing permeable strata is a well-known geotechnical hazard. Ward (1957) describes a case in which high water pressures in a stratum of laminated sand caused uplift at the bottom of a trench excavation in stiff clay. The problem was solved by the use of simple gravel-filled relief boreholes.

Various methods were considered for relieving the water pressures immediately beneath the final basement level of the car park, including the use of relief wells. Detailed flow net calculations were carried out to assess the upward seepage gradients. However, there were considerable doubts about the short- and long-term effectiveness of relief wells. Moreover, the author was very unwilling to rely on a seepage analysis, knowing that even a small undetected permeable region could completely nullify the analysis. A robust solution was adopted, which involved taking the diaphragm retaining walls down into the intact clay layer at a depth of 30 m, thereby cutting off all horizontal seepage along the sand partings.

The choice and depth of the foundations was also profoundly influenced by the soil profile. The indications were that below a depth of 34 m (i.e. below the 4 m thick intact clay layer) there was the possibility of significant erosive seepage during excavation of the piles due to the silty nature of the clay at that depth. At a much shallower depth the significant reductions in effective stress immediately beneath the excavation would give rise to long-term reductions in stiffness and strength. With these considerations in mind, it was decided to use under-reamed bored piles founded on the intact clay layer at a depth of about 30 m, thereby making use of the higher shear strength beneath that level but avoiding difficulties due to seepage. A trial pile was constructed ahead of the main job to check on the groundwater and soil conditions and to evaluate the proposed foundation construction procedure. In situ inspection of the soil in the trial pile confirmed all the earlier deductions.

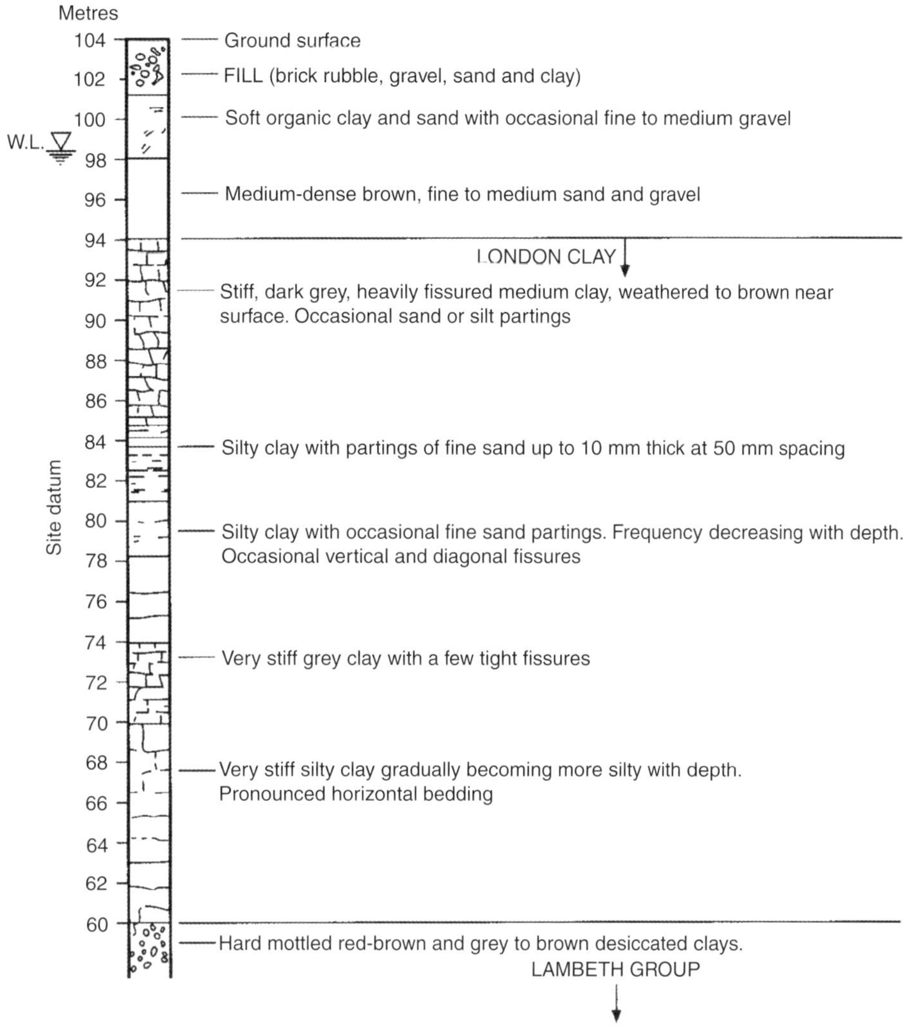

Figure 4.4 The ground profile at the Palace of Westminster

4.7.3 Modelling the ground movements

A major task in the design of the car park was to estimate the effects of the excavation on the surrounding buildings and on the structure itself. Accordingly, a detailed finite element analysis was undertaken. The analysis was based on a number of idealisations, and it is important to be absolutely clear about these.

Regarding the mechanical properties of the ground, the London Clay was assumed to behave as a linear elastic isotropic porous material having a stiffness that varies with depth. (See Chapter 17 *Strength and deformation behaviour of soils* for a discussion on the strength and deformation behaviour of soils.) This assumption was felt to be reasonable as the best research data available at the time suggested that at small strains, samples of London Clay tested in the laboratory exhibited linear behaviour. The variation of the undrained Young's modulus, E_u, with depth used in the analysis is given by the line in **Figure 4.5**.

The stiffness values for the London Clay and Lambeth Group were based on values obtained by back-analysis of measurements of the movements of the retaining walls for the deep basement excavation for Britannic House in the City of London (Cole and Burland, 1972). Thus, the choice of stiffness values was based on previous relevant case histories and observations in the London area – a clear example of the application of 'well-winnowed experience'. The adopted values of E_u for the basement beds of the London Clay and the Lambeth Group are somewhat lower than those obtained at Britannic House. It was felt necessary to be conservative in view of the lack of knowledge about the Lambeth Group in the Westminster area. Nevertheless, it is worth noting that the values of E_u used in the analysis were three to five times larger than the values deduced from careful laboratory tests. The author would never have chosen such high values without the benefit of experience from the analysis of previous relevant case histories.

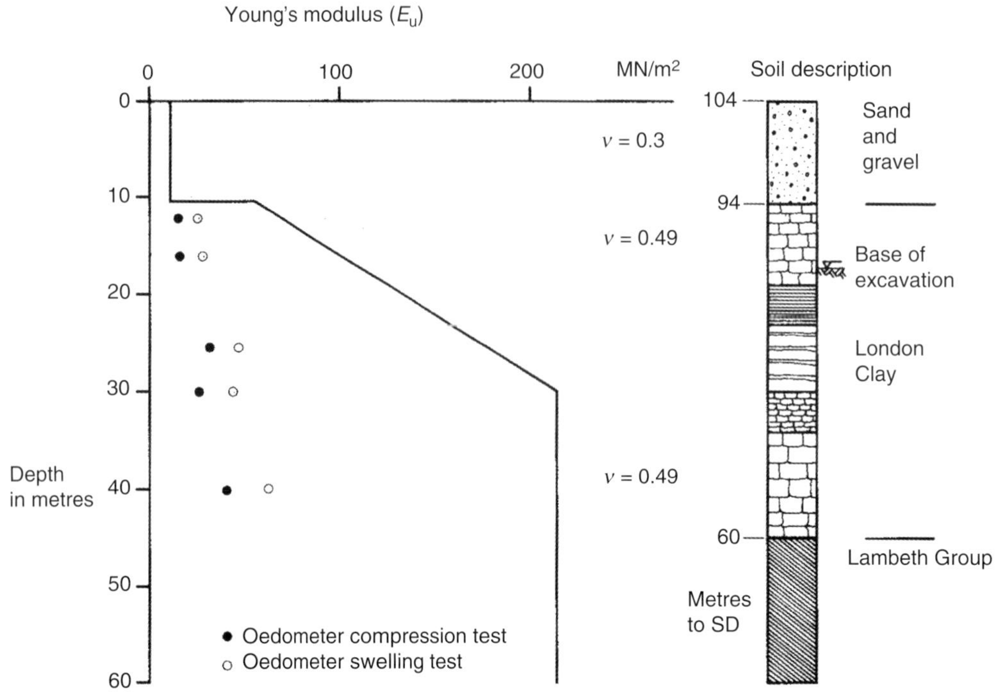

Figure 4.5 Values of undrained Young's modulus used in the finite element analysis

A key aspect of the modelling involved assessing the magnitude of the initial horizontal stresses in the clay, since these had to be progressively released during excavation. Once again the author made use of values deduced by others for various sites in London Clay. An assessment had to be made for the effects of erosion of the surface of the London Clay and the subsequent deposition of the gravel. This illustrates the importance of understanding the genesis of the ground profile – see Section 4.2.

Even though the problem is highly three-dimensional, a plane strain analysis was adopted as being approximately representative of the centre of the north and south walls, respectively. The finite element mesh extended to the base of the Lambeth Group (60 m depth) and laterally to a distance of 80 m from the edge of the excavation.

Numerical analysis was carried out in a step-by-step process simulating the excavation and installation of the floor props at each stage, as described by Burland and Hancock (1977). **Figure 4.6** shows the predicted wall displacements at the various stages of excavation, the maximum inward displacement being about 22 mm. The predicted short-term horizontal and vertical movements of the ground surface outside the excavation are given in **Figure 4.7**. These predicted ground surface movements gave a maximum change of gradient of 1/800 and a maximum extension strain of 0.02%. These strains were felt to be unlikely to cause any significant damage to the surrounding buildings.

Estimates were also made of the movements of the Clock Tower using an axisymmetric analysis, which was felt to be more appropriate for conditions close to the north-east corner of the car park. The Clock Tower was predicted to rotate away from the excavation by about 1/6000, while its foundations were predicted to move in towards the excavation by about 3 mm. The finite element model was also used to estimate the long-term movements and the swelling of the clay at the base of the excavation. On the basis of these calculations it was decided to use a suspended basement floor with a drained void beneath, rather than a solid slab, so that the clay could swell without interacting with the rest of the structure. It is very important to note that the predicted movements were published prior to work starting on the project (Ward and Burland, 1973).

The numerical model was also used to estimate the vertical movement of the under-reamed piles during and subsequent to excavation. This information was used to assess the approximate magnitudes of the differential movements across the floor slabs and the associated bending and shear forces within them due to ground movements.

4.7.4 Field monitoring

In view of the depth of excavation, the close proximity of such priceless historic buildings and the sensitivity of the project, a comprehensive programme of monitoring was undertaken during all phases of excavation and construction. Briefly, the monitoring can be considered under three broad headings: precision surface surveying, ground movements at depth and pore water pressures. Approximately 60 movement points were established by grouting Building Research Station levelling

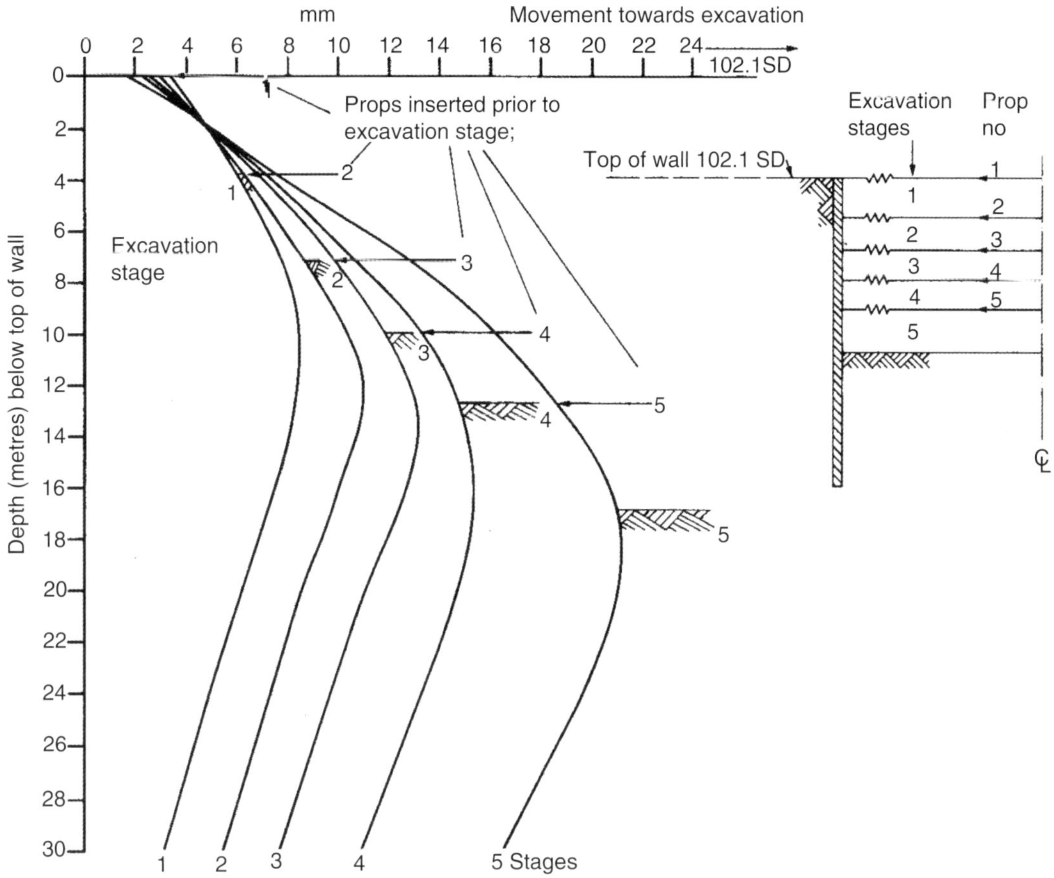

Figure 4.6 Predicted horizontal wall movements during excavation

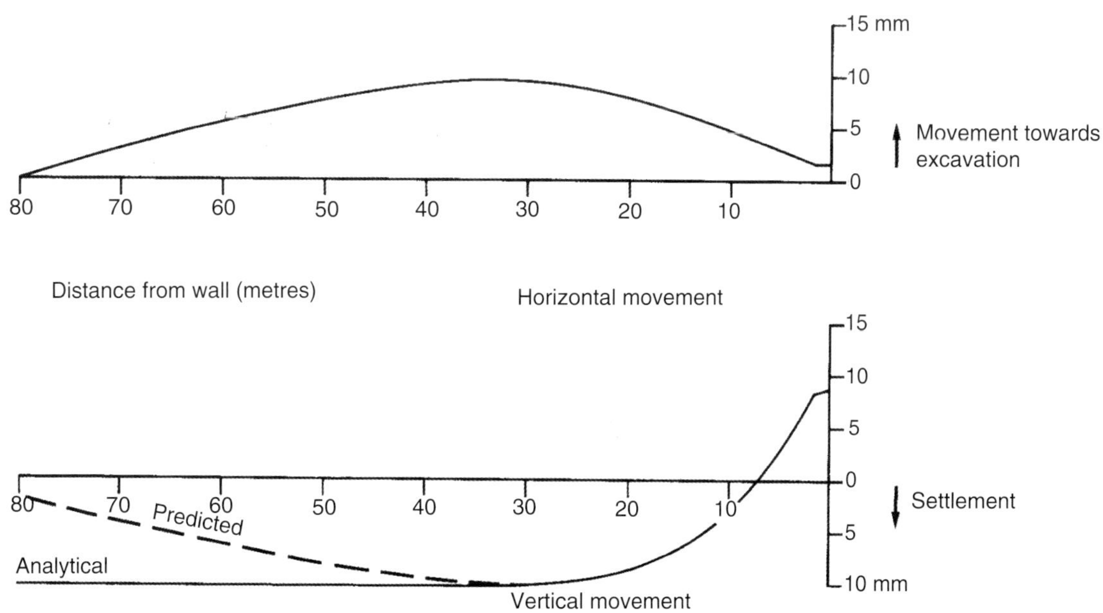

Figure 4.7 Predicted horizontal and vertical ground surface movements outside the excavation

stations into the masonry of the surrounding buildings (Cheney, 1973). Precision settlement surveys were carried out continuously during construction. The frequency of observations was increased close to areas where construction activity was concentrated. Complete geodetic surveys in both plan and elevation were made at intervals of approximately 2 months over the construction period of about 2 years. Changes in the verticality of the Clock Tower were monitored daily by means of a Hilger and Watts 'autoplumb'.

The subsurface horizontal deflections of the diaphragm retaining walls were measured by means of inclinometers. Special demountable targets were fitted to the tops of the inclinometer tubes, and their positions were measured in plan and elevation at 2-monthly intervals. An important aspect of the below-ground monitoring was to measure the heave at various depths during excavation as a check on the effectiveness of the design measures. Two magnet extensometers (Burland *et al.*, 1972) were installed at the centre of the excavation for this purpose.

A number of standpipes and pneumatic piezometers were installed within the area of the excavation and around its perimeter. They were to be used, in conjunction with the magnet extensometers, to check that hydraulic uplift was not developing. Unfortunately the pneumatic piezometers and many of the standpipes within the excavation were destroyed, but the standpipes outside the excavation all functioned satisfactorily.

4.7.5 Observed behaviour

Figure 4.8 shows the observed deflected shapes of the centre of the south diaphragm wall (inclinometer 8) at the various stages of excavation and support. These may be compared with the predicted behaviour given in **Figure 4.6**, where it is evident that the agreement is very satisfactory. The final predicted shape of the wall is shown as a broken line in **Figure 4.8(a)**. The major difference between the predicted and observed movement is the overestimation of the inward movement beneath the final excavation level. This is a direct result of the deliberate choice of conservative values for the undrained Young's modulus, E_u, for the basement beds of the London Clay and the Lambeth Group.

A large range of deflected shapes were observed for the various wall panels. **Figure 4.8(b)** shows the two extreme cases given by inclinometers 3 and 10. The results demonstrate that, even for relatively uniform ground conditions, widely differing deflected shapes can be anticipated. These differences may be due in part to the precise manner in which the excavation was carried out at various locations. Nevertheless, the differences shown in **Figure 4.8(b)** are a salutary reminder that there are limits to the precision with which predictions can be made.

In **Figure 4.9** the measured total horizontal and vertical surface movements behind the south wall are compared with the predictions given in **Figure 4.7**. The predicted horizontal movements are somewhat less than the observed values within

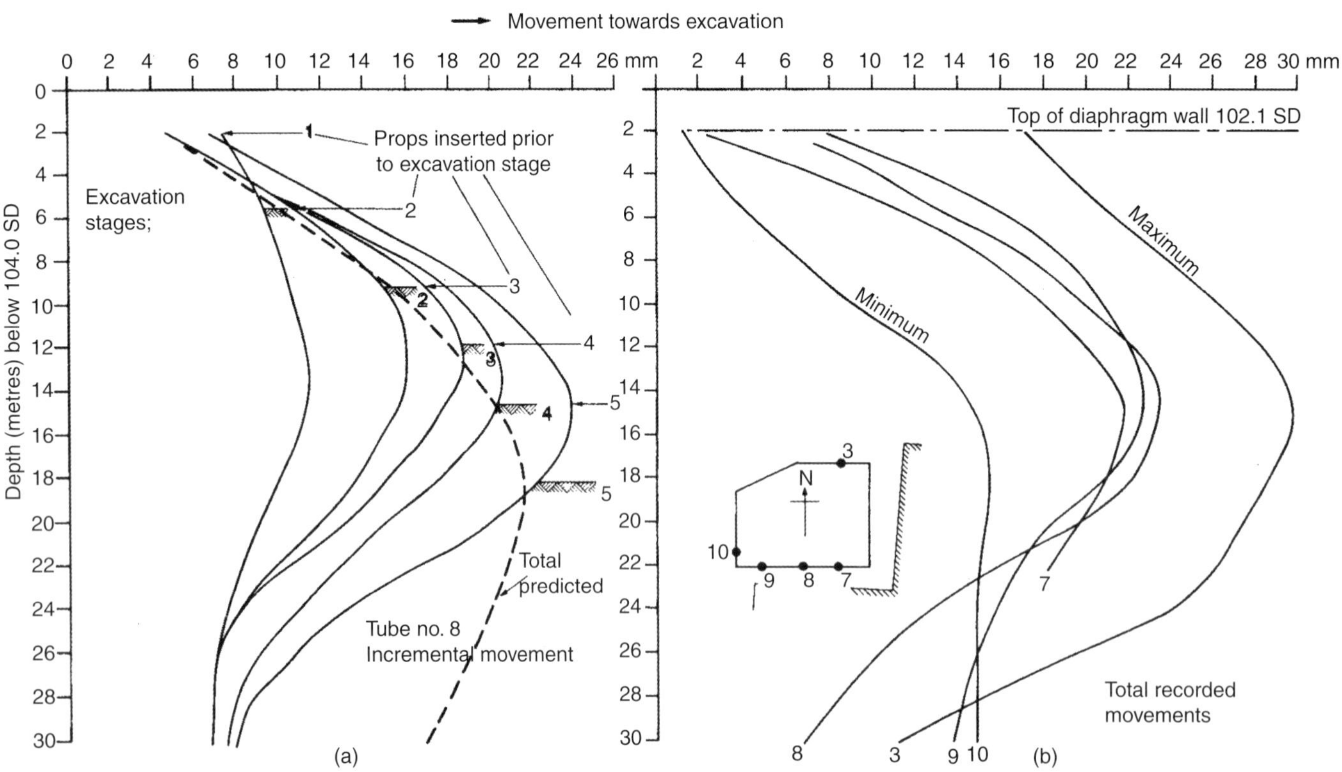

Figure 4.8 Observed movements of the diaphragm walls. (a) Inclinometer tube 8 at various stages of excavation. (b) Range of measured movements at the end of excavation

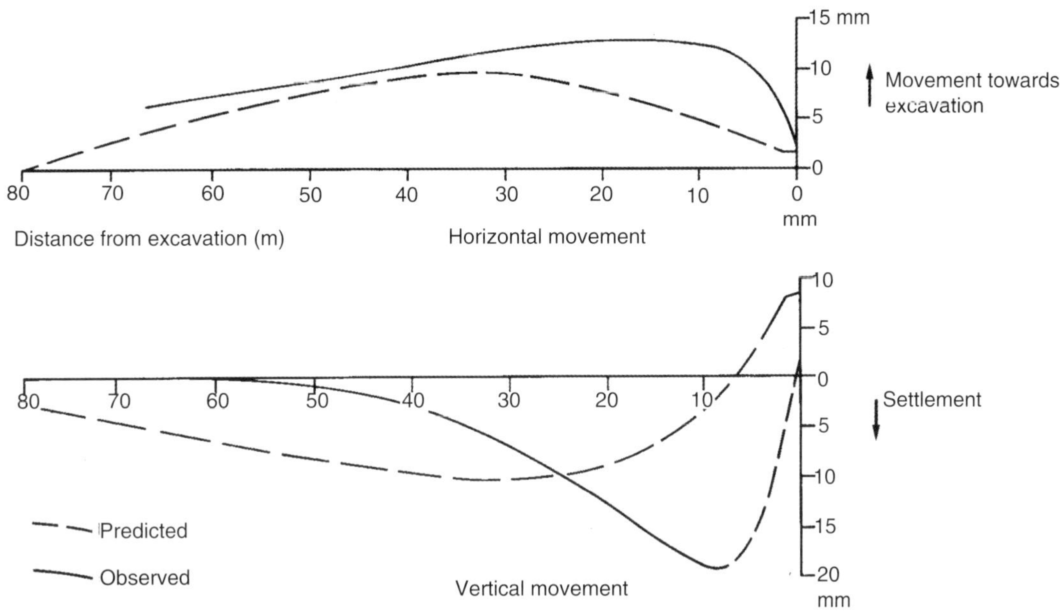

Figure 4.9 Comparison of predicted and measured ground surface movements away from the south wall

25 m of the wall, but further away the agreement is good. In general the overall form and magnitude of the prediction is reasonable. However, agreement between the observed and predicted vertical movements is not nearly as good. The upward displacement of the wall has been overestimated, while the maximum settlement has been underestimated. The 'settlement trough' is much nearer the wall than predicted. A consequence was that, whereas the Clock Tower was predicted to tilt away from the excavation by about 1/6000, it actually tilted towards the excavation by about 1/7000. The predicted and observed horizontal displacements were much more satisfactory, being 3 and 5 mm, respectively, towards the excavation.

In summary: The detailed ground profile together with experience of previous similar constructions led to the key conceptual design decisions – top-down construction to minimise ground movements, diaphragm walls cutting off horizontal seepage through the layer containing silt and sand partings, and founding the under-reamed piles in the intact clay layer. The stiffness parameters used in the finite element analysis were based on a back-analysis of other deep excavations in the London area. Numerical modelling was used primarily to assess the movements in the surrounding Palace of Westminster, but it was also used in the structural design of the diaphragm walls. The monitoring of the movements and groundwater pressures was carried out as a check that the behaviour was within acceptable limits. The car park was completed without significant damage to the surrounding buildings.

4.7.6 Subsequent refinements of the ground model

The difference between the predicted and measured vertical ground surface displacements just described proved very puzzling. However, shortly after the measurements were published by Burland and Hancock (1977), Simpson *et al.* (1979) showed that, by using a bilinear stress–strain law with a high initial stiffness, the agreement between observations and predictions could be greatly improved – particularly with respect to the vertical movements, as shown by the broken lines in **Figure 4.10**. At the same time as this theoretical work was being carried out, laboratory studies began at Imperial College London in which axial strains were measured locally on soil samples instead of between the end plattens, as was traditional. These measurements gave highly *nonlinear* stress–strain behaviour, with stiffnesses at small strains which were much larger than those inferred from traditional measurements. It now became clear that the pattern of ground surface movements observed at New Palace Yard, in which the vertical movements are concentrated close to the edge of the

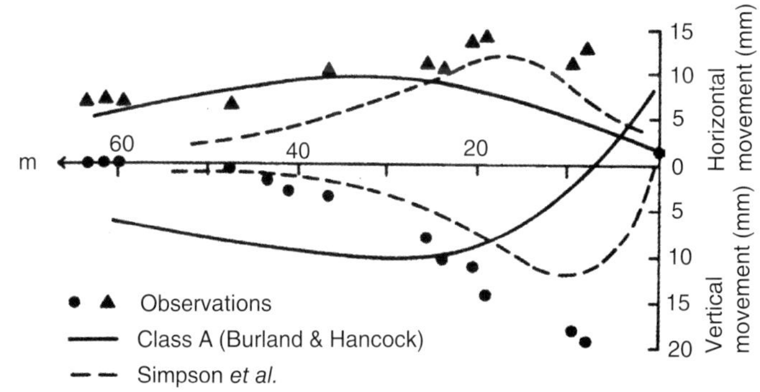

Figure 4.10 Improved predictions of the shape of the settlement trough (shown as a broken line) due to inclusion of small-strain nonlinearity in the soil model

excavation, is due to the *nonlinear* nature of the stress–strain behaviour of the soil at small strains.

This process of prior publication of predictions, although uncomfortable at the time, has proved highly beneficial as it caused the author to ponder long and hard as to the explanation for the discrepancies. Without such public disclosure it would have been all too tempting to quietly ignore the discrepancies and move on to other things. The work at New Palace Yard, and the measured response of the Clock Tower, has spawned a whole new important area of study of the behaviour of the ground at small strains – indeed whole international conferences are now devoted to the subject. These studies are proving very important for modelling the key mechanisms governing the interaction effects between ground and structure. Understanding these mechanisms is particularly important in the urban environment, where underground construction is a vital part of infrastructure development.

This case study is a clear demonstration of the importance of publishing well-documented case histories, as they serve to contribute to the profession's collective experience and can be used to calibrate new numerical modelling techniques for many years to come.

4.8. Concluding remarks

This chapter outlines a coherent approach to the key aspects of geotechnical engineering by making use of the *geotechnical triangle* (**Figure 4.1**). The distinct and rigorous activities of (*a*) ground investigation, (*b*) measurement and observation and (*c*) appropriate modelling are represented as the apexes of the triangle with (*d*) 'well-winnowed experience' (an equally rigorous activity) linked to all three at the centre of the triangle.

The application of the geotechnical triangle has been illustrated by revisiting the well-known case history of the underground car park at the Palace of` Westminster, London. A deep excavation was chosen as a case history since this type of problem demands all the traditional skills of the engineer, including: reliance on observation and measurement; a deep understanding of both geotechnical and construction materials; the effects of groundwater and seepage; the development of appropriate conceptual and analytical models; and, above all, judgement based on a knowledge of case histories and construction methods.

It should be noted that the key conceptual design decisions for the underground car park at Westminster were taken on the basis of a thorough knowledge of the ground profile obtained from the visual and tactile description of the soil samples and measurements of groundwater pressures from standpipes.

The choice of stiffness properties for the London Clay was based on the careful and rigorous analysis of other appropriate case histories of deep excavations in the London area, illustrating the importance of case histories and 'well-winnowed experience'. The field measurements made during the excavation of the car park have proved most valuable in gaining an understanding of the influence of small-strain *nonlinear* behaviour of the ground. Even more importantly, these measurements have shown that precise prediction of behaviour is not possible and that successful design requires that inherent uncertainties of response must be taken into account.

References

Burland JB (1987) Kevin Nash lecture: The teaching of soil mechanics – a personal view. *Proceedings of the 9th European Conference on Soil Mechanics and Foundation Engineering*, Dublin, Ireland, vol. 3, pp. 1427–1447.

Burland JB and Hancock RJR (1977) Underground car park at the House of Commons, London: geotechnical aspects. *The Structural Engineer* **55(2)**: 87–100.

Burland JB, Moore JFA and Smith PDK (1972) A simple and precise borehole extensometer. *Géotechnique* **22(1)**: 174–177.

Cheney JE (1973) Techniques and equipment using the surveyor's level for accurate measurement of building movement. *Proceedings of the Symposium on Field Instrumentation, London*. British Geotechnical Society, London, UK, pp. 85–99.

Cole KW and Burland JB (1972) Observations of retaining wall movements associated with a large excavation. *Proceedings of the 5th European Conference on Soil Mechanics and Foundation Engineering*, Madrid, Spain, vol. 1, pp. 445–453.

Peck RB (1962) Art and science in subsurface engineering. *Géotechnique* **12(1)**: 60–66.

Simpson B, O'Riordan NJ and Croft DD (1979) A computer model for the analysis of ground movements in London Clay. *Géotechnique* **29(2)**: 149–175.

Standing JR and Burland JB (2006) Unexpected tunnelling volume losses in the Westminster area, London. *Géotechnique* **56(1)**: 11–26.

Ward WH (1957) The use of simple relief wells in reducing water pressure beneath a trench excavation. *Géotechnique* **7(3)**: 134–139.

Ward WH and Burland JB (1973) The use of ground strain measurements in civil engineering. *Philosophical Transactions of the Royal Society, London* **A274**: 421–428.

All chapters within Sections 1 *Context* and 2 *Fundamental principles* together provide a complete introduction to the Manual and no individual chapter should be read in isolation from the rest.

doi: 10.1680/icemge.66816.0029

Chapter 5
Structural and geotechnical modelling

John B Burland Imperial College London, London, UK

CONTENTS

During his career the author has encountered profound differences in the approaches taken by structural and geotechnical engineers, often leading to a lack of understanding and difficulties in communication. This chapter explores these differences and the reasons for them.

The term *modelling* is used extensively. It is defined as the process of idealising a structure and the underlying ground, including the geometry, material properties and loading, in order to make the problem amenable to analysis and, hence, assessment for fitness of purpose. It is demonstrated that traditional structural modelling is very different from geotechnical modelling and these differences need to be appreciated if ground–structure interaction problems are to be realistically modelled. It is concluded that concepts such as *ductility* and *robustness* underpin the success of both structural and geotechnical modelling and more explicit recognition of these is needed.

5.1. Introduction

There is always an interaction between a structure and its foundation … whether or not the designers allow for it. In some situations the interaction can be minimised, such as by founding the building on rigid piles. Such an approach can be very costly and is often not feasible. If structure–ground interaction is to be taken into account, structural and geotechnical engineers themselves have to interact.

There have been many occasions when the author has both witnessed and experienced difficulties in communications between structural and geotechnical engineers, and it has been a continuing source of interest to him as to why this should be so (Burland, 2006). It is a matter of outstanding importance because poor communication and lack of understanding can lead to poor engineering and even failures.

At the heart of the problem there are differences in the day-to-day approach to *modelling* structural and geotechnical behaviour. By identifying these differences it is possible to move towards more integrated and realistic approaches to designing for ground–structure interactions.

The term *modelling* is used extensively in this chapter. It is intended to describe the process of idealising a real-life structure, foundations and ground (or parts of them) including the geometry, material behaviour and loading, in order to make them amenable to analysis, carrying out that analysis, reviewing it and then assessing the design for 'fitness for purpose'. Thus, the process of modelling entails very much more than the process of carrying out the analysis itself.

5.2. Structural modelling

The late Edmund Hambly, a most innovative structural engineer who obtained his PhD in soil mechanics, published a little book *Structural Analysis by Example* (Hambly, 1994). It was intended as a handbook for undergraduates and professionals who like to use physical reasoning and off-the-shelf computer modelling to understand structural form and behaviour. Fifty examples are given of structural problems of increasing complexity, ranging from simple frames, through beams, columns and slabs, to shear lag and torsion, and then on to whole structures such as offshore platforms and spiral staircases. The book represents an admirable summary of the range of problems and types of analysis that a structural engineer encounters in day-to-day practice.

In terms of *modelling*, it is evident that the geometry of most structures is well defined and reasonably easy to idealise. Rather simple linear elastic material behaviour is usually assumed with a limiting stress imposed. The major idealisations in the modelling process are in the loading, although in Hambly's book this is specified, as is usually the case in codes of practice and standards. It is evident that the process of routine structural modelling mainly consists of idealising the structural form (often termed the 'conceptual model') and carrying out analysis – usually on the computer. Recently, the Institution of Structural Engineers (IStructE, 2002) and MacLeod (2005) set out more formal procedures for the structural modelling process.

5.2.1 Limitations to structural modelling

Using Hambly's book as an exemplar, it is evident that structural engineers think primarily in terms of the forces and stresses (or equivalent elastic displacements) that are the outputs from the structural models that they use in day-to-day practice. Yet most studies on real whole structures (especially buildings) show that the measured strains and displacements bear little resemblance to the calculated ones (Walley, 2001). Classic experiments on large-scale steel structures, carried out in the early 1930s under the direction of Lord Baker, revealed that the measured stresses under working loads bore very little relation to the calculated values (Baker *et al.*, 1956). National Building Studies Research Paper 28 records that the biggest change of strain in the steel beams of the Ministry of Defence building in Whitehall was caused by the shrinkage of the concrete after the floors had been cast – this was larger than that induced by the subsequent loading of the floors (Mainstone, 1960). In his own work, the author has found that the thermal and seasonal movements of buildings are often as large as the movements induced by tunnelling, and yet these are overlooked when assessing the response of buildings to foundation movements (Burland *et al.*, 2001). These limitations have been appreciated for a long time but are easily forgotten. The fact is that the uncertainties of modelling real projects are dealt with in good practice not only by factors of safety, but also by incorporating appropriate levels of ductility and robustness.

5.2.2 Ductility and robustness

Ductility may be defined as *the ability to undergo inelastic deformations without significant loss of strength*, while robustness can be defined as *the ability to absorb damage without collapse*.

On page 1 of Hambly's book, reference is made to the *safe design theorem*, which, of course, derives from the lower bound theorem of plasticity. This theorem states that:

A structure should be able to carry its design loads safely if:

1. The calculated system of forces is in equilibrium with the loads and reactions throughout the structure.
2. Each component has the strength to transmit its calculated force and the ductility to retain its strength while deforming.
3. The structure has sufficient stiffness to keep deflections small and to avoid buckling before design loads are reached.

If the real structure deforms under load with a different flow of forces from that calculated, it should still be safe as long as the materials are ductile (like steel) and not brittle (like glass). The equilibrium check ensures that any underestimate of the force flowing through one part of the structure is balanced by an overestimate in the force in another part. Then ductility ensures that the component that is overstressed retains its strength while deforming, and sheds the excess force to the parts that have available capacity as a result of the equilibrium check. It is the ductility of structures that ensures the success of current design methods. This is frequently overlooked and, in the author's experience, many structural engineers still seem to believe that their structures behave as calculated – a belief that powerful computer programs and prescriptive codes of practice tend to reinforce. Both Heyman (2005) and Mann (2005) have written on this topic and stressed the need to ensure that the structural details are ductile – also a prerequisite for earthquake design. The inherent ductility and robustness of steel structures is well known and gave rise to modern methods of plastic analysis. Beeby (1997, 1999) summarised the developments in the understanding of ductility for traditional reinforced-concrete design and stressed the importance of designing for robustness.

It is important to note that the majority of structural failures result from defects at joints and connections (Chapman, 2001). Engineers do, of course, carry huge responsibility for ensuring that local instabilities cannot develop – particularly in temporary works and propping systems. For steel structures Burdekin (1999) drew attention to how important it is for designers to be aware of the factors that give rise to brittle fracture and fatigue.

5.2.3 The three- and four-legged stools (Hambly's paradox)

A simple but profound physical model can be used to illustrate what has been discussed so far. **Figure 5.1** shows two stools, one with three legs and one with four (Hambly, 1985). Imagine that each must support a person milking a cow who weighs 60 kg, and who always sits with their centre of gravity directly over the middle of the stool. The problem is to determine how much load must be carried by each leg of the three-legged stool and the four-legged stool.

The three-legged stool is straightforward in that one-third of the person's weight must go down each leg (i.e. 20 kg). For the four-legged stool the answer of 15 kg is wrong! Careful inspection of **Figure 5.1** shows that one of the four legs does not quite touch the ground, either because the leg is slightly short or because the ground is uneven; consequently the leg is not carrying any load. The opposite leg will also not be carrying any load because its load must balance that in the short leg when equilibrium of moments is considered about the diagonal through the other two legs. Thus, all of the weight is carried by two legs (i.e. 30 kg per leg), instead of being shared by the four legs. Hence, the paradox (Heyman, 1996) – the

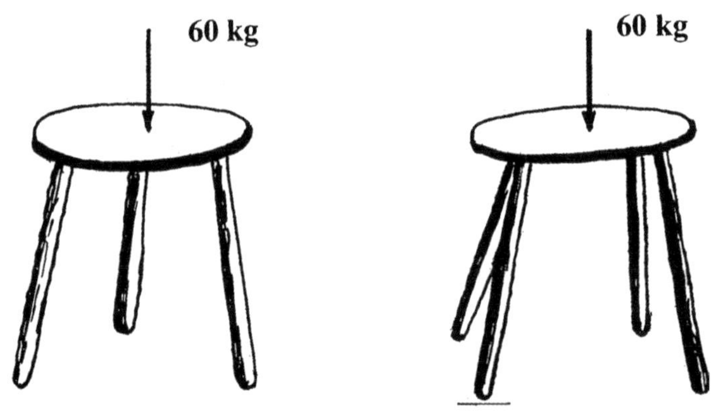

Figure 5.1 Three- and four-legged stools – Hambly's paradox

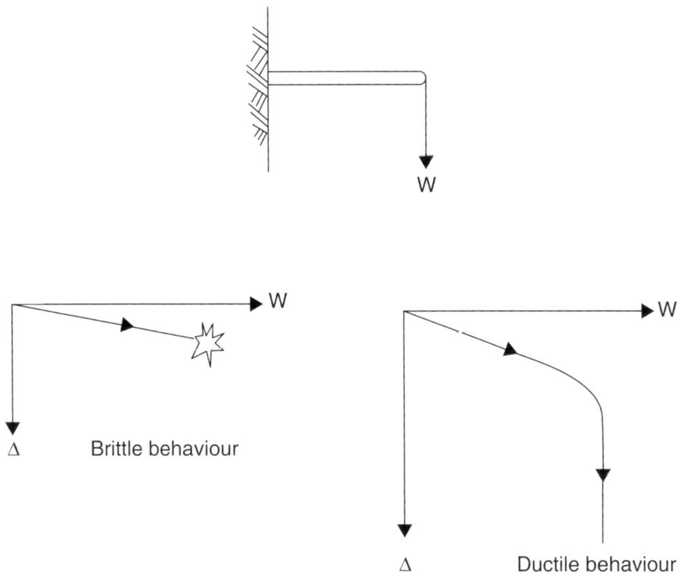

Figure 5.2 Brittle and ductile behaviour of a stool leg

addition of a fourth leg to a three-legged stool can increase, rather than decrease, the force likely to be carried by each leg. So what load *should* the legs be designed to carry?

It is here that concepts of ductility and robustness come in and the illustrative model can be extended to include material properties. **Figure 5.2** shows the load–displacement properties of a leg, either when acting as a cantilever or in axial compression. In one case, very brittle behaviour is illustrated such that at a certain load the strength of the leg is lost completely, as it would be if it were made of glass. In the second case, the member exhibits ductile behaviour and retains its strength once yield has occurred.

If brittle material is used for the three-legged stool, then accidental overload, due perhaps to a very heavy person or the cow kicking out at the stool, can easily result in total collapse. Clearly high factors of safety are required to deal with this design. It may be decided to opt for four legs, but this may be of little help. The design load for each leg would have to be 1.5 times higher than for the three-legged design. Moreover, accidental overload may cause loss of one member and there is then a risk of progressive collapse. In other words, the structure is *fragile*.

If ductile properties are chosen, there is little likelihood of catastrophic collapse if one of the three legs is damaged. Moreover, with four legs there is scope for redistribution of the load once the carrying capacity of a leg is reached. Even accidental removal or serious damage to one member is unlikely to give rise to progressive collapse.

This simple example is very profound and can be extended to other aspects of structural behaviour and design, including buckling and ground–structure interaction. Above all, it illustrates the importance of ductility, robustness and redundancy.

It is useful to quote Heyman's (1996) conclusions to his study of Hambly's paradox:

> Hambly's four-legged stool stands, of course, for the general problem of design of any redundant structure. It has long been recognised that, in order to calculate the 'actual' state of a structure under specified loading, all three of the basic structural statements must be made – equilibrium, material properties and deformation (compatibility and boundary conditions). However, the calculations do not in fact lead to a description of the actual state. Boundary conditions are, in general, unknown and unknowable; an imperfection in assembly, or a small settlement of a footing, will lead to a state completely different from that calculated. This is not a fault of the calculations, whether elastic or not – it is a result of the behaviour of the real structure. … There is no correct solution to the equations, but one solution that will lead to the greatest economy in material. This is the solution sought by the simple plastic designer, and it is safe and valid provided that no instability is inherent in the structure.

The important message is that, in the process of structural modelling, the inherent uncertainties are such that the precise state of the structure cannot usually be calculated. The art of strutural engineering is to use the process of modelling to produce a design that is robust enough to safely cope with the uncertainties, at reasonable cost, and that is fit for purpose. Difficulties arise in interacting with geotechnical engineers when the inherent uncertainties in the structural analysis are not recognised – that is, when the output from a computer program is believed to represent the true state of a structure.

5.3. Geotechnical modelling

Soil mechanics is a difficult subject and is regarded by many structural engineers as a kind of black art. It is helpful to discuss the reasons for some of the difficulties.

One difficulty is that the soil, unlike concrete or steel, is a particulate material with little or no bonding between the particles. It is made up of an infinite variety of shapes and sizes of particles. This material is usually modelled as a continuum – but it is important never to forget that its properties are determined by its particulate nature.

Because soil is particulate, the water pressures acting within the soil pores are just as important as the stresses applied to its boundaries. This means that changes in the groundwater regime can be crucial in stabilising, or destabilising, a slope, retaining wall or foundation.

In summary, whereas the strength of structural materials, such as steel and concrete, is primarily cohesive and well defined, soil is primarily frictional, so that the confining pressure and pore water pressure determine its strength and stiffness.

5.3.1 The geotechnical triangle

But it is not just the complexity of soil as a material which causes difficulty. As pointed out in Chapter 4 *The geotechnical triangle* there are at least four distinct but interlinked aspects of any geotechnical problem

- the ground profile – what is there and how it got there
- the measured behaviour of the ground
- prediction using appropriate models
- empirical procedures; judgement based on precedent and 'well-winnowed experience'.

It is the author's experience that the major difficulty in soil mechanics lies not so much in the complexity of the material as such, but in the fact that, all too often, the boundaries between the above four aspects become confused (Burland, 1987).

The first three of these may be depicted as forming the apexes of a triangle with empirical procedures occupying the centre, as shown in **Figure 5.3**. Associated with each of these aspects is a distinct and rigorous activity or discipline.

Ground profile: This is the visual description in simple engineering terms of the successive strata making up the ground together with the groundwater conditions. The importance of handling and describing the soil so as to establish the ground profile cannot be overemphasised. It is also important to understand how the ground got there and what might have happened to it during its formation – what is termed its *genesis*. As described in Chapter 13 *The ground profile and its genesis*, establishing the ground profile is a key outcome of carrying out a site investigation – often far more important than measuring the material properties.

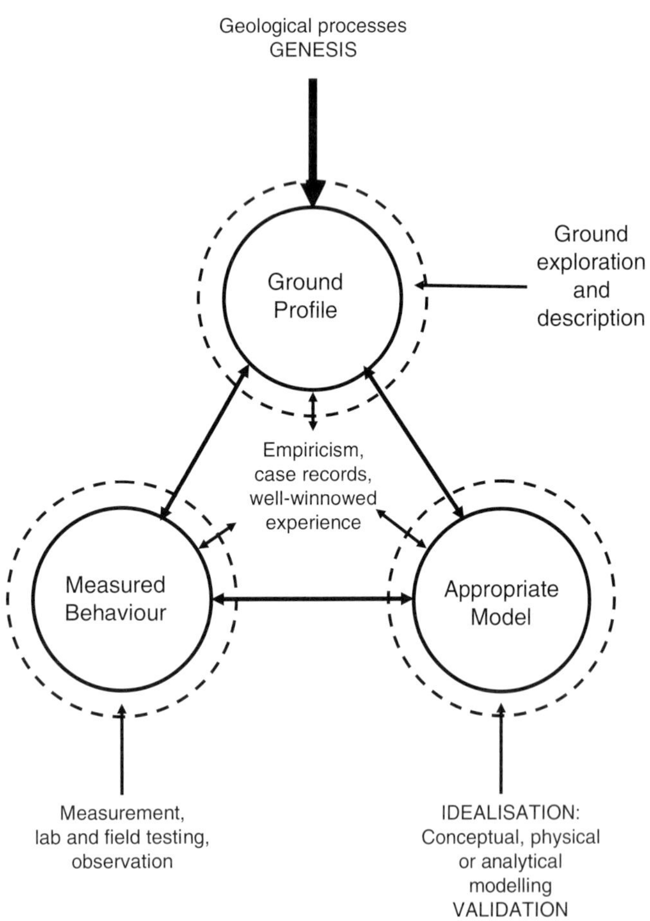

Figure 5.3 The geotechnical triangle: each activity has its own distinct methodology and its own rigour

Behaviour of the ground: This is established from laboratory or in situ tests and usually requires very careful and experienced interpretation. The properties may also be inferred from field measurements of behaviour, such as the back analysis of a landslip or of settlement observations. Section 2 *Fundamental principles* of this manual provides a practical introduction to the fundamentals of soil and rock behaviour.

Appropriate modelling: This can be carried out at all sorts of levels. It may be purely conceptual and instinctive, it might be physical modelling or it can involve extremely sophisticated mathematical or numerical work. At whatever level, modelling first involves identifying what is to be modelled, and then follows a process of idealisation (to analyse is first to idealise), after which the analysis itself is undertaken. This should then be followed up by review and assessment – it often isn't! Frequently in soil mechanics, as in structural engineering, it is understanding the basic mechanisms of behaviour that is the key rather than the fine quantitative detail. It may be of interest to note that, whereas the success of much structural modelling relies on the *lower bound theorem of plasticity*, geotechnical modelling has traditionally made use of both the *lower* and *upper bound theorems of plasticity* for the study of limiting equilibrium using stress characteristics and slip-line fields. Thus, both disciplines are firmly grounded in plasticity. For both disciplines, if the material or structure is brittle in its response, extreme caution is required, as the lower and upper bound theorems no longer strictly apply.

Empirical procedures: With a material as complex as the ground, empiricism is inevitable and it is (and will always remain) an essential aspect of ground engineering. Many of our design and construction procedures are the product of what the author has termed *well-winnowed experience*.

Each activity in the geotechnical triangle has its own distinct methodology and rigour. Geotechnical modelling requires that each aspect be considered and that the geotechnical triangle remains 'in balance'. See Chapter 4 *The geotechnical triangle* for a more detailed discussion of the geotechnical triangle.

5.4. Comparisons between structural and geotechnical modelling

It is immediately obvious from the above that the processes of geotechnical modelling usually involve much greater explicit uncertainties and complexities in idealising both the geometry and the material properties than in structural modelling, where both of these are usually specified by the engineer.

In introducing the subject of soil–structure interaction to students, the author starts by asking them to imagine the utopian situation of having unlimited computational power in which, given the *geometry* of the problem, the *material properties* and the *loading*, any problem can be analysed. With this unlimited computational power, how much better would the results be? As soon as the idealisations that are involved in most ground–structure interaction problems and the uncertainties of the boundary conditions are examined, it becomes obvious that the results will not be much better than they are today. Any design has to make allowance for these uncertainties.

5.4.1 The geotechnical triangle revisited

Hopefully, the above discussion will serve to help structural engineers to understand why the process of modelling geotechnical problems is inherently less certain than it appears to be for most structural problems. This is because much of the time the structural engineer is working with materials that are specified and manufactured under strict control. Usually the structural form can be idealised in a reasonably straightforward manner. The major uncertainties lie in the loading, in inevitable imperfections and in the connections. In geotechnical engineering both the geometry (ground profile) and the properties (ground behaviour) of the 'structure' are laid down by nature and not specified. Precise analysis is usually not possible and the key requirement is to understand the dominant mechanisms of behaviour and their likely bounds.

Perhaps the differences between routine structural and geotechnical modelling can best be illustrated by comparing the approach of the structural engineer working on an existing building to that of the geotechnical engineer. In this situation the structural engineer is no longer able to specify the material and the structural form is often difficult to idealise (Burland, 2000). **Figure 5.4** shows an isometric of the West Tower of Ely Cathedral, which was strengthened during 1973 and 1974 as described by Heyman (1976). Also shown in **Figure 5.4** is the geotechnical triangle of **Figure 5.3** but with some descriptions changed to show the key activities undertaken by a structural engineer. For the *soil profile* at the top of the triangle, the *structure of the building and its materials* has been inserted. To establish these requires the most careful examination and investigation. As with the ground, small discontinuities and weaknesses can play a major role in determining the overall response. It is also vital to establish the way the building was constructed and the changes that have taken place historically – this might be called the *genesis of the building* and it is analogous to the geological processes that have formed the ground profile. At the bottom left of the triangle are the *properties* of the materials and the *observed behaviour* of the building. This aspect requires observation, field measurement, sampling and testing. At the bottom right of the triangle there is the need to develop *appropriate predictive models* that take account of the form and structure of the building, its history, its material properties and known behaviour – an almost identical requirement to that for the ground. There is a whole spectrum of models that can be developed, ranging from the intuitive and conceptual right through to highly sophisticated numerical models. The key is to appreciate the inevitable idealisations that have to be made and the limitations that they impose. Finally, as in ground engineering, well-winnowed experience is of supreme importance and well-documented case records are invaluable. It is of interest to note that, in developing his structural understanding of the behaviour of the West Tower of Ely Cathedral, Heyman (1976) drew extensively on the *safe design theorem* and did not attempt to model the 'actual' stress distributions within the structure.

It is evident from the foregoing that, even if engineers were in possession of unlimited analytical power, the uncertainties in both the ground and the structure are so great that precision in the prediction of behaviour would be unlikely to improve significantly. As in so many fields of engineering, modelling is only one of many tools required in designing for the soil–structure interaction. In most circumstances the real value of modelling will be in assisting the engineer to place bounds on the likely overall behaviour, in understanding the mechanisms of behaviour and in beneficially modifying that behaviour if necessary.

5.5. Ground–structure interaction

Both structural engineering and geotechnical engineering adopt a variety of procedures for modelling a wide range of problems. Section 2 *Fundamental principles* of this manual describes the basic approaches to a variety of geotechnical problems, such as bearing capacity, settlement prediction and slope stability. The book by Hambly (1994), mentioned previously, summarises the range of problems and types of analysis that a structural engineer encounters in day-to-day practice. Difficulties emerge when it is necessary to combine structural and geotechnical modelling.

Structural engineers are tempted to model the ground as a series of discrete springs since this allows conventional structural analysis software to be used. Sometimes sophisticated springs are used incorporating bilinear or nonlinear load–deflection characteristics to simulate the onset of plastic behaviour. Under certain circumstances it is reasonable to adopt springs to model the interaction with the ground (e.g. in some retaining wall problems) but in the majority of situations their use is inappropriate and leads to very misleading results.

The simple case of a uniformly loaded flexible slab resting on clay can be used to illustrate the inappropriateness of adopting springs to model the ground – see **Figure 5.5(a)**. It is well known that for the majority of types of ground the slab will settle more in the middle than at the edges and settlement will also take place outside the loaded area, as shown in **Figure 5.5(b)**. Methods of calculating these settlements are described in Chapter 19 *Settlement and stress distributions*.

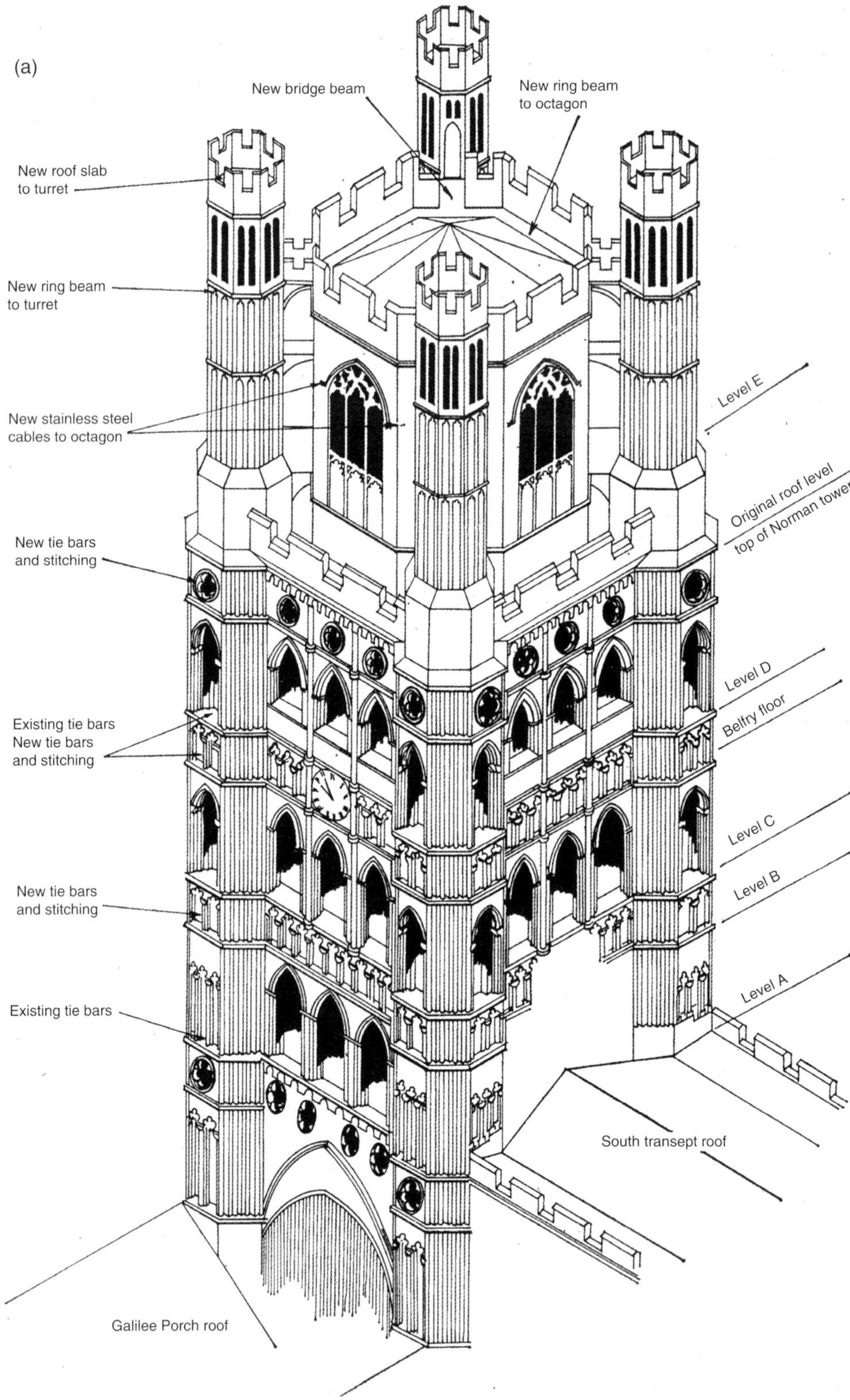

Figure 5.4 (a) The strengthening of the West Tower of Ely Cathedral and associated structural activities. (b) The geotechnical triangle changed to show the key activities undertaken by a structural engineer (Derived from Heyman, 1976) (continued on next page)

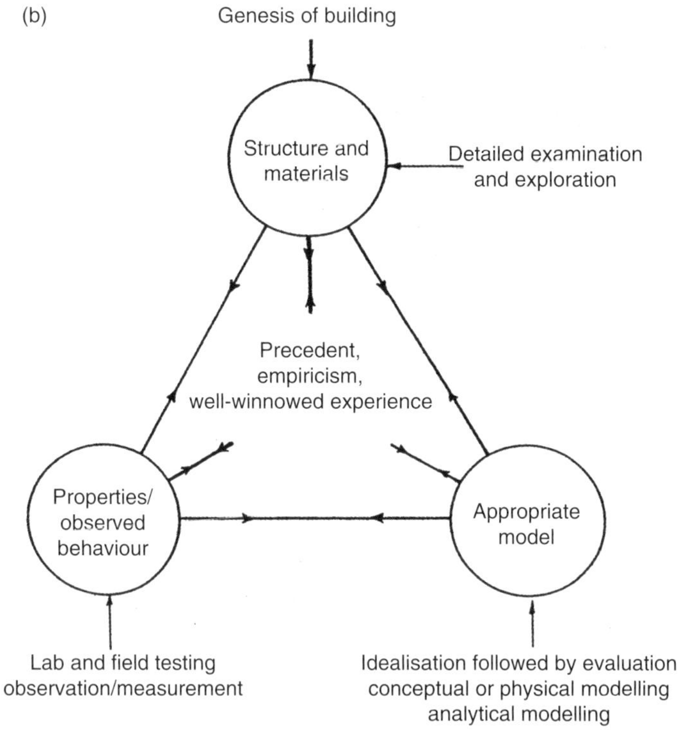

Figure 5.4 Continued

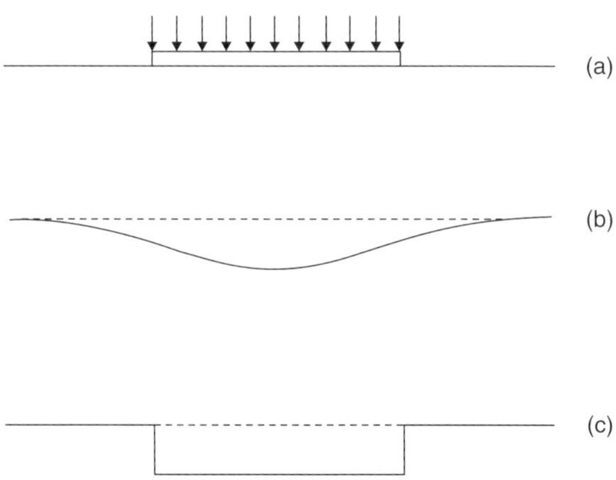

Figure 5.5 (a) Uniformly loaded flexible slab. (b) Settlement profile for slab founded on a clay soil. (c) Settlement profile for soil modelled as springs

If the soil is represented by a series of springs, then, as shown in **Figure 5.5(c)**, it is evident that the slab will settle uniformly and there will be no settlement outside the loaded area. Moreover, there is no simple standard way of calculating the appropriate spring stiffness from the measured compressibility or stiffness of the ground. Thus, the use of springs to represent the ground in this case will not be capable of capturing the differential settlements across a building founded on the slab or of simulating possible interaction effects with adjacent structures or services.

In cases where the ground–structure interaction is likely to be important it is essential that the structural engineer and geotechnical engineer interact so that an appropriate approach to modelling can be developed. It is hoped that this chapter will help each to understand the other so as to lead to positive and constructive interactions.

5.6. Conclusions

This chapter explores the differences in the philosophy of modelling and analysis between structural and geotechnical engineers. The idealisations that are involved in both disciplines are compared. It has been demonstrated that the approach that has to be adopted by a structural engineer working on an existing historic building is very similar to that of a geotechnical engineer working with the ground.

Superficially it might seem that the idealisations adopted by a structural engineer in routine practice are more certain than those that have to be adopted by a geotechnical engineer. However, on closer examination it has to be concluded that the calculation of the 'actual' state of a structure or building under a known set of loads is very uncertain.

The success of structural design calculations owes much to the inherent ductility of the structural elements, so that the *safe design theorem* applies. Geotechnical engineers, too, owe much to the plastic ductile behaviour of their materials and foundations for the success of their designs. In both cases it is vital to identify brittle behaviour as this invalidates the *safe design theorem* and can lead to progressive collapse.

Structural engineers tend to work and think in terms of forces and stresses. Geotechnical engineers are much more used to working with strain and deformation. This difference becomes particularly apparent for situations in which elements of the building or ground approach their maximum resistance. Structural engineers brought up on concepts of limiting stress find it difficult to accept behaviour that implies the full mobilisation of resistance (which is not the same as 'failure'). Reference to Hambly's three- and four-legged stool (*Hambly's paradox*) greatly aids in the understanding of the above ideas.

Of overriding importance is the need to gain a clear understanding of the mechanisms of behaviour of a soil–structure interaction system. If the analytical, computer or conceptual model does not capture the key mechanism, no amount of sophisticated computation will help solve the problem.

In summary, understanding and designing for the ground–structure interaction requires all the traditional skills of the engineer: reliance on observation and measurement; a deep understanding of materials, both ground and structural; the development of appropriate physical and analytical models to reveal the underlying mechanisms of behaviour; and *well-winnowed experience* based on a discerning knowledge of precedents and case histories.

References

Baker JF, Horne MR and Heyman J (1956) *The Steel Skeleton,* vols. 1 and 2. Cambridge University Press, Cambridge, UK.

Beeby AW (1997) Ductility in reinforced concrete: why is it needed and how is it achieved? *The Structural Engineer* **75(18)**: 311–318.

Beeby AW (1999) Safety of structures, and a new approach to robustness. *The Structural Engineer* **77(4)**: 16–19.

Burdekin FM (1999) Size matters for structural engineers. *The Structural Engineer* **77(21)**: 23–29.

Burland JB (1987) The teaching of soil mechanics: a personal view. *Proceedings of the 9th European Conference on Soil Mechanics and Foundation Engineering, Dublin, Ireland*, vol. 3, pp. 1427–1447.

Burland JB (2000) Ground–structure interaction: does the answer lie in the soil? *The Structural Engineer* **78(23/24)**: 42–49.

Burland JB (2006) Interaction between structural and geotechnical engineers. *The Structural Engineer* **84(8)**: 29–37.

Burland JB, Standing JR and Jardine FM (eds) (2001) *Building Response to Tunnelling,* vols 1 and 2. CIRIA and Thomas Telford, London, UK.

Chapman JC (2001) *Learning from Construction Failures* (Campbell P (ed)). Whittles, Caithness, UK, pp. 71–101.

Hambly EC (1985) Oil rigs dance to Newton's tune. *Proceedings of the Royal Institution* **57**: 79–104.

Hambly EC (1994) *Structural Analysis by Example.* Archimedes, Berkhamsted, UK.

Heyman J (1976) The strengthening of the West Tower of Ely Cathedral. *Proceedings of the Institution of Civil Engineers* **60**: 123–147.

Heyman J (1996) Hambly's paradox: why design calculations do not reflect real behaviour. *Proceedings of the Institution of Civil Engineers, Civil Engineering* **114**: 161–166.

Heyman J (2005) Theoretical analysis and real-world design. *The Structural Engineer* **83(8)**: 14–17.

IStructE (Institution of Structural Engineers) (2002) *Report: The Use of Computers for Engineering Calculations.* IStructE, London, UK.

MacLeod IA (2005) *Modern Structural Analysis: Modelling Process and Guidance.* Thomas Telford, London, UK.

Mainstone RJ (1960) Tests on the new government offices, Whitehall Gardens. Studies in Composite Construction, Part III. National Building Studies Research Paper 28. BRS, HMSO, London, UK.

Mann A (2005) Contribution to Verulam on 'Are structures being repaired unnecessarily?'. *The Structural Engineer* **83(20)**: 20.

Walley F (2001) Introduction. In *Learning from Construction Failures* (Campbell P (ed)). Whittles, Caithness, UK.

Further reading

IStructE (1989) *Report: Soil–Structure Interaction. The Real Behaviour of Structures.* IStructE, London, UK.

All chapters within Sections 1 *Context* and 2 *Fundamental principles* together provide a complete introduction to the Manual and no individual chapter should be read in isolation from the rest.

doi: 10.1680/icemge.66816.0037

Chapter 6
Computer analysis principles in geotechnical engineering

David Potts Imperial College London, London, UK
Lidija Zdravkovic Imperial College London, London, UK

This chapter categorises various forms of computational analysis in geotechnical engineering and explains both positive and negative aspects of their performance. The conventional methods of analysis are considered first, followed by some of the main aspects of advanced numerical analysis.

CONTENTS

6.1. General

Geotechnical analysis is part of the geotechnical design process, performed with the aim of assessing (*a*) the stability and serviceability of the new construction and (*b*) the effects that the new construction imposes on existing adjacent structures and services. With respect to this, there are several forms of stability to be checked by geotechnical calculations. First, it must be checked that the structure and its support system are stable as a whole, with no danger of promoting either local failure of the structure itself, or global failure involving the surrounding soil domain. The loads on any structural elements must be calculated so that these can be designed to carry them safely. Finally, movements of both the structure and the ground must be evaluated, in particular if there are structures and services nearby. It may also be necessary to calculate any structural forces induced in the existing structures and services, to assess whether they can be safely sustained.

As part of the design process, it is necessary for an engineer to perform calculations to obtain estimates of the above quantities. Analysis provides the mathematical framework for such calculations. However, the usefulness of an analysis depends on its input. Only a simulation involving realistic material properties and loading conditions enables a proper understanding of the problem.

6.2. Theoretical classification of analysis methods

For the analysis of a geotechnical problem to be theoretically exact, it has to satisfy the following four requirements: equilibrium, compatibility, constitutive behaviour and boundary conditions.

6.2.1 Equilibrium

Equilibrium under static conditions implies that the applied external loads to the soil domain are fully counteracted by the internal stresses generated within the soil domain. External loads are any concentrated forces and distributed loads on the boundaries of the soil domain, as well as any body forces such as the soil's self-weight. In general three-dimensional (3D) stress space the internal stresses have six components, three direct and three shear stresses, in the three coordinate directions. This theoretical requirement can be written in terms of the following equilibrium equations, for each of the coordinate directions:

$$\begin{aligned}\sum F_x &= \frac{\partial \sigma_x}{\partial x} + \frac{\partial \tau_{yx}}{\partial y} + \frac{\partial \tau_{zx}}{\partial z} + \gamma = 0 \\ \sum F_y &= \frac{\partial \tau_{xy}}{\partial x} + \frac{\partial \sigma_y}{\partial y} + \frac{\partial \tau_{zy}}{\partial z} = 0 \\ \sum F_z &= \frac{\partial \tau_{xz}}{\partial x} + \frac{\partial \tau_{yz}}{\partial y} + \frac{\partial \sigma_z}{\partial z} = 0\end{aligned} \qquad (6.1)$$

where *x*, *y* and *z* are the coordinate directions; σ_x, σ_y and σ_z are the direct stresses in the three coordinate directions; and τ_{xy}, τ_{xz} and τ_{yz} are the three shear stresses. The only external load in Equation 6.1 is the self-weight, γ, which is assumed to act in the *x*-direction.

6.2.2 Compatibility

Compatibility is a requirement related to the soil's deformations. If the applied loads do not result in rupture or overlapping of the soil, such deformations are considered to be compatible. Otherwise, they are non-compatible. Compatible deformations can be ensured by defining consistent strains. These strains are calculated from the displacements of the soil element, which have three components, *u*, *v* and *w*, in

Method of analysis		Theoretical requirements				
					Boundary conditions	
		Equilibrium	Compatibility	Constitutive behaviour	Force	Displacement
Closed form		✓	✓	Linear elastic	✓	✓
Limit equilibrium		✓	✗	Rigid with a failure criterion	✓	✓
Stress field		✓	✗	Rigid with a failure criterion	✓	✗
Limit analysis	LB	✓	✗	Ideal plasticity	✓	✗
	UB	✗	✓		✗	✓
Beam-spring approach		✓	✓	Soil modelled by springs	✓	✓
Full numerical analysis		✓	✓	Any	✓	✓

Table 6.1 Theoretical requirements satisfied by various methods of analysis

the three coordinate directions, x, y and z, respectively. They are:

$$\varepsilon_x = \frac{\partial u}{\partial x}\,;\ \varepsilon_y = \frac{\partial v}{\partial y}\,;\ \varepsilon_z = \frac{\partial w}{\partial z} \tag{6.2}$$

$$\gamma_{xy} = \frac{\partial v}{\partial x} + \frac{\partial u}{\partial y}\,;\ \gamma_{yz} = \frac{\partial v}{\partial z} + \frac{\partial w}{\partial y}\,;\ \gamma_{xz} = \frac{\partial w}{\partial x} + \frac{\partial u}{\partial z} \tag{6.3}$$

Similar to stresses, there are three direct strains, ε_x, ε_y and ε_z, and three shear strains, γ_{xy}, γ_{yz} and γ_{xz}.

6.2.3 Constitutive behaviour

For analysis purposes the behaviour of the soil, in terms of its load–deformation characteristics, is expressed through a set of constitutive equations. These provide a link between stresses and strains, and hence a link between equilibrium and compatibility. Depending on the stress–strain relationship, a soil can be characterised as ductile or brittle, or strain hardening or strain softening.

For calculation purposes the constitutive behaviour can be expressed in the mathematical form

$$\begin{Bmatrix} \sigma_x \\ \sigma_y \\ \sigma_z \\ \tau_{xy} \\ \tau_{yz} \\ \tau_{xz} \end{Bmatrix} = \begin{bmatrix} D_{11} & D_{12} & D_{13} & D_{14} & D_{15} & D_{16} \\ D_{21} & D_{22} & D_{23} & D_{24} & D_{25} & D_{26} \\ D_{31} & D_{32} & D_{33} & D_{34} & D_{35} & D_{36} \\ D_{41} & D_{42} & D_{43} & D_{44} & D_{45} & D_{46} \\ D_{51} & D_{52} & D_{53} & D_{54} & D_{55} & D_{56} \\ D_{61} & D_{62} & D_{63} & D_{64} & D_{65} & D_{66} \end{bmatrix} \cdot \begin{Bmatrix} \varepsilon_x \\ \varepsilon_y \\ \varepsilon_z \\ \gamma_{xy} \\ \gamma_{yz} \\ \gamma_{xz} \end{Bmatrix} \tag{6.4}$$

or

$$\{\sigma\} = [D] \times \{\varepsilon\} \tag{6.5}$$

where D_{ij} ($i = 1,6; j = 1,6$) are the components of the constitutive matrix $[D]$ and $\{\sigma\}$ and $\{\varepsilon\}$ are vectors of stress and strain components, respectively. The elements D_{ij} of the constitutive matrix represent soil properties (e.g. Young's modulus, Poisson ratio, angle of shearing resistance) and depend on the complexity of soil behaviour and the constitutive model applied to simulate it.

6.2.4 Boundary conditions

Boundary conditions are any restrictions applied on the boundaries of the soil domain, which have to be satisfied in order that equilibrium, compatibility and constitutive behaviour are not violated. They can be applied in terms of forces and displacements.

6.2.5 Categories of analysis methods

As stated at the beginning of this section, all four theoretical requirements have to be satisfied for the analysis solution to be exact. It is therefore useful to review the broad categories of analysis currently in use against these theoretical requirements.

Current methods of analysis can be grouped into the following categories: closed form, simple (or classical) and numerical analysis. The ability of each method to satisfy the fundamental theoretical requirements and provide design information is summarised in **Tables 6.1** and **6.2**.

Method of analysis		Design requirements		
		Stability	Movements	Adjacent structures
Closed form		✗	✓	✗
Limit equilibrium		✓	✗	✗
Stress field		✓	✗	✗
Limit analysis	Lb	✓	✗	✗
	Ub	✗	Crude estimate	✗
Beam-spring approach		✓	✗	✗
Full numerical analysis		✓	✓	✓

Table 6.2 Design requirements satisfied by various methods of analysis

6.3. Closed form solutions

A closed form solution is in many ways the ultimate method of analysis. If it is possible for a particular geotechnical structure to establish a realistic constitutive model for material behaviour and appropriate boundary conditions, and to combine these with equations of equilibrium and compatibility, then an exact solution can be obtained which satisfies all the above theoretical requirements.

However, as soil is a highly complex multiphase material which behaves nonlinearly when loaded, a closed form solution to realistic geotechnical problems is normally not possible. Closed form solutions are only possible for two very simple situations.

(*a*) Where the soil is an isotropic linear elastic material. The solution provides an estimate of movements and structural forces, but cannot be used for assessing stability. However, comparison with observed behaviour of geotechnical structures indicates that even the predictions of movement are often not realistic.

(*b*) Elasto-plastic problems where the geometry of the problem contains sufficient geometric symmetries such that it can be reduced to being essentially one-dimensional. Examples of this are solutions for the expansion of spherical and infinitely long cylindrical cavities in an infinite elasto-plastic continuum.

6.4. Classical methods of analysis

Classical methods of analysis enable more realistic solutions to geotechnical problems; however, they also involve approximations. Whereas mathematics is still used to obtain a solution, one of the theoretical requirements may not be satisfied. Therefore, the solution is not exact in a theoretical sense. This approach was adopted by the pioneers of geotechnical engineering, when computer power was non-existent and it was therefore impossible to obtain a complete solution utilising only hand calculations.

This group of analysis methods consists of the limit equilibrium, stress fields and limit analysis methods. They consider only the ultimate limit states, when the soil is at failure, but differ in the manner in which they arrive at a solution.

6.4.1 Limit equilibrium

This method of analysis involves the following steps.

- An arbitrary failure surface is adopted for the problem – that is, its position and shape, which can be either planar or curved, or a combination of both.
- The failure criterion is assumed to hold everywhere along the failure surface.
- Only the global equilibrium of the rigid blocks of soil between the failure surfaces and the boundaries of the problem is considered; external forces acting on blocks include the self-weight; internal forces (stresses) are generated only along the failure surfaces.
- The internal stress distribution within the blocks of soil is not considered (i.e. the blocks are assumed to be rigid).

The Coulomb's wedge analysis for retaining walls and the method of slices for slopes are examples of limit equilibrium calculations. These calculations do not consider the compatibility requirement and it is therefore not possible to obtain any information about the possible movements of the soil.

6.4.2 Stress field solution

As can be seen in **Table 6.1**, the stress field method satisfies the same theoretical requirements as the limit equilibrium method of analysis and adopts the same assumption for soil behaviour (i.e. rigid with a failure criterion). However, there is an important difference in the assumptions of this method: while the limit equilibrium method assumes failure in the soil only along the failure surface, the stress field method assumes failure everywhere in the soil domain. Therefore, the equilibrium equations in the stress field method have to be written for an infinitesimal soil element, in the form of partial differential equations. This complicates the solution process as it leads, when combined with the equations of the failure criterion, to a set of hyperbolic partial differential equations. These equations can only be solved analytically if the soil is assumed to be weightless. While a weightless soil is clearly unrealistic, there are some scenarios where the soil's weight does not affect the solution and consequently where analytical solutions are applicable. For example, appropriate analytical solutions are available for some boundary value problems where the soil behaves in an undrained manner with a Tresca (S_u) failure criterion (i.e. bearing capacity of surface footings). For boundary value problems where weight is important and/or the soil behaves in a drained manner with a Mohr–Coulomb failure criterion, then the governing partial differential equations must be solved using an approximate finite difference approach. Again, because compatibility is not considered, the method does not provide any information about ground movements and considers only the ultimate limit states (i.e. failure).

Examples of stress field solutions are the Rankine active and passive stress fields for earth pressure calculations and the earth pressure tables of Sokolovski (1960, 1965). This method is also used for solving bearing capacity problems. The Prandtl (1920) and Hill (1950) stress fields for a rigid strip footing are examples. Computer programs are also available for performing stress field calculations for problems with complex geometries (Martin, 2005).

6.4.3 Limit analysis

Limit analysis calculations provide bounds to the exact solution. The upper bound (UB) approach results in an unsafe estimate of the exact solution and arrives to the solution by ignoring equilibrium. The lower bound (LB) approach provides a safe estimate of the exact solution by ignoring the compatibility requirement. If the two solutions for the same problem are the same, then an exact solution to the problem is obtained.

As opposed to the limit equilibrium and stress field methods, the soil behaviour is assumed to be ideal elasto-plastic instead of rigid.

6.4.3.1 Safe theorem

> If a statically admissible stress field covering the whole soil mass can be found, which nowhere violates the yield condition, then the loads in equilibrium with the stress field are on the safe side or equal to the true collapse loads.

This theorem is often referred to as the lower bound theorem. A statically admissible stress field consists of an equilibrium distribution of stress which balances the applied loads and body forces. As compatibility is not considered, an infinite number of stress fields can be postulated. The accuracy of the solution depends on how close the assumed stress field is to the real one for a given problem.

6.4.3.2 Unsafe theorem

> An unsafe solution to the true collapse loads can be found by selecting any kinematically possible failure mechanism and performing a work rate calculation. The loads so determined are either on the unsafe side or equal to the true collapse loads.

This theorem is also known as the upper bound theorem. A kinematically possible failure mechanism is one that satisfies the compatibility requirements in the shear zones between blocks of soil in the assumed mechanism. Equilibrium of the work done by external and internal forces is then considered, instead of the equilibrium of the external and internal forces. Because the latter is not considered, an infinite number of failure mechanisms can be assumed. The accuracy of the solution depends on how close the assumed mechanism is to the real one in a given problem.

In practice it is often easier to postulate a realistic failure mechanism than an admissible stress field, and consequently more use has historically been made of the unsafe theorem. It is rare to be able to obtain the same solution from both safe and unsafe calculations, and when two such solutions are obtained they usually provide differing answers which bracket the true solution, the true solution being that applicable to the ideal soil behaviour assumed to develop the limit theorems. To the authors' knowledge there are only two boundary value problems in which both safe and unsafe calculations have yielded the same result. These are the short-term lateral load that can be sustained by an infinitely long cylindrical pile in an infinite clay deposit and the bearing capacity of surface strip footings.

Recently considerable research effort has been expended in combining limit analysis with finite element techniques to develop an analysis tool that can deal with complex boundary value problems involving mixed soils (Sloan, 1988, 1989). Such a tool looks promising and should provide an alternative method for analysing ultimate limit states.

6.5. Numerical analysis

6.5.1 Beam-spring approach

This approach is used to investigate soil–structure interactions. For example, it can be used to study the behaviour of axially and laterally loaded piles, raft foundations, embedded retaining walls and tunnel linings. The major approximation is the assumed soil behaviour and two approaches are commonly used. The soil behaviour is either approximated by a set of unconnected vertical and horizontal springs (Borin, 1989) or by a set of linear elastic interaction factors (Pappin *et al.*, 1985). Only a single structure can be accommodated in the analysis. Consequently, only a single pile or retaining wall can be analysed. Further approximations must be introduced if more than one pile, retaining wall or foundation interact. Any structural support, such as props or anchors (retaining wall problems), is represented by simple springs (**Figure 6.1**).

To enable limiting pressures to be obtained, for example on each side of a retaining wall, 'cutoffs' are usually applied to the spring forces and interaction factors representing soil behaviour. These cutoff pressures are usually obtained from one of the simple analysis procedures discussed above (e.g. limit equilibrium, stress fields or limit analysis). It is important to appreciate that these limiting pressures are not a direct result of the beam-spring calculation, but are obtained from separate approximate solutions and then imposed on the beam-spring calculation process.

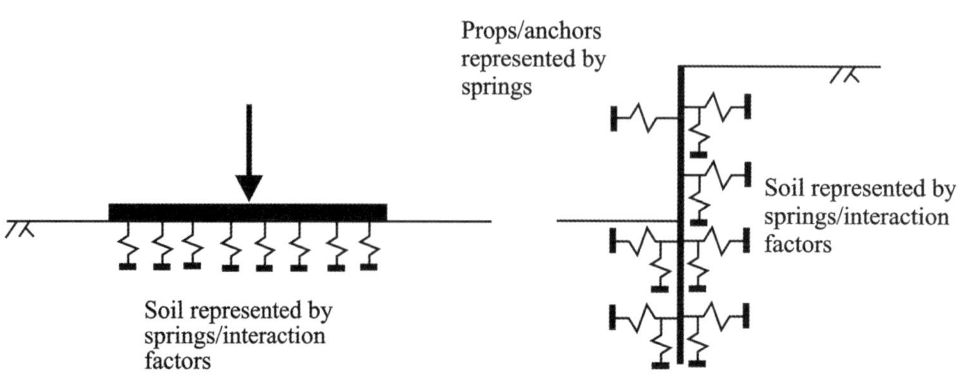

Figure 6.1 Examples of the beam-spring approach

Having reduced the boundary value problem to studying the behaviour of a single isolated structure (e.g. a pile, a footing or a retaining wall) and made gross assumptions about soil behaviour, a complete theoretical solution to the problem is sought. Due to the complexities involved, this is usually achieved using a computer. The structural member (e.g. pile, footing or retaining wall) is represented using either finite differences or finite elements and a solution that satisfies all the fundamental solution requirements is obtained by iteration.

Sometimes computer programs which perform such calculations are identified as finite difference or finite element programs. However, it must be noted that it is only the structural member that is represented in this manner and these programs should not be confused with those that involve full discretisation of both the soil and structural members by finite differences or finite elements (see Section 6.5.2).

As solutions obtained in this way include limits to the earth pressures that can develop adjacent to the structure, they can provide information on local stability. This is often indicated by a failure of the program to converge. However, numerical instability may occur for other reasons and therefore a false impression of instability may be given. Solutions from these calculations include forces and movements of the structure. They do not provide information about global stability or movements in the adjacent soil. They do not consider adjacent structures.

It is difficult to select appropriate spring stiffnesses and to simulate some support features. For example, when analysing a retaining wall it is difficult to account realistically for the effects of soil berms. Retaining wall programs using interaction factors to represent the soil can have problems in dealing with wall friction and often neglect shear stresses on the wall, or make further assumptions to deal with them. For the analysis of retaining walls a single wall is considered in isolation and structural supports are represented by simple springs fixed at one end (grounded). It is therefore difficult to account for realistic interaction between structural components such as floor slabs and other retaining walls. This is particularly so if 'pin-jointed' or 'full moment' connections are appropriate. As only the soil acting on the wall is considered in the analysis, it is difficult to model realistically the behaviour of raking props and ground anchors which rely on resistance from soil remote from the wall.

6.5.2 Full numerical analysis

This category of analysis includes methods which attempt to satisfy all theoretical requirements, include realistic soil constitutive models and incorporate boundary conditions that realistically simulate field conditions. Because of the complexities involved and the nonlinearities in soil behaviour, all methods are numerical in nature. Approaches based on finite difference and finite element methods are those most widely used. These methods essentially involve a computer simulation of the history of the boundary value problem from green field conditions, through construction and in the long term.

Their ability to accurately reflect field conditions essentially depends on (*a*) the ability of the constitutive model to represent real soil behaviour and (*b*) the correctness of the boundary conditions imposed. The user has to define the appropriate geometry, construction procedure, soil parameters and boundary conditions. Structural members may be added and withdrawn during the numerical simulation to model field conditions. Retaining structures composed of several retaining walls, interconnected by structural components, can be considered and, because the soil mass is modelled in the analysis, the complex interaction between raking struts or ground anchors and the soil can be accounted for. The effect of time on the development of pore water pressures can also be simulated by including coupled consolidation. No postulated failure mechanism or mode of behaviour of the problem is required, as these are predicted by the analysis. The analysis allows the complete history of the boundary value problem to be predicted and a single analysis can provide information on all design requirements.

The methods can solve 1D, 2D or 3D problems and suffer none of the limitations discussed previously for the other methods.

It is often claimed that these approaches have limitations. Usually these relate to the fact that detailed soils information or knowledge of the construction procedure is needed. In the authors' opinion, neither of these are limitations. If a numerical analysis is anticipated during the design stages of a project, it is then not difficult to ensure that the appropriate soil information is obtained from the site investigation. It is only if a numerical analysis is an afterthought, once the soil data have been obtained, that this may present difficulties. In this respect it is noted that site investigations are not always performed to an adequate standard to obtain certain parameters, and that not all parameters for advanced constitutive models can be directly obtained from the site investigation.

If the behaviour of the boundary value problem is not sensitive to the construction procedure, then any reasonable assumed procedure is adequate for the analysis. However, if the analysis is sensitive to the construction procedure then, clearly, this is important and it will be necessary to simulate the field conditions as closely as possible. Therefore, far from being a limitation, numerical analysis can indicate to the design engineer where, and by how much, the boundary value problem is likely to be influenced by the construction procedure. This will enable adequate provision to be made within the design.

Full numerical analyses are complex and should be performed by qualified and experienced staff. The operator must understand soil mechanics and, in particular, the constitutive models that the software uses, and be familiar with the software package to be employed for the analysis. Nonlinear numerical analysis is not straightforward and at present there are several algorithms (i.e. solution procedures) available for solving the nonlinear system of governing equations. Some of these are more accurate than others and some are increment-size-dependent. There are approximations within these algorithms and errors associated with discretisation. However, these can be controlled by the experienced user so that accurate predictions can be obtained.

Full numerical analysis can be used to predict the behaviour of complex field situations. It can also be used to investigate the fundamentals of soil–structure interaction and to calibrate some of the simple methods discussed above. However, as noted above, numerical analyses involve approximations and there are many pitfalls that can be encountered. Some of these approximations and pitfalls are considered below. In doing so, the finite element method is considered; however, it should be noted that similar considerations apply to finite difference approaches.

6.6. Overview of the finite element method

The finite element method involves following steps.

(*a*) **Element discretisation:** This is the process of modelling the geometry of the problem under investigation by an assemblage of small regions, termed finite elements. These elements have nodes defined on the element boundaries, or within the element.

(*b*) **Primary variable approximation:** A primary variable must be selected (e.g. displacements, stresses) and rules as to how it should vary over a finite element established. This variation is expressed in terms of nodal values. In geotechnical engineering it is usual to adopt displacements as the primary variable.

(*c*) **Element equations:** Use of an appropriate variational principle (e.g. minimum potential energy) to derive element equations

$$[\boldsymbol{K}_{\mathrm{E}}] \times \{\Delta \boldsymbol{d}_{\mathrm{E}}\} = \{\Delta \boldsymbol{R}_{\mathrm{E}}\} \quad (6.6)$$

where $[\boldsymbol{K}_{\mathrm{E}}]$ is the element stiffness matrix, $\{\Delta \boldsymbol{d}_{\mathrm{E}}\}$ is the vector of incremental element nodal displacements and $\{\Delta \boldsymbol{R}_{\mathrm{E}}\}$ is the vector of incremental element nodal forces. It should be noted that the element stiffness matrix $[\boldsymbol{K}_{\mathrm{E}}]$ is derived from the constitutive matrix $[D]$ discussed in Section 6.2.3.

(*d*) **Global equations:** Combine element equations to form global equations

$$[\boldsymbol{K}_{\mathrm{G}}] \times \{\Delta \boldsymbol{d}_{\mathrm{G}}\} = \{\Delta \boldsymbol{R}_{G}\} \quad (6.7)$$

where $[\boldsymbol{K}_{\mathrm{G}}]$ is the global stiffness matrix, $\{\Delta \boldsymbol{d}_{\mathrm{G}}\}$ is the vector of all incremental nodal displacements and $\{\Delta \boldsymbol{R}_{\mathrm{G}}\}$ is the vector of all incremental nodal forces.

(*e*) **Boundary conditions:** Formulate boundary conditions and modify global equations. Loadings (e.g. line and point loads, pressures, body forces, construction and excavation) affect $\{\Delta \boldsymbol{R}_{\mathrm{G}}\}$, while the prescribed displacements affect $\{\Delta \boldsymbol{d}_{\mathrm{G}}\}$.

(*f*) **Solution of global equations:** The above global equations are in the form of a large number of simultaneous equations. These are solved to obtain the displacements $\{\Delta \boldsymbol{d}_{\mathrm{G}}\}$ at all the nodes. From these nodal displacements secondary quantities, such as stresses and strains, are evaluated.

Several approximations are involved in the above process. Some of these are considered in detail in the following sections.

6.7. Element discretisation

The geometry of the boundary value problem under investigation must be defined and quantified. Simplifications and approximations may be necessary during this process. For example, restraints on computational resources might impose that plane strain conditions have to be assumed. This geometry is then replaced by an equivalent finite element mesh which is composed of small regions called finite elements. For 2D problems, the finite elements are usually triangular or quadrilateral in shape (**Figure 6.2**). Their geometry is specified in terms of the coordinates of key points on the element, called nodes. For elements with straight sides these nodes are usually located at the element corners. If the elements have curved sides then additional nodes, usually at the midpoint of each side, must be introduced. Further nodes may be added to improve numerical accuracy (e.g. 15-noded triangular elements). The set of elements in the complete mesh are connected together by the element sides and a number of nodes.

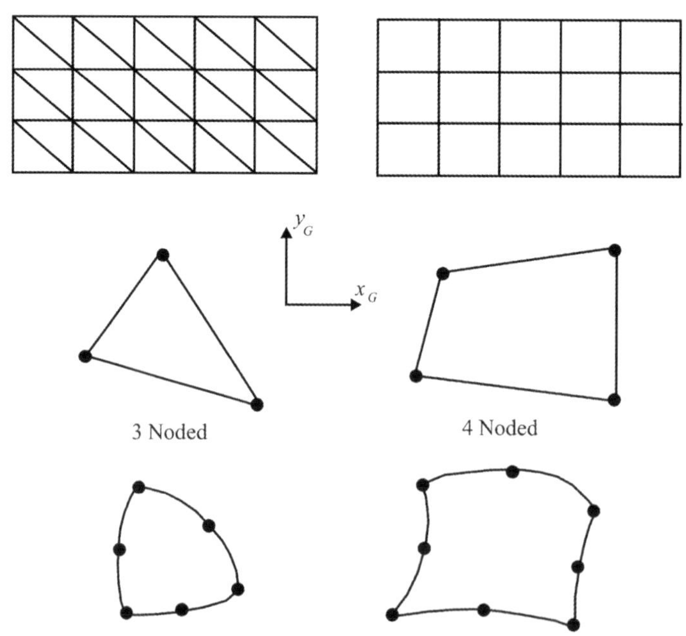

Figure 6.2 Typical 2D finite elements

When constructing the finite element mesh the following should be considered.

- The geometry of the boundary value problem must be approximated as accurately as possible.
- If there are curved boundaries or curved material interfaces, the higher-order elements with mid-side nodes should be used (**Figure 6.3**).
- In many cases geometric discontinuities suggest a natural form of subdivision. For example, discontinuities in boundary gradient, such as re-entrant corners or cracks, can be modelled by placing nodes at the discontinuity points. Interfaces between materials with different properties can be introduced by element sides (**Figure 6.4**).
- Mesh design may also be influenced by the applied boundary conditions. If there are discontinuities in loading, or point loads, these can again be introduced by placing nodes at the discontinuity points (**Figure 6.5**).

In combination with the above factors, the size and the number of elements depend largely on the material behaviour, since this influences the final solution. For linear material behaviour the procedure is relatively straightforward and only the zones where unknowns vary rapidly need special attention. In order to obtain accurate solutions, these zones require a refined mesh of smaller elements and/or the inclusion of higher-order elements. The situation is more complex for general nonlinear material behaviour, since the final solution may depend, for example, on the previous loading history. For such problems the mesh design must take into account the boundary conditions, the material properties and, in some cases, the geometry, which all vary throughout the solution process. In all cases a mesh of regularly shaped elements will give the best results. Elements with widely distorted geometries or long thin elements should be avoided.

To illustrate the error that can arise from the discretisation of the problem geometry into finite elements, the behaviour of smooth rigid strip and circular surface foundations under vertical loading will be considered. The soil is assumed to be linear elastic–perfectly plastic, with a Tresca yield surface, with $E = 10\,000$ kPa, $\mu = 0.45$ and $S_u = 100$ kPa.

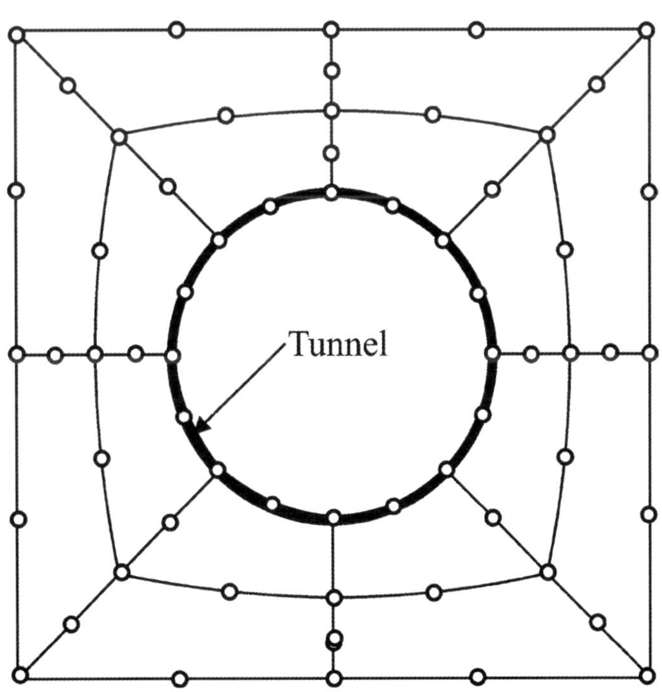

Figure 6.3 Use of higher-order elements

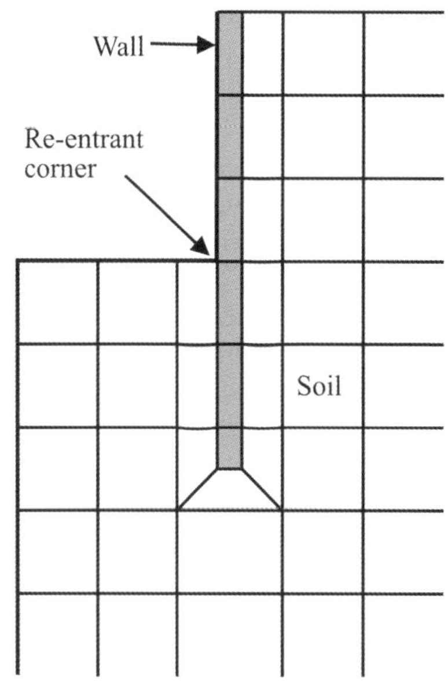

Figure 6.4 Effect of boundary conditions

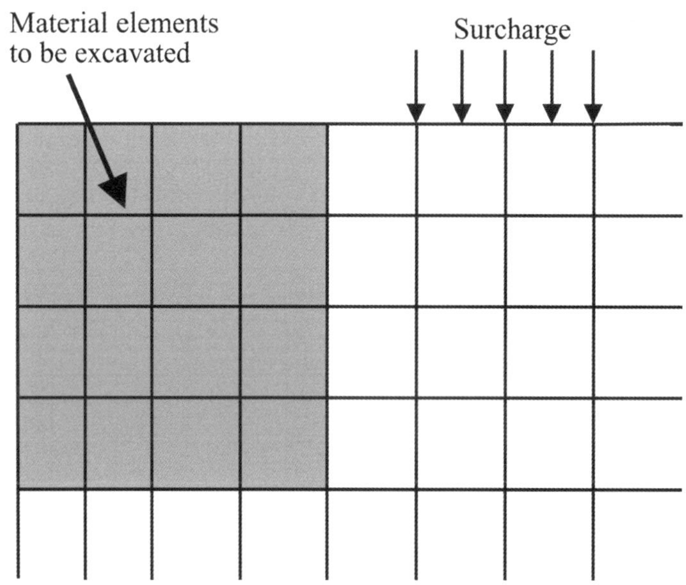

Figure 6.5 Geometric discontinuities

Two alternative finite element meshes are shown in **Figures 6.6** and **6.7**. The mesh shown in **Figure 6.6** has 110 8-noded isoparametric elements, whereas the mesh shown in **Figure 6.7** has only 35 8-noded elements. Both meshes have been used to perform plane strain (i.e. strip footing) and axisymmetric (circular footing) analyses, in which the foundation is displaced vertically downward to obtain the complete load–displacement curve. From a cursory glance at the two meshes, most users would probably conclude that the 110-element mesh was likely to produce the more accurate predictions due to its larger number of elements and their more even spatial distribution.

Although there is no analytical solution for the complete load–displacement curve for either the strip or circular surface footings, there is an analytical solution for the ultimate load for the strip footing which gives $Q_{max} = q_f \times 2B$ per metre length, where $q_f = N_c S_u$, $N_c = (2 + \pi)$ and $2B$ is the width of the footing. Often, textbooks quote an equivalent analytical solution for the circular footing of $Q = q_f \times S_u \times \pi R^2$, where $q_f = N_c S_u$, $N_c = 5.69$ and R is the radius of the footing. However, this solution is not strictly an analytical solution as it involves the numerical integration of some stress field equations. It is, however, thought to be reasonably accurate.

Results from finite element analysis using the meshes shown in **Figures 6.6** and **6.7** are presented in **Figure 6.8**. Also marked on this figure are the theoretical limit loads discussed above. Assuming that the theoretical limit loads are correct, then the analyses with the 110-element mesh overestimate the collapse loads by 3.8% for the strip footing and 8.8% for the circular footing. However, the errors are much smaller, 0.5% for the strip and 2% for the circular footing, for the analyses performed using the 35-element mesh. The reason for the relatively poor performance of the 110-element mesh can be seen from **Figure 6.9**, which shows vectors of incremental nodal displacements for the last increment of the 110-element strip footing analysis. Each vector indicates the magnitude (by its length) and the direction (by its orientation) of the associated incremental nodal displacement. While the absolute magnitude of the incremental displacements is not important, the relative magnitude of each of the vectors and their orientation clearly indicate the mechanism of failure. Attention should, however, be focused on the corner of the footing, where a rapid change in the direction of the incremental vectors is indicated. The direction changes by approximately 120° between the vector immediately below the corner of the footing and the adjacent node on the soil surface.

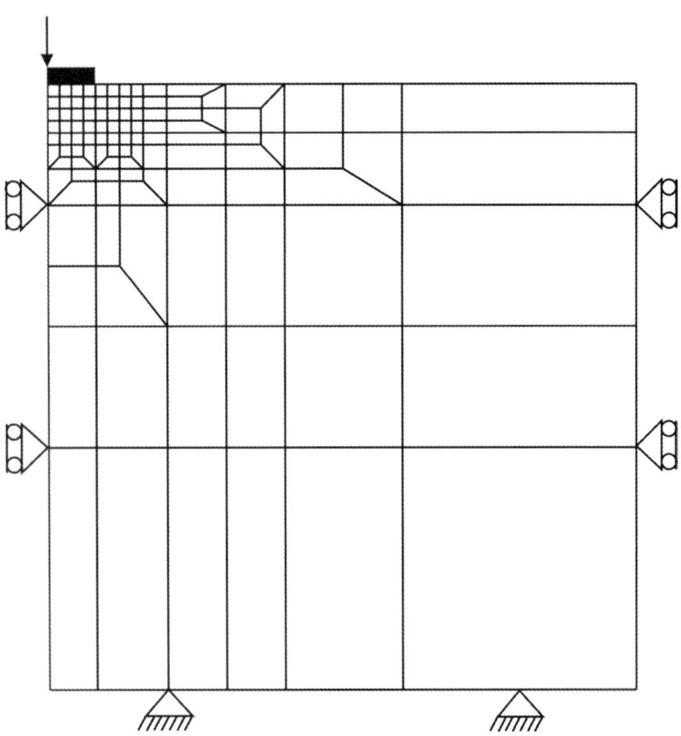

Figure 6.6 Smooth footing: 110-element mesh

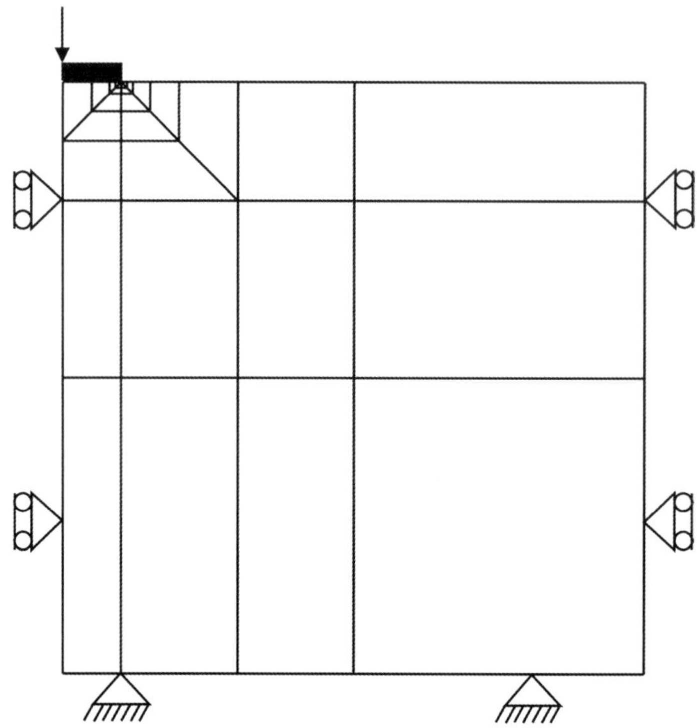

Figure 6.7 Smooth footing: 35-element mesh

Such behaviour clearly indicates that under the corner of the footing there are large stress and strain gradients. In the 110-element mesh the elements in the vicinity of this corner are too large and of insufficient number to accurately reproduce this concentration of behaviour. In comparison, the 35-element mesh, which has smaller elements and a graded mesh under the corner of the footing (**Figure 6.10**) can accommodate this behaviour more easily and therefore produces more accurate solutions.

Clearly it is important that smaller and therefore more elements are placed where there are rapid changes in stresses

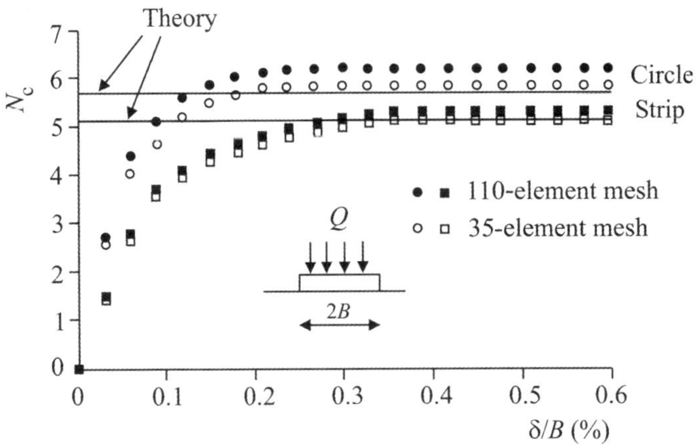

Figure 6.8 Load–displacement curves for smooth footing

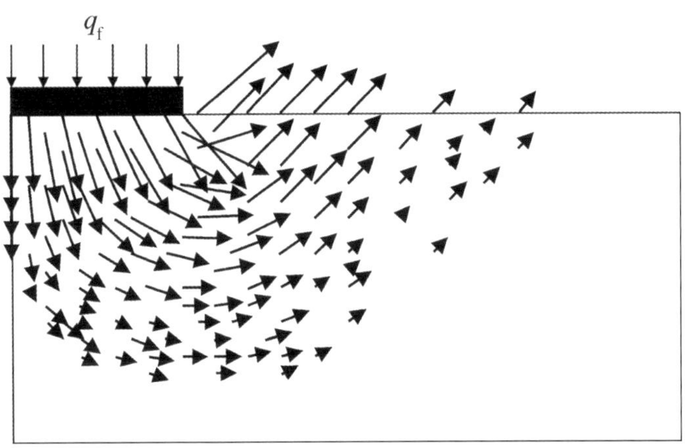

Figure 6.9 Vectors of incremental displacements at failure (110-element mesh)

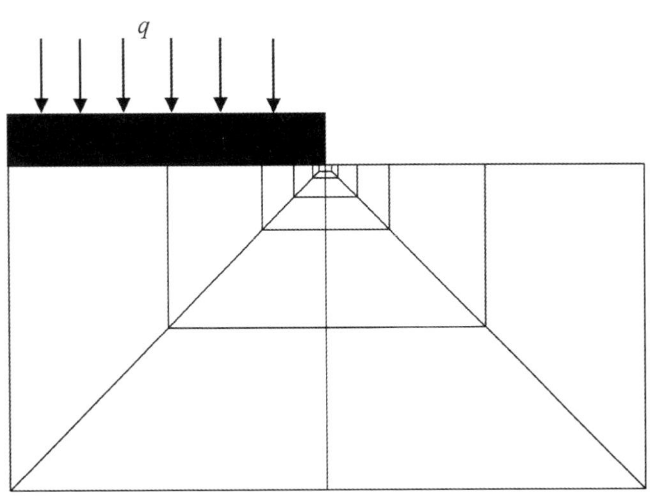

Figure 6.10 Detail from the 35-element mesh

and strains and therefore directions of movement. Some computer programs will automatically regenerate the finite element mesh to improve the accuracy of the solution. They work by first performing a solution with a relatively even mesh. Based on the results of this analysis a new mesh is generated with more elements in the zones of greatest stress and strain changes. The analysis is then repeated. In principle, this process of refinement could be repeated until the results were unaffected by further mesh refinement.

6.8. Nonlinear finite element analysis

If the soil is nonlinear elastic and/or elasto-plastic, the equivalent constitutive matrix is no longer constant, but varies with stress and/or strain. It therefore changes during a finite element analysis. Consequently, a solution strategy is required that can account for this changing material behaviour. This strategy is a key component of a nonlinear finite element analysis, as it can strongly influence the accuracy of the results and the computer resources required to obtain them.

As noted above, four basic solution requirements need to be satisfied: equilibrium, compatibility, constitutive behaviour and boundary conditions. Nonlinearity introduced by the constitutive behaviour causes the governing finite element equations to be reduced to the following incremental form

$$[\boldsymbol{K}_{\mathrm{G}}]^i \times \{\Delta \boldsymbol{d}\}^i_{\mathrm{nG}} = \{\Delta \boldsymbol{R}_G\}^i \tag{6.8}$$

where $[\boldsymbol{K}_{\mathrm{G}}]^i$ is the incremental global system stiffness matrix, $\{\Delta \boldsymbol{d}\}^i_{\mathrm{nG}}$ is the vector of incremental nodal displacements, $\{\Delta \boldsymbol{R}_{\mathrm{G}}\}^i$ is the vector of incremental nodal forces and i is the increment number. To obtain a solution to a boundary value problem, the change in boundary conditions is applied in a series of increments, and the above equation must be solved for each increment. The end solution is obtained by summing the results of each increment. Due to the nonlinear constitutive behaviour, the incremental global stiffness matrix $[\boldsymbol{K}_{\mathrm{G}}]^i$ is dependent on the current stress and strain levels and therefore is not constant, but varies over an increment. Unless a very large number of small increments are used, this variation should be accounted for. Hence, the solution of the above equation is not straightforward and different solution strategies exist. The objective of all such strategies is that the solution satisfies the four basic requirements listed above. Three different categories of solution algorithm, which are most popular in existing computer codes, are briefly considered below. For detailed information the reader is referred to Potts and Zdravkovic (1999).

6.8.1 Tangent stiffness method

The tangent stiffness method, sometimes called the variable stiffness method, is the simplest solution strategy. In this approach, the incremental stiffness matrix $[\boldsymbol{K}_{\mathrm{G}}]^i$ is assumed to be constant over each increment and is calculated using the current stress state at the beginning of each increment. This is equivalent to making a piecewise linear approximation to the nonlinear constitutive behaviour. To illustrate the application of this approach, the simple problem of a uniaxially loaded bar of nonlinear material is considered (**Figure 6.11**). If this bar is loaded, the true load–displacement response is shown in **Figure 6.12**. This might represent the behaviour of a strain-hardening plastic material which has a very small initial elastic domain.

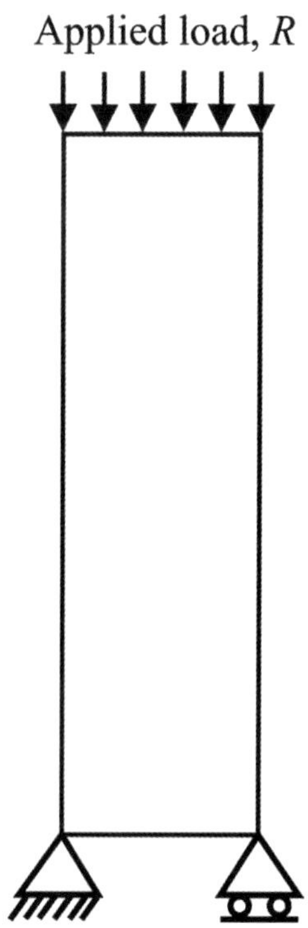

Figure 6.11 Uniaxially loaded bar

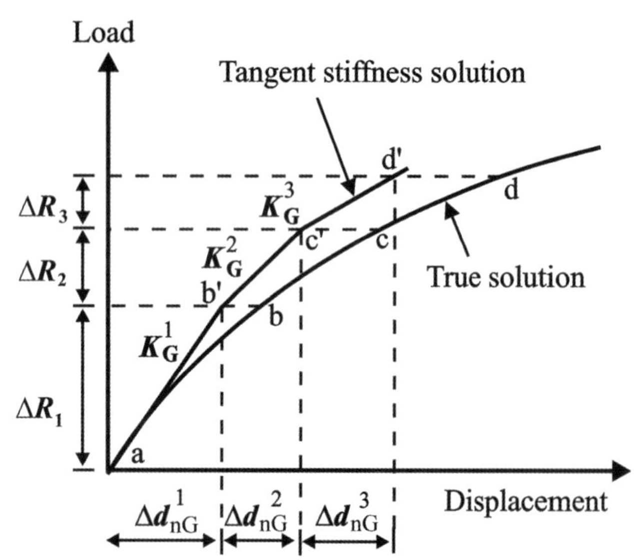

Figure 6.12 Application of the tangent stiffness algorithm

In the tangent stiffness approach the applied load is split into a sequence of increments. In **Figure 6.12** three increments of load are shown, $\Delta \boldsymbol{R}_1$, $\Delta \boldsymbol{R}_2$ and $\Delta \boldsymbol{R}_3$. The analysis starts with the application of $\Delta \boldsymbol{R}_1$. The incremental global stiffness matrix $[\boldsymbol{K}_G]^1$ for this increment is evaluated based on the unstressed state of the bar corresponding to point **a**. For an elasto-plastic material this might be constructed using the elastic constitutive matrix. The first incremental equation is then solved to determine the nodal displacements $\{\Delta \boldsymbol{d}\}^1_{nG}$. As the material stiffness is assumed to remain constant, the load–displacement curve follows the straight line **ab′** in **Figure 6.12**. In reality, the stiffness of the material does not remain constant during this loading increment and the true solution is represented by the curved path **ab**. There is therefore an error in the predicted displacement equal to the distance **b′b**; however, in the tangent stiffness approach this error is neglected. The second increment of load, $\Delta \boldsymbol{R}_2$, is then applied, with the incremental global stiffness matrix $[\boldsymbol{K}_G]^2$ evaluated using the stresses and strains appropriate to the end of increment 1 (i.e. point **b′** on **Figure 6.12**). Solution of the incremental governing equation then gives the nodal displacements $\{\Delta \boldsymbol{d}\}^2_{nG}$. The load–displacement curve follows the straight path **b′c′** in **Figure 6.12**. This deviates further from the true solution, the error in the displacements now being equal to the distance **c′c**. A similar procedure now occurs when $\Delta \boldsymbol{R}_3$ is applied. The stiffness matrix $[\boldsymbol{K}_G]^3$ is evaluated using the stresses and strains appropriate to the end of increment 2 (i.e. point **c′** in **Figure 6.12**). The load–displacement curve moves to point **d′** and again drifts further from the true solution. Clearly, the accuracy of the solution depends on the size of the load increments. For example, if the increment size was reduced so that more increments were needed to reach the same accumulated load, the tangent stiffness solution would be nearer to the true solution.

From the above simple example it may be concluded that in order to obtain accurate solutions to strongly nonlinear problems, many small solution increments are required. The results obtained using this method can drift from the true solution and the stresses can fail to satisfy the constitutive relations. Thus the basic solution requirements may not be fulfilled. It can be shown (Potts and Zdravkovic, 1999) that the magnitude of the error is problem-dependent and is affected by the degree of material nonlinearity, the geometry of the problem and the size of the solution increments used. Unfortunately, in general, it is impossible to predetermine the size of solution increment required to achieve an acceptable error.

The tangent stiffness method can give particularly inaccurate results when soil behaviour changes from elastic to plastic or vice versa. For instance, if an element is in an elastic state at the beginning of an increment, it is assumed to behave elastically over the whole increment. This is incorrect if during the increment the behaviour becomes plastic and results in an illegal stress state which violates the constitutive model. Such illegal stress states can also occur for plastic elements if the increment size used is too large – for example, a tensile stress state could be predicted for a constitutive model which cannot sustain tension. This can be a major problem with critical state type models, such as modified Cam Clay, which employ a $v - \ln p'$ relationship (v = specific volume, p' = mean effective stress), since a tensile value of p' cannot be accommodated. In that case either the analysis has to be aborted or the stress state has to be modified in some arbitrary way, which would cause the solution to violate the equilibrium condition and the constitutive model.

6.8.2 Visco-plastic method

This method uses the equations of visco-plastic behaviour and time as an artifice to calculate the behaviour of nonlinear, elasto-plastic, time-independent materials (Owen and Hinton, 1980; Zienkiewicz and Cormeau, 1974).

The method was originally developed for linear elastic visco-plastic (i.e. time-dependent) material behaviour. Such a material can be represented by a network of the simple rheological units shown in **Figure 6.13**. Each unit consists of an elastic and a visco-plastic component connected in series. The elastic component is represented by a spring and the visco-plastic component by a slider and dashpot connected in parallel. If a load is applied to the network, then one of two situations occurs in each individual unit. If the load is such that the induced stress in the unit does not cause yielding, the slider remains rigid and all the deformation occurs in the spring. This represents elastic behaviour. Alternatively, if the induced stress causes yielding, the slider becomes free and the dashpot is activated. As the dashpot takes time to react, initially all deformation occurs in the spring. However, with time the dashpot moves. The rate of movement of the dashpot depends on the stress it supports and its viscosity. As time progresses the dashpot moves at a decreasing rate because some of the stress the unit is carrying is dissipated to adjacent units in the network, which as a result suffer further movements themselves. This represents visco-plastic behaviour. Eventually, a stationary condition is reached where all the dashpots in the network stop moving and are no longer sustaining stresses. This occurs when the stress in each unit drops below the yield surface and the slider becomes rigid. The external load is now supported purely by the springs within the network, but, importantly, straining of the system has occurred not only due to compression or extension of the springs, but also due to movement of the dashpots. If the load was now removed, only the displacements (strains) occurring in the springs would be recoverable, the dashpot displacements (strains) being permanent.

Application to finite element analysis of elasto-plastic materials can be summarised as follows, using the example of a uniaxially loaded bar in **Figure 6.11**. On application of a solution increment the system is assumed to instantaneously behave linear elastically (**Figure 6.14**). If the resulting stress state lies within the yield surface, the incremental behaviour is elastic and the calculated displacements are correct. If the resulting stress state violates yield, the stress state can only be sustained momentarily and visco-plastic straining occurs. The magnitude of the visco-plastic strain rate is determined by the value of the yield function, which is a measure of the degree by which the current stress state exceeds the yield condition. The visco-plastic strains increase with time, causing the material to relax with a reduction in the yield function and hence the visco-plastic strain rate. A marching technique is used to step forward in time until the visco-plastic strain rate is insignificant. At this

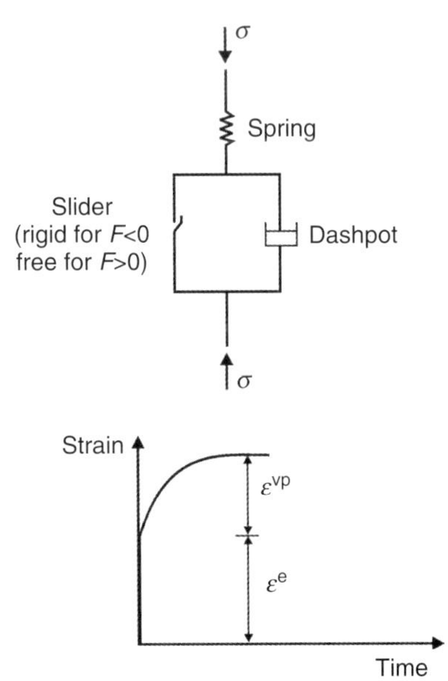

Figure 6.13 Rheological model for visco-plastic material

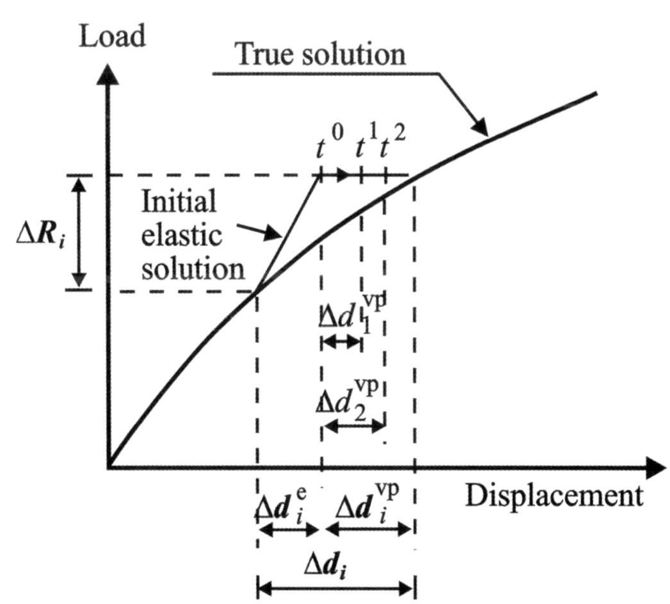

Figure 6.14 Application of the visco-plastic algorithm

point, the accumulated visco-plastic strain and the associated stress change are equal to the incremental plastic strain and stress change, respectively.

In order to use the procedure described above, a suitable time step, Δt, must be selected. If Δt is small many iterations are required to obtain an accurate solution. However, if Δt is too large numerical instability can occur. The most economical choice for Δt is the largest value that can be tolerated without causing such instability. For simple constitutive models, such as Tresca and Mohr–Coulomb, it can be shown that the value of the critical time step is dependent only on the elastic stiffness and strength parameters. As these parameters are constant, the critical time step has to be evaluated only once during an analysis. However, for more complex constitutive models the critical time step is also dependent on the current state of stress and strain and therefore is not constant. It must therefore be continuously evaluated throughout the analyses.

Due to its simplicity, the visco-plastic algorithm has been widely used. However, in the authors' opinion the method has severe limitations for geotechnical analysis. First, the algorithm relies on the fact that for each increment the elastic parameters remain constant. The simple algorithm cannot accommodate elastic parameters that vary during the increment because, for such cases, it cannot determine the true elastic stress changes associated with the incremental elastic strains. The best that can be done is to use the elastic parameters associated with the accumulated stresses and strains at the beginning of the increment to calculate the elastic constitutive matrix, $[\boldsymbol{D}]$, and assume that this remains constant for the increment. Such a procedure only yields accurate results if the increments are small and/or the elastic nonlinearity is not great.

A more severe limitation of the method arises when the algorithm is used as an artifice to solve problems involving non-viscous material (i.e. elasto-plastic materials). Here, calculations to determine the plastic strains and elasto-plastic stress changes are evaluated at illegal stress states which lie outside the yield surface. This is theoretically incorrect and results in failure to satisfy the constitutive equations. The magnitude of the error depends on the constitutive model (see Potts and Zdravkovic, 1999).

6.8.3 Modified Newton–Raphson method

In both the tangent stiffness and visco-plastic algorithms errors can arise because the constitutive behaviour is based on illegal stress states. The modified Newton–Raphson (MNR) algorithm attempts to rectify this problem by only evaluating the constitutive behaviour in, or very near to, legal stress space.

The MNR method uses an iterative technique to solve the system of equations. The first iteration is essentially the same as the tangent stiffness method. However, it is recognised that the solution is likely to be in error and the predicted incremental displacements are used to calculate the residual load, a measure of the error in the analysis. The system of equations is then solved again with this residual load, $\{\psi\}$, forming the incremental right-hand side vector. The system of equations can be then rewritten as

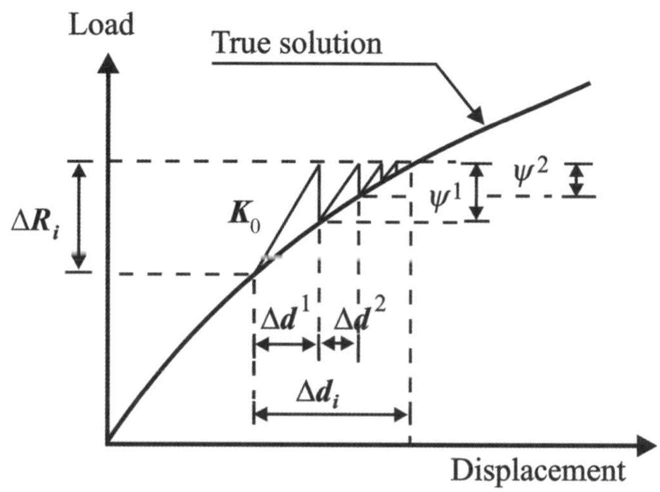

Figure 6.15 Application of the modified Newton–Raphson algorithm

$$[K_G]^i \times \left(\{\Delta d\}^i_{nG}\right)^j = \{\psi\}^j \quad (6.9)$$

The superscript j refers to the iteration number and $\{\psi\}^0 = \{\Delta \boldsymbol{R}_G\}^i$. This process is repeated until the residual load is small. The incremental displacements are equal to the sum of the iterative displacements. This approach is illustrated in **Figure 6.15** for the simple problem of a uniaxially loaded bar of nonlinear material. In principle, the iterative scheme ensures that for each solution increment the analysis satisfies all solution requirements.

A key step in this calculation process is determining the residual load vector. At the end of each iteration the current estimate of the incremental displacements is calculated and used to evaluate the incremental strains at each integration point. The constitutive model is then integrated along the incremental strain paths to obtain an estimate of the stress changes. These stress changes are added to the stresses at the beginning of the increment and used to evaluate consistent equivalent nodal forces. The difference between these forces and the externally applied loads (from the boundary conditions) gives the residual load vector. A difference arises because a constant incremental global stiffness matrix $[\boldsymbol{K}_G]^i$ is assumed over the increment. Due to the nonlinear material behaviour, $[\boldsymbol{K}_G]^i$ is not constant but varies with the incremental stress and strain changes.

Since the constitutive behaviour changes over the increment, care must be taken when integrating the constitutive equations to obtain the stress change. Methods of performing this integration are termed *stress point algorithms* and both explicit and implicit approaches have been proposed in the literature. There are many of these algorithms in use and, as they control the accuracy of the final solution, users must verify the approach used in their software (see Potts and Zdravkovic, 1999).

The process described above is called a Newton–Raphson scheme if the incremental global stiffness matrix $[\boldsymbol{K}_G]^i$ is recalculated and inverted for each iteration, based on the latest estimate of the stresses and strains obtained from the previous iteration. To reduce the amount of computation, the MNR method only calculates and inverts the stiffness matrix at the beginning of the increment and uses it for all iterations within the increment. Sometimes the incremental global stiffness matrix is calculated using the elastic constitutive matrix, $[\boldsymbol{D}]$, rather than the elasto-plastic matrix, $[\boldsymbol{D}^{ep}]$. Clearly, there are several options here and many software packages allow the user to specify how the MNR algorithm should work. In addition, an acceleration technique is often applied during the iteration process (Thomas, 1984).

6.8.4 Comparison of the solution strategies

A qualitative comparison of the three solution strategies presented above suggests the following. The tangent stiffness method is the simplest, but its accuracy is influenced by increment size. The accuracy of the visco-plastic approach is also influenced by increment size (and time step size) if complex constitutive models are used. The MNR method is potentially the most accurate and is likely to be the least sensitive to increment size. However, considering the computer resources required for each solution increment, the MNR method is likely to be the most expensive in terms of time required for an analysis, the tangent stiffness method is the cheapest and the visco-plastic method is probably somewhere in between. It may be possible to use larger and therefore fewer increments with the MNR method to obtain a similar accuracy. Thus, it is not obvious which solution strategy is the most economic for a particular solution accuracy.

To illustrate the difference between the solution strategies, two examples will be considered. First, a smooth rigid strip footing subjected to vertical loading, as depicted in **Figure 6.16** is analysed. The soil has been modelled using a form of modified Cam Clay and has been assumed to behave undrained. The distribution of undrained strength with depth is shown in **Figure 6.16**. For further information on the constitutive model and the input parameters, see Potts and Zdravkovic (1999). The finite element mesh is shown in **Figure 6.17**. Note that due to symmetry about the vertical line through the centre of the footing, only half of the problem needs to be considered in the finite element analysis. Plane strain conditions are assumed. Before loading the footing, the coefficient of earth pressure at rest, K_o, was assumed to be unity and the vertical effective stress and pore water pressure were calculated using a saturated bulk unit weight of the soil of 20 kN/m^3 and a static water table at the ground surface. The footing was loaded by applying a series of equally sized increments of vertical displacement until the total displacement was 25 mm.

The load–displacement curves for the tangent stiffness, visco-plastic and MNR analyses are presented in **Figure 6.18**. For the MNR method, analyses were performed using 1, 2, 5, 10, 25, 50 and 500 increments to reach a footing settlement of 25 mm. With the exception of the analysis performed with only a single increment, all analyses gave very similar results and plot as a single curve, marked MNR on this figure. The MNR results are therefore insensitive to increment size and show a well-defined collapse load of 2.8 kN/m.

For the tangent stiffness approach, analyses using 25, 50, 100, 200, 500 and 1000 increments have been carried out. Analyses with a smaller number of increments were also attempted, but illegal stresses (negative mean effective stresses, p') were predicted. As the constitutive model is not defined for such stresses, the analyses had to be aborted. Some finite element packages overcome this problem by arbitrarily resetting

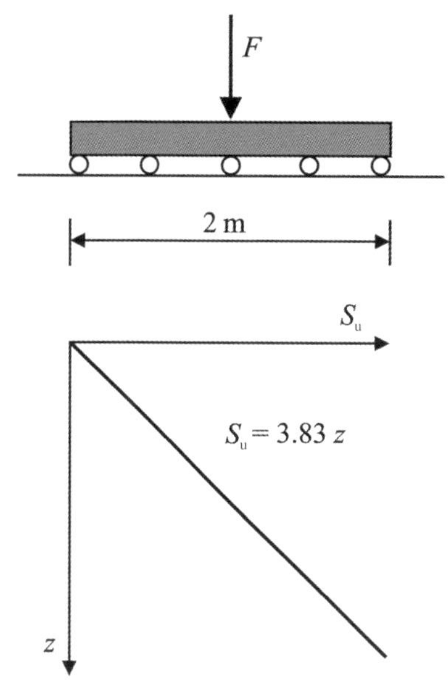

Figure 6.16 Geometry of footing

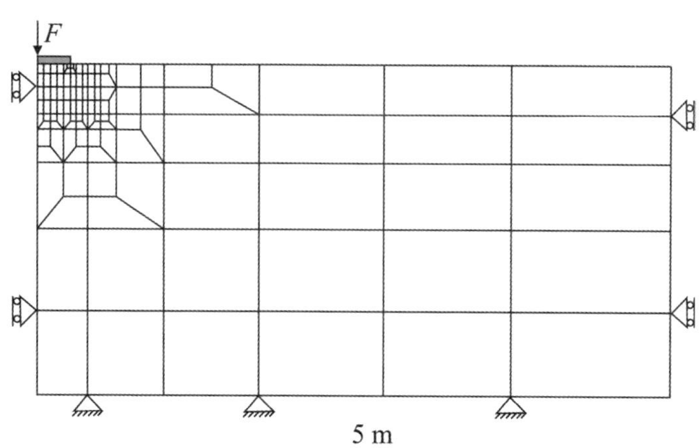

Figure 6.17 Finite element mesh for footing analyses

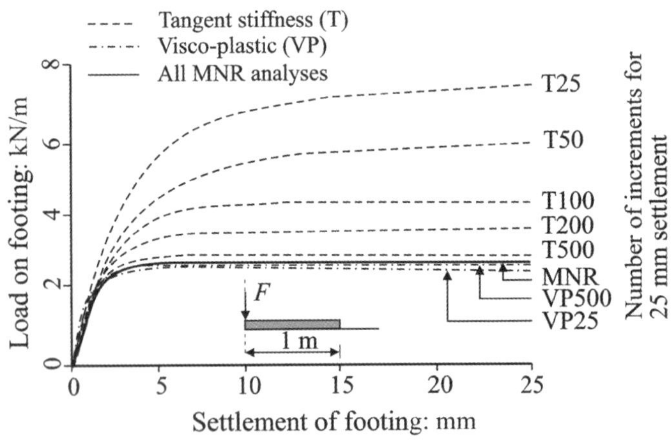

Figure 6.18 Footing load–displacement curves

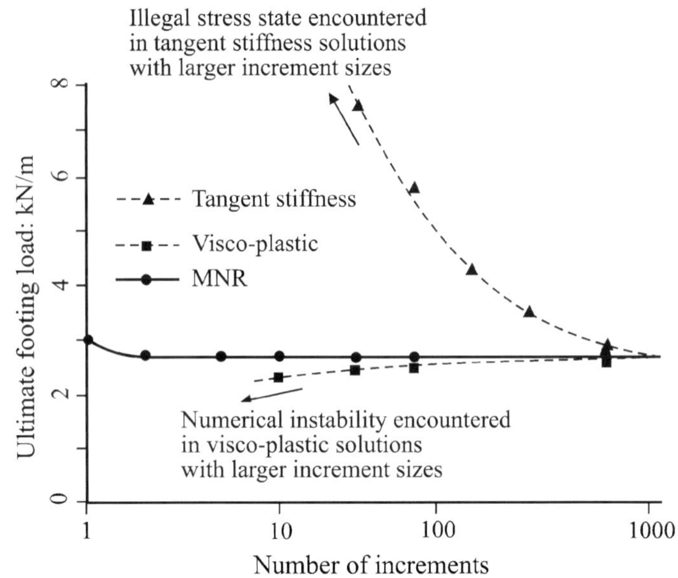

Figure 6.19 Ultimate footing load against number of increments

the offending negative p' values. There is no theoretical basis for this and it leads to violation of both the equilibrium and the constitutive conditions. Although such adjustments enable an analysis to be completed, the final solution is in error.

Results from the tangent stiffness analyses are shown in **Figure 6.18**. When plotted, the curve from the analysis with 1000 increments is indistinguishable from those of the MNR analyses. The tangent stiffness results are strongly influenced by increment size, with the ultimate footing load decreasing from 7.5 kN/m to 2.8 kN/m with reduction in the size of the applied displacement increment. There is also a tendency for the load–displacement curve to continue to rise and not reach a well-defined ultimate failure load for the analysis with large applied displacement increments. The results are unconservative, overpredicting the ultimate footing load. There is also no indication from the shape of the tangent stiffness load–displacement curves as to whether the solution is accurate, since all the curves have similar shapes.

Visco-plastic analyses with 10, 25, 50, 100 and 500 increments were also performed. The 10-increment analysis had convergence problems in the iteration process, which would initially converge but then diverge. Similar behaviour was encountered for analyses using still fewer increments. Results from the analyses with 25 and 500 increments are shown in **Figure 6.18**. The solutions are sensitive to increment size, but to a lesser degree than the tangent stiffness approach.

The load on the footing at a settlement of 25 mm is plotted against number of increments, for all tangent stiffness, visco-plastic and MNR analyses, in **Figure 6.19**. The insensitivity of the MNR analyses to increment size is clearly shown. In these analyses the ultimate footing load only changed from 2.83 kN/m to 2.79 kN/m as the number of increments increased from 2 to 500. Even for the MNR analyses performed with a single increment, the resulting ultimate footing load of 3.13 kN/m is still reasonable and is more accurate than the value of 3.67 kN/m obtained from the tangent stiffness analysis with 200 increments. Both the tangent stiffness and visco-plastic analyses produce ultimate failure loads which approach 2.79 kN/m as the number of increments increases. However, tangent stiffness analyses approach this value from above and therefore overpredict, while visco-plastic analyses approach this value from below and therefore underpredict. The CPU times and results for selected analyses are shown in **Table 6.3**. It should be noted that these analyses were completed many years ago. Consequently the absolute time values are not indicative of analyses performed on modern computers. However, the relative times are still valid.

The second example considers the mobilisation of the stresses in the soil immediately adjacent to a pile shaft during drained loading. The behaviour of a segment of an incompressible 2 m dia. pile, well away from the influence of the soil surface and pile tip, is examined (see **Figure 6.20**). This boundary value problem has been discussed in detail by Potts and Martins (1982) and the alternative methods of representing the problem in finite element analysis have been explored by Gens and Potts (1984). The soil is assumed to be normally consolidated, with initial stresses $\sigma'_v = \sigma'_h = 200$ kPa. The parameters for the modified Cam Clay model are the same as used for the footing problem

Solution type	Number of increments	CPU time: s	Relative CPU times	Collapse load: kN
MNR	10	15 345	1	2.82
Tangent stiffness	500	52 609	3.4	3.01
Tangent stiffness	1000	111 780	7.3	2.82
Visco-plastic	25	70 957	4.6	2.60
Visco-plastic	500	1 404 136	91.5	2.75

Table 6.3 CPU times and failure loads for the footing analyses

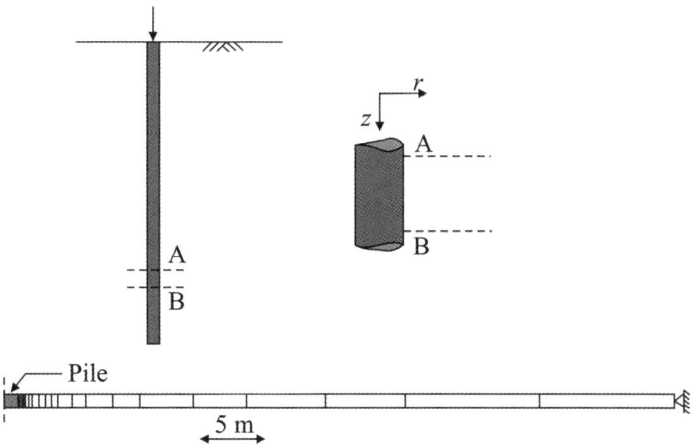

Figure 6.20 Geometry and finite element mesh for pile problem

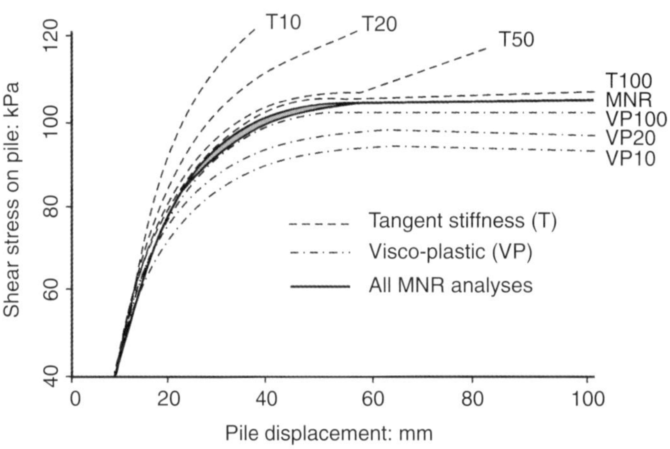

Figure 6.21 Mobilised pile shaft resistance against pile displacement

and are discussed in detail in Potts and Zdravkovic (1999). Axisymmetric conditions are applicable to this problem and the finite element mesh is shown in **Figure 6.20**. Loading of the pile has been simulated in the finite element analyses by imposing a series of equal-sized increments of vertical displacement to the pile shaft to give a total displacement of 100 mm. The soil was assumed to behave in a drained manner throughout.

Results from analyses using all three solution strategies and with varying numbers of increments are presented in **Figure 6.21** in the form of mobilised shaft resistance, τ, against vertical pile displacement. The MNR analyses with 20 increments and above plot as a single curve on this figure and are represented by the upper solid line. MNR analyses with smaller numbers of increments (i.e. with larger applied displacement increments) showed minor differences with the above results between pile displacements of 25 mm and 55 mm. The MNR analysis with 5 increments is shown by the lower solid line in the figure. The spread of results between the MNR analyses with 5 and 20 increments is therefore given by the shaded band. Because this band is small it may be concluded that, as with the footing problem, the MNR results are insensitive to increment size.

Results from some of the tangent stiffness and visco-plastic analyses are also shown in **Figure 6.21**. If small increments of applied displacement (large number of increments) are used, both of these solution strategies give results which approach those of the MNR analyses. However, analyses using both of these approaches are sensitive to increment size, and analyses with larger increments of applied displacement are inaccurate. The tangent stiffness overpredicts and the visco-plastic underpredicts the mobilised shaft resistance for any given pile displacement. The visco-plastic analyses predict that the mobilised pile shaft resistance increases to a peak value and then slowly reduces with further pile displacement. This behaviour is not predicted by the MNR analyses, which show that once the maximum shaft resistance is mobilised it remains constant. For the tangent stiffness analyses with fewer than 100 increments the mobilised pile shaft resistance continually climbs and no peak value is indicated.

Values of the mobilised pile shaft resistance, τ_f, at a pile displacement of 100 mm (the end point on the curves in **Figure 6.21**) are plotted against number of increments used in **Figure 6.22**. As noted above, the MNR results are insensitive to increment size, with τ_f only increasing from 102.44 kPa to 103.57 kPa as the number of increments increases from 1 to 500. The results from the tangent stiffness and visco-plastic analyses are much more dependent on the size of the applied displacement increment, and over 100 increments for the tangent stiffness and 500 increments for the visco-plastic analyses are needed to give reasonably accurate results. As both the tangent stiffness and visco-plastic solutions approach those of the MNR when the applied displacement increment reduces, the 500-increment MNR analysis may be assumed 'correct'. The error in any analysis may therefore be expressed as

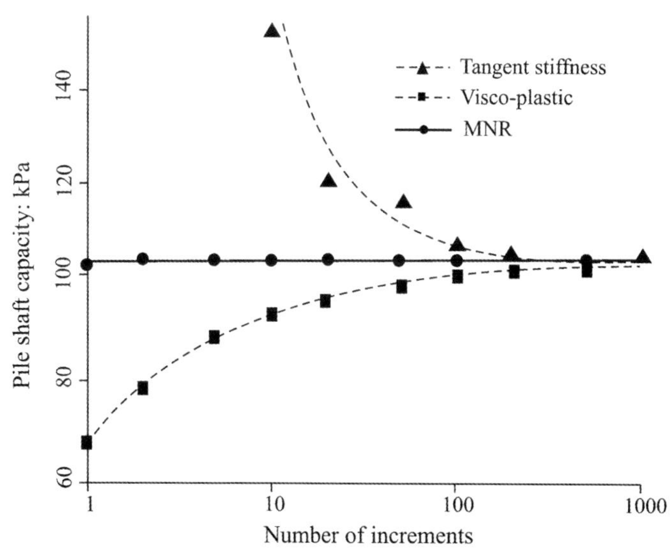

Figure 6.22 Pile capacity against number of increments

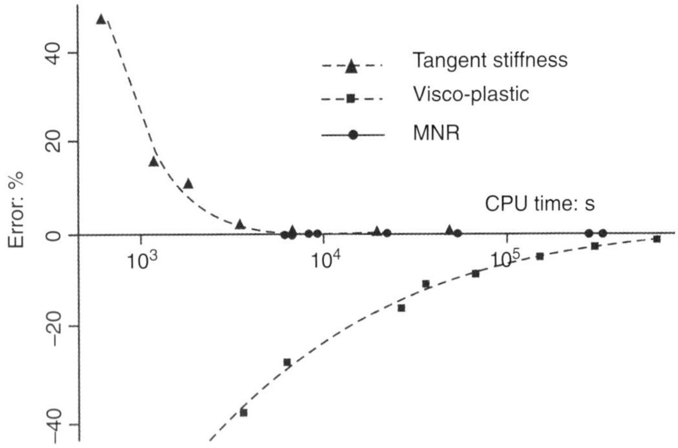

Figure 6.23 Error in pile analysis against CPU time

$$\text{Error} = \frac{\tau_f - 103.6}{103.6} \tag{6.10}$$

where τ_f for the 500-increment MNR analysis equals 103.6 kPa. This error is plotted against CPU time in **Figure 6.23** for all analyses. The results clearly show that as well as being the least accurate, the visco-plastic analyses are also the most expensive in terms of CPU time consumed. As the MNR analyses give accurate solutions with small error values, typically less than 0.2%, the results plot on the CPU time axis in this figure. These CPU times compare favourably with those of the tangent stiffness analyses.

The relatively poor performance of the visco-plastic strategy should be noted. This contrasts to its better performance for the footing problem. However, the footing problem involved undrained soil behaviour, whereas for the pile problem the soil is assumed to remain drained. Results from the analysis of a range of boundary value problems presented in Potts and Zdravkovic (1999) also indicate that the relative performance of the visco-plastic strategy is worse for drained than undrained problems. It may be concluded that for the analyses with the modified Cam Clay model the visco-plastic approach is likely to be better behaved and less sensitive to the size of the solution increment when soil behaviour is undrained than when the soil is drained.

6.8.5 Comments

Results from the tangent stiffness analyses of both boundary value problems are strongly dependent on increment size. The error associated with the tangent stiffness analyses usually results in unconservative predictions of failure loads and displacements in most geotechnical problems. For the footing problem large overpredictions of failure loads are obtained unless a very large number of increments (>1000) is employed. Inaccurate analyses based on too large an increment size produced ostensibly plausible load–displacement curves. Analytical solutions are not available for most problems requiring a finite element analysis. Therefore it is difficult to judge whether a tangent stiffness analysis is accurate on the basis of its results. Several analyses must be carried out using different increment sizes to establish the likely accuracy of any predictions. This could be a very costly exercise, especially if there was little experience in the problem being analysed and no indication of the optimum increment size.

Results from the visco-plastic analyses are also dependent on increment size. For boundary value problems involving undrained soil behaviour these analyses are more accurate then tangent stiffness analyses with the same increment size. However, if soil behaviour is drained, visco-plastic analyses are only accurate if many small solution increments are used. In general, the visco-plastic analyses use more computer resources than both the tangent stiffness and MNR approaches. For both the footing and pile problems, the visco-plastic analyses underpredict failure loads if insufficient increments are used and are therefore conservative in this context.

Close inspection of the results from the visco-plastic and tangent stiffness analyses indicates that a major reason for their poor performance is their failure to satisfy the constitutive laws. This problem is largely eliminated in the MNR approach, where a much tighter constraint on the constitutive conditions is enforced.

The results from the MNR analyses are accurate and essentially independent of increment size. For the boundary value problems considered, the tangent stiffness method requires considerably more CPU time than the MNR method to obtain results of similar accuracy (e.g. over seven times more for the foundation problem). Similar comparisons can be found between the MNR and visco-plastic solutions. Thus not withstanding the potentially very large computer resources required to find the optimum tangent stiffness or visco-plastic increment size, the tangent stiffness or visco-plastic method with an optimum increment size is still likely to require more computer resources than an MNR analysis of the same accuracy. Although it may be possible to obtain tangent stiffness or visco-plastic results using fewer computer resources than with the MNR approach, this is usually at the expense of the accuracy of the results. Alternatively, for a given amount of computing resources, an MNR analysis produces a more accurate solution than either the tangent stiffness or visco-plastic approaches.

The study has shown that the MNR method appears to be the most efficient solution strategy for obtaining an accurate solution to problems using critical state type constitutive models for soil behaviour. However, it should be noted that the MNR method is highly sensitive to the stress point algorithm (see Section 6.9.3) that it uses, which is code-dependent. The large errors in the results from the tangent stiffness and visco-plastic algorithms in the present study emphasise the importance of checking the sensitivity of the results of any finite element analysis to increment size.

Many of the above concerns also apply when developing and/or using user-defined constitutive models in a commercial software. Care must be taken by the user to ensure that the nonlinear integration algorithms are fit for purpose and not excessively increment-size-dependent.

6.9. Modelling of structural members in plane strain analysis

With many geotechnical problems, while the main feature might be adequately represented by a plane strain idealisation, other, smaller components of the same problem might not be. For example, consider a typical urban excavation problem as shown in **Figure 6.24**. Here is an embedded retaining wall supported by ground anchors and a foundation slab tied down by tension piles. This could represent a cross-section of an excavation for a road. Consider the structural members in turn.

6.9.1 Walls

If the wall is of the concrete diaphragm type, then it is best modelled using solid finite elements with the appropriate geometry and material properties. There is no serious modelling problem here as the wall satisfies the plane strain assumption. However, if the wall is made from secant or contiguous concrete piles, steel sheet piles or some combination of steel columns and sheeting, then the properties and geometry of the wall will vary in the out-of-plane direction and therefore will not satisfy the requirements of plane strain. Modelling of these components in a plane strain analysis will therefore involve some approximations. Usually it is possible to estimate the average axial (*EA*) and bending stiffness (*EI*) per meter length of wall. If beam elements are used to model the wall, then these parameters can be input directly as the material properties. However, if solid elements are to be used, then the average axial and bending stiffness must be converted into an equivalent thickness *t* of the solid elements and an equivalent Young's modulus, E_{eq}. This is done by solving the following two simultaneous equations.

$$\text{Axial stiffness: } t \times E_{eq} = E \times A \quad (6.11)$$

$$\text{Bending stiffness: } (E_{eq} \times t^3)/12 = E \times I \quad (6.12)$$

where E is the Young's modulus of the wall and A and I are cross-sectional area and moment of inertia per meter length, respectively. It may also be necessary to calculate some form of average strength, but this will depend on the constitutive model employed to represent the wall. Clearly the above procedure only treats the wall in an approximate manner and it will not be possible to accurately estimate the details of the stress distribution within the separate components of the wall.

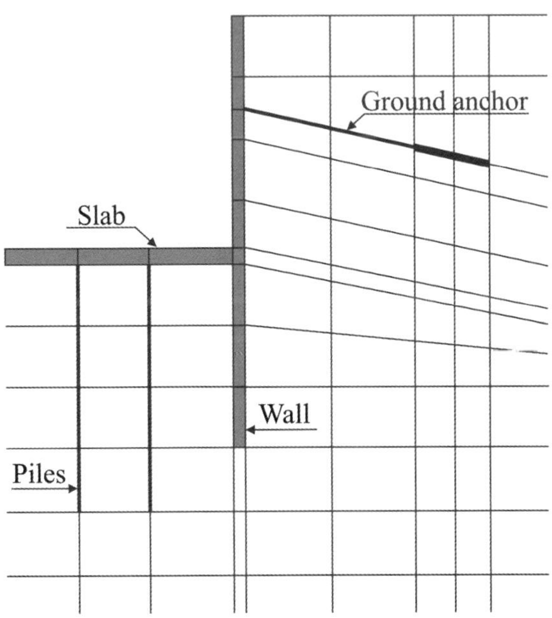

Figure 6.24 Typical urban excavation

6.9.2 Piles

Modelling the piles below the base slab involves additional assumptions as the piles are not continuous in the out-of-plane direction, but are separated by relatively large expanses of soil. While it is again possible to estimate average axial and bending stiffnesses per unit length in the out-of-plane direction and calculate equivalent parameters as was shown for the wall above, modelling the piles with solid or beam elements implies that the soil is not able to freely move between the piles, as the piles are essentially modelled as a wall in the out-of-plane direction (**Figure 6.25**). As a consequence, the lateral movements of the soil below the excavation will be restricted. It may therefore be more realistic to neglect the bending stiffness of the piles so that they provide no resistance to lateral soil movement. A spring or a series of membrane elements can then be used to represent the piles.

If a linear spring is used to represent the pile only, elastic behaviour of the pile can be represented. In addition, if the spring is connected between a node on the concrete slab and

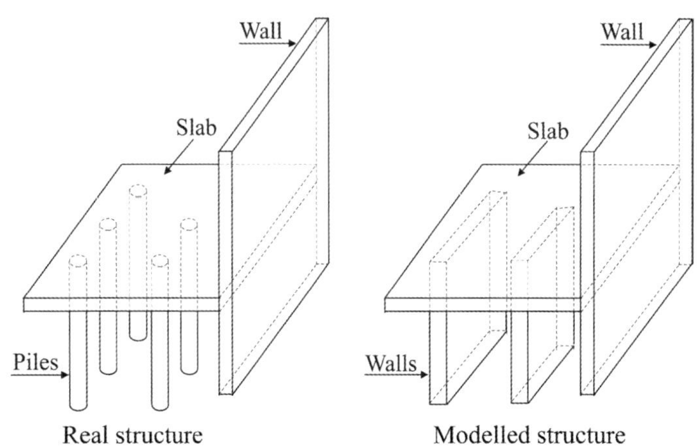

Figure 6.25 Comparison between real and simulated conditions for piles

a node in the soil (i.e. nodes A and B in **Figure 6.26**), only forces at these two nodes will be accounted for (i.e. the pile and soil are not connected between A and B). It will then not be possible to account for shaft friction nor set a limit to the end-bearing resistance of the pile (i.e. at node B).

If a series of membrane (or beam) elements is used to model the pile (**Figure 6.27**), then the pile will be connected to the soil along its shaft and consequently mobilisation of shaft resistance will be simulated to some limited extent. However, the zero thickness of the elements in the plane of analysis implies that base resistance will be mobilised through a single node; it is therefore difficult to impose a limit to the end-bearing resistance.

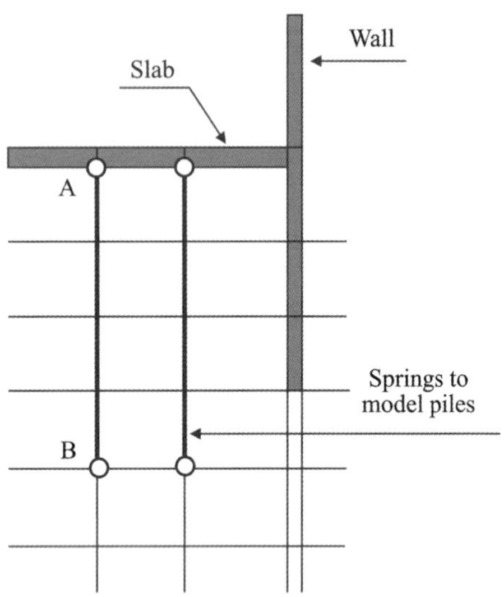

Figure 6.26 Springs used to model piles

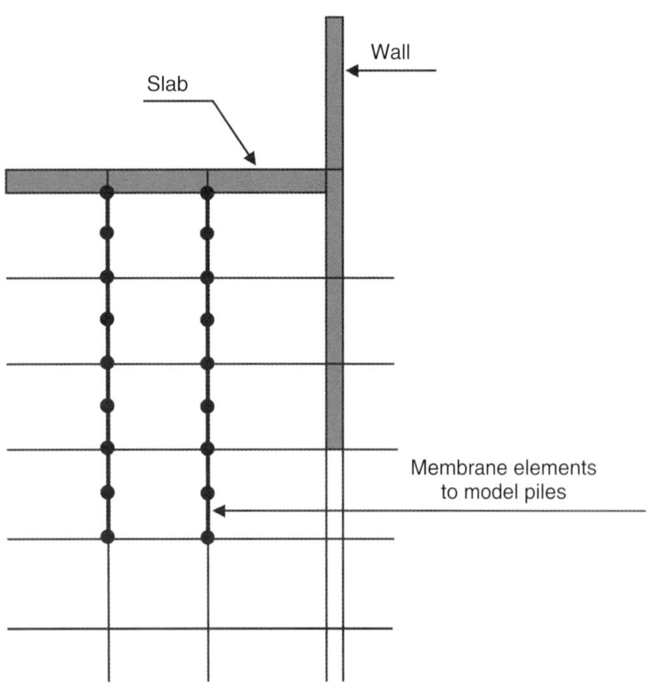

Figure 6.27 Membrane elements used to model piles

6.9.3 Ground anchors

Ground anchors located behind a wall can be conveniently divided into two parts: the *fixed* and *free* anchor length (**Figure 6.28**). As with the piles, equivalent axial and bending stiffnesses per unit length in the out-of-plane direction can be calculated for both components of a row of anchors. Account must be taken of the spacing of the anchors in the out-of-plane direction when calculating these equivalent values. Again, it is probably best to ignore the bending stiffness and model the two anchor components with either springs, membrane elements or a combination of both (**Figure 6.29**). The situation shown in **Figure 6.29(c)**, where a spring is used to model the free anchor length, and a series of membrane elements the fixed length, is probably the more realistic approach. It is usual to neglect any shear stresses mobilised between the soil and the free anchor length; consequently, the springs are connected to the wall at one end (point A in **Figure 6.29(c)**) and the soil (and fixed anchor length) at the other end (point B in **Figure 6.29(c)**). As with the piles, modelling the anchors in this way is rather crude as it is not possible to accurately account for the limiting end and shaft capacity of the fixed anchor length.

An alternative way of modelling the fixed anchor length is shown in **Figure 6.30** and involves the use of solid and interface elements. It is now possible to simulate the mobilisation of both shaft and end-bearing resistance and to apply limits to these quantities. However, it is necessary to assign equivalent properties to the solid and interface elements to take account of the spacing of the anchors in the out-of-plane direction. This is often not straightforward. By assigning an equivalent Young's modulus to the solid elements, the fixed anchor length will have bending stiffness. This will limit soil movements perpendicular to the anchor. In the real problem the soil will be able to deform between the anchors.

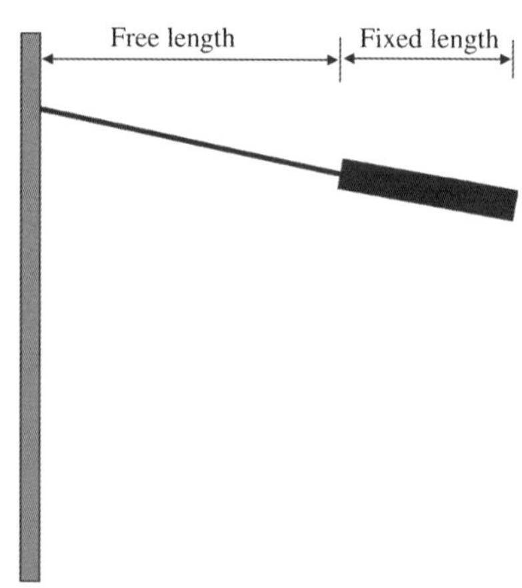

Figure 6.28 Ground anchor

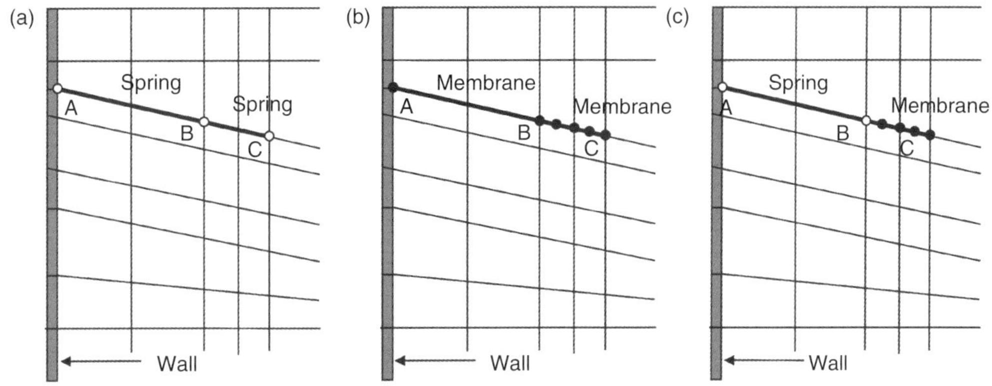

Figure 6.29 Modelling ground anchors using (a) springs, (b) membrane elements and (c) a combination of spring and membrane elements

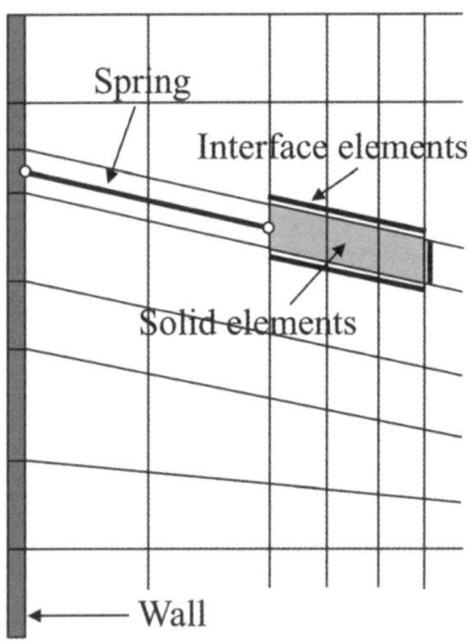

Figure 6.30 Modelling a ground anchor with solid elements

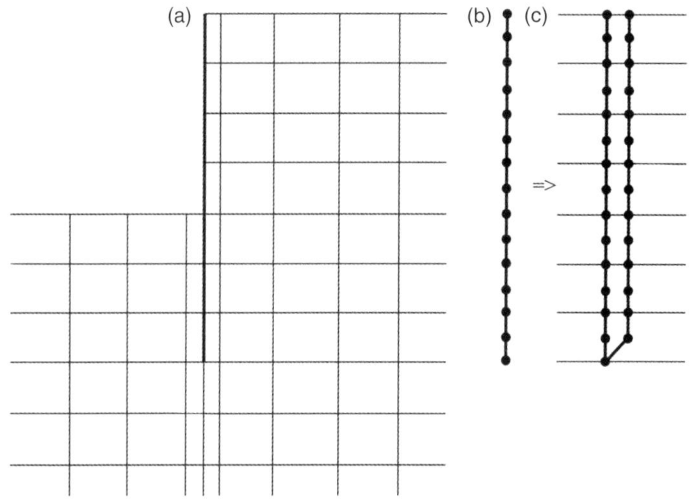

Figure 6.31 Dealing with impermeable line elements in consolidation problems (use of interface elements)

The above discussion indicates that while it is possible to approximate the wall behaviour in a realistic manner, much greater approximations are involved with the modelling of both the piles and ground anchors. This arises because these structural elements are not continuous in the out-of-plane direction.

6.9.4 Structural members in coupled analyses

A further pitfall can occur when using membrane or beam elements to model structural members in coupled consolidation analyses. The problem is best explained by considering the example of the retaining wall shown in **Figure 6.31**. The wall is modelled using beam elements and therefore shares a common set of nodes with both the elements in front of and behind it (**Figure 6.31(b)**). As the solid elements on either side of the wall have common nodes at the wall, the wall will be implicitly assumed to be permeable in a consolidation analysis. Water will therefore flow freely through the wall. This may not be the desired outcome. If the wall is to be assumed impermeable, this can be achieved by placing interface elements along one side of the wall as shown in (**Figure 6.31(c)**). If these elements are non-consolidating, they will provide an impermeable break between the solid elements on either side of the wall.

6.9.5 Structural connections

Potential problems can arise when modelling connections between different structural components – for example, the connection between props and a retaining wall. Such connections can be modelled as either *simple*, *pin-jointed* or *full moment* connections. These conditions and the possible alternative ways that can be used to model them in plane strain analyses are shown in **Figures 6.32–6.34**. It is often all too easy to use the options given in **Figure 6.34** when in reality the connection is pin-jointed.

6.9.6 Segmental tunnel linings

Many tunnel linings are constructed from segments. Usually 8–12 segments are required to form a complete circular ring

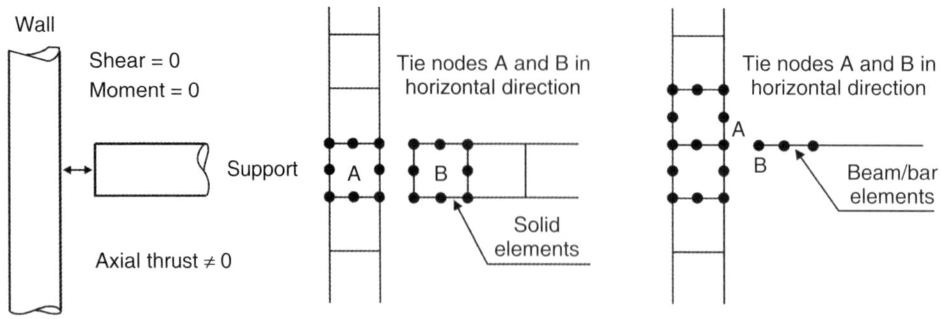

Figure 6.32 Simple connection between wall and beam

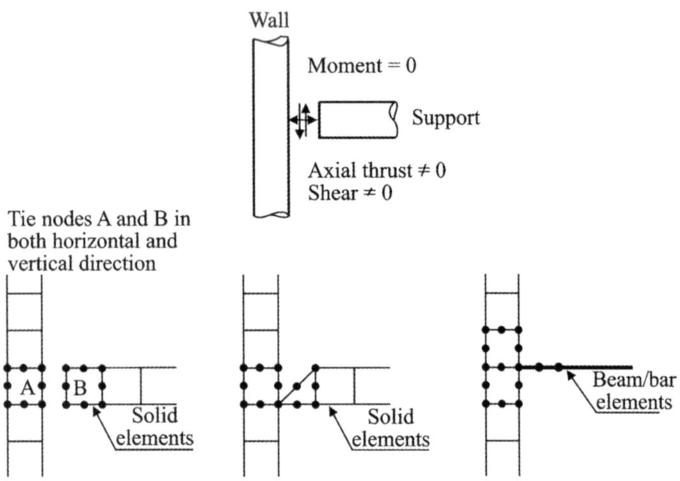

Figure 6.33 Pin-jointed connection between wall and prop

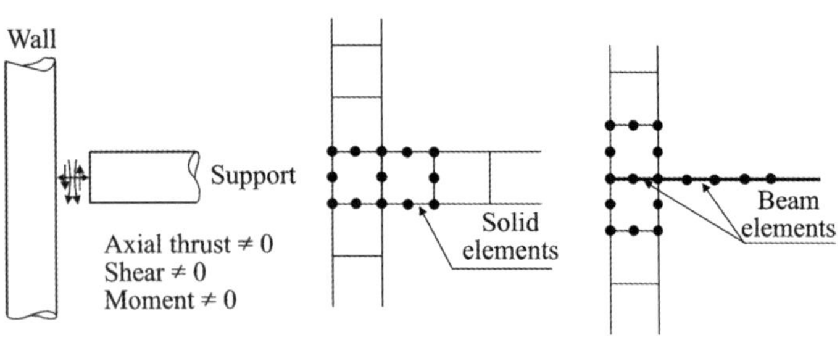

Figure 6.34 Full connection between wall and prop

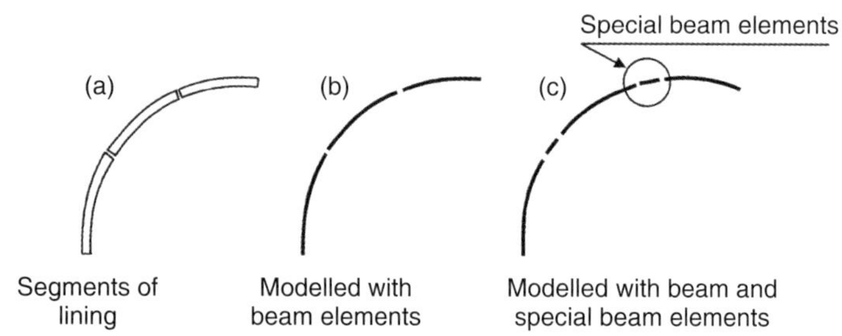

Figure 6.35 Modelling tunnel lining

(**Figure 6.35(a)**). Sometimes the segments are bolted to each other, whereas sometimes they are not. In either event there is little bending resistance between the segments. It is often convenient to model the lining using beam or solid elements. However, if these elements are placed as a sequential ring (**Figure 6.35(b)**) a full moment connection between the segments will be implied. If this is not what is required, then an alternative approach must be used. For example, special beam elements which represent the interface between the segments can be used. Each segment is modelled as a group of beam elements and these are then joined together by these small special beam elements (**Figure 6.35(c)**).

6.10. Some pitfalls with the Mohr–Coulomb model

The Mohr–Coulomb constitutive model is the simplest representation of soil behaviour and forms part of most geotechnical software packages. It is also the most widely used constitutive model in geotechnical engineering practice. This is a linear elastic–perfectly plastic model, requiring only a few input parameters that can be obtained from standard laboratory testing: Young's modulus E and Poisson ratio μ to describe the elastic part of the model, and cohesion c' and angle of shearing resistance ϕ' to describe the plastic (failure) part of the model. If no other input parameter is required, this implies associated plasticity for the model and negative (dilative) plastic volumetric strains. Problems with such a formulation of the model are twofold: (*a*) soil dilation is usually smaller than that implied by associated plasticity (i.e. angle of dilation ν equal to ϕ'); and (*b*) once the soil starts to dilate, it will dilate forever, without reaching a limit load for volumetrically constrained problems.

6.10.1 Drained loading

An example of the performance of the Mohr–Coulomb model in drained conditions is given in **Figure 6.36**, which shows load–displacement curves for a vertically loaded pile, 1.0 m dia. and 20 m long, in a drained soil with $c' = 0$ and $\phi' = 25°$ (Potts and Zdravkovic, 2001). If the angle of dilation, ν, is equal to ϕ' (i.e. associated plasticity), the pile never reaches a

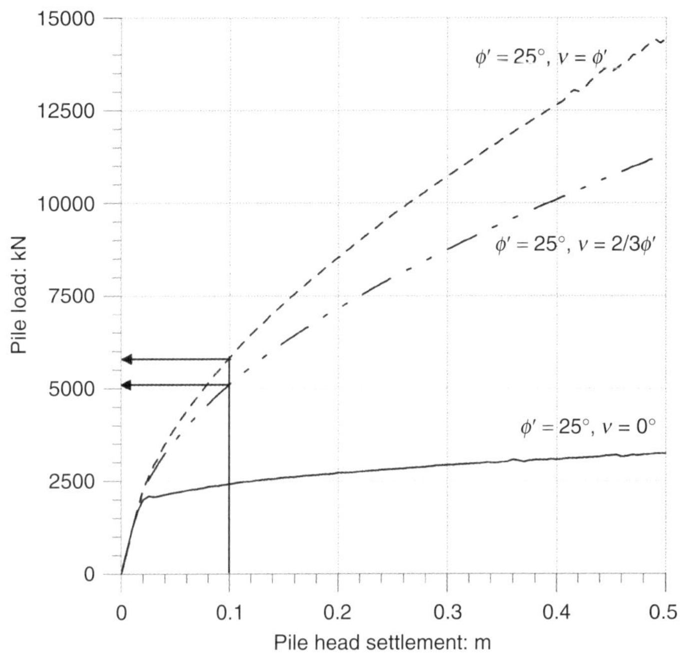

Figure 6.36 Load–displacement curves for vertically loaded pile

limit load, no matter how far it is pushed into the ground. Even if the model has the flexibility to input a value of ν smaller than ϕ', for any value of ν that is greater than zero the model still does not predict a limit load. The magnitude of ν makes a difference only in the magnitude of the dilative volumetric plastic strains, which reduce with a reduction of ν and hence reduce the magnitude of the pile load for a particular displacement of the pile head. Consequently, a practical approach in determining a limit load from such analysis has been to arbitrarily adopt the load value that corresponds to the displacement equal to 10% of the pile diameter (0.1 m in this case). This can result in a significant overestimate in the load capacity of the pile in the case of associated plasticity, as soil dilation is generally overpredicted. Only if $\nu = 0$ can the pile reach a limit load. This is a conservative prediction as most soils dilate to some extent, but it is at least a theoretically correct value obtained without any arbitrary decisions from the user.

6.10.2 Undrained loading

As noted above, the Mohr–Coulomb model can be used with a dilation angle ranging from $\nu = 0$ to $\nu = \phi'$. This parameter controls the magnitude of the plastic dilation (plastic volume expansion) and remains constant once the soil is on the yield surface. This implies that the soil will continue to dilate indefinitely if shearing continues. Clearly such behaviour is not realistic as most soils will eventually reach a critical state condition, after which they will deform at constant volume if sheared any further. While such unrealistic behaviour does not have a great influence on boundary value problems which are unrestrained (i.e. the drained surface footing problem), it can have a major effect on problems which are constrained, as noted above for the drained pile problem, due to the restrictions on volume change imposed by the boundary conditions. In particular, unexpected results can be obtained in undrained analysis in which there is a severe constraint imposed by the zero total volume change restriction associated with undrained soil behaviour. To illustrate this problem two examples will now be presented.

The first example considers ideal (no end effects) undrained triaxial compression ($\Delta\sigma_v > 0$, $\Delta\sigma_h = 0$) tests on a linear elastic Mohr–Coulomb plastic soil with parameters $E' = 10\,000$ kPa, $\mu = 0.3$, $c' = 0$, $\phi' = 24°$. As there are no end effects, a single finite element is used to model the triaxial test with the appropriate boundary conditions. The samples were assumed to be initially isotopically consolidated with $p' = 200$ kPa and zero pore water pressure. A series of finite element runs were then made, each with a different angle of dilation, ν, in which the samples were sheared undrained. Undrained conditions were enforced by setting the bulk modulus of water to be 1000 times larger than the effective elastic bulk modulus of the soil skeleton, K' (see Potts and Zdravkovic, 1999). The results are shown in **Figures 6.37(a)** and **6.37(b)** in the form of deviatoric stress (J) – mean effective stress (p') and deviatoric stress (J) – axial strain (ε_z) plots. It can be seen that in terms of $J - p'$ all analyses follow the same stress path. However, the rate at which the stress state moves up the Mohr–Coulomb failure line differs for each analysis. This can be seen from **Figure 6.37(b)**. The analysis with zero plastic dilation, $\nu = 0$, remains at a constant J and p' when it reaches the failure line. However, all other analyses move up the failure line, with those with the larger dilation moving up more rapidly. They continue to move up this failure line indefinitely with continued shearing. Consequently, the only analysis that indicates failure (i.e. a limiting value of J) is the analysis performed with zero plastic dilation.

The second example considers the undrained loading of a smooth rigid strip footing. The soil was assumed to have the same parameters as those used for the triaxial tests above. The initial stresses in the soil were calculated on the basis of a saturated bulk unit weight of 20 kN/m^3, a groundwater table at the soil surface and a $K_o = 1 - \sin\varphi'$. The footing was loaded by applying increments of vertical displacement and undrained conditions were again enforced by setting the bulk modulus of the pore water to be 1000 times K'. The results of two analyses, one with $\nu = 0°$ and the other with $\nu = \varphi'$, are shown in **Figure 6.38**. The difference is quite staggering: while the analysis with $\nu = 0°$ reaches a limit load, the analysis with $\nu = \varphi'$ shows a continuing increase in load with displacement. As with the triaxial tests, a limit load is only obtained if $\nu = 0°$.

It can be concluded from these two examples that a limit load will only be obtained if $\nu = 0°$. Consequently, great care must be exercised when using the Mohr–Coulomb model in undrained analysis. It could be argued that the model should not be used with $\nu > 0$ for such analysis. However, reality is not that simple and often a finite element analysis involves both an undrained and a drained phase (i.e. undrained excavation followed

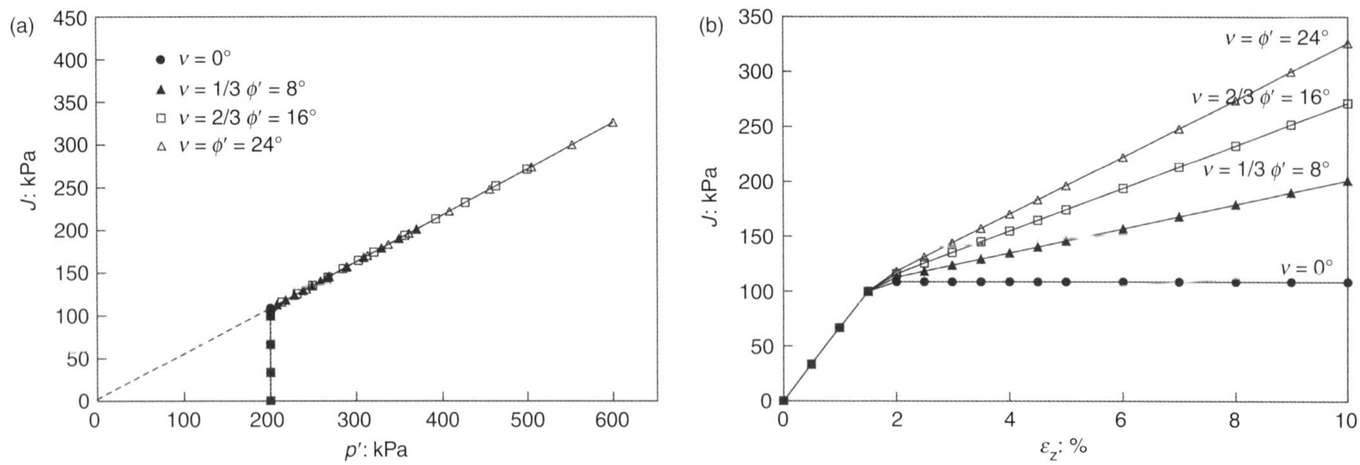

Figure 6.37 Prediction of (a) stress paths and (b) stress–strain curves in undrained triaxial compression using a Mohr–Coulomb model with different angles of dilation

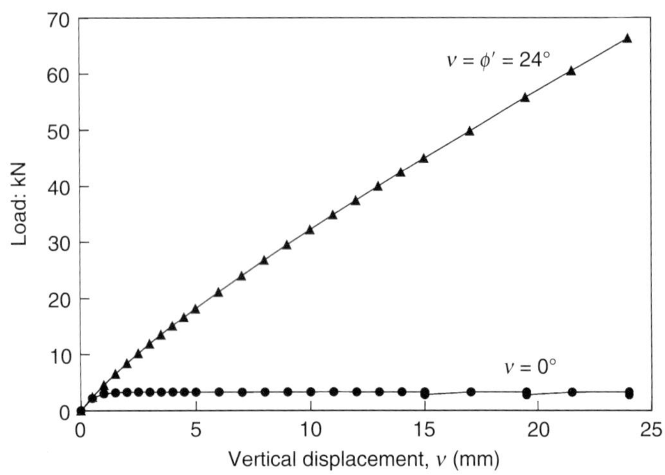

Figure 6.38 Load–displacement curves for a strip footing, using the Mohr–Coulomb model with different angles of dilation

by drained dissipation). Consequently, it may be necessary to adjust the value of ν between the two phases of the analysis. Alternatively, a more complex constitutive model which better represents soil behaviour may have to be employed.

6.11. Summary

This chapter identifies the main design requirements for a geotechnical structure, for which calculations have to be performed in order to confirm the stability and safety of the design. These requirements are

- local stability
- overall stability
- structural forces
- ground movements.

Several methods of analysis are available for the above geotechnical calculations, and they are classified into three main groups

- closed form solutions
- classical analysis (limit equilibrium, stress fields and limit analysis)
- numerical analysis (beam-spring approach and full numerical analysis).

In terms of satisfying the main theoretical requirements, it is important to note the following.

- Closed form solutions are possible to obtain in some cases only if the soil is assumed to be linear elastic. However, this is a significant simplification of real soil behaviour and may result in the wrong predictions of ground movements. Additionally, these solutions cannot be used to assess stability of geotechnical structures.
- All classical methods of analysis do not satisfy at least one of the theoretical requirements, hence in general they do not produce an exact solution. An exception is when calculations using both the 'safe' and 'unsafe' theorems of limit analysis result in the same answer. However, such cases are rare.
- The beam-spring approach can analyse only one structural element, with the soil represented by a number of springs. The main issue in this approach is to assign the appropriate spring stiffness.
- Full numerical analysis, in terms of finite elements or finite differences, is the ultimate analysis method which satisfies all theoretical requirements, discretises both the soil and the structure in the same way and can apply a constitutive behaviour which models the soil in a realistic manner.

However, it is important to recognise that full numerical analysis should not be performed blindly, without understanding the numerical methods (algorithms) and software used in the analysis. Some of the main causes of error in finite element analysis discussed in this chapter are

- element discretisation (i.e. the type and size of finite element)
- choice of the nonlinear solver (i.e. solution of the system of non-linear finite element equations)
- modelling of structural elements within the soil continuum
- choice of appropriate constitutive models.

The number of potential errors increases as the analysis becomes more advanced, involving, for example, coupled problems (i.e. both mechanical behaviour and fluid flow, where the choice of correct permeability models and hydraulic boundary conditions become important), as well as soil dynamics and earthquake engineering problems (where appropriate constitutive models, boundary conditions and time integration schemes become important).

If all the above aspects of numerical analysis are applied properly, then the numerical analysis has an enormous potential to explain the engineering behaviour of geotechnical structures and produce an optimal and safe design.

References

Borin DL (1989) WALLAP: computer program for the stability analysis of retaining walls. *GEOSOLVE.*

Gens A and Potts DM (1984) Formulation of quasi axi-symmetric boundary value problems for finite element analysis. *Engineering Computations* **1**: 144–150.

Hill R (1950) *The Mathematical Theory of Plasticity.* Clarendon Press, Oxford, UK.

Martin CM (2005) Exact bearing capacity calculations using the method of characteristics. *Proceedings of the 11th International Conference IACMAG, Torino, Italy*, vol. 4 (Barla G and Barla M (eds)). Patron Editore, Bologna, Italy, pp. 441–450.

Owen DRJ and Hinton E (1980) *Finite Moments in Plasticity: Theory and Practice.* Peneridge Press, Swansea, UK.

Pappin JW, Simpson B, Felton PJ and Raison C (1985) Numerical analysis of flexible retaining walls. *Conference on Numerical Methods in Engineering Theory and Application, Swansea*, pp. 789–802.

Potts DM and Martins JP (1982) The shaft resistance of axially loaded piles in clay. *Géotechnique* **32(4)**: 369–386.

Potts DM and Zdravkovic L (1999) *Finite Element Analysis in Geotechnical Engineering: Theory.* Thomas Telford, London, UK.

Potts DM and Zdravkovic L (2001) *Finite Element Analysis in Geotechnical Engineering: Application.* Thomas Telford, London, UK.

Prandtl L (1920) Uber die Harte Plastischer Korper. *Nachrichten von der Koniglichen Gasellschaft der Wissenschaften, Gottingen, Math. Phys. Klasse* **1920**: 74–85.

Sloan SW (1988) Lower bound limit analysis using finite elements and linear programming. *International Journal for Numerical and Analytical Methods in Geomechanics* **12**: 61–67.

Sloan SW (1989) Upper bound limit analysis using finite elements and linear programming. *International Journal for Numerical and Analytical Methods in Geomechanics* **13**: 263–282.

Sokolovski VV (1960) *Statics of Soil Media.* Butterworth, London, UK.

Sokolovski VV (1965) *Statics of Granular Media.* Pergamon Press, Oxford, UK.

Thomas JN (1984) An improved accelerated initial stress procedure for elasto-plastic finite element analysis. *International Journal for Numerical and Analytical Methods in Geomechanics* **8**: 359–379.

Zienkiewicz OC and Cormeau IC (1974) Visco-plasticity, plasticity and creep in elastic solids: a unified numerical solution approach. *International Journal for Numerical Methods and Engineering* **8**: 821–845.

All chapters within Sections 1 *Context* and 2 *Fundamental principles* together provide a complete introduction to the Manual and no individual chapter should be read in isolation from the rest.

doi: 10.1680/icemge.66816.0061

Chapter 7
Geotechnical risks and their context for the whole project

Tim Chapman Arup, London, UK
Marianne Walsh Arup, London, UK

This chapter provides an understanding of how project developers consider construction projects and their huge vulnerability to project delays, and when it can be profitable to proceed with a development. By example, it shows the proportion of geotechnical-related costs for a typical London office project and compares those costs to the tiny investments in geotechnical risk mitigation by desk study and ground investigation. It explains that geotechnical risks can be 'wildly random' instead of 'mildly random' and so the consequences of a problem can be hugely disproportionate to the initial cost. It compares typical frequencies of ground-related problems and finds that even the average cost of mildly random problems have consequences that far outweigh the very modest cost of a good ground investigation. It concludes that management of geotechnical risk by contingency is an approach fraught with peril and that management by mitigation is much more likely to produce the desired outcome.

CONTENTS

7.1. Introduction

The aim of this chapter is to provide data to enable geotechnical engineers to communicate simply and clearly with their clients, and to persuade them of the merits of timely geotechnical investigations and other interventions.

It is well known that

- construction problems occur too frequently
- the consequential delays and their associated costs are huge
- the problems could often have been avoided.

The aim of this chapter is to quantify (to some extent) the first and second points, and to offer some potential solutions to the third.

This chapter considers financial risks only and not safety ones – it examines the overall design and construction process; safety risks are usually triggered by errors in discrete operations, such as a missing toe-board. Projects where the overall design and construction process is skimped may share characteristics with those that suffer the sorts of organisational breaches that lead to accidents. Delays due to injuries and scrutiny by the Health and Safety authorities can have a devastating effect on programme and out-turn costs, not to mention the human impact on all involved. Safety in geotechnical engineering is discussed in more detail in Chapter 8 *Health and safety in geotechnical engineering.*

The examples provided in this chapter are based on the financial climate in 2009. There have been multiple changes to the financial and political climate in the intervening period, including the COVID-19 pandemic, Brexit and the Russian invasion of Ukraine, which have resulted in significant impacts on inflation and interest rates. While the particular costs have changed, the overall relationship between elements of geotechnical projects remains the same. For example, in November 2022 the costs for a ground investigation provided in Section 7.7 have (based on data from sites in London) changed to

Desk study		£7 000
Ground investigation		
5 CP boreholes 20–40 m	£30 000	
10 machine-dug trial pits	£7 000	
Contamination testing	£7 000	
Soils laboratory testing	£7 000	£51 000
		£58 000

7.2. Motivation of developers

> Heathrow's new terminal is on time and on budget. How odd.
> Headline in *The Economist*, 20 August 2005

To understand the critical issues that affect developments, it is worth examining the motives of developers, who fall into three categories.

- Commercial developers create developments essentially to make a profit, either speculatively where the eventual occupier is not yet known or pre-let to a bespoke tenant.
- Self-developers intend the development to facilitate other operations, such as building an office or a manufacturing function for a large company. Sometimes assistance from a commercial developer is sought on a consultancy basis.
- State developers create the national infrastructure to allow the rest of the economy to function (e.g. roads, railways, schools and hospitals).

In each case, a rational cost–benefit analysis can be carried out to decide whether a particular development should proceed.

To assist in understanding the sensitivity of commercial drivers on a development, it is helpful to study a hypothetical example. The example created by Chapman and Marcetteau (2004) has been altered and extended to show the main issues that influence the commercial success of a development. The building under consideration is

- a commercial office building in the City of London
- a currently occupied 30-year-old building with a gross area of 16 000 m^2 (172 000 ft^2)
- occupying a site which could be replaced by a new building with a gross area of about 20 000 m^2 (215 000 ft^2); with 85% efficiency the net lettable area would be 17 000 m^2 (183 000 ft^2). The site value, based on net lettable area, is estimated to be £376/ft^2 (a typical range would be £250–£500/ft^2) – so the land on which it is to be built would cost some £68.8 million.

The developer first needs to decide whether it is worth maintaining the existing asset, or proceeding with a new development. The type of development contemplated is shown in **Figure 7.1**, and its details are summarised in Appendix A of this chapter. The relative costs for the development are summarised in **Figure 7.2**. The basis for the choice to proceed with the development is given in **Table 7.1**. The new building will be attractive to a broader range of occupiers, including those able and willing to pay higher rents.

Figure 7.1 Typical new six-storey London office development

From the calculations in **Table 7.1**, proceeding with the development makes commercial sense based on the increase in demand (and hence rent) for the newer and bigger building. The table omits the prospect of capital growth, which has also made property development a popular investment in good times. It also omits any requirement to provide affordable housing as a proportion of the extra space created, which can be a tax that inhibits new developments. **Table 7.2** shows a comparison of typical investment returns in 2001, given by RIBA (2003), showing why investment in property was favoured at that time.

By 2005, in a stronger property market, most income yields had remained the same but capital growth for equities had accelerated to 8%, giving a total return of 11%, while capital growth for commercial property peaked at 11%, giving a total return of 18%. Both had declined by early 2008, making the choice to invest in commercial property considerably more marginal. *The Economist* (Anon 2008) reported that office rents in the City of London had risen by 22% from September 2006 to September 2007, and predicted that they may fall by 5% over the following year. At the time, it was predicted to lead to a fall in property value of up to 30% over the next three years, potentially the biggest collapse in commercial property prices since the Second World War. Of the speculative offices under construction now in the City of London, not all are forecast to find tenants in the short term. By 2009, banks had become more cautious about lending to this sector and the drop in collateral prices was not encouraging further lending to developers for this purpose. By later that year, with returns on all assets at long-term lows and liquidity re-entering the financial system, some signs of banking investment interest were returning. This persisted with the start of a full recovery in London by mid-2011 but the property market elsewhere in the UK remained in recession.

A developer can make a simple profit by using their own funds for the whole development, as shown in **Table 7.3** (RIBA, 2003).

Developers can leverage their success by use of a bank loan, but it can also magnify failure. This is illustrated in a simplified form in **Table 7.4**.

Thus the developer can make a much greater profit by using a loan, but can also make a much greater loss if the project costs more than expected. Developers therefore crave certainty even above cheapness. Once a site has been bought (potentially with an outlay of over half the development cost) and demolished (so no further income will be received until the new building is occupied), the developer is very vulnerable. They are particularly vulnerable to

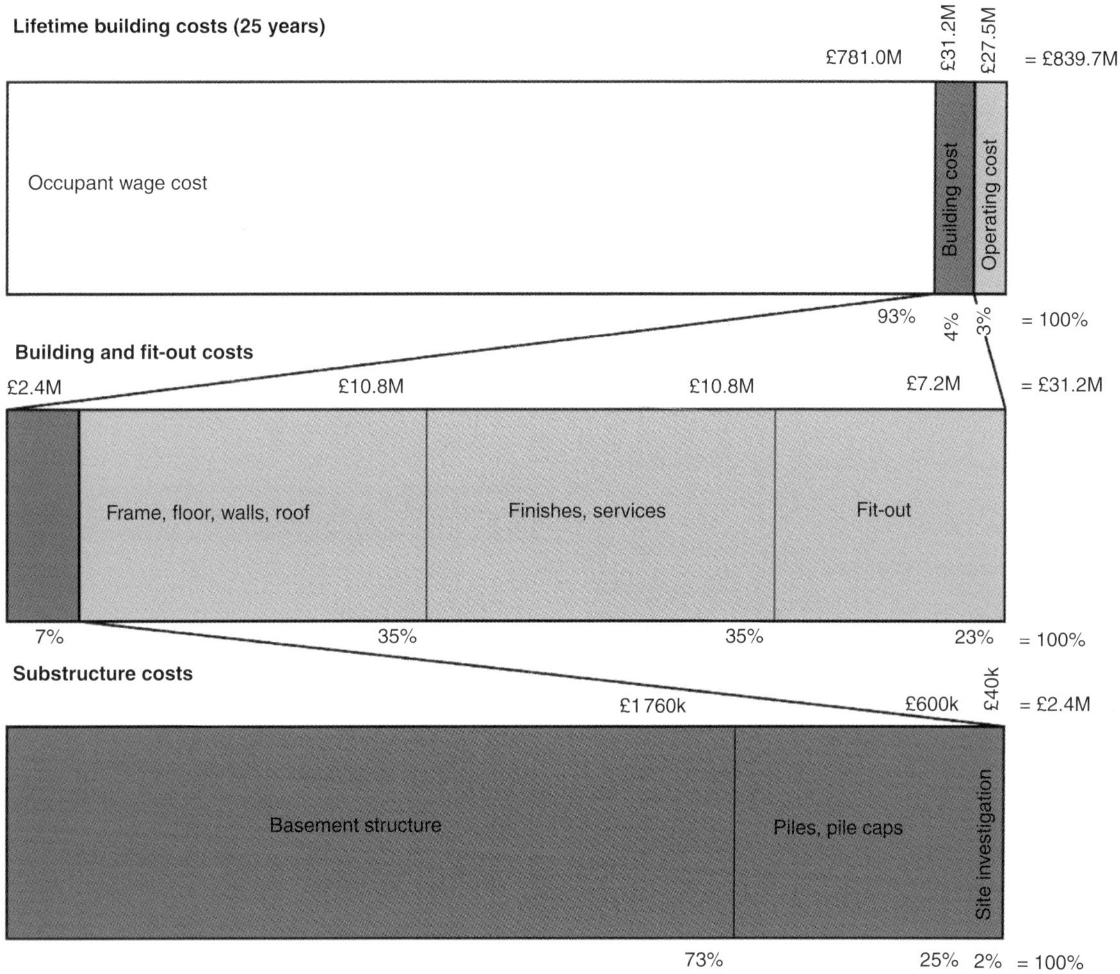

Figure 7.2 Breakdown cost for a typical building (Reproduced from Chapman, 2008; all rights reserved)

	No new development	New building
Annual rent per square foot (£)	39	55
Floor area	172 000ft² (16 000 m²)	215 000 ft² (20 000 m²)
Annual rent (£)	6.7 million	11.8 million
Payback period considered (years)	25	25 – 3 = 22 (3-year construction/rent-free period)
Total rent (£)	168.0 million	259.6 million
Less construction cost (£)	–	31.2 million
New revenue (£)	168.0 million	228.4 million (i.e. 60.4 million extra)
Notes	Probably longer fallow periods between tenants; possible shorter rental periods	Probably reduced maintenance and operating costs

Table 7.1 Comparison of value from new development to refurbishment of the building

	Building society	Gilts (10 year bond)	Equities (FTSE 500)	Commercial property
Income	4%	5%	3%	7%
Capital	0%	0%	6%	3%
Total	4%	5%	9%	10%

Table 7.2 Typical investment returns in 2001 (Data taken from RIBA, 2003)

- increase in costs – distorting the investment basis
- delay in completion – hence delaying the start of the income stream to pay for it; this is particularly painful when the development funds are borrowed and interest has to be paid over the interim period until rental cash starts coming in
- increase in interest rates, particularly if unexpected
- reduction in standard of the completed building – making it less attractive to potential clients and hence lowering its rental value.

Costs	£ million
1. Land, including fees and demolition	68.8
2. Design and construction costs	31.2
3. Leasing – at 20% of rent	2.0
4. Interest and financing	3.9
5. Void – 1 year rent free	11.8
Total cost	**117.7**
Value	
6. 215 000 ft² at £55/ft²	11.8/annum
7. Investment yield 7% (income from commercial property)	165.2
Dividing (7) by (6) gives a multiplier of 14	
Surplus	
8. Value less total cost (165.2 – 117.7)	47.5
Surplus as % of cost (47.5 × 100/117.7)	40%

Table 7.3 Overall consideration of costs and value for a new development (Data taken from RIBA, 2003)

Costs (as % of cost of a successful project)	Successful project		A failure	
	Without a loan	With a loan	Without a loan	With a loan
Total cost	100	100	160	160
Loan		–70		–70
Developer's equity outlaid	100	30	160	90
Sale proceeds	140	140	140	140
Loan repayment		–70		–70
Outcome	140	70	140	70
Surplus at completion	40	40	–20	–20
Return on developer's equity (surplus/outlay)	40%	133%	–13%	–22%

Table 7.4 Illustration of the effect of a loan on project success or failure

This is illustrated in **Figure 7.3**; this shows how rental rates need to develop rapidly to catch up with the mounting bank interest changes.

These calculations have been carried out for a City of London development, where demand generally tends to be strong and potential rental rises may offset extra costs due to any problems. In other locations the calculations underpinning a new development are often more marginal and the prospect of a favourable rate rise is more remote.

7.3. Government guidance on 'optimism bias'

> 229 If a builder builds a house for someone, and does not construct it properly, and the house which he built falls in and kills its owner, then that builder shall be put to death.
>
> Hammurabi's Code of Laws 1700 BC, Mesopotamia

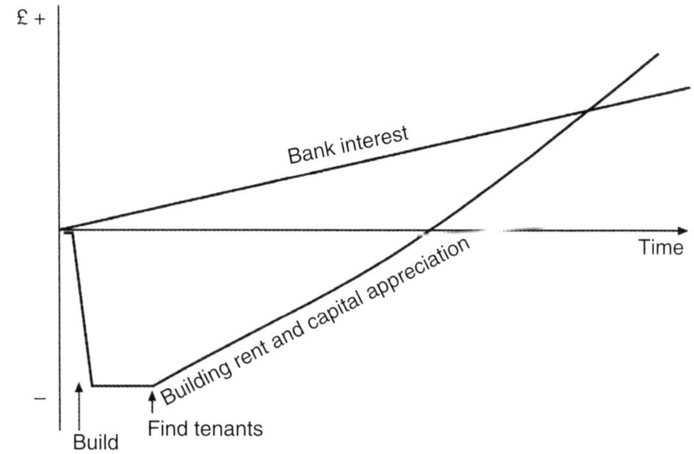

Figure 7.3 Comparison of building investment and bank interest (Reproduced from Chapman, 2008; all rights reserved)

The UK government has studied the tendency of project costs and works durations to increase and has come up with advice in its Green Book (HM Treasury, 2022) to redress the systematic optimism (which it terms 'optimism bias') that has historically afflicted the appraisal process for major projects.

It suggests that the two main causes for this bias in capital cost estimates are:

- poor definition of the scope and objectives of projects in the business case due to poor identification of stakeholder requirements
- poor management of projects during implementation, so that schedules are not adhered to and risks are not mitigated.

In *Supplementary Green Book Guidance – Optimism Bias* (HM Treasury, 2022) values are suggested for the extent of optimism bias typically found and are recommended for use unless robust evidence is available to support other values. The ranges are given in **Table 7.5**. These are based on various contributory factors, listed in **Table 7.6**. Unfortunately, these contributory factors are not broken down into the project phase in which the problems occurred, so the proportion of problems originating in the ground is not discernible. **Table 7.7** shows an extract from the *Supplementary Guidance* illustrating the sorts of steps taken to mitigate risk for an example project.

From the guidance the following can be deduced.

- Non-standard projects carry more risk, and non-standard building projects carry more risk than non-standard civil engineering projects. The reason for this is not obvious, but may be due to building projects involving a greater mix of different skills or perhaps because civil engineering projects procured by the state have more rigorous stage reviews and design checks (such as Category 3 checking).
- Building and civil engineering projects seem to be less risky than government projects to procure equipment, which includes software projects.

	Optimism bias (%)*			
	Works duration		Capital expenditure	
Project type	Upper	Lower	Upper	Lower
Standard buildings	4	1	24	2
Non-standard buildings	39	2	51	4
Standard civil engineering	20	1	44	3
Non-standard civil engineering	25	3	66	6
Equipment/development	54	10	200	10
Outsourcing	N/A	N/A	41**	0**

* These are indicative starting values recommended and not highest or lowest possible values.
**The optimism bias for outsourcing projects is measured for operating expenditure.

Table 7.5 Range of expected values for optimism bias for different sorts of projects (Reproduced from HM Treasury, 2013; © Crown Copyright)

- Building projects' greatest failings have, as prime causes, inadequacy of the business case by the client and poor procurement, leading to disputes and claims (presumably due to poor specification).
- Civil engineering projects suffer from environmental impact (presumably mitigation too late in the process), poor contractor capabilities and poor procurement leading to disputes and claims.

It is worth noting that better geotechnical design is an excellent way to mitigate the uncertainties that the allowance for optimism bias is meant to counter. By doing so, the allowance for optimism bias can be reduced – making the project more affordable. This service is of significant advantage to the project promoter.

7.4. Typical frequency and cost of ground-related problems

> When anyone asks me how I can best describe my experience in nearly forty years at sea, I merely say, uneventful. Of course there have been winter gales and storms and fog and the like but, in all my experience, I have never been in an accident … of any sort worth speaking about. I have seen but one vessel in distress in all my years at sea. I never saw a wreck and have never been wrecked nor was I ever in any predicament that threatened to end in disaster of any sort.
>
> Captain Edward John Smith, 1907 (five years before he became captain of the *RMS Titanic* for her maiden voyage)

A review of 5000 industrial buildings by the National Economic Development Office (NEDO, 1983) found 50% of the projects overran by at least 1 month. A study of a representative group of these projects showed that 37% of the overruns were due to ground problems. Of 8000 commercial buildings, one-third overran by more than a month and a further one-third by up to a month (NEDO, 1988). Half of a representative group of these had suffered delays due to unforeseen ground conditions.

The National Audit Office (2001) reported little improvement by 2001: it cites an Office of Government Commerce study which found that 70% of a range of public projects were delivered late and 73% were over the tender price. Contemporary reports from the author's experience of legal disputes and the construction press suggest that incidences of unforeseen ground conditions still occur all too frequently. The UK airport operator BAA, quoted in an article in *The Economist* in 2005, estimated the industry average for significant delays at 40% (with their own record at just under 20% due to their use of partnering and other ways of aligning their objectives with those of their project collaborators). These data are presented graphically in **Figure 7.4** – roughly one-third of the projects studied were delayed by up to a month, and a further one-third suffered more serious delays. Where data on the cause are available, about one-third to half of delays were blamed on troublesome ground conditions. Thus significant delays due to ground conditions probably occur some 17–20% of the time. It is notable from **Figure 7.4** that the incidence of delays does not seem to have improved with time.

Van Staveren and Chapman (2007) examined figures on the financial impact of project failures, which were huge. In the Netherlands alone, failure figures are assessed as 5–13% of the yearly expenditure of the Dutch construction industry (for all failures, not only in the ground). As Dutch spending on construction totals some €70 billion per year, €3.5–9 billion are spent to respond to failures (van Staveren, 2006). Chapman and Marcetteau (2004) showed that in the UK about one-third of construction projects are significantly delayed, and of those half of the delays are caused by problems in the ground. Applying this ratio to the overall Dutch construction failure rate would suggest that an average of some €2.5 billion are due to ground-related causes in the Netherlands, or probably equal to about €50 billion across the whole of the European Union (the extrapolation is based on the size of the Dutch GDP in 2006 (US $613) billion to the whole of the European Union (US $13 600 billion) (CIA Factbook, 2007)).

These ground-related problems often originate in an earlier phase than the phase in which they occur, as highlighted by Chowdhury and Flentje (2007), based on the extensive work of Sowers (1993). This latter study showed that in 57% of the projects studied, the geotechnical problems originated from flaws in the design phase and in 38% of the cases from the construction phase. However, these geotechnical problems actually materialised 41% of the time in the construction phase and 57% in the operational phase of the project.

An important part of the structural and geotechnical design processes should be for the whole project team to work together to reduce risks in the ground. Ground risks are one of the major causes of project delay, and when they occur they are seldom resolved easily or quickly. As the installation of foundations and construction of basements is usually on the project's critical path, any delay affects all follow-on activities, and can impact directly on completion date.

		Non-standard buildings		Standard buildings		Non-standard civil engineering		Standard civil engineering	
Upper bound optimism bias (%)*		39	51	4	24	25	66	20	44
		Work duration	Capital expenditure	Works duration	Capital expenditure	Works duration	Capital expenditure	Works duration	Capital expenditure
Contributory factors to upper bound optimism bias (%)**		Non-standard buildings		Standard buildings		Non-standard civil engineering		Standard civil engineering	
Procurement	Complexity of contract structure	3	1	1		4			
	Late contractor involvement in design	6	2	3	2	<1			3
	Poor contractor capabilities	5	5	4	9	2		16	
	Government guidelines								
	Dispute and claims occurred	5	11	4	29	16			21
	Information management								
	Other (specify)					1	2		
Project specific	Design complexity	2	3	3	1	5	8		
	Degree of innovation	8	9	1	4	13	9		
	Environmental impact						5	46	22
	Other (specify)	5	5			3			18
Client specific	Inadequacy of the business case	22	23	31	34	3	35	8	10
	Large number of stakeholders			6					
	Funding availability	3		8			5	6	
	Project management team	5	2		1		2		
	Poor project intelligence	5	6	6	2	3	9	14	7
	Other (specify)	1	2		<1				
	Public relations			8	2				9
Environment	Site characteristics	3	1	5	2		5	10	3
	Permits/consents/ approvals	3	<1	9					
	Other (specify)	1	3						
External influences	Political	13				19			
	Economic		13		11	24	3		7
	Legislation/regulations	6	7	9	3		8		
	Technology	4	5			6	8		
	Other (specify)		2			<1	1		

* Indicative starting values – optimism bias profile for a project will change during its life cycle.
** Contributions from each area as a percentage of the recorded optimism bias. Rounding errors may result in the sum of values not adding up to 100%.

Table 7.6 Optimism bias upper bound guidance (Reproduced from HM Treasury, 2013; © Crown Copyright)

Contributory factor	Project cost % contribution to optimism bias	Mitigation factor	Example cost of risk management
Poor contractor capabilities	5	1.0	£0
Design complexity	3	1.0	£140 000
Inadequacy of the business case	23	0.4	£700 000
Poor project intelligence	6	1.0	£10 000
Site characteristics, e.g. ground investigation*	1	1.0	£40 000

* In the example, it is suggested that a trust has owned a site for at least 20 years with a comprehensive site investigation carried out within the last 5 years; therefore only desk study and limited site investigation are needed.

Table 7.7 Main contributory factors to optimism bias for project cost and works duration for an example project given by HM Treasury (Reproduced from HM Treasury, 2013; © Crown Copyright)

7.5. Expect the unexpected

> … as we know, there are **known knowns**; there are things we know we know. We also know there are **known unknowns**; that is to say we know there are some things we do not know. But there are also **unknown unknowns** – the ones we don't know we don't know.
>
> Former US Secretary of Defense, Donald Rumsfeld, explained a state of intelligence at a US Defense Department Briefing on 12 February 2002

To a large extent, this awareness of what might happen underpins successful geotechnical design. Managing risks for the 'unknown unknowns' is difficult as it is tricky to identify a likelihood and severity for unimagined events, which are not amenable to statistical understanding.

The concept of the 'Black Swan' was introduced by Nassim Nicholas Taleb (2007) – the name based on the unassailable belief by people in the Old World that all swans were white, confirmed by centuries of empirical evidence, until black swans were encountered in Australia in January 1697 by Dutch explorer Willem de Vlamingh. Taleb has created a whole theory based on the Black Swan event, to which he attributes the following characteristics.

- As an outlier event, it lies outside the realm of regular expectations because no prior experience can point to the possibility.
- It has an extreme impact.
- Despite being an outlier, once it has happened human nature can concoct explanations for its occurrence – it is therefore obvious, but only with the benefit of hindsight.

Professor Taleb identifies many examples, ranging from the 9/11 events to the 'winner takes all' phenomenon for the wages of pop stars, footballers and artists. His thesis is that Black Swan events happen all too frequently and tend to dominate the success of many endeavours. He distinguishes between 'mild randomness' and 'wild randomness'. Dealing with ground risks is potentially a wildly random type of Black Swan because the consequences can be so severe, and so it is important that the possibility of unexpected events should not be discounted in geotechnical assessments.

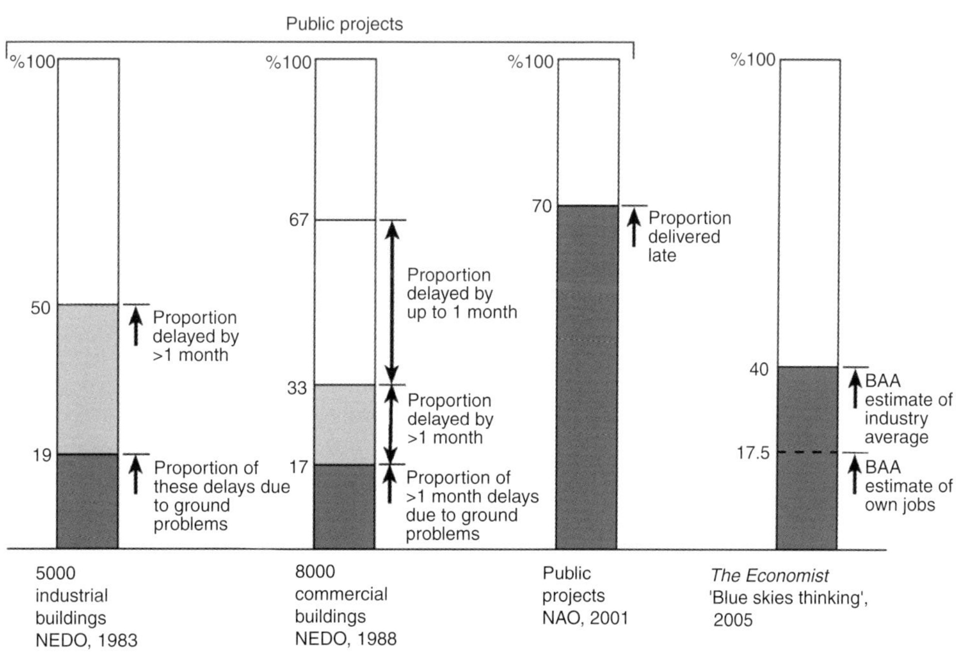

Figure 7.4 Typical UK delays for projects against time

It is possible by good planning to minimise the likelihood of a Black Swan. However, by the very nature of a Black Swan, their incidence can never be entirely eliminated. Those who manage complex geotechnical projects need to appreciate the possibility that they may be subject to Black Swans – a sense of humility is needed, plus an understanding of the range of catastrophic discoveries that can arise on any apparently innocuous site. Often Black Swans arise because engineers rely purely on previous experience and do not consider what might happen. This point is considered in greater detail in the example towards the end of this chapter.

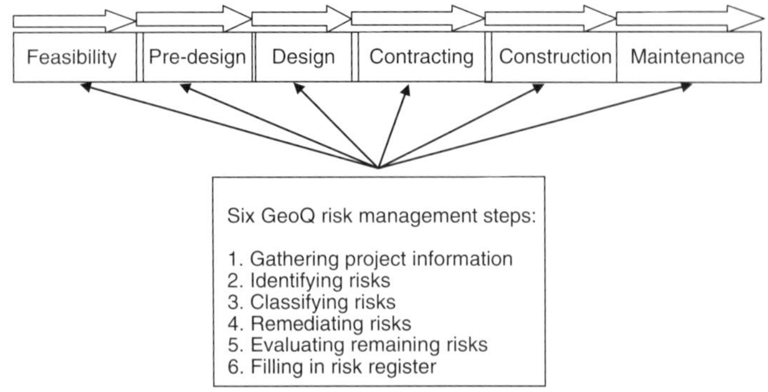

Figure 7.5 The GeoQ risk management model

7.6. Importance of site investigation

> Without site investigation, ground is a hazard.
>
> Site Investigation Steering Group, 1993

As explained in Section 4 *Site investigation*, site investigation consists of two main activities, as discussed in BS 5930 (BSI, 1999).

- Desk study, also known as *Preliminary Investigation* by Eurocode 7 (BSI, 2004) – the gathering of information about a site from documentary sources, made easier by free online aerial photography and other easily available resources, such as Envirocheck in the UK. A very valuable part of the desk study process is often to summarise all the ground-related hazards onto one site plan which can be used throughout the whole design and construction process. For larger (and more complex) sites, a geographic information system (GIS) can fulfil the same function.
- Ground investigation – the physical investigation of a site by boreholes and trial pits, with laboratory testing and reading of piezometers, to establish the ground and groundwater conditions. The ground investigation should be driven by the hazards identified in the desk study, together with an assessment of how those hazards may affect the intended development proposals and construction processes.

In addition to normal geotechnical considerations like stratigraphy and stratum properties, it is important that the following aspects of a site are also investigated

- potential presence of valuable archaeological remains
- potential presence of contaminated ground or groundwater in the vicinity
- potential presence of obstructions or other impediments to foundation installation, as well as neighbouring foundations, tunnels and so on
- buried cables and other services
- potential presence of UXO (unexploded ordnance).

Often these man-made issues are a bigger risk for the project than normal geological variability. Desk studies that only consider, for instance, easily available mapping and not mapping data that go back to the start of the site's development, should be seen as potentially deficient.

The site investigation should also address constructability – it is therefore important that likely foundation types and processes needed on the site should be considered before the site investigation takes place. The risk of delay will be considerably increased if the contractors cannot derive useful information on the efficacy of installation for different types of foundation from the site investigation reports.

It is worth remembering that a site investigation is a careful but selective sampling of the ground; therefore the vast majority of the ground will not be encountered until the main construction takes place. Using the soil volumes described in Appendix A of this chapter, a typical ground investigation will only recover some 0.03% of the soil beneath the site; only a fraction of that will be examined by a professional geologist or geotechnical engineer.

The GeoQ process, introduced by van Staveren in 2006, is a formalisation of normal good geotechnical practice, emphasising the need for early data gathering. **Figure 7.5** shows a conventional contract, with the tendering stage between design and construction. A simplified example of applying the GeoQ steps during a number of normal project stages is shown in **Table 7.8**.

7.7. Costs and benefits of site investigation

> Quality is never an accident; it is always the result of intelligent effort.
>
> John Ruskin (1819–1900)

Geotechnical risks can be managed and mitigated. Sound geotechnical practices – for example, hazard recognition and exploration by means of the desk study and ground investigation – can reduce risks substantially. The value of these investments needs to be seen in the context of the risks reduced.

For the building considered earlier and described in Appendix A of this chapter, a site investigation could consist of the following

GeoQ step	Normal project stage	Typical example
1. Gathering project information	Desk study	Collecting historical and geological maps and other information
2. Risk identification		Discovery of old maps showing a filled-in watercourse crossing the site
3. Risk classification	Normally done during the **ground investigation** stage, when the magnitude and extent of the hazard can be quantified	Prove that the channel is up to 10 m in depth and infilled with soft organic alluvium
4. Risk remediation	Normally done during **design**, when the development can protect itself against the main risks	Design of different foundations for that part of the site; inclusion of methane protection measures for basement
5. Risk evaluation	**Prior to tender**; collection of unmitigated risks so they can be managed and/or transferred to the contractor as part of the contract documents	Assess remaining extent of issues connected with the channel
6. Risk mobilisation – filling in the risk register		Highlight to contractor that their plant may get bogged down in alluvial formation; care of piling rig stability

Table 7.8 A simplified example of GeoQ steps during a few normal project stages

Desk study		£5 000
Ground investigation		
5 CP boreholes 20–40 m	£25 000	
10 machine-dug trial pits	£5 000	
Contamination testing	£5 000	
Soils laboratory testing	£5 000	£40 000
		£45 000

This needs to be seen in the context of the £24 million building cost excluding fit-out costs (or £13.2 million structural cost or £2.4 million substructure cost). This site investigation is equal to 0.19% of the building cost, 0.34% of the structure cost and 1.9% of the substructure cost. Compare this with Rowe's (1972) estimate for site investigations of 0.05–0.22% of building cost.

The relative costs for delay can be compared to the savings that may have contributed to them. Using the statistics given earlier, it can be assumed that some 20% of projects are significantly delayed by ground conditions, by perhaps 1 month or more. From the perspective of the developer, the cost of 1 month's delay can be calculated as

Land purchase cost	£68.8 million
Building cost (with fit-out)	£31.2 million
Total cost	£100 million

Taking 7% annual return, the annual cost to service the loan is £7 million, or £583 000 per month. If the delay prevents the occupier from working in the building then the costs will be much higher, as they will also have to include the lost productive effort of the people working there, which is assumed to exceed the cost of employing them – a further £31.25 million per year (equivalent to £2.6 million per month) or nearly five times more than the building finance costs.

Of course the loss of use of the building could be much higher as these figures do not include

- missing the top of a property rental cycle, so that the achieved rent over subsequent years is lower
- missing an important season, such as Christmas for a shop or summer for a holiday facility
- having to pay any dispute costs, which can be several times more than the original loss being argued over.

The developer's art involves predicting, potentially several years ahead, when there will be most demand for property and initiating the planning, design and construction process in sufficient time to provide property that meets the demand. This is illustrated in **Figure 7.6**. Thus, they are already vulnerable to uncertainties in the state of the property market when the facility is completed, as well as uncertainties in construction cost and programme.

If the developer skimps on the site investigation, a saving of half of the investigation cost might translate into £22 500 'saved' (but often a poorly thought-out site investigation makes very little saving).

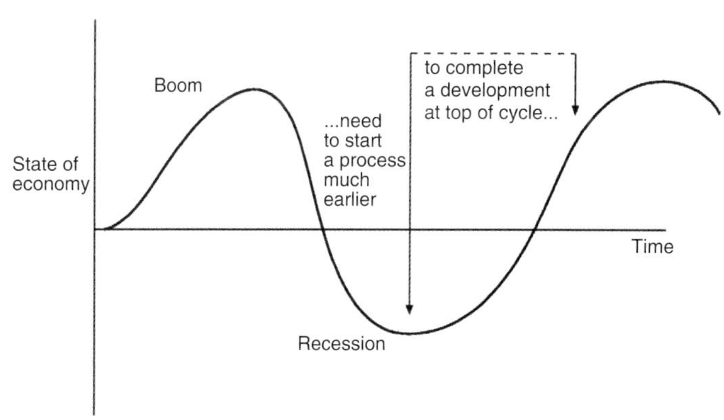

Figure 7.6 Graph showing how the start point of a development process is key to hitting the top of the economic cycle (Reproduced from Chapman, 2008; all rights reserved)

Omitting dispute and occupier costs (as well as contractor liquidated damages), the delay costs for 1 month might be

Client – delay interest		£583 000
Main contractor		
Establishment	£100 000	
Extra investigation	£30 000	
Redesign	£30 000	£160 000
		£743 000

There may also be increased construction costs due to a less efficient foundation design caused by the inadequate site investigation. As an example, the London District Surveyors Association (2000) allows a design basis for piles with a good site investigation that is 20% more efficient than one that is deficient. Thus a poor site investigation may result in piles that are some 20% less efficient, which may cost 10% more to construct. Thus, for the example under consideration, the extra foundation costs may be £60 000 more, increasing the £743 000 to £803 000.

The extra 'average' cost for deficient ground investigation can be calculated as £209 000, spread over all developments. This figure includes an average extra delay cost due to unexpected ground conditions of £149 000 based on 20% of the calculated typical delay cost of £743 000, and an allowance for less efficient design of £60 000.

Thus a saving of half of the site investigation budget of £22 500 would have to be balanced against a conservative estimate for a delay of more than £800 000 on the one-fifth of projects where the risks materialise, or an average extra cost over all projects of £209 000. In many cases the consequences are much more severe than the fairly optimistic assumptions made in this illustration.

7.8. Mitigation not contingency

> It is unwise to pay too much, but worse to pay too little; when you pay too much, you lose a little money, that's all. When you pay too little, you sometimes lose everything, because the thing you bought was incapable of doing the things it was bought to do. The common law of business balance prohibits paying a little and getting a lot. It can't be done. If you deal with the lowest bidder, it is as well to add something for the risk you run. And if you do that, you will have enough to pay for something better. There is hardly anything in the world that someone can't make a little worse and sell a little cheaper – and people who consider price alone are this man's lawful prey.
>
> John Ruskin (1819–1900)

Often when faced with a low tender price, clients concentrate on the direct saving and not on the consequential losses to which they may be subjected. They may believe that a contingency will provide them with protection, rather than choosing to resolve the issues directly. This can be valid when the consequences are 'mildly random' but not when they may cause the sort of severe 'wildly random' impact on project finances that were illustrated earlier.

Many developers bear the financial risk themselves, and if things go wrong in a way that invalidates the assumptions in their financial model they are left to sort out the consequences. The main problems are usually delays, resulting in late handover (delaying receipt of income streams), or higher costs. Often a disappointed developer will wish to recover losses from their design and construction team, for instance by liquidated damages for delay from the contractor, or by way of the designers' professional indemnity insurance policies. These are not reliable methods for recovering losses, and especially consequential losses. Better ways of covering risks are to insure them explicitly by project insurance that covers the whole team, or by latent defects insurance (LDI) that covers the final building (although LDI does not always cover consequential losses). While these insurance products encourage a shared approach to resolving problems, they do not let designers and constructors completely off the hook. For instance, the insurer may later decide to try to recover some of the losses (by a subrogated claim) if they suspect negligence was a cause.

LDI should not be seen as a solution by itself, in the same way that ever tighter liabilities on contractors and designers are unlikely to resolve a deteriorating position to the benefit of a developer. The best way to avoid the sort of issues that result in delays and losses is to avoid the problems in the first place. Thus, spending more on proper investigation of ground issues is likely to be more fruitful than engaging lawyers to devise more onerous contracts to impose on project participants.

7.9. Mitigation steps

> The rain came down, the streams rose, and the winds blew and beat against that house; yet it did not fall, because it had its foundation on the rock. But everyone who hears these words of mine and does not put them into practice is like a foolish man who built his house on sand.
>
> Matthew 7: 25–26

The standard project work stages (based on RIBA's *Plan of Work* – RIBA, 2020) are shown in **Table 7.9**, along with typical geotechnical design activities.

Timely intervention with appropriate advice at the best time is needed for all projects. Geotechnical advisers need to be aware of the timescale over which the rest of the project is being designed and the potential implications of late advice, particularly for more complex projects where there should be much early interaction with other parts of the design team, especially the structural designers.

Often geotechnical elements will present a major constraint on other parts of the development – for instance, on a tight site a complete retaining wall system might be up to 2 m wide (e.g. 0.15 m for guide wall, 0.9 m for hard/soft secant wall, 0.15 m for tolerances and protuberances to specification, reinforced-concrete facing wall of 0.30 m, void of 0.15 m, blockwork wall of 0.15 m, giving a total thickness of 1.80 m), yet sometimes architects hope that a 0.3 m reinforced-concrete wall

Project phase	RIBA work stage	Corresponding geotechnical activities
Preparation	0. Strategic Definition Identification of client's needs and objectives, business case and possible constraints on development	• Preliminary desk study, identifying major hazards that may profoundly affect development proposals
	1. Preparation and Briefing 1a. Preparation of feasibility studies and assessment of options to enable the client to decide whether to proceed 1b. Development of initial statement of requirements of the design brief by or on behalf of the client, confirming key requirements and constraints 1c. Identification of procurement method, procedures, organisational structure and range of consultants and others to be engaged for the project 1d. Development of Project Programme and Responsibility Matrix	• Full desk study, identifying all key hazards
Design and pre-construction	2. Concept Design 2a. Implementation of design brief and preparation of additional data Preparation of concept design including outline proposals for structural and building services systems, outline specifications and preliminary cost plan Review of procurement route 2b. If contractor led or two-stage D&B procurement route, prepare Employer's Requirements	Ground investigation • Geotechnical interpretative report • Initial designs for various options • Advice on preferred options • Agree requirement for site supervision
	3. Spatial Coordination 3a. Development of concept design to include structural and building services systems, updated outline specifications and cost plan Completion of project brief Submit application for detailed planning permission 3b. If one-stage D&B procurement route, prepare Employer's Requirements and tender documents	• Design development • Detailed calculations for preferred option • Agree requirement for site supervision if design relies on it • Final calculation package and meet building control standards
	4. Technical Design 4a. Production information i Preparation of technical design(s) and specifications, sufficient to coordinate components and elements of the project; confirm that they meet statutory standards and construction safety ii Preparation of detailed information for construction including manufacturing and construction information and final specifications iii Application for statutory approvals iv Preparation of further information for construction required under the building contract v Review of information provided by specialists 4b. If traditional procurement route, tender documents prepared and tendered mid-stage	• Complete tender documents • Construction design and management submission • Identify preferred foundation contractors • Contractors to identify principal construction hazards • Tender reviews • Agreement to winning contractors – name, design, proposals, etc.
Construction	5. Manufacturing and Construction 5a. Administration of the building contract to practical completion 5b. Provision to the contractor of further information as and when reasonably required 5c. Review of information provided by contractors and specialists 5d. Preparation of building manual including Health and Safety File and Fire Safety Information	• Brief resident engineer (RE) • RE supervision • Act as geotechnical adviser to the contractor • Assist with claims • Piling close-out report
Handover	6. Handover 6a. Comply with planning conditions as required 6b. Feedback on project performance 6c. Final certificate	
Use	7. Use 7a. Administration of the building contract after practical completion and making final inspections 7b. Assisting building user during initial occupation period 7c. Review of project performance in use	

Table 7.9 RIBA work stages with normal corresponding geotechnical activities (Data taken from RIBA, 2020 and ACE, 2004)

(implicitly expecting an open-cut solution) will be possible. This will have severe implications for space planning around the complete perimeter of the building, and may have worse consequences if a parking layout has been derived for the smaller wall size. It is therefore important for the geotechnical team to make sure that their indicative solutions are provided early. So, for instance, not providing a retaining wall design until the results of some advanced laboratory tests are known (and hence relatively late in the overall design process) – while protecting the geotechnical designer from possibly having to change their solution – may cause consternation among the rest of the design team.

Whenever a test is carried out on a site it carries the risk of not conforming to the specification. It is therefore wise to test earlier (and ideally off the project's critical path) instead of leaving important test results to be measured later and at a time when a non-conformance will bring greatest disruption. The following provides some examples.

- A geotechnical baseline report (GBR), as recommended by the British Tunnelling Society (2003), is a good way of ensuring clarity on risk sharing for all project types. Rather than giving comprehensive coverage of the ground conditions as in a geotechnical interpretative report, a GBR sets out the expected conditions simply, so that deviations from them are clear. It therefore seeks to provide clarity regarding when extra payment to a contractor is due, thereby reducing the likelihood of disputes.
- A ground investigation should concentrate on the greatest uncertainties usually revealed by the desk study. A ground investigation based on a poor desk study will inevitably be hit-and-miss. Considering that less than 0.03% of the soil beneath is seen, it will tend to miss more than it hits. A good ground investigation will target the most likely hazards, which will often be the near-surface stratum boundaries as man's activities will tend to be more erratic (in most cases) than geological processes.
- The initial choice of foundations can have a profound effect on the processes that follow; piles in silty soils with a high water table tend to go wrong more frequently. Choosing shallow foundations on dense sandy silts above the water table can obviate the hazard. However, if piles have been selected, and particularly if they are specified with the design to be done by the piling contractor, then a shallow foundation solution is unlikely to be selected later.
- Early testing of a preliminary test pile provides confidence in both the constructability and capacity of a chosen pile type, leaving sufficient time for the design or pile choice to be amended if problems develop. The same principle applies to other forms of early testing; for instance, testing for the efficacy of a ground improvement method or for problems during trial boring or trial drives.
- Late working pile tests (pile testing during the works), in addition to obstructing progress, also raise the possibility of adverse results with no time to react. Completion of pile load testing after installation of all of the works piles means that every pile may need to be augmented if an unexpectedly low pile capacity is measured, which can be very difficult to resolve.
- As well as reducing risks, doing more pile tests before and fewer tests during works pile installation is often cheaper and causes less disruption to a construction programme.

The process should be to consider all risks from the outset, then mitigate them according to their likelihood and potential severity. Risks should be owned and managed by those best able to do so, not just delegated or passed on. The stages for doing so are summarised against RIBA workstages in **Table 7.9**.

7.10. Example

> He who has not first laid his foundations may be able with great ability to lay them afterwards, but they will be laid with trouble to the architect and danger to the building.
>
> Extract from *The Prince*, Niccolo Machiavelli (1469–1527)

For an example of how the occurrence and severity of a geotechnical Black Swan event can be mitigated by sound engineering processes for various levels of study and investigation by a design team, see **Table 7.10**.

In the case of poor investigation of the risks before starting, there would be no inkling of the possibility of a potentially catastrophic feature; the risk would initially be evaluated based on an expected range of mildly pessimistic assumptions. It is unlikely that any allowance would be made in the cost planning for such a catastrophic feature with potentially dire consequences. Its occurrence would cost heavily – it could have the potential to cause serious injuries to site workers, especially if the collapse was exacerbated by the large construction equipment used, the possible disposal of water or by significant vibrations from construction plant. In the first case there is also a possibility that the feature might not even be discovered during construction. Often such features can collapse subsequently, resulting in the closure of an operational building with huge financial costs – and possibly with a risk of injuries or fatalities.

7.11. Conclusions

> Everybody, sooner or later, sits down to a banquet of consequences.
>
> Robert Louis Stevenson (1850–1894)

A frequent cause of problems on projects is a failure to resolve issues at an early stage. There are many issues where the associated risk can be effectively mitigated during

- initial data gathering
- design
- procurement of construction.

Many clients procure their design consultants on the assumption that all will provide design products of similar quality and that price should be a major basis for differentiating between

Site status – amount of investigatory work undertaken	Example of what might be known with the stated extent of investigation	Likely engineering approach given state of knowledge	Potential consequence if foundations were to be installed with state of knowledge	Rumsfeldian classification of feature when its presence was detected	Validity of approach
No desk study with hit-and-miss ground investigation	No knowledge of a large old mineshaft on the site	Evaluate site as normal	Major safety risk for site workers (e.g. collapse failure of a foundation during installation)	Unknown unknown	Low known costs but high risk of subsequent failure. Also likely to have long dispute about costs and responsibility
Only a desk study, no ground investigation	Knowledge there could possibly be a mineshaft somewhere on the site	Know of possible presence but not its poor condition	Feature's location and condition to be discovered so it is likely that there will be site delay while condition of feature is investigated	Known unknown	Possibly low known costs but still risk of significant delay and dispute
Desk study and ground investigation	Know there is an old mineshaft, its location and dimensions. Also know that its condition is poor and it is at risk of collapse	All aspects known – best foundation solution chosen at the start; contractor knows how to cope with issue at tender time	Condition known in advance so low risk of extra direct costs or delay. Therefore unlikely to be a dispute	Known known	Potentially higher known costs but much lower risk. Development should be on time and to budget without debilitating row

Table 7.10 Illustration of risk mitigation and consequences

them. For simpler projects with easily defined deliverables and where external influences are more easily controlled, price can be a valid comparator, especially if the design consultants have been pre-qualified for their skill and relevant abilities. For more complex projects, what matters most is the quality of the output, such as

- collection of data into a very coherent and useful desk study report
- consideration of buildability and ways to resolve such issues
- better integration of temporary and permanent works solutions.

The mineshaft example in **Table 7.10** illustrates the very different consequences from various levels of preparation and study by the design team prior to the start on site.

If total design fees are 5% of the project cost, then an improvement in output quality with 20% more effort (more time or more experienced input) can translate into a 1% increase in project cost. Hopefully this will translate into much bigger savings in out-turn project cost in terms of

- a lower incidence of hazards causing problems
- a more deliverable programme
- smarter ideas resulting in lower capital cost.

For projects where viability is very tight, a greater investment up-front is even more important, otherwise the project is more likely to encounter unexpected events that will turn it into a financial failure. Perversely, it is often on such projects where most effort is made to minimise design costs that the likelihood of a Black Swan event occurring is inadvertently increased.

This chapter has sought to demonstrate the following.

- While foundations are a relatively small part of the overall project cost (perhaps 7%), their problems can be responsible for half of significant project delays, occurring some 17–20% of the time.
- Typically only 0.05–0.22% of the project's building cost are spent investigating those risks, even though 'average' consequences are many times more.
- Developers' financial models are very vulnerable to unexpected shocks such as project delays, especially when funded by loans, which is usually the case.
- Geotechnical uncertainties can be 'wildly random' rather than 'mildly random' – the effects of a problem can quickly magnify and be disproportionately expensive to solve. Therefore management by mitigation is far more effective than management by contingency.
- A coherent geotechnical risk reduction process can significantly reduce risks. A comprehensive desk study is one of the most cost-effective steps.
- The design, by admitting the possibility (or even probability) of some uncertainty, can go further in reducing the incidence of risk.
- Employment of experienced contractors who make proper allowance for uncertainty will generally provide a more reliable solution than a cheap solution by a less experienced contractor. Poor procurement and poor contractor capabilities were highlighted by the UK Treasury as major contributors to cost and programme overruns on government-funded projects.
- A testing strategy that confronts risks at an early stage by early testing (e.g. load testing of piles) and which reduces the incidence of unexpected or anomalous results during the main construction works will reduce programme vulnerability.

- Clarity of risk ownership is important, with the best strategy being ownership of the risk by those best able to manage it – not just dumping of risks on those at the bottom of the supply chain. Early consideration and mitigation at the opportune time are better than late reaction to a deteriorating situation.
- Reliance on damages and negligence is not an effective way for developers to manage their risks with their constructors and designers – they will seldom recover all their consequential losses, which will usually dwarf any direct losses.
- Only very seldom are ground engineering problems caused by gaps in engineering knowledge. Most problems are caused by a failure to apply known processes adequately.
- Communication and information sharing will often result in earlier identification of problems, enabling them to be solved earlier and more easily.

This chapter illustrates that there are no shortcuts to gathering pertinent information, digesting it into a form that can be shared across the whole team and using it throughout the whole design and construction process. The main aim should therefore be for robust solutions and overall reliability. The reduction of ground risk needs better application of existing processes, not necessarily better science.

References

ACE (Association of Consulting Engineers) (2004) *Conditions of Engagement*. ACE, London, UK.

Anon (2005) Heathrow's new terminal is on time and on budget. How odd. *The Economist*, 20 August.

Anon (2008) Commercial property: dominoes on the skyline. *The Economist*, 5 January.

British Tunnelling Society (2003) *The Joint Code of Practice for Risk Management of Tunnel Works in the UK*. BTS, London, UK.

BSI (British Standards Institution) (1999) BS 5930:1999: British standard code of practice for site investigations. BSI, London, UK.

BSI (2004) BS EN 1997-1:2004: Eurocode 7: Geotechnical design – General rules. BSI, London, UK.

Chapman T (2008) The relevance of developer costs in geotechnical risk management. *Proceedings of the Second BGA International Conference on Foundations, Dundee, UK* (Brown MJ, Bransby MF, Brennan AJ and Knappett JA (eds)). BRE Press, London, UK.

Chapman T and Marcetteau A (2004) Achieving economy and reliability in piled foundation design for a building project. *The Structural Engineer*, 2 June: pp. 32–37.

Chowdhury R and Flentje P (2007) Perspectives for the future of geotechnical engineering. *Proceedings of the International Conference on Civil Engineering For the New Millennium: Opportunities and Challenges, Shibpur, India*, pp. 59–75.

CIA Factbook (2007) https://www.cia.gov/the-world-factbook (accessed 02/05/2023).

HM Treasury (2013) *The Green Book: Appraisal and Evaluation in Central Government*. TSO, London, UK.

HM Treasury (2022) *Supplementary Green Book Guidance: Optimism Bias*. TSO, London, UK.

London District Surveyors Association (2000) Guidance notes for the design of bored straight shafted piles in London Clay. Foundations Note No 1. LDSA, London, UK.

National Audit Office (2001) *Modernising Construction, HC87*. TSO, London, UK.

NEDO (National Economic Development Office) (1983) *Faster Building for Industry*. NEDO, London, UK.

NEDO (1988) *Faster Building for Commerce*. NEDO, London, UK.

RIBA (Royal Institute of British Architects) (2003) *The Commercial Offices Handbook* (Battle T (ed)). RIBA Publishing, London, UK.

RIBA (2020) *Plan of Work*. RIBA Publishing, London, UK.

Rowe PW (1972) The relevance of soil fabric to site investigation practice: 12th Rankine Lecture. *Géotechnique* **22(2)**: 195–300.

Site Investigation Steering Group (1993) *Without Site Investigation, Ground is a Hazard*. Thomas Telford, London, UK.

Sowers GF (1993) Human factors in civil and geotechnical engineering failures. *Journal of Geotechnical Engineering* **119(2)**: 238–256.

Taleb NN (2007) *The Black Swan: The Impact of the Highly Improbable*. Allen Lane, London, UK.

van Staveren M Th (2006) *Uncertainty and Ground Conditions: A Risk Management Approach*. Butterworth Heinemann, Oxford, UK.

van Staveren M Th and Chapman T (2007) Complementing code requirements: managing ground risk in urban environments. *Proceedings of the XIV European Conference on Soil Mechanics and Geotechnical Engineering, Madrid, Spain* (Cuéllar V, Dapena E, Alonso E, *et al.* (eds)). Millpress Science Publishers, Rotterdam, the Netherlands.

All chapters within Sections 1 *Context* and 2 *Fundamental principles* together provide a complete introduction to the Manual and no individual chapter should be read in isolation from the rest.

Appendix A
Example development

This example is an idealisation to provide indicative total costs for a typical building development. It is acknowledged that the total and relative costs will vary enormously for different developments. Assumptions are based on a site slightly smaller than 1 ha (1 hectare = 10 000 m^2) in a central area of London. As the site will accommodate 17 000 m^2 (183 000 ft^2) of net lettable area, the land price was set at £68.8 million (the site value, based on net lettable area, is assumed to be £376/ft^2 – mid-way in the 2009 typical range of £250–500/ft^2). On the site is placed a 20 000 m^2 gross area high-quality air-conditioned office building, six storeys high with a single-level basement for cycle and car parking and plant. The net lettable area is therefore 85% of gross floor area. A generous allowance of 17 m^2 per person gives a total building population of about 1000. An average salary for the occupants of this building has been taken as £25 000 per annum. Employer's payroll costs are taken as an extra 25%, giving an average employee cost of £31 250. The total salary roll per year is therefore £31.25 million. Over a 25-year building life, the total employment cost will have been about £781 million (expressed in constant 2009 £).

A building cost has been calculated as typically £1200/m^2, giving a total of £24 million. The costs have been apportioned as about 10% for substructure (foundations and basement), 45% for frame, floors, walls and roof, and 45% for finishes and services. Of the £2.4 million for substructure, it has been assumed that £600 000 went on piles and pile caps, with the rest for the basement structure. Category A fit-out of the building has been assumed to cost some 30% of the total building cost (£7 million). The tenant's fit-out has been taken as £10/m^2 (£200 000), giving a total of about £7.2 million for fit-out. The total building costs for construction and fit-out are therefore £31.2 million, which equates to £1560/m^2 or £145/ft^2. Professional fees are assumed to be included in the construction cost figures. The annual operating costs are assumed to be approximately £55/m^2 per year, giving a total of £1.1 million per year, or a total of £27.5 million over a 25-year design life. The site investigation costs have been taken as £5000 for a desk study and £40 000 for a ground investigation, including interpretation. These yield a total site investigation cost of £45 000. This represents about 0.19% of the building cost excluding fit-out (which was £24 million); this is typical for the site investigation proportion of structural costs which vary between 0.05% and 0.22% for building projects collected by Rowe in 1972.

Based on the hazards identified in the desk study, it has been assumed that a site less than 1 ha will need five boreholes to depths ranging from 20 to 40 m, ten trial pits and associated testing for geotechnical properties and contamination, both on site and in the laboratory. Assuming a site of 90 × 90 m and considering the relevant soil depth that might influence the development down to 40 m, the total volume of soil for which geotechnical representation is needed is 324 000 m^3. The soil actually seen on site will be as follows:

- boreholes: $5 \times (0.15\ m)^2\ \pi/4 \times 30$ m average depth = 2.65 m^3
- trial pits: $10 \times 3\ m \times 1\ m \times 3\ m = 90\ m^3$.

This gives a total of around 93 m^3. The proportion of soil recovered by the ground investigation is therefore some 0.03% of the total. The proportion examined by a professional engineer or geologist will be a much smaller proportion again.

doi: 10.1680/icemge.66816.0077

Chapter 8
Health and safety in geotechnical engineering

Pauline Sloan COWI, London, UK
Delwynne Cuttlian previously of Balfour Beatty Ground Engineering, Basingstoke, UK
Tony P Suckling A-squared Studio Engineers Ltd, London, UK

CONTENTS

Construction is a hazardous occupation, with 30 workers killed in work-related accidents in Great Britain in 2021/22. This is in comparison with the average number of 36 fatalities for 2017/18–2021/22. In addition, there were 59 000 non-fatal work-related injuries each year averaged over the three-year period 2019/20–2021/22. There were an estimated 78 000 work-related ill-health cases (new or long-standing), estimated as the annual average for 2019/20–2021/22. It is important to note that the ill-health cases are estimated as there could be many more cases which were not reported. Safety is everyone's responsibility, including clients, designers, contractors and general workers, and everyone must work towards reducing the human cost of our profession. Every construction project is unique, with a particular combination of clients, customers, end-users, contractors, subcontractors, processes, materials, hazards and risks. Maintaining safety is a primary focus of any site; there are legal obligations to be met by all directly involved in any project and all associated with it. Moreover, there is a moral and ethical obligation to ensure that no one is harmed by our site activities. It is the responsibility of all to identify risks, eliminate them where possible and to mitigate against the residual risk. Geotechnical engineering has its own specific hazards and risks. Safety on every project must be planned, managed, continuously monitored and improved by competent personnel as part of the 'way we work' every day. Responsibility for safety cannot be delegated.

8.1. Introduction

Every construction project will have a unique combination of particular contractors, subcontractors, processes, materials, hazards and risks. Maintaining health, safety and well-being should be a primary focus of any site. There are legal obligations to be met by all directly involved in any project and all associated with it. Clients must ensure that competent people are appointed to manage the project, designers must ensure that risks are identified and eliminated during the design phase (pre-construction phase), the principal contractor and contractors on site must manage and monitor the processes to ensure the construction work is completed safely. It is everyone's responsibility to identify risks, to eliminate them where possible and to mitigate against residual risks. The accident record of the construction industry is improving; however, in relation to most other industries construction remains dangerous, with 30 deaths reported in 2021/2022 in Great Britain. Everyone, from the most senior to the most junior on any project, has the right to expect to work in a safe place and to return home safe and well at the end of the working day. Geotechnical engineering has its own specific hazards and risks, and these are generally described. Safety on each and every project must be planned, continuously monitored and improved by competent personnel as part of the 'way we work' every day. Responsibility for safety cannot be delegated.

8.2. An introduction to the legislation

Health and safety has gained great importance in the construction industry over the years, partly due to increased health and safety legislation, partly due to pressure exerted on the industry because of its poor previous safety record and more importantly due to increased levels of health and safety awareness. This chapter gives a brief overview of some of the liabilities under such legislation and is not intended to be a detailed reference on the subject. It should be noted that much of the legislation described applies only to England and Wales; however, the legislation applicable to Scotland and Northern Ireland is almost identical in most respects. For projects outside the UK different legislation will apply.

Most health and safety legislation is goal setting: the legislation describes broad principles which are expected to be achieved; the detail of how to achieve these principles is primarily left to the employer. Any breach of health and safety legislation is a criminal offence and as such both fines and/or imprisonment can be imposed; the levels of penalties are set out in the Health and Safety (Offences) Act 2008. Conviction can result in a criminal record. Health and safety law places a reverse burden of proof on any defendant, by the nature of the fact that there is an injured party and an accident has taken place. It is up to the defendant to demonstrate that they have complied with the legislation rather than the prosecutor having a responsibility to demonstrate that they have not. The employer must demonstrate that they have taken all reasonably practicable steps to prevent the foreseeable accident occurring.

The Health and Safety at Work etc. Act 1974 (HSAWA) imposes duties on all parties (including the employer and the employee) with the purpose of ensuring the health and safety of all people at work. The HSAWA is an enabling act, which means there are many regulations, often accompanied by approved codes of practice (ACOPs) and guidance notes that take effect under the act. The HSAWA gives appointed inspectors wide-ranging powers of investigation and enforcement, including the serving of improvement and prohibition notices.

The Management of Health and Safety at Work Regulations (HSE, 1999) places duties on employers and the self-employed in relation to others affected by their work. The main duties placed on employers are the assessment, elimination and reduction of risk, the need to establish emergency procedures and health surveillance, and to provide information and training for employees. The duties in respect of risk assessment in this legislation are the basis of all risk assessment at work.

The Construction (Design and Management) Regulations (CDM) were first enacted in 1994 and were substantially revised in 2007 (HSE, 2007) and then further revised in 2015 (HSE, 2015). CDM places duties on everyone involved in the construction process and is one of the most significant pieces of legislation governing them. The 'dutyholders' are divided into six key categories: client (including domestic clients), principal designer, designer, principal contractor, contractor and worker. Each of these dutyholders have responsibilities. A summary of the main responsibilities of the dutyholders is provided in **Table 8.1**.

CDM requires that anyone who wishes to appoint or engage someone to do a construction activity, or undertake design work, must be satisfied that the person they want to appoint has sufficient knowledge, skills and experience to carry out the tasks and their duties under the regulations. Accidents and major incidents often occur because the people carrying out potentially complex tasks have been appointed because they are the cheapest and can do the job the quickest. This is as true for geotechnical works as for any other construction activity. There have been many incidents where contractors have not understood the correct sequences (see Chapter 79 *Sequencing of geotechnical works*) or have not been experienced enough to undertake the job – thereby being ignorant of the hazards and risks involved. All persons have this duty; so a structural designer who needs to engage the services of a geotechnical engineer must ensure that the person has the required skills, knowledge and experience and not appoint on the basis of cost alone.

CDM requires that a designer must not begin work on a project unless satisfied that the client is aware of their duties under the regulations. The client must make suitable arrangements for managing their project, including allocating sufficient time and other resources to ensure the construction work can be carried out, so far as reasonably practicable, without risk to the health and safety of anyone affected by the project. The client is responsible for appointing in writing the principal designer and the principal contractor. A client should ultimately be satisfied that the whole design, procurement and implementation team have sufficient skills, knowledge and relevant experience. The client should therefore ensure that if members of the project team appoint subcontractors, they have rigorous procedures in place for similar assessment. The client must also provide any relevant pre-construction information to every designer and contractor appointed or being considered for appointment to the project. In addition, before the construction phase begins the client must ensure that the principal contractor/contractor has drawn up the construction phase plan.

A project is 'notifiable' if the construction work is scheduled to last more than 30 days and have more than 20 workers or exceed 500 person-days. The client is responsible for notifying the Health and Safety Executive (HSE) using online form *F10 – Notification of construction project*. If the client is a 'commercial' client, then they must notify the HSE. If the client is a 'domestic' client then they can arrange for the principal contractor, contractor or principal designer to notify the HSE on their behalf.

Designers from all disciplines have a contribution to make in reducing and avoiding health and safety risks which are inherent in the whole life of the project, including construction process and subsequent work, maintenance, usage and demolition. The most important contribution a designer can make to improve health and safety will often be at the concept and feasibility stages. For geotechnical works, maintenance and demolition are often ignored as being irrelevant, but they should always be considered. In addition to focusing on the physical health and safety risks associated with the construction phase, designers must equally consider the occupational health risks associated with the whole life of the project.

The main duty of designers of 'structures' is to assess the implications of their design on the health and safety of people affected by the building when in use, during maintenance of the 'structure' and at the demolition stage. This is a significant departure from previous safety legislation where most responsibility was placed on contractors. The definition of a 'structure' is wide-ranging and includes items such as buildings, railway lines, tunnels, bridges, roads, airfields, pipelines, sewers, drainage works, earthworks, lagoons, caissons and earth-retaining structures. It also refers to any formwork, scaffold or falsework which is designed to provide support or access during construction.

The client must also ensure that the principal designer prepares a Health and Safety File for the project; if the principal designer's duties end before the end of construction, then the principal contractor should complete the file. Once construction is complete, CDM requires the client to retain the Health and Safety File. This file also commonly includes or makes reference to the operations and maintenance manuals (O&M Manual). The focus of the O&M Manual is to provide information on the 'structure' relevant to the health and safety of those using, maintaining, repairing or renovating the structure.

CDM dutyholders:* Who are they?	Summary of role/main duties
Clients are organisations or individuals for whom a construction project is carried out.	Make suitable arrangements for managing a project. This includes making sure • other dutyholders are appointed • sufficient time and resources are allocated • relevant information is prepared and provided to other dutyholders • the principal designer and principal contractor carry out their duties • welfare facilities are provided.
Domestic clients are people who have construction work carried out on their own home, or the home of a family member that is not done as part of a business, whether for profit or not.	Domestic clients are in the scope of CDM 2015, but their duties as a client are normally transferred to • the contractor, on a single-contractor project or • the principal contractor, on a project involving more than one contractor. However, the domestic client can choose to have a written agreement with the principal designer to carry out the client duties.
Designers are those who, as part of a business, prepare or modify designs for a building, product or system relating to construction work.	When preparing or modifying designs, to eliminate, reduce or control foreseeable risks that may arise during • construction • the maintenance and use of a building once it is built. Provide information to other members of the project team to help them fulfil their duties.
Principal designers** are designers appointed by the client in projects involving more than one contractor. They can be an organisation or an individual with sufficient knowledge, experience and ability to carry out the role.	Plan, manage, monitor and coordinate health and safety in the pre-construction phase of a project. This includes • identifying, eliminating or controlling foreseeable risks • ensuring designers carry out their duties. Prepare and provide relevant information to other dutyholders. Provide relevant information to the principal contractor to help them plan, manage, monitor and coordinate health and safety in the construction phase.
Principal contractors are contractors appointed by the client to coordinate the construction phase of a project where it involves more than one contractor.	Plan, manage, monitor and coordinate health and safety in the construction phase of a project. This includes • liaising with the client and principal designer • preparing the construction phase plan • organising cooperation between contractors and coordinating their work. Ensure • suitable site inductions are provided • reasonable steps are taken to prevent unauthorised access • workers are consulted and engaged in securing their health and safety • welfare facilities are provided.
Contractors are those who do the actual construction work and can be either an individual or a company.	Plan, manage and monitor the construction work under their control so that it is carried out without risk to health and safety. For projects involving more than one contractor, coordinate their activities with others in the project team – in particular, comply with directions given to them by the principal designer or principal contractor. For single-contractor projects, prepare a construction phase plan.
Workers are the people who work for or under the control of contractors on a construction site.	They must • be consulted about matters which affect their health, safety and welfare • take care of their own health and safety and others who may be affected by their actions • report anything they see which is likely to endanger either their own or others' health and safety • cooperate with their employer, fellow workers, contractors and other dutyholders.

* Organisations or individuals can carry out the role of more than one dutyholder, provided they have the skills, knowledge, experience and (if an organisation) the organisational capability to carry out those roles in a way that secures health and safety.
** Principal designers are not a direct replacement for CDM coordinators. The range of duties they carry out is different to those undertaken by CDM coordinators under the CDM 2007.

Table 8.1 A summary of roles and duties under CDM 2015 (Reproduced from *Managing Health and Safety in Construction, Construction (Design and Management) Regulations 2015, Guidance on Regulations*, L153, HSE (2015) © Crown Copyright)

Information on the safe demolition of the 'structure' should also be included. The Health and Safety File should include any residual risks and detailed as-built drawings, providing valuable information for the next client.

CDM also makes provisions relating to welfare. Both the client and the principal contractor have responsibilities for ensuring that suitable welfare facilities are available prior to construction works starting.

There is no specific health and safety legislation for geotechnical engineering. British Standards such as 'BS 8008:1996: Safety precautions and procedures for the construction and descent of machine-bored shafts for piling and other purposes' (BSI, 1996) provide accepted standard practice; while they have no legal status they are often referred to in order to establish if legal duties have been adequately discharged. However, the specific requirements of geotechnical design and construction must be considered within the legislative requirements for the whole industry.

Other legislation which may need to be considered includes, but is not limited to

- Control of Substances Hazardous to Health (COSHH) Regulations (2002)
- Manual Handling Operations Regulations (1992)
- The Provision and Use of Work Equipment Regulations (1998)
- The Lifting Operations and Lifting Equipment Regulations (1998)
- Personal Protective Equipment Regulations (1992)
- The Work at Height Regulations (2005)
- The Control of Vibration at Work Regulations (2005)
- The Control of Noise at Work Regulations (2005)
- The Confined Spaces Regulations (1997)
- The Workplace (Health, Safety and Welfare) Regulations (1992).

For more detailed information, refer to the *ICE Manual of Health and Safety in Construction* (McAleenan and Oloke, 2015).

8.3. Hazards

A hazard is anything that has the potential to cause harm to someone or something. This could apply to almost any item if used in a particular way; combining two or more items could give numerous hazards that may need to be considered on any project.

Ground investigation works are hazardous as the personnel are often the first onto the site as part of the development. The principal contractor may not have been appointed at this stage, so the specialist ground investigation organisation will be required to perform these duties while on site. As these works are, by their very nature, an investigation, it should always be anticipated that there will be hazards which will not have been identified or even anticipated by the desk study. Also, if these works are to be undertaken inside an abandoned building or structure, then consideration must be given to particular hazards (e.g. asbestos, lead), and a structural assessment must be undertaken prior to the ground investigation.

Most geotechnical works are specialist and are normally undertaken by specialist subcontractors. The subcontractor often undertakes both the design and the construction of their element of the works and so has the best understanding of the particular hazards and risks associated with the product or process. The principal contractor and the designer/engineer are most likely to have the best understanding of the specific risks and hazards of the overall project. Therefore, all parties must work together, both before and during the works, to identify the hazards and introduce appropriate mitigation and control measures.

A list of general hazard areas that must be considered on any construction project is provided below, with more detailed discussion of some of the hazards associated with geotechnical works provided in the clauses which follow. This list is not exhaustive and the unique combination of factors which go to make up any project means that each project must be reviewed on its own merits. Reviews should take place as necessary to reflect significant changes. Some areas for consideration are

- access and egress to the place of work for vehicles and pedestrians
- the movement of vehicles on site, including the need to separate vehicles and pedestrians wherever practical
- protection of the public (almost every construction project will have some interaction with the public; consideration must be given as to how to ensure they remain safe and that the project inconveniences them as little as possible)
- protection of the environment (watercourses, groundwater, habitat, species, noise, vibration, dust, etc.)
- exposure to biological agents or zoonotic diseases (e.g. leptospirosis, Lyme disease)
- exposure to harmful plants (e.g. giant hogweed)
- provision of welfare for workers (this can be a particular challenge on geotechnical projects which may be small projects of a transient nature or large projects which stretch over many miles)
- the effects of weather (provision must be made for ensuring adequate lighting during winter months, shelter and protection in hot weather, workers must be able to dry wet clothes, the effect of wind speed on tall plant may need to be considered, as well as how to maintain safe ground conditions in heavy rain or snow)
- site security (not only must provision be made to ensure that members of the public do not inadvertently access the works areas, but consideration should also be given to maintaining records of who is on site and what they are doing)
- exposure to buried services and overhead power lines
- designing for safety (consideration must be given to designing out risks wherever practical; this should include any temporary works structures and geotechnical works in the temporary state)
- emergency situations (consideration needs to be given to dealing with incidents on site such as the provision of first aiders, the location of the closest hospital and mechanisms for workers to report emergency situations)
- management of works on site (including the programming of works to ensure that risks are minimised both from a design point of view and taking into account the physical working space)
- the elimination of the need to work at height, or the control of the risk of falling (this should include the risk of falling into excavations and falling objects)
- the elimination of the need to work in a confined space
- management of fatigue (including working hours, travel time and rest periods); fatigue can increase the risk of accidents and mistakes

- protection of workers' health (workers' short- and long-term health can be affected by a number of things such as noise, dust, biological hazards, psychosocial/mental health issues, manual labour and vibration; these hazards must be identified and managed).

8.3.1 Shallow excavations

The stability of the sides of excavations depends on the ground conditions, including the effects of groundwater, as well as the depth of excavation and the surcharge placed on the sides of the excavation. It is important to note that stability is likely to decrease with time and can reduce very quickly by the introduction of water from precipitation or elsewhere. Unfortunately, every year people are killed or seriously injured, having been trapped in unsupported excavations which have collapsed or by falling materials within the excavation. CDM states that all practicable steps must be taken to prevent danger to persons in or near excavations, including, where necessary, the provision of supports or battering. No one should enter any unsupported excavations, regardless of depth. CDM does not specify any requirement for the types of support or battering, but they must be inspected by a competent person before any construction work can take place within the excavations. An assessment of stability should be based on sound engineering design principles and not judgement. Excavation stability must be assessed by a competent person, taking into account items such as ground conditions, weather conditions, activities adjacent to the excavation and the movement of groundwater. The excavation should be inspected before each shift and following significant changes such as heavy rainfall.

Consideration must also be given to access/egress to excavations (for people, plant, equipment and materials), maintaining the safety of the excavation with regards to water ingress, underfoot conditions and the possible build-up of gases. Measures must be taken to ensure that people, equipment and/or vehicles do not fall into the excavation and that people can egress the excavation safely in the event of an emergency.

Care must also be taken to ensure that when an excavation has been completed it is backfilled with appropriate material and compacted where required. This is of particular importance where other activities will be using the area. Plant tracking over a poorly reinstated excavation is at significant risk of overturning. This not only applies to shallow excavations, but also to pile bores and deep urban excavations.

8.3.2 Piling works

Piling works use specialist plant which can weigh more than 140 tonnes, can be over 30 m high and is front-heavy; the driver usually does not have full visibility around the machine. A safe exclusion zone must be maintained around the piling plant, the public must be protected from risk and vehicle and plant movement around the piling plant must be properly controlled. Piling rigs require a significant amount of space to operate safely and the specialist contractor should be consulted as early as possible to ensure that the project is sequenced in a way which ensures that this space can be provided and maintained.

Three particular hazards associated with piling works are as follows.

- Design, maintenance and repair of the working platform (**Figure 8.1** and Chapter 81 *Types of bearing piles*). The platform must be adequate for the full duration of its use. Soft or hard areas beneath the platform, either existing before or created after construction of the platform, can cause rig instability.
- Failure of a pile load test reaction system (**Figure 8.2**). Significant and real forces are applied through the test pile and reaction system, and each and every component must be designed and constructed rigorously.

Figure 8.1 A piling rig which overturned due to the platform not having sufficient capacity to support the bearing pressure of the machine

Figure 8.2 A reaction test set-up which collapsed when a component part failed

- Site personnel getting caught up in rotating parts such as drill strings, augers or other rotating parts of the specialist machinery. The risk is much greater when operators need to work closer to the machine when adding/removing sections. The Provision and Use of Work Equipment Regulations (PUWER) 1998 and associated ACOP set out a hierarchy for safeguarding. 'Appendix 4: The hierarchy of protection under PUWER and SMSR' (HSE, n.d.), sets out the hierarchy that should be applied in selecting safeguarding for drilling and piling rigs. Where fixed guarding is appropriate and achievable it should be used, before selecting moveable interlocked guards, trip wires, pressure-sensitive mats and exclusion zones.

8.3.3 Deep urban excavations

Figure 8.3 shows an example of a congested city centre project which requires the excavation of a deep basement.

The particular hazards in deep urban excavations include

- the effect of the works on adjacent structures
- proximity to underground infrastructure and services beneath the basement, particularly in city centres
- sequencing of works is of particular concern where ground movement could occur as excavation of basement structures progresses
- protection of members of the public
- planning and sequencing the works to ensure that there is sufficient space for each of the contractors to work safely
- use of tower cranes – these have a unique set of hazards and must be appropriately planned, designed, registered, maintained and operated by competent persons.

8.3.4 Contaminated ground

Short- and long-term health problems can arise from contact with contaminated ground. Potentially dangerous substances need to be identified, the related risks assessed and safe systems of work put in place in accordance with the Control of Substances Hazardous to Health Regulations (COSHH).

In 1992, the British Drilling Association (BDA) published *Guidance Notes for the Safe Drilling of Landfills and Contaminated Land*, and these were subsequently adopted by the Institution of Civil Engineers Site Investigation Steering Group and republished in 1993 as Part 4 of the publication *Site Investigation in Construction*. In 2008, the British Drilling Association revised edition of the 1992 BDA publication was introduced, *Guidance For Safe Intrusive Activities on Contaminated or Potentially Contaminated Land*, introducing a site classification system (green, yellow, red). Anyone involved with drilling into the ground, whether thought to be contaminated or not, should refer to such specialist documentation.

Specialist advice must be sought when dealing with contaminated land.

Figure 8.3 A typical congested city centre project

8.4. Risk assessment

Risk assessments are used as a means of documenting, defining and managing risk as a response to the Management of Health and Safety at Work Regulations. Risk identification and evaluation is the process of looking at the hazards identified, calculating the severity of the outcome if their potentials were realised, and checking who might be affected. Risk control is ensuring that the hazard's potential cannot be realised. Therefore the risk assessment process, put simply, is figuring out how to keep people safe and then recording what has been decided, and why – and then revisiting this regularly to ensure it is still relevant and appropriate.

All geotechnical processes will have associated hazards and risks; these are identified and defined in the risk assessment form. Mitigation measures (controls) are introduced to hopefully eliminate the hazard. If complete elimination of the hazard is not possible then measures are introduced to reduce the risk of harm to an acceptable level. What defines an acceptable level of risk depends on a number of factors, including the legislative requirements. Competent advice should be sought from geotechnical specialists when reviewing to check that proposed controls measures are adequate and realistic.

The best mechanisms are to avoid or eliminate hazards at source; however, if this is not possible then consideration must first be given to substituting the hazard for a safer alternative which does not introduce a new hazard. The next option is isolating the individuals from the hazard. If none of these options are viable, then consideration should be given to engineering controls. These are generally controls which work for a large number of people at the same time and do not require human input (e.g. using a fixed guard rail around an excavation). If preventive measures cannot be adopted then protective measures should be considered next. Protective measures will generally only work for one person at a time and will require the input of that person – for instance, they may be required to wear a safety harness to prevent a fall into an excavation. This whole process is documented in the risk assessment form. All control measures must be put in place, monitored and maintained, so that the risk to health and safety remains at a low level. Project-specific hazards and risks are dealt with as part of the same process.

The risk assessment process cannot be programmed to be undertaken at a particular point in a project, as the hazards and risks associated with a project will change daily. The process of identifying hazards, assessing risk and implementing appropriate controls is an ongoing one which should involve everyone who is associated with the project. Many hazards can be effectively eliminated and/or controlled long before the work on site starts, provided the right people, with the right skills, are involved in the planning process.

The other vital part of the risk assessment process is ensuring that the people who are going to do the work have the right information, instruction and training in order to keep themselves safe. Any significant risks on a project must be briefed to the workers. This is usually done in an induction. Inductions do not need to be long and complex; they should simply explain the key site risks and controls, the emergency procedures and the site rules. The inductor should confirm that everyone has understood the briefing before the activity begins. Finally, after exhausting all the other control measures, there is the introduction of personal protective equipment (PPE).

Most construction companies have their own processes and procedures in relation to identifying and recording hazards and risks. Specialist geotechnical organisations are the same. These processes are usually based around the principles described in the Management of Health and Safety at Work Regulations, in particular Schedule 1 – General Principles of Prevention. Further guidance is freely available on the HSE's website.

8.5. Well-being

Although the construction industry can be very rewarding it can also be very challenging. Many projects require long working hours, irregular shift working, very tight programmes, limited budget and challenging environmental conditions. All of these can have an impact on our mental health and well-being and can lead to stress, anxiety, depression and other mental health disorders. In construction, there were an estimated 20,000 work-related cases of stress, depression or anxiety detailed in the *Construction Statistics in Great Britain, 2022* (HSE, 2022).

Many companies and construction sites will have dedicated mental health first aiders available to support and signpost people to help. In addition, many companies will have employee assistance programmes available 24/7 to offer specialist support. We all have a responsibility to break the stigma associated with mental ill health to encourage people to reach out and get the help and support they need. We must raise awareness of mental ill health and train our colleagues in the signs and symptoms to enable them to identify changes in themselves and others and to offer support. As an industry we must train our people in positive well-being, resilience and the importance of self-care. Together we can help to ensure that everyone goes home safe and well each day.

References

BDA (British Drilling Association) (1992) *Guidance Notes for the Safe Drilling of Landfills and Contaminated Land*. BDA, Brentwood, UK (withdrawn).

BDA (2008) *Guidance for Safe Intrusive Activities on Contaminated or Potentially Contaminated Land*. BDA, Brentwood, UK.

BSI (British Standards Institution) (1996) BS 8008:1996+A1:2008: Safety precautions and procedures for the construction and descent of machine-bored shafts for piling and other purposes. BSI, London, UK.

HSE (Health and Safety Executive) (2007) *Managing Construction for Health and Safety, Construction (Design and Management) Regulations 2007. Approved Code of Practice*. HSE Books, London, UK (withdrawn).

HSE (2015) *Managing Health and Safety in Construction, Construction (Design and Management) Regulations 2015, Guidance on Regulations*. HSE, London, UK. See https://www.hse.gov.uk/pubns/books/l153.htm (accessed 15/05/2023).

HSE (2022) *Construction Statistics in Great Britain, 2022*. HSE, London, UK. See http://www.hse.gov.uk/statistics/industry/construction.pdf (accessed 15/05/2023).

HSE (n.d.) 'Appendix 4: The hierarchy of protection under PUWER and SMSR'. HSE, London, UK. See http://www.hse.gov.uk/foi/internalops/og/og-00055-appendix4.pdf (accessed 15/05/2023).

Institution of Civil Engineers Site Investigation Steering Group (1993) *Site Investigation in Construction. Part 4: Guidelines for the Safe Investigation by Drilling of Landfills and Contaminated Land*. Thomas Telford, London, UK.

McAleenan C and Oloke D (eds) (2015) *ICE Manual of Health and Safety in Construction*, 2nd edn. Thomas Telford, London, UK.

Further reading

BDA (2007) *Guidance for the Safe Operation of Dynamic Sampling Rigs and Equipment*. BDA, London, UK.

BDA (2008) *Guidance for Safe Intrusive Activities on Contaminated or Potentially Contaminated Land*. BDA, London, UK.

BDA (2015) *Health and Safety Manual for Land Drilling: 2015 – A Code of Safe Drilling Practice*. BDA, London, UK

BDA (2020) *Guidance for the Operation of Cable Percussion Rigs and Equipment, v2.2*. BDA, London, UK.

BSI (2009) BS 5228-1:2009+A1:2004: Code of practice for noise and vibration control on construction and open sites – Part 1: Noise. BSI, London, UK.

BSI (2011) BS 22475-3:2011: Geotechnical investigation and testing. Sampling methods and groundwater measurements. Conformity assessment of enterprises and personnel by third party. BSI, London, UK.

BSI (2011) BS 22475-2:2011: Geotechnical investigation and testing. Sampling methods and groundwater measurements. Qualification criteria for enterprises and personnel. BSI, London, UK.
BSI (2017) ISO 18400-103:2107: Soil quality sampling. Part 103: Safety. BSI, London, UK.
BSI (2019) BS EN 50104:2019: Electrical equipment for the detection and measurement of oxygen. performance requirements and test methods. BSI, London, UK.
BSI (2022) BS EN 60079-29-1:2016+A1:2022: Explosive atmospheres, Part 29-1: Gas detectors – Performance requirements of detectors for flammable gases. BSI, London, UK.
Building Research Establishment (2004) *Working Platforms for Tracked Plant*. Building Research Establishment, Watford, UK.
CIRIA (Construction Industry Research and Information Association) (1995) *Contaminated Land Risk Assessment: A Guide to Good Practice*. CIRIA Report C552. CIRIA, London, UK.
CIRIA (1996) *A Guide for Safe Working on Contaminated Sites*. CIRIA Report R132D. CIRIA, London, UK.
CIRIA (1999) *Environmental Issues in Construction: A Desk Study*. CIRIA Report 73. CIRIA, London, UK.
CIRIA (2008) *Site Health Handbook*, 2nd edn. CIRIA Report C670. CIRIA, London, UK.
CIRIA (2008) *Site Safety Handbook*, 4th edn. CIRIA, London, UK.
CIRIA (2015) *Environmental Good Practice on Site Guide*, 4th edn. CIRIA, London, UK.
Clayton CRI (2001) *Managing Geotechnical Risk: Improving Productivity in UK Building and Construction*. Thomas Telford, London, UK.
Engineering Council, The (2021) *Guidelines on Risk for the Engineering Professional*. The Engineering Council, London, UK. See http://www.engc.org.uk/media/3824/guidance-on-risk.pdf (accessed 15/05/2023).
Engineering Council, The (2021) *Guidelines on Sustainability for the Engineering Profession*. The Engineering Council, London, UK. See http://www.engc.org.uk/media/3555/sustainability-a5-leaflet-2021-web_pages.pdf (accessed 15/05/2023).
Environment Agency (2014) *Technical Guidance WM3: Waste Classification – Guidance on the Classification and Assessment of Waste*. EA, London, UK. See http://www.gov.uk/government/publications/waste-classification-technical-guidance (accessed 15/05/2023).
Environment Agency (2020) *Land Contamination Risk Management (LCRM)*. EA, London, UK. See http://www.gov.uk/government/publications/land-contamination-risk-management-lcrm (accessed 15/05/2023).
Environment Agency (2020) *Dispose of Waste to Landfill*. EA, London, UK. See http://www.gov.uk/guidance/dispose-of-waste-to-landfill (accessed 15/05/2023).
Environmental Services Association (2012) *Landfill Gas, Industry Code of Practice, The Management of Landfill Gas*. ESA, London, UK.
HSE (1999) *Reducing Error and Influencing Behaviour*, 2nd edn. HSE, London, UK. See https://www.hse.gov.uk/pubns/books/hsg48.htm (accessed 15/05/2023).
HSE (2006) *Health and Safety in Construction*, 3rd edn. HSE, London, UK. See https://www.hse.gov.uk/pubns/books/hsg150.htm (accessed 15/05/2023).
HSE (2012) *Working with Substances Hazardous to Health, A Brief Guide to COSHH*. HSE, London, UK. See https://www.hse.gov.uk/pubns/indg136.htm (accessed 15/05/2023).
HSE (2013) *The Health & Safety (First-Aid) Regulations 1981. Guidance on Regulations*, 3rd edn. HSE, London, UK. https://www.hse.gov.uk/pubns/books/l74.htm (accessed 15/05/2023).
HSE (2013) *Managing for Health and Safety*, 3rd edn. HSE, London, UK. See https://www.hse.gov.uk/pubns/books/hsg65.htm (accessed 15/05/2023).
HSE (2013) *Respiratory Protective Equipment at Work, A Practical Guide*, 4th edn. HSE, London, UK. See https://www.hse.gov.uk/pubns/books/hsg53.htm (accessed 15/05/2023).
HSE (2014) *Safe Work in Confined Spaces. Confined Spaces Regulations 1997. Approved Code of Practice and Guidance*, 3rd edn. HSE, London, UK. See https://www.hse.gov.uk/pubns/books/l101.htm (accessed 15/05/2023).
HSE (2014) *Safe Use of Lifting Equipment, Lifting Operations and Lifting Equipment Regulations 1998, Approved Code of Practice and Guidance*, 2nd edn. HSE, London, UK. See https://www.hse.gov.uk/pubns/books/l113.htm (accessed 15/05/2023).
HSE (2014) *Avoiding Danger from Underground Services*, 3rd edn. HSE, London, UK. See https://www.hse.gov.uk/pubns/books/hsg47.htm (accessed 15/05/2023).
HSE (2018) *Work with Ionising Radiation, Ionising Radiation Regulations 2017, Approved Code of Practice and Guidance*, 2nd edn. HSE, London, UK. See https://www.hse.gov.uk/pubns/books/l121.htm (accessed 15/05/2023).
HSE (2018) *Safe Use of Work Equipment, Provision and Use of Work Equipment Regulations 1998, Approved Code of Practice and Guidance*, 4th edn. HSE, London, UK. See https://www.hse.gov.uk/pubns/books/l22.htm (accessed 15/05/2023).
HSE (2019) *Hand–Arm Vibration*, 2nd edn. HSE, London, UK. See https://www.hse.gov.uk/pubns/books/l140.htm (accessed 15/05/2023).
HSE (2020) *EH40/2005 Workplace Exposure Limits*, 4th edn. HSE, London, UK. See https://www.hse.gov.uk/pubns/books/eh40.htm (accessed 15/05/2023).
HSE (2023) *Health and safety statistics*. www.hse.gov.uk/statistics/index.htm (accessed 15/05/2023).
Hughes P and Ferrett E (2020) *Introduction to Health and Safety at Work*, 7th edn. Routledge, Oxon, UK.

Useful websites

British Standards Institution: http://www.bsigroup.com/en-GB
Environment Agency: http://www.gov.uk/government/organisations/environment-agency
Health and Safety Executive: http://www.hse.gov.uk
Institute of Occupational Safety and Health: http://www.iosh.com
Institution of Civil Engineers website: http://www.ice.org.uk
Sentencing Guidelines Council: http://www.gov.uk/government/organisations/the-sentencing-council-for-england-and-wales

All chapters within Sections 1 *Context* and 2 *Fundamental principles* together provide a complete introduction to the Manual and no individual chapter should be read in isolation from the rest.

doi: 10.1680/icemge.66816.0085

Chapter 9
Foundation design decisions

Tony O'Brien Mott MacDonald, UK
John B Burland Imperial College London, London, UK
Imran Farooq Mott MacDonald, UK

CONTENTS

The objective of this chapter is to dispel the widely held view that foundation engineering is simply a matter of designing and then constructing a foundation to carry a given load. In practice, many key decisions have to be made in developing the conceptual design before detailed design calculations can be undertaken. An approach to tackling these key decisions is given. For foundations to be constructed economically and sustainably it is necessary for the geotechnical and structural engineer to work closely together from an early stage. Examples are given of the benefits associated with close cooperation across the design team. Great emphasis is placed on recognising that foundation engineering is a 'process' involving a number of interlinked operations. Success depends on the skill of the engineers and operators and the way these operations are monitored and modified as work progresses. The chapter contains a number of case histories describing problems that emerged during construction and how these were solved. They illustrate how important it is to keep the 'geotechnical triangle' in balance.

9.1. Introduction

In 1951 Professor Karl Terzaghi, in a lecture given at the Building Research Congress in London, said (Terzaghi, 1951):

> On account of the fact that there is no glory attached to the foundations and that the sources of success or failure are hidden deep in the ground, building foundations have always been treated as stepchildren and their acts of revenge for lack of attention can be very embarrassing.

The aim of this chapter is to describe the decisions and procedures that have to be undertaken in arriving at a foundation design and in successfully constructing the foundation and the structure it is intended to support. It is important to understand right from the start that foundation engineering is essentially a process. The process involves a number of interlinked operations, and success depends on the skill of the engineers and operators and the way these operations are monitored and modified as work progresses. Thus, the design and construction of a foundation is critically dependent not only on the choice of foundation, but on numerous other factors including the ground and groundwater conditions and the way they are influenced by the construction method, the skill and experience of the contractor, the supervision and quality control of each operation and a clear understanding of the performance requirements of the structure resting on the foundations. Furthermore, a good understanding of ground-related risk and early engagement between designers, contractors and owners/operators can lead to practical and sustainable risk mitigation measures being incorporated in the process.

A number of case histories are included which are chosen to emphasise the importance of the geotechnical engineer, the contractor and the structural engineer working closely together to ensure that the process as a whole is successful.

9.2. Foundation selection

Many textbooks cover the topic of foundation design; however, the engineer reading these textbooks could be left with the impression that foundation design is solely concerned with the calculation of bearing capacity and settlement. In practice, many of the key decisions have been made before bearing capacity and settlement calculations are carried out. A useful mnemonic for the foundation designer is the 'five Ss'.

- *Site:* What are the nature, topography and size of the site? Are there major constraints, such as limited headroom or major underground services? Are there settlement-sensitive structures adjacent to the site? What is the site history – for example, is it underlain by historic mining, old buried foundations?
- *Soil:* What is the soil (or ground) profile and groundwater regime? This should be understood in the context of the local geology and hydrogeology. Do competent soils exist at shallow depth? Will groundwater seriously affect foundation construction?
- *Structure:* The size of structure, its structural form and layout, the type, magnitude and direction of applied loads, together with the allowable foundation movements, will be major factors in foundation selection.
- *Safety:* Numerous ground-related hazards may affect the site. These may be due to the intrinsic hazards associated with the soil/rock types which underlie the site and the way they may have been affected by post-depositional geological processes (such as landslips), or with human activities, such as historical industrial processes which have left a legacy of ground/groundwater contamination or hazardous gases, or mining, quarrying or non-engineered materials (dumped fills or landfills) which are potentially

unstable. Reference should be made to Chapters 8 *Health and safety in geotechnical engineering*, 40 *The ground as a hazard* and 41 *Man-made hazards and obstructions*, in addition to guidance provided by the Health and Safety Executive (e.g. HS(G)66 (HSE, 1991)) and CIRIA guides (e.g. CIRIA Reports 132 (Steeds *et al.*, 1996), 552 (Rudland *et al.*, 2001) and 807 (Davis *et al.*, 2023)).

- *Sustainability:* This requires a consideration of environmental and social factors and the economic implications of different foundation options. The aim should be to minimise the environmental and social impact of foundation construction, providing the costs are reasonable. Environmental impacts include an assessment of embodied energy, and the carbon footprint of different options. Social factors would include potential disruptions to adjacent communities from construction traffic, noise, vibration and so on.

The 'five Ss' are discussed in more detail in Chapter 52 *Foundation types and conceptual design principles*. At the end of this chapter a case history is given which provides a practical example of the factors that influence foundation selection. Foundation types are commonly classified into two types (**Figure 9.1**).

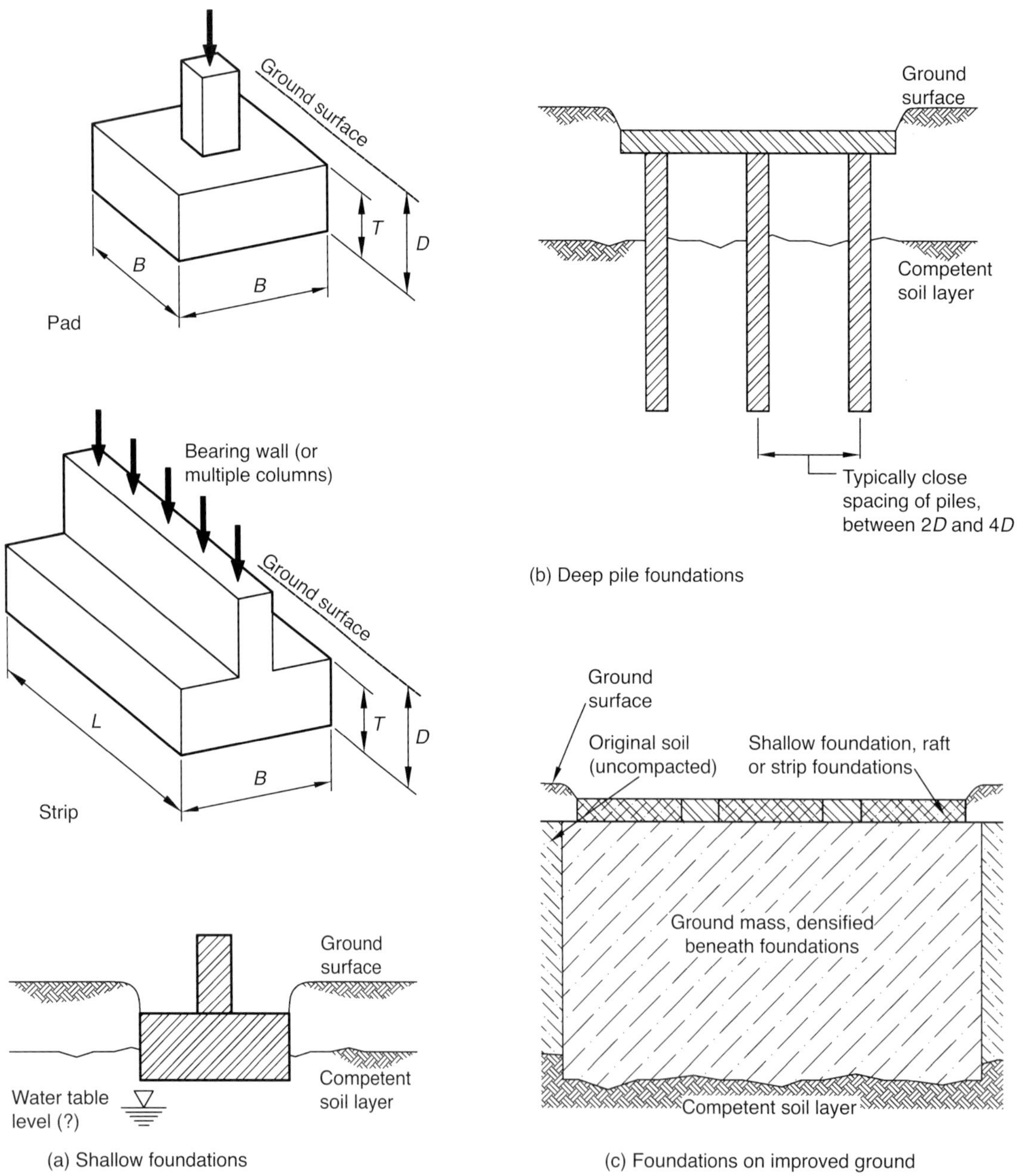

Figure 9.1 Foundation types

- *Shallow foundations:* these include discrete pad footings which usually support a single superstructure column; strip footings which may support a wall; and raft foundations which support the entire superstructure. These are discussed in more detail in Chapter 53 *Shallow foundations.*
- *Deep foundations:* most commonly these include piled foundations, either as single piles below superstructure columns/piers (refer to Chapter 54 *Single piles*) or as several relatively closely spaced piles acting together as a pile group to support either the entire superstructure or heavily loaded parts of the structure (refer to Chapter 55 *Pile-group design*). More specialist deep foundations include deep shafts, caissons and barrettes.

It is also useful to consider a third generic foundation type (**Figure 9.1**).

- *Hybrid foundations:* this type comprises elements common to both shallow and deep foundations. Examples include: deep ground improvement (to increase the bearing capacity and/or reduce the compressibility of soils below the structure), with shallow foundations constructed on the improved ground; piled rafts where the structure is supported by a raft and a few widely spaced piles. Chapter 59 *Design principles for ground improvement* discusses ground improvement design principles and Chapter 56 *Rafts and piled rafts* discusses piled rafts.

The ground profile and the allowable foundation movement are two key factors in selecting the most appropriate foundation type. If competent founding strata are close to the ground surface then shallow foundations should be considered, since they are usually relatively simple and cheap to construct. However, as foundation loads increase and/or allowable movements are reduced, then either shallow foundations with ground improvement or the use of piled foundations will become more appropriate. Often, the decision is fairly simple – for example, if there is a deep layer of heterogeneous made ground covering the project site, and the foundations are heavily loaded (say, for a multistorey public building or large bridge), then the use of piles will be obvious. However, if the proposed structure applies modest loads and is relatively tolerant of ground movement (say, a steel-frame single-storey warehouse), then it may be possible to found on the made ground (or non-engineered fill), especially if a deep ground improvement technique can be used to strengthen/stiffen the made ground. The important topic of foundation engineering within non-engineered fills is considered in Chapter 58 *Building on fills.*

For some sites the foundation designer may be faced with a difficult decision in deciding upon the most appropriate foundation type. The 'five Ss' will then need to be considered in great detail and the designer will have to juggle with a wide range of factors, such as ground strength/compressibility, allowable movement, available space for construction equipment, potential impacts on neighbours and the local community, and the ease, speed and safety of different construction options. Once the advantages/disadvantages are considered and balanced for each option, then a final selection of foundation types, and their dimensions, can be made. Some of the factors that will affect these decisions are subjective, and will be affected by perceptions of how to best minimise risks, especially during construction. This inevitably varies between individuals and organisations and will also reflect local construction practices (which vary markedly around the world). Common factors in properly understanding ground risks are

- understanding the site history and geology – this requires a comprehensive desk study to be carried out (refer to Chapter 43 *Preliminary studies*)
- understanding the site's groundwater regime and how it impacts upon the proposed construction and permanent works
- hazards that may be encountered during construction that may affect different construction techniques (e.g. encountering running sands may limit the possibility of using CFA piling, and alternative foundation options may need to be considered).

In the vast majority of cases when foundation problems occur it is because the above have not been properly understood by the foundation designer. A coherent approach to proper understanding of ground risk is provided by the 'geotechnical triangle' (introduced in Chapter 4 *The geotechnical triangle*), and practical examples are provided in Section 9.4.

Figure 9.2(a) provides examples of the wide range of pad and pile sizes and shapes for the support of a nominal 9000 kN load on a deep layer of stiff overconsolidated clay. Although all the options are 'safe', in the sense that they all have an acceptable factor of safety against bearing capacity failure, their expected settlement and the issues which need to be considered for construction all vary markedly.

- Pad foundation: Is the expected settlement (at 100 mm) too large? If yes, can the pad be made larger to reduce settlement to a tolerable amount? Is there sufficient space to form the pad? If the pad is excavated, will it undermine adjacent structures or slopes? If settlements will be relatively quick then can an extended hold period with staged construction allow criteria to be met?
- Can shallow dig-and-replace or deeper ground improvement improve the performance and meet the movement criteria?
- Pier foundations: If deep excavations are required, can an open excavation be formed (at a stable slope angle) or will temporary retaining walls be needed? Or will specialist caisson/shaft sinking methods be more appropriate? Again available space across the site will be a critical consideration. Will excavation need to be carried out below the water table? If the groundwater table is to be lowered to facilitate excavation, what settlement will be induced? What volume of water will be extracted and can it be disposed of? Can a caisson be constructed under water?
- Under-reamed piles: Will the ground be stable enough to form an under-ream? This will often depend on the detailed 'fabric' of the ground, together with the groundwater regime. Do silt/sand layers exist within the clay (which may lead to the under-ream becoming unstable prior to concreting), or is the clay uniform and of low permeability?
- Large-diameter (1.2 or 1.5 m dia.) piles: These will require a large heavy piling rig to be mobilised to site, and a thick temporary working platform, which will be expensive. Will a lot of

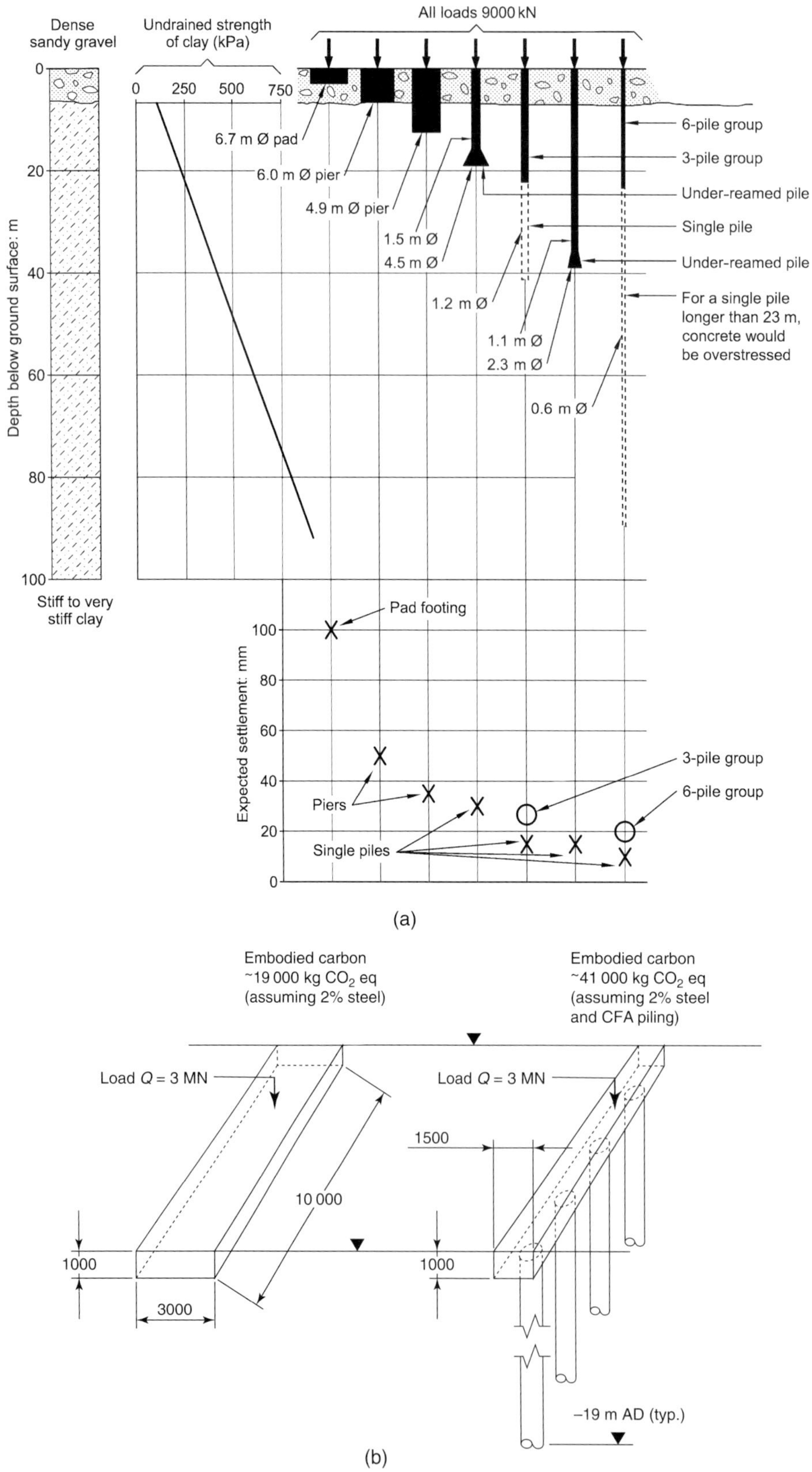

Figure 9.2 (a) Methods of supporting a 9000 kN load (Reproduced from Cole, 1988); (b) Embodied carbon comparison for shallow and piled foundations founded on firm clay

large-diameter piles be needed, or is this just for a few isolated foundations (hence the overall foundation requirements need to be considered for the whole project)? Are there significant weight restrictions for heavy plant, either due to weak surface soils or weak bridges to gain entry into the site?

- Medium-diameter (0.6 m dia.) piles: Is there sufficient headroom across the site to construct piles using conventional plant? For bored piles, will any temporary support be needed, such as temporary casing through surface gravels or support fluids at greater depth (if the clay is interbedded with silts/sands/gravels which lead to pile bore instability, especially if groundwater pressures are high)? Headroom and pile bore stability will also need to be considered for large-diameter piles.

For smaller and less powerful piling rigs an important consideration will be the risk of hitting obstructions – for example, if old foundations are buried beneath the site, or if hard rock layers need to be penetrated. In these circumstances, alternative methods may be preferred, such as rotary bored methods with core barrel, reverse circulation drilling, roller bit ODEX methods or others.

Clearly when practical construction issues are being considered, early discussions with specialist contractors will be helpful. However, it is important for the foundation designer to develop a broad knowledge of construction issues so that discussions with specialist contractors can be meaningful.

Often designers may not fully appreciate or check the embodied carbon for the foundation options being considered. Designers should, where possible, consider the sustainability aspects as part of their option selection. **Figure 9.2(b)** illustrates an example of a piled solution compared with a shallow foundation for exactly the same ground condition and loading. The calculated embodied carbon for the shallow foundation solution is approximately 19 000 kg CO_2 eq (assuming 2% steel), whereas the embodied carbon is approximately 41 000 kg CO_2 eq (assuming 2% steel and CFA piling) for the piled solution. Clearly there is a sustainability benefit of using a shallow foundation in this scenario.

The predicted settlements for the different foundation types, shown in **Figure 9.2(a)**, vary by an order of magnitude; hence an understanding of allowable foundation movement is clearly important when selecting the most appropriate and cost-effective foundation type. Often the limits set for allowable total and differential settlement are arbitrary and over-conservative. It is important for the foundation designer to have early discussions with the structural engineer on this matter. Usually total settlement is used as an indicator of what is 'allowable'; however, structural damage will be a function of *differential* rather than total settlement. It is also important to note that it is the differential settlement which occurs *after* the structure is built (and structural connections formed, and sensitive finishes completed) that will lead to damage to the structure. Hence, a considerable percentage of the total settlement that will occur during construction (due to the dead weight of the major structural elements) will be built out and have no impact on the risk of structural damage. This topic is considered in further detail in the context of foundation design in Chapters 26 *Building response to ground movements and tunnelling* and 52 *Foundation types and conceptual design principles*. Finally, it should be noted that the main opportunity for economic foundation design is to set realistic (rather than over-conservative) limits on allowable foundation movement, especially differential movements.

9.3. A holistic approach to foundation engineering

Most textbooks tend to focus on the behaviour of a foundation element (say a strip footing or a pile) acting in isolation. However, practical foundation engineering requires the foundation designer to consider the overall site redevelopment strategy, and in particular how different foundation types may be used in order to ensure the foundations are safe, simple and economic to construct.

Foundations are assemblages of different components, which may include various ancillary processes to facilitate design or construction (such as groundwater control, grouting, ground anchors, etc.) together with the use of retaining walls (either as temporary or permanent works). The interrelationship between foundation components, ancillary processes and assemblages is illustrated in **Figure 9.3**. A simple example would be the construction of a bridge abutment adjacent to a river bank (**Figure 9.4**).

(*a*) The *bridge abutment* is a composite structure. The stem of the inverted-T retaining wall retains the approach embankment fill, and the base of the T acts as a pad footing to support the bridge deck and traffic loads. The settlement of the abutment is affected by the loads and settlement induced by the adjacent embankment, as well as those directly applied from the bridge.

(*b*) *Temporary excavations* will be required for construction and may involve a combination of temporary sheet piling and/or groundwater control.

(*c*) On the *adjacent river bank* the stability of the slope will need to be checked (under varying river levels), and the river bank will need erosion protection, so scour does not lead to the bridge abutment being eroded and undermined.

(*d*) The interaction between the above elements also needs to be considered – for example, will the temporary excavation or temporary surcharge loads for the bridge abutment construction destabilise the river bank? Will the ground movements associated with the approach embankment adversely affect the bridge abutment? During a flood event there may be seepage under the inverted-T wall, reducing the compressive load and increasing the eccentricity and frictional resistance to sliding.

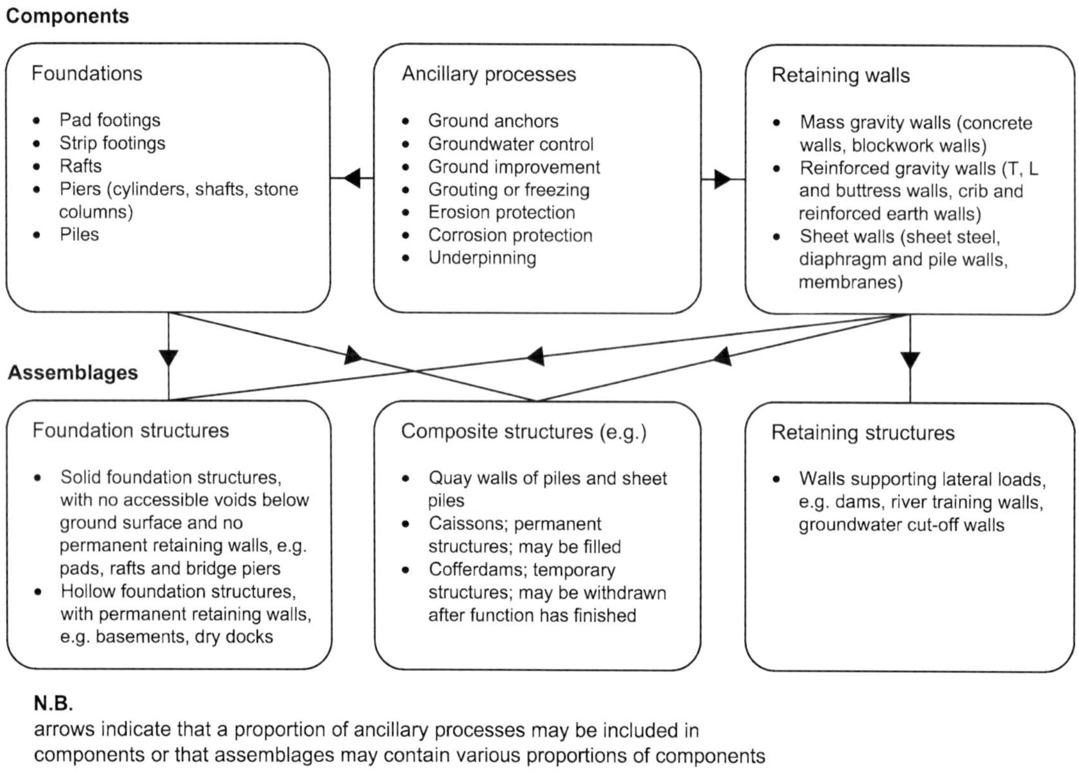

Figure 9.3 Relationships between components, ancillary processes and assemblages (Reproduced from Cole, 1988)

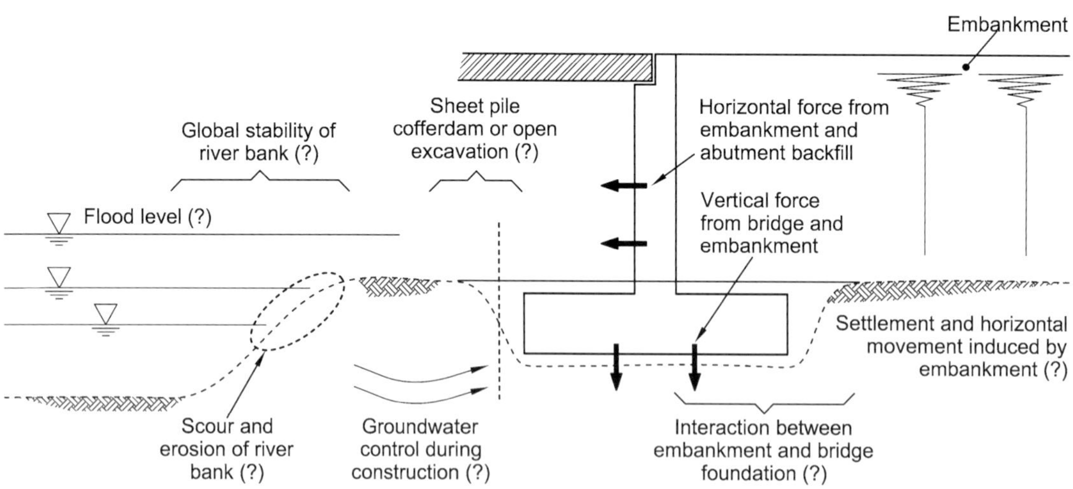

Figure 9.4 Foundation components, ancillary processes and assemblages: a bridge abutment adjacent to a river

Sometimes these different components of a project (say tasks (*a*), (*b*) and (*c*) above) are designed by different teams within an organisation, or by different organisations. It is important that potential interactions are considered by a single team and preferably this should be the lead project designer; otherwise serious problems can occur.

9.3.1 Examples of circumstances that influence foundation design

Because of the huge variety of ground and groundwater conditions which may be encountered and the widely different circumstances in which foundations are constructed, foundation design may involve a wide range of components, ancillary processes and assemblages. The main factors which influence design decisions are: ground shape and location relative to other natural features (rivers, banks, slopes); adjacent site use and structures (surface and underground) and their sensitivity to ground movement/vibration; the site history; proposed superstructure requirements (allowable movements) and loads; and the site geology and hydrogeology. A few practical examples are discussed below and illustrated in **Figures 9.5** and **9.6**.

- *Ground shape and location*: The overall shape of the ground (its geomorphology) both within and around the project site can have a major influence on foundation design and construction. A walkover survey by an experienced geomorphologist may allow the risk of, say, historic landslides being encountered to be assessed (Fookes *et al*., 2007; Phipps, 2003). Other examples are: sites in mining areas, which may be subject to subsidence and local instability; sites affected by large-scale erosion processes and/or flooding. The overall shape of the site may need to be changed to give a more satisfactory layout for the project development, for example by cutting and/or filling to form terraces, or to increase/decrease the overall site level. Although this may be necessary for the project, it may substantially increase the complexity of the foundation engineering because it will cause significant 'global' ground movements to develop (for example, settlement below deep areas of fill, or heave/swelling beneath deep excavations). If cut/fill is carried out on sites underlain by clays, these global ground movements may take several years to fully develop. Hence, additional forces (both vertical and horizontal) and ground movements would need to be considered in the design of individual foundation components. Chapter 57 *Global ground movements and their effects on piles* considers the effects of global ground movements on pile design.
- *Site history and constraints*: Nowadays, 'greenfield' sites – that is, sites which have had no previous structures or industrial use – are increasingly rare. Obstructions, constraints and various hazards are commonly found on the site of a new project. Some sites may contain remains of archaeological value. Carrying out a comprehensive desk study at an early stage of the project is by far the best way of identifying these potential problems. Guidance on how to discover as much as possible about a site prior to carrying out ground investigation is given in Chapter 43 *Preliminary studies*. Some sites affected by industrial activities may be underlain by chemically contaminated materials (these may have originated at the site under consideration, or have 'leaked' in from neighbouring sites, or have been spread across an area due to poorly controlled demolition). These contaminated materials may have impacted the soil or groundwater, or be in the form of noxious/explosive gases. If contaminated soils impact a site then the foundation designer will need to work closely with a contaminated land specialist and the relevant regulatory authorities, and it is likely to have a significant impact on foundation design/construction. Contaminated land guidance is given in a wide range of publications, including those by the Department of the Environment and the Environment Agency; CIRIA Reports such as C552 (Rudland *et al*., 2001) provide a good introduction.
- *Site geology and hydrogeology*: The composition and structure of the ground and the groundwater conditions, both now, during construction, and in the long term, during the design life of the structure, affect the choice of foundations. For example, consider the foundations for a multistorey building. At one extreme, the site may be underlain by deep deposits of very soft alluvial clays/silts and peat, over an irregularly weathered and fractured rock mass affected by historic coal mining (a relevant case history is discussed in Chapter 52 *Foundation types and conceptual design principles*). The foundations may require numerous deep large-diameter piles. Prior to piling, an ancillary process, involving grouting of the mine workings, will be required to stabilise the site and avoid collapse of the mine workings in the future. At the other extreme, a massive layer of a strong rock (say, granite) with only a few widely spaced discontinuities may be encountered close to the ground surface. This rock could support the superstructure frame (columns/walls, etc.) with little enlargement to form the footings (this rock type would be stronger than the foundation concrete). Between these extremes lies a vast range of different combinations of ground conditions. Perhaps less obvious than the ground types which underlie the site are the potential influence of groundwater conditions on foundation engineering and how they may fluctuate seasonally and in the long term and how these may affect foundation behaviour. **Table 9.1** provides a checklist of some groundwater-related factors which may affect foundation design, and an associated commentary. An additional challenge is how rapidly the groundwater regime will lead to changes in ground behaviour, for example softening of clays adjacent to an excavation. This rate of change is affected by the mass (or bulk) permeability of the ground. At one extreme are highly permeable soils, such as 'clean' gravels (i.e. with negligible silt/clay content) and at the other extreme are low-permeability homogeneous clays. This range covers a change in permeability of about seven

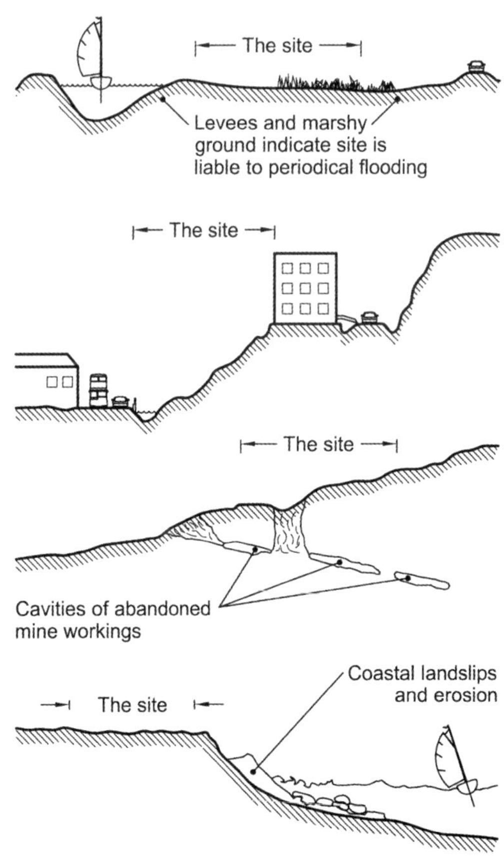

Figure 9.5 Influence of shape of ground on foundation design and construction (Reproduced from Cole, 1988)

orders of magnitude, between 10^{-3} m/s and 10^{-10} m/s. Most natural soils comprise varying amounts of clay, silt and sands; and the 'fabric' – that is, the arrangement of silt/sand layers/lenses within clays, or vice versa – has a major influence on the ground's mass permeability. Laboratory measurements of permeability are usually misleading, and in situ permeability testing is preferable (refer to CIRIA C750 (Preene *et al.*, 2016)). It is important to note that the groundwater effects outlined in **Table 9.1** may lead to substantial foundation movements which are independent of the bearing pressures applied by the structure. These effects will often be overlooked by a structural or general civil engineer, and it is important for the geotechnical engineer to carefully consider groundwater-related effects and associated risks.

- *Ground-related construction hazards:* Understanding ground hazards and their impact on foundation construction techniques is key to selection of the foundation type by the designer (e.g. piled foundation over a contaminated site would be preferred to a pier-type foundation constructed in an open cut; conversely, where frequent deeper obstructions are suspected a raft foundation may be preferred to piled foundation). CIRIA C807 (Davis *et al.*, 2023) provides guidance for use of geotechnical baselines to allocate ground-related risks during construction. Key takeaways for designers are (*a*) a conservative design may seem a 'safe' approach, but increasing size and depths of foundations often increases the ground-related risks during construction and therefore may increase project and CDM risks, carbon footprint, construction cost and duration; and (*b*) early engagement with the contractor and owners is optimal to understand, allocate and manage ground-related risk.

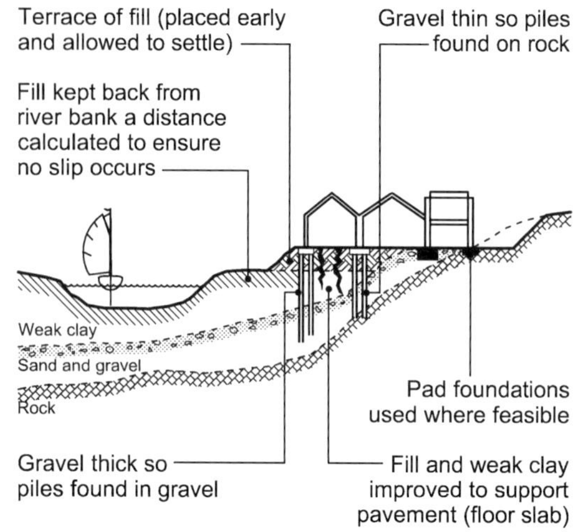

Figure 9.6 Example of how site re-levelling requires several foundation engineering activities (Reproduced from Cole, 1988)

9.4. Keeping the geotechnical triangle in balance: ground risk management

9.4.1 Background

During the last 50 years or so, research developments in soil mechanics and the related engineering sciences have led to immense improvements in knowledge and understanding. Enormous developments in computer technology and

Swelling and softening of strata below excavation	For clays, as effective stresses reduce, the strength reduces and compressibility increases. In rocks, joints may open (or joint infill materials may soften), and overall rock mass strength/stiffness is reduced. Repeated wetting/drying of weak rocks may lead to breakdown of rock structure and large loss of strength.
Dewatering during construction and re-establishment after dewatering switched off	In clays/silts: dewatering of adjacent more permeable soils (such as sands/gravel) can lead to a reduction in pore water pressure in the clay/silt, leading to consolidation settlements. After dewatering is switched off, 'rebound' will occur, potentially leading to swelling (depending on foundation bearing pressure and net increase/decrease in effective stresses). In sands/gravel: dewatering may lead to a loss of fines and substantial differential settlement, and ultimately ground instability. Hence, appropriate filters and observations are essential to avoid loss of fines.
Excavation, basal instability	Due to excessive groundwater pressures or hydraulic gradients. Even if ground failure does not occur, there may be substantial loss of ground strength/stiffness.
Erosion and scour	Surface water flows may lead to undermining of shallow foundations, or loss of lateral support to shallow and deep foundations. Erosion protection is essential in high-risk situations. Silts and fine sands are especially vulnerable.
Frost action	Leads to volumetric expansion due to ice lens formation and/or freezing of water in cracks. Local codes of practice usually specify minimum foundation depths to avoid adverse effects.
Groundwater level changes (low-lying areas may be vulnerable to flooding)	Natural variations in groundwater table level may occur, or man-made changes may occur (e.g. due to burst water main). These may lead to uplift forces on foundations, and a requirement for anchors and/or drains/relief wells to avoid buoyancy failure.
Collapsible soils	These soils are either very loose unsaturated soils, or (more commonly in the UK) loose non-engineered fills. If these soils wet up due to a water table rise, then a large 'collapse'-type settlement may occur.
Expansive soils	Desiccated clays, located above the water table, will be prone to substantial swelling pressures and volumetric expansion (heave or swelling) if the clay wets up. High-plasticity clays are most vulnerable. In the UK, this problem is usually associated with vegetation effects and seasonal wetting/drying (see below).
Vegetation effects and seasonal wetting/drying	Large trees (especially high-water-demand varieties), together with high-plasticity clays, can induce substantial seasonal foundation movements due to winter wetting/summer drying of the clays. Chapter 53 *Shallow foundations* discusses this problem in more detail. If trees have been removed, then the desiccated clay will slowly wet up over several years, as pore water pressures re-equilibrate, leading to large swelling and uplift forces on foundations.

Table 9.1 Groundwater-related issues which may affect foundation design and cause large foundation movements

numerical modelling have enabled engineers to analyse highly complex geotechnical problems, which would have been unthinkable a few decades ago. Alongside these theoretical and technical developments have been major improvements in practical construction techniques which enable a much wider range of ground conditions to be safely engineered and extremely large foundation loads to be resisted. Despite all these improvements, major geotechnical difficulties still regularly occur, causing costly delays to projects, and sometimes more severe failures that cause damage to adjacent infrastructure and injury or death to members of the public and construction workers. There are numerous reasons for these failures, including the fragmented nature of the industry, deficiencies in procurement of specialist services and competitive and time-related pressures. At a technical level a major problem can be the lack of a coherent approach to geotechnical engineering and a systematic approach to ground risk management. In Chapter 4 *The geotechnical triangle* the geotechnical triangle was described. The four main components of the geotechnical triangle (**Figure 9.7**) are as follows.

(*a*) *Understanding the ground profile* (and the associated geological processes and man-made activities that formed the ground profile) *including groundwater conditions.*

(*b*) *The measured or observed mechanical behaviour of the ground* – this includes carrying out laboratory and field (or in situ) testing, and careful interpretation of the test data in the context of the proposed foundation requirements.

(*c*) *Appropriate modelling* – this is much more than carrying out an analysis. It involves: idealisation, or simplification, of the real soil/structure; identifying and carrying out appropriate analyses; and validation and assessment of the output. Several iterations may be needed, and a particularly difficult part of the process is selecting appropriate input parameters (based on (*a*) and (*b*)).

(*d*) *Empirical procedures and 'well-winnowed' experience* – empiricism is an inevitable and essential aspect of geotechnical engineering. It is also important to understand the limitations of available empirical procedures. Experience is important in order to make the difficult judgements which are often required; however, the term 'well-winnowed' emphasises the point that the experience needs to be founded on a proper understanding of the facts and key behavioural mechanisms associated with past activities and observations.

For practical applications of the 'geotechnical triangle' it has been helpful to consider a fifth component to the triangle (the 'outer loop' shown in **Figure 9.7**).

(*e*) *Construction processes and design verification* – the information which is available to the foundation designer is seldom perfect (e.g. contradictory information about the site's history, variable data from ground investigations, etc.). Hence, several important assumptions often have to be made to develop a design. Therefore, it is important that these assumptions are verified during the construction process. There are a wide range of activities which may have to be carried out, and a few examples are provided in **Table 9.2**. At a basic level, it may be simply a visual check during construction that the geology of the founding stratum for, say, a shallow footing, and its strength, is the same as that assumed for design. For more demanding projects it may involve sophisticated tests of deep piles. Different organisations/design teams may be involved in the foundation engineering activities at different stages of the project. Throughout this process it is important that key assumptions are communicated to other parties, so that ultimately these assumptions can be checked when the foundations are being constructed.

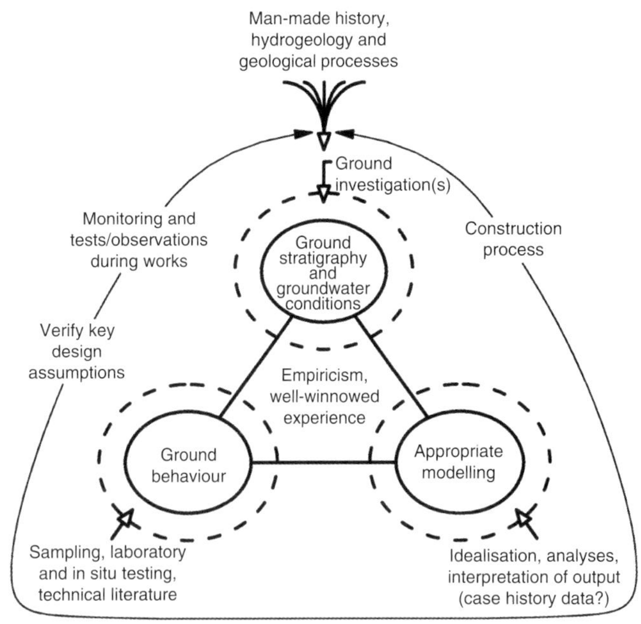

N.B.
This geotechnical triangle is modified by way of the outer circle for 'construction process'; this emphasises the importance of construction processes in foundation engineering, in the context of the potential impact of various processes on key design assumptions

Figure 9.7 The geotechnical triangle, foundation engineering applications

All five components are important, and problems are likely to occur if one component is forgotten or poorly implemented and the triangle goes 'out of balance'. Outlined below are several short case histories, which briefly summarise some foundation engineering problems that have occurred on UK construction projects.

Foundation engineering issue	Possible tests/observations[a]
Confirm founding strata	Inspect excavated surface and compare with design assumptions, simple visual check. For layered strata and/or deep foundations, may need extra boreholes (continuous sampling) if GI boreholes too widely spaced.
Assess variability of strata	Usually simple in situ tests (static cone CPT, or dynamic probing) appropriate to check *relative* hardness or density of strata across site, if heterogeneous conditions expected.
Groundwater regime	Important if construction below water table required. Install piezometers in key strata. If dewatering necessary, large-scale pumping tests may be required, prior to works beginning.
Impact on adjacent structures	Precise levelling of affected structures from regular array of survey pins. Subsurface measurements may be needed in some circumstances, especially if lateral movements expected (inclinometers).
Foundation load–deformation	Large-scale load tests on key foundation elements, especially if sensitive to construction/installation, such as deep bored piles. Tests will be more complex if lateral load behaviour needs to be checked.
Deep ground improvement	Verification testing is usually critically important for controlling the effectiveness of the works. Type of test dependent on improvement mechanism (e.g. reinforcement, densification or providing increased cohesion, by addition of cementing agents, or reduction of permeability).

[a] The list is not intended to be comprehensive and merely highlights some common examples.

Table 9.2 Examples of verification tests/observations which may be required during foundation engineering

9.4.2 Shallow foundations, adjacent to a landslip

Figure 9.8 provides a plan and cross-section of the site. Shallow foundations, comprising pad and strip footings, were planned for an out-of-town retail development. A ground investigation had been carried out, which included five boreholes and several trial pits. These indicated a thin layer of made ground (typically less than 2 m thick) over rockhead which comprised strong limestone. In one corner of the site a shallow retaining wall (about 1–2 m high) was required to create extra space to construct the shallow foundations. The low-height retaining wall was formed from proprietary precast elements with a shallow foundation. While the excavations were being carried out for the retaining wall and shallow foundations at the toe of the slope, a large landslip was triggered. This caused significant delay to the project and public concern, since the landslip caused the closure of a footpath and adjacent public areas. What had gone wrong?

- A geological desk study had not been carried out. The positioning of the boreholes and trial pits by the structural engineers had been based mainly on ease of access, rather than technical considerations.
- Inspection of geological maps indicated that a fault passed through the site; unfortunately all the boreholes/trial pits were located on one side of the fault. On the other side of the fault the ground conditions were quite different, with interbedded mudstone and siltstone.

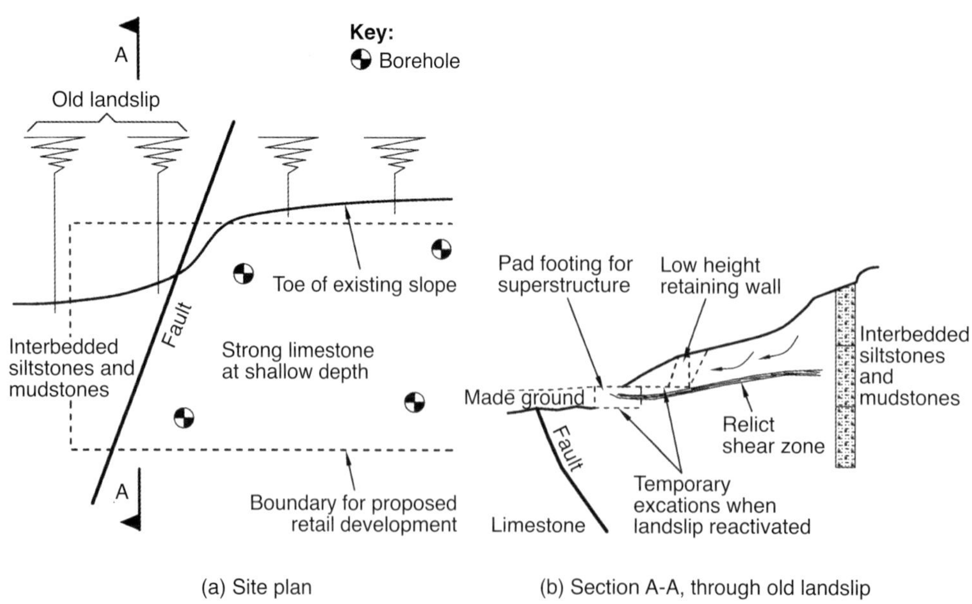

Figure 9.8 Shallow foundation adjacent to a landslip

- Based on a subsequent geomorphological inspection of the area, it was clear that the area to the west of the fault had suffered from past landslips, and the natural slope was at a factor of safety close to 1.0. Even quite minor excavations were sufficient to reactivate the landslip and cause further instability.

This is a stark reminder that even simple foundation engineering needs to be done in a careful and systematic way. A desk study and an understanding of the site's geology is a fundamental requirement, and is essential in order to properly plan and design a ground investigation.

At this site, the first construction activity should have been to stabilise the landslip area. The landslip was stabilised by placing a temporary berm (at the landslip toe) and then installing a 'shear key' through the shear zone of the landslip. The shear key involved large-diameter heavily reinforced bored piles, constructed as a contiguous embedded retaining wall. Following this the toe of the slope could be safely removed and excavations for the shallow foundations constructed.

9.4.3 Water treatment works, differential swelling

The design of a water treatment works required varying excavation and founding levels across the facility. The operation of the works meant that water levels in the various tanks would fluctuate over time. The construction of the works went well; however, just before the works opened cracking of floor slabs was observed during commissioning tests. The cracks were monitored, and they became progressively worse over time. At this stage, many of the areas which had suffered the worst damage were also relatively lightly loaded.

Figure 9.9 shows a cross-section through part of the works. The overall configuration of the substructure is quite complex, comprising an irregular array of reinforced-concrete boxes of varying size. The site geology is quite simple, comprising a deep layer of heavily overconsolidated Gault clay. The site was in a semi-rural area and was a greenfield site. A simple open non-retained excavation was formed since there were no space constraints. The excavation was carried out during a relatively dry summer and there were no problems with ground or surface water. The concrete substructure 'boxes' were then built within the excavation and conventional granular backfill used (**Figure 9.9**).

When the cracking became apparent, survey points were fixed and monitored and these indicated differential vertical movements across the structure, and net upward swelling movement beneath the deepest part of the substructure. A standpipe piezometer in the backfill indicated about 1 m head of water. Subsequent analysis highlighted the following issues.

- Owing to the different founding levels, the different types of equipment and water levels in certain parts of the works, the *net* change in bearing pressure is extremely variable, from small net *increases* in bearing pressure, to large net *decreases* in bearing pressure.
- Because of the high plasticity index of the Gault clay and its high degree of overconsolidation, it is prone to significant long-term swelling, and if movement is restrained large swelling pressures are developed against the structure in contact with the Gault clay.
- The trigger for the swelling movements and structural damage was a readily available source of water. Over winter, rainwater seeped into the granular backfill, and along the (relatively poor and cracked) blinding concrete. During the summer period it is likely that desiccation cracking of the exposed Gault clay would have created a relatively high mass permeability and allowed relatively quick infiltration of water into the clay.
- The fundamental flaw, however, was the lack of awareness on the part of the designers that differential movements would develop, and the structural design was vulnerable to differential movement. The combination of relatively flexible slabs and stiff walls/columns meant that the substructure was particularly sensitive to differential movement.

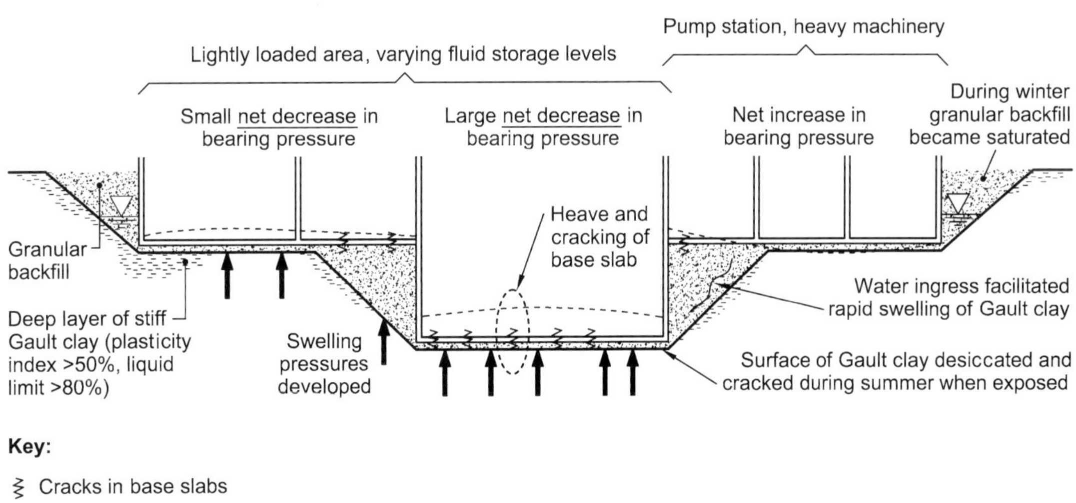

Figure 9.9 Water treatment works, differential swelling

- An important cause of the problem was poor communication between the geotechnical and structural engineers. The structural engineer asked the geotechnical engineer to provide 'allowable bearing pressures'. There was no discussion about the nature of the structure or the variation in applied loads imposed on the foundations. As a result, the structural engineer did not realise there was a risk of swelling or differential movements across the structure. The geotechnical engineer did not realise there would be net unloading across large parts of the structure. This lack of communication led to the design of an intrinsically flawed foundation and structural design.

For this project the mechanical behaviour of the ground was not assessed *in the context* of the proposed foundation requirements. A more appropriate design would have created more uniform loading, or reduced the net unloading beneath the deepest part of the substructure, by reconfiguration of the plant layout, or if impractical, provided a structural design which would have minimised differential movement (e.g. a thicker, stronger raft) or allowed differential movement to occur without structural damage (provision of movement joints, rocker slabs, etc.).

9.4.4 Driven piling in mudstone, plugging of open-ended tubular piles

A highway bridge was to be constructed across a small river. The bridge and foundations were designed and let under a conventional ICE contract to a local civil engineering contractor. The foundations were bored pile groups constructed in the Coal Measures rocks, which comprised interbedded mudstone and sandstone. A high-quality ground investigation was carried out; **Figure 9.10** summarises the ground conditions. The upper Coal Measure rocks were relatively poor quality, and predominantly comprised mudstones with an unconfined compressive strength of about 10 MN/m^2, and RQD values of 0–10%, with frequent reference to open or clay-filled joints in the rock mass. The rock quality improved at depth, with more frequent layers of sandstone with unconfined compressive strengths of about 30 MN/m^2 and RQD values of 30–60%. SPT 'N' values were typically in excess of 100; however, in these ground conditions the SPT was meaningless and of little value.

The contractor offered an alternative foundation design which comprised driven open-ended tubular piles. The driven piles were expected to be quicker to install, and provide equivalent overall performance (capacity and settlement) to the original foundation design. Pile driving proved to be difficult, with some piles experiencing premature refusal during driving (probably due to local blocks of sandstone). The project specification required the contractor to carry out pile load tests on selected piles. These tests on working piles were meant to impose loads up to 1.5 times the working load with pile head settlement of less than 10 mm. Unexpectedly, at the working load the pile settlement was excessive and, for practical purposes, the pile had failed. Several other pile load tests were carried out and these also failed. A detailed review was carried out of both the pile installation and the pile design. A number of problems were apparent both in terms of pile construction and design. The pile design had made the following assumptions.

- The base of the tubular pile would 'plug' – that is, the end-bearing resistance would be mobilised over the full pile diameter. The end-bearing resistance was assumed equal to 4.5 times the unconfined compressive strength of the rock (i.e. equivalent to 9 times the undrained shear strength, as commonly assumed for clay).
- The shaft resistance was based on a mobilised adhesion of twice the SPT 'N' value (i.e. in excess of 200 kN/m^2); the assumption that shaft shear resistance $\tau = 2$ 'N' is commonly used when designing driven piles in sands and gravels.

Based on these assumptions the geotechnical capacity of the pile was expected to be extremely high, and the factor of safety was expected to be well in excess of 3.0 at the working load. Unfortunately, as shown by the pile tests, the actual factor of safety was close to 1.0.

The above design assumptions were based on a published design guide for driven piles. However, the guidance and associated equations for calculating pile capacity were actually based on experience of driven piles in soils, not rocks. Hence, the design assumptions were inappropriate. A careful back-analysis of the pile test data indicated the following.

- The open-ended tubular piles did not plug – that is, end-bearing resistance was only mobilised across the annulus of the pile.
- Pile resistance was predominantly mobilised by shaft friction, both internally and externally; the average shaft adhesion varied between about 60 and 100 kN/m^2, with a median value of about 85 kN/m^2. The intense remoulding of the mudstone due to pile driving had effectively created a thin zone of very stiff clay along the sides of the steel pile.

The pile tubes had not been strengthened to facilitate hard driving (normally a tubular pile would have a specially designed thick-walled driving shoe if hard driving was expected). The pile-driving hammer was also of inadequate capacity. Hence, driving the piles deeper to the higher-quality sandstone at depth was not a viable option. The most effective solution to these design and construction problems was therefore to drill out the materials inside the pile tubes and infill the void with concrete. This enabled the full end-bearing resistance to be mobilised.

Subsequent pile load tests indicated that the modified piles had adequate capacity and settlement characteristics (**Figure 9.11**). Because of the fractured nature of the rock mass, the end-bearing resistance was 1.5–2.5 times the unconfined compressive strength (i.e. 3–5 times S_u, considerably lower than soil mechanics theory would indicate).

This project experience clearly emphasises the need to understand the ground behaviour in the context of the proposed construction process (in this case the rock mass behaviour during pile driving, which is quite different from that of soil), and understand the limitations of empirically based design guidance. 'Well-winnowed' experience was clearly *not*

Sampling			Strata
Depth (m)	TCR	(RQD)	Description
10.40	75%	(0%)	Weak grey slightly carbonaceous SILTSTONE. Locally weathered to slightly sandy fine to coarse angular gravel. Discontinuities horizontal to subhorizontal very closely to closely spaced planar closed to slightly open (<3 mm) clean. At 9.80 m, clay smeared discontinuity. At 10.02 m, 10° clay smeared discontinuity.
11.40	93%	(0%)	At 11.10 to 11.20 m, weathered to clayey fine to coarse angular gravel.
12.90	93%	(0%)	At 13.10 to 13.70 m, dark grey and carbonaceous, tending to silty MUDSTONE. From 13.70 m, moderately weak to moderately strong sandy siltstone.
14.40	100%	(8%)	At 14.40m and 14.50 m, open (10 mm) slightly sandy clay infilled discontinuities. At 15.27 to 15.30 m, open (30 mm) slightly clayey fine to medium angular gravel infilled discontinuity. At 15.30 to 16.45, subvertical irregular rough closed to slightly open (<3mm) clean discontinuity.
15.90	84%	(40%)	At 16.20 to 16.70 m, dark grey and carbonaceous. At 16.55 m, open (10 mm) gravelly sand infilled discontinuities. At 16.70 to 16.75 m and 16.92 to 16.95 m, open slightly sandy fine to medium angular gravel infilled discontinuities. At 16.92 to 18.30 m, slightly sandy. From 17.00 m, moderately strong to strong. From 17.40 m, discontinuities very closely to medium spaced.
17.40	100% (93%)	(40%)	At 18.50 to 18.80 m, subvertical irregular smooth closed clean discontiuity.
18.90	100% (93%)	(35%)	
Equipment: Rotary coring, mist flush		Borehole dia. (mm): 92 to 25.40 m depth	Casing dia. (mm): 120 to 6.00 m depth

Figure 9.10 Extract from borehole log, upper Coal Measures

used either in terms of design or construction. When designing open-ended tubular piles, the end-bearing resistance of the pile annulus plus the internal shaft resistance should be compared against the bearing resistance across the full pile area, and the lesser of the two values should be used for design. In addition, the influence of the construction process, in this case the pile driving, on the mobilised shear resistance along the pile shaft was not appreciated. Estimating the capacity of driven tubular piles in weak rocks is actually quite challenging. Even if the design assumptions had been more appropriate, it would have been wise to carry out preliminary pile tests (to failure) on non-working piles, prior to the start of the main piling works. This would have significantly reduced the risk of unexpected pile failures, although the cost and duration of the piling works would have been increased.

9.4.5 The geotechnical triangle revisited

Reflecting upon the above case histories, the value of the geotechnical triangle can be appreciated. All aspects of the triangle should be considered and kept in balance – that is, one element should not be given excessive attention while ignoring other aspects. There is a high risk of failures occurring if one or more aspects are forgotten.

Table 9.3 provides a commentary on the problems encountered in the above case histories in the context of the five aspects of the triangle. The errors and omissions which caused the foundation problems may seem obvious; however, all the engineers involved were experienced. In some cases there was an assumption that because the engineering was perceived as 'simple and routine', shortcuts to normal good practice could

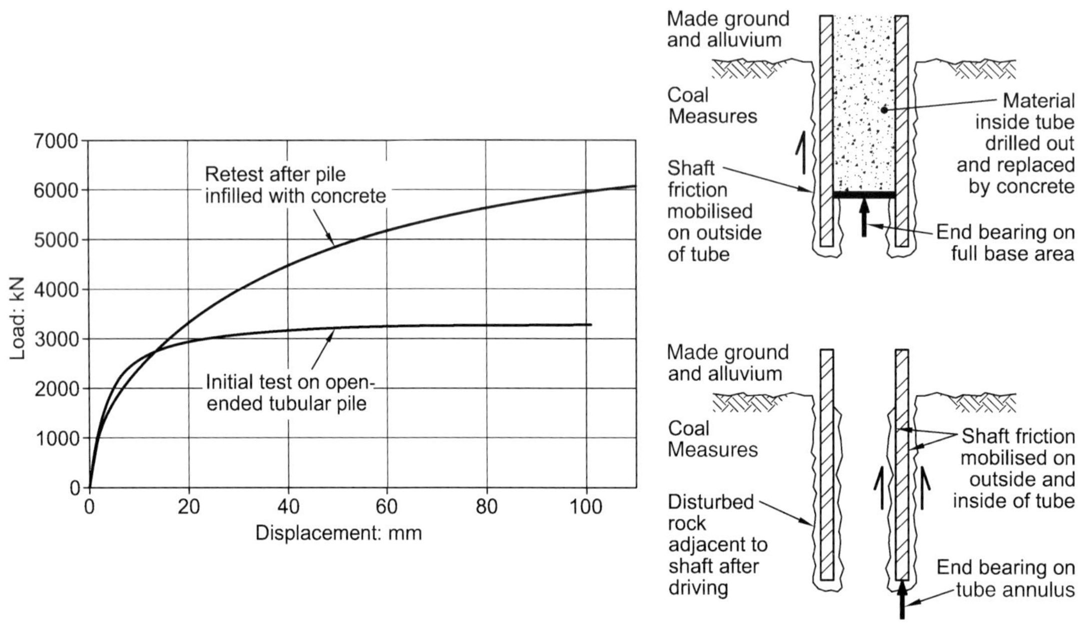

Figure 9.11 Driven tubular piles in mudstone, pile test data: open-ended pile vs concrete-infilled piles

Case history	Ground profile and groundwater conditions	Ground behaviour	Appropriate modelling	Empiricism, case records, well-winnowed experience	Construction process test/observations and verifications
Shallow foundations, adjacent to landslip	No desk study. Geology *not* understood. Poor positioning of boreholes relative to fault.	Since geology not understood, no awareness of adjacent landslip.	Global stability of site not assessed, only local stability of retaining walls/ foundations considered.	Geological maps and memoirs highlighted the risk of landslips in this area (but not reviewed by designers).	Because hazard not identified, construction process was inappropriate. Should have stabilised the slope *prior* to excavations.
Water treatment works, differential swelling	Geology and ground profile understood. Deep layer of Gault clay.	Poor communication between geotechnical and structural engineers. Hence, net unloading of O/C plastic clay not identified.	Only bearing capacity and settlement checks for net increases in bearing pressure. Consequences of swelling not assessed.	The swelling potential of high-plasticity overconsolidated clays, such as Gault clay, is well known. But poor communication meant that potential risk for this structure was not identified.	Since design was flawed, construction process, tests, etc. would not have mitigated the risk.
Driven piling in mudstone, plugging of tubular piles	Good desk study and high-quality ground investigation carried out.	Appropriate testing and observations; ground behaviour was well established. But not understood by designer/ contractor in the context of proposed construction, i.e. driven piling.	Poor modelling. Key assumptions were incorrect. Open-ended tubular piles were unlikely to plug.	Inappropriate design methods used (not relevant for weak rock). Poor construction methods (piles not strengthened for hard driving, underpowered piling hammer), so unable to drive to competent strata.	Pile tests on working piles identified problem but too late, so overall project was delayed. Tests on preliminary piles would (technically) be preferable but cost/time penalty.

Table 9.3 Keeping the geotechnical triangle in balance: case history examples

be made without risk. Clearly, with the benefit of hindsight, the engineers involved would now have a different opinion!

9.5. Foundation applications

9.5.1 Introduction

In this section two case histories are given which describe the foundation engineering for two very different projects. The first describes the selection of an appropriate foundation type for a multi-span bridge across an important railway line. The site was underlain by estuarine sands of variable density. The second discusses groundwater problems during construction of foundations for a building, which was underlain by chalk. Common to both case histories are the important observations and tests which were carried out during the construction works (i.e. the fifth key component of the geotechnical triangle), and the critical role that these played in successful completion of the works.

9.5.2 Vibrocompaction of estuarine sands

9.5.2.1 Site location and ground conditions

The site is located in North Wales, adjacent to the River Dee estuary. The project involved the construction of an access road into a new business park, which was to be located between an existing main road and the estuary. A bridge was required to carry the new road over a railway line which traverses the centre of the site. This railway line is a strategically important route connecting the UK network to the Irish Sea ferries. The bridge comprised a two-span deck, each span being about 30 m long. To attain adequate clearance to the railway, the approach embankments were relatively high at about 12 m. This case history is described in detail by O'Brien (1997).

The site is located on a flat estuarine plain and the geological sequence is shown in **Figure 9.12**. The groundwater table level was about 2.5–3.0 m beneath ground level, and pore water pressures varied hydrostatically with depth. Of concern, for bridge foundation design, was the variability in relative density of the estuarine sand. Although typically described as loose to medium dense, there were also very loose and dense layers. SPT 'N' values varied between 5 and 35, but the repeatability of the test was poor due to the difficulty of maintaining borehole base stability beneath the water table. CPT q_c values tended to increase with depth, although there was significant variability in the CPT q_c profiles.

9.5.2.2 Foundation selection

Several different foundation types were considered. The net increase in bearing pressure was about 300 kN/m^2. The bearing capacity of the sands was adequate, and the main concern was ensuring settlement, particularly differential settlement, was sufficiently low. For the multi-span bridge a limit on differential settlement of about 15 mm was targeted. Founding the bridge on shallow foundations, beneath the clay/silt capping layer, would have been feasible if the approach embankments (which imposed the majority of the bearing pressure) could have been constructed before placing the bridge deck. Unfortunately the bridge deck beams had to be placed within a narrow time frame, when a track possession (for construction over the railway line) had been granted. This possession period was fixed and critical to the whole project, and all other activities had to be as flexible as possible. If the deck was placed before construction of the approach embankments, then differential settlement was expected to be higher than the 15 mm allowable (perhaps double this value, due to the variable density of the estuarine sands). The main foundation options are summarised in **Table 9.4**. Pile integrity and durability was a particular concern, since previously the site

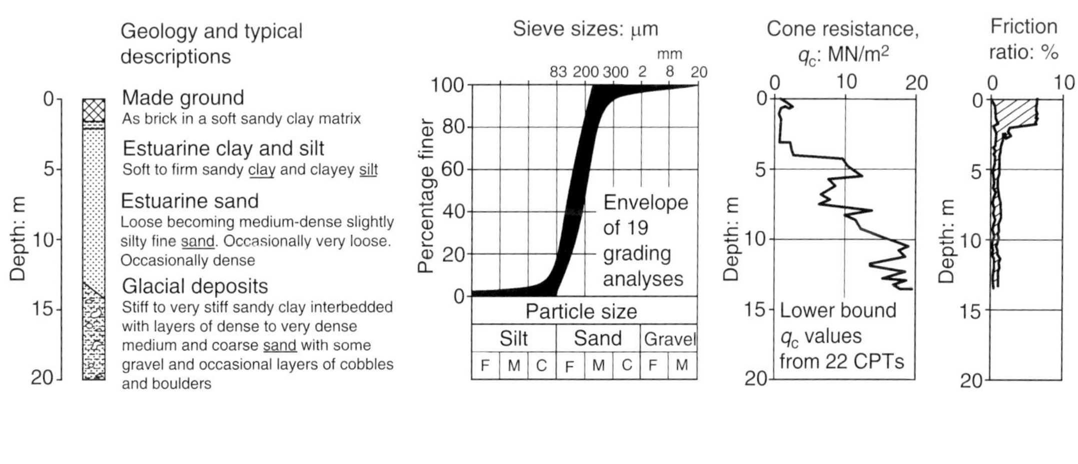

(a) Typical soil profile

(b) Particle size distribution estuarine sand

(c) CPT, cone resistance and friction ratio in estuarine deposits and made ground

Figure 9.12 Vibrocompaction of estuarine sands, ground conditions had been used as a chemical works, and the ground conditions were aggressive to reinforced concrete. At the time the works were implemented (mid-1990s), CFA piling rigs were much less powerful than today. Large-diameter (>1.0 m) CFA rigs were not available, and installing heavy pile reinforcement, especially to accurate verticality tolerances, was judged to be impractical. Heavy pile reinforcement would have been required because the bridge abutment and embankment backfill imposed significant lateral loads onto the piles. Because of the reinforced-concrete durability concerns, it would have been important to ensure the reinforcement had adequate concrete cover; hence accurate reinforcement installation was necessary. Driven piles would probably have caused damaging vibrations and settlement to the adjacent railway. Large-diameter bored piles (due to the need to use bentonite slurry support) were expected to be very expensive. Vibrocompaction was believed to be the most appropriate foundation engineering option compared with the other available options. Deep ground improvement, using vibrocompaction, would increase the relative density of the sands, and in particular eliminate the zones of loose sand that existed. This would enable simple foundations to be constructed on the improved ground, and reduce the risk of differential settlement. Vibrocompaction is discussed in more detail in Chapter 59 *Design principles for ground improvement.* The vibrocompaction method used at this site used a 'wet top feed' technique, which involves using a vibroflot (a large vibrating poker) to compact the ground. Water is used as a jetting medium to facilitate insertion of the vibroflot, and gravel is fed into the hole around the vibroflot from the ground surface and compacted by the vibroflot.

The ground conditions (fine estuarine sands, with a low silt/clay content) were believed to be ideal for vibrocompaction. Discussions with local contractors were held, but locally vibrocompaction was generally only carried out for lightly loaded industrial buildings founded on made ground. Typically, the performance criteria were less onerous, with net bearing pressures being of the order of 100–150 kN/m^2. Nevertheless, a review of the technical literature and discussions with specialist contractors indicated that vibrocompaction would be feasible, although possibly challenging. The specification was developed carefully in order to control the anticipated project risks.

- A preliminary trial was required, using different compaction spacings, and compaction effectiveness was checked by carrying out CPT tests before and after compaction (during this trial, remote from the railway line, ground vibration and settlement were also monitored).
- The specification was an 'end product' type: a target CPT cone resistance profile had to be met by the specialist ground improvement contractor (**Figure 9.13**).
- A limit on allowable ground vibration and settlement at the railway line (based on discussions with the railway authorities) was stated. It was recognised that this constraint would prevent the use of the most powerful rigs for vibrocompaction, but lower-powered rigs could be used and would still be able to densify the sands.

9.5.2.3 Observations and tests during the works

The successful tenderer used a vibroflot with a power rating of 50–80 kW, and medium/coarse gravel was used as backfill

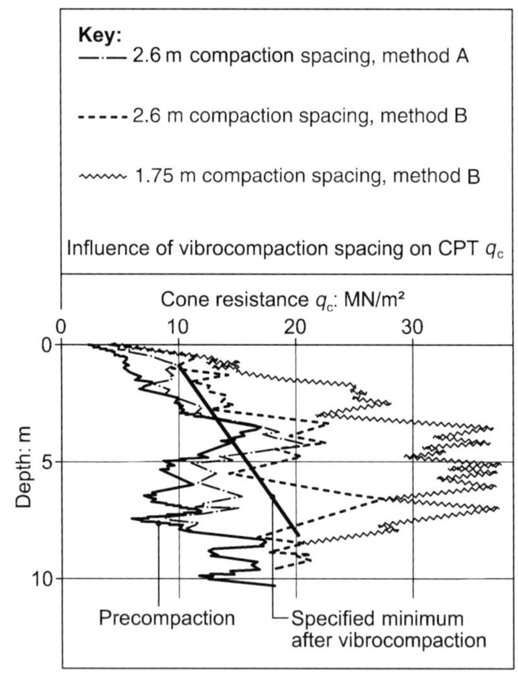

Figure 9.13 Vibrocompaction preliminary trial: influence of different compaction methods (method A vs method B)

Option	Potential advantages	Potential disadvantages
1. Driven piles	Suited to ground and groundwater conditions. Good vertical capacity.	Difficult driving due to dense sand layers. Large vibrations and damage to railway. Limited lateral/bending capacity unless raking.
2. CFA piles	Low vibrations. Relatively inexpensive.	Inadequate lateral/bending capacity. Poor integrity/durability.
3. Large-diameter bored piles	Good lateral/bending capacity. Low vibrations.	Need temporary casing and slurry support. Very expensive.
4. Vibrocompaction	Quick and inexpensive (less than 50% of option 3). Suited to ground and groundwater conditions. Allows simple substructure.	Vibrations and possible damage to railway. Inadequate degree of improvement. Lack of precedent in UK for heavily loaded bridges.

Table 9.4 Case history: vibrocompaction of estuarine sands, advantages/disadvantages of different foundation options

at each compaction point. The initial trial using a compaction spacing of 2.6 m was very disappointing (**Figure 9.13** (method A)), since the post-compaction cone resistance profile failed to meet the specification requirement. There were a range of possible reasons discussed by the designer and contractor (ground conditions different from expected? specified target cone resistance profile reasonable?), although the discussions quickly focused on how the vibroflot was being used by the rig operator. Method A involved withdrawal of the vibroflot in 0.6–0.9 m depth increments, compaction duration per increment of 20–30 seconds and high water flow rates from side jets during withdrawal. This method had been used successfully to treat made ground, when the improvement mechanism was mainly 'reinforcement' by construction of relatively large-diameter stone columns. In contrast, the prime requirement for this project was to densify the sands. An alternative construction method, 'method B', was proposed, which involved withdrawal in 0.3 m depth increments, compaction continuing for a longer period until the power consumption peaked, and water flow during withdrawal kept to a minimum (to avoid washing out excessive quantities of sand). From **Figure 9.13** it can be seen that post-compaction CPT q_c tests showed that method B was far more effective than method A in terms of compacting the sands. Method B was used for the remainder of the trial (to check the influence of different compaction spacings on CPT cone resistance) and for the main works.

The measured ground vibrations were towards the lower bound of those reported in the literature (a resultant peak particle velocity of less than 15 mm/s) and ground surface settlements between 3 mm and 7 mm. Following completion of the trial the railway authorities were satisfied that vibrocompaction could take place up to 2 m from the railway.

For the main works a compaction spacing of 2.2 m was selected since at this spacing there was a negligible risk of the post-compaction CPT q_c values falling below the target profile. The plan area of compaction extended about 3 m beyond the bridge pad footings, with a minimum of one compaction point beyond the footing. The pad foundations were founded at about 2.5 m below ground level, to ensure the foundations were below a surface layer of silt and clay, and above the groundwater table level. Pre- and post-compaction CPT tests were carried out to verify that the vibrocompaction was adequate. CPT profiles for one of the bridge abutments are shown in **Figure 9.14**. The post-compaction CPT q_c values exceed the target profile by a comfortable margin, and are typically about two to three times higher than the pre-compaction values across the main depth of interest. The subsequent performance of the bridge foundations has been satisfactory. Based on the post-compaction CPT profiles, routine calculations indicated a settlement of about 15 mm, while more sophisticated analyses indicated a settlement of about 10 mm. Measured settlements were between about 5 and 10 mm, with differential settlements of less than 5 mm. These were comfortably within the allowable limits for the bridge deck.

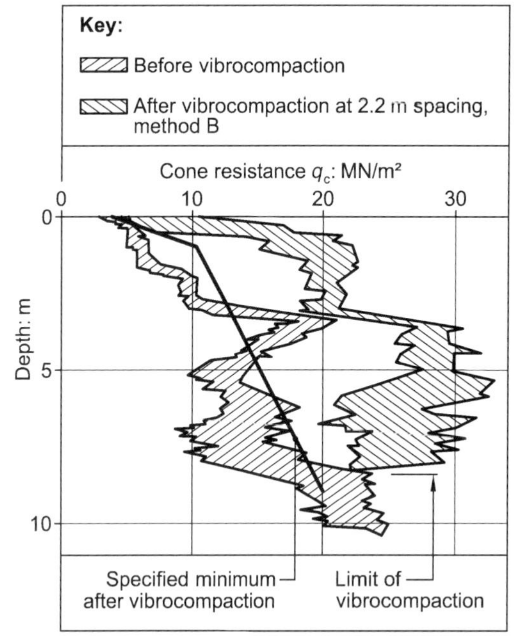

Figure 9.14 Vibrocompaction, before and after CPT tests, 2.2 m spacing, method B

9.5.2.4 Concluding remarks

This case history describes some of the factors which need to be considered when deciding upon an appropriate type of foundation. The foundation type, deep ground improvement with shallow foundations, performed extremely well. However, successful use of the technique drew heavily on a balanced application of all five components of the geotechnical triangle. In particular, the fifth component, 'understanding the construction process and verifying the design assumptions', was important. The initial vibrocompaction method (method A) was inappropriate; observations and tests during a preliminary trial identified the problem and enabled a more effective construction method to be identified. This was critical for the successful completion of these works.

9.5.3 Shallow foundation on soft chalk

9.5.3.1 The ground conditions and construction problems

The construction of a telephone exchange in the centre of Salisbury came to an abrupt halt when unexpected ground conditions were encountered. Work was not to resume for some 6 months. The geological map shows the site to be covered by valley gravel overlying Upper Chalk. Four boreholes were sunk on the site using a shell and auger rig. The borehole logs showed the profile to consist of 0.9–1.4 m of made ground overlying 4.9–6.4 m of medium-dense flint gravel with layers of silty clay. All the boreholes were reported as entering hard fissured chalk below the flint gravel. The water table was about 2 m below ground level. This case history is discussed in detail by Burland *et al.* (1983).

The telephone exchange is a four-storey building of reinforced-concrete flat slab construction. **Figure 9.15** shows a plan and cross-section through the building. Most of the structure to the east is founded on pad footings bearing on gravel with a pressure of about 200 kN/m^2. A notable feature is a chamber running along the western edge of the building. The base of the cable chamber was to be founded in chalk some 6.6 m below the original ground surface at a gross bearing pressure of approximately 185 kN/m^2. The cable chamber was constructed in a cofferdam 9 m wide and 47 m long. The cofferdam was constructed mainly of sheet piles but in the north-west corner a contiguous bored pile retaining wall was used to minimise vibration alongside an existing building. **Figure 9.16** is a cross-section through the cofferdam.

Work progressed smoothly until excavation within the cofferdam penetrated the chalk. Thereafter progress was difficult. A close visual examination of the chalk in the base of the excavation revealed that, far from being hard and fissured, it was very soft, consisting of small angular pieces of chalk in a silt-like putty chalk matrix. At this stage two-thirds of the pad foundations had been completed. The cofferdam area had been partially excavated. At the south end in the base of the excavation the chalk had turned into a slurry, making it impossible to obtain a firm bottom. The cofferdam was being kept partly free of water by pumping from two sumps, and a considerable quantity of water was flowing along the bottom of the cofferdam.

Since the ground conditions within the excavation had proved to be so different from those assumed, the whole basis of the foundation design had to be questioned. In particular

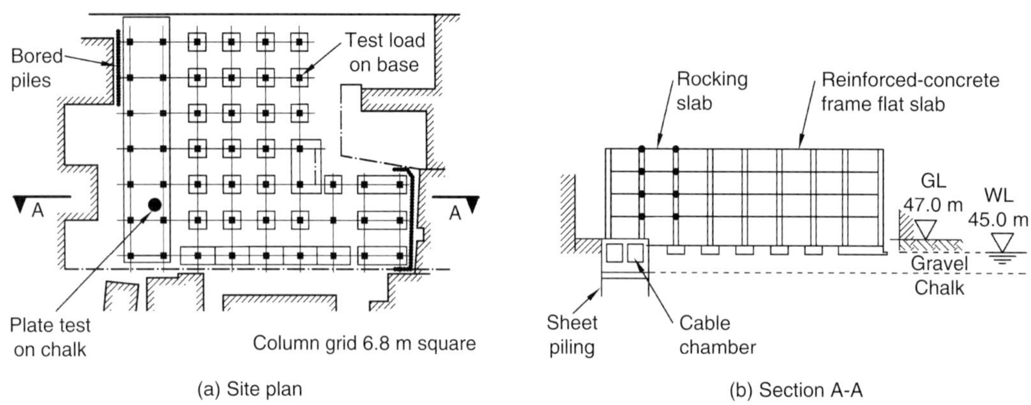

Figure 9.15 Shallow foundations on soft chalk; plan and cross-section through building

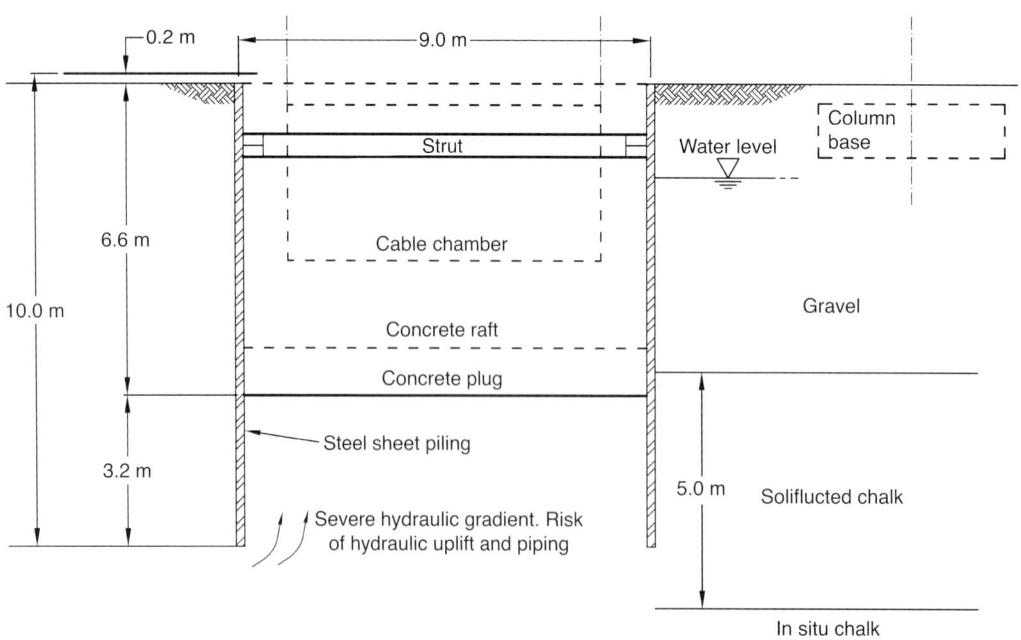

Figure 9.16 Shallow foundations on soft chalk; cross-section through cofferdam

the use of pad footings and a raft foundation beneath the cable chamber were thrown into doubt. Before reaching any conclusions it was felt that a further site investigation should be carried out. The soil strata, as they affect the foundations, may be divided into three categories.

- *Medium-dense sandy gravel – soil type G:* This flint gravel contains very sandy pockets and also pockets or lenses of a more clayey nature. Near the bottom of the stratum there is an increasing proportion of cobble-size flints. At the base of the stratum is a layer of about 0.3 m thickness of gravel and sand-sized rounded particles of chalk.
- *Soliflucted chalk – soil type Ch 1:* This stratum consists of fragments of hard white chalk in a soft matrix of remoulded chalk with flints. The matrix is mostly silt size but the materials exhibit cohesion and plasticity. Samples of the material broke open easily and were similar to soft or firm clayey silts in appearance and feel. During drilling the chalk was found to flow up into the borehole and the casing sank under its own weight. It appears therefore that the material is very sensitive and loses much of its strength when it is remoulded without drainage. The SPT values for this material are very low, ranging from 1 to 10, with an average of about 7.
- *In situ chalk – soil type Ch 2:* It proved difficult to accurately define the changeover between Ch 1 and Ch 2 chalk by examination of samples. There is, however, a noticeable increase in the SPT values when moving from Ch 1 into Ch 2 material. Moreover, on reaching the in situ chalk, the borehole casing stopped sinking. As a rough guide a level of about 35 m OD was taken as the change from Ch 1 to Ch 2.

The soliflucted Ch 1 material had probably been redeposited, and it was felt that it would behave like a silt. Four 100 mm dia. U100 samples of soliflucted chalk were selected for consolidated drained triaxial testing. The results of the tests gave $\Phi' = 40°$ and $c' = 0$. Values of m_v, measured during oedometer tests, ranged from 0.04 to 0.11 m^2/MN. The measured permeability was about 4×10^{-8} m/s, which is consistent with a silt. Pumping tests in the underlying Ch 2 chalk gave a permeability of about 2×10^{-5} m/s, reflecting the fissured nature of the chalk. Hence, the natural chalk has a permeability which is about 1000 times higher than the overlying soliflucted chalk.

A simple calculation of vertical equilibrium of the chalk in the base of the cofferdam (during the excavation stage) showed that the total vertical stress due to the chalk (at the sheet pile toe level) was about 69 kN/m^2, whereas the groundwater pressure at that level was 76 kN/m^2. There was therefore a high risk of failure by hydraulic uplift and piping in the base of the cofferdam.

9.5.3.2 The pad footings

The first problem to deal with was the adequacy of the existing pad footings. Based on the ground investigation data for the soliflucted chalk, preliminary calculations indicated settlements of 75–175 mm, which were deemed to be too large. An expensive solution was removal of the pads and installation of piles. Before taking this drastic action, an alternative means of assessing the pad footing settlement was required. Laboratory testing was believed to be unrealistic due to disturbance when sampling the chalk. In situ plate load tests would have been difficult below the water table. The simple and direct approach which was chosen was to load two of the existing pad footings up to the working load of 4000 kN (i.e. 400 tonnes) and measure the settlement. Although expensive, it was relatively quick, and all uncertainties (in terms of interpretation of alternative indirect methods) were eliminated.

The footings are 4.2 m square, 0.9 m thick and founded at about 44.5 m OD. In order to measure soil deformation below the footing, holes were drilled through the pad: one hole to the gravel–chalk interface (about 3.5 m below founding level) and the other extended 8 m into the chalk. The holes were sleeved and rods embedded in mortar at the base of the hole. Precision levelling was used to measure the settlement of the rods and the four corners of the pad. The footing was loaded in 25 tonne increments up to 200 tonnes, and left for 42 hours. It was then loaded to 400 tonnes and left for 70 hours. Settlement was measured frequently throughout the test. Under full load the settlement was only 2.5 mm, with about one-third of this in the chalk. The time-dependent settlement was negligible. Because of these very small settlements, a second test was unnecessary. The full-scale test demonstrated unambiguously that the pad footings were more than adequate.

9.5.3.3 The cable chamber

In view of the very soft nature of the chalk it was felt that the stability of the sheet pile walls was marginal. Pumping was therefore stopped and the cofferdam was allowed to flood. The next task was to investigate the chalk at the base of the cofferdam and assess its degree of disturbance. It was felt that a simple method of doing this was to use a cone penetrometer. A number of tests carried out within the cofferdam and outside it should provide a means of assessing the variability of conditions within the cofferdam and the degree of disturbance in relation to the surrounding undisturbed chalk.

Eleven tests were carried out alongside the existing pad footings. The tests were made from the bottom of new boreholes drilled by shell and auger methods through the gravel and lined with rigid 74 mm dia. UPVC pipe. The results of five typical tests are given in **Figure 9.17(a)**. As is to be expected, the results show considerable scatter but there is a well-defined lower limit as shown by the broken line. Below a level of 39 m OD the penetration resistance increases steadily with depth.

Seventeen tests were carried out in the base of the cofferdam from a rigid working platform spanning the excavation. At the north end of the cofferdam the depth of excavation was approximately 3.5 m below original ground level and was still in the gravel. The results of five cone penetration tests are shown in **Figure 9.17(b)**. The gravel extends to approximately

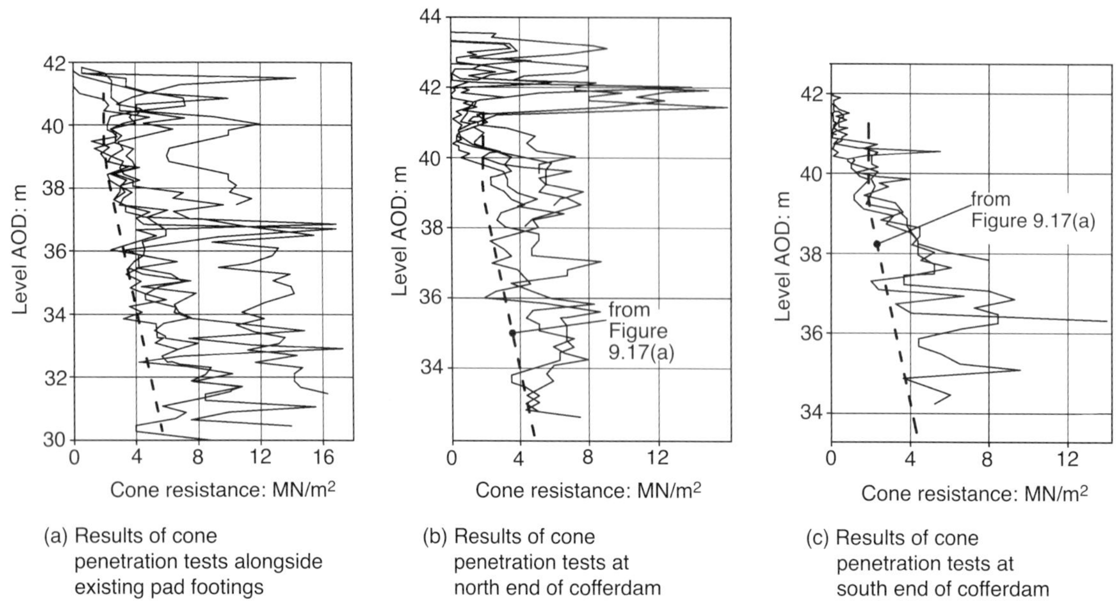

Figure 9.17 Shallow foundations on soft chalk; cone penetration tests

41.5 m OD, which is confirmed by an adjacent borehole log. For comparison the broken line is taken from **Figure 9.17(a)**. It can be seen that the cone resistances lie well below the broken line down to about 40 m OD, below which there is reasonable agreement.

At the southern end of the cofferdam, excavation was approximately 5 m below the original ground level. **Figure 9.17(c)** shows the results of three cone penetration tests, and it can be seen that very low penetration resistances were encountered down to approximately 40.5 m OD. Below a level of about 38 m OD the results are in good agreement with those given in **Figure 9.17(a)**. The results of the cone penetration tests point strongly to the fact that the soliflucted chalk in the base of the cofferdam had become loosened and disturbed by the upward flow of water, causing piping. On the basis of the cone test results it was felt that it would be unwise to found the raft for the cable chamber above a level of 40.4 m OD.

Before reaching a final decision to found the cable chamber on a raft it was necessary to make an estimate of the settlement. A simple large-scale in situ loading test was devised. The aim of the test was to reproduce as realistically as possible the actual construction conditions. The test was sited at the southern end of the cofferdam, where the cone resistances were least. A 2 m square pit was excavated underwater to a level of 40.4 m OD (i.e. approximately 1.5 m below the existing bottom). A diver was used to guide and control the excavation. A 1.5 m dia. concrete cylinder was then constructed and loaded by kentledge. There were a number of practical problems with carrying out this test; in particular, the cylinder was not vertical and tilted during loading. Nevertheless, load–settlement measurements were made up to a load of 51 tonnes, and the test data showed two important features.

- Up to working load, settlement was approximately proportional to load.
- Time-dependent settlement under constant load was very small.

Back-analysis of the test, using linear elastic theory, allowed the settlement of the proposed cable chamber raft to be predicted; the results of these predictions are shown in **Table 9.5** for various construction stages. Construction up to ground floor level was to be carried out with groundwater pumping from wells in the natural (Ch 2) chalk. Dewatering would then be stopped, leading to a decrease in *net* bearing pressure (and during the net decrease and subsequent increase in pressure, settlement was assumed to be zero until the previous maximum net pressure was exceeded). The total predicted measurable settlement is 62 mm. About half of this takes place during construction up to ground floor level. During superstructure construction the calculated settlement is 26 mm, with a further 10 mm due to live load. The large-scale test indicated that most of the settlement would be immediate. Hence, most of the settlement during superstructure construction would be 'built out'. Provision was made for residual differential settlement between the cable chamber raft foundation and adjacent pad footings by incorporating 'rocker' slabs. These were omitted until *after* the superstructure was completed.

9.5.3.4 Cable chamber construction and foundation settlement

Concreting of the concrete plug in the bottom of the excavation was carried out underwater in a series of transverse bays so as not to impair the stability of the cofferdam. Before the concrete was placed, a layer of filter fabric was placed on the chalk type Ch 1 and covered with a layer, 100 mm thick, of coarse granular fill to form a drainage layer. Three 450 mm

Construction stage	Bearing pressure: kN/m²		Foundation settlement: mm	
	Incremental	Net increase	Incremental	Total
Concrete plug	22	22	11*	–
Construction up to ground floor	52	74	26	26
Cease dewatering	−40	34	0	26
Construction to roof level	60	94	10	36
Internal walls and finishes	32	126	16	52
Live load	20	146	10	62

Note: * Not measurable

Table 9.5 Case history: shallow foundations on soft chalk – cable chamber, predicted settlements

dia. sleeves were cast in the plug through which sleeved wells were installed to a depth of 8 m below the excavation level and into chalk Ch 2. In addition, two 75 mm dia. pressure relief wells were installed through the plug in each bay and down into the chalk Ch 2. Pumping from the 8 m deep wells using a 150 mm dia. pump in each well lowered the water level to the level of the drainage layer beneath the concrete plug. Pumping continued until the raft and basement walls were complete, when the structure had sufficient weight to overcome buoyancy. Concrete for the complete raft was placed in a day.

The maximum settlement recorded for any of the pad footings was 5 mm and the average settlement was 2–3 mm. The time-dependent settlement was negligible. These results are in excellent agreement with the results of the full-scale loading test on one of the pad footings. The maximum settlement of the cable chamber near the south end was 15 mm, while near the north end it was 8 mm. Most of the settlement took place during the construction of the raft. Thereafter the settlements were small and very uniform, ranging from 3 to 5 mm.

9.5.3.5 Concluding remarks

It is evident from the measured settlements that the pad and raft foundations adopted for the telephone exchange were entirely satisfactory. Yet the decision to continue with these foundations was not an easy one. The top 5 m of chalk was very soft and all the available evidence pointed towards large settlements. It is very doubtful whether a traditional site investigation with sampling and laboratory testing would have resolved the issue because of the difficulty of obtaining representative undisturbed samples. In this case a direct approach to the problem was adopted by carrying out large-scale loading tests. The full-scale loading test on the pad footing was carried out quickly and simply and provided conclusive evidence of the adequacy of the pad foundations. The loading test in the cofferdam was less satisfactory in that the results led to a significant overestimate of the settlements (62 mm predicted and 8–15 mm measured). Nevertheless, the large-scale test played a very important part in arriving at the final foundation solution.

This case history highlights the problems that can occur during construction, especially when the groundwater regime and groundwater effects during construction are not understood. Careful observations and tests during the works were of vital importance, in order to successfully complete the works. Construction of the cable chamber below the water table required a proper dewatering and depressurisation system to be installed. This was only switched off once the weight of the permanent works was sufficient to avoid buoyancy.

9.6. Conclusions

- A useful mnemonic for foundation design is the 'five Ss': site, soil, structure, safety and sustainability.
- The ground profile and the allowable foundation movement are two key factors in selecting the most appropriate foundation type.
- Common factors in properly understanding ground risks are
  - understanding the site history and geology
  - understanding the site's groundwater regime and how it impacts upon the proposed construction and permanent works
  - understanding potential ground hazards and their impact on the foundation construction techniques.

In the vast majority of cases when foundation problems occur it is because the above has not been properly understood by the foundation designer. A coherent approach to proper understanding of ground risk is provided by the geotechnical triangle (introduced in Chapter 4 *The geotechnical triangle*).

- Often the limits set for allowable total and differential settlement are arbitrary and over-conservative. It is important for the foundation designer to have early discussions with the structural engineer on this matter. Note that it is the *differential* settlement which occurs *after* the structure is built and finishes applied that has the potential to cause damage. The main opportunity for economic foundation design is to set realistic limits on allowable foundation movement, especially differential movements.
- The foundation engineering associated with a project can involve the assemblage of a number of components and a variety of processes. It is important that a holistic approach be adopted. When different components of a project are designed by different teams the potential interactions should be considered by a single team – preferably the lead project designer.

- Examples are given in this chapter illustrating the importance of keeping the geotechnical triangle 'in balance'. In addition to the four activities included in the triangle, it has been found helpful to consider a fifth aspect which involves *construction processes and design verification*. This aspect recognises the fact that foundation engineering is essentially a process requiring monitoring and verification of key design assumptions.

References

Burland JB, Hancock RJ and May J (1983) A case history of a foundation problem on soft chalk. *Géotechnique* **33(4)**: 385–395.

Cole KW (1988) *Foundations*. Thomas Telford, London, UK.

Davis J, Essex R, Farooq I and Drake A (2023) *Geotechnical Baseline Reports: A Guide to Good Practice*. CIRIA Report C807. CIRIA, London, UK.

Fookes PG, Lee EM and Griffiths JS (2007) *Engineering Geomorphology: Theory and Practice*. Taylor and Francis, Boca Raton, FL.

HSE (Health and Safety Executive) (1991) *Protection of Workers and the General Public during the Development of Contaminated Land*. HS(G)66. HSE, London, UK.

O'Brien AS (1997) Vibrocompaction of loose estuarine sands. *Proceedings of the 3rd International Conference on Ground Improvement Geosystems*, London, UK. Thomas Telford, London, UK, pp. 120–126.

Phipps PJ (2003) Geomorphological assessment for transport infrastructure projects. *Proceedings of the ICE – Transport* **156(3)**: 131–143.

Preene M, Roberts TOL and Powrie W (2016) *Groundwater Control: Design and Practice*, 2nd edn. CIRIA Report C750. CIRIA, London, UK.

Rudland DJ, Lancefield RM and Mayell PN (2001) *Contaminated Land Risk Assessment: A Guide to Good Practice*. CIRIA Report C552. CIRIA, London, UK.

Steeds JE, Shepherd E and Barry DL (1996) *A Guide for Safe Working Practices on Contaminated Sites*. CIRIA Report R132. CIRIA, London, UK.

Terzaghi K (1951) The influence of modern soil studies on the design and construction of foundations. *Building Research Congress, London, UK*, vol. 1, pp. 139–145.

All chapters within Sections 1 *Context* and 2 *Fundamental principles* together provide a complete introduction to the Manual and no individual chapter should be read in isolation from the rest.

Chapter 10
Codes and standards and their relevance

Trevor Orr Trinity College, Dublin, Ireland

This chapter explains the relevance and benefits of geotechnical codes and standards, serving as a means to achieve the safety, serviceability and economy required by society, as a means of communication, providing a common understanding between all involved in a project, and as a means of ensuring the quality and reliability of investigations, testing, construction and materials. The former separate standards for different geotechnical design situations, based on the permissible stress method with overall factors of safety, have been replaced by a single standard, Eurocode 7, based on the limit state method with partial factors. The 10 soil mechanics or features causing geotechnical design to differ from structural design and Eurocode 7 to differ from the other Eurocodes are explained. The four main components of a geotechnical design: geometric data, loads and geotechnical parameters, verification method and complex nature of design situations are presented in a geotechnical design triangle, and related to the aspects of the geotechnical triangle for the teaching of soil mechanics. Details of the BSI codes and standards for geotechnical investigation and testing, geotechnical design, geotechnical construction, earthworks and geotechnical materials are presented. In addition the changes planned for the second generation of Eurocode 7 are summarised.

doi: 10.1680/icemge.66816.0107

10.1. Introduction

As society grows more complex, with additional laws and regulations, and as people become more litigious, those involved in geotechnical engineering need to be confident that their designs are reliable and have the level of safety required by society. Reliable and safe geotechnical designs may be achieved through possession of a good understanding of soil behaviour, through adopting appropriate models for geotechnical design situations and, in view of the complex nature of soil, through taking appropriate account of previous comparable experience. Design codes of practice and standards aim to provide recommendations for good design practice, setting out the processes and procedures to achieve geotechnical designs that meet the needs of society with regard to health and safety. In addition, they aim to provide designs that meet the requirements of the owners and users with regard to serviceability – that is, with regard to acceptable settlement and deformation and to economy. As Simpson *et al.* (2009) noted, codes have a responsibility to balance economy with safety and serviceability in a world in which limited resources are increasingly recognised.

10.2. Statutory framework, objectives, status and relevance of codes and standards

The Building Regulations 2010, published by the UK Department for Communities and Local Government (DCLG) and made under the powers of the Building Act 1984, provide the statutory framework for building projects by setting out the functional requirements to ensure the health and safety of people in and around all types of buildings, the welfare and convenience of disabled persons, as well as the requirements for energy conservation. Practical guidance on ways to comply with the functional requirements of the Building Regulations is contained in Approved Documents, which are published in a number of parts by the National Building Specification (NBS) for the DCLG. For example, Approved Document A (NBS, 2013), which is concerned with structural safety, refers to two parts of the geotechnical British Standard, published by the British Standards Institution (BSI): BS EN 1997-1:2004+A1:2013 and BS EN 1997-2:2007. These are 'Part 1: General rules' and 'Part 2: Ground investigation and testing' of Eurocode 7, 'Geotechnical design', which are standard parts providing practical guidance for meeting the structural safety requirements of the Building Regulations.

The prefix BS indicates a British Standard and the prefix EN indicates a European Norm (i.e. a standard). References to Eurocode 7 in this chapter are to these two parts of Eurocode 7, and where references are made to particular text or paragraphs, they are to those in Part 1. A second generation of Eurocode 7 is due to be published by the European Committee for Standardization (CEN) by March 2026 for subsequent publication by BSI by September 2027. Some comments on the objectives of this version of Eurocode 7 and how it will differ from the existing version are given in Section 10.9.4.

It should be noted that the guidance given in the Approved Documents is not a set of statutory requirements and does not have to be followed. Compliance with the codes and standards referred to in the Approved Documents is just one way of fulfilling (i.e. being deemed to satisfy) the functional requirements in the Building Regulations for some of the more common building situations. Hence, the referenced standards in an Approved Document are not mandatory and alternative methods may be used, although the designer may need to prove that the alternative methods satisfy all the relevant functional requirements and that the resulting design still complies with the Building Regulations. There is a legal presumption that if a designer has followed the guidance in an Approved Document, this is evidence that the design has complied with the Building Regulations.

The status and relevance of Eurocode 7, and the other Structural Eurocodes, as stated in the forewords to all the Eurocodes, are that they are to serve as reference documents for the following purposes

- as a means to prove compliance of building and civil engineering works with the essential requirements of Council Directive 89/106/EEC, particularly Essential Requirement No. 1 – mechanical resistance and stability – and Essential Requirement No. 2 – safety in case of fire
- as a basis for specifying contracts for construction works and related engineering services
- as a framework for drawing up harmonised technical specifications for construction.

With regard to those who use codes and standards in geotechnical engineering, Eurocode 7 makes the following important assumptions, which should not be overlooked.

(a) Data required for design are collected, recorded and interpreted by appropriately qualified personnel.

(b) Structures are designed by appropriately qualified and experienced personnel.

(c) Adequate continuity and communication exist between the personnel involved in data collection, design and construction.

(d) Adequate supervision and quality control are provided in factories, in plants and on site.

(e) Execution is carried out according to the relevant standards and specifications by personnel having the appropriate skill and experience.

(f) Construction materials and products are used as specified in Eurocode 7, or in the relevant material or product specifications.

(g) The structure will be adequately maintained to ensure its safety and serviceability for the designed service life.

(h) The structure will be used for the purpose defined for the design.

Ensuring compliance with all of these assumptions is more important for ensuring the safety of geotechnical structures and reducing the risk of failure than is precision in the design calculations, or the values adopted for the partial factors. Regarding compliance with *(a)*, *(b)* and *(e)*, Eurocode 7 does not define what is meant by people having appropriate qualifications, skills and experience. This will depend on the nature of the project and the ground conditions, and has been left to national determination. A formal measure to address this issue has been undertaken by the Institution of Civil Engineers together with the Geological Society and the Institute of Materials, Minerals and Mining with the establishment in June 2011 of the UK Register of Ground Engineering Professionals (UK RoGEP). Details about the Register and how to become a registered ground engineering professional are given in document 3009(4) (ICE, 2011), prepared by the British Geotechnical Association under the auspices of the Ground Forum. The UK RoGEP provides external stakeholders, including clients and other professionals, with a means to identify individuals who are suitably qualified and competent in ground engineering.

10.3. Benefits of codes and standards

According to the BSI website (www.bsigroup.com), the benefits of using standards are that they fine-tune one's performance and manage the risks faced while operating in more efficient and sustainable ways; they allow one to demonstrate to customers the quality of what one does; and they help one see how to embed best practice into one's organisation. Thus, standards help to make life simpler and increase the reliability and effectiveness of the goods and services we use. In the case of geotechnical design, their benefits are that they establish the procedures and requirements, including the safety elements, for the design process to achieve the levels of safety, serviceability and economy required by society. Codes and standards are created by those involved in the industry and hence they bring together important collective understanding and relevant experience. Another benefit is that they provide a common reference and means of communication for all those involved in geotechnical projects: geotechnical designers, ground investigators, those working in geotechnical laboratories, structural engineers, contractors, clients, users, public authorities and regulators. Simpson *et al.* (2009) have elaborated on the communication aspect of codes and how they provide *(a)* a link between analysis and design, where analysis is a tool in the design process; and *(b)* a link between the activities of the designer and the requirements of society regarding safety and serviceability.

In the case of the Eurocodes, the European Commission has published Guidance Paper L (EC, 2003), in which the intended benefits and opportunities of the Eurocodes are listed as

- providing common design criteria and methods to fulfil the specified requirements for mechanical resistance, stability and resistance to fire, including aspects of durability and economy

- providing a common understanding regarding the design of structures between owners, operators and users, designers, contractors and manufacturers of construction products
- facilitating the exchange of construction services between member states
- facilitating the marketing and use of structural components and kits in member states
- facilitating the marketing and use of materials and constituent products, the properties of which enter into design calculations, in member states
- being a common basis for research and development in the construction sector
- allowing the preparation of common design aids and software
- increasing the competitiveness of European civil engineering firms, contractors, designers and product manufacturers in their worldwide activities.

In the case of geotechnical design, the Eurocodes have the additional benefit and relevance that, being based on the same limit state design method with consistent partial factors, they harmonise geotechnical design with structural design and thus provide a 'common language' for geotechnical design in the different European countries. Until the publication of Eurocode 7, geotechnical design codes and standards in the European countries varied greatly.

10.4. Characteristics and development of codes and standards for geotechnical engineering

10.4.1 Characteristics of codes and standards

The British Standard BS 0:2021 'A standard for standards: Principles of standardization' defines a standard as a document that is 'established by consensus and approved by a recognised body, that provides, for common and repeated use, rules, guidelines or characteristics for activities or their results, aimed at the achievement of the optimum degree of order in a given context'. It also notes that standards are based on the consolidated results of science, technology and experience, and aimed at the promotion of optimum community benefits. All BSs

(*a*) represent good practice as defined by experts

(*b*) are developed by balanced and broadly representative standing committees that retain responsibility for them indefinitely and that reach agreement by consensus

(*c*) are subjected to unrestricted open consultation

(*d*) remain subject to periodic review as to their continuing validity.

BSI publishes documents concerned with geotechnical engineering that are sometimes termed *codes of practice* and sometimes *standards*. To understand what exactly a code of practice is and what a standard is and how they are developed, it is helpful to refer to BS 0:2021, where the different types of standardisation publication, published by BSI as BSs with the prefix BS, are defined and their characteristics are listed.

- A *specification* gives a coherent set of absolute requirements, each objectively verifiable. The result is a non-negotiable set of criteria for products, materials, services, processes, or systems. It is particularly suited to giving the performance criteria demanded of a product, or the fundamental elements of a service or management system.
- A *management system standard* provides a model to follow when setting up and operating a management system. It is usually written in the style of a specification.
- A *code of practice* contains recommendations and supporting guidance, where the recommendations relevant to a given user have to be met in order to support a claim of compliance. Users may also justify the substitution of any of the recommendations in a code of practice with practices of equivalent or better outcome. A code of practice usually reflects current good practice as employed by competent and conscientious practitioners. An example of a geotechnical code of practice is Eurocode 7, BS EN 1997-1.
- A *guide* primarily contains information and guidance. It may also include recommendations where appropriate, but these are generally of a nature that would not support reliable claims of compliance.
- A *method of test* provides repeatable and reproducible procedures with consistent outcomes for the assessment of material, product, or process performance. An example of a geotechnical method of test is BS EN ISO 17892-1 'Geotechnical investigation and testing – Laboratory testing of soil – Part 1: Determination of water content'.
- A *method of specifying* gives characteristics of a material, product, process, or system so that a customer can select the values needed before agreement with a supplier.
- A *vocabulary* is a compendium of terms and definitions which helps to harmonise the use of language within a given sector, field, or discipline. A useful vocabulary for those involved in civil engineering is BS ISO 6707-1:2020 'Building and civil engineering works – Vocabulary – Part 1: General terms'.
- A *classification* is an ordering of items or grading system for use across a given sector, field, or discipline.

Thus, a code of practice is used for design as it aims to reach an acceptable technical level of quality and is expressed as functional requirements based on scientific/technical principles. It normally avoids standardising particular design methods or detailed construction procedures, leaving these choices to the designer (Krebs Ovesen and Orr, 1991). For example, in Eurocode 7 all the design methods are presented as application rules, which are optional, rather than as Principles, which are requirements and hence mandatory; this means the designer can choose another design method but needs to demonstrate that it provides as least an equivalent level of safety as the method given in Eurocode 7 so as to satisfy the Principle. A method of test or a method of specifying, on the other hand, aims to achieve a certain degree of order in a given context through specification, such as through specifying a particular test method or execution procedure to be used.

According to BS 0, the purposes of BSs, some of which are similar to the intended benefits and opportunities of the Eurocodes noted in Section 10.3, include

- facilitating trade, particularly in reducing technical barriers and artificial obstacles to international trade
- supporting public policy objectives and, where appropriate, offering effective alternatives to regulation
- providing a framework for achieving economies, efficiencies and interoperability
- enhancing consumer protection and confidence
- supporting and defining good environmental, social and governance practices
- addressing climate change
- supporting innovation
- enabling equality and inclusion.

BS 0 states that it is the main document by which BSI sets out how BSs and related documents are developed and maintained. The majority of documents now published by BSI as BSs have their origin in European or international standards, developed by CEN or the International Organization for Standardization (ISO). The remainder are developed exclusively by BSI to meet particular needs in the UK. The standards development processes of BSI, CEN and ISO all operate on the basis of technical committees which are appointed to be responsible usually for a range of standards, both published and in development, that fall within their scope. Technical committees report to the managing bodies of their respective organisations, and these are responsible for monitoring the overall work and arbitrating disputes about committees' terms of reference and spheres of activity. The BSI technical committee responsible for all the BS geotechnical codes and standards is B/526, *Geotechnics* while the CEN technical committee responsible for all the Eurocodes is CEN/TC 250: *Structural Eurocodes*.

An important feature of BSs, pointed out in BS 0, is that they 'are voluntary' in that there is no obligation to apply them or to comply with them, except in those few cases where their applicability is directly demanded by regulatory instruments. They are tools devised for the convenience of those who wish to use them. In certain circumstances the actions of a third party might have the effect of making the application of a standard a commercial necessity, for example in a contract, but BSI has no control over these actions and is not a party to them. The forewords to many BSs (e.g. BS 1377-1 and BS EN 1997-1) draw attention to the consequence of this by including the important statement, which is presented in bold text: 'Compliance with a British Standard cannot confer immunity from legal obligations.'

10.4.2 Development of UK codes and standards

The first codes and standards for geotechnical engineering in the UK were prepared in the 1940s and 1950s by committees convened by the engineering institutions on behalf of the Codes and Practice Committee for Civil Engineering, Public Works, Building and Construction Work, under the aegis of the former Ministry of Works, for publication in the Civil Engineering and Public Works Series. These design codes of practice were based on the permissible stress method with overall factors of safety. More details about the evolution of British codes of practice and standards have been provided by McWilliam (2002), who has reported on the issues that arose during the development of the codes of practice in relation to the traditional role of professional engineers – who regarded design rules as the individual choice for the engineer – and their developing relationships with government agencies and international trade.

Work on 'Code of Practice no. 1 (CP 1): Site investigations' started in 1943 and was published by the Institution of Civil Engineers in 1950; 'Code of Practice no. 2 (CP 2): Earth retaining structures' was published by the Institution of Structural Engineers in 1951; and 'Code of Practice no. 4 (CP 4): Foundations' was published by the Ministry of Works in the late 1940s. The first standard for the testing of soils was published by BSI in 1948 as BS 1377 'Method of test for soil description and compaction', which consisted of just one part when first published, but has since increased to nine parts. In 1949 the responsibility for the preparation and issue of the codes of practice of the Civil Engineering and Public Works Series in the UK was handed over to the following four professional engineering institutions: the Institution of Civil Engineers, the Institution of Municipal Engineers, the Institution of Water Engineers and the Institution of Structural Engineers.

In 1954, while the drafting of codes of practice remained with the professional engineering institutions, the responsibility for publishing the codes of practice was transferred from the engineering institutions to BSI, where it has remained. Revised versions of CP 1 and CP 4 were published by BSI as CP 2001 in 1957 and CP 2004 in 1972, respectively. While they still had the phrase 'code of practice' in their titles, subsequent revisions of these codes of practice were published during the 1980s as British Standards with the prefix BS rather than CP. For example, the revised version of CP 2001 was published as BS 5930 in 1981 and the revised version of CP 2004 was published as BS 8004 in 1986. The first three revisions of BS 1377, in 1961, 1967 and 1975, were published as single documents. In the 1961 version the title was changed to 'Methods of testing soil for civil engineering'. The 1975 version was published as a metricated version with SI rather than Imperial units and included procedures for a greater range of soil tests. Subsequently, in 1990, the next revision was published in nine separate parts with the slightly modified umbrella title 'Method of test for soils for civil engineering purposes'.

10.4.3 Development of European and international codes and standards

With the move towards harmonisation in Europe through the formation of the European Economic Community and then the European Union, the European Commission (EC) in 1975 initiated a programme to prepare harmonised standards for the design of buildings and civil engineering works, known as the Structural Eurocodes. As noted in Section 10.2, the Structural Eurocodes were prepared to serve as reference documents

- as a means to prove compliance of building and civil engineering works with the essential requirements of Council Directive 89/106/EEC (EC, 1989), particularly Essential Requirement No. 1 – mechanical resistance and stability – and Essential Requirement No. 2 – safety in case of fire
- as a basis for specifying contracts for construction works and related engineering services
- as a framework for drawing up harmonised technical specifications for construction.

Council Directive 89/106/EEC (EC, 1989) is the Construction Products Directive (CPD) of 21 December 1988 on the approximation of laws, regulations and administrative provisions of the member states relating to construction products. This has since been repealed and superseded by the Construction Products Regulation (CPR) (EU) No. 305/2011 of the European Parliament and Council (EU, 2011) of 9 March 2011, laying down harmonised conditions for the marketing of construction products in the EU. The CPR provides a common technical language to assess the performance of construction products and in this way also enables EU countries to ensure the safety of construction works. Although the UK left the EU on 1 January 2021 (Brexit), the essential requirements for construction products, including buildings, in the UK are the same as those in the EU and hence the Eurocodes continue to be recognised in the UK as reference documents to meet the UK Building Regulations.

After 15 years' work on the Eurocodes, it was decided in 1990 to transfer this work from the EC to CEN. The national standards organisations of all the countries in Europe, including the UK post-Brexit, are members of CEN (BSI is the UK member). Indeed, BSI provides the secretariat for CEN/TC 250: *Structural Eurocodes*, the CEN technical committee responsible for the drafting of all the Eurocodes. CEN publishes the European harmonised standards, which are those standards officially recognised by reference in the *Official Journal of the European Union* as meeting the essential requirements of a relevant European directive. In the case of the Eurocodes, the relevant directive is the CPD.

Since establishing the required level of safety for structures in a particular country is a national responsibility, Eurocode 7 and each of the other Eurocodes is published by BSI with a National Annex for designs in the UK. The National Annexes provide the values of the partial factors, referred to as Nationally Determined Parameters (NDPs), and other safety elements to be used with a particular Eurocode. The National Annexes also provide references to non-contradictory complementary information (NCCI) – for example, national standards that have been prepared or revised so as not to conflict with Eurocodes, which may be used to complement the Eurocodes.

Each Eurocode has been prepared by a subcommittee (SC) of CEN/TC 250, with Eurocode 7 prepared by SC7 *Geotechnical design*, hosted by the Royal Netherlands Standardization Institute (NEN). As well as preparing the Eurocodes for structural and geotechnical design, CEN has also in recent years been preparing, through its technical committees, a large number of standards relating to other aspects of geotechnical engineering; most of these have now been published by CEN as European Standards (i.e. as ENs). Some standards relating to geotechnical engineering have also been prepared in recent years by technical committees of ISO and published as ISO standards (these have the prefix ISO). To avoid duplication of work, it was decided in 1991, under the Vienna Agreement, that when CEN and ISO both required a standard in a particular area, they would cooperate in the preparation of the standard and exchange technical information. Examples of this include the collaboration between CEN and ISO in the preparation of the standards for geotechnical investigation and testing, with some standards being prepared by CEN/TC 341 *Geotechnical investigation and testing* in collaboration with ISO/TC 182 *Geotechnics*, and some being prepared by ISO/TC 182 in collaboration with CEN/TC 341.

As a member of CEN, BSI is obliged to adopt identical versions of all European Standards as BSs, and to withdraw any national standards that might conflict with them. Hence, since the Eurocodes and other European Standards have been approved by CEN as European Standards with the prefix EN, they have been adopted and published by BSI as BSs with the prefix BS EN, and the conflicting BSs have been superseded and withdrawn or revised so as to provide only NCCI. If an ISO standard is adopted as a BS, it is published by BSI with the prefix BS ISO, unless it has been prepared by an ISO or CEN committee working in collaboration under the Vienna Agreement, in which case once it is approved by CEN as a European Standard, with the prefix EN ISO, it is then published by BSI with the prefix BS EN ISO. Examples of such standards are provided in **Tables 10.3–10.7**.

10.5. Why geotechnical and structural codes and standards differ

Geotechnical design codes and standards, such as Eurocode 7, differ in both structure and content from structural design codes. This is because soil has ten properties or special features that the materials used for structural design (e.g. steel) either do not have or have only to a much lesser extent. The consequences for geotechnical design and Eurocode 7 of each of these properties or special features of soil is explained in the following paragraphs and summarised in **Table 10.1**.

- *Soil being a natural material.* The principal property or special feature of soil which affects geotechnical design and causes geotechnical codes to differ from structural codes is that it is a natural material, unlike the materials used in structural design, such as concrete and steel, which are manufactured. The consequence of this is that the properties of the soil need to be determined as part of the geotechnical design process rather than being specified by the designer, as in the case of structural design. It is for this reason that Eurocode 7 has two parts to cover all the components of geotechnical design. These are 'Part 1: General rules', with the principles and requirements for the geotechnical aspects of the design of buildings and civil engineering works, and 'Part 2: Ground investigation and testing', with the design requirements relating to the planning and reporting of ground investigations, the general requirements for a number of commonly used laboratory and field tests for soil, the interpretation of test results and the derivation of geotechnical parameters and coefficients.

- *The particulate and frictional nature of soil.* Since soil is a particulate material, it is frictional and hence has shear strength, which is a function of the normal stress and the coefficient of friction. This is unlike coherent non-frictional manufactured materials, such as steel or concrete, where direct tensile or compressive strengths, which are not functions of the normal stress, are usually relevant. Since soil strength is frictional, depending on the normal stress as well as the coefficient of friction, it is necessary to be careful when factoring permanent loads in geotechnical design calculations to ensure that normal loads which are unfavourable (i.e. those tending to cause failure), but which are also contributing to the soil strength and resistance through providing the normal stress on the failure surface, are not increased. It is for this reason that a partial factor of 1.0 is normally used on permanent loads, including the self-weight of soil, in geotechnical design, unlike in structural design where the partial factor on permanent loads is 1.35. Further discussion of the frictional nature of soil is given in Chapter 17 *Strength and deformation behaviour of soils.*

- *The two- or three-phase nature of soil.* Since soil consists of mineral particles, which have shear strength, and water and air, which have no shear strength, the principle of effective stress needs to be taken into account in geotechnical design. This principle states that all effects of a change of stress on soil are due to changes in effective stress, where the effective stress is the total stress minus the pore water pressure – that is, it implies that all changes of soil strength, compression and distortion are due to changes in the stresses acting on the soil particles, not due to changes in the pore water pressure. It is for this reason that the principle of effective stress is so fundamental to geotechnical design and why, in order to predict how the soil will behave in a geotechnical design situation, it is normally necessary to determine the pore water pressures and understand how the pore water is behaving. Eurocode 7 takes account of this by constantly referring to the need in geotechnical design to consider pore water pressures and the effects of changes in them (e.g. Paragraph 2.4.2(9)P). Further discussion of this behaviour is given in Chapter 17 *Strength and deformation behaviour of soils.*

- *The wide range of permeability of soil.* As a consequence of its wide range of particle sizes and void ratios, soil has a wide range of permeability from very low (i.e. effectively impermeable) to very high. A consequence of the wide range of permeability is that the rate at which pore water can flow through soil, and hence the rate at which pore water pressures can dissipate, varies greatly. Furthermore, since changes in pore water pressure take a significant length of time to occur in fine-grained, low-permeability soils and since the properties and behaviours of soils change as the pore water pressures dissipate, it is necessary in geotechnical design, as stated in Paragraph 2.2(1)P of EN 1997-1, to consider both short- and long-term design situations. These are analysed using undrained and drained strengths and immediate and consolidation deformation parameters. This time-dependent behaviour of soil is a unique feature of geotechnical engineering that does not occur in structural design. Further discussion of this time-dependent behaviour of soil is given in Chapter 17 *Strength and deformation behaviour of soils.*

- *The inherent variability of soil.* Soil is inherently more variable than manufactured materials such as steel and concrete. Hence, larger partial material factor values are generally chosen for geotechnical design than for structural design (e.g. 1.25 on $\tan\phi'$ compared with 1.15 on the strength of steel, where ϕ' is the effective angle of friction). As a result of the greater variability of soil, there is often less uncertainty in the loads compared to the soil parameter values in geotechnical design, particularly when the loads are due to the self-weight of the soil, or compared with the uncertainty in the loads in structural design; for this reason, lower partial load factors are often chosen for geotechnical design. For example, 1.0 on permanent loads and 1.3 on variable loads, compared with the partial load factors of 1.35 and 1.5, respectively, chosen for structural design.

- *The spatial variability and non-homogeneous nature of soil.* Due to spatial variability and since the volume of the zone of ground governing the behaviour of a geotechnical structure is usually very large compared with the volume that is sampled, and since the number of test results is limited and the types of samples may vary, it is often not appropriate to select the 5% fractile of the test results as the characteristic value of a geotechnical parameter for use in geotechnical designs, unlike in the case of structural designs where an unlimited series of test results is, in principle at least, available and the material does not vary spatially. According to Paragraph 2.5.4.2(2)P of Eurocode 7, 'the characteristic value of a geotechnical parameter shall be selected as a cautious estimate of the value affecting the occurrence of the limit state', which is often the mean of a range of values covering a large surface or volume of the ground. Although it is not anticipated that statistics will normally be employed in the selection of the characteristic value, Eurocode 7 states that a cautious estimate of the mean value is a selection of the mean value of the limited set of geotechnical parameter values, with a confidence level of 95%. Due to the spatial variability and non-homogeneous nature of soil, it is important that the characteristic value should take account of the extent of the zone of ground governing the behaviour of the geotechnical structure and experience of similar soil conditions and the behaviour of similar structures in such soil, and also to identify any weak zones or layers whose strength might determine and control the failure mechanism. When the zone of influence is large – for example, in the case of a slope failure – then, provided there are no weak layers, the characteristic strength value will be a cautious estimate of the mean strength along the failure surface. However, in the case of a local failure involving a small volume of soil – for example, the failure of a footing – the characteristic value should be chosen as a more cautious estimate. An example of this is given in Paragraph 2.4.5.2(9), which notes that, in the case of a flexible structure on isolated footings, where the building is unable to transfer loading to resist local failure, the characteristic strength should be the mean strength under each footing. The need to consider the extent of the zone of influence in geotechnical design is indicated by the note to Paragraph 2.4.5.2(10), which states 'a cautious estimate of the mean is a selection of the mean of the limited set of geotechnical parameter values, with a confidence level of 95%; where local failure is concerned, a cautious estimate of the value is a 5% fractile'.

- *The ductile behaviour of soil.* Soil is usually very ductile – it can undergo large deformations when sheared and still retain significant shear strength, even if, on shearing, it first reaches a peak strength and then reduces to a residual strength. Although some structural materials, such as steel, are ductile, the extent of their ductility is normally less than for soil, so unless special measures are included to improve the ductility, structures are not able to deform as much as soil before failing. A consequence of the ductile behaviour of soil is that when ground supports a redundant (stiff) structure, those areas of the ground that are more heavily loaded will yield, allowing load to be transferred through the structure from the more heavily loaded to the less heavily loaded areas. Hence, the unplanned redistribution process due to soil's ductile behaviour is favourable and thus beneficial for geotechnical design. However, the situation in structural design is different because manufactured materials are normally homogeneous, without weak and strong zones, so that any unplanned load redistribution in structures is usually unfavourable and thus necessitates higher partial load factors in structural design. This is another reason why the partial factors on permanent and variable loads in geotechnical design are less than those in structural design.

- *The dilatant behaviour of soil.* Some soils – for example dense, coarse-grained soils – dilate when sheared, exhibiting a peak angle of friction, ϕ'_p. After sufficient shearing, they reach the critical state or constant volume angle of friction, ϕ'_{cv} when the soil is completely remoulded. In the case of other soils, such as clays, the soil particles may become aligned along the failure surface after large strains, with the result that a residual angle of friction is reached which is less than the critical state angle. Generally what is important in geotechnical design is the soil strength that is available to prevent failure occurring in a particular situation. Eurocode 7 states in Paragraph 2.4.3(2)P that the geotechnical parameter 'values obtained from test results and other data shall be interpreted appropriately for the limit state considered' and in Paragraph 2.4.3(4)P notes that 'account shall be taken of the possible differences between the ground properties and geotechnical parameters obtained from test results and those governing the behaviour of the geotechnical structure', giving soil and rock structure and brittleness as two of the factors that may cause these differences. Further discussion of the dilatant behaviour of soil is given in Chapter 17 *Strength and deformation behaviour of soils.*

  Eurocode 7 generally does not state if the soil shear strength to be used in design calculations is the peak, critical state or residual value. Paragraph 3.3.6(1)P notes that, when assessing the shear strength, the influence of features including the stress level, anisotropy of strength, strain rate effects, very large strains where these may occur in a design situation, pre-formed slip surfaces and degree of saturation shall be considered. Since structures will generally be considered to have failed well before the deformations required to get to the critical state have been reached (geotechnical designs have traditionally been carried out using the peak strength), and since most field tests are correlated to the peak strength, this is the strength that is normally used in designs to Eurocode 7. This is indicated by Paragraph 6.5.3(5), which, referring to the strength used to determine the resistance to sliding, states that 'for large movements, the possible relevance of post-peak behaviour should be considered'. The only paragraphs in Eurocode 7 that refer to the use of a specific angle of friction are Paragraphs 6.5.3(10) and 9.5.1(6). Paragraph 6.5.3(10) states that 'the design (structure–ground interface) friction angle δ_d may be assumed equal to the design value of the effective critical state angle of shearing resistance, $\phi'_{cv,d}$, for cast *in situ* concrete foundations and equal to 2/3 $\phi'_{cv,d}$ for smooth pre-cast foundations'. Paragraph 9.5.1(6) states that 'A concrete wall or sheet pile wall supporting sand or gravel may be assumed to have a design wall–ground interface parameter $\delta_d = k\phi_{cv,d}$. k (the wall–ground interface friction angle factor) should not exceed 2/3 for precast concrete or sheet piling.' Further discussion of the dilatant behaviour of soil is given in Chapter 17 *Strength and deformation behaviour of soils.*

- *The compressible nature and low shear stiffness of soil.* Since soil, particularly fine-grained soil, is compressible with a low shear stiffness, and is normally very much more compressible and less stiff than structural materials, it can undergo large deformations exceeding the limiting values for the structure. Hence the controlling criterion in many geotechnical design situations is limiting the deformations rather than preventing failure. Geotechnical designs are therefore often controlled by serviceability limit state considerations rather than by ultimate limit state requirements.

- *The nonlinear and complex stress–strain behaviour of soil.* The nonlinear and complex stress–strain behaviour of soil, and the difficulties in determining appropriate parameter values and modelling soil deformations, sometimes make it difficult to obtain reliable predictions of the ground deformations. Consequently, for simple design situations geotechnical designs may be carried out using ultimate limit state calculations with appropriate partial or overall safety factor values, chosen to limit the mobilised shear strength and hence the ground deformations. An example of this in Eurocode 7 is the application rule in Paragraph 6.6.2(16), which states that 'For conventional structures founded on clays, the ratio of the bearing capacity of the ground, at its initial undrained shear strength, to the applied serviceability loading should be calculated. If this ratio is less than 3, calculations of settlements should always be undertaken. If the ratio is less than 2, the calculations should take account of nonlinear stiffness effects in the ground.'

10.6. The geotechnical design triangle

Burland (1987) proposed that the teaching of soil mechanics should focus on four distinct aspects. He originally presented these four aspects as the soil mechanics triangle, a concept he has developed and expanded into the geotechnical triangle shown in **Figure 10.1** and discussed in Chapter 4 *The geotechnical triangle.* The geotechnical triangle has the first three aspects at its apexes: the ground profile, the soil behaviour and

Soil	Steel	Consequences for geotechnical design
Natural material	Manufactured	Properties are determined, not specified, so ground investigation and testing are part of the design process.
Particulate and frictional	Non-frictional	Soil has shear strength, which depends on normal stress, hence loads affect resistances and so care is needed when factoring permanent loads and hence partial factor of unity is usually appropriate for soil weight.
Two or three phases	Single phase	Soil particles have shear strength but water has none, so in geotechnical design it is necessary to take account of the principle of effective stress and the effect on soil behaviour of effective stress changes due to changes in pore water pressure.
Wide range of permeability	Impermeable	Soil has a wide range of particle sizes and hence permeability, from very low to very high, so that the rate at which pore water pressures dissipate can vary greatly and therefore it is necessary to consider time effects (i.e. both short- and long-term design situations) by examining both undrained and drained conditions.
Inherently variable	Low variability	Due to soil's inherent variability, there is greater uncertainty in soil strength than in the strength of structural materials and hence higher partial factors on soil strength than on steel strength (e.g. 1.25 on $\tan\phi'$).
Spatially variable and non-homogeneous	Homogeneous	Characteristic value is not 5% fractile of test results but defined as a cautious estimate of value affecting the occurrence of a limit state. It is necessary to take account of the zone of influence and use comparable experience when selecting the characteristic value. The estimate should be more cautious for local rather than global failures. It is also necessary to identify and take account of any weak layers that may affect a limit state.
Ductile	Most manufactured materials are not as ductile as soil	Soil ductility enables unplanned load redistribution in the ground supporting redundant structures so that strong areas attract load and weak areas shed load. Such load redistribution is favourable and beneficial in geotechnics. However, in structures such unplanned load redistribution is usually unfavourable. This is another reason for lower partial factors on loads in geotechnical design than structural design.
Dilatant	Non-dilatant	Dilation on shearing and the magnitude of the shear strains need to be considered when deciding whether the peak, critical state or residual strength parameters are appropriate.
Compressible and with a low shear stiffness	Non-compressible	Design is often controlled by serviceability limit state, not by ultimate limit state.
Nonlinear with a complex stress–strain behaviour	Linear and simple	Serviceability limit state design situations are usually difficult to model and parameters may not be available, so that for simple situations geotechnical design may be carried out using ultimate limit state calculations.

Table 10.1 Properties or special features of soil compared with steel and consequences for geotechnical design

an appropriate model. The fourth aspect: precedent, empiricism and well-winnowed experience, lies at its centre and is linked to the other three.

In a similar way, Orr (2008) has shown that the different components of geotechnical design can be presented by the geotechnical design triangle shown in **Figure 10.2**, the aim of which is to clarify and improve understanding of the geotechnical design process in Eurocode 7. The components of geotechnical design that form the apexes of the geotechnical design triangle are the geometrical data, the loads and geotechnical parameters, and the verification method. Linking these three components is the fourth component, the design complexity. How the design complexity affects the other geotechnical design components is an important design consideration and hence design complexity has been placed at the centre of the geotechnical design triangle, linked to each of the other three design components in the same way as precedent, empiricism and well-winnowed experience are linked to the components at the apexes of the geotechnical triangle.

The constituents and design procedures involved in the four geotechnical design components are listed as bullet points in the geotechnical design triangle in **Figure 10.2**. The constituents involved in the first component, the data describing the geometry of the design situation, include the level and slope of the ground surface, water levels, levels of interfaces between strata, excavation levels and the dimensions of the geotechnical structure. With regard to the second component, the loads involved in geotechnical designs are usually either provided by the structural designer or determined from the weight density (unit weight) of the soil, while the geotechnical parameters, such as the soil strength and stiffness, are determined as derived values from geotechnical investigations involving field and laboratory tests.

Regarding the third component, the verification method, with the introduction of Eurocode 7 the limit state design method has replaced the permissible stress method for geotechnical design. This design method requires that all the relevant limit states are considered and shown to be sufficiently improbable. This is achieved by adopting one or more of the following: calculations, prescriptive measures, load or model tests, an observational method. The constituents that make up the fourth component, the design complexity, are the nature of the ground and the groundwater conditions, the nature of the structure, the nature of the surroundings, the environmental

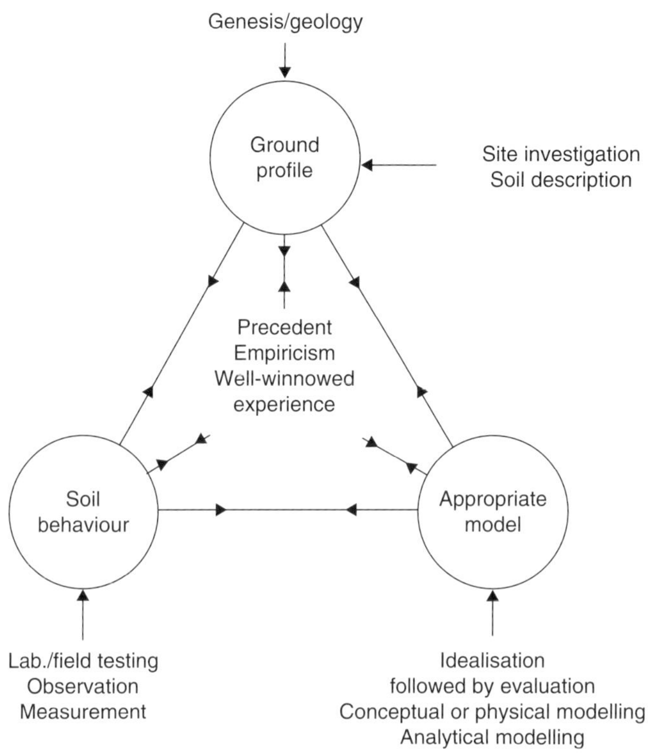

Figure 10.1 Geotechnical triangle

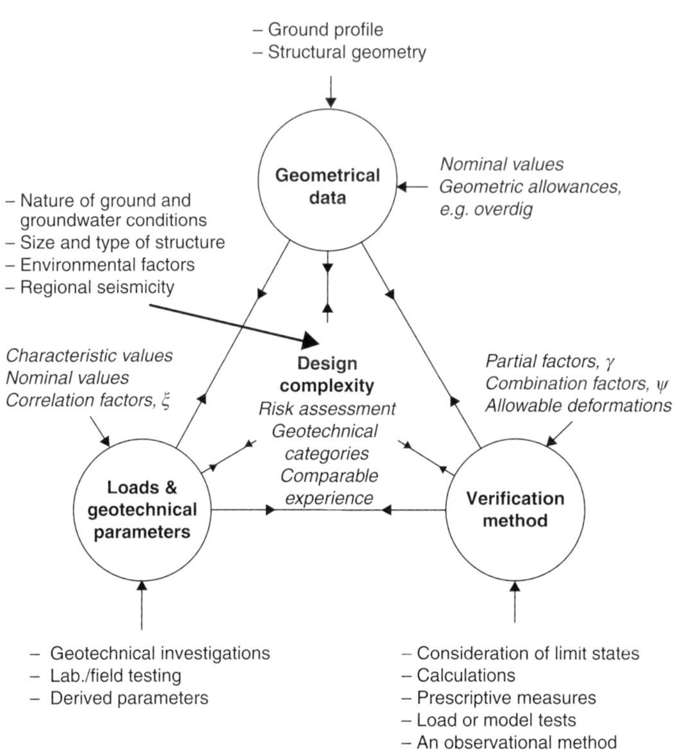

Figure 10.2 Geotechnical design triangle

factors and, where relevant, the regional seismicity. Each of these factors and their level of complexity needs to be considered at the start of a geotechnical design as they will affect how each of the other three components is treated.

10.7. Safety elements adopted in Eurocode 7

In geotechnical designs to Eurocode 7, each component of the geotechnical design triangle is treated in an appropriate manner with appropriate safety elements to ensure the required level of safety is achieved. The treatment and safety elements adopted in the case of each component are explained in the following paragraphs and are shown in italics in **Figure 10.2**.

In the case of the geometrical data, the values used for design are measured or chosen nominal values. Where appropriate, geometrical allowances are included as the measures to ensure safety; examples of such allowances in Eurocode 7 are the overdig allowance in the design of retaining structures (Paragraph 9.3.2.2(2)) and the additional constructional tolerance on the foundation width in the case of spread foundations with highly eccentric loads (Paragraph 6.5.4(2)).

In the case of the loads and geotechnical parameters, the uncertainty and variability in these are taken into account and safety is achieved through the selection of suitably cautious values, termed characteristic values. Where sufficient data are available, the characteristic value of a load is defined as the 95% fractile value of the load distribution, while in the case of a favourable material parameter, the characteristic value is determined as the 5% fractile value. Often sufficient data are not available to obtain the characteristic value of a load statistically and hence a 'nominal' value is used instead. When combinations of loads are involved in a design situation, combination factors are applied to the characteristic loads to obtain 'representative' values of loads. These are the values to which the partial action (load) factors are applied to obtain the design values that are used in design calculations to Eurocode 7. Regarding the characteristic value of a geotechnical parameter, as a consequence of the variability and limited test results, noted in Section 10.5 in connection with a property or special feature of soil, Paragraph 2.5.4.2(2)P of Eurocode 7 states that it 'shall be selected as a cautious estimate of the value affecting the occurrence of the limit state'. When the pile resistance is determined from pile load tests or profiles of field tests, Eurocode 7 provides correlation factors, ξ, that are applied to the test results or profiles to determine the characteristic pile resistance. The ξ values used to determine the characteristic pile resistance depend on the number of pile load tests or test profiles.

With regard to the verification method, the required level of safety when using the limit state design method in Eurocode 7 is achieved by applying appropriate partial factors to the loads and/or the geotechnical parameter values or resistances in the ultimate or serviceability limit state calculation models. Partial model factors may also be introduced to account for uncertainty in the calculation model, although this uncertainty is normally accounted for in the values chosen for the partial action and material factors. The values of the partial factors provided in the UK National Annex to Eurocode 7 are all for ultimate limit states, since the partial factors for serviceability limit states are all equal to unity.

The way that safety is introduced in Eurocode 7 with regard to the complexity of a design situation is by the requirement in Paragraph 2.1(8)P that the complexity of each design situation shall be identified together with the associated risks so that minimum requirements for the extent and content of the geotechnical investigations, calculations and construction and control checks can be established. When identifying and addressing the complexity of a design situation, it is important to take account of comparable experience, which is defined in Paragraph 1.5.2.2 as 'documented or other clearly established information related to the ground being considered in design, involving the same types of soil and rock and for which similar geotechnical behaviour is expected, and involving similar structures. Information gained locally is considered to be particularly relevant.' An example of a risk assessment procedure in Eurocode 7 to address the design complexity is the three Geotechnical Categories given in Paragraph 2.1(10) to establish the requirements for designs to Eurocode 7. These Geotechnical Categories are presented in Eurocode 7 as an application rule, not as a principle – hence their use is optional for geotechnical designs carried out in the UK. The three Geotechnical Categories – GC1, GC2 and GC3 – relate to the following design situations.

GC1 small and relatively simple structures with ground conditions known from local experience to be straightforward and with negligible risk in terms of overall stability

GC2 conventional types of structures and foundations with no exceptional risk or difficult ground or loading conditions

GC3 Structures or parts of structures that fall outside the limits of GC1 and GC2 – examples include very large or unusual structures and structures involving abnormal risk or unusual or exceptionally difficult ground or loading conditions.

10.8. Relationship between the geotechnical design triangle and the geotechnical triangle

The relationships between the four components of the geotechnical design triangle and the four aspects of the geotechnical triangle, which were identified as being essential for the understanding of soil behaviour, are shown in **Table 10.2**. This table shows that the components of the geotechnical design triangle are related to and cover the same areas as Burland's aspects of soil mechanics. However, the safety elements of the geotechnical design components, which are a fundamental feature of geotechnical designs, are not included in the first three of the four aspects of the geotechnical triangle, which were identified as being essential for the understanding of soil behaviour, but the fourth aspect includes precedent and well-winnowed experience, which are essentially the comparable-experience safety element of the fourth component of geotechnical design, accounting for the complexity of geotechnical designs.

Geotechnical triangle aspects	Geotechnical design components	Treatment manner and safety elements
Ground profile – Genesis/geology – Site investigation – Soil description	Geometrical data – Level and slope of ground surface – Water levels – Levels of strata interfaces – Excavation levels – Structural dimensions	Nominal values Foundation tolerance Overdig allowance
Soil behaviour – Laboratory/field tests – Observation – Measurement	Loads and geotechnical parameters – Geotechnical investigations – Field testing – Laboratory testing – Derived parameters	Characteristic values Nominal values Correlation factors, ξ
Appropriate model – Idealisation followed by evaluation – Conceptual or physical modelling – Analytical modelling	Verification method – Consideration of all relevant limit states Verification by – Calculations – Prescriptive measures – Load or model tests – An observational method	Partial factors, γ Combination factors, ψ Design values Limiting movements
– Precedent – Empiricism – Well-winnowed experience	Design complexity – Soil conditions – Groundwater situation – Influence of the environment – Regional seismicity – Nature of structure – Nature of surroundings	Risk assessment Geotechnical Categories Comparable experience

Table 10.2 Relationship between geotechnical triangle aspects and geotechnical design triangle components and safety elements

With regard to complexity, Burland (2006) stated that he considered the main difficulty in soil mechanics was not so much in the complexity of the material, but in the fact that, all too often, the boundaries between the aspects become confused. He also noted the need for precedent, empiricism and well-winnowed experience to moderate and account for uncertainties in the aspects of soil mechanics. A similar situation exists in geotechnical design whereby it is not so much the complexity of the situation that affects the safety of a design, but the need to identify the uncertainties involved in each of the design components and adopt appropriate treatment procedures and safety elements for each component. This involves using comparable experience in geotechnical design. An important statement in Eurocode 7 related to ensuring safety in complex geotechnical design situations is Paragraph 2.4.1(2), which notes that 'knowledge of the ground conditions depend on the extent and quality of the geotechnical investigations. Such knowledge and the control of workmanship are usually more significant to fulfilling the fundamental requirements than is precision in the calculation models and partial factors'.

10.9. Codes and standards for geotechnical engineering

10.9.1 Groups of geotechnical codes and standards

The codes and standards in the UK for geotechnical engineering may be divided into five groups, presented in **Tables 10.3–10.7**, covering: geotechnical investigation and testing, geotechnical design, the execution of geotechnical works, earthworks and materials used in geotechnical engineering. These tables provide, for each standard, its prefix and reference number, title, scope/description and status in 2023. Standards that have been published by BSI are shown in the tables with their date of publication after the reference number. Standards that are being or have been prepared by CEN in collaboration with ISO, but which have not yet been approved by CEN or ISO, are shown with no date and the expected prefix when published by BSI is shown in brackets, for example BS EN ISO.

10.9.2 Codes and standards for geotechnical investigation and testing

The BSs for geotechnical investigation and testing are listed in **Table 10.3**. The first standards listed in **Table 10.3** are the current versions of the original BSs for soil tests and for ground investigations:

- 1377 Methods of test for soils for civil engineering purposes
  – Part 1: General requirements and sample preparation
  – Part 2: Classification tests and determination of soil properties
  – Part 3: Chemical and electro-chemical testing
  – Part 9: In situ tests
- 5930 Code of practice for ground investigations.

These standards have been revised so as not to conflict with or duplicate the large number of European Standards for geotechnical investigation and testing that have been prepared by CEN and ISO. In addition, both parts of Eurocode 7 have sections that cover the design aspects of geotechnical investigations – for example, the planning and location of investigations. As a result, BS 1377 now only has four parts (1, 2, 3 and 9) that are current BSs. The other parts of BS 1377 have all been withdrawn as they have been superseded by CEN/ISO standards; much of 'Part 9 In situ tests' has also been superseded.

In recent years CEN has published, and in a few cases is still preparing or planning, the very comprehensive range of standards for geotechnical investigation and testing listed in **Table 10.3** (60 standards or standard parts are listed on the CEN/TC 341 website). The following standards and their parts, all under the umbrella title 'Geotechnical investigation and testing', have been prepared collaboratively with ISO, under the Vienna Agreement, with sometimes CEN/TC 341 *Geotechnical investigation and testing* taking the lead and sometimes ISO TC 182 *Geotechnics* taking the lead:

- 14688 Identification and classification of soil
  – Part 1: Identification and description
  – Part 2: Principles for a classification
- 14689 Identification, description and classification of rock
  – Part 1: Identification and description
- 16383 Laboratory testing of rock
  – Part 1: Determination of water content
- 17628 Geothermal testing
- 17892 Laboratory testing of soil
  – Part 1: Determination of water content
  – Part 2: Determination of density of fine-grained soil
  – Part 3: Determination of particle density
  – Part 4: Determination of particle size distribution
  – Part 5: Incremental loading oedometer test
  – Part 6: Fall cone test
  – Part 7: Unconfined compression test
  – Part 8: Unconsolidated undrained triaxial test
  – Part 9: Consolidated triaxial compression tests on water-saturated soils
  – Part 10: Direct shear tests
  – Part 11: Permeability tests
  – Part 12: Determination of liquid and plastic limits
- 18674 Geotechnical monitoring by field instrumentation
  – Part 1: General rules
  – Part 2: Measurement of displacements along a line: extensometers
  – Part 3: Measurement of displacements across a line: inclinometers
  – Part 4: Measurement of pore water pressure: piezometers
  – Part 5: Stress change measurements by total pressure cells (TPC)
  – Part 7: Measurement of strains: strain gauges
  – Part 8: Measurement of forces: load cells
- 22282 Geohydraulic testing
  – Part 1: General rules

- Part 2: Water permeability test in a borehole using open systems
- Part 3: Water pressure tests in rock
- Part 4: Pumping tests
- Part 5: Infiltrometer tests
- Part 6: Water permeability test in a borehole using closed systems

- 22475 Sampling methods and groundwater measurements
  - Part 1: Technical principles for execution
  - Part 2: Qualification criteria for enterprises and personnel
  - Part 3: Conformity assessment of enterprises and personnel by third party
- 22476 Field testing
  - Part 1: Electrical cone and piezocone penetration test
  - Part 2: Dynamic probing
  - Part 3: Standard penetration test
  - Part 4: Prebored pressuremeter test by Ménard procedure
  - Part 5: Prebored pressuremeter test
  - Part 6: Self-boring pressuremeter test
  - Part 7: Borehole jack test
  - Part 8: Full displacement pressuremeter test
  - Part 9: Field vane test (FVT and FVT-F)
  - Part 10: Weight sounding test
  - Part 11: Flat dilatometer test
  - Part 12: Mechanical cone penetration test
  - Part 13: Plate loading test
  - Part 14: Borehole dynamic probing
  - Part 15: Measuring while drilling
  - Part 16: Borehole shear test
- 22477 Testing of geotechnical structures
  - Part 1: Testing of piles: static compression load testing
  - Part 2: Testing of piles: static tension load testing
  - Part 3: Testing of piles: static transverse load testing
  - Part 4: Testing of piles: dynamic load testing
  - Part 5: Testing of grouted anchors
  - Part 6: Testing of nailing
  - Part 7: Testing of reinforced fill
  - Part 10: Testing of piles: rapid load testing
- 24283 Qualification criteria and assessment (3 parts).
  - Part 1: Qualified technician
  - Part 2: Responsible expert
  - Part 3: Qualified enterprise

The two parts of 14688 on the identification and classification of soil and the one part of 14689 on the identification, description and classification rock have been published as BSs.

Standard 16383 on the laboratory testing of rock, with just one part for determining the water content, is being drafted in 2023. It is anticipated that additional parts on other tests to determine rock properties will be prepared.

Standard 17628 on the geothermal testing of soil, with one part, has been published as a BS.

All 12 parts of 17892 on laboratory testing of soil to determine the most common soil properties have been published as BSs. All parts have a note in their scopes which states that the document fulfils the requirements of the determination of the properties of soils for geotechnical investigation and testing in accordance with EN 1997-1 and EN 1997-2.

Standard 18674 on geotechnical monitoring by field instrumentation has seven parts. In 2023 Parts 1–5 were published as BSs; Part 7, 'Measurement of strains: strain gauges', is being drafted; and Part 8, 'Measurement of forces: load cells', is under approval. No information is available on Part 6. Each part has a statement that it fulfils the requirements for the performance monitoring of the ground, of structures interacting with the ground and of geotechnical works by the means of inclinometers as part of the geotechnical investigation and testing in accordance with EN 1997-1 and EN 1997-2.

The six parts of 22282 on geohydraulic testing have been published as BSs. It is stated in the scope for each part that they specify the requirements for the determination of particular geohydraulic properties of soils and rock as part of the geotechnical investigation services according to EN 1997-1 and EN 1997-2. As of 2023 it is 11 years since Parts 1, 2, 3, 5 and 6 were published; they are under review in 2023.

The three parts of 22475 on sampling methods and groundwater measurements have been published as BSs.

Standard 22476 on field testing has Parts 1–16. Parts 1–12, 14 and 15 have been published as BSs with a broader Part 5 published in 2023 titled 'Prebored pressuremeter test' that covers the flexible dilatometer test. As of 2023, Part 13, 'Plate loading test', is planned and Part 16, 'Borehole shear test', is under development. The parts state that they fulfil the requirements for field tests as part of geotechnical investigation and testing according to EN 1997-1 and EN 1997-2.

Standard 22477 on testing of geotechnical structures has Parts 1–10. Parts 1, 4, 5 and 10 have been published as BSs and all have the note that they are intended to be used in conjunction with EN 1997-1. In 2023, Part 2, 'Testing of piles: static tension load testing', has been issued for public comment. Parts 3, 6 and 7, 'Testing of piles: static transverse load testing', 'Testing of nailing' and 'Testing of reinforced fill', are planned. No information is available on Parts 8 and 9.

The three parts of 24283 on qualification criteria and assessment are concerned with the qualification criteria for technicians and operators, responsible experts and enterprises performing sampling, testing, measuring, monitoring and installation of equipment in the framework of a geotechnical investigation. They were approved by CEN on 24 December 2021 for provisional application and published by CEN as Technical Specifications (TSs) in 2022, rather than as ENs. Hence they have been published by BSI as Published Documents (PDs) rather than as BSs. Their period of validity is limited initially to 3 years. After 2 years the CEN members will be requested to submit their comments, particularly on the question of whether the TSs can be converted into ENs.

One standard on soil testing, the standard for measuring the shear wave velocity profile in the ground, that has been published by ISO TC 182 without any involvement of or cooperation with CEN, is

- 24057 Array measurement of microtremors to estimate shear wave velocity profile.

Reference	Title	Scope/description	Status
BS 1377-1:2016	Methods of test for soils for civil engineering purposes – Part 1: General requirements and sample preparation	This part of BS 1377 specifies general requirements for the testing of soils for civil engineering purposes and describes methods for the calibration of equipment and the preparation of test samples.	Current BS
BS 1377-2:2022	Methods of test for soils for civil engineering purposes – Part 2: Classification tests and determination of soil properties	This part of BS 1377 specifies methods of test for the classification of soil and for the determination of geotechnical properties of soils in the laboratory. It includes – common laboratory tests required for the classification of soils – determination of compaction characteristics of soils for earthworks, permeability, compressibility and erodibility – determination of shear strength of soils in terms of both total and effective stresses.	Current BS
BS 1377-3:2018+A1:2021	Methods of test for soils for civil engineering purposes – Part 3: Chemical and electro-chemical testing	This part of BS 1377 describes test methods for determining the amount of chemical substances in samples of soil and extremely weak and very weak rocks, as defined by BS 5930, and groundwater including organic matter content, total carbon content, loss on ignition, sulfur content and seven other substances. The chemical tests in this part may be used on other rocks if required. It also describes test methods for the determination of some electrochemical and resistivity properties of solid samples. These tests provide data to assess the potential of the ground and solutes to damage construction materials, including cementitious materials and metals in the ground. They can also be used in assessment of the potential for volume change of the ground due to chemical reaction. Resistivity test results can also be used to assess in situ resistivity results.	Current BS
BS 1377-9:1990	Methods of test for soils for civil engineering purposes – Part 9: In situ tests	This part of BS 1377 describes in situ methods of test on soils for civil engineering purposes – that is, tests made directly on the soil in place.	Current, but largely superseded by parts of BS EN ISO 22476
BS 5930:2015 +A1:2020	Code of practice for ground investigations	This document gives recommendations for the investigation of sites for the purposes of assessing their suitability for the construction of civil engineering and building works and of acquiring knowledge of a site's characteristics that could affect the design and construction of such work and the security of neighbouring land and property. It provides guidance on the application of BS EN 1997-1 and BS EN 1997-2 and related test standards cited therein.	Current BS
BS EN ISO 14688-1:2018	Geotechnical investigation and testing – Identification and classification of soil – Part 1: Identification and description	This document specifies the rules for the identification and description of soils and is intended to be read in conjunction with ISO 14688-2, which outlines the basis of classification of those material characteristics most commonly used for soils for engineering purposes. This document specifies procedures for the identification and description of soils based on a flexible system for use by experienced persons, covering both material and mass characteristics by visual and manual techniques. Details are given of the individual characteristics for identifying soils and the descriptive terms in regular use, including those related to the results of hand tests carried out in the field as part of the descriptive process.	Current BS
BS EN ISO 14688-2:2018	Geotechnical investigation and testing – Identification and classification of soil – Part 2: Principles for a classification	This document specifies the basic principles for classification of those material characteristics most commonly used for soils for engineering purposes. It is intended to be read in conjunction with ISO 14688-1, which gives rules for the identification and description of soils. The classification principles established in this document allow soils to be classified into groups of similar composition and geotechnical properties, based on the results of field and laboratory tests with respect to their suitability for geotechnical engineering purposes. This document is applicable to natural soil in situ, natural soil reworked artificially and synthetic materials. A more detailed classification specific to use in earthworks is given in EN 16907-2 .	Current BS
BS EN ISO 14689:2018	Geotechnical investigation and testing – Identification, description and classification of rock – Part 1: Identification and description	This document specifies the rules for the identification and description of rock material and mass on the basis of mineralogical composition, genetic aspects, structure, grain size, discontinuities and other parameters. It also provides rules for the description of other characteristics as well as for their designation. This document applies to the description of rock for geotechnics and engineering geology in civil engineering. The description is carried out on cores and other samples of rock and on exposures of rock masses.	Current BS

Table 10.3 Codes and standards for geotechnical investigation and testing (continued on next page)

Reference	Title	Scope/description	Status
(BS EN ISO) 16383-1:	Geotechnical investigation and testing – Laboratory testing of rock – Part 1: Determination of water content	No information available.	Under development in 2023
BS EN ISO 17628:2015	Geotechnical investigation and testing – Geothermal testing – Determination of thermal conductivity of soil and rock using a borehole heat exchanger	This document specifies requirements for the geothermal response test (GRT). This test comprises the in situ determination of the thermal conductivity in saturated and unsaturated soil and rock in a heat exchanger installed in a borehole. For this test, liquid heat transfer media not subjected to phase changes are used. The thermal conductivity is an important parameter used in the design of thermal storage and thermal exchange systems. A GRT measures the temperature response to a thermal energy forcing of a borehole heat exchanger (BHE) or the extraction of thermal energy from a borehole. The temperature response is related to the thermal parameters of the ground and borehole filling material, such as thermal conductivity and borehole resistivity, and is therefore used to obtain estimated or derived values of these parameters. This document applies to heat exchangers installed in vertical or inclined boreholes with length up to 400 m and with a diameter of up to 200 mm.	Current BS
BS EN ISO: 17892-1:2014 +A1:2022	Geotechnical investigation and testing – Laboratory testing of soil – Part 1: Determination of water content	This standard specifies a method of determining the water content of soils. It is applicable to the laboratory determination of the water (also known as moisture) content of a soil test specimen by oven-drying. The practical procedure for determining the water content is to determine the mass loss on drying the specimen to a constant mass in a drying oven controlled at a given temperature. The mass loss is assumed to be due to free water and is referenced to the remaining dry mass of solid particles.	Current BS
BS EN ISO: 17892-2:2014	Geotechnical investigation and testing – Laboratory testing of soil – Part 2: Determination of bulk density	This standard specifies three methods for determining the bulk density of soils: (*a*) linear measurement method; (*b*) immersion in fluid method; and (*c*) fluid displacement method. The linear measurement method is suitable for determining the bulk density of a soil specimen of regular shape. The specimens used are either rectangular prisms or cylinders with circular cross-sections. The immersion in fluid method covers determining the bulk density of a specimen of natural or compacted soil by measuring its mass in air and apparent mass when suspended in fluid. The fluid displacement method covers determining the bulk density of a soil specimen by measuring its mass in air and the mass of fluid displaced by immersion.	Current BS
BS EN ISO: 17892-3:2015	Geotechnical investigation and testing – Laboratory testing of soil – Part 3: Determination of particle density	This standard specifies methods for the laboratory determination of the particle density of soil within the scope of geotechnical investigations, and describes two methods, a pycnometer method by fluid displacement and a pycnometer method by gas displacement. The fluid pycnometer method described in this part of ISO 17892 applies to soil types with particle sizes under about 4 mm, or soils crushed to meet this requirement. Larger pycnometers are used for coarser materials. The particle size of soils suitable for testing in the gas pycnometer is limited by the dimensions of the specimen container of the particular gas pycnometer being used.	Current BS
BS EN ISO: 17892-4:2016	Geotechnical investigation and testing – Laboratory testing of soil – Part 4: Determination of particle size distribution	This document specifies a method of determining the particle size distribution of soils by sieving, or sedimentation, or a combination of both. Classification of soils is mainly based on the particle size distribution. The particle size distribution provides a description of soil based on a subdivision in discrete classes of particle sizes. The size of each class can be determined by sieving and/or sedimentation. Coarse soils are usually tested by sieving, but fine and mixed soils are usually tested by a combination of sieving and sedimentation, depending on the composition of the soil. The sieving method described is applicable to all non-cemented soils with particle sizes less than 125 mm. Two sedimentation methods are described: the hydrometer method and the pipette method.	Current BS
BS EN ISO: 17892-5:2017	Geotechnical investigation and testing – Laboratory testing of soil – Part 5: Incremental loading oedometer test	This document specifies methods for determining the compression and deformation characteristics of soils by incremental loading in an oedometer. The oedometer test is carried out on a cylindrical test specimen confined laterally by a rigid ring. The specimen is subjected to discrete increments of vertical axial loading or unloading and is allowed to drain axially from the top and bottom surfaces. Tests may be carried out on undisturbed, remoulded, recompacted or reconstituted specimens. The small specimen size generally does not adequately represent the fabric features in natural soils. Analysis of consolidation tests is generally based on the assumption that the soil is saturated. In case of unsaturated soils, some derived parameters may not be appropriate.	Current BS

Table 10.3 Continued

Reference	Title	Scope/description	Status
BS EN ISO 17892-6:2017	Geotechnical investigation and testing – Laboratory testing of soil – Part 6: Fall cone test	This document specifies a method of undrained strength index testing of both undisturbed and remoulded specimens of fine-grained soils by the fall cone method to estimate the undrained shear strength. A cone is allowed to fall with its tip towards a soil specimen, and the resulting penetration of the cone into the soil is measured. The penetration values are used to estimate the undrained shear strength by correlation to some estimate of undrained shear strength determined in the laboratory by other methods. As the test is performed on a small specimen, the result may not agree with tests on larger specimens and can be regarded as an estimation of undrained shear strength rather than a true measurement of it. The ratio of the remoulded shear strength to the undisturbed shear strength may be used to estimate the sensitivity of a soil specimen.	Current BS
BS EN ISO: 17892-7:2018	Geotechnical investigation and testing – Laboratory testing of soil – Part 7: Unconfined compression test	This document specifies a method for the unconfined compression test to determine the unconfined compressive strength for a homogeneous specimen of undisturbed, recompacted, remoulded or reconstituted soil under compression loading. This test method is useful to estimate the undrained shear strength of soil. It is noted that drainage is not prevented during this test. The estimated value for undrained shear strength is, therefore, only valid for soils of low permeability, which behave sufficiently undrained during the test.	Current BS
BS EN ISO 17892-8:2018	Geotechnical investigation and testing – Laboratory testing of soil – Part 8: Unconsolidated undrained triaxial test	This document specifies a method for unconsolidated undrained triaxial compression tests to determine the undrained triaxial shear strength of soil under compression loading. The cylindrical specimen, which can comprise undisturbed, recompacted, remoulded or reconstituted soil, is subjected to an isotropic stress under undrained conditions and thereafter is sheared under undrained conditions. The test allows the determination of shear strength and stress–strain relationships in terms of total stresses.	Current BS
BS EN ISO 17892-9:2018	Geotechnical investigation and testing – Laboratory testing of soil – Part 9: Consolidated triaxial compression tests on water-saturated soils	This document specifies a method for consolidated triaxial compression tests on water-saturated soils to determine the triaxial shear strength under compression loading. The cylindrical specimen, which can comprise undisturbed, recompacted, remoulded or reconstituted soil, is subjected to an isotropic or an anisotropic stress under drained conditions and thereafter is sheared under undrained or drained conditions. The test allows the determination of shear strength, stress–strain relationships and effective stress paths. This document provides a test for a single specimen. A set of at least three relatable tests are required to determine the shear strength parameters from these tests.	Current BS
BS EN ISO 17892-10:2018	Geotechnical investigation and testing – Laboratory testing of soil – Part 10: Direct shear tests	This document specifies two laboratory test methods to determine the effective shear strength parameters of soils in direct shear under consolidated drained conditions using either a shearbox or a ring shear device. The tests included in this document are for undisturbed, remoulded, recompacted or reconstituted soils. The procedure describes the requirements of a determination of the shear resistance of a specimen under a single vertical (normal) stress. Generally three or more similar specimens from one soil are prepared for shearing under three or more different vertical pressures to allow the shear strength parameters to be determined.	Current BS
(BS EN ISO) 17892-11:2019	Geotechnical investigation and testing – Laboratory testing of soil – Part 11: Permeability tests	This document specifies methods for the laboratory determination of the water flow characteristics in soil. The permeability test is carried out on a cylindrical test specimen that is either confined laterally by a rigid container or by a flexible membrane. The specimen is subjected to differential hydraulic head and the water flow is measured under either a constant or falling head. The results are used to determine the coefficient of permeability of the soil specimen. Tests can be carried out on undisturbed, remoulded, compacted or reconstituted specimens.	Current BS
BS EN ISO 17892-12:2018 +A1:2022	Geotechnical investigation and testing – Laboratory testing of soil – Part 12: Determination of liquid and plastic limits	This document specifies methods for determining the liquid and plastic limits of a soil. These comprise two of the Atterberg limits for soils. The liquid limit is the water content at which a soil changes from the liquid to plastic state. This document describes the determination of the liquid limit of a specimen of natural soil, or of a specimen of soil from which material larger than about 0.4 mm has been removed. This document describes two methods: the fall cone method and the Casagrande method. The fall cone method in this document should not be confused with that of BS EN ISO 17892-6. The plastic limit of a soil is the water content at which a soil ceases to be plastic when dried further.	Current BS

Table 10.3 Continued

Reference	Title	Scope/description	Status
BS EN ISO 18674-1:2015	Geotechnical investigation and testing – Geotechnical monitoring by field instrumentation – Part 1: General rules	This document lays out the general rules and fulfils the requirements for the performance monitoring of the ground, of structures interacting with the ground, of geotechnical fills and of geotechnical works as part of the geotechnical investigation and testing according to EN 1997-1 and EN 1997-2. Specifically, this document applies to field instrumentation and measurements carried out – in connection with site investigations of soils and rocks – in connection with observational design procedures – in connection with the performance of geotechnical structures before, during and after construction – for ground behaviour evaluation (e.g. unstable slopes, consolidation) – for the proof or follow-up of a new equilibrium within the ground after disturbance of its natural state by construction measures (e.g. foundation loads, excavation of soil, tunnelling) – for the proof or follow-up of the stability, serviceability and safety of structures and operations which might be influenced by geotechnical construction – for perpetuation of evidence – for the evaluation and control of geotechnical works.	Current BS
BS EN ISO 18674-2:2016	Geotechnical investigation and testing – Geotechnical monitoring by field instrumentation – Part 2: Measurement of displacements along a line: extensometers	This document specifies the measurement of displacements along a line by means of extensometers carried out for geotechnical monitoring. If applied in conjunction with ISO 18674-3, this document allows the determination of displacements acting in any direction. It is applicable to – monitoring the behaviour of soils, fills and rocks – checking geotechnical designs in connection with the observational design procedure – deriving geotechnical key parameters (e.g. from results of pile load tests or trial tunnelling) – evaluating stability ahead of, during or after construction (e.g. stability of natural slopes, slope cuts, embankments, excavation walls, foundations, dams, refuse dumps, tunnels).	Current BS
BS EN ISO 18674-3:2017+ A1:2020	Geotechnical investigation and testing – Geotechnical monitoring by field instrumentation – Part 3: Measurement of displacements across a line: inclinometers	This document specifies the measurement of displacements across a line by means of inclinometers carried out for geotechnical monitoring. This document also refers to deflectometers to supplement inclinometers for the determination of horizontal displacements across horizontal measuring lines. In general, there are two independent displacement components acting across measuring lines. Inclinometers allow the determination of the two components for vertical measuring lines. For horizontal lines, inclinometers are limited to the determination of the vertical component only. If applied in conjunction with ISO 18674-2, this document allows the determination of displacements acting in any direction. This document is applicable to – checking geotechnical designs in connection with the observational design procedure – monitoring of geotechnical structures prior to, during and after construction (e.g. natural slopes, slope cuts, embankments, excavation walls, foundations, dams, refuse dumps, tunnels) – deriving geotechnical key parameters (e.g. from results of pile load tests or trial tunnelling) – identification and monitoring of active shear planes in the ground.	Current BS
BS EN ISO 18674-4:2020	Geotechnical investigation and testing – Geotechnical monitoring by field instrumentation – Part 4: Measurement of pore water pressure: piezometers	This document specifies the measurement of pore water pressures and piezometric levels in saturated ground by means of piezometers, which are installed in the ground for geotechnical monitoring. If applied in conjunction with ISO 18674-5, the procedures described in this document allow the determination of effective stresses acting in the ground. This document is applicable to – monitoring of water pressures acting on and in geotechnical structures (e.g. quay walls, dykes, excavation walls, foundations, dams, tunnels, slopes, embankments) – monitoring of consolidation processes of soil and fill (e.g. beneath foundations and in embankments) – evaluating stability and serviceability of geotechnical structures – checking geotechnical designs in connection with the observational design procedure. For pore water pressure measurements carried out in connection with cone penetration tests, see ISO 22476-1.	Current BS

Table 10.3 Continued

Reference	Title	Scope/description	Status
BS EN ISO 18674-5:2019	Geotechnical investigation and testing – Geotechnical monitoring by field instrumentation – Part 5: Stress change measurements by total pressure cells (TPC)	This document specifies the measurement of stress changes by means of total pressure cells. If applied in conjunction with ISO 18674-4 it allows the determination of effective stress acting in the ground. It is applicable to – monitoring changes of the state of stress in the ground and in geoengineered structures (e.g. in earthfill dams or tunnel lining) – monitoring contact pressures at the interface between two media (e.g. earth pressure on retaining wall; contact pressure at the base of a foundation) – checking geotechnical designs and adjustment of construction in connection with the observational design procedure – evaluating stability during or after construction. Guidelines for the application of TPC in geotechnical engineering are presented.	Current BS
(BS EN ISO) 18674-7	Geotechnical investigation and testing – Geotechnical monitoring by field instrumentation – Part 7: Measurement of strains: strain gauges	The methods of measuring strain are described and rules for measurement of strain in structural members are given, from which stresses can be computed in geotechnical engineering or more generally in foundation engineering. Strains and stresses in geotechnical structural members are needed to judge the loading of engineered construction in the ground.	Under development in 2023
(BS EN ISO) 18674-8	Geotechnical investigation and testing – Geotechnical monitoring by field instrumentation – Part 8: Measurement of forces: load cells	This document specifies the measurement of forces by means of load cells carried out for geotechnical monitoring. This document is applicable to – performance monitoring of geotechnical structures such as anchors, end-anchored rock bolts, tiebacks, piles, struts, props and steel linings – checking geotechnical designs and adjustment of construction in connection with the observational design procedure – evaluating stability during or after construction.	Draft under approval in 2023
BS EN ISO 22282-1:2012	Geotechnical investigation and testing – Geohydraulic testing – Part 1: General rules	This document establishes the general rules and principles for geohydraulic testing in soil and rock as part of the geotechnical investigation services in accordance with EN 1997-1 and EN 1997-2. It defines concepts and specifies requirements relating to permeability measurement in soil and rock. The different purposes of geohydraulic testing are to obtain information on the permeability of soil or rock in natural or treated states, transmissivity and storage coefficient, and hydrodynamic parameters of aquifers.	Current BS, under review in 2023
BS EN ISO 22282-2:2012	Geotechnical investigation and testing – Geohydraulic testing – Part 2: Water permeability tests in a borehole using open systems	This document specifies requirements for the determination of the local permeability in soils and rocks below and above groundwater level in an open hole by water permeability tests as part of the geotechnical investigation services according to EN 1997-1 and EN 1997-2.	Current BS, under review in 2023
BS EN ISO 22282-3:2012	Geotechnical investigation and testing – Geohydraulic testing – Part 3: Water pressure tests in rock	This document specifies the requirements for water pressure tests (WPTs) carried out in boreholes drilled into rock as part of geotechnical investigation and testing according to EN 1997-1 and EN 1997-2. The tests are used to investigate – hydraulic properties of the rock mass, which are mainly governed by discontinuities – absorption capacity of the rock mass – tightness of the rock mass – effectiveness of grouting – geomechanical behaviour (e.g. hydrofracturing, hydrojacking). Many effects of geohydraulic tests are not only influenced by the ground itself, but stem from the testing procedure. Historically, the WPT was evaluated based on the assumption that stationary behaviour was achieved. Recent advances in geohydraulics have shown that transient phenomena are often present.	Current BS, under review in 2023
BS EN ISO 22282-4:2021	Geotechnical investigation and testing – Geohydraulic testing – Part 4: Pumping tests	This document establishes requirements for pumping tests as part of the geotechnical investigation service in accordance with EN 1997-1 and EN 1997-2. This document applies to pumping tests performed on aquifers whose permeability is such that pumping from a well can create a lowering of the piezometric head within hours or days, depending on the ground conditions and the purpose. It covers pumping tests carried out in soils and rock.	Current BS

Table 10.3 Continued

Reference	Title	Scope/description	Status
BS EN ISO 22282-5:2012	Geotechnical investigation and testing – Geohydraulic testing – Part 5: Infiltrometer tests	This document establishes requirements for ground investigations by means of infiltrometer tests. It applies to the in situ determination of the water permeability of an existing geological formation or of treated or compacted materials. The infiltrometer test is used to determine the infiltration capacity of the ground at the surface or shallow depth. It is a simple test for determining the permeability coefficient. The method can be applied using either steady-state or transient conditions in saturated or unsaturated soils. The principle of the test is based on the measurement of a surface vertical flow rate of water which infiltrates the soil under the influence of a positive hydraulic head. Surface infiltration devices include single- and double-ring infiltrometer designs of the open or closed type.	Current BS; under review in 2023
BS EN ISO6 22282-6:2012	Geotechnical investigation and testing – Geohydraulic testing – Part 6: Water permeability tests in a borehole using closed systems	This document specifies requirements for the determination of the local permeability in soils and rocks below or above the groundwater table in a closed system by the water permeability tests. The tests are used to determine the permeability coefficient *k* in low-permeability soil and rock lower than 10^{-8} m/s. They can also be used to determine the transmissivity *T* and the storage coefficient *S*.	Current BS under review in 2023
BS EN ISO 22475-1:2021	Geotechnical investigation and testing – Sampling methods and groundwater measurements – Part 1: Technical principles for the sampling of soil, rock and groundwater	This document deals with principles of sampling of soil, rock and groundwater as part of the programme of geotechnical investigation and testing. The aims of such ground investigations are (*a*) to recover soil, rock and water samples of a quality appropriate to assess the general suitability of a site for geotechnical engineering purposes and to determine the required ground characteristics in the laboratory; (*b*) to obtain information on the sequence, thickness and orientation of strata and discontinuities; (*c*) to establish the type, composition and condition of strata; (*d*) to obtain information on groundwater conditions and recover water samples for assessment of the interaction of groundwater, soil, rock and construction material. Water sampling for the purposes of quality control, quality characterisation and identification of sources of pollution of water, including bottom deposits and sludges, is not covered.	Current BS
BS 22475-2:2011	Geotechnical investigation and testing – Sampling methods and groundwater measurements – Part 2: Qualification criteria for enterprises and personnel	This standard specifies the qualification criteria for an enterprise and for personnel performing sampling and groundwater measurement services in accordance with BS EN ISO 22475-1. Enterprises shall be able to demonstrate adequate competence and have suitably qualified and experienced personnel, conforming equipment, a health and safety system and a quality assurance system.	Current BS
BS 22475-3:2011	Geotechnical investigation and testing – Sampling methods and groundwater measurements – Part 3: Conformity assessment of enterprises and personnel by third party	This document specifies requirements for the conformity assessment by third-party control of enterprises, personnel and qualified operators performing specified parts of sampling and groundwater measurements in accordance with BS EN ISO 22475-1 and complying with the technical qualification criteria specified in BS 22475-2.	Current BS
BS EN ISO 22476-1:2023	Geotechnical investigation and testing – Field testing – Part 1: Electrical cone and piezocone penetration test	This document deals with equipment requirements, the execution of and reporting on electrical cone and piezocone penetration tests. Within the electrical cone and piezocone penetration test, two subcategories of the cone penetration test are considered: the electrical cone penetration test (CPT), which includes measurement of cone resistance and sleeve friction; and the piezocone test (CPTU), which is a cone penetration test with the additional measurement of pore pressure. The CPTU is performed like a CPT with measurement of the pore pressure at one or several locations on the penetrometer surface. This document specifies the following features: type of CPT, application class, penetration length or penetration depth, elevation of the ground surface or the underwater ground surface at the test location with reference to a datum, location of the test relative to a reproducible fixed reference point, pore pressure dissipation test.	Current BS

Table 10.3 Continued

Reference	Title	Scope/description	Status
BS EN ISO 22476-2:2005+A1:2011	Geotechnical investigation and testing – Field testing – Part 2: Dynamic probing	This document deals with the equipment requirements for, execution of and reporting on dynamic probing. It covers the determination of the resistance of soils and soft rocks in situ to the dynamic penetration of a cone. A hammer of a given mass and given height of fall is used to drive the cone. The penetration resistance is defined as the number of blows required to drive the cone over a defined distance. A continuous record is provided with respect to depth, but no samples are recovered. Four procedures are included, covering a wide range of specific work per blow: dynamic probing light, dynamic probing medium, dynamic probing heavy and dynamic probing super heavy. The test results are specially suited for the qualitative determination of a soil profile together with direct investigations (e.g. sampling according to ISO 22475-1) or as a relative comparison of other in situ tests. They may also be used for determining the strength and deformation properties of soils, generally cohesionless but also possibly in fine-grained soils, through appropriate correlations. The results can also be used to determine the depth to very dense ground layers (e.g. to determine the length of end-bearing piles) and to detect very loose, voided, backfilled or infilled ground.	Current BS
BS EN ISO 22476-3:2005+A1:2011	Geotechnical investigation and testing – Field testing – Part 3: Standard penetration test	This document deals with the equipment requirements for, execution of and reporting on the penetration test to complement direct investigations (e.g. ISO 22475-1). The standard penetration test aims to determine the resistance of soils at the base of a borehole to the dynamic penetration of a split barrel sampler and the recovering of disturbed samples for identification purposes (SPT). In gravelly soils and in soft rocks a solid cone is also used (SPT(C)). The standard penetration test is used mainly to assess the strength and deformation parameters of cohesionless soils, but some valuable data may also be obtained in other soil types. The basis of the test consists in driving a sampler by dropping a hammer of 63.5 kg mass onto an anvil or drive head from a height of 760 mm. The number of blows (***N***) necessary to achieve a penetration of the sampler of 300 mm (after its penetration under gravity and below a seating drive) is the penetration resistance.	Current BS
BS EN ISO 22476-4:2021	Geotechnical investigation and testing – Field testing – Part 4: Prebored pressuremeter test by Ménard procedure	This document specifies equipment requirements, the execution of and reporting on the Ménard pressuremeter test. It describes the procedure for conducting a Ménard pressuremeter test in natural grounds, treated or untreated fills, either on land or offshore. The test is performed by the radial expansion of a cylindrical probe placed in the ground. Pressure applied to and the associated radial expansion of the probe are measured and recorded to obtain the stress–strain relationship of ground. The pressuremeter test results are suited to a quantitative determination of ground strength and deformation parameters. They can yield lithological information in conjunction with measuring while drilling performed when creating the borehole (according to ISO 22476-15). They can also be combined with direct investigation (e.g. sampling according to ISO 22475-1) or compared with other in situ tests (see EN 1997-2).	Current BS
BS EN ISO 22476-5:2023	Geotechnical investigation and testing – Field testing – Part 5: Prebored pressuremeter test	This document is applicable to pressuremeter tests using cylindrical flexible probes placed in pre-existing boreholes using testing procedures other than the Ménard procedure. Pressuremeter tests following the Ménard procedure are provided in ISO 22476-4. A high-pressure flexible pressuremeter probe which contains transducers for the measurement of radial displacements is also known as a flexible dilatometer probe or a high-pressure dilatometer probe. This document applies to tests performed in any kind of grounds, starting from soils, treated or untreated fills, hard soils and soft rocks, up to hard and very hard rocks. The parameters derived from this test can include stiffness, strength, initial in situ stress state and consolidation properties.	Current BS
BS EN ISO 22476-6:2018	Geotechnical investigation and testing – Field testing – Part 6: Self-boring pressuremeter test	This document specifies the equipment requirements, execution of and reporting on self-boring pressuremeter (SBP) tests. Tests with the self-boring pressuremeter cover the measurement in situ of the deformation of soils and weak rocks by the expansion and contraction of a cylindrical flexible membrane under pressure.	Current BS

Table 10.3 Continued

Reference	Title	Scope/description	Status
BS EN ISO 22476-7:2012	Geotechnical investigation and testing – Field testing – Part 7: Borehole jack test	This document specifies the equipment requirements, execution of and reporting on borehole jack tests. It specifies the procedure for conducting a borehole jack test in ground stiff enough not to be adversely affected by the drilling operation. Two diametral cylindrical steel loading plates are placed in the ground and opened by pressure. Pressure applied to, and associated opening of the probe are measured and recorded so as to obtain a stress–displacement relationship of the ground for the range of the expected design stress. This document applies to test depths ≤100 m and to testing either on land or offshore.	Current BS
BS EN ISO 22476-8:2018	Geotechnical investigation and testing – Field testing – Part 8: Full displacement pressuremeter test	This document specifies the equipment requirements, execution of and reporting on full displacement pressuremeter (FDP) tests. Tests with the FDP cover the measurement in situ of the deformation of soils and weak rocks by the expansion/contraction of a cylindrical flexible membrane under pressure.	Current BS
BS EN ISO 22476-9:2020	Geotechnical investigation and testing – Field testing – Part 9: Field vane test (FVT and FVT-F)	This document deals with the equipment requirements, execution and reporting of field vane tests for the measurement of peak and remoulded vane shear strength together with the sensitivity of fine-grained soils. In addition, post-peak shear strength behaviour can be evaluated. It covers onshore and nearshore field vane testing. Two types of field vane test are described: the ordinary field vane test (FVT) and the fast field vane test (FVT-F). The uncertainties of the vane test result are described in Annex D.	Current BS
BS EN ISO 22476-10:2017	Geotechnical investigation and testing – Field testing – Part 10: Weight sounding test	This document specifies the equipment, execution and reporting requirements of the weight sounding test. It specifies the procedure for conducting a test with the weight sounding device in natural soils, made ground, and fill either on land or on water. It is applicable to the determination of the resistance of soil to the static load or the static load and the specified turning of the sounding point. It gives guidelines for the use of the weight sounding test to give a continuous soil profile and an indication of the layer sequence. The use includes the estimation of the density of cohesionless soils and the depth to very dense ground layers indicating the length of end-bearing piles.	Current BS
BS EN ISO 22476-11:2017	Geotechnical investigation and testing – Field testing – Part 11: Flat dilatometer test	This document establishes guidelines for the equipment requirements, execution of and reporting on flat dilatometer tests. The basic flat dilatometer test consists of inserting vertically into the soil a blade-shaped steel probe with a thin, expandable, circular steel membrane mounted flush on one face and determining two pressures at selected depth intervals: the contact pressure exerted by the soil against the membrane when the membrane is flush with the blade and, subsequently, the pressure exerted when the central displacement of the membrane reaches 1.10 mm. Results of flat dilatometer tests are used mostly to obtain information on soil stratigraphy, in situ state of stress, deformation properties and shear strength. It is also used to detect slip surfaces in clays. The flat dilatometer test is most applicable to clays, silts and sands, where particles are small compared to the size of the membrane.	Current BS
BS EN ISO 22476-12:2009	Geotechnical investigation and testing – Field testing – Part 12: Mechanical cone penetration test (CPTM)	This document specifies a mechanical cone penetration test (CPTM), including equipment requirements, execution and reporting. The results from such geotechnical testing are especially suited to the qualitative and/or quantitative determination of a soil profile – together with direct investigations – or as a relative comparison with other in situ tests. The results from a CPT can in principle be used to evaluate stratification, soil type, and geotechnical parameters such as soil density, shear strength parameters and deformation and consolidation characteristics. This document specifies the following features: type of CPT, application class, penetration length or penetration depth, elevation of ground surface or underwater ground surface at the location of the test with reference to a datum, location of test relative to a reproducible fixed reference point.	Current BS
(BS EN ISO) 22476-13	Geotechnical investigation and testing – Field testing – Part 13: Plate loading test	No details available.	Planned

Table 10.3 Continued

Reference	Title	Scope/description	Status
BS EN ISO 22476-14:2020	Geotechnical investigation and testing – Field testing – Part 14: Borehole dynamic probing	This document specifies the equipment requirements, execution of and reporting on borehole dynamic probing. It specifies technical requirements in respect of equipment and implementation in order to extensively prevent incorrect appraisals of the subsoil conditions and to limit scatter in the probing results due to equipment and implementation.	Current BS
BS EN ISO 22476-15:2016	Geotechnical investigation and testing – Field testing – Part 15: Measuring while drilling	This document specifies the technical principles for measuring equipment requirements, the execution and reporting on the parameters of an investigation drilling process for geotechnical purposes. It is applicable to top-driven, destructive drilling methods performed by a fully hydraulically powered drill rig and driving device. It is commonly used with destructive drilling techniques but can also be used with core drilling. The recording of the drilling parameters during soil grouting, drilling of nails, anchors or piles are beyond the scope of this document.	Current BS
(BS EN ISO) 22476-16	Geotechnical investigation and testing – Field testing – Part 16: Borehole shear test	No information available.	Under development in 2023
BS EN ISO 22477-1:2018	Geotechnical investigation and testing – Testing of geotechnical structures – Part 1: Testing of piles: static compression load testing	This document establishes the specifications for the execution of static pile load tests in which a single pile is subjected to an axial static load in compression in order to define its load–displacement behaviour. This document is applicable to vertical piles as well as raking piles. All types of piles are covered by this document. The tests considered are limited to maintained load tests. Pile load tests with constant penetration rate and cyclic load tests are not covered by this document. EN 1997-1 provides numerical values of partial factors for limit states and of correlation factors to derive characteristic values from static pile load tests to be taken into account in design. This document provides specifications for the execution of static axial pile load tests: (*a*) checking that a pile will behave as designed, and (*b*) measuring the resistance of a pile.	Current BS
(BS EN ISO) 22477-2	Geotechnical investigation and testing – Testing of geotechnical structures – Part 2: Testing of piles: static tension load testing	This document establishes the specifications for the execution of static pile load tests in which a single pile is subjected to an axial static load in tension in order to define its load–displacement behaviour. This document is applicable to vertical piles as well as raking piles. All types of piles are covered by this document. The tests considered in this document are limited to maintained load tests. Pile cyclic load tests are not covered by this document. Numerical values of partial factors for limit states and of correlation factors to derive characteristic values from static pile load tests to be taken into account in design are provided in EN 1997-1. This document provides specifications for the execution of static axial pile load tests – checking that a pile will behave as designed – measuring the resistance of a pile.	Draft issued for public comment in 2023
(BS EN ISO) 22477-3	Geotechnical investigation and testing – Testing of geotechnical structures – Part 3: Testing of piles: static transverse load testing	No details available.	Planned
BS EN ISO 22477-4:2018	Geotechnical investigation and testing – Testing of geotechnical structures – Part 4: Testing of piles: dynamic load testing	This document establishes the specifications for the execution of dynamic load tests in which a single pile is subject to an axial dynamic load in compression. It outlines the methods of testing required to allow assessment of pile resistance to be determined from the following methods and procedures described in EN 1997-1:2004+A1:2013: (*a*) dynamic impact testing, (*b*) pile-driving formulae, (*c*) wave equation analysis and (*d*) multi-blow dynamic testing. This document is applicable to all pile types mentioned in EN 1536, EN 12699 and EN 14199. The tests considered in this document are limited to dynamic load tests on piles only. Numerical values of partial factors for limit states from pile load tests to be taken into account in design are provided in EN 1997-1. For design to EN 1997-1 the results from dynamic load tests will be considered equivalent to the measured compressive resistance $R_{c,m}$ after being subject to appropriate analysis. This document provides specifications for investigation tests and control tests.	Current BS

Table 10.3 Continued

Reference	Title	Scope/description	Status
BS EN ISO 22477-5:2018	Geotechnical investigation and testing – Testing of geotechnical structures – Part 5: Testing of grouted anchors	This document establishes specifications for the execution of tension tests to be carried out on an anchor grouted in the ground, as defined in EN 1997-1 and EN 1537. Three methods of testing are recognised by this document. Test method 1 involves cyclic tension loading with measurement of displacement at the load stages; test method 2 involves cyclic tension loading with measurement of load loss at the load stages; and test method 3 involves step-loading with measurement of displacement under successive maintained tension loads. This document provides specifications for the experimental devices, the measurement apparatus, the test procedures, the definition and presentation of the test results and the content of records.	Current BS
(BS EN ISO) 22477-6	Geotechnical investigation and testing – Testing of geotechnical structures – Part 6: Testing of nailing	No details yet available.	Planned
(BS EN ISO) 22477-7	Geotechnical investigation and testing – Testing of geotechnical structures – Part 7: Testing of reinforced fill	No details yet available.	Planned
BS EN ISO 22477-10:2016	Geotechnical investigation and testing – Testing of geotechnical structures – Part 10: Testing of piles: rapid load testing	This document establishes the specifications for the execution of rapid load pile tests in which a single pile is subject to an axial load in compression to measure its load–displacement behaviour under rapid loading and to allow an assessment of its measured compressive resistance ($R_{c,m}$) and corresponding load–displacement behaviour. This document is applicable to piles loaded axially in compression. All pile types mentioned in EN 1536, EN 12699 and EN 14199 are covered by this document. The tests in this document are limited to rapid load pile tests only. It can be used in conjunction with EN 1997-1. For design to EN 1997-1, the results from rapid load pile testing will be considered equivalent to the measured compressive resistance, $R_{c,m}$, after being subject to appropriate analysis. This document provides specifications for – investigation tests, whereby a sacrificial test pile is loaded up to ultimate limit state (ULS) – control tests, whereby a pile is loaded up to a specified load in excess of serviceability limit state (SLS). Generally, an investigation test focuses on general knowledge of a pile type; a control test focuses on one specific application of a pile.	Current BS
BS ISO 24057:2022	Array measurement of microtremors to estimate shear wave velocity profile	This document specifies requirements for equipment, survey procedure, data analysis and reporting of array measurement of microtremors, which is one of the non-destructive testing methods with an array of sensors deployed on the ground surface. This document applies to the array measurement of microtremors to estimate a 1D shear wave velocity profile. Especially, this document describes array measurement of microtremors using vertical ground vibration to estimate an S-wave velocity profile by processing microtremor records based on the fundamental mode of Rayleigh waves.	Current BS
(BS EN ISO) 24283-1	Geotechnical investigation and testing – Qualification criteria and assessment – Part 1: Qualified technician and qualified operator	This document specifies the qualification criteria for a person performing sampling, testing, measuring, monitoring and installation of equipment (e.g. piezometers, BHEs, inclinometers and extensometers) in the framework of a geotechnical investigation.	Published by CEN as TS and published by BSI as PD
(BS EN ISO) 24283-2	Geotechnical investigation and testing – Qualification criteria and assessment – Part 2: Responsible expert	This document specifies the qualification criteria for the person who is responsible for the performance of sampling, testing, measuring, monitoring and installation of equipment (e.g. piezometers, BHEs, inclinometers and extensometers) in the framework of geotechnical investigation.	Published by CEN as TS and published by BSI as PD
(BS EN ISO) 24283-3	Geotechnical investigation and testing – Qualification criteria and assessment – Part 3: Qualified enterprise	This document specifies the qualification criteria for enterprises performing sampling, testing, measuring, monitoring and installation of equipment (e.g. piezometers, BHEs, inclinometers and extensometers) in the framework of a geotechnical investigation.	Published by CEN as TS and published by BSI as PD

Table 10.3 Continued

This standard has been published by BSI as a BS ISO since CEN has not been involved and it has not been published as an EN.

10.9.3 Codes and standards for geotechnical design

The BSs concerned with geotechnical design are listed in **Table 10.4**. The first codes of practice listed in this table are the following, which were originally prepared solely by BSI prior to the publication of EN 1997-1 and EN 1997-2.

- 8002 Code of practice for earth retaining structures
- 8004 Code of practice for foundations
- 8006 Code of practice for strengthened/reinforced soils
  – Part 1: Code of practice for strengthened/reinforced soils
  – Part 2: Soil nail design
- 8081 Code of practice for grouted anchors
- 8103 Structural design of low-rise buildings
  – Part 1: Code of practice for stability, site investigation, foundations, precast concrete floors and ground floor slabs for housing

These standards all have the prefix BS and are all called codes of practice. The original versions of these standards, prior to the publication of BS EN 1997, were broad in scope and descriptive documents, providing not only requirements for design but also guidance on construction operations, construction materials, occupational safety and legislation, and also a bibliography. However, since BS EN 1997 has been adopted by BSI, the design of geotechnical structures in the UK should be carried out in accordance with BS EN 1997, and hence these codes of practice have been revised to provide only NCCI to BS EN 1997 while still retaining the guidance information that was considered helpful. For example, the scope of BS EN 8002 states that this standard 'gives recommendations for the design and construction of earth retaining structures' and that 'it provides non-contradictory, complementary information for use in conjunction with BS EN 1997 and its UK National Annex'.

The standard BS 8103-1 is a code of practice with requirements for the geotechnical design of low-rise buildings, not exceeding 15 m and constructed using traditional methods. It has provisions, including prescriptive measures such as the minimum widths in its Table 8 for foundations on different types of rock or soil, for what in Paragraphs 2.1(14) and (15) of BS EN 1997-1 are termed GC1 structures.

The standard for geotechnical design in the UK is the European Standard, prepared by CEN/TC 250/SC7: *Geotechnical design* and published by BSI with the prefix BS EN:

- 1997 Eurocode 7: Geotechnical design
  – Part 1: General rules
  – Part 2: Ground investigation and testing.

BS EN 1997 has two parts: Part 1 is intended to be used as a general basis for the geotechnical aspects of the design of buildings and civil engineering works, and is concerned with the requirements for strength, stability, serviceability and durability of geotechnical structures. It has sections covering the design of spread foundations, pile foundations, anchors, retaining structures, hydraulic failure, overall stability and embankments. Part 2 provides the requirements for the planning, performance and reporting of field and laboratory tests and the evaluation of test results and derivation of geotechnical parameter values.

Geotechnical designs in the UK using BS EN 1997-1 should be carried out using the National Annexes to Parts 1 and 2. The National Annexes provide the NDPs, including the values of the partial factors, and information on NCCI to be used with the two parts of BS EN 1997 in the case of geotechnical designs in the UK. The UK National Annex to EN 1997-1 lists the following publications as providing NCCI: BS 1377, BS 5930, BS 6031, BS 8008, BS 8081, PD 6694-1 and CIRIA C580, while the National Annex to BS EN 1997-2 only refers to BS 5930. The NCCI publication 'PD 6694-1: Recommendations for the design of structures subject to traffic loading to BS EN 1997-1', included in **Table 10.4**, covers some geotechnical aspects of the design of bridges and other structures subject to traffic loading designed to BS EN 1997-1 and other relevant Eurocodes. However, as it is published by BSI as a PD it does not have the status of a BS. The listed publication CIRIA C580 (Gaba *et al.*, 2003) is a guidance document providing NCCI for the economic design of embedded retaining walls. Another Eurocode part, included in **Table 10.4**, that is relevant to geotechnical design is Part 5 of EN 1993, 'Design of steel structures: piling'.

BS EN 1997-1 provides no requirements for designs involving ground reinforcement and states in Paragraph 8.1.1(5) that the design of soil nails and rock bolts is not covered, hence designs for ground reinforcement and soil nailing should be carried out in accordance with BS 8006-1 and BS 8006-2. The forewords of these standards state that they are 'in line with' and 'follow the principles of' BS EN 1997-1. In the scope of BS EN 8006-1 it is noted that 'the values of the partial factors and load factors given in BS EN 1997-1 have not been calibrated for reinforced soil structures' and hence 'BS EN 1997-1 is not for use in the design and execution of reinforced soil'. It states that 'the partial factors set out in BS 8006-1 cannot be replaced by similar factors in BS EN 1997-1'. In the design of soil nailing, BS EN 8006-2 states that this 'standard is intended to harmonise the design approach of soil nailing with other structures designed using BS EN 1997-1'.

The following standard for the thermal design of foundations to avoid frost heave has been prepared by the CEN/TC 89 *Thermal performance of buildings and building components* in collaboration with ISO/TC 163 *Thermal insulation* and published by BSI as a BS

- 13793 Thermal performance of buildings: Thermal design of foundations to avoid frost heave

Reference	Title	Scope/description	Status
BS 8002:2015	Code of practice for earth-retaining structures	This document gives recommendations for the design and construction of earth-retaining structures to support ground at slopes steeper than the ground would naturally assume. It provides NCCI for use in conjunction with BS EN 1997 and its UK National Annex. It gives general recommendations for the design and construction of all types of earth-retaining structures and gives specific recommendations for the design and construction of gravity walls, semi-gravity walls and embedded walls, and for the design and construction of cofferdams, basements, strutted excavations and deadman anchors. It gives information about specific geological formations encountered in the UK.	Current BS
BS 8004: 2015+A1:2020	Code of practice for foundations	This document gives recommendations for the design and construction of foundations for the normal range of buildings and engineering structures. It provides NCCI for use in conjunction with BS EN 1997 and its UK National Annex. It gives general recommendations for the design and construction of all types of foundations with specific recommendations for the design and construction of spread foundations and pile foundations. It gives specific recommendations for the design and construction of helical steel piles and underpinning. It gives information about specific geological formations encountered in the UK and about the UK government's policy regarding archaeological finds.	Current BS
BS 8006-1:2010+ A1:2016	Code of practice for strengthened/ reinforced soils and other fills	This document contains recommendations and guidance for the application of reinforcement techniques to soils, as fill or in situ, and to other fills. The standard is written in a limit state format and guidelines are provided in terms of partial material factors and load factors for various applications and design lives. This code is to be read in conjunction with BS EN 1997-1 and BS EN 14475. BS EN 1997-1 does not cover the design and execution of reinforced soil structures. The values of partial factors and load factors given in BS EN 1997-1 have not been calibrated for reinforced soil structures. BS EN 1997-1 is not for use in the design and execution of reinforced soil. The partial factors set out in BS 8006-1 cannot be replaced by similar factors in BS EN 1997-1. In line with current practice, the design methods described are based on limit state principles. The clauses are supplemented by a substantial list of references to enable the user to consider in greater depth the applications of the technique.	Current BS
BS 8006-2:2011+ A1:2017	Code of practice for strengthened/ reinforced soil – Part 2: Soil nail design	This document gives recommendations and guidance for stabilising soil slopes and faces using soil nails. Other methods of stabilisation using reinforced soil methods are given in BS 8006-1:2010 and both parts might be needed for complex structures. Additional considerations might be required for unusually loaded or high soil nailed slopes, or where they interface with other structures. While BS EN 1997-1:2004 specifically excludes soil nailing, this standard is intended to harmonise the design approach of soil nailing with other geotechnical structures designed using BS EN 1997-1:2004. The principal purpose of this standard is to provide design guidance; however, where knowledge of construction methodology is required for design purposes then appropriate paragraphs have been included. Construction guidance is given in execution standard BS EN 14490:2010. At the time of preparation of this standard, CEN TC 341 is drafting a standard covering the testing of soil nails – see 22477-6 in **Table 10.3**.	Current BS
BS 8081: 2015+A2:2018	Code of practice for grouted anchors	This document gives recommendations for the design, construction, stressing, testing, monitoring and maintenance of grouted anchors as defined in BS EN 1997-1, BS EN 1537 and BS EN ISO 22477-5. Further general recommendations for corrosion hazards and protective measures, construction techniques and quality controls, stressing procedures and the testing of grouted anchor components and complete installations are provided. Information supporting the practical implementation of these recommendations is provided in annexes to this document.	Current BS
BS 8103-1:2011	Structural design of low-rise buildings – Part 1: Code of practice for stability, site investigation, foundations, precast concrete floors and ground floor slabs for housing	This document gives recommendations for the design of low-rise housing and covers the site investigation and foundations and ground floor slabs used in the construction. Foundations comprising strip footings or trench fill founded in normal ground are the only type described. Low-rise housing comprises detached, semi-detached and terraced houses and flats (with not more than four self-contained dwelling units per floor accessible from one staircase), of not more than three storeys above ground intended for domestic occupation and of traditional masonry construction with timber roofs and floors of timber or concrete.	Current BS

Table 10.4 Codes and standards for geotechnical design (continued on next page)

Reference	Title	Scope/description	Status
		For the purposes of this standard, additional habitable accommodation in the roof space constitutes a storey of the house. The recommendations also apply to certain single-storey, non-residential buildings (e.g. domestic garages and annexes to residential buildings) with a floor area not exceeding 36 m^2.	
BS EN 1997-1:2004+A1:2013 incorporating Corrigendum February 2009	Eurocode 7: Geotechnical design – Part 1: General rules	This document is intended to be used in conjunction with EN 1990, which establishes the principles and requirements for safety and serviceability, describes the basis of design and verification and gives guidelines for related aspects of structural reliability. It is intended to be applied to the geotechnical aspects of the design of buildings and civil engineering works. It is concerned with the requirements for strength, stability, serviceability and durability of structures. The following subjects are dealt with in separate sections: basis of geotechnical design, geotechnical data, supervision of construction, monitoring and maintenance, fill, dewatering, ground improvement and reinforcement, spread foundations, pile foundations, anchors, retaining structures, hydraulic failure, overall stability and embankments.	Current BS to be used with NA+A2:2020 to BS EN 1997-1:2004 +A1:2013
NA+A2:2022 to BS EN 1997-1:2004 +A1:2013 incorporating Corrigendum No. 1	UK National Annex to Eurocode 7: Geotechnical design – Part 1: General rules	This National Annex gives – the UK decisions for the NDPs described in the different subclauses in the body and Annexes of BS EN 1997-1:2004+A1:2013 – the procedure to be used where alternative procedures are given – the UK decisions on the status of BS EN 1997-1:2004+A1:2013 informative annexes – references to NCCI.	Current NA to BS EN 1997-1:2004+A1:2013
BS EN 1997-2:2007 incorporating Corrigendum October 2010	Eurocode 7 Geotechnical design – Part 2: Ground investigation and testing	This document is intended to be used in conjunction with EN 1997-1 and provides rules supplementary to EN 1997-1 related to: planning and reporting of ground investigations; general requirements for a number of commonly used laboratory and field tests; interpretation and evaluation of test results; and derivation of values of geotechnical parameters and coefficients. In addition, examples of the application of field test results to design are given. This document gives no specific provisions for environmental ground investigations. Only commonly used geotechnical laboratory and field tests are covered. These were selected on the basis of their importance in geotechnical practice, availability in commercial geotechnical laboratories and existence of an accepted testing procedure in Europe. The laboratory tests on soils are mainly applicable to saturated soils. The provisions of this document apply primarily to projects of GC2, as defined in Section 2.1 of EN 1997-1.	Current BS
NA to BS EN 1997-2:2007	UK National Annex to Eurocode 7: Geotechnical design – Part 2: Ground investigation and testing	There are no NDPs in BS EN 1997-2:2007. This document contains all the information concerning the application of BS EN 1997-2:2007 in the UK. It gives – the UK decisions on the status of BS EN 1997-2:2007 informative annexes – references to NCCI.	Current NA to BS EN 1997-2:2007
BS EN 1993-5:2007 incorporating Corrigendum May 2009	Eurocode 3 Design of steel structures – Part 5: Piling	This document provides principles and application rules for the structural design of bearing piles and sheet piles made of steel. It also provides examples of detailing for foundation and retaining wall structures. The field of application includes: steel piled foundations for civil engineering works on land and over water; temporary or permanent structures needed to carry out steel piling work; and temporary or permanent retaining structures composed of steel sheet piles, including all kinds of combined walls. The field of application excludes offshore platforms and dolphins. It includes application rules for steel piles filled with concrete and design provisions for walings, bracing and anchorages.	Current BS
BS EN ISO 13793:2001	Thermal performance of buildings – Thermal design of foundations to avoid frost heave	This document gives simplified procedures for the thermal design of building foundations so as to avoid the occurrence of frost heave. It applies to foundations on frost-susceptible ground, and includes buildings with both slab-on-ground floors and suspended floors. It covers heated and unheated buildings, but other situations requiring frost protection (e.g. roads, water pipes in the ground) are not included. The standard applies in climates where the annual average air temperature is above 0°C, but does not apply in permafrost areas where the annual average air temperature is below 0°C.	Current BS

Table 10.4 Continued

Reference	Title	Scope/description	Status
PD 6694-1: 2011+A1:2020	Recommendations for the design of structures subject to traffic loading to BS EN 1997-1:2004+A1:2013	BS EN 1997-1 sets out principles and requirements for the geotechnical design of structures, but in many cases it does not identify specific requirements for particular structure types. This document provides NCCI relating to the design of structures subject to traffic loading to accompany BS EN 1997-1. This document covers some geotechnical aspects of bridges and other structures subject to traffic loading designed to BS EN 1997-1 and other relevant Eurocodes. The document includes information relating to the design of spread foundations, piled foundations, gravity-retaining walls, embedded retaining walls, integral bridges and buried concrete structures.	Current PD, not a BS

Table 10.4 Continued

This standard gives simplified procedures for the thermal design of building foundations so as to avoid the occurrence of frost heave. It has the prefix BS EN ISO, and some details are given in **Table 10.4**. It is referred to in Application Rule 6.4(3) of EN 1997-1.

10.9.4 Second-generation version of Eurocode 7

CEN regularly reviews and updates its standards. In 2015–2016 a systematic review was carried out of all the Eurocodes. It was decided to prepare a second generation of the Eurocodes with the following priorities identified

- improved ease of use, particularly for day-to-day calculations
- increased harmonisation through a reduction in NDPs.

With regard to Eurocode 7, the following additional priorities were identified for the second generation

- simplification of design approaches leading to design cases with combinations of partial factors
- reorganisation of structure and presentation – more consistent sections
- better alignment with EN 1990
- improved reliability discrimination
- improved rules for groundwater pressures
- improved rules for common geotechnical structures and more calculation models
- improved rules for numerical methods
- improved rules for design involving rock and dynamic design.

Work on the revision and development of EN 1997 has been ongoing since 2017 by subcommittees of CEN/TC 250/SC7 and is being reviewed by subcommittees of the national standards bodies. The second-generation EN 1997 will have three parts – 'Part 1: General rules', 'Part 2: Ground properties' and 'Part 3: Geotechnical structures' – rather than two parts, as in the current version. More guidance will be provided on the selection of geotechnical property values, with the present cautious estimate renamed as the nominal value rather than the characteristic value, and the term characteristic value reserved for when statistics are used to determine the value. Partial material factors will be applied to representative values, which may be either nominal or characteristic values, to obtain the design values. This is closer to the way design actions are obtained at present. The concept of the ground model will be included. There will be more guidance on the use of geotechnical calculation models and numerical methods. To improve the reliability of geotechnical designs, the risk assessment procedure involving the use of the Geotechnical Categories will be developed by defining the Geotechnical Categories as a combination of the complexity of the structure – that is, the Consequence Class – and the complexity of the ground – that is, the Geotechnical Complexity Class. Hence the Geotechnical Categories will become an integral part of geotechnical design rather than being optional, as at present. The scope of Eurocode 7 will be broadened to include requirements for rock engineering, pile groups and piled rafts, reinforced fill structures, ground reinforced by bolts, nails and mesh, ground improvement including stone columns, and groundwater control, in addition to the requirements for the types of geotechnical structures covered by the existing Part 1. All parts of the second-generation version of EN 1997 (and the other Eurocodes) are due to be published by CEN by March 2026, with BSI to publish its National Annexes by September 2027. There will then be a coexistence period in which either version can be used, before the existing version has to be withdrawn. Withdrawal of the existing Eurocodes is scheduled for 30 March 2028.

10.9.5 Codes and standards for geotechnical construction

The BSs concerned with the construction aspects of geotechnical engineering, or, to use the CEN terminology, the execution of geotechnical works, are listed in **Table 10.5**. The first standards listed in this table are the following that have been prepared by BSI and not by CEN.

- 8008 Safety precautions and procedures for the construction and descent of machine-bored shafts for piling and other purposes
- 8102 Protection of below ground structures against water ingress – Code of practice.

Reference	Title	Scope/description	Comment
BS 8008:1996+ A1:2008	Safety precautions and procedures for the construction and descent of machine-bored shafts for piling and other purposes	This document provides guidance on the safety precautions and procedures involved where it is required to lower personnel into a shaft that is machine-bored and not less than 750 mm in diameter. It does not cover the descent of hand-dug shafts lined progressively as excavation proceeds, for which reference ought to be made to BS 6164:2019 'Health and safety in tunnelling in the construction industry'.	Current BS
BS 8102:2022	Protection of below ground structures against water ingress – Code of practice	This document gives recommendations and provides guidance on methods of dealing with and preventing the entry of water from external sources into structures that are partly or wholly below ground. It covers the use of – waterproofing barrier materials applied to the structure – structurally integral watertight construction – drained cavity construction. It refers to the evaluation of groundwater conditions and consideration of harmful ground gases, risk assessment and how to manage these risks in below-ground structures. It includes guidance on drainage outside the structure and recognises the risk of water entering a structure through openings.	Current BS
BS EN 1536:2010+ A1:2015	Execution of special geotechnical works – Bored piles	This document establishes general principles for the execution of bored piles. It covers piles or barrettes which are formed in the ground by excavation and are structural members used to transfer actions and/or limit deformations. It applies to bored piles with – uniform cross-section (straight shaft) – telescopically changing shaft dimensions – excavated base enlargements – excavated shaft enlargements.	Current BS, under review in 2023
BS EN 1537:2013	Execution of special geotechnical works – Ground anchors	This document covers ground anchors grouted into the ground which are stressed and tested. They can be used for permanent or temporary applications. The anchors are designed in accordance with EN 1997-1 and tested in accordance with EN ISO 22477-5.	Current BS
BS EN 1538:2010+ A1:2015	Execution of special geotechnical works – Diaphragm walls	This document establishes general principles for the execution of diaphragm walls as either retaining walls or cut-off walls. It covers only structures constructed in a trench excavated with a support fluid or in dry conditions, where soil is removed and replaced by concrete or slurry and with wall thickness $B \geq 40$ cm. The following types of structure are considered – retaining walls: usually constructed to support the sides of an excavation in the ground – cut-off walls: usually constructed to prevent migration of groundwater, clear or polluted, or of other contaminants present in the ground.	Current BS, under review in 2023
BS EN 12063:1999	Execution of special geotechnical work – Sheet pile walls	This document specifies requirements, recommendations and information concerning the execution of permanent or temporary sheet pile wall structures. It applies only to steel sheet pile walls, combined walls and wooden sheet pile walls. Composite structures such as Berliner walls and sheet pile walls in combination with shotcrete are not the subject of this document.	Current BS, under review in 2023
BS EN 12699:2015	Execution of special geotechnical works – Displacement piles	This document establishes general principles for the execution of displacement piles, that means piles which are installed in the ground without excavation or removal of material from the ground except for limiting heave and/or limiting vibration as well as removal of obstructions or to assist penetration. Piles are driven into the ground using impact, vibration, pressing, screwing or a combination of these methods. The material of displacement piles covered by this document can be: steel, cast iron, concrete, mortar, timber, grout or a combination of these. This document covers prefabricated, cast in situ or a combination of these methods to form displacement piles of regular shape. Displacement piles may be installed in soils enhanced by ground improvement techniques.	Current BS, revision underway in 2023
BS EN 12715:2020	Execution of special geotechnical work – Grouting	This document is applicable to the execution, testing and monitoring of geotechnical grouting work. Grouting for geotechnical purposes is a process in which the remote placement of a pumpable material in the ground is indirectly controlled by adjusting its rheological characteristics and by the manipulation of the placement parameters. The following principles and methods of geotechnical grouting are covered by this document – displacement grouting – grouting without displacement of the host material.	Current BS
BS EN 12716:2018	Execution of special geotechnical works – Jet grouting	This document establishes general principles for the execution, testing and monitoring of jet grouting work.	Current BS

Table 10.5 Codes and standards for the execution of geotechnical works (continued on next page)

Reference	Title	Scope/description	Comment
BS EN 14199:2015	Execution of special geotechnical works – Micropiles	This document establishes general principles for the execution of micropiles. They are for drilled piles constructed using a drilling tool with a diameter less than 300 mm. It is not applicable to driven piles, the execution of which is governed by EN 12699.	Current BS
BS EN 14475:2006	Execution of special geotechnical works – Reinforced fill	This document establishes general principles for the construction of reinforced fill. It covers engineered fills that are reinforced by the inclusion of horizontal or sub-horizontal reinforcement placed between layers of fill during construction. The scope of reinforced fill applications considered in this document includes – earth-retaining structures with a facing to retain fill placed between the reinforcing layers – reinforced steep slopes with a facing, either built-in or added or wrap-around, reinforced shallow slopes without a facing, but covered by some form of erosion protection without a facing, reinstatement of failed slopes – embankments with basal reinforcement and embankments with reinforcement against frost heave in the upper part.	Current BS
BS EN 14490:2010	Execution of special geotechnical works – Soil nailing	This document establishes general principles for the execution, testing, supervision and monitoring of soil nailing. Soil nailing is a construction technique used to enhance/maintain the stability of a soil mass by installation of reinforcing elements (soil nails). The scope of soil nailing applications considered in this document includes the installation and testing of soil nails and associated operations, required when stabilising existing and newly cut slopes and faces in soil, existing earth-retaining structures, embankments, existing tunnels and the excavated facing of new tunnels in soil. Soil nailing may be used to form part of a hybrid construction. This document is relevant only to the soil nailing aspect of such constructions.	Current BS, revision underway in 2023
BS EN 14679:2005	Execution of special geotechnical works – Deep mixing	This document specifies general principles for the execution, testing, supervision and monitoring of deep mixing works carried out by two different methods: dry mixing and wet mixing. Deep mixing considered in this document is limited to methods which involve – mixing by rotating mechanical mixing tools where the lateral support provided to the surrounding soil is not removed – treatment of the soil to a minimum depth of 3 m – different shapes and configurations – treatment of natural soil, fill, waste deposits, slurries, etc. Guidance on practical aspects of deep mixing is given. Methods of testing, specification and assessment of design parameters, which are affected by execution, are presented.	Current BS, under review in 2023
BS EN 14731:2005	Execution of special geotechnical works – Ground treatment by deep vibration	This document is applicable to the planning, execution, testing and monitoring of ground treatment by deep vibration achieved by depth vibrators and compaction probes. The following types of treatment are covered – deep vibratory compaction to densify the existing ground – vibrated stone columns to form a stiffened composite ground structure by the insertion of granular material which itself shall be densified. Generally, stone columns have a diameter greater than 0.6 m and less than 1.2 m. The following treatment methods are covered by this document – methods in which depth vibrators, containing oscillating weights which cause horizontal vibrations, are inserted into the ground – methods in which compaction probes are inserted into the ground using a vibrator which remains at the ground surface and in most cases oscillates in a vertical mode.	Current BS, under review in 2023
BS EN 15237:2007	Execution of special geotechnical works – Vertical drainage	This document establishes general principles for the execution, testing, supervision and monitoring of vertical drain projects. It includes the application of prefabricated vertical drains and sand drains and deals with requirements to be placed on design, drain material and installation methods. It applies to the improvement of low-permeability, highly compressible soils by vertical drainage and preloading. Information regarding loading and preloading is given. Vertical drainage is used for both on-land and marine constructions for the following purposes – (pre-) consolidation and reduction of post-construction settlements – speeding up the consolidation process by decreasing the path lengths for pore water dissipation – increase of stability (by increasing effective stresses in the soil) – groundwater lowering – mitigation of liquefaction effects.	Current BS

Table 10.5 Continued

CEN/TC 288, *Execution of special geotechnical works*, has prepared the following comprehensive range of 12 standards for the construction aspects of geotechnical engineering, all with the umbrella title 'Execution of special geotechnical works'

- 1536 Bored piles
- 1537 Ground anchors
- 1538 Diaphragm walls
- 12063 Sheet pile walls
- 12699 Displacement piles
- 12715 Grouting
- 12716 Jet grouting
- 14199 Micropiles
- 14475 Reinforced fill
- 14679 Deep mixing
- 14731 Ground treatment by deep vibration
- 15237 Vertical drainage.

These standards have all been published by BSI as BSs with the prefix BS EN. Details of these standards, including their scopes and status, are given in **Table 10.5**. The standards 1536, 1537, 12063, 12699 and 14199 are referred to in principles and application rules of Sections 7 'Pile foundations' and 8 'Anchors' of EN 1997-1. In 2023, as indicated in **Table 10.5**, the standards 1536, 1538, 12063, 14679 and 14731 were under review and the standards 12699 and 14490 were being revised.

10.9.6 Codes and standards for earthworks

Earthworks is a particular aspect of geotechnical engineering for which specific codes of practice, standards and specifications have been published. Earthworks are a civil engineering process that involves selection of suitable fill material and modification of the geometry of the ground surface to create stable and durable structures. Earthworks are often associated with transport infrastructure, but are also involved in other types of structures such as industrial, commercial and residential buildings, water facilities, flood defences and coastal protection works (see the chapters in Section 7 *Design of earthworks, slopes and pavements*). The BSs for earthworks are listed in **Table 10.6**. The first standard is

- 6031 Code of practice for earthworks.

The introduction to this standard states that 'The overall aim is for BS 6031 to be non-conflicting complementary information (NCCI) to BS EN 1997-1:2004'. It should be noted that the geotechnical aspects of the design of earthworks should be in accordance with BS EN 1997-1.

CEN/TC 396: *Earthwork,* has prepared, and is still preparing without the involvement of ISO, the following 20 standard parts/standards for the execution of earthworks and for testing and treatment of soils and rocks for use in earthworks, all with the umbrella title *Earthworks* (detailed in **Table 10.6**)

- 16907 Earthworks
  - Part 1: Principles and general rules
  - Part 2: Classification of material
  - Part 3: Construction procedures
  - Part 4: Soil treatment with lime and/or hydraulic binder
  - Part 5: Quality control
  - Part 6: Land reclamation earthworks using dredged hydraulic fill
  - Part 7: Hydraulic placement of extractive waste
  - Part X: Use of alternative materials in earthworks
  - Part Y: Sustainable earthworks
- 17006 Earthworks – Continuous compaction control (CCC)
- 17542 Earthworks – Geotechnical laboratory tests
  - Part 1: Degradability test standard
  - Part 2: Fragmentability test standard
  - Part 3: Methylene blue value VBS on soils and rocks
  - Part 4: Collapse potential test
  - Part 5: Swelling potential test
- 17685 Earthworks – Chemical tests
  - Part 1: Determination of organic matter content by loss on ignition
  - Part 2: Organic matter content test by potassium permanganate method
- 17693 Earthworks – Soil treatment tests
  - Part 1: pH-test for determination of the lime requirement of soils for stabilization (lime fixation point LFP, lime modification optimum LMO)
  - Part 2: Test of evaluation of the aptitude of a dry material to emit dust
  - Part 3: Non-destructive seismic test method for mechanical performance.

The first seven parts of 16907 'Earthworks' were all published by BSI as standards with the prefix BS EN. A further two parts, 'Use of alternative materials in earthworks' and 'Sustainable earthworks' are at a preliminary stage of development in 2023.

Standard 17006 'Continuous compaction control' was approved and published by CEN in 2016 as a TS for provisional application, initially for 3 years; hence, it was published by BSI as a PD. It has not yet in 2023 been approved by CEN as an EN, so it remains a PD.

Standard 17542 'Geotechnical laboratory tests' has five parts. The first three have been published by CEN as ENs and hence published by BSI as BSs. The fourth and fifth parts, 'Collapse potential test' and 'Swelling potential test', are at a preliminary stage of development in 2023.

Reference	Title	Scope/descriptors	Status
BS 6031:2009	Code of practice for earthworks	This document gives recommendations and guidance for unreinforced earthworks forming part of general civil engineering construction, with the exception of dams. This standard also gives recommendations and guidance for temporary excavations such as trenches and pits. Reinforced earthworks are covered in BS 8006-1 and BS 8006-2. This document applies to earthworks classified as GC1, 2 and 3 structures as defined in BS EN 1997-1:2004.	Current BS
BS EN 16907-1:2018	Earthworks – Part 1: Principles and general rules	This document gives definitions, principles and general rules for the planning, design and specification of earthworks. It introduces the other parts of the standard, which will be used together with Part 1. Earthworks are a civil engineering process aimed at creating earth structures by changing the geometry of the earth surface for construction or other activities. They are characterised by the need to use available natural or recycled materials and to handle them in a way appropriate to yield prescribed properties. This document is applicable to all types of earth structures, except – some specific types of works such as the execution of trenches and small earthworks may be organised using simplified or specific rules – some structures, such as dykes and dams, need earthworks which have specific design and construction requirements: these may extend beyond the rules of this standard.	Current BS
BS EN 16907-2:2018	Earthworks – Part 2: Classification of materials	This document defines a common basis for description and classification for use by all parties involved in the design, planning and construction of earthworks. It specifies the processes and properties to be used in the description and classification of earthworks materials. It specifies soil and rock groups as a basis of material specifications for earth structure elements. This classification relates to the physical and chemical properties of the soil and rock materials. The approach to the description of soil and rock set out in EN ISO 14688-1 and EN ISO 14689, respectively, and the approach to the classification of soil set out in EN ISO 14688-2 are applicable to earthworks, but the range and scope of classification for earthworks given here is more detailed and orientated to the specific demands of earthwork procedures and earth structure elements. Informative examples of existing experience-based classification systems and their use in the UK are presented in Annex H to EN 16907-1:2018.	Current BS
BS EN 16907-3:2018	Earthworks – Part 3: Construction procedures	This document provides execution procedures for excavating, transporting and placing soils and rocks for the construction of earth structures and guidance for the work. Additionally, it includes excavation and placement of rock materials underwater. Dredging of soils and the associated hydraulic placement of fills are covered by EN 16907-6 and EN 16907-7. Execution of earthworks follows the conclusions of the earthworks design and optimisation phase (EN 16907-1), which should anticipate soil and rock specificities and their suitability. In case some events could not be foreseen, additional design is performed during the execution of works.	Current BS
BS EN 16907-4:2018	Earthworks – Part 4: Soil treatment with lime and/or hydraulic binders	This document applies to the treatment with binders of natural soils, weak rocks, intermediate rocks, chalk, recycled materials and artificial materials for the execution of earthworks during the construction and maintenance of roads, railways, airfields, platforms, dykes, ponds and any other type of earth structure. It relates only to the treatment in layers, produced for earthworks in situ or from a mixing plant, as opposed to the treatment by columns, for example. It specifies the requirements for the constituents of the mixtures, the preliminary laboratory testing methodology, the laboratory performance classification, the execution and control. The informative annexes also give example of good practices for execution and control. The laboratory performance classification specified covers the two types of treatment: improvement and stabilisation. For improvement, the classification relates to the short-term performance. For stabilisation, the classification relates to the medium- to long-term performance.	Current BS
BS EN 16907-5:2018	Earthworks – Part 5: Quality control	This document provides recommendations and guidance on the quality assurance and quality control of earthworks construction forming part of general civil engineering and building works. It provides guidance on the techniques to be used to give clients, contractors and designers confidence that the earthworks have been constructed in accordance with their requirements.	Current BS

Table 10.6 Codes and standards for earthworks (continued on next page)

Reference	Title	Scope/descriptors	Status
BS EN 16907-6:2018	Earthworks – Part 6: Land reclamation earthworks using dredged hydraulic fill	This document deals with underwater excavation and hydraulic placement of fill material for land reclamation projects. The scope is limited to soils that exhibit free-draining behaviour during and after placement. The main purpose of this document is to ensure that functional requirements and specifications for such projects are in harmony with site boundary conditions and construction methods. This document specifies minimum requirements for site-related data to be acquired before the procurement and execution stage of a dredging and land reclamation project. It gives guidance on how the selection of the dredging equipment shall be undertaken. It also gives guidance on the selection of a borrow area and on the judgement regarding the suitability of the fill material for the project. It offers the general principles on how to design the actual execution of a dredged hydraulic fill project and offers guidelines for monitoring and quality control of that execution in order to guarantee that the fill mass exhibits the behaviour as intended by the designer.	Current BS
BS EN 16907-7:2021	Earthworks – Part 7: Hydraulic placement of extractive waste	This document gives recommendations for the hydraulic placement of extractive wastes and may be applicable to – all stakeholders engaged in the deposition of extractive wastes using hydraulic placement techniques with respect to geotechnical and geochemical aspects of the investigation, engineering design, construction and operation of a mine waste facility and all associated monitoring activities – those extractive industries involving the production of fine particulate wastes which, in the course of industrial processing, require to be stored in a safe, stable and environmentally acceptable location – practitioners in non-extractive industries in fields where similar techniques may be applicable and for which no other European guidance exists. The scope of this document includes all aspects of a dam, confining embankment or other structure serving to contain, retain, confine or otherwise support such wastes on the surface in a terrestrial environment. It addresses the characterisation of the extractive waste for the purposes of hydraulic placement in the mine waste facility, both as part of the confining embankment and for safe storage.	Current BS
(BS EN) 16907-X	Earthworks – Part X: Use of alternative materials in earthworks	No information available.	Preliminary stage in 2023
(BS EN) 16907-Y	Earthworks – Part Y: Sustainable earthworks	No information available.	Preliminary stage in 2023
PD CEN/TS 17006:2016	Earthworks – Continuous compaction control (CCC)	This technical specification provides guidance, specifications and requirements on the use of continuous compaction control (CCC) as a quality control method in earthworks by means of roller integrated dynamic measuring and documentation systems. The CCC method is suitable for soils, granular materials and rockfill materials which can be compacted using vibratory rollers. A CCC technology based on the measure of propel energy necessary to overcome the rolling resistance is also available and can be used as a quality control method in earthworks. The propelling power of the compactor provides an indication of the material stiffness and it is measured as a function of the machine ground speed, slope angle and rolling resistance.	Current PD, not a BS
BS EN 17542-1:2022	Earthworks – Geotechnical laboratory tests – Part 1: Degradability test standard	This document defines the principle and the methods for the determination of the 'degradability coefficient' of rocky material. The degradability coefficient I_{DG} distinguishes the behaviour of certain rocky material and is used to show the change in the geotechnical characteristics (particle size, clay content, plasticity, etc.) in relation to the characteristics seen immediately following excavation. Changes in the particle size occur due to the combined action of climatic or geohydrological elements (frost, soaking–drying cycles) and mechanical stress to which it is subjected. In the case of degradable rocky material, this leads to a fairly significant and continuous reduction in the mechanical and geometric characteristics of the works in which they are used. The two methods developed in this document for the determination of I_{DG} are not equivalent. The results obtained by this document refer to the method used.	Current BS

Table 10.6 Continued

Reference	Title	Scope/descriptors	Status
BS EN 17542-2:2022	Earthworks – Geotechnical laboratory tests – Part 2: Fragmentability test standard	This document defines the principle and the methods for the determination of the 'fragmentability coefficient' of rocky material. The fragmentability coefficient I_{FR} distinguishes the behaviour of certain rocky material and is used to show the change in particle size from the moment that the material is excavated through to its subsequent implementation and in certain cases during its whole service life. Changes in the particle size occur due to the structural resistance of the rock being unable to support the mechanical stress to which it is subjected during its implementation and use.	Current BS
BS EN 17542-3:2022	Earthworks – Geotechnical laboratory tests – Part 3: Methylene blue value VBS on soils and rocks	This document describes the reference method for the determination of the methylene blue value (V_{BS}) in soils and rocks for earthworks. The test is based on measuring the quantity of methylene blue that can be adsorbed by the material suspended in water. This quantity of adsorbed methylene blue is reported by direct proportionality to the 0/50 mm ground. The soil blue value is directly related to the specific surface area of the soil particles or rocky material. The V_{BS} test uses common equipment and calibration as the methylene blue test (MB) for aggregates (EN 933-9), but it applies to another granular fraction (5 mm for V_{BS} and 2 mm for MB, respectively). Thus, the results obtained between the two tests cannot be compared in the general case.	Current BS
(BS EN) 17542-4	Earthworks – Geotechnical laboratory tests – Part 4: Collapse potential test	No information available.	Preliminary stage in 2023
(BS EN) 17542-5	Earthworks – Geotechnical laboratory tests – Part 5: Swelling potential test	No information available.	Preliminary stage in 2023
BS EN 17685-1:2023	Earthworks – Chemical tests. Part 1: Determination of loss on ignition	This document describes a method for the determination of the loss on ignition (w_{LOI}) of fine, intermediate, composite and coarse soils, organic soils and anthropogenic materials (according to EN 16907-2) after ignition under air at 550°C. The loss of mass suffered by these materials at 550°C is usually due to the release of volatile compounds, water (absorbed, crystallised or structural) and gases from decomposition of organic matter and inorganic substances such as sulfur, sulfides or hydroxides (e.g. H_2O, CO_2, SO_2). A method is given in order to estimate the organic matter content (C_{OM}) from the value of w_{LOI} for clayey soils.	Current BS
(BS EN) 17685-2	Earthworks – Chemical tests. Part 2: Organic matter content test by potassium permanganate method	No information available.	Preliminary stage in 2023
PD CEN/TS 17693-1:2021	Earthworks – Soil treatment tests – Part 1: pH-test for determination of the lime requirement of soils for stabilization (lime fixation point LFP, lime modification optimum LMO)	This document describes the reference method for the determination of the lime fixation point (LFP) in soil treatment for earthworks. The test consists in measuring the lowest quantity of lime to be added in a soil suspension in water that will result in a pH value of the soil–lime mix suspension of 12.4, measured at 25°C $\pm$ 1°C. This test method cannot be used to provide information about soil reactivity with lime, or other performance values (mechanical characteristics of soil–lime mixes) applicable for improvement or stabilisation purposes. Those performance tests will be conducted in a laboratory from a specific study, the lime dosage to be applied being indicated from this method.	Current PD, not a BS
PD CEN/TS 17693-2:2021	Earthworks – Soil treatment tests – Part 2: Test of evaluation of the aptitude of a dry material to emit dust	This document describes the reference method for the determination of the index of dust emission (I_{DE}) in soil treatment for earthworks. This test concerns more particularly – limes in conformity with EN 459 'Building lime' – cements in conformity with EN 197 'Cement' – road binders in conformity with EN 13282 'Hydraulic road binders' – fly ashes in conformity with EN 14227 'Hydraulically bound mixtures – Specifications' – siliceous fly ashes in conformity with EN 450-1 'Fly ash for concrete' – ground granulated blast furnace slag in conformity with EN 15167-1 'Ground granulated blast furnace slag for use in concrete, mortar and grout'.	Current PD, not a BS
(BS EN) 17693-3	Earthworks – Soil treatment tests – Part 3: Non-destructive seismic test method for mechanical performance	No information available	Preliminary stage in 2023

Table 10.6 Continued

Standard 17685 'Chemical tests' has two parts. As of 2023 the first part, 'Determination of organic matter content by loss on ignition', has been published by BSI as a BS, while the second part 'Organic matter content test by potassium permanganate method', is at a preliminary stage of development.

Standard 17693 'Soil treatment tests' has three parts. The first part is 'pH-test for determination of the lime requirement of soils for stabilization (lime fixation point LFP, lime modification optimum LMO)' and the second part is 'Test of evaluation of the aptitude of a dry material to emit dust'. Both parts were approved and published by CEN in 2021 as TSs for provisional application and hence published by BSI as PDs. Their period of validity is limited initially to 3 years. After 2 years the members of CEN will be requested to submit their comments, particularly on the question of whether the CEN/TS can be converted into a European Standard. A third part, 'Non-destructive seismic test method for mechanical performance', is at a preliminary stage of development.

10.9.7 Codes and standards for geotechnical materials

BSI has published many standards for the materials used in geotechnical engineering and for their testing. In recent years most of these have been CEN standards prepared either just by a CEN TC or by a CEN TC in collaboration with an ISO TC. An example of a standard prepared just by CEN is BS EN 12794 'Precast concrete products – Foundation piles', which was prepared by CEN/TC 299: *Precast concrete products*. Three examples of standards prepared by CEN/TC 189 *Geosynthetics* in collaboration with ISO/TC 221 *Geosynthetics* are: BS EN ISO 13433 'Geosynthetics – Dynamic perforation test (cone drop test)', BS EN ISO 13437 'Geosynthetics – Installing and retrieving samples in the field for durability assessment' and BS EN ISO 13438 'Geosynthetics – Screening test method for determining the resistance of geotextiles and geotextile-related products to oxidation'. Some details of these standards are given in **Table 10.7**. Since the CEN/TC 189 website lists 85 standards on geosynthetics that have either been published or are at different stages of preparation, it has not been possible to include all of those standards in **Table 10.7**.

Reference	Title	Scope/description	Status
BS EN 12794:2005	Precast concrete products – Foundation piles	This document specifies the terminology, requirements, basic performance criteria, test methods and evaluation of conformity that will be applied to precast concrete foundation piles, factory-produced for building and civil engineering works and installed at the site. This document may also be applied to products manufactured in temporary plants on site where production is controlled and is protected against adverse weather conditions as necessary. This document applies to foundation piles produced in a plant as reinforced or prestressed concrete elements. The cross-section may be solid or provided with a hollow core, either prismatic or cylindrical. The cross-section may be constant over the full length or tapered partly or wholly along the pile or pile segment length. This document deals with foundation piles manufactured either in one length or in segments with cast-in pile joints. The foundation piles may have an enlarged toe or a pile shoe. This document applies to normal weight concrete compacted so as to retain no appreciable amount of entrapped air other than entrained air.	Current BS
BS EN ISO 13433:2006	Geosynthetics – Dynamic perforation test (cone drop test)	This document specifies a method to determine the resistance of geosynthetics to penetration by a steel cone dropped from a fixed height. The degree of penetration is an indication of the behaviour of the geosynthetic when sharp stones are dropped on its surface. The method is generally applicable to geosynthetics. However, the validity of this test for some types of products should be considered carefully, as the test principle may not be applicable.	Current BS
BS EN ISO 13437:2019	Geosynthetics – Installing and retrieving samples in the field for durability assessment	This document specifies a method for the on-site installation and retrieval of geosynthetic samples, irrespective of the particular degradation mechanisms to which they are exposed. The method is also appropriate to test for mechanical damage, much of which occurs during installation, and to provide an owner with information about the state of the geosynthetic product in their structure.	Current BS
BS EN ISO 13438:2018	Geosynthetics – Screening test method for determining the resistance of geotextiles and geotextile-related products to oxidation	This document specifies a screening test method for determining the resistance of geotextiles and geotextile-related products to oxidation. The test is applicable to products as follows – Method A for material consisting solely in polypropylene (PP), polyethylene (PE), polyamide (PA), aramide (AR) – Method B for material consisting solely in polyvinyl alcohol (PVAL). The data are suitable for screening purposes but not for deriving performance data such as lifetime, unless supported by further evidence.	Current BS

Table 10.7 Codes and standards for materials used in geotechnical engineering

10.10. Conclusions

Geotechnical design in the UK is now dominated by codes and standards prepared by CEN and/or ISO and published by BSI. Those involved in geotechnical design need to know these codes and standards, in particular Eurocode 7 (i.e. BS EN 1997) for geotechnical design, and its associated standards, their relevance and how to apply them. The move from the use of the traditional permissible stress design method with overall factors of safety, and from codes specific to the design of a particular geotechnical structure, to the broad limit state design method that not only harmonises all geotechnical designs with each other but also harmonises geotechnical designs with structural designs, has presented challenges and opportunities for the geotechnical designer. How the properties and special features of soil have influenced the development of Eurocode 7 and caused its design components to differ from those for structural design is explained. The components of geotechnical design, as embodied in Eurocode 7, have been presented in the form of the geotechnical design triangle, a concept with the aim of clarifying and improving understanding of the geotechnical design process.

The reason for the formation first of ISO and then CEN in the past century and the large number of standards prepared in recent years by these two international organisations is because of the importance and benefits of international standards in facilitating trade, including in the construction industry and geotechnical engineering. The benefits of international standards are that they: provide common and agreed design criteria and test methods; provide a common understanding regarding the design and construction of structures; facilitate the exchange of construction services between different states; facilitate the marketing and use of structural components in different states; provide a common basis for research and development in the construction sector; allow the preparation of common design aids and software; increase the worldwide competitiveness of civil engineering firms, contractors, designers and product manufacturers; and, particularly in the case of Eurocode 7 and the other Eurocodes, provide engineers with a common language for geotechnical and structural design.

The reference number, titles, scope and current status of the standards published by BSI, including those prepared by BSI and the standards prepared by CEN and ISO for geotechnical investigation and testing, geotechnical design, geotechnical construction, earthworks and some of the standards for materials used in geotechnical engineering, are given in five tables. The linkage of these standards to Eurocode 7 is explained. Also the aims, new features and broader scope of the second generation of Eurocode 7, which include improved ease of use, improved reliability discrimination and improved rules for common geotechnical structures and designs involving rock, are summarised. In addition, the publication schedule is outlined, with 30 March 2028 given as the date for withdrawal of the existing Eurocodes and replacement by the second generation versions.

References

Burland JB (1987) The teaching of soil mechanics: a personal view, groundwater effects in geotechnical engineering. *Proceedings of the IX European Conference on Soil Mechanics and Foundation Engineering, Dublin, Ireland* (Hanrahan ET, Orr TLL and Widdis TB (eds)). Balkema, London, UK, pp. 1427–1447.

Burland JB (2006) Interaction between structural and geotechnical engineers. *The Structural Engineer*, Apr.: pp. 29–37.

EC (European Community) (1989) Council Directive 89/106/EEC of 21 December 1988 on the approximation of laws, regulations and administrative provisions of the Member States relating to construction products. *Official Journal of the European Communities* **L40**.

EC (2003) Guidance Paper L (concerning the Construction Products Directive 89/106/EEC): Application and use of Eurocodes (edited version: April 2003). European Commission Enterprise Directorate General, Brussels, Belgium.

EU (2011) Regulation No. 305/2011 (Construction Products Regulation, or CPR) of the European Parliament and of the Council of the European Union. *Official Journal of the European Union*, **L88**.

Gaba AR, Simpson B, Powrie W and Beadman DR (2003) *Embedded Retaining Walls: Guidance for Economical Design*. CIRIA Report C580. CIRIA, London, UK.

ICE (Institution of Civil Engineers) (2011) UK register of ground engineering professionals. *Document ICE* 3009(4). The Geological Society, The Institute of Materials, Minerals and Mining and Institution of Civil Engineers, London.

Krebs Ovesen N and Orr TLL (1991) Limit states design: the European perspective. *Geotechnical Engineering Congress 1991*, vol. 2 (McLean FG, Campbell DA and Harris DW (eds)). American Society for Civil Engineers, Reston, VA, pp. 1341–1352.

McWilliam RC (2002) *The Evolution of British Standards*. PhD thesis, University of Reading, Reading, UK.

NBS (National Building Specification) (2013) *Approved Document A: Structure*. National Building Specification, London, UK.

Orr TLL (2008) Geotechnical education and Eurocode 7. *Proceedings of International Conference on Geotechnical Education, Constanta, Romania*, pp. 87–99.

Simpson B, Morrison P, Yasuda S, Townsend B and Gasetas G (2009) State of the art report: analysis and design. *Proceedings of the XVII International Conference on Soil Mechanics and Geotechnical Engineering, The Academia and Practice of Geotechnical Engineering, Alexandria, Egypt*, vol. 4 (Hamza M, Shabien M and El-Mossallamy Y (eds)). IOS Press, Amsterdam, the Netherlands, pp. 2873–2929.

All chapters within Sections 1 *Context* and 2 *Fundamental principles* together provide a complete introduction to the Manual and no individual chapter should be read in isolation from the rest.

Chapter 11
Sustainable geotechnics

doi: 10.1680/icemge.66816.0141

CONTENTS

Alice Berry Arup, London, UK
Matthew Free Arup, London, UK
Heleni Pantelidou Arup, London, UK
Cameron Hughes Arup, London, UK
Duncan Nicholson previously of Arup, London, UK
Asim Gaba previously of Arup, London, UK

This chapter describes how geotechnical engineers can contribute to sustainable development. Climate change and environmental degradation, together with social and economic inequalities are among the biggest threats to humankind and the planet. Ensuring that sustainable development is at the heart of every decision made as geotechnical engineers and that design and construction contributes towards environmental regeneration and social equity is undoubtedly the biggest challenge facing engineering in the twenty-first century. Geotechnical engineers need to play their part to make geotechnical practice sustainable and actively question whether their projects deliver outcomes that satisfy the triple bottom line: environmental, social and economic sustainability.

This chapter cannot cover the full field of sustainability and geotechnics in any significant depth, but rather signposts the reader to where to find more information. Sustainable development in the context of geotechnical engineering includes minimising carbon emissions and other environmental impacts from construction; preserving and restoring biodiversity through use of regenerative and nature-based solutions; maintaining water quality and minimising its use; preserving topsoils and enabling carbon sequestration in soil; minimising use of natural resources and greater use of circular economy; and developing resilient geotechnical solutions.

11.1. Context

11.1.1 Background

The Intergovernmental Panel on Climate Change (IPCC) *Sixth Assessment Report* states that 'It is unequivocal that human influence has warmed the atmosphere, ocean and land' (IPCC, 2021). The report warns that 'The magnitude and rate of climate change and associated risks depend strongly on near-term mitigation and adaptation actions, and projected adverse impacts and related losses and damages escalate with every increment of global warming.' In 2022 the IPCC estimated that by 2021 the increase in global surface temperatures was already 1.09°C (IPCC, 2022). There is an urgent need to take significant action to mitigate against the effects of climate change, and to adapt to the expected impacts of climate change.

The Intergovernmental Science-Policy Platform on Biodiversity and Ecosystem Services (IPBES) states that 'Biodiversity and nature's benefits to people underpin almost every aspect of human development … Yet they are being depleted and degraded faster than at any other point in human history' (IPBES, 2022). The 2021 Dasgupta Review of the economics of biodiversity notes that 'Our economies, livelihoods and well-being all depend on our most precious asset: Nature' (UK Government, 2021) and highlights the many linkages between nature and climate change. Meanwhile, the Planetary Boundaries framework (Stockholm Resilience Centre, 2009) indicates that we are already past the safe operating space and in the zone of high risk for several variables, including biogeochemical flows (e.g. flows of nitrogen) and biosphere integrity.

In 2015, the United Nations (UN) set out their 17 Sustainable Development Goals (SDGs), which every UN country has signed up to aiming to achieve by 2030 (UN General Assembly, 2015). The SDGs cover all areas of sustainable development, and have become a common framework to communicate sustainability.

National legislation and international commitments are rapidly changing, as the environmental crises (climate change and biodiversity loss) have become a planetary emergency. Some particularly notable recent UK changes include the following.

- In 2008, the UK was the first country to introduce the Climate Change Act, a legal national requirement to reduce greenhouse gas (GHG) emissions in 2050 by 80% against 1990 emissions and increase resilience to the changing climate.
- In 2019 the UK updated the Climate Change Act with a legal requirement to transition to net-zero carbon by 2050 (Climate Change Act 2008 (2050 Target Amendment); HMG, 2019).
- There are now net-zero requirements for all publicly funded projects over £5 million revenue (Cabinet Office, 2021).
- Under the 2021 Environment Act, 10% biodiversity net gain will become a legal requirement in 2023 (Environment Act 2021; HMG, 2021).

EU legislation is also forging ahead (EFFC, 2023) and other countries worldwide may have different legislative requirements. At the same time, large, listed companies in the UK and Europe are being asked to quantify their responses to the climate and biodiversity crises. The Taskforce for Climate-related Financial Disclosures (TCFD, 2022) established common requirements for companies to address climate change, introducing the dual, and mutually related, concepts of

- transition risk, related to the transition to a net-zero carbon economy – this transition may entail extensive policy, legal, technology and market changes to address mitigation requirements
- physical risk, related to the physical impacts of climate change – these may be acute, event-driven risks such as extreme weather events, or chronic risks such as sustained higher temperatures. These could manifest as safety issues or as operational issues, such as delays to projects from supply chain disruption.

The Taskforce for Nature-related Financial Disclosures (TNFD) also seeks to replicate these disclosures for biodiversity and natural capital. Projects will need to stand up to scrutiny from private investors in order to access 'green finance', also known as Environmental, Social and Governance (ESG) funding.

There is growing public awareness of the severity of the climate and biodiversity emergencies, as evidenced by recent high-profile protests. Some major UK projects, such as Heathrow's third runway, have been successfully legally challenged over failure to take due consideration of climate and sustainability legislation, such as compliance with the 2015 Paris Climate Agreement (UN, 2015). More than 150 UK civil engineering organisations are signatories of UK Civil Engineers Declare a Climate and Biodiversity Emergency (Civil Engineers Declare, 2019), which also commits them to minimum undertakings with relation to sustainability.

11.1.2 Putting project work into context

Buildings and infrastructure projects ought to positively contribute to the economy and society, but at present this is often at the expense of the environment. At a time when climate change and biodiversity loss are threatening the very integrity of the planet that supports our societies and economies, engineers must urgently transform their traditional practice to not cause net environmental harm (sustainable), or better still, reverse the damage already done to the environment (restorative) or regenerate it. The environmental impacts of an engineering intervention most often reach far beyond the physical site boundary. There are likely implications across the complex man-made and natural systems with which they interact.

Figure 11.1 illustrates a range of project approaches. Projects on the left-hand side of the scale can be considered conventional, where technical needs are addressed, but there is a net detrimental effect on the environment. Current geotechnical engineering practice is largely focused in this area. Moving further to the right are 'green' projects, which make an effort to minimise environmental impacts but still fall short of being environmentally neutral (i.e. environmentally sustainable). Projects need to move beyond this, providing net environmental gain: repairing and reversing ecological damage (restorative) or enabling a 'snowball' effect of environmental improvements (regenerative).

Figure 11.2 shows illustratively that sustainability needs to be considered right from the earliest stages of projects. This is when the most impactful decisions are taken, and it enables

Figure 11.1 Engineering projects relative to their impact on the environment. Best practice today (green) still accepts environmental damage but seeks to minimise the negative impacts. We need to move to the right on this diagram, producing restorative and regenerative outcomes (Adapted from Arup Better Way Strategy (Arup, 2020))

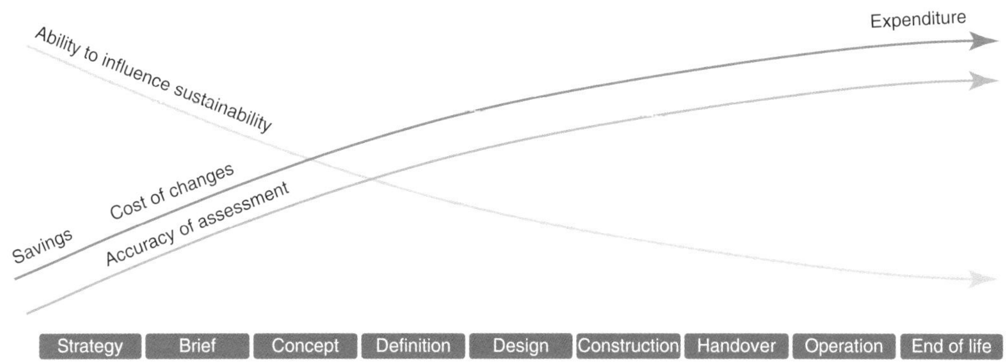

Figure 11.2 Conceptual diagram showing the ability to influence sustainability across the different work stages of project delivery (Adapted from PAS 2080:2016; BSI, 2016)

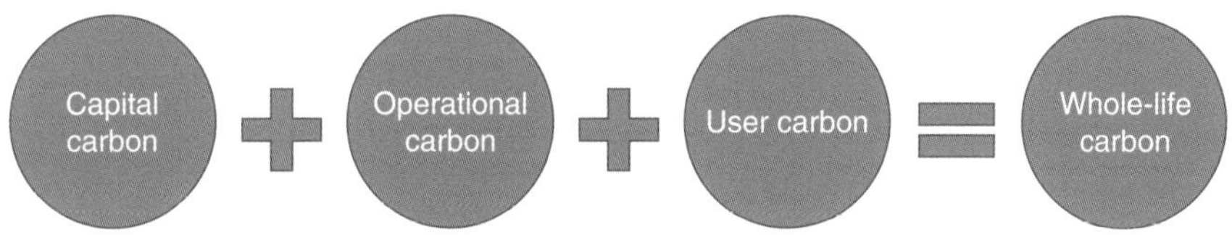

Figure 11.3 Whole-life carbon as a sum of its parts

sustainability to be 'baked in' to the project, rather than 'retrofitted' later on, vulnerable to being cut out during value engineering. Notwithstanding, there are still sustainability interventions that can be taken at all stages of projects, and these should be pursued.

Geotechnical design and construction of the 'right' projects can significantly contribute to social and economic improvement. Understanding which projects are 'right' can be aided by reviewing against the SDGs. The Value Toolkit also provides a useful framework (Construction Innovation Hub, 2022).

The rest of this chapter sets out how we, as geotechnical engineers, can influence the environmental impact of our projects to move them towards the right of the diagram in **Figure 11.1**. Social and economic sustainability, while clearly an important impact for our sector, are not addressed directly in this engineering-focused section. Environmental sustainability themes have been broadly grouped into four areas: decarbonisation (also known as climate change mitigation), circular economy, climate resilience and ecology and biodiversity.

11.2. Sustainability and geotechnics

11.2.1 Decarbonisation (or climate change mitigation)

11.2.1.1 Whole-life carbon and systems thinking

GHG emissions (e.g. carbon dioxide and methane) cause climate change. Reducing GHG emissions is essential for mitigating the impacts of climate change, and is mandated by UK law. GHGs are measured in carbon dioxide equivalent, usually in tCO_2e or $kgCO_2e$, with 'carbon' used as a shorthand for all GHGs. PAS2080:2023 (BSI, 2023) provides a framework for the management of whole-life carbon reduction. When aiming to reduce carbon, it is important to consider the whole-life carbon impact in the control and influence of any particular project.

As illustrated in **Figure 11.3**, whole-life carbon is the sum of the following.

- **Capital carbon:** Emissions associated with building an asset, and with subsequently decommissioning it at the end of its life. For a bridge this would include sourcing materials and construction of the bridge, for a residential building it would also include fixtures and fittings.
- **Operational carbon:** Emissions associated with running an asset for its intended life. For a bridge this might be electricity for lighting, for a residential building it would include replacing equipment and heating.
- **User carbon:** Emissions associated with use of the asset for the intended life, as influenced by the presence of the asset. For a bridge this would be the energy used by vehicles running over the bridge and the transport network of which it is part.

All projects should reduce whole-life carbon. It may be necessary to provide a capital carbon investment to enable this, and there will be an associated 'carbon payback period' (**Figure 11.4**). For example, a new railway branch could facilitate lower-carbon and shorter- distance journeys, and a windfarm would enable transition to lower-carbon electricity. Simply minimising the capital carbon invested is not enough. For example, **Figure 11.5** shows the operational energy required for a high-speed train running through a tunnel – while minimising the tunnel diameter would reduce the capital carbon of

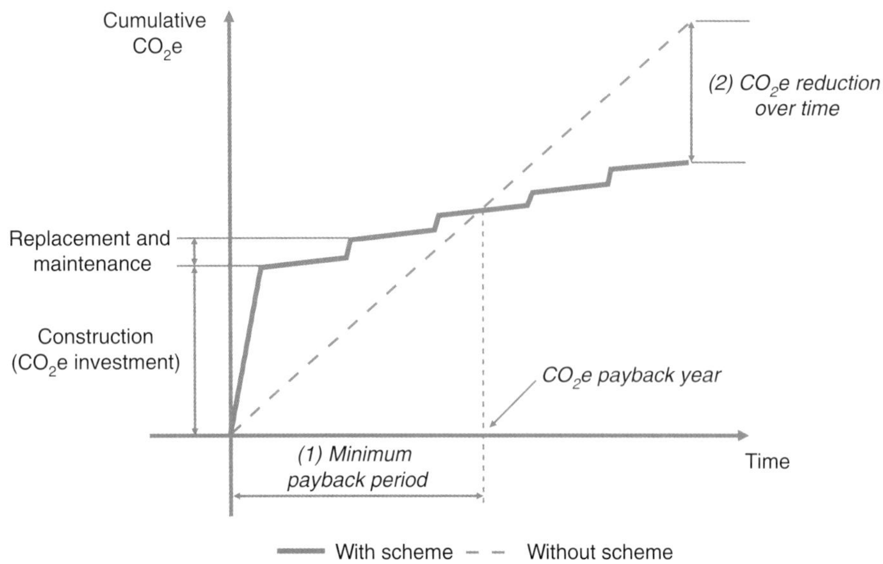

Figure 11.4 Conceptual diagram illustrating carbon payback of a carbon investment during an asset life cycle

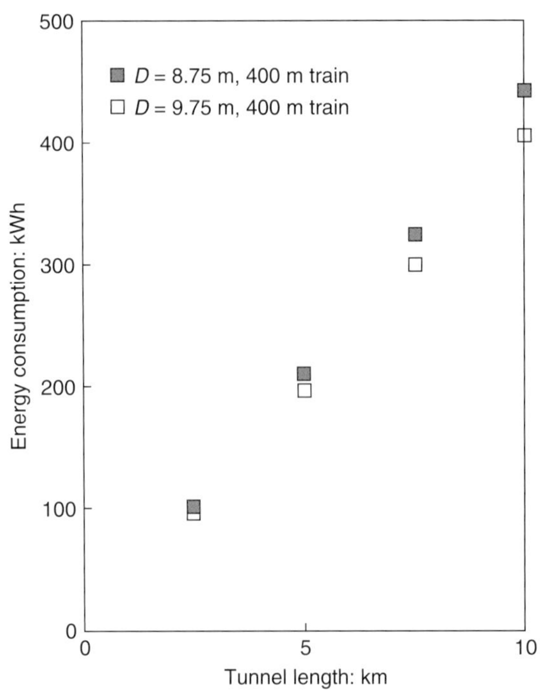

Figure 11.5 Operational energy consumption with change in tunnel diameter (From Pantelidou *et al.*, 2017)

the tunnel, doing so would be detrimental to whole-life carbon by increasing the energy that the train must expend to get through it (due to the aerodynamic 'piston effect').

Buildings and infrastructure assets cannot be viewed in isolation – they all form part of a bigger system, and the beneficial or detrimental impact of the project on carbon emissions beyond the physical boundaries of the project must be considered.

11.2.1.2 Carbon-reduction hierarchy

PAS 2080:2023 (BSI, 2023) provides a framework for managing and reducing whole-life carbon. When aiming to minimise whole-life carbon, it helps to follow a hierarchy of actions, so that the decisions with greatest ability to reduce carbon are targeted first. **Figure 11.6** shows this hierarchy.

- **Avoid:** Find ways to achieve the desired outcomes without adding carbon. For example, by examining the programme and taking a risk-based approach to fire safety, an escape shaft could be eliminated between two underground stations. Or, foundations from an existing building could be reused.
- **Shift:** Shift to a lower-carbon alternative. For example, by providing upstream natural flood storage, the need for flood defence walls can be reduced or eliminated. Earthworks or ground improvement could also be considered as an alternative to 'hard' concrete and steel solutions.
- **Improve:** Use design or technological means to reduce the carbon intensity of an element. For example, use of the observational method to decrease temporary works, management of residual risk through ongoing monitoring, novel pile types such as ribbed or impression piles, use of preliminary pile load testing to reduce partial factors, and lower-carbon concretes.

11.2.1.3 Setting Scope 1, 2 and 3 emissions in context for engineering

Globally, companies tend to classify their carbon emissions as Scope 1 (direct emissions, such as fuel use), Scope 2 (indirect emissions from energy/electricity use) and Scope 3 (all other indirect emissions, mostly associated with the supply chain). While this thinking about Scope 1 and 2 emissions can be a useful framework, a geotechnical engineer has influence over a disproportionate amount of Scope 3 carbon emissions, as illustrated by **Figures 11.7** and **11.8**. **Figure 11.7** shows the

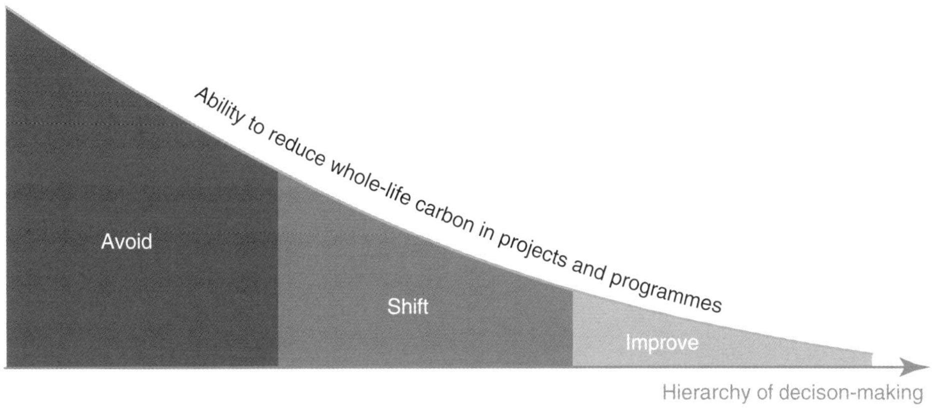

Figure 11.6 Carbon-reduction hierarchy (Adapted from PAS 2080:2023)

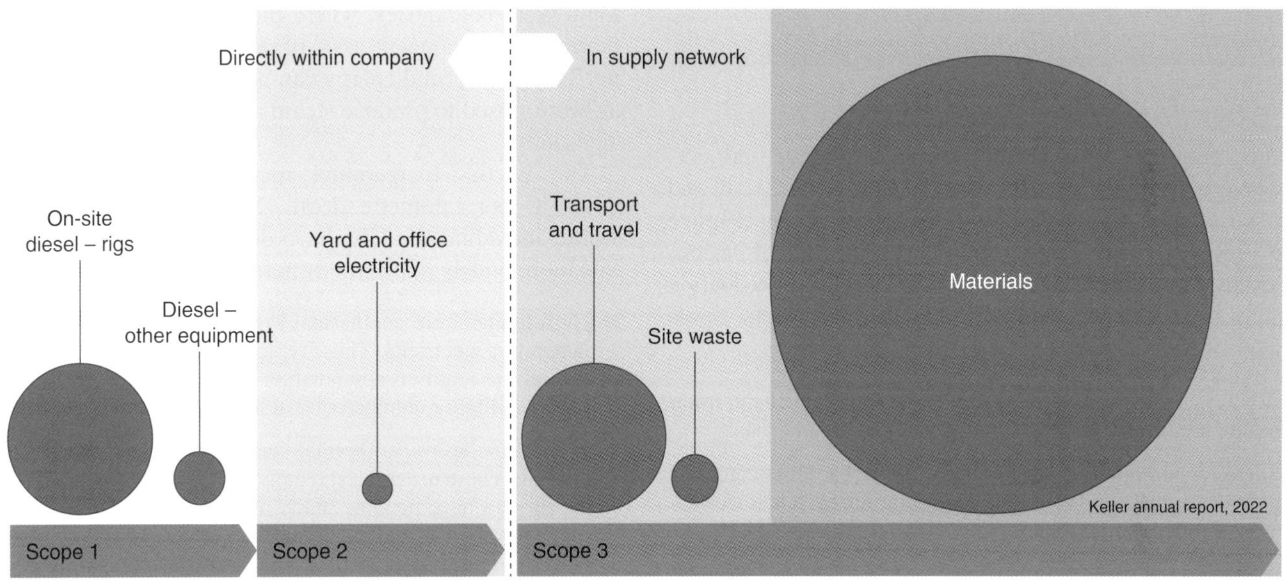

Figure 11.7 Relative emission Scope sizes on a heavy foundations project (Adapted from Keller's *Annual Report*, 2022: p. 58)

approximate size of Scope 1, 2 and 3 emissions for a geotechnical contractor, while **Figure 11.8** shows the approximate size of Scope 1, 2 and 3 emissions for an engineering consultancy (or contractor design house) set alongside the emissions associated with the projects they work on. Geotechnical engineers' influence on the supply chain and on our projects should not be underestimated, and the following sections provide advice on how to leverage this.

11.2.1.4 Common worries

- **'We'll do ourselves out of a job.'** Application of the 'avoid' principle can make engineers worry that society will stop building, but by considering whole-life carbon it becomes clear that this is not the case. Sustainable geotechnical engineering is essential for buildings, infrastructure and nature as we transition to a low-carbon future, resilient to the effects of climate change.
- **'What real influence can I have? I'm not a masterplanner or a policymaker.'** As a geotechnical engineer, it can feel like the scope of what we can influence is very small. We can have an impact through consistent good design and efficient construction, but we should not stop there. Through challenging our project teams and engaging with wider stakeholders, we can catalyse change and influence well beyond our own direct area of design or construction. Geotechnical engineers are well placed to influence a project to incorporate sustainable and resilient solutions from its earliest stages.
- **'The project can't afford it.'** Carefully considered sustainable designs can bring about cheaper, more efficient and more inclusive design, particularly when integrated from the outset.
- **'The client isn't interested.'** Most major clients have their own sustainability targets. They need help from engineering experts to accomplish these goals. Geotechnical advice on carbon reduction may be unexpected but can lead to significant savings in cost and carbon, and wider benefits.

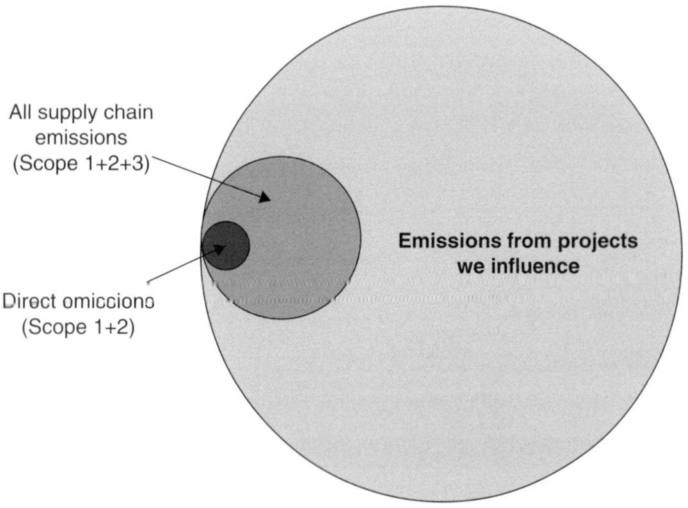

Figure 11.8 Relative emission Scope sizes for a typical engineering consultancy (Arup internal document, 2017)

11.2.1.5 Carbon reduction through geotechnical design

As stated above, it is important to seek to influence projects outside our own area of control – this falls under the 'avoid' and 'shift' parts higher up the carbon-reduction hierarchy (**Figure 11.6**). However, there is also plenty that geotechnical designers can do within the 'improve' part of the carbon-reduction hierarchy to reduce the whole-life carbon of projects. Some examples of this include

- undertaking sufficient site investigation and monitoring to minimise the use of overdesign to compensate for ground risk
- challenging design criteria – for example, by allowing more settlement or by specifying long-term monitoring, a less carbon-intensive foundation option may be selected
- consideration of soil–structure interaction in analysis
- adoption of the observational method (Chapter 100 *Observational method*)
- using a performance-based approach, such as for seismic design
- considering ground improvement instead of heavy foundations, and earthworks instead of hard engineering
- integration of earthworks with landscaping and habitat generation to minimise waste and increase potential for long-term sequestration
- undertaking instrumented preliminary pile load testing to enable back-analysis of pile design parameters and adoption of lower partial factors
- specifying, or enabling the use of, lower-carbon materials – for example by setting concrete strength requirements at beyond 28 days to enable greater use of cement replacement, or by allowing 'or equivalent' steel grades to enable reuse of waste steel from other industries.

Other chapters in this manual contain good advice to ensure better design.

It is also important that sufficient time is allowed for geotechnical design. Where geotechnical packages are to be constructed in advance of completion of the rest of a design, this can lead to conservative assumptions and overdesign. Similarly, a rushed design will not have time to rigorously undertake more refined analysis.

11.2.1.6 Geothermal energy and ground source heat

Geothermal energy is heat that is generated within the Earth by the decay of radioactive isotopes within the Earth's core, mantle and crust. It provides a significant opportunity for sustainable power generation. High temperatures at relatively shallow depth, and the associated hydrothermal systems caused by heated groundwater, are typically located along tectonic plate boundaries such as the Pacific margin. Geothermal energy is also present beneath regions located away from tectonic plate boundaries, where the Earth's crust is thinner or there is a large concentration of radioactive bedrock like granite. This geothermal energy can be captured and used directly as heat or used to produce steam and subsequently to generate electricity.

The Earth's temperature increases gradually, at about 25–30°C per kilometre depth. Many countries around the world, including the UK, have developed methods to access this more widely distributed energy source.

- High-temperature geothermal systems are used for heating and generating electricity. These typically involve drilling deep wells to find the required combination of temperature and rock mass permeability (or conductivity) at significant financial cost.
- Low-temperature geothermal systems are used for heating and cooling. These use modest temperatures in the ground at shallow depth, and have become much more common because of their much lower costs and installation times. These systems are typically combined with heat pumps to raise the temperatures to the required levels for heating purposes. Although ground source heat pumps have a much lower capital cost, they typically have higher operating costs, which need to be included in the assessment of the system economics.

11.2.1.7 Underground space

In 2007, the world's urban population exceeded the rural population for the first time; by 2050 it is predicted that nearly 7 in 10 people will live in cities (World Bank, 2022). Dense living enables lower-carbon lifestyles and leaves space for biodiversity, 'ecosystem services' (see Section 11.2.4) and food production. This 'land-sparing' for nature also provides the potential for natural carbon sequestration, an essential part of mitigating climate change (the 'net' of the net-zero carbon transition). Therefore, despite the more capital-carbon-intensive nature of underground construction compared to above ground, geotechnical engineers will continue to need to facilitate use of underground space, for example for transport corridors and appropriate basement uses. Decisions relating to use of underground space should be taken considering all areas of sustainability.

11.2.1.8 Underground storage of carbon dioxide

Effective use of underground space for the storage of carbon dioxide is an area of significant and prioritised interest and research in response to the climate emergency and associated sustainability requirements. Lagesse *et al.* (2022) note the contribution of underground storage and the alignment of these activities to the UN SDGs.

Carbon capture and storage (CCS) are initiatives to capture CO_2 at source or from the atmosphere and to store it away in the long term with the intention of reducing the concentration of CO_2 in the atmosphere. Carbon capture utilisation and storage (CCUS) can be more controversial, using the captured CO_2 for various purposes, including increasing extraction from oil wells. The range of methods developed to capture the CO_2 are described by Bui *et al.* (2018). Two main techniques are being considered for the long-term storage of the captured CO_2.

- **Geological storage** involves injecting the captured CO_2, generally in supercritical liquid form, into underground geological formations. Depleted oil or gas fields or saline aquifers are typically proposed as alternatives. The injection of CO_2 is already used to enhance the extraction of oil and gas reserves. It is important that the storage formation has characteristics such as highly impermeable caprock that prevent the CO_2 from escaping to the surface.
- **Mineral storage** involves injecting CO_2 into naturally occurring rock masses which react with the CO_2 to produce stable carbonate minerals (e.g. calcite) in the rock mass discontinuities at depth. This process occurs naturally over geological time and can be enhanced to provide the same carbon capturing benefits over shorter timescales. The organisation 44.01, which plans to capture CO_2 in olivine-rich rocks in Oman, was one of the EarthShot prize winners in 2022 (EarthShot Prize, 2022). Basalt rock mineral storage is already being applied by the organisation CarbFix successfully in Iceland to capture and store CO_2 (CarbFix, 2022).

Underground space is also being assessed for storage of hydrogen, and for longer-term storage of nuclear waste.

11.2.1.9 Construction carbon

As stated in Section 11.1.2, it is important to seek to influence projects outside our own area of control – this would fall under the ‘avoid’ and ‘shift’ parts higher up the carbon-reduction hierarchy (**Figure 11.6**). However, there is also plenty that geotechnical engineers can do within the ‘improve’ part of the carbon-reduction hierarchy to reduce the whole-life carbon of construction. Some examples of this include

- engaging and incentivising the supply chain to provide lower-carbon solutions, including lower-carbon materials
- minimising journey distances for materials by using local yards and suppliers
- maintaining equipment and minimising engine idling
- connecting sites to the electric grid where possible
- investing in electric and more efficient equipment.

The European Federation of Foundation Contractors (EFFC) has published guidance suggesting ways to reduce whole-life carbon for foundation contractors (EFFC, 2022).

11.2.1.10 Measuring carbon

Carbon is emitted from extracting and fabricating materials, transporting them to site and constructing the materials into a project. Carbon is emitted from using, running and maintaining the project, from replacing parts or equipment and from end-of-life processing. Carbon may also be emitted or sequestered by changes of land use and changes to soil chemistry and biology. Applying a consistent breakdown of whole-life carbon into constituent contributions can help to ensure that important emissions contributions are not overlooked (for example, following RICS, 2017).

The maths for estimating whole-life carbon is simple: quantities (of materials or of energy use) are multiplied by appropriate emissions factors, and summed up. The difficulties come in selecting appropriate emission factors, deciding the system boundary for the carbon estimate, making appropriate assumptions and providing consistency of calculations. Various carbon measurement standards exist, such as the ISO 14040 series. Emission factors and carbon benchmarks can also be found from various sources – some commonly used UK sources are the Inventory of Carbon and Energy (ICE) database (Circular Ecology, 2019), DEFRA and BEIS factors (NAEI, 2022) and the RICS guidance (RICS, 2017). Carbon estimates can be undertaken using hand calculations, spreadsheets, benchmarks, BIM add-ons or specialist carbon assessment software tools.

Carbon estimates should be based on the best information available at the time, but should not be delayed by waiting for better information – timely estimates allow decision-making to happen at the most impactful time.

11.2.1.11 Cost of carbon

While the effects of global warming can be difficult to quantify, it can sometime be useful to assign carbon emissions a monetary value. The Department for Business, Energy and Industrial Strategy (BEIS) publishes recommended monetary values for carbon (BEIS, 2021). The cost of purchasing certified, quality, carbon offsets could also be taken into account. At the time of publication of this edition, carbon offset costs are generally lower than the BEIS carbon values; however, they are expected to rise as demand increases. Certified carbon offsetting should be considered a last resort after all other measures to reduce carbon have been taken, and carbon offsetting schemes should be chosen carefully to avoid unintended detrimental impacts to other areas of sustainability.

11.2.2 Circular economy

11.2.2.1 What is the circular economy?

Over the last few centuries the world’s economy has been based on a linear model, where resources are extracted, used and then

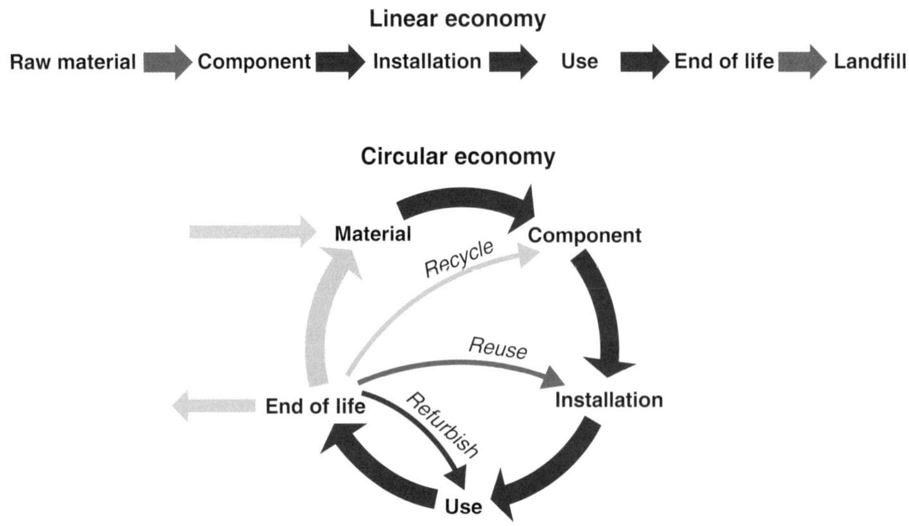

Figure 11.9 Linear economy against circular economy (From Luke Deamer)

disposed of. Due to this unsustainable linear model, in 2020 the amount of anthropogenic mass exceeded the weight of all global living biomass for the first time (New Capital, 2021). There are finite resources on the planet, so we should aim for a circular economy, where resources are kept in circulation at their highest value for as long as possible (**Figure 11.9**).

The circular economy is based on three main principles (Ellen MacArthur Foundation, 2013)

(*a*) eliminate waste and pollution

(*b*) circulate products and materials (at their highest value)

(*c*) regenerate nature.

Item (*c*) is covered in more detail in Section 11.2.4. Further information can be found at the Ellen MacArthur Foundation website (Ellen MacArthur Foundation, 2022) and in the Arup Circular Buildings Toolkit (Arup, 2022).

11.2.2.2 Eliminate waste and pollution

For geotechnical engineering, eliminating waste and pollution includes ground and groundwater remediation (Chapter 48 *Geo-environmental testing*). Eliminating pollution also encompasses reducing whole-life carbon (Section 11.2.1), and avoiding the use of hazardous materials. Waste reduction hierarchies, such as the '9Rs', can help to reduce and appropriately handle waste. As with all elements of sustainability, a holistic view needs to be taken – for example, materials used for retaining the ground could be minimised by way of ground freezing, but energy use would be high compared to installing a conventional retaining wall.

11.2.2.3 Earthworks strategy and soil reuse

The volume of material to be excavated on a project should be minimised at all stages as this, self-evidently, minimises carbon and waste. The earthworks strategy should be recorded in a Preliminary Earthworks Strategy Report and regularly updated as the project develops (Phear and Fraser, 2023). Sustainability opportunities could include

- reducing overall volumes of cut and fill and aiming to achieve a net balance, including consideration of ground improvement in situ
- maximising the on-site reuse of excavated materials
- minimising waste
- minimising mass-haul (earth-moving) distances
- increasing the resilience of the earthworks design
- adopting landscape and other regenerative environmental earthworks for carbon sequestration and net increase in biodiversity.

Excavated materials can be used to blend infrastructure projects into the local landscape terrain; for sound and visual screening; improvement of agricultural land; and creating the potential for ecological gain. The land-take required for environmental mitigation earthworks needs to be carefully justified (Hooton *et al.*, 2020). Topsoils and growing soils are particularly valuable, as explained in Section 11.2.4, so should be handled with care. Care should also be taken to ensure contaminated ground is not included in reuse without treatment.

When reuse of earthworks materials requires significant handling and processing, for example adding carbon-intensive quicklime to clay soils, then a holistic decision on sustainability needs to be taken (Hughes *et al.*, 2011). Reuse on site also significantly reduces landfill costs and reduces use of primary aggregates.

Quicklime with much lower embodied carbon is being developed and this has the potential to change the choices of earthwork solutions (including lime treatment) and their competitiveness. Research also shows that clays may be calcined for use as a cement replacement project for concrete. While still innovative at present, this is expected to come into mainstream use over the next few years.

11.2.2.4 Foundation reuse

When sites are reused there may be foundations left in situ from previous uses, or entire assets may be refurbished and adapted for different uses. On urban sites these foundation 'obstructions' can cause significant difficulties in locating space for new foundations. Where previous foundations exist, they can be considered for reuse in situ. This can save cost, carbon, materials and construction time compared to building new foundations. However, more resources may need to be allocated at early project stages for thorough investigation of existing foundations. The client's and insurer's attitudes to risk need to be considered.

Technical guidance on foundation reuse from early feasibility study, through record assessment, to site investigation, design and construction of any strengthening can be found in *Reuse of Foundations for Urban Sites* (Butcher *et al.*, 2006), CIRIA C653 *Reuse of Foundations* (Chapman *et al.*, 2007) and the IStructE 'Short Guide to Reusing Foundations' (Tayler, 2020). The guidance advocates

- a traffic-light system of red (high), amber (medium) and green (low) to represent risk
- the foundation reuse strategy, agreed with the client and insurers at an early project stage
- clear communication on allocation of risks
- development of a risk matrix and a risk register
- risk progressively reduced to low through design strategies, investigation and potentially strengthening works
- residual risk progressively monitored and closed out through inspection and testing as well as load/movement monitoring during construction
- latent defects insurance taken out.

There is ongoing work to standardise and store piling information to enable future foundation reuse. Sheet piles and steel pipe piles can also be extracted and reused or repurposed. Some sheet pile manufacturers offer a rental or buy-back schemes. Hollow piles can be extended for future reuse in situ.

11.2.2.5 Circular economy for geotechnical construction

The EFFC has published a guide to the circular economy for foundation contractors (EFFC, 2023). Actions include refurbishing and upgrading equipment, keeping meticulous construction records and constructing with reused materials.

11.2.3 Resilience and climate change adaptation

11.2.3.1 Providing resilience to climate change

The effects of climate change are already being felt around the globe. Existing infrastructure and buildings, and our future projects, will need to be made resilient to both acute weather, such as storms and floods, as well as chronic events such as drought and desertification. This will lead to demand for geotechnical designs to retrofit climate resilience to existing systems, and will also affect the assumptions needed for geotechnical projects. Future climate and weather data are available from the UK Met Office (Met Office, 2022a, 2022b).

As highlighted in Section 11.2.4.2, water is highly influential in how the ground behaves, and in the forces experienced by geotechnical structures. This means that climate change is expected to lead to displacements from swelling/desiccation, changing effective stresses, increased water flows (scour, piping, etc.), surface flooding and more. Stronger winds and greater temperature extremes may lead to higher loads acting on foundations. These factors will need to be taken into account in design. The expected future amount of global warming is not known, and local effects may be significantly different to global trends. Therefore judgement, and risk-based analysis, needs to be used in determining the appropriate scenarios to provide resilience for. Using a performance-based approach to designing mitigation may be appropriate.

11.2.3.2 Resilience and carbon

The whole-life carbon impacts of providing climate change resilience (and resilience to other factors such as earthquakes and other geohazards) also need to be considered. Not providing flood resilience, for example, could lead to greater carbon expenditure by triggering significant repairs after a flooding event – this is shown illustratively in **Figure 11.10** (dashed line). Alternatively, 'hard' flood defences could be built, requiring considerable investment of capital carbon but avoiding the possible future carbon from flood repairs (**Figure 11.10**, solid line). Or, it may be possible to provide flood resilience by other means, such as relocating critical infrastructure and providing nature-based solutions with 'soft' flood storage areas upstream that might even enable carbon sequestration (**Figure 11.10**, dotted line).

11.2.3.3 Resilience and earthworks

Always a principal consideration in the design and operational performance of earthworks and earth structures, resilience is increasingly important because of the vital need for infrastructure that can withstand the impacts of climate change (Phear and Fraser, 2023). Recent studies found that most earthworks failures are due to heavy or extreme rainfall and drainage or porewater problems (Lane *et al.*, 2020; Mair *et al.*, 2021). For earthworks and earth structures, resilience needs to be considered holistically together with drainage, vegetation management and whole-life carbon.

The main stages in a typical risk-based resilience assessment framework for earth structures are set out by Codd *et al.* (2023). Many infrastructure owners, such as National Highways (2020) and Network Rail (Mair *et al.*, 2021), have comprehensive earthworks asset management systems. Further guidance is available in the references.

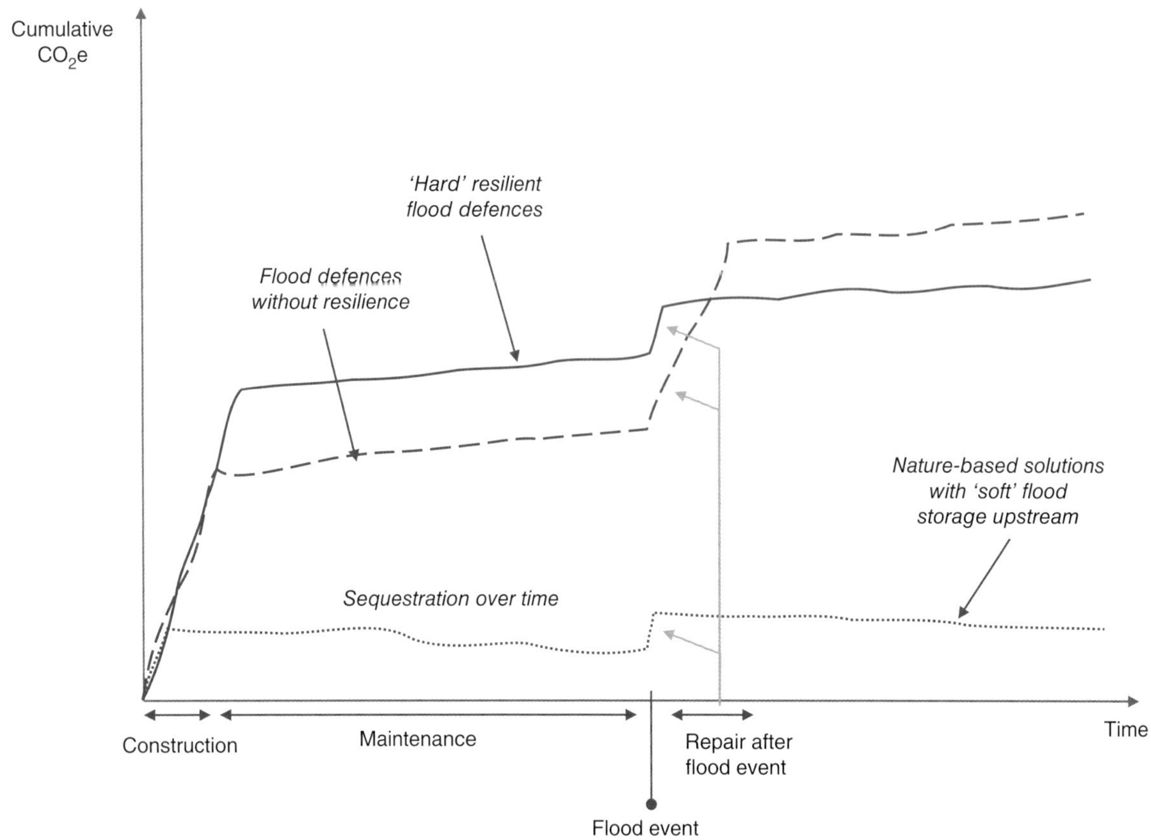

Figure 11.10 Conceptual diagram illustrating the relationship between resilience and carbon for different flood protection schemes, with different approaches to resilience

11.2.4 Ecology, ground health and biodiversity

11.2.4.1 Restore and regenerate

With their understanding of, and influence on, the ground and groundwater, geotechnical engineers are uniquely placed to be able to protect, restore and regenerate nature. **Figure 11.11** illustrates various areas of influence.

11.2.4.2 Water

Water is an integral part of geotechnical engineering. Changes in groundwater result in ground movements and impacts to geotechnical structures. One of the main concerns for a geotechnical project should be its impact on the wider aquatic system and the balance of that system, and should consider the following.

- **Sustainable drainage systems (SuDs):** Water drainage methods that replicate natural infiltration systems rather than conventional hard covering that inhibits natural water infiltration into the underlying aquifer.
- **Aquifer storage and recovery:** A method of supplementing or replacing natural infiltration into an aquifer by storing water in an aquifer at times of surplus, for use at times of scarcity (Jones *et al.*, 1998).
- **Water in construction:** Sustainable construction practices should aim to minimise water consumption by optimising water usage during construction and operation, and ensuring that the water quality for discharge is appropriate for the receiving aquatic environment.
- **Design for reduced operational energy:** For example, use sustainable alternatives to eliminate the need for permanent dewatering.
- **Design for reduced water consumption in a design:** For example, consider 'dry' solutions such as surcharge consolidation, dry soil mixing or ground improvement over water-intensive solutions such as jet grouting.

11.2.4.3 Earthworks

Earthworks, if planned, designed and constructed appropriately, can assist in achieving biodiversity net gain on infrastructure projects. To achieve sustainable development requirements, planners of infrastructure projects will increasingly need to look to landscape-scale solutions. For example, the Keyn Glas A30 trunk road upgrade featured individual environmental enhancement projects up to 3 km from the road, which resulted in up to 250% net gain in biodiversity (Oakman and Wood, 2022).

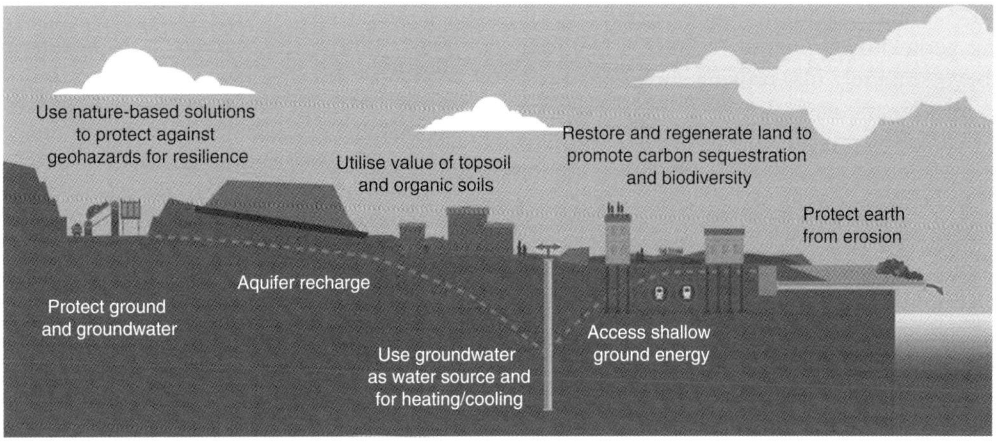

Figure 11.11 Some activities geotechnical engineers can do to aid sustainable development (Courtesy of Arup)

11.2.4.4 Soils health

Life on Earth depends on soil. Healthy soil provides a growing medium for food, it aids resilience, it conducts, stores, filters and retards flow of water and it is the foundation of terrestrial biodiversity. Soil has a vast potential to store carbon, both in trapped organic material (organic carbon) and through chemical reactions with soil minerals (inorganic carbon). Understanding soil health and how to promote and protect healthy soils is an area of growing research. This includes

- soil science, focused on interactions between ecosystems and soils
- the carbon sequestration capacity of soils
- the importance of land use and land-use change
- regenerative land management (e.g. peatland restoration or resurfacing culverted watercourses)
- how to quantify and test soil health
- the role of soil in flood, drought and fire resilience
- creation of topsoil (e.g. through composting or animal husbandry)
- the role of soil in water quality improvement
- maintaining and improving nutrient balance and biogeochemical flows.

11.2.4.5 Topsoils

Recent European legislation has recognised the importance of topsoils and this will change the way they are used in civil engineering and construction. Careful soil management is an essential element of both sustainable earthworks and sustainable agriculture. The UK Department for the Environment, Food and Rural Affairs (DEFRA) code of practice for the sustainable use of soils on construction sites (DEFRA, 2009) summarises good practice on this subject. Stripping, stockpiling and replacing topsoils needs to be carried out carefully and in a way that enables them to continue their main function as growing mediums.

Growing soils hold much more carbon than the atmosphere (Environment Agency, 2019). Disturbing them releases a lot of carbon into the atmosphere and efforts should be made to limit this. The ability of growing soils to sequester carbon from the atmosphere will also become a function that will need to be increasingly considered on projects.

11.3. What can I do, personally?

Sometimes it can feel that as a geotechnical engineer, rather than a masterplanner or policymaker, our ability to influence the sustainability of our projects can be limited, and it can feel that the constraints of our projects do not allow us to influence the sustainability beyond creating an efficient design/construction.

However, the following are within the control of geotechnical engineers, and can be the catalyst for important change.

- Ask questions about project decisions, and seek to influence outside your official role/remit.
- Find out what your client has publicly said they want to achieve in terms of sustainable development and then question how your work contributes to that.
- Estimate whole-life carbon impacts so that decisions can be properly informed.
- Consider resilience in your work.
- Obtain sufficient data to manage risk without oversizing.
- Share good practice and lessons learned, and keep up to date with developments.
- Engage with stakeholders.
- Aim for the most efficient design and construction possible.
- But think of the big picture, including appropriate future-proofing, rather than just concentrating on shrinking your own part.

The climate and biodiversity crises are some of the greatest challenges to face society to date. Unchecked, they will have a detrimental effect on many of the other UN SDGs, including those relating to social and economic improvement. Civil engineers, including geotechnical engineers, have a key role to play in moving beyond merely maximising social and economic benefits and minimising environmental impact within their project scope, to creating projects that have a net positive, regenerative effect.

References

Arup (2020) A better way: our net zero carbon strategy. https://www.arup.com/-/media/arup/files/policies/a-better-way-our-net-zero-carbon-strategy.pdf (accessed 01/12/2022).

Arup (2022) Circular Buildings Toolkit. https://www.arup.com/services/climate-and-sustainability-services/circular-economy-services/circular-buildings-toolkit (accessed 01/12/2022).

BEIS (2021) Valuation of energy use and greenhouse gas: supplementary guidance to the HM Treasury Green Book on Appraisal and Evaluation in Central Government. Guidance document and dataset. See https://www.gov.uk/government/publications/valuation-of-energy-use-and-greenhouse-gas-emissions-for-appraisal (accessed 01/12/2022).

BSI (British Standards Institution) (2016) PAS 2080:2016: Carbon management in infrastructure. BSI, London, UK.

BSI (2023) PAS 2080:2023: Carbon management in infrastructure. BSI, London, UK.

Bui M, Adjiman C, Bardow A, *et al.* (2018) Carbon capture and storage: the way forward. *Energy & Environmental Science* **11(5)**: 1062–1176.

Butcher AP, Powell JJM and Skinner HD (2006) *Reuse of Foundations for Urban Sites: A Best Practice Handbook*. IHS BRE Press, Bracknell, UK.

Cabinet Office (2021) PPN 06/21: Technical standard for completion of carbon reduction plans. Cabinet Office, London, UK.

CarbFix (2022) https://www.carbfix.com (accessed 01/12/2022).

Chapman T, Anderson S and Windle J (2007) *Reuse of Foundations*. CIRIA Report C653. CIRIA, London, UK.

Circular Ecology (2019) ICE Carbon Database V3.0. https://circularecology.com/embodied-carbon-footprint-database.html#.XKX_oJhKhPY (accessed 01/12/2022).

Civil Engineers Declare (2019) https://www.civilengineersdeclare.com (accessed 01/12/2022).

Codd J, Pritchard O and Mian J (2023) A resilience assessment framework for England's strategic road network. *Proceedings of the BGA Geo-resilience Conference, Cardiff, UK*.

Construction Innovation Hub (2022) https://constructioninnovationhub.org.uk/value-toolkit (accessed 01/12/2022).

DEFRA (2009) Construction code of practice for the sustainable use of soils on construction sites. DEFRA, London, UK. See https://www.gov.uk/government/publications/code-of-practice-for-the-sustainable-use-of-soils-on-construction-sites (accessed 20/04/2023).

EarthShot Prize (2022) https://earthshotprize.org/winners-finalists/44-01 (accessed 01/12/2022).

EFFC (European Federation of Foundation Contractors) (2022) *Guide No.1 Carbon Reduction*. Sustainability Guide. https://www.effc.org/content/uploads/2022/05/EFFC-Carbon-Reduction-Guide_FINAL.pdf (accessed 20/04/2023).

EFFC (2023) *Circular Economy*. Sustainability Guide. https://www.effc.org/effc-circular-economy-guide/ (accessed 05/06/2023).

Ellen MacArthur Foundation (2013) Restorative and regenerative: exploring the concepts in the circular economy. *Journal of Industrial Ecology* **24(4)**: 763–773.

Ellen MacArthur Foundation (2022) *What Is a Circular Economy?* https://ellenmacarthurfoundation.org/topics/circular-economy-introduction/overview (accessed 01/12/2022).

Environment Agency (2019) *The State of the Environment: Soil*. Environment Agency, Bristol, UK, Report.

HMG (Her Majesty's Government) (2019) *Climate Change Act 2008 (2050 Target Amendment) Order 2019*. The Stationery Office, London, UK.

HMG (2021) *Environment Act 2021*. The Stationery Office, London, UK.

Hooton M, Doughty L, Newton L and Hutchison S (2020) *Minimisation of Spoil to Landfill from a Collaborative Approach to Re-Use on Site*. https://learninglegacy.hs2.org.uk/document/minimisation-of-spoil-to-landfill-from-a-holistic-approach-to-re-use-on-site (accessed 20/04/2023).

Hughes L, Phear A, Nicholson D, *et al.* (2011) Carbon dioxide from earthworks: a bottom-up approach. *Proceedings of the Institution of Civil Engineers, Civil Engineering* **164**: 1–7.

IPBES (Intergovernmental Science-Policy Platform on Biodiversity and Ecosystem Services) (2002) https://ipbes.net (accessed 01/12/2022).

IPCC (Intergovernmental Panel on Climate Change) (2021) Summary for policymakers. In *Climate Change 2021: The Physical Science Basis. Contribution of Working Group I to the Sixth Assessment Report of the Intergovernmental Panel on Climate Change*. Cambridge University Press, Cambridge, UK, pp. 3–32.

IPCC (2022) Summary for policymakers. In *Climate Change 2022: Impacts, Adaptation and Vulnerability. Contribution of Working Group II to the Sixth Assessment Report of the Intergovernmental Panel on Climate Change*. Cambridge University Press, Cambridge, UK, pp. 3–33.

Jones HK, MacDonald DMJ and Gale IN (1998) *The Potential for Aquifer Storage and Recovery in England and Wales*. Environment Agency, London, UK, Report.

Keller (2022) *Annual Report*. https://investors.keller.com/~/media/Files/K/Keller-IR/reports/2022rep/ar2022/keller-ar22.pdf (accessed 10/05/2023).

Lagesse R, Hambling J, Gill J, Lim C and Ingvorsen P (2022) The role of engineering geology in delivering the United Nations Sustainable Development Goals. *Quarterly Journal of Engineering Geology and Hydrogeology* **55(3)**: 1–16.

Lane M, Halstead K, Power C, Spink T and Bailey A (2020) Establishing and quantifying the causal linkage between drainage and earthworks performance for Highways England. *Quarterly Journal of Engineering Geology & Hydrogeology* **53**: 333–338.

Mair R, *et al.* (2021) *A Review of Earthworks Management*. Network Rail, London, UK.

Met Office (2022a) UK Climate Projections (UKCP). https://www.metoffice.gov.uk/research/approach/collaboration/ukcp (accessed 01/12/2022).

Met Office (2022b) UK Climate Projections User Interface. https://ukclimateprojections-ui.metoffice.gov.uk/ui/home (accessed 01/12/2022).

NAEI (National Atmospheric Emissions Inventory) (2022) Emission Factors for Transport. https://naei.beis.gov.uk/data/ef-transport (accessed 01/12/2022).

National Highways (2020) *Managing the Maintenance of Highway Geotechnical Assets*. National Highways, Bristol, UK, Report CS 641.

New Capital (2021) *Visualizing the Accumulation of Human-Made Mass on Earth*. https://www.newcapitalmgmt.com/news/visualizing-the-accumulation-of-human-made-mass-on-earth (accessed 01/12/2022).

Oakman B and Wood P (2022) Keyn Glas, Cornwall, UK: a green legacy. *The Arup Journal* **1/2022**: 28–33.

Pantelidou H, Stephenson S, Alexander J and Sturt R (2017) Designing tunnels for whole life value. *Railway Engineering*, 10.25084/raileng.2017.0124.

Phear A and Fraser N (2023) Developing a new European standard on sustainable earthworks. *Proceedings of the BGA Georesilience Conference, Cardiff, UK*.

RICS (Royal Institution of Chartered Surveyors) (2017) *Whole Life Carbon Assessment for the Built Environment*. RICS, London, UK.

Stockholm Resilience Centre (2009) Planetary boundaries: exploring the safe operating space for humanity. *Ecology and Society* **14(2)**: 32. https://www.ecologyandsociety.org/vol14/iss2/art32/ (accessed 20/04/2023).

Tayler H (2020) A short guide to reusing foundations. *The Structural Engineer*, 10.56330/GYPR7761

TCFD (Task Force on Climate-related Financial Disclosures) (2022) *2022 Status Report*. TCFD, Basel, Switzerland. https://assets.bbhub.io/company/sites/60/2022/10/2022-TCFD-Status-Report.pdf (accessed 20/04/2023).

UK Government (2021) *The Economics of Biodiversity: The Dasgupta Review*. HM Treasury, London, UK.

UN General Assembly (2015) Resolution adopted by the General Assembly on 25 September 2015: Transforming Our World: The 2030 Agenda for Sustainable Development. https://www.refworld.org/docid/57b6e3e44.html (accessed 01/12/2022).

United Nations (2015) Adoption of the Paris Agreement. 21st Conference of the Parties. United Nations, Paris, France.

World Bank (2022) Urban Development. https://www.worldbank.org/en/topic/urbandevelopment/overview (accessed 01/12/2022).

Further reading

Alonso I, Weston K, Gregg R and Morecroft M (2012) Carbon Storage by Habitat: Review of the Evidence of the Impacts of Management Decisions and Condition on Carbon Stores and Sources. Natural England, York, UK, Research Report NERR043.

Armour T (2003) Channel Tunnel Rail Link section 1: landscape design. *Proceedings of the Institution of Civil Engineers, Civil Engineering* **156**: 54–59.

CL:AIRE (2011) Definition of Waste: Development Industry Code of Practice. Version 2. http://www.claire.co.uk/cop (accessed 20/04/2023).

Highways Agency (1995) *Earthworks: Design and Preparation of Contract Documents*. Departmental Advice Note. Highways Agency, Bedford, UK.

IEA (International Energy Agency) (2019) *2019 Global Status Report for Buildings and Construction*. IEA, London, UK.

IEA (2022a) https://iea-gia.org/ (accessed 01/12/2022).

IEA (2022b) *2021 Geothermal Annual Report*. IEA, London, UK.

McMillan P, Harber A and Nettleton I (2011) *Rock Engineering Guides to Good Practice: Rock Slope Remedial and Maintenance Works*. Transport Research Laboratory, Wokingham, UK.

Mellings L and Limna G (2017) *Crossrail: Excavated Materials Story*. Crossrail Learning Legacy. https://learninglegacy.crossrail.co.uk/documents/excavated-materials-story (accessed 20/04/2023).

Noakes AJ, Mason-Jarvis LF, Taylor GR and Evans E (2020) Geospatial assessment methods for geotechnical asset management of legacy railway embankments. *Quarterly Journal of Engineering Geology and Hydrogeology* **53**: 339–348.

Perry J, Pedley M and Brady K (2003a) *Infrastructure Cuttings: Condition Appraisal and Remedial Treatment*. CIRIA Report C591. CIRIA, London, UK.

Perry J, Pedley M and Reid M (2003b) *Infrastructure Embankments: Condition Appraisal and Remedial Treatment*. CIRIA Report C592. CIRIA, London, UK.

Spink T, Duncan I, Lawrence A and Todd A (2013) *Transport Infrastructure Drainage: Condition Appraisal and Remedial Treatment*. CIRIA Report C714. CIRIA, London, UK.

All chapters within Sections 1 *Context* and 2 *Fundamental principles* together provide a complete introduction to the Manual and no individual chapter should be read in isolation from the rest.

Section 2: Fundamental principles

Section editors: **William Powrie and John B Burland**

doi: 10.1680/icemge.66816.0157

Chapter 12
Introduction to Section 2

William Powrie University of Southampton, Southampton, UK
John B Burland Imperial College London, London, UK

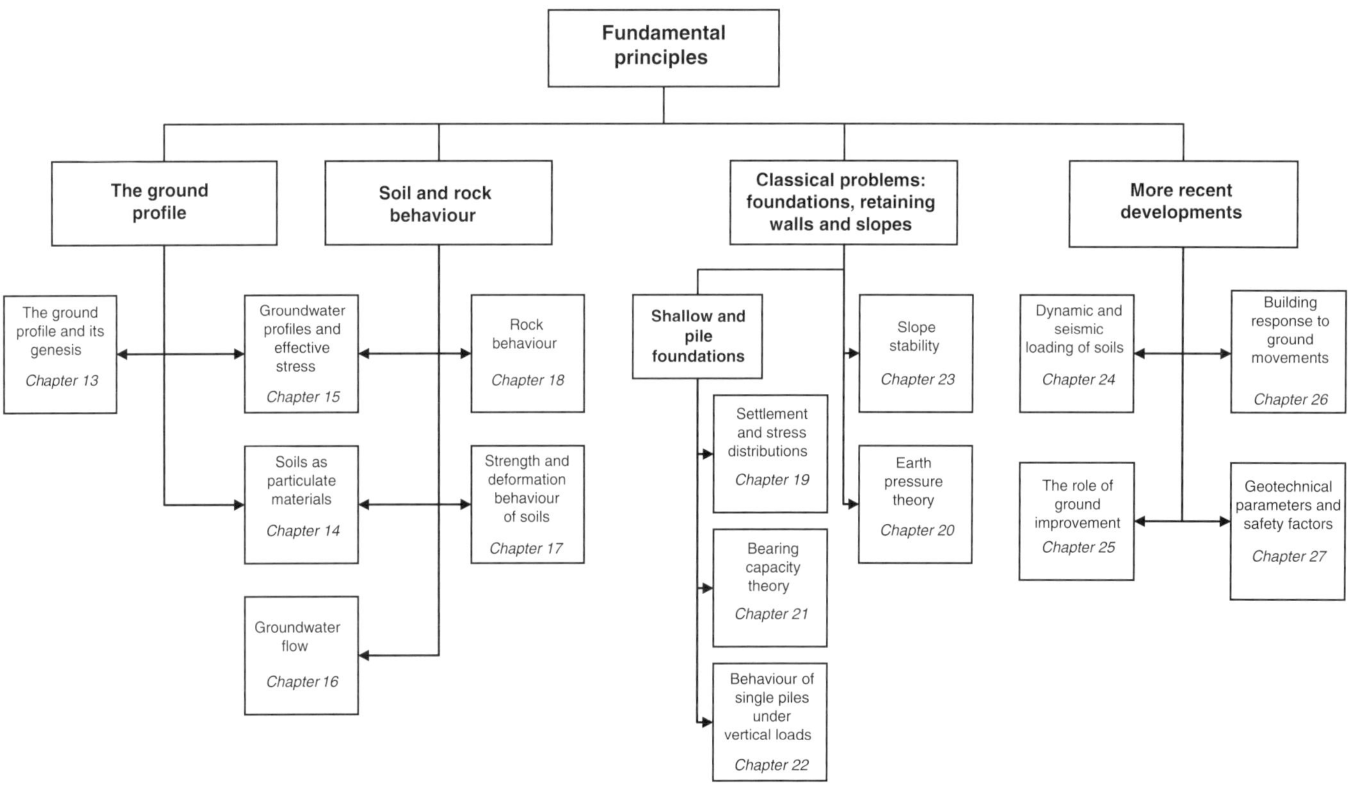

Figure 12.1 Layout of chapters in Section 2

Figure 12.1 outlines the layout and contents of Section 2 *Fundamental principles*.

Section 2 serves as an aide-memoire to the fundamental principles and theories underlying geotechnical engineering, from characterising the ground, through the behaviour of geomaterials to the basic conceptual models used in the analysis and design of geotechnical engineering structures.

The first six chapters (Chapters 13 *The ground profile and its genesis* to 18 *Rock behaviour*) are concerned with characterising the ground and the fundamental aspects of material behaviour. Chapter 13 *The ground profile and its genesis* describes and discusses the development of the ground profile from a geological perspective, and how this knowledge may be used to inform the development of a conceptual ground model for use in geotechnical engineering. Chapter 14 *Soils as particulate materials* explains the factors that make soil, as a particulate material, fundamentally different from many other civil engineering materials such as steel and concrete.

Without groundwater, geotechnical engineering would be much simpler – but also less of a challenge and much less interesting. In many parts of the world, the soils of interest to the geotechnical engineer are saturated or nearly so; thus an understanding of pore water pressures and groundwater flow is essential. Chapter 15 *Groundwater profiles and effective stresses* describes the estimation of pore water pressure and effective stress profiles in the ground, while Chapter 16 *Groundwater flow* discusses the calculation of flow rates and changes in pore water pressure induced by construction activities such as excavation. Chapters 17 *Strength and deformation behaviour of soils* and 18 *Rock behaviour* summarise the important aspects of soil and rock behaviour, and the conceptual and mathematical models commonly used to characterise and quantify them.

The next five chapters (Chapters 19 *Settlement and stress distributions* to 23 *Slope stability*) address the theory underlying the three classic problems in geotechnical engineering: foundations (Chapters 19 *Settlement and stress distributions*, 21 *Bearing capacity theory* and 22 *Behaviour of single piles under vertical loads*), retaining structures (Chapter 20 *Earth pressure theory*) and slopes (Chapter 23 *Slope stability*).

The last four chapters (Chapters 24 *Dynamic and seismic loading of soils* to 27 *Geotechnical parameters and safety factors*) address a range of topics representing arguably more recent developments in geotechnical engineering. The dynamic and seismic loading of soils (Chapter 24 *Dynamic and seismic loading of soils*) is generally considered more difficult than the static behaviour, but for certain applications and in many parts of the world cannot be ignored. Ground improvement (Chapter 25 *The role of ground improvement*) encompasses a range of techniques that geotechnical engineers use to improve certain properties of the ground. These often require an understanding of chemical or electrochemical phenomena in addition to mechanics. Given the uncertainties involved in applying these to the ground, they represent to many an area still closer to an art than a science. Understanding and predicting the response of buildings to ground movements (Chapter 26 *Building response to ground movements*) has become increasingly important as more and more ambitious construction projects involving open excavations and tunnels are undertaken in highly built-up areas. This need, combined with the development of computing power enabling complex and sophisticated three-dimensional numerical analysis to be carried out, has driven a major area of advance in geotechnical knowledge and understanding in the last decade or so. Finally, Chapter 27 *Geotechnical parameters and safety factors* addresses the choice of design parameters and factors of safety. This is a topic of huge importance, about which there is often a lack of coherence and clarity, partly because it is difficult to revise customs and codes of practice in a way that keeps pace with developments in geotechnical engineering knowledge and understanding.

Section 2 sets out to cover a huge range of soil and rock mechanics and geotechnical engineering knowledge. Inevitably, it can serve as no more than a reminder for those already familiar with a topic, or an introduction for those who are not. The following bibliography may be helpful to those wishing to find out more about any of the variety of subjects covered.

Further reading

Engineering geology (Chapter 13)
Waltham AC (2009) *Foundations of Engineering Geology*, 3rd edn. Spon Press, London, UK.

Fundamental soil mechanics (Chapters 14–17 and 19–22)
Atkinson JH (2007) *The Mechanics of Soils and Foundations*, 2nd edn. Spon Press, London, UK.
Muir Wood D (1990) *Soil Behaviour and Critical State Soil Mechanics*. Cambridge University Press, Cambridge, UK.
Muir Wood D (2009) *Soil Mechanics: A One-dimensional Introduction*. Cambridge University Press, Cambridge, UK.
Powrie W (2018) *Soil Mechanics: Concepts and Applications*, 3rd edn. CRC Press, Boca Raton, FL.

Rock mechanics (Chapter 18)
Jaeger JC, Cook NGW and Zimmerman RW (2007) *Fundamentals of Rock Mechanics*, 4th edn. Wiley-Blackwell, Oxford, UK.

Slope stability (Chapter 23)
Bromhead EN (1992) *The Stability of Slopes*, 2nd edn. Blackie Academic & Professional, London, UK.

Soil dynamics and seismic loading (Chapter 24)
Jefferies M and Been K (2006) *Soil Liquefaction: A Critical State Approach*. Taylor & Francis, London, UK.
Verruijt A. (2010) *An Introduction to Soil Dynamics*. Springer, Dordrecht, the Netherlands.

Ground improvement (Chapter 25)
Moseley MP and Kirsch K (2004) *Ground Improvement*, 2nd edn. Spon Press, London, UK.

doi: 10.1680/icemge.66816.0159

Chapter 13
The ground profile and its genesis

Michael H de Freitas Imperial College London, London, UK, and Director of First Steps Ltd, London, UK

A ground profile, or vertical profile, is the starting place for developing an understanding of the ground for design, analyses and construction. It is a record of the ground at depth and an interpretation of its formation up to the present. The correct identification of the materials within a profile and interpretation of their history indicates how ground is likely to respond to changes arising from construction or from natural processes such as rainfall and earthquakes. Decisions based on a profile that is incorrect are unlikely to be corrected later by subsequent calculations, no matter how sophisticated these calculations may be, because they will be quantifying something that does not exist. Mistakes made with the ground profile can delay works and even stop contracts and may result in claims. Profiles enable cross-sections to be constructed and these are the templates upon which the distribution and magnitude of all other geotechnical data should be superimposed. That is the surest way to obtain a relationship between the geotechnical properties for the materials present and their geology: many case histories demonstrate this. Case histories are part of the inherited wisdom of geotechnical engineering and remove any excuse for either ignoring a ground profile or for obtaining one but failing to use it appropriately in design and construction.

CONTENTS

13.1. Overview

A ground profile, also called a vertical profile, describes the ground beneath a particular location. It lists the materials encountered with depth, as would be intersected by a borehole or trial pit, and then uses this and other data to interpret how the profile was formed and how it might behave in response to engineering within it. This response will reflect the character of the materials present, their amounts and the environment under which they exist in that profile, by which is meant the vertical and horizontal in situ magnitudes of effective stress, and chemical quality of the liquid and gas phases present. A ground profile therefore has three basic components: (*a*) the vertical sequence, which can be sampled, seen and touched and provides evidence of material types and quantities; (*b*) the environment, which is not so easy to see and touch, as stress is invisible and water may be invisible too, and generally has to be observed using instruments such as piezometers and standpipes, and devices for measuring stress or strain; and (*c*) the interpretation of these data.

The vertical sequence usually starts at ground level and extends to whatever depth is deemed necessary for the geotechnical work for which it was requested. The age of the materials encountered increases with depth, so a profile represents the history of the ground, with the most recent part of that history being at ground level. The correct interpretation of the materials and history of a profile and its environment can indicate how the ground is likely to respond to engineering.

There are many forms of profile; all are crucial to decisions taken for geotechnical design, analyses and construction. If the description of the materials with depth is wrong or is missing important data, then decisions based on it will be wrong no matter how correct and accurate subsequent calculations may be. This is because these calculations will be quantifying something that does not exist. Mistakes made with the ground profile can be so seriously wrong that they may result in groundwork being stopped and contracts being delayed, sometimes even being abandoned.

Having established the profile at one location, it is usually necessary to compare it with profiles at other locations nearby, so that a series of profiles can be related to each other. There are hazards associated with such a seemingly simple task. The outcome of such comparisons is a depiction of the site geology in 3D. Rarely can this be done without considering how each profile has been formed – that is, its genesis – because design, analysis and construction come from understanding what has happened to the ground in the past.

Thus, a ground profile is not just the sequence of soils and rocks, but incorporates an understanding of what has happened to these during their formation, their subsequent history to the present day and the environment they are now under. Aspects of this are explained in this chapter because understanding the formation, history and environment of a profile is one of the key tasks for an engineering geologist to complete. Working with appropriate material descriptions is critical, and using a scale appropriate for relating the ground to the intended engineering project is necessary. The sections, maps and 3D depictions so created become the basis upon which all other geotechnical data should be superimposed; only in this way can the distribution of geotechnical properties reflect the materials present and their geological history (de Freitas, 2009).

The acquisition and use of geotechnical data that has not been related to the materials and history to which it relates can easily result in substandard geotechnical engineering; there are many case histories that demonstrate this. Case histories are part of the inherited wisdom of geotechnical engineering, and they remove any excuse for either ignoring a ground profile or for obtaining it but failing to use it appropriately.

13.2. The ground profile

Consider the excavation and design for a basement foundation, 10 × 15 m in plan, with its formation level 7 m below ground level, the ground for which has been investigated using five boreholes each taken to a depth of 15 m and each used to recover a 100 mm dia. core. If the core recovery is 100%, the volume of ground seen is less than 0.01% of the ground to be excavated and considerably less than that of the material relied upon to carry the foundation loads beneath the excavation for the next 60–100 years. Thus, the commercial investment that has facilitated this construction is based on virtually no knowledge. Such simple calculations demonstrate the value that must be placed on these tiny pieces of information and the need for this information to be gathered with care; this is the subject of ground investigation and the reason why ground investigation now has protocols that should be followed (e.g. Eurocode 7; BSI, 2004). Retrieving a sample of ground and describing it so that it reflects the condition of the ground from which it came are fundamental steps towards creating a ground profile; they require skill, knowledge and experience. However, such a log, by itself, is insufficient; its data must be used for design and analyses, and the first step in using it this way is to generate a ground profile.

Two logs are shown in **Figure 13.1**, one from London and the other from Hong Kong; they illustrate that a description of a vertical sequence is the starting point for a ground profile. Note that they provide little direct information about either the origin or the history of the ground represented by the log; this comes later. Thus, ground profiles require either a log of borehole cores or a log of the boreholes themselves, or their equivalent from a quarry, trial pit or other excavation. Such a profile refers to the ground beneath the point on the surface of the earth below which it is made, but may also be relevant to locations nearby if the ground does not vary significantly from place to place.

Such logs are used for many purposes (e.g. basic design of subsurface structures, geotechnical analyses, deciding methods of construction, installation of monitoring equipment, assessment of material volumes). Thus, logs must provide as accurate a record as possible of the ground as it exists in situ. This can be considered the most basic of ground profiles.

Logs are the starting point for developing a fuller ground profile because the information they contain is used to

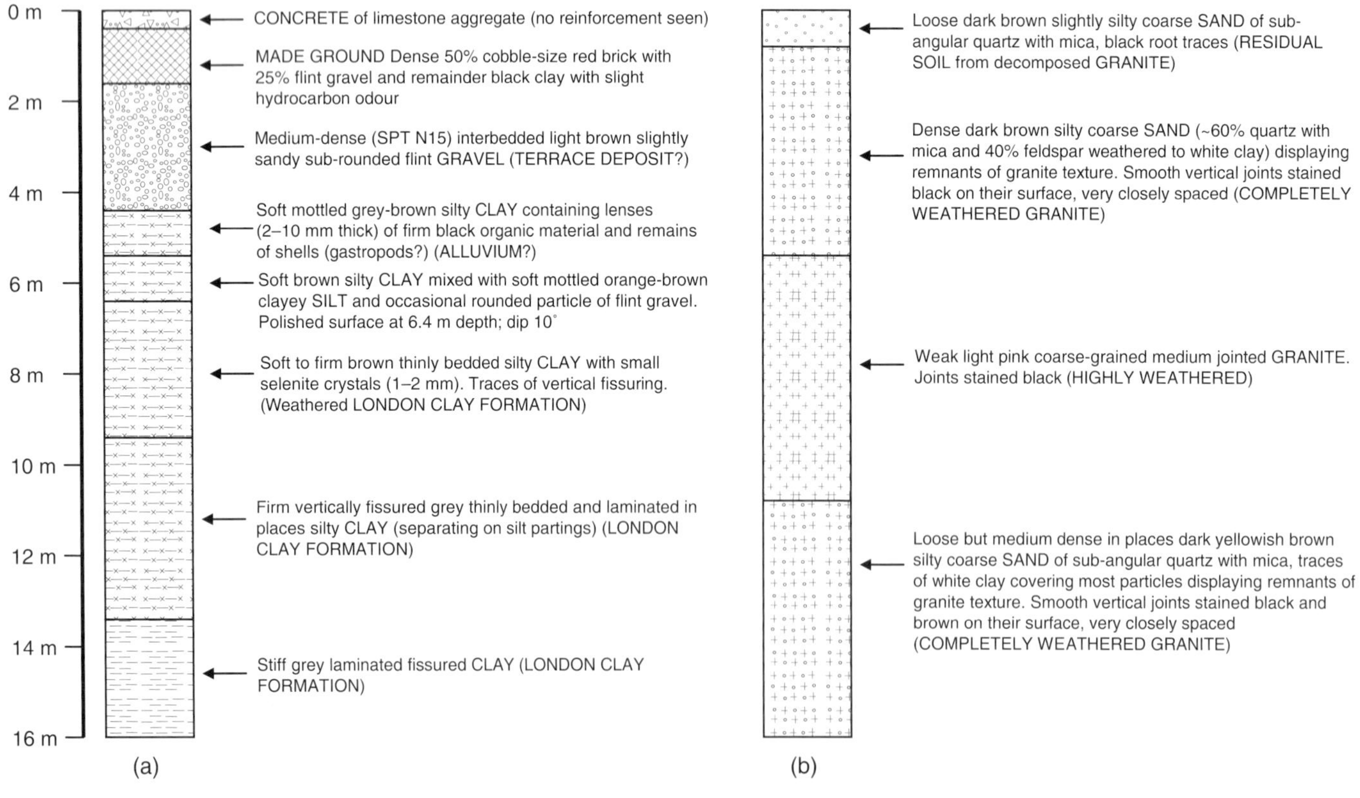

Figure 13.1 Borehole logs from (a) London and (b) Hong Kong illustrating vertical sequences formed mainly by sedimentation and in situ weathering of pre-existing granite. Details of recovery, water returns, etc., are omitted

understand how that vertical sequence will respond to the particular changes that engineering work is likely to create. Such logs also indicate where further studies of the ground are required (e.g. water levels, strengths and permeabilities). The interpretation of logs therefore requires care and attention, and this improves with experience. Profiles that must be created from the logs of others present different problems from those created by personal observation. The usefulness of a profile created from the records of others depends upon the extent to which those records are a relevant, complete and correct representation of the ground – that is one of the major reasons why protocols are now provided and should be followed, because they enable a common language of descriptors to be used.

The most obvious characteristics to observe are the sequence that is present in the log and the materials the sequence contains, as shown in **Figure 13.1**. The less obvious, yet sometimes crucial, characteristics to note are the fabric of the materials, the structure of their layers and the nature of their boundaries (i.e. their interfaces). Even more subtle is recognising what is missing; this is not usually a problem when a vertical exposure exists, as in the face of a quarry, cliff or trial pit, but becomes of crucial importance when core recovery is not 100% and the core recovered is no longer representative of the condition in situ. All these aspects are static in character – they are either present or not present in the profile. However, most profiles have a component that is not static but can come and go with time. Liquids and gases are the most obvious of these; the free face that exposes a profile and enables it to be observed in the side of a trial pit, shaft, quarry or cliff can affect the presence of liquids and gases and their location in the ground.

Water will drain towards a free face if its level on the free face side is lower than that in the ground behind the face, so the level of water recorded on the free face may not be that further back in the ground. Air will replace the voids drained of water, and as this is rich in oxygen it will tend to chemically change the ground with which it comes into fresh contact. Soil and rock, and surfaces cutting them which may be black in ground devoid of oxygen, may become yellow-brown, brown or even whitish in the presence of oxygen-rich water and atmospheres. Drainage can turn dark browns to light browns, and turn surfaces that are solid when wet, powdery when dry. Other gases, such as methane, and other organic-rich volatile material, which are all pertinent to the geotechnical character of the ground, may escape if present. All of these are changes that occur with time, but they need not be the only ones to occur. Any excavation, natural or man-made, removes the vertical and lateral loads that were on the ground and the ground will respond to their loss by relaxing towards the free face that has been created. This relaxation can itself create many changes to the original profile, notably the amount of fissuring and jointing, and a general loss of in situ density, and the magnitude and extent of these changes depends on the time they have had to develop.

A good log is thus essential for developing a ground profile for use in design and analysis; some aspects of the ground can be captured by a photograph, and it is now good practice to obtain an image of the core or exposure that has been logged. Some aspects cannot be captured; they need to be deduced from considering what can be seen, what has changed and what has been lost. The geographical location of the profile also needs to be recorded (to the nearest decimetre) on the surface of the Earth. The elevation of the datum from which the profile has been measured is also required, at least to the nearest decimetre, and if possible to the nearest centimetre. The age of the exposure or borehole core upon which the profile is to be based is also relevant and should be recorded, or if unknown, this should be noted. Finally, the profile needs a complete description. Should any one of these pieces of information be missing the profile is incomplete; that means the geotechnical engineering intended at the site will begin in ignorance of essential information. Its absence will call into question the geotechnical competence of those involved.

A checklist of basic requirements for achieving a vertical profile is provided in **Table 13.1**. The task has been divided for convenience into two phases: an initial vertical profile which is then augmented by reference to other related data. The initial profile essentially captures the 'static' aspects of what is below ground level, and the augmented profile essentially captures the 'dynamic' aspects – that is, how the ground has responded to unloading, and in the case of boreholes how its fluids have responded to the creation of a hydraulic conduit.

13.3. Importance of a profile

Section 1 *Context* has set geotechnical engineering in context; it is a branch of engineering that came from the need to engineer in soil and rock. Experience in this field is recorded as case histories which describe what happened in certain circumstances, even if the reason for the ground's response was not understood at the time. Geotechnical engineering uses the knowledge of the ground and of its mechanics in the way described in Chapter 4 *The geotechnical triangle* – see **Figure 4.1**. The ground profile forms the apex of the triangle and describes the material and its history to which the 'well-winnowed' case histories of the triangle refer. It is easy to appreciate why this is so. Ground has four attributes of direct relevance to geotechnical work: its strength (the loads it can carry), deformability (the deformations those loads will generate), conductivity (the flows that will be permitted) and durability (the alterations that will occur as a result of changing the environment under which the ground existed before engineering began). All this information comes from a ground profile, either directly by testing and monitoring the ground, or indirectly by calculation and analogy. The success with which this is done is a function of the quality of the ground profile. A basic log is the minimum required; one that has been put into context and interpreted is better (**Table 13.1**).

Boreholes: initial vertical profile	
Geologist's log	Materials and boundaries
Driller's log	Method, progress, casing depth, sampling, strikes for groundwater/gas, drilling fluid returns, in situ tests (e.g. SPT)
Geographical location	Essential
Vertical or inclined	Angle of inclination and direction relative to north
Date of drilling	Essential
Date of logging	Essential
Assessment of above data	
• Core recovery	Compare logs with photos of core and drilling progress
• What may have been lost	Consider core loss
• Larger geological setting	Compare with maps of geology and topography
• Presence of groundwater	Dampness or visible water
Boreholes: augmented vertical profile	
BH observations	
• Instrumental	Water levels (standpipe or piezometer), gas
• Other logs	Geophysical/geochemical
Assessment of above data by considering	
• Time-dependent changes with drilling	Results from in situ testing, BH logging and lab testing
• Time-dependent changes since drilling	Water level fluctuations, gas flows
• Relationship of present condition with initial assessment	Compare with initial assessment
Excavated or natural exposures: initial vertical profile	
Geologist's log	Materials and boundaries
Excavator's log	
• TBM	Method, progress, samples and groundwater/gas
• Open tunnel face advance	Support, groundwater, gas
• Surface excavator	Stability, water
Geographical location	Essential – may be chainage
Orientation of exposure	Relative to north
Vertical or inclined	Inclination of exposure
Date of drilling	Essential
Date of logging	Essential
Excavated or natural exposures: augmented vertical profile	
Assessment of above data by considering	
• Time-dependent changes	Compare logs with photos of exposure and results from in situ testing and lab testing
• What may have been unseen	Orientation of structures relative to exposure
• Larger geological setting	Compare with maps of geology and topography
• Presence of groundwater	Damp or visible water

This table lists the checks required to achieve a basic profile. It is not exhaustive; other subjects and aspects can be required, depending upon the details of the geology, the natural environment and the requirements for engineering design. This list pre-supposes that the logs have been acquired by a geologist trained in logging and using an approved code for description (e.g. Eurocode 7).

Table 13.1 A checklist of basic requirements for achieving a vertical profile

The profile illustrated in **Figure 13.2** comes from investigations for the underground car park at the Palace of Westminster in London, otherwise known as the Houses of Parliament (Burland and Hancock, 1977; see also Chapter 4 *The geotechnical triangle*). The site is adjacent to the bell tower of Big Ben and the north embankment of the River Thames. The works required an excavation to 20 m below ground level into the London Clay and the creation of a box supported on piles cast in situ and under-reamed at their base. Diaphragm walls were considered for lateral support, but their depth gave cause

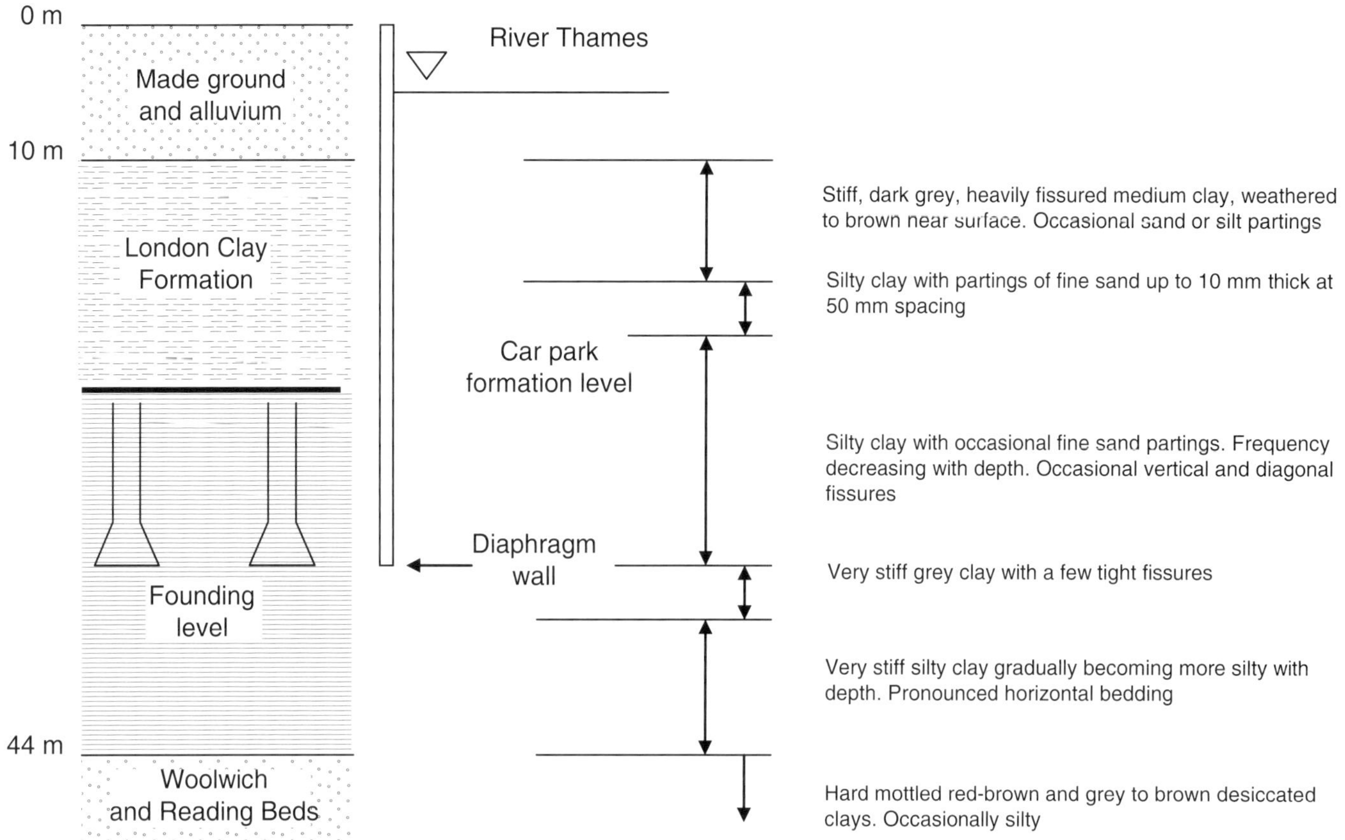

Figure 13.2 Portion of a log from the underground car park at the Palace of Westminster in London (Burland and Hancock, 1977)

for concern once the ground profile for the site was properly studied. The core was carefully taken and logged in detail by splitting it along its length to reveal the fabric and grain size of the material. Thin layers of silt were found within the clay, and these were identified as a serious potential hazard because they could transmit water from the alluvium of the River Thames, if hydraulically connected to it, to the excavation for the basement, possibly causing softening and erosion of the adjacent clay. Such silt layers were a source of drainage for the interlayers of clay, so facilitating consolidation; their presence would also jeopardise the efficiency of any under-ream executed within that zone. None of this was considered an acceptable risk adjacent to so important a national monument. The design was changed to deepen the diaphragm walls to cut off all water access through the 10 m of clay interbedded with silt, and the piles were lengthened so that their under-reamed base started below the 10 m section containing the silt horizons.

It is important to appreciate that it was the profile that made the case for design change self-evident and unquestionable; analyses of the changes followed but no calculation had been required to see the need to change the design. Interestingly, the problems arising from silts seem not to have been considered until the succession had been studied as a profile. Prior to that, emphasis had been placed on assessing strength; over 100 unconfined compression tests had already been conducted.

The profile in **Figure 13.3** comes from an area where soft marine sediments overlie very stiff clays that cover limestone bedrock. An excavation to bedrock for cut-and-cover tunnelling was required with diaphragm walling for permanent support. **Figure 13.3(a)** illustrates the character of the ground and **Figure 13.3(b)** as it appeared in the borehole logs together with its description. The contractor had priced on the basis of the pictorial log and found that soil conditions terminated at higher levels, so upsetting both progress and contracted price; delays followed and a claim resulted.

These examples highlight why the correct interpretation of a profile is important and illustrate some of the difficulties in creating one. In the case of the underground car park, the first mistake came from not appreciating the subtleties of geological nomenclature. 'London Clay' is shorthand for the name of a geological formation (the London Clay Formation) and not a lithology. A lithology is a sediment type whereas a Formation is a deposit of a certain age usually named after the dominant lithology within it – clay, in this case. Other lithologies, notably silt, can also be present. Always ask: '*What is this material?*' and request the answer in plain English, not in technical geological terms. The answer should describe

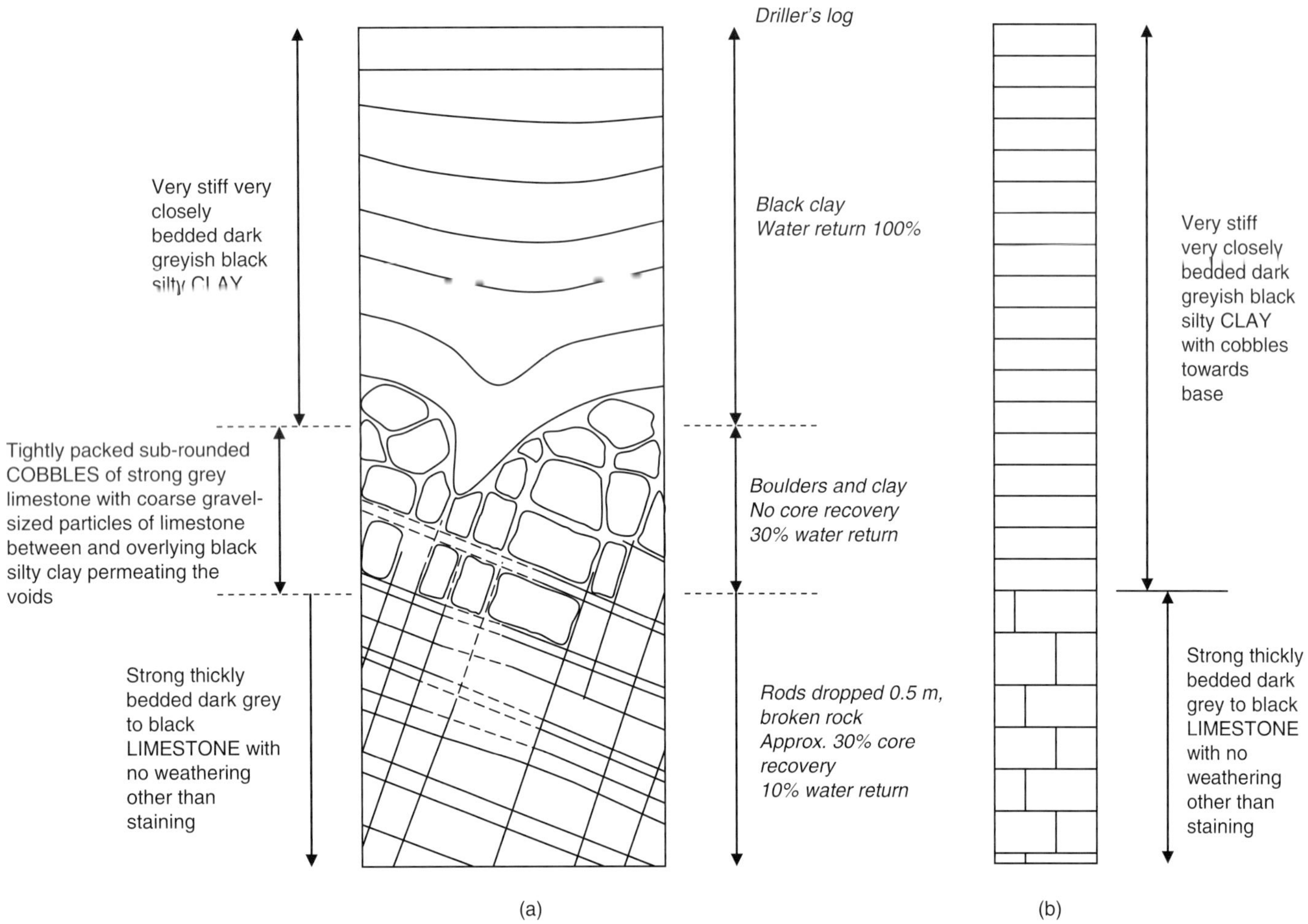

Figure 13.3 The borehole log of soft marine sediment over glacial deposits blanketing karstic limestone in Eire; (a) sketch of vertical profile as exposed by excavation and (b) as described by the driller and in the final logs

both the character of the material and its condition in situ. For sediments and sedimentary rocks, the description of character should be in terms of grain size, mineralogy and fabric; for igneous rocks it will be in terms of crystal size, mineralogy and fabric (sometimes called texture). Metamorphosed equivalents of these, created by the geological action of heat and pressure, can also be described in these terms, depending on whether they are now more like sedimentary or igneous rocks. The in situ condition of the material will describe its fissuring, jointing or faulting (i.e. its discontinuities) and its weathering.

The second mistake with the underground car park followed from the first because the problem was initially perceived to be one of strength and deformation. Obviously strength and deformation were relevant, but they were not the only issue of substance; much damage to both the ground and the surrounding structures could have resulted during excavation, long before the 'strength' of the ground in response to significant changes in either lateral or vertical loads were mobilised. In other words, the design and an important part of the ground investigation to support it had overlooked a key component governing ground response to the engineering work.

Not appreciating what the ground profile implied is also how the second profile caused problems for the excavation of diaphragm walls. A number of contractors had declined to bid for this work, sensing something was 'odd' about the site; the pictorial representation did not seem to reflect the written description. To understand the evidence it was necessary to appreciate the origins of the materials on site and their history. The limestone was of Carboniferous age (~360 million years old) and much deformed by folding and faulting; further, there had been plenty of time and opportunity for dissolution and karst to develop within it. The clay above was hard, because it had been deposited beneath a glacier, and contained large pieces of limestone picked up by glaciers as they travelled over the surrounding country. Glaciers had covered the area in the geologically recent past (~10 000 years ago) and any debris they deposited or disturbance they created to the rock surface over which they travelled would have filled karstic voids within the underlying limestone. None of this appears in the description; it has to be deduced from the profile and the geological history of the area. The problem here was that a zone of karstic rock with its voids infilled (as described above) had been recovered during the borehole investigation as loose debris with fragments of limestone and interpreted as 'soil' above more solid bedrock. The cutters for the diaphragm walling machine encountered the top of the karstic limestone at depths where this so-called 'soil' had been expected.

These examples, two of many, illustrate that a ground profile is more than just the log of a borehole core, trial pit or similar exposure as it requires the factual records such logs provide to be interpreted in terms of the response of that ground to engineering within it; such interpretations bring the profiles 'to life'. In this way, a qualitative ground profile leads to the quantitative means for predicting ground response to engineered change, namely mechanical properties and analyses, and these are the two other points of the geotechnical triangle. Selecting the mechanical properties to be considered and the analyses to be used needs the direction and focus provided by a considered ground profile as this can prevent serious mistakes being made in the quantitative aspect of the work.

13.4. The formation of a profile

Figure 13.4 illustrates four ways a ground profile can form. In **Figure 13.4(a)** layers of sediment have been deposited one on top of the other, which would be the situation in many areas where sediments accumulate more quickly than they weather and erode. A profile of normally consolidated sediment would be expected. In **Figure 13.4(b)** weathering has changed the material near ground level so that the fresh parent material is now found at depth. In soils this would typically result in desiccation near ground level and a crust where the sediments are slightly overconsolidated. In rocks such weathering commonly occurs in regions where tropical weathering is either operating now or has operated in the past and alters material more quickly than it erodes. Such weathering can be well developed in crystalline materials, especially in granite and gabbro, where it can leave pseudomorphs of the original minerals. When that happens, shapes that were clearly original crystals of feldspar, amphibole and pyroxene are now found composed almost entirely of clay. Where these have remained undisturbed their structure and fabric remain intact and produce a zone of remarkable character known as saprolite.

Profiles often contain mixtures of weathered and unweathered material because the rates of weathering, accumulation and erosion during their formation have changed with time. Many of the shallow profiles encountered in ground engineering through natural ground (as distinct from made ground) will have been formed over at least 10 000 years (i.e. from the start of the Pleistocene and the beginning of the Quaternary). Profiles through Tertiary sediments are likely to encounter sediments whose ages can go back 65 million years from the present and these would have experienced tropical conditions when the Earth was a much warmer place than it is at present. For example, deep excavations near Aberdeen in Scotland (latitude $\sim 57°$ N) penetrated stiff glacial clay that was laid down in the Pleistocene over much older granite that had been tropically weathered, probably during the Tertiary.

Figure 13.4(c) illustrates a variation of **Figure 13.4(b)**, as both are associated with the transport of material in a dissolved form. In **Figure 13.4(b)** the movement is predominantly downwards as infiltrating rainwater percolates through the profile, carrying solids it has dissolved out of the profile – typical of weathering in wet climates. In **Figure 13.4(c)** dissolved solids in groundwater close to ground level are drawn up by evaporation and capillarity, and precipitate in the pores of material and at ground level as a mineral cement. This binds particles together and fills pores to produce a natural 'concrete' called a pedocrete; this is further named after its predominant constituents: ferricrete, also known as laterite (iron and aluminium), calcrete, also known as caliche (calcium), silcrete (silica), gypcrete (after gypsum; calcium and sulfur), alucrete also known as bauxite (alumina) and salcrete (after salt; sodium and chlorine). The durability they impart to the materials they indurate causes them to resist weathering and form a more durable crust to the landscape, hence they are also known as duricrusts. The layers beneath these crusts are often leached (i.e. weathered) by such upward migration of atoms and molecules, leaving them weaker than they were originally. Such weathering occurs mainly in hot climates, especially in dry ones.

Figure 13.4(d) illustrates one of the consequences of disturbing the ground. In **Figure 13.4(d1)** the commonly occurring effects of overconsolidation are illustrated; fractures and bedding separation increase in number and become closer to each other towards ground level. This occurs when the ground has carried a greater effective stress than it does at present, the reduction normally being due to erosion over geological time and relaxation following the retreat of the glaciers at the end of the Ice Age. Relaxation will also occur because of excavation either by natural erosion or by man. Natural erosion produces the slopes of landscape and in so doing reduces the stresses below and behind their profile, causing overconsolidated deposits and rock to relax, so enabling fissures, fractures and joints to form. Engineered excavations (e.g. below opencast mines, quarries, foundations especially deep ones, and behind cut slopes) cause relaxation while these excavations are in progress. An example of relaxation associated with excavation is illustrated in **Figure 13.4(d2)** – in this case the collapse of a mine below the base of the profile being studied. As layers of strata drop into the void at depth, bedding opens starting with the beds immediately above the void; the effect decreases with distance from the point of collapse. Thus, the profile detects an increase in fractures per metre depth, interpreted (correctly) as a progressively deteriorating ground condition with depth – rather the opposite of what is normally encountered.

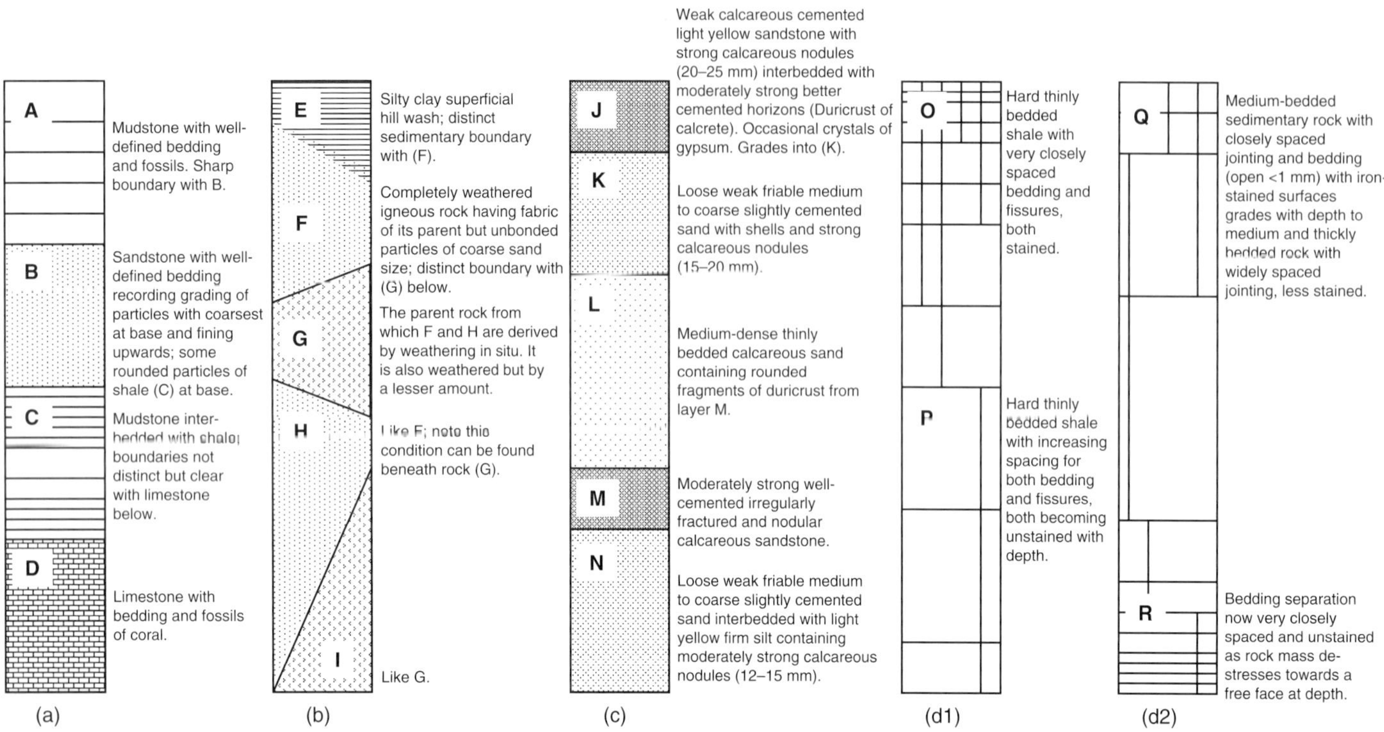

Figure 13.4 Examples of profile formation by (a) sedimentation, (b) weathering, (c) evaporation and (d) stress relief

Ground adjusting to the excavation of tunnels, shafts, inclines, basements and the like can create similar changes, causing the intensity of fissuring and jointing to increase approaching the free face of the excavation.

Although profiles can form in a variety of ways, two features of their formation are unchanging and referred to in geology as the principles of superposition and uniformitarianism.

Any layer found above another is younger than the layer it covers – this is the principle of superposition: **Figure 13.5** illustrates how care needs to be taken even when applying this simplest of laws. Mass movement has transported clay of greater age (layer B) downhill over younger beds (layer A) to form the deposit (layer E); detailed analyses of the included particles, fossil content and possibly carbon dating of layer E all confirm the original age of the material making up layer E is greater than that of layer A. However, the crucial point is that layer E came after layer A and is stratigraphically younger than layer A. The fabrics of the layers should also be studied as these may differ and would differ in the example illustrated in **Figure 13.5**, as should the boundaries between the layers – the boundary below layer E is a palaeo-shear surface. Such a situation is commonplace on many slopes, and excavation across such a junction as that below layer E can reactivate slope instability (Skempton and Weeks, 1976).

If the laws of physics and chemistry permit an action to occur naturally, then that action can be expected to have happened in the geological past. Thus, in **Figure 13.5**, the log of an excavation or borehole that crosses layers E, A and B would be correctly interpreted as younger over older if the observer makes a connection between the properties of the profile, the existence of a slope and the likely presence of mass transport on the slope (i.e. landslides). This connection between a profile and the geological processes that generated it requires the interpreter either to have experience with landslides or to have read case histories describing them. In other words, actions experienced at present, in this case landslides, are relevant to interpreting similar actions that occurred in the past. This is the principle of uniformitarianism, often oversimplified as 'the present is the key to the past'.

The log of a vertical section or borehole records geological products as a sequence of materials arranged in layers and zones, with their depth, and by applying the principles of superposition and uniformitarianism this record of geological products can be used to reveal the likely processes that formed them. All this helps to secure a better understanding of the implications that the profile has for engineering work that is to come. It also means that care must be taken when investigating and interpreting such profiles.

13.5. Investigating a vertical profile

From what has been said in Section 13.2, it is evident that if resources and circumstances permit, it is best for the person who logs the ground to also provide a version of the profile.

By far the greatest number of profiles will consist of natural geological sequences covered by a layer of made ground, and it is likely that at the start of an investigation the processes

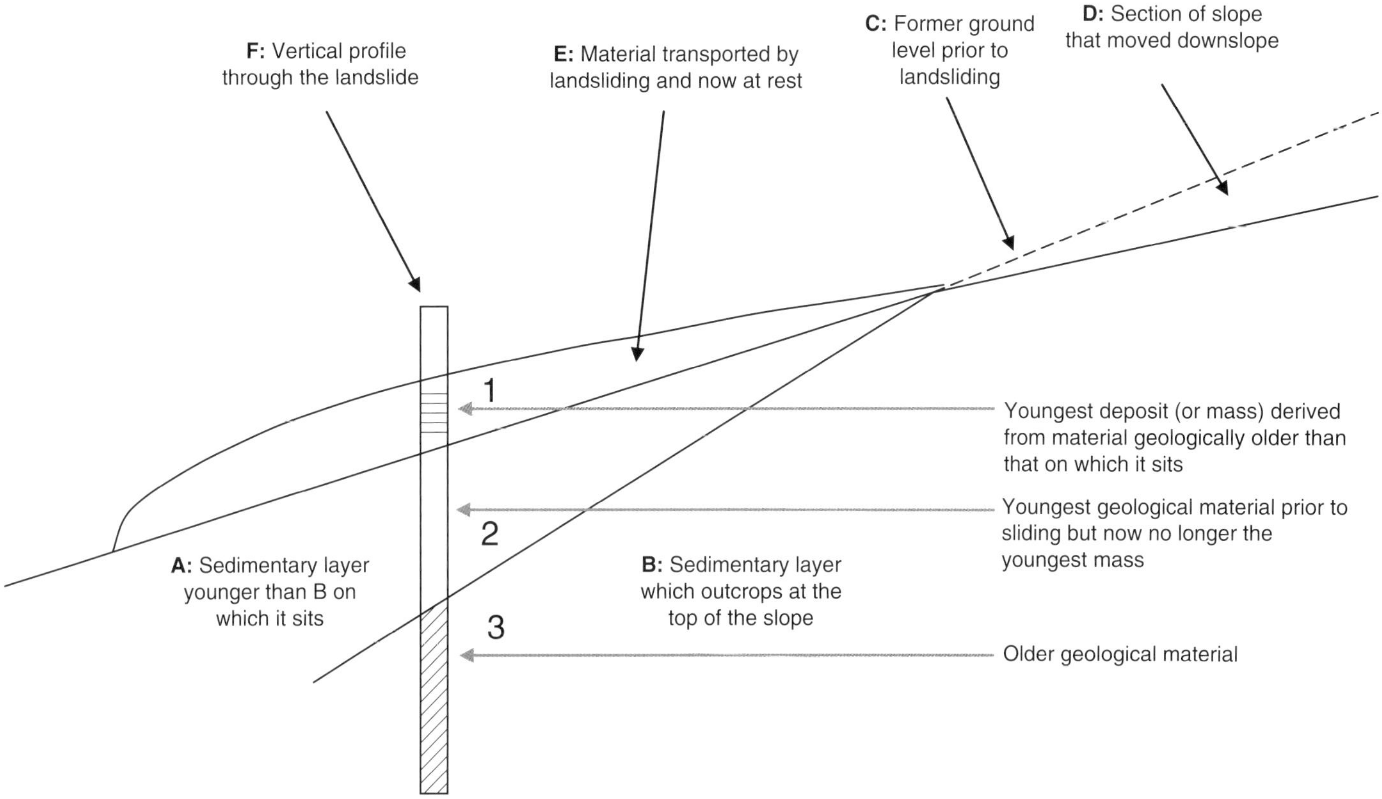

Figure 13.5 A profile from ground covered by landsliding

involved in forming these would be known only from the information provided by a desk study. Thus, any investigation into the ground and any description of it usually take place against this background of some ignorance. The ground, even most made ground, is not manufactured to specification; there is no manufacturing standard against which it can be compared, natural processes have provided the vertical profile; it is what it is, and its investigation is assisted by considering as much of the ground and as many aspects of the ground as is economically possible.

The easiest profile to investigate is one which is totally exposed, as in the face of a quarry or a trench. Compare that with a good-quality borehole core from the same profile; the profile obtained from an exposure can be seen in context – that is, it can be seen relative to the ground to the left and right of it, and possibly below it too. No such context exists for the borehole core, regardless of how good it may be. Context is important when considering any data and there are three reasons for considering context when generating a vertical profile.

- The first is that any investigation is limited; only when the ground is opened by excavation can it be properly seen. The unseen volume of ground that carries the reaction to applied loads can only be seen directly through small excavations such as boreholes, shafts and pits; thus the context of their very limited amounts of data is important.
- The second reason is that soils and rock are complex composite materials and descriptions of them have, of necessity, to be idealised summations of features considered relevant to engineering and the environment. Seeing a profile in context makes a description of its components easier because it helps the significance of those features around which the idealisation and summation are made to be better appreciated; thus in the case of the car park at the Palace of Westminster (**Figure 13.2**) the stiffness and strength of the profile were considered secondary to the presence of silty laminations.
- The third reason is that the scale used for recording has to be decided. Relevant features in a profile do not have to be large; Terzaghi records the importance of small-scale features to the stability of dams, in this case discontinuities measured in millimetres (Terzaghi, 1929). If recording had been at a scale of 1:1 the discontinuities would have been recorded on a pictorial log; however, logging at such a scale is impractical for most purposes and for that reason it is standard to capture an image of cores and exposures for later reference. If the scale of a log reduces to 1:10 the fractures can barely be shown pictorially. Logs presented at around a 1:50 scale result in small-scale features being recorded entirely by words in the written description, and seen only in a good-quality photograph, or image, of the core or exposure.

It is normal not to know the full context of a single ground profile and this should be remembered when interpreting the log of a succession. That is why it helps to refer to other related information, especially reports from a geological survey, geological maps, aerial photographs and historical land-use maps, and to visit the site and its surroundings. It also helps to have a number of profiles from locations close to each other because they help to put each other into context.

The descriptive log upon which the profile is based uses words, and here codes of good practice and norms provide rules and guidance on the words to be used for description and their meaning; sound guidance is provided by Norbury (2020). There is a tendency to believe that it is unnecessary to go beyond what a standard requires. That thinking is wrong – the standards outline the *minimum* that is required and it is for this reason that Burland was instrumental in recommending the following descriptors (BRE, 1993) – only some of which are used in standards – be given particular attention when logging for the purposes of creating a ground profile.

Moisture condition: this can be indicated using simple tactile descriptors such as *dry*, *moist* (i.e. neither wet nor dry), *slightly moist*, *moist*, *very moist* and *wet* when water is visible. It is the state of the material as inspected and a condition that can change with time; it may be the only record of moisture content in a profile where samples are not sent to the laboratory for it to be measured. That is why it is so valuable to use the simple test of rolling a sample between the fingers when logging fine-grained soils to assess whether the soil is wetter or drier than its plastic limit. Many profiles penetrate partially saturated soils whose properties can change dramatically on wetting; clays that have shrunk can swell and sands that cohere through capillarity lose that bond when saturated.

Colour: this is the most obvious characteristic of sediments and rocks, regardless of their particle size. Although required for most standard descriptions, it is more than just an identifier because it is often related to moisture content and length of exposure to oxygen. In fissured and jointed materials, the colour on the surface of discontinuities can differ from that of the material they bound. In this way the relative ages of discontinuities can be determined, those in situ being stained and those formed later (e.g. by coring and/or stress relief) being less stained. A change in the colour of the sample between the field and the laboratory is the first sign that the sample has changed chemically since its collection. A colour chart, such as the Munsell charts, should be used to provide uniformity for those logging and for those interpreting the colours recorded.

Consistency: this can be considered as a tactile assessment of hardness or density and thus reflects stiffness and strength. Such assessments are a normal requirement of standard descriptions and use the terms *very soft*, *soft*, *firm*, *stiff* and *very stiff* in clayey plastic materials; these describe the ease with which material can be moulded. It is important to handle a soil: to roll it in the fingers, squeeze it and push it to see how it holds together and whether it is on the wet or dry side of its plastic limit, and to record that as this may be the only record made of the soil's condition if samples are not sent for testing. These simple tactile tests become even more important when dealing with silt, especially when the content of silt begins to alter the behaviour of a soil. The presence of silt can be detected by its grittiness, and provided the sample has not been contaminated this can be readily detected on the tongue and between the teeth. An alternative and safer method comes from the ease with which fine soil can be brushed from hands and clothing. If your clothing can be cleaned under the blows of a good beating from your hands (which produce clouds of fine dust) then the soil is largely silt. Failure to remove the soil this way suggests the presence of clay; see further comments below under *Soil and rock type*. Silts and sands exhibit little in the way of plasticity and crumble under finger pressure; this has led to use of the terms *loose* (easy to excavate or to penetrate), *dense* (difficult to do so) and *slightly cemented* (strong enough to be abraded). It is always worth attempting to test silts and sands this way as the origin of resistance within them may be no more than the capillary bond of water and easily lost with the addition of water; mineral cement provides something that can be much stronger, as can fabric, and not lost by the simple addition of water.

Although consistency is not a word associated with rock, weathered rock can share some of the characteristics described for soils, especially the silty-sandy soils. Weathered rock is by no means easy to describe and tends to grade from the surface of joint and bedding-bounded blocks to their interior. It is extremely valuable to describe the materials present, their locations and the volumetric percentage they occupy within the mass. All this is best done qualitatively by careful observation, aided by picking and prodding the material (Hencher and McNicholl, 1995; Martin and Hencher, 1986; Norbury, 2020).

Structure: there are arbitrary divides between what is called *structure*, *fabric* and *texture*. In geology these terms have been used synonymously; here, *structure* will be used for features that extend for some distance within a mass and are easily visible to the naked eye. In profiles it is important to observe the structure of the mass – *bedding*, including *laminations*, *fissuring* (a term used for steeply inclined fractures in clayey soils), *jointing* (the term used for such fractures in rock) and *shear surfaces*, especially those from landsliding and faulting. These elements of the mass become important when in situ properties are considered, as clearly seen when paper-thin shear surfaces govern the overall strength of a mass and reduce it to residual values. Structure is easy to describe using terms such as *thinly laminated* and *very thickly bedded* for the range of bedding and *extremely closely spaced* to *very widely spaced* for the range in fissures and joints. Between these surfaces the material will possess a fabric, which can be thought of as being the spatial distribution of materials (like the pattern in a cloth woven from different threads) and texture, which is the morphological result of their size and distribution (like the surface of such a woven material, which could be rough or smooth). Fabric and texture determine how the unfissured and unjointed material will behave.

Structure and fabric can have a profound effect upon the properties of a soil mass even when small, and many case histories exist showing the effect small-scale structures can have, which seem out of proportion to their size and extent (Rowe, 1972). They are often revealed by inclusions of different material, such as rounded chalk pellets and angular gravel-sized fragments of flint in a homogeneous matrix of putty-like chalk, or thin partings of silt in clay. Such small-scale features and subtle changes are best seen in the roughness of a freshly formed surface produced by breaking the soil in tension and allowing it to air dry; this is why it is useful to split soil core along its length. Sometimes clayey sediment separates easily on bedding, as partings, as happens when bedding has traces of silt, and this should be noted. It is extremely difficult to describe much of what can be seen at a small scale and use has to be made of concepts such as *structureless*, *heterogeneous* and *mottled* (where colour and heterogeneity combine). The objective at this scale must be to record as much as can be seen using the vocabulary available, and to support these descriptions with a scaled photograph. The difficulties here are well reflected in the problems associated with describing weathering (Geological Society Engineering Group Working Party, 1995). Small-scale structures and fabric can also have a profound effect upon the behaviour of rock en masse (Terzaghi, 1929).

Texture is something perceived on surfaces and should be considered, because if the surfaces have formed by shearing they will almost certainly contain grooves and striations that betray *slickensides*; these indicate that their shear strength will be at a residual value in their direction of grooving. If surfaces are rough, their shear strength will be a function of friction and dilation that comes from overriding any roughness in the direction of shear.

Soil and rock type: every description requires some judgement to be made on what to call the soil or rock being described, and care needs to be taken because description, which records what is present, can easily slip into classification, which organises materials of similar character into groups.

For soils, grain size and plasticity dominate the decision of what to call them; grain size determines the name in materials where individual particles can be seen, and plasticity provides crucial clues on what name to use for materials where individual particles are indiscernible. The boundary between soils where grains are discernible from those where they are not is where mistakes are easily made, especially in the silt/clay transition of the soil classification. Here, tactile evidence is of great help; it is important to know whether silt is present in clay and to provide a warning while logging that laboratory tests may be needed to adequately determine what a soil should be called. A particularly sensitive tactile test appropriate for revealing the presence of significant contents of silt requires a sample of the soil to be placed in the palm of the hand and gradually dispersed with water until a paste is produced which can then be squeezed by closing the palm and relaxed by reopening the palm. The paste can behave in different ways depending on its silt content. If dominantly silt, the paste, on opening the palm, will appear dry and possibly brittle, but tapping the palm will soon bring its free water to the surface as its brittleness melts away and it reverts to a paste again. This behaviour reflects the response of pore water in the sample to dilation of its particles, and to capillary tensions between them; it indicates that the paste dilates readily, that its particles are larger than clay and that the porosity between them permits the easy movement of water; this would not happen in a clay. In contrast, a soil that is dominantly clay will produce a paste which is still wet and creamy after the palm is squeezed and reopened. The name eventually given (*silt*, *clay*, *silty clay*, etc.) must conform to the specifications of the standards and codes that govern these things and can normally only be achieved with the aid of laboratory tests.

For rocks the task is rather different. Rock names, unlike soil names, are traditionally based on their geological origin, petrological fabric and mineralogical composition (e.g. granite, gabbro, slate, limestone, mudstone and sandstone). Their mechanical properties are linked to material and fabric, and there are no rules that can be used to relate rock names to their mechanical properties; the link for this is empirical. It is therefore important to name rocks correctly, as their name will be used to link with archives of behaviour for rocks of similar character. Serious difficulties for both description and naming arise when rock is weathered, and for this reason weathered material should always be described as well as possible. This can take considerable time and for that reason classifications of weathering (e.g. Grade 1, 2, 3, etc.) become a desirable and practical necessity to speed logging. However, classifications should only be used where they can be applied unambiguously, and for this they need to be based on good descriptions (Geological Society Engineering Group Working Party, 1995); the rule is 'first describe then classify'. Although some standards and codes may do things differently, the advice from the Geological Society Engineering Group Working Party (1995) is scientifically sound and worth recording as it can be applied to rocks in all climatic zones.

Origin: until now, the work of recording a profile has mainly involved description with some classification when dealing with soil and rock type, but to deduce the origin of a material is to venture into interpretation, unless the formation of the deposit or rock has been seen. The many ways in which geological materials may form can be grouped into two categories: either they were transported to where they are now by wind, water or ice or as a viscous downslope flow driven by gravity or by man; or they originated in place and are the remains, or residues, of what was there before. Geological experience is required to decide on origin especially as the materials within a profile may have formed in a number of ways. In Section 13.4 the four basic routes that can result in a profile are described; however, one route does not preclude the operation of another later. Thus, a saprolite formed in place may have its upper levels disturbed by hill wash and be covered by transported material that is later indurated to create a duricrust – the whole profile being subsequently disturbed by adjacent excavation which relaxes the ground. In other words, deducing the origin of a profile need not be straightforward. Allocating a correct name to a sediment, particularly a stratigraphical name (e.g. Oxford Clay Formation), is a helpful indicator to origin but by itself insufficient. Stratigraphic names refer to the accumulation of the material when it was formed. However, the engineering behaviour of the profile will also reflect what has happened to it since then – and in geology that can be a long time ago.

Groundwater: groundwater is an intrinsic component of soil and rock and can govern their strength, stiffness and permeability, yet it is often not appreciated that groundwater is truly an independent variable because the level of groundwater can rise and fall as a result of infiltration, recharge and discharge. The presence or absence of groundwater should be recorded, and where the descriptive log fails to mention either the presence or absence of groundwater, standpipe and piezometric data need to be obtained later to complete the profile (see **Table 13.1**).

13.6. Joining vertical profiles

Individual profiles from specific locations can be combined as a vertical cross-section to acquire an insight into the character of the ground on site and thus a means for revealing what may lie elsewhere on site at locations that have not been investigated. Cross-sections enable the tiny pieces of information represented by a profile to be used to the best advantage. However, there are three hazards associated with this exercise.

The first hazard is the quality of the information: if a cross-section uses data from different sources (e.g. logs from a variety of sources) then it is important to calibrate the data so that like can be joined with like. Different schemes of logging may have been used and observers with different experiences might have been logging. Similarly, descriptions might have been acquired at different times of the year when the profile may have exhibited seasonal changes linked to moisture content and water level. The second hazard is scale: it is necessary to consider the scale at which data have been recorded, as mentioned in Section 13.5; this is necessary to know what has been combined and is now presented as a 'single' unit. The third is origin: here, attempts at understanding the origins of a profile are rewarded because that knowledge constrains the data that can be joined. Units of different origin, even if they are at the same elevation, should not be joined. The fact that they cannot be joined suggests there is some other reason on site, perhaps not yet revealed, that prevents this (e.g. weathering, erosion, faulting, landsliding, subsidence).

Figure 13.6 illustrates a vertical cross-section created from individual ground profiles. Three boreholes intersect firm, thinly bedded clay that becomes soft towards ground level and stiff above a sand layer at depth. The only clear marker horizon is the junction between the clay and the sand; with no other information that would be the only line on the section. However, when standard penetration test results obtained in the clay are plotted against the profile (**Figure 13.6(a)**) the change in consistency described qualitatively can be seen to have quantitative limits which divide the profile into three, so making four zones (**Figure 13.6(b)**). These can be detected across the site and, when used, show that although the site has horizontal strata there is superimposed upon this original structure a second profile that has sloping boundaries. Estimates of mechanical parameters and material response based on depth alone might therefore be in error. Such profiles can result from weathering and erosion.

13.7. Interpreting profiles

Interpretation is applied both consciously and subconsciously throughout the assembly of a profile; every description requires an interpretation to convert what has been seen and recorded into a decision on what it is. Photographs are taken because they are considered to involve none of the subjective interpretation required with logging cores, yet in taking them decisions have been made, possibly unconsciously, on the subject matter to be photographed, scale and colour; thus, what photographs record must be interpreted when incorporated into a profile. All this is a prerequisite to an interpretation of the response of the ground to engineering.

The interpretation of core and borehole logs and of photos might be possible without any further information, as was partly the case with the excavation at the Palace of Westminster (**Figure 13.2**). However, in many cases there is other information of a more mechanical character that can be superimposed on the basic cross-section(s) that have been constructed from the individual vertical profiles. Core recovery, fluid returns, water levels, penetration rates, borehole instability, in situ tests, laboratory tests and similar information can all be superimposed upon the vertical profiles to assist their interpretation and augment the cross-section of possible conditions on site (**Figure 13.6**).

When using ground profiles, it is helpful to consider the ground response that can be anticipated in terms of strength, deformability, permeability and durability, and to review the profile in relation to each, selecting whichever is pertinent to the work in mind; often this will be three of the four. From this will come an holistic view of the likely response of the ground: one that is qualitative but can be justified and its magnitude approximated to give sufficient focus to more quantitative methods for analyses that will have to idealise the ground for the practical purposes of calculation. In this way the basis for that idealisation is placed on reasoned judgement and good geotechnical practice. In 2014 the International Association of Engineering Geology (IAEG) published their Commission 25 on *Engineering Geological Models*. This publication illustrates how basic data such as that gathered in constructing vertical profiles can be used in practice, and categorises the models for the ground that result as *conceptual models*

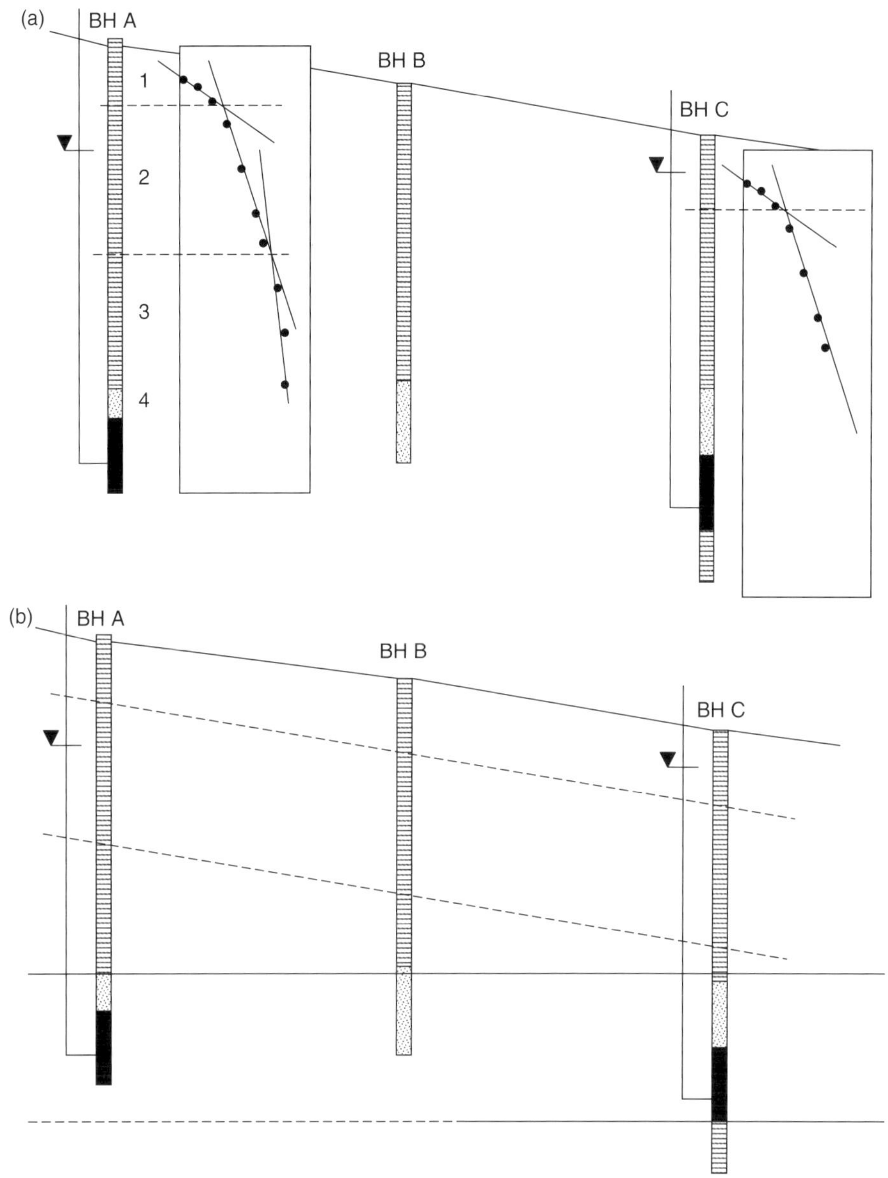

Figure 13.6 A vertical cross-section created by joining ground profiles

to begin with, and later as *observational models* as ground truth from ground investigation is used to constrain the former concepts, and eventually as *analytical models* where quantification of amounts and magnitudes of ground response to engineering can be made and justified (Parry *et al.*, 2014). The IAEG has since published further guidelines illustrating these models in practice (Baynes and Parry, 2022).

13.8. Conclusions

- A ground profile relies on an accurate and complete description of the ground, usually in the form of a log.
- The primary purpose of that log is to describe the ground as it exists in the ground, in situ; it forms the most basic of ground profiles.
- To that basic data should be added the observations and interpretations that put the log into context in terms of geological history, especially in terms of characters that are time dependent, such as groundwater, gas, in situ stress and stress relief.
- This allows some thought to be given to the history of the ground, from its formation to the present day, and especially over the past 10 000 years: essentially since the end of the Ice Age.
- These data should then be considered at a scale that is appropriate for the engineering work to which they will be applied, and joined to give cross-sections from which a conceptual model of the ground can be gained.
- Such a model enables the grouping of materials that are sufficiently similar to be allocated characteristic values to be justified on the basis of available evidence, and then verified as the ground

is exposed, and adjusted as needs be, with the implications of such adjustments being implemented directly into the mechanical properties of the ground and its analysis.

- Ground profiles are therefore both the starting point for and the justification of quantitative analyses of the ground. Securing a good set of profiles is the best investment a ground investigation can make.

References

Baynes FJ and Parry S (2022) *Guidelines for the Development and Application of Engineering Geological Models on Projects.* International Association for Engineering Geology and the Environment (IAEG), Toulouse, France.

BRE (Building Research Establishment) (1993) *Site Investigation for Low Rise Building: Soil Description*. BRE, Watford, UK.

BSI (British Standards Institution) (2004) *BS EN 1997:2004: Eurocode 7: Geotechnical design*. BSI, London, UK.

Burland JB and Hancock RJR (1977) Underground car park at the House of Commons, London: geotechnical aspects. *The Structural Engineer* **55(2)**: 87–100.

de Freitas MH (2009) Geology; its principles, practice and potential for Geotechnics. 9th Glossop lecture. *Quarterly Journal of Engineering Geology and Hydrogeology* **42**: 397–441.

Geological Society Engineering Group Working Party (1995) The description and classification of weathered rocks for engineering purposes. Geological Society Engineering Group Working Party Report. *Quarterly Journal of Engineering Geology* **28**: 207–242.

Hencher SR and McNicholl DP (1995) Engineering in weathered rock. *Quarterly Journal of Engineering Geology* **28**: 253–266.

Martin RP and Hencher SR (1986) Principles for description and classification of weathered rock for engineering purposes. In *Investigation Practice: Assessing BS 5930* (Hawkins AB (ed)). Geological Society of London, London, UK, pp. 299–308.

Norbury D (2020) *Soil and Rock Description in Engineering Practice*, 3rd edn. Whittles Publishing, Dunbeath, UK.

Parry S, Baynes FJ, Culshaw MG, *et al.* (2014) Engineering geological models: an introduction – IAEG Commission 25. *Bulletin of Engineering Geology and the Environment* **73(3)**: 689–706, 10.1007/s10064-014-0576-x.

Rowe PW (1972) The relevance of soil fabric to site investigation practice. *Géotechnique* **27**: 195–300.

Skempton AW and Weeks AG (1976) The Quaternary history of the Lower Greensand escarpment and Welad Clay vale near Sevenoaks, Kent. *Philosophical Transactions of the Royal Society A* **203**: 493–526.

Terzaghi K (1929) Effect of minor geologic details on the safety of dams. Originally published in American Institute of Mining and Metallurgical Engineers Technical Publication 215. American Institute of Mining and Metallurgical Engineers, Wilkes-Barre, PA, pp. 31–44. Also in: Bjerrum L, Casagrande A, Peck RB and Skempton AW (eds) (1960) *From Theory to Practice in Soil Mechanics: Selections from the Writings of Karl Terzaghi*. Wiley, New York, NY.

All chapters within Sections 1 *Context* and 2 *Fundamental principles* together provide a complete introduction to the Manual and no individual chapter should be read in isolation from the rest.

doi: 10.1680/icemge.66816.0173

Chapter 14
Soils as particulate materials

John B Burland Imperial College London, London, UK

Soil comprises countless particles of different shapes and sizes which are in frictional contact with each other and are acted on by gravity. This chapter illustrates the key features of mechanical behaviour that stem directly from the particulate nature of soils. A simple base friction apparatus is described that demonstrates some of these features. In particular the importance of self-weight and microfabric are illustrated, together with the phenomena of contractancy and dilatancy. The apparatus is used to study particle movements associated with active and passive earth pressures, settlement and bearing capacity. A simple mechanistic approach is presented to illustrate the importance of pore water pressure, leading to a demonstration of Terzaghi's effective stress principle. A similar mechanistic approach is used to introduce some key features of unsaturated soil behaviour.

CONTENTS

14.1. Introduction
It was pointed out in the introduction to Chapter 4 *The geotechnical triangle* that soil is a particulate material. Thus a sample of soil is made up of countless particles of a variety of shapes and sizes. The particles are in contact with each other, and the arrangement of particles is often referred to as the 'soil skeleton'. Loads are transmitted through this skeleton at points of contact between the particles. Frequently the contacts between the particles are essentially frictional. However, in many natural soils there is a small amount of bonding between the particles either due to cementation or physico-chemical effects (Mitchell, 1993). The presence of even a small amount of bonding between the particles can have an important influence on both the stiffness and strength of a soil.

The purpose of this chapter is to illustrate in a very simple mechanistic way the key principles governing the behaviour of particulate materials. To carry out calculations of ground displacement and stability it is necessary to idealise the soil as a continuum with certain stiffness and strength properties. The danger is that it is possible to get so used to thinking in terms of the stress–strain response of idealised continua (e.g. porous elastic materials or elastic–perfectly plastic materials) that it is all too easy to forget that the soil is actually particulate and its behaviour is controlled by this fact.

14.2. Phase relationships
Figure 14.1 illustrates an element of soil made up of a number of discrete particles. The spaces between the particles are referred to as the 'voids'. The voids may be filled with water or a mixture of water and air. Completely dry soils whose voids are filled with air are not often encountered in nature, but can be reproduced in the laboratory.

The closeness of packing of the particles has a dominant influence on the mechanical behaviour of a soil. The more densely packed the particles, the greater will be the stiffness and strength of the soil and the lower will be its permeability.

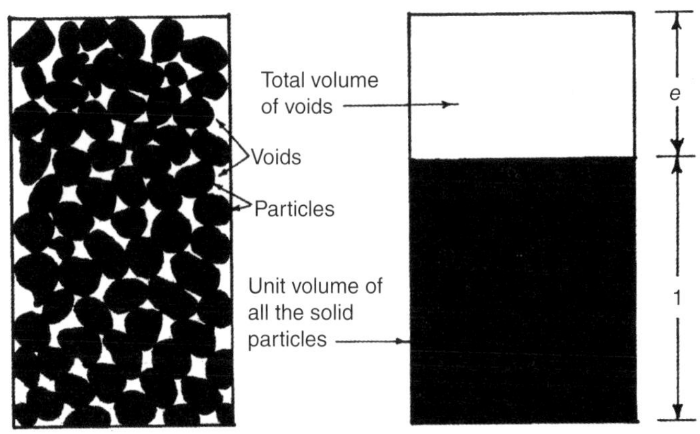

Figure 14.1 Soil is a particulate material. Void ratio e = volume of voids per unit volume of solids

A widely used measure of the 'closeness of packing' of the particles is the volume of the voids in an element of soil divided by the volume of all the particles (solids). This is termed the 'void ratio' of the soil and is denoted as e. Thus the void ratio e is the volume of the voids per unit volume of the solids – see **Figure 14.1**. Another less widely used measure of the closeness of packing of the particles is *porosity* (n), which is defined as the ratio of the volume of voids to the total volume of the soil element, so that, from **Figure 14.1**, $n = e/(1 + e)$. Because of its ease of determination, the closeness of packing of the particles in fully saturated soils (particularly clayey soils) is frequently specified in terms of the mass of water divided by the mass of the solids and is defined as the moisture content w. The various relationships between the volumes and masses of solids, voids, water and air are termed *phase relationships* and these can be found in any basic textbook on soil mechanics (e.g. Knappett and Craig, 2012; Powrie, 2018). These relationships are not complicated, but remembering them can be tedious!

14.3. A simple base friction apparatus

It is possible to illustrate many of the basic mechanisms of behaviour of particulate materials using a simple physical model. **Figure 14.2** shows a photograph of what is termed a 'base friction apparatus'. It consists of a Perspex base across which a standard acetate strip is drawn by means of a small variable-speed electric motor. The arrow in **Figure 14.2** shows the direction of movement of the sheet. In this case the model particles consist of short lengths of copper tube of three different diameters. When the electric motor is switched on the acetate sheet carries the particles with it until they come up against a boundary, after which the acetate sheet 'drags' against each particle, simulating the force of gravity acting on it. The behaviour of the particles can be projected onto a vertical screen by means of an overhead projector so that they appear to be 'dragged' downwards under the action of gravity.

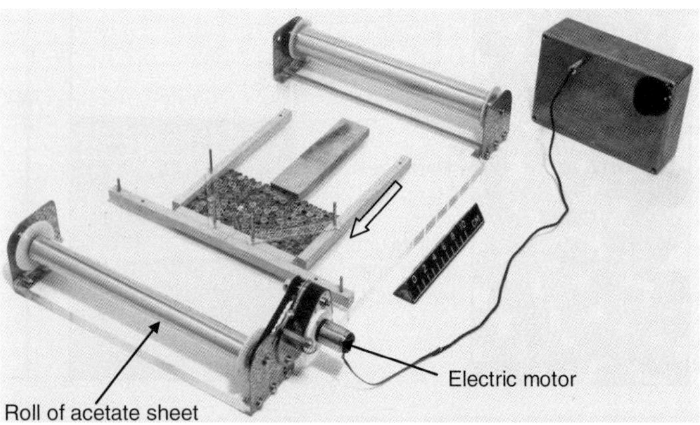

Figure 14.2 Base friction model; the arrow shows the direction of movement of the acetate sheet (Reproduced from Burland, 1987)

The key feature of a base friction apparatus is that it simulates the forces of gravity which give rise to the all-important effects of self-weight. In soil mechanics, the self-weight of the material generates the confining pressures that govern its strength and stiffness, and it also provides the dominant stabilising and destabilising mechanisms for most stability problems.

14.3.1 The process of deposition

Figure 14.3 illustrates the process of vertical deposition of a dispersed suspension of particles acted on by gravity into a container with a horizontal base and vertical sides. **Figure 14.3(c)** shows the particle arrangements on completion of deposition and illustrates a number of important features. The small particles tend to form into clusters. There are a number of voids around which the particles tend to 'arch' so that beneath an arch the particles will not be transmitting much load. If a top plate is placed on the assemblage and gravity is 'switched off', then a gentle up-and-down movement of the whole assemblage shows that there are numerous 'loose' particles that are not transmitting any load – that is, there are many more 'arches' than was at first apparent. The whole assemblage is clearly in a very loose state with a high void ratio.

Another key observation is that it is possible to trace numerous vertical and sub-vertical columns of particles showing well-defined preferred arrangements or 'structures'. Thus the particles have arranged themselves so as to resist the dominant vertical gravity forces and the assemblage is stiffer and stronger in the vertical direction than in the horizontal direction. In other words, the assemblage has anisotropic properties which are inherent to the mode of deposition under the action of gravity.

The assemblage of particles shown in **Figure 14.3(c)** is termed 'normally consolidated' because the vertical stresses acting on it are those imposed as a result of deposition. If the vertical stresses are reduced subsequent to deposition, the

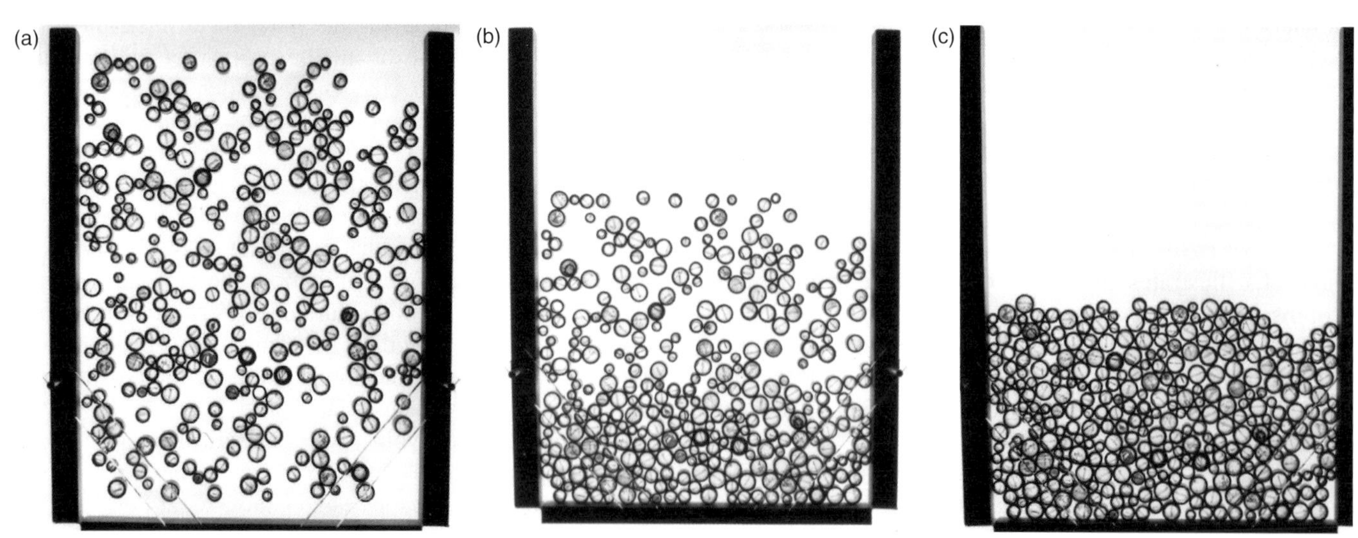

Figure 14.3 Successive stages of the deposition of an initially dispersed suspension of particles (Reproduced from Burland, 1987)

material is defined as 'overconsolidated'. The 'overconsolidation ratio' is the maximum previous vertical stress divided by the present vertical stress. Normally consolidated soils are invariably in a loose state.

14.3.2 Contractant and dilatant behaviour

If a top plate is placed on the surface of the assemblage of particles, gravity is 'switched on' and the sides of the container are then slowly rotated by equal amounts in the same sense, then the assemblage will be deformed under what is termed 'simple shear'. During this process of shearing it will be observed that the various 'arches' of particles are broken down and the particles move to take on a closer packing – that is, the void ratio *decreases*. This process of void ratio reduction during pure shearing is known as 'contractancy'.

The assemblage of particles can be compacted to a closer pack by 'switching on' gravity and tamping the top plate. If, after compaction, the assemblage is again subjected to simple shear it can then be observed that the particles ride over one another and the overall volume *increases*. This process of void ratio increase during pure shearing is known as 'dilatancy'.

The phenomenon of contractant and dilatant behaviour during shearing is almost unique to particulate materials and is an extremely important property with a profound influence on the shearing resistance of the material.

14.3.3 Active and passive earth pressures

The base friction apparatus can be used to illustrate the development of active and passive regions behind and in front of retaining walls. Refer to Chapter 20 *Earth pressure theory*, which discusses active and passive earth pressures. **Figure 14.4(a)** shows what happens if gravity is 'switched on' and one of the sides is then rotated away from the retained assemblage of particles. It can be seen that the movement of the particles is confined to a relatively narrow wedge-shaped region close to the wall. This is known as the 'active' wedge. At a relatively small rotation the earth pressure reaches a steady minimum value, termed the 'active earth pressure'.

Figure 14.4(b) illustrates the movement of the particles when the wall is rotated inwards towards the retained material. It can be seen that the region of movement of the particles extends much further than for the active wedge; it is known as the 'passive' region – in this case it is still wedge-shaped. Much larger rotations are required to fully mobilise full 'passive earth pressure' than for active conditions.

14.3.4 Foundation settlement and bearing capacity

The mechanisms involved in foundation settlement and bearing capacity can be studied by placing a thin rectangular strip of wood on the acetate sheet of the apparatus and then 'switching on' gravity. The foundation moves slowly down until it is in contact with the ground surface. By placing a relatively light weight on the wooden strip the frictional drag on the strip

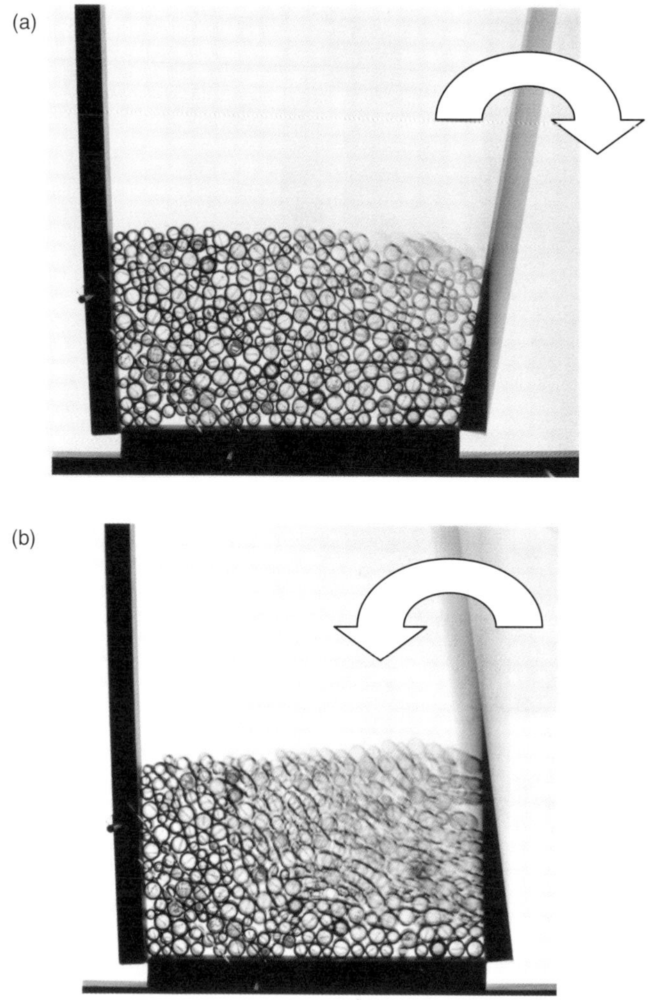

Figure 14.4 (a) Development of an active earth pressure region. (b) Development of a passive earth pressure region (Reproduced from Burland, 1987)

is increased (simulating a small increase in foundation loading) and the foundation settles due to the compaction of some of the particles immediately beneath the foundation. The zone of compaction is confined to a depth equal to about half the width of the foundation.

The load can be increased further by replacing the wooden strip with a brass strip. When gravity is 'switched on' the foundation settles significantly and the zone of particle movement extends further beneath the foundation and outwards on either side, as shown in **Figure 14.5**. The foundation has reached bearing capacity failure and the shearing resistance of the assemblage of particles has been fully mobilised within the observed zone of movement. Further loading of the brass strip causes it to plunge and the zone of particle movement extends around the base and upwards to the surface alongside the strip. This behaviour is modelling the driving of a pile into granular material, and it is possible to visualise the development of high end resistances and also frictional resistance along the shaft of the pile.

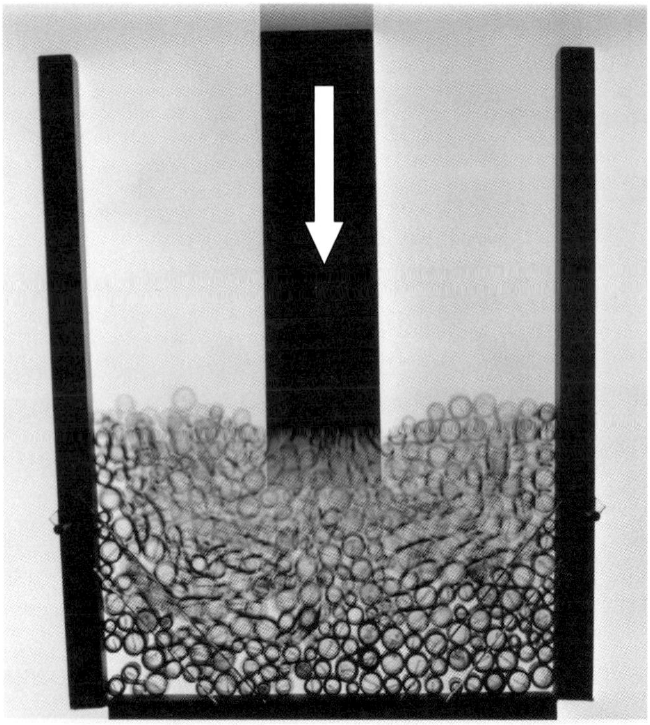

Figure 14.5 Development of bearing capacity failure (Reproduced from Burland, 1987)

The base friction apparatus can be used to illustrate mechanisms of ground behaviour for other geotechnical problems as well. For example, it can be used to illustrate subsidence above tunnels and arching around them as they are excavated. It is also possible to illustrate the mechanisms of deep-seated movement around propped excavations.

14.4. Soil particles and their arrangements

The previous section explored the mechanical response of a relatively simple assemblage of round particles of three different sizes. It was evident that, even with an assemblage composed of uniformly shaped and sized particles, relatively complex arrangements can occur. Most natural soils are made up of much more complex arrangements of particles.

14.4.1 Particle size

The range of sizes of particles that make up a soil is huge, varying from boulders, through gravels, sands and silts down to clay sizes. **Table 14.1** shows the approximate sizes of these particles and some important visual and tactile descriptions. **Table 14.2** illustrates the amazing range of sizes, numbers and surface areas of typical grains in a gram of soil. **Table 14.3** describes dimensions of three typical clay minerals. The engineering importance of grain size lies primarily in the drainage characteristics of the soil, thereby influencing the permeability and rate at which the volume changes and strength responds to changes in loading.

Soil type		Size (mm)	Visual or tactile description
Boulders			**Visible to the naked eye**
Gravel	Coarse	>60	**Particle shape:**
	Medium	20	Angular; subangular, rounded, flat, elongated
	Fine	36	**Texture**: Rough, smooth, polished
Sand	Coarse	2	**Grading**: Well graded (wide range of sizes)
	Medium	0.6	
	Fine	0.2	Poorly (uniformly) graded
			Gap graded
Silt		0.01	**Not visible to naked eye**
			Gritty to hand or teeth
			Disintegrates quickly in water
			Exhibits dilatancy when squeezed in hand
Clay		0.002	Feels soapy when rubbed with water in hand
			Sticks to fingers and dries slowly
			When spread with knife, leaves a shiny surface
Organic soils			Contains substantial amounts of organic matter
Peats			Predominantly plant remains, dark brown or black
			Low bulk density

Table 14.1 Soil types

Particle	Diameter (mm)	Mass (g)	No. per gram	Surface area m²/g
Boulder	75	590	1.7×10^{-3}	3×10^{-5}
Coarse sand	1	1.4×10^{-3}	720	2.3×10^{-3}
Fine sand	0.1	1.4×10^{-6}	7.2×10^{5}	2.3×10^{-2}
Medium silt	0.01	1.4×10^{-9}	7.2×10^{8}	0.23
Fine silt	0.002	5.6×10^{-12}	9×10^{11}	1.1

Table 14.2 Typical grain dimensions (assumed spherical)

Mineral	Width (mm)	Width/thickness	Surface area (m²/g)
Kaolinite	0.1–4.0	10:1	10
Illite	0.1–0.5	20:1	100
Smectite	30.1–0.5	100:1	1000

Table 14.3 Typical clay mineral dimensions

Very few soils in nature fall entirely within one grain size classification. Mostly soils are mixtures (e.g. silty clay, very sandy, fine to coarse gravel). The distribution of grain size in a soil is termed its 'grading'. Thus the 'soil type' is primarily based on grain size and grading. There is an unfortunate tendency in practice to refer to 'cohesive' and 'non-cohesive' soils to distinguish between soils with a high or a low clay content. This can be very misleading as many normally consolidated clay soils do not exhibit cohesion and many granular soils can be bonded. Moreover, all soils that are sheared without allowing drainage exhibit undrained strength, which is often treated as an equivalent cohesion in analysis. True cohesion in a soil is a very difficult property to determine and its precise definition is far from clear. It is much better simply to refer to 'clayey soils' and 'granular soils' without implying anything about their cohesion (sometimes the terms 'fine-grained' and 'coarse-grained' are used).

14.4.2 Particle shape

The particles used in the simple base friction model illustrated in Section 14.3 are circular in shape. Soil particles vary in shape. For granular soils (boulders down to silt-size particles) the variety of shapes may be embraced by the following descriptors: angular, subangular, rounded, flat, elongated – see **Table 14.1**. The texture of the particle surfaces also varies, ranging from rough, through smooth, to polished. The shape and angularity of the particles have a profound influence on its shearing resistance. This was impressed firmly on the author when he went for a walk on Chesil Beach in Dorset. The beach is made up of smooth, rounded gravel-sized particles – see **Figure 14.6**. As a consequence it is extremely difficult to walk on the beach because the grains move around so much underfoot!

Clay particles are generally much smaller than their granular counterparts and are usually plate-like in shape with plan dimensions much greater than the thickness. **Table 14.3** gives some typical dimensions of three common clay minerals: kaolinite, illite and smectite (montmorillonite). The clay content in a soil, together with stress history, has a dominant influence on its compressibility.

14.4.3 Fabric and micro-structure

The above sizes and shapes of particles combine to form a bewildering array of particle arrangements – termed the 'fabric' of the soil. **Figures 14.7** and **14.8** show, respectively, some typical schematic representations of elementary particle arrangements and particle assemblages (Collins and McGown, 1974).

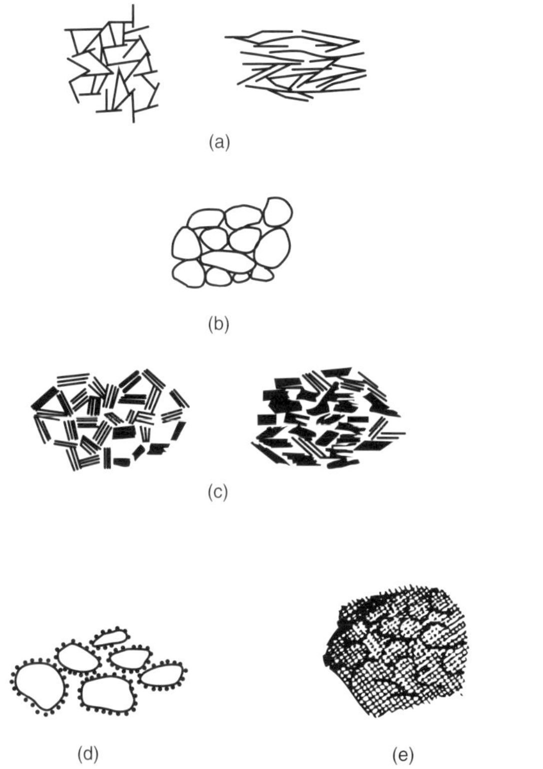

Figure 14.7 Schematic representation of elementary particle arrangements. (a) Individual clay platelet interaction. (b) Individual silt or sand particle interaction. (c) Clay platelet group interaction. (d) Clothed silt or sand particle interaction. (e) Partly discernible particle interaction (Reproduced from Collins and McGown, 1974)

Figure 14.6 (a) Chesil Beach, Dorset, is made up of (b) smooth, rounded gravel with particle matrix

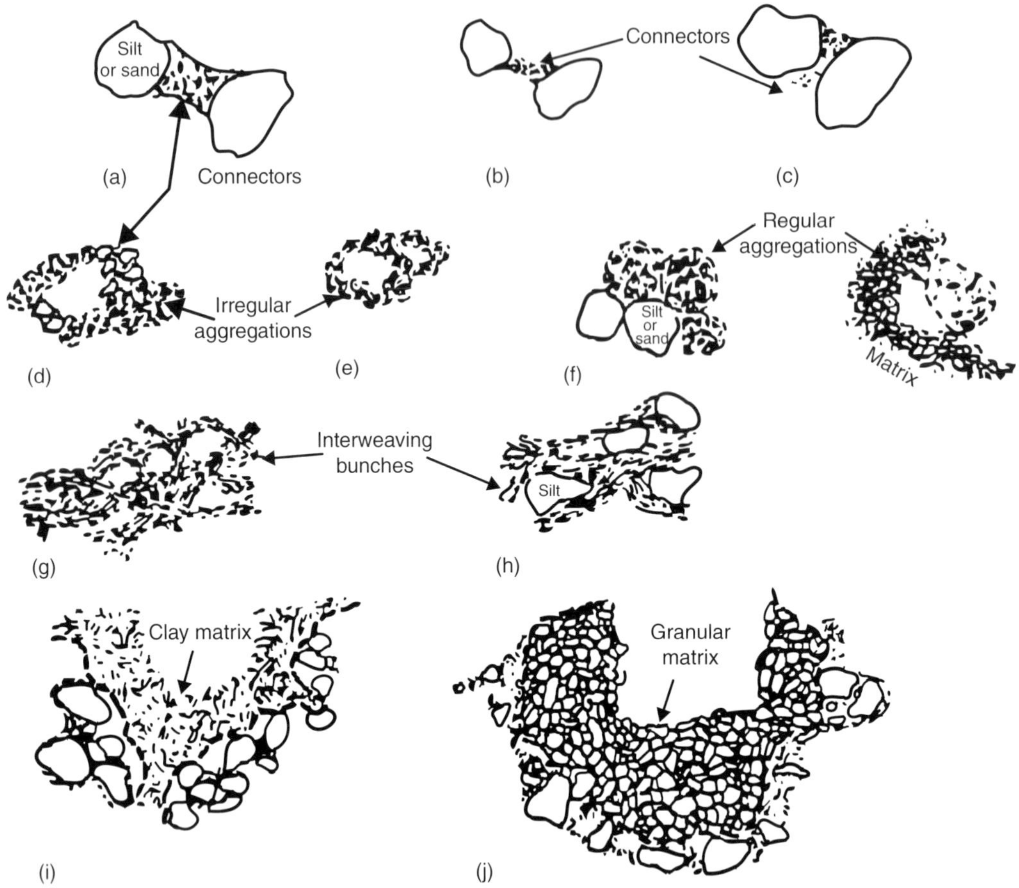

Figure 14.8 Schematic representation of particle assemblages. (a), (b) and (c) Connectors. (d) and (e) Irregular aggregations. (f) Regular aggregations interacting with particle matrix. (g) and (h) Interweaving bunches of clay. (i) Clay particle matrix. (j) Granular particle matrix (Reproduced from Collins and McGown, 1974)

As mentioned previously, the particles of soil are usually in mechanical contact with each other, and resistance to sliding at the contact results from friction and interlocking. Frequently in natural soils the particles are lightly bonded at their grain contact points due to cementation or physicochemical effects.

The term 'micro-structure' means the combination of 'fabric' (the arrangement of the particles) and interparticle 'bonding' (Mitchell, 1993).

14.4.4 Macro-structure

It is important to appreciate that many soils have discontinuities through the mass of the soil as a result of geological or man-made processes. For example, joints and fissures can form due to deformations during deposition or tectonic activity. These give rise to important planes of weakness, particularly if clay particles on the joint surface have realigned themselves to give very low frictional resistance. Similarly, pre-existing shear planes can exist in clayey soils due to previous landslipping or geological processes. The weathering of rocks containing relic joints can also lead to important planes of weakness in the soil mass. The presence of such discontinuities can have a dominant influence on the strength of the ground.

Varying depositional environments can lead to alternating layers of clays and granular materials. These more permeable layers usually need to be taken into account in the design and construction of excavations, slopes and embankments. An example of this is given in Chapter 4 *The geotechnical triangle*, where it is shown that the design of the underground car park at the Palace of Westminster was dominated by the discovery of thin silt and sand partings in the clay just beneath excavation level. It is also important to note that the presence of joints and fissures can profoundly influence the mass permeability of the ground, particularly if any bonding or cementation is present in the soil mass, because there will be a tendency for these discontinuities to remain open.

14.5. The concept of effective stress in fully saturated soils

Appreciation that soils are essentially particulate materials, albeit very complicated ones, leads to a simple mechanistic understanding of the underlying factors that govern the

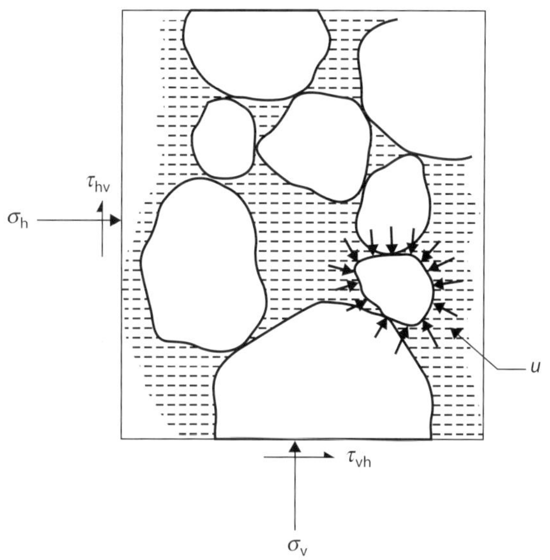

Figure 14.9 Mechanistic model of a loose granular material (Reproduced from Burland and Ridley, 1996)

behaviour of fully saturated soils. **Figure 14.9** shows an element of soil containing an aggregation of particles where the voids between the particles are full of water. The sides of the element are acted on by the stresses σ_v, σ_h and τ_{vh} as shown, and there is a positive pressure u acting in the water. The stresses acting on the boundaries of the element (called total stresses) are made up of two parts: (*a*) the water pressure u, which acts in the water and in the solids in every direction with equal intensity, as shown; and (*b*) the balances ($\sigma_v - u$), ($\sigma_v - u$), and τ_h which are carried by the soil skeleton. Terzaghi (1936) defined the balances ($\sigma - u$) and τ as 'effective' stresses. He then stated his *effective stress principle*, which is that 'all measurable effects of a change of stress, such as compression, distortion and a change of shearing resistance, are exclusively due to changes in effective stress'. Thus the two variables, the total normal stress σ and the pore pressure u, have been replaced by a single variable $\sigma' = (\sigma - u)$. This is a tremendous advantage when measuring the mechanical properties of a soil, as only the relevant effective stresses have to be applied and not the absolute values of the total stress and pore pressure. On the other hand, the effective stress principle requires that, in carrying out a soil investigation, it is essential to know the in situ pore pressures as well as the total stresses acting on the soil.

It is important to note that the effective stress principle says nothing about the way that the effective stresses are transmitted through the soil skeleton – that is, it gives no information about the stresses at the particle contact points. Thus an effective stress is *not* an intergranular stress even though it is sometimes called that. The effective stress is simply that part of the total stress that is carried by the soil skeleton. Note that a shear stress is always an effective stress because water cannot carry shear.

The validity of Terzaghi's effective stress principle can be demonstrated mechanistically as follows. **Figure 14.10(a)** schematically illustrates an element of loose granular material which in **Figure 14.10(b)** is compressed by the application of an all-round stress σ with the pore pressure u held

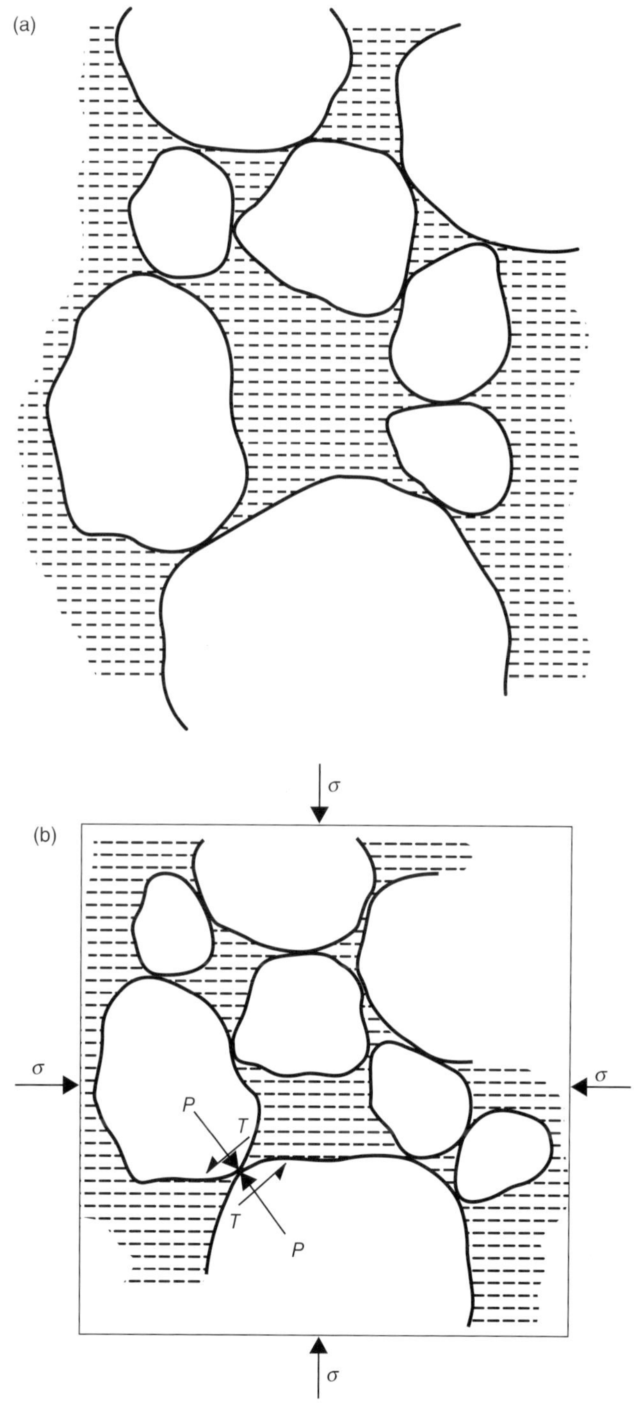

Figure 14.10 Mechanistic model of a loose granular material compressing under equal all-round effective stress: (a) uncompressed; (b) compressed (Reproduced from Burland and Ridley, 1996)

constant. At each grain contact point is acting a normal force P and a shear force T, as illustrated in **Figure 14.10(b)**. In view of the large number of grain contact points for each grain, relative displacement between grains can only take place as a result of slip at the contact points. For a grain to be in equilibrium under its contact forces, the ratio T/P must be less than or equal to μ at every contact point, where μ is the coefficient of friction of the material composing the grains.

An increase in all-round stress $\Delta\sigma$ at constant u will cause a small decrease in volume. This occurs as a result of grain slip at some contact points at which $T/P = \mu$ giving rise to a closer pack of the particles. Therefore an increment in all-round stress induces increases in shear force as well as normal force at the grain contact points. If, instead of an *increase* in total all-round stress σ, the pore pressure u is *decreased* by the same amount, the assemblage of grains would behave in exactly the same way. Clearly a decrease in u is equivalent to an increase in σ in its effects on the grain contact forces. This is a simple mechanistic demonstration of Terzaghi's principle of effective stress for fully saturated soils. It provides a clear demonstration that knowledge of the pore pressure in a soil is just as important as a knowledge of the applied stresses.

14.6. The mechanistic behaviour of unsaturated soils

When a soil is unsaturated the voids contain both air and water. If the air voids are continuous then the water tends to migrate towards the grain contact points due to surface tension effects. **Figure 14.11** shows two spherical particles in contact, with a lens of water around the contact point. It can be shown that surface tension within the curved surface of the water, called the meniscus, generates a compressive force F between the particles. This phenomenon is of profound importance to the behaviour of an unsaturated assemblage of particles. The pressure within the lens of water is less than atmospheric – that is, the water pressure is negative and is often termed 'suction'.

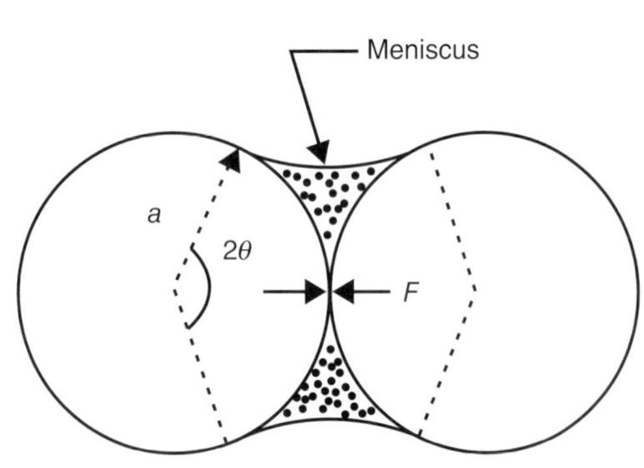

Figure 14.11 Inter-particle force due to surface tension (Reproduced from Burland and Ridley, 1996)

14.6.1 The process of drying a soil

Figure 14.12 illustrates mechanistically the process of drying for a loose granular soil. In **Figure 14.12(a)** the assemblage of grains is fully saturated and the pore pressure is positive. Evaporation, or drying, then starts to take place, reducing the pore pressure. In **Figure 14.12(b)** the assemblage is still fully saturated, but menisci have formed at the boundary. In this situation the pore pressure is less than atmospheric and the tension in the boundary menisci induces compression of the assemblage, resulting in some grain slip and a small reduction of volume.

At a limiting value of suction (known as the 'air entry value') air will enter the pores of the assemblage and the pore water will form lenses around the grain contact points, as illustrated in **Figure 14.12(c)**. These water lenses will generate inter-particle contact forces which are essentially normal to the planes of contact. It is evident that the inter-particle forces generated in this way will tend to inhibit grain slip, thereby stabilising the aggregation of grains. Water at the grain contact points can be thought of as acting rather like blobs of glue or bonds which increase both the stiffness and the strength of the soil. This is why a sandcastle constructed from damp sand is so much more stable than one constructed from either perfectly dry sand or very wet sand.

In clayey soils the process of drying can give rise to considerable shrinkage of the soil, often resulting in the formation of shrinkage cracks and fissures. Tree roots are a prime cause of drying out of soils in periods of drought, and in clayey soils

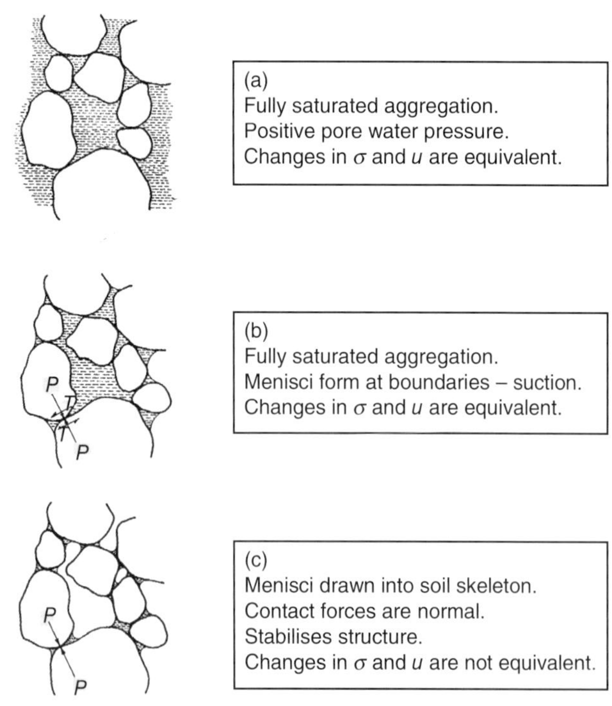

Figure 14.12 Mechanistic model showing the drying of a loose granular material (Reproduced from Burland and Ridley, 1996)

the resulting shrinkage, and consequent subsidence of the foundations, can be very damaging to buildings.

14.6.2 The process of wetting up of a dry soil

The previous section showed that the process of drying an initially saturated granular soil stabilises the grain contact points and increases both the stiffness and strength. The converse is that if an initially dry soil is wetted up, the stabilising forces at the grain contact points are removed, resulting in a loss of strength and, for loose soils, a rapid reduction in volume. This process of rapid volume reduction during wetting is known as 'collapse of grain structure' or simply 'collapse'. The phenomenon of 'collapse' on wetting is well known for dry, loose sandy soils in arid climates. This phenomenon is also well known for fill materials that have been inadequately compacted during placement and which subsequently wet up, often due to a rising groundwater table.

The wetting up of initially dry clay soils can also give rise to serious problems. Such soils are often fissured and cracked due to shrinkage. If water becomes available it can penetrate the clay rapidly through the cracks and cause the clay to swell, which can be very damaging to buildings founded on such soils. An excavation with steep sides in a dry clay soil may become unstable during heavy rain due to the rapid penetration of surface water down shrinkage cracks.

See Burland (1965) and Burland and Ridley (1996) for a detailed discussion of the mechanistic behaviour of unsaturated soils.

14.7. Conclusions

The key conclusions that can be drawn from this chapter are listed below. Some of them are oversimplified but they form a good basis for gaining an understanding of the complex behaviour of soils.

- All soils, even high-plasticity clays, consist of countless particles of a variety of shapes and size. The spaces between the particles are referred to as the 'voids'. The void ratio is a measure of the closeness of pack of the particles and is an important parameter in determining the stiffness, strength and permeability of a soil.
- The particles are in contact with each other and may be thought of as forming the soil 'skeleton'. The arrangement of the particles is referred to as the 'microfabric'.
- The contacts between the particles are largely frictional, but in most natural soils there is frequently a certain amount of bonding between the particles as well.
- A fundamental feature of the ground is that it has weight. The self-weight of the soil generates the confining pressures that govern its strength and stiffness. Moreover, the self-weight provides the dominant stabilising and destabilising mechanisms that operate in most stability problems.
- A simple base friction apparatus is described which simulates that action of gravity on the soil particles. The apparatus is used to introduce the formation of soil fabric during deposition. It is demonstrated that deposition under vertical gravity leads to preferred orientation of the particle arrangements and this leads to inherent anisotropy of the soil skeleton.
- When sheared, loosely packed particulate materials contract and densely packed particulate materials dilate. The base friction apparatus is used to illustrate these important phenomena.
- A number of standard soil mechanics problems are simulated in the base friction apparatus to illustrate the behaviour of particulate materials under gravity, including active and passive pressures, settlement and bearing capacity.
- The importance of particle grading, shape and fabric are discussed, as well as the dominant influence that macro-structure can exert on the behaviour of a mass of soil.
- A simple mechanistic approach is presented to illustrate the importance of pore water pressure leading to the demonstration of Terzaghi's effective stress principle.
- This mechanistic approach is then used to illustrate some of the basic features that control the behaviour of unsaturated soils during the process of drying and wetting. In unsaturated soils the pore water forms lenses at particle contact points which provide an important stabilising effect on the aggregation of particles. During wetting up, this stabilising effect is reduced, leading to collapse of grain structure and reduction in strength.

References

Burland JB (1965) Some aspects of the mechanical behaviour of partly saturated soils. *Symposium in Print on Moisture Equilibria and Moisture Changes in Soils Beneath Covered Areas* (Aitchison GD (ed)). Butterworth, Victoria, Australia, pp. 270–278.

Burland JB (1987) Kevin Nash lecture: The teaching of soil mechanics – a personal view. *Proceedings of the 9th European Conference on Soil Mechanics and Foundation Engineering, Dublin, Ireland*, vol. 3, pp. 1427–1447.

Burland JB and Ridley AM (1996) Invited special lecture: The importance of suction in soil mechanics. *Proceedings of the 12th Southeast Asian Conference, Kuala Lumpur, Malaysia*, vol. 2, pp. 27–49.

Collins K and McGown A (1974). The form and function of microfabric features in a variety of natural soils. *Géotechnique* **24(2)**: 223–254.

Knappett J and Craig RF (2012) *Craig's Soil Mechanics*, 8th edn. CRC Press, Boca Raton, FL.

Mitchell JK (1993) *Fundamentals of Soil Behaviour*, 2nd edn. Wiley, New York, NY.

Powrie W (2018) *Soil Mechanics: Concepts and Applications*, 3rd edn. CRC Press, Boca Raton, FL.

Terzaghi K (1936) The shearing resistance of saturated soils. *Proceedings of the 1st International Conference on Soil Mechanics and Ground Engineering, Cambridge, MA*, vol. 1, pp. 54–56.

All chapters within Sections 1 *Context* and 2 *Fundamental principles* together provide a complete introduction to the Manual and no individual chapter should be read in isolation from the rest.

doi: 10.1680/icemge.66816.0183

Chapter 15
Groundwater profiles and effective stresses

William Powrie University of Southampton, Southampton, UK

This chapter covers the generation of profiles of effective stresses and pore water pressures within the ground for use in geotechnical engineering calculations. Total stresses and pore water pressures are usually determined independently, with the effective stress calculated as the difference between them. The calculation of geostatic and hydrostatic vertical total stresses and pore water pressures arising from the self-weight of the soil and the water, respectively, is described. Non-hydrostatic pore pressure profiles (corresponding to artesian conditions of higher pore pressures in an underlying aquifer and underdrainage into an underlying, more permeable stratum) are discussed. Finally, the difficulties of estimating pore water pressures above the water table, and lateral soil stresses, are considered.

CONTENTS

15.1. Importance of pore pressure and effective stress profiles

The nature of soil as a particulate material, generally with at least two phases present, was described in Chapter 14 *Soils as particulate materials*. If the soil is saturated, the two phases present are liquid (i.e. water in the pores) and solid (i.e. the soil grains). These two phases have very different stress–strain–strength responses – in particular, the pore water cannot carry any of the shear stress τ. However, the presence of a pore water pressure u reduces the normal effective stress σ' to below the total stress σ, and hence also the ability of the soil skeleton to resist shear. According to Terzaghi's *principle of effective stress*

$$\sigma' = \sigma - u \tag{15.1}$$

and the effective stress failure criterion for soils is

$$\tau = \sigma' \tan \phi' \tag{15.2}$$

where ϕ' is the effective angle of shearing resistance or the effective angle of soil friction. Effective stresses, which govern the behaviour of all soils, cannot generally be calculated unless both the total stresses and pore water pressures are known. Thus, in geotechnical engineering it is essential to be able to define the distributions of vertical total stress and pore water pressure with depth as part of the initial ground model for the site.

15.2. Geostatic vertical total stress

It is usually relatively straightforward to estimate the profile of vertical total stress within the ground. Imagine a column of soil of unit weight γ, cross-sectional area A and height z. The force resulting from vertical total stress acting on the base, $\sigma_v(z)A$, must balance the total weight of the column, γAz, giving $\sigma_v(z) = \gamma z$. If the column of soil has a number of different layers having different unit weights, the calculation of vertical total stress must take this into account, so that more strictly

$$\sigma_v(z) = \int_0^z \gamma(z)\,dz \tag{15.3}$$

Normally, Equation 15.3 is discretised into finite layers. The effect of any surface surcharge q must also be taken into account.

The resulting distribution of vertical total stress with depth corresponds to natural equilibrium conditions in the ground, and is often termed *geostatic*.

When applied to a body of soil, Equation 15.3 gives the average total vertical stress over a cross-section. Chapter 13 *The ground profile and its genesis* already showed how the stress distribution around a cavity (void) or around a stiff inclusion will be altered, and Chapter 14 *Soils as particulate materials* showed that the structures set up during soil deposition may result in some grain contacts being relatively unstressed.

15.3. Hydrostatic conditions for pore water pressures

If the groundwater is static, the water pressure is unaffected by the presence of the soil skeleton as long as the pore network is continuous. Thus, in the absence of groundwater flow, pore water pressures below the water table (defined as the level at which the pore water pressures are zero) can be calculated in a similar way to the total vertical stress, by considering a column of water of depth z below the water table. The pore water pressure would therefore be

$$u(z) = \gamma_w z \tag{15.4}$$

where γ_w is the unit weight of water. These conditions, and the profile represented by Equation 15.4, are known as *hydrostatic*.

(The presence of the soil grains makes no difference to the pressure in the pore water column because each soil grain experiences an upthrust equal to the weight of water displaced, while the residual weight of the soil grains – that is, the actual weight minus the upthrust – is carried by contact forces between the grains.)

15.4. Artesian conditions

Most ground profiles will comprise a succession of strata of differing permeabilities (see Chapter 16 *Groundwater flow*). For example, two high-permeability strata through which water may flow relatively easily may be separated by a low-permeability stratum through which water flows only slowly and with difficulty. The relatively permeable strata are known as *aquifers* and the relatively impermeable stratum is known as an *aquiclude*. It is quite possible for the water tables (or, more strictly, the *piezometric levels*) in the two aquifers to be different. In this case, vertical flow will occur through the separating aquiclude in response to the head difference across it (Chapter 16 *Groundwater flow*). The upper aquifer is then termed an *unconfined* or *water table aquifer* (because it is not overlain or confined by a low-permeability layer), and the lower aquifer is *confined* (because it is).

A well-known example of this is in the London basin, England, where a near-surface aquifer comprising alluvial sands and gravels (the Thames Gravels) is separated from the underlying chalk aquifer by the relatively impermeable London Clay. As the chalk is in hydraulic continuity with higher ground outside London, the piezometric level within it is naturally higher than (*a*) ground level in the valley and (*b*) the water table in the surface aquifer, which is hydraulically connected to the River Thames. The natural flow of groundwater will be upward through the London Clay, and the pore water pressures in this stratum will be greater than hydrostatic relative to the upper surface of the clay (**Figure 15.1**, in which the upper aquifer is omitted for clarity). These conditions are known as *artesian*. A borehole drilled through the clay into the chalk would overflow, acting as an *artesian well*. Originally, the fountains in Trafalgar Square, London, operated on this principle and did not need to be pumped.

The pore water pressure at the upper surface of the underlying aquifer cannot exceed the vertical total stress arising from the weight of the overlying strata without the latter becoming unstable as the vertical effective stress is reduced to zero. Excessive pore water pressures at depth in nature may result in *boiling* or *piping* of water up through the overlying soil, or *quicksand* if the overlying material is more permeable. The creation of unstable uplift conditions by the removal of overlying material to form an excavation is discussed in Chapter 16 *Groundwater flow*. *Sub-artesian* conditions, in which the piezometric level in the underlying aquifer is above the water table in the overlying material (but below ground level), are perhaps more common than true artesian conditions.

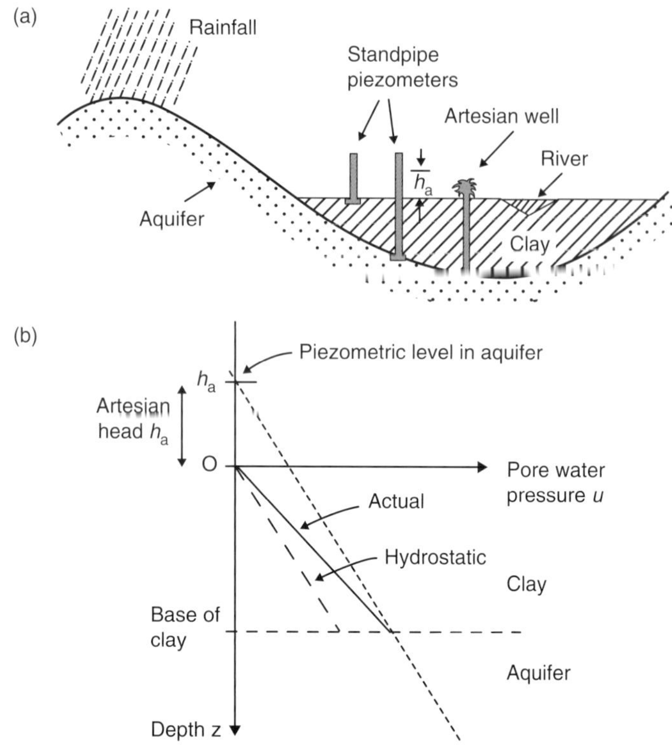

Figure 15.1 (a) Artesian groundwater conditions and (b) associated pore water pressures in the confining clay aquiclude (Reproduced from Powrie, 2018)

15.5. Underdrainage

If the piezometric level in the underlying aquifer is below the water table in the upper aquifer, the direction of natural groundwater flow through the intervening aquiclude is reversed and is slowly downward. Pore pressures in the clay will be below the hydrostatic line drawn from the upper aquifer water table, as shown in **Figure 15.2**. This is known as *underdrainage*, and represents current conditions in London because the historic abstraction of groundwater by pumping from the chalk for industrial and domestic use has lowered the piezometric level within the chalk to below the water table in the alluvial gravels.

The water table in the upper aquifer is sometimes described as *perched*: this term is also more commonly applied to zones of higher pore water pressure resulting from extensive but discontinuous lenses of clay or other low-permeability material.

The simple linear distributions of pore water pressure with depth in the clay aquicludes indicated in **Figures 15.1(b)** and **15.2(b)** tacitly assume that the permeability of these layers is uniform with depth. For a clay, it is more likely that the permeability will decrease with vertical effective stress and hence with depth, giving an equilibrium pore pressure distribution more like that shown in **Figure 15.3**.

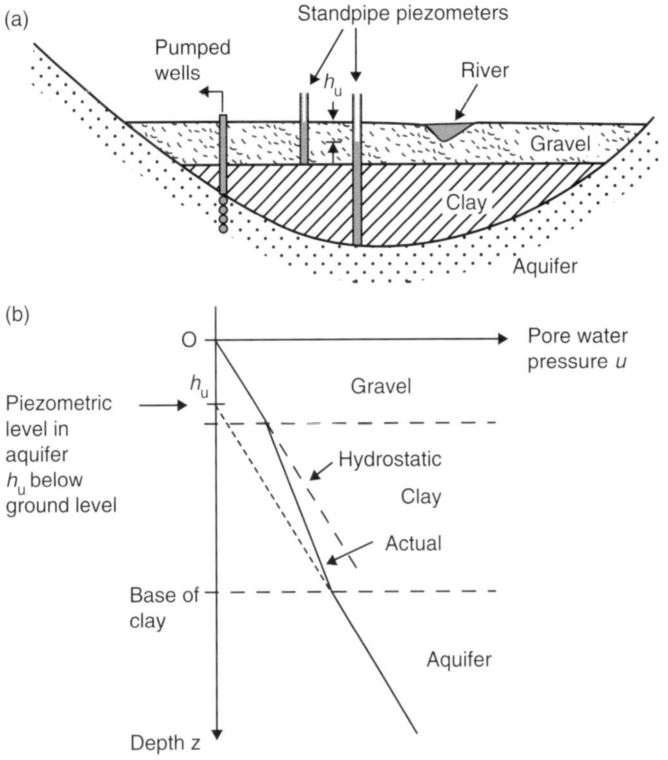

Figure 15.2 (a) Underdrainage and (b) associated pore water pressures in the intervening clay aquiclude (Reproduced from Powrie, 2018)

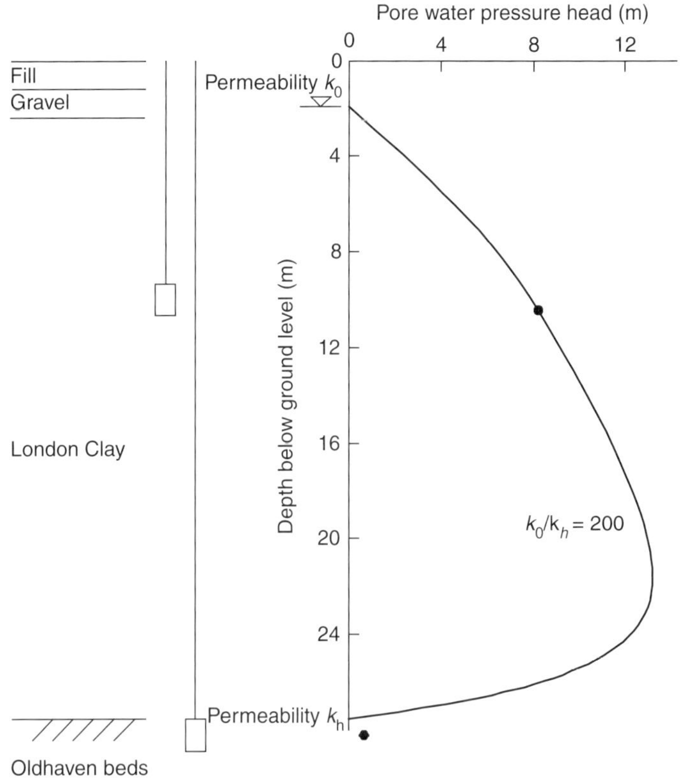

Figure 15.3 Variations of pore water pressure head with depth in a clay aquiclude in which the permeability decreases with depth (Modified from Vaughan, 1994)

15.6. Conditions above the water table

Above the water table in an unconfined aquifer, pore water will be retained in the soil by capillary action. Pore water pressures will fall (i.e. they will become negative) at the hydrostatic gradient with height above the water table, until a suction is reached at which sufficient air has entered the soil to disrupt the continuity of the water phase within the pores. This suction will depend on a variety of factors, including the particle size and the particle size distribution, the density and whether the soil is wetting or drying, but will increase dramatically with decreasing particle size. Thus gravel, with its large particle size, would become sufficiently unsaturated to disrupt the continuity of the liquid phase at a very low suction. A fine sand would hold a suction in the order of one to a few tens of kPa before becoming unsaturated; silt, several tens of kPa; and a clay, tens to a few hundred kilopascal. While soil suction can cause unwelcome and damaging shrinkage and settlements, in terms of the stability of a geotechnical structure (such as a slope) it is usually a good thing. However, because it is difficult to calculate and may be unreliable (e.g. varying between seasons and with climate and vegetation cover), suctions are often ignored in most engineering calculations, with the pore pressures above the water table taken as zero. An exception may need to be made in the calculation of settlements, when suctions can have a significant effect.

15.7. In situ horizontal effective stresses

In situ horizontal stresses are difficult to estimate. They may lie anywhere between the active and passive limits for the soil, which is too wide a range to be useful (see Chapter 20 *Earth pressure theory*). One possibility is to calculate the horizontal effective stress using the theory of elasticity and the fact that during deposition of a soil the lateral strains are zero. However, this requires estimates to be made of the elastic parameters. This is difficult for soils because of the dependence of soil elastic parameters on stress history, stress state, stress path and stress change.

The most common approaches to estimating the in situ horizontal stresses in a soil deposit are empirical. The in situ horizontal effective stress is characterised by the earth pressure coefficient $K_0 = \sigma'_h/\sigma'_v$. For a normally consolidated soil (i.e. one that has not previously been subjected to a vertical effective stress greater than that which currently acts), Jaky's equation

$$\frac{\sigma'_h}{\sigma'_v} = K_0 = (1 - \sin\phi') \qquad (15.5)$$

is frequently used (Jaky, 1944).

For overconsolidated clays that have previously been subjected to a maximum vertical effective stress of $\sigma'_{v,max}$, some of the horizontal stress caused on first loading tends to remain 'locked in' on unloading, resulting in an increase in K_0. There are several empirical equations that can be used to capture this – for example, Mayne and Kulhawy (1982) suggest

$$\frac{\sigma'_h}{\sigma'_v} = K_0 = \left(1 - \sin\varphi'\right)(\text{OCR})^{\sin\varphi'} \qquad (15.6)$$

where OCR is the overconsolidation ratio ($\text{OCR} = \sigma'_{v,max}/\sigma'_{v,current}$).

15.8. Summary

- Profiles of pore water pressure and effective stress with depth are generally needed for use in geotechnical calculations.
- Vertical total stresses will normally increase geostatically with depth at a gradient equal to the bulk unit weight of the soil.
- If the groundwater is static, pore water pressures will increase hydrostatically with depth at a gradient equal to the unit weight of water, provided the water phase is continuous (e.g. the soil is saturated).
- Vertical effective stresses must be calculated by subtracting the pore water pressure from the vertical total stress.
- The horizontal effective stress profile may be estimated from the vertical effective stress profile, using an empirical relationship appropriate to the stress history of the deposit.

References

Jaky J (1944) The coefficient of earth pressure at rest. *Journal of the Union of Hungarian Engineers and Architects* **1944**: 355–358 [in Hungarian].

Mayne PW and Kulhawy FH (1982) K_0–OCR relationships for soils. *Journal of the Geotechnical Engineering Division, American Society of Civil Engineers* **108(6)**: 851–872.

Powrie W (2018) *Soil Mechanics: Concepts and Applications*, 3rd edn. CRC Press, Boca Raton, FL.

Vaughan PR (1994) Assumption, prediction and reality in geotechnical engineering. 34th Rankine Lecture. *Géotechnique* **44(4)**: 573–609.

All chapters within Sections 1 *Context* and 2 *Fundamental principles* together provide a complete introduction to the Manual and no individual chapter should be read in isolation from the rest.

doi: 10.1680/icemge.66816.0187

Chapter 16
Groundwater flow

William Powrie University of Southampton, Southampton, UK

This chapter covers the fundamental principles of steady-state groundwater flow in a saturated soil, based on Darcy's Law and the concept of soil permeability or hydraulic conductivity. Simple flow regimes including radial flow to single wells are described, and the use of plane flownets in plan and cross-section to calculate flow rates and pore water pressures is discussed. Flow in strata of anisotropic permeability and across a boundary between two soils having different permeabilities is briefly considered. The importance of controlling pore water pressures in the vicinity of an excavation is illustrated with reference to a case study, and some common methods of groundwater control are summarised. Finally, the concepts of transient flow and consolidation are introduced.

CONTENTS

16.1. Darcy's Law

For almost all practical purposes, the flow of groundwater through a soil is governed by Darcy's Law

$$q = Aki \tag{16.1}$$

where q is the volumetric flowrate in m^3/s; A is the gross area through which flow occurs (m^2); k is a parameter known as the *hydraulic conductivity* or the *permeability* (m/s), which is a measure of the ease with which water can flow through the soil; and i is the hydraulic gradient in the direction of flow being considered. i is defined as

$$i = \frac{\mathrm{d}h}{\mathrm{d}x} \tag{16.2}$$

that is, the rate of loss of total head, h (m), with distance x (m) along the direction of flow. The hydraulic gradient is therefore dimensionless. The head used is always the total head, measured above an arbitrary datum which, once chosen, remains fixed for the problem being analysed. In **Figure 16.1** the total head for the point at the centre of the left-hand cross-section, A, above the datum indicated, is h_A. This is the level above datum to which water will rise in a standpipe piezometer with its tip at the point of interest A. Other types of head, which are not relevant for the purposes of Darcy's Law, are

- *the elevational head* – that is, the elevation of the point of interest above the chosen datum, z_A
- *the pressure head*, which is defined as the pore water pressure at the point of measurement divided by the unit weight of water, and is indicated by the height to which water in the piezometer rises above the measurement point – in the case shown in **Figure 16.1** this would be $h_A - z_A$
- *the velocity head*, defined as $v^2/2g$ where v is the velocity of groundwater flow and g is acceleration due to gravity; v and the associated velocity head are usually negligibly small in groundwater flow problems.

Darcy's Law is illustrated schematically in **Figure 16.1**. This figure also shows the basis of the simplest form of hydraulic conductivity measurement test, which involves subjecting a specimen of known cross-sectional area A to a known head drop h (from which the hydraulic gradient h/x may be calculated) and measuring the corresponding volumetric flowrate.

The main caveats in relation to Darcy's Law are as follows.

(*a*) It applies only if the flow is laminar. However, unless the ground is very highly permeable, this will almost always be the case in groundwater flow applications.

(*b*) The flowrate q is divided by the cross-sectional area A to obtain an *apparent*, *superficial* or *Darcy* flow velocity

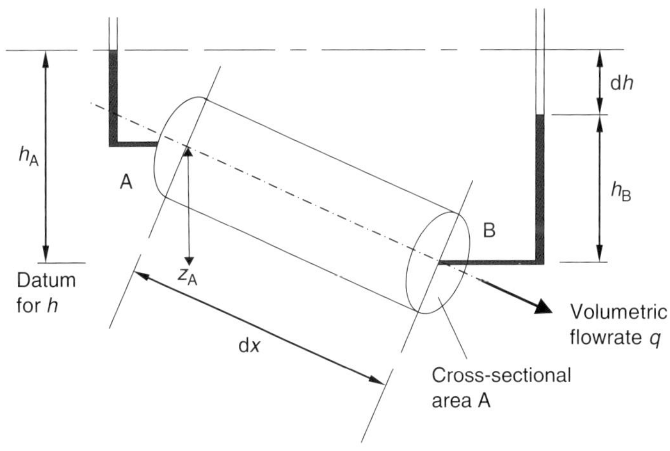

Figure 16.1 Schematic representation of hydraulic gradient and Darcy's Law

$v_D = q/A$. This is not the true average fluid flow velocity, because in reality flow takes place only through the soil pores. The pores occupy a portion nA of the total area, where n is the soil porosity defined as the volume of pores divided by the total volume, or $n = e/(1+e)$ where e is the void ratio. Thus the true average fluid flow velocity v_{true} is $(1+e)/e$ times the superficial or Darcy flow velocity v_D

$$v_{true} = \frac{(1+e)}{e} v_D \quad (16.3)$$

(*c*) The head loss h used in calculating the hydraulic gradient is in terms of total head, determined as the rise in water level above an arbitrary datum in an imaginary standpipe piezometer driven into the soil, with its tip at the point at which the head is to be measured.

(*d*) The hydraulic conductivity depends on both the nature of the soil and the physical properties of the permeant fluid, in this case water.

(*e*) Even in a uniform soil, the hydraulic conductivity depends hugely on the particle size ($k \propto D_{10}^2$) and to a lesser extent on the void ratio. In the ground it can vary greatly from point to point because of natural inhomogeneities, and be strongly anisotropic (with the horizontal permeability k_h being much greater than the vertical permeability k_v) owing to soil fabric and structure. In finer-grained and compressible materials such as clays and landfilled wastes, the hydraulic conductivity is especially likely to vary – often significantly – with stress and stress history, and hence with depth (Chapter 15 *Groundwater profiles and effective stresses*). In construction, what is usually sought is a reliable estimate of the effective bulk hydraulic conductivity of the volume of ground contributing to the flow of interest. This can be difficult to estimate, even to within an order of magnitude, especially from small-scale or laboratory element tests.

(*f*) Darcy's Law forms the basis of nearly all groundwater flow calculations, whether they are carried out by hand or numerically. Darcy's Law can generally be applied to fissure flow in rocks (e.g. chalk), in which case the hydraulic conductivity used is that representative of a volume of rock whose linear dimensions are large in relation to the typical fissure spacing.

Items (*d*), (*e*) and (*f*) are discussed in Section 16.2.

16.2. Hydraulic conductivity (permeability)

The hydraulic conductivity, k, used in Darcy's Law depends on both the soil and the properties of the permeant fluid – in this case water – according to Equation 16.4

$$k = \frac{K\gamma_f}{\eta_f} \quad (16.4)$$

where K (m^2) is the intrinsic permeability of the soil matrix, γ_f is the unit weight of the permeant fluid (kN/m^3), and η_f is its dynamic viscosity (kNs/m^2). In geotechnical engineering the hydraulic conductivity is often known simply as the permeability (although the term *Darcy permeability* is sometimes used to avoid ambiguity).

The practical implications of this are for the most part fairly limited. The Darcy permeability will change if the permeant fluid is not water, but the dependence of the dynamic viscosity on temperature could also be significant in some cases (η_f reduces, and hence k increases by a factor of about 2 over a temperature rise from 20 to 60°C).

The hydraulic conductivity of a granular material is governed by the smallest 10% of the particles, and varies roughly with the square of the D_{10} particle size. Hazen's empirical formula

$$k(\text{m/s}) \approx 0.01[D_{10}(\text{mm})]^2 \quad (16.5)$$

is often used to obtain a rough estimate of the hydraulic conductivity (even though it was originally intended for use only with clean filter sands). Thus the potential range of hydraulic conductivity even for uniform soils is very wide indeed, varying over perhaps ten orders of magnitude from 1 m/s for a clean, open gravel to 10^{-10} m/s for an unfissured clay (**Table 16.1**). This degree of variation is far greater than that of almost any other engineering parameter for natural materials – for example, the shear strength of a soft clay is a factor of about 10^5 less than that of high-tensile steel.

Natural inhomogeneities such as lenses or zones of different materials (e.g. an infilled river channel) may cause the hydraulic conductivity in the ground to vary from point to point. Even in uniform soils, the in situ hydraulic conductivity may be anisotropic owing to the natural fabric or structure of the soil. In clays with no obvious layered structure, the hydraulic conductivity in the horizontal direction might be expected to be up to around ten times that in the vertical. If the soil displays obvious layering (e.g. interlaminated bands of clay/silt and silt/ sand) the ratio of horizontal to vertical hydraulic conductivity could easily be in the order of 10^2 or 10^3.

Some typical values of hydraulic conductivity are given in **Table 16.1**.

In ground engineering it is usual to use a fairly simple flow model based on a single value of hydraulic conductivity, at least in an initial calculation. It therefore becomes important that the value used is reasonably representative of the whole volume within which flow is taking place. Large-scale, in situ tests that involve pumping on one or more wells, with an array of piezometers to monitor the resulting drawdown, are likely to give the

Indicative soil type	Degree of permeability	Typical hydraulic conductivity range: m/s
Clean gravels	High	$>1 \times 10^{-3}$
Sand and gravel mixtures	Medium	1×10^{-3} to 1×10^{-5}
Very fine sands, silty sands	Low	1×10^{-4} to 1×10^{-7}
Silt and interlaminated silt/sand/clays	Very low	1×10^{-6} to 1×10^{-9}
Intact clays	Practically impermeable	$<1 \times 10^{-9}$

Table 16.1 Range of hydraulic conductivity for different soil and ground types (Data from CIRIA C750, Preene *et al.*, 2016; www.ciria.org)

most realistic estimates. Building a reliable picture from smaller-scale tests can be difficult, not only because of the small volume of soil involved in the test, but also because of the potential for disturbance during sample retrieval (for laboratory tests) or borehole installation (for rising or falling head tests in single boreholes). A review of the various methods of determining hydraulic conductivity for the purpose of large-scale groundwater control system design is given by Preene *et al.* (2016).

16.3. Calculation of simple flow regimes

Hand calculations based on Darcy's Law, combined with the condition of continuity of flow (i.e. that the volumetric flow of water into an element of soil must be equal to the volumetric flow of water out over a given increment of time if the soil element is not changing in volume), can be carried out for simple flow regimes that are essentially two-dimensional in nature. Common situations include the following.

(*a*) Radial flow to a fully penetrating well in a confined aquifer (illustrated, with the terms defined, in **Figure 16.2**)

$$q \frac{2\pi Dk\left(H-h\right)}{\ln\left(R_o/r_w\right)} \tag{16.6}$$

(*b*) Radial flow to a fully penetrating well in an unconfined aquifer (illustrated in **Figure 16.3**)

$$q \frac{\pi k\left(H^2-h^2\right)}{\ln\left(R_o/r_w\right)} \tag{16.7}$$

Full details of the analyses leading to Equations (16.6) and (16.7) are given in Powrie (2018).

A large excavation of plan dimensions a (length) $\times\, b$ (breadth) surrounded by a ring of dewatering wells can be analysed approximately as a single well of equivalent radius r_e, provided that the aspect ratio of the excavation a/b is between 1 and 5, and the distance from the edge of the excavation to the recharge boundary, L_o, is greater than $3a$. The equivalent radius r_e is given by $(a+b)/\pi$, and the equivalent radius of influence R_o is $(L_o + a/2) \approx L_o$. Further details are given by Preene and Powrie (1992).

(*c*) Plane flow: for excavations with close recharge boundaries, flowrates can be calculated by means of a flownet or numerical analysis representing the flow within one-half of the cross-sectional plane. This gives the flowrate per metre run, which is then multiplied by the perimeter of the excavation to give the overall flowrate. If the excavation is long

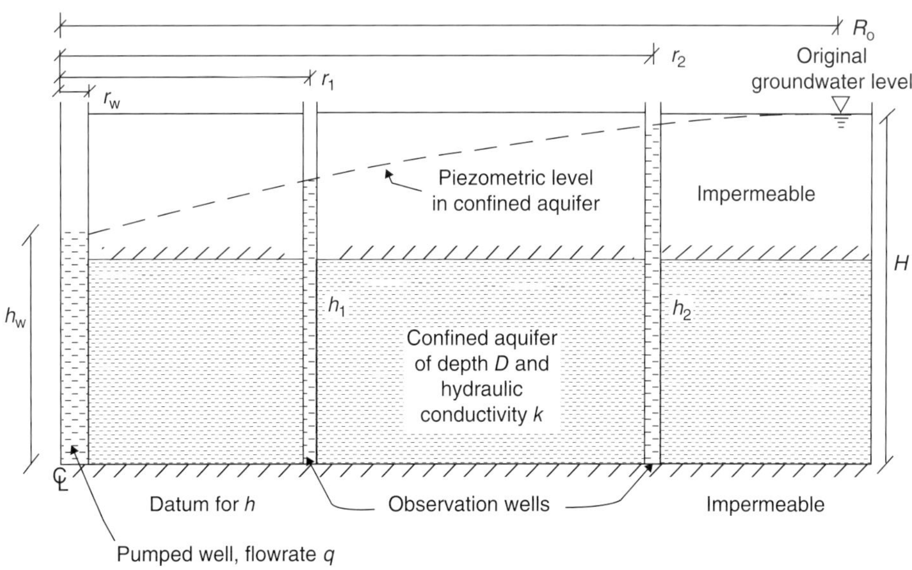

Figure 16.2 Radial flow to a fully penetrating well in a confined aquifer (Reproduced from Powrie, 2018)

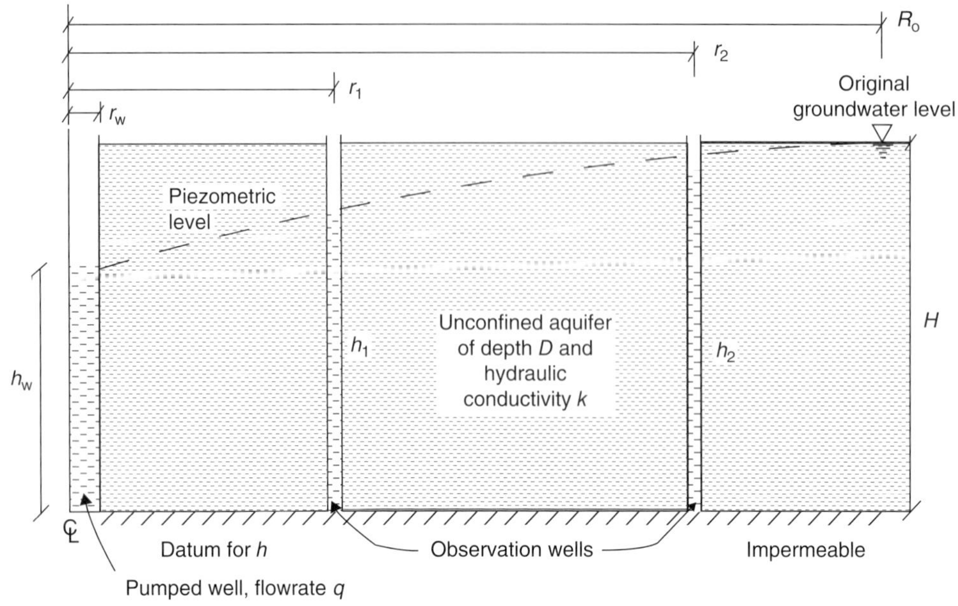

Figure 16.3 Radial flow to a fully penetrating well in an unconfined aquifer (Reproduced from Powrie, 2018)

($a/b > 10$) with close recharge boundaries ($L_o/a < 0.1$), plane flow to the long sides will dominate and end effects may be neglected. If the excavation is rectangular ($1 < a/b < 5$) with close recharge boundaries ($L_o/a < 0.3$), plane flow to the four sides dominates and corner effects may be neglected.

Drawing a flownet is a useful and highly effective graphical technique for solving the Laplace equation, which governs planar flow of groundwater through a soil of uniform or anisotropic hydraulic conductivity. (The technique is generally termed ‘flownet sketching’, although the more careful the ‘sketch’ the better the result). The derivation of the Laplace equation, and a comprehensive introduction to flownet sketching, is given by Powrie (2018).

In summary, plane flownet sketching involves determining the upper and lower equipotentials (i.e. lines of equal head) representing the source of water and the sink, and two bounding flowlines. The space so defined is then filled with a network of intermediate flowlines, which intersect intermediate equipotentials at 90° to form elements known as ‘curvilinear squares’, whose length is equal to their breadth. (In practice, this means that a circle can be fitted inside each of them, although the circles will vary in diameter depending on the position of the element in the flownet.) In an unconfined aquifer, the need to determine the position of the top flowline (which represents the drawn down water table) as part of the general flownet sketching process represents a small additional complication. The flowrate per metre run is then calculated as

$$q = kH \, {}^{N_F}\!/_{N_H} \qquad (16.8)$$

where H is the overall head drop between the upper and lower equipotentials (the source and the sink), N_F is the number of flowtubes (which is equal to the number of flowlines minus 1) and N_H is the number of equipotential drops (i.e. the number of equipotentials minus 1). A full description of the basic technique, which progresses by trial and error, is given by Powrie (2018).

An example of a plane flownet in the vertical plane is shown in **Figure 16.4**.

In cases where there is no similarity of vertical cross-sections and flow is primarily horizontal, flownets may be sketched in the horizontal plane as shown in **Figure 16.5**. The principles are the same, but the total flowrate is calculated by multiplying

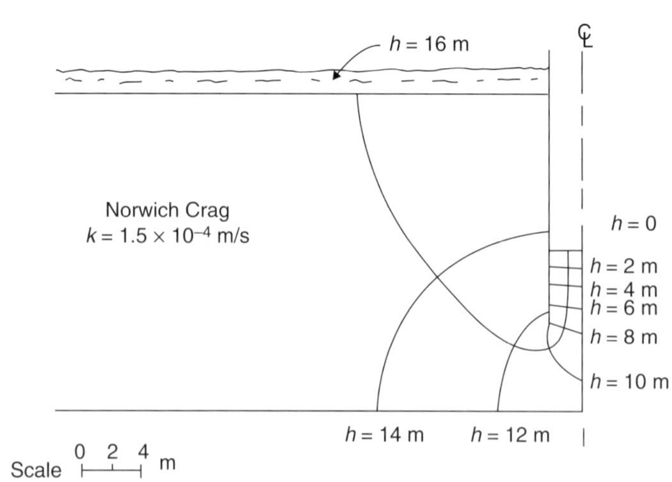

Figure 16.4 Example of a plane flownet in one-half of the vertical cross-sectional plane for flow into a trench excavation (Reproduced from Powrie, 2018)

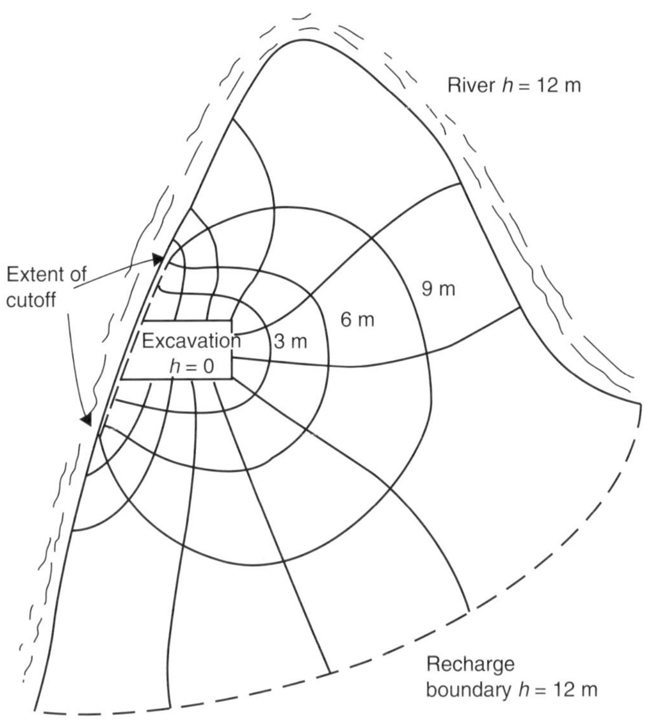

Figure 16.5 Example of a horizontal plane flownet for flow to an excavation in a deep chalk aquifer on the Medway peninsula (Reproduced from Powrie, 2018)

the flownet result (Equation 16.8) by the depth of the aquifer. An inherent assumption is that regions of vertical flow, for example up into the excavation, are small and do not significantly affect the basic premise of substantially horizontal flow.

Although the technique could be viewed as having been superseded by computer-based numerical methods, flownet sketching is quick and simple and has the great advantage of giving strong insights into the physicality of, and the main factors governing, the flow regime.

A finite difference solution to a plane flow problem can be obtained by establishing a grid of nodes and using Darcy's Law to write an expression for the flowrate into each node in terms of the head difference between it and the adjacent nodes, the implied area of flow and the soil hydraulic conductivity. At the steady state, the net flowrate into or out of each node is zero (except of course for the nodes on the constant head or equipotential boundaries). To obtain values of head that satisfy the condition of zero net flow into or out of each internal node, and to give a sum of flowrates *into* the boundary nodes on the upper equipotential (the source) equal to the sum of the flowrates *out* of the boundary nodes on the lower equipotential (the sink), solution is by trial and error and is greatly facilitated by the use of a spreadsheet.

16.4. More complex flow regimes

Flownets can be sketched for anisotropic soils following the same rules as for isotropic soils, provided that the real geometry of the cross-section has first been 'transformed' to account for the difference between the vertical and the horizontal hydraulic conductivities. The true horizontal dimensions, x, must be multiplied by $\sqrt{(k_v / k_h)}$, where k_v and k_h are the hydraulic conductivities in the vertical and horizontal directions, respectively. As k_v is usually less than k_h, this generally involves shrinking the horizontal distances on the transformed section. Equation 16.8 for the calculation of flowrates holds, except that the effective hydraulic conductivity of the transformed section, k_t, that must be used for this purpose is the geometric mean of k_v and k_h – in other words, $k_t = \sqrt{(k_v k_h)}$. Derivations and further details are given in Powrie (2018).

Flownet sketching can also accommodate the presence of two or more strata of different hydraulic conductivities. If the ratio of hydraulic conductivities is on the order of a few hundred, one will act as a reservoir compared with the other (i.e. any head drop within it will be negligible compared with that in the less permeable stratum), and this is the appropriate approximation to make. If the ratio of hydraulic conductivities is on the order of 1 to 100, the flownet should be sketched so that the flowlines defect through an angle ($\beta_1 - \beta_2$) as they pass from a soil with hydraulic conductivity k_1 into a soil with smaller hydraulic conductivity k_2, such that

$$\frac{\tan \beta_1}{\tan \beta_2} = \frac{k_1}{k_2} \tag{16.9}$$

This is analogous to the refraction of light as it passes between media of different optical densities. The derivation is given in Powrie (2018).

A comprehensive guide to flownet sketching for these more complex cases and even transient flow is given by Cedergren (1989), but given the ready availability of spreadsheets and low-cost personal computers it is probably easier to solve such problems using the finite difference approach and a spreadsheet, or one of the commercial computer programs available for modelling groundwater flow. The underlying principles, however, remain Darcy's Law and continuity of flow, and success still depends on understanding the physical constraints to the flow domain and the judicious selection of numerical values of soil hydraulic conductivity.

16.5. Groundwater control for stability of excavations

Whenever an excavation is made below the natural water table, at least in a non-clay soil, groundwater and pore water pressures must be controlled to prevent flooding and potential instability of the excavation. This can be achieved by physical means, such as grouting or a cutoff wall, pumping from dewatering wells placed in or around the excavation or a combination of both. Even when the sides of the excavation are supported by retaining walls, pumping may still be necessary to prevent instability of the base. Pumping will also often reduce the pore water pressures acting on the outside of the retaining walls, which may enable them to be constructed more economically. In built-up areas where compressible strata (e.g. overlying layers of clay or peat) are present, consideration should be given to mitigating the potential for settlement damage to buildings resulting from ground settlements associated with reducing the pore water pressures outside the excavation. Powrie and Roberts (1995) provide a case history on this subject.

As outlined earlier in this chapter, flow calculations are required to assess (*a*) the likely flowrate that a dewatering system will have to pump, and (*b*) the reductions in pore water pressure that a dewatering system will achieve. The pore water pressures will feed in turn into calculations relating to the stability of sheet pile and other retaining walls, excavation side slopes and the excavation base. Most of these calculations are discussed elsewhere in this manual, but the importance of maintaining stability of the base will be discussed briefly here.

The base of an excavation will become unstable if the pore water pressures within a certain depth of the excavated surface become equal to the vertical total stress due to the weight of soil. The 'certain depth' will depend to some extent on the width of the excavation. If the sidewalls are of the embedded type, they may offer some additional frictional resistance to uplift, but the effect will diminish towards the centre of the excavation, especially if the excavation is wide. In any case, once the effective stress has reduced to zero there will be no friction on the sidewalls. Ignoring sidewall friction, it can easily be shown that instability or uplift will occur when the hydraulic gradient associated with upward flow to the base of the excavation is approximately 1 (see Powrie, 2018) or when the pore water pressure on the base of a plug of low-permeability soil exceeds the vertical total stress due to its weight.

The perils of ignoring the potential for unrelieved pore water pressures at depth to cause major damage, and possibly even loss of life, should not be underestimated. They are illustrated by the case study summarised in **Figure 16.6**. This relates to an excavation for a pumping station near Weston-super-Mare which the contractor attempted to make without taking steps to reduce the pore water pressures in the underlying silty sand. As formation level was approached, the base became unstable and the excavation rapidly filled with water. This was quite predictable: at full depth, the pore water pressure at point A on the base of the 6 m depth of soft clayey silt that remains below the excavation floor is 13 m × 10 kN/m^3 = 130 kPa, which is substantially greater than the total vertical stress due to the weight of the clayey silt of 6 m × 18 kN/m^3 = 108 kPa. The remedial works necessitated by the failure were substantially more expensive than the cost of a pre-emptive groundwater control scheme, quite apart from the damage to plant and the potential for personal injury or even loss of life.

Methods of groundwater control by pumping are summarised in **Table 16.2**. Further details are given in Powrie

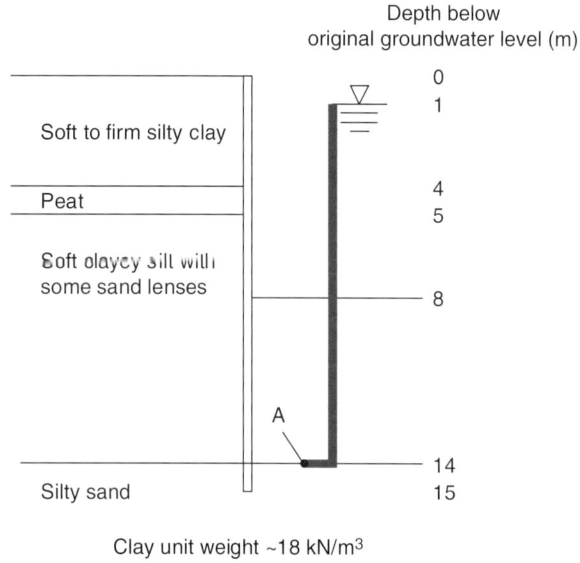

Figure 16.6 Case study demonstrating the importance of controlling pore water pressures to an adequate depth below the excavation floor

(2018), Cashman and Preene (2021), Preene *et al.* (2016) and Powers (1992). The suitability of each method depends on the pumped flowrate, and hence the soil hydraulic conductivity and the drawdown, as shown schematically in **Figure 16.7**.

16.6. Transient flow

So far this chapter has considered steady-state flow, in which the volume of water flowing into a soil element in a given time is the same as the volume of water flowing out (i.e. there is no change in the volume of the soil element). This is not always the case; if the volume of the soil element is changing over time (e.g. in response to a change in the external loading regime), the flow into the element will be greater than the flow out if the soil volume is increasing (swelling), or less than the flow out if the soil volume is decreasing (compression). Non-steady-state flow is known as *transient flow*, and the time-dependent process of volume change associated with it is known as *consolidation* (or *swelling* if it is a volume increase).

For the effective stress to increase, the soil skeleton must compress: this requires water to flow from the pores, which cannot occur instantaneously because the rate of flow is limited by the hydraulic conductivity of the soil. Thus an increase in total stress results initially in an increase in the pore water pressure. The increase in pore water pressure causes local hydraulic gradients, in response to which water will flow out of the soil. As water flows from the pores, the pore water pressure falls back towards its equilibrium value, the soil compresses and the effective stress increases to make up for the reduction in pore water pressure. Eventually, the pore water pressures return to the equilibrium values, and transient flow stops. The soil has compressed, the effective stress has

Method	Thumbnail description	Comments and case studies
Open sump	Literally an open sump into which groundwater is allowed to flow under gravity, and pumped away by a standalone diesel or electric site pump	Subject to limitations on space and accessibility, this method is generally only suitable for modest drawdowns (up to about 2 m) in well interlocked, open gravels
Vacuum wellpoints	Closely spaced small wells (typically in a line 1–3 m apart) approximately 50 mm dia. connected by risers to a common header main, pumped by a surface vacuum pump	Maximum drawdown (owing to limitations of vacuum lift) is about 6 m, but can be used in multiple stages. Comfortable with flowrates typical of sandy soils
Deep wells	Large-diameter wells (e.g. 200 mm dia. screens in 350 mm dia. bores) pumped individually by slimline electrical submersible pumps. Widely spaced so recovery of water table between wells is significant and well depth must allow for this	Can deal with higher flowrates and drawdowns than wellpoints. Restriction on depth of drawdown likely to be economic rather than physical. Vacuum can be applied by means of a separate system in lower-permeability ground (Bevan *et al.*, 2010; Powrie and Roberts, 1995)
Ejectors (sometimes called eductors)	Nozzle and venturi devices installed at depth in deep but small-diameter boreholes (minimum 50 mm), driven by high-pressure water pumped from the surface	Will draw air as well as water so will create a vacuum automatically if the well is sealed. Particularly useful for lower-permeability soils. Their low energy efficiency makes them expensive for higher flowrates (Powrie and Roberts, 1990; Roberts *et al.*, 2007)

Table 16.2 Methods of groundwater control by pumping

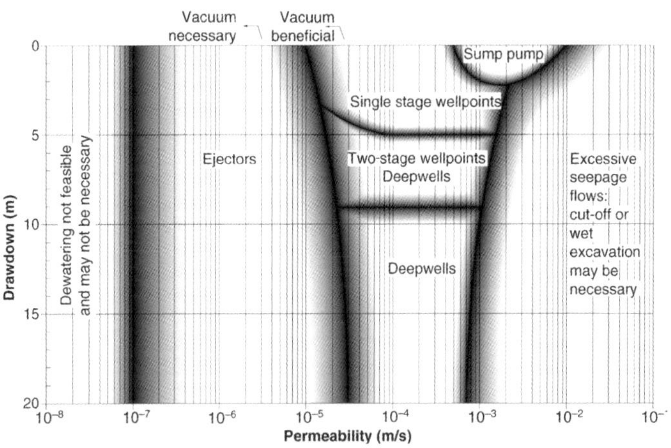

Figure 16.7 Range of application of dewatering techniques (Reproduced with permission from CIRIA C750, Preene *et al.*, 2016; www.ciria.org. A slightly revised version of this diagram is proposed by Roberts *et al.*, 2023)

increased to mirror the increase in total stress and consolidation is complete.

The rate of consolidation increases with the soil hydraulic conductivity k. The amount of compression associated with the eventual increase in effective stress decreases as the soil stiffness in compression, E'_o, increases. The parameters k and E'_o are combined to give the consolidation coefficient, c_v

$$c_v = \frac{kE'_o}{\gamma_w} \qquad (16.10)$$

where γ_w is the unit weight of water. The consolidation coefficient c_v will vary with soil stiffness and hydraulic conductivity, hence with stress history, stress state and stress path. For stiff soils subjected to small changes in stress, it might be reasonable to assume an equivalent single value of c_v in analysis. However, for problems involving softer sediments and larger changes in stress, an approach that allows c_v to vary with both depth and stress will probably be required.

The consolidation time also depends on the distance to the drainage boundary, d, and the flow geometry. For simple flow geometries approximating to one-dimensional flow, consolidation is usually considered to be substantially complete after a time t such that the dimensionless group T defined in Equation 16.11 is approximately equal to 1

$$T = \frac{c_v t}{d^2} \qquad (16.11)$$

In principle, all soils will undergo transient flow in response to a change in total stress or boundary pore water pressure. However, consolidation is normally only of practical importance in saturated, low-permeability, fine-grained soils. The reason for this, as illustrated in **Table 16.3**, is that sandy

Soil parameter	Medium sand	Fine sand	Silt	Clay
Hydraulic conductivity k (m/s)	10^{-3}	10^{-4}	10^{-6}	10^{-8}
Stiffness in one-dimensional compression E'_o (MPa)	100	50	10	2
Time t for completion of transient flow with drainage path length $d = 50$ m	4 min	1.4 h	1 month	40 years

Table 16.3 Indicative times for completion of transient flow with a drainage path length of 50 m (Adapted from CIRIA C750, Preene *et al.*, 2016; www.ciria.org)

soils are so relatively stiff and permeable that the timescale for transient flow is very short – perhaps a matter of minutes rather than months or even years.

While much soil mechanics theory is relatively straightforward, huge simplifications are involved both in idealising the complex behaviour of the soil and in choosing often a single value to characterise highly variable parameters such as hydraulic conductivity, soil stiffness and consolidation coefficient. This is why experience, judgement and a respect for reality are so important in translating soil mechanics theory into geotechnical engineering practice.

16.7. Summary

- The flow of water through the ground is governed by Darcy's Law, and controlled by a parameter known as the *hydraulic conductivity* or *Darcy permeability*.
- Hydraulic conductivity varies widely across the range of soils and can be difficult to determine with confidence, especially on the basis of tests involving small volumes of soil, owing to scale and fabric effects.
- Simple flow regimes can be analysed mathematically or by the use of plane flownet sketching.
- Anisotropy, and the deflection of flowlines as they pass between soil strata of differing hydraulic conductivity, can be taken into account if needed.
- Steps must be taken to lower groundwater levels and pore water pressures in the vicinity of an excavation below the natural water table, to prevent instability of the excavation sides and base. This is usually achieved by pumping from wells in a process known as construction dewatering.
- Transient flow occurs as a soil is changing in volume in response to a change in boundary stress. This leads to the process known as consolidation, which is usually only significant for low-permeability (clay) soils.

References

Bevan MA, Powrie W and Roberts TOL (2010) Influence of large scale inhomogeneities on a construction dewatering system in chalk. *Géotechnique* **60(8)**: 635–649, 10.1680/geot.9.P.010.

Cashman PM and Preene M (2021) *Groundwater Lowering in Construction: A Practical Guide*, 3rd edn. CRC Press, London, UK.

Cedergren H (1989) *Seepage, Drainage and Flownets*, 3rd edn. Wiley, New York, NY.

Powers JP (1992) *Construction Dewatering: New Methods and Applications*, 2nd edn. Wiley, New York, NY.

Powrie W (2018) *Soil Mechanics: Concepts and Applications*, 3rd edn. CRC Press, Boca Raton, FL.

Powrie W and Roberts TOL (1990) Field trial of an ejector well dewatering system at Conwy, North Wales. *Quarterly Journal of Engineering Geology* **23(2)**: 169–185.

Powrie W and Roberts TOL (1995) Case history of a dewatering and recharge system in chalk. *Géotechnique* **45(4)**: 599–609.

Preene M and Powrie W (1992) Equivalent well analysis of construction dewatering systems. *Géotechnique* **42(4)**: 635–639.

Preene M, Roberts TOL and Powrie W (2016) *Groundwater Control: Design and Practice*, 2nd edn. CIRIA Report C750. CIRIA, London, UK.

Roberts TOL, Roscoe H, Powrie W and Butcher D (2007) Controlling clay pore pressures for cut-and-cover tunnelling. *Proceedings of the Institution of Civil Engineers, Geotechnical Engineering* **160(GE4)**: 227–236.

Roberts TOL, Preene M and Powrie W (2023) The changing nature of groundwater control for temporary works. *Proceedings of the Institution of Civil Engineers, Civil Engineering* **176(5)**: 11–20, 10.1680/jcien.22.00128.

All chapters within Sections 1 *Context* and 2 *Fundamental principles* together provide a complete introduction to the Manual and no individual chapter should be read in isolation from the rest.

Chapter 17
Strength and deformation behaviour of soils

John B Burland Imperial College London, London, UK

This chapter explains the basic strength and stiffness properties of soils by frequent reference to their particulate nature. The results of simple shear box tests on sand and clay are presented and compared with the Coulomb strength criterion: $\tau_f = c' + \sigma'_n \tan \varphi'$. It is demonstrated that the angle of shearing resistance φ' depends on a number of factors, including the effective confining pressure, the initial relative density and the stress history. The key roles played by contractancy and dilatancy in controlling undrained strength are explained. The phenomena of critical state strength and residual strength are described, and the extension of the Coulomb strength criterion to the widely used Mohr–Coulomb strength criterion is explained, together with its limitations. The latter part of the chapter deals with the deformation properties of soil and its behaviour under one-dimensional compression. The important influence of inter-particle bonding is introduced, leading onto the concept of yield. In ground engineering, extensive use is made of elasticity in the solution of boundary value problems – such as the settlement of foundations. The behaviour of an ideal isotropic porous elastic material is explored in detail and compared with real soils. The important phenomenon of nonlinear stress–strain behaviour is introduced briefly.

doi: 10.1680/icemge.66816.0195

CONTENTS

17.1. Introduction

This chapter deals exclusively with the behaviour of fully saturated soils. Discussion of the behaviour of unsaturated soils may be found in Chapters 14 *Soils as particulate materials*, 32 *Collapsible soils*, 33 *Expansive soils*, 30 *Tropical soils* and 29 *Arid soils*. As stressed in Chapter 14, a saturated soil of whatever type may be considered as a skeleton of solid particles in contact with each other and with the surrounding voids, which are filled with water. Nevertheless, for the purposes of analysis, it is necessary to idealise the soil as a continuum in which the sizes of the particles are neglected in relation to the sizes of the elements of soil that are being considered. It is essential in making this idealisation to remember that it is the particulate nature of soil that controls its behaviour.

Some of the key controlling features of a particulate material listed in the conclusions to Chapter 14 are as follows.

- The effective stress is that part of the total stress that is transmitted through the particle contact points and controls soil behaviour.
- The closeness of pack of the particles (i.e. the void ratio e) plays an important role in determining the stiffness, strength and permeability of a soil.
- The self-weight of the soil generates the effective confining pressures that govern its strength and stiffness.
- When sheared, loose soils tend to contract and dense soils tend to dilate.
- Particle grading, shape and micro-fabric are important in determining soil properties.
- The presence of discontinuities through the mass of the soil, such as shear and bedding planes, can exert a key influence on the behaviour of the mass.

The element of soil shown in **Figure 17.1(a)**, with a pore pressure u acting in the voids, is idealised as an element of a continuum in **Figure 17.1(b)**. When considering the normal stress σ and shear stress τ acting on any arbitrary plane through the continuum, it is also necessary to consider the pore pressure u acting within the plane. Therefore the effective stresses acting on the plane are $\sigma' = (\sigma - u)$ and τ. In this chapter, strength and stiffness properties of soil are introduced and explained in simple mechanistic terms by referring to their particulate nature. References are given to aid more advanced study.

17.2. Analysis of stress

The analysis of stress is thoroughly covered by most textbooks on applied mechanics and strength of materials, and will be dealt with only briefly here. It is important to appreciate the sign conventions that are adopted in soil mechanics. In particular, compressive stress is taken as positive as tensile stresses seldom occur in practical problems. **Figure 17.2(a)** shows a two-dimensional element referred to Cartesian coordinates (x, z) acted on by compressive stresses σ'_x, σ'_z and τ_{xz} ($=\tau_{zx}$). A convenient sign convention is that, when standing at the origin and looking at the two nearest faces of the element, positive stress is in the direction of increasing x and increasing z. Thus the positive stress σ'_x is acting normal to the constant x-plane directed towards increasing x. Similarly the positive shear stress τ_{xz} is acting parallel to the constant x-plane directed towards increasing z. Note that for equilibrium,

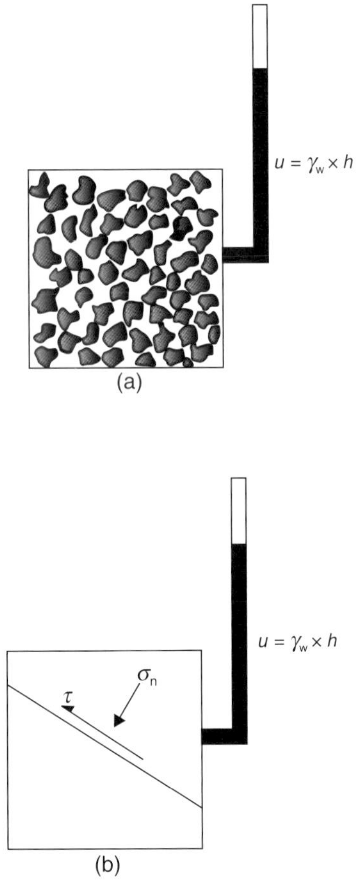

Figure 17.1 (a) An element of particulate material; (b) the element idealised as a continuum

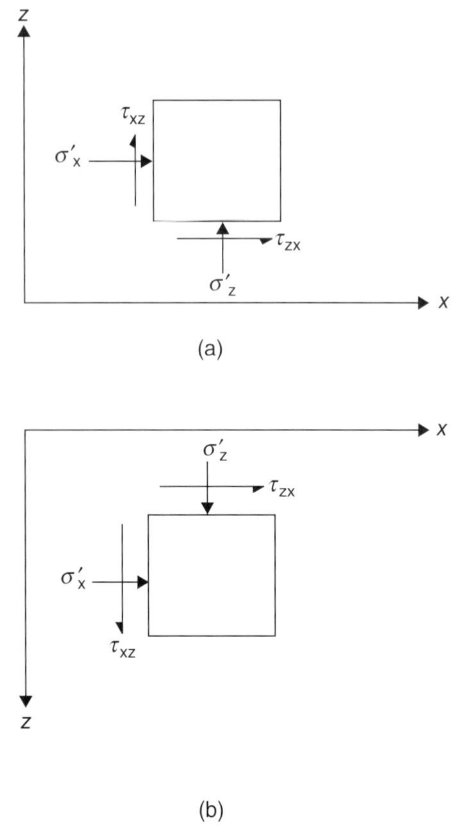

Figure 17.2 Sign convention for positive stresses (a) *z* positive upwards; (b) *z* positive downwards

$\tau_{xz} = \tau_{zx}$, which satisfies the above sign convention. Also, for equilibrium, there are stresses acting on the other two faces in opposite directions from their counterparts. If z is taken as increasing downwards, which is common in geotechnical problems, then the direction of positive stresses acting on the element are as shown in **Figure 17.2(b)**.

17.2.1 Mohr's circle of stress

The stresses shown in **Figure 17.2** may be resolved to find the associated stresses acting on any plane through the element. The equations for such transformation of stress are given in most textbooks. In soil mechanics, extensive use is made of the well-known graphical procedure for transformation of stress by means of the Mohr's circle geometric construction – see, for example, Atkinson (1993). This is a very important tool in soil mechanics as it relates to the Mohr–Coulomb strength criterion, which is universally used in soil mechanics. When using the Mohr's circle construction it is often not appreciated that it is necessary to make a small change to the sign convention for the shear stresses shown in **Figure 17.2**. For the geometric construction to work, it is necessary to plot anticlockwise shear stresses as positive and clockwise shear stresses as negative. Thus the shear stress τ_{zx} shown in **Figure 17.2(a)** is plotted as positive in a Mohr diagram, whereas the shear stress τ_{xz} is plotted as negative. It is important to stress that this sign convention only applies to Mohr's circles and not to stress analysis in general.

Figure 17.3(a) shows an element acted upon by major and minor principal stresses σ'_{I} and σ'_{III}. The stress state can be represented by a circle on a Mohr diagram of shear stress τ against normal stress σ, as shown in **Figure 17.3(b)**. The points representing the major and minor principal stresses plot on the σ' axis since there are no shear stresses on the planes on which they act.

An important feature of a Mohr's circle is the pole (sometimes called the 'origin of planes'). If the pole can be located, then the stresses acting on any plane can be found very simply. The pole is located by drawing a line parallel to the plane on which the stress acts through a known stress on the circle. Where the line intersects the circle again locates the pole. Thus, in **Figure 17.3(b)**, the line I–I is parallel to the principal plane on which σ'_{I} acts and passes through the point on the Mohr's circle representing σ'_{I}. Where this line intersects the circle again represents the pole.

Having found the pole, the stresses acting on any other plane may be found by drawing a line through the pole, parallel to

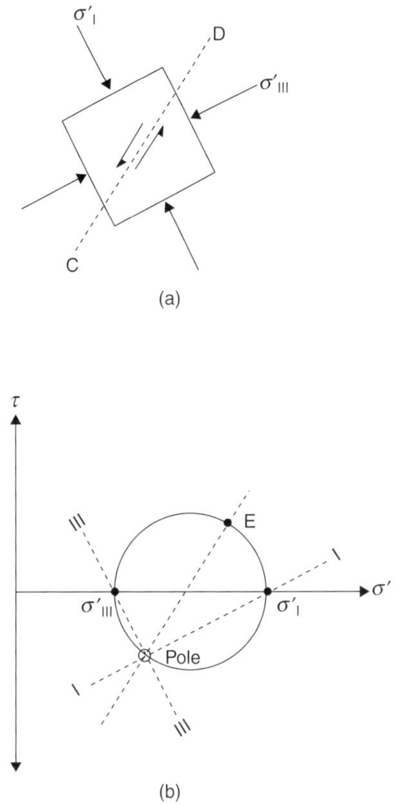

Figure 17.3 Application of Mohr's circle for the transformation of stress. For this construction anticlockwise shear stresses are plotted as positive

the plane. Where the line intersects the circle again gives the required stress state. For the case shown in **Figure 17.3(b)**, the point E represents the stresses acting on the plane CD through the element. It can be seen that the shear stress is positive so that the shear stresses acting on either side of the plane CD are in an anticlockwise direction, as shown.

17.3. The drained strength of soils

17.3.1 Coulomb's equation

In 1773 Coulomb read a paper to the French Royal Academy of Science with the title 'On the application of the rules of maximum and minimum to some statical problems, relevant to architecture' (Heyman, 1997). Besides dealing with the bending of beams and the stability of arches, Coulomb discussed the strength of masonry columns and pressures of 'fresh earth' against retaining walls. This was a most remarkable contribution which is still widely used in geotechnical practice.

For this work, Coulomb assumed that sliding takes place along an arbitrary plane on which the material has both friction and cohesion. By varying the orientation of the sliding plane he was able to work out the minimum value of the column load to give failure and the maximum thrust against a gravity-retaining wall. Although he never explicitly stated it as such, the strength criterion adopted by Coulomb, expressed in terms of effective stresses, is given by the equation

$$\tau_{\text{f}} = c' + \sigma'_{\text{n}} \tan \varphi' \tag{17.1}$$

where τ_{f} is the shear stress at failure on the plane, σ'_{n} is the effective stress normal to the plane, c′ is the effective cohesion and φ' is the effective angle of shearing resistance (frequently referred to as the effective friction angle). However, the author prefers the term 'effective angle of shearing resistance' as its value depends on other (more dominant) factors than inter-particle friction, such as particle shape and grading – see Chapter 14 *Soils as particulate materials*. Equation 17.1 is known as Coulomb's equation; it plots as two straight lines in a Mohr diagram, as shown in **Figure 17.4** for positive and negative shear stress. The following sections examine the results of laboratory tests using a simple shear box and compare them with the Coulomb equation.

17.3.2 Results of drained shear box tests on granular materials

For drained shearing, the shear stress is applied slowly so that the pore water can drain into or out of the soil sample as it dilates or contracts. Roscoe (1953), working at Cambridge University, developed a sophisticated shear box within which a soil sample is caused to deform uniformly in simple shear, as shown in **Figure 17.5**. Drained tests were carried out on a variety of granular soils and clear patterns of behaviour emerged. **Figure 17.6** shows the results for some simple shear tests carried out on samples of sand at various initial relative densities ranging from very loose (Test 1) to very dense (Test 4). All the tests were carried out at the same normal effective stress σ'_{n}. The results are plotted in **Figures 17.6(a–c)**.

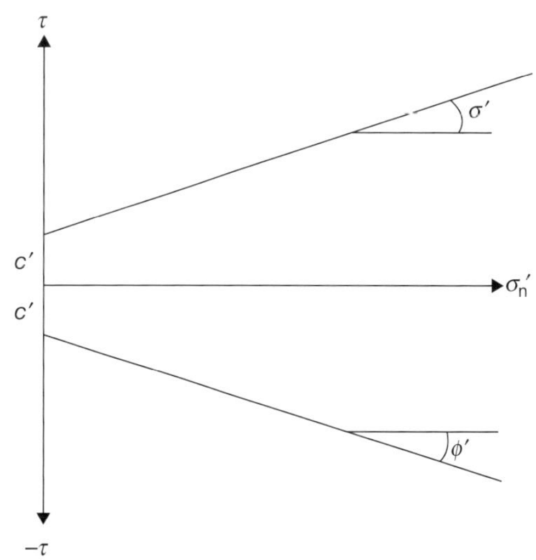

Figure 17.4 The Coulomb strength criterion

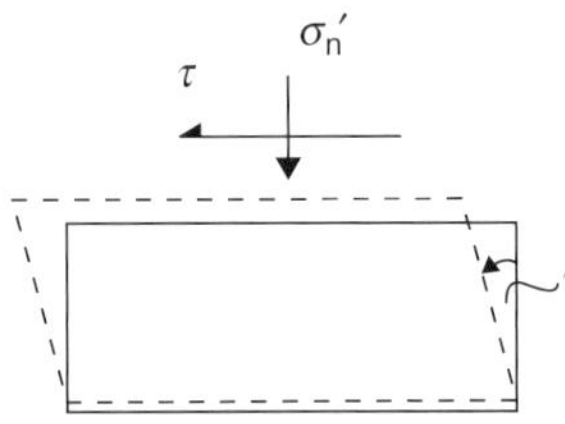

Figure 17.5 The mode of deformation of the Cambridge simple shear apparatus

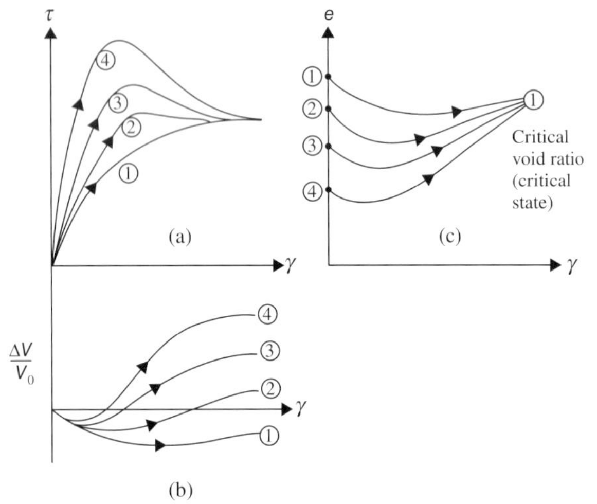

Figure 17.6 Results of drained tests on sand at various initial relative densities in the simple shear apparatus, carried out at the same value of σ_n'

Figure 17.6(a) shows graphs of shear stress τ against the shear strain γ. For the initially very loose sample (Test 1) it can be seen that the stress–strain curve rises at a decreasing rate until a maximum shear stress is reached, which then remains approximately constant as further deformation takes place. **Figure 17.6(b)** shows the corresponding volumetric strains during the test. It can be seen that as shearing starts the sample reduces in volume (contracts) initially, then increases very slightly and thereafter the volume remains constant for further deformation. Overall the sample can be seen to be contractant during shearing.

The above behaviour can be contrasted with that of an initially very dense sample (Test 4). In **Figure 17.6(a)** it can be seen that the stress–strain curve rises steeply and reaches a maximum (peak) value, after which the strength reduces. Ultimately, with continued deformation, the shear stress reaches the same value of strength as the initially very loose sample. In **Figure 17.6(b)** it can be seen that after an initial small contraction, the very dense sample begins to dilate, reaching a maximum rate of dilation at peak strength. Thereafter, the rate of dilation decreases until ultimately there is no further change in volume as shearing continues. Two samples of intermediate relative density show behaviour which lies between that of the initially very loose (Test 1) and very dense (Test 4) samples.

One of the key features of the Cambridge simple shear apparatus is that uniform deformations are imposed on the boundaries of the sample so that changes of void ratio can be measured throughout the test, right through to ultimate conditions at quite high shear strains. **Figure 17.6(c)** shows the changes of void ratio for each sample as the shear strains increase. The initially very loose sample starts at a high void ratio which reduces during most of the test. In contrast, the initially very dense sample starts at a low void ratio which then increases during the test. The important observation was made that all four samples ultimately ended up at the same void ratio, termed the 'critical void ratio'. It was further concluded that when a sample of soil reaches a condition in which it continues to shear at constant shear stress and constant void ratio, it has reached what is termed the 'critical state'.

All the tests described above were carried out at the same normal effective stress. **Figure 17.7** shows the results of tests at the four initial void ratios, but with different values of vertical effective stress. **Figure 17.7(a)** shows that the value of the critical void ratio reduces as the vertical effective stress increases, giving what has been termed the 'critical void ratio line'. In **Figure 17.7(b)** the peak strengths and ultimate strengths are plotted for each of the samples in a Mohr diagram. It can be seen that the ultimate strength line is straight and passes through the origin, having a slope defined as φ_{cs}' – the critical state angle of shearing resistance. However, the peak strength lines, particularly for the denser samples, are curved and lie above the critical state strength line. It can be shown that this upward displacement of the strength envelopes, together with their curvature, is due to the additional work done during shearing as a result of dilation.

The following broad conclusions can be drawn from these results.

- The effective confining pressure plays a dominant role in controlling the strength of a particulate material.
- The initial density influences the peak strength. The greater the initial density, the steeper and more curved is the peak strength envelope in a Mohr diagram of τ against σ_n'.
- After a large shear strain, the shearing resistance of loose and dense samples under the same vertical effective stress converge to the same value. Similarly, the void ratios of each sample converge to the same value. This ultimate state is known as the 'critical state' in which a sample at a given value of vertical effective stress will continue to shear at constant shear stress and constant void ratio (i.e. at the critical state, the rate of dilation is zero).
- On a Mohr diagram, the critical state strength line is straight and passes through the origin with an angle of shearing resistance φ_{cs}'.
- The initial density and the magnitude of σ_n' influence the amount of dilation during shear. The greater the initial density and the lower the magnitude of σ_n', the greater is the dilation. It is noted that the maximum rate of dilation corresponds approximately to peak strength. The curvature of the strength envelope is related to the work done during dilation against the confining pressure.

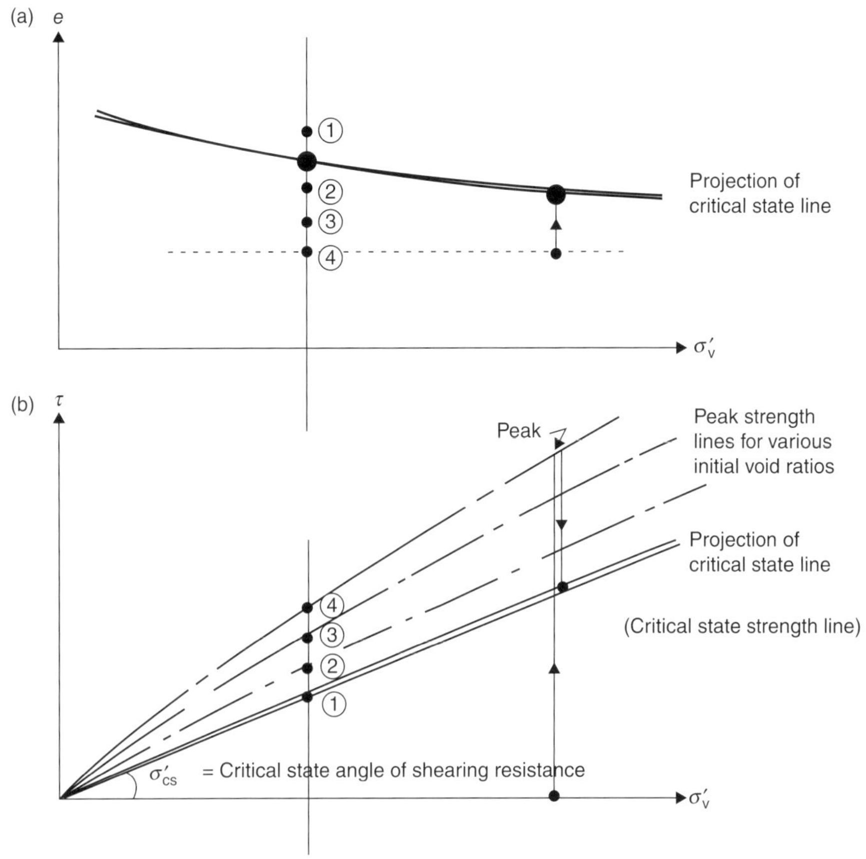

Figure 17.7 Results of drained tests on sand at various initial densities in the simple shear apparatus, carried out at various values of σ'_n

- At a given normal effective stress σ'_n, the peak strength τ_p is greater than the critical state strength τ_{cs}. The strength difference $(\tau_p - \tau_{cs})$ is due to the rate of dilation and is sometimes termed the 'enhanced strength'.

17.3.3 Results of drained shear box tests on clay soils

The basic strength properties of clay soils have been researched using samples that have first been formed with high moisture content and then consolidated to various values of effective normal stress in a shear box, giving what is called 'the virgin compression line'. Such samples are described as remoulded or 'reconstituted'. **Figure 17.8(a)** shows the virgin compression line of void ratio *e* against σ'_v for a clay soil that has been formed in this way. At given values of effective normal stress σ'_v (points A and B), drained shear box tests are carried out. The behaviour is very similar to that of loose sand. **Figure 17.8(c)** shows the stress–strain curves of τ against γ, which rise at a decreasing rate until a maximum shear stress is reached, which then remains approximately constant as further deformation takes place (points A′ and B′). It can be seen from **Figure 17.8(c)** that as shearing starts, the samples contract at a rate that progressively decreases until at maximum strength there is no further volume reduction and void ratios remain constant (points A′ and B′ in **Figure 17.8(a)**). Thus the samples appear to reach a critical state condition. When plotted on a Mohr diagram of τ against σ'_v, the strengths lie on a straight line through the origin, as shown in **Figure 17.8(b)**, having an angle of shearing resistance φ'_{cs} – the critical state strength line.

Overconsolidated reconstituted clay soils behave in a rather similar manner to dense sands. **Figure 17.9(a)** shows the relationship between void ratio *e* and σ'_v for a sample of reconstituted clay that has been consolidated to a high value of σ'_v (point B), after which the effective stress has been progressively decreased, causing the sample to swell and come to equilibrium at a much lower value of σ'_v (point C). If a drained shear box test is now carried out with σ'_v held constant, the stress–strain curve (see **Figure 17.9(c)**) will reach a peak value of τ at *C*′, which then gradually reduces as shearing continues. At the same time, the sample will tend to dilate so that the void ratio increases. When plotted on a Mohr diagram, the peak strength *C*′ lies above the corresponding critical state value and the strength envelope for the soil is curved as shown in **Figure 17.9(b)**.

17.3.4 Residual strength of clay soils

As described in Section 17.3.2, for dense granular soils, tests show that after the strength reaches a peak value it gradually reduces to a steady-state value corresponding to a critical state condition. It is often assumed that overconsolidated clay

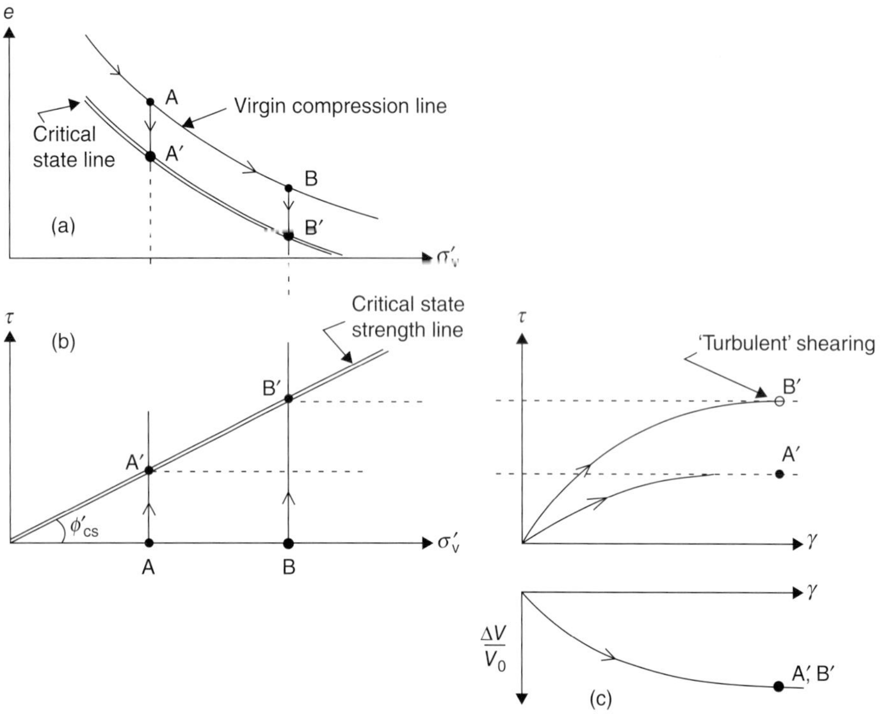

Figure 17.8 Results of drained simple shear tests on a reconstituted normally consolidated clay

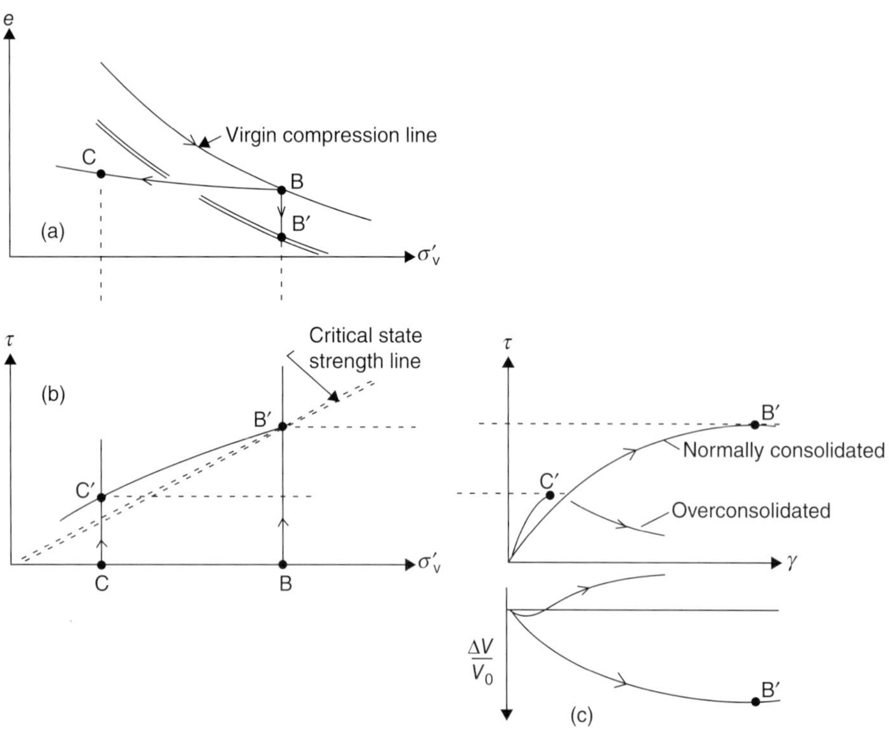

Figure 17.9 Results of drained simple shear tests on a reconstituted overconsolidated clay

soils behave in a similar way. However, the behaviour of clay soils is complicated by the fact that the shearing process can dramatically change the fabric (arrangement of the particles). Ring shear box tests on soils with a high clay content show that after peak strength is reached, the clay particles progressively align themselves in the zone of shearing until a polished shear surface is formed. In so doing, the strength steadily decreases with increasing displacement across the shear plane until the clay particles are fully aligned and a low 'residual strength' is reached – see **Figure 17.10**. Lupini *et al.* (1981) termed this 'sliding shearing'. They contrasted this with shearing at the critical state, which they termed 'turbulent shearing' to convey the idea that during continued shearing at the critical state, the particles are tumbling over one another in a turbulent manner.

Figure 17.11 shows the comparison between the critical state angle of shearing resistance φ'_{cs} and the residual friction angle φ'_r for a sand–bentonite mixture in which the bentonite content was steadily increased. It can be seen that for a clay fraction of less than 20% there is little difference between φ'_{cs} and φ'_r. However, as the clay fraction increases the values of φ'_r become progressively less than φ'_{cs}. For an overconsolidated clay tested in a ring shear apparatus, the strength drops smoothly from peak to residual, as shown in **Figure 17.10**, and it is not possible to distinguish an obvious critical state strength. It seems that the simplest way of measuring the critical state strength is on normally consolidated reconstituted samples of the soil which contract during shearing.

17.3.5 The influence of bonding on the drained strength of clay soils

In the previous section, the influence of changes in microfabric on the drained shearing resistance of clay soils was described. Another important influence is the effect of interparticle bonding, which most natural clays exhibit to a greater or lesser extent. Bonding has the effect of increasing the strength and stiffness of the material. Tests on initially intact, natural, stiff, overconsolidated clays show that the breaking of the bonds results in a very rapid post-peak reduction of strength with the formation of well-defined rupture surfaces – see **Figure 17.12**. The strength along the rupture surface, known as the post-rupture strength, has an effective angle of shearing resistance φ'_{pr}. This is remarkably similar to the critical state value for the reconstituted material, but often with a small effective cohesion c'. In view of the very brittle post-peak behaviour of such soils, it would be unwise for the design to rely on strengths exceeding the post-rupture values.

17.4. The undrained strength of clay soils

When a shear stress is applied to a clayey soil of relatively low permeability, it takes a considerable time for the pore water to drain into or out of the pores. When shearing takes place with little or no drainage, the process is defined as 'undrained shearing', and the measured strength is defined as the 'undrained strength' S_u. The tendency for a soil to contract or dilate during drained shearing has a profound effect on the undrained strength of the material. This behaviour can be illustrated by considering elements of soil sheared undrained in a shear box, in which the applied normal stress σ_n is kept constant during the test.

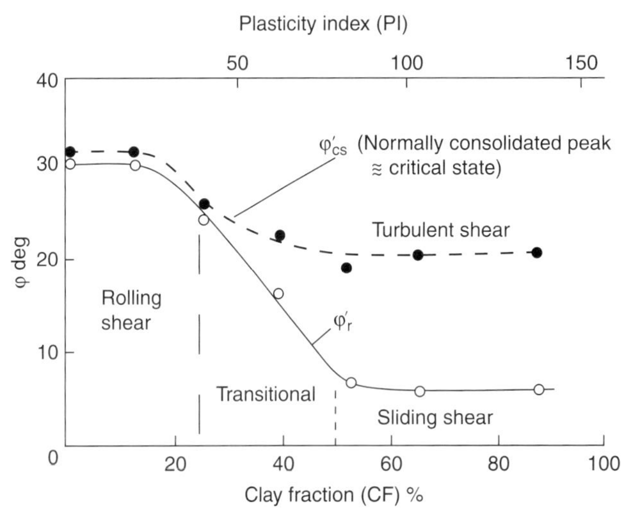

Figure 17.11 Ring shear tests on sand–bentonite mixtures. Normally consolidated at $\sigma' = 350$ kPa; PI/CF = 1.55 (Data taken from Lupini *et al.*, 1981)

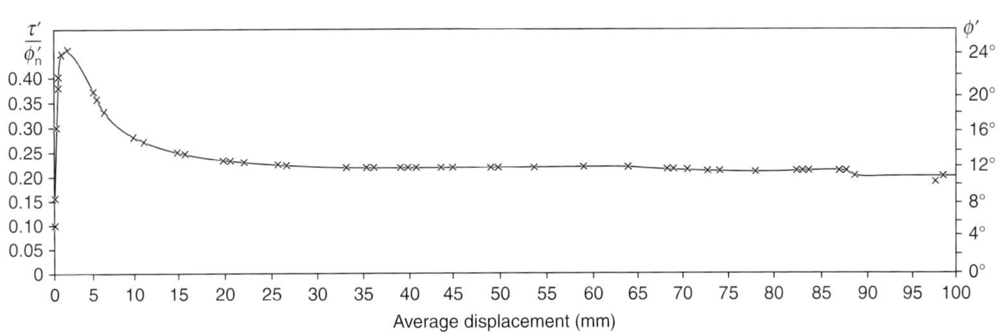

Figure 17.10 Results of drained ring shear tests on an undisturbed sample of Blue London Clay. Nominal normal stress $\sigma'_n = 207$ kPa; nominal shearing rate 0.0076 mm/min (Redrawn with data taken from Bishop *et al.*, 1971)

For a loose (normally or lightly overconsolidated) sample, as τ is increased, the element will try to contract – but is prevented from doing so. As a consequence, stress is thrown onto the pore water so that the pore pressure u increases and, since σ_v is constant, σ'_v decreases. The resulting effective stress path and stress–strain behaviour are shown by the full lines in **Figure 17.13**. It can be seen that the undrained strength S_u (given by point A) is less than the corresponding drained strength (given by point B). If, at a given value of τ, drainage now takes place with τ and σ_v kept constant, the positive pore

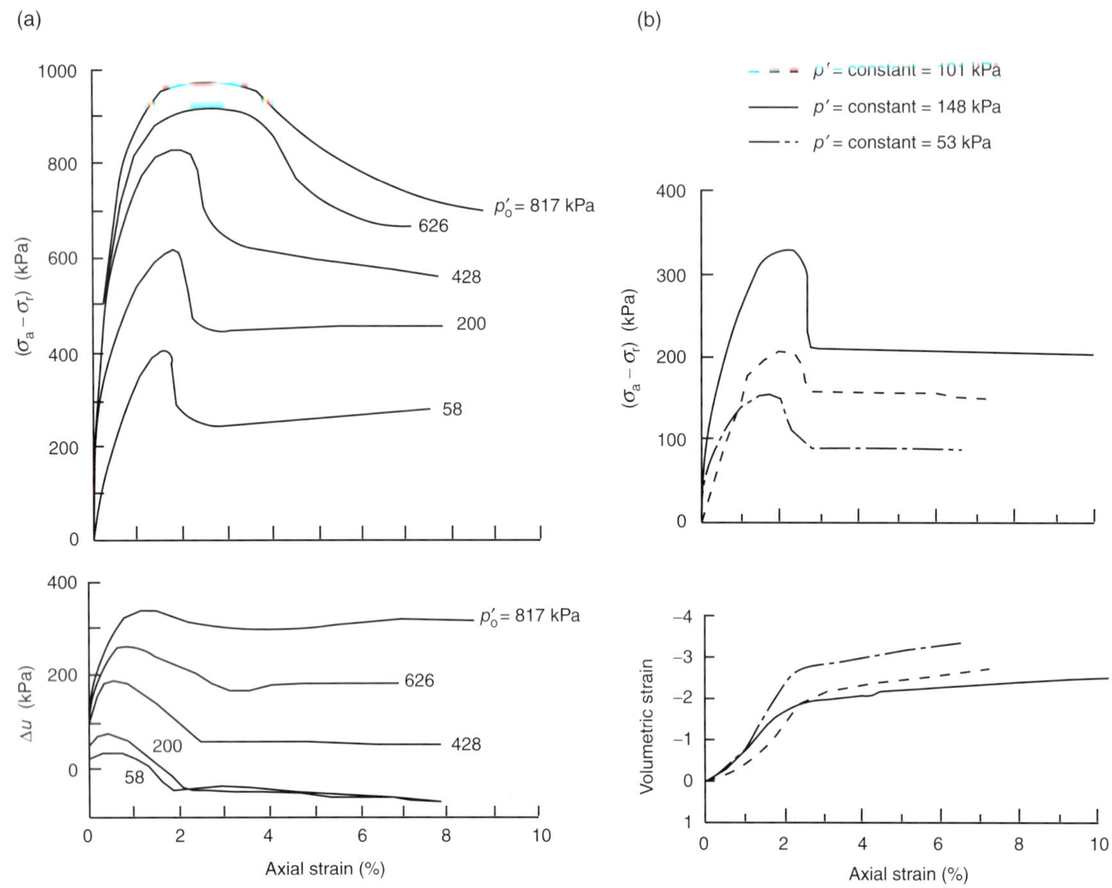

Figure 17.12 Results of (a) drained and (b) undrained triaxial tests on intact Vallerica Clay showing very brittle post-peak strength reductions (Reproduced from Burland *et al.*, 1996)

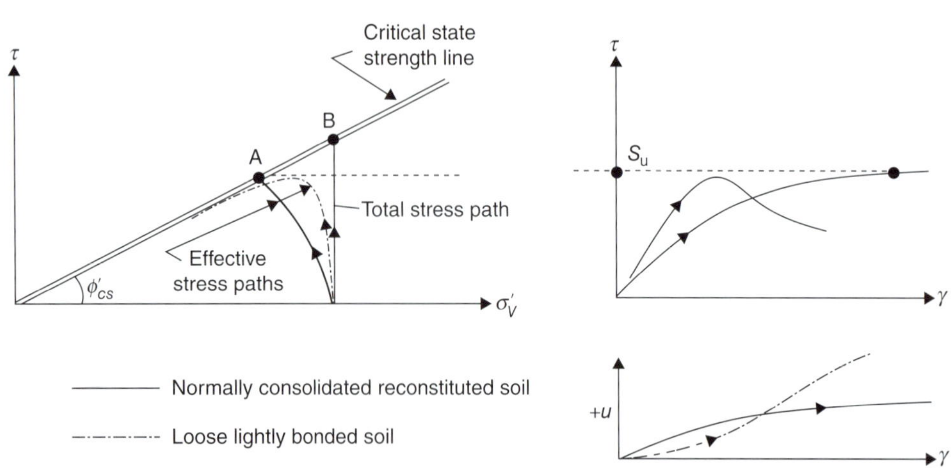

Figure 17.13 Undrained shearing of a normally consolidated clay

pressure dissipates, the volume decreases and the strength of the sample will increase. *Hence contractant soils tend to get stronger as they drain after the undrained application of a shear stress.*

For an initially dense (overconsolidated) sample, as τ is increased the element will try to dilate but is prevented from doing so. In contrast to the loose sample, the pore pressure u decreases as the soil tries to dilate and hence σ_v' increases. The resulting stress path and stress–strain behaviour are shown by the full lines in **Figure 17.14**. It can be seen that the undrained strength S_u (point A) is greater than the drained strength (point B). If, at a given value of τ, drainage now takes place with τ and σ_n kept constant, the negative pore pressure dissipates, the volume increases and the strength of the sample will decrease. Clearly if the drained strength is less than the applied shear stress τ, failure will occur as drainage is taking place. This is the cause of long-term (or delayed) slope failures in overconsolidated cut slopes – see Chapter 23 *Slope stability*. Hence dilatant soils tend to get weaker as they drain after the undrained application of a shear stress.

The chain dotted lines in **Figure 17.13** represent the undrained behaviour of an initially very loose soil with some bonding between the particles. The material is highly contractive and positive pore pressures continue to develop after peak undrained strength is reached, and the bonds are progressively broken down. As a consequence, the strength reduces and the effective stress path continues to travel to the left, moving down the critical state strength line. Hence loose soils, such as normally and lightly overconsolidated natural clays, can exhibit very brittle undrained shearing behaviour with significant loss of strength after the peak has been reached, particularly if they are lightly bonded or cemented.

The degradation of undrained strength upon remoulding is termed 'sensitivity', which is defined as the undrained strength of a clay divided by its remoulded strength at the same moisture content.

17.5. The Mohr–Coulomb strength criterion

So far this chapter has only considered the strength of a mass of soil in terms of the classical Coulomb equation (shown in Equation 17.1) applied to a shear box in which the normal effective stress σ_n' and the shear stress τ are known. This equation gives no information about the principal stresses or their orientation. Mohr showed that when a material fails according to the Coulomb criterion, the Mohr's circle of stress corresponding to the state of failure is tangential to the Coulomb strength line, as shown in **Figure 17.15**.

Figure 17.16 shows the Mohr's circle of failure for a drained triaxial compression test for which the axial effective stress σ_a' is the major principal effective stress $(=\sigma_I')$ and the radial effective stress σ_r' is the minor principal effective stress $(=\sigma_{III}')$. From simple geometry, it can be shown that σ_{III}' is related to σ_I' by the expression:

$$\sigma_{III}' = \sigma_I' \tan^2(\pi/4 - \varphi'/2) - 2c' \tan(\pi/4 - \varphi'/2) \qquad (17.2)$$

which is known as the Mohr–Coulomb strength criterion.

In **Figure 17.16**, since σ_a' $(=\sigma_I')$ acts normal to a horizontal plane, the pole coincides with the σ_{III}' point as shown. The lines connecting the pole to the points of tangency of the Mohr's circle of failure are often termed 'stress characteristics' or 'failure planes'. It can be seen that the failure planes make an angle of $\sigma_n' \pm (\pi/4 + \varphi'/2)$ to the major principal plane.

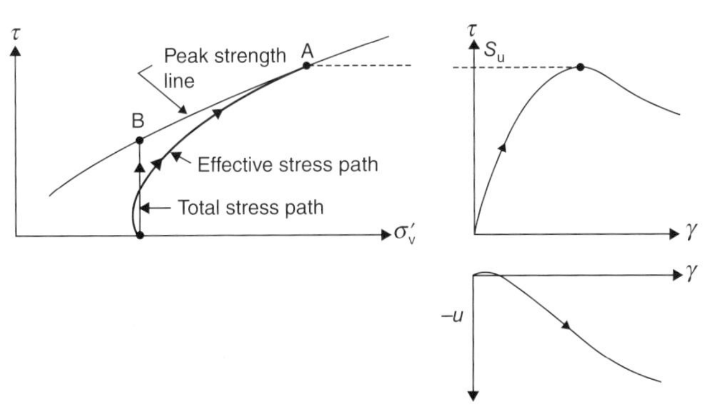

Figure 17.14 Undrained shearing of an overconsolidated clay

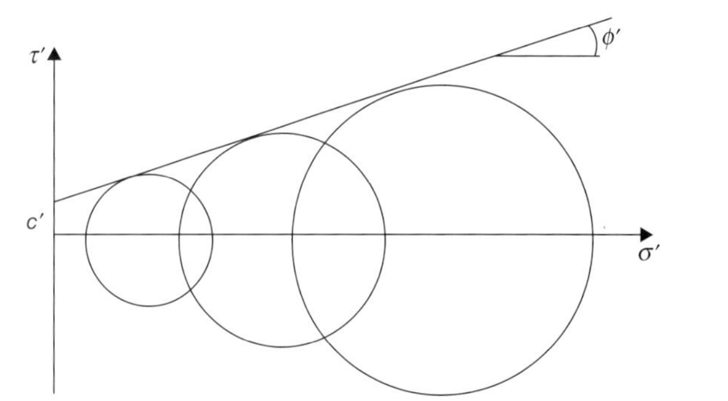

Figure 17.15 The Mohr–Coulomb strength criterion

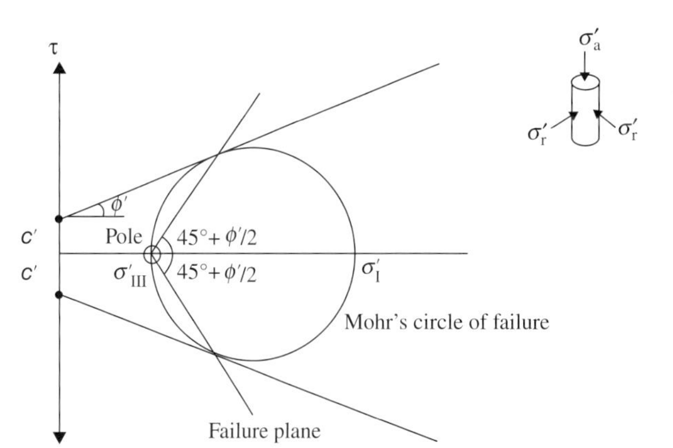

Figure 17.16 Failure in a triaxial compression test

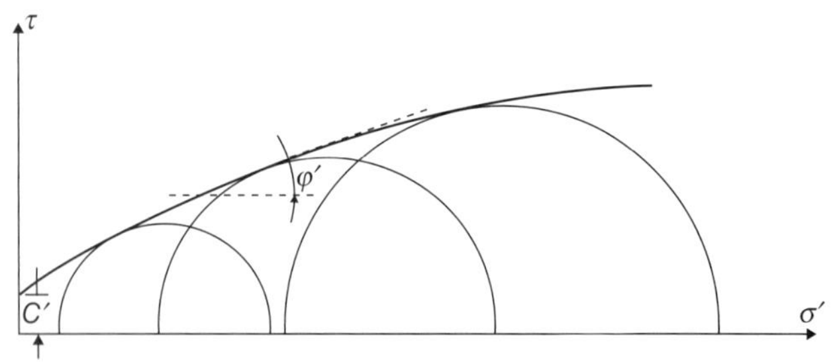

Figure 17.17 Curved strength envelope drawn tangential to Mohr's circles of failure

The measured strength envelope from a series of shear box tests is often curved. In this situation the Mohr's circles are still tangential to the envelope. By carrying out a number of drained or undrained triaxial tests with pore pressure measurements, it is possible to construct the strength envelope by constructing the Mohr's circles of effective stress corresponding to failure. The best fit failure envelope is then sketched in tangential to these circles, as shown in **Figure 17.17**. In this particular case, an effective cohesion intercept c' is indicated which will only exist if there is some inter-particle bonding at very low confining pressures.

17.6. Choice of strength parameters for analysis and design

The previous sections identified at least three common definitions of strength: peak strength, critical state strength and residual strength. Peak and critical state strength may be applied to either drained or undrained conditions. Residual strength is usually a drained condition.

The Coulomb equation is a simplification. First, the peak value of φ' is not unique to a given soil but depends on its initial void ratio or density. Second, the strength envelope is often curved and therefore does not strictly satisfy the Coulomb equation. As a consequence of these observations, which effective strength parameters to use in an analysis that makes use of the Mohr–Coulomb strength criterion is not straightforward. A simple practical approach is to choose a linear failure envelope defined by values of φ' and c' that are appropriate to the range of normal effective stresses encountered in the problem. In these circumstances it is important to appreciate that the chosen design value of c' is an analytical convenience rather than a true measure of the cohesion of the material.

The choice of a design value of undrained strength S_u is also not straightforward. The key point to remember about undrained strength is that it is not a unique quantity for a given deposit of soil or soil sample. Its value depends on at least three factors: (*a*) the rate at which the sample is tested (i.e. S_u is rate-dependent – the higher the rate the greater the strength); (*b*) the undrained strength of most soils is anisotropic and depends on the orientation of the principal stresses; and (*c*) for stiff fissured clays, the undrained strength depends on the size of the sample being tested – the mean strength of 100 mm dia. samples of stiff fissured clays will usually be significantly lower than that of 35 mm dia. samples. The important subject of choice of strength parameters is covered in Chapter 27 *Geotechnical parameters and safety factors.*

17.7. The compressibility of soils

This section moves from discussing the strength of soils to their deformability. Because of their particulate nature, soils are compressible – especially clay soils. As discussed in Chapter 14 *Soils as particulate materials*, increases in all-round effective stress cause slip to take place between particles which then take on a closer pack; as a consequence, water is squeezed out of the voids. The time taken for the water to be expelled is a function of the permeability and compressibility of the soil. Perhaps the earliest, and still most widely used, apparatus for measuring the compressibility of soils is the oedometer, in which a sample of soil is compressed one-dimensionally as shown in **Figure 17.18**. Detailed descriptions of the test and its application are given in most textbooks on the subject (e.g. Knappett and Craig, 2020; Powrie, 2018).

17.7.1 The coefficient of volume compressibility (m_v)

Explored here are the one-dimensional (confined) compression characteristics of an initially very loose sample of soil that is placed in an oedometer, subjected to a number of increments of increasing vertical effective stress with full drainage between increments, and then subsequently unloaded. **Figure 17.19(a)** shows a graph of void ratio e against vertical effective stress σ'_v for the sample, which is termed the 'virgin compression line'. It can be seen that this compression line ABC is concave upwards, showing that as σ'_v increases the soil becomes increasingly less compressible. The coefficient of volume compressibility m_v is defined as the increment in volumetric strain $\Delta V/V_0$ divided by the corresponding increment in vertical effective stress σ'_v. Consider a state B on the compression line for the sample corresponding to a void ratio e_0 and a vertical effective stress σ'_{v0}. An increment $\Delta\sigma'_v$ causes a decrease in void ratio Δe. Since the initial volume of the sample per unit volume of solid is $(1 + e_0)$ and the corresponding change of volume is equal to Δe, it follows that

$$m_v = \frac{\Delta e \,/\, (1 + e_0)}{\Delta \sigma'_v} \quad (17.3)$$

Clearly m_v is not a constant, but it has to be determined for the appropriate stress range. It is usually expressed in the units m^2/MN – the inverse of stress.

17.7.2 Simple mechanistic explanation of compression and swelling

The soil sample is loaded to a vertical effective stress σ'_{vp} (point C in **Figure 17.19(a)**) and then unloaded in increments.

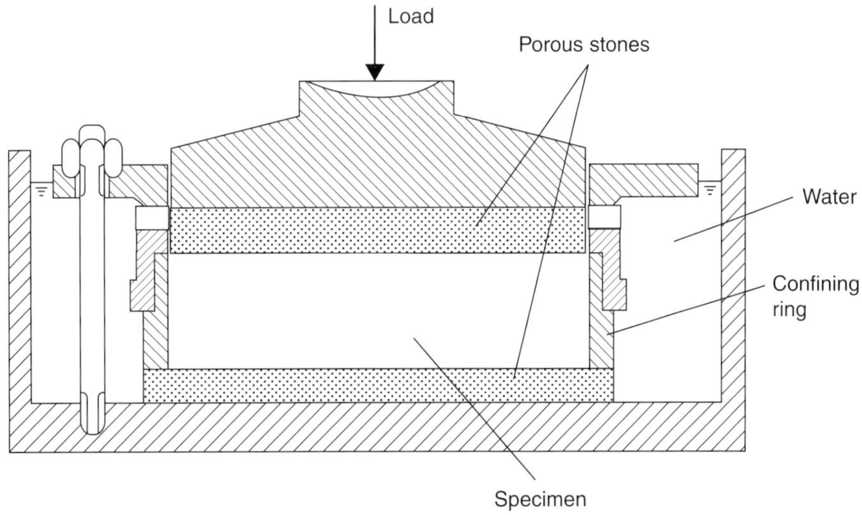

Figure 17.18 The oedometer

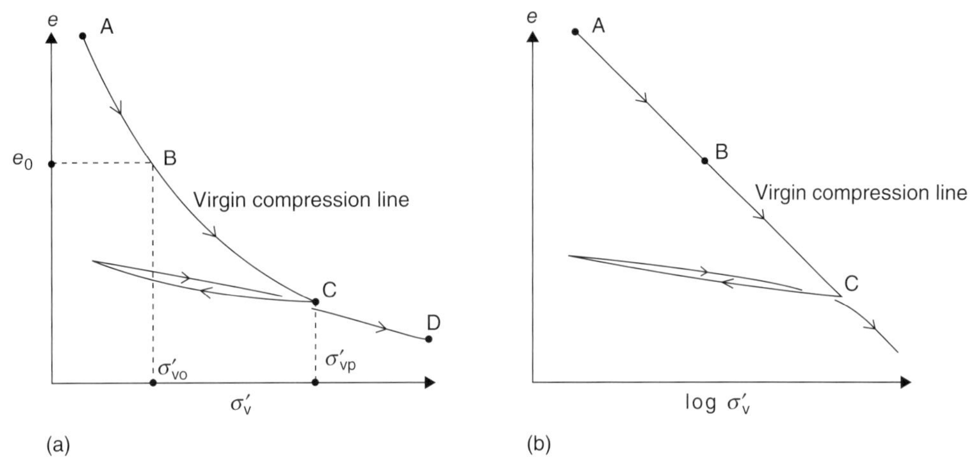

Figure 17.19 One-dimensional (confined) compression characteristics of an initially very loose soil subject to loading, unloading and re-loading: (a) σ'_v plotted to a natural scale; (b) σ'_v plotted to a logarithmic scale

It can be seen that the sample increases in volume as unloading takes place. The unloading (or swelling) curve is much flatter than the initial compression line. A simple mechanistic explanation of this behaviour is that during initial loading, much of the volume reduction results from slip at the particle contact points as the particles take on a closer pack. On unloading, little 'reverse' slip occurs and much of the recovery results from elastic unloading of the particles and, in the case of clays, absorption of water into the clay particle lattice structure.

The sample is now re-loaded to an effective stress corresponding to point D which is greater than σ'_{vp}. Initially, the re-loading compression line lies a little above the unloading curve – until the stress approaches the maximum previous stress σ'_{vp} where it crosses the unloading curve and rejoins the virgin compression line. The simple mechanistic explanation of this behaviour is that initially, on re-loading, the material is responding approximately reversibly, and the particles deform with little slip at the grain contact points. However, once the effective stress approaches the previous maximum value, slip at the contact point begins to restart and the re-loading compression line rejoins the original 'virgin compression line' as shown. Thus, the maximum previous effective stress acts as a kind of 'preloading' such that, at lower stresses, the compressibility of the sample is much less than on first loading. Once the 'preloading' is exceeded, significant slip at the particle contact points restarts. This concept of the onset of non-reversible particle contact slip is very important and is analogous to 'yield' of a ductile metal, which signals the onset of irrecoverable (plastic) strains.

17.7.3 The compression curve plotted on semi-logarithmic axes

It is common practice to plot the compression and swelling lines illustrated in **Figure 17.19(a)** using a log scale for the stress axis, as shown in **Figure 17.19(b)**. When plotted in this manner, the virgin compression and swelling lines approximate to straight lines. The slope of the virgin compression line in a semi-logarithmic plot is defined as the compression index, C_c, and for any two points on the approximately linear portion, it is given by

$$C_c = \frac{e_0 - e_1}{\log(\sigma'_{v1}/\sigma'_{v2})} \tag{17.4}$$

The slope of the unloading part of the $e - \log \sigma'_v$ plot is referred to as the swelling index C_s.

For the re-loading line, the sharp downward curvature as it rejoins the virgin compression line has traditionally been used to determine the maximum past pressure σ'_{vp}. As will be explained in the following sections, some care is needed in interpreting oedometer curves in this way, as the sharp downward curvature also indicates the breaking of inter-particle bonds.

17.7.4 The influence of inter-particle bonding on compression and yield

Most natural soils have some inter-particle bonding (due to cementation or physico-chemical effects) which has an important influence on the compression characteristics. Physico-chemical bonds can develop over quite a short period of time in clay soils. Their influence was demonstrated by Leonards and Ramiah (1959) in some classic experiments, as reported by Burland (1990). A clay soil was reconstituted and compressed in an oedometer for which special precautions were taken to minimise wall friction. At a given value of vertical effective stress, the load was held constant for 12 weeks. After this ageing period, the loading process was then continued.

The results are illustrated schematically in **Figure 17.20** on a graph of e against log σ'_v. As σ'_v was incremented daily, the soil compressed along the virgin compression line. At a stress σ'_{v0}, the load was maintained constant for 12 weeks, during which time some creep compression took place and the void ratio of the soil dropped below the virgin compression line. When loading was resumed, the resulting compression curve crossed to the right of the virgin compression line, then bent downwards sharply and rejoined the virgin compression line from above. Some additional tests were carried out in which creep was prevented during the ageing process. Once again, the subsequent compression line moved out to the right of the virgin compression line before bending down sharply and rejoining it as before. The downward 'kink' in the compression curve results from the breaking of the particle bonds and the onset of inter-particle slip. In the case of Leonards and

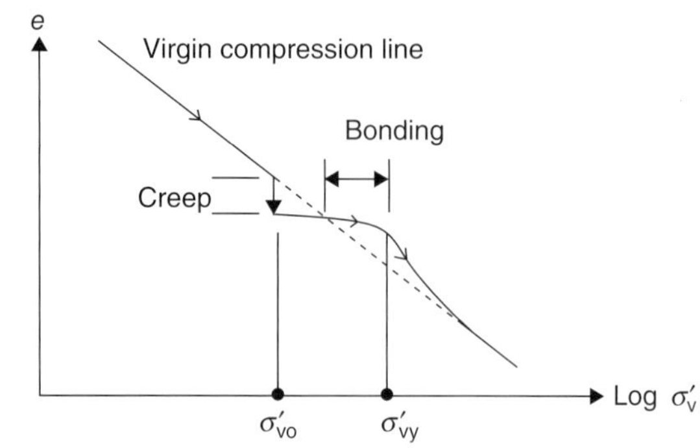

Figure 17.20 Influence of ageing and inter-particle bonding on the compression characteristics of a reconstituted normally consolidated clay

Ramiah's experiments, this 'kink' occurred at a stress equal to about 1.5 times the value of σ'_{v0}.

The important conclusion from the above work is that care is needed in adopting the classical interpretation of the point of maximum curvature of an oedometer compression curve as indicating the maximum past effective overburden pressure. What it actually indicates is the stress at which significant particle slip begins to take place, often due to the breaking of inter-particle bonds. For this reason, Burland (1990) recommended that this point be referred to as the vertical yield stress σ'_{vy} rather than the pre-consolidation pressure. The latter term should be reserved for situations in which the magnitude of such a pre-consolidation pressure can be established by geological means. It is the norm rather than the exception that normally consolidated natural clays have values of $\sigma'_{vy}/\sigma'_{v0}$ (termed the yield stress ratio) as high as 1.5. In cases where cementation has taken place, the yield stress ratio can be very much higher.

The accurate determination of σ'_{vy} is of great practical importance – if vertical effective stresses can be kept below this value, then settlements will generally be small. If, for soft clays, the vertical yield stress is exceeded, very large movements can take place. The vertical yield stress is also useful as an empirical indicator of undrained compressive strength S_{uc}. It has been found that the ratio S_{uc}/σ'_{vy} is roughly in the range 0.28–0.32 for a wide variety of natural clays (Burland, 1990).

Sometimes the use of an e against log σ'_v plot can be misleading when attempting to evaluate σ'_{vy}. For example, a linear line on an e against σ'_v plot becomes a convex upwards curve on an e against log σ'_v plot. It is not unknown for such a line to be interpreted as exhibiting a pre-consolidation pressure or yield stress! A more reliable approach is to plot the data on double log axes as suggested by Butterfield (1979). **Figure 17.21** is taken from Butterfield (1979), who plotted $\ln(1 + e)$ against log σ'_v for four sets of data given by Taylor (1948). It is evident that the double logarithmic plots give a clearer indication of the vertical yield stress than do the semi-logarithmic plots.

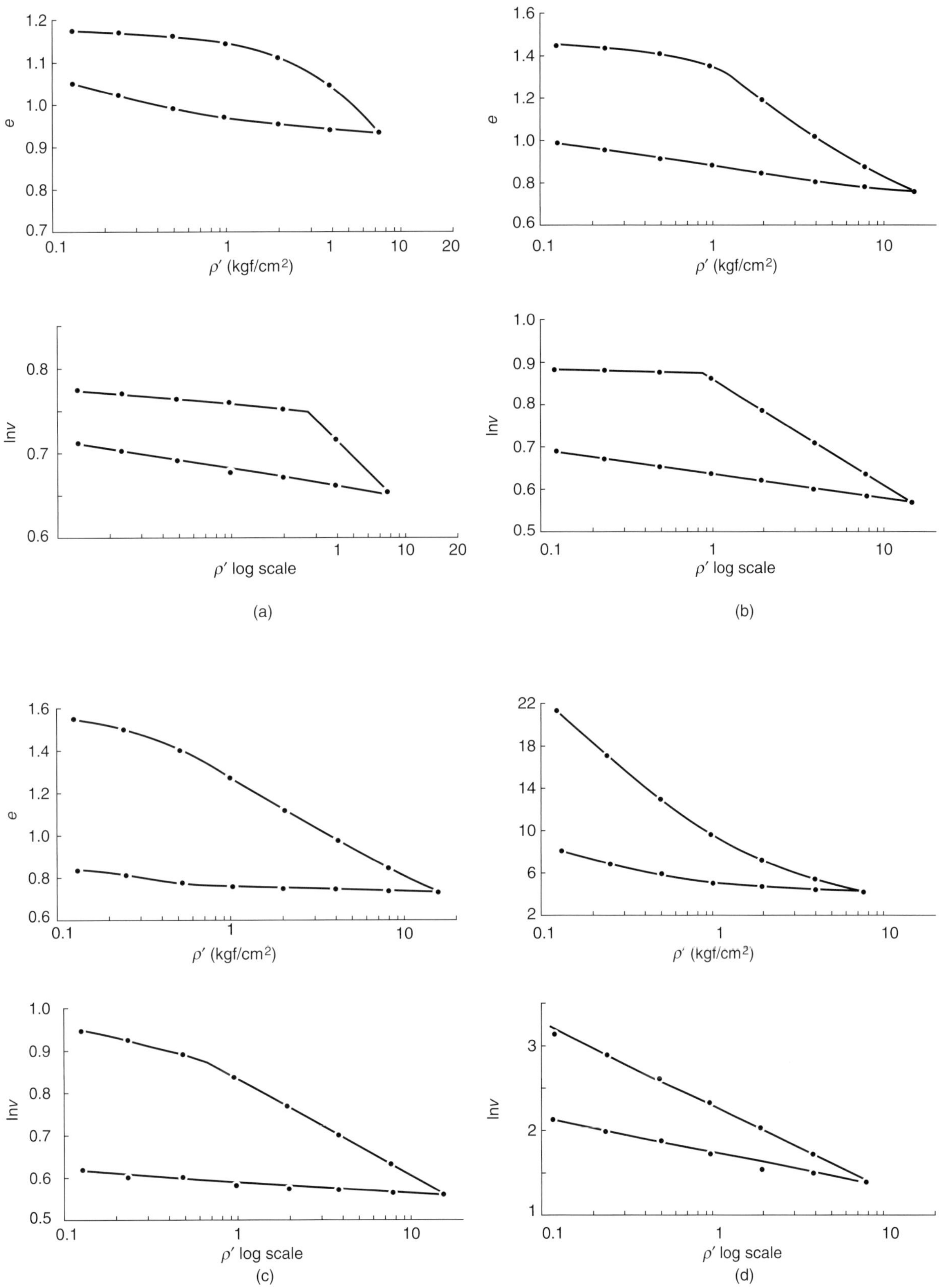

Figure 17.21 The use of double logarithmic plots to identify σ'_{vy} for various soils (a) Boston Blue Clay; (b) Chicago Clay; (c) Newfoundland silt; (d) Newfoundland peat (Reproduced from Butterfield, 1979)

17.8. The stress–strain behaviour of soils

Section 17.3 began with a description of the most commonly used idealisation of the strength of soils – namely, the Coulomb equation. Thereafter, the results of soil tests were examined to assess the equation's accuracy and limitations. A similar approach is adopted for the study of the stress–strain behaviour of soils. The most commonly used model of the stress–strain behaviour of soil is that of an ideal porous elastic solid.

17.8.1 Ideal isotropic porous elastic solid

A basic property of an elastic material is that when it is subjected to a closed cycle of stress, the deformations are reversible and fully recoverable. It is not necessary for the stress–strain behaviour to be linear. It is helpful to have a physical image of an isotropic porous elastic material: imagine a material which consists of very fine metal filings (elongated particles) arranged in a random manner and spot-welded at their contact points (to ensure that the assemblage is elastic). The particles are so fine and the pores so small that the material has a low permeability. Hence, when loaded quickly, pore pressure changes can take place within the voids.

An isotropic elastic material is fully defined by two parameters. The most commonly used ones are Young's modulus E and Poisson's ratio v. The properties of interest are those of the *skeleton* of this ideal soil. These are the properties that control its drained behaviour; they are called the 'effective' properties and denoted by a prime. The two most commonly used effective parameters are

- the effective Young's modulus E'
- the effective Poisson's ratio υ'.

The corresponding undrained parameters are denoted as E_u and v_u.

17.8.2 The elastic equations

As is usually the case in soil mechanics, compressive stresses are taken as positive – so reductions in length or volume give rise to positive strains. In developing the elastic equations, *stress changes* and their corresponding *strain changes* are considered. Consider a vertical rectangular prism of isotropic porous elastic material as shown in **Figure 17.22**. An increment of principal vertical effective stress $\Delta\sigma_1'$ is applied allowing full drainage. The resulting increment in vertical strain $\Delta\varepsilon_1$ is $\Delta\sigma_1'/E'$. It also follows that the increments in horizontal strain are given by $\Delta\varepsilon_2 = \Delta\varepsilon_3 = -\upsilon'\ \Delta\sigma_1'/E'$.

Independent applications of $\Delta\sigma_2'$ and $\Delta\sigma_3'$ in the horizontal plane will cause increments of strain in the vertical direction as follows

$$\Delta\varepsilon_1 = -\upsilon'\Delta\sigma_2'/E' \quad (17.5)$$

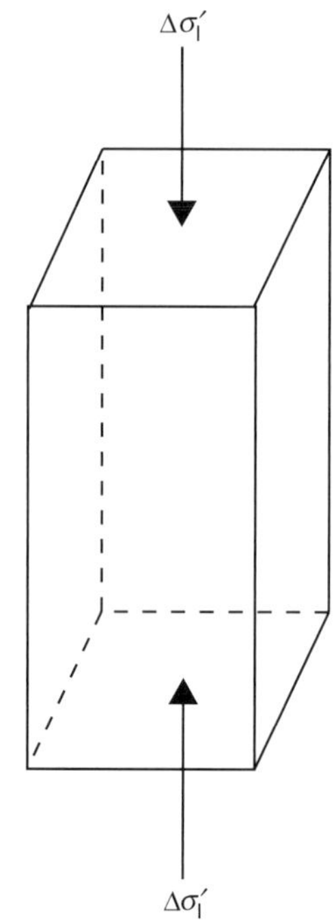

Figure 17.22 Element of porous isotropic elastic material subjected to an increment in vertical effective stress

and

$$\Delta\varepsilon_1 = -\upsilon'\Delta\sigma_3'/E' \quad (17.6)$$

By superposition, for the general case of changes in all three principal effective stresses, the increment of vertical strain is given by

$$\Delta\varepsilon_1 = \{\Delta\sigma_1' - \upsilon'\Delta\sigma_2' - \upsilon'\Delta\sigma_3'\}/E' \quad (17.7)$$

Similarly

$$\Delta\varepsilon_2 = \{\Delta\sigma_2' - \upsilon'\Delta\sigma_1' - \upsilon'\Delta\sigma_3'\}/E' \quad (17.8)$$

and

$$\Delta\varepsilon_3 = \{\Delta\sigma_3' - \upsilon'\Delta\sigma_1' - \upsilon'\Delta\sigma_2'\}/E' \quad (17.9)$$

These three simple equations can be used to derive a number of important properties for an ideal isotropic porous elastic material.

17.8.3 Volumetric strain

An increment in volumetric strain Δv ($=\Delta V/V_0$) is equal to the sum of the three principal increments of strain (neglecting second-order terms). Adding Equations 17.7–17.9 gives

$$\Delta v = \frac{1-2\upsilon'}{E'}\{\Delta\sigma_1' + \Delta\sigma_2' + \Delta\sigma_3'\} \tag{17.10}$$

Therefore

$$\Delta v = \frac{3(1-2\upsilon')}{E'}\Delta p' = \frac{\Delta p'}{K'} \tag{17.11}$$

where $\Delta p' = (\Delta\sigma_1' + \Delta\sigma_2' + \Delta\sigma_3')/3$, the effective mean normal stress, and $K' = E'/3(1 - 2v')$, the effective bulk modulus.

It follows from Equation 17.11 that if there is no volume change during the application of a general stress increment, then there will be no change in mean normal effective stress – that is, if $\Delta v = 0$, then $\Delta p' = 0$, and vice versa. This result is an inevitable consequence of the idealisation of isotropic porous elastic behaviour. It has important consequences for undrained behaviour, as discussed later.

17.8.4 Shear strain

By definition, an increment in maximum shear strain $\Delta\gamma_{max} = \Delta\varepsilon_1 - \Delta\varepsilon_3$. Substituting from Equations 17.7 and 17.9, and noting that $\Delta\tau_{max} = (\Delta\sigma_1' - \Delta\sigma_3')/2$, it follows that

$$\Delta\gamma_{max} = \frac{2(1+\upsilon')}{E'}\Delta\tau_{max} = \frac{\Delta\tau_{max}}{G'} \tag{17.12}$$

where G' is the effective shear modulus $= E'/2(1 + \upsilon')$. It follows from Equation 17.12 that the application of an increment, in shear stress only, causes no volume change. Therefore this ideal material is non-dilatant. Moreover, the effects of applying $\Delta p'$ and $\Delta\tau_{max}$ are entirely separate, in that the former only causes volume change and the latter only causes shear distortion. This is termed 'uncoupled' behaviour.

17.8.5 One-dimensional compression

Section 17.7 showed that the coefficient of one-dimensional volume compressibility $m_v = \Delta v/\Delta\sigma'_v$ where $\Delta\varepsilon_2 = \Delta\varepsilon_3 = 0$. It can be shown from the elastic equations that

$$m_v = \frac{1 - \dfrac{2\upsilon'^2}{1-\upsilon'}}{E'} \tag{17.13}$$

It should also be noted that if $\Delta\varepsilon_3 = 0$ and $\Delta\sigma_2' = \Delta\sigma_3'$ in Equation 17.9

$$\frac{\Delta\sigma_3'}{\Delta\sigma_1'} = \frac{\upsilon'}{1-\upsilon'} \tag{17.14}$$

which gives the value of the coefficient of earth pressure at rest K_0 for the one-dimensional compression of an isotropic porous elastic material (see Chapter 15 *Groundwater profiles and effective stresses* for a discussion on the important topic of the coefficient of earth pressure at rest, K_0). The following limiting values of m_v from Equation 17.13 should be noted

- when $\upsilon' = 0$, $m_v = 1/E'$ – that is, m_v is the reciprocal of E'
- when $\upsilon' = 0.5$, $m_v = 0$ – that is, the material is incompressible.

17.8.6 The measurement of E' and υ' in an ideal drained triaxial test

For a standard drained triaxial test, increments of axial effective stress are applied without change in effective radial stress – that is, $\Delta\sigma_a' = \Delta\sigma_1' > 0$ and $\Delta\sigma_r' = \Delta\sigma_3' = 0$. Both the axial strain $\Delta\varepsilon_a = \Delta\varepsilon_1$ and the volumetric strain $\Delta V/V_0$ are measured and plotted as shown in **Figure 17.23**.

From elastic Equation 17.7

$$E' = \Delta\sigma_1'/\Delta\varepsilon_1 \tag{17.15}$$

By definition $\upsilon' = -\Delta\varepsilon_3/\Delta\varepsilon_1$ and $\Delta V/V_0 = \Delta\varepsilon_1 + 2\Delta\varepsilon_3$. Therefore

$$\Delta\varepsilon_3 = \frac{\left[\dfrac{\Delta V}{V_0} - \Delta\varepsilon_1\right]}{2}$$

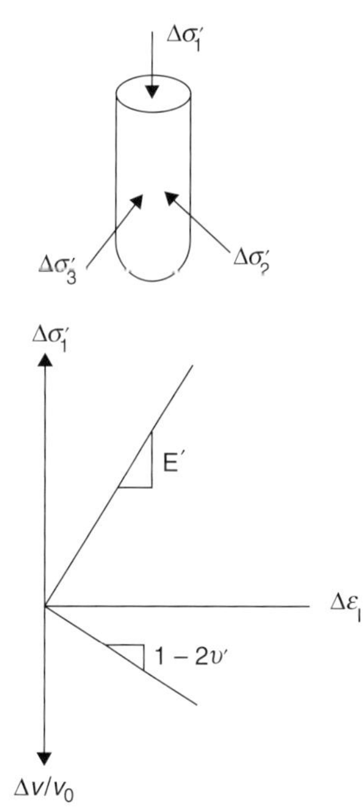

Figure 17.23 Determination of E' and υ' from a drained triaxial test on an ideal isotropic porous elastic material

and hence

$$\upsilon' = \frac{\left[1 - \dfrac{\Delta V/V_0}{\Delta \varepsilon_1}\right]}{2} \quad (17.16)$$

Note from **Figure 17.23** that the slope of the $\Delta\sigma_1'$ against $\Delta\varepsilon_a$ line is equal to E' and the downward slope of the $\Delta V/V_0$ against $\Delta\varepsilon_1$ line is equal to (1–2υ').

17.8.7 Ideal undrained triaxial test

For an undrained test, the pore water cannot drain out of or into the sample. Hence 'excess' pore pressures develop which can be positive or negative. Unless the pore pressures are measured, only the applied total stresses are known. The undrained Young's modulus E_u is given by

$$E_u = \Delta\sigma_1/\Delta\varepsilon_1 \quad (17.17)$$

As shown in the next section, the undrained Young's modulus is not usually equal to the drained Young's modulus E'. For an undrained test, full saturation is usually ensured so that there is no volume change – the water can be assumed to be incompressible. Hence υ_u (the undrained Poisson's ratio) is 0.5. When the excess pore pressure Δu is measured, as is usually the case for consolidated undrained tests, its value may be used to calculate the pore pressure coefficient A, which is referred to later.

17.8.8 Relationship between drained and undrained Young's modulus

Since water cannot carry shear, an increment in shear stress $\Delta\tau$ is an effective stress, irrespective of drainage. For an ideal isotropic porous elastic material the application of $\Delta\tau_{max}$ will not cause any volume change in a drained test (or any pore pressure change in an undrained test). Hence the only deformation is a shear strain $\Delta\gamma_{max}$ which must be identical for both drained and undrained conditions. Hence the effective shear modulus G' is identical to the undrained shear modulus G_u. From Equation 17.12 and its undrained equivalent

$$G' = \frac{E'}{2(1+\upsilon')} = G_u = \frac{E_u}{2(1+\upsilon_u)} \quad (17.18)$$

Setting $\upsilon_u = 0.5$ and solving for E_u

$$E_u = \frac{3}{2(1+\upsilon')}E' \quad (17.19)$$

This result is an inevitable consequence of the idealisation of isotropic elastic behaviour. It has particular significance when calculating undrained and drained settlements of foundations.

17.8.9 Pore pressure changes during undrained loading

Equation 17.11 shows that, for an isotropic porous elastic material, if there is no volume change during general loading $(\Delta\sigma_1 + \Delta\sigma_2 + \Delta\sigma_3)/3$ there will be no change in mean normal effective stress p' – that is, $\Delta p' = 0 = (\Delta\sigma'_1 + \Delta\sigma'_2 + \Delta\sigma'_3)/3$.

Since, from the effective stress principle $\Delta\sigma'_1 = \Delta\sigma_1 - \Delta u$ and so on, it follows that

$$0 = (\Delta\sigma_1 - \Delta u + \Delta\sigma_2 - \Delta u + \Delta\sigma_3 - \Delta u)/3 \quad (17.20)$$

and

$$0 = \Delta p = \Delta u \quad (17.21)$$

that is, during undrained loading of a fully saturated porous isotropic elastic material, $\Delta u = \Delta p$ (assuming the pore fluid to be incompressible in comparison with the skeleton). For an undrained triaxial compression test on a fully saturated sample of the ideal material in which both the axial and the radial total stresses are changed, $\Delta\sigma_a = \Delta\sigma_1$ and $\Delta\sigma_r = \Delta\sigma_2 = \Delta\sigma_3$. Hence, from Equation 17.21

$$\Delta u = \Delta p = \frac{(\Delta\sigma_1 + 2\Delta\sigma_3)}{3} = \Delta\sigma_3 + \frac{(\Delta\sigma_1 - \Delta\sigma_3)}{3} \quad (17.22)$$

Equation 17.22 is written in the form of Skempton's pore pressure equation

$$\Delta u = B[\Delta\sigma_3 + A(\Delta\sigma_1 - \Delta\sigma_3)] \quad (17.23)$$

where A and B are experimentally determined coefficients. When the soil is fully saturated, $B = 1$. It can be seen by inspection of Equations 17.22 and 17.23 that for an isotropic porous elastic soil, the value of A for an undrained triaxial test is fixed at 1/3. (It can be shown that under plain strain conditions the value of $A = 1/2$. Hence A is dependent on the total stress changes and is not a fundamental property, even for the ideal case of a porous isotropic elastic material.)

Further details on the determination and application of Skempton's pore pressure coefficients may be found in most basic soil mechanics textbooks (e.g. Knappett and Craig, 2020; Powrie, 2018). Note that the coefficient B can be measured by applying an increment of equal all-round stress $\Delta\sigma_3$ in an undrained triaxial test and measuring the change of pore pressure Δu. If B is less than unity then the sample is not fully saturated.

17.8.10 Summary of the properties of an ideal isotropic porous elastic material

Many soil mechanics problems are modelled using an ideal isotropic porous elastic material. This section has shown that this idealised material possesses some important and rather restrictive results which practising engineers should be aware of.

- Only changes in stress and strain are considered.
- Volume change only takes place if there is a change in mean normal effective stress: $\Delta v = \Delta p'/K'$.
- A change in shear stress only causes distortion and not volume change: $\Delta\gamma_{\max} = \Delta\tau_{\max}/G'$.
- The undrained Young's modulus E_u is uniquely related to the effective Young's modulus E' and the effective Poisson's ratio υ'.
- In undrained triaxial compression, the pore pressure coefficient $A = 1/3$.
- E' and υ' can be measured in the standard drained triaxial test.
- The one-dimensional volumetric compressibility m_v can be expressed in terms of E' and υ'.

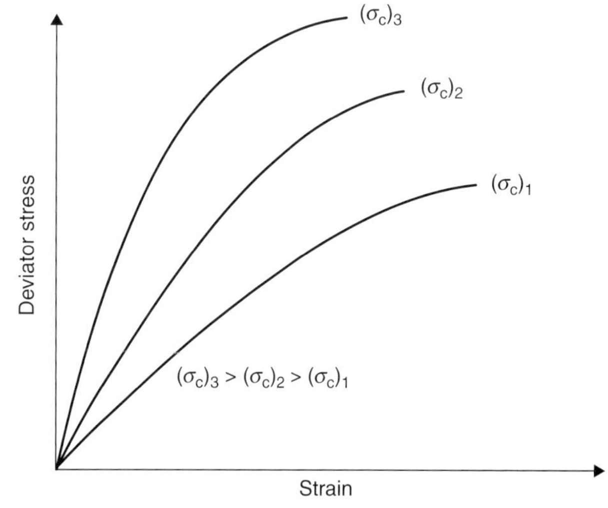

Figure 17.24 Influence of confining pressure on the drained stress–strain behaviour of soils

17.8.11 The stress–strain behaviour of real soils

There are many useful solutions available for practical geotechnical problems based on ideal isotropic elastic materials – see, for example, Poulos and Davis (1974). In making use of such solutions, the engineer has to make allowances for the fact that the stress–strain behaviour of real soils differs from that of the ideal isotropic elasticity in a number of ways. The three most obvious differences are nonlinear behaviour, the dependence of stiffness on confining pressure and the effects of anisotropy. All three stem directly from the particulate nature of soils.

It was shown in Section 17.7 that the compressibility of a soil reduces with increasing pressure. Similarly, research has shown that, in the same way that strength is significantly dependent on the initial effective confining pressure p_0', so is the stiffness – as shown in **Figure 17.24**. Since p_0' increases with depth, it can be anticipated that the value of E' will usually increase with depth.

In Chapter 14 *Soils as particulate materials*, it was shown that the process of deposition of a sediment leads to preferred arrangements of the particles so that their stiffness and strength properties will be anisotropic. The geotechnical engineer needs to be aware of this departure from the ideal of isotropic behaviour. It is relatively straightforward to develop the elastic equations for a cross-anisotropic porous elastic material in which the properties in the horizontal direction differ from those in the vertical direction. For such a material, five elastic constants are required: $E_v', \upsilon_{vh}', E_h', \upsilon_{hh}'$ and G_{vh}. The determination of all these constants is far from straightforward experimentally, and the application of a cross-anisotropic model should be carried out by an expert.

It is evident from the above that real soils depart from ideal isotropic elastic behaviour in a number of respects. The application of simple elastic theory can be very instructive in solving practical geotechnical problems. However, the key to its successful application is to understand clearly the properties of such a material and how the real material departs from these properties. The application of elastic theory to the calculation of foundation settlement and stress distributions is discussed in more detail in Chapter 19 *Settlement and stress distributions*.

Most soils exhibit nonlinear stress–strain behaviour of the form shown in **Figure 17.25**. A value of Young's modulus E' or E_u has to be selected which is appropriate for the range of stress changes involved. Frequently a value is chosen corresponding to about 0.1% axial strain. In sophisticated applications, numerical analysis is carried out using nonlinear elastic formulations. The equations used are the same as those given in this chapter, but the elastic constants (E', G, K', etc.) are altered as the analysis progresses to reflect the reduction in stiffness of the soil as the strains increase. More advanced nonlinear models make use of elastic–plastic formulations. A large number of nonlinear constitutive models are available for undertaking the numerical analysis of geotechnical problems. When embarking on such complex analysis, it is important to remember that such models are idealisations and always carry with them a number of assumptions and limitations. Every effort should be made to understand these. For a detailed discussion of the benefits and limitations of the various soil models that are used in practice, reference should be made to Potts and Zdravković (2001). Nonlinear analysis should only be undertaken with the advice and guidance of a specialist geotechnical engineer with expertise in numerical analysis.

17.9. Conclusions

The purpose of this chapter is to summarise the basic strength and stiffness properties of soils, emphasising the particulate

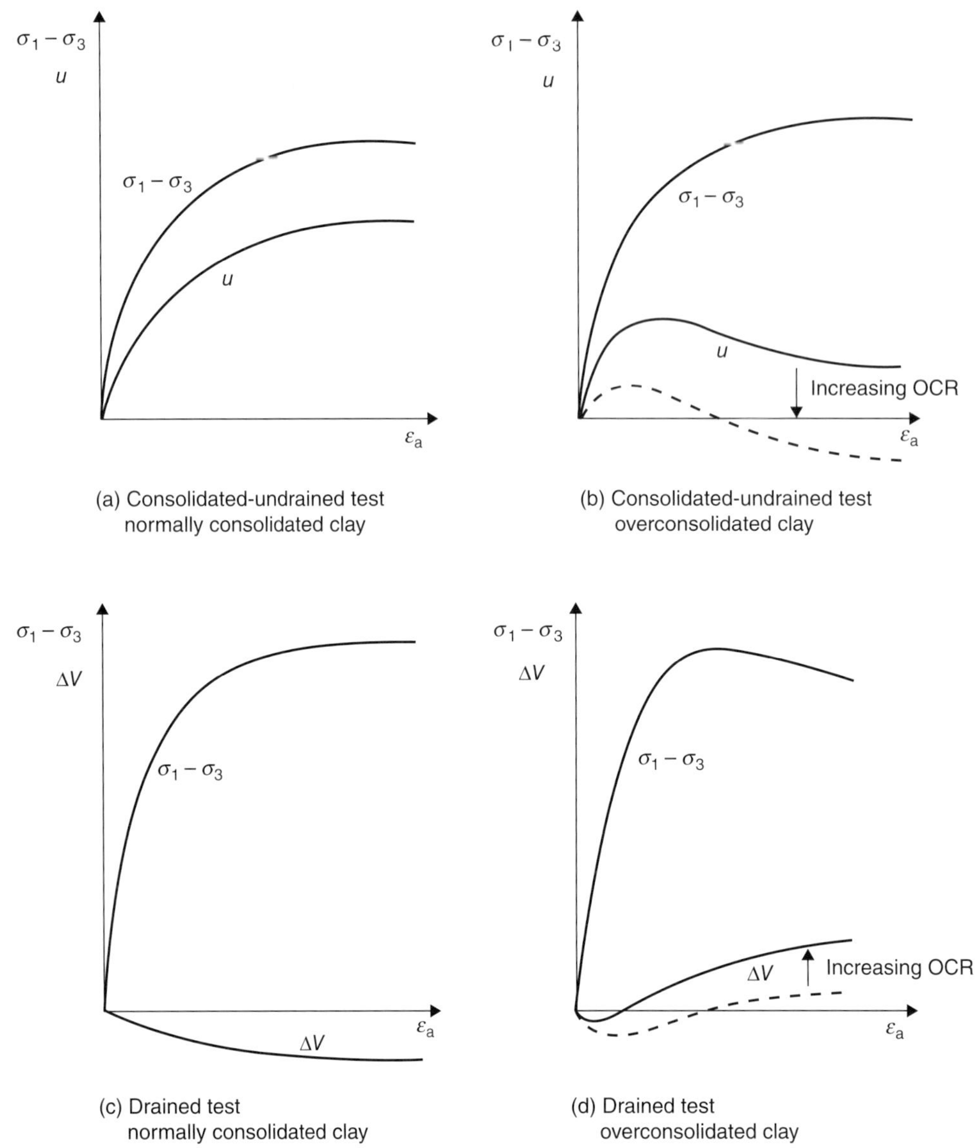

Figure 17.25 Typical stress–strain curves from consolidated undrained and drained triaxial tests

nature of the material. In doing so, it has been necessary to relate the behaviour to simple idealised frameworks. Reference to the geotechnical triangle in **Figure 4.1** of Chapter 4 *The geotechnical triangle* shows that there is a link between 'measured behaviour' in the bottom left-hand corner of the triangle and 'appropriate model' in the bottom right-hand corner. Simple models and idealisations are necessary to aid in prediction, but they need to be tempered with a clear understanding of real behaviour.

In the case of strength, the Coulomb criterion is universally used and the results of experiments on real soils have been compared with it. It is apparent that at least three measures of strength are used: peak strength, critical state strength and residual strength. In the case of peak strength, it has been shown that for dense or overconsolidated soils, the strength envelope is curved and depends on the initial relative density of the material. The engineer has to decide what measure of strength is relevant for a given problem and what values of φ' and c' to adopt in an analysis. This subject is covered in Chapter 27 *Geotechnical parameters and safety factors*; it is essential to have a clear understanding of the idealisations that are being made and their limitations.

Most clayey soils are very compressible compared with other structural materials, and classical soil mechanics has focused particularly on the one-dimensional (confined) compressibility. In this chapter, discussion of the one-dimensional compressibility of soils follows traditional approaches. Particular emphasis is placed on the importance of inter-particle bonding which is present in most natural soils. In particular, the existence of the vertical yield pressure σ'_{vy} is introduced. This yield pressure represents the vertical effective stress at which the inter-particle bonds begin to break down and slip between the grains. In the past this pressure has been confused with the pre-consolidation pressure, but this can be misleading as most natural normally consolidated soils exhibit a value of σ'_{vy} which is greater than the effective overburden pressure σ'_{v0}. The ratio $\sigma'_{vy}/\sigma'_{v0}$ is known as the 'yield stress ratio'. A most useful empirical finding is that the undrained strength in triaxial compression S_{uc} is closely related to σ'_{vy} such that the ratio S_{uc}/σ'_{vy} usually lies roughly in the range 0.28–0.32.

Many practical solutions exist for deformation problems using simple isotropic elasticity. In this chapter, the properties of an ideal isotropic porous elastic material have been described in detail. This ideal behaviour has then been compared with the known stress–strain behaviour of real soils. It is hoped that this discussion will assist the engineer both in the choice of appropriate stiffness parameters and in assessing the limitations of predictions based on the simple elastic models. If a more complex analysis is required employing nonlinear constitutive relations, it is important to remember that such models are idealisations and always carry with them a number of assumptions and limitations. Such an analysis should only be undertaken with the advice and guidance of a specialist geotechnical engineer with expertise in numerical analysis.

References

Atkinson J (1993) *An Introduction to the Mechanics of Soils and Foundations*. McGraw-Hill, New York, NY.

Bishop AW, Green GE, Garga VK, Andresen A and Brown JD (1971) A new ring shear apparatus and its application to the measurement of residual strength. *Géotechnique* **21(4)**: 273–328.

Burland JB (1990) On the compressibility and shear strength of natural clays. *Géotechnique* **40(3)**: 329–378.

Burland JB, Rampello S, Georgiannou VN and Calabresi G (1996) A laboratory study of the strength of four stiff clays. *Géotechnique* **46(3)**: 491–514.

Butterfield R (1979) A natural compression law for soils (an advance on e–log p'). *Géotechnique* **29(4)**: 469–480.

Heyman J (1997) *Coulomb's Memoir on Statics: An Essay in the History of Civil Engineering*. Imperial College Press, London, UK.

Knappett J and Craig RF (2020) *Craig's Soil Mechanics*, 9th edn. CRC Press, Boca Raton, FL.

Leonards GA and Ramiah BK (1959) Time effects in the consolidation of clay. *ASTM Special Technical Publication No. 254*. ASTM, Philadelphia, PA, pp. 116–130.

Lupini JF, Skinner AE and Vaughan PR (1981) The drained residual strength of cohesive soils. *Géotechnique* **31(2)**: 181–213.

Potts DM and Zdravković L (2001) *Finite Element Analysis in Geotechnical Engineering: Application*. Thomas Telford, London, UK.

Poulos HG and Davis EH (1974) *Elastic Solutions for Soil and Rock Mechanics*. Wiley, New York, NY.

Powrie W (2018) *Soils Mechanics: Concepts and Applications*, 3rd edn. CRC Press, Boca Raton, FL.

Roscoe KH (1953) An apparatus for the application of simple shear to soil samples. *Proceedings of the 3rd International Conference on SMFE, Guimarães, Portugal*, vol. 1, pp. 186–191.

Taylor DW (1948) *Fundamentals of Soil Mechanics*. Wiley, New York, NY.

All chapters within Sections 1 *Context* and 2 *Fundamental principles* together provide a complete introduction to the Manual and no individual chapter should be read in isolation from the rest.

doi: 10.1680/icemge.66816.0215

Chapter 18
Rock behaviour

David J Sanderson University of Southampton, Southampton, UK

Rocks are naturally occurring, polycrystalline materials that play a wide range of roles in civil engineering. Their behaviour can be understood in terms of their intrinsic textures (grains, cement, voids and discontinuities) and through the application of standardised testing methods. Rocks and soils display a wide range of rheological properties, even under near-surface conditions. The void space in rocks is generally filled with water, at some pore pressure. For many engineering applications rocks may be considered as poroelastic, with a proportion of the applied load being supported by the pore pressure. Thus, their deformation is related to the effective stress (= total stress – pore pressure) as is generally found for soils. Rocks have a wide range of stiffness, strength and permeability, all of which will determine their suitability for different engineering applications. Most importantly, rocks are heterogeneous, displaying various forms of layering or grain fabrics, and are almost ubiquitously fractured. Thus the rock mass properties vary over a range of scales (heterogeneity) and orientations (anisotropy).

CONTENTS

18.1. Rocks

Rocks are naturally occurring polycrystalline materials that play a wide range of roles in civil engineering, ranging from their support for structures to their use as construction materials.

All rocks can be considered as being made up of (**Figure 18.1**)

- **grains:** discrete elements of either individual crystals or aggregates of crystals, usually having a distinct composition and shape
- **cement:** usually crystalline material that binds the grains together
- **voids:** spaces between grains, usually in the form of interconnected pores, often filled with water
- **discontinuities:** macroscopic surfaces that separate the rock mass into blocks or layers. They may take the form of fractures (or joints), movement surfaces (faults), bedding planes or other surfaces. The discontinuities usually have markedly different physical properties from the rest of the rock material.

The composition and arrangement of these components gives rise to a variety of different rock types (which geologists describe with a bewildering range of names), which have a wide range of engineering properties. The nature of the grains, cement, voids and discontinuities is important in evaluating the behaviour of rocks and rock masses. Some understanding of the nature of rocks is necessary in order to discuss their physical behaviour.

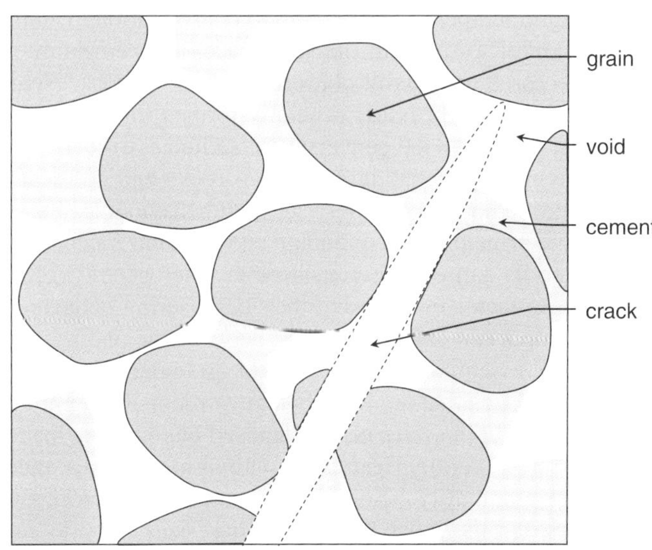

Figure 18.1 Diagrammatic representation of the components that make up a typical rock

18.2. Classification of rocks

Geologists recognise three broad categories of rocks, based on their mode of formation:

- **Sedimentary rocks** form at or near the surface of the Earth from grains, usually eroded from previously formed rock. The grains are then transported across the surface and deposited to form new *sediments*. The main agents for these processes are water (in the form of rivers and seas), ice and wind. Soils can be regarded as a form of sediment.

- **Igneous rocks** form by crystallisation of grains from molten rock either deep within the Earth (magma) or at the Earth's surface (lava).
- **Metamorphic rocks** form deep within the Earth due to solid-state recrystallisation of other rock types.

Igneous and metamorphic rocks share many characteristics and consist of an interlocking aggregate of mineral grains, usually resulting in strong and stiff materials. The boundaries between grains are usually very narrow, with no cement, and it is the interlocking nature of these boundaries and the lattice-scale forces between grains that impart their strength. Void space is small, with the porosity (ratio of void space to total volume) generally being $<1\%$.

Sedimentary rocks are generally formed as loose aggregates of grains, initially with a large amount of void space (porosities of 30% or greater). In the Earth this porosity is usually saturated with water (and occasionally other fluids). The water may contain dissolved materials that can precipitate in the pore space to form cement. At an initial stage, when porosities are high, the sediments resemble soils in their mechanical behaviour (see Chapter 17 *Strength and deformation behaviour of soils*). As the sediments are buried they experience elevated pressures and temperatures, leading to consolidation (usually by expulsion of water from the void space) and cementation. These processes essentially convert the 'soil' to rock. Near-surface chemical processes, generally termed diagenesis by geologists, accelerate the conversion of sediment to rock.

There is no clear boundary between rocks and soils, and many of the methods used to describe their behaviour are based on the same principles of continuum and granular mechanics. Both materials can exhibit considerable heterogeneity, often an important factor in assessing their engineering behaviour, but some important differences exist between the two.

Soils are essentially particulate (or granular) materials, with particle sizes that are usually many orders of magnitude less than the length scales of the imposed engineering loads. Thus they are generally treated as continuum materials, their microscale granularity being approximated by macroscopic parameters. On the other hand, the block sizes of rocks are often of a similar length scale to the applied loads and the discrete nature of rock masses is usually of greater importance in rocks than for soils (e.g. Hudson and Harrison, 1997).

Generally, both rocks and soils contain water, the void space often being saturated. Thus in both materials the effects of fluid pressure and the principles of effective stress are of great importance. However, the fundamental behaviour of soil changes dramatically with water content, changing from solid → plastic → liquid at the plastic and liquid limits, respectively. Rock properties are also affected by water, but not to anything like the same extent; for example, the strengths of dry and saturated rocks rarely differ by more than a factor of two.

18.3. Rock composition

An important distinction between the crystalline (igneous and metamorphic) and sedimentary rocks is the composition of the grains. The bulk composition of the Earth's crust is dominated by the elements silicon (Si) and oxygen (O) that readily combine to form the silica ion (SiO_4^{4-}). The next most abundant elements are the metals aluminium (Al), iron (Fe), magnesium (Mg), calcium (Ca), potassium (K) and sodium (Na), which form positively charged ions and combine with the SiO_4^{4-} to form a group of minerals known as rock-forming silicates. These include the common minerals quartz and feldspar, together with a complex array of other silicates (pyroxenes, amphiboles, micas, etc.). Most igneous and metamorphic rocks consist of a small number, typically three or four, of these rock-forming silicates, the composition of which depends on the chemistry of the patent magma or rock, and the pressure–temperature conditions under which the rock formed. The details of these minerals need not concern us here; what is important is that they generally have moderate to high strengths, contributing to the strength of igneous and metamorphic rocks.

Many of these rock-forming silicates weather on exposure to the Earth's surface, and are converted to various types of clay minerals and other salts; the latter often dissolve in water. An exception is the mineral quartz, which is highly resistant to weathering. Weathering leads to a reduction in strength of igneous and metamorphic rocks, and is facilitated by fracturing of the rock mass. Thus a key issue in the behaviour of such rocks is the rock mass quality.

Weathering leads to the development of sediment – hence, it can be expected to be composed mainly of quartz and clay minerals. Sediment is also generated by the precipitation of dissolved salts, either through evaporation (e.g. rock salt) or more commonly by biochemical action, mainly involving fixing of dissolved salts in the shells of invertebrate organisms. The nature of the sediment determines the type of sedimentary rock that is subsequently produced.

Quartz sand	→	sandstone
Clay minerals	→	claystone (also called mudstone or shale)
Calcite shells	→	limestone
Dissolved salts	→	rock salt (and other 'evaporates')
Organic matter	→	coal, oil, etc.

18.4. Porosity, saturation and unit weight

From the previous description of rock in terms of grains, cement and voids, the grains and cement together can be considered as a solid phase (**Figure 18.2**), with the voids being liquid (pore water) and/or gas (air). The volume of the rock (V_R)

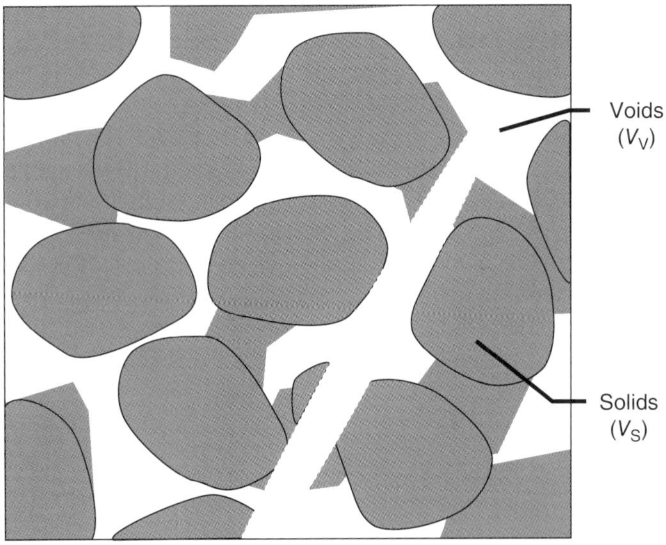

Figure 18.2 Rock components (as Figure 18.1) classified into solids (V_s) and voids (V_v)

is made up of solids (V_S) and voids (V_V), where $V_R = V_S + V_V$. The relative proportions of these define two parameters

$$\text{Porosity } (n) = V_v / V_R \text{ or } V_V / (V_S + V_V) \quad (18.1a)$$

$$\text{Void ratio } (e) = V_V / V_S \quad (18.1b)$$

where $$e = n / (1 + n) \text{ and } n = e / (1 + e) \quad (18.1c)$$

Since the volume of voids can be filled with water and/or air, $V_V = V_W + V_A$, which can be expressed in terms of the **saturation** S, where:

$$S = V_W / V_V \quad (18.2)$$

Porosity, void ratio and saturation are usually expressed as a fraction (0–1), but can be expressed as a percentage.

Knowing the densities (ρ) of the solid, water and air phases, and using $g = 9.81$, the weights of the phases may be calculated, where $W = \rho g$. For example,

$$\text{unit weight } (\gamma) = (W_S + W_W + W_A) / (V_S + V_W + V_A) \quad (18.3a)$$

$$\text{unit dry weight } (\gamma_d) = W_S / (V_S + V_W + V_A) \quad (18.3b)$$

These parameters are discussed more fully in Chapter 17 *Strength and deformation behaviour of soils*, but can provide a useful basis for characterising the multi-component nature of rocks and, hence, assessing rock behaviour. For example, the uniaxial compressive strength is related empirically to the porosity for different rock types, which provides a useful guide to rock strength in the absence of tests.

18.5. Stresses and loads

Stress exists throughout the Earth's crust and may be changed by surface and subsurface construction. There are four universal components responsible for this stress:

- the weight of the overlying column of rock, which is a relatively simple function of depth and rock density (unit weight), and is usually referred to as **overburden stress (σ_v)**
- **fluid (or pore) pressure (u)**, which arises from the fact that most rocks are basically two-phase materials consisting of mineral grains (which may or may not be cemented together) and fluid-filled pores and cracks
- **thermal stresses** arising from heating or cooling of rock, which tends to cause rocks to expand or contract
- **externally applied loads** that may be imposed by geological processes (tectonics, topography, etc.) or by construction.

These four components interact in different ways, but their combined effect is to act on or load materials to induce strain (change in shape and/or volume). Civil engineering is mainly concerned with how the ground responds to externally applied loads, but should not overlook the possible effects of other sources of stress.

Most rocks contain pores and cracks that are generally saturated by water. The externally applied loads create both a stress in the framework of grains and cement, and a pressure in the fluid, which is referred to as fluid pressure or pore pressure. Terzaghi (1943) suggested that the applied loads may be supported by both the stress in the solids and the fluid pressure in the pores. At the grain boundaries, these oppose one another and create an **effective stress**, such that

$$\text{Effective stress} = \text{total stress} - \text{pore pressure} \quad (18.4a)$$

$$\sigma' = \sigma - u \quad (18.4b)$$

In a granular material it is this effective stress that promotes deformation, with the relationship between effective stress and strain being determined by the rheology and properties of the material. The effective stress principle is applied almost universally in soil mechanics and widely in rock mechanics. This is discussed further in Section 18.8 on poroelasticity.

18.6. Rock rheology

There are three basic responses of materials to applied stress that are easily recognised from plots of stress against strain or strain rate (**Figure 18.3**).

- **Elasticity** is where the strain is linearly proportional to the stress (**Figure 18.3(a)**). This typifies the behaviour of solid materials and the ratio of stress to strain is referred to as the stiffness (Young's modulus, rigidity, etc.). In the ideal case, the deformation is completely recovered on removal of the stress and the rock exhibits no significant change in structure. Many crystalline rocks approximate this behaviour and are fairly rigid; that is, they exhibit a high stiffness, with Young's modulus being measured in gigapascals (GPa). This is one reason for their widespread use as construction materials.

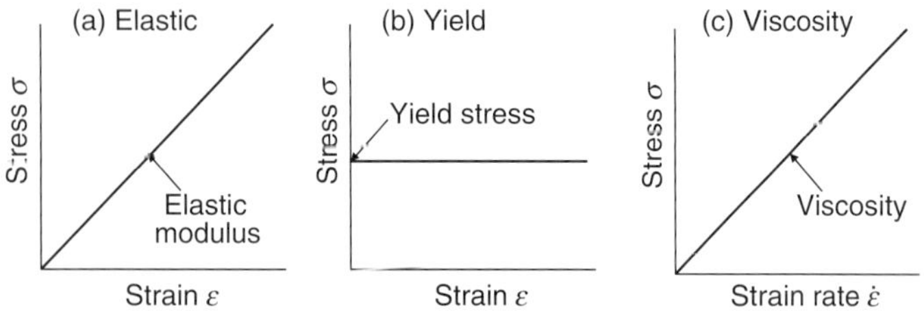

Figure 18.3 Idealised rheological behaviour showing (a) linear elasticity, (b) plastic yield and (c) Newtonian viscosity

- **Viscosity** is where the material flows and is the basic characteristic of liquids. In rock mechanics *creep* experiments are performed where a sample is allowed to deform under a constant applied stress and the strain plotted against time, the slope being the strain rate (**Figure 18.3(c)**). If the strain rate is proportional to stress, the material exhibits linear or Newtonian viscosity, where the viscosity is the ratio of stress to strain rate. Some rocks exhibit Newtonian viscosity, but generally the behaviour is more complex (nonlinear viscosity).
- **Yield** occurs when a material behaves elastically at low stress, but is ductile (i.e. flows) at higher stresses. The stress at which this transition takes place is the yield stress (**Figure 18.3(b)**). This behaviour is characteristic of plastic materials.

Rocks, in common with most other materials, exhibit all of these basic rheologies, the stiffness, viscosity and yield stress often being complex (nonlinear) and dependent on the temperature, confining pressure and strain rate.

It is useful, at least from a conceptual basis, to model rocks by combining the three basic rheological elements in various ways. For example, the stress–strain curves for rock (**Figure 18.4(a)**) at room temperature are elastic, whereas high confining pressures show a close correspondence to an elastic–plastic rheology. Confining pressure effectively inhibits failure and the rock exhibits a reasonably clearly defined yield stress at ~500 MPa (**Figure 18.4(a)**). Yield stress and the stiffness decrease with increasing temperature (**Figure 18.4(b)**), as does the viscosity (not shown in **Figure 18.4**).

The rheology of a material is described by a *constitutive law* – an equation relating the deformation (strain) induced in the rock to the applied stress (or vice versa). Simple examples were given in the previous section. However, to define a constitutive law more rigorously there is a need to consider the relationships between different components of stress and strain. In general, the deformation can be described by a second-order strain tensor, ε_{ij}, which is related to the displacements (u):

$$\varepsilon_{ij} = \tfrac{1}{2}\left(\frac{\delta u_i}{\delta x_j} + \frac{\delta u_j}{\delta x_i}\right) \tag{18.5}$$

Where one subscript (for example, i or j) is used, it denotes the direction of the main quantity. For example, u_i is the displacement in the i direction and x_j is distance in the j direction. Where two subscripts are used (for example, ij or 11), as is often the case for stresses or strains, the first subscript denotes the direction of the normal to the plane on which the stress acts, and the second the direction of the stress or strain. For example, σ_{xx} is the stress acting in the x-direction on the plane whose normal is also in the x-direction; it is therefore a direct or normal stress. σ_{xy} is the stress acting in the y-direction on the plane whose normal is in the x-direction; it is therefore a shear stress. The planes or directions may be denoted by numerals 1, 2 and 3 rather than ordinates x, y and z, especially (but not exclusively) where these are principal planes on which the shear stress is zero.

Three commonly encountered deformation types (unconfined compression, simple shear and volumetric strain) are illustrated in **Figure 18.5**.

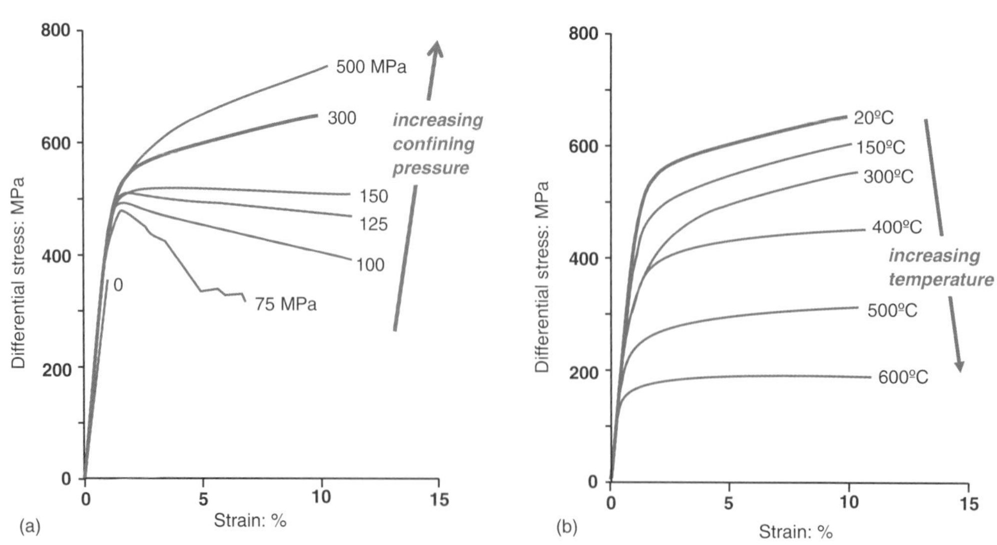

Figure 18.4 Stress–strain curves from triaxial tests on Solenhofen Limestone (After Heard, 1960). All tests at strain rate of 2×10^{-4} s^{-1}; (a) changing confining pressure at constant temperature of 25°C; (b) changing temperature at 300 MPa confining pressure

18.7. Elasticity and rock stiffness

For elastic behaviour, the relationship between stress and strain is linear and, for small deformations, can be described by a series of stiffness (or elastic) constants, defined by the ratio of different stress-to-strain components. The following elastic constants are widely used and relate to the three common deformation types (**Figure 18.5**) as follows.

- **Young's modulus** (*E*) where $E = \sigma_{11}/\varepsilon_{11}$. This is the stiffness in unconfined (uniaxial) compression (**Figure 18.5(a)**), where σ_{11} is the only non-zero stress, although there will be lateral strains in the other principal directions (i.e. $\varepsilon_{33} = \varepsilon_{22} \neq 0$).
- **Poisson's ratio** (ν) where $\nu = -\varepsilon_{33}/\varepsilon_{11}$ is the ratio of lateral to axial strains in uniaxial compression.
- **Shear modulus** (*G*) where $G = ½(\sigma_{31}/\varepsilon_{31})$; it is easily determined from simple shear experiments (**Figure 18.5(b)**).
- **Bulk modulus** (*K*) where $K = p/\varepsilon_{vol}$, *p* is the mean stress or uniform confining pressure and ε_{vol} is the volumetric strain (**Figure 18.5(c)**). The **compressibility** (*c*) is simply 1/*K*.

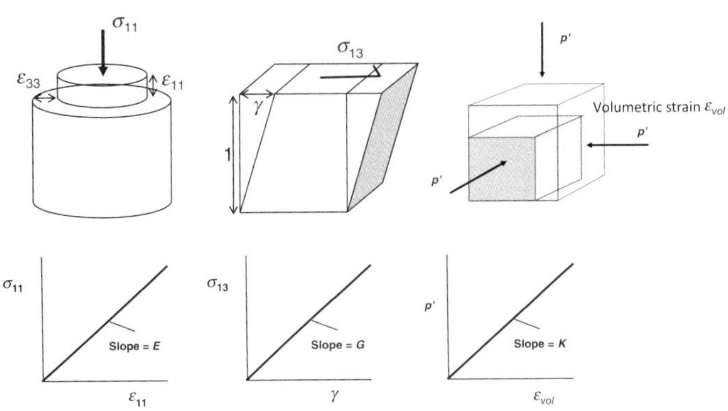

Figure 18.5 Three types of deformation involving different components of stress and strain and leading to different descriptions of rock stiffness

More generally, the **constitutive equation** for linear elasticity is given by

$$\sigma_{ij} = \lambda\delta_{ij}\varepsilon_{kk} + 2G\,\varepsilon_{ij} \tag{18.6}$$

where δ_{ij} is the Kronecker delta, which has a value 1 if $i = j$ and 0 if $i \neq j$. The term λ is a fifth elastic constant known as **Lame's constant** and, together with *G*, provides a more concise mathematical formulation of the constitutive equation in the theory of elasticity. The other, more practical, stiffness constants can be related to λ and *G* as

$$K = \lambda + 2G/3 \tag{18.7a}$$

$$E = (3\lambda + 2G)/(\lambda/G + 1) \tag{18.7b}$$

$$\nu = \lambda/[2(\lambda + G)] \tag{18.7c}$$

	Young's modulus *E*	Poisson's ratio ν	Bulk modulus *K*	Rigidity (shear modulus) *G*	Lame constant X
E, ν			$\frac{E}{3(1-2\nu)}$	$\frac{E}{2(1+\nu)}$	$\frac{E\nu}{(1+\nu)(1-2\nu)}$
E, G		$\frac{E}{2G} - 1$	$\frac{EG}{3(3G-E)}$		$G\frac{E-2G}{3G-E}$
E, K		$\frac{3K-E}{6K}$		$\frac{3KE}{9K-E}$	$\frac{3K(3K-E)}{9K-E}$
ν, G	$2G(1+\nu)$		$\frac{2G(1+\nu)}{3(1-2\nu)}$		$\frac{2G\nu}{(1-2\nu)}$
ν, K	$3/K(1-2\nu)$			$\frac{3K(1-2\nu)}{2(1+\nu)}$	$\frac{3K\nu}{(1-\nu)}$
ν, λ	$\frac{\lambda(1+\nu)(1-2\nu)}{\nu}$		$\frac{\lambda(1+\nu)}{3\nu}$	$\frac{\lambda(1-2\nu)}{2\nu}$	
K, G	$\frac{9KG}{(3K+G)}$	$\frac{3K-2G}{2(3K+G)}$			$K - \frac{2G}{3}$
λ, K	$9K\frac{K-\lambda}{3K-\lambda}$	$\frac{\lambda}{3K-\lambda}$		$3\frac{K-\lambda}{2}$	
λ, G	$G\frac{3\lambda+2G}{\lambda+G}$	$\frac{\lambda}{2(\lambda+G)}$	$\lambda + \frac{2G}{3}$		

Table 18.1 Conversion of elastic constants

	Crystalline rock	Soil	Water
Youngs's modulus (MPa)	4×10^4	10^1–10^2	0
Bulk modulus (MPa)	2×10^4	10^{-1}	2.2×10^3
Rigidity (MPa)	2×10^4	10^1–10^2	0
Yield stress (MPa)	$\sim 2 \times 10^2$	10^{-3}–10^{-1}	0
Poisson's ratio	0.1–0.25	0.2–0.45	0.5
Viscosity (Pas)	$\sim 10^{19}$	$\sim 10^4$	10^{-4}

Table 18.2 Simplified material properties of rock, soil and water (liquid)

Any two elastic constants can be used to define the material and, hence, all other constants (**Table 18.1**). The range of elastic properties in natural materials is very large, especially when water is included, as illustrated in **Table 18.2**.

The discussion of elasticity presented above assumes that rocks are homogeneous and isotropic. In reality a rock mass is inhomogeneous (properties vary with location) and anisotropic (properties vary with direction). A major contribution to anisotropy comes from the grain fabric due to depositional processes (commonly parallel to bedding) and ductile deformation. Fractures also contribute significantly to anisotropy, both on a macroscopic scale and due to micro-cracks (e.g. Goodman, 1989; Hudson and Harrison, 1997). This elastic anisotropy is manifest in directional variation in seismic wave velocities, but discussion of this is beyond the scope of this study. Anisotropy of rock mass strength is addressed in Section 18.16.

18.8. Poroelasticity

Poroelasticity is an attempt to describe the deformation of a solid material with connected voids (pores) that are saturated with fluid, and is a useful description for many soils or rocks (e.g. Biot, 1941; Wang, 2000).

Consider a rock under an isotropic confining pressure σ_C, with pores that are subject to a fluid pressure u. The volumetric strain ε will be affected by both pressures σ_C and u. The material can be conceptualised as comprising a solid 'framework' and a series of pores, shown schematically in **Figure 18.6(a)**. Based on the law of superposition, this stress system can be considered as the addition of (*a*) a confining stress ($\sigma_C - u$) acting on the outer boundary, with no pore pressure (**Figure 18.6(b)**); and (*b*) the pore pressure u acting on all boundaries (**Figure 18.6(c)**).

The total volumetric strain $\varepsilon_{vol,a}$ is simply

$$\varepsilon_{vol,a} = \varepsilon_{vol,b} + \varepsilon_{vol,c} \tag{18.8}$$

but

$$\varepsilon_{vol,b} = \frac{(\sigma_C - u)}{K} \text{ and } \varepsilon_{vol,c} = \frac{u}{K_G} \tag{18.9}$$

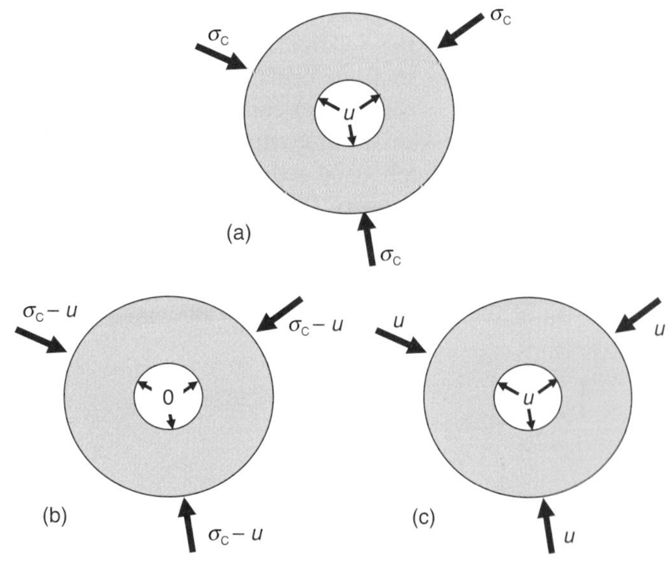

Figure 18.6 Decomposition of confining stress (σ_C) and pore pressure (u) – for explanation see text

where K and K_G are the bulk moduli of the rock and mineral grains, respectively and σ_c and u are, strictly, the increments in total stress and pore water pressure giving rise to the various strains.

Combining Equations 18.8 and 18.9 and rearranging gives

$$\varepsilon_{vol,a} = \frac{(\sigma_C - \alpha u)}{K} \tag{18.10}$$

Thus, deformation is produced by an effective stress

$$\sigma' = K\varepsilon_{vol,a} = \sigma_C - \alpha u \tag{18.11}$$

where $\alpha = (1 - K/K_G)$ is a dimensionless constant, usually referred to as the Biot constant. Note that Equation 18.11 is identical to Equation 18.4b, and hence Terzaghi's effective stress principle, for the case $\alpha = 1$.

For unconsolidated materials, where $K \ll K_G$, the Biot constant is $\alpha \approx 1$, as was originally proposed for soils (Terzaghi, 1943). For unfractured crystalline rock the grain framework may be stiffened by cement and/or interlocking grains, thus $K \to K_G$ and $\alpha \to 0$. Partly cemented sediments and fractured rock would be expected to have $0 < \alpha < 1$. For example, the bulk modulus for quartz is ~38 GPa, but that of sandstones is around 5–10 GPa; therefore $\alpha = 0.75$–0.9 would be the expected range.

18.9. Failure and rock strength

In simple terms there are two broad types of failure in rocks.

- Brittle failure is where the rock undergoes some sort of fracture, usually accompanied by a volume increase (as cracks develop) and preceded by an essentially elastic behaviour of the rock mass.

- Ductile failure is where the rock undergoes some form of plastic yielding, usually accompanied by a volume decrease (collapse of pore structure) and preceded by some form of elastic compression or compaction.

Traditionally rock mechanics has focused on the brittle type of failure and soil mechanics on ductile yielding, but in reality both types of failure are seen in both rocks and soils.

The strength of a rock can be considered as the stress at which some sort of failure occurs. In rocks, strength is a complex concept and will depend not only on the properties of the rock components (grains, cement, voids and discontinuities) and their interactions, but on the type of failure and the conditions under which it occurs. Thus any specification of rock strength must include a careful description of the test conditions, or have been obtained under standard (i.e. prespecified) conditions.

A simple view of rock failure is represented in **Figure 18.7**, which incorporates three widely recognised types of failure criteria.

Tensile failure occurs when one of the principal effective stresses is negative (tensile) and cracks form normal to the minimum principal stress (σ_3), the magnitude of which is the tensile strength (T_O); that is

$$T_O = -\sigma_3. \tag{18.12}$$

The concept is simple, but direct tensile testing is difficult to achieve and interpret, and most tensile strengths are measured indirectly, such as from a Brazilian test. For most rocks tensile strengths are low, generally 0–30 MPa.

Following the work of Griffith (1921), materials fail in tension due to the growth of micro-cracks. For a uniaxial far field tensile stress (σ_r), failure will occur when the stress concentration at the crack tip attains a critical value (K_C – the critical stress intensity factor or fracture toughness)

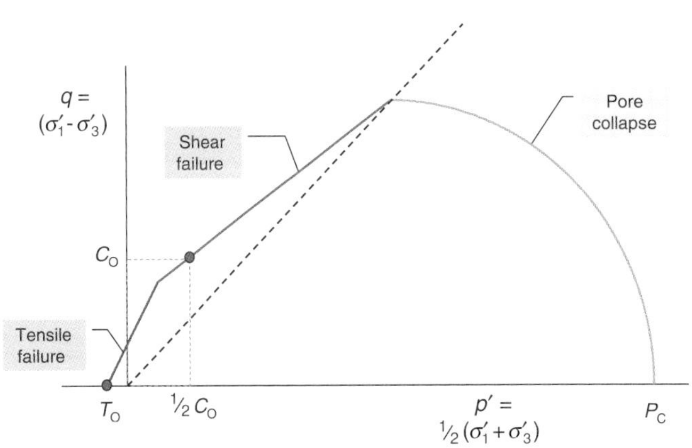

Figure 18.7 Plot of differential stress (q) against mean effective stress (p') showing the three main types of failure mechanism; T_O – tensile strength, C_O – uniaxial compressional stress, P_C – pre-consolidation pressure

$$K = Y\sigma_r(\pi a)^{\frac{1}{2}} \tag{18.13}$$

where a is the half-length of the crack and Y is a factor (usually ~1) that depends on the geometry of the crack and sample. Combining Equations 18.12 and 18.13, and using $Y = 1$, gives

$$T_O \approx K_c(\pi a)^{\frac{1}{2}} \tag{18.14}$$

Thus the tensile strength is a function of fracture toughness (a material property) and crack length (a textural property). The micro-cracks and other defects in an unfractured rock generally approximate to the size of the grains or pores (10^{-3} to 10^{-4} m) and K_C is in the range 0.3–3 MPa/m^{½}. Thus, a tensile strength of 5–170 MPa would be expected for most rocks, as is observed.

Shear failure occurs on planes oblique to the principal stresses that, hence, experience a shear stress. A simple and widely used criterion was developed by Coulomb, which states that the shear stress (τ) must exceed a linear function of the normal stress (σ_n), such that

$$\tau \geq S_O + \mu\sigma_n \tag{18.15}$$

where S_O is the cohesion – the shear stress required for failure in the absence of a normal stress – and μ is the coefficient of internal friction. This behaviour is analogous to frictional sliding (see Section 18.11).

Coulomb failure is represented by a straight line on a plot of shear stress against normal stress – the Mohr diagram (**Figure 18.8**). Failure will occur on planes where 2θ is normal to the failure envelope – that is, $2\theta = 90 + \varphi$, where φ is the angle of friction and $\mu = \tan(\varphi)$. For most rocks $\varphi = 20–50°$, hence $\theta = 55–70°$, with shear fractures forming at 20–35° to the maximum compressive stress (σ_1) – a commonly observed orientation in tests and nature.

When triaxial test data are plotted on either the Mohr diagram or on a plot of σ_1 against σ_3 they tend to show a nonlinear

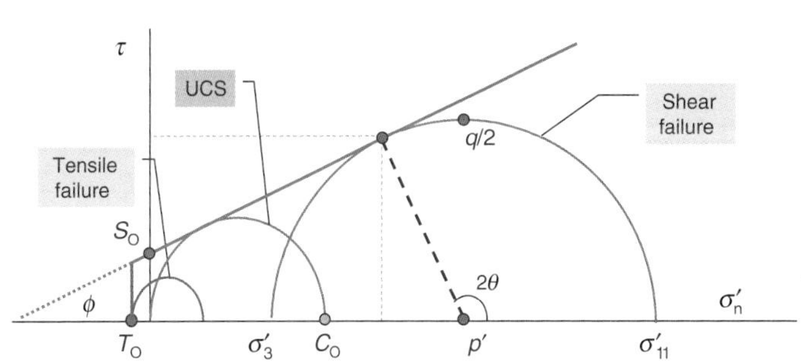

Figure 18.8 Mohr diagram (plot of shear stress, τ, against normal effective stress, σ_n') showing conditions for tensile and shear failure; T_O, tensile strength; C_O, uniaxial compressional stress; S_O, cohesion; ϕ, angle of friction; p', mean effective stress; $q/2$, maximum shear stress = ½ (differential stress)

relationship. Hoek (1966) originally proposed an empirical failure criterion for rocks, which was modified by Hoek and Brown (1980) to

$$\sigma_1 = \sigma_3 + \left[mC_0\sigma_1 + sC_0^2\right]^{1/2} \quad (18.16)$$

where C_0 is the uniaxial compressive strength and the parameters m and s are chosen to best fit the data. For more discussion of this criterion see Chapter 49 *Sampling and laboratory testing*.

Tensile and shear failure both involve the development of cracks and are generally accompanied by small volume increases (dilatancy). At high confining pressures, many porous rocks show a form of plastic yielding associated with localised **pore collapse** and compaction. This behaviour is typical of soils (Chapter 17 *Strength and deformation behaviour of soils*). This is shown schematically in **Figure 18.7** by the addition of an 'end cap' which meets the p' axis at a pressure equivalent to the pre-consolidation pressure in soils. Stress paths, such as that produced by hydrostatic compression ($q = 0$), that lead to failure by pore collapse have a lower ratio of q/p than that for shear failure. Such deformation is best analysed using the techniques of critical state soil mechanics (see Chapter 17 *Strength and deformation behaviour of soils*).

18.10. Strength testing

The field and laboratory testing of rock samples is described in detail elsewhere (Chapters 47 *Field geotechnical testing* and 49 *Sampling and laboratory testing*). Tests are usually performed on rock cylinders of standard shape and size, and some are briefly introduced below.

- **Uniaxial (unconfined) compressive test:** A simple and widely used test in which a cylinder or cube of rock is compressed between platens with no confining stress. The stress producing failure is the unconfined compressive strength (UCS).
- **Brazilian test:** A cylinder of rock is loaded between two platens transverse to its axis. This test is used to determine the tensile strength (T).
- **'Triaxial' test:** Axial compression (σ_1) of a cylinder under a radial confining stress ($\sigma_2 = \sigma_3$) – geometry similar to UCS. This test is usually run at several confining stresses and the failure envelope constructed, usually by plotting on a Mohr diagram (as in **Figure 18.8**).
- **Hydrostatic compression test:** This is usually carried out in a triaxial rig and involves increasing the confining pressure, in the absence of an axial load, until there is a volumetric collapse of the pore space. This is similar to the pre-consolidation pressure in soil mechanics and is used to define the plastic yield surface (**Figure 18.7**).
- **Shear tests:** The direct shear test involves a simple shear loading of a rectangular prismatic sample in a shear box. It is widely used to measure the shear strength (S_S) of soils, but is not suitable for anything but the weakest of rocks. A **torsional ring shear test** has been developed that involves the twisting of a hollow disc between rigid end-discs to impart a shear on the specimen, which can be applied to a wider range of rock strengths.
- **Point load test:** This is a widely used test involving transverse loading of a cylinder of rock between two conical 'points' of standard shape (60° conical angle and tip radius of 5 mm). It can be performed with portable apparatus and applied to borehole core and irregular rock samples.

Other field-based methods of measuring rock strength have been developed, which usually rely on an empirical calibration of some strength measure (usually UCS) to some physical response measured in the apparatus. A good example of this approach is the widely used **Schmidt hammer**, which measures the rebound of a spring-loaded rod propelled against a rock surface to estimate UCS. A **scratch test** determines the normal and tangential forces required to attain a constant depth of scratch on a rock surface, which can be related empirically to UCS.

A scheme for characterising rock strength based on simple field classification is outlined in **Table 18.3**. Essentially this is a subjective scheme based on the response of the rock to a series of simple physical tests using one's hand or a hammer. The scheme conforms broadly with the Working Party Report

Description	UCS: MPa	φ	Cohesion: MPa	Field test	Rock type(s)
Very strong rock	300	50	20	Repeated hammering to break	Most igneous rocks
Strong rock	100	45	9	Breaks with hammer	Greywacke, quartzite, gneiss
Average rock	30	40	3.3	Dented by hammer	Sandstone, limestone
Moderately weak rock	10	35	1.35	Cannot be broken by hand	Shale, claystone
Weak rock	3	32	0.46	Crumbles under hammer blows	Soft chalk
Very stiff soil/very weak rock	1	30	0.17	Easily broken by hand	Sand
Stiff soil	0.3	28	0.054	Indented by fingernail	Marl
Firm soil	0.1	25	0.020	Moulded by fingers	Clay
Soft soil	0.03	22	0.007	Easily moulded by fingers	Clay

Table 18.3 Field estimation of rock strength (UCS, internal friction angle for intact rock (φ) and cohesion)

(Geological Society Engineering Group Working Party, 1977) and is found in Clayton *et al.* (1995) and Waltham (2009).

In terms of using these tests to describe rock behaviour, it is important to remember that they are usually carried out on small pieces of the rock mass, typically at the centimetre-scale. Such tests may be useful in the characterisation of pieces of aggregate and building stone, but do not directly characterise the resultant structures (concrete, wall or foundation). Nor do the tests relate directly to the behaviour of the rock mass as a whole in groundworks, slopes, tunnels, etc. The following considerations are important in utilising laboratory estimates of rock strength.

- **Sampling:** Is the small specimen used in a test representative? All rocks are heterogeneous, usually at a wide range of scales – does the sampling capture this heterogeneity? This is particularly important in evaluating many sedimentary rock units that are made up of layers of different rock types and textures, with samples of a suitable size often being much easier to obtain from the thicker and stronger layers. These effects may be less of a problem in igneous and metamorphic rocks, although the fabrics in the latter produce anisotropy, which requires careful treatment (see Section 18.16).
- **In situ conditions** are difficult to replicate in a laboratory test for a number of reasons. Changes in boundary stress and fluid pressure and saturation occur during the extraction, transport and storage of samples that can lead to permanent physical and chemical damage, which cannot be reversed under test conditions. This is common in soils and poorly consolidated rocks, but is less important in most rocks.
- Virtually all routine tests are on intact rock and *do not measure the contribution of the rock fractures.*

18.11. Behaviour of discontinuities

Discontinuities are macroscopic surfaces that separate the rock mass into blocks or layers. They include fractures (or joints), movement surfaces (faults) and bedding planes or other surfaces. The discontinuities usually have markedly different physical properties from the rest of the rock matrix and may fail in tension or shear.

Experiments of sliding on planar surfaces cut in rock show a simple linear relationship between the shear stress (τ) and normal effective stress (σ'_n) at low normal stresses (**Figure 18.9**), such that

$$\tau = \mu_S \, \sigma'_n \qquad (18.17)$$

where μ_S is the coefficient of sliding friction. Byerlee (1978) proposed an average of $\mu_S = 0.85$, but at higher normal effective stresses ($\sigma_n > 200$ MPa) he found

$$\tau \approx 100 + 0.65\sigma'_n (\text{MPa}) \qquad (18.18)$$

These results appear to be largely independent of rock type (at least for well-consolidated and crystalline rocks). Based on

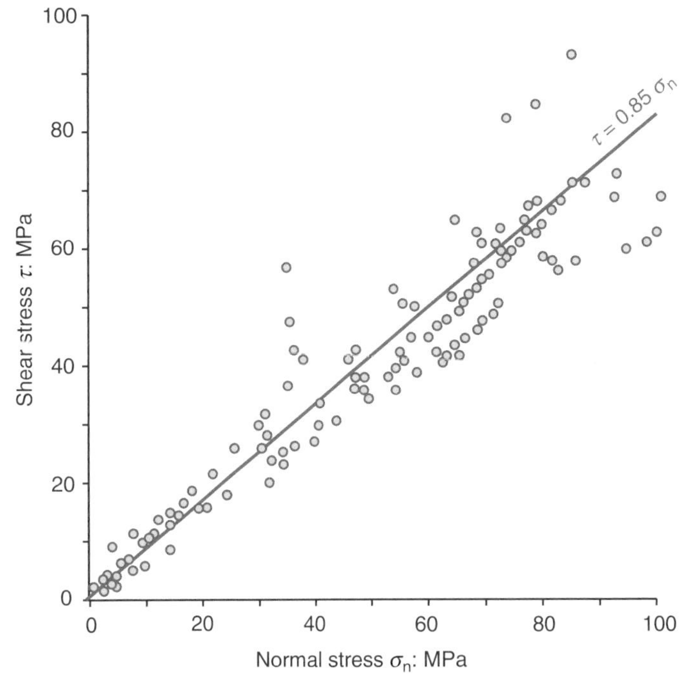

Figure 18.9 Data obtained by Byerlee (1978) for frictional sliding of planar rock surfaces at low normal stress

Byerlee's results, it is widely assumed that cohesionless friction characterises many fractured rocks, with coefficients of sliding friction generally being between 0.65 and 0.85 – this has become known as Byerlee's Law. The presence of weak material (e.g. clay fault gouge) in the fault plane can considerably reduce the coefficient of sliding friction. Water in the fractures mainly supports a pore pressure that controls the effective stress.

Since both shear and normal stress depend on the orientation of the surface in relation to the principal stress axes, failure by frictional sliding on rock fractures will be strongly controlled by the orientation of the fractures. If, however, sufficient variation in fracture orientation exists in a rock mass, then shear failure on optimally orientated fractures may be the dominant failure mechanism (e.g. Zhang and Sanderson, 2001).

18.12. Permeability

Rocks consist of solid phases (grains and cement) separated by voids. Flow of fluid between connected voids creates permeability in the porous medium. Flow is possible only through the connected void fraction, but for many rocks this is most of the void space. An exception is a rock such as pumice, which is a volcanic rock with isolated, unconnected gas bubbles; these make the rock light enough to float in water, but the rock is virtually impermeable – hence it does not get 'waterlogged' and sink.

For most porous rocks, the flow is laminar (Reynolds number $R_e \ll 1000$) and the flow rate is linearly proportional to the pressure gradient. This was originally demonstrated by Henry Darcy in 1856, who showed that the flux (ql – m^3s^{-1}) of water in a pipe full of sand and subjected to a uniform drop in head (Δh) was proportional to its cross-sectional area (A) and inversely proportional to its length (L). Thus

$$q = A \cdot k \cdot \frac{\Delta h}{L} \tag{18.19}$$

The negative sign indicates that flow is from high to low pressure and the constant (k) is known as the hydraulic conductivity. k has SI units of m/s and depends on the nature of both the rock and the fluid.

$$k = \frac{K\rho g}{\mu} \tag{18.20}$$

where ρ and μ are the density and dynamic viscosity of the fluid, g is the gravitational acceleration and K the intrinsic permeability of the material. The intrinsic permeability has units of square metres (m^2), with rocks typically having values in the range 10^{-18} to 10^{-10} m^2. The Darcy ($\approx 10^{-12}$ m^2) is often a more convenient unit as many rocks have permeabilities in the Darcy to milliDarcy range.

Civil engineering is normally concerned only with groundwater, whose properties change relatively little. For water, at room temperature, $\rho g/\mu \approx 10^7$ in SI units; hence, $k \approx 10^7 K$. Again, because of the low values k is often measured in m/day or m/year. Another common practice in civil engineering is to refer to k as a 'coefficient of permeability' (or even just 'permeability'), a situation that can lead to confusion, and one that is both unnecessary and unhelpful.

A porous medium can be modelled as a system of pores, with porosity (n), connected by much finer tubes or throats of radius (r). Models can be used to relate the pore structure to the intrinsic permeability. For example, a simple model of a bundle of cylindrical tubes gives

$$K \approx r^2\, n/8 \tag{18.21}$$

Thus for $r = 10^{-5}$ m and $n = 0.1$ this yields an intrinsic permeability of ~1 Darcy, equivalent to a hydraulic conductivity of a little less than 1 m/day.

The size and distribution of rock components (grains, voids, cement and fractures) have a major influence on the hydraulic conductivity of rocks.

18.13. Fracture-controlled permeability

Many rocks have very low porosities, such as most igneous and metamorphic rocks, and many cemented limestones. Others have very small grain size (and hence pore-throat radius), such as claystones, and this results in very low intrinsic permeability. Small specimens, when tested in the laboratory, confirm this. On the other hand when tested in the field many of these rocks have permeability in the milliDarcy to Darcy range that can be attributed to flow in fractures. Thus it is the rock mass, rather than grain-scale, properties that control flow.

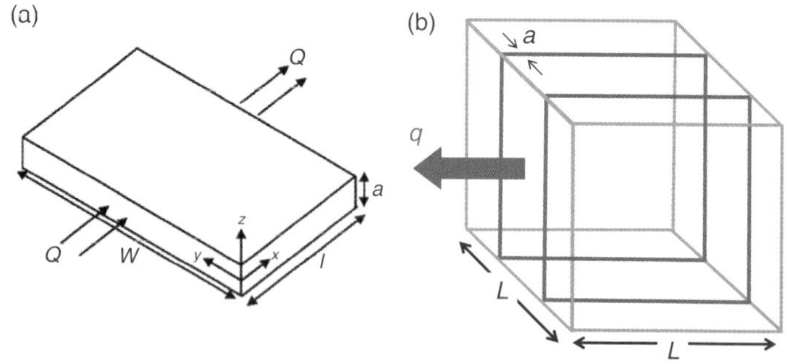

Figure 18.10 Flow of fluid through a set of parallel fractures

Fracture flow can be modelled by laminar flow between two parallel plates (**Figure 18.10(a)**) where

$$q = \frac{w \cdot a^2}{12} \times \frac{\rho g}{\mu} \times \frac{\Delta h}{L} \tag{18.22}$$

where, in addition to the symbols already defined, W is the width of the aperture and a is its height.

This is the well-known 'cubic flow law'. Since the cross-sectional area $A = Wa$, it follows that the intrinsic permeability of the fracture is $K = a^2/12$; this would be the permeability of a layer assigned to model the fracture.

For a set of fractures orientated parallel to the pressure gradient (**Figure 18.10(b)**), with aperture (a) and density (d = number per unit length), then from Equation 18.22 it is clear that the intrinsic permeability of the rock mass as a whole is proportional to the fracture density (d) and the cube of the fracture aperture (a) and is

$$K = da^3/12 \tag{18.23}$$

Since d has SI units m^{-1}, it follows that K has units m^2.

Fractures are an important component of the permeability of many rocks. A single fracture of aperture $a = 100$ microns (10^{-4} m) in a cubic metre of rock (i.e. $d = 1$ m^{-1}) would provide a permeability of ~10^{-13} m^2 (or 100 milliDarcys), from Equation 18.23. This is equivalent to the permeability of many sandstones used as oil or water reservoirs and would represent a hydraulic conductivity $k \approx 1$ m/day.

18.14. Rock mass characterisation

Rock mass classification schemes have been around for over a century. They have generally been developed for assessment of behaviour in different engineering environments (tunnelling, slope stability, etc.). Early schemes were simple and qualitative (e.g. Terzaghi, 1946); more recent schemes use standardised measures and an algorithmic approach that can be implemented using spreadsheets.

Terzaghi's descriptions (simplified from his original paper) are as follows.

- *Intact* rock contains no discontinuities and, hence, breaks across the rock matrix, either across grains or along grain boundaries.
- *Stratified* rock consists of layers separated by boundaries that may represent discontinuities across which the rock may have less resistance to separation and/or shear.
- *Moderately jointed* rock contains a network of joints that generally separate the rock into blocks, which interact along a large proportion of their surface and may be intimately interlocked.
- *Blocky and seamy* rock consists of grains and other rock fragments that are imperfectly interlocked and interact through small areas of contact.
- *Crushed* rock consists of small fragments and grains with no re-cementation that generally are surrounded by voids and interact at point contacts. The voids may be saturated with fluid.
- *Squeezing* rock flows into large voids (tunnel and other excavations) without perceptible volume increase. Such rocks typically have microscopic particles of mica or clay minerals with a low swelling capacity.
- *Swelling* rock flows into large voids on account of expansion. Such rocks contain clay minerals such as montmorillonite, with a high swelling capacity.

This scheme is entirely qualitative, but has the advantage of focusing attention on the main characteristic of the mass behaviour. It is rarely used in modern engineering design.

Rock quality designation (RQD) is a single parameter used widely to describe the degree of rock fracturing in borehole core and was developed by Deere *et al.* (1967). RQD is simply the percentage of intact core pieces longer than 0.1 m (4 inches) in the total length of core. The core should be at least 50 mm (2 inches) dia. RQD is dependent on the spacing and direction of fractures relative to the borehole axis, and should exclude any fractures induced by drilling or handling of the core.

RQD clearly depends on spacing of fractures – the inverse of the number of fractures per unit length (λ). For a negative exponential distribution of fracture (i.e. one obtained from random placement of fractures) Priest and Hudson (1976) showed that

$$\text{RQD} = 100\, e^{-\lambda t}(1 + \lambda t) \qquad (18.24)$$

where t is the threshold length used to determine RQD. Using the conventional value $t = 0.1$ m, RQD varies between 5 and 95% in a range of 3–50 fractures/m and thus is insensitive to variations outside this range.

Rock mass rating (RMR) was designed for the estimation of rock strength (Bieniawski, 1973, 1989). It uses six quantitative measures to classify a rock mass

(*a*) uniaxial compressive strength of rock material
(*b*) RQD
(*c*) spacing of discontinuities
(*d*) condition of discontinuities
(*e*) groundwater conditions
(*f*) orientation of discontinuities.

Values of each of these measures are used to define 'ratings'. For measures (*a*)–(*e*) these ratings are numbers usually between 0 and 15–30, which sum to a total between 0 (good rock) and 100 (poor rock). The final factor (*f*) is expressed as a negative rating reflecting the favourable (low negative value) or unfavourable (high negative value) of the orientation of fractures to the structure (tunnel wall, slope, etc.). The magnitude of the negative value also varies with the type of excavation, being greater for slopes than for tunnels.

Factors (*b*) and (*c*) are closely related, and together account for 40% of the RMR, indicating that the 'degree of fracturing' is a key factor in this rock mass classification. RMR also incorporates the condition of the fractures (alteration and water content as well as the strength of the rock matrix.

The classification system is generally applied to different regions within the rock mass, usually based on the distribution of different rock types and major discontinuities.

18.15. Rock tunnelling quality index, *Q*

This was developed by the Norwegian Geotechnical Institute for underground excavations (Barton *et al.*, 1974) and is based on six quantitative measures

- RQD (as defined above)
- Jn, the number of joint sets
- Jr, the joint roughness number
- Ja, the joint alteration number
- Jw, the joint water reduction factor
- SRF, a stress reduction factor.

These parameters are used to define three ratios.

- RQD/Jn represents the structure of the rock mass. As RQD is assessed relative to a 10 cm length of intact core, the ratio is crudely related to the size of fracture blocks.
- Jr/Ja is a ratio that accounts for the roughness and frictional characteristics of the fracture walls and/or filling materials.
- Jw/SRF is a ratio that consists of two stress parameters and attempts to describe the in situ loading conditions within the rock

mass. Jw is a measure of water pressure, which reduces the effective normal stress, hence the shear strength of fractures. SRF is a measure of the total stress or load applied to the rock mass, but also incorporates effects such as loosening due to excavation and squeezing loads in plastic materials.

For a simple introduction to the *Q* system, see Waltham (2009: pp. 86–87).

Both the RMR and *Q* schemes use geological and engineering parameters to provide a quantitative assessment of rock mass quality. Many of the parameters used are similar, but the calculations and weightings differ. Both schemes consider the degree of fracturing and the conditions of the fractures (groundwater, roughness, alteration, etc.). A significant difference is that RMR uses compressive strength directly, whereas *Q* only considers this in relation to in situ stress (through the SRF parameter). The RMR also incorporates the orientation of the fractures relative to the structure. Some estimate of orientation can be incorporated into *Q* using guidelines presented by Barton *et al.* (1974).

Both the RMR and Q schemes are best implemented using standardised procedures, which can be easily and consistently executed using spreadsheets, and details of the precise implementation of the methods are not given here.

18.16. Anisotropy

As well as the presence of discontinuities, there are often marked differences in rock properties with orientation within the rock mass. The major causes of such anisotropy are related to depositional layering in sedimentary rocks and flow fabrics in metamorphic and some igneous rocks; generally the latter have the greatest homogeneity. The anisotropy of rocks can be of three fundamentally different types.

- Intrinsic anisotropy at the grain-scale is caused primarily by the preferred orientation of mineral grains. The mineral lattice properties may be strongly isotropic, hence alignment produces a macroscopic anisotropy in the rock sample. Alternatively the mineral grains may have a shape fabric and, hence, even for isotropic minerals the alignment of grain boundaries produces anisotropy. Preferred orientation of micro-cracks can produce similar anisotropy.
- Many rock masses consist of layers of different rock materials and these produce a composite material, whose properties are not simply the average of the two (or more) materials involved. An example of this sort would be thin clay layers in sandstones, where the lower stiffness, strength and permeability of the clay may dominate the physical behaviour of the material. In such cases there is usually a strong directional dependence on the material properties.
- The presence of fractures (see Section 18.11).

Recognition of one or other of these types of anisotropy necessitates careful design of procedures to characterise the material – sampling different layers, testing large representative samples, testing in different orientations and so on. Conversely, one must be very careful in applying test results to construction in such anisotropic and/or composite materials.

Anisotropy is a major consideration in many claystones (shales). Stiffness is generally greater for compression normal to layering. Strength (e.g. UCS) is generally greater for compression normal to layering, but for shear failure in fissile shales the strength may be less when loaded in directions oblique to layering.

References

Barton NR, Lien R and Lunde J (1974) Engineering classification of rock masses for the design of tunnel support. *Rock Mechanics* **6**: 189–239.

Bieniawski ZT (1973) Engineering classification of jointed rock masses. *Transactions of the South African Institute of Civil Engineers* **15**: 335–344.

Bieniawski ZT (1989) *Engineering Rock Mass Classifications*. Wiley, New York, NY.

Biot MA (1941) General theory of three-dimensional consolidation. *Journal of Applied Physics* **12**: 155–164.

Byerlee JD (1978) Friction of rocks. *Pure and Applied Geophysics* **116**: 615–626.

Clayton CRI, Matthews MC and Simons NE (1995) *Site Investigation*, 2nd edn. Blackwell Science, Oxford, UK.

Deere DU, Hendron AJ, Patton FD and Cording EJ (1967) Design of surface and near surface construction in rock. *Proceedings of the 8th U.S. Symposium Rock Mechanics*, Minneapolis, MN, pp. 237–302.

Geological Society Engineering Group Working Party (1977) The description of rock masses for engineering purposes: report by the Geological Society Engineering Group Working Party. *The Quarterly Journal of Engineering Geology* **10(4)**: 355–388.

Goodman RE (1989) *Introduction to Rock Mechanics*, 2nd edn. Wiley, Chichester, UK.

Griffith AA (1921) The phenomena of rupture and flow in solids. *Philosophical Transactions of the Royal Society* **A221**: 163–198.

Heard HC (1960) Transition from brittle to ductile flow in Solenhofen limestone as a function of temperature, confining pressure, and interstitial fluid pressure. *Geological Society of America Memoir* **79**: 193–226.

Hoek E (1966) Rock mechanics: an introduction for the practical engineer. *Mineralogical Magazine* **114**: 236–243.

Hoek E and Brown ET (1980) Empirical strength criterion for rock masses. *Journal of Geotechnical Engineering Division, ASCE* **106**: 1013–1035.

Hudson JA and Harrison JP (1997) *Engineering Rock Mechanics*. Pergamon, Oxford, UK.

Priest SD and Hudson JA (1976) Discontinuity spacing in rock. *International Journal of Rock Mechanics and Mining Sciences* **13**: 134–153.

Terzaghi K (1943) *Theoretical Soil Mechanics*. Wiley, New York, NY.

Terzaghi K (1946) Rock defects and loads on tunnel supports. In *Rock Tunneling with Steel Supports*, vol. 1 (Proctor RV and White TL (eds)). Commercial Shearing and Stamping Company, Youngstown, OH, pp. 17–99.

Waltham T (2009) *Foundations of Engineering Geology*, 3rd edn. Spon, London, UK.

Wang HF (2000) *Theory of Linear Poroelasticity with Applications to Geomechanics and Hydrogeology*. Princeton University Press, Princeton, NJ.

Zhang X and Sanderson DJ (2001) Evaluation of instability in fractured rock masses using numerical analysis methods: effects of fracture geometry and loading direction. *Journal of Geophysical Research* **106(B11)**: 26671–26688.

Further reading

Barton N, Løset F, Lien R and Lunde J (1980) Application of the Q-system in design decisions. In *Subsurface Space*, vol. 2 (Bergman M (ed)). Pergamon, New York, NY, pp. 553–561.

Deere DU (1989) *Rock Quality Designation (RQD) after 20 Years.* U.S. Army Corps Engineers, Vicksburg, MS, Contract Report GL-89-1.

Farmer I (1983) *Engineering Behaviour of Rocks*, 2nd edn. Chapman and Hall, London, UK.

Jaeger JC and Cook NGW (1969) *Fundamentals of Rock Mechanics.* Chapman and Hall, London, UK.

All chapters within Sections 1 *Context* and 2 *Fundamental principles* together provide a complete introduction to the Manual and no individual chapter should be read in isolation from the rest.

doi: 10.1680/icemge.66816.0229

Chapter 19
Settlement and stress distributions

John B Burland Imperial College London, London, UK

This chapter describes the basic steps involved in calculating both the undrained and drained settlement of foundations on clay soils. It is demonstrated that the vertical stress changes beneath loaded areas, derived from simple elasticity, can be used with confidence for this purpose. In contrast, the changes in horizontal and shearing stresses are very sensitive to the assumed stress–strain behaviour and the theory of elasticity is much less reliable for calculating these. The four most common methods for calculating the final settlement are described. The somewhat surprising conclusion is reached that the simple one-dimensional method of analysis is even better than the more sophisticated methods, provided the appropriate soil stiffness parameters have been obtained. Settlements on granular soils are usually small and much of it takes place during construction. Numerous empirical methods have been proposed for predicting the settlement of foundations on granular materials, which lead to a wide range of estimates. A straightforward analysis of the measured settlements of over 200 records has led to a remarkably simple approach relating settlement to bearing pressure, breadth and average standard penetration test blow count, or cone resistance, over the depth of influence.

CONTENTS

19.1. Introduction

This chapter describes in simple terms the most widely used methods of settlement analysis. These range from the simple classical one-dimensional (1D) method through to methods that make use of numerical analysis. It is demonstrated that the simple traditional methods of analysis are usually adequate for practical purposes, provided that the appropriate soil stiffness parameters have been measured. Thus, the main emphasis should be placed on high-quality sampling and testing rather than sophisticated analysis. For granular soils, these procedures are not usually possible because of the difficulties of sampling them. Therefore, for granular soils a completely different approach is usually adopted.

19.2. Total, undrained and consolidation settlement

When a load is applied to a foundation and then kept constant, a time–settlement graph of the form shown in **Figure 19.1** is observed. Some settlement takes place almost immediately during loading. If the soil is saturated and of low permeability, there will be no time for drainage; this is termed the undrained (or immediate) settlement ρ_u.

As the excess pore water pressures (set up during loading) drain away under approximately constant load, consolidation (or time-dependent compression) of the soil takes place until all the excess pore water pressure has been dissipated. The settlement occurring during this process is termed the consolidation settlement ρ_c, and is shown in **Figure 19.1**.

For some soils, settlement continues after the excess pore water pressure has been dissipated due to creep of the soil skeleton. This is termed 'secondary' or 'creep' settlement ρ_s.

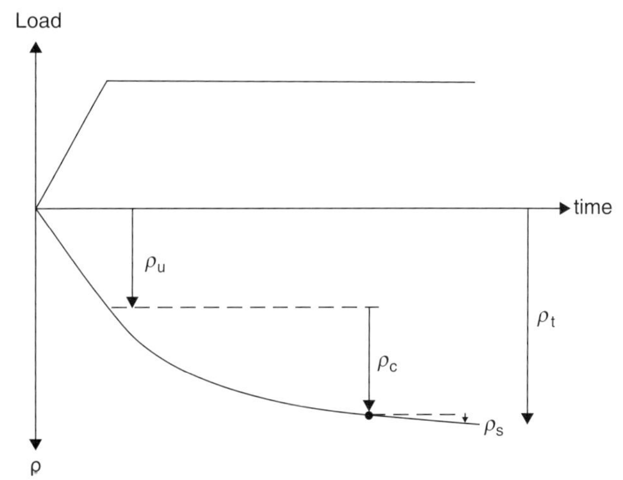

Figure 19.1 Idealised time–settlement curve for a foundation showing the undrained (ρ_u), consolidation (ρ_c) and secondary (ρ_s) components

Thus the total settlement ρ_t at a given time following completion of consolidation settlement is made up of three components

$$\rho_t = \rho_u + \rho_c + \rho_s \quad (19.1)$$

For rapidly draining or partly saturated soils, ρ_u and ρ_c may, for all practical purposes, be indistinguishable. Little attention will be devoted to ρ_s as it is not well understood and is frequently, although not invariably, small – see Simons and Menzies (2000) for a more detailed discussion on this topic.

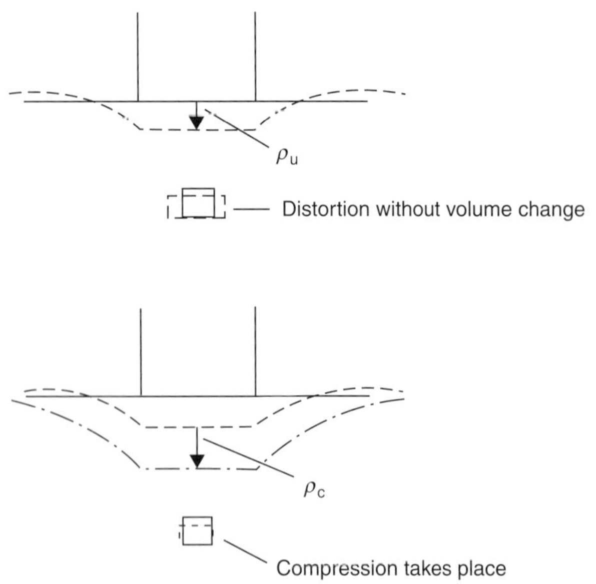

Figure 19.2 Schematic deformations of a soil element beneath a foundation during undrained loading and subsequent consolidation

Figure 19.2 shows schematically the ground surface movements that take place owing to (*a*) undrained settlement and (*b*) subsequent consolidation settlement. The associated deformations of an element of soil beneath the centre line of the foundation are shown. It is evident that during undrained settlement only distortion without volume change occurs, whereas during consolidation volume reduction of the soil skeleton takes place.

19.3. Stress changes beneath loaded areas

To carry out a settlement calculation it is necessary to derive the vertical strains ε_v at various depths beneath the foundation. The vertical compressions of the various layers of thickness Δz are then summed up over the depth of influence such that

$$\rho = \sum \varepsilon_v \Delta z \tag{19.2}$$

The calculation of ε_v at any depth requires knowledge of two things: (*a*) the stiffness or deformability of the soil at that depth; and (*b*) the stress changes at that depth induced by the loading of the foundation.

The theory of elasticity is invariably used to calculate the stress changes. Most textbooks outline methods of obtaining the vertical stress changes beneath simple loaded areas on an elastic, isotropic homogeneous half-space. For some particularly simple problems other stress changes, such as the principal stress changes or the horizontal stress changes, are given as well. All these results are obtained by integrating Boussinesq's famous solution for a point load on the surface and are often referred to as 'Boussinesq stresses'.

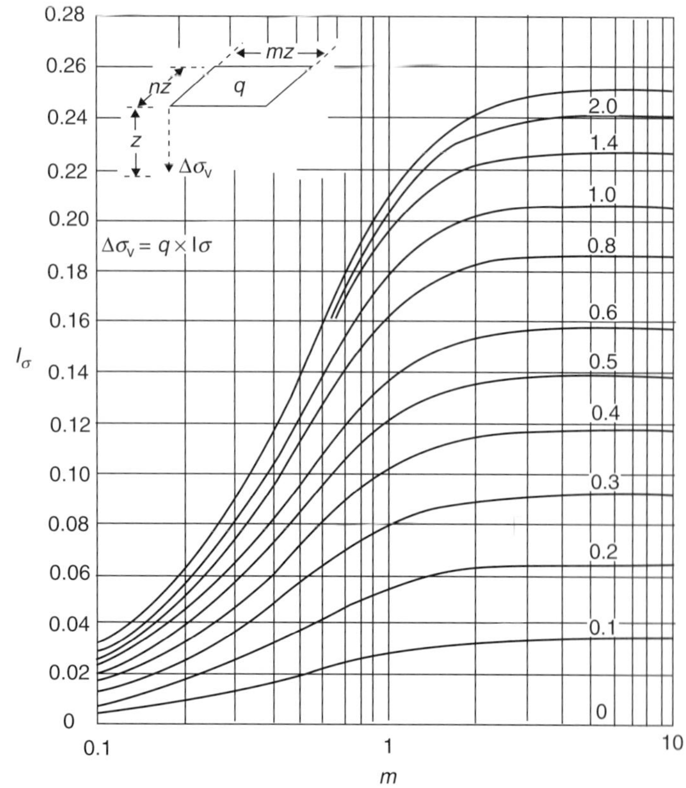

Figure 19.3 Vertical stress under the corner of a rectangular area carrying a uniform pressure (Reproduced from Simons and Menzies, 2000); data originally taken from Fadum, 1948)

The chart given in **Figure 19.3** is the well-known one for deriving the vertical stress change at depth z beneath the corner of a rectangular uniformly loaded area. Superposition can be used to calculate the vertical stress change at any location beneath or outside the rectangle, as explained in most textbooks. The expressions for the vertical and horizontal stress changes beneath the centre of a uniformly loaded circular area of radius r are as follows

$$\Delta\sigma_v = q\left\{1 - \left[\frac{1}{1+\left(\frac{r}{z}\right)^2}\right]^{3/2}\right\} \tag{19.3}$$

$$\Delta\sigma_h = \frac{q}{2}\left\{(1+2\nu) - \frac{2(1+\nu)}{\left[1+(r/z)^2\right]^{1/2}} + \frac{1}{\left[1+(r/z)^2\right]^{3/2}}\right\} \tag{19.4}$$

where ν is Poisson's ratio. **Figure 19.4** is derived from Equation 19.3 and shows the contours of increase in vertical stress Δq_v beneath a uniformly loaded circular area expressed as a proportion of the applied pressure at the surface Δq_s. The contour of $\Delta q_v/\Delta q_s = 0.1$ encloses what is termed the 'bulb of

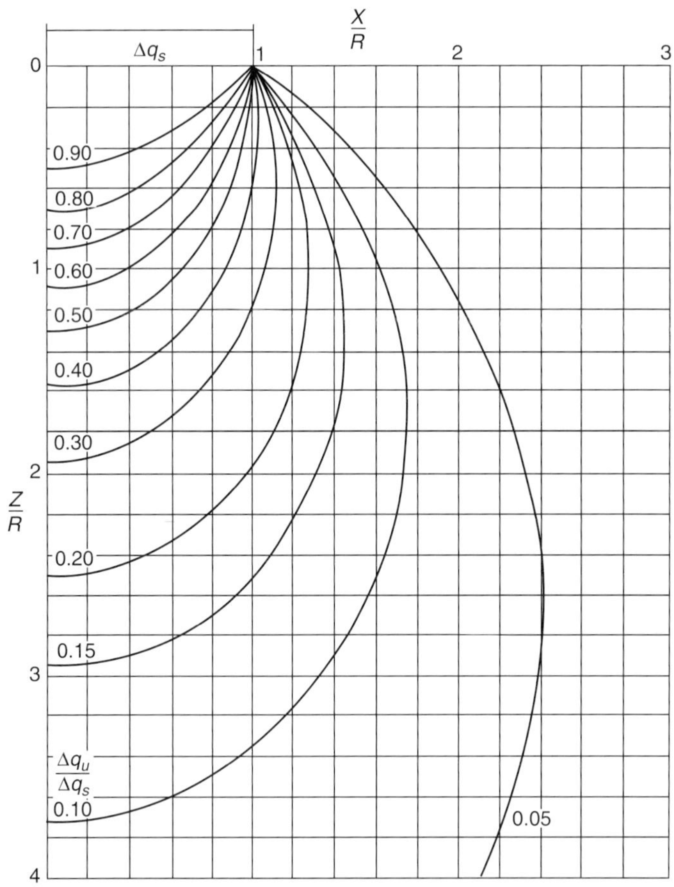

Figure 19.4 Vertical stress changes beneath a uniformly loaded circular area of radius *R* (Reproduced from Lambe and Whitman, 1969; © John Wiley & Sons, Inc.)

pressure', which is taken as an approximate but useful guide to the zone of influence beneath and around a loaded area. For a circular loaded area, the bulb of pressure is seen to extend to a depth of about twice its diameter. It is important to note that for an elastic homogeneous, isotropic half-space, the changes in vertical stress are always independent of Poisson's ratio ν. However, for most problems, the other stresses are dependent on it. Poulos and Davis (1974) have assembled a very complete set of elastic solutions for a wide range of problems.

An obvious and important question that is not addressed in most textbooks is: how valid is it to use the Boussinesq theory to predict stress changes in soils beneath foundations? Most soils are patently not elastic, not isotropic and not homogeneous, so at face value the theoretical basis for calculating settlements looks pretty flimsy! As a result of recent analytical and numerical work using more 'soil-like' stress–strain properties, more is known about the answer to this question than a few years ago. In particular, the effects of nonlinearity, nonhomogeneity and anisotropy have been studied.

19.3.1 Nonlinear stress–strain behaviour

A number of theoretical studies have been carried out to assess the influence of nonlinear behaviour on the stress changes beneath uniformly loaded areas. These studies have included elastic–perfectly plastic materials and continuously nonlinear stress–strain behaviour up to failure (see Burland *et al.*, 1977). All these studies have shown that the changes in vertical stress are remarkably similar to those given by the Boussinesq theory. However, the other stresses, in particular the horizontal and shear stresses, depart significantly from Boussinesq. **Figure 19.5** shows the stress changes beneath the centre of

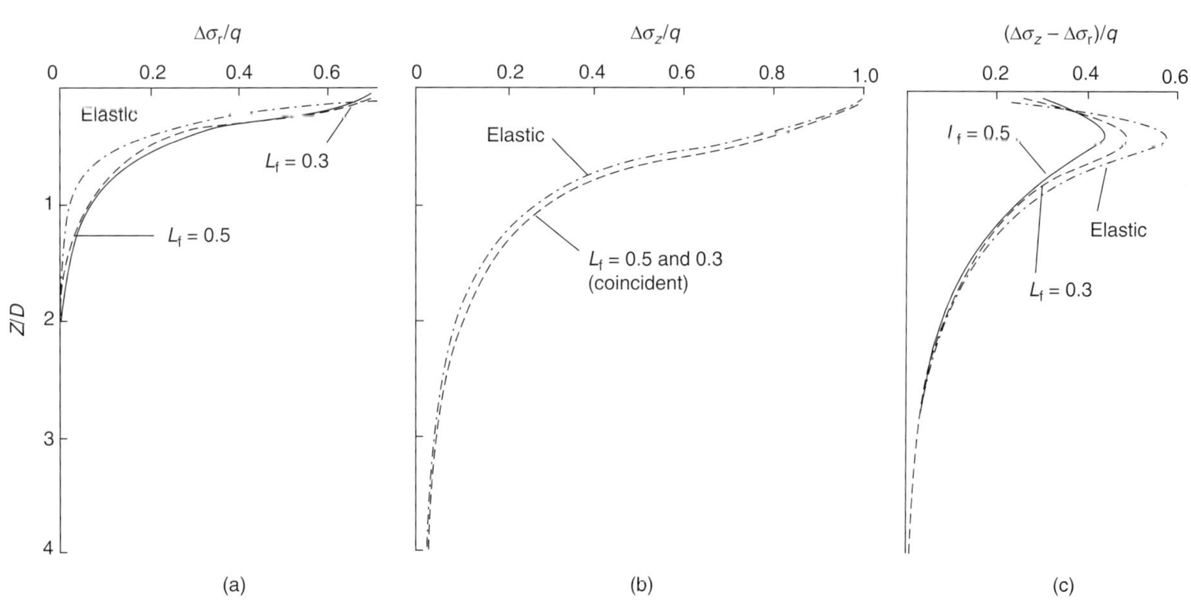

Figure 19.5 Distribution of stress increments beneath the centre of a flexible circular footing for various load factors *L*: (a) radial stress changes; (b) vertical stress changes; (c) deviator stress changes (Reproduced from Jardine *et al.*, 1986)

a uniform circular surface load for an undrained nonlinear plastic material. It can be seen that the vertical stress changes are almost coincident with the Boussinesq values, but that the radial and deviator stress changes differ significantly from the Boussinesq values and depend on the stress level.

19.3.2 Non-homogeneity

There is of course an infinite variety of types of non-homogeneity, but it is instructive to examine some of the more common types. One of the most common situations is where stiffness increases with depth, sometimes termed a Gibson soil. Since, for a given soil type, stiffness is approximately proportional to effective confining pressure, this increase in stiffness with depth is to be expected. **Figure 19.6** shows the distribution of vertical and horizontal stress changes beneath the centre of a uniform strip load on the surface of an elastic half-space, in which Young's modulus increases linearly with depth from zero at the surface (Gibson and Sills, 1971). The full lines give the Boussinesq solution. Remarkably, for Poisson's ratio $\nu = 0.5$, the vertical and horizontal stress changes are identical to those given by Boussinesq. For $\nu < 0.5$ the vertical stress changes are only slightly larger than those given by Boussinesq, but the horizontal stress changes are much less – being very sensitive to the value of Poisson's ratio.

Another common form of non-homogeneity is that of a compressible layer of soil overlying rock. **Figure 19.7** shows the distribution of vertical stress change beneath the centre line of a uniform circular load resting on a compressible layer overlying a rigid layer. It can be seen that the vertical stress changes increase only slightly as the depth of the rigid layer decreases. It can be shown that the horizontal stress changes are very significantly affected by the depth of the rigid layer.

Unlike the previous two situations, the presence of a stiff layer overlying a soft layer has a significant effect on the vertical stress changes. **Figure 19.8** shows the vertical stress changes beneath the centre of a uniform circular load resting on a stiff layer overlying a more compressible layer. It can be seen that the presence of a stiff overlying layer significantly reduces the magnitudes of the vertical stress changes in comparison with the Boussinesq stresses. Thus the presence of a stiff overlying layer is very beneficial in reducing the stresses in an underlying softer material.

19.3.3 Anisotropy

In Chapter 14 *Soils as particulate materials* it was demonstrated that deposition under vertical gravity leads to preferred orientation of the particle arrangements, and this leads to inherent anisotropy of the soil skeleton. Hence anisotropy is a common feature of most soils. In Chapter 17 *Strength and deformation behaviour of soils* readers were introduced to the cross-anisotropic porous elastic model of soil, in which the

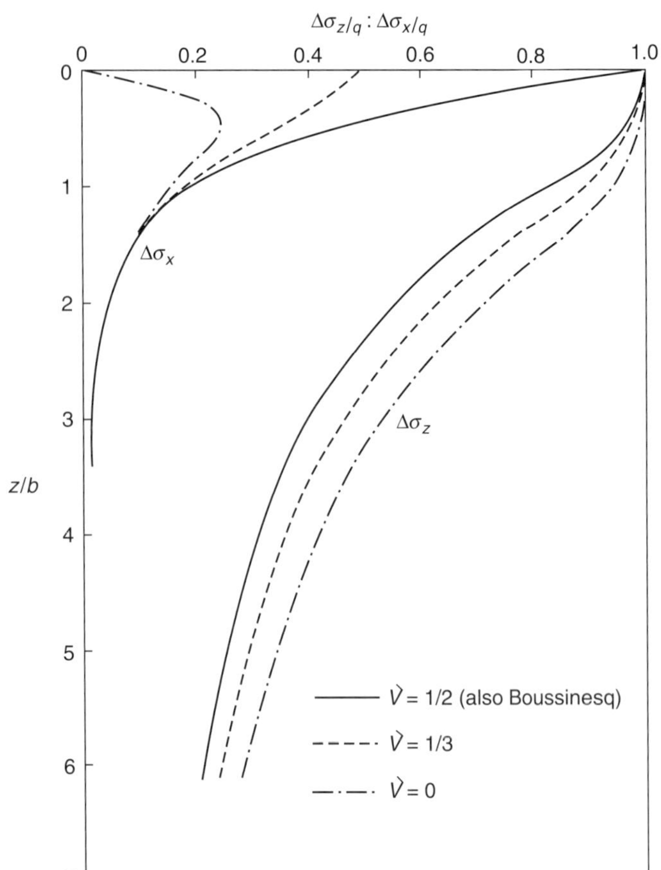

Figure 19.6 Stress changes beneath centre line of a uniform strip load on a Gibson half-space (Reproduced from Gibson and Sills, 1971; all rights reserved)

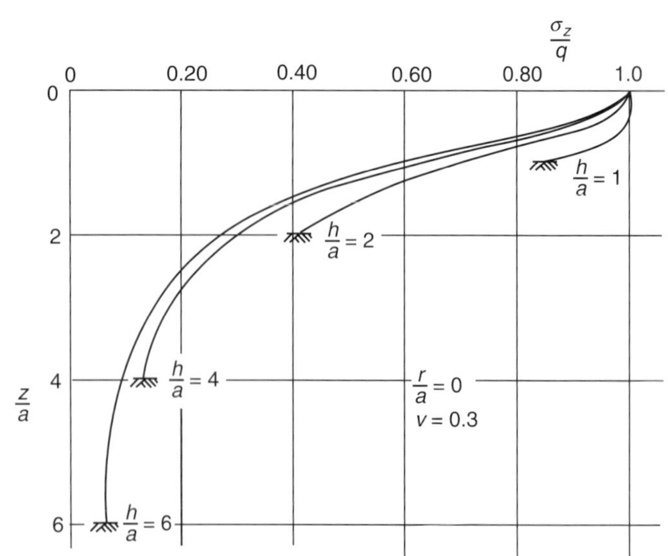

Figure 19.7 Vertical stress changes beneath the centre of a uniform circular load of a radius *a* on a compressible layer of various thicknesses *h* (Reproduced from Davis and Poulos, 1963; all rights reserved)

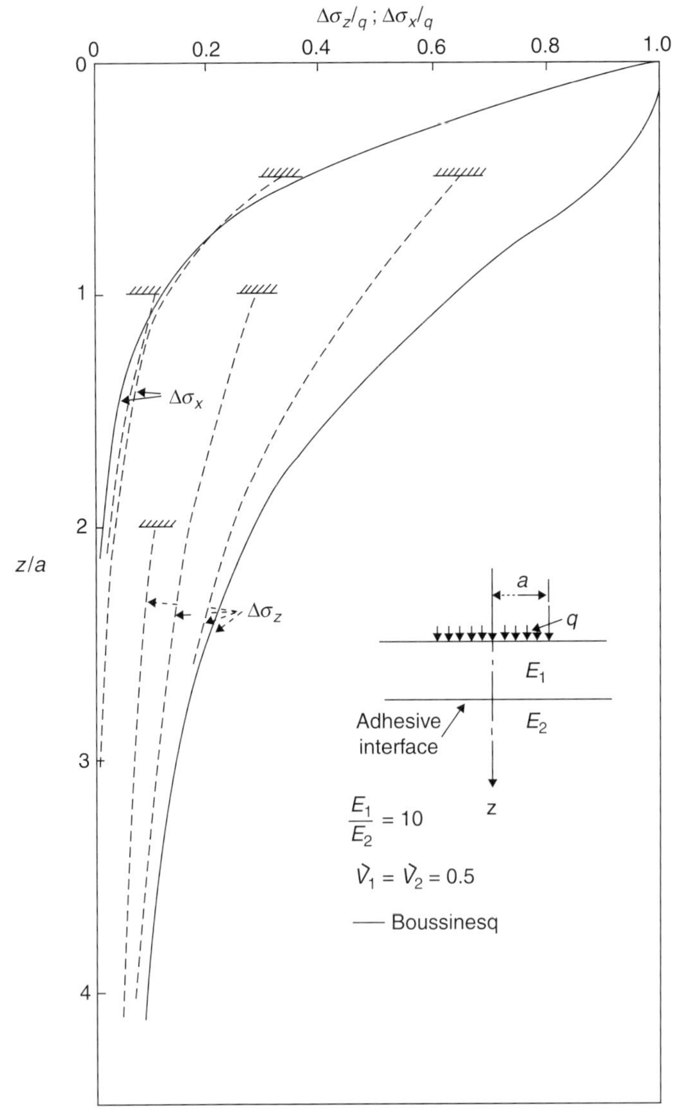

Figure 19.8 Influence of a stiff upper layer of various thicknesses on the vertical and horizontal stress changes beneath the centre of a uniformly loaded circular area (Reproduced from Burland *et al.*, 1977)

horizontal stiffness properties differ from the vertical ones. This model can be fully defined by the five parameters

E'_v = effective Young's modulus in the vertical direction
E'_h = effective Young's modulus in the horizontal direction
ν'_{vh} = effective Poisson's ratio for the effect of vertical stress change on horizontal strain
ν'_{hh} = effective Poisson's ratio for the effect of horizontal stress change on orthogonal horizontal strain
G'_{vh} = shear modulus in the vertical plane.

It is convenient to define the following ratios

$$E'_h/E'_v = n' \text{ and } G'_{vh}/E'_v = m'$$

Figure 19.9 shows the vertical stress changes beneath the centre of a uniform circular load on a porous elastic anisotropic soil for (*a*) varying n', with m' fixed at its equivalent isotropic value of $0.5(1 + \nu'_{vh})$ and (*b*) varying m', with n' fixed at its isotropic value (—1). It can be seen from **Figure 19.9(a)** that varying n' has little effect on the vertical stress changes. However, varying G'_{vh} (i.e. the value of m') has a larger effect.

19.3.4 Conclusions on stress changes

The following important conclusions can be drawn from the examples given in this section.

- For many conditions which depart from the ideal of linear, isotropic, homogeneous elasticity, the Boussinesq theory gives very reasonable estimates of the changes in vertical stress beneath areas of known pressure distribution.

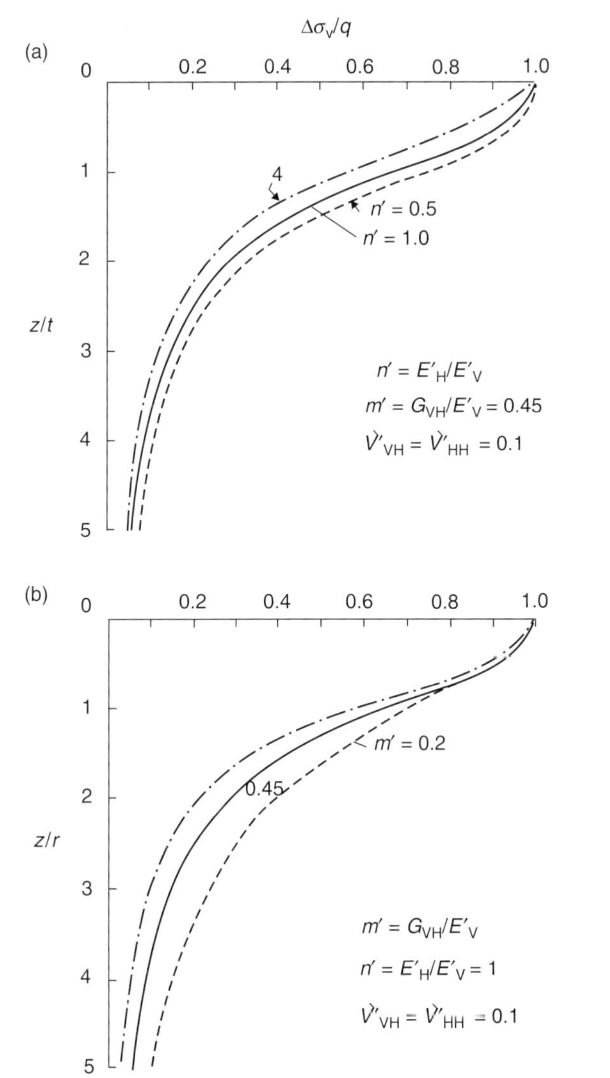

Figure 19.9 Influence of anisotropy on the vertical stress changes beneath the centre of a uniformly loaded circular area: (a) influence of change of E'_h ; (b) influence of change of shear modulus G'_{vh}

- Two exceptions to this are (*a*) for a stiff layer overlying a softer layer where large errors on the conservative side result from using the Boussinesq theory; and (*b*) for anisotropic properties, whereas the vertical stress changes are relatively insensitive to the ratio $n' = E'_h/E'_v$, they are more sensitive to the ratio $m' = G'_{vh}/E'_v$. It is of interest to note that the static value of G'_{vh} is seldom measured in a routine manner because it is difficult to do so.
- In contrast to the vertical stress changes, the horizontal (and shear) stress changes are very sensitive to all the variables studied. Hence the Boussinesq equations are unlikely to give accurate estimates of these stress changes.

Although only a limited number of examples are given here, these conclusions are widely supported by numerous studies.

19.4. Summary of methods of settlement prediction for clay soils

This section gives a very brief résumé of the most commonly used methods of settlement prediction on soils that can be reasonably sampled and tested to give the deformation properties.

19.4.1 Conventional 1D method

This method was proposed by Terzaghi and was originally used for estimating the compression of a layer of clay situated at some depth. It later became used for thick layers of clay extending right to the surface. The procedure is as follows, referring to **Figure 19.10**.

(*a*) The compressible stratum is divided into a number of layers of thickness Δz (as shown at the bottom of **Figure 19.10**).

(*b*) The initial vertical effective stress σ'_{vi} is calculated for the centre of each layer.

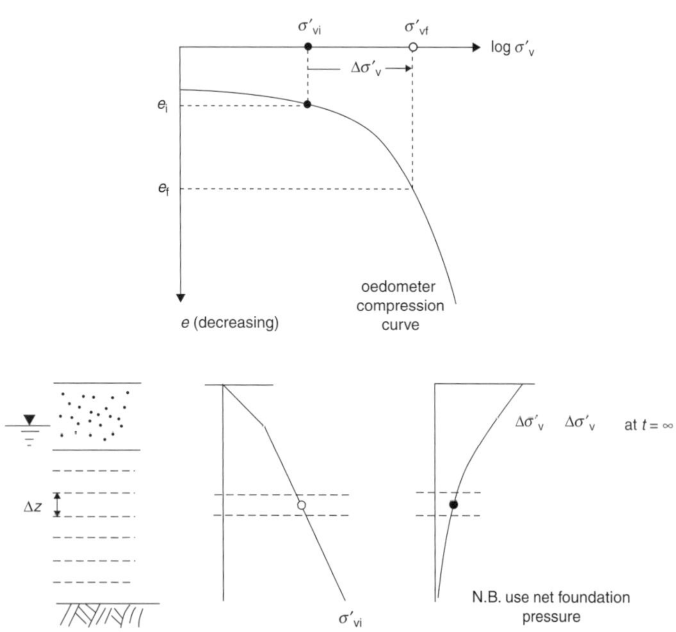

Figure 19.10 Classical one-dimensional method of settlement analysis

(*c*) Using Boussinesq stresses, the increase in vertical stress $\Delta\sigma'_v$ is calculated for the centre of each layer. At the end of consolidation, when all excess pore water pressures have been dissipated, the stress increases will be effective stresses. Note that if the foundation is founded in an excavation, it is the *net* increase in vertical effective stress that gives rise to compression.

(*d*) The vertical strain ε_v for each layer may be obtained directly from a plot of *e* against σ'_v from one-dimensional (oedometer) compression tests (see the top of **Figure 19.10**) such that

$$\varepsilon_v = \frac{e_i - e_f}{1 + e_i} \quad (= m_v \Delta\sigma'_v) \tag{19.5}$$

where the subscript i refers to the initial value of *e* and the subscript f refers to the final value of *e*. If quoted values of m_v are used, it is important to ensure that they have been measured over the appropriate stress range – see Chapter 17 *Strength and deformation behaviour of soils.* It should also be noted that in Equation 17.13 it was shown that m_v is related to the equivalent effective Young's modulus E' and Poisson's ratio ν' by the expression

$$m_v = \left\{1 - \frac{2\nu^2}{1-\nu}\right\}/E' \tag{19.6}$$

(*e*) The compression of each layer $= \varepsilon_v \Delta z$ and, summing over the total thickness of the compressible stratum, the total 1D settlement is given by

$$\rho_{1XD} = \sum \varepsilon_v \Delta z \tag{19.7}$$

Note that Terzaghi assumed that ρ_{1XD} is equal to the total settlement ρ_t. The accuracy of this assumption will be evaluated later in this chapter.

19.4.2 Skempton and Bjerrum method

In 1955, Skempton, Peck and MacDonald used the 1D method to calculate the total settlement of a number of buildings, but they recognised that the undrained settlement ρ_u could be a significant proportion of the total settlement ρ_t (see Equation 19.1). They used elastic displacement theory, described in Section 19.5, to evaluate ρ_u. This led them to conclude that the consolidation settlement ρ_c is given by $\rho_c = \rho_{1XD} - \rho_u$. In the discussion of the above paper, various writers suggested that it was more logical to assume that $\rho_c = \rho_{1XD}$, so that

$$\rho_t = \rho_u + \rho_{1XD} \tag{19.8}$$

With this background, Skempton and Bjerrum (1957) put forward their seminal method for calculating the consolidation settlement ρ_c They argued that the consolidation settlement resulted from the dissipation of excess pore pressure Δu. The key assumption was made that consolidation takes place *one-dimensionally* so that the vertical consolidation strain ε_{vc} at a given depth is given by $\varepsilon_{vc} = \Delta u \times m_v$ where Δu can be calculated knowing the values of the pore pressure coefficient A (see Equation 17.23 in Chapter 17 *Strength and deformation behaviour of soils*), $\Delta\sigma_v$ and $\Delta\sigma_h$. Hence the consolidation settlement ρ_c for the whole layer is obtained by summation

$$\rho_c = \sum \Delta u \times m_v \Delta z \tag{19.9}$$

Skempton and Bjerrum defined a correction factor μ to be applied to ρ_{1XD} such that

$$\rho_c = \mu \times \rho_{1XD}$$

Hence the total settlement is given by

$$\rho_t = \rho_u + \mu\rho_{1XD} \tag{19.10}$$

Simple elastic Boussinesq stresses were used to evaluate μ for a range of values of A and foundation geometries. The well-known Skempton and Bjerrum chart for values of μ is given in **Figure 19.11**.

It is important to note that the key assumption of the method is that consolidation takes place *one-dimensionally*. For soils which are approximately elastic in their response, this assumption can be significantly in error. Moreover, the values of μ are based on values of $\Delta\sigma_v$ and $\Delta\sigma_h$ obtained from isotropic, homogeneous, semi-infinite elastic theory. Values of $\Delta\sigma_h$ obtained in this way are unreliable.

Thus, although the method is widely used and taught, caution is needed in applying it.

19.4.3 Stress path method

Although attributed to Lambe (1964), the stress path method was suggested earlier by Skempton (1957) and developed by Davis and Poulos (1963). In essence, the method consists of the following.

(*a*) Take undisturbed samples from appropriate depths.

(*b*) Set up each sample in the triaxial apparatus under the appropriate initial in situ effective stresses.

(*c*) Without allowing drainage, apply the calculated values of $\Delta\sigma_v$ and $\Delta\sigma_h$. Measure the vertical strain and the excess pore pressure.

(*d*) Allow drainage to occur – if necessary, adjust the applied stresses to the final calculated values. Measure the vertical strain and volume change.

(*e*) The measured strains for stages (*c*) and (*d*) are summed to obtain the undrained settlement and consolidation settlement.

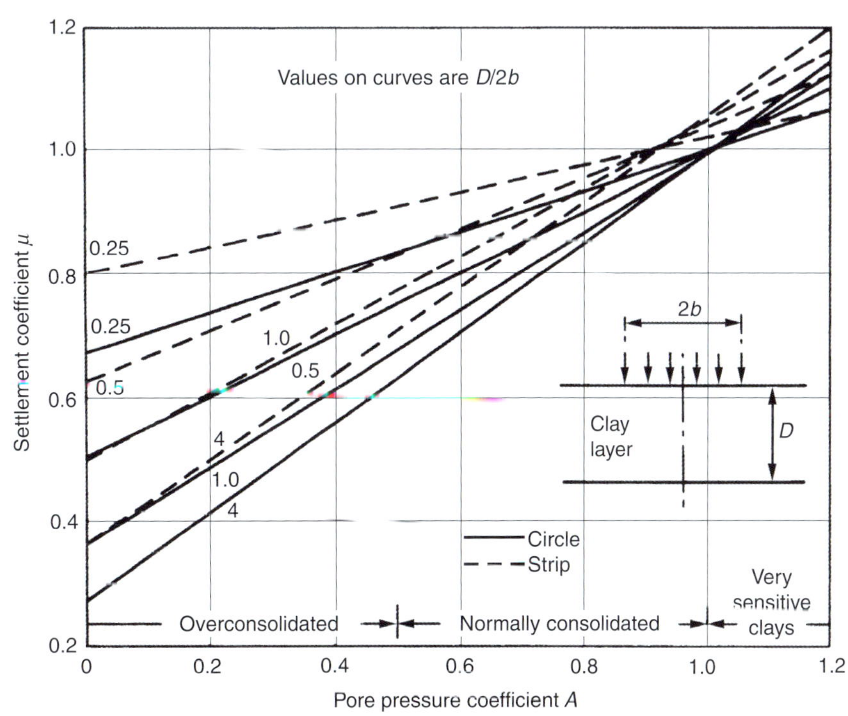

Figure 19.11 Skempton and Bjerrum's settlement coefficient μ for circular and strip foundations (Reproduced from Simons and Menzies, 2000; data originally taken from Skempton and Bjerrum, 1957)

The method appears to be logical but carries with it serious difficulties. The initial vertical and horizontal effective stresses must be known. A high degree of experimental skill is required. Isotropic, homogeneous elasticity is used to calculate the stress changes, and uncertainties in the accuracy of $\Delta\sigma_h$ therefore exist. The testing is expensive and time-consuming, and a change in the foundation size or loading could render the whole testing programme worthless.

A more realistic and flexible approach was suggested by Davis and Poulos (1963). Samples are subjected to appropriate stress levels and stress changes, and the associated equivalent elastic constants E_u, E' and ν' are deduced from the measured strains. The undrained and total settlements can then be obtained by calculating the corresponding equivalent elastic strains and summing them over the appropriate depth

$$\rho = \sum \varepsilon_v \times \Delta z = \sum \frac{1}{E}\left\{\Delta\sigma_v - \nu(\Delta\sigma_x + \Delta\sigma_y)\right\}\Delta z \qquad (19.11)$$

Alternatively, average values of E and ν can be used in conjunction with elastic displacement theory – see Section 19.5.

19.4.4 Finite element method

In recent years, powerful numerical methods have become available which can be used for settlement prediction. Most finite element (FE) packages can handle non-homogeneous anisotropic elastic materials. Increasingly more packages can handle nonlinear stress–strain behaviour of varying complexity. It is perhaps a truism to say that the accuracy of the predictions depends on the quality of the input data. However, engineers are easily seduced by the apparent power and versatility of FE packages and forget, or are not aware of, the many idealisations and assumptions that are built into them.

It is important to examine some of the limitations of routine testing, however well conducted, in providing input data for a numerical analysis.

(*a*) For triaxial and oedometer tests, only axisymmetric stress conditions can be applied: σ_a and σ_r.

(*b*) The principal stresses and stress changes remain vertical and horizontal.

(*c*) Usually only the vertical stiffness E'_v and Poisson's ratio ν'_{vh} are determined.

(*d*) Any nonlinear behaviour that is measured is strictly for the highly restricted stress conditions referred to in (*b*).

(*e*) The in situ stresses and, in particular, stress history of the natural soil are usually open to considerable speculation.

These restrictions in the test data should be contrasted with the real situation in the ground.

- In real problems, axisymmetric stress conditions almost never occur.
- Rotations of principal stress directions take place during loading.
- The stiffness both in direct loading and shear is almost certainly anisotropic and related in a complex way to the rotating stresses.
- The previous stress history, and in particular whether the soil is in a loading or unloading cycle, profoundly affects the stress–strain behaviour.

In view of the above, any experimental data used in an FE analysis usually entails profound implicit idealisations.

19.4.5 Comments on methods of settlement prediction

It is evident from the methods presented above that the methods of settlement prediction have progressively become more sophisticated, moving from the classical 1D method through to numerical methods. This has been accompanied by the need for increasingly more soil parameters and more advanced testing. From a purely practical point of view, the engineer is entitled to ask whether such sophistication is necessary and whether greater accuracy is in fact achieved. This question is addressed in Section 19.6 by comparing the various approximate methods with known rigorous solutions. First it is necessary to briefly discuss elastic displacement theory.

19.5. Elastic displacement theory

At the beginning of Section 19.2 it was pointed out that a settlement calculation involves deriving the vertical strains ε_v at various depths beneath the foundation. For an ideal isotropic elastic material the vertical strain at any depth below a loaded area is given by

$$\varepsilon_v = \frac{1}{E}\left\{\Delta\sigma_v - \nu(\Delta\sigma_x + \Delta\sigma_y)\right\} \qquad (19.12)$$

Note that the two horizontal stress changes and Poisson's ratio are involved.

Elastic displacement theory is based on integrating the vertical strains from $z = 0$ to infinity, such that

$$\rho = \int_0^\infty \varepsilon_v dz \qquad (19.13)$$

Solutions for a wide range of problems are usually expressed in the form

$$\rho = \left\{\frac{qB}{E}(1-\nu^2)\right\} I_\rho \qquad (19.14)$$

where:

q = average foundation pressure

B = a characteristic dimension of the foundation (e.g. breadth)

$I\rho$ = a factor to take account of geometry, depth of founding, depth of rigid layer and so on.

Examples of some well-known solutions are

- Settlement of the centre of a uniform circular load of diameter D

$$\rho = \frac{qD}{E}(1-\nu^2) \tag{19.15}$$

- Settlement of the centre of a uniform circular load at great depth

$$\rho = \frac{qD}{E}(1-\nu^2)\frac{1}{2} \tag{19.16}$$

- Settlement of a rigid circular load

$$\rho = \frac{qD}{E}(1-\nu^2)\frac{\pi}{4} \tag{19.17}$$

It is of interest to note that the settlement of a rigid circular load given by Equation 19.17 is only very slightly less than the average settlement of a uniformly loaded (flexible) circular area (Equation 19.15). Thus, while stiffening a foundation significantly reduces the differential settlements, it has little effect on the average total settlements.

Numerous charts are available for calculating the settlement of various types of loaded area. For example, **Figure 19.12** is the well-known settlement chart for estimating the average undrained settlement of uniformly loaded rectangular areas at depth D with a rigid layer at a depth H beneath foundation level (Christian and Carrier, 1978). These charts are only for Poisson's ratio = 0.5 and are strictly only relevant for estimating undrained settlement. Moreover, an approximation was used in deriving them in that it was assumed that the rigid layer has no influence on the stress changes (known as the Steinbrenner approximation).

Section 19.3 showed that simple elastic theory gives a good estimate of the vertical stress changes for a range of soil-like properties and can therefore be used with considerable confidence. The same cannot generally be said of elastic displacement theory, which must be used with considerable caution.

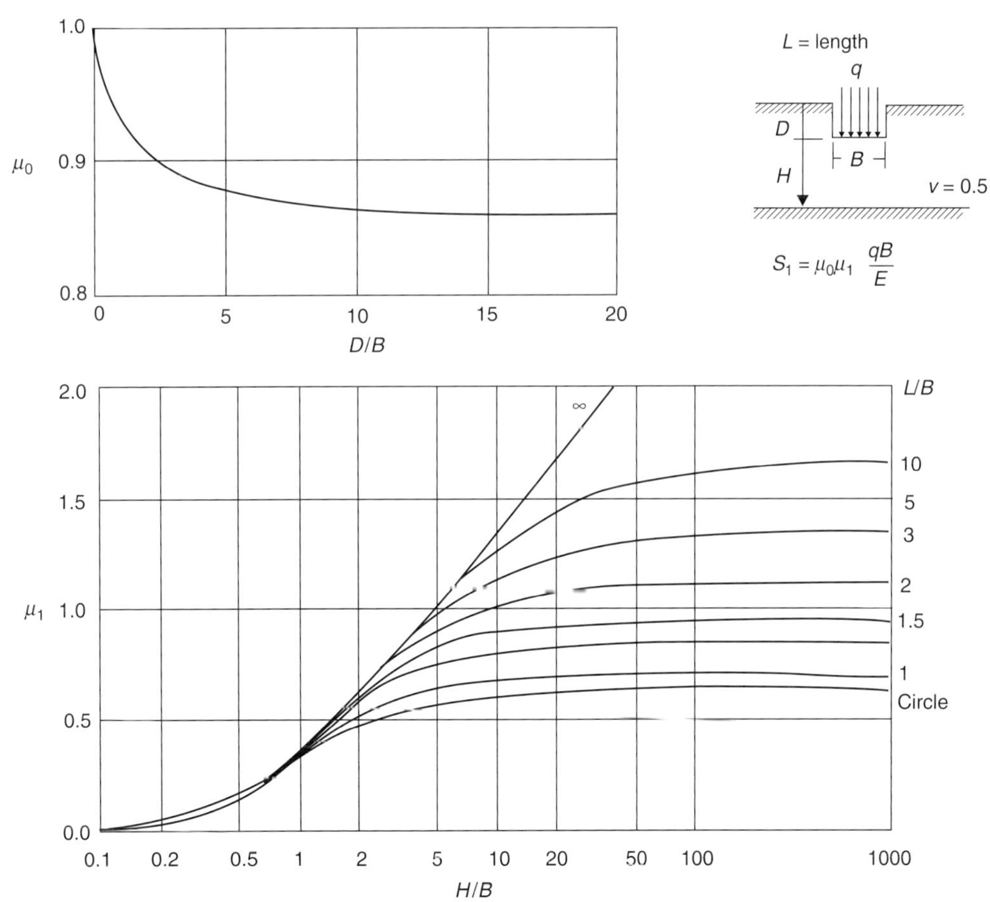

Figure 19.12 Average undrained settlement of uniformly loaded rectangular areas with a foundation depth *D* and a rigid layer at depth *H* beneath founding level (Reproduced from Christian and Carrier, 1978; © Canadian Science Publishing or its licensors)

In particular, the calculation of the strains in Equation 19.12 involves horizontal stress changes which are very sensitive to many factors. Moreover, the integration of the strains in Equation 19.13 involves the assumption that the stiffness properties are constant with depth (homogeneity).

When using solutions based on elastic displacement theory, it is no easy matter to make allowance for the influence of such factors as anisotropy, non-homogeneity and nonlinearity on stress–strain behaviour. These influence the stress changes (in particular the horizontal ones) as well as the stiffness. However, when used with discrimination, elastic displacement theory can be very useful, particularly for estimating undrained settlements.

19.6. Theoretical accuracy of settlement predictions

In the past, the difficulty of assessing the accuracy of methods of settlement prediction was that the methods of testing were intimately linked with the analysis. This has meant that it has not been possible to isolate inaccuracies in sampling and testing from the limitations of the analysis.

With the advent of the FE method, it has been possible to obtain rigorous solutions using realistic soil-like constitutive relationships, and compare them with predictions based on the various current approximate methods of settlement prediction. The startling conclusion is reached that there is little to choose between the methods and that in most cases the simple 1D method gives the best predictions of total settlement ρ_t. Hence all the effort should be devoted to measuring as accurately as possible the straightforward stiffness parameters such as m_v, E'_v and ν'. It is demonstrated that the vertical shear modulus G_{vh} is also a key parameter, but is very seldom determined from static tests.

The above important conclusions are briefly illustrated for a few cases in the following sections. The approach is to compare the rigorous solution with the predictions of the various methods.

19.6.1 Accuracy of the one-dimensional method for homogeneous isotropic elastic material

As explained in Chapter 17 *Strength and deformation behaviour of soils*, a homogeneous porous elastic material is completely defined by the two parameters E' and ν'. It was further shown in Equation 17.13, and repeated as Equation 19.6, that

$$m_v = \left\{1 - \frac{2\nu'^2}{1-\nu'}\right\}/E'$$

For m_v constant with depth, Equation 19.7 for the 1D settlement is given by

$$\rho_{1XD} = m_v \int \Delta\sigma'_v dz \tag{19.18}$$

where $\Delta\sigma'_v$ is obtained from Boussinesq stresses. For a uniform circular load of radius r, the values of $\Delta\sigma'_v$ are given by Equation 19.3. It turns out that $\int \Delta\sigma'_v dz$ for this case is particularly simple and is equal to $2qr$. Hence, Equation 19.18 for the 1D settlement becomes

$$\rho_{1XD} = 2 \times q \times r \times m_v = \frac{q \times D}{E'}\left\{1 - \frac{2\nu^2}{1-\nu'}\right\} \tag{19.19}$$

This is the analytical solution for the 1D settlement of the centre of a uniform circular load on an isotropic semi-infinite porous elastic material. The exact total settlement for this problem is given by Equation 19.15, so

$$\rho_{t(\text{exact})} = \frac{qD}{E'}(1-\nu'^2) \tag{19.20}$$

Hence the ratio of the one-dimensional settlement to the exact settlement, derived from Equation 19.16 is

$$\frac{\rho_{1XD}}{\rho_{t(\text{exact})}} = \frac{1-2\nu'}{(1-\nu')^2} \tag{19.21}$$

Figure 19.13 shows a graph of Equation 19.16. It can be seen that when $\nu' = 0$, the 1D settlement is exactly equal to the total settlement. As ν' increases, the ratio $\rho_{1XD}/\rho_{t(\text{exact})}$ decreases slowly at first, and then more rapidly until it equals zero when $\nu' = 0.5$. Most soils and weak rocks have values of $\nu' < 0.25$ over the range of working stress changes. It can therefore be concluded that ρ_{1XD} will usually be within 10% of the total

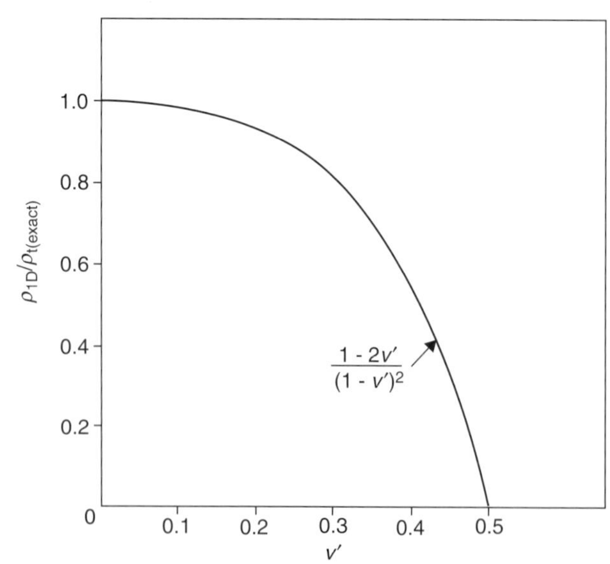

Figure 19.13 The influence of effective Poisson's ratio on the accuracy of the 1D method for a uniform circular load on an isotropic homogeneous elastic half-space

settlement on the low side. If the compressible layer is of limited depth, the 1D settlement will be even closer to the exact total settlement.

19.6.2 Accuracy of the one-dimensional method for homogeneous cross-anisotropic elastic material

In Section 19.3.3, the five elastic constants that define a cross-anisotropic material were listed. It was pointed out in Chapter 17 *Strength and deformation behaviour of soils* that such a material is generally more representative of a stiff soil or a soft sedimentary rock than an isotropic elastic material. Using the example of a uniform circular load, this section compares the settlements predicted by the four methods described in Section 19.4 with the solutions obtained by a rigorous FE analysis (for practical purposes it can be assumed that these solutions are exact).

It must be remembered that the engineer does not necessarily know that the material is anisotropic. It is assumed that the appropriate routine tests are carried out on vertical samples to measure parameters such as m_v, E', ν', E_u and A (the pore pressure parameter). Moreover, as is normal practice, Boussinesq stress changes are used in carrying out the settlement predictions.

Figure 19.14 shows graphs of $\rho_{(\mathrm{predicted})}/\rho_{\mathrm{t(exact)}}$ against the ratio $E'_h/E'_v = n'$ for the four methods of settlement prediction described in Section 19.4. For the purposes of this analysis, the value of $G_{vh}/E'_v = 0.45$ was adopted (the isotropic value) and with $\nu'_{vh} = \nu'_{hh} = 0.1$. It is found that the solutions are insensitive to small changes in ν'.

The results for the 1D method are shown by the full line in **Figure 19.14**. It can be seen that for $n' < 1.2$, the exact total settlement is slightly underpredicted. As n' increases to 4, the method tends to overpredict slightly, reaching a maximum of about 6%. None of the other three methods give better predictions of the total settlement than the 1D method, despite its simplicity. The greatest discrepancy is given by the Skempton and Bjerrum method, which for $n' > 4$ overpredicts the total settlement by 22%. Even the FE method, which makes use of the measured values of E' and ν', overpredicts the total settlement by as much as 20% for $n' > 4$.

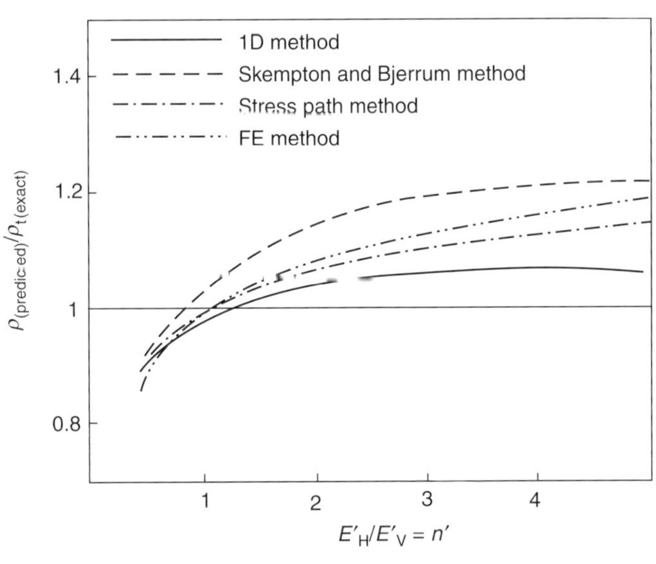

Figure 19.14 The accuracy of four methods of predicting the total settlement of the centre of a uniformly loaded circular area resting on a homogeneous, anisotropic elastic soil. All four methods make use of stiffness parameters obtained from oedometer or triaxial tests

19.6.3 Accuracy of the one-dimensional method for cross-anisotropic elastic material: stiffness increasing with depth

As explained in Section 19.3.2, increasing stiffness with depth is one of the most common situations in soil mechanics – far more common than uniform stiffness with depth. If cross-anisotropy is also included, this is a very realistic and reasonably general condition that can be used for evaluating the accuracy of various prediction methods. Using methods identical to those described in Section 19.6.2, the four common methods of settlement prediction have been compared with exact values obtained using FE analysis. It was found that the 1D method always gives predictions of the total settlement which are within 10% of the exact solution – significantly better than the other methods.

19.6.4 The influence of vertical shear modulus G_{vh}

For all of the above comparisons a value of $G_{vh}/E'_v = 0.45$ was used. This corresponds to the equivalent isotropic value given by $G = E'/2(1 + \nu')$ with $\nu' = 0.1$. The vertical shear modulus G_{vh} is an independent variable which is seldom measured other than by dynamic methods. In principle, there seems to be no reason why the value of G_{vh}/E'_v should not vary over quite wide limits. Rigorous FE analyses have been carried out for a uniform circular load varying the value of G_{vh}/E'_v, and in **Figure 19.15** the results are compared with the 1D and the Skempton and Bjerrum methods, which cover the range given by all four methods. It can be seen that for values of $G_{vh}/E'_v > 0.45$ all of the methods overpredict the settlement, whereas for values of $G_{vh}/E'_v < 0.45$ they underpredict. It can be concluded from this analysis that the vertical shear modulus is a key parameter in determining the magnitude of settlement, yet it is seldom measured by static means and traditional methods of settlement prediction take no account of it.

19.6.5 Accuracy of the one-dimensional method for normally consolidated clay

In recent years, considerable developments have been made in developing mathematical models for the stress–strain behaviour of soft normally consolidated clays. A detailed discussion of these developments is outside the scope of this chapter. However, enough has been done to demonstrate that, for factors of safety on undrained bearing capacity greater than about

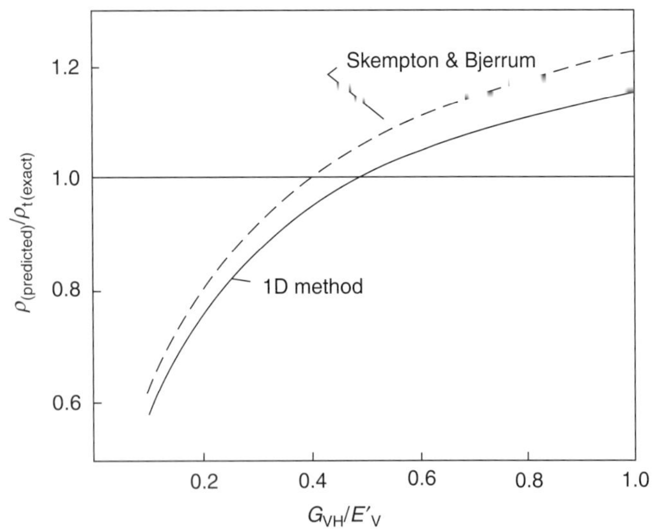

Figure 19.15 Results of rigorous finite element analyses, illustrating the important influence of the vertical shear modulus G_{vh} on settlement

3, the measured and theoretical values of the consolidation settlement ρ_c are only slightly larger than the 1D settlement. **Figure 19.16** shows the results of two model footing tests on normally consolidated clays compared with the predictions of the 1D method and the modified Cam Clay model for each model footing test (Burland, 1971). It can be seen that for bearing pressures up to about one-third of the undrained bearing capacity the measured settlements are only slightly larger than the predicted 1D values.

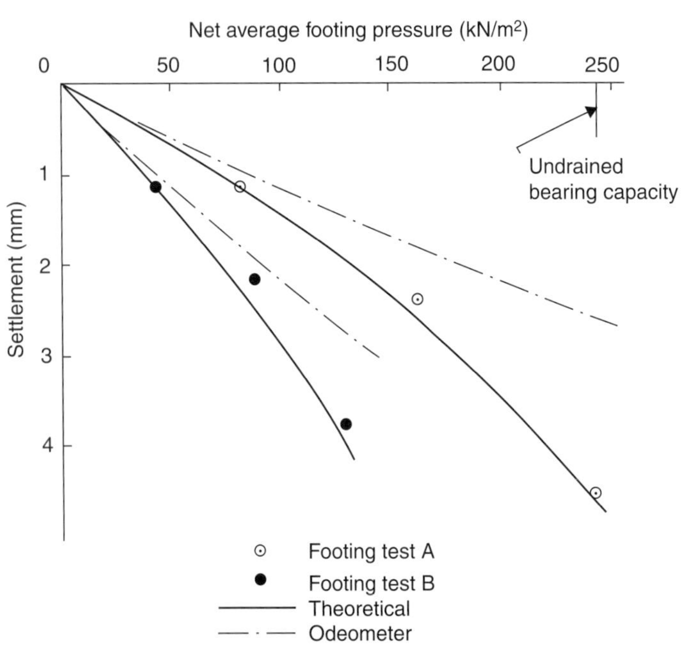

Figure 19.16 Observed consolidation settlements for two model footings on clay with predictions using the modified Cam Clay model (Reproduced from Burland, 1971)

19.7. Undrained settlement

The undrained settlement ρ_u is usually calculated by making use of elastic displacement theory (see Section 19.5). Most available charts and formulae are based on homogeneous elastic theory and are therefore of limited value. Butler (1974) developed some useful charts for E_u increasing linearly with depth. A major problem in predicting undrained settlement is the accurate measurement and choice of appropriate values of E_u.

A simple pragmatic empirical approach is to estimate the undrained settlement as a proportion of the total settlement ρ_t. Simons and Som (1970) have reported on the settlement of 12 buildings on overconsolidated clays. They show that ρ_u/ρ_t varies from 0.3 to 0.8, with an average of 0.58. This figure is in agreement with elastic theory and many recent case histories.

For normally consolidated clays, Simons and Som (1970) reviewed nine case histories and showed that ρ_u/ρ_t varies from 0.08 to 0.2, with an average of 0.16. Undoubtedly some consolidation would have taken place during construction and ρ_u/ρ_t will usually be less than 0.1.

19.8. Settlement on granular soils

Numerous methods of predicting settlement of foundations on sands and gravels have been published – many more than for clays. The reason lies in the difficulty of obtaining undisturbed samples for the laboratory determination of compressibility under appropriate conditions of stress and stress history. Hence resort has been made to the empirical interpretation of field in situ tests, such as standard penetration tests (SPTs) and cone penetration tests (CPTs). This extensive literature has been adequately covered by Sutherland (1974) and Simons and Menzies (2000). The latter authors apply the six most up-to-date methods to a simple illustrative example and show that the predictions differ by a factor of five, even when the representative values of the penetration results are given. Presumably the range would be even wider in practice, where the engineer has, in addition, to interpret the penetration data.

The practical significance of the problem was put into perspective by Terzaghi (1956) when he stated that all buildings resting on sand, which were known to him, had settled less than 75 mm, whereas the settlement of buildings on clay foundations quite often exceeded 500 mm!

In view of the unsatisfactory state of the art, and mindful of Terzaghi's assertion, Burland *et al.* (1977) went back to the original databases on which the various empirical methods are based, to see whether a simpler picture emerges which is less dependent on quantitative correlations with erratic penetration tests.

Indeed, a much simpler picture does emerge, as shown in **Figure 19.17**, where measured settlement per unit pressure (ρ/q) is plotted against foundation breadth B. In each case, the sand is broadly classified as 'loose', 'medium dense' and 'dense', based on average SPT blow counts N. In this correlation no account has been taken of such factors as the water

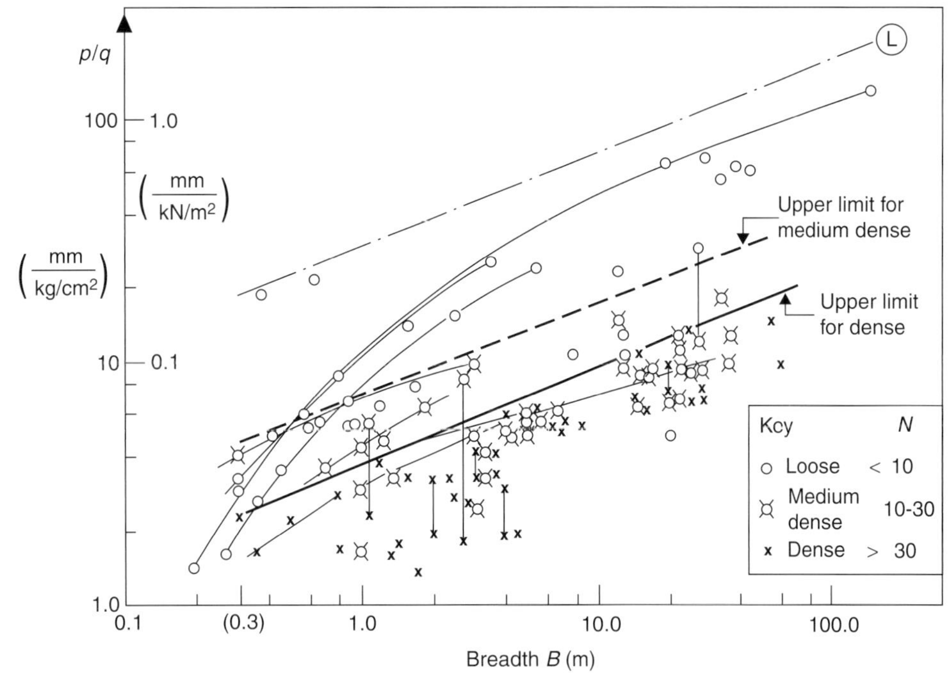

Figure 19.17 Measured settlement of foundations on sand of various relative densities (Reproduced from Burland *et al.*, 1977)

table, depth of loaded area and geometry. Reasonably well-defined upper limits for dense, medium-dense and loose sands are shown by the full, dotted and chain dotted lines, respectively. When assessing the most probable settlement, the engineer may elect to work to about 50% of the upper limit values. In most cases the maximum settlement would be unlikely to exceed about 75% of the upper limit. **Figure 19.17** is an example of an empirical approach in which, as far as possible, the measurements of foundation performance are allowed to speak for themselves with a minimum of indirect interpretation; see Chapter 4 *The geotechnical triangle*.

In **Figure 19.17**, the points which are connected by thin full lines are for different sizes of foundation at the same site. If plate loading tests are to be used to extrapolate to larger foundation dimensions, it is clear that the trends are not established at $B = 0.3$ m and that tests with $B = 1$ m are more likely to be successful.

Figure 19.17 is useful for making a preliminary assessment of the probable settlements on granular materials. Subsequent to the publication of this figure, Burland and Burbidge (1985) carried out a detailed statistical analysis of 200 case records of the settlement of foundations on sands and gravels. This resulted in a simple but more refined approach, which leads to an assessment of the likely range of settlements of a given foundation. It was also shown that time-dependent settlements can take place which, after 30 years, can reach 1.5 times the immediate settlements for static loads, and as high as 2.5 times for fluctuating loads. A brief summary of this more refined approach is given by Knappett and Craig (2012). In concluding this section, it is appropriate to bear in mind the following remarks by Sutherland (1974):

> Before a designer becomes entangled in the details of predicting settlement (in sand) he must satisfy himself whether a real problem actually exists and ascertain what advantages and economies can result from refinements in settlement prediction.

19.9. Summary

- Vertical stress changes beneath loaded areas are insensitive to such factors as nonlinear stress–strain behaviour, anisotropy and increasing stiffness with depth. Therefore the Boussinesq equations can be used to calculate values of $\Delta\sigma_v$ with confidence.
- Two important limitations to the above are where a stiff layer overlies a soft layer and where the vertical shear modulus G_{vh} deviates significantly from the equivalent isotropic value. In the former case, the value of $\Delta\sigma_v$ will be overestimated by using Boussinesq.
- In contrast to the vertical stress changes, the horizontal stress changes are very sensitive to all the variables studied and are therefore likely to depart significantly from the Boussinesq theory.
- Elastic displacement theory is based on very restrictive idealisations and should be used with great caution.
- As methods of settlement prediction have become more complex, increasingly more sophisticated types of testing are required. The limitations and cost of laboratory testing must be recognised.
- A careful comparison between exact solutions and four common prediction methods covering a wide range of properties shows

that the simple classical 1D method is at least as good as any other method for calculating the total settlement ρ_t. For normally consolidated clays it appears that the 1D method gives a good estimate of the consolidation settlement ρ_c.

- For overconsolidated clays $\rho_u \approx 0.5\rho_t$ and $\rho_t \approx \rho_{1XD}$.
- For normally consolidated clays $\rho_u \approx 0.1\rho_t$ and $\rho_t \approx 1.1\rho_{1XD}$.
- A most important conclusion is that most effort must be directed towards the measurement of simple properties such as m_v and E'_v and their variation with depth. Errors in these are much more significant than differences in analytical methods.
- The settlement of foundations on granular soils will usually be small and much of it takes place during construction. Numerous empirical methods have been proposed for predicting the settlement of foundations on granular materials which lead to a wide range of estimates. A straightforward analysis of the measured settlements of over 200 records has led to a remarkably simple approach relating settlement to bearing pressure, breadth and average SPT blow count (or cone resistance) over the depth of influence.

References

Burland JB (1971) A method of estimating the pore pressures and displacements beneath embankments on soft natural clay deposits. *Proceedings of Roscoe Memorial Symposium, Cambridge, UK*, pp. 505–536.

Burland JB and Burbidge MC (1985) Settlement of foundations on sand and gravel. *Proceedings of the Institution of Civil Engineers, Part 1* **78**: 1325–1381.

Burland JB, Broms BB and de Mello VFB (1977) Behaviour of foundations and structures: state of the art review. *9th International Conference on SMFE, Tokyo, Japan*, vol. 2, pp. 495–546.

Butler FG (1974) Heavily overconsolidated cohesive materials: state of the art review. *Conference on Settlement of Structures*. Pentech Press, London, UK, pp. 531–578.

Christian JT and Carrier WD (1978) Janbu, Bjerrum and Kjaernsli's chart reinterpreted. *Canadian Geotechnical Journal* **15**: 123–128.

Davis EH and Poulos HG (1963) Triaxial testing and three-dimensional settlement analysis. *Proceedings of the 4th Australia–New Zealand Conference on SM & FE, Adelaide, Australia*, pp. 233–243.

Fadum RE (1948) Influence values for estimating stresses in elastic foundations. *Proceedings of the 2nd International Conference on Soil Mechanics and Foundation Engineering, Rotterdam, the Netherlands*, vol. 3, pp. 77–84.

Gibson RE and Sills GC (1971) Some results concerning the plane deformation of a non-homogeneous elastic half-space. *Proceedings of Roscoe Memorial Symposium, Cambridge, UK*, pp. 564–572.

Jardine RJ, Potts DM, Fourie AB and Burland JB (1986) Studies of the influence of nonlinear stress–strain characteristics in soil–structure interaction. *Géotechnique* **36(3)**: 377–396.

Knappett, J and Craig RF (2012) *Craig's Soil Mechanics*, 8th edn. Spon Press, Oxford, UK.

Lambe TW (1964) Methods of estimating settlement. *Journal of Soil Mechanics and Foundation Division, ASCE*, **90(SM5)**: 43–67.

Lambe TW and Whitman RV (1969) *Soil Mechanics*. Wiley, New York, NY.

Poulos HG and Davis EH (1974) *Elastic Solutions for Soil and Rock Mechanics*. Wiley, New York, NY.

Simons N and Menzies B (2000) *A Short Course in Foundation Engineering*, 2nd edn. Thomas Telford, London, UK.

Simons N and Som NN (1970) *Settlement of Structures on Clay, with Particular Emphasis on London Clay*. Construction Industry Research and Information Association, London, UK.

Skempton AW (1957) Discussion: session 5. *Proceedings of the 4th International Conference SM & FE, London, UK*, vol. 3, pp. 59–60.

Skempton AW and Bjerrum L (1957) A contribution to the settlement analysis of foundations on clay. *Géotechnique* **7(4)**: 168–178.

Skempton AW, Peck RB and MacDonald DH (1955) Settlement analyses of six structures in Chicago and London. *Proceedings of the Institution of Civil Engineers* **4(4)**: 525–542.

Sutherland HB (1974) State of the art review on granular materials. *Proceedings of the Conference on the Settlement of Structures*. Pentech Press, Cambridge, UK, pp. 473–499.

Terzaghi K (1956) Discussion on paper by Skempton and MacDonald 'The allowable settlement of buildings'. *Proceedings of the Institution of Civil Engineers, Part 3* **5**: 775–777.

All chapters within Sections 1 *Context* and 2 *Fundamental principles* together provide a complete introduction to the Manual and no individual chapter should be read in isolation from the rest.

Chapter 20
Earth pressure theory

William Powrie University of Southampton, Southampton, UK

doi: 10.1680/icemge.66816.0243

CONTENTS

This chapter describes the classical calculation of limiting lateral earth pressures for use in the analysis of retaining walls. The concept of simple active and passive zones of soil at failure is developed for total stress and effective stress failure criteria, for adhesionless and frictionless walls. The effects of dry and flooded tension cracks for the total stress (undrained) condition are considered. The way in which the simple active and passive zones can be modified to take account of the effects of wall adhesion or friction is described, and finally the possible in-service stresses acting on a retaining wall are discussed.

20.1. Introduction

The need to be able to calculate lateral earth pressures arises particularly in the design of vertical or near-vertical retaining walls. Owing to the difficulties of quantifying the detailed stress–strain behaviour of a soil, the approach to the calculation of lateral earth pressures in geotechnical engineering has traditionally been based on the concepts of plasticity. By considering the stress states of soil elements on the verge of failure, it is possible to deduce limits between which the lateral earth pressures must lie.

20.2. Simple active and passive limits

20.2.1 Mohr circle of stress

A state of plane stress within a soil (or indeed any solid material) can be characterised by means of the Mohr circle of stress, plotted on a diagram with axes of normal stress σ (axis runs in the x-direction) and shear stress τ (axis runs in the y-direction). The Mohr circle shows how the combination of τ and σ on a given line (direction) within the plane varies with the orientation of that line. The Mohr circle is in effect a locus defining all possible combinations of τ and σ, acting on lines at different orientations, that make up a particular state of stress within a plane. In the current context, the plane is usually the cross-sectional plane of a long retaining wall. The circle is symmetrical about the σ-axis: the two points at which it crosses the σ-axis represent *principal directions*, lines on which the shear stress τ is zero. The corresponding normal stresses are termed the *major* (greater) and *minor* (lesser) *principal stresses*. The stress state on a line at an angle θ anticlockwise from the line on which the major principal stress σ_1 acts is given by the point on the circle whose radius subtends an angle 2θ at the centre of the circle with the radius through the major principal stress point. This is illustrated in **Figure 20.1**. The angle 2θ between the radii corresponding to the major and the minor principal stresses is 180°, which means that in reality their directions (or the directions of the lines on which they act) are at 90°.

In soils, the Mohr circle may be plotted in terms of either total stresses (σ, τ) or effective stresses (σ', τ), depending on the form of analysis being used.

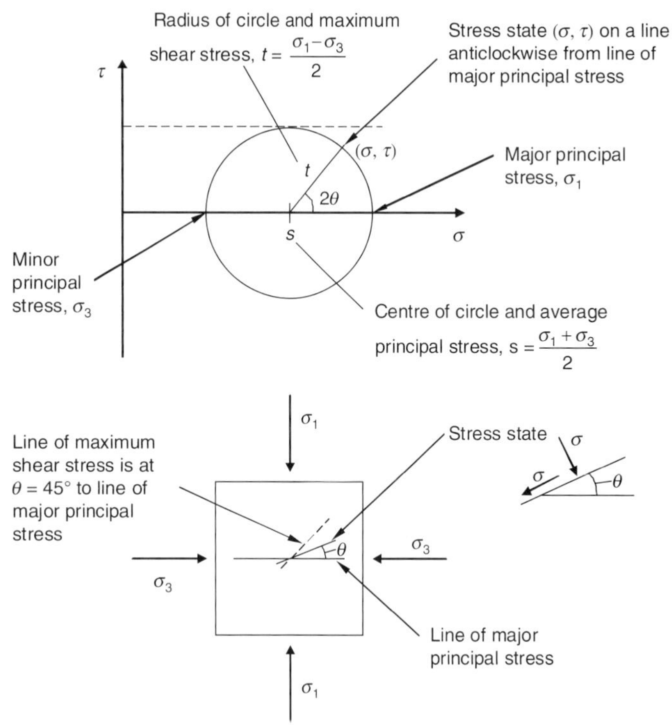

Figure 20.1 Mohr circle representation of the stress state within a principal plane

If the soil is in a state of limiting equilibrium, there will be some orientation within the plane for which the combination of stresses (σ, τ) or (σ', τ) lies on the line defining states of stress at failure, which is known as the *failure envelope*. In soils, two alternative failure criteria or failure envelopes are used; the first is discussed in Section 20.2.2, the second in Section 20.2.3.

20.2.2 Total stress failure criterion

The total stress failure criterion specifies the maximum shear stress possible anywhere within the plane, τ_{max}, in terms of the undrained shear strength of the soil, c_u (the symbols s_u and τ_u

may also be used). This failure criterion applies only to clay soils sheared rapidly to failure. The undrained shear strength c_u is not a soil property, but depends on the water content of the clay as sheared. The undrained shear strength failure criterion is written as

$$\tau_{max} = c_u \tag{20.1}$$

A Mohr circle of total stress that touches the undrained shear strength failure envelope is shown in **Figure 20.2**. Consideration of the geometry of the Mohr circle shows that the difference between the major and minor principal total stresses, $\sigma_1 - \sigma_3$, is equal to twice the radius of the circle, and the radius is equal to c_u. Thus at failure

$$\sigma_1 - \sigma_3 = 2c_u \tag{20.2}$$

Assuming for the moment that the principal total stress directions are vertical and horizontal, there are two ways in which the soil could fail. In the first, the major principal total stress is vertical, and the minor principal total stress is horizontal. This could correspond to the withdrawal of lateral support at constant vertical stress until failure is reached. Such conditions might occur behind a retaining wall, and are known as *active*. In active conditions, the horizontal total stress has fallen to its minimum possible value for a given vertical total stress. If the horizontal total stress were reduced any further (with the vertical total stress remaining constant), the Mohr circle of stress would expand beyond the failure envelope, which is not permissible. In active conditions the (minimum) horizontal total stress is given by

$$\sigma_{h,min} = \sigma_v - 2c_u \tag{20.3}$$

In the second mode of failure, the major principal total stress is horizontal and the minor principal total stress is vertical.

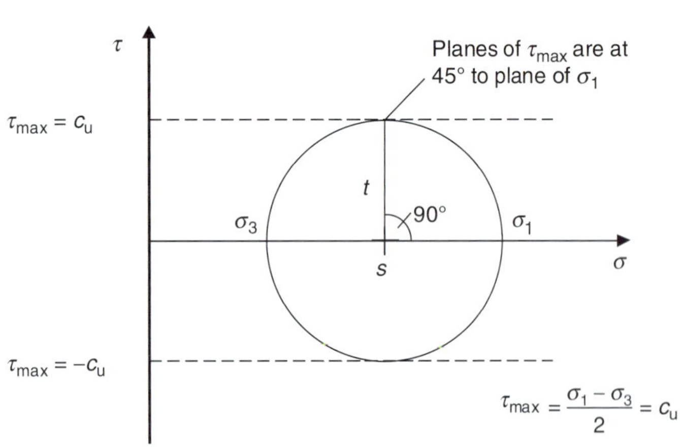

Figure 20.2 Mohr circle of total stress at failure for a clay soil obeying the undrained shear strength failure criterion, $\tau_{max} = c_u$

This could correspond to an increase in lateral stress and a constant or possibly reducing vertical stress, as might occur in the soil in front of an embedded retaining wall. These conditions ($\sigma_h > \sigma_v$) are known as *passive*. In passive conditions, the horizontal total stress has increased to its maximum possible value for a given vertical total stress. If the horizontal total stress were increased any further, the Mohr circle of stress would expand beyond the failure envelope, which is again not permissible. In passive conditions, the (maximum) horizontal total stress is given by

$$\sigma_{h,max} = \sigma_v + 2c_u \tag{20.4}$$

As discussed in Chapter 15 *Groundwater profiles and effective stresses*, the vertical total stress is usually taken as $\int \gamma dz$ ($= \gamma z$ in a uniform soil), plus the effect of any surface surcharge q.

When calculating active earth pressures behind a retaining wall using Equation 20.3, there are two important provisos. First, negative (tensile) stresses are not permitted, so a no-tension cutoff at $\sigma_h = 0$ comes into play where the uncritical application of Equation 20.3 would otherwise give a negative value for $\sigma_{h,min}$. This is equivalent to the development of a tension crack penetrating to a depth z_{tc} given by

$$\sigma_{h,min} = \left(q + \gamma z_{tc} - 2c_u\right) = 0 \tag{20.5}$$

or

$$z_{tc} = \frac{\left(2c_u - q\right)}{\gamma} \tag{20.6}$$

which, in the absence of a surface surcharge, reduces to the more familiar $z_{tc} = 2c_u/\gamma$.

The second proviso is that if there is any possibility that the tension crack could flood, the total lateral stress acting on the soil in the active zone becomes equal to the hydrostatic pressure of water. A flooded tension crack may remain open to a greater depth, z_{ftc}, given by

$$\sigma_{h,min} = \left(q + \gamma z_{ftc} - 2c_u\right) = \gamma_w z_{ftc} \tag{20.7}$$

or

$$z_{ftc} = \frac{\left(2c_u - q\right)}{\left(\gamma - \gamma_w\right)} \tag{20.8}$$

where γ_w is the unit weight of water. As $\gamma_w \approx 0.5\gamma$, if $q = 0$ the depth to which a flooded tension crack may remain open (and, more importantly, hydrostatic pressures are exerted on the retaining wall) is $z_{ftc} \approx 4c_u/\gamma$. The tendency of the undrained shear strength c_u to increase with depth means that the depth z_{ftc} could well be more than twice the depth to which a dry tension crack can remain open.

20.2.3 Effective stress failure criterion

The more fundamental effective stress failure criterion specifies the maximum possible ratio of shear to normal effective stress anywhere within the plane, $(\tau/\sigma')_{max}$, in terms of the tangent of the angle of shearing resistance or the effective friction angle of the soil, φ'. This applies to all soils at all times, but the pore water pressures must be known to calculate the normal effective stress. Also, there are different possible angles of shearing resistance that might be used. The critical state angle of shearing resistance is a soil property, while the peak angle of shearing resistance is not. This is because the latter depends on the potential of the soil to dilate, which in turn depends on a combination of the relative density and the stress state, and must therefore be expected to vary with depth.

The effective stress failure criterion is written as

$$\left(\frac{\tau}{\sigma'}\right)_{max} = \tan\phi' \tag{20.9}$$

A Mohr circle of effective stress that touches the effective stress failure envelope is shown in **Figure 20.3**. The radius t of the Mohr circle is equal to half the difference between the major and minor principal effective stresses

$$t = \frac{(\sigma_1' - \sigma_3')}{2} \tag{20.10}$$

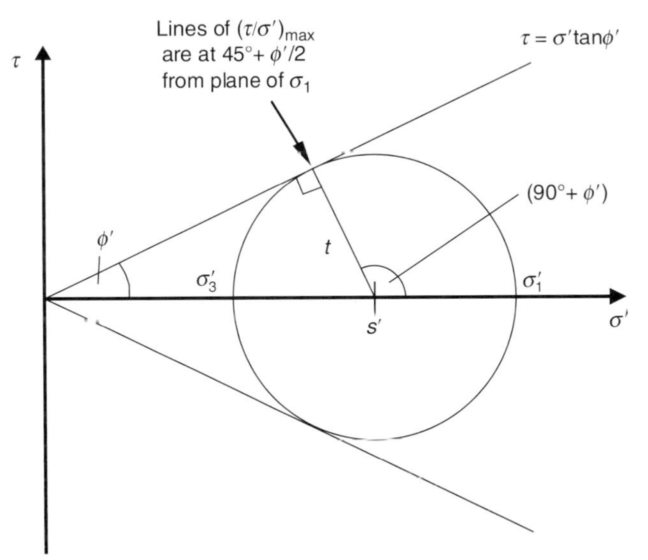

Figure 20.3 Mohr circle of effective stress at failure for a soil obeying the effective stress (frictional) failure criterion $(\tau/\sigma')_{max} = \tan\phi'$

while the distance of its centre from the origin, s', is given by half the sum of the principal effective stresses

$$s' = \frac{(\sigma_1' + \sigma_3')}{2} \tag{20.11}$$

Further consideration of the Mohr circle geometry shows that $t = s' \sin\phi'$. Substitution of the expressions for t and s' in terms of the major and minor principal effective stresses, σ_1' and σ_3', into this equation, followed by a little manipulation, gives the ratio of the major and minor principal total stresses, σ_1'/σ_3', at failure as

$$\frac{\sigma_1'}{\sigma_3'} = \frac{(1+\sin\phi')}{(1-\sin\phi')} \tag{20.12}$$

Again assuming that the principal effective stress directions are vertical and horizontal, there are two ways in which the soil could fail. The first is again active, corresponding potentially to conditions behind a retaining wall with the major principal effective stress vertical and the minor principal effective stress horizontal. In active conditions, the (minimum) horizontal effective stress is given by

$$\sigma'_{h,min} = \frac{(1-\sin\phi')}{(1+\sin\phi')}\sigma_v' = K_a \times \sigma_v' \tag{20.13}$$

where K_a is the well-known active earth pressure coefficient. From Equation 20.13, $K_a = (1-\sin\phi')/(1+\sin\phi')$ – this is sometimes written as the trigonometric identity $\tan^2(45° - \phi')$.

The second failure mode is again passive, corresponding potentially to conditions in front of an embedded retaining wall, with the major principal total stress horizontal and the minor principal total stress vertical. In passive conditions, the (maximum) horizontal total stress is given by

$$\sigma'_{h,max} = \frac{(1+\sin\phi')}{(1-\sin\phi')}\sigma_v' = K_p\sigma_v' \tag{20.14}$$

where K_p is the passive earth pressure coefficient. From Equation 20.14, $K_p = (1+\sin\phi')/(1-\sin\phi')$ – this is sometimes written as $\tan^2(45° + \phi')$.

As discussed in Chapter 15 *Groundwater profiles and effective stresses*, the vertical effective stress is usually taken as $\int\gamma dz - u$ ($= \gamma z - u$ in a uniform soil), plus the effect of any surface surcharge q.

In the case of the total stress (undrained shear strength) failure criterion, a difference between the principal stresses at failure is calculated, whereas the effective stress (frictional) failure criterion leads to a ratio of principal effective stresses at failure.

Of course, real walls are not frictionless, and the presence of a shear stress on the back or front of a retaining wall will invalidate the assumption that the principal stress directions are vertical and horizontal, at least adjacent to the retaining wall. This usually works to the designer's advantage in that wall friction or adhesion, provided it is in the 'right' direction (see below), reduces the active stresses and increases the passive stresses. The calculation of earth pressures taking wall friction into account is discussed in Section 20.3.

20.3. Effects of wall friction or adhesion

It is relatively straightforward to calculate revised earth pressure parameters that take into account a known degree of soil/wall friction (in the case of an effective stress analysis) or undrained soil/wall adhesion (in the case of a total stress analysis). This is done by considering the effect on the average effective or total stress (s' or s) of the rotation in the direction of principal stresses that must occur between (*a*) a zone of soil near the free surface (within which the principal stresses are horizontal and vertical, as assumed in Section 20.2 for the whole of the soil behind or in front of the retaining wall); and (*b*) the soil adjacent to the wall (in which the presence of a shear stress on the vertical surface of the wall means that this cannot be a principal direction).

Details of this calculation may be found in Powrie (2018). In practice, it is usual to use the resulting general formulae, charts or tables based upon them, rather than deriving the required result from first principles every time. The key decisions facing the designer then concern the magnitude of the soil/wall adhesion or friction (usually expressed as a proportion of the design soil strength, c_u or ϕ'), and the direction in which the adhesion or friction acts.

A discussion on soil/wall friction is given in Powrie (2018). Eurocode 7 (EC7) (BSI, 2014) gives limiting values of soil/wall friction δ, where $\delta = \phi'_{crit}$ for rough (cast in situ) concrete, and $\delta = 2/3\,\phi'_{crit}$ for smooth (precast) concrete or sheet piling supporting sand and gravel. EC7 (BSI, 2014) is silent with regard to values of soil/wall adhesion in an undrained analysis, but Gaba *et al.* (2017) recommend a maximum value of half the design undrained shear strength for walls in stiff clays.

If the soil behind a retaining wall settles relative to the wall and the soil in front heaves, then the directions of soil/wall adhesion or friction act so as to reduce the active earth pressures and increase the passive. Both of these tend to enhance the stability of the wall. However, there are circumstances in which these conventional directions of relative soil/wall movement (and hence adhesion or friction) could be reversed. For example, if the wall carries a vertical load – such as the perimeter wall for a building basement – then the wall may settle relative to the retained soil and the direction of active side adhesion or friction will be reversed. Dewatering inside an excavation could cause relative settlement of the soil in front of the wall, but this is less likely.

For undrained conditions, a stress analysis taking into account the effects of a favourable soil/wall adhesion τ_w gives the following expressions for horizontal components of (*a*) the active stresses

$$\sigma_{h,min} = \sigma_v - \tau_u\left(1 + \Delta + \cos\Delta\right) \tag{20.15}$$

and (*b*) the passive stresses

$$\sigma_{h,max} = \sigma_v + \tau_u(1 + \Delta + \cos\Delta)_{h,max} \tag{20.16}$$

where $\sin\Delta = \tau_w/\tau_u$.

If $\tau_w = 0.5\tau_u$, the term in brackets in Equations 20.15 and 20.16 becomes numerically equal to 2.39, compared with 2 in the case of the frictionless wall in Section 20.2. The same comments regarding the possibility and depth of a dry or flooded tension crack behind the wall apply as in Section 20.2. However, there is a minor intellectual complication concerning the depth below the bottom of the tension crack at which soil/wall adhesion starts to be mobilised (no adhesion can be mobilised within the tension crack because the wall and the soil are not in contact).

In effective stress terms, active and passive earth pressure coefficients, defined now as the horizontal component of the effective stress on the wall at depth z, $\sigma'_{h(z)}$, divided by the notional vertical effective stress at the same level, $(q + \int\gamma dz - u)$, where q is any surface surcharge, with an angle of soil/wall friction δ acting favourably are

$$K_a = \frac{\sigma'_{h(z)}}{\left(q + \int\gamma dz - u\right)} = \frac{\left[1 - \sin\varphi'\cos(\Delta - \delta)\right]}{\left[1 + \sin\varphi'\right]}e^{-[(\Delta-\delta)\tan\varphi']} \tag{20.17}$$

and

$$K_p = \frac{\sigma'_{h(z)}}{\left(q + \int\gamma dz - u\right)} = \frac{\left[1 + \sin\varphi'\cos(\Delta + \delta)\right]}{\left[1 - \sin\varphi'\right]}e^{[(\Delta+\delta)\tan\varphi']} \tag{20.18}$$

respectively, where $\sin\Delta = \sin\delta/\sin\varphi$.

Values of K_a and K_p for various effective friction angles and relative soil/wall friction δ/ϕ' are given in **Tables 20.1** and **20.2** respectively. These, together with Equations 20.17 and 20.18, are consistent with the equations and charts given in EC7 (BSI, 2014) and Gaba *et al.* (2017).

So far, no judgement has been made as to the value of soil strength used. If conditions at collapse are under investigation, then the soil strength used would be the best estimate of the soil and interface strengths relevant to that limit state. As discussed by Powrie (1996), for embedded retaining walls these

ϕ': degrees	K_a with $\delta = 0$	K_a with $\delta = \phi'/2$	K_a with $\delta = 2\phi'/3$	K_a with $\tan\delta = 0.75 \times \tan\phi'$	K_a with $\tan\delta = \phi'$
12	0.6558	0.6133	0.6038	0.5999	0.5931
13	0.6327	0.5892	0.5794	0.5734	0.5683
14	0.6104	0.5660	0.5560	0.5519	0.5446
15	0.5888	0.5437	0.5336	0.5293	0.5219
16	0.5678	0.5223	0.5120	0.5076	0.5002
17	0.5475	0.5017	0.4914	0.4869	0.4793
18	0.5279	0.4894	0.4715	0.4669	0.4593
19	0.5088	0.4629	0.4525	0.4478	0.4402
20	0.4903	0.4446	0.4341	0.4294	0.4218
21	0.4724	0.4269	0.4165	0.4117	0.4041
22	0.4550	0.4100	0.3996	0.3947	0.3872
23	0.4381	0.3956	0.3833	0.3784	0.3709
24	0.4217	0.3778	0.3676	0.3627	0.3552
25	0.4059	0.3626	0.3380	0.3330	0.3257
26	0.3905	0.3480	0.3380	0.3330	0.3257
27	0.3755	0.3339	0.3240	0.3189	0.3118
28	0.3610	0.3202	0.3105	0.3054	0.2984
29	0.3470	0.3071	0.2976	0.2924	0.2855
30	0.3333	0.2944	0.2851	0.2799	0.2731
31	0.3201	0.2822	0.2730	0.2678	0.2612
32	0.3073	0.2704	0.2614	0.2562	0.2497
33	0.2948	0.2590	0.2502	0.2450	0.2387
34	0.2827	0.2479	0.2394	0.2342	0.2280
35	0.2710	0.2373	0.2289	0.2238	0.2177
36	0.2596	0.2270	0.2189	0.2137	0.2078
37	0.2486	0.2171	0.2092	0.2040	0.1983
38	0.2379	0.2075	0.1998	0.1947	0.1891
39	0.2275	0.1983	0.1908	0.1856	0.1803
40	0.2174	0.1893	0.1820	0.1770	0.1718

Table 20.1 Active earth pressure coefficients K_a consistent with EC7 (BSI, 2014), calculated using Equation 20.17 for various values of effective friction angle ϕ' and relative soil/wall friction δ/ϕ' (Data taken from Powrie, 2018)

ϕ': degrees	K_p with $\delta = 0$	K_p with $\delta = \phi'/2$	K_p with $\delta = 2\phi'/3$	K_p with $\tan\delta = 0.75 \times \tan\phi'$	K_p with $\tan\delta = \phi'$
12	1.525	1.6861	1.724	1.739	1.763
13	1.580	1.7657	1.809	1.826	1.855
14	1.638	1.8500	1.900	1.920	1.953
15	1.698	1.9393	1.996	2.020	2.057
16	1.761	2.0341	2.099	2.126	2.168
17	1.826	2.1347	2.209	2.240	2.287
18	1.894	2.2417	2.326	2.361	2.415
19	1.965	2.3556	2.451	2.492	2.552
20	2.040	2.4770	2.584	2.631	2.699
21	2.117	2.6066	2.728	2.782	2.857
22	2.198	2.7449	2.881	2.943	3.028
23	2.283	2.8930	3.047	3.117	3.212
24	2.371	3.0515	3.225	3.305	3.411
25	2.464	3.2215	3.416	3.509	3.627
26	2.561	3.4042	3.623	3.729	3.861
27	2.663	3.6006	3.847	3.969	4.116
28	2.770	3.8123	4.090	4.229	1.393
29	2.882	4.0407	4.353	4.512	4.695
30	3.000	4.2877	4.639	4.822	5.026
31	3.124	4.5550	4.951	5.162	5.389
32	3.255	4.845	5.291	5.534	5.788
33	3.392	5.160	5.664	5.944	6.227
34	3.537	5.504	6.072	6.395	6.712
35	3.690	5.879	6.522	6.895	7.250
36	3.852	6.289	7.017	7.449	7.847
37	4.023	6.738	7.564	8.066	8.512
38	4.204	7.232	8.170	8.754	9.255
39	4.395	7.777	8.844	9.525	10.088
40	4.599	8.378	9.595	10.390	11.026

Table 20.2 Passive earth pressure coefficients K_p consistent with EC7 (BSI, 2014), calculated using Equation 20.17 for various values of effective friction angle ϕ' and relative soil/wall friction δ/ϕ' (Data taken from Powrie, 2018)

may be the critical state soil strength, ϕ'_{crit}, with full wall friction, $\delta = \phi'_{crit}$. In design, an appropriate partial factor would need to be applied to the soil and interface strengths, together with other measures including potentially an unexpected surcharge on the retained side, an unexpected overdig in front of the wall and the most onerous foreseeable pore water pressure conditions, which might (in the case of an undrained analysis) include a flooded tension crack between the wall and the retained soil.

The same techniques can also be used to calculate earth pressure differences and coefficients in cases where the backfill slopes and/or the back of the wall is battered (see, for example, Powrie, 2018).

20.4. In-service conditions

Modern design codes, including EC7 (BSI, 2014), generally require the designer of a retaining wall to guard against failure in the ground by carrying out an ultimate limit state (ULS) calculation with the actual soil strength reduced by a (partial) factor. Certain other provisions may also apply, such as an allowance for an unexpected surcharge load on the retained side, an unexpected overdig in front of the wall, and the most onerous foreseeable groundwater conditions. The factored or 'design' strength, being lower than the estimated actual strength of the soil, will result in increased active earth pressure coefficients, reduced passive earth pressure coefficients and an increased depth of embedment in the case of an embedded retaining wall.

It will be possible to draw a free body diagram for the wall that is in equilibrium under the action of the active and passive lateral stresses calculated using the factored soil strength. In fact, this is the basis on which the required depth of embedment (for an embedded wall) is calculated. However, the idealised earth pressures obtained from the limit equilibrium analysis with factored soil strengths are unlikely to correspond to those really acting on the wall, and are likely to overestimate wall bending moments. The reasons for this are varied and complex; for an embedded wall they may include

- relative wall/soil flexibility, which will influence local strains and hence mobilised strengths along the depth of the wall (see Rowe, 1952)
- different relationships between wall movement and shear strain on either side of an embedded wall (see Bolton and Powrie, 1988; Diakoumi and Powrie, 2013)
- different rates of mobilisation of soil strength with strain or wall movement on either side of an embedded wall, arising from the different recent stress histories relative to the stress paths followed during excavation (see Powrie *et al.*, 1998).

These effects are most likely to affect the linearity and magnitude of the lateral stresses with embedded walls propped at the crest, and multi-propped structures such as building basements. For these types of construction, a numerical soil–structure interaction analysis may lead to some economies in design and specification (Gaba *et al.*, 2017).

Concern has also been expressed regarding the possibility of high lateral stresses on embedded in situ retaining walls in overconsolidated soils, as a result of the stress history of these deposits (a discussion of which is given in Powrie, 2018). However, field studies of these walls have produced no evidence for this (e.g. Richards *et al.*, 2007), probably as a result of the factors listed above, together with the effects of wall installation in reducing the lateral stresses prior to excavation in front of the wall to below the in situ values (Richards *et al.*, 2006). In many cases, the use of active earth pressure coefficients behind the wall based on the full soil strength and steady-state pore water pressures will probably give a reasonable estimate of the bending moments likely to occur in service (Batten and Powrie, 2000; Gaba *et al.*, 2017; Powrie and Batten, 2000). A useful discussion of the effects of concrete diaphragm and bored pile wall installation is given in Gaba *et al.* (2017: Appendix A4.6).

20.5. Summary

- Classical earth pressure theory is based on calculating limiting lateral earth pressures, corresponding to conditions of failure in the soil, and in its simplest form is for frictionless or adhesionless retaining walls.
- Active conditions correspond to failure with the major principal stress vertical, represent the minimum possible lateral stress for a given vertical stress and are usually associated with conditions behind a retaining wall. Passive conditions correspond to failure with the major principal stress horizontal, represent the maximum possible lateral stress for a given vertical stress and are usually associated with conditions in front of an embedded retaining wall.
- Active and passive pressures can be calculated using either the total stress failure criterion (for clay soils, undrained only) or the effective stress failure criterion. In the latter case, the pore water pressures must be considered explicitly as well.
- In an undrained calculation, consideration must be given to the possibility of a flooded tension crack developing between a retaining wall and the soil behind it.
- The simple expressions for active and passive lateral stresses developed for a frictionless or adhesionless wall are usually modified for use in practice to take account of the effects of wall friction or adhesion. However, consideration should then be given to the likely direction of relative soil/wall movement to ensure that the beneficial effects of wall adhesion or friction can be relied on.
- The in-service stresses acting on a retaining wall are likely to be different from those at failure in the ground, owing to the effects of wall flexibility and differential rates of mobilisation of soil strength (soil stiffness) behind and in front of the wall.

References

Batten M and Powrie W (2000) Measurement and analysis of temporary prop loads at Canary Wharf underground station, east London. *Proceedings of the Institution of Civil Engineers (Geotechnical Engineering)* **143(3)**: 151–163.

Bolton MD and Powrie W (1988) Behaviour of diaphragm walls in clay prior to collapse. *Géotechnique* **38(2)**: 167–189.

BSI (British Standards Institution) (2014) BS EN 1997-1: Eurocode 7: Geotechnical design – Part 1: General rules. BSI, London, UK.

Diakoumi M and Powrie W (2013) Mobilisable strength design for flexible embedded retaining walls. *Géotechnique* **63(2)**: 95–106.

Gaba AR, Hardy S, Doughty L, Powrie W and Selemetas D (2017) *Guidance on Embedded Retaining Wall Design*. CIRIA Report C760. CIRIA, London, UK.

Powrie W (1996) Limit equilibrium analysis of embedded retaining walls. *Géotechnique* **46(4)**: 709–723.

Powrie W (2018) *Soil Mechanics: Concepts and Applications*, 3rd edn. CRC Press, Boca Raton, FL.

Powrie W and Batten M (2000) Comparison of measured and calculated temporary prop loads at Canada Water station. *Géotechnique* **50(2)**: 127–140.

Powrie W, Pantelidou H and Stallebrass SE (1998) Soil stiffness in stress paths relevant to diaphragm walls in clay. *Géotechnique* **48(4)**: 483–494.

Richards DJ, Clark J and Powrie W (2006) Installation effects of a bored pile retaining wall in overconsolidated deposits. *Géotechnique* **56(6)**: 411–425.

Richards DJ, Powrie W, Roscoe H and Clark J (2007) Pore water pressure and horizontal stress changes measured during construction of a contiguous bored pile multi-propped retaining wall in Lower Cretaceous clays. *Géotechnique* **57(2)**: 197–205.

Rowe PW (1952) Anchored sheet pile walls. *Proceedings of the Institution of Civil Engineers*, Part 1 **1**: 27–70.

All chapters within Sections 1 *Context* and 2 *Fundamental principles* together provide a complete introduction to the Manual and no individual chapter should be read in isolation from the rest.

Chapter 21
Bearing capacity theory

William Powrie University of Southampton, Southampton, UK

This chapter discusses the theory behind the well-known formulae for the bearing capacity of soils. Theoretical solutions for an infinitely long strip foundation on a weightless soil, with the soil above the founding plane acting simply as a surcharge, are presented for both total stress and effective stress failure criteria. The use of empirical adjustment factors to take account of foundation shape, depth and the self-weight of the soil below the founding plane is described. Finally, the effect on bearing capacity of an inclined and/or eccentric load is discussed.

doi: 10.1680/icemge.66816.0251

CONTENTS

21.1. Introduction

The bearing capacity of a foundation is one of the classical problems in geotechnical engineering. The solution for a long foundation on a soil obeying the undrained shear strength failure criterion $\tau_{max} = c_u$ was published by Prandtl (1920), in the context of the indentation of a metal by a flat die. The corresponding solution for a weightless soil of effective friction angle ϕ' obeying the effective stress (frictional) failure criterion $(\tau/\sigma')_{max} = \tan\phi'$ was given by Reissner (1924).

For the total stress failure criterion, the solution relates the difference between the vertical total stress on the foundation at failure, σ_f, and the vertical total stress in the ground on either side at the level of the bottom of the foundation (the *founding plane*), σ_o, to the undrained shear strength, c_u

$$\frac{(\sigma_f - \sigma_o)}{c_u} = N_c = (2 + \pi) = 5.14 \qquad (21.1)$$

The ratio $(\sigma_f - \sigma_o)/c_u$ is known as a *bearing capacity factor* and is given the symbol N_c in an undrained (total stress) analysis. The side surcharge, σ_o, will usually arise from the weight of the soil above the base of the foundation: $\sigma_o = \int \gamma dz = \gamma D$ for a foundation of depth D in a soil of uniform unit weight γ. The weight of the soil does not otherwise affect the calculation, because below the founding plane it adds equally to both σ_f and σ_o, and therefore cancels out.

For the effective stress failure criterion, the solution is in terms of the ratio of the vertical effective stress on the foundation at failure, σ'_f, to the vertical effective stress in the ground on either side at the level of the foundation base σ'_o and the soil frictional strength ϕ'

$$\frac{\sigma'_f}{\sigma'_o} = N_q = K_p e^{(\pi \tan\phi')} \qquad (21.2)$$

where ϕ' is the effective angle of friction and $K_p = (1 + \sin\phi')/(1 - \sin\phi')$ is the passive earth pressure coefficient (see Chapter 20 *Earth pressure theory*). In this case, the bearing capacity factor N_q is the ratio of the vertical effective stresses on the foundation and in the soil to either side, at the level of the foundation base. As before, the side surcharge σ'_o will usually arise from the weight of the soil above the base of the foundation, $\sigma'_o = \int \gamma dz - u = (\gamma D - u)$ for a foundation of depth D in a soil of uniform unit weight γ, where u is the pore water pressure at the level of the foundation base.

The undrained shear strength and effective stress failure criteria are discussed in Chapter 20 *Earth pressure theory*, and the analyses leading to Equations 21.1 and 21.2 are given in full in, for example, Powrie (2018).

21.2. Bearing capacity equation for vertical load: empirical adjustments for shape and depth

The fact that the effective stress solution is developed for a weightless soil is, in practice, an important potential limitation in the use of Equation 21.2 in unmodified form. For every 10 kPa increase in the vertical effective stress σ'_o on either side of the foundation, the limiting effective stress underneath the foundation, σ'_f, increases by N_q times this, which for $\phi' = 30°$ would be 184 kPa (with $\phi' = 30°$, $N_q = 18.4$ according to Equation 21.2). Failure of the foundation will in reality not occur just at the level of the founding plane, but will involve a mechanism extending to a depth approximately equal to the foundation width B below the foundation base. Thus the use of Equation 21.2 on its own would significantly underestimate the true bearing capacity, because the increase in σ'_f with depth is neglected. The depth of soil below the founding plane involved in the eventual failure mechanism may be taken into account empirically by adding a further term to Equation 21.2, known as the N_γ term, given by

$$N_\gamma \times (0.5\gamma B - \Delta u) \qquad (21.3)$$

where N_γ is a factor analogous to N_q, B is the foundation width, γ is the unit weight of the soil, Δu is the increase in pore water pressure between the founding plane and a depth of $0.5B$ below it, and $(0.5\gamma B - \Delta u)$ is then the increase in vertical effective stress over the same depth, in the soil on either side of the foundation. In effect, Equation 21.3 gives the bearing capacity of a surface footing for which the depth of embedment D (and hence σ'_o) is zero. This is added to the basic Equation 21.2 to estimate the bearing capacity of a long footing embedded to a depth D.

Two further problems with Equations 21.1–21.3 are that (*a*) most real foundations are not infinitely long and (*b*) the soil on either side of the foundation above the founding plane, which is assumed in the analyses to act simply as a surcharge, has a strength that will contribute to the bearing capacity of the foundation, especially as the depth of the founding plane D is increased. These are overcome by the use of empirical factors, usually based on the results of laboratory tests on model footings, to modify the basic theoretical equations.

In effective stress terms, the bearing capacity equation is written in the form

$$\sigma'_f = \left(N_q \times s_q \times d_q \times \sigma'_o\right) + [N_\gamma \times s_\gamma \times d_\gamma \times r_\gamma \times (0.5\gamma B - \Delta u)] \qquad (21.4)$$

where, in addition to the symbols already defined

- s_q and s_γ are shape factors, which account for the fact that the footing is not infinitely long
- d_q and d_γ are depth factors, which account for the fact that the soil above the founding plane has some strength and therefore does more than act simply as a surcharge
- r_γ is a reduction factor, to prevent the N_γ term increasing indefinitely with the footing width B.

N_q, the basic bearing capacity factor, is generally taken as $K_p e^{\pi \tan\phi'}$ for a vertical load applied through the centre of the footing. Values of s_q, d_q, N_γ, s_γ, d_γ and r_γ suggested by Meyerhof (1963), Brinch Hansen (1970) and Bowles (1996) are given in **Table 21.1**.

For foundations on clay soils loaded quickly, the undrained bearing capacity equation may be written as

$$(\sigma_f - \sigma_o) = [N_c \times s_c \times d_c] \times c_u \qquad (21.5)$$

For a vertical load acting at the centre of the foundation, the basic bearing capacity factor N_c is $(2 + \pi) = 5.14$, and s_c and d_c are enhancement factors accounting for the shape and depth of the footing, respectively. In this case, there is no N_γ effect because the difference between the total stresses σ_f and σ_o is unaffected by the self-weight of the soil.

Parameter	Meyerhof (1963)	Brinch Hansen (1970)
Shape factor s_q	$1 + 0.1K_p(B/L)$	$1 + (B/L)\tan\phi'$
Depth factor d_q	$1 + 0.1\sqrt{(K_p)}(D/B)$	$1 + 2\tan\phi'(1 - \tan\phi')k$
N_γ	$(N_q - 1)\tan(1.4\phi')$	$1.5(N_q - 1)\tan\phi'$
Shape factor s_γ	$= s_q$	$1 - 0.4(B/L)$
Depth factor d_γ	$= d_q$	1

Meyerhof's expressions apply for $\phi' > 10°$. $k = D/B$ if $D/B \le 1$. $k = \tan^{-1}(D/B)$ (in radian) if $D/B > 1$. $K_p = (1 + \sin\phi')/(1 - \sin\phi')$. $r_\gamma = 1 - 0.25\log_{10}(B/2)$ for $B \ge 2$ m (Bowles, 1996). Foundation length L, breadth B and depth D.

Table 21.1 Bearing capacity enhancement factors, effective stress failure criterion

Parameter	Skempton (1951)	Meyerhof (1963)
Shape factor s_c	$1 + 0.2(B/L)$	$1 + 0.2(B/L)$
Depth factor d_c	$1 + 0.23\sqrt{(D/B)}$, up to a maximum of 1.46 ($D/B = 4$)	$1 + 0.2(D/B)$

Foundation length L, breadth B and depth D.

Table 21.2 Bearing capacity enhancement factors, undrained shear strength failure criterion

Values of s_c and d_c suggested by Skempton (1951) and Meyerhof (1963) are given in **Table 21.2**.

Brinch Hansen (1970) suggests an undrained bearing capacity equation of a slightly different form, in which the enhancement factors are additive rather than multiplicative. He also gives adjustment factors that take account of inclined loads and founding planes.

Annex B of BSI (2014) gives equations of a similar form to Equations 21.4 and 21.5, together with factors i_q and i_c intended to account for the effect of inclined loads but with the depth factors d_q and d_c omitted. However, a completely general bearing capacity equation in which parameter values are inserted to suit particular circumstances can lead to confusion. In any case, inclined loads, non-horizontal footing bases and sloping ground on one or both sides of the footing can be taken into account analytically in the calculation of N_c or N_q as shown, for example, in Powrie (2018).

21.3. Inclined loading

The bearing capacity of a shallow foundation subjected to an inclined load acting through its centre can be investigated by modifying the stress analysis used to obtain the basic bearing capacity Equations 21.1 and 21.2. An inclined load acting through the centre of the foundation may be split into its horizontal and vertical components, H and V expressed in kN per m run of the foundation.

For a foundation on a clay soil obeying the undrained shear strength (maximum shear stress) failure criterion, the loads are characterised by a normal stress $\sigma_f = V_f/B$ and a shear stress $\tau_f = H_f/B$, where B is the foundation width and V_f and H_f

are the vertical and horizontal loads (in kN/m) at failure. The modified bearing capacity is then given by

$$\frac{(\sigma_f - \sigma_o)}{c_u} = N_c = (1 + \pi - \Delta + \cos\Delta) \qquad (21.6)$$

where σ_o is the surcharge on either side of the foundation at the level of the founding plane, and $\Delta = \sin^{-1}(\tau_f / c_u)$.

For a foundation on soil obeying the effective stress (maximum stress ratio or frictional) failure criterion, the loads are characterised by a normal effective stress $\sigma'_f (= V_f / B)$, assuming that the pore water pressure on the base of the foundation is zero, and a shear stress τ_f ($= H_f / B$), where B is the foundation width and V_f and H_f are the vertical and horizontal loads (in kN/m) at failure. The modified bearing capacity is then given by

$$\frac{\sigma'_f}{\sigma'_o} = N_q = \frac{[1 + \sin\phi' \cos(\Delta + \delta)]}{[1 - \sin\phi]} \times e^{(\pi - \Delta - \delta)\tan\phi'} \qquad (21.7)$$

where σ'_o is the effective stress surcharge on either side of the foundation at the level of the founding plane, $\sin\Delta = \sin\delta / \sin\phi'$, and $\delta = \tan^{-1}\left(\tau_f / \sigma'_o\right)$.

21.4. Offset loading

A vertical load V acting at a distance e from the geometrical centre of the foundation is statically equivalent to a vertical load V acting through the centre, together with a moment of Ve about the centre. Foundations subjected to combined vertical and moment loading are considered in Section 21.5.

Alternatively, a simple approach (suggested by Meyerhof, 1963) to calculate the bearing capacity of a long shallow foundation subjected to an offset or eccentric vertical load is to reduce the width of the foundation to $B' = (B - 2e)$ so that the load acts through the geometric centre. If this reduced width is used with the limiting vertical stresses given by the appropriate equations presented in Sections 21.1 and 21.2, a conservative (i.e. safe) estimate of the reduced load-carrying capacity of the foundation will be obtained.

BSI (2014) requires 'special precautions' to be taken 'where the eccentricity of loading exceeds 1/3 of the width of a rectangular footing or 0.6 of the radius of a circular footing', because this brings the point of application outside the equivalent of the 'middle third' of the foundation, leading to a tendency to tension or uplift at the edge away from the load.

21.5. Combined vertical, horizontal and moment (*V–H–M*) loading interaction diagram for a surface foundation

In general, a foundation may be subjected to a combination of vertical (V), horizontal (H) and moment (M) loads, either as a result of the application of an inclined point load offset from the centre of the footing (**Figure 21.1(a)**), or directly (**Figure 21.1(b)**). These are statically equivalent, with $V = R\cos\alpha$, $H = R\sin\alpha$ and $M = R\cos\alpha\, e$.

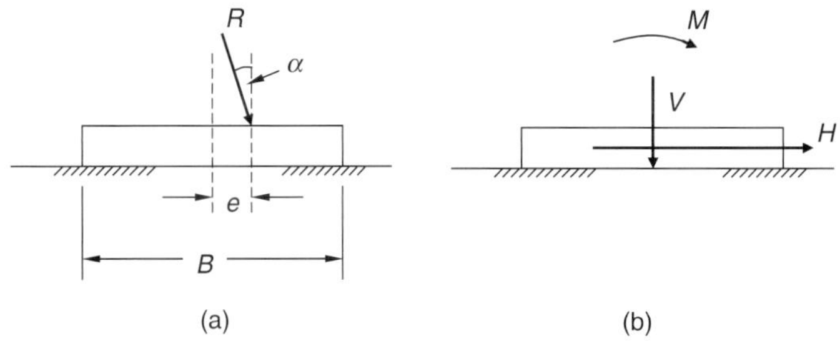

Figure 21.1 Statical equivalence of (a) an eccentric, inclined point load and (b) combined vertical, horizontal and moment loading through the centroid of a shallow foundation (Butterfield and Gottardi, 1994)

An alternative approach to calculating the bearing capacity of a surface footing subjected to simultaneous vertical, horizontal and moment loading is given by Butterfield and Gottardi (1994). On the basis of a large number of small-scale (model) tests, they showed that combinations of *V*, *H* and *M* at failure lay on a unique three-dimensional surface or failure envelope when plotted on a three-dimensional graph with axes *V*, *H* and *M/B*, where *B* is the breadth of the footing. The failure surface may be described as cigar-shaped, and is parabolic when viewed in either the *V–H* or the *V–M/B* plane (**Figure 21.2**).

Butterfield and Gottardi (1994) normalised their experimental values of *V*, *H* and *M/B* with respect to the vertical load V_{max} that would cause failure of the footing when acting on its own. For a foundation placed directly on the soil surface, the bearing capacity in symmetrical vertical loading σ'_f may be calculated using Equation 21.4 with the surcharge on either side due to the weight of soil above the founding plane $\sigma'_o = 0$. The depth factor d_γ and the associated reduction factor r_γ are also omitted because the depth of burial of the foundation D is zero, giving

$$\sigma'_f(\text{kPa}) = N_\gamma \times s_\gamma \times (0.5_\gamma B - \Delta u) \qquad (21.8)$$

If the foundation is of width B and length L, the vertical load at failure, acting through the centroid, is

$$V_{max}(\text{kN}) = \sigma'_f BL = N_\gamma s_\gamma (0.5\gamma B - \Delta u) BL \qquad (21.9)$$

The parabolae representing the projections of the three-dimensional failure surface in the *V–H* and *V–M/B* planes have equations

$$\frac{H}{t_h} = \frac{V \cdot (V_{max} - V)}{V_{max}} \qquad (21.10a)$$

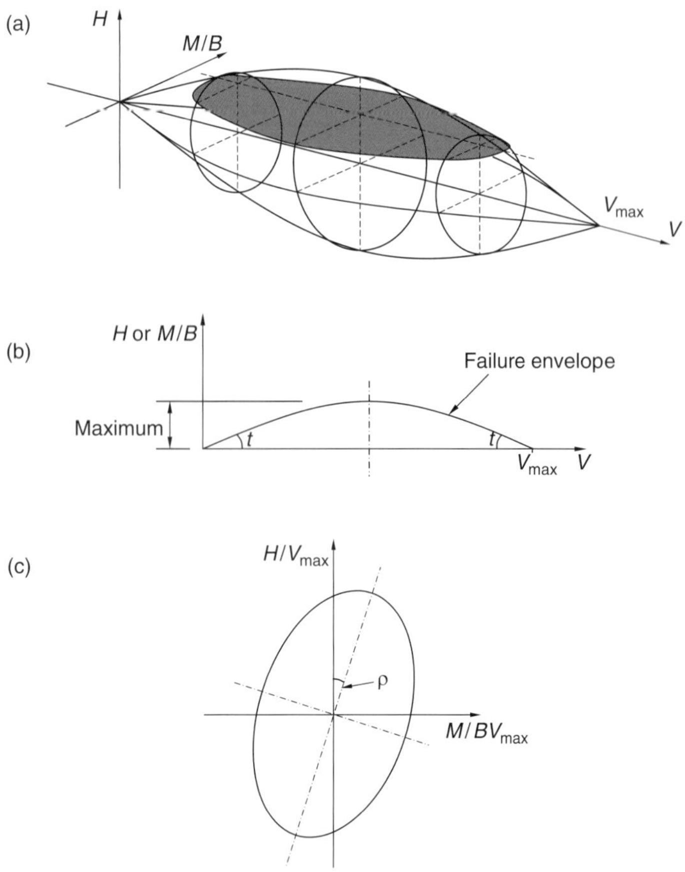

Figure 21.2 Failure envelopes for a shallow footing: (a) three-dimensional view in (V–M/B–H) space; (b) H or M/B against V; (c) normalised cross-section perpendicular to the V axis at $V = V_{max}/2$ ((a) and (b) from Butterfield and Gottardi, 1994)

and

$$\frac{\left(\frac{M}{B}\right)}{t_m} = \frac{V(V_{max} - V)}{V_{max}} \tag{21.10b}$$

where t_h and t_m are the slopes of the parabolae at the origin in the V–H and V–M/B planes, respectively (**Figure 21.2(b)**). For Butterfield and Gottardi's experiments, $H_{max} \approx V_{max}/8$ and $(M/B)_{max} \approx V_{max}/11$, corresponding to $t_h = 0.5$ and $t_m = 0.36$. H_{max} is defined at $M/B = 0$, $(M/B)_{max}$ is defined at $H = 0$, and both occur at $V = V_{max}/2$. Sections through the failure surface in the M/B–H plane (i.e. when viewed along the V axis) are ellipses, rotated through an angle ρ from the H axis towards the positive M/B axis (**Figure 21.2(c)**). The reason for this rotation is that the H and M/B loads can act either together or in opposition: if H and M/B act in the directions shown in **Figure 21.1**, the horizontal load at failure will be smaller than if either H or M/B is reversed.

The three-dimensional failure surface may be represented by an equation of the form

$$\left(\frac{H/V_{max}}{t_h}\right)^2 + \left(\frac{M/BV_{max}}{t_m}\right)^2 - \left(\frac{2C\frac{M}{BV_{max}}\frac{H}{V_{max}}}{t_h t_m}\right) = \left[\frac{V}{V_{max}}\left(1 - \frac{V}{V_{max}}\right)\right]^2 \tag{21.11}$$

where the constant C is a function of t_h, t_m and the inclination ρ to the H axis of the major axis of the cross-sectional ellipse. Data from three series of plane strain tests on foundations of different widths on different types and densities of sand have a best fit failure surface represented by Equation 21.11, with $t_h = 0.52$, $t_m = 0.35$, $\rho = 14°$ and $C = 0.22$.

The main practical implications of the above are as follows.

- Surface footings are particularly vulnerable to horizontal and moment loading, with $H \approx V_{max}/8$ or $M/B \approx V_{max}/11$ being sufficient to cause failure. For this reason, foundations that are required to carry significant horizontal or moment loads are generally buried or possibly piled. Truly unembedded surface footings are very unusual.
- The remoteness from the failure surface of a point on the (V–M/B–H) diagram representing a general load will depend on the load path followed to failure. Hence the use of an old-fashioned 'factor of safety' defined purely as a factor on the collapse load is ambiguous and unadvisable.
- If a surface footing is to remain remote from failure when subjected to a general load increment, the design vertical load should be about $V_{max}/2$. It is shown in an example given by Powrie (2018) that load paths involving an increase in H or M coupled with a reduction in V are particularly onerous for surface footings.

21.6. Summary

- Classical theory can be used to calculate the bearing capacity of an infinitely long strip foundation on a weightless soil, with the soil above the founding plane acting simply as a surcharge, for both total stress and effective stress failure criteria.
- Empirical adjustment factors are then applied to take account of foundation shape, depth and the self-weight of the soil below the founding plane.
- An inclined and/or eccentric load will affect the bearing capacity of a foundation: either a theoretical or an empirical approach can be used to take this into account, as appropriate.

References

Bowles JE (1996) *Foundation Analysis and Design*, 5th edn. McGraw-Hill, New York, NY.

Brinch Hansen J (1970) *A Revised and Extended Formula for Bearing Capacity*. Danish Geotechnical Institute, Copenhagen, Denmark.

BSI (British Standards Institution) (2014) BS EN 1997-1: Eurocode 7: Geotechnical design – Part 1: General rules. BSI, London, UK.

Butterfield R and Gottardi G (1994) A complete three-dimensional failure envelope for shallow footings on sand. *Géotechnique* **44(1)**: 181–184.

Meyerhof GG (1963) Some recent research on the bearing capacity of foundations. *Canadian Geotechnical Journal* **1(1)**: 16–26.

Powrie W (2018) *Soil Mechanics: Concepts and Applications*, 3rd edn. CRC Press, Boca Raton, FL.

Prandtl L (1920) Uber die härte plastisher körper (On the hardness of plastic bodies). *Nachrichten kon gesell. der wissenschaffen, Göttingen, Mathematisch-Physikalische Klasse* **1920**: 74–85.

Reissner H (1924) Zum erddruckproblem. *Proceedings of the 1st International Congress for Applied Mechanics, Delft, the Netherlands* (Bienzo CB and Burgers JM (eds)), pp. 295–311.

Skempton AW (1951) The bearing capacity of clays. *Proceedings of the Building Research Congress* **1**: 180–189.

All chapters within Sections 1 *Context* and 2 *Fundamental principles* together provide a complete introduction to the Manual and no individual chapter should be read in isolation from the rest.

doi: 10.1680/icemge.66816.0257

Chapter 22
Behaviour of single piles under vertical loads

John B Burland Imperial College London, London, UK

This chapter concentrates on the prediction of the axial capacity of piles in clay and granular materials. It is stressed that piling is a geotechnical process, controlled by an operator, the outcome of which depends on numerous factors. It is therefore not reasonable to expect to be able to predict pile behaviour with great precision. The fundamental mechanisms of load–settlement behaviour are described. The traditional approach to pile design in clay is presented using undrained strength as the controlling variable. The limitations of this approach are stressed and particular attention is drawn to the wide scatter of results on which design guidance is based. Because of these limitations, much of the chapter is devoted to the application of the effective stress method for pile shaft design. It is shown to give reasonable predictions of the shaft resistance of piles in soft clays with negative friction. The effective stress approach is also used in correlating the results of pile tests on overconsolidated clays. It is shown that these correlations are much better defined than by using undrained strength. The traditional approaches to the design of driven piles in granular soils are described.

CONTENTS

22.1. Introduction

The installation of a pile is rather like other geotechnical processes, such as grouting and tunnelling. In essence, a geotechnical process is controlled by an operator or contractor and the outcome of the process depends on numerous factors – including the ground profile and groundwater conditions (see Chapter 13 *The ground profile and its genesis*), the type of pile (see Chapter 54 *Single piles*), the type of piling plant used, the skill and experience of the operator and the supervision and quality control. Thus, successful delivery of the project depends not only on the design and specification of the piles, but on the process of installation.

It is often said that there are as many theories of pile behaviour as there are piling engineers! What is abundantly clear as one studies the vast literature on the subject is that much of it is highly empirical. As pointed out in Chapter 4 *The geotechnical triangle*, empiricism is a key activity but it must be used rigorously with a clear understanding of the basis of the empiricism.

Even though piling engineering is operator-dependent and so inevitably empirical, nevertheless the behaviour of piles has been shown to depend on basic soil mechanics principles. Understanding these principles provides the key to understanding pile behaviour and the many factors that control it. The aim of this chapter is therefore to describe the basic soil mechanics principles that control the carrying capacity of piles.

This chapter does not cover the specialist subject of piles in rock. For an introduction to this topic the reader is referred to Fleming *et al.* (2009).

22.2. Basic load–settlement behaviour

Figure 22.1(a) shows a straight-shafted pile which has been installed in the ground in some manner and is then loaded by applying an axial load P to the top of the pile. This load is transmitted into the ground by the shaft resistance Q_s and the base resistance Q_b. For simplicity, it can be assumed that the pile is rigid and does not compress under load.

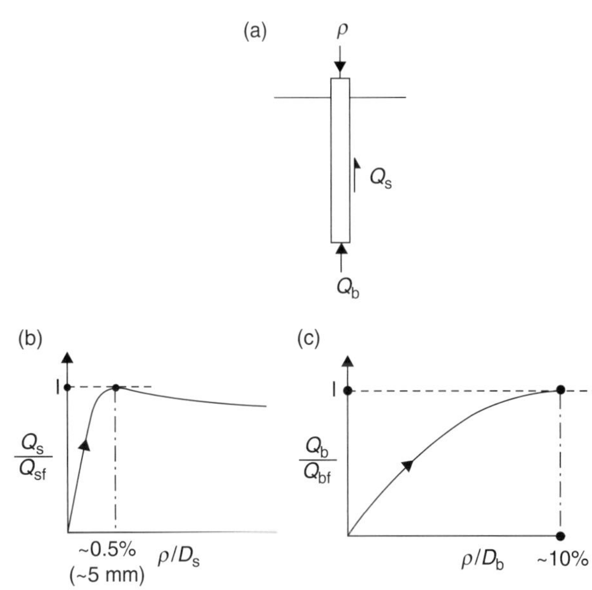

Figure 22.1 Contrasting development of shaft and base resistance with increasing settlement for straight-shafted pile in clay

As the applied load P is steadily increased, settlement ρ of the pile takes place. The development of the shaft resistance and the base resistance as the settlement increases are considered separately.

22.2.1 Development of the shaft resistance

Figure 22.1(b) shows a graph of the development of the shaft resistance with increasing settlement. The axes are expressed non-dimensionally as Q_s/Q_{sf} against ρ/D_s, where Q_{sf} is the maximum shaft resistance and D_s is the diameter of the shaft. Numerous instrumented pile tests in a variety of ground types have shown that the shaft resistance is fully developed at small settlements. This settlement is usually about 0.5% of the shaft diameter or, alternatively, somewhere between 5 and 10 mm for routine pile diameters. Sometimes the shaft resistance reaches a maximum value and then slowly decreases with further settlement; on other occasions it may continue to increase slowly with increasing settlement. It is extremely unusual for the shaft resistance to reduce rapidly and by a large amount after the maximum value has been reached. In this sense, the shaft resistance usually exhibits ductile load–settlement behaviour.

22.2.2 Development of the base resistance

Figure 22.1(c) shows a graph of the development of the base (or end) resistance of a pile with increasing settlement. As for the shaft, the axes are expressed non-dimensionally as Q_b/Q_{bf} against ρ/D_b, where Q_{bf} is the maximum base resistance and D_b is the diameter of the base. In contrast to the shaft resistance, the base resistance (both for bored and driven piles) requires much larger movements to develop fully – of the order of 10–20% of the base diameter.

22.2.3 Development of the combined resistance

The load–settlement behaviour of a pile is determined by the combined behaviour of the shaft and the base. **Figure 22.2(a)** shows the resultant load–settlement behaviour of a straight-shafted pile in clay, where the base only contributes a small proportion of the ultimate carrying capacity of the pile. The contributions of the shaft and base are shown as broken lines. Such a pile is often referred to as a *friction pile*. This behaviour can be contrasted with the load–settlement behaviour of a pile with an enlarged base, as shown in **Figure 22.2(b)**. In this case, the base makes a much larger contribution to the ultimate carrying capacity of the pile. For such piles, the shaft resistance is often fully developed under working load. Hence a pile with an enlarged base must settle more than a friction pile under working load if it is to operate efficiently.

In summary, it is important to have a clear appreciation of the two very different mechanisms of load transfer – the shaft resistance develops at relatively small settlements, whereas the base resistance requires much larger settlements to fully develop. It is also evident from **Figures 22.2(a)** and **22.2(b)** that the proportion of the applied load, carried by the shaft and the base, changes as the applied load increases. Thus, for a pile with an enlarged base, whereas under working load the shaft might carry, say, 80% of the applied load, at failure the shaft might only be carrying 20% of the applied load. Having a clear understanding of these two mechanisms of load transfer is very helpful when interpreting the results of pile loading tests.

22.2.4 Measured load–settlement behaviour of an under-reamed bored pile

Whitaker and Cooke (1966) carried out a seminal series of load tests on bored piles in London Clay at Wembley, in which the load carried by the base was measured by means of a sophisticated load cell. Knowing the applied and base loads at any stage of the test, it was possible to deduce the load carried by the shaft.

Figure 22.3 shows the results of one of Whitaker and Cooke's tests. In this case, the diameters of the shaft and base were 0.8 and 1.7 m, respectively, and the depth of the pile was 12 m. For the first part of the test the loads on the pile were applied in increments, with the load at each stage being

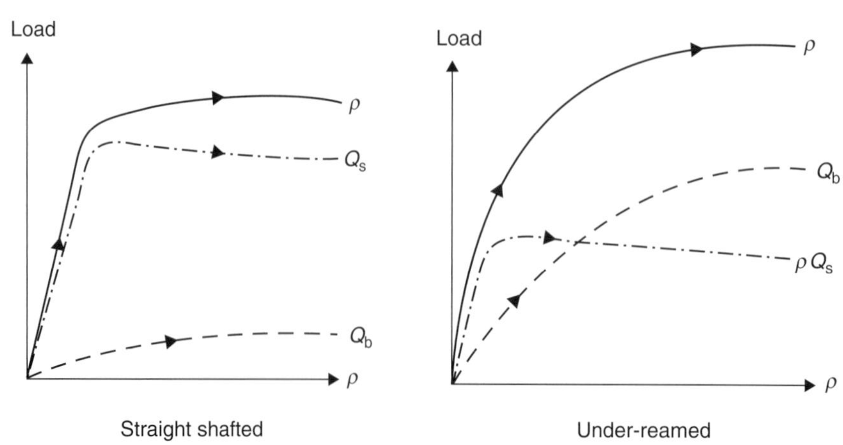

Figure 22.2 Contrasting load–settlement behaviour of (a) straight-shafted and (b) under-reamed piles in clay

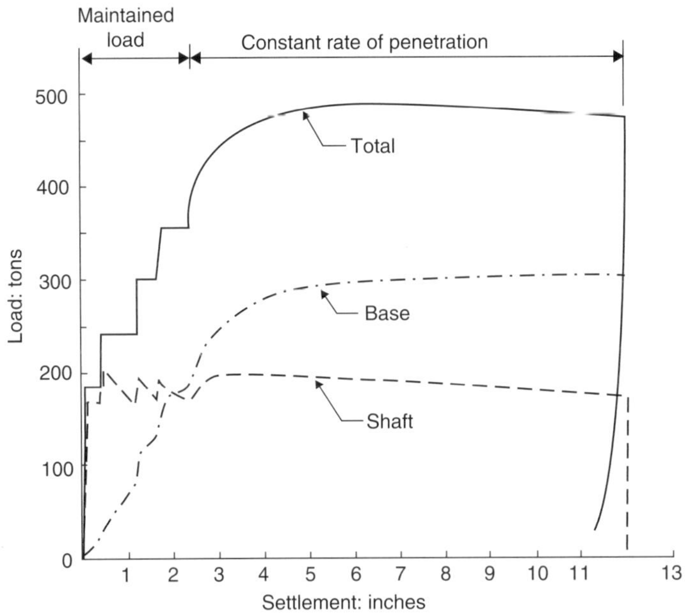

Figure 22.3 Measured load–settlement relationships for instrumented test on an under-reamed pile in London Clay (Reproduced from Whitaker and Cooke, 1966)

maintained constant until settlement had practically ceased. In some cases the load was held constant for 30 or more days. The second part of the test consisted of carrying out a constant rate of penetration test on the pile so as to determine its ultimate carrying capacity. The full line in **Figure 22.3** shows the load–settlement behaviour of the pile as a whole. Note that, for clarity, the measurements made during the unload–reload cycle that took place in changing from incremental loading to constant rate of penetration loading have been omitted from the figure.

The broken line labelled 'Shaft' in **Figure 22.3** shows the measured load–settlement behaviour of the shaft. It is clear that the carrying capacity of the shaft is fully mobilised at very small settlements. The jagged 'sawtooth' shape of the load–settlement curve is of interest. When an additional increment of load is applied to the pile, the load carried by the shaft initially increases and then reduces as settlement takes place. The temporary increase of load carried by the shaft is a 'rate effect' due to the initial high rate of settlement with time (Burland and Twine, 1988). As the rate of settlement steadily reduces, so the load carried by the shaft reduces back to its previous value – with the reduction in shaft load being shed to the base. It can be seen that the equilibrium fully mobilised carrying capacity of the shaft remains remarkably constant with increasing settlement, confirming the ductile nature of the shaft behaviour as described in Section 22.2.1. It is evident from **Figure 22.3** that the measured load–settlement behaviour of the base is dramatically different from that of the shaft, being much 'softer' and with full mobilisation taking place at significant settlements.

It is important to note that for working loads in excess of about 170 tons the pile will be operating with full mobilisation of the shaft-carrying capacity. There must be tens of thousands of large-diameter bored piles operating satisfactorily in this way worldwide. As will be described in Section 22.4, shaft resistance is essentially a frictional phenomenon, and as such is permanent and reliable – provided the effective stresses acting on the pile do not change significantly.

22.3. Traditional approach to estimating the axial capacity of piles in clay

It has become a well-established tradition to relate the carrying capacity of a pile in clay to undrained strength S_u. This is because undrained strength data are usually included with pile test results and it has proved very convenient to attempt to establish correlations between measured pile capacity and undrained strength. Unfortunately this empirical convenience is often overlooked and many piling engineers have come to regard the undrained strength as a fundamental controlling factor. In Section 22.4 it is shown that this is not the case for shaft resistance. It is necessary to understand that undrained strength is not a straightforward parameter; this is discussed below.

22.3.1 Undrained strength of clay

The undrained strength of a soil depends on many factors, the following being the principal ones.

- **Stress path:** For many soils, particularly normally or lightly overconsolidated clays, the undrained strength in plane strain or triaxial extension can differ significantly from the undrained strength measured in triaxial compression, S_{utc}. For soft clays, the strength in plane strain or triaxial extension can be less than the value in triaxial compression and may be as low as half this value.
- **Orientation:** The undrained strength usually depends on the orientation of the principal stresses – that is, the undrained strength is anisotropic.
- **Size:** The undrained strength is often critically dependent on the volume of soil being tested. In stiff fissured clays the undrained strength in the mass may be much less than the average strength of small samples due to the dominant influence of the presence of fissures.
- **Rate:** The undrained strength depends on the rate of testing. Usually, the slower the rate, the lower the strength.
- **Sample disturbance:** This can be a major factor and can operate either way – sometimes reducing strength and sometimes increasing it. A particularly important problem occurs with stiff clays containing saturated silt or sand partings. The stress relief during sampling causes the samples to take up moisture from the partings and soften.

When the above factors are appreciated, it is obvious that different types of test and different sizes of samples give differing values of S_u. In particular, the values of S_u derived from in situ tests, such as vane, cone and pressuremeter, will frequently differ from the values measured in triaxial compression. For these reasons, when pile behaviour (in particular shaft resistance) is correlated with undrained strength, it is necessary to be absolutely clear as to how the strength was determined.

22.3.2 Ultimate capacity of the whole pile

Figure 22.4 depicts a pile of weight W in a state of failure under a load P_f such that the total shaft resistance is Q_{sf} and the total base resistance is Q_{bf}. The average undrained strength over the length L of the pile is $S_{u(shaft)}$ and the undrained strength appropriate to the base is $S_{u(base)}$. For the whole pile at failure

$$P_f + W = Q_{bf} + Q_{sf} \tag{22.1}$$

The base resistance can be derived from basic bearing capacity theory given in Chapter 21 *Bearing capacity theory* (Equation 21.5)

$$Q_{bf} = A_b[\{N_c \times s_c \times d_c\} \times S_{u(base)} + p_o] \tag{22.2}$$

where A_b is the area of the base and p_o is the overburden pressure ($=\gamma_{bulk} \times L$). For a surface strip foundation, the basic bearing capacity factor $N_c = 5.14$, and for a deep circular or square foundation, the shape factor $s_c = 1.2$ and the depth factor $d_c = 1.46$. Hence, Equation 22.2 can be rewritten as

$$Q_{bf} = A_b[9 \times S_{u(base)} + \gamma_{bulk} \cdot L] \tag{22.3}$$

Undrained strength is used for assessing the base resistance of piles in clay because the undrained bearing capacity is usually much less than the drained bearing capacity, and the end resistance is at its lowest during the application of most of the working load. For the shaft

$$Q_{sf} = A_s \times \bar{\tau}_{sf} \tag{22.4}$$

where A_s is the surface area of the shaft and τ_{sf} is the average shaft friction over the length of the shaft. Methods of estimating the value of τ_{sf} are discussed later.

The ultimate capacity of the whole pile is obtained by substituting Equations 22.3 and 22.4 into Equation 22.1 as

$$P_f + W = A_b[9 \times S_{u(base)} + \gamma_{bulk} \cdot L] + A_s \times \bar{\tau}_{sf} \tag{22.5}$$

The assumption is frequently made that the weight of the pile is equal to the weight of the soil it displaces, so that $W = A_b \times \gamma_{bulk} \times L$ and Equation 22.5 simplifies to:

$$P_f = A_b \times 9 \times S_{u(base)} + A_s \times \bar{\tau}_{sf} \tag{22.6}$$

For piles bearing on a strong stratum or with enlarged bases, much of the load at failure will be carried by the base. Hence, care is needed in selecting the appropriate value of $S_{u(base)}$ to insert into Equation 22.6. When the clay is fissured, the design strength should be close to the lower limit of the scatter of unconsolidated undrained triaxial compressive strengths (Burland, 1990).

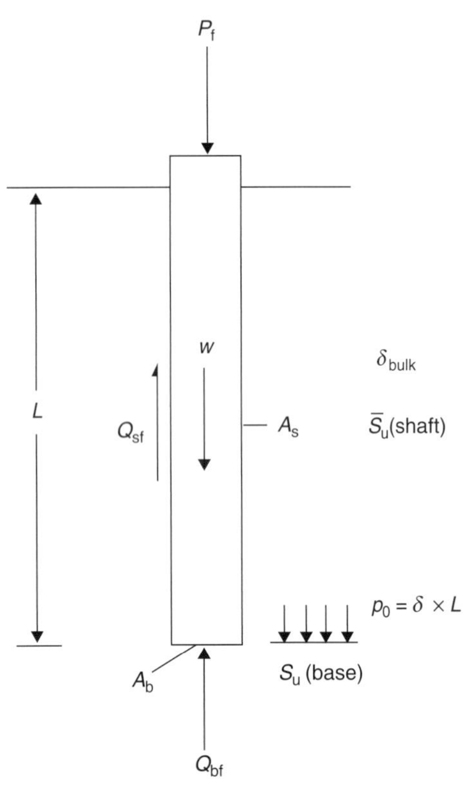

Figure 22.4 Ultimate capacity of a whole pile: a pile of weight W in a state of failure under a load P_f such that the total shaft resistance is Q_{sf} and the total base resistance is Q_{bf}

22.3.3 Estimation of shaft resistance from undrained strength

Fleming *et al.* (2009) point out that most piles in clay develop a high proportion of their overall capacity along the shaft. Hence the reliable determination of the shaft friction is important. For reasons that were explained at the beginning of this section, it has become traditional to relate the average shaft friction τ_{sf} to the average undrained strength $S_{u(shaft)}$ over the length of the shaft such that

$$\bar{\tau}_{sf} = \alpha \times \bar{S}_{u(shaft)} \tag{22.7}$$

The factor α is known as the 'shaft adhesion factor' and is a purely empirical quantity which has been evaluated from the results of pile tests. It is not a constant but depends on a number of variables.

One of the most important considerations when adopting values of α in design is to be very clear about how the values of $S_{u(shaft)}$ were determined in the original empirical correlation. This point can be illustrated by referring to the classic work of Skempton (1959), who analysed a number of load tests on cast in situ bored piles in London Clay and obtained an average value of $\alpha = 0.45$. He began his study by plotting the results of undrained triaxial tests on 1.5 inch (38 mm) dia. samples against depth, as shown in **Figure 22.5**. Each point represents the average of two to four samples at a given depth. Even so, it can be seen that there is a considerable scatter and this is common for London Clay, due largely to its fissured nature. Skempton drew a mean line through the strength results as shown by the full line. He used this basic line to obtain values of $S_{u(shaft)}$ and $S_{u(base)}$ for each of the piles in the study. For each test pile, the value of Q_{bf} was estimated from bearing capacity theory and was subtracted from the measured ultimate load to obtain the value of Q_{sf} and hence the value of τ_{sf} ($=Q_{sf}/A_s$). **Figure 22.6** shows Skempton's plot of τ_{sf} (*y*-axis) against $S_{u(shaft)}$ (*x*-axis). It can be seen that the points correspond to values of α ($\tau_{sf}/S_{u(shaft)}$) ranging between 0.3 and 0.6, with an average value of 0.45. Hence, the recommended value of $\alpha = 0.45$ is based on two averaging processes of widely scattered data, first of the measured strength–depth results and second of the correlation between τ_{sf} and $S_{u(shaft)}$.

In recent years, the practice has developed of measuring the undrained strength in triaxial compression using 98 mm rather than 38 mm dia. samples. It has been found that the average strength of 98 mm dia. samples of stiff fissured clay is significantly lower than for 38 mm dia. samples because the larger samples contain more fissures (Burland, 1990; Marsland, 1974). Hence it is no longer appropriate to adopt a value of $\alpha = 0.45$; a higher value should be used to account for the lower average measured strengths.

Tomlinson (1995) has published design curves of α against $S_{u(shaft)}$ for driven piles in clay, based on the analysis of published and unpublished records of pile tests. In general, the suggested design values of α reduce from about 1 (at low undrained strengths) to as low as 0.25 (for strengths greater than about 150 kPa). The data points on which the suggested design curves are based are illustrated in **Figure 22.7** – which can be seen to show a very wide scatter (Tomlinson, 1971). Similarly, Weltman and Healy (1978) propose curves of α against undrained strength for bored and driven piles in glacial till, based on data which show a wide scatter. Hence, great care is needed in the application of such design curves, which should be regarded as indicative only.

In view of the uncertainties associated with the correlations between the shaft adhesion factor α and the average undrained strength, it is clearly advisable to carry out pile loading tests for specific projects. It is of course vital that the test piles are installed in the same way as the proposed working piles.

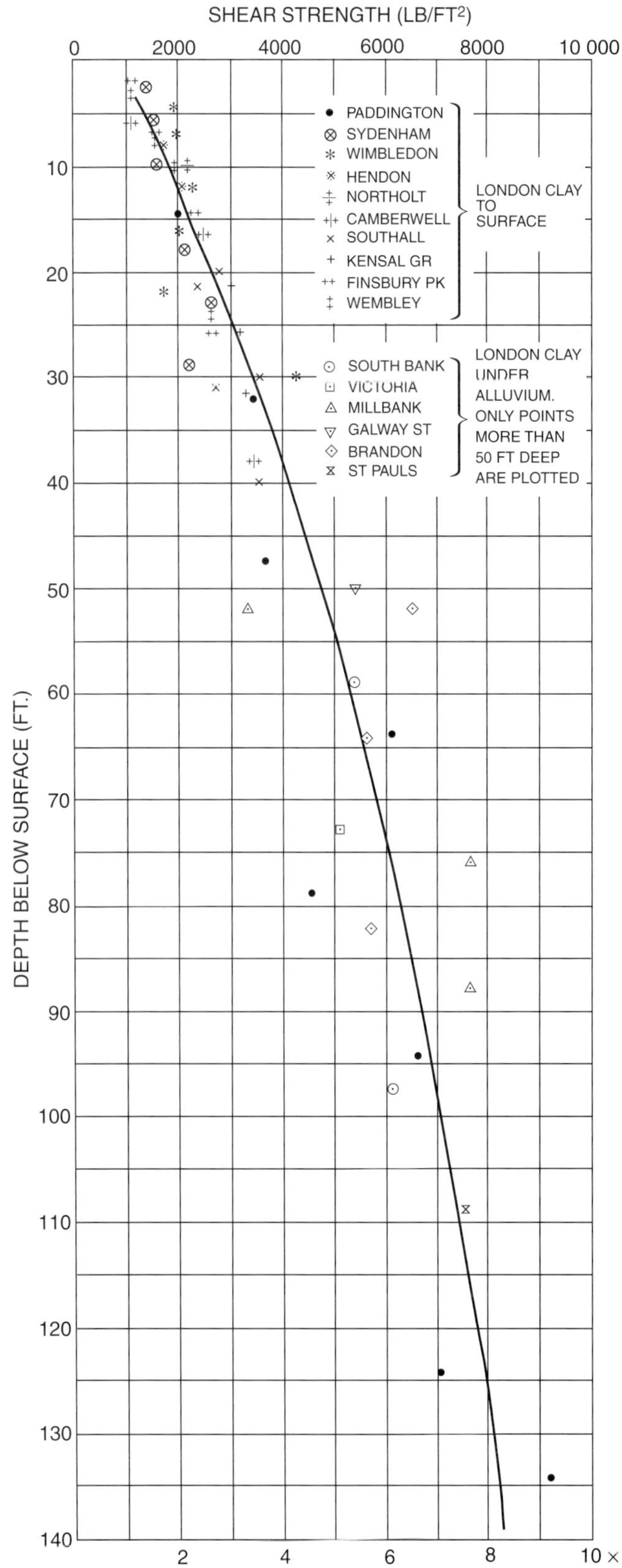

Figure 22.5 Values of average undrained strength S_u plotted against depth from various sites in London (Reproduced from Skempton, 1959)

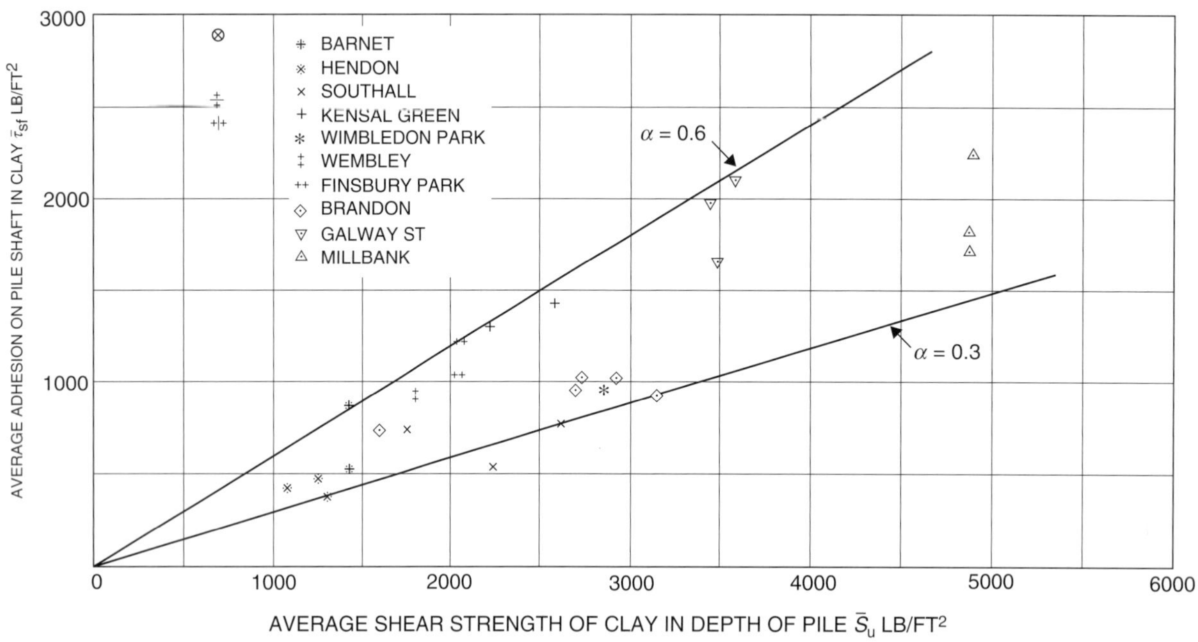

Figure 22.6 Range of measured α values for bored piles in London Clay (Reproduced from Skempton, 1959)

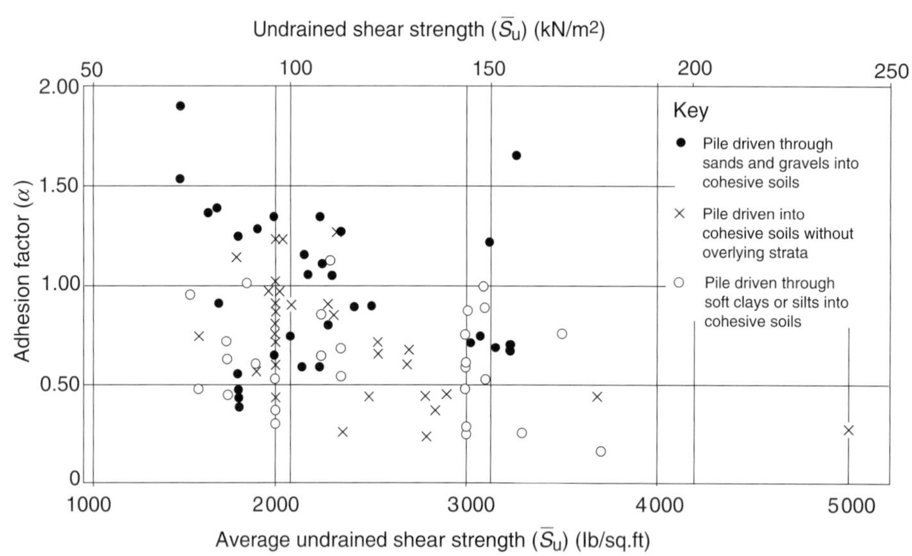

Figure 22.7 Relationship between adhesion factor α and average undrained strength over length of piles embedded in stiff to very stiff clays (Reproduced from Tomlinson, 1971)

22.4. Shaft friction of piles in clay, in terms of effective stress

Section 22.3.3 describes the traditional approach of evaluating shaft friction as a proportion α of the average undrained strength $S_{u(shaft)}$ over the length of the shaft. It was pointed out that the undrained strength does not have a unique value for a soil and is dependent on the test method and the quality of sampling. In stiff clays, the scatter of the undrained strength results can be very large due to the fissured nature of the clay. Examination of the pile test data on which suggested design values of α are based shows that the scatter is also very large. In addition to these uncertainties, a major problem with the traditional approach to evaluating shaft friction is that it is wholly empirical; there is no fundamental underpinning theoretical framework as an aid to understanding the key factors controlling shaft friction. The effective stress approach also has its limitations, but it is soundly based on fundamental soil mechanics principles.

22.4.1 Shaft friction in terms of the effective stress parameter β

The basis of the effective stress approach was first suggested by Johannessen and Bjerrum (1965) in relation to assessment of negative friction. Chandler (1968) suggested it could be applied to the interpretation of the results of tests on bored piles in London Clay, and Burland (1973) applied it more generally. Other papers followed, notably by Meyerhof (1976) and Flaate and Selnes (1977).

Research on instrumented piles (e.g. Lehane *et al.*, 1994) shows that limiting shaft friction is controlled by the simple Coulomb effective stress interface sliding law

$$\tau_{sf} = \sigma'_{rf} \tan\delta'_f \tag{22.8}$$

where σ'_{rf} is the radial effective stress acting on the shaft at failure and δ'_f is the effective angle of interface friction at failure. Thus, the problem of evaluating shaft friction at any depth on a pile requires the determination of these two quantities.

The value of the radial effective stress at failure σ'_{rf} depends on the following factors, in approximate order of significance: (*a*) the vertical effective overburden stress σ'_v; (*b*) the overconsolidation history of the deposit, which can be simplified to the evaluation of the overconsolidation ratio (OCR); (*c*) the effects of installing the pile; and (*d*) the change in σ'_r during loading. With regard to (*c*), it is normally assumed that any excess pore pressures set up during installation are completely dissipated. Regarding (*d*), it is reasonable to assume that loading takes place under fully drained conditions because the zone of major distortion adjacent to the pile is relatively thin and drainage will take place rapidly.

Because of the dominant influence of the effective self-weight of the surrounding ground and its ease and reliability of determination, the vertical effective overburden stress σ'_v can be incorporated into Equation 22.8 as

$$\frac{\tau_{sf}}{\sigma'_v} = \frac{\sigma'_{rf}}{\sigma'_v}\tan\delta'_f = K_{sf}\tan\delta'_f = \beta \tag{22.9}$$

where K_{sf} is the earth pressure coefficient on the shaft at failure. Burland (1973) pointed out that the parameter β can be obtained empirically by direct measurement of τ'_{sf} in pile tests (usually giving mean values β). When determined in this way, β is analogous to the adhesion factor *a* in Equation 22.7 but is much more rigorously defined, since σ'_v can be determined accurately and unambiguously, which is not the case for the undrained strength S_u. However, not only is β more rigorously defined empirically, but it can be evaluated by measuring δ'_f and determining K_{sf} from soil mechanics principles. Thus the parameter β is a much more clearly defined and fundamental parameter than α is.

22.4.2 Predicted shaft friction of an 'ideal' pile in normally consolidated clay

This section investigates the shaft resistance of an 'ideal' pile with a view to establishing a basic frame of reference for comparison with the measured behaviour of actual piles. An 'ideal' pile is one which has been installed in the ground without changing the in situ stresses. Moreover, it is assumed that shearing takes place through the remoulded ground next to the pile shaft so that $\delta'_f = \phi'_{cv}$, the critical state angle of shearing resistance. Hence, for an 'ideal' pile Equation 22.9 becomes

$$\frac{\tau_{sf}}{\sigma'_v} = K_o \tan\varphi'_{cv} = \beta_{ideal} \tag{22.10}$$

where K_o is the earth pressure coefficient at rest – see Chapter 15 *Groundwater profiles and effective stresses.* For a normally consolidated clay, the value of K_o is accurately given by the Jaky expression $(1 - \sin\varphi'_{cv})$. It follows that for a normally consolidated clay, Equation 22.10 for an 'ideal' pile becomes

$$\beta_{ideal} = (1 - \sin\varphi'_{cv})\tan\varphi'_{cv} \tag{22.11}$$

As shown in **Figure 22.8**, whereas K_o and tan ϕ'_{cv} vary widely as ϕ'_{cv} changes, their product (β_{ideal}) changes very little, varying between 0.21 to 0.29 for the range of practical ϕ'_{cv} values. Equation 22.11 provides a basic framework for ideal pile behaviour in normally consolidated clays. Its validity can

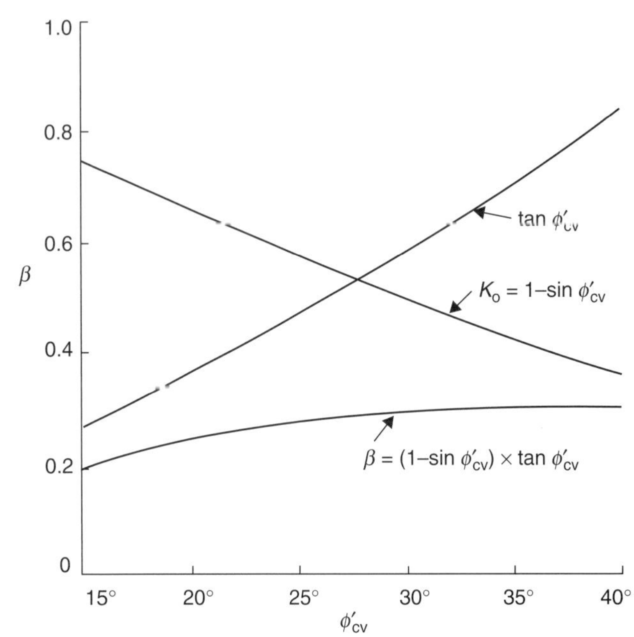

Figure 22.8 Relationship between the effective stress shaft friction parameter β and ϕ'_{cv} (Reproduced from Burland, 1973; Emap)

be tested against real pile behaviour and as a consequence it is possible to judge the extent to which the assumptions that $K_{sf} = K_o$ and $\delta'_{sf} = \phi'_{cv}$ need to be adjusted to account for the measured behaviour.

22.4.3 Comparison between measured and ideal values of β for normally consolidated clays

Burland (1973) compared this 'ideal' behaviour with careful experiments on some driven piles in two very different normally consolidated clays. **Figure 22.9** shows a plot of deduced average shaft friction τ_{sf} against average effective overburden pressure σ'_{vo}. All except one of the results plotted in this figure are from tests described by Hutchinson and Jensen (1968) on a number of concrete, steel and timber piles driven into deep estuarine clay in the Port of Khorramshahr, Iran. The average liquid and plastic limits (*LL* and *PL*) for the clay are 48% and 23%, respectively, and it has a sensitivity of 2.5–3.0. The deduced values of α ranged from 0.43 to 0.79. It can be seen from **Figure 22.9** that the deduced average values of β lie between 0.25 and 0.4, with an average value of 0.32. Eide *et al.* (1961) published the results of some long-term tests on a timber pile driven into Drammen clay. The average liquid and plastic limits for this material are 35% and 15%, respectively, and the clay has a sensitivity of 4–8. The deduced value of α was 1.6. The black point in the figure corresponds to a deduced value of β equal to 0.32.

In spite of the clay at the two sites having very different properties and the deduced values of α for the two sites having extreme upper and lower limits for soft clay ranging from 1.6 to 0.42, the average values of β are the same. This average measured value of 0.32 is slightly larger than the average value for an 'ideal' pile of about 0.27 (corresponding to $\varphi'_{cv} = 25°$ in Equation 22.10). Burland (1973) went on to examine a large number of results for driven piles in normally consolidated soils and found that the results lay between $\beta = 0.25$ and 0.4. He suggested that a reasonable average design value would be 0.3, a little higher than the value of 0.27 given by an 'ideal' pile.

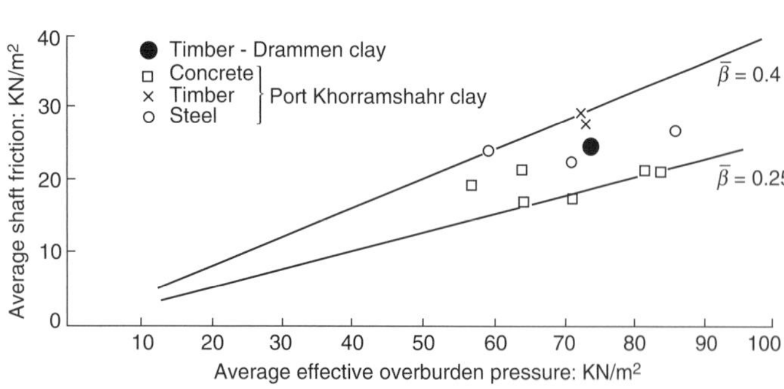

Figure 22.9 Comparison between results of pile tests on Port Khorramshahr clay (*LL* = 48; *PL* = 23; sensitivity = 2.5–3.0; α = 0.43–0.79) and Drammen clay (*LL* = 39; *PL* = 20; sensitivity = 4–8; α = 1.6) (Reproduced from Burland, 1973; Emap)

For a real pile driven into a normally consolidated sediment, it is reasonable to expect that the radial effective stresses would be somewhat larger than the 'at rest' pressures given by K_o, leading to somewhat higher measured values of β. Hence, for the case of most normally consolidated clays, the simple ideal effective stress approach is seen to give very reasonable results. Results from fully instrumented tests on driven piles in a normally consolidated clayey silt give values of β of less than 0.2 (Burland, 1993). Such exceptionally low values are strongly suggestive of arching around the pile due to the highly contractive behaviour of the material.

22.4.4 Negative friction on piles in soft clay

Negative friction, or 'downdrag', can occur on the shaft of a pile installed through a compressible layer of clay which undergoes consolidation. Such consolidation may result from the placing of fill on the surface, groundwater lowering or even the dissipation of excess pore water pressures set up during pile driving. The consolidating layer exerts a downward drag on the pile and therefore the direction of shaft friction in the layer is reversed. Negative skin friction usually develops gradually as consolidation of the clay proceeds and the vertical and horizontal effective stresses increase with time due to the dissipation of pore water pressure.

Burland and Starke (1994) analysed ten case records in which negative friction had been measured. The results were analysed in terms of the effective shaft friction factor β. The causes of negative friction included surface surcharge, underdrainage and disturbance during pile driving. Both steel and concrete driven piles were considered, some of which were end bearing and others 'floating'. The period of measurement varied from a few months up to 17 years.

In spite of the wide range of conditions, average values of β for soft compressible sediments were found to lie within relatively narrow limits of 0.15–0.35. A careful study of the soil types revealed that for low-plasticity marine clays the values of β were generally in the range 0.15–0.25, whereas for higher-plasticity clays and silty clays β tended to lie between 0.2 and 0.35. It therefore appears that the values of β operating for negative friction are similar but somewhat lower than those associated with driven friction piles in soft clays.

22.4.5 Predicted shaft friction of an 'ideal' pile in overconsolidated clay

There can be no doubt that the effective stress approach to shaft friction for piles in soft clays has major advantages over the traditional α approach, with all its uncertainties. This is because the coefficient of earth pressure at rest K_o in Equation 22.10 is well defined for normally consolidated clays. For overconsolidated clays the determination of the in situ horizontal effective stresses requires very sophisticated in situ or laboratory testing, and the effective stress approach is therefore not nearly so straightforward. Nevertheless, establishing the behaviour of an 'ideal' pile in overconsolidated clay provides very useful insights into the factors controlling its behaviour.

As pointed out in Chapter 15 *Groundwater profiles and effective stresses*, for overconsolidated clays which have previously been subjected to a maximum vertical effective stress $\sigma'_{v,max}$, the horizontal effective stress caused on first loading tends to remain 'locked in' on unloading, resulting in an increase in K_o. A widely used empirical equation for capturing this was developed by Mayne and Kulhawy (1982)

$$\frac{\sigma'_h}{\sigma'_v} = K_o = (1 - \sin\varphi'_{cv}) \times \text{OCR}^{\sin\varphi'_{cv}} \tag{22.12}$$

where OCR is the overconsolidation ratio ($= \sigma'_{v,max}/\sigma'_{v,current}$). Note that when OCR=1, Equation 22.12 reduces to the Jaky expression $K_o = (1 - \sin\phi'_{cv})$. Substituting Equation 22.12 into Equation 22.10

$$\frac{\tau_{sf}}{\sigma'_v} = \beta_{ideal} = (1 - \sin\varphi'_{cv}) \times \text{OCR}^{\sin\varphi'_{cv}} \times \tan\varphi'_{cv} \tag{22.13}$$

Hence it would seem appropriate to investigate the relationship between β and OCR. However, since the determination of OCR requires high-quality sampling and oedometer testing, there is little available pile test data with which to test Equation 22.13. What is needed is some simple indirect measure of OCR. It is well known (Jamiolkowski *et al*., 1985) that the ratio (S_u/σ'_v) is related to OCR by an expression of the form

$$S_{utc}/\sigma'_v = (S_{utc}/\sigma'_v)_{nc} \times \text{OCR}^m \tag{22.14}$$

where S_{utc} is the undrained strength measured in an unconsolidated undrained triaxial compression test and $(S_{utc}/\sigma'_v)_{nc}$ refers to normally consolidated conditions where m is an empirical coefficient. Solving for OCR in Equation 22.14 and substituting in Equation 22.13 gives

$$\beta_{ideal} = (1 - \sin\varphi'_{cv}) \times \tan\varphi'_{cv} \times \left[\frac{S_{utc}/\sigma'_v}{\left(S_{utc}/\sigma'_v\right)_{nc}}\right]^{\frac{\sin\varphi'_{cv}}{m}} \tag{22.15}$$

Although Equation 22.15 looks rather complicated, it suggests that there should be a relationship between β_{ideal} and S_{utc}/σ'_v. To explore this relationship, Equation 22.15 may be considerably simplified as follows.

The term $(1 - \sin\varphi'_{cv}) \times \tan\varphi'_{cv}$ equals β_{ideal} for a normally consolidated clay, and β_{ideal} changes very little with changes in φ'_{cv}. It is reasonable to set it equal to 0.27, corresponding to $\varphi'_{cv} = 25°$. It is well known (Burland, 1990) that for normally consolidated natural clays the ratio $(S_{utc}/\sigma'_v)_{nc}$ is remarkably well defined, generally varying between about 0.28 and 0.32; a mean value of 0.3 can reasonably be adopted. A major uncertainty exists over the likely range of m and hence Equation 22.15 can be simplified to

$$\beta_{ideal} = 0.27\left[\frac{S_{utc}/\sigma'_v}{0.3}\right]^E \tag{22.16}$$

where E is an empirically determined exponent. This work suggests that there is likely to be a relationship between the effective shaft friction factor β and the undrained compressive strength normalised by the effective overburden pressure S_{utc}/σ'_v.

22.4.6 Measured values of β for driven piles in overconsolidated clays

Flaate and Selnes (1977) published the results of a number of pile tests on timber and concrete piles driven into normally and lightly overconsolidated clays. Semple and Rigden (1984) assembled an extensive database for piles driven into overconsolidated clays which formed the basis of guidance by the American Petroleum Institute for many years. After following the reasoning described in the previous section, Burland (1993) plotted these two sets of data on a graph of β against S_{utc}/σ'_v and demonstrated that there is a very satisfactory correlation. Chow (1996) added significantly to these two databases and Patrizi and Burland (2001) showed that the effective shaft friction factor could be accurately represented by the expression

$$\beta = 0.1 + 0.4\left(S_{utc}/\sigma'_v\right) \tag{22.17}$$

The solid line in **Figure 22.10** shows the comparison between the results from the three databases and Equation 22.17. The insert shows the data points for values of $S_{utc}/\sigma'_v < 1$ plotted to a larger scale. (It is of interest to note that the broken line plotted in this figure is given by Equation 22.16 for an 'ideal' pile with the empirical exponent E = 0.72.) The open points in **Figure 22.10** are for low-plasticity clayey silts and it can be seen that there is a tendency for the values of β to be somewhat lower for these materials – a result referred to previously in Sections 22.4.3 and 22.4.4.

It is necessary to mention that the values of S_{utc} are determined for intact samples of clay. If the clay is fissured, the values of undrained strength will not be representative of the intact material and are likely to be significantly lower than the intact strengths. When plotting results from tests, this leads to points lying significantly above the line given by Equation 22.17, and when using this equation for predicting design values of β they are likely to be significantly underestimated.

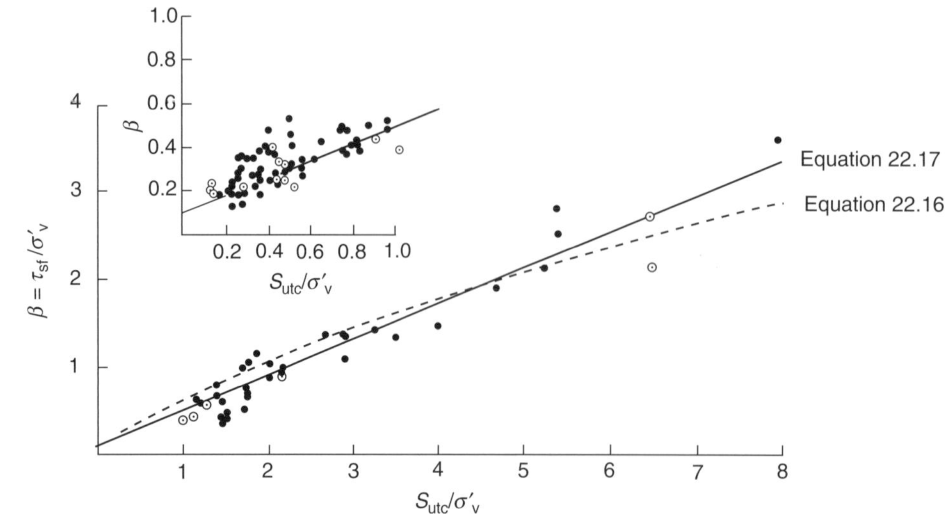

Figure 22.10 Driven piles in clay – the relationship between the effective stress shaft friction parameter β and S_{utc}/σ'_v, where S_{utc} is the *intact* unconsolidated undrained strength in triaxial compression (this correlation is not appropriate for fissured clays for which S_{utc} is not representative of intact strength)

Patrizi and Burland (2001) demonstrated that the correlation between Equation 22.17 and the results of the database is very good. An even better correlation can be obtained using the Imperial College Pile (ICP) design methods developed by Jardine *et al.* (2005), which are based on instrumented pile tests driven into both sands and clays. These approaches have been developed for the offshore industry and require advanced laboratory and in situ tests.

22.4.7 The effective stress behaviour of bored piles in stiff fissured clays

Burland (1973) developed the following expression for the average shaft friction $\overline{\tau}_{sf}$ for an 'ideal' pile in overconsolidated clay, which takes account of the variation of K_o with depth:

$$\overline{\tau}_{sf} = \frac{1}{L}\sum_{0}^{L} \sigma'_v \times K_o \times \tan\delta'_f \times \Delta L \qquad (22.18)$$

For London Clay, the variation of K_o with depth was determined by Skempton (1961) and Bishop *et al.* (1965) and an average distribution adopted. The assumption for an 'ideal' pile that $\delta'_f = \phi'_{cv} = 21.5°$ was also made. **Figure 22.11** shows a plot of average shaft friction τ_{sf} against average depth in clay. The full line was obtained from Equation 22.18 and represents the predicted values of average shaft friction of 'ideal' piles in London Clay.

The open points plotted in **Figure 22.11** represent the average measured shaft friction deduced from maintained load tests on instrumented piles carried out at Wembley by Whitaker and Cooke (1966), previously referred to in Section 22.2.4. It can be seen that, apart from one outlier, the results show relatively little scatter and give measured values of average shaft friction which are less than the 'ideal' values. Burland (1973) attributed this to the effect that the boring of the pile had on reducing the equilibrium value of the horizontal effective stress to below the 'at rest' K_o condition assumed for the 'ideal' pile.

Burland and Twine (1988) re-examined the Wembley data and concluded that the main reason for the measured values of τ_{sf} being less than the 'ideal' values was that the angle of interface friction δ'_f was close to the residual angle of shearing resistance φ'_r, having values as low as 12°. Various causes for this were suggested, the two most probable being (*a*) the upward and downward passage of the auger causing a smooth surface of highly orientated clay particles; and (*b*) thermal shortening of the shaft of only a few millimetres would be sufficient to induce a residual shear surface over much of the length of the pile. Burland and Twine also concluded that constant rate of penetration (CRP) testing must be interpreted with considerable caution, as current rates of testing can lead to a significant overestimation of the shaft resistance under static load.

Patel (1992) analysed a number of load tests on bored piles from 28 different sites in London. Maintained load tests were carried out at 16 of these sites and the shaft resistances were evaluated by subtracting the calculated base resistances from the ultimate resistance of the pile. The full points in **Figure 22.11** show the deduced values of $\overline{\tau}_{sf}$. Once again, apart from two outliers at shallow depth, the scatter is remarkably small. The chain dotted line representing a value of $\beta = 0.7$ is close to the lower limit of the scatter and would

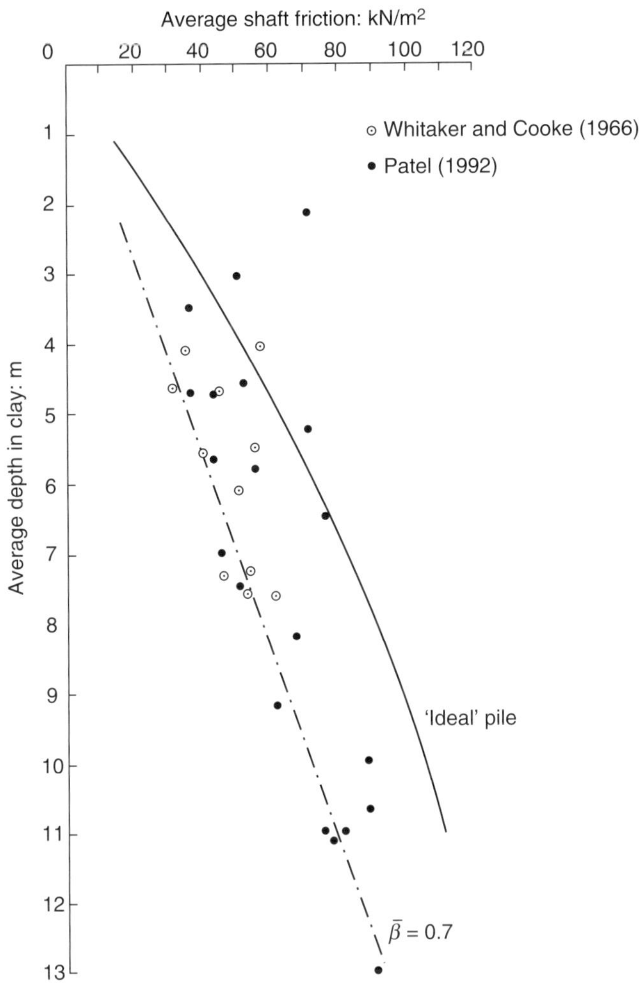

Figure 22.11 The relationship between measured average shaft friction and average depth in clay for bored piles in London Clay

be appropriate for design purposes for bored piles in London Clay (i.e. $\bar{\tau}_{sf} = 0.7 \times \sigma'_v$). Patel (1992) also confirmed the conclusion of Burland and Twine (1988) that the use of CRP tests can overestimate the true values of shaft friction by up to 20% due to rate effects.

The use of **Figure 22.11** for assessing shaft friction in terms of effective stress for bored piles in London Clay offers a very simple and dependable method which is based on highly reliable load tests. It can provide a valuable check when using the traditional method described in Section 22.3.3, or when interpreting the results of pile tests. The approach is particularly appropriate for assessment of shaft resistance where long-term changes in effective stress will occur, as is the case beneath a deep excavation or where changes in groundwater level are taking place. The traditional α method of assessing shaft friction is completely inappropriate for such conditions.

More generally, when analysing the results of pile tests in clay, it is worthwhile plotting the deduced values of τ_{sf} against average depth in clay and evaluating the associated values of β ($=\tau_{sf}/\sigma'_v$). In this way, the basic data are allowed to speak for themselves without the confusion caused by correlating the results with average undrained strength and all the uncertainties that are associated with it. Bown and O'Brien (2008) have shown that when reliable methods are available for measuring σ'_{ho} (namely the in situ horizontal effective stress), Equation 22.8 can be applied directly. They proposed that the average shaft friction of a bored pile in stiff plastic clay can be expressed as

$$\bar{\tau}_{sf} = c \cdot \bar{\sigma}'_{ho} \tan \delta'_f \tag{22.19}$$

where c is a reduction factor for installation effects with values varying typically from 0.9 to 0.8, and decreasing with depth. For London Clay, a value of effective angle of interface friction $\delta'_f = 16°$ was found to be appropriate.

22.5. Piles in granular materials

Piles in granular soils usually act primarily in end bearing and, unlike many piles in clay, shaft friction usually contributes only a small proportion of the load-carrying capacity. It is extremely difficult to obtain reasonably undisturbed samples of granular soils from boreholes. Moreover, the process of driving the pile will inevitably change the density of soil, which will significantly affect the carrying capacity of both the base and the shaft. These facts, in addition to the inherent variability of granular deposits, make the prediction of pile behaviour by analytical methods both difficult and unreliable. It is therefore the usual practice to assess the ultimate bearing capacities of piles in these soils from the results of in situ tests. Carrying capacity can be assessed on the basis of

- bearing capacity theory aided by standard penetration tests (SPTs)
- cone penetration tests (CPTs)
- pile-driving formulae.

22.5.1 Bearing capacity theory

Equation 21.4 in Chapter 21 *Bearing capacity theory* gives the basic bearing capacity equation for a granular soil. For the tip of a pile the self-weight term involving $N\gamma$ becomes negligibly small. Moreover, the shape and depth factors, s_q and d_q respectively, may be subsumed into the bearing capacity factor for surcharge N_q. Hence the ultimate base resistance of a pile embedded in a granular soil becomes

$$q_{bf} = N_q \times \sigma'_v \tag{22.20}$$

A wide range of values for N_q are quoted in the literature, but perhaps the most generally accepted are those given by Berezantsev *et al.* (1961). **Figure 22.12** shows their proposed relationships between N (which include s_q and d_q) and the effective angle of shearing resistance ϕ' adopted by Tomlinson (1995). If the depths of penetration of the pile into the granular

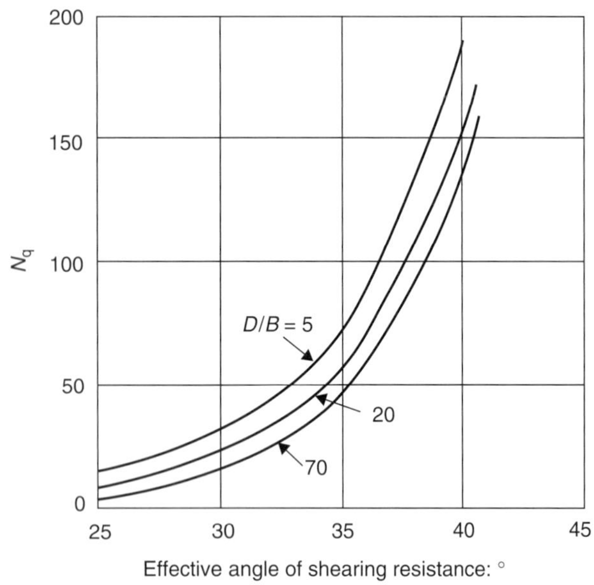

Figure 22.12 Relationship between N_q and ϕ' (Reproduced from Berezantsev *et al.*, 1961; all rights reserved)

material are less than about five times the pile width, values of N corresponding to a surface footing should be used – see Chapter 21 *Bearing capacity theory*.

Evaluation of N_q in Equation 22.20 requires knowledge of the effective angle of shearing resistance of the soil ϕ'. This is often done by making use of the well-known relationship between SPT N-value and ϕ' proposed by Peck *et al.* (1967), illustrated in **Figure 22.13**.

The expression for the ultimate shaft friction τ_{sf} is identical to Equation 22.8 and can be rewritten as

$$\tau_{sf} = \sigma'_v \times K_{sf} \times \tan \delta'_f \quad (22.21)$$

where K_{sf} is the earth pressure coefficient against the pile shaft at failure and δ'_f is the angle of interface friction along the shaft.

There are various rules for evaluating δ'_f in Equation 22.21. For a concrete or rough steel interface it appears that $\delta'_f \approx \phi'_{cv}$. For smoother interfaces, Jardine *et al.* (1992) conclude that δ'_f is a function of mean particle size (D_{50}) and varies from about 34° to 22° as D_{50} increases from 0.05 to 2.0 mm.

The evaluation of K_{sf} in Equation 22.21 presents a real problem as it depends largely on the initial density of the soil, the pile installation method and the volume of the displaced soil. Broms (1965) suggested the following values of K_{sf} in granular soils

Type of pile	Loose	Dense
Steel	0.5	1.0
Concrete	1.0	2.0
Timber	1.5	3.0

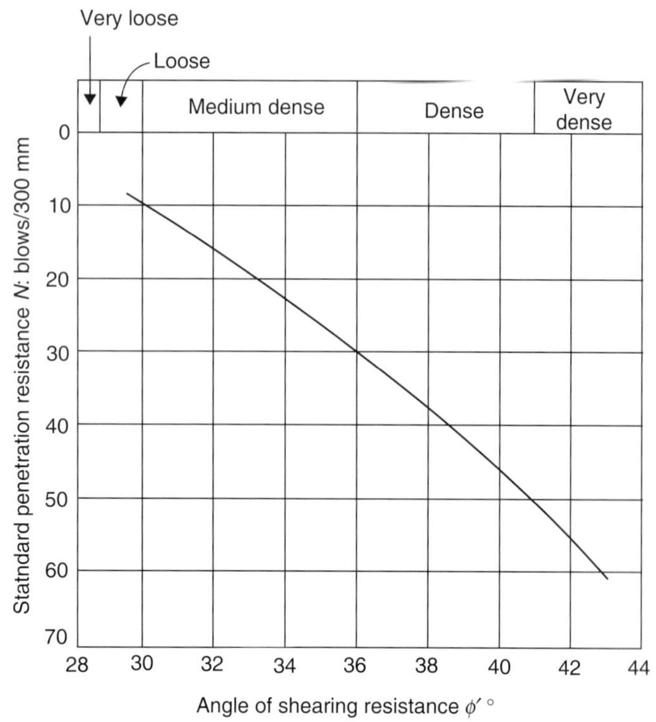

Figure 22.13 Relationship between SPT N values and ϕ' (Reproduced from Peck *et al.*, 1967; Wiley)

Fleming *et al.* (2009) argue that the value of K_{sf} varies in a similar fashion to N_q and may be estimated from the expression

$$K_{sf} = N_q / 50 \quad (22.22)$$

22.5.2 Methods based on static cone penetration tests

Methods based on CPTs are widely used for the design of offshore piles, and their use on land is increasing. The electric cone provides a profile with depth of end resistance q_c and sleeve friction f_s which can be used for the design of driven piles. A brief description of the equipment is given by Simons *et al.* (2002).

Where the pile is deeply embedded in the granular layer, the tip of the cone may be regarded as a model of the base of the pile. Hence the base resistance of the pile q_{bf} is taken as equal to the cone resistance q_c. However, the cone normally gives a fluctuating resistance with depth and requires careful interpretation. The currently accepted method of taking account of these fluctuations in deriving q_{bf} for a deeply embedded pile is given by Thorburn and Buchanan (1979) as

$$q_{bf} = 0.25q_{c0} + 0.25q_{c1} + 0.5q_{c2} \quad (22.23)$$

where

- q_{c0} is the average static cone resistance over a distance of two pile diameters below the pile base

- q_{c1} is the minimum cone resistance over the same distance below the pile base
- q_{c2} is the average of the minimum cone resistances over a distance above the pile base, neglecting any values greater than q_{c1}.

In situations where the cone resistance decreases to depths below the pile base greater than or equal to 3.5 pile diameters, the following expression gives a reasonable prediction of the ultimate base resistance

$$q_{bf} = 0.5q_{cb} + 0.5q_{ca} \qquad (22.24)$$

where q_{cb} is the average cone resistance over a distance below the pile base equivalent to 3.5 pile diameters and is determined by

$$q_{cb} = \frac{q_{c1} + q_{c2} + q_{c3} + \ldots + q_{cn}}{2n} + n \cdot q_{cn} \qquad (22.25)$$

where

- q_{ca} is the average cone resistance over a distance above the pile base equivalent to 8 pile diameters, neglecting any values greater than q_{cn}
- q_{c1}, q_{c2}, q_{c3}, … q_{cn}, are the cone resistances at regular intervals to a depth of 3.5 pile diameters below the pile base, q_{cn} being the lowest resistance at this depth.

The compressed rules given by Beringen *et al.* (1979) are broadly in agreement with the above.

De Beer *et al.* (1979) present some case histories of piles penetrating only a small distance into dense sands. They show that for small penetrations the end resistance of a pile can be significantly less than that given by the cone resistance. Various methods have been proposed to make allowance for this important scale effect. The method proposed by De Beer (1963) is widely used but is believed to be very conservative.

The results of CPTs are regularly used for predicting shaft friction τ_{sf}. This is usually taken as some fraction of the cone tip resistance q_c. Thorburn and Buchanan (1979) recommend that for sands $\tau_{sf} = q_{c(ave)}/200$ where $q_{c(ave)}$ is the average point resistance over the embedded length of the pile. For silts, he recommends that $\tau_{sf} = q_{c(ave)}/150$. For driven H-piles in granular materials, Meyerhof (1976) recommends $\tau_{sf} = q_{c(ave)}/400$.

22.5.3 Methods based on standard penetration tests

In Section 22.5.1 reference was made to the use of SPTs for evaluating the angle of shearing resistance ϕ' for use in the bearing capacity equation (Equation 22.20). Some approaches to pile design are based on direct empirical correlations between SPT blow count N and allowable end bearing and shaft friction – see, for example, GEO (1996) for bored piles in completely decomposed granite in Hong Kong. It is most important that the basis of the correlations is understood when using them, and that the piling method and ground conditions are appropriate.

When CPT tests are not available, a reasonable approach is to convert the SPT N values to equivalent cone resistance q_c using the graph in **Figure 22.14** (Burland and Burbidge, 1985). The methods outlined in Section 22.5.2 can then be applied.

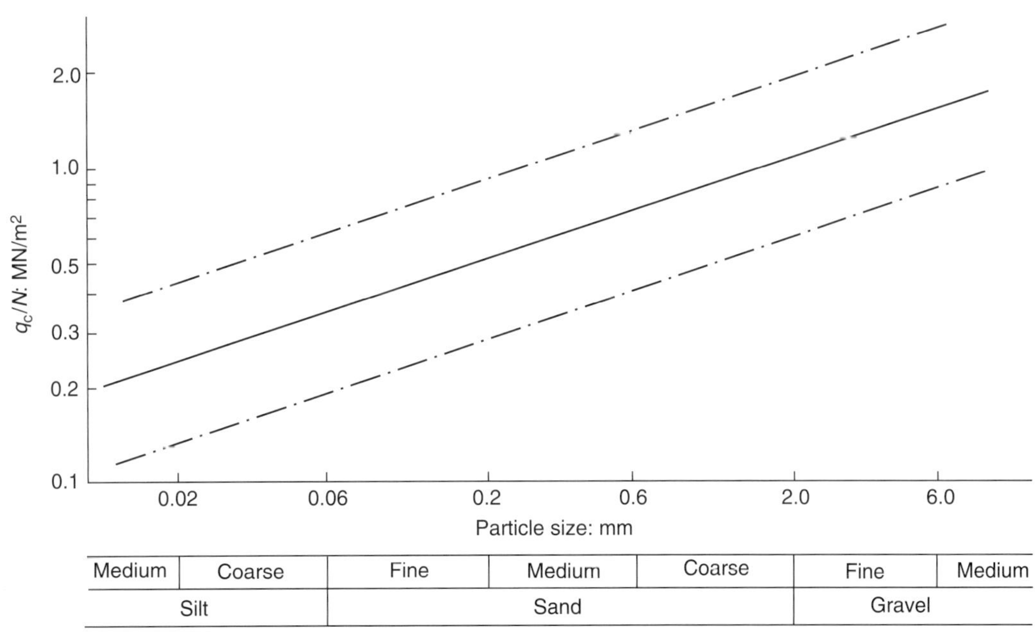

Figure 22.14 Relationship between q_c/N and grain size (Reproduced from Burland and Burbidge, 1985)

22.5.4 Pile-driving formulae

Because of the uncertainties in predicting the carrying capacity of piles driven into granular materials, it has long been recognised that the driveability of a pile should provide a useful measure of its static carrying capacity. Moreover, in variable ground conditions the driving record of each pile should be a good indicator of its quality. It is therefore not surprising that many attempts have been made to determine the relationship between the dynamic resistance of a pile during driving and the static load-carrying capacity of the pile.

Whitaker (1975) has carried out a detailed review of a number of the better-known pile-driving formulae. The purpose of this section is to provide a brief introduction to the subject. Driving formulae assume an energy balance between the dynamic input energy ($=\eta \times W \times h$) of the hammer and the work ($=R(s + c/2)$) required to displace the pile permanently by a small amount s, where

- η is the efficiency of the hammer
- W is the weight of the hammer
- h is the drop height
- R is the pile resistance
- s is the permanent set of the pile
- c is the elastic, or recoverable movement of the pile.

Hence

$$R = \frac{\eta \times W \times h}{(s + c/2)} \tag{22.26}$$

According to the widely used Hiley formula of 1925 the efficiency is given by

$$\eta = \frac{k\left(W + e^2 W_p\right)}{W + W_p} \tag{22.27}$$

where

- k is the output efficiency of the hammer
- W_p is the weight of the pile
- e is the coefficient of restitution between the hammer and the pile cushion.

Typical values of k and e are given by Whitaker (1975), and Fleming *et al.* (2009). In applying any pile-driving formula, the fundamental assumption is made that the driving resistance R is equal to the ultimate static bearing capacity of the pile. This assumption can be significantly in error, particularly in clay soils, fine granular soils and cemented soils such as chalk. Whitaker (1975) and Tomlinson (1995) give valuable guidance on the practical application of pile-driving formulae and more sophisticated forms of dynamic analysis.

22.5.5 Concluding remarks on piles in granular materials

The present state of the art on the prediction of pile capacity in granular soils is such that pile loading tests should be carried out if at all possible. Driving resistance records on all piles should be kept (blows per 0.2 m or 0.5 m penetration). These can be useful, particularly when correlated with loading tests at the site. Beware of pile heave due to adjacent pile driving – this is particularly important with saturated fine-grained granular soils (i.e. silts). It may be necessary to re-drive some of the piles.

22.6. Overall conclusions

- The installation of a pile is a geotechnical process, controlled by an operator, the outcome of which depends on numerous factors. In these circumstances it is not reasonable to expect to be able to predict pile behaviour with great precision.
- Methods of prediction of pile behaviour are inevitably highly empirical. It is therefore important to seek to understand the basis of any empirical factor or correlation that is used, and the likely range of prediction.
- The load–settlement behaviour of a pile results from the summation of two very different mechanisms. The shaft resistance is essentially frictional and is fully mobilised at small settlements – typically about 0.5% of the shaft diameter or 5–10 mm. The behaviour of the shaft is usually ductile in the sense that little reduction in carrying capacity takes place, with continued settlement after full mobilisation. The full mobilisation of the base resistance requires much larger settlements than the shaft – typically 10–20% of the base diameter. Having a clear understanding of these two mechanisms of load transfer is very helpful when interpreting the results of pile loading tests.
- Basic undrained bearing capacity theory can be used to calculate the base resistance of a pile in clay. Usually the average undrained strengths from unconsolidated undrained triaxial compression tests (S_{utc}) are used for this purpose. In stiff fissured clays, for which the undrained strengths usually show considerable scatter, the in situ operational strength is close to the lower limit of the scatter and should be used for design purposes.
- Traditionally, the measured shaft friction τ'_{sfh} of piles in clay has been correlated with undrained strength (i.e. $\tau_{sf} = \alpha \times S_u$) for the simple reason that undrained strength data are usually included with pile test results. This approach has serious limitations. It has been shown that the undrained strength of a particular soil is not a unique quantity and can show considerable scatter. The value of α has been shown to vary over wide limits. Frequently published design curves relating α to undrained strength do not include the enormous scatter on which they are based. It is essential that the engineer is fully aware of these limitations as they are frequently glossed over.
- It is because of these serious limitations that much of this chapter has been devoted to the determination of shaft friction in terms of effective stress. It is now clear beyond all reasonable doubt that shaft friction is controlled by the simple effective stress Coulomb sliding law given by Equation 22.8. This equation can be normalised by the vertical effective stress and rewritten as

$$\frac{\tau_{sf}}{\sigma'_v} = \frac{\sigma'_{rf}}{\sigma'_v} \tan \delta'_f = K_{sf} \tan \delta'_f = \beta \tag{22.28}$$

The effective stress shaft friction parameter β can be obtained empirically from pile tests and is much more rigorously defined than α since σ'_v can usually be accurately determined. The value of β can also be obtained from fundamental soil mechanics principles to give 'ideal' values, which can be used for judging the influence of various installation and operational processes.

- It has been shown that for driven piles in normally consolidated clays, measured values of β are well defined, lying between 0.25 and 0.4 – a little larger than the 'ideal' values.
- The effective stress approach is really the only viable approach to evaluating negative friction or 'down drag'. It has been shown that local values of β lie between 0.15 and 0.25 for low-plasticity clays, whereas for high-plasticity clays the values lie between 0.2 and 0.35.
- An approach is outlined for applying the effective stress method to overconsolidated clays when values of K_o are not known. It is shown that a theoretical relationship exists between β and S_{utc}/σ'_v, where S_{utc} is the *intact* undrained strength in unconsolidated triaxial compression. Three large databases for driven piles in clay have been used to show that there exists a simple approximately linear relationship between β and S_{utc}/σ'_v.
- The effective stress method has been applied to bored piles in London Clay for which the distribution of K_o with depth is approximately known. The 'ideal' curve has been compared with a number of incremental load tests from various parts of London. The measured values are significantly less than the ideal values – the reason is believed to be that the angle of interface friction for traditional bored piles in London Clay operates close to residual strength rather than critical state strength, which was assumed in the 'ideal' case.
- For bored piles in London Clay, the simple effective stress relationship $\tau_{sf} = 0.7 \times \sigma'_v$ is close to the lower limit of the scatter in **Figure 22.11** and is appropriate for design purposes. More generally, it is argued that there is a strong case for presenting the results of tests on bored piles in stiff clays in terms of average shaft resistance against depth, with the appropriate values of β drawn in.
- Piles in granular material act primarily in end bearing. Because of the difficulty of obtaining undisturbed samples of granular soils, considerable reliance is placed on the use of in situ tests for evaluating the end and shaft resistance. The standard empirical methods of applying the results of CPTs and SPTs to pile design are described.
- A brief introduction is given to the application of pile-driving formulae and their limitations.

References

Berezantsev VC, Khristoforov V and Golubkov V (1961) Load bearing capacity and deformation of piles foundations. *Proceedings of the 5th International Conference on SMFE, Paris, France*, vol. 2, pp. 11–15.

Beringen FL, Windle D and van Hooydonk WR (1979) Results of load tests on driven piles in sand. *Proceedings of the Conference on Recent Developments in the Design and Construction of Piles, London, UK*. ICE, London, UK, pp. 213–225.

Bishop AW, Webb DC and Lewin PI (1965) Undisturbed samples of London Clay from Ashford Common shaft: strength/effective stress relationships. *Géotechnique* **15(1)**: 1–31.

Bown AS and O'Brien AS (2008) Shaft friction in London Clay: modified effective stress approach. *Proceedings of the 2nd BGA International Conference on Foundations, Dundee, UK* (Brown MJ, Bransby MR, Brennan AJ and Knappett JA (eds)). IHS BRE Press, Watford, UK.

Broms BB (1965) Methods of calculating the ultimate bearing capacity of piles: a summary. *Sols-Soils* **5(18–19)**: 21–32.

Burland JB (1973) Shaft friction of piles in clay: a simple fundamental approach. *Ground Engineering* **6(3)**: 30–42.

Burland JB (1990) On the compressibility and shear strength of natural clays. *Géotechnique* **40(3)**: 329–378.

Burland JB (1993) Closing address. In *Large Scale Pile Tests*. Thomas Telford, London, UK, pp. 590–595.

Burland JB and Burbidge M (1985) Settlement of foundations on sand and gravel. *Proceedings of the Institution of Civil Engineers, Part 1* **78**: 1325–1381.

Burland JB and Starke W (1994) Review of negative pile friction in terms of effective stress. *Proceedings of the 13th International Conference on SMFE, New Delhi, India*, pp. 493–496.

Burland JB and Twine D (1988) The shaft friction of bored piles in terms of effective stress. *Conference on Deep Foundations on Bored and Auger Piles, London, UK* (Van Impe WF (ed)). Balkema, Rotterdam, the Netherlands, pp. 411–420.

Chandler RJ (1968) The shaft friction of piles in cohesive soils in terms of effective stress. *Civil Engineering and Public Works Review* **63**: 48–51.

Chow F (1996) *Investigation into the Displacement Pile Behaviour or Offshore Foundations*. PhD thesis, Imperial College London, London, UK.

De Beer EE (1963) The scale effect in the transposition of the results of deep sounding tests on the ultimate bearing capacity of piles and caisson foundations. *Géotechnique* **13(1)**: 39–75.

De Beer EE, Lousberg E, de Jonghe A, Carpenter R and Wallays M (1979) Analysis of the results of loading tests performed on displacement piles of different types and sizes penetrating at a relatively small depth into a very dense sand layer. *Proceedings of the Conference on Recent Developments in the Design and Construction of Piles, London, UK*. ICE, London, UK, pp. 199–211.

Eide O, Hutchinson JN and Landa A (1961) Short and longterm test loading of a friction pile in clay. *Proceedings of the 5th International Conference on SMFE, Paris, France*, vol. 2, pp. 45–53.

Flaate K and Selnes P (1977) Side friction of piles in clay. *Proceedings of the 9th International Conference on SMFE, Tokyo, Japan*, vol. 1, pp. 517–522.

Fleming WGK, Weltman AJ, Randolph MF and Elson WK (2009) *Piling Engineering*, 3rd edn. Taylor & Francis, Oxford, UK.

GEO (Geotechnical Engineering Office) (1996) *Pile Design and Construction*. Geotechnical Engineering Office, Civil Engineering Department, Hong Kong.

Hiley A (1925) A rational pile-driving formula and its application in piling practice explained. *Engineering, London*, **119**: 657, 721.

Hutchinson JN and Jensen EV (1968) *Loading Tests on Piles Driven into Estuarine Clays of Port of Khorramshahr, and Observations on the Effect of Bitumen Coatings on Shaft Bearing Capacity.* NGI, Oslo, Norway.

Jamiolkowski M, Ladd CC, Germaine JT and Lancellotta R (1985) New developments in field and laboratory testing of soils. *Proceedings of the 11th International Conference on SMFE, San Francisco, CA*, vol. 1, pp. 57–153.

Jardine RJ, Lehane BM and Everton SJ (1992) Friction coefficients for piles in sands and silts. *Proceedings of the International Conference on Offshore Site Investigation and Foundation Behaviour, London, UK.* Society of Underwater Technology, London, UK, pp. 661–680.

Jardine RJ, Chow F, Overy R and Standing J (2005) *ICP Design Methods for Driven Piles in Sands and Clays.* Thomas Telford, London, UK.

Johannessen IJ and Bjerrum L (1965) Measurements of the compression of a steel pile to rock due to settlement of the surrounding clay. *Proceedings of the 6th International Conference on SMFE, Montreal, QC*, vol. 2, pp. 261–264.

Lehane BM, Jardine RJ, Bond AL and Chow FC (1994) The development of shaft resistance on displacement piles in clay. *Proceedings of the 13th International Conference on SMFE, New Delhi, India*, vol. 2, pp. 473–476.

Marsland A (1974) Comparison of the results from static penetration tests and large in situ plate tests in London Clay. *Proceedings of the European Symposium on Penetration Testing, Stockholm, Sweden.*

Mayne PW and Kulhawy FM (1982) K_0–OCR relationships for soils. *Journal of Geotechnical Engineering Division, ASCE* **108(6)**: 851–872.

Meyerhof GG (1976) Bearing capacity and settlement of pile foundations. *Journal of Geotechnical Engineering Division, ASCE* **102(GT3)**: 197–228.

Patel DC (1992) Interpretation of results of pile tests in London Clay. *Piling: European Practice and World-wide Trends.* Thomas Telford, London, UK, pp. 100–110.

Patrizi P and Burland JB (2001) Developments in the design of driven piles in clay in terms of effective stresses. *Revista Italiana di Geotecnica* **35(3)**: 35–49.

Peck RB, Hanson WE and Thornburn TH (1967) *Foundation Engineering*, 2nd edn. Wiley, New York, NY.

Semple RM and Rigden WJ (1984) Shaft capacity of driven piles in clay. *Proceedings of ASCE National Convention, San Francisco, CA*, pp. 59–79.

Simons N, Menzies B and Matthews M (2002) *A Short Course on Geotechnical Site Investigation.* Thomas Telford, London, UK.

Skempton AW (1959) Cast in situ bored piles in London Clay. *Géotechnique* **9(4)**: 153–173.

Skempton AW (1961) Horizontal stresses in overconsolidated Eocene clay. *Proceedings of the 5th International Conference on SMFE, Paris, France*, vol. 1, pp. 351–357.

Thorburn S and Buchanan NW (1979) Pile embedment in fine-grained non-cohesive soils. *Proceedings of the Conference on Recent Developments in the Design and Construction of Piles, London, UK.* ICE, London, UK, pp. 191–198.

Tomlinson MJ (1971) Some effects of pile driving on skin friction. *Proceedings of the Conference on Behaviour of Piles, London, UK.* ICE, London, UK, pp. 107–114.

Tomlinson MJ (1995) *Foundation Design and Construction*, 6th edn. Longman Scientific & Technical, Harlow, UK.

Weltman AJ and Healy PR (1978) *Piling in Boulder Clay and Other Glacial Tills.* CIRIA Report PG5. CIRIA, London, UK.

Whitaker T (1975) *The Design of Piled Foundations*, 2nd edn. Pergamon, Oxford, UK.

Whitaker T and Cooke RW (1966) An investigation of the shaft and base resistance of large bored piles in London Clay. *Proceedings of the Symposium on Large Bored Piles, London, UK.* ICE, London, pp. 7–49.

All chapters within Sections 1 *Context* and 2 *Fundamental principles* together provide a complete introduction to the Manual and no individual chapter should be read in isolation from the rest.

Chapter 23
Slope stability

Edward N Bromhead Kingston University, London, UK

doi: 10.1680/icemge.66816.0273

CONTENTS

This chapter explores the basics of slope stability and how relative stability is calculated using a limit equilibrium approach. It discusses the principles of slope stabilisation as they impact on the analysis required, and makes recommendations regarding the management of slope instability risks within the context primarily of the analysis.

23.1. Factors affecting the stability and instability of natural and engineered slopes

23.1.1 Introduction

It is customary to consider the problems of natural slopes as being distinct from those of engineered slopes, and in the latter to discriminate between cut (excavated) slopes and filled slopes, despite many of the technical issues being the same. Filled slopes give the engineer the option of controlling the material properties of the soils involved as well as controlling the shape of the resulting slope, whereas in both cut and natural slopes the geological succession and resulting properties are whatever nature presents. In cut slopes, one at least has the possibility of forming them to a desired shape, but in natural slopes, again, the engineer has often to deal with what is there.

Engineered slopes are created in connection with urban development, transportation routes, dams, mining and quarrying, municipal waste disposal and numerous other construction activities requiring the formation of excavations, and in building mounds. Where the available space does not permit stable slopes to be created, retaining structures are required (Section 6 *Design of retaining structures*).

Natural slopes present stability problems associated with contemporary coastal and river erosion; the erosion processes cause oversteepening. In addition, the range of processes that form hills and valleys commonly leave hill slopes in a state where they can be readily destabilised. For example, extremes of rainfall or modest earthworks may cause hitherto dormant soil masses that were at one time unstable to move again.

Many engineers would further distinguish between rock slopes and soil slopes, where in the former the behaviour is characterised largely by failure primarily along pre-existing discontinuities such as joints and faults, and much less so through the intact material and vice versa. However, many soil slopes fail along bedding, or weak layers constructed into a fill, so that such a distinction is somewhat arbitrary. The rest of this chapter deals primarily with soils.

23.1.2 Factors giving rise to slope failure

Factors giving rise to slope failure comprise external and internal geometric factors, intrinsic properties of the materials involved, the presence of discontinuities and other structural weaknesses (which may in part be due to geological causes and reflect the processes by which the soils were formed or emplaced), the presence of fluids (air and water) in the soil and processes of weathering or ageing – some of which relate to long-term fluid pressure changes. Much of the loading in a soil comes from self-weight, but some slopes may be subjected to external loads from foundations and anchorages, and stresses induced dynamically, of which the most obvious is seismic shaking.

External geometric factors include its height and slope angle, and its shape, both in section and plan. Internal geometric factors include the disposition of different material types inside the slope. It is extremely common to find that failure in an engineered slope occurs because a weak layer was left in place in an embankment within its foundation, or in a cut slope within the body of the soil mass. Natural deposits, particularly of sedimentary origin, often contain strata which are weaker than the surrounding soils, and failures occur along such weaker layers.

23.1.3 The soil strength model

In the majority of slopes, the most significant loading comes from the self-weight of the soil mass, and therefore its density (unit weight γ, or *weight density* according to Eurocode 7) is a significant factor. It is convenient to model the resistance to shear s using a simple c–ϕ effective stress-based relationship:

$$s = c' + (\sigma - u)\tan\phi' \qquad (23.1)$$

For a small range of stresses, and with appropriately selected parameters, this permits the analysis to consider the self-weight and pore fluid pressure effects in the soil (these are usually due to water, so the phrase *pore water pressure* is taken as equivalent in the following), both from steady and unsteady seepage, and caused or modified by undrained stress changes in the soil mass. Experience shows that laboratory-measured peak parameters are often poor choices for c' and ϕ', and reduced ϕ' values (largely based on empiricism or experience) may be required.

Soils vary so much in their composition that it is rarely possible to identify values for c' and ϕ' merely from a soil description, although some rules of thumb are useful. For example, in granular soils such as sands and gravels, ϕ' is almost invariably greater than 30° and increases with density and particle interlock, so that granular soils with angular particles and a range of particle sizes are normally much stronger than those with rounded, non-interlocking, similar-sized particles. Micaceous sands may not exhibit the interlock phenomenon, and may be weaker. Mineral cements such as iron compounds and calcite may further increase ϕ', but their effect is greatest on c', which would otherwise (in granular soils) be negligible. Cements are unreliable if present to a small degree – decalcification, for example, can occur with weak acids from vegetation or acid rain.

Clay soils commonly exhibit ϕ' in a range from just below 20° to about 25°. Factors that increase ϕ' are, again, the content of cementitious minerals and the type of clay minerals present (which may be apparent through their plasticity). Most clays lose significant amounts of strength when sheared, leading to the formation in the soil mass of slickensided shear surfaces. The amount of strength degradation or brittleness varies widely, and is usually highest where the soils contain smectites and lowest where they contain kaolinites. The strength along slickensided shear surfaces may also be represented by a c–ϕ, effective stress-based relationship, but using different c–ϕ values for the sheared or *residual strength* condition. This is denoted by a subscript 'r'. Commonly, illite clays exhibit ϕ'_r in the range 9–15°, but smectites may be much lower, down to around 4°. Effective residual cohesion is usually negligible. Silty soils fall in between the two limits, depending on whether they are devoid of, or rich in, clay.

The stability of a saturated clay mass is dominated not by the mean strength, but by the strength of the weakest path through the mass. Such weaker zones may be created at the time of deposition or placement of the soil and are often the result of having less effective cohesion. Therefore, notwithstanding the existence of effective cohesion in many clays (and it is almost invariably shown up in effective stress shear strength tests), stability analyses routinely employ little or no effective cohesion.

All assessments of stability using effective stress-based shear strength rely on the correctness of the pore water pressures used in the analyses. Soils with low densities, particularly organic soils such as peat and municipal solid waste, are extremely sensitive to fluctuations in the pore water pressure: in Equation 23.1 the direct stress σ (which arises from self-weight) may not be much greater than the pore water pressure u.

23.1.4 Undrained strength

Under some loading conditions – notably the application of loads without change in the moisture content of the soil, and for clay fills that are placed in an unsaturated state – it may be convenient to employ an undrained strength soil model. Usually, this is nominally a c–ϕ model, but ϕ_u is commonly small or zero, and is ignored. Strictly, the undrained strength is s_u, but the term c_u is used and the two are interchangeable.

When using the undrained strength, pore pressure effects (some of which are highly complex) are lumped in with the other strength parameters, and the analysis is done as though the soil is devoid of pore pressures. Needless to say, that simplifies the analysis.

Undrained strengths in some saturated soils are subject to degradation upon remoulding or reworking. This is termed sensitivity. Unlike drained brittleness, which is a phenomenon of the soil particles, sensitivity is commonly a pore water pressure effect, although that in turn comes from disruption to the soil fabric. Unsaturated soils obtain much of their undrained strength from capillarity, and this may be lost upon soaking or saturation.

Slopes with clay soils or clay strata tend to fail in the clays, and subsequent to failure slickensided shear surfaces are left in the ground. These surfaces dominate the subsequent behaviour and stability of the slope as a whole, although local slope facets may remain where the peak or unsheared strength continues to control local stability.

23.2. Modes and types of failure commonly encountered

There are a number of classification and descriptive schemes for slope failures (Cruden and Varnes, 1996; Dikau *et al.*, 1996). The critical distinction between slope deformation and slope failure is that in the latter a mass of soil or rock is detached from the parent mass and moves independently, whereas slope deformation is movement of a part of the parent mass with no, or incomplete, detachment. The detached mass may stay in contact as it moves, or it may lose contact. If the moving mass breaks into a few pieces and it loses contact (the pieces may bounce or roll, or simply free fall until impact), then it is usual to describe the slope failure as a *fall* (e.g. rock fall). Where a fall breaks up into many pieces during movement, the mass tends to hug the topographic surface and exhibit characteristics associated with fluid flow, and it is then described as a *flow* (e.g. debris flow, mud flow).

Slides occur when one (or a few) discrete blocks of soil and rock stay in contact with, but are clearly detached from, the parent mass. Slides may also degenerate into flows if they break up into many parts that move independently.

From the perspective of engineered slopes, falls are most commonly experienced from rock slopes, and result from weathering or water ingress into the slope – causing blocks to loosen and become detached along pre-existing discontinuities, with little or no failure through the intact rock mass. Falls of soil can, and do, occur from oversteepened temporary excavations such as trenches and pits.

Flows occur where the soil loses strength en masse. In loose granular soils this can occur by vibration, rapid loading creating pore pressures, movement of water into or through the soil mass and deformation during sliding. Loose fills on sidelong ground are particularly susceptible to this mode of failure, as are soils placed hydraulically (e.g. from dredgings) where the failure may be termed *liquefaction*.

Slide failures are the most common type of failure, as they can occur on a range of slope angles, including extremely low-angled slopes, and are often the first stage of an event that turns into a flow or fall. Slide failures are particularly common where there is a pre-existing appropriately orientated weakness in the slope. This can range from a fault, bedding or joints in a rock mass, particularly where they dip out of the slope, to a weak (possibly organic) horizon in the foundation of an earth embankment. Such inappropriately orientated weaknesses are sometimes incorporated in earthworks by accident or neglect, including clay capping layers in municipal solid waste (MSW) facilities, or poorly prepared interfaces in clay fills.

Other modes of failure have been described in the relevant literature, but they occur chiefly in natural slopes, and are not commonly encountered.

23.3. Methods of analysis for slopes, exploring their limitations of applicability

23.3.1 Sliding analyses

Since many failures of slopes initiate with sliding, slope stability is usually assessed by means of a sliding model, employing a shear strength relationship of the type described above. The simplest of possible models employs a planar sliding surface, although this can be elaborated to cope with curved or compound (partly planar, partly curved) shapes, and is most often applied to a plane section through the slope. Occasionally, the three-dimensional shape of the slope and transverse curvature of potential sliding surfaces are taken into account. In rare cases, where the geology of a slope has demanded it, a kinked or dogleg section has been adopted. An analysis considering the collapse or ultimate limit state (giving rise to the epithet of 'limit equilibrium analysis') is usually sufficient for the analysis and design of routine engineered slopes and for the analysis of many natural slopes.

Continuum methods of analysis are used to explore the stress, strain and deformation conditions in slopes under their expected conditions of operation when the financial value of the slope itself (or the consequences of failure or underperformance) merits the additional costs of analysis and determination of parameters. Sometimes, such analyses are used to explore the conditions that led to failure in a *post hoc* or forensic investigation, as these methods must follow the correct stress path (or sequence of events).

Applied external forces such as live loads and anchors are in addition to, and often much less significant than, self-weight and pore pressure forces. More detailed treatment of the available methods is given by Bromhead (1992) and Duncan and Wright (2005).

23.3.2 Planar sliding

Figure 23.1 shows a cross-section of a soil slope. If the mass of soil of weight W above a planar surface A–A inclined at an angle to the horizontal of α, the component of this weight D acting down the plane A–A is

$$D = W \sin\alpha \tag{23.2}$$

The component normal to the plane is $W \cos \alpha$, so that the maximum possible resistance R to sliding in the absence of water is

$$R = W \cos\alpha \tan\phi' + c'A \tag{23.3}$$

where A is the area of contact between the block and the parent soil or rock mass. If there is a pore water pressure acting over the area of contact, then the force from the average pressure u multiplied by the area A must be taken into account, thus

$$R = (W \cos \alpha - uA)\tan\phi' + c'A \tag{23.4}$$

Since the resistance is a reaction, and must therefore be equal to the applied force

$$D = W \sin\alpha = \frac{R}{F} = (W \cos\alpha - uA)\frac{\tan\phi'}{F} + \frac{c'A}{F} \tag{23.5}$$

Note that if R is less than D, the mass of soil would accelerate under the unbalanced force $D-R$.

This equation could be used to determine the actual proportion of the shear strength parameters mobilised (i.e. to compute a factor of safety) or, by employing an inequality (<), to assure that the scenario is adequately stable when employing factored parameters.

23.3.3 Sliding on curved and compound surfaces

As soil slopes commonly fail on slip surfaces that combine curves and straights in a plane section, it is useful to consider the sliding mass to be made up from a series of vertical-sided slices, each sliding down its own plane. (Non-vertical-sided slices can also be employed, but vertical slices are much more common.) The precise number of slices is taken to ensure that the base of each is approximately planar. Since the above

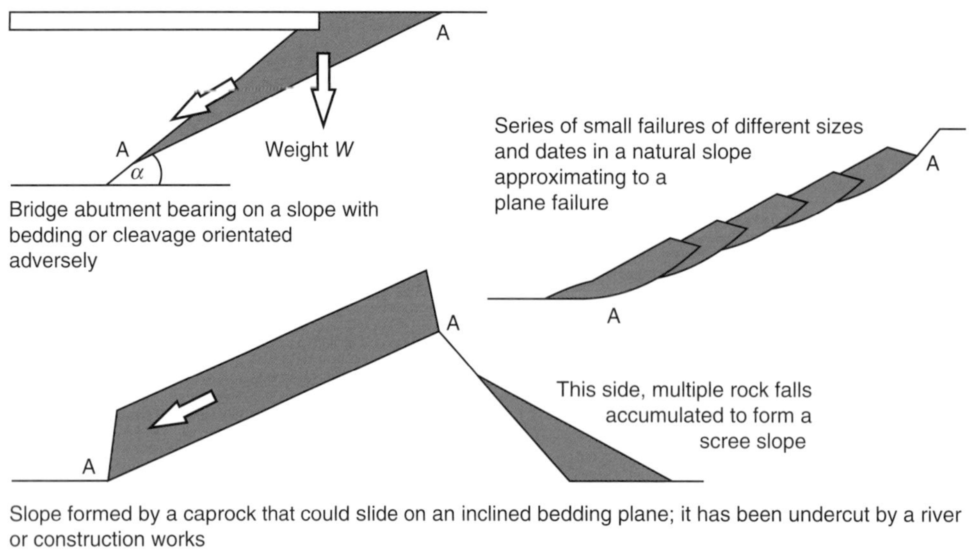

Figure 23.1 Examples of plane failure; the mass slides along plane A–A

equation could be applied to each slice in turn to calculate a 'local' factor of safety, the factor of safety of the whole mass must lie between the maximum and minimum values so computed. Indeed, it is likely that the factor of safety of the whole is some sort of average of the individual values. An average would also take into account the tendency of some slices, if isolated, to slide in the opposite sense to the mass as a whole, especially in a toe zone that breaks out upwards.

A convenient averaging technique is to compute the sums of the driving (D) and resisting (R) terms, before dividing the latter by the former, since this takes into account implicitly the relative sizes and importance of the slices, and gives the overall factor of safety F as

$$F = \frac{\sum_{i=1}^{n}\left\{(W\cos\alpha - uA)\tan\phi' + c'A\right\}_i}{\sum_{i=1}^{n}\left\{W\sin\alpha\right\}_i} \qquad (23.6)$$

Note that this definition of F continues to be identical to the mobilisation factor on the shear strength parameters.

Consideration of the form of this equation shows that it reverts to the planar slide formula under the appropriate circumstances and is useful for shallow slip surfaces. However, if the slip surface is markedly curved, or descends deeply in the slope (especially when pore pressures are variable and high), the lack of computed interaction between the slices (i.e. the absence of inter-slice forces) makes the stress analysis change from merely approximate to highly dubious, and introduces errors which lead to a conservative estimate of F. Starting with an iterative procedure for dealing solely with the horizontal component of inter-slice forces published by Bishop (1955), efforts have been made to incorporate – as realistically as possible – the effects of inter-slice forces. One outcome of this is the publication of a multiplicity of named methods for the analysis of curved and compound sliding surfaces. Fortunately, provided that some attempts are made to model the inter-slice forces, the results do not appear to depend greatly on the precise technique used. Methods include using horizontal inter-slice forces (e.g. Bishop, 1955), inclined but parallel inter-slice forces (Spencer, 1967) and positioning the inter-slice force line of action at, say, one-third height on the slice. Morgenstern and Price (1965) introduced a procedure where the distribution of inter-slice forces is decided a priori by the analyst, but the precise magnitudes are determined during the equilibrium computation. This approach is followed by Sarma (1979).

The computational procedures almost always involve an iterative solution method (except where the soils are purely cohesive) once inter-slice forces are considered, however approximately. The analyst is usually faced with selecting one of a small range of options in whatever software they have, and does not have a free choice among the various methods available at large.

Since even Bishop's procedure is laborious to compute by hand, slope stability calculations became an early application for computers. To summarise, the essential differences between the methods lie in the treatment of inter-slice forces – and this choice is largely immaterial, and even the once-important issue of how long it took to produce results is of little significance, given the speed of modern computing hardware and software.

23.3.4 Slip circles

It is found that in many cases an arc of a circle (**Figure 23.2**) on the two-dimensional slope section provides an adequately good first approximation to the shape of a failure. Indeed, some case examples show that under particular combinations of conditions it can be a *good* approximation. These surfaces

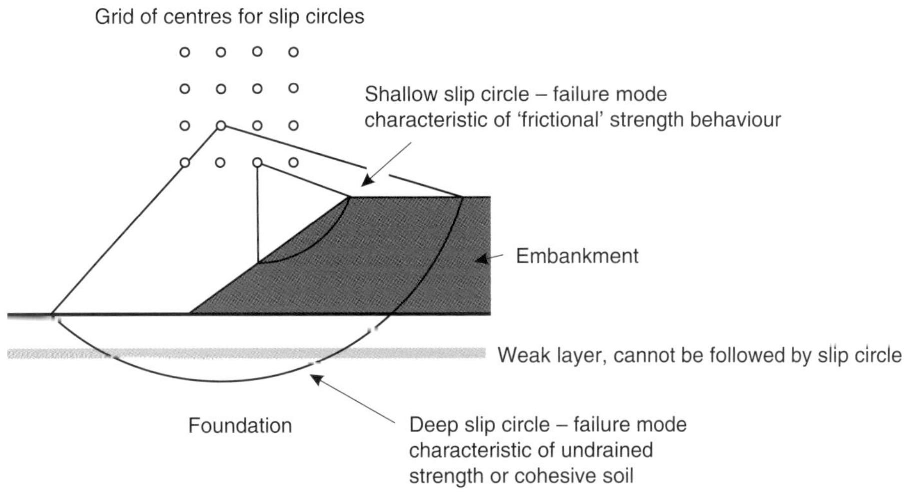

Figure 23.2 The slip circle, commonly employed where the critical clip surface is unknown

are called slip circles. Since all of the geometric factors can be calculated simply from the position of the centre of a slip circle together with its radius, the slip circle lends itself to repetitive analyses using many slip circles to find the one or ones with the least factors of safety. Most analysis software uses a regular rectangular grid of centres together with multiple circles emanating from each centre. Refinements may include constraints on where the circles may intersect the slope, and grids of centres of various types and orientations. The particular method of indicating the critical slip circle (i.e. the one with the least factor of safety) and how it relates to other possible slip circles is not standardised, but depends in part on the imagination of the programmer and, to an extent, that of the analyst.

Stability analyses of slopes with soil strengths that are dominated by their angles of shearing resistance (ϕ') tend to indicate that shallow slip surfaces are the most critical. In the undrained case, and where strength is dominated by a cohesion parameter, the critical slip circles are deep, akin to foundation failures. A fundamental defect of the slip circle approach is that no circle can follow a thin stratum of weak soil (**Figures 23.2** and **23.3**), and where slip circles do indicate the weaker zones of a slope (low shear strength parameters, but also where high pore pressures exist) these zones are discoverable by visual inspection anyway. Since it is not possible to generate compound slip surfaces with quite the facility with which slip circles can be generated (from a grid of centres), it is useful to undertake a slip circle-based analysis in the first instance, and use the most critical slip circles as a basis for extending into compound and other shapes to include the weak zones.

Many cases of slope failure in engineered slopes occur because at the time of construction a weak layer in an adverse orientation was incorporated into the fill or left in the embankment foundation. Similarly, in cuts and natural slopes, a weak horizon again provides a path for a failure surface. It is therefore essential in these cases not to rely simply on slip circle approaches.

23.3.5 Three-dimensional effects

Three-dimensional analyses are rarely done in routine engineering earthworks design, simply because there is no 3D analogue of the slip circle search procedure. In forensic (post-failure) analyses, the 3D approach may convey some benefits. However, an alternative approach is to make 2D analyses and combine the results. This is often sufficient.

A determined analyst will find that 2D calculations often yield a lower factor of safety than the 3D calculation. This is sometimes the result of failing to consider adequately the variation in pore water pressure in the out-of-plane direction: field failures all have a 3D element.

23.3.6 Pore water pressure changes during the 'lifetime' of a slope

In the foundation of an earth fill, undrained loading usually gives rise to substantial pore pressures. Provided that failure does not occur, the dissipation of excess foundation pore pressures leads to a gain in foundation strength and therefore to increased stability with time, although the effect of pore pressure dissipation in the foundation decreases with time. In some cases, the migration of pore water laterally in horizontal layers of higher permeability may, early in the process, lead to a transfer of pressures from under the embankment crest, where they do little harm, to the toe, where they do damage. The critical point for stability, therefore, in the lifetime of a filled slope on a weak foundation is either at the end of construction, or shortly thereafter if pore pressures do migrate out to the toe. Some benefit is obtained by filling slowly, or by allowing rest periods (often throughout the rainy season when fill cannot be placed), during which time the foundation consolidates and is therefore capable of bearing the following season's lift of fill.

In the case of a cut slope, the total stresses decrease during excavation, and the undrained pore pressure change in

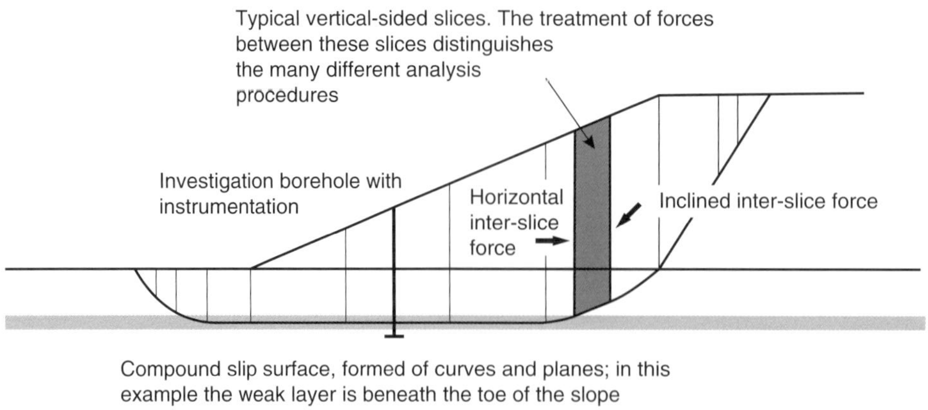

Figure 23.3 Compound failure with sub-horizontal bedding-controlled shear surface

response decreases the pore pressure state over the lower part of the cut slope to a state of suction. Loss of this suction occurs with infiltration of water from the ground surface or from depth, and this leads to loss of strength and in some cases to delayed failure. The equilibration of pore pressures may take decades. The process is most marked in stiff clays.

Clay fills are inevitably placed incompletely saturated and manifest strength due to capillary effects in whatever pore water there is. As in the case of a cut slope, loss of suction may occur due to water ingress, and this can, in clay fill slopes, also lead to delayed failure. However, suctions near the surface can be reinstated by dry weather, or by moisture abstraction through the roots of vegetation. The progress towards delayed failure is neither as certain nor as monotonic as the gain in strength of an embankment foundation as it consolidates. Although recent research work has been directed to understanding and quantifying these effects, they are not routinely considered in slope stability analyses, which proceed on the basis of pessimistically assessed positive pore pressure conditions.

Partly submerged slopes gain support from the external water load, but lose strength as water enters the slope and exerts pore water pressures. These two effects act in opposite senses, and the result may increase or decrease the overall stability relative to the slope prior to submergence. In many cases, it is found that there is a particular water level – termed the 'critical pool level' – at which point the stability is least. Care on first impounding water-retaining earthworks is therefore called for. Stability is often severely affected if the support from the water is rapidly withdrawn ('rapid drawdown'), especially if there is insufficient time for water in the slope to flow out. This, as in all the cases of partly submerged slopes, is modelled in slope stability analyses through the use of appropriate distributions of pore water pressures.

23.3.7 Seismic stability

During an earthquake, slopes may be accelerated and decelerated multiple times in a variety of directions, but usually for very short durations in each. A simple approach to dealing with this is to take the inertial forces as some fraction of gravity, and to apply the resulting forces in a standard slope stability analysis as a horizontal body force. In some cases, vertical body forces are applied in addition to weight. There may also be undrained pore pressure changes. Inevitably, slopes with a high static factor of safety and moderate to low inertial forces appear to perform well, and those with low initial stability or where the inertial forces exceed some threshold appear to be in line for failure. The body forces are calculated in the horizontal direction from $k_h W$, where k_h is the proportion of gravitational acceleration assumed to act horizontally.

An alternative procedure is to calculate the value of k_h that causes F to drop to 1. This factor is termed k_{crit} or k_c. Then, whenever k_h exceeds k_c, the slope will begin to move. However, if the times involved are in milliseconds, the displacements per seismic pulse are tiny, although they increment pulse by pulse through each design model earthquake. Thus, $F < 1$ is not necessarily a disaster, provided that it occurs for a short time, and the soil properties are not left in a damaged state that will permit failure even when the shaking stops (e.g. by way of increased pore water pressures).

Slopes that continue to move after the earthquake vibrations have ceased are of course the primary concern. This may be the result of high earthquake-generated pore pressures. Loose fine granular soils and saturated cohesive soils with a porous fabric are most prone to loss of strength on shaking. Stiff clays form slip surfaces with diminished residual strength parameters, but this is often associated with dilation and lowered undrained pore pressures, and may not be significant.

23.3.8 Errors in slope stability analyses

Errors made in slope stability analyses are usually the result of

(a) incorrect assessment of design values for shear strength, including a failure to consider whether drained, consolidated drained or consolidated undrained conditions are operating

(b) incorrect assessment of pore water pressures in the ground

(*c*) inadequate assessment of the effect of weak layers in the slope which may be present in the soil for stratigraphic reasons, or have been created by construction activities

(*d*) selection of inadequate mobilisation factors in a particular case, especially where there is uncertainty surrounding the soil parameters and pore water pressures, but also, less obviously, where there is the risk of progressive failure.

Item (*c*) in this list sometimes manifests itself by a weak layer not having been recognised during investigation or construction, and also by the analyst not considering slippage along such weak layers. The use of slip circles alone may result in this situation arising in the analysis.

The technique most commonly employed in finding the critical (least safe) slip surface is a 'brute force' approach of analysing often thousands of slip circles. It is easy to forget that the critical surface may not be the arc of a circle.

Errors in the computation itself are much rarer than in the above technique, probably the result of the widespread deployment of quality-controlled computer software.

23.3.9 Continuum methods

Continuum methods (such as finite element and finite difference stress analyses) are occasionally employed in slope stability problems, but they need parameters that are more difficult to obtain, and an even more skilled analyst, so are expensive to apply in routine situations. They provide estimates of deformations, absent from a limit equilibrium analysis (including the pattern of deformations) and they do not require analyst intervention to show the most probable mode of deformation leading to failure.

However, the problem of dealing with thin weak bands in the soil is perhaps more intractable than in limit equilibrium analyses.

23.4. Rectification of unstable slopes

23.4.1 Introduction

Unstable, or potentially unstable, slopes may pose threats to human interests, and intervention of some sort is often called for. The baseline case is that nothing is done. Value judgements can then be made on the cost of intervention against the anticipated losses.

Slope stabilisation methods are described in Chapter 72 *Slope stabilisation methods*, with the generalities of design in Section 7 *Design of earthworks, slopes and pavements* as a whole; here, the analysis issues with those methods are discussed. Retaining structures in particular are described in Section 6 *Design of retaining structures*.

One form of intervention that may be highly successful is to ensure that no valuable assets lie in areas of high risk. Many of the best examples of the application of such controls are adjacent to the coast. They include prohibition of development by statutory ordinances, realignment inland of coastal roads and, in one celebrated case, the historic Belle Tout lighthouse at Beachy Head, East Sussex, was shifted inland by about 50 m to protect it from coastal cliff erosion.

Where the commonest form of instability is that of small to moderate flows (e.g. debris flows in mountainous areas) or of occasional rock falls, assets may be protected by various forms of barrier and catch structures, or deflection barriers and channels. Barriers range from earth or rockfill bunds, through reinforced-concrete walls, to steel or occasionally timber post walls. Catch structures may be bunded enclosures or fences and netting. Where the failures are frequent, the catch structures need periodic clearing to maintain their effectiveness, combined with some repairs (especially where a catch structure has been called upon to resist loading beyond its design envelope) to compensate for corrosion, rot, vandalism or changed environmental conditions.

The final alternative is to reduce the probability of failure by engineering works on the slope itself.

23.4.2 Earthworks solutions

Slopes may be reprofiled to an overall flatter gradient. Where the likely mode of failure is relatively shallow, this is the optimum approach. However, there are numerous cases where grading to a flatter angle is not only useless, but may provoke failures. For example, a coastal slope with a large, pre-existing landslide may exhibit a coastal cliff with small failures. Flattening this cliff may solve the small-scale local stability problem, while at the same time destabilising the deeper slide system as it constitutes an unloading of the toe zone of the deep-seated slide. Many slopes with pre-existing adversely orientated weaknesses, or with pore water pressure conditions independent of the regraded slope facet, are subject to this effect.

Deeper-seated failure modes tend to be improved with the construction of toe-weighting fills and berms. However, such earthworks may cause local slope stability issues if inappropriately founded – for example, on weak compressible subsoils or on sidelong ground. Toe fills may not be fully mobilised to support a slope until movements have taken place and compressed the fill laterally. A further effect of some concern to geotechnical engineers is the problem that arises in slope stability analysis if the toe of a slip surface rises through a toe fill with a high angle of shearing resistance ϕ'. This tends to make the computed factor of safety excessively high as a result of the assumptions in the analysis method. Care is needed in interpretation of these results.

Unloading the head of a slip failure is a useful method of stabilising it, but at the risk of affecting the head scar of the failure, which may require more significant interventions to keep it stable.

In general terms, when attempting to stabilise a slip, fills are highly effective above parts of the slip surface where it rises to ground level at the toe, and cuts are highly effective where the slip surface is steeper than ϕ' (where the slip surface rises towards the head of the slope). In between these locations, the relative effectiveness of cuts and fills depends on the pore pressure response in the slope to stress change.

The analysis of the stabilisation solution is a re-run with the modified slope profile and soil properties.

23.4.3 Drainage solutions

Slope drainage systems fall naturally into two main groups: preventing water from penetrating the slope in the first instance, and removing it once it is there.

A good tree canopy is effective in controlling rainwater ingress, but deciduous trees lose this capability annually. In any case, the canopy is established slowly, and can be breached by the loss of individual trees or wholesale felling. Creation of impermeable coverings inhibits the movement outwards of water already in the slope, and creates problems of peak discharge during heavy storms, as well as potential aesthetic issues. Very shallow drainage in a slope, such as ditch networks and land drains formed from clay pipes, tile drains and modern perforated plastic pipes, has the effect of controlling infiltration. Buried systems have the additional benefit of slowing down the release of captured water. Land drainage systems are easily blocked and are difficult to maintain. When blocked, they back up water, overpressuring the drainage system and the soil, causing local failures.

Deep drainage systems may be installed to release pore pressures, such as undrained loading-induced pore pressures that are then unlikely to be replenished, or to control seepage-induced pore pressures that are continuously replenished. In the former case, filtration is unlikely to be of major significance, but in the latter case it is. Filters prevent loss of fines into and through the drainage system that would result in internal erosion in the soil, and possible blockages in the drainage system.

Pore pressures induced by undrained loading are most often associated with the construction of embankments over weak and deformable foundations, where the load from the embankment fill compresses the foundation. Release systems for these pore water pressures usually take the form of vertical filter-fabric wicks or sand-filled drain wells, with hybrid systems sometimes employed. They may connect to permeable strata at depth, or to a collector drainage layer at their top, or both. Accelerated consolidation of the ground can be caused by applying suction, through the use of electrical systems, or where stability permits, by applying surcharge loads. Where the pore water pressures are induced in high fills, the drainage system may be laid at the time of construction. In this case, it is often a drainage layer formed from appropriately selected granular material with a high permeability. The cost of drainage layers in a very large earthwork can be high, and to minimise this the layers are formed from a grid of interconnecting 'strips'. In earth dams, the drainage layers to control construction-induced pore pressures may form part of the permanent drainage system, in which case filters are obligatory. Filters may take the form of further granular layers or be made from thin sheets of manufactured and non-degradable fabric.

Permanent drainage systems may also be incorporated in a slope by drilling sub-horizontal or other drains. These are kept open in soils with perforated plastic (or other) liners. Filter protection of these employs proprietary systems of variable longevity, most of which prove difficult to maintain. Bored drains can be installed from the surface, or from shafts and tunnels, both of which operate as drains. Shafts collect water at their bases and ideally are interconnected by a tunnel or series of borings to permit the water to drain under gravity; if not, then pumps must be employed. These pumps are subject to breakdown, and have running and replacement costs.

The efflux from drainage systems may contain contaminants, requiring treatment before it can be discharged. These contaminants may include microbial contaminants (e.g. from MSW, or where septic tank foul drainage is employed locally), or mineral content such as the iron-rich oxygen-demanding acid mine drainage water from underneath some types of mine waste tips. In some cases, discharge of water from a slope drainage system may not be permitted, or may be permitted only subject to severe constraints, by the appropriate local authority or environmental protection agency. Some forms of drainage water, most notably calcite-rich or iron-rich water, can cause drain blockage due to precipitation, in the latter case assisted by bacteriological action.

Drainage trenches of several metres or more in depth are often employed to stabilise failures in cut slopes. Regardless of their depth, they can reduce pore water pressures. However, when they are deep enough to penetrate sheared surfaces or other zones of weakness, they also have a reinforcing or buttressing effect in the slope, in which case they are termed counterfort drains. Drainage trenches infilled with gravel and open to the surface may permit infiltration of water, and if the efflux is obstructed, they can operate in the opposite sense to which they were intended. Although drains may accelerate the equilibration of lowered excavation-induced pore pressures to their final state, that state is usually better for stability than if the drains were not there, and the net effect is beneficial overall.

Significant drainage actions lead to consolidation in the ground and surface deflections and strains. Drainage water collected in a pipework system has the capacity to do damage if it is released in the wrong place. Transporting drainage water across or through unstable slopes is not usually without risk of promoting failure should the drains leak.

Analysis of slope stability with drains is simply a question of modifying the pore pressures used in the analysis.

23.4.4 Anchorages and reinforcements

Active systems of restraint can be installed using pre-tensioned ground anchors. These employ cable anchorages running in boreholes through a slope into stable ground underneath where they are anchored, with tensioning loads applied at the ground surface. Pre-tensioned anchors may need periodic restressing if the ground consolidates, and in the case of some types of ground movement. They are highly susceptible to corrosion, and require elaborate protection systems. Furthermore, the existence of large cable tensions in ground anchors may give rise to long-term safety concerns – for example, if they are forgotten and the site is redeveloped in ignorance. It is rarely possible to install pre-tensioned ground anchors in their mathematically optimum orientation and position (their length, access to anchor head positions for installation and restressing, etc., being key constraints). They are often installed to control ground movement in their vicinity even if they do not stabilise the slope as a whole. It is preferable to spread the load from anchors into the ground with individual pads, as this minimises the risk of progressive failure of the whole anchorage system (**Figure 23.4**).

Plastic and metal grids or strips can be incorporated into fill slopes to retain the face, where this is oversteep, or to prevent lateral spreading (sliding on sub-horizontal planes). They are also used to restrain the sliding tendency of the face of a fill on sidelong ground. Since the capacity and durability of these systems depends on their manufacturer, it is conventional to employ manufacturer-specific design tools and construction specifications. Reinforcement grids may be installed in multiple layers, but then care is needed to ensure that the layers do not slide over each other. Loads are induced in these reinforcement systems by deformations during construction, which makes them a passive system.

Passive restraints can be induced in other reinforcing elements in the slope and are usually inserted into drill holes. Where they are of comparatively small diameter and often low angle they are called 'soil nails'. Soil nails are essentially small-diameter steel-reinforced concrete 'piles', which act in tension. Piles may also be employed to reinforce the soil mass in a slope. They may take the form of dowels (rows or grids of medium-diameter piles reinforced with traditional reinforcement cages of bars), or larger-diameter piles reinforced with steel tubes (**Figure 23.5**), or structural steel I- or H-sections. Bored piling techniques are frequently preferred to driven piling techniques, as they prevent dynamic loading of the ground and employ larger sections than can be conveniently driven.

Trenches that go deep enough to penetrate sheared surfaces or other zones of weakness and which are subsequently filled with concrete or compacted rockfill provide shear keys. As noted above, if they are permeable, they can provide some drainage.

Sheet piles may prove effective to stabilise small slides, but they are intrinsically flexible, susceptible to corrosion, and come apart at the interlocks (**Figure 23.6**).

Figure 23.4 Failed slope with planar slide which occurred largely as a result of failure of the ground anchor system (Courtesy of Professor J. J. Hung; all rights reserved)

Figure 23.5 Riverbank stabilisation in the Ironbridge Gorge, Shropshire: construction of piles reinforced by steel tubes by Birse Civils (contractor) and Jacobs (engineer)

Figure 23.6 Failed sheet pile retaining wall, originally installed in association with grading, drainage and sea defences, to stabilise compound slip with basal shear surface at approximately mid-slope height

In analyses of slopes with reinforcing elements, the necessary forces and reactions can be considered in the equilibrium calculation. It is usually a serious error to incorporate soil zones with 'concrete' or 'steel' properties into a slope stability analysis dataset and then run slip surfaces through it.

23.5. Factors of safety in slope engineering

23.5.1 Risk

Risk is defined as the potential loss from a hazard. It is the product of the probability of the damaging occurrence (in a given place and timescale) multiplied by the consequences when it happens. *When* is used, not *if*, as any probability of failure other than zero implies that sooner or later it *will* happen. Those consequences are most easily expressed in financial terms, and reflect the number and value of elements at risk and their vulnerability, or the proportion of their value lost in the damaging event(s). Fortunately, in the UK, fatalities and injuries from slope instability are comparatively rare. Apart from one small event, the entire record shows that financial effects far outweigh personal injury. This makes most risk analyses simply a matter of money. Other parts of the world are less fortunate and must find ways of dealing with the two aspects concurrently.

The entire gamut of engineering slope stabilisation measures deals with risk by reducing the probability of occurrence of damaging events (slope failures). Relocation of assets and control of access to sites reduces the number and value of elements at risk. Barriers and control structures may be seen as reducing the vulnerability of those elements. Hence, the treatment of slope instability (actual or potential) is a matter of risk reduction (Bromhead, 1997, 2005).

23.5.2 Factor of safety approach

In an ideal world, every slope might be approached with the intention to reduce the risk from failure to tolerable levels. In practice, engineering calculations rely on factors of safety, and these (as articulated above) represent a factoring down of shear strength parameters. Thus the traditional methods of slope stability engineering fit neatly with Eurocode 7's philosophy. The recommended factors in Eurocode 7 are believed to result (in most everyday practice) in tolerable levels of risk, provided that appropriate assessments of parameters and mechanisms are made in the first instance. There are almost certain to be cases where larger factors are required – for example, where the consequences of failure are dire, or where uncertainties in soil behaviour or properties demand it. Natural slopes of large scale may require smaller factors of safety (or those recommended by Eurocode 7 may simply be unachievable) for an equivalent level of security. For example, residual strength on a very deep sheared surface is unlikely to be subject to further reduction, and this may not require factoring in the same way as the unsheared strength that may be sensitive or subject to degradation with strain. However, a highly sensitive soil may be subject to progressive failure at a mobilisation factor that is satisfactory where the subsoil is insensitive.

23.5.3 Serviceability limits

In the case of the routine design of earthworks of moderate height with conventional fills and placement methods, the Eurocode 7 factors also provide some control over deflections associated with the mobilisation of shear resistance in the ground (but not, of course, to consolidation settlement and related mechanisms). In these types of earthwork, adoption of lower factors of safety gives rise to ground deformations as shear strength is mobilised. The surface manifestations include toe bulging and cracking at the embankment crest. If the earthwork consolidates and gains strength, it may be possible to give the embankment a cosmetic 'makeover' without permanent impairment of its performance. However, such deformations are more commonly experienced as a precursor to collapse, probably because the soils involved are susceptible to progressive brittle failure, and as a result should be treated with caution. This is especially so if the embankment fill or subsoil is clayey and can develop slickensided shear surfaces at residual strength.

In-slope geotechnical instrumentation can detect the onset of ground deformations before they are apparent to the naked eye. Careful inspection and interpretation of geotechnical instrumentation readouts is therefore an essential adjunct to construction, especially of particularly high or steep earthworks, those on weak subsoils and where the consequences of collapse are critical (from the perspective of collateral damage) and affect the construction programme.

23.6. Post-failure investigations

Post-failure investigations are conducted from a variety of motivations, from facilitating remediation through to attributing blame. In clay soils, an essential element of investigation is to ascertain the location of the basal shear surface of the slide. This may be done by visual examination of borehole samples, or, when the slide is still moving, by the use of appropriate instrumentation. Inferences on slip surface position and depth may be drawn from the surface morphology and surface deformations, but logged boreholes and instrumentation provide the best information. Shafts, adits and pits are used to expose shear surfaces in situ, but they are potentially hazardous, requiring proper support at all times, and they should only be entered by trained operatives acting within an appropriate safety plan.

Many failures occur under climate-driven or other transient pore water conditions; using post-failure instrumentation it may still not be possible to recover the situation.

In operation, failures in infrastructure earthworks and in both temporary and permanent slopes may prove disruptive and inconvenient to leave for later forensic investigation. It is therefore essential that proper investigations take place concurrently with the clear-up operations. During construction works the costs of remediation of some slope failures may be significantly less than after completion of the works – as plant is on site and access is usually better. However, as failures are not normally contemplated at the design stage, it may be necessary to implement remedial works with additional care; they may well contain activities not catered for in the initial site health and safety plan.

It is sometimes the case that an engineer investigating a failed slope needs to form an opinion on the cause or causes of failure. In this situation, the failure is often the result of several preparatory factors, combined with a final trigger. It is naive to single out the final trigger as the causative factor. Instead, the individual preparatory factors should be investigated and ranked in decreasing order of impact on the slope. It is then usually clear that several factors in combination have led to the failure.

References

Bishop AW (1955) The use of the slip circle in the stability analysis of earth slopes. *Géotechnique* **5**: 7–17.

Bromhead EN (1992) *Stability of Slopes*, 2nd edn. Blackies, London, UK.

Bromhead EN (1997) The treatment of landslides. *Proceedings of the ICE, Geotechnical Engineering* **125(2)**: 85–96.

Bromhead EN (2005) Geotechnical structures for landslide risk reduction. In *Landslide Hazard and Risk* (Glade T, Anderson M and Crozier M (eds)). Wiley, New York, NY, pp. 549–594.

Cruden DM and Varnes DJ (1996) Landslide types and processes. In *Landslides: Investigation and Mitigation* (Turner AK and Schuster RL (eds)). Transportation Research Board, Washington, DC, pp. 36–75.

Dikau R, Brunsden D, Schrott L and Ibsen M-L (1996) *Landslide Recognition*. Wiley, New York, NY.

Duncan JM and Wright SG (2005) *Soil Strength and Slope Stability*. Wiley, New York, NY.

Morgenstern NR and Price VE (1965) The analysis of the stability of general slip surfaces. *Géotechnique* **15**: 79–93.

Sarma SK (1979) Stability analysis of embankments and slopes. *Proceedings of ASCE, Journal of the Geotechnical Engineering Division* **105**: 1511–1524.

Spencer EE (1967) A method of the analysis of the stability of embankments assuming parallel inter-slice forces. *Géotechnique* **17**: 11–26.

All chapters within Sections 1 *Context* and 2 *Fundamental principles* together provide a complete introduction to the Manual and no individual chapter should be read in isolation from the rest.

doi: 10.1680/icemge.66816.0285

CONTENTS

Chapter 24
Dynamic and seismic loading of soils

Jeffrey A Priest University of Calgary, Canada

There are many instances where soils experience dynamic loading, either through natural forces or human-induced forces. Although the magnitudes of these dynamic loads are often much smaller than static loads, the cyclic nature of the loading introduces inertial forces that become important and must be considered in geotechnical design. This chapter introduces the concept of dynamic loading of soils and the main characteristics that differentiate dynamic and static loading: loading frequency and loading cycles, and the strain-dependent behaviour of soil. The various methods that can be employed for characterising the dynamic behaviour of the soil are outlined, differentiating those that employ low- or high-strain cyclic loading, both in field measurements and laboratory testing. The theory that captures the dynamic response of the soil to cyclic loading is briefly examined with the main features highlighted. The main factors that influence the dynamic behaviour of soil are explored, with reference to results from the extensive laboratory testing that has been conducted in this area. The phenomenon of 'soil liquefaction' is also highlighted, along with the factors that influence it.

24.1. Introduction

In geotechnical engineering design the engineer is predominantly concerned with soil strength and stiffness so that they can define the failure state of the soil and/or control excessive deformation of the soil/structure. In these circumstances, the loading applied to the ground is considered static and the strain imposed within the soil may vary from about 0.1% (in service) to a few per cent (at failure).

There are many situations, however, where the loads applied to the soil are dynamic (cyclic) in nature and can arise through natural forces such as earthquakes (seismic), wind and water, or from human-induced forces such as bomb blasts, traffic loads (planes, trains and automobiles) and machine foundations. The magnitudes of these dynamic loads are generally much smaller than the static loads considered in design and can generate strains within the soil as low as 10^{-4}%. However, during cyclic loading the soil is subject to inertial forces, which increase proportionally with the square of the loading frequency and may become important. It is therefore imperative for the geotechnical design engineer to understand the dynamic behaviour of soils in circumstances where the soil may be subject to these loading events.

This chapter introduces the dynamic behaviour of soils and outlines some of its main features. References are given where more detailed information can be obtained if required.

24.1.1 Dynamic loading

Dynamic loading can be defined by two main characteristics, the loading frequency and the number of load cycles (**Figure 24.1**). Loading is generally considered dynamic when the load is applied at frequencies greater than around 1 Hz (one cycle per second). Depending on the loading event, the number of cycles can vary from single loads, such as those generated by bombs or blasting, up to many thousands of cycles as in machine foundation vibration. Problems with dynamic load can be classified into three general areas based on the number of load cycles: (*a*) shock loading, such as bomb blasts where a single impulse of short duration occurs; (*b*) vibration or wave propagation, generally associated with earthquake events and construction

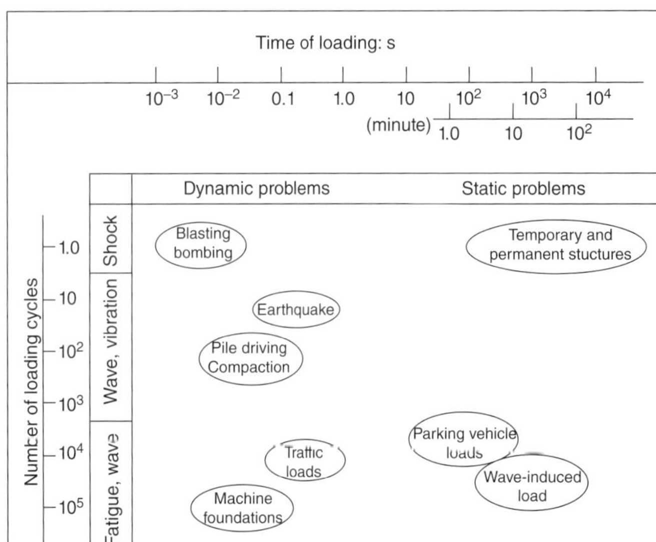

Figure 24.1 Classification of dynamic problems (Reproduced with permission from Ishihara, 1996; © Oxford University Press)

activities where tens to thousands of loading cycles occur at frequencies of 1–100 Hz; (*c*) where the number of cycles is inordinately large, such as those associated with machine foundations and traffic loading (cars and trains). In this case the soil behaviour can be considered as potentially *fatigue* related.

24.1.2 Strain-dependent behaviour of soils

Since the 1970s it has been shown that the stiffness and therefore the deformation behaviour of soils is heavily dependent on the shear strain (γ) to which the soil is subjected. **Figure 24.2** shows the general trend for the shear strain-dependent behaviour of soils. Typically, for soils subject to very small strains (10^{-4} to 10^{-2}%), such as those induced by wave propagation and vibration, the soil response is considered elastic and recoverable (i.e. there is no permanent deformation). At these strains, both the number of loading cycles and the rate of loading have negligible effect on the soil behaviour. In the intermediate strain range (10^{-2} to 1%) the soil is elasto-plastic and irrecoverable permanent deformations occur, usually associated with cracks and differential settlement of structures supported on the soil. At these strains, the number of loading cycles and the rate of loading both have an effect. These strain levels are typical for geotechnical structures such as retaining walls, tunnels and deep excavations (Burland, 1989; Mair, 1993). At higher strains (>1%) the material exhibits failure, where the strains increase with no further increase in shear stress. These are typically associated with slope failure, compaction and liquefaction of non-clay soils.

Also listed in **Figure 24.2** are typical test methods that can be employed for measuring dynamic soil properties along with an indication of the strains that each method can impose on the soil. Historically, there was little crossover between dynamic tests and those defined as static (<1 Hz frequency) with respect to the shear strain applied. Historical differences in stiffness measured using dynamic and static tests are now generally assumed to have resulted from differences in the degree of strain imposed on the soil. With field measurements using vibration or wave propagation techniques, large amounts of energy are required to generate strains in excess of 0.1% due to the inertial forces involved in mobilising the soil at the applied frequency. Therefore, for tests at higher strains, low-frequency repeated load tests (where inertial forces are negligible) are more commonly used. In these circumstances, the loading becomes a repeated static load (Ishihara, 1996). Large-strain vibration tests are similarly difficult to achieve in the laboratory. Although sufficient energy may be available to induce large strains in a soil sample during vibration, the governing equations become invalid (see Richart *et al.*, 1970). With repeat load tests at large strains, the determination of elastic constants is straightforward. It is therefore important to specify an appropriate test method to derive accurate parameters for use in the design process.

24.2. Wave propagation in soil

The test methods highlighted in **Figure 24.2** can be broadly grouped into low-strain and high-strain measurements. At high strains, the stiffness and damping are calculated directly from the stresses and strains measured in the soil as in a static test (see Chapter 49 *Sampling and laboratory testing*). However, difficulties in measuring small deformations within the soil to accurately compute strains have led to vibration/wave propagation techniques being employed for testing at small strains. Although recent advances in measuring small deformations in soils have allowed direct measurements of soil properties at small strains (Clayton and Heymann, 2001; Jardine *et al.*, 1984;

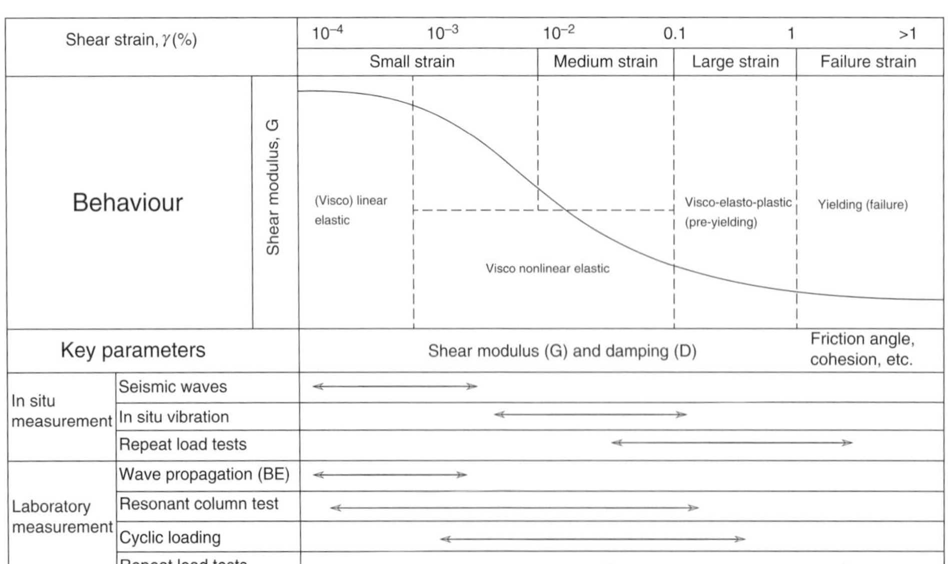

Figure 24.2 Strain-dependent properties of soils. BE, bender element, a laboratory-based wave propagation technique (Modified with permission from Ishihara, 1996; © Oxford University Press)

Witowski, 2018; Xu *et al.*, 2007), they are not routinely undertaken in a commercial soil testing environment.

24.2.1 Wave velocity

Soil parameters are computed from wave propagation techniques based on the theoretical response of the soil during the propagation of a stress wave through it. A variety of waves can propagate in an elastic material (see Richart *et al.* (1970) or Kramer (1996) for a detailed theoretical background to ground-borne wave propagation). For soils, three types of waves are readily encountered and are of importance. The first two are body waves that propagate within the soil and comprise the compressional wave (P-wave) and shear wave (S-wave), while the third wave is the Rayleigh wave, which travels along the ground surface.

For near-surface saturated granular soils (which is likely to be the type of soil encountered for a large number of geotechnical problems), the compressibility of the soil matrix is orders of magnitude greater than the compressibility of the pore fluid. This leads to the P-wave velocity (V_p) of the saturated soil being close to that of the pore fluid. If the soil is unsaturated, V_p can range from that of the soil matrix (with no pore fluid) to that for the saturated case. Therefore, the use of V_p to determine the properties of soils is problematic, even with detailed knowledge of the hydrological regime. In contrast, the velocity of S-waves (V_s) is only influenced by the soil matrix since the pore fluid cannot carry shear stresses. From the theory of propagating waves, it can be shown that the shear wave velocity through the soil is related to the stiffness of the soil by

$$G = \rho V_s^2 \qquad (24.1)$$

where G is the shear modulus of the soil and ρ is the soil density.

Therefore, measuring changes in V_s allows the shear stiffness of a soil to be determined, and measurement of S-waves is a commonly employed technique for determining the dynamic behaviour of soils at small strains.

Rayleigh wave measurements have gained in importance in recent years since ground surface measurements can be readily and easily undertaken without the need for any intrusive investigations, such as those required for measuring P-waves or S-waves. There is no direct link between the velocity of Rayleigh waves (V_R) and soil stiffness (Young's modulus, E or G); however, it has been shown that for nearly all values of Poisson's ratio, $V_s \approx 1.09V_R$ (Richart *et al.*, 1970). Therefore, measurements of Rayleigh waves can give a good correlation with shear wave velocities (Hiltunen and Woods, 1988), from which G can be derived.

24.2.2 Damping-wave attenuation

The theory of wave propagation is usually considered in terms of linear elastic materials, such that the waves would travel indefinitely without loss of energy (wave amplitude). However, in real soils the propagating waves are attenuated due to inherent material damping of the soil and geometric damping, where geometric damping relates to the loss of energy as the wavefront increases in length as it radiates out from the source. Material damping is associated with dissipation of energy during the passage of the stress wave because of friction, heat generation and so on. Thus, material damping is also a key parameter in determining the dynamic response of a soil.

24.3. Dynamic measurement techniques

As previously highlighted, the key parameters for understanding the dynamic behaviour of soils are shear modulus (G) and damping (D). Although other parameters, such as Poisson's ratio and density, influence the dynamic behaviour of soils, these are generally secondary to G and D. A variety of field and laboratory techniques have been developed for measuring G and D, each with its own advantages and limitations. As the dynamic properties of soils are highly nonlinear with strain, the choice of which technique (or range of techniques) to adopt needs to be carefully considered with respect to the existing stress conditions within the soil and those that may be imposed.

Detailed descriptions of soil investigation techniques and the data reduction methods used to determine soil properties are given in Chapter 45 *Geophysical exploration and remote sensing* (field measurements) and Chapter 49 *Sampling and laboratory testing* (laboratory measurements). Section 24.4 lists the techniques relevant to determining dynamic soil properties.

24.3.1 Field measurement techniques

There are a number of well-known advantages in performing dynamic field tests. These include: the ability to measure soil properties under in situ stress conditions; no sample disturbance to be accounted for during testing; and the ability to survey large volumes of soil rather than measuring properties at discrete locations (laboratory sampling). However, these must be considered in the light of certain disadvantages, such as: working stress conditions cannot easily be applied to the soil; hydrological conditions cannot be controlled (pore water drainage); and for most tests, material properties are not measured directly but inferred (either theoretically or empirically).

24.3.1.1 Low-strain tests

A number of low-strain field measurement techniques can be used to determine dynamic soil properties and all utilise the measurement of velocity of wave propagation in the soil. These include seismic reflection profiling and seismic refraction profiling – typically used to measure V_p and V_s, and continuous surface wave (CSW) or spectral analysis of surface waves (SASW) – used to measure V_R from which V_s is calculated. In these methods the instruments to measure wave velocity are placed on the ground surface. Instruments can also be inserted into pre-drilled boreholes to enable seismic cross-hole and down-hole tests to be conducted – which can be used to measure both V_p and V_s and respective damping parameters (with multiple receivers). In recent years the seismic cone penetration test (SCPT) has grown in popularity. This test is similar to the down-hole test but does not require a separate borehole since the cone is mechanically pushed into the ground with wave velocity measurements conducted as the cone advances.

24.3.1.2 High-strain tests

For most high-strain field tests, the soil stiffness is usually obtained through empirical correlation with other non-related parameters from these tests. Such tests include the standard penetration test (SPT), the cone penetration test (CPT) and the dilatometer test (DMT) (see Chapter 47 *Field geotechnical testing*). The Ménard pressuremeter test (PMT) is the exception; it can measure the soil stiffness directly through the stress–strain response of the soil.

24.3.2 Laboratory measurement techniques

In a similar manner to field measurement techniques, laboratory tests can be classified as either small- or large-strain tests. Also important in choosing a test method is an understanding of the stress regime within the soil and the likely stresses that may be imposed, since each laboratory test can represent certain aspects of the required stress path.

24.3.2.1 Low-strain tests

As highlighted above, advances in transducer accuracy and measurement techniques have allowed local strain measurements to be conducted on soil samples under static loading to determine small-strain soil stiffness and damping. However, the most common methods to determine small-strain stiffness parameters of soils are those that utilise wave propagation. The three most popular methods are the *resonant column*, *ultrasonic pulse transmission* and *bender element* tests. These tests are generally considered to be non-destructive in that the soil remains unchanged during the dynamic test.

The resonant column allows stiffness to be determined for applied strains ranging from $\sim10^{-4}$ to $\sim0.1\%$. Different modes of operation have allowed different soil parameters to be determined, such as G and E, as well as their respective damping (see Cascante *et al.*, 1998; Clayton *et al.*, 2005). The resonant column method utilises the theory of vibration of a linearly visco-elastic cylindrical rod to quantify the velocity of propagation. In a test the soil specimen is vibrated over a range of frequencies to identify the resonant frequency of the soil from which the wave propagation velocity can be derived. In contrast, the ultrasonic pulse transmission and bender element tests directly record the travel time of a wave that is generated by a source, travels through the soil and is recorded by a receiver. Typically the ultrasonic pulse transmission test will generate P-waves, while the bender element test generates S-waves. For these two tests the strain imposed upon the soil is reasonably constant at $\sim10^{-4}\%$. Both of these tests can be incorporated into conventional soil testing apparatus such as the conventional triaxial apparatus (Viggiani and Atkinson, 1995), the simple shear device, oedometers (Ayala *et al.*, 2017; Lee *et al.*, 2020) and others.

24.3.2.2 High-strain tests

For laboratory samples subject to high dynamic shear strains, the stiffness of soils is derived from the stresses and strains measured during a test. Soils tested in drained conditions are likely to exhibit changes in volumetric strain, while those tested in undrained conditions (where volumetric strains are prevented) exhibit changes in pore pressure (thereby changing the effective stress). Thus, for high-strain tests, measurements of pore pressure and/or volume change are required to accurately determine soil stiffness and damping.

A number of different devices are used to measure stiffness at high strains, including the *cyclic triaxial*, the *cyclic direct simple shear* and the *hollow cylinder cyclic torsional* tests. All tests measure the same basic parameters; however, there are differences between the tests which relate to how well they represent the shear stresses within the ground. Although the cyclic triaxial is the most commonly used apparatus for measuring the dynamic soil properties at high strain, it is the least accurate in reproducing the stress conditions that exist during dynamic loading. This is because the principal stresses are constrained to act in the vertical and horizontal directions. The cyclic direct simple shear test overcomes this shortfall by deforming the soil in a similar way to that during propagation of shear stress in the ground. However, shear stresses are only applied on the top and bottom surfaces of the sample. The hollow cylinder cyclic torsional test overcomes these shortcomings, and the stress conditions reproduce those under dynamic loading more closely.

24.4. Dynamic soil properties

24.4.1 Theoretical behaviour of soils

Complex geotechnical design problems focusing on soil–structure interactions can now be solved using a variety of numerical modelling techniques. In addition, advances in desktop computing power have allowed these problems to be routinely solved by most practising engineers if a model of the main characteristics of soil behaviour under cyclic loading is provided (see Chapter 6 *Computer analysis principles in geotechnical engineering*).

As discussed previously, at very small strains ($10^{-4}\%$) the behaviour of the soil is generally considered to be elastic in nature, such that shear modulus is a key parameter. At intermediate strains, from 10^{-3} to 0.1%, the soil behaviour becomes elasto-plastic, with the shear modulus reducing with increasing strain. In addition, energy is dissipated during each cycle of stress due to slippage of particles at grain contacts. The energy loss (damping) is rate-dependent and hysteretic in nature. Thus, the dynamic behaviour of soils can be fairly accurately described using nonlinear visco-elastic theory (it should be noted that the material damping within the soil is not viscous in nature, but it is mathematically convenient to assume so; see Thomson (1988) for a mathematical derivation of viscous damping). The general hysteretic behaviour of a nonlinear visco-elastic material during a stress cycle is illustrated in **Figure 24.3**. The stress–strain curve can be considered as being composed of two curves: (*a*) the backbone/skeleton curve which represents the montonic loading, and (*b*) the hysteretic loop. The shear modulus at small strains, G_{max} can be obtained from the initial gradient of the backbone curve, while the secant shear modulus at any given strain is given by

$$G = \frac{\tau_a}{\gamma_a} \tag{24.2}$$

where τ_a and γ_a are the shear stress and shear strain, respectively, at the given point in the loading cycle.

The inherent damping of the soil is related to the energy dissipated during a loading cycle and is defined as

$$D = \frac{\Delta W}{4\pi W} \tag{24.3}$$

where D is the damping ratio, W is the total stored energy (given by the area under the stress–strain curve) and ΔW is the energy loss per cycle.

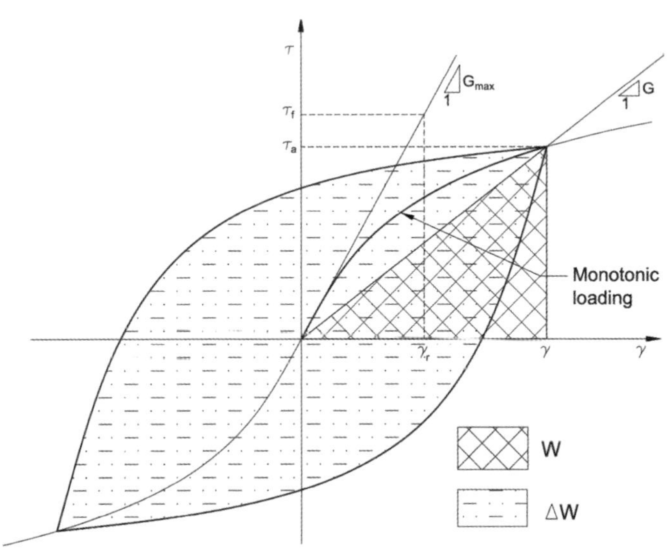

Figure 24.3 Stress–strain relationship for hysteretic soils

Numerous constitutive models have been developed to describe the dynamic nonlinear stress–strain response of a soil, idealised in **Figure 24.3**. The simplest model considers a two-parameter hyperbolic function to relate shear stress to shear strain (Hardin and Drnevich, 1972a), given by

$$\frac{G}{G_{max}} = \frac{1}{1 + \frac{\gamma_a}{\gamma_r}} \tag{24.4}$$

$$G = \frac{\tau_a}{\gamma_a} \tag{24.5}$$

where γ_r is a reference strain derived from the stress at failure from monotonic loading, τ_f, and assuming the soil behaves elastically. A drawback of the hyperbolic model is that the two parameters G_{max} and the shear strength, τ_f, can prevent the determination of damping ratio since the strain-dependent damping ratio for a soil is determined automatically, and therefore cannot be adjusted to fit experimental data. Therefore, a four-parameter Ramberg–Osgood (R-O) model can be considered for expressing the strain-dependent shear modulus of a soil under cyclic loading (Ishihara, 1996), given by

$$\frac{G}{G_{max}} = \frac{1}{1 + \alpha\left[\frac{G}{G_{max}} \cdot \frac{\gamma_a}{\gamma_r}\right]^{r-1}} \tag{24.6}$$

To ensure that τ_a does not exceed τ_f the constant α can be defined by the relation

$$\alpha = \frac{\gamma_f}{\gamma_r} - 1 \tag{24.7}$$

where γ_f is the shear strain at failure derived from monotonic loading. The constant r can be determined by taking into account the damping characteristics of the soil given by

$$r = \frac{1 + \frac{\pi D_{min}}{2} \frac{1}{1 - G_f/G_{max}}}{1 - \frac{\pi D_{min}}{2} \frac{1}{1 - G_f/G_{max}}} \tag{24.8}$$

where G_f is the shear modulus at failure given by

$$G_f = \frac{\tau_f}{\gamma_f} \tag{24.9}$$

The parameters highlighted above, namely G_{max}, τ_f, α and r, can be derived by fitting the respective models to experimental data.

24.4.2 Shear modulus and damping of soils

It has been previously highlighted that cyclic shear strain amplitude, γ_c, is a very important parameter for determining the dynamic behaviour of soils. Extensive laboratory testing, which has been undertaken to determine this behaviour, has shown that shear modulus (G) and corresponding damping ratio of soils (D) are influenced by a range of parameters, such that

$$G;D = f(\gamma_c, \sigma_0', e, N, S_r, \dot{\gamma}, OCR, C_m, t) \quad (24.10)$$

where

γ_c = cyclic shear strain amplitude
σ_0' = mean effective principal stress
e = void ratio
N = number of loading cycles
S_r = saturation ratio
$\dot{\gamma}$ = strain rate (frequency of cyclic loading)
OCR = overconsolidation ratio
C_m = grain characteristics – shape, size, mineralogy
t = time of confinement under applied load.

When γ_c is around $10^{-4}\%$ the dynamic behaviour of unconsolidated granular soils, such as sands, depends mostly on σ_0' and e, while for clayey materials OCR and t also become important. At higher strains, both clays and non-clay soils are also affected by $\dot{\gamma}$ and N.

24.4.2.1 Small-strain shear modulus, G_{max}

When soils are subjected to very small strains (generally taken as $\leq 10^{-4}\%$), the shear modulus for a given soil reaches a maximum value (see **Figure 24.2**) and is termed G_{max} or G_0. As shown in Equation 24.10 the shear modulus of a soil depends on the confining stress. **Figure 24.4** shows the typical increase in shear modulus for a range of dry granular materials, with increasing isotropic effective stress (σ'). Empirically, a simple exponential relationship is found between G_{max} and σ' of the form

$$G = A\sigma'^b \quad (24.11)$$

where A and b are constants (Hardin and Black, 1968; Hardin and Drnevich, 1972a). By fitting Equation 24.11 to the data in **Figure 24.4**, the b-value for the soil can be obtained. For non-clay soils the b-value is generally found to vary from 0.4 to 0.6 and represents both the nature of the contact stiffness and fabric change (Cascante *et al.*, 1998) during isotropic compression.

Based on Hertzian contact theory, the b-value for a closely packed array of elastic spheres is 0.33 (Duffy and Mindlin, 1957), while a value of 0.5 is achieved for a random packing of elastic particles with non-spherical contacts (Goddard, 1990).

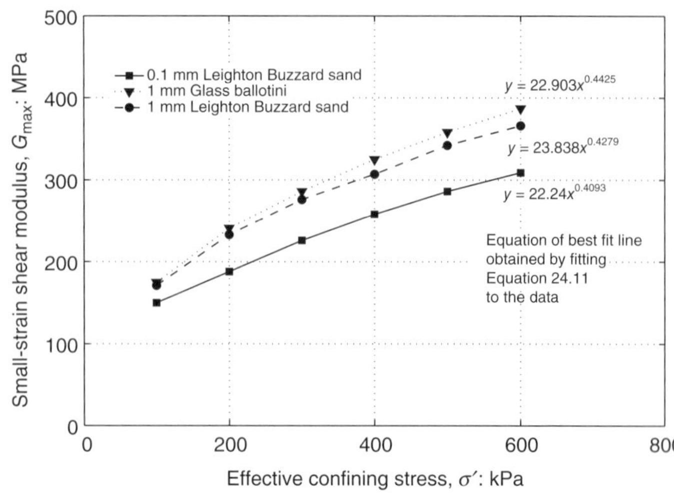

Figure 24.4 Influence of effective confining pressure (σ') on small-strain shear modulus of a range of dry granular geomaterials (Data taken from Bui, 2009)

Thus, for non-clay soils, where changes in σ_0' produce minor changes in fabric, particle shape and size (which affect e) and the contact geometry influence the relationship between G_{max} and σ_0'. However, for clay soils where changes in σ_0' can produce large changes in fabric (especially for soils with a high void ratio), the b-value may be appreciably higher than those for non-clay soils (Rampello *et al.*, 1997).

The influence of void ratio can be more clearly observed in **Figure 24.5**, which highlights changes in G_{max} with e reported in the literature for a range of geomaterials. It can be seen that as the void ratio reduces, G_{max} increases. For non-clay soils the coordination number, N_c (the number of grain-to-grain contacts per grain), increases with reducing void ratio, leading to a reduction in average contact force under a given stress regime. Thus, during dynamic loading higher void ratio soils are likely to exhibit higher contact forces, giving rise to localised yielding at particle contacts, which reduces G_{max}. To account for the influence of void ratio on shear modulus, G_{max} is often normalised using a void ratio function, F(e) to eliminate the influence of void ratio. Two empirically derived void ratio functions are commonly used, including

$$F(e) = \frac{(e_r - e)^2}{1 + e} \quad (24.12)$$

where e_r is a soil constant with values of 2.97 for angular sands and 2.17 for rounded sands (Hardin and Drnevich, 1972b; Iwasaki and Tatsuoka, 1977) for soils e in the range 0.57 to 0.98; and

$$F(e) = e^c \quad (24.13)$$

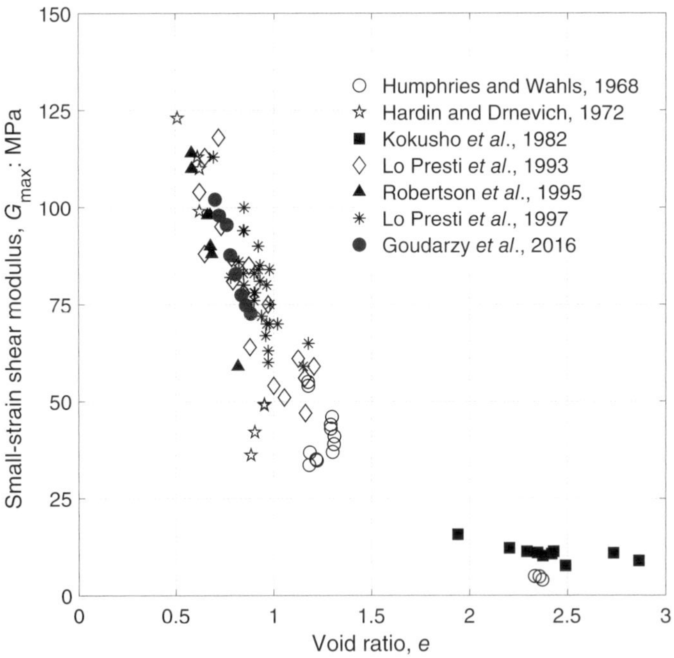

Figure 24.5 Changes in G_{max} with e for a range of soils (Data obtained for samples subjected to $\sigma_0' \sim 100$ kPa)

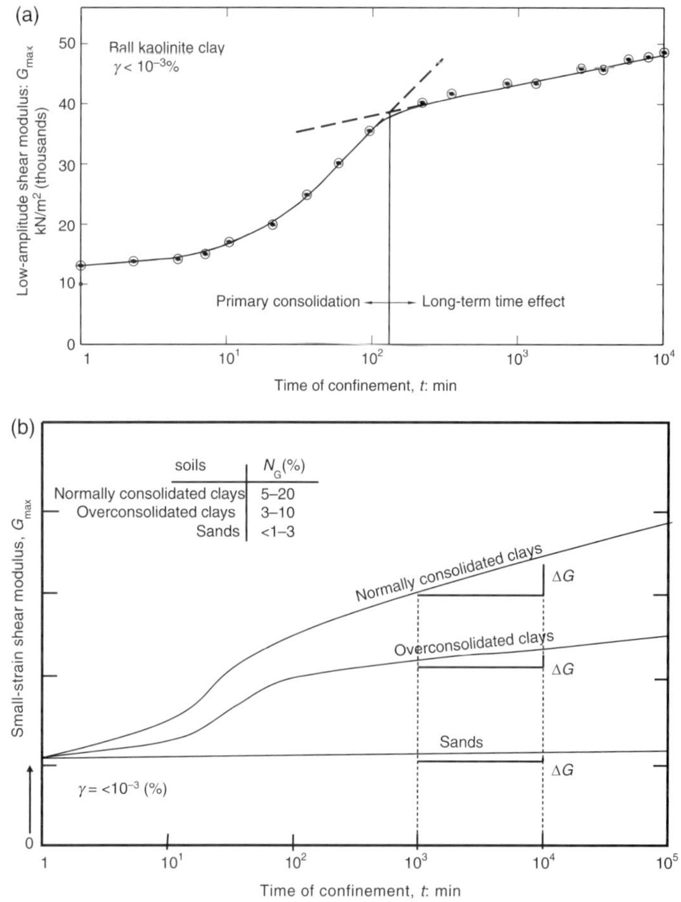

Figure 24.6 Variation in small-strain shear modulus with time of confinement for (a) ball kaolinite clay sample showing primary and secondary consolidation, and (b) a range of geomaterials (Reprinted, with permission, from ASTM STP 654 Dynamic Geotechnical Testing; Anderson and Stokoe, 1978 © ASTM International)

where c is a fitting function that was found to be -1.3 for values of e between 0.81 and 1.18 (Lo Presti *et al.*, 1997). These functions were derived over a narrow void ratio range and there are no universal void ratio functions that can account for changes in G_{max} over the wide range of void ratios usually encountered. A further factor that may need to be considered is the effect of fines content (f_c) on the behaviour of sands, where increasing f_c leads to a reduction in G_{max} even though void ratio may reduce (Iwasaki and Tatsuoka, 1977; Salgado *et al.*, 2000; Wichtmann *et al.*, 2015). To account for this, an equivalent granular void ratio, e^*, can be considered to replace e, which is dependent on the ratio of f_c to sand grains within the soil (Goudarzy *et al.*, 2016).

Increasing *OCR* at a given value of σ_0', primarily for clay soils, will increase G_{max}. This can be related to irrecoverable reduction in void ratio for a given soil as *OCR* increases, leading to increases in G_{max}. In addition, rearrangement of particle orientation inherently changes the structure of the soil, causing an increase in G_{max} above that suggested by simply a reduction in void ratio (Kokusho *et al.*, 1982).

The influence of time on the measured shear modulus soils is related to the consolidation of the soil under applied stress (Humphries and Wahls, 1968), and can be characterised by two phases: (*a*) the initial phase which results from the primary consolidation of the soil, and (*b*) a secondary phase which results from the long-term (creep) effects. **Figure 24.6** shows the effect of confinement time on G_{max} for a variety of soils. It can be seen that for clay soils both an initial phase and a secondary phase occur (**Figure 24.6**(a)), while for sands only a secondary phase is evident (**Figure 24.6**(b)). In addition, changes in G_{max} during the secondary phase for clay soils may be up to 15 times greater than that for sand. This observed behaviour has major implications for the testing of soils in the laboratory. For clay soils, testing must be conducted at the end of primary consolidation for each stress increment applied to the specimen. In addition, for all soils, testing must be conducted at equal intervals during the secondary phase for each sample.

24.4.2.2 Strain-dependent shear modulus, *G*

As has been highlighted above, the dynamic behaviour of soils is highly nonlinear, and the material properties that are important, such as shear modulus and damping, change significantly with increasing strain during cyclic loading. Clays and sands have historically been treated individually. Seed and Idriss (1970) presented results from a wide range of tests conducted on sands compiled from the literature. They showed that when the ratio of G/G_{max} is plotted against strain, the results tended to fall within a narrow band for a given confining stress.

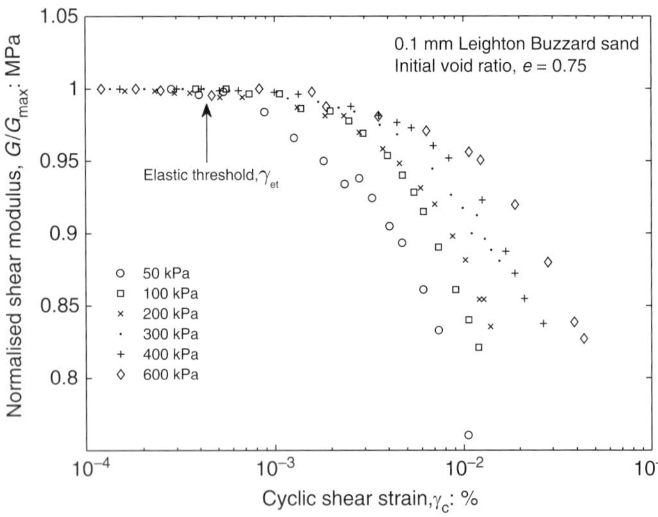

Figure 24.7 Changes in normalised shear modulus, G/G_{max}, as a function of applied cyclic shear strain at different effective confining pressures

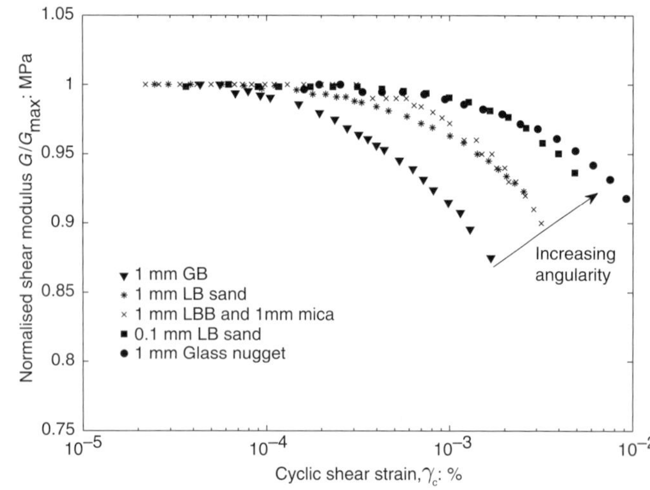

Figure 24.8 Changes in normalised shear modulus, G/G_{max}, as a function of applied cyclic shear strain at σ'_0 = 100 kPa for a range of geomaterials; increasing particle angularity and surface roughness increase the elastic threshold strain, γ_{et} (Data taken from Bui, 2009)

Figure 24.7 shows the results of resonant column tests on 0.1 mm dia. Leighton Buzzard sand. The results show a similar trend to that identified by Seed and Idriss (1970) and others, where shear modulus starts to reduce once a given strain is exceeded. The value of strain at this point is referred to as the linear elastic threshold shear strain, γ_{et}. Another observation is that the shear modulus reduction curve is influenced by effective confining pressure. At higher confining pressures, γ_{et} increases – that is. the linear elastic region occurs at higher strain and can be considered as shifting the whole curve to the right. Numerous studies (e.g. Bui, 2009; Cho *et al.*, 2006; Liu and Yang, 2018; Otsubo *et al.*, 2015) have shown that soil attributes such as particle shape and surface roughness influence G_{max} and γ_{et}. Typically, increasing particle angularity and surface roughness increased γ_{et}, while more rounded particles with low roughness such as glass ballotini gave a value of γ_{et} of around 10^{-6} (**Figure 24.8**) such that there was little or no observable G_{max}. Vucetic and Dobry (1991) reviewed data from a broad range of materials, especially those for high-plasticity clays. Their results, as presented in **Figure 24.9**, showed that γ_{et} was dependent on the plasticity of the soil and not its void ratio. They showed that as soil plasticity increased, γ_{et} increased. In addition, it was shown that the influence of effective confining pressure was reduced as soil plasticity

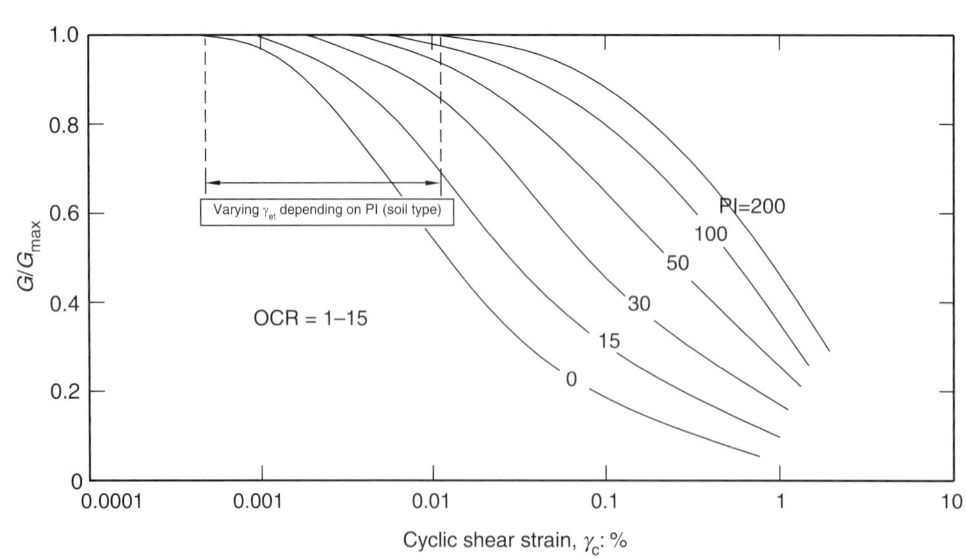

Figure 24.9 Small-strain response of clays with different values of PI (Reproduced from Vucetic and Dobry, 1991, with permission from ASCE)

increased. It was also observed that there was a high degree of similarity between the curves for soils with a plasticity index (PI) of zero, and those obtained from Seed and Idriss (1970) on sands. Coarse gravels (Seed *et al.*, 1986) show a similar response to sands, although the shear modulus reduction curve was slightly flatter.

24.4.2.3 Damping ratio

Theoretically, for soils subject to shear strain within the elastic region ($\sim 10^{-4}\%$) no hysteretic energy is dissipated, since frictional losses through grain slippage and yielding are thought not to occur (Winkler *et al.*, 1979). However, even within the elastic limit a minimum value of damping is observed; this is often termed as D_{min} and may result from fluid squirt flow at grain contacts, or from Bauschinger effects (plastic yielding of individual grains at grain contacts). Equipment damping due to the apparatus is also present and can affect measurements of material damping, hence careful calibration is required to remove its effect from measured data (Priest *et al.*, 2006). This is important at low strains where the equipment damping may be the same order of magnitude as that of the material being tested.

Studies of the factors affecting damping ratio, D, are not as extensive as those for G. In Equation 24.10, several factors were highlighted as having an effect on damping. It has been shown that D is strongly influenced by γ_c and σ_0'. **Figure 24.10** highlights the measured damping ratios as a function of both γ_c and σ_0' for the same samples whose normalised shear modulus was presented in **Figure 24.7**. D reduces for increasing σ_0' and increases for increasing γ_c. As σ_0' increases, slippage at particle contacts is restricted thereby leading to a reduction in D, while at higher strains the shear stress at particle contacts is increased, leading to increased frictional losses.

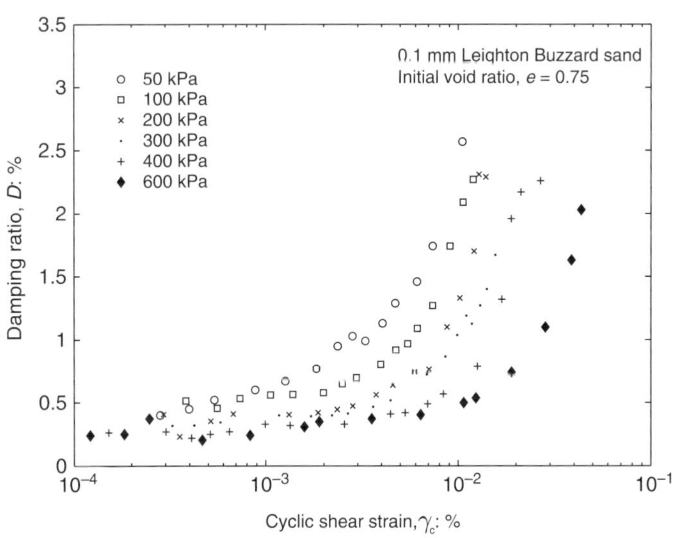

Figure 24.10 Influence of confining stress on the strain-dependent damping ratio for Leighton Buzzard sand

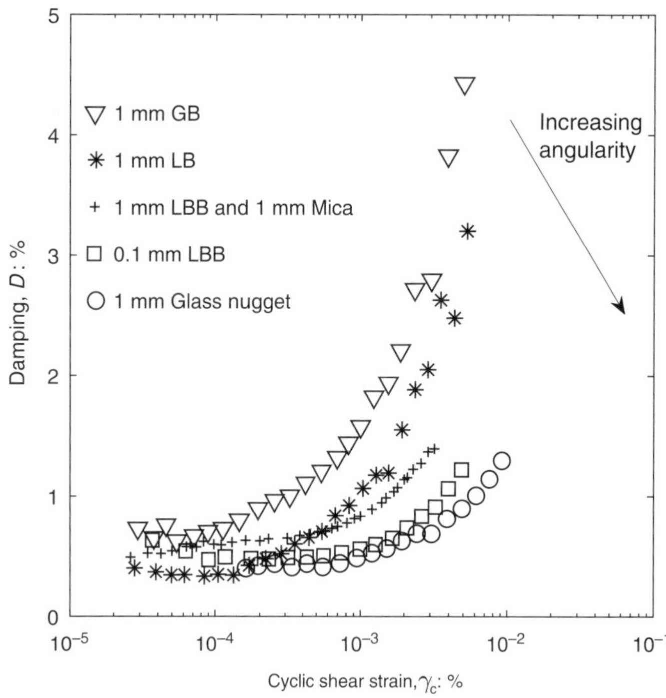

Figure 24.11 Influence of particle shape and roughness on the strain-dependent damping ratio for a range of geomaterials at σ_0' = 100 kPa (Data taken from Bui, 2009)

Figure 24.11 highlights the influence of particle angularity on the measured damping ratio for a range of geomaterials. The results show that increased angularity leads to reduced damping for the range of geomaterials tested. γ_c and σ_0' have major effects on D, but the time of confinement and e have only minor influences. The number of loading cycles is important at high-strain amplitudes, but becomes less important as the induced strains become small.

24.5. Liquefaction of soils

For soils subject to small cyclic strains, in the range 10^{-4} to $\sim 0.1\%$, the deformation of the soil is negligible during cyclic loading. However, at higher strains, a phenomenon called soil liquefaction can occur in soils, and in particular in sands. During cyclic loading (such as that caused by an earthquake), the large shear strains developed in the soil can lead to frictional sliding and yielding at grain contacts. These lead to changes in soil volume, which for loose sandy soils can produce a build-up of pore pressure and subsequent reduction in effective stress. In extreme cases the effective stress can become zero (suggesting no grain-to-grain contacts), transforming the soil from a solid to a liquefied state (Youd and Idriss, 2001). Thus, large shear deformations can be induced over short timescales and can lead to catastrophic failure of structures located on these soils (**Figure 24.12**). This dramatic change in behaviour is usually manifested in loose to moderately dense sands capped by impermeable layers, or in silty sands that have poor drainage

Figure 24.12 Aerial view of leaning apartment houses in Niigata, Japan, produced by soil liquefaction during the 1964 earthquake (Photograph in the public domain, courtesy of NOAA/NGDC; www.ngdc.noaa.gov)

properties; both prevent the high pore pressures induced in the soil from dissipating rapidly. For moderately dense to dense soils, liquefaction causes transient softening of the material and increases in shear strain. However, the tendency for dense soils to dilate during shearing inhibits major strength losses and restricts the development of large ground deformations. Although liquefaction has been predominantly reported in sands and silty sands, significant ground deformations have also been induced in fine-grained soils (Martin *et al.*, 2004; Tsai *et al.*, 2022). Boulanger and Idriss (2006) suggest that fine-grained soils can be classified as 'sand-like' (PI < 7) where the soil is liable to suffer liquefaction, or 'clay-like' (PI > 7) where the soil is liable to suffer cyclic softening. The terms 'liquefaction' and 'cyclic softening' do not imply strong differences in the observed stress–strain response of the soil during cyclic loading. They are used to identify soils where the mechanical behaviours are very different from each other and so require different procedures to be adopted to evaluate their seismic behaviours (Boulanger and Idriss, 2004, 2007).

The capacity of a soil to resist liquefaction is quantified by the cyclic resistance ratio (*CRR*). From laboratory tests the *CRR* is defined as the point where the ratio of shear (τ) and mean effective (P') stresses produces a 5% peak-to-peak amplitude shear strain, or when the build-up of pore pressure in the soil equals that of the mean effective stress (Ishihara, 1996). To constrain the tests, the number of load cycles to achieve liquefaction is required. An arbitrary value can be taken if relevant correction factors are applied, but given the number of significant cycles present in actual time histories, a value of 10–20 cycles is usually appropriate.

Figure 24.13 shows the typical response of saturated loose/medium-dense sand during cyclic loading under a constant shear stress. For each successive stress cycle a corresponding increase in pore pressure occurs. As the pore pressure increases and approaches that of the applied mean effective stress, large strains are manifested. Factors that influence the onset of liquefaction include initial confining stress, cyclic shear stress (intensity of shaking) and initial void ratio (relative density). In the application of constant shear stress, factors such as soil fabric are inherently important with regard to the pore pressure response. Therefore, tests using constant shear strain amplitude are preferred for remoulded sands, since the pore pressure response under constant strain is not significantly affected by the disturbance. **Figure 24.14** highlights the response from a cyclic triaxial test for medium-dense sand

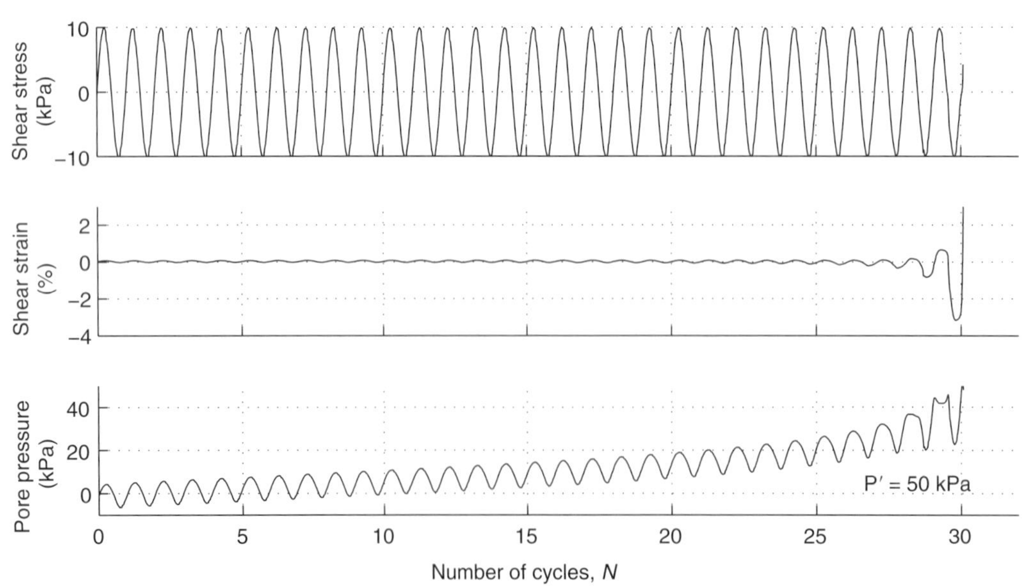

Figure 24.13 Shear stain and pore pressure response from cyclic triaxial test for applied constant stress for a loose/medium-dense sand (Data courtesy of Fugro GeoConsulting Ltd)

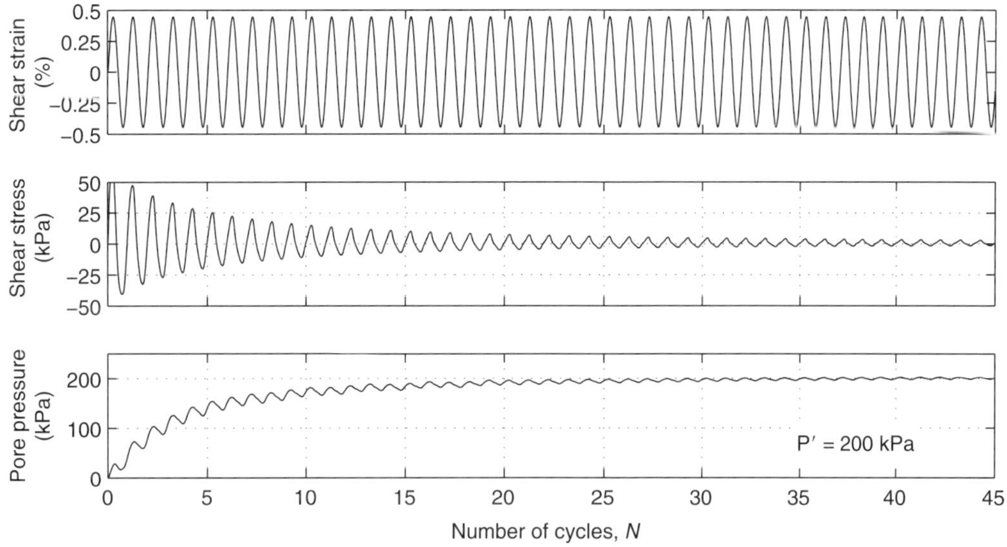

Figure 24.14 Shear stress and pore pressure response from cyclic triaxial test for applied constant strain (Data courtesy of Fugro GeoConsulting Ltd)

with constant strain amplitude. As for constant shear stress test, pore pressure increases for each cycle, with the shear stress reducing to maintain the constant shear strain amplitude. The stiffness response of the soil during constant strain amplitude can easily be incorporated into earthquake response programmes.

Due to the inherent difficulty in obtaining undisturbed granular soil samples for testing and these soils being the most likely to liquefy, field tests are routinely carried out to assess liquefaction potential using techniques such as the SPT and the CPT, as well as measurement of shear wave velocity, V_s. Correlation curves have been developed over the years to relate field results to the laboratory-derived *CRR*. Various corrections are required for these curves to take account of factors such as fines content, soil plasticity, soil type and overburden pressure, as well as test-specific correction factors. A detailed discussion of these can be found in Youd and Idriss (2001), Idriss and Boulanger (2004) and Boulanger and Idriss (2016).

24.6. Summary of key points

- The ground on which structures are founded can be subject to dynamic loads from both natural (earthquake, wave, water) and man-made sources (vibrating machinery, traffic, bomb blasts), which generate strains in the soil (10^{-4} to 0.1%) that are generally much lower than most static loads.
- Dynamic loading can be characterised by the frequency of the applied load (generally greater than 1 Hz) and the number of load cycles, which can vary from 1–2 (bomb blasts) up to possibly millions (traffic loading and vibrating machinery).
- All soils exhibit nonlinear stress–strain behaviour where permanent, non-recoverable deformation of the soil occurs. At very low strains (10^{-4} to 10^{-2}%) soil behaviour could be described as elastic, with recoverable deformations.
- Dynamic soil properties can be ascertained using field or laboratory techniques through either determining the velocity of wave propagation through the soil, or by direct measurement of the stress–strain response of the soil. The measurement techniques adopted will depend on what soil parameters are required and at what strains the measurements can be determined.
- The nonlinear behaviour of soils can be described using several nonlinear constitutive models, which can be adopted in numerical analysis. The key parameters for these models, such as very small-strain shear modulus (G_{max}), and shear strength (τ_f) can be derived from experimental data.
- The shear modulus and damping measured in soils are influenced by a range of parameters, with the most important parameters being cyclic shear strain amplitude (g_c (%)), effective confining stress (σ') and void ratio (e). For clay soils both *OCR* and time of confinement (t) have also been shown to be important.
- At strains greater than 0.1%, frictional sliding and yielding at grain contacts can lead to changes in volume of the soil producing a build-up of pore pressures and subsequent reduction in effective stress, allowing large shear deformations to occur in the soil. This behaviour is usually observed in loose to moderately loose sands and is termed liquefaction. In dense sands the tendency to dilate during shearing inhibits major strength loss.
- Fine-grained soils can be classified as having 'sand-like' behaviour and being liable to liquefy, or exhibit 'clay-like' behaviour, where cyclic softening is likely to occur.

References

Anderson DG and Stokoe KHII (1978) Shear modulus: A time dependent soil property. *Dynamic Geotechnical Testing, ASTM* **654**: 66–90.

Ayala JL, Villalobos FA and Alvarado G (2017) Study of the elastic shear modulus of Bio Bio sand using bender elements in an oedometer. *Geotechnical Testing Journal* **40**: 673–682.

Boulanger RW and Idriss IM (2004) *Evaluating the Potential for Liquefaction or Cyclic Failure of Silts and Clays.* University of California, Davis, CA, Report UCD/CGM-04/01.
Boulanger RW and Idriss IM (2006) Liquefaction susceptibility criteria for silts and clays. *ASCE Journal of Geotechnical and Geoenvironmental Engineering* **132(11)**: 1413–1426.
Boulanger RW and Idriss IM (2007) Evaluation of cyclic softening in silts and clays. *ASCE Journal of Geotechnical and Geoenvironmental Engineering* **133(6)**: 641–652.
Boulanger RW and Idriss IM (2016) CPT-based liquefaction triggering procedure. *ASCE Journal of Geotechnical and Geoenvironmental Engineering* **142(2)**: 04015065.
Bui MT (2009) *Influence of Some Particle Characteristics on the Small Strain Response of Granular Material.* PhD thesis. University of Southampton, Southampton, UK.
Burland JB (1989) Small is beautiful: the stiffness of soils at small strains, 9th Laurits Bjerrum Memorial Lecture. *Canadian Geotechnical Journal* **26**: 499–516.
Cascante G, Santamarina C and Yassir N (1998) Flexural excitation in a standard torsional-resonant column. *Canadian Geotechnical Journal* **35**: 478–490.
Cho GC, Dodds J and Santamarina JC (2006) Particle shape effects on packing density, stiffness, and strength: natural and crushed sands. *ASCE Journal of Geotechnical and Geoenvironmental Engineering* **132(5)**: 591–602.
Clayton CRI and Heymann G (2001) Stiffness of geomaterials at very small strains. *Géotechnique* **51(3)**: 245–255.
Clayton CRI, Priest JA and Best AI (2005) The effects of disseminated methane hydrate on the dynamic stiffness and damping of a sand. *Géotechnique* **55(6)**: 423–434.
Duffy J and Mindlin RD (1957) Stress–strain relations and vibrations of a granular medium. *Journal of Applied Mechanics, Transactions of ASME* **24**: 585–593.
Goddard JD (1990) Nonlinear elasticity and pressure-dependent wave speeds in granular media. *Proceedings of the Royal Society A* **430**: 105–131.
Goudarzy M, Rahman MM, König D and Schanz T (2016) Influence of non-plastic fines content on maximum shear modulus of granular soils. *Soils and Foundations* **56(6)**: 973–983.
Hardin BO and Black WL (1968) Vibration modulus of normally consolidated clay. *ASCE Journal of the Soil Mechanics and Foundations Division* **94(SM2)**: 353–369.
Hardin BO and Drnevich VP (1972a) Shear modulus and damping in soils: design equations and curves. *ASCE Journal of the Soil Mechanics and Foundations Division* **98(SM6)**: 667–692.
Hardin BO and Drnevich VP (1972b) Shear modulus and damping in soils: measurement and parameter effects. *ASCE Journal of the Soil Mechanics and Foundations Division* **98(SM6)**: 603–624.
Hiltunen DR and Woods RD (1988) SASW and crosshole test results compared. *Proceedings of the Earthquake Engineering and Soil Dynamics II: Recent Advances in Ground Motion Evaluation, Park City, UT.* ASCE, New York, NY, pp. 279–289.
Humphries WK and Wahls EH (1968) Stress history effects on dynamic modulus of clay. *ASCE Journal of the Soil Mechanics and Foundation Division* **94(SM2)**: 371–389.
Idriss IM and Boulanger RW (2004) Semi-empirical procedures for evaluating liquefaction potential during earthquakes. *Proceedings of the 11th ICSDEE 3rd ICEGE, Berkeley, CA*, pp. 32–56.
Ishihara K (1996) *Soil Behaviour in Earthquake Geotechnics.* Clarendon Press, Oxford, UK.
Iwasaki T and Tatsuoka F (1977) Effects of grain size and grading on dynamic shear moduli of sands. *Soils and Foundations* **17(3)**: 19–35.
Jardine RJ, Symes MJ and Burland JB (1984) The measurement of soil stiffness in the triaxial apparatus. *Géotechnique* **34(3)**: 323–340.
Kokusho T, Yoshida Y and Esashi Y (1982) Dynamic properties of soft clay for wide strain range. *Soils and Foundations* **22(4)**: 1–18.
Kramer SL (1996) *Geotechnical Earthquake Engineering.* Prentice Hall, Upper Saddle River, NJ.
Lee J-S, Park G, Byun Y-H and Lee C (2020) Modified fixed wall oedometer when considering stress dependence of elastic wave velocities. *Sensors* 20(**21**): 6291.
Liu X and Yang J (2018) Shear wave velocity in sand: effect of grain shape. *Géotechnique* **68(8)**: 742–748.
Lo Presti DCF, Pallara O, Cavallaro A, *et al.* (1993) Monotonic and cyclic loading behaviour of two sands at small strains. *Geotechnical Testing Journal* **16(43)**: 409–424.
Lo Presti DCF, Jamiolkowski M, Pallara O, Cavallaro A and Pedroni S (1997) Shear modulus and damping of soils. *Géotechnique* **47(3)**: 603–617.
Mair RJ (1993) Developments in geotechnical engineering: application to tunnels and deep excavations. Unwin Memorial Lecture. *Proceedings of the ICE , Civil Engineering* **97(1)**: 27–41.
Martin IJR, Olgun CG, Mitchell JK and Durgunoglu HT (2004) High-modulus columns for liquefaction mitigation. *ASCE Journal of Geotechnical and Geoenvironmental Engineering* **130(6)**: 551–561.
Otsubo M, O'Sullivan C, Sim WW and Ibraim E (2015) Quantitative assessment of the influence of surface roughness on soil stiffness. *Géotechnique* **65(8)**: 694–700.
Priest JA, Best A and Clayton CRI (2006) Attenuation of seismic waves in methane gas hydrate-bearing sand. *Geophysical Journal International* **164(1)**: 149–159.
Rampello S, Viggiani GMB and Amorosi A (1997) Small-strain stiffness of reconstituted clay compressed along constant triaxial effective stress ratio paths. *Géotechnique* **47(3)**: 475–489.
Richart FE, Hall JR and Woods RD (1970) *Vibration of Soils and Vibrations.* Prentice Hall, Englewood Cliffs, NJ.
Robertson PK, Sasitharan S, Cunning JC and Sego DC (1995) Shear-wave velocity to evaluate in-situ state of Ottawa sand. *ASCE Journal of Geotechnical Engineering* **121(3)**: 262–273.
Salgado R, Bandini P and Karim A (2000) Shear strength and stiffness of silty sand. *ASCE Journal of Geotechnical and Geoenvironmental Engineering* **126(5)**: 451–462.
Seed HB and Idriss IM (1970) *Soil Moduli and Damping Factor for Dynamic Response Analyses.* Earthquake Engineering Research Center, University of California, Berkeley, CA, Report EERC 70-10.
Seed HB, Wong RT, Idriss IM and Tokimatsu K (1986) Moduli and damping factors for dynamic analyses of cohesionless soil. *ASCE Journal of Geotechnical Engineering* **112(11)**: 1016–1032.
Thomson WT (1988) *The Theory of Vibrations with Applications.* Prentice Hall, Englewood Cliffs, NJ.
Tsai C-C, Yang Z-X, Chung M-H and Shang S-Y (2022) Case study of large-scale levee failures induced by cyclic softening of clay during 2016 Meinong earthquake. *Engineering Geology* **297**: 106518.
Viggiani G and Atkinson JH (1995) Stiffness of fine-grained soil at very small strains. *Géotechnique* **45(2)**: 249–265.

Vucetic M and Dobry R (1991) Effect of soil plasticity on cyclic response. *ASCE Journal of Geotechnical Engineering* **117(1)**: 89–107.

Wichtmann T, Navarrete Hernández MA and Triantafyllidis T (2015) On the influence of a non-cohesive fines content on small strain stiffness, modulus degradation and damping of quartz sand. *Soil Dynamics and Earthquake Engineering* **69**: 103–114.

Winkler K, Nur A and Gladwin M (1979) Friction and seismic attenuation on rocks. *Nature* **277**: 528–531.

Witowski M (2018) Local displacement transducer with miniature position encoder. *Geotechnical Testing Journal* **41(6)**: 1147–1154.

Xu M, Bloodworth AG and Clayton CRI (2007) The behaviour of a stiff clay behind embedded integral abutments. *ASCE Journal of Geotechnical and Geoenvironmental Engineering* **133(6)**: 721–730.

Youd TL and Idriss IM (2001) Liquefaction resistance of soils: summary report from the 1996 NCEER and 1998 NCEER/NSF workshops on evaluation of liquefaction resistance of soils. *ASCE Journal of Geotechnical and Geoenvironmental Engineering* **127(4)**: 297–313.

Further reading

Jeffries M and Been K (2006) *Soil Liquefaction: A Critical State Approach*. Taylor & Francis, Oxford, UK.

Johnson KL (1987) *Contact Mechanics*. Cambridge University Press, Cambridge, UK.

All chapters within Sections 1 *Context* and 2 *Fundamental principles* together provide a complete introduction to the Manual and no individual chapter should be read in isolation from the rest.

Chapter 25
The role of ground improvement

Chris DF Rogers University of Birmingham, Birmingham, UK

doi: 10.1680/icemge.66816.0299

CONTENTS

Ground improvement can take many forms to cause temporary or permanent change, usually with some specific engineering purpose in mind. In common with all geotechnical design, specific engineering purposes (e.g. stable foundation) must be defined and the ground's current state in the zone of influence fully established. Analysis of the problem in the light of this information usually yields a range of design options, some of which might involve altering the ground's properties. Techniques are available to strengthen or stiffen soft or weak ground, either by physically altering its structure or by changing its chemical properties. Similarly, groundwater flow regimes can be altered to improve stability or to stop flow of contaminants. Any analysis of ground treatment must therefore encompass the engineering purpose(s), the ground's current state and its potential to be altered, the proposed means of its alteration, and any side effects of treatment. Physical ground improvement includes application of static, vibratory or dynamic loading; water can be drained using gravity, increased stresses or an electrical gradient; and chemicals can be introduced as solids or liquids. Hybrids of these processes and approaches such as ground freezing complete a range of alternatives to hard structural solutions and often provide more sustainable outcomes.

25.1. Introduction

Ground improvement covers any method by which the ground, whether natural or disturbed in some way by anthropogenic processes, has its performance for any specific (geotechnical) purpose enhanced. Typically the goal is strengthening and/or stiffening of soft or weak soil or fill, either temporarily or permanently, although other specific improvements (e.g. reduction or increase of hydraulic conductivity) might bring about the desired change or be the goal of the exercise. Equally the goal might be to make the properties worse (e.g. slurries introduced at the face of tunnelling machines to facilitate excavation and transportation of soil), while some techniques have unwanted side effects (e.g. vertical stone columns provide a route for venting gases), so the term 'improvement' should be treated with caution. Nevertheless, it can be generally stated that a ground improvement process has some associated desired physical or chemical consequence, such as

- settlements of overlying or embedded structures being reduced, speeded up or rendered unimportant
- the risk of failure being removed or reduced
- the need for excessive or unnecessary use of natural resources (import of virgin material to a site) and/or waste generation (poor ground disposed of in landfill) being obviated
- the movement of chemicals being enhanced (for extraction) or inhibited (for containment)
- the excavation of contaminated ground for treatment, thereby exposing it to the environment, being avoided.

Ground improvement for physical purposes is generally an alternative to structural solutions which either bypass the soils concerned (e.g. deep foundations) or act in spite of the soils concerned (e.g. rigid raft foundations). Ground improvement for chemical purposes might be used instead of permanent physical solutions (e.g. barriers to stop chemical contaminant migration), novel semi-permanent processes (e.g. permeable reactive barriers; see Boshoff and Bone, 2005) or costly temporary operations (such as pumping for groundwater lowering to generate an inward hydraulic gradient). Phear and Harris (2008) track developments in this subject.

There are many reasons for carrying out some form of ground improvement. As with all engineering undertakings, the costs and benefits of the alternative engineering solutions must be examined, and this should be done using the broad three-pillar (environmental, social and economic) or four-pillar (adding natural resources; see Braithwaite, 2007) sustainability models (Fenner *et al.*, 2006, 2022). This is a most important aspect of an engineer's role due to the wider impact of geotechnical processes (Jowitt, 2004) – in the category of resource use falls energy (embodied and directly used) and hence carbon dioxide emissions, water and similar concerns. Sustainability arguments are highly germane to ground improvement; if what is in place can be made adequate by some means, then the environmental and social (and resource) benefits are usually very substantial, while economic benefits are also often favourable. A good example of rather too focused – or siloed – thinking concerns the adoption of low-carbon concrete and steel as the primary route to decarbonisation of the construction industry.

The aim is inherently worthy, of course, but the application of systems thinking would include analysis of all the consequences – positive and negative – of each alternative design option (Rogers, 2018). In this case the carbon cost of ground improvement added to reduced quantities of carbon and steel could yield a lower embodied carbon outcome for the project in question. As long as the sum of the social, environmental and direct and indirect economic outcomes is favourable, ground improvement as a route to decarbonisation demands consideration (Rogers *et al.*, 2022). The civil engineering industry has now moved to a state of global awareness that requires it to go beyond the traditional 'cost, quality and time' approach to engineering design. Ground improvement alternatives to 'structural' solutions represent an excellent example of how these considerations play out.

The aim of this chapter is to introduce the fundamental principles of ground improvement techniques, the detailed manifestations of which can be seen routinely in academic or professional journals. The scope of this chapter is limited to hydraulic, physical and chemical means of improvement. Soil reinforcement by way of the introduction of tensile elements in the form of grids, strips, bars, fibres or similar is covered elsewhere (see Chapter 86 *Soil reinforcement construction*). Uncommon techniques, such as the thermal treatment of clays and the use of bitumen, fly ash or other 'non-standard' additives to improve soil properties, lie outside the scope of this manual. For a more complete review of ground treatment techniques, refer to Mitchell and Jardine (2002). Helpful guidance on implementation in practice is provided by Han (2015). Hereafter the term 'soil' will be used to represent undisturbed natural, disturbed natural and artificial ground (such as fills).

25.2. Understanding the ground

Common with all geotechnical design, the specific geotechnical engineering purposes for which the site is to be used must first be defined. These purposes in turn will define what is expected of the ground. So, for example, if the site in question is for a multistorey building, the expectation of the ground is to provide a stable foundation; whereas if a river is to be protected from contamination migrating from an adjacent industrial site, the expectation of the ground is to provide a barrier to groundwater flow. In the light of these expectations, the current state of the ground in the appropriate zone of influence (the influences of the stresses imposed by the building in the case of the foundation, or the potential routes through which the contamination might flow to the river) must then be established. Moreover, the future state of the ground for the design life of the expectation must equally be anticipated.

A useful tool for this analysis is presented by the 'Burland Triangle' (**Figure 25.1**), which provides a framework in which to explore whether the expectation can be met by the ground as it exists and is likely to continue to exist on the site in question. A site investigation will be conducted to establish the

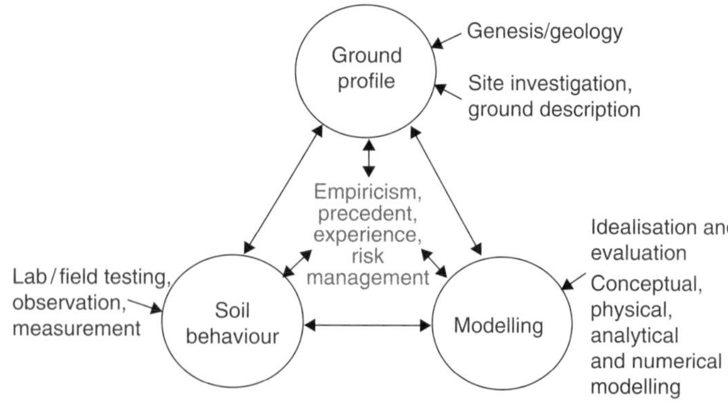

Figure 25.1 The 'Burland Triangle' for analysing the ground and how it might react for a geotechnical engineering purpose (Burland, 1987). Ground improvement processes can be analysed in the same manner

ground profile and provide clues to its geological history, and this provides the first pointer to the future likely state of the ground; for example, whether exposure might lead to further weathering of the natural soil and a change in its properties. If the building is to be founded on a clay soil at a certain depth below the ground surface, for example, then the net stress increase (maximum likely stress imposed by the building minus the stress relief due to excavation of soil above the foundation depth) must be determined, although of course it could be a stress reduction were it a lightweight structure founded at depth. The makeup of the soil (see Chapter 14 *Soils as particulate materials*) will usually be established by the site investigation, while the properties of the soil (strength and deformation behaviour; see Chapter 17 *Strength and deformation behaviour of soils*) will likewise be established by laboratory testing; for a clay soil the deformation may take several years before it is complete.

Some form of modelling will determine whether, in general terms, the ground can fulfil the geotechnical expectation placed on it, taking cognisance of the ground properties, while an understanding of the geotechnical context of the wider area will provide clues as to whether the ground properties are likely to remain stable in the long term. Groundwater flow regimes (see Chapter 16 *Groundwater flow*) are important here, as they are for the second example of contaminant transport. Similarly, changes to the groundwater profile are important since they will alter the effective stress regimes operating on the site (see Chapter 15 *Groundwater profiles and effective stresses*), which in turn will alter the soil properties. The important point here is that sites are dynamic – the ground properties can change with time, and potential future changes must be accounted for in any geotechnical design, whether it is a conventional 'structural' solution (concrete foundation or cutoff wall) or a ground improvement process that is adopted. Analysis of the problem in the light of the totality of the above information will usually yield a range of design options, some of which might be to alter the ground's properties. The remainder of this chapter is devoted to outlining the ways in which ground properties can be altered to obviate the need for more elaborate 'structural' solutions, and thereby provide additional weapons in the geotechnical engineer's arsenal.

25.3. Removal of water

25.3.1 Introduction

Water in soil is usually a nuisance (it gets in the way of construction operations) and, more importantly, the strength of a soil is usually very sensitive to both water content and water pressure (see Chapter 15 *Groundwater profiles and effective stresses*). Water is added to soil in certain situations (for example, to aid compaction of a granular soil or to weaken a clay soil to aid excavation) but on the whole it is a hindrance. The problems posed by water in the context of ground improvement are generally that there is too much of it and/or it cannot escape quickly enough when additional, superimposed stresses are applied to the soil. Moreover, if water flows it can strengthen or weaken the soil according to the direction of flow and the pore water pressure distribution generated by the flow, which can be determined from a flow net (see Chapter 16 *Groundwater flow*). Only saturated soil (soil mineral particles and water) will be considered in terms of the need for water removal, which is the primary focus of the discussion hereafter. Natural groundwater flow is excluded from this discussion, whereas techniques that induce the flow of water out of the soil are central to ground improvement.

From Terzaghi's law of effective stress

$$\sigma = \sigma - u \text{ or more generally } p' = p - u \qquad (25.1)$$

where σ' is the effective normal stress on any given plane in a soil mass, σ is the equivalent total normal stress, u is the pore water pressure, p' is the mean normal effective stress in three dimensions and p is the equivalent mean normal total stress (all measured in kiloPascals (kPa)). The parameter σ' (or p') is the factor which determines the strength of the soil on any given plane (or in three dimensions). If the pore water pressure increases above its original level, σ' (or p') reduces, and vice versa.

Given that the pore water pressure increases linearly with depth below the groundwater table in the majority of cases (see **Figure 15.3** and the associated text for exceptions to this rule) and water is held by capillary suction such that the suction (or negative pore water pressure) increases with height above the water table until the suction can no longer be maintained (which can be many metres in fine-grained soil having very small pores), one obvious means of ground improvement is to lower the groundwater table by some means of drainage. This principle stood the Romans in good stead when they were creating their roads, or (literally) their highways, in areas with a naturally high water table which adversely influenced the properties of fine-grained soil. Of course, the same principle informs our current drainage specification for roads, railways, airport runways and other such traffic-carrying infrastructure. By digging deep ditches on either side of their roads the Romans depressed the water table, thereby increasing the suction in the soil above the water table – that is, making u (the pore water pressure) more negative in Equation 25.1. This in turn increased σ' (or p') and made the soil effectively stronger (**Figure 25.2**). Not only this, but by creating a cambered surface and paving it with stone flags, they minimised the possibility of water entering the road pavement structure from above and thereby maintained the suction regime. Drainage is therefore an assured means of ground improvement and is used widely: behind retaining walls, in slope faces and adjacent to dam cores, for example.

Fine-grained soil formation typically involves soil particles transported by water and settling in slow flowing conditions, such as in the base of lakes or the sea. The soil starts off as very wet indeed, but as more soil particles are progressively

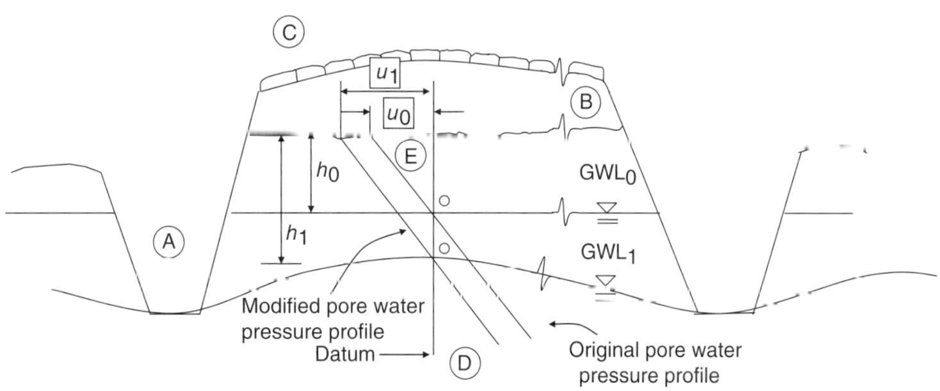

Figure 25.2 The effects on the strength of soil due to groundwater table lowering, as illustrated by a Roman road construction. A, deep ditches on either side of the road lower groundwater level (from GWL_0 to GWL_1); B, excavated soil creates the highway; C, stone flags on the cambered surface limit surface water ingress; D, pore water pressure increases linearly with depth below GWL; E, suction, or negative pore water pressure, increases linearly with height h above GWL such that suction $u_1 > u_0$

laid down above it, the soil densifies as water is progressively removed from the soil – it is squeezed out as the vertical (and hence mean normal) stress builds up by a process known as consolidation. The relationship between volume and mean normal effective stress is described by the normal consolidation line (NCL) and is shown using critical state theory terminology in **Figure 25.3**. A natural soil may undergo sequences of erosion and deposition, whereby the effective stress it experiences reduces and increases. One such sequence is shown in **Figure 25.3**, in which the soil is progressively consolidated as p' increases past points A and B until point C is reached. After this, p' reduces (due to erosion) until point D is reached. The degree to which the soil swells back relative to the projection of the critical state line (CSL) is important in determining the future response of the soil to applied stress regimes. The point on the NCL at which the soil at point D (and at any point on the swellback line, which is idealised as a straight line here for simplicity; in truth it is curved) has experienced its maximum p' is termed the *pre-consolidation pressure*. If the soil at point D is thereafter reloaded, it will recompress and its state will move down the path of the recompression line (idealised here as linear and coincident with the swellback line) to reach point C. It will then follow the NCL downwards once more.

In practice, if the soil is at point D in equilibrium ($p' = p_0'$) and a superimposed stress is applied to it of $\delta p'$, say due to a new building, according to the idealised graph the soil state would move to point E and the amount of volume change would be δv_2 once consolidation due to the imposed stress was complete – that is, once water had been caused to flow out of the soil and a new equilibrium at $p' = p_1'$ had been reached. If, however, the soil encountered was at point A, also in equilibrium at $p' = p_0'$, and a similar stress of $\delta p'$ was imposed on it due to construction of the same building, then the far greater volume change of δv_1 would result once a new equilibrium at $p' = p_1'$ had been established as the soil state moved to point B; far more water would have been driven out of the soil to achieve this state. For practical purposes, therefore, it is desirable to ensure that the stresses imposed by construction activities remain below the pre-consolidation pressure. One form of ground improvement is to create this situation – that is, to induce artificial overconsolidation.

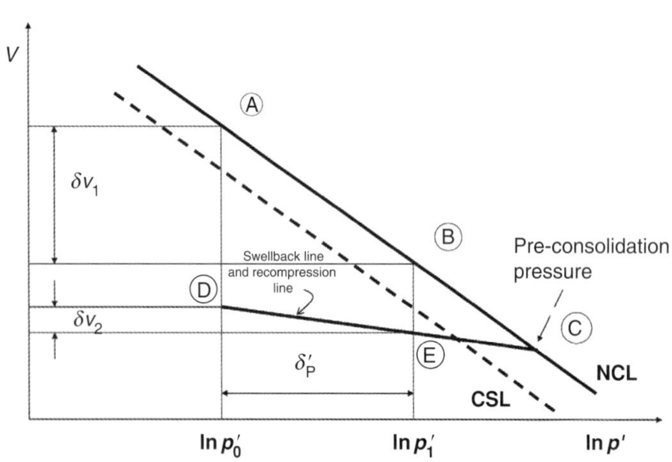

Figure 25.3 Graph of volume against mean normal effective stress plotted on a log scale

The velocity (v in m/s) at which water flows through a saturated soil, which is a particulate material with interconnected pore spaces, is governed by the driving force (the hydraulic gradient, i, which is dimensionless) and the resistance to flow (the hydraulic conductivity, k in m/s) according to Darcy's Law

$$v = ki \tag{25.2}$$

When an increment of total stress (δp) is applied to a soil, this stress is instantaneously carried by the pore water and creates an excess pore water pressure of $\delta u = \delta p$. Water flows from high to low pressure and therefore away from this area of high pressure, thereby progressively reducing δu to zero, causing $\delta p'$ to rise progressively until it increases by δp. The driving force (i.e. the hydraulic gradient) is therefore due to the excess pore water pressure induced by the increment of stress (δp). In effect there is now a seepage problem, for which isochrones could be drawn to illustrate the direction of flow and progression of consolidation within the soil. In fine-grained soil, with hydraulic conductivities ranging from $\sim 10^{-5}$ m/s for clean silts to $\sim 10^{-11}$ m/s for heavy clay soils, the rate of flow is small, while the distance that the water has to flow clearly governs the time required for consolidation to occur.

25.3.2 Preloading

Returning to the example illustrated in **Figure 25.3**, if a soil is at point A and a stress is temporarily applied equal to (applying $\delta p = \delta p'$ would take the state to point B) or greater than (say to point C) the stress subsequently to be imposed by the building, and were to allow the stress to remain in place for a sufficient length of time for consolidation of the soil to occur fully – that is, for the water to flow out and equilibrium to be regained – then removal of the temporary stress and imposition of the new stress due to the building would cause only a small amount of volume change. Unloading (by removal of the temporary stress at points B or C) would cause the soil to swell – that is, move upwards along parallel swellback lines – while the addition of the building would recompress the soil – that is, move downwards along the relevant recompression line (being coincident with the swellback line). If this process of stress removal and replacement by the building were to take place quickly, and/or if the soil had a small coefficient of hydraulic conductivity (k), or the flow distance for the water to exit the soil is large, then the heave caused by the unloading and the compression caused by the building could be very small, and for buildings or foundations that are very sensitive to settlement this process can be used to advantage to create negligible settlements.

This process is effective, especially for soils in which secondary compression (creep) is significant, but it is slow. One way to speed up the process is to increase the temporarily applied stress by increasing the driving force element. However, for thick soil deposits, especially where the direction of flow is upwards only (e.g. the soil overlies an impermeable rock), it can prove impractically slow; the rate of flow is proportional to the square of the drainage path. The temporary load can be earth or rubble fill, although water tanks or vacuum preloading (by pumping from beneath an impervious membrane) have been used. Placement of fill, ensuring continued drainage, is effected in layers as quickly as possible (staged loading is commonly needed to avoid overstressing the soft, weak soil being treated and avoiding failure); likewise, fill is removed as quickly as possible (staged removal is not necessary). The advantages are that this is a simple method that is easy to effect and certain to work, and is suitable for soils that undergo large volume decreases. It can, however, require a long time and much material, which has to be double-handled.

25.3.3 Vertical drains

Although it adds to the cost and complicates the construction process somewhat, the time required for consolidation using preloading can be greatly reduced by installing vertical drains. The rate of consolidation depends upon the length of the drainage path and the hydraulic conductivity of the soil associated with this path. In general, the water will flow vertically on loading to exit the soil; k_{vertical}, which is governed by the lowest coefficient of hydraulic conductivity of the substrata (e.g. interlayered fine sands, silts and clays in alluvium), will determine the rate of flow. By permitting horizontal drainage to vertical drains installed at predetermined intervals, not only is the length of the drainage path greatly reduced, but the rate of flow is governed by $k_{\text{horizontal}}$ rather than k_{vertical}. In general, the horizontal coefficient of hydraulic conductivity is far greater due to the nature of sedimentary soil particle deposition, and since it is the higher of the values of the substrata that dominate, consolidation occurs more quickly. For interlayered sands, silts and clays this can mean a few orders of magnitude. This is why normally consolidated or lightly overconsolidated alluvial soils and stratified soils are most effectively improved using vertical drains. Types of vertical drains include sand-filled boreholes (although horizontal shearing would compromise their performance), sand-filled stockings that are inserted into vertical boreholes and proprietary band drains consisting of a strong, flexible core (manufactured from polyethylene or similar) containing drainage channels and covered by a filter fabric. A horizontal free-draining layer is constructed at the top of the vertical drains prior to application of the preload. Connection with an underlying permeable layer (e.g. sand or gravel) is also achieved where possible. Smearing of the borehole sides should be avoided so that horizontal drainage is not compromised (Hird and Moseley, 2000).

25.3.4 Electro-osmosis

When a voltage gradient is induced across a soil, the water in the pore structure of the soil moves from the anodes (often metal rods) to the cathodes (e.g. perforated metal pipes creating a series of filter wells), which are water abstraction points. The electro-osmotic flow of water through soil is expressed by

$$v_e = k_e i_e \tag{25.3}$$

where v_e is the electro-osmotic flow rate, k_e is the coefficient of electro-osmotic permeability (measured in m^2/Vs) and i_e is the potential gradient $= V/L$ (where V is the potential difference and L is distance between the electrodes). This is analogous to Darcy's Law for hydraulic flow (Equation 25.2), yet there is a fundamental difference since k_e is commonly considered to be independent of the size of the pore spaces, while k is very strongly influenced by pore size. The technique is therefore most effective in fine-grained soils, in which k_e is typically 10^{-8} to 10^{-9} m^2/Vs (Mitchell and Soga, 2005: p. 275) whereas k can vary by orders of magnitude down to $\sim 10^{-11}$ m/s (example comparisons are provided by Mitchell, 1993: p. 270). Electro-osmotic flow is primarily caused by cations in the water surrounding negatively charged clay particle surfaces being transported towards the cathode, drawing water (a polar molecule that is attached to the cations) with them. (Very occasionally, positively charged surfaces surrounded by anions will result in flow towards the anode.)

There are many phenomena acting (see Liaki *et al.*, 2008), and some can bring about additional improvements.

- Degradation of a metallic anode and transport of metal ions by electromigration can cause improvement of clay soils by way of cation exchange on the exchange sites of clay minerals.
- Transport of stabilising agents added as aqueous solutions (electrolytes) at the anode (anolyte) and cathode (catholyte) can result in stabilising reactions.
- Electrokinetic remediation can be used to draw contaminants to the electrodes where they can be removed, hence remediating chemically contaminated sites (Ottosen *et al.*, 2008), thus strongly contributing to more sustainable practices if designed sensitively (Vocciante *et al.*, 2021).

Electro-osmosis can be used for temporary dewatering in soils not readily treated by well pumping, or for permanent improvement, especially when accompanied by chemical changes. Moreover, it is fast and can be used in a confined area, although it remains little used in practice. The soil immediately adjacent to the anodes tends to dry out, causing increased resistance and current requirements, but can be countered by periodic short-duration reversals of polarity or by an appropriate anolyte solution. Mitchell and Soga (2005) is a particularly helpful reference on the subject of ion movements, while the use of the techniques in association with geosynthetics is described by Jones *et al.* (2017).

25.3.5 Groundwater lowering

Well pumping is conceptually simple, being more of an accepted construction technique than a geotechnical process, and works best in soils with relatively high coefficients of hydraulic conductivity such as sands and gravels (see Chapter 80 *Groundwater control*).

25.4. Improvement of soils by mechanical means

25.4.1 Introduction

Compaction is widely used to increase the density of soils, and thereby increase their strength and stiffness, notably in road construction where surface compaction technology has reached a considerable level of sophistication. The principal goal is to cause the soil particles to pack together as closely as possible so as to minimise voids and to maximise 'particle interlock' – so that when the soil is sheared, the soil in the sheared region must dilate (i.e. particles must do work against the normal, or confining, force) as well as doing work against the frictional force. The types of rollers available include smooth-wheel dead weight rollers, vibrating rollers and sheepsfoot rollers (see Chapter 75 *Earthworks material specification, compaction and control*). Compaction technology has developed into deep compaction for more general ground improvement based on both vibratory and impact methods. Explosives have also been used to induce deep densification (Hausmann, 1990), but this approach lies outside the scope of the standard approaches and is not covered. A static form of compaction is achieved by preloading, which is covered in Section 25.3.2.

25.4.2 Dynamic compaction

Compaction of soil by the repeated dropping of weights on the ground has been practised for centuries with the aim of improving the bearing capacity. The extension of this technique to dynamic compaction, involving heavy weights and large drop heights, was first used in the early 1970s to provide a stable foundation for structures of large plan area and for reclaimed land development. The principle of operation is simply based on the application of large vertical stresses to a considerable depth, thereby compressing the soil skeleton directly where the degree of saturation (S_r) is low (Gu and Lee, 2002). Where S_r is not low, high excess pore water pressures are generated; if they are sufficiently high they will cause failure and fissures (or cracks) to be created extending from the surface to significant depths. As the excess pore water pressures progressively dissipate, aided greatly by the fissures providing convenient drainage paths, the bearing capacity of the soil increases such that it exceeds the previous maximum value.

Dynamic compaction is effective in coarse-grained fills (including decomposed domestic waste and industrial fills) and soils, but relatively ineffective in soft clays and peats that absorb the applied energy. The weights, or pounders, typically have a mass of 10–20 tonnes (100–200 kN) and are typically dropped, using a crane, from heights of 10–20 m at wide intervals initially, the spacing being progressively reduced. Repeated application of the pounder causes craters, which are backfilled with coarse fill prior to the next series of load applications. The surface layers require rolling at the end of the process. It is a simple, rapid, relatively inexpensive method of treatment involving large equipment (although it cannot be used in confined spaces) and considerable ground vibration (hence it cannot be used adjacent to sensitive buildings or structures). A depth of compaction of 5–6 m can be easily achieved; a 'rule of thumb' for the treatment depth given by Hausmann (1990) is $0.5\sqrt{(WH)}$, where W is the mass of the weights and H is the height from which they are dropped. The same principle is adopted by *rapid impact compactors*, which were developed to provide a more controlled, rapid and thus readily applied form of the technique (see Serridge and Synac, 2006) and which typically deliver an energy input of around 8 tonne metres at a rate of 40 blows per minute. Guided or leader dynamic compaction rigs are also available.

25.4.3 Vibrocompaction

Vibrocompaction (sometimes termed vibroflotation) aims to cause deep densification, and hence strengthening and stiffening, of a specifically designed volume of ground by means of lateral vibration using a poker lowered by a crane. The poker contains an eccentric weight mounted on a vertical shaft which is rotated by a motor. The poker vibrates laterally, thus compacting the ground while sinking to the design depth under its self-weight. This is known as the 'dry method' of operation; the 'wet method' involves forcing water through the end of the poker while it is lowered into the ground, thereby producing what is stated to be a 'flushing effect'. The principles that underlie the standard compaction curve (described in Section 25.5.2 and **Figure 25.4**) also apply here. A void is thereby formed within an area of dense, compacted soil. The poker is withdrawn and a granular fill is dropped into the void and is compacted in 250–500 mm lifts by lowering the poker to refusal. The purpose of the granular fill is to compensate for the reduction in volume due to compaction of the existing soil. Vibrocompaction thus aims to create a uniformly dense, uniformly stiff block of soil onto which to locate some type of foundation. The size of each treated block of ground is designed to accommodate the bulb of pressure beneath the foundation – that is, the plan area of the treated block is square for a pad foundation (using typically 4, 9, 16 or 25 insertions) and linear for a strip foundation. The poker typically compacts the soil to a radius of 1–2 m from the insertion point, the volume decrease being generally about 5–10%, and occasionally up to 15% (see Chapter 85 *Embedded walls*).

Loose natural soils and fills ranging from fine sand to coarse gravel are most effectively treated, while coarser material is difficult to penetrate unless it is well graded. A high fines content prevents rapid dissipation of the excess pore water pressures that are generated by vibration in soils with a high degree of saturation. It also serves to dampen the vibration, hence reducing the area of influence of compaction. Cohesionless soils with less than 15% fines are consequently commonly quoted as being suitable; see Slocombe *et al.* (2000). The technique has also been widely applied to fills consisting of brick rubble, ash and general demolition debris, the voids in these materials being broken down by regular insertion of the poker and granular fill forced into the remaining spaces.

25.4.4 Stone columns

The same equipment and construction methods are used as for vibrocompaction, although a single-sized crushed rock is used as the granular fill. The end result is a 0.9–1.2 m dia. column of dense stone within a less competent stratum – typically a saturated weak/compressible fine-grained soil, although stone columns can in principle be installed in any soil type. In this case, the vibration does not necessarily influence the soil between the columns significantly, especially when the soil has a high fines content and a high water content. This will lead to largely undrained shearing displacements since there is no time for the water to flow out of the soil (although any air within the soil voids would, of course, be compressed). The process of adding stone and compacting it using the poker, however, will inevitably cause stone to be forced sideways into the soil until it is resisted. Three zones of material might be considered once the process is complete: largely undisturbed soil that has undergone some displacement due to undrained shearing, a soil–stone intermediate zone and the compacted stone column itself. This technique is sometimes known as vibroreplacement or vibrodisplacement.

The stone columns serve to reduce settlement of overlying foundations by acting as bearing piles and to strengthen the soil by virtue of zones of granular material having a much higher shear strength than the soil. It is also suggested that the stone columns act as vertical drains for the rapid dissipation of excess pore water pressures, although this remains a matter of debate associated with the degree to which the displaced soil fills the voids between the compacted stone. The depth of stone columns rarely exceeds 12 m, although lengths approaching 30 m have been recorded. The column fill typically consists of uniform 20–50 mm crushed stone, gravel or slag. Surface disturbance (due to a lack of confinement to resist the compaction energy) means that surface rolling is required after the columns have been formed (see Chapter 85 *Embedded walls*). The technique can be enhanced by encasing the stone columns in geosynthetics (Tang *et al.*, 2015).

25.4.5 Micropiles or root piles

Micropiles were originally introduced for underpinning operations. They are now used for slope stability problems, retaining structures and underground construction. Micropiles are constructed primarily to carry tensile and compressive forces and thereby increase the strength of soil for a particular purpose. They have only a limited degree of flexural (or bending) resistance and are therefore generally constructed at appropriate angles to carry direct forces. They can be used in most types of soil (Cadden *et al.*, 2004).

To create a micropile, a casing is jetted, driven or drilled to the desired depth, reinforcement is inserted into the casing and cement mortar is pumped in to fill the hole while the casing is withdrawn. In the case of root piles, the cement mortar is forced into the holes under higher pressure using a grouting technique so that it penetrates cracks or soft layers around the void. This creates a rough pile surface, thereby gaining maximum skin friction.

25.4.6 Soil nails, ground anchors and soil reinforcement

These techniques do not really constitute ground improvement and they are mentioned here only for completeness. Their purpose is to give the soil into which they are installed or buried an enhanced strength, either by providing tensile elements, increasing the mean normal effective stress, holding in place some form of structural facing to retain soil, or a mixture of all three. Soil nails are described in Chapters 72 *Slope stabilisation methods*, 73 *Design of soil reinforced slopes and structures*, 74 *Design of soil nails* and 88 *Soil nailing construction*. Ground anchors are described in Chapters 64 *Geotechnical design of retaining walls*, 87 *Rock stabilisation* and 89 *Grouted anchor construction*. Soil reinforcement is described in Chapters 62 *Types of retaining walls*, 64 *Geotechnical design of retaining walls*, 72 *Slope stabilisation methods*, 73 *Design of soil reinforced slopes and structures* and 86 *Soil reinforcement construction*.

25.5. Improvement of soils by chemical means

25.5.1 Introduction

The addition of chemicals to clay soils to bring about improvements in strength, stiffness and volume stability (i.e. to inhibit shrinkage and swelling), or for other purposes, is hardly new – the Romans were adept at using the technique. The fact that lime has been used for this purpose in stabilising clays is equally unsurprising since limestone and shale (which is highly compressed clay) are the two primary ingredients of cement; cement factories being located where limestone and shale outcrop together. Other than the use of chemicals to dry very wet soils and make sites workable following heavy rain (quicklime is particularly effective for this), chemical soil stabilisation can be divided into two categories: (*a*) soil improvement as a result of the combined action of the soil constituents and the stabiliser to create a cemented product (as occurs by mixing lime and clay); and (*b*) addition of a stabiliser that binds the soil particles together (as is the case with a sand–cement mix). The traditional manner of introducing the stabiliser is either as a powdered solid or a liquid suspension that is mixed with the soil at the surface, laid and compacted.

However, more modern techniques have evolved to mix the stabilisers at depth or to pump stabilising solutions, such as cement grouts, into the ground. There are several applications (see, for example, Rogers *et al*., 1996) and while all have different requirements, the basic principles remain the same. Problems can be encountered when stabilising soils having a high plasticity index due to difficulties in achieving a thorough mix and ensuring a sufficiently large volume of stabiliser to chemically modify the clay minerals, leading to some concern over long-term softening due to water uptake.

25.5.2 Lime and cement stabilisation

Traditionally, lime has been used to improve the properties of clay soils by surface mixing, wetting where necessary, laying and compaction. Mixing lime with a clay soil alters its physical properties by fundamentally changing its nature as a result of cation exchange – a process termed *modification*. If sufficient lime has been added, and the modified material is compacted and allowed to cure for a significant period of time, the material cements by a process termed *stabilisation* (or *solidification*). The technique is most commonly used as a general ground improvement process, essentially to improve soft, weak clay soils, but it can also be used to treat contaminated soils.

Lime is a general term used to describe either quicklime (calcium oxide, CaO, created by heating hydrated lime to 450°C), hydrated lime (calcium hydroxide, $Ca(OH)_2$, which is formed when CaO comes into contact with water or water vapour) or lime slurry (a mixture of hydrated lime with water). Agricultural lime refers to calcium carbonate ($CaCO_3$), which is unreactive and will form if lime comes into contact with the atmosphere; carbonation must be avoided. In its simplest application, quicklime is used as a construction expediency, to dry out sites that are unworkable due to water, by way of a strongly exothermic reaction that turns CaO into $Ca(OH)_2$, with dehydration taking place due to the reaction and to steam being released. Lime is simply spread evenly on the surface to effect this improvement, perhaps with some limited surface mixing.

Lime modification occurs as a result of cation exchange, the calcium ions exchanging with the cations on the clay mineral sites (usually monovalent sodium or potassium ions), which causes

- a reduction in the thickness of the adsorbed water layer, and hence a reduced susceptibility of the clay to water
- flocculation of the clay particles due to greater electrical attraction (as the water layer has thinned)
- an increased internal angle of friction (i.e. greater shear strength)
- a textural change from a plastic clay to a friable material that appears granular in nature
- reduced plasticity as a result primarily of the plastic limit (PL) rising considerably.

To bring about this improvement the lime is spread evenly on the surface, mixed thoroughly with the clay to the required depth using a rotavator and allowed to mellow (or cure) for 24–72 hours. This time span will ensure that all quicklime is fully hydrated (it hydrates by way of an expansive reaction and this must not be allowed to happen post-compaction), and that the cation exchange reactions are complete – the surface of the mixed material having been lightly rolled to seal it from rain and to inhibit carbonation. Lime modification can be achieved with the addition of a modest quantity of lime that is determined using the initial consumption of lime (ICL) test, which establishes the quantity of lime required to fully satisfy the cation exchange reactions. The ICL test is based on measuring pH changes in the soil, the ICL value being the lime addition that causes the pH to rise to 12.4 (or failing that, 12.3). This result can be confirmed by Atterberg limit tests of clay mixed with lime at different quantities.

To bring about the lime stabilisation reactions, sufficient additional lime is needed to maintain the pH at a high level (>12) and, once the cation exchange demand of the clay has been met, provide the additional calcium ions which are needed to react with constituents of the clay minerals (which are dissolved in the high pH environment) to achieve stabilisation by way of pozzolanic reactions (Boardman *et al*., 2001). It is important that, as the reactions are taking place, there is a close proximity of the clay flocs so that the reaction products can bind the particles together in a coherent cemented mass, and that this process takes place in an environment that is free from chemical or organic contaminants that adversely affect the reaction processes. The reactions can be superficially stated as

$$Ca^{2+} + OH^- + Al_2O_3 + H_2O \rightarrow \text{calcium aluminate hydrate gel}$$
$$Ca^{2+} + OH^- + SiO_2 + H_2O \rightarrow \text{calcium silicate hydrate gel}$$

and are the result of alumina (Al_2O_3) and silica (SiO_2) dissolving from the clay mineral under conditions of high pH. The two gels crystallise over time to yield strong, brittle bonds. This causes a considerable increase in strength and results in a material that is subject to brittle fracture (i.e. is non-plastic).

The construction process is similar to that of modification, except that the mellowing process should not be as long (24 hours should be ample, and even this length of time has been considered to be detrimental to the long-term performance of the lime-stabilised clay). The material should then be remixed, adding water if necessary to bring the water content up to the right level (between 0% and 2% above the optimum water content so that density is maximised while air voids are minimised – see **Figure 25.4**). It can then be placed and compacted, in layers of appropriate thickness for the compaction energy applied, to ensure a density close to the maximum dry density throughout the full depth of the layer. The material should then be rolled with a smooth-wheeled roller if necessary to seal the surface while ensuring that there is a slight fall to avoid water

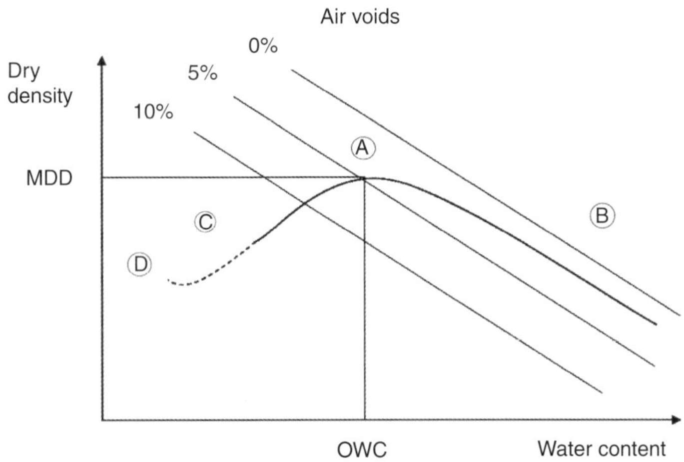

Figure 25.4 A standard compaction curve for fine-grained soil. Compaction seeks to maximise density and minimise air voids; for this application both of these compaction objectives are important. A, the maximum dry density (MDD) is achieved at the optimum water content (OWC) at which point the soil has minimum resistance to shear, hence greatest densification occurs; B, additional water inhibits densification and undrained shearing occurs at progressively greater water content (hence lower dry densities); C, as soil dries below the OWC, suctions progressively increase thereby inhibiting shear under compaction stresses and the dry density falls; D, where there is insufficient water, suction is lost and the dry density increases

ponding. Finally, it needs to be left to cure, free from frost and traffic until the reactions have advanced. The sulfate content must be sufficiently low (some specify <0.5%, others <1.0%) to avoid the creation of ettringite, which forms by way of a highly expansive reaction, destroying the compacted state and compromising long-term strength. Organic material in the clay can be detrimental because it tends to be acidic, and hence lowers the pH, but more importantly calcium reacts preferentially with the organic material and thus less is available for the stabilisation reactions. The most assured method of design is to carry out laboratory trials of compacted mixes with different lime contents above the ICL value and allow them to cure for different periods before being tested for strength.

Cement stabilisation can be used in any soil (there is no requirement for the alumina and/or silica from the clay mineral). The principles behind the technique and the processes of design and construction are similar. A larger quantity of cement than lime is usually required to provide sufficient binding agent to create a matrix sufficient to encompass the soil particles. Cement contains a significant proportion of quicklime and, like lime stabilisation, the use of cement causes a large rise in pH. This effect can be reduced if a cement replacement material, such as ground granulated blast furnace slag (GGBS), is used.

25.5.3 Lime, lime–cement and cement columns

The modern techniques of lime column construction were developed in Sweden and Japan, although there is now a far greater use of cement-only and lime–cement mixes rather than lime-only. This technique, commonly referred to as 'deep mixing', has generally only been carried out in soft/weak soils due to the difficulty of operating the mixing plant in stiff/strong soils. The principles listed above remain valid for the deeper process, notably the need for sufficient water for the reactions to take place, the mixing to be achieved thoroughly and the resulting material to be adequately densified. The mixing takes place with a purpose-designed tool that is augured into the ground; the densification occurs by counter-rotation of this tool such that there is downward pressure on the mixed material as the tool is withdrawn. The correct water content can be achieved by the inclusion of quicklime in a powder mix (nowadays consisting primarily of cement to guarantee the reaction products) to dry out wet soils, or, by using a cement slurry in cases where additional water is needed (see Chapter 84 *Ground improvement*).

25.5.4 Lime piles

An alternative stabilisation technique, commonly restricted to clay slope stabilisation, is to create lime piles, which are columns of compacted quicklime. The quicklime hydrates by the exothermic hydration reaction, causing water content reduction and immediately reducing the pore water pressure and creating suctions; these have been found to arrest slope movement (Rogers *et al.*, 2000a). The lime hydration releases calcium and hydroxide ions into the ground and they migrate from the quicklime pile into the surrounding soil, which is thus stabilised, to a distance that is dependent on the water content of the soil and the reactivity of the clay (Rogers and Glendinning, 1996). This distance is usually a few tens of millimetres, although clay cracking due to the rapid dehydration causes fissures that radiate from the pile and these provide conduits along which the lime migrates and stabilises the clay further away from the pile. The process can be enhanced by the inclusion of different additives, GGBS being particularly effective (Rogers *et al.*, 2000b).

25.5.5 Lime slurry pressure injection

Lime slurry pressure injection is another alternative that borders on the use of cement grout but remains distinct since the aim of the process is to cause chemical changes in the clay soils into which the slurry is injected. A lime–water mix is injected at different depths into the clay in a grid pattern with the aim of creating an approximately uniform distribution of slurry. Migration of ions into the soil adjacent to the slurry seams results in the classical lime stabilisation reactions occurring. The technique has been used for clay embankment stabilisation in the USA but has been little used in the UK and requires further research before its efficacy can be proved.

25.5.6 Grouting

Grouting has been practised for roughly 100 years and can be effected in many ways. The aim is generally to strengthen the soil by introducing a cementitious grout and/or by causing densification as a result of introducing the grout, or to prevent the flow of water by filling the voids in the soil. Essentially a fluid is injected into the soil and hardens at some later stage. The 'split spacing' method of group injection is generally used, whereby predetermined quantities of grout are injected through holes spaced at two or three times the final spacing, the hole spacing gradually being reduced. The voids become progressively smaller and the ground progressively tighter. Grouting is covered in Chapters 84 *Ground improvement* and 90 *Geotechnical grouting and soil mixing*, and so will be covered only briefly here.

25.5.6.1 Permeation grouting

Permeation grouting refers to the process in which no volume change or soil structure change occurs as the grout permeates the soil through the pore spaces. Once the grout is in position it hardens with time to strengthen the soil. Particulate grouts can be used where the soil is relatively coarse-grained. Chemical grouts can penetrate smaller voids (medium silts or coarser), but are unsuitable for soils having more than 15% fines. Electrochemical injection can be used in silts and silty clays.

25.5.6.2 Displacement grouting

Displacement grouting refers to the process by which high-viscosity grouts are pumped into the soil under high pressures with the aim of causing lateral or radial compaction of the soil. The grout usually consists of mixtures of cement, soil and/or clay and water. Either spherical bulbs of grout are formed at the end of the grout pipe or columns of grout are formed by withdrawing the pipe. The process was first developed in the 1940s in France but was of limited use until more recently developed pumping capabilities enabled the method to be more effective. The technique is applicable to soils which are readily compactable.

25.5.6.3 Jet grouting

Jet grouting is a variation that was introduced in Japan in the 1970s, the technique being based on the use of high-speed water jets. The native soil is either mixed with a stabiliser (grout) in situ or, in the case of poor soils, the soil can be removed and grout columns formed in their place.

25.5.7 Artificial ground freezing

Artificial ground freezing is a construction expedient only. It is a temporary measure used to stabilise an area or zone of saturated soil while a particular construction operation is performed. It is a potentially dangerous construction process (significant precautions are necessary) and an expensive last resort, but it is a useful technique in certain cases.

The water within a soil mass is frozen by installing closely spaced freezing pipes through which a continuous supply of coolant is passed. The frozen soil has a greatly increased bearing capacity, a far higher strength and a permeability approaching zero. Excavation (for tunnels, underpinning, etc.) can proceed safely once the ground is frozen.

Freezing causes a volume increase that is dependent on the water content of the soil, although a significant heave of the soil can be expected and this can cause problems similar to those of expansive clays if precautions are not taken. Conversely, adverse settlements occur during thaw and these too can create unwanted side effects. Sands and gravels are little affected because they are free-draining, whereas movement can be large in silts and clays.

References

Boardman DI, Glendinning S and Rogers CDF (2001) Development of stabilisation and solidification in lime–clay mixes. *Géotechnique* **51(6)**: 533–543.

Boshoff GA and Bone BD (2005) *Permeable Reactive Barriers*. International Association of Hydrological Sciences, Wallingford, UK.

Braithwaite P (2007) Improving company performance through sustainability assessment. *Proceedings of the Institution of Civil Engineers – Engineering Sustainability* **160(ES2)**: 95–103.

Burland JB (1987) Nash Lecture: the teaching of soil mechanics – a personal view. *Proceedings of the 9th European Conference on Soil Mechanics and Foundation Engineering, Dublin, Ireland*, vol. 3, pp. 1427–1447.

Cadden A, Gómez J, Bruce D and Armour T (2004) Micropiles: recent advances and future trends. In *Current Practices and Future Trends in Deep Foundations* (ASCE (ed)). ASCE, Reston, VA, pp. 140–165.

Fenner RA, Ainger CM, Cruickshank HJ and Guthrie PM (2006) Widening engineering horizons: addressing the complexity of sustainable development. *Proceedings of the Institution of Civil Engineers – Engineering Sustainability* **159(ES4)**: 145–154.

Fenner RA, Sykes J and Ainger CM (2022) *Sustainable Infrastructure: Principles into Practice*, 2nd edn. ICE, London, UK.

Gu Q and Lee F-H (2002) Ground response to dynamic compaction of dry sand. *Géotechnique* **52(7)**: 481–493.

Han J (2015) *Principles and Practice of Ground Improvement*. Wiley, Chichester, UK.

Hausmann MR (1990) *Engineering Principles of Ground Modification*. McGraw-Hill, New York, NY.

Hird CC and Moseley VJ (2000) Model study of seepage in smear zones around vertical drains in layered soil. *Géotechnique* **50(1)**: 89–97.

Jones C, Lamont-Black J, Huntley D, Alder D and Glendinning S (2017) Electrokinetic geosynthetics: from research to hype to practice. *Proceedings of the Institution of Civil Engineers – Civil Engineering* **170(3)**: 127–134.

Jowitt PW (2004) Sustainability and the formation of the civil engineer. *Proceedings of the Institution of Civil Engineers – Engineering Sustainability* **157(ES2)**: 79–88.

Liaki C, Rogers CDF and Boardman DI (2008) Physicochemical effects on uncontaminated kaolinite due to electrokinetic treatment using inert electrodes. *Journal of Environmental Science and Health Part A* **43(8)**: 810–822.

Mitchell JK (1993) *Fundamentals of Soil Behavior*, 2nd edn. Wiley, New York, NY.
Mitchell JK and Soga K (2005) *Fundamentals of Soil Behavior*, 3rd edn. Wiley, New York, NY.
Mitchell JM and Jardine FM (2002) *A Guide to Ground Treatment. CIRIA Report C573.* CIRIA, London, UK.
Ottosen LM, Christensen IV, Rörig-Dalgård I, Jensen PE and Hensen HK (2008) Utilisation of electromigration in civil and environmental engineering: processes, transport rates and matrix changes. *Journal of Environmental Science and Health Part A* **43(8)**: 795–809.
Phear AG and Harris SJ (2008) Contributions to *Géotechnique* 1948–2008: ground improvement. *Géotechnique* **58(5)**: 399–404.
Rogers CDF (2018) Engineering future liveable, resilient, sustainable cities using foresight. *Proceedings of the Institution of Civil Engineers – Civil Engineering* **171(6)**: 3–9.
Rogers CDF and Glendinning S (1996) The role of lime migration in lime pile stabilisation of slopes. *Quarterly Journal of Engineering Geology* **29(4)**: 273–284.
Rogers CDF, Glendinning S and Dixon N (1996) *Lime Stabilisation.* Thomas Telford, London, UK.
Rogers CDF, Glendinning S and Holt CC (2000a) Slope stabilisation using lime piles: a case study. *Ground Improvement* **4(4)**: 165–176.
Rogers CDF, Glendinning S and Troughton VM (2000b) The use of additives to improve the performance of lime piles. *Proceedings of the 4th International Conference on Ground Improvement Geosystems*, Kuala Lumpur, Malaysia. Building Information Ltd, Helsinki, Finland, pp. 127–134.
Rogers CDF, Makana LO, Leach JM and the UKCRIC Community (2022) *The Little Book of Theory of Change for Infrastructure and Cities.* University of Birmingham, Birmingham, UK.
Serridge CJ and Synac O (2006) Application of the rapid impact compaction (RIC) technique for risk mitigation in problematic soils. *Proceedings of the International Association of Engineering Geology Congress, Nottingham, UK, paper* 294.
Slocombe BC, Bell AL and Baez JL (2000) The densification of granular soils using vibro methods. *Géotechnique* **50(6)**: 715–725.
Tang L, Cong S, Ling X, Lu J and Elgamal A (2015) Numerical study on ground improvement for liquefaction mitigation using stone columns encased with geosynthetics. *Geotextiles and Geomembranes* **43(2)**: 190–195.
Vocciante M, Dovì VG and Ferro S (2021) Sustainability in electrokinetic remediation processes: a critical analysis. *Sustainability* **13(2)**: 770.

Further reading

Chai J-C and Miura N (1999) Investigation of factors affecting vertical drain behavior. *ASCE Journal of Geotechnical and Geoenvironmental Engineering* **125(3)**: 216–225.
Davies MCR (ed) (1997) *Ground Improvement Geosystems: Densification and Reinforcement.* Thomas Telford, London, UK.
Heibrock G, Kessler S and Triantafyllidis T (2006) *On modelling vibro-compaction of dry sands. In Numerical Modelling of Construction Processes in Geotechnical Engineering for Urban Environment* (Triantafyllidis T (ed)). Taylor & Francis, London, UK, pp. 126–140.
Horpibulsuk S, Miura N and Nagaraj TS (2003) Assessment of strength development in cement-admixed high water content clays with Abram's law as a basis. *Géotechnique* **53(4)**: 439–444.
Larsson S, Stille H and Olsson L (2005) On horizontal variability in lime–cement columns in deep mixing. *Géotechnique* **55(1)**: 33–44.
Lee FH, Lee CH and Dasari GR (2006) Centrifuge modelling of wet deep mixing processes in soft clays. *Géotechnique* **56(10)**: 677–691.
Moseley MP and Kirsch K (2004) *Ground Improvement*, 2nd edn. Spon, Abingdon, UK.

All chapters within Sections 1 *Context* and 2 *Fundamental principles* together provide a complete introduction to the Manual and no individual chapter should be read in isolation from the rest.

Chapter 26
Building response to ground movements

John B Burland Imperial College London, London, UK

Ground movements affecting buildings can result from the loads applied by the building itself to the subsoil or from a variety of external causes, including swelling or shrinkage of the subsoil, subsidence due to tunnelling or mining, and groundwater lowering. This chapter summarises an approach to assessing the response of buildings to ground movement and draws together in a coherent way a number of related studies including definitions of foundation movements, the description and classification of damage, and the factors controlling limiting distortions of buildings. The latter part of the chapter is devoted to applying the above principles to the important problem of assessing the impacts on buildings of subsidence induced by tunnelling. Methods of mitigating these impacts are described.

doi: 10.1680/icemge.66816.0311

CONTENTS

26.1. Introduction

Ground movements affecting buildings can result from the loads applied by the building itself to the subsoil or from external causes such as swelling or shrinkage of the subsoil, subsidence due to tunnelling or mining and groundwater lowering. Whatever the cause of the ground movements, they are likely to have an impact on the building and its function. In almost all cases, it is the differential movements (both vertical and horizontal) that are of importance – these can lead to tilting, distortion and possible damage to the building.

Compared with the literature on the prediction of ground movements, the question of allowable ground movements and their influence on the performance and serviceability of structures has received relatively little attention. Burland and Wroth (1974) listed some possible reasons for this.

- Serviceability is very subjective and depends both on the function of the building and the reactions of the users.
- Buildings vary so much one from another, both in broad concept and in detail, that it is very difficult to lay down general guidelines as to allowable movements.
- Buildings, including foundations, seldom perform as designed because construction materials display different properties from those assumed in design. Moreover, a 'total' analysis including the ground and the cladding would be impossibly complex and would still contain a number of questionable assumptions.
- As well as depending on loading and settlement, movements in buildings can be attributed to a number of factors such as creep, shrinkage and temperature. As yet, there is little quantitative understanding of these factors and there is a lack of careful measurements of the performance of actual buildings.

There is tendency among foundation engineers, indeed structural engineers as well, to believe that the movements of the foundations are the major cause of distress in buildings, and that by controlling these the satisfactory performance of the building is guaranteed. At a conference on *Design for movement in buildings* hosted by The Concrete Society in 1969, many cases were quoted of damage to finishes which resulted from movements of the structural members rather than the foundations. Another aspect of the problem which engineers may overlook is that a certain amount of cracking is unavoidable if the building is to be economic (Peck *et al.*, 1956). It is said that it is impossible to build a structure that does not crack due to shrinkage, creep and so on. Little (1969) has estimated that in the case of one particular type of building, the cost of preventing any cracking could exceed 10% of the total building cost. This chapter outlines the fundamental principles associated with assessing the response of buildings to ground movement. Two key aspects are: (*a*) the importance of clear definitions of ground and foundation movement; and (*b*) the objective description and classification of building damage. These are required for an objective approach to the assessment of impacts of ground movement on buildings and are discussed in the following two sections.

26.2. Definitions of ground and foundation movement

A study of the literature reveals a wide variety of confusing symbols and terminology describing foundation movements. Burland and Wroth (1974) proposed a consistent set of definitions based on the displacements (either measured or calculated) of a number of discrete points on the foundations of a building.

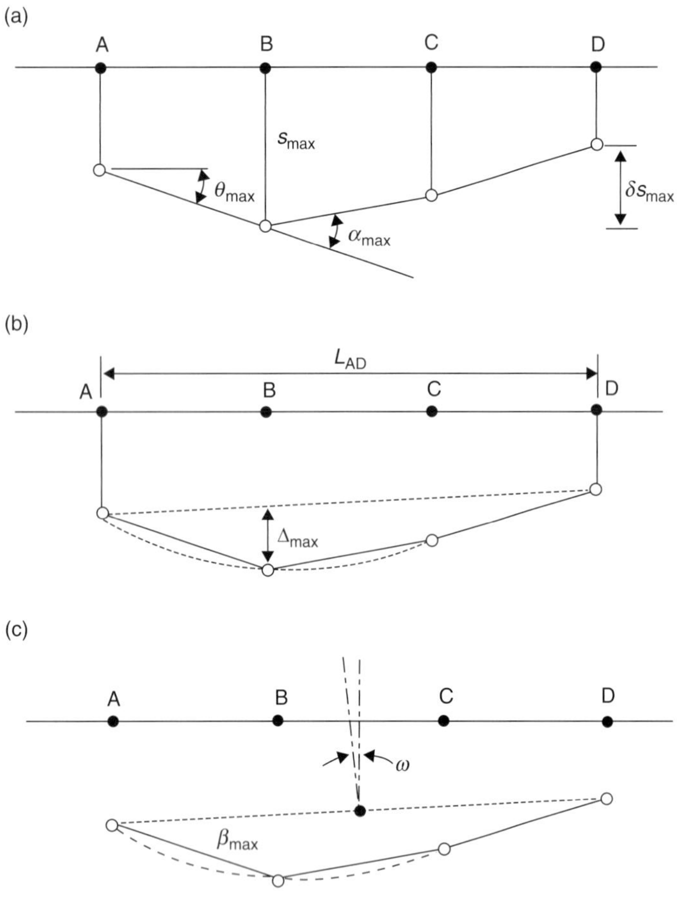

Figure 26.1 Definitions of ground and foundation movement (a) rotation or slope θ and angular strain α; (b) relative deflection Δ and defection ratio Δ/L; (c) tilt ω and relative rotation β (Reproduced from Burland, 1997; all rights reserved)

Care was taken to ensure that the terms did not prejudice any conclusions about the distortions of the superstructure itself, since these depend on a large number of additional factors. The definitions appear to have been widely accepted and are illustrated in **Figure 26.1**. A few points to note are as follows.

(*a*) *Rotation* (or slope θ) is the change in gradient of a line joining two reference points (e.g. A, B in **Figure 26.1(a)**).

(*b*) The *angular strain* α is shown in **Figure 26.1(a)**. It is positive for upward concavity (sagging) and negative for downward concavity (hogging).

(*c*) *Relative deflection* Δ is the displacement of a point relative to the line connecting two reference points on either side (see **Figure 26.1(b)**). The sign convention is as for (*b*).

(*d*) *Deflection ratio* (sagging ratio or hogging ratio) is denoted by Δ/L where L is the distance between the two reference points defining Δ. The sign convention is as for (*b*).

(*e*) *Tilt* ω describes the rigid body rotation of the structure or a well-defined part of it – see **Figure 26.1(c)**.

(*f*) *Relative rotation* (angular distortion) β is the rotation of the line joining two points, relative to the tilt ω – see **Figure 26.1(c)**. It is not always straightforward to identify the tilt and the evaluation of β can sometimes be difficult. It is also very important not to confuse relative rotation β with angular strain α. For these reasons, Burland and Wroth (1974) preferred the use of Δ/L as a measure of building distortion.

(*g*) *Average horizontal strain* ε_h is defined as the change of length δL over the length L. In soil mechanics it is customary to take a reduction of length (compression) as positive.

The above definitions only apply to in-plane deformations and no attempt has been made to define three-dimensional behaviour.

26.3. Classification of damage

26.3.1 Introduction

Assessment of the degree of building damage can be a highly subjective, and often emotive, matter. It may be conditioned by a number of factors such as local experience, the attitude of insurers, the cautious approach of a professional engineer or surveyor who might be concerned about litigation, market value and 'saleability' of the property. In the absence of objective guidelines based on experience, extreme attitudes and unrealistic expectations towards building performance can develop. It is worth stressing that most buildings experience a certain amount of cracking, often unrelated to foundation movement, which can be dealt with during routine maintenance and decoration.

Clearly, if an assessment of the risk of damage due to ground movement is to be made, the *classification* of damage is a key issue. In the UK, the development of an objective system of classifying damage is proving to be very beneficial in creating realistic attitudes towards building damage, and also in providing logical and objective criteria for designing for movement in buildings and other structures. This classification system will now be described.

26.3.2 Categories of damage

Three broad categories of building damage can be considered. These affect: (*a*) visual appearance or 'aesthetics'; (*b*) serviceability or function; and (*c*) stability. As foundation movements increase, damage to a building will progress successively from (*a*) through to (*c*).

It is only a short step from the above three broad categories of damage to the six (more detailed) classifications given in **Table 26.1**. Normally, categories 0, 1 and 2 relate to aesthetic damage, 3 and 4 relate to serviceability damage and 5 represents damage affecting stability. It was first put forward by Burland *et al.* (1977), who drew on the work of Jennings and Kerrich (1962), the National Coal Board (1975) and MacLeod

Category of damage	Normal degree of severity	Description of typical damage[a] Ease of repair is underlined
0	Negligible	Hairline cracks less than about 0.1 mm
1	Very slight	Fine cracks which are easily treated during normal decoration. Damage generally restricted to internal wall finishes. Close inspection may reveal some cracks in external brickwork or masonry. Typical crack widths up to approximately 1 mm
2	Slight	Cracks easily filled. Redecoration probably required. Recurrent cracks can be masked by suitable linings. Cracks may be visible externally and some repointing may be required to ensure weathertightness. Doors and windows may stick slightly. Typical crack widths 2–3 mm but may be up to approximately 5 mm locally
3	Moderate	The cracks require some opening up and can be patched by a mason. Repointing of external brickwork and possibly a small amount of brickwork to be replaced. Doors and windows sticking. Service pipes may fracture. Weathertightness often impaired. Typical crack widths are approximately 5–15 mm or several closely spaced cracks >3 mm
4	Severe	Extensive repair work involving breaking-out and replacing sections of walls, especially over doors and windows. Windows and door frames distorted, floor sloping noticeably.[b] Walls leaning[b] or bulging noticeably, some loss of bearing in beams. Service pipes disrupted. Typical crack widths are 15–25 mm, depending on the number of cracks
5	Very severe	This requires a major repair job involving partial or complete rebuilding. Beams lose bearing, walls lean badly and require shoring. Windows broken with distortion. Danger of instability. Typical crack widths are >25 mm, depending on the number of cracks

(a) Crack width is only one factor in assessing category of damage and should not be used on its own as a direct measure of it.
(b) Local deviation of slope, from the horizontal or vertical, of more than 1/100 will normally be clearly visible. Overall deviations in excess of 1/150 are undesirable.

Table 26.1 Classification of visible damage to walls with particular reference to ease of repair of plaster and brickwork or masonry

and Littlejohn (1974). Since then the classification has been adopted with only slight modifications by BRE (1981) and the Institution of Structural Engineers (1978, 1994).

The system of damage classification in **Table 26.1** is based on 'ease of repair' of the visible damage. Thus, in order to classify visible damage it is necessary, when carrying out the survey, to assess what type of work would be required to repair the damage both externally and internally. The following important points should be noted.

- The classification relates only to the visible damage at a given time and not to its cause or possible progression, which are separate issues.
- The strong temptation to classify the damage solely on crack width must be resisted. It is the ease of repair which is the key factor in determining the category of damage.
- The classification was developed for brickwork, blockwork and stone masonry. It can be adapted for other forms of cladding. It is not intended to apply to reinforced-concrete structural elements.
- More stringent criteria may be necessary where damage may lead to corrosion, penetration or leakage of harmful liquids and gases or structural failure.

Besides defining numerical categories of damage, **Table 26.1** also describes the 'normal degree of severity' associated with each category. These descriptions of severity relate to standard domestic and office buildings and serve as a guide to building owners and occupiers. In special circumstances, such as for a building with valuable or sensitive finishes, these descriptions of severity of damage may not be appropriate.

26.3.3 The division between categories 2 and 3 damage

The dividing line between categories 2 and 3 damage is particularly important. Studies of many case records show that damage up to category 2 can result from a variety of causes, either from within the structure itself (e.g. shrinkage or thermal effects) or associated with the ground. Identification of the cause of damage is usually very difficult as it frequently results from a combination of causes. If the damage exceeds category 2, the cause is usually much easier to identify and it is frequently associated with ground movement. Thus the division between categories 2 and 3 damage represents an important threshold which will be referred to later.

26.4. Routine guides on limiting deformations of buildings

Perhaps the two best-known studies leading to recommendations on allowable deformations of buildings are those of Skempton and MacDonald (1956) and Polshin and Tokar (1957). Skempton and MacDonald summarised settlement and damage observations on 98 buildings, 40 of which showed signs of damage. The study was limited to traditional steel and reinforced-concrete frame buildings and to a few load-bearing wall buildings.

As their measure of building distortion, they used the ratio of the differential settlement δ and the distance l between two points after eliminating the influence of tilt of the building. They defined the ratio δ/l as the 'angular distortion' which corresponds to the 'relative rotation' β defined in Section 26.2.

It was concluded that the limiting value of β to cause cracking in walls and partitions is 1/300 and that values of β greater than 1/150 will cause structural damage. It was recommended that values of β greater than 1/500 should be avoided. Subsequently, Bjerrum (1963) supplemented these recommendations by relating the magnitude of β to various types of impacts and structures. It is important to appreciate that Skempton and MacDonald's conclusions relate to traditional frame buildings and that the evidence relating to load-bearing walls was not reliable. Also, no classification of damage was used other than 'architectural', 'functional' and 'structural'. The limitations of the data and the tentative nature of the conclusions were emphasised by Skempton and MacDonald in their paper, but these qualifications are seldom emphasised in textbooks and design recommendations.

Polshin and Tokar (1957) discussed the question of allowable deformations and settlements, and defined three criteria: relative rotation β; deflection ratio Δ/L; and average settlement. The limiting values of these, adopted by the 1955 Building Code of the USSR, were then listed. It is of particular interest to note that frame structures were treated separately from continuous load-bearing walls. Recommended maximum relative rotations vary from 1/500 for steel and concrete frame infilled structures to 1/200 where there is no infill and no danger of damage to cladding.

Much stricter criteria were laid down for load-bearing brick walls. For L/H ratios (where L is the length and H the height) of less than 3, the maximum values of Δ/L are 0.3×10^{-3} for sand and 0.4×10^{-3} for soft clay. For L/H ratios greater than 5, the corresponding maximum values of Δ/L are 0.5×10^{-3} and 0.7×10^{-3} respectively. In their paper, Polshin and Tokar (1957) introduced two important variables: (*a*) the L/H ratio of the building or wall; and (*b*) the concept of limiting tensile strain before cracking. Using a limiting tensile strain of 0.05%, the limiting relationship between L/H and Δ/L was presented and was shown to be in good agreement with a number of cracked and uncracked brick buildings. The above recommendations for load-bearing masonry walls are based on a requirement for no cracking; if adhered to, the degree of damage would be unlikely to exceed category 1 (very slight) in **Table 26.1**.

The reader is referred to Chapter 52 *Foundation types and conceptual design principles* for a more detailed discussion on routine guides to allowable building movements.

26.5. Concept of limiting tensile strain

26.5.1 Onset of visible cracking

Cracking in masonry walls and finishes usually, but not always, results from tensile strain. Following the work of Polshin and Tokar (1957), Burland and Wroth (1974) investigated the idea that tensile strain might be a fundamental parameter in determining the onset of cracking. A study of the results from numerous large-scale tests on masonry panels and walls carried out at the UK Building Research Establishment (BRE) showed that, for a given material, the onset of visible cracking is associated with a reasonably well-defined value of average tensile strain which is not sensitive to the mode of deformation. They defined this as the critical tensile strain ε_{crit}, which is measured over a gauge length of 1 m or more.

Burland and Wroth (1974) made the following important observations.

- The average values of ε_{crit} at which visible cracking occurs are very similar for a variety of types of brickwork and blockwork and are in the range 0.05–0.1%.
- For reinforced-concrete beams the onset of visible cracking occurs at lower values of tensile strain in the range 0.03–0.05%.
- The above values of ε_{crit} are much larger than the local tensile strains corresponding to tensile failure.
- The onset of visible cracking does not necessarily represent a limit of serviceability. Provided the cracking is controlled, it may be acceptable to allow deformations well beyond the initiation of visible cracking.

Burland and Wroth (1974) showed how the concept of critical tensile strain could be used in conjunction with simple elastic beams to develop deflection criteria for the onset of visible damage. This work will be discussed in more detail later.

26.5.2 Limiting tensile strain: a serviceability parameter

Burland *et al.* (1977) replaced the concept of *critical tensile strain* with that of *limiting tensile strain* (ε_{lim}). The importance of this development is that ε_{lim} can be used as a serviceability parameter which can be varied to take account of differing materials and serviceability limit states.

Boscardin and Cording (1989) developed this concept of differing levels of tensile strain. Seventeen case records of damage due to excavation-induced subsidence were analysed. A variety of building types were involved and they showed that the categories of damage given in **Table 26.1** could be broadly related to ranges of ε_{lim}. These ranges are tabulated in **Table 26.2**, which is important as it provides the link between estimated building deformations and the possible severity of damage.

26.6. Strains in simple rectangular beams

This section describes how Burland and Wroth (1974) and Burland *et al.* (1977) used the concept of limiting tensile strain to study the onset of cracking in simple weightless elastic beams undergoing sagging and hogging modes of deformation. This simple approach gives considerable insight into the mechanisms controlling cracking. Moreover, it is shown that the criteria for initial cracking of simple beams are in very good agreement with the case records of damaged and undamaged buildings undergoing settlement. Therefore, in many circumstances, it is both reasonable and instructive to

Category of damage	Normal degree of severity	Limiting tensile strain (ε_{lim}) (%)
0	Negligible	0–0.05
1	Very slight	0.05–0.075
2	Slight	0.075–0.15
3	Moderate	0.15–0.3
4–5	Severe to very severe	>0.3

Boscardin and Cording (1989) describe the damage corresponding to ε_{lim} in the range 0.15–0.3% as 'moderate to severe'. However, none of the cases quoted by them exhibit severe damage for this range of strains. There is therefore no evidence to suggest that tensile strains up to 0.3% will result in severe damage.

Table 26.2 Relationship between category of damage and limiting tensile strain (ε_{lim}) (Data taken from Boscardin and Cording, 1989)

represent the facade of a building by means of a simple rectangular beam.

26.6.1 The relationship between limiting values of Δ/L and limiting tensile strain

The approach adopted by Burland and Wroth (1974) is illustrated in **Figure 26.2**, where the building is represented by a rectangular beam of length L and height H. The problem is to calculate the tensile strains in the beam for a given deflected shape of the building foundations and hence obtain the sagging or hogging ratio Δ/L at which cracking is initiated. It is immediately obvious that little can be said about the distribution of strains within the beam unless its *mode* of deformation is known. Two extreme modes are: (*a*) bending only about a neutral axis at the centre (**Figure 26.2(d)**); and (*b*) shearing only (**Figure 26.2(e)**). In the case of bending only, the maximum tensile strain occurs in the top fibre and that is where cracking will initiate as shown. In the case of shear only, the maximum tensile strains are inclined at 45°, giving rise to diagonal cracking. In general, both modes of deformation will occur simultaneously and it is necessary to calculate both bending and diagonal tensile strains to ascertain which type is limiting.

The expression for the total mid-span deflection Δ of a centrally loaded beam having both bending and shear stiffness is given by Timoshenko (1957) as

$$\Delta = \frac{PL^3}{48EI}\left(1 + \frac{18EI}{L^2HG}\right) \quad (26.1)$$

where E is Young's modulus, G is the shear modulus, I is the second moment of area and P is the point load. Equation 26.1 can be rewritten in terms of the deflection ratio Δ/L and the maximum extreme fibre strain $\varepsilon_{b\,max}$ as

$$\frac{\Delta}{L} = \left(\frac{L}{12t} + \frac{3I}{2tLH}\frac{E}{G}\right)\varepsilon_{b\,max} \quad (26.2)$$

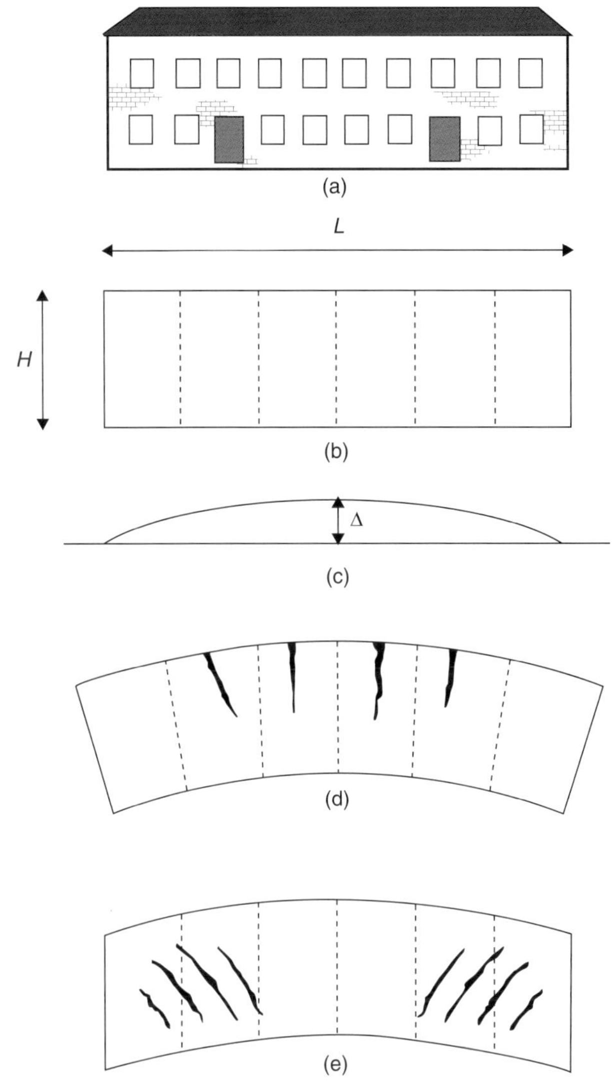

Figure 26.2 Cracking of a simple beam in bending and in shear: (a) actual building; (b) equivalent deep beam; (c) deflected shape of foundation; (d) bending deformation; (e) shear deformation (Reproduced from Burland, 1997; all rights reserved)

where t is the distance of the neutral axis from the edge of the beam in tension. Similarly for the maximum diagonal strain $\varepsilon_{d\,max}$, Equation 26.1 becomes

$$\frac{\Delta}{L} = \left(1 + \frac{HL^2}{18I}\frac{G}{E}\right)\varepsilon_{d\,max} \quad (26.3)$$

Similar expressions are obtained for the case of a uniformly distributed load with the diagonal strains calculated at the quarter points of the span (i.e. at $L/4$ and $3L/4$). Burland and Wroth (1974) drew the important conclusion that the maximum tensile strains are much more sensitive to the value of Δ/L than to the precise distribution of loading.

By setting $\varepsilon_{max} = \varepsilon_{lim}$, Equations 26.2 and 26.3 define the limiting values of Δ/L for the deflection of simple beams.

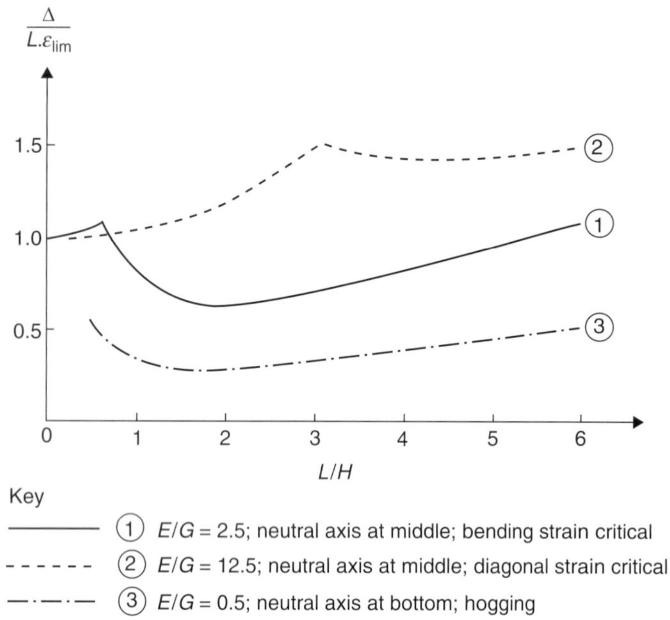

Figure 26.3 The influence of *E/G* and mode of deformation on the cracking of a simple rectangular beam (Reproduced from Burland, 1997; all rights reserved)

It is evident that, for a given value of ε_{lim}, the limiting value of Δ/L (whichever is the lowest in Equations 26.2 and 26.3) depends on *L*/*H*, *E*/*G* and the position of the neutral axis. For an isotropic beam ($E/G \approx 2.5$) with a neutral axis in the middle, the limiting relationship between $\Delta/L \times \varepsilon_{\text{lim}}$ and *L*/*H* is given by curve (1) in **Figure 26.3**. For a beam which has relatively low stiffness in shear ($E/G = 12.5$), the limiting relationship is given by curve (2). A particularly important case is that of a beam which is relatively flexible in bending, and is subjected to hogging such that its neutral axis is at the bottom. Curve (3) shows the limiting relationship for such a beam having $E/G = 0.5$. These curves serve to illustrate that even for simple beams, the limiting values of Δ/L causing cracking can vary over wide limits.

26.6.2 Limiting values of Δ/L for very slight damage

Burland and Wroth (1974) carried out a preliminary survey of data for cracking of infill frames and masonry walls. They concluded that the range of values of average tensile strain at the onset of visible cracking for a variety of common building materials was remarkably small. For brickwork and blockwork set in cement mortar ε_{lim} lies in the range 0.05–0.1%, while for reinforced concrete having a wide range of strengths, the values lie in the range 0.03–0.05%.

In order to assess the potential value of the limiting tensile strain approach in estimating the onset of visible cracking in buildings, Burland and Wroth compared the limiting criteria obtained from the analysis of simple beams with observations of the behaviour of a number of buildings – many of them of modern construction. For this comparison, a value of $\varepsilon_{\text{lim}} = 0.075\%$ was used. As can be seen from **Table 26.2**, this value corresponds to the threshold between category 1 (very slight) and category 2 (slight) damage. The buildings were classified as frame, load-bearing walls undergoing sagging and load-bearing walls undergoing hogging. **Figures 26.4(a)–26.4(c)** show the comparison with curves (2), (1) and (3), respectively, from **Figure 26.3**, with $\varepsilon_{\text{lim}} = 0.075\%$. Also shown is the criterion of limiting relative rotation $\beta = 1/300$ proposed by Skempton and MacDonald (1956) and the limiting relationship proposed by Polshin and Tokar (1957) for load-bearing walls. In spite of its simplicity, the analysis reflects the major trends in the observations. In particular, the prediction is borne out that load-bearing walls, especially when subjected to hogging, are more susceptible to damage than are frame buildings, which are relatively flexible in shear.

26.6.3 The relationship between Δ/L and levels of damage

At this point it is necessary to emphasise that limiting tensile strain is not a fundamental material property like tensile strength. It should be regarded as a measure of serviceability (as set out in **Table 26.2**) which aids the engineer in deciding whether a building is likely to develop visible cracks, their severity and where the critical locations might be. Equations 26.2 and 26.3 can be used in this way to assess what levels of damage might be expected for a given (perhaps calculated) value of Δ/L.

The advantages of this approach over traditional empirical rules limiting deformations are as follows.

- It can be applied to complex structures employing well-established stress analysis techniques.
- It makes explicit the fact that damage can be controlled by paying attention to the modes of deformation within the building structure and its finishes.
- The limiting values can be varied to take account of differing materials and serviceability limit states – for example, the use of soft bricks and lime mortar can substantially reduce cracking as it raises the value of ε_{lim}.

It is also important to appreciate that the onset of visible cracking does not necessarily represent a limit of serviceability. Provided cracking is controlled, it may be acceptable to allow deformations to continue well beyond the initiation of cracking. Cases where the propagation of initial cracks may be fairly well controlled are framed structures with panel walls, and also reinforced load-bearing structures. Unreinforced load-bearing walls undergoing sagging under the restraining action of the foundations may also fall into this category. An important mode of deformation where uncontrolled cracking can occur is that of hogging of unreinforced free-standing load-bearing walls. Once a crack forms at the top of the wall there is nothing to stop it propagating downwards. Low-rise

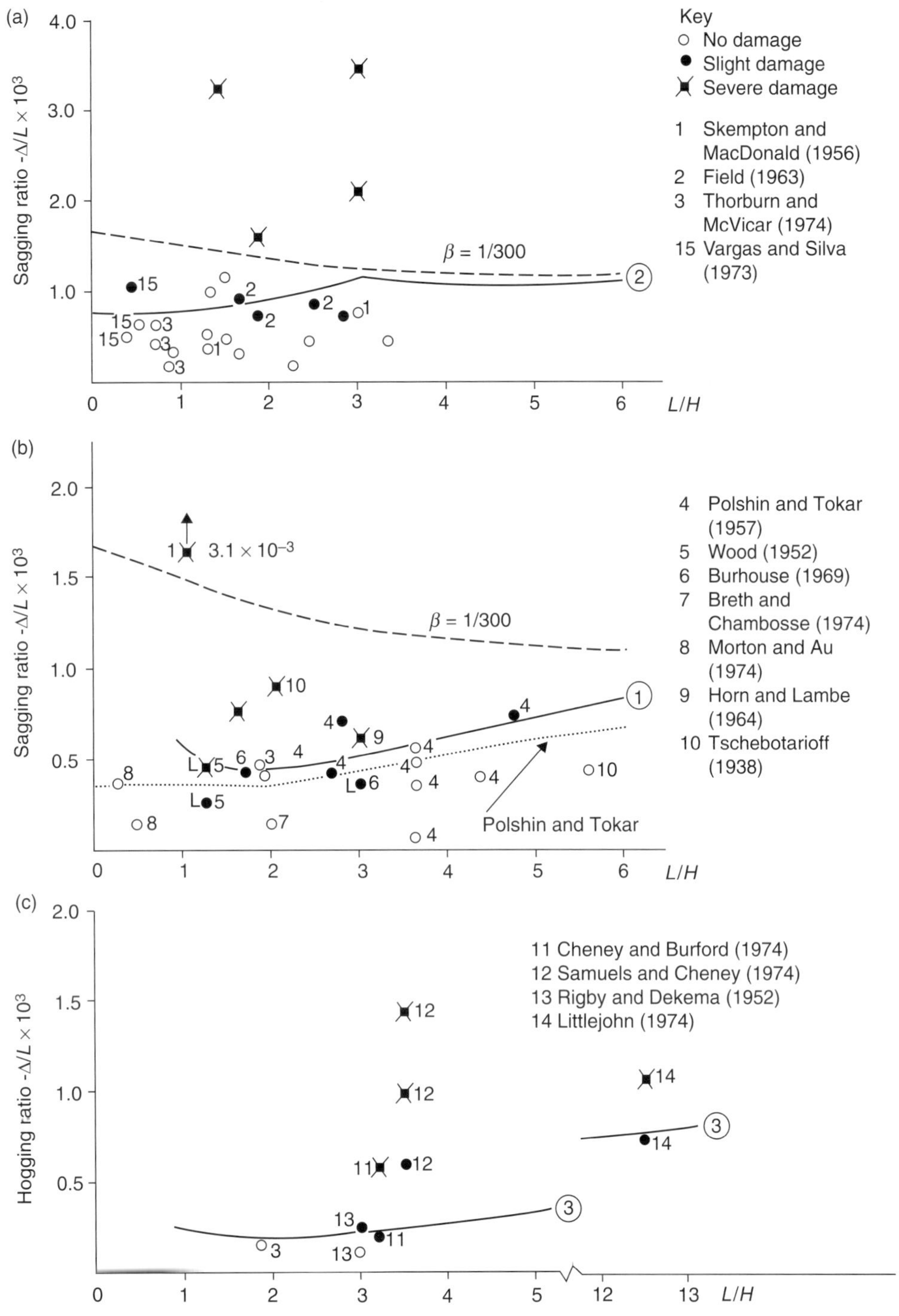

Figure 26.4 Relationship between Δ/*L* and *L*/*H* for buildings showing various degrees of damage: (a) frame buildings; (b) load-bearing walls; (c) hogging of load-bearing walls (Reproduced from Burland *et al.*, 1977; all rights reserved)

houses with shallow foundations on swelling or shrinking clays are particularly prone to this form of damage.

The remainder of this chapter is devoted to applying the principles set out above to the important case of assessing the impacts on buildings of subsidence induced by tunnelling and excavations.

26.7. Ground movement due to tunnelling and excavation

26.7.1 Introduction

In recent years, assessment of the environmental impact of major construction projects has become a normal and required procedure. The construction of tunnels in urban areas, while

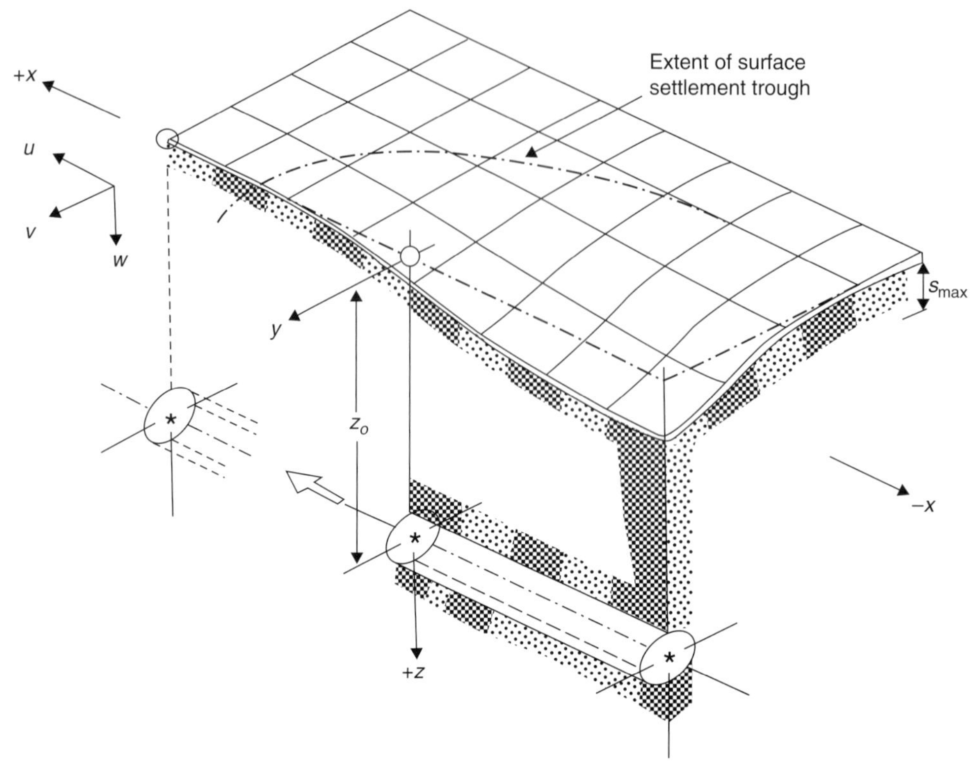

Figure 26.5 Settlement trough above an advancing tunnel (Reproduced from Attewell *et al.*, 1986; Blackie and Sons Ltd. All rights reserved)

having many long-term environmental benefits, can also create significant environmental impacts. During construction, such impacts would include construction traffic, noise, vibration and dust, as well as temporary restrictions on access to certain roads and other public areas. Longer-term impacts would include land and building acquisition, traffic and ventilation noise and vibration levels, and other impacts such as pollution, groundwater changes and effects on ecology.

An environmental impact which is causing increasing public awareness and concern is that of subsidence and its effects on structures and services due to tunnelling and deep excavations. Construction of tunnels and deep excavations is inevitably accompanied by ground movements. It is necessary, both for engineering design and for planning and consultation, to develop rational procedures for assessing the risks of damage. Coupled with such assessments is, of course, the requirement for effective protective measures which can be deployed when predicted levels of damage are judged to be unacceptable.

The following section summarises the approach that was originally developed in the 1990s for assessing the risks of subsidence damage for the London Underground Jubilee Line Extension project, which involved tunnelling under densely developed areas of central London. The experience gained from this project has been summarised in a two-volume book (Burland *et al.*, 2001a, 2001b). The approach to assessing the risks of subsidence damage, which is presented here, draws on the results of the studies described previously, including the description and classification of damage and limiting distortions of brickwork and masonry walls.

The construction of tunnels or surface excavations will inevitably be accompanied by movement of the ground around them. At the ground surface, these movements manifest in what is called a 'settlement trough'. **Figure 26.5** shows diagrammatically the surface settlement trough above an advancing tunnel. For greenfield sites, the shape of this trough transverse to the axis of the tunnel approximates closely to a normal Gaussian distribution curve – an idealisation which has considerable mathematical advantages.

26.7.2 Settlements caused by tunnel excavation

Figure 26.6 shows such an idealised transverse settlement trough. Attewell *et al.* (1986) and Rankin (1988) have summarised the current widely used empirical approach to the prediction of immediate surface and near-surface ground displacements. The settlement s is given by

$$s = s_{max} \exp\left(\frac{-y^2}{2i^2}\right) \qquad (26.4)$$

where s_{max} is the maximum settlement and i is the value of y at the point of inflection. It has been found that, for most

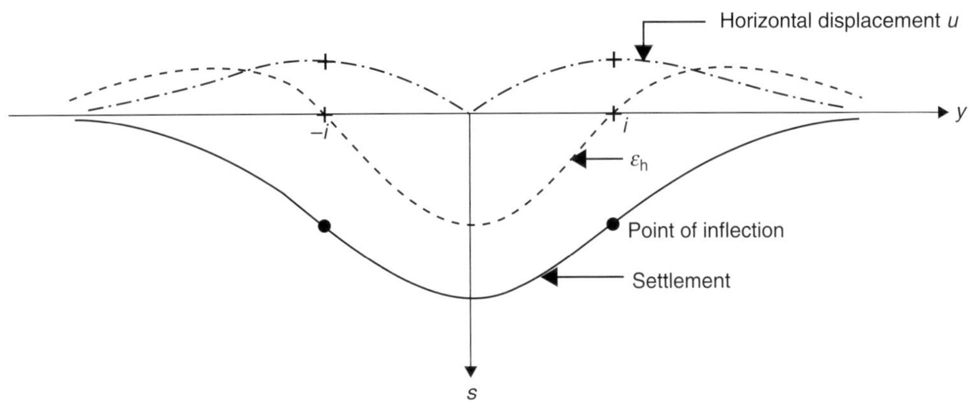

Figure 26.6 Transverse settlement trough and horizontal displacement profiles

purposes, i can be related to the depth of the tunnel axis z_o by the linear expression

$$i = Kz_o \tag{26.5}$$

The trough width parameter K depends on the soil type. It varies from 0.2 to 0.3 for granular soils, 0.4 to 0.5 for stiff clays and to as high as 0.7 for soft silty clays. As a general rule, the width of the surface settlement trough is about three times the depth of the tunnel for tunnels in clay strata. It is important to note that, although the value of K for *surface* settlements is approximately constant for various depths of tunnel in the same ground, Mair *et al.* (1993) have shown that its value increases with depth for *subsurface* settlements.

The immediate settlements caused by tunnelling are usually characterised by the 'volume loss' V_L which is the volume of the surface settlement trough per unit length V_s expressed as a percentage of the notional excavated volume of the tunnel.

Integration of Equation 26.4 gives

$$V_s = \sqrt{2\pi}\, i\, s_{max} \tag{26.6}$$

so that

$$V_L = \frac{3.192 i\, s_{max}}{D^2} \tag{26.7}$$

where D is the diameter of the tunnel. Combining Equations 26.4, 26.5 and 26.7 gives the surface settlement s at any distance y from the centre line as

$$s = \left(\frac{0.313 V_L D^2}{Kz_o}\right)\exp\left(\frac{-y^2}{2K^2 z_o^2}\right) \tag{26.8}$$

26.7.3 Horizontal displacements due to tunnelling

Building damage can also result from horizontal tensile strain, and therefore predictions of horizontal movement are required. Unlike settlements, there are few case histories where horizontal movements have been measured. The data that do exist show reasonable agreement with the assumption of O'Reilly and New (1982) that the resultant vectors of ground movement are directed towards the tunnel axis. It follows that the horizontal displacement u can be related to the settlement s by the expression

$$u = \frac{s \times y}{z_o} \tag{26.9}$$

Equation 26.9 is easily differentiated to give the horizontal strain ε_h at any location on the ground surface.

Figure 26.6 shows the relation between the settlement trough, the horizontal displacements and the horizontal strains occurring at ground level. In the region $-i < y < i$, the horizontal strains are compressive. At the points of inflection the horizontal displacements are a maximum and $\varepsilon_h = 0$. For $y > | i |$ the horizontal strains are tensile.

26.7.4 Assessment of surface displacements due to tunnelling

The above empirical equations provide a simple means for estimating the near-surface displacements due to tunnelling, assuming greenfield conditions – that is, ignoring the presence of any building or structure.

A key parameter in this assessment is the volume loss V_L. This results from a variety of effects which include movement of ground into the face of the tunnel and radial movement towards the tunnel axis due to reductions in supporting pressures. The magnitude of V_L is critically dependent on the type of ground, the groundwater conditions, the tunnelling method, the length of time in providing positive support, and the quality of supervision and control. The selection of an appropriate value of V_L for design requires experience and is greatly aided by well-documented case histories in similar conditions.

A number of other assumptions are involved in the prediction of ground displacements due to tunnelling. For example, in ground containing layers of clay and granular soils, there is uncertainty about the value of the trough width parameter *K*. When two or more tunnels are to be constructed in close proximity, the assumption is usually made that the estimated ground movements for each tunnel acting independently can be superimposed. In some circumstances this assumption may underestimate the movements and allowance needs to be made for this.

It is clear from the above that, even for greenfield conditions, precise prediction of ground movements due to tunnelling is not realistic. However, it is possible to make reasonable estimates of the likely range of movements provided tunnelling is carried out under the control of suitably qualified and experienced engineers.

26.7.5 Ground movements due to deep excavations

Ground movements around deep excavations are critically dependent both on the ground conditions (e.g. stratigraphy, groundwater conditions, deformation and strength properties) and the method of construction (e.g. sequence of excavation, sequence of propping, rigidity of retaining wall and supports). In general, open excavations and those supported by cantilever retaining walls give rise to larger ground movements than strutted excavations and those constructed by top-down methods. In the urban situation, the latter are clearly to be preferred if building damage is to be minimised.

The calculation of ground movements is not straightforward and much experience is required to make any sensible use of complex analyses. It is therefore essential that optimum use is made of previous experience and case histories in similar conditions. Peck (1969) presented a comprehensive survey of vertical movements around deep excavations which was updated by Clough and O'Rourke (1990). Burland *et al.* (1979) summarised the results of over ten years of research into the movements of ground around deep excavations in London Clay. The Norwegian Geotechnical Institute have published a number of case histories of excavations in soft clay in the Oslo area (e.g. Karlsrud and Myrvoll, 1976). For well-supported excavations in stiff clays, Peck's settlement envelopes are generally very conservative as settlements rarely exceed 0.15% of the depth of excavation. However, movements can extend to three or four times the excavation depth behind the basement wall. Horizontal movements are generally of similar magnitude and distribution to vertical movements, but may be much larger for open and cantilever excavations in stiff clays.

Advanced methods of numerical analysis, based on the finite element method, are widely used for prediction of ground movements around deep excavations. Such analyses can simulate the construction process, modelling the various stages of excavation and support conditions. However, comparison with field observations shows that successful prediction requires high-quality soil samples with the measurement of small-strain stiffness properties using local strain transducers mounted on the sides of the samples (Jardine *et al.*, 1984).

As for tunnelling, it is essential that deep excavation work be carried out under the close supervision of an experienced engineer. Unless positive support is provided rapidly and groundwater is properly controlled, large unexpected ground movements can develop.

26.7.6 The influence of horizontal strain

It was shown in Section 26.7.3 that ground surface movements associated with tunnelling and excavation not only involve sagging and hogging profiles, but significant horizontal strains as well – see **Figure 26.6**. Boscardin and Cording (1989) included the influence of horizontal extension strain ε_h in Equations 26.2 and 26.3 by using simple superposition – that is, it is assumed that the deflected beam is subjected to uniform extension over its full depth. The resultant extreme fibre strain ε_{br} is given by

$$\varepsilon_{br} = \varepsilon_{b\,max} + \varepsilon_h \tag{26.10}$$

where $\varepsilon_{b\ max}$ is obtained from Equation 26.2. In the shearing region of the beam the resultant diagonal tensile strain ε_{dr} can be evaluated using Mohr's circle of strain. The value of ε_{dr} is then given by

$$\varepsilon_{dr} = \varepsilon_h\left(\frac{1-\nu}{2}\right) + \sqrt{\varepsilon_h^2\left(\frac{1+\nu}{2}\right)^2 + \varepsilon_{d\,max}^2} \tag{26.11}$$

where $\varepsilon_{d\ max}$ is obtained from Equation 26.3 and ν is Poisson's ratio. The maximum tensile strain is the greater of ε_{br} and ε_{dr}. Thus, for a beam of length *L* and height *H*, it is a straightforward matter to calculate the maximum value of tensile strain ε_{max} for a given value of Δ/L, and ε_h, in terms of *t*, *E*/*G* and ν. This value of ε_{max} can then be used in conjunction with **Table 26.2** to assess the potential associated damage.

The physical implications of Equations 26.2, 26.3, 26.10 and 26.11 can be illustrated by considering the previous case of the isotropic beam undergoing hogging, with its neutral axis at the bottom edge and $\nu = 0.3$. By combining Equations 26.2 and 26.10, the influence of ε_h on the limiting values of Δ/L can be examined for bending strains only by setting $\varepsilon_{b\,max} = \varepsilon_{lim}$. **Figure 26.7(a)** shows the normalised relationship between Δ/L and ε_h. For $\varepsilon_h = 0$, the limiting values of Δ/L are the same as given in **Figure 26.3** for various values of *L*/*H*. It can be seen that as ε_h increases towards the value of ε_{lim}, the limiting values of Δ/L for a given *L*/*H* reduce linearly, becoming zero when $\varepsilon_h = \varepsilon_{lim}$.

Similarly, **Figure 26.7(b)** has been derived from Equations 26.3 and 26.11 for the diagonal strains only. Once again, for $\varepsilon_h = 0$ the limiting values of Δ/L are recovered from

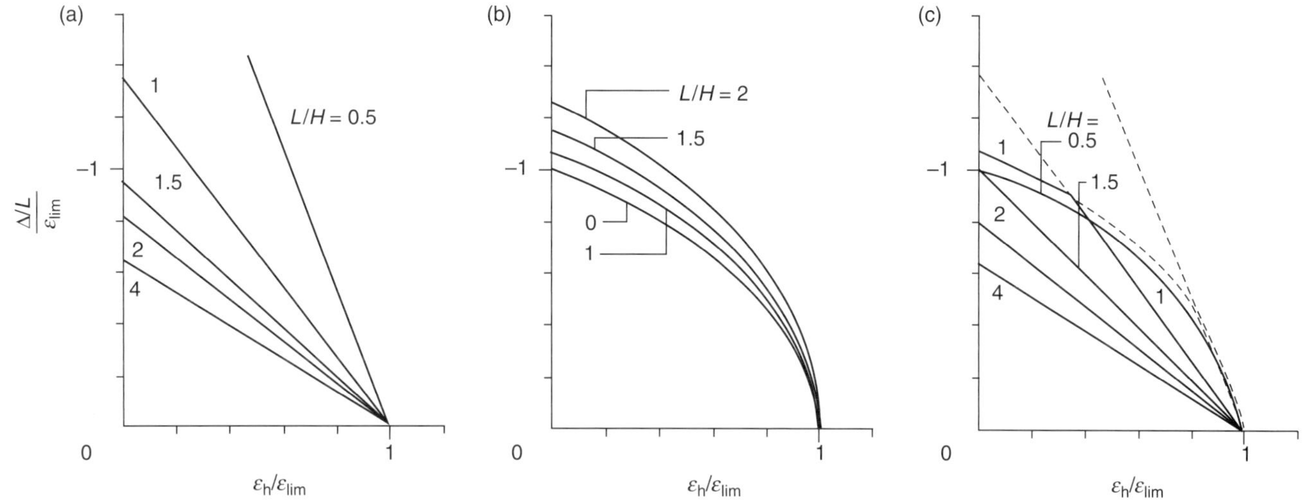

Figure 26.7 Influence of horizontal strain on $(\Delta/L)/\varepsilon_{lim}$ for (a) bending strain controlling; (b) diagonal strain controlling; and (c) limiting combinations of (a) and (b). Note that $E/G = 2.5$ and the mode of deformation is hogging (Reproduced from Burland, 1997; all rights reserved)

Figure 26.3. As ε_h increases, the limiting values of Δ/L decrease nonlinearly at an increasing rate towards zero. It is of interest to note that the values of Δ/L are not very sensitive to the values of L/H between 0 and 1.5.

Figures 26.7(a) and **26.7(b)** can be combined to give the resultant normalised limiting relationships between Δ/L and ε_h for various values of L/H, as shown in **Figure 26.7(c)**. It can be seen that, for $L/H > 1.5$, the bending strains always control. Also, for lower values of L/H, as ε_h increases, the controlling strain changes from diagonal to bending. It must be emphasised that **Figure 26.7** relates to the specific case of hogging with the neutral axis at the lower face and with $E/G = 2.5$.

By adopting the values of ε_{lim} associated with the various categories of damage given in **Table 26.2**, **Figure 26.7(c)** can be developed into an interaction diagram showing the relationship between Δ/L and ε_h for a particular value of L/H. **Figure 26.8** shows such a diagram for $L/H = 1$.

There are similarities between **Figure 26.8** and the well-known Boscardin and Cording (1989) chart of angular distortion β against ε_h. The latter chart has the following limitations.

- It only relates to $L/H = 1$.
- Maximum bending strains $\varepsilon_{b\,max}$ are ignored.
- β was assumed to be proportional to Δ/L, whereas Burland *et al.* (2004) have shown that the relationship is in fact very sensitive to the load distribution and to the value of E/G.
- As mentioned in Section 26.2, the evaluation of β is not always straightforward.

26.7.7 Relevant building dimensions

An important consideration is the definition of the relevant height and length of the building. A typical case of a building

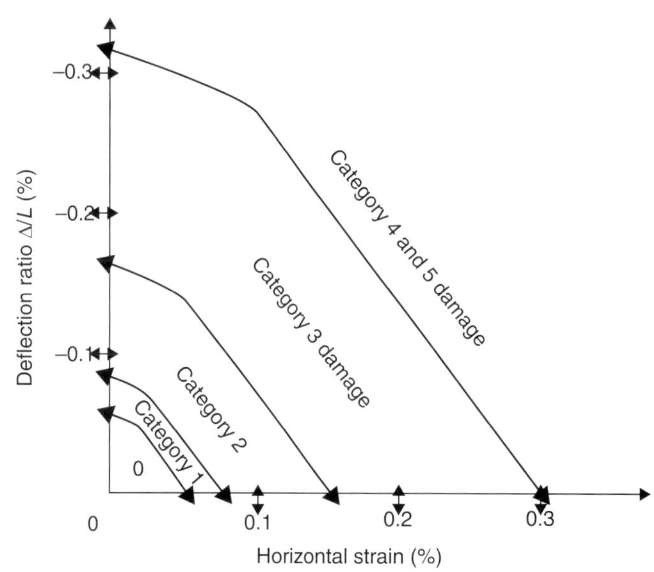

Figure 26.8 Relationship of damage category to deflection ratio and horizontal strain for hogging ($L/H = 1$, $E/G = 2.5$) (Reproduced from Burland, 1997; all rights reserved)

affected by a single tunnel settlement trough is shown in **Figure 26.9**. The height H is taken as the height from foundation level to the eaves. The roof is usually ignored. It is assumed that a building can be considered separately either side of a point of inflexion – that is, points of inflexion of the settlement profile (at foundation level) will be used to partition the building. The length of building is not considered beyond the practical limit of the settlement trough, which for a single tunnel can be taken as $2.5i$ (where $s/s_{max} = 0.044$). In a calculation of building strain, the building span length is required; it is defined as the length of building in a hogging or sagging

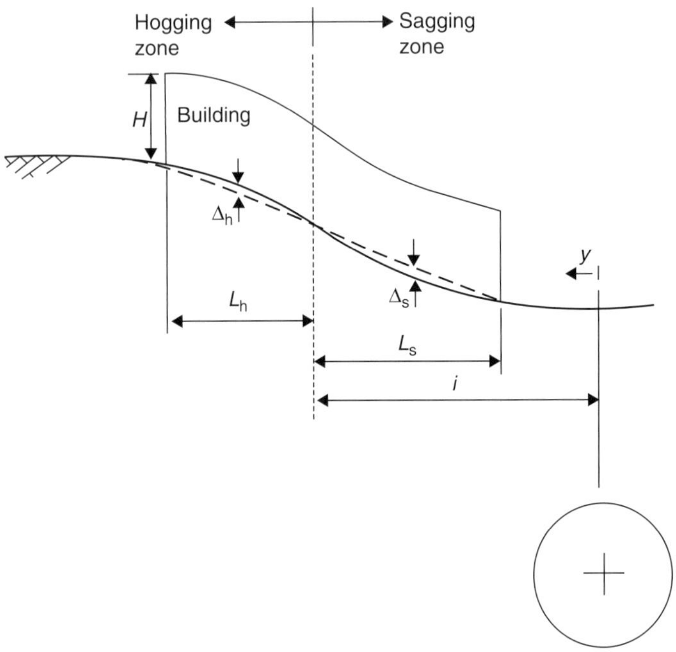

Figure 26.9 Building deformation: partitioning between sagging and hogging (Reproduced from Mair *et al.*, 1996; all rights reserved)

zone (shown as L_h or L_s in **Figure 26.9**) and limited by a point of inflexion or extent of settlement trough.

26.8. Evaluation of risk of damage to buildings due to subsidence

The various concepts discussed in the previous sections can be combined to develop a rational approach to the assessment of risk of damage to buildings due to tunnelling and excavation. The following broadly describes the approach that was adopted during the planning and enquiry stages of the Jubilee Line Extension underground railway in London, and which is now widely used internationally with minor variations (Burland, 1997).

26.8.1 Level of risk

The term 'the level of risk', or simply 'the risk', of damage refers to the *possible degree of damage* as defined in **Table 26.1**. Most buildings are considered to be at low risk if the predicted degree of damage falls into the first three categories 0 to 2 (i.e. negligible to slight). At these degrees of damage, structural integrity is not at risk and damage can be readily and economically repaired. It will be recalled from Section 26.3.3 that the threshold between categories 2 and 3 damage is a particularly important one. A major objective of design and construction is to maintain the level of risk below this threshold for all buildings. It should be noted that special consideration must be given to buildings judged to be of particular sensitivity, such as those in poor condition, containing sensitive equipment or of particular historical or architectural significance.

Because of the large number of buildings that are usually involved, the method of assessing risk is a staged process as follows: preliminary assessment; second stage assessment; detailed evaluation. These three stages will now be described briefly.

26.8.2 Preliminary assessment

So as to avoid a large number of complex and unnecessary calculations, a very simple and conservative approach is adopted for the preliminary assessment. It is based on a consideration of both maximum slope and maximum settlement of the ground surface at the location of each building. According to Rankin (1988), a building experiencing a maximum slope θ of 1/500 and a settlement of less than 10 mm has negligible risk of any damage. By drawing contours of ground surface settlement along the route of the proposed tunnel and its associated excavations, it is possible to eliminate all buildings having negligible risk. This approach is conservative because it uses ground surface displacements rather than foundation-level ones. It also neglects any interaction between the stiffness of the buildings and the ground. For particularly sensitive buildings it may be necessary to adopt more stringent slope and settlement criteria.

26.8.3 Second stage assessment

The preliminary assessment described above is based on the slope and settlement of the ground surface and provides a conservative initial basis for identifying those buildings along the route requiring further study. The second stage assessment makes use of the work described in the previous sections of this chapter. In this approach, the facade of a building is represented by a simple beam whose foundations are assumed to follow the displacements of the ground in accordance with the greenfield site assumption mentioned in Section 26.7. The maximum resultant tensile strains are calculated from the pairs of Equations 26.2, 26.10 and 26.3, 26.11. The corresponding potential category of damage, or level of risk, is then obtained from **Table 26.2**.

The above approach, although considerably more detailed than the preliminary assessment, is usually still very conservative. Thus the derived categories of damage refer only to *possible* degrees of damage. In the majority of cases, the likely *actual* damage will be less than the assessed category because in calculating the tensile strains the building is assumed to have no stiffness – so that it conforms to the greenfield site subsidence trough. In practice, however, the inherent stiffness of the building will be such that its foundations will interact with the supporting ground and tend to reduce both the deflection ratio and the horizontal strains.

Potts and Addenbrooke (1996, 1997) carried out a parametric study of the influence of building stiffness on ground movements induced by tunnelling using finite element

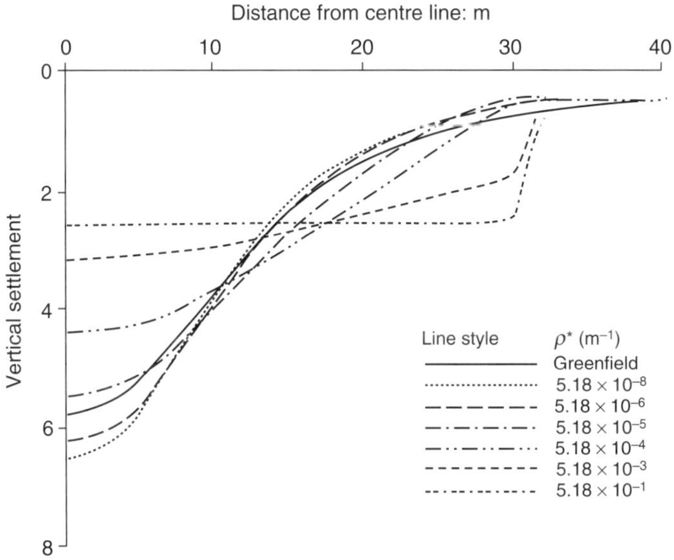

Figure 26.10 Influence of relative bending stiffness on settlement profile (Reproduced from Potts and Addenbrooke, 1997)

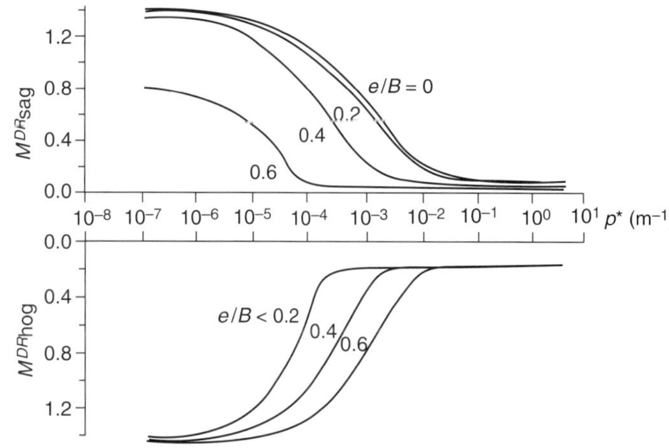

Figure 26.11 Bending stiffness modification factors (Reproduced from Potts and Addenbrooke, 1997)

methods, incorporating a nonlinear elastic–plastic soil model. The building was represented by an equivalent beam having axial and bending stiffness EA and EI (where E is Young's modulus, A the cross-sectional area and I the moment of inertia of the beam). The relative axial stiffness α^* and bending stiffness ρ^* are defined as

$$\alpha^* = EA/E_s H \text{ and } \rho^* = EI/E_s H^4 \qquad (26.12)$$

where H is the half-width of the beam (=B/2) and E_s is a representative soil stiffness.

Figure 26.10 shows the influence of relative bending stiffness on the settlement profile for a 20 m deep tunnel excavated beneath a 60 m wide building with zero eccentricity. **Figure 26.11** shows the modification factors MDR that are applied to the greenfield site deflection ratios (Δ/L) for sagging and hogging modes of deformation at different e/B ratios, where e is the eccentricity of the tunnel centre line. It can be seen that, for hogging in particular, the building changes from relatively very flexible to relatively very stiff over quite a small range in relative bending stiffness.

Potts and Addenbrooke's assessment of the influence of global stiffness of a building is a most valuable addition to the existing methodology for assessing the risk of damage. The results can be used to make more realistic assessments of relative deflection, and hence average strains within a building.

26.8.4 Detailed evaluation

Detailed evaluation is carried out on those buildings that, as a result of the second stage assessment, are classified as being at risk of category 3 damage, or above (see **Table 26.1**).

The approach is a refinement of the second stage assessment in which the particular features of the building and the tunnelling and/or excavation scheme are considered in detail. Because each case is different and has to be treated on its own merits, it is not possible to lay down detailed guidelines and procedures. Factors that are taken more closely into account are discussed below.

26.8.4.1 Tunnelling and excavation

The sequence and method of tunnel and excavation construction should be given detailed consideration, with a view to reducing volume loss and minimising ground movements as far as is practical.

26.8.4.2 Structural continuity

Buildings possessing structural continuity, such as those of steel and concrete frame construction, are less likely to suffer damage than those without it, such as load-bearing masonry and brick buildings.

26.8.4.3 Foundations

Buildings on continuous foundations such as strip footings and rafts are less prone to damaging differential movements (both vertical and horizontal) than those on separate individual foundations, or where there is a mixture of foundations (e.g. piles and spread footings).

26.8.4.4 Orientation of the building

Buildings orientated at a significant skew to the axis of a tunnel may be subjected to warping or twisting effects. These may be accentuated if the tunnel axis passes close to the corner of the building.

26.8.4.5 Soil–structure interaction

The predicted greenfield displacements will be modified by the stiffness of the building. The detailed analysis of this problem

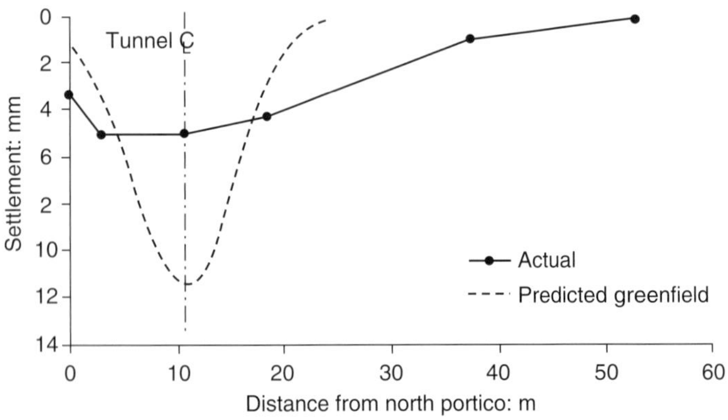

Figure 26.12 Comparison of observed and greenfield site settlements of the Mansion House, London, from driving a 3.05 m dia. tunnel at 15 m depth (Reproduced from Frischman *et al.*, 1994)

is exceedingly complex and resort is usually made to simplified procedures – some of which are described in the report published by the Institution of Structural Engineers (1989). The beneficial effects of building stiffness can be considerable, as demonstrated by some measurements on the Mansion House in the City of London during tunnelling beneath and nearby – see **Figure 26.12** from a paper by Frischman *et al.* (1994).

26.8.4.6 Previous movements

The building may have experienced movements due to a variety of causes such as construction settlement, groundwater lowering and nearby previous construction activity. It is important that these effects be assessed as they may reduce the tolerance of the building to future movements. If the building is already cracked, an assessment should be made as to whether the tunnelling-induced movement will focus on these cracks. Sometimes this can be an advantage, as other parts of the building are then at less risk of damage.

As many factors are not amenable to precise calculations, the final assessment of possible degree of damage requires engineering judgement based on informed interpretation of available information and empirical guidelines. Because of the inherently conservative assumptions used in the second stage assessment, the detailed evaluation will usually result in a reduction in the possible degree of damage. Following the detailed evaluation, consideration is given as to whether protective measures need to be adopted. These will usually only be required for buildings remaining in damage categories 3 or above (see **Table 26.1**).

26.9. Protective measures

Before considering near-surface measures, consideration should be given to measures that can be applied from within the tunnel to reduce the volume loss. There are a variety of such measures – for instance, increasing support at or near the face, reducing the time to provide such support, the use of forepoling, soil nailing in the tunnel face and the use of a pilot tunnel. These approaches tackle the root cause of the problem and may prove much less costly and disruptive than near-surface measures.

If, for a particular building, tunnelling protective measures are considered either technically ineffective or too expensive, then it will be necessary to consider applying protective measures near the surface or to the building itself. However, it must be emphasised that such measures are generally disruptive and can have a significant environmental impact. The main forms of protective measures currently available fall into the following six broad groups.

26.9.1 Strengthening of the ground

This can be achieved by means of grout injection (cement or chemical) or by ground freezing. It is normally undertaken in granular water-bearing soils. Its primary purpose is to provide a layer of increased stiffness below foundation level or to prevent loss of ground at the tunnel face during excavation.

26.9.2 Strengthening of the building

Strengthening measures are occasionally undertaken in order that the building may safely sustain the additional stresses or accommodate deformations induced by ground movements. Such measures include the use of tie rods and temporary or permanent propping. Caution is needed when adopting such an approach as the work can be very intrusive and may lead to greater impacts than simply allowing some cracking to take place which can subsequently be repaired.

26.9.3 Structural jacking

In some special situations (e.g. sensitive equipment) it may be possible to insert jacks and use them to compensate for tunnelling-induced settlement as it takes place.

26.9.4 Underpinning

There may be circumstances in which the introduction of an alternative foundation system can be used to eliminate or minimise differential movements caused by tunnelling. However, because the ground movements are usually deep-seated, such an approach may not be possible. If the existing foundations are inadequate or in a poor condition, underpinning may be used to strengthen them and provide a more robust and stiffer support system.

26.9.5 Installation of a physical barrier

Occasionally, consideration has been given to the installation of a physical barrier between the building foundation and the tunnel. This might take the form of a slurry trench wall or a row of secant bored piles. Such a barrier is not structurally connected to the building's foundation and therefore does not provide direct load transfer. The intention is to modify the shape of the settlement trough and reduce ground displacements adjacent to and beneath the building.

26.9.6 Compensation grouting

Compensation grouting is the controlled injection of grout between the tunnel and the building foundations in response to observations of ground and building movements during tunnelling. As its name implies, the purpose is to compensate for ground loss. The technique requires detailed instrumentation to monitor the movements of the ground and the building. Experience of compensation grouting is reported by, among others, Harris *et al.* (1994) and Harris (2001). The technique was successfully used on the Jubilee Line Extension Project in London for the protection of many historic buildings, including the Big Ben clock tower at the Palace of Westminster (Harris *et al.*, 1999). More recently, Mair (2008) described the successful application of compensation grouting in granular soils and the innovative application of directional drilling for installation of the grout tubes.

It cannot be emphasised too strongly that all of the above measures are expensive and disruptive, and should not be regarded as a substitute for good-quality tunnelling and excavation practice aimed at minimising settlement.

26.10. Conclusions

Damage due to foundation movements is only one aspect of the wider problem of serviceability of buildings. The problem of coping with differential settlement, as with creep, shrinkage and structural deflection, may frequently be solved by designing the building, and in particular the cladding and partitions, to accommodate movement rather than resist it. Successful and economic design and construction of the total structure requires cooperation between the geotechnical engineer, the structural engineer and the architect from the earliest stages of planning.

In this chapter, clear definitions have been given of foundation movement; it is vital that these are understood and used. Also, a widely accepted classification of damage based on ease of repair is described. These two aspects form the basis of a coherent and objective approach to assessing limiting distortions of buildings. The concept of limiting tensile strain as a serviceability parameter is introduced as a means of gaining insight into some of the factors influencing limiting distortions in buildings.

The latter part of the chapter summarises briefly a rational and coherent approach to the assessment of risk of damage to buildings due to tunnelling and excavation. A three-stage approach is recommended for assessing the potential for damage due to subsidence. The first two stages neglect the stiffness of the building in assessing the magnitude of the induced strains within it. The assessments are therefore very conservative. The advantages of this staged approach are that it is simple to use and the buildings which require detailed analysis are identified without recourse to a large amount of unproductive complex analysis.

A wide variety of protective and mitigation measures are described to minimise impacts of ground movement. In general, it is much better to control ground movements by controlling the tunnelling operation (so as to minimise ground loss) than to embark on expensive building protective measures which can themselves be disruptive and damaging.

References

Attewell PB, Yeates J and Selby AR (1986) *Soil Movements Induced by Tunnelling and Their Effects on Pipelines and Structures.* Blackie, Oxford, UK.

Bjerrum L (1963) Discussion. *Proceedings of the European Conference on SM&FE*, Wiesbaden, Germany, vol. 2, p. 135.

Boscardin MD and Cording EG (1989) Building response to excavation-induced settlement. *Journal of Geotechnical Engineering, ASCE* **115(1)**: 1–21.

BRE (Building Research Establishment) (1981). *Assessment of Damage in Low Rise Buildings with Particular Reference to Progressive Foundation Movements.* BRE, Garston, UK.

Burland JB (1997) Invited special lecture: assessment of risk of damage to buildings due to tunnelling and excavation. *1st International Conference on Earthquake Geotechnical Engineering, Tokyo, Japan*, vol. 3, pp. 1189–1201.

Burland JB and Wroth CP (1974) Settlement of buildings and associated damage: SOA Review. *Conference on Settlement of Structures, London, UK.* Pentech Press, Cambridge, UK, pp. 611–654.

Burland JB, Broms BB and de Mello VFB (1977) Behaviour of foundations and structures: SOA Report, Session 2. *Proceedings of the 9th International Conference on SMFE, Tokyo, Japan*, vol. 2, pp. 495–546.

Burland JB, Simpson B and St John HD (1979) Movements around excavations in London Clay: invited national paper. *Proceedings of the 7th European Conference on SM&FE, Brighton, UK*, vol. 1, pp. 13–29.

Burland JB, Standing JR and Jardine FM (eds) (2001a) *Building Response to Tunnelling: Case Studies from the Jubilee Line Extension, London, vol. 1, Projects and Methods.* CIRIA and Thomas Telford, London, UK.

Burland JB, Standing JR and Jardine FM (eds) (2001b) *Building Response to Tunnelling: Case Studies from the Jubilee Line Extension, London, vol. 2, Case Studies.* CIRIA and Thomas Telford, London, UK.

Burland JB, Mair RJ and Standing JR (2004) Ground performance and building response to tunnelling. *Proceedings of the International Conference on Advances in Geotechnical Engineering, The Skempton Conference* (Jardine RJ, Potts DM and Higgins KG (eds)). Thomas Telford, London, UK, vol. 1 pp. 291–342.

Clough GW and O'Rourke TD (1990) Construction induced movements of in-situ walls. In *Design and Performance of Earth Retaining Structures* (ASCE (ed)). ASCE, Reston, VA, pp. 439–470.

Frischman WW, Hellings JE, Gittoes S and Snowden C (1994) Protection of the Mansion House against damage caused by ground movements due to the Docklands Light Railway Extension. *Proceedings of the Institution of Civil Engineers, Geotechnical Engineering* **102(2)**: 65–76.

Harris DI (2001) Protective measures. In *Building Response to Tunnelling: Case Studies from Construction of the Jubilee Line Extension, London* (Burland JB, Standing JR and Jardine FM (eds)). CIRIA and Thomas Telford, London, UK, vol. 1, pp. 135–176.

Harris DI, Mair RJ, Love JP, Taylor RN and Henderson TO (1994) Observations of ground and structure movements for compensation grouting during tunnel construction at Waterloo Station. *Géotechnique* **44(4)**: 691–713.

Harris DI, Mair RJ, Burland JB and Standing J (1999) Compensation grouting to control the tilt of Big Ben clock tower. In *Geotechnical Aspects of Underground Construction in Soft Ground* (Kushakabe O, Fujita K and Miyazaki Y (eds)). Balkema, Rotterdam, the Netherlands, pp. 225–232.

Institution of Structural Engineers (1978/1989) *State of the Art Report: Structure–Soil Interaction*. Institution of Structural Engineers, London, UK. (Revised and extended in 1989.)

Institution of Structural Engineers (1994) *Subsidence of Low Rise Buildings*. Institution of Structural Engineers, London, UK.

Jardine RJ, Symes MJ and Burland JB (1984) The measurement of soil stiffness in the triaxial apparatus. *Géotechnique* **34(3)**: 323–340.

Jennings JE and Kerrich JE (1962) The heaving of buildings and the associated economic consequences, with particular reference to the orange free state goldfields. *The Civil Engineer in South Africa* **5(5)**: 122.

Karlsrud K and Myrvoll F (1976) Performance of a strutted excavation in quick clay. *Proceedings of the 6th European Conference on SM&FE, Vienna, Austria*, vol. 1, pp. 157–164.

Little MER (1969) Discussion, Session 6. *Proceedings of the Symposium on Design for Movement in Buildings, London, UK*. The Concrete Society, London, UK.

MacLeod IA and Littlejohn GS (1974) Discussion on Session 5. *Conference on Settlement of Structures, London, UK*. Pentech Press, Cambridge, UK, pp. 792–795.

Mair RJ (2008) 46th Rankine Lecture: tunnelling and geotechnics – new horizons. *Géotechnique* **58(9)**: 695–736.

Mair RJ, Taylor RN and Bracegirdle A (1993) Subsurface settlement profiles above tunnels in clays. *Géotechnique* **43(2)**: 315–320.

Mair RJ, Taylor RN and Burland JB (1996) Prediction of ground movements and assessment of risk of building damage due to bored tunnelling. *International Symposium on Geotechnical Aspects of Underground Construction in Soft Ground, London, UK*, pp. 713–718.

National Coal Board (1975) *Subsidence Engineers' Handbook*. National Coal Board Production Dept, London, UK

O'Reilly MP and New BM (1982) Settlements above tunnels in the United Kingdom: their magnitude and prediction. *Tunnelling '82, London, UK*, pp. 173–181.

Peck RB (1969) Deep excavations and tunnelling in soft ground: SOA Report. *7th International Conference on SM&FE, Mexico City, Mexico*, State of the Art Volume, pp. 225–290.

Peck RB, Deere DU and Capacete JL (1956) Discussion on paper by Skempton, A. W. and MacDonald, D. H. 'The allowable settlement of buildings'. *Proceedings of the Institution of Civil Engineers, Part 3* **5**: 778.

Polshin DE and Tokar RA (1957) Maximum allowable non-uniform settlement of structures. *Proceedings of the 4th International Conference on SM&FE, London, UK*, vol. 1, p. 402.

Potts DM and Addenbrooke TI (1996) The influence of existing surface structure on ground movements due to tunnelling. *Geotechnical Aspects of Underground Construction in Soft Ground, Tokyo, Japan* (Mair RJ and Taylor RN (eds)). Balkema, Rotterdam, the Netherlands, pp. 573–578.

Potts DM and Addenbrooke TI (1997) A structure's influence on tunnelling-induced movements. *Proceedings of the Institution of Civil Engineers, Geotechnical Engineering* **125**: 109–125.

Rankin WJ (1988) Ground movements resulting from urban tunnelling: predictions and effects. In *Engineering Geology of Underground Movement* (Geological Society (ed)). Geological Society, London, UK, pp. 79–92.

Skempton AW and MacDonald DH (1956) Allowable settlement of buildings. *Proceedings of the Institution of Civil Engineers, Part 3* **5**: 727–768.

Timoshenko S (1957) *Strength of Materials: Part I*. D van Nostrand Co, Inc., London, UK.

All chapters within Sections 1 *Context* and 2 *Fundamental principles* together provide a complete introduction to the Manual and no individual chapter should be read in isolation from the rest.

doi: 10.1680/icemge.66816.0327

CONTENTS

Chapter 27
Geotechnical parameters and safety factors

Paul Morrison, Paul Bailie and Tim Chapman Arup, London, UK

A key aspect of geotechnical design involves the choice of appropriate parameters for characterising the ground. This chapter describes how the engineer should set about this important aspect of design. Great emphasis is placed on the fact that the selection of the appropriate geotechnical parameters must be undertaken in the context of overall considerations of risk. A number of examples are given to illustrate how the characteristic values and design values of various geotechnical parameters are selected for a variety of problems. Much of the discussion relates to the requirements of Eurocode 7, but reference is also made to other more traditional approaches.

27.1. Introduction

27.1.1 The design process

This chapter presents the process that should be followed for choosing the appropriate geotechnical parameter for a particular safety factor regime – the choice of parameter should always be made in the context of the design situation and the safety factor regime. The chapter generally follows the approach proposed by Part 1 of Eurocode 7 (BS EN 1997-1:2004a). Other design codes are also referenced to illustrate how different approaches to design can be used to achieve a similar result.

Eurocode design is based on a cautious estimate of the parameters affecting the occurrence of a limit state; these characteristic parameter values are used along with partial factors to calculate design values. The design values are then used to assess whether a limit state is being satisfied (e.g. the design axial load applied to a pile is less than the design resistance to axial load of the pile). Note that design to the Eurocode uses partial factors applied to the individual variables under consideration rather than a combined factor of safety. Within geotechnical design to Eurocode 7 partial factors are applied to both material strengths (or resistances in piles) and applied actions, in order to factor each element in the design according to the degree of uncertainty to which the parameter/action is known and according to the variability in the value of the parameter/action.

Regardless of the code being used, the choice of geotechnical parameters and safety factors is a balance between economy, sustainability and reliability. A cautious approach to parameter selection is safer and the resulting structure will be more robust with a lower risk of under-performing than if a less cautious approach is taken. However, few clients are prepared to pay for an excessive level of redundancy resulting in design being the art of making elements safe and functional as well as economical. In geotechnical design especially, where the risk of failure is very real and the ability to predict failure accurately is not easy, the art of design is to select a design basis that is robust, but not excessively so. The cautious choice of parameter must reflect the design situation or limit state being considered; in some situations, a high strength of the ground could be 'cautious' (e.g. the strength of the ground when it is moving laterally compared to a restrained pile within it).

Design is usually carried out within the constraints of a code, for example Eurocode 7. The preparation of a code allows the experience of others to be encapsulated in a concise manner. The choice of appropriate ground parameters must therefore be compatible with the intent of the code writers for the values of partial factors that are to be used. Furthermore, the choice of geotechnical parameters must be made in the context of the complete structural design being embarked upon. For geotechnical parameters to be correctly identified and assessed through the design and construction process the actions listed in **Table 27.1** are recommended.

In addition to the actions detailed in **Table 27.1**, the following considerations are important in ensuring good design and execution.

- Consideration of how a design can be robust to accommodate unforeseen or unexpected loading events is important to both ground and structural elements.
- It is important to utilise a procurement process in which the allocation of risk is carried out appropriately and which rewards experienced contractors who make proper allowance for those risks for which they are responsible. To deal with unexpected conditions the contract should place an onus on the contractor to identify such conditions at as early a stage as possible and should thereafter allow for reasonable compensation (time/money).

The confidence that errors in design or construction have been avoided can always be increased by additional third-party checking, both of the design process (such as Category 3 checks proposed by the Highways Agency standards) or on construction sites (such as independent supervision, full-time on-site auditing of the processes and verification of records). These factors are explored by Chapman and Marcetteau (2004).

Action	Output (Eurocode 7)
A well-planned desk study should be carried out; this may include a site walkover (recommended) as well as a preliminary site investigation – see Section 27.4.	Desk study report with preliminary ground model
Carrying out well-planned ground investigation that responds to the risks identified in the desk study. Interpretation of the findings of the investigation, allowing sufficient time for monitoring of groundwater and environmental gases to be completed. Part 2 of Eurocode 7 'Ground investigation and testing' (BS EN 1997-2:2007; BSI, 2007a) gives suggested scopes.	Factual and ground investigation report with refined ground model
Use of experienced and knowledgeable designers who take part in a coherent design process using proven models which include consideration of geotechnical, structural and performance requirements.	Geotechnical design report with recommendations for construction and monitoring
Implementation of construction trials/tests where necessary to validate the design method.	Update/revision of the geotechnical design report
Clear design documentation (geotechnical design report) must state the assumptions made in the design in a clear manner to allow the professional involved during the construction (execution in Eurocode terminology) phase to assess ground conditions against the design assumptions. Revaluation of the geotechnical parameters/conditions may be necessary at any stage of design and construction. In the design of geotechnical elements, it is important for the design to anticipate the construction processes to be used, as those processes may affect the magnitude of the design parameter that is used – for instance, different methods of piling cause different values of skin friction.	Feedback loop from contractor to designer to ensure that design intent and assumptions are correct
Where appropriate, validation testing of parts of the construction to verify that design assumptions and construction methods are compatible (e.g. working pile load tests or ground anchor tests) to validate the design models and geotechnical parameters adopted. Testing samples of the works (e.g. working pile load tests) may also be necessary to verify that the constructed elements do not differ from the trial constructions or design intent.	Feedback to designer for revision of the design; use in subsequent designs to enable continuous development

Table 27.1 Recommended actions for identifying and assessing geotechnical parameters

The management of temporary works risks during construction can include use of the observational method – this permits the adoption of less pessimistic parameters and replaces the otherwise lost confidence by more robust management processes, supported by construction monitoring and a set of feasible predetermined contingency measures. This is covered by Nicholson *et al.* (1999).

Within the stricture of modern design codes there is limited opportunity to deviate from the partial factors (or factors of safety) stated in design codes. It is only for design of the most basic of structures that design based on comparable experience without the direct incorporation of factors of safety is permissible. (It is noted that for structures of high importance, such as hospitals in earthquake regions, there is a logical need for increased safety in these structures due to the consequences of failure being so great, and this may require an approach to assessment of ground model and parameters that are beyond conventional codes.) As such, it is vitally important that the choice of ground parameter values is appropriate to the level of safety provided for in the code's stated partial factors; the designer must maintain control of the design process by informed choice of the geotechnical parameter to use within a framework of partial factors.

27.1.2 Design codes

In design to Eurocode 7 the value of the partial factors (e.g. in the UK NA to BS EN 1997-1:2004; BSI, 2004b) is based on the assumption that the designer will adopt a ground parameter value that is a cautious estimate of the parameter's value relevant to the limit stage being considered (Clause 2.4.5.2 of BS EN 1997-1:2004; BSI, 2004a). Failure to adopt a cautious estimate of the ground strength would result in a potentially unsafe construction. In contrast to a cautious estimate approach, the London District Surveyors Association (LDSA) note for pile design in London (LDSA, 2017) provides an alternative approach to pile design (in addition to a Eurocode 7 compliant design) which stipulates that the undrained strength of London Clay should be taken as an average value (and not a cautious estimate). Such an approach is reasonable based on the great knowledge from long history of pile testing to failure in the area and due to the size of factor of safety applied to pile resistance; the combination of mean ground strength and factor of safety provides a level of confidence deemed reasonable. Both of these approaches – Eurocode 7 and the LDSA – are valid alongside many others when applied consistently across a design; they must not, however, be mixed. It is recommended that when the rest of the structure is being designed to the Structural Eurocodes that a Eurocode 7 compliant foundation design is adopted for the geotechnical components, this results in a unified design where there is reduced risk of introducing errors when factors applied to actions are transferred from one part of the design to another.

The list of codes in **Table 27.2**, including Eurocode 7 and the LDSA (2017) note, shows how the degree of conservatism used in the geotechnical parameter is balanced by the value of the partial factor on the geotechnical parameter. The methods are all different but aim at achieving similar end constructions.

27.2. Overall consideration of risk

27.2.1 An example

When considering the choices for partial factors and ground parameters, it is necessary to consider the overall risk regime. Eurocode 7 Part 1, Clause 2.1 introduces the concept of

Reference	Caution applied to soil parameters	Partial factor/factor of safety
LDSA (2017) – Option 2	Mean value	High value of partial factor/factor safety
NA to BS EN 1997-1:2004 (BSI, 2004b)	Characteristic line or a cautious estimate	NA to BS EN 1997-1 partial factors
CIRIA R185 *The Observational Method in Ground Engineering* (Nicholson *et al.*, 1999)	Moderately conservative line or conservative best estimate Allows less conservatism applied to soil strengths through monitoring of structure	
BS 8002:2015 Code of practice for earth retaining walls (BSI, 2015b)	Minimum = when combined with partial factor of: • cautious estimate of peak strength • constant volume	• NA to BS EN 1997-1 values • 1.0

Table 27.2 Code comparison

geotechnical safety/risk categories for design, where Category 1 is for small and relatively simple structures with negligible risk, Category 2 is for normal structures and Category 3 is for special structures where the consequences of failure are worse than normal. Some of these concepts are illustrated in **Figure 27.1**.

The idealised example of a 'project' in **Figure 27.1** is provided to highlight the need to design not only for the 'project' but also for the surrounding context of the project. It is suggested that the design of the foundations for the project building need have no more than typical safety requirements (as supplied by the combination of caution in choice of geotechnical parameters and partial factors) on the assumption that local failure of the building is not anticipated to impact on the nuclear power plant below. However, the design of the retaining wall above the nuclear plant will need to have enhanced safety due to the consequences of its failure. This enhanced safety would logically include larger partial factors on the building loads from the 'project' than would be used to design the building itself.

The design for the retaining wall supporting the building will be subservient to the design of the wall protecting the nuclear plant. The implication of this is that the combination of geotechnical parameter and partial factor used to design the wall to protect the plant will be the controlling feature and will result in lower *design soil strength* than would be used for design of the 'project' foundations (same ground with different levels of safety).

In the context of geotechnical safety/risk categories, the 'project' could be a Category 2 structure, while the wall should be a Category 3 structure. It must be recognised that the stability of the building depends crucially on that of the retaining wall, so the building designer should be satisfied that the retaining wall is both sufficiently stable and has a design life commensurate with that intended for the building.

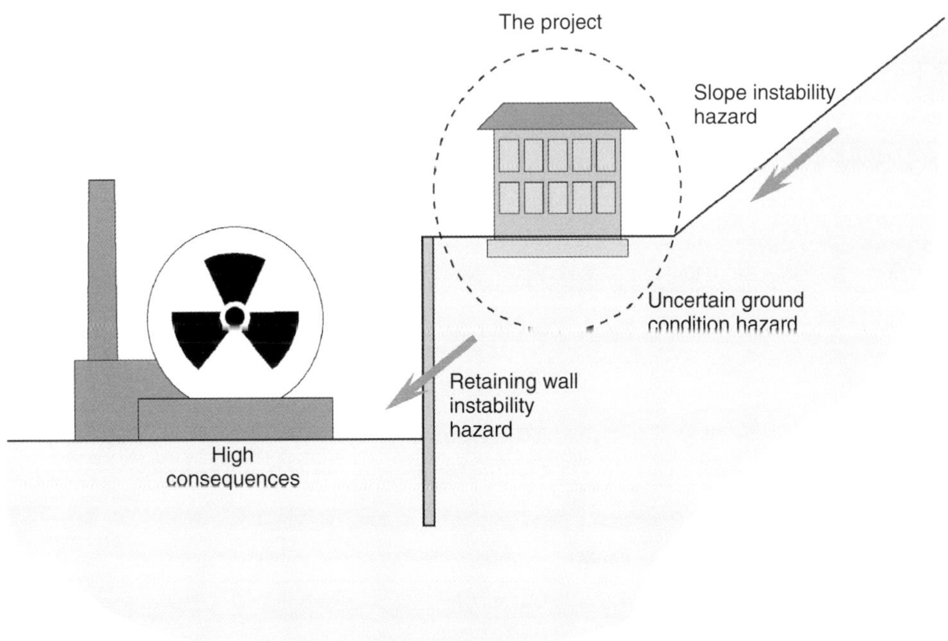

Figure 27.1 The context of a design problem

As a parting comment on this example, it is interesting to consider the geotechnical category that should be placed on assessment of the slope uphill of the project building. If a slope failure could impact on the nuclear plant, then it would be Category 3; if not, then Category 2.

27.2.2 Risk factors

The factors that may affect the level of caution applied in design are summarised in **Table 27.3**.

Precautionary measures to reduce the risk for geotechnical elements can be applied by

- robust design processes including desk study and ground investigations
- checks on design
- checks on construction
- some redundancy in the structures (e.g. many piles per column).

27.2.3 Geotechnical categories to Eurocode 7

The concept of geotechnical category to take account of risk as presented in Eurocode 7, Part 1 (see Section 27.2.1) is useful in defining the level of caution that is appropriate when considering the design of a structure. These geotechnical categories are summarised in **Table 27.4**. An analogous assessment of risk is presented in the Eurocode (BS EN 1990:2002+A1: 2005; BSI, 2005).

27.3. Geotechnical parameters

27.3.1 General

Geotechnical conditions are represented in design by the ground model and the geotechnical parameters for the ground units within the ground model, along with groundwater conditions such that they can be modelled for the purposes of that design. Geotechnical parameters can be measured directly or obtained through the use of correlation, theory and empiricism, or they can be sourced from published data; it is often best to obtain geotechnical parameters in more than one way to allow confidence in the variability of the parameter to be developed.

The terms 'geotechnical parameters' and 'design parameters' are often used interchangeably. However, in the context of this chapter and in the Eurocodes, a design value is the particular value of a geotechnical parameter selected for a specific design purpose. The numeric value of the design value will vary depending on the limit state being considered as a function of

Factors due to the supported structure	Factors due to the project or its situation
Structural form very vulnerable to damage or collapse (e.g. statically determinate)	Difficult or complex ground, or poorly understood ground conditions, for example susceptible to progressive failure
Columns supported by single piles	High or very variable groundwater conditions and groundwater flows
High economic consequences of failure (e.g. the structure is vital to the cash flow or financial well-being of a major organisation)	Seismic conditions or ground otherwise unstable (e.g. on a hillside suspected as prone to failure)
High safety consequences of a failure (e.g. risk that many people may be injured or killed by a geotechnical failure, such as a concert hall or sports stadium)	High live loads give a greater level of uncertainty of the maximum load likely to be applied. Cyclic loading results in large settlement
High safety consequences of a failure due to the processes carried out within, such as a chemical production or electricity generation from nuclear power	Ground conditions where accurate prediction is difficult. Complex ground, or poorly understood ground conditions; attention to settlement at working loads and how these vary over time
Monitoring and maintenance regimes are important	
Choice of foundations which are unproven in the circumstances or where comparable experience suggests a high rate of defects	Appointment of an inexperienced or poorly equipped foundation contractor

Table 27.3 Consideration of risk

	Category 1	Category 2	Category 3
Risk level	Negligible	Not exceptional	Abnormal
Groundwater level	Construction above the water table only	Construction above or below the water table	Construction above or below the water table
Ground	Straightforward with good record of performance for type of structure proposed	Routine	Exceptionally difficult
Type of analysis	Qualitative	Quantitative	Special (quantitative)
Typical structures	Small and relatively simple	Normal spread footings, piles, rafts, retaining walls, etc.	Large or unusual; or in highly seismic areas, etc.

Table 27.4 Geotechnical categories (Data taken from BS EN 1997-1:2004)

the particular level of safety required in the specific calculation being carried out and the manner in which safety is achieved in the calculation (factoring of actions or factoring of resistances or a combination of both). For ultimate limit state (ULS) calculations, the design parameter is typically the factored version of the *characteristic* geotechnical parameter. For serviceability limit state (SLS), the design and characteristic geotechnical parameters are typically equal. In all cases the partial factor is applied to adjust the characteristic geotechnical parameter. Sometimes an upper value of a parameter may result in a less safe condition than a lower one – hence just picking a lower bound value is not always appropriate without understanding the calculation to be carried out. This may mean it is necessary to increase the 'cautious upper' estimate of the geotechnical parameter upwards rather than downwards to obtain the value of the design parameter to use in ULS assessments.

In certain circumstances, rather than determining a unique value of a geotechnical parameter, it may be appropriate to consider a justifiable range of parameters to see how these influence the design being carried out; such a situation may be applicable to ground strength, but is more typically applicable to ground stiffness. The estimate may have varying levels of caution, depending on the limit state being considered, and this should accommodate the balance between caution in choice of parameter and caution inherent in the magnitude of factor of safety or partial factor being applied and the implication of failure of the limit state being considered.

A geotechnical parameter is the culmination of an interpretation process in which geotechnical data are critically examined through a process that considers code requirements and issues such as

- ground model
  - geological history
  - uncertainty with respect to the ground conditions
  - scope of ground investigation – is it adequately thorough?
  - groundwater conditions and variation with time
  - treatment of anomalous data
  - anisotropic or heterogeneous ground
  - brittle or strain softening ground
- risks associated with proposed structure and design simplifications
  - type and complexity of the intended structure
  - required design life
  - geotechnical design models
  - scale/rate/strain effects
  - comparable experience
- construction-related factors
  - anticipated construction methods and workmanship
  - adjacent site activities
  - construction monitoring and feedback loop involving the designer
- social considerations
  - level of acceptable risk/importance of the intended structure.

27.3.2 Eurocode 7

Eurocode 7 uses the following four terms when referring to geotechnical data

- measured data
- derived values
- characteristic values
- design values.

Measured data are, as the name suggests, data collected in their most basic format – for example, standard penetration test (SPT) *N* values.

Derived values are ground properties interpreted from measured data directly or indirectly (e.g. correlation, theory and empiricism), and from other sources such as published data. An example of an indirectly derived ground property could be undrained shear strength based upon SPT data.

Characteristic values are the basic form of geotechnical parameters used in design and are defined in Eurocode 7 as 'a cautious estimate of the value affecting the occurrence of the limit state'. They are selected from derived values.

Design values are characteristic values that have been modified by a partial factor determined according to the limit state being considered and the design approach being adopted. It is noted that in some instances (e.g. when dealing with water pressures) design values may also be selected directly without the use of partial factors (albeit the partial factors give guidance on the level of caution required).

27.3.3 Considerations for parameter choice

27.3.3.1 General

The partial factors in Eurocode 7 are based on a geotechnical parameter being a cautious estimate of the value controlling the limit state (referred to as a 'characteristic value' as noted above). Clearly this cautious estimate of the geotechnical parameter implies a value that is offset from what is deduced to be the most probable value by a margin to provide a level of caution.

Choosing a characteristic value is not about putting a line through a set of data. It is about understanding the issues that lead to the choice. Once these are understood, the selection of a characteristic value or line should be straightforward, and while it could be curved, straight, stepped, constant with depth or increasing with depth, an overly complex line is generally not necessary or helpful. An overly complex choice more frequently leads to errors in calculations.

Box 27.1 Case study 1

This example is one of a pile embedded in a stiff clay formation which is subjected to a fixed lateral displacement at the top of the pile 1 m above ground level, as shown in **Figure 27.2**.

The undrained shear strength with depth, measured on 100 mm dia. samples tested in quick undrained triaxial compression, is shown in **Figure 27.3**.

The measurement of undrained shear strength has been carried out by means of undrained triaxial compression tests. The clay stratum was logged as being uniform with the exception of a sand layer at 16.5 m depth. Prior to lines being drawn on the data it is necessary to review the data to isolate any results that are not representative of the ground. Notes are provided in **Figure 27.3** for three of the data points which are considered to be atypical of the in situ state. The high value at approximately 6.5 m depth is noted in the laboratory test log as having dried out; such data may be excluded from further assessments. The two data points at 16–17 m depth appear to be soft and coincide with a sand layer; it is likely that sand and water from the sand layer have been included in the samples obtained, allowing the clay to swell within the sample tube prior to testing. While the measurement of undrained strength is therefore not representative of in situ conditions, the observation that the sand layer may have implications for pile construction should be carried through to the design process. The end product of this review is that three data points have been removed from the ground data prior to assessing the characteristic strength of the clay.

It is usual to assume that a characteristic design line (cautious estimate) through the data would be in keeping with the 'cautious lower' line. By adopting this line the strength of the soil will be towards a lower bound. However, when considering the limit state relevant to the structural design of the pile, adopting this line as the characteristic property of the ground will likely underestimate the bending moment and shear force in the pile as it would result in a low assessment of mobilised resistance of the ground to imposed pile head movement (δh). Hence for this particular design case the correct characteristic design line through the undrained shear strength data will be the 'cautious upper' line as only this will provide the appropriate resistance to the pile head movement for carrying forward to structural design of the pile.

The lines shown for the 'cautious lower' and 'cautious upper' are open to discussion and tweaking. For the dataset used the lines are close to being parallel; this is clearly not a requirement and the splay from the average line should be a function of the scatter in the data and the confidence that the designer has in the data.

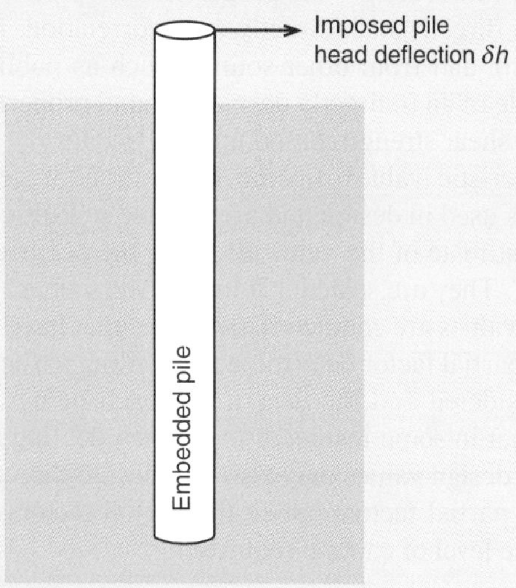

Figure 27.2 Laterally loaded pile

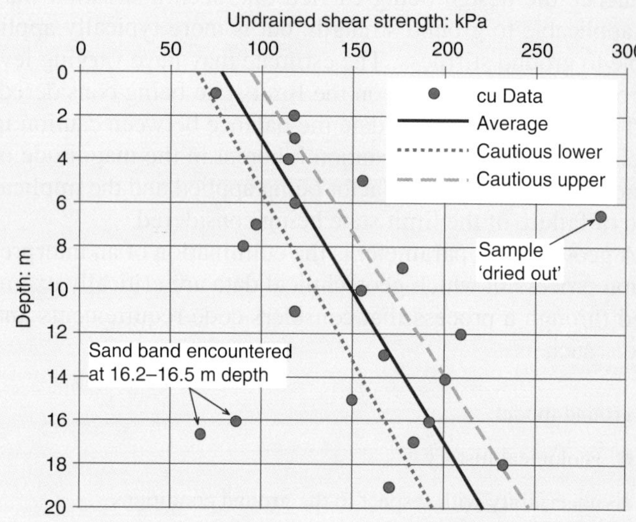

Figure 27.3 Selection of characteristic line

While the use of statistical methods to determine a characteristic line, by attributing a statistical definition to the characteristic value, has been promoted by various authors, they are often a trap. It should be kept in mind that the ground is not a well-behaved material (e.g. like concrete), datasets are often small and not of normal distribution, and they can contain significant scatter – not an ideal situation for the application of rigorous statistical approaches. This issue is given further attention in the proposed update to Eurocode 7 (also see Section 27.5).

Rather than try to describe how a characteristic line could be selected, it seems more appropriate to communicate the idea using an example. The case studies in **Boxes 27.1–27.5** provide a simple illustration of how such an assessment could be made in the context of Eurocode 7.

As a final comment on the choice of value of the characteristic parameter, it is necessary to also consider how it is used in design. BS EN 1997-1 provides partial factors for the unconfined strength (UCS) of a rock; however, the behaviour of rock is rarely related linearly to the rock's UCS, hence factoring in the UCS is not necessarily helpful – the designer needs to consider how their design uses the data and the implications of how partial factors are applied (this topic is anticipated to be addressed in the next version of BS EN 1997-1).

27.3.3.2 Illustrative considerations for shallow foundations

For shallow foundations, design parameters are very dependent on the construction process – in particular, the preparation of the bearing surface. They also typically engage far less ground than, for example, a pile, and can be affected by localised changes in ground properties or by non-typical soils such as sensitive clays, loose sands or collapsible soils. Sometimes those more onerous conditions may lurk just below the exposed formation and may not be observed in the formation preparation. Geotechnical parameters need to be selected with these issues in mind.

27.3.3.3 Illustrative considerations for piled foundations

For piles, empirical design parameters/coefficients are very dependent on the construction processes used to form the piles and a clear separation between design and construction responsibilities is not as simple as in other branches of engineering. For instance, in the event of a pile failure it may be because the geotechnical parameters/coefficients were too optimistic for the likely construction methods, or because the construction methods failed to live up to the standards expected by the designer.

Hence, geotechnical parameters/coefficients need to be selected in full cognisance of

- the likely construction methods
- the range of results of load tests carried out for the particular pile type and ground conditions
- the specification to be employed, to ensure that processes known to be detrimental are not used, or, where it would be impractical for them not to be used, that the design parameters are modified accordingly.

The detailed choice of construction methodology needs to be made in full cognisance of the design assumptions and parameters.

27.3.3.4 Illustrative considerations for slopes

It is often a thin layer of soil that is most critical for the stability of a slope. For example, the back-analysis of the failure at the Carsington dam by Potts *et al.* (1990) showed that a thin layer of soil had been neglected in the analyses. Understanding the groundwater regime is equally important and, in particular, accidental events such as burst water pipes should not be forgotten.

The selection of design parameters is absolutely critical for slopes and it may be that just one data point (perhaps an outlier) is most representative for the behaviour of the overall slope. This is where the 'art' of engineering is particularly important – mere statistical analysis will not suffice.

For slopes, too often a computer is used to plot results of several drained triaxial tests, and the choice of what should be characteristic values for effective cohesion (c') and effective angle of internal friction (ϕ') is not made with sufficient understanding of the true soil behaviour. c' is unlikely to persist for shallow cut slopes as it will be lost due to weathering, and ϕ' should be chosen over a stress range that is representative of that which will occur in practice.

Box 27.2 Case study 2

An example of how the construction process can lead to problems is seen in CFA piles. A good CFA pile may corkscrew (or almost corkscrew) its way into the ground, thereby creating a pile with a ribbed shape. This creates a high-capacity pile where the shaft resistance is a combination of a failure interface with elements of intact soil to soil and intact soil and remoulded soil to concrete. Continued auger rotation during the construction process can, however, quite quickly remove the beneficial ribs by a low rate of penetration per revolution. Hence parameters derived from a test on a 'perfect' pile may not apply to a pile built by the same equipment but with less control, or where circumstances have led to greater auger rotations. Hence, where empiricism is used in design, the basis of the empiricism needs to be traced from the investigation stage, through the design stage to the construction stage (Morrison *et al.*, 2020).

27.3.3.5 Illustrative considerations for retaining walls

For retaining walls, critical geotechnical parameter choices can include

- appropriate selection of water pressures and cognisance of the effect of seepage throughout construction stages and design life
- allowance for locked-in earth pressures caused by compaction plant or integral bridges
- knowing when undrained behaviour can be relied upon
- allowing for construction control such as overdig allowance.

27.4. Factors of safety, partial factors and design parameters

27.4.1 General

Geotechnical design falls into two main categories: design against failure and design against unacceptable movement. In limit state design theory, these two concepts are known as breaching an ULS and breaching a SLS.

In many instances the historical choice of the value of a factor of safety allowed for settlements or lateral movements of a structure that were acceptable for typical performance requirements. Use of such all-encompassing factors of safety is acceptable (even in Eurocode – for example, Geotechnical Category 1) where there is well-established experience allowing 'rules of thumb' to be used. Where such rules do not exist, or where they are not suitable for the level of rigour required, then the designer must assess serviceability explicitly.

In other situations the factor of safety has simply been a factor to create a margin against a real failure, such as rupture or constant movement. Given the provenance of historic factors it is necessary for the designer to be aware of the implications of choosing a particular value to ensure that the resulting performance is satisfactory in terms of SLS and ULS for the design being undertaken.

Often geotechnical factors are composite and are created so as to be sufficiently high to prevent both ultimate and

Box 27.3 Case study 3

In the following example a situation is described along with discussion of how a designer may accommodate the ground conditions in the calculation process.

Situation: Open-cut excavation in a stiff high-plasticity clay stratum (**Figure 27.4**).

Context: Long-term design case critical.

Consideration: Case history records of open excavations in stiff overconsolidated high-plasticity clays exhibit progressive failure whereby a range of mobilised strength (c', ϕ') values are mobilised at different parts of the resulting failure mechanism.

- What design parameters should be used in design (long-term conditions)? How is progressive failure accommodated?
- Measurements of shear strength parameters may be made in the laboratory (triaxial, shear box, ring shear, etc.). These typically return values of peak resistance and may also show strength reduction with increasing shear strain. How can these results be carried into the analysis of slope stability?
- One option would be to incorporate in a sophisticated numerical analysis model of the slope and ground a soil model which fully accounted for the strain softening of the ground – in this manner a geotechnical parameter model could be populated with appropriately cautious ground parameters. In subsequent calculations partial factors can then be applied to achieve the appropriate level of risk. This is not a standard approach and not one advocated for general design.
- An alternative solution would be to look at the worst possible strength parameters (e.g. residual strength) that are considered possible and to use them with reduced partial factors, possibly equal to unity, to provide an equally coherent design approach.
- A third approach would be to use case history data from similar slopes to derive 'safe' parameters to bring forward into the design process. Clearly for this approach to be valid a large set of data for slopes with age similar to the design life of the project slope would be necessary; extrapolation for time-dependent failure is not viable.
- Adjustments will be necessary for any design input to allow for environmental changes in the future that are not replicated by past experience (e.g. global warming and associated changes in vegetation, precipitation and temperature variations).
- Clearly, consideration of strength parameters is only one part of the stability assessment. Equally important is the consideration of water pressures in the slope. Such considerations need to look at anisotropy of permeability and recharge boundaries. Any drainage provisions that are put in place will need to be maintained if they are to be relied upon in design (it is sometimes the case that drainage provisions are assumed to fail due to risk of not carrying out maintenance).

In this example three possible approaches to the slope stability are proposed. They are quite different in approach (cautious estimate of strength to worst possible estimate of strength), but by using an appropriate value of partial factors in each case the design can be shown to be coherent. As always, water pressures need to be chosen correctly.

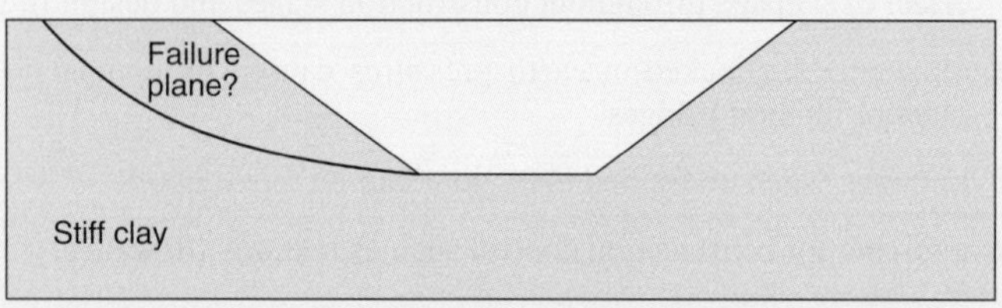

Figure 27.4 Open-cut excavation

serviceability failure. This is one reason why geotechnical factors are sometimes higher than in other branches of engineering. Another is that ground engineering is an inexact subject and utilises materials with wide natural variation and hence the extent of uncertainty in any parameter is much larger than for manufactured materials.

27.4.2 Eurocode basis

The Eurocode use of partial factors is significantly different from a factor of safety approach, in that they are written in terms of limit state principles rather than those of working stress design. The creation of Eurocode 7 brings geotechnical engineers into line with the longer established structural engineering limit state approach to design, which will hopefully lead to a more efficient and robust design process and better understanding between geotechnical and structural engineers.

In limit state design the designer is obliged to think about the problem at hand in terms of a particular limit mode and level of risk. Safety is applied in the form of partial factors on individual parameters, actions and action effects in a way that more rigorously takes account of uncertainty compared with the lumped factor of safety approach. The combination of partial factors on strength and actions requires clear communication when moving results from one code to another (e.g. between concrete elements designed to BS EN 1992 and geotechnical elements designed to BS EN 1997).

The Eurocodes divide limit state design into ULS and SLS.

- ULS are those concerned with collapse and safety; they are not intended to be realistic in terms of the day-to-day performance of a structure. Accidental cases are also covered by the ULS condition.
- SLS are those concerned with the day-to-day requirements for a structure, such as comfort of people and cracking; while the limit state is not anticipated to be passed, it is possible that it may be occasionally.

In order to verify that potential limit states will not occur, Eurocodes typically adopt the method of partial factors to achieve a suitable level of safety. Eurocode 7 goes one step further to define the following ways by which limit states may be verified

- by prescriptive methods ('rules of thumb')
- using calculation methods

Box 27.4 Case study 4

In the following example a situation is described along with discussion of how a designer may accommodate the ground conditions in the calculation process.

Situation: Basement formed in a retained excavation in stiff clay with layers of water-bearing sand (**Figure 27.5**).
Context: Design for short- and long-term conditions.
Consideration: The presence of water-bearing sand layers suggests that the overall behaviour of the ground to a retained excavation will be dominated by the presence of the sand layers.

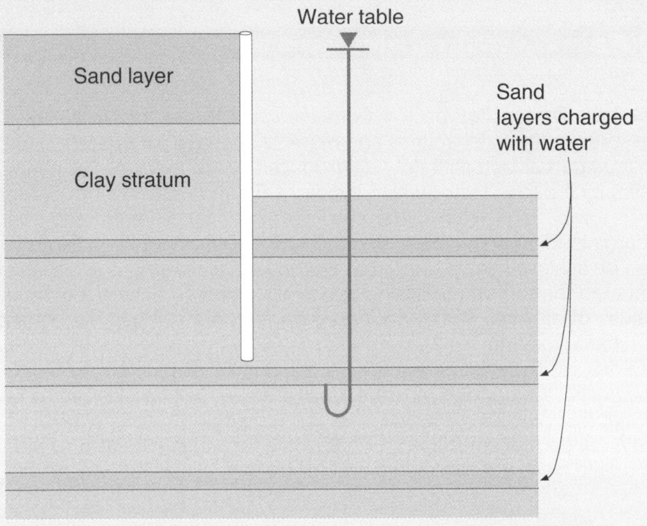

Figure 27.5 Retaining wall in layered soil

What design parameters should be used in temporary and permanent design?

- Typically in stiff clay undrained parameters are assumed for temporary works design. This is possible due to the drainage path length preventing all but the exposed formation from swelling. However, in this situation, what impact do the sand layers have? The designer must assess the effect that these layers have on the rate that the clay layers can swell and thus the length of time that the undrained strength can be relied upon. Is the geotechnical parameter a drained or undrained parameter? Caution would indicate a drained parameter in the absence of other case histories or additional data.
- The presence of a sand layer close to the toe of the wall requires consideration of uplift of the soil below the excavation. Could the base of the excavation 'blow'? Have water levels and supply rate both been investigated to allow such a design decision to be made? What design conditions should be attributed to the sand layer (permeability and recharge potential)? How can the water level (piezometric pressure) be managed?
- The assumption of the retaining wall forming a cutoff below formation level cannot be taken for granted without careful construction and trial dewatering. If the wall is not continuous then pore pressure in the upper sand band may be recharged through the retaining wall and cause loss of passive resistance in the zone immediately below formation level (this is the zone that controls retaining wall movement the most). What design assumptions are made and how can these be validated?

While design parameters are usually considered in terms of strength and stiffness, in this case in order for these to be useful they need to be understood in the context of permeability of the ground, the anisotropy of the ground and the hydrogeological boundary conditions. It is as important to make cautious assessment of these other factors as it is for strength and stiffness.

- through models and load tests
- by observational methods.

In the partial factor method, actions, materials, resistances and geometry are factored using prescribed, partial (γ) and combination (ψ) factors. The majority of factors contained within the Eurocodes have been determined by calibration to well-established experience, so a design undertaken using them should be similar to conventional designs undertaken with a single lumped factor of safety. Some factors have, however, been determined in a probabilistic manner from laboratory and field data.

27.4.3 Traditional code basis

The traditional, and now superseded, standards for geotechnical design (note: the Eurocodes are British Standards) have been replaced by Eurocode 7 in the UK. However, the traditional standards are often revised and redated in the form of non-contradictory complementary information (NCCI) referenced in the Eurocodes. This means that the very useful information and guidance contained within these 'codes of practice' can be used in conjunction with Eurocodes; examples include 'BS 8004:2015: Code of practice for foundations' and 'BS 8002:2015: Code of practice for earth retaining structures' (BSI, 2015a, 2015b).

27.4.4 Consideration of safety

27.4.4.1 General

Historically, factors of safety in geotechnical engineering have incorporated both ULS and SLS considerations without being explicit about these limit states. For foundations, as per Eurocode, it is better practice for these to be considered separately. Factors only providing a margin against failure can be lower than those seeking to control movements, for instance.

27.4.4.2 Illustrative considerations for shallow foundations

For shallow foundations designed to Eurocode 7, the design bearing resistance is calculated using ground parameters modified by partial factors. This design bearing resistance is then compared with the design effect of actions which are also factored (note: in traditional design the ultimate bearing resistance would be factored and referred to as the safe, or allowable, bearing pressure).

Box 27.5 Case study 5

For design against failure the factor of safety has historically been defined as the ratio of resisting actions to disturbing actions. The units of these actions can be force, moment or indeed pressure or stress, depending on the calculation being carried out. In order to best understand the factor of safety, the following example is presented.

In **Figure 27.6** the factor of safety for the gravity wall in sliding is

$$\text{FOS} = \Sigma_{\text{resisting}}/\Sigma_{\text{disturbing}}$$
$$= (P_{P1} + T_{S1})/P_{A1}\,(\text{typically a value of} \geq 2 \text{ would be required}) \qquad (27.1)$$

where all actions are based on unfactored parameters rather than factored parameters.

It is noted that the 'resisting actions' should be taken as the sum of all resisting actions and the 'disturbing actions' the sum of all disturbing actions. When calculating a bulked factor of safety it is not permissible to consider the net disturbing or restoring action as part of the calculation.

With the introduction of limit state design, the term 'factor of safety' has been dispensed with and the level of safety is now accommodated through independent partial factors on actions, materials, geometry and resistances to arrive at design values. The resulting verification of safety for a particular ultimate limit state calculation becomes

$$\text{Design value of the effect of actions } (E_d) \leq \text{design value of the resistance } (R_d) \qquad (27.2)$$

Considering the loading conditions in **Figure 27.6** and using Eurocode 7 calculation requirements for the GEO/STR condition of the ULS (Design Approach 1 Combination 2 as per the UK NA to BS EN 1997-1:2004), all actions and material strength parameters attract unique partial factors. Safety is introduced by reducing the soil parameters by a partial factor which increases P_{A1} and reduces both T_{S1} and P_{P1} as well as increasing the detrimental influence of transient surcharge loading. Verification that the limit state has not been exceeded can be achieved by showing that the design value of P_{A1} is less than or equal to the sum of the design values of P_{P1} and T_{S1}, where the design values of all three forces are based on factored soil parameters and surcharge loading.

Note: for Design Approach 1 Combination 2 of BS EN 1997-1:2004 the partial factors that must be applied are: $\gamma_{\varphi'} = 1.25$ (applied to tan φ'); and $\gamma_{c'} = 1.25$; $\gamma_{cu} = 1.4$.

For all permanent loads $\gamma_G = 1.0$, while for variable loads only unfavourable loads are considered and for these $\gamma_{Q;unfav} = 1.3$ ($\gamma_{Q;fav} = 0.0$).

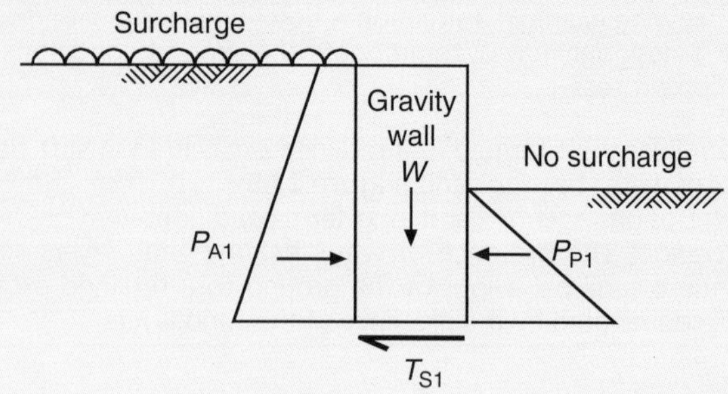

Figure 27.6 Factor of safety against sliding

	Permanent actions γ_G	Variable actions γ_Q	Tanϕ' $\gamma_{\varphi'}$	c' $\gamma_{c'} = 1.25$	c_u γ_{cu}
Combination 1	Favourable = 1.0 Unfavourable = 1.35	Favourable = 0.0 Unfavourable = 1.5	1.0	1.0	1.0
Combination 2	Favourable = 1.0 Unfavourable = 1.0	Favourable = 0.0 Unfavourable = 1.3	1.25	1.25	1.4

Table 27.5 Partial factors for GEO and STR limit states for NA to BS EN 1997-1:2004

The values of the partial factors for the Geotechnical (GEO) and Structural (STR) considerations encapsulated in Design Approach 1 (Combinations 1 and 2) of BS EN 1997-1 are shown in **Table 27.5**.

In many instances the STR limit state will be controlled by Design Approach 1 Combination 1 and GEO will be controlled by Design Approach 1 Combination 2. However, BS EN 1997-1 requires that both GEO and STR are checked against Design Approach 1 Combination 1 *and* Combination 2.

For Combination 2 the ground strength values are factored, leaving density or surcharge effects largely unfactored or with small partial factors, while for Combination 1 the actions or effects of the actions are factored (e.g. surcharge loading and calculated shear stress or bending moment in an element). By splitting into these two assessments it is possible to be confident that both the geotechnical stability (usually this would be assessed first and controlled by Combination 2) and then the structural forces are addressed correctly with both combinations.

Settlements (SLS) are considered separately and computed using partial factors of unity for the actions imposed. They are compared to the values of settlement that it is thought the structure, or the function of the structure, can withstand without impact on its use. If the calculated settlement is too high, then the design is modified (note: in traditional design the safe bearing resistance would be reduced to an allowable bearing pressure if settlements controlled the design).

27.4.4.3 Illustrative considerations for piled foundations

In traditional design for piles, two criteria are often applied

- an overall factor on the combined shaft and base resistances, giving a margin against failure

- a reduced factor on the stiffer and normally more reliable shaft capacity alone, to limit movements to the stiffer part of the pile load–settlement curve, giving control on movements of the supported structure.

There are cases when this approach can be too conservative, such as when the pile base performance is known to be both stiff and reliable; in these cases the factor on shaft capacity is not used.

Under Eurocode 7 the capacity and settlement of a pile are usually dealt with separately; the design resistance is calculated by modifying the characteristic resistance by partial factors. However, where serviceability is not considered explicitly, an alternative set of partial factors on resistance is used.

For temporary works, sometimes lower safety factors are considered. However, their use is only appropriate where the likely consequences justify it, for example the higher resulting risks can be accommodated/mitigated and where larger displacements are acceptable.

27.4.4.4 Illustrative considerations for slopes

For slope stability, overall moment and vertical stability are the key ULS concerns. Serviceability is generally related to the effect that slope displacement has on surrounding structures rather than internal displacements, unless there are aesthetic constraints to consider.

27.4.4.5 Illustrative considerations for retaining walls

For retaining walls, the geotechnical modes of failure are more varied and typically require separate calculations – for example, rotation about the top, rotation about the base, vertical displacement and horizontal displacement. While safety is typically accommodated in partial factors on actions and ground parameters, geometrical variation and progressive failure checks are also important.

Serviceability is typically addressed in terms of lateral movement and the effect this has on supported structures, although excessive displacement of a wall into a basement, for

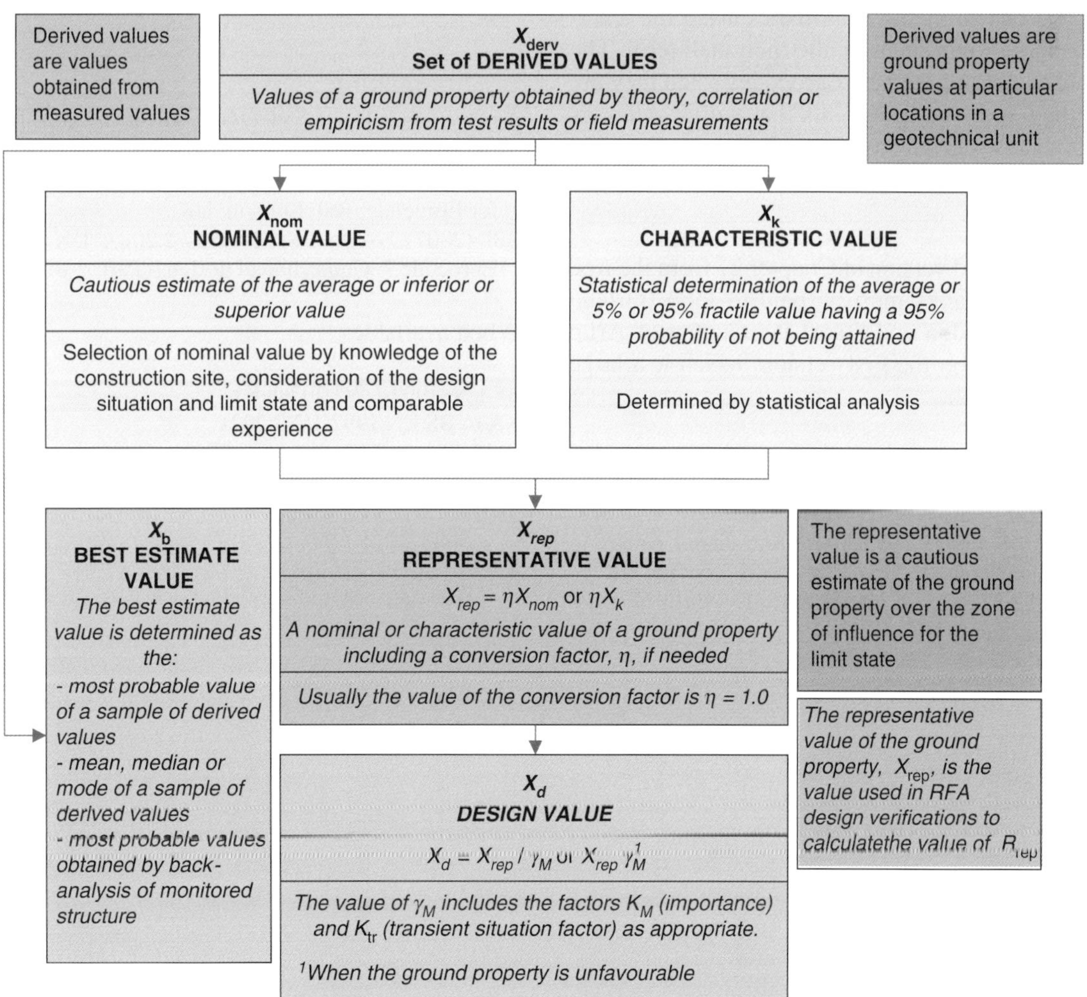

Figure 27.7 Possible revisions to BS EN 1997-1 terminology currently in review (see Guideline document to EN1997-1:202x when published)

example, may impact upon its amenity despite it not causing problems for structures that it supports.

27.5. Concluding remark

Geotechnical parameters and safety factors must be viewed holistically: the manner by which one is chosen needs to mirror the level of risk in the other. Beyond ground strength and designing against failure, the assessment of soil–structure interaction, movement and performance of structures associated with the ground elements need also to be assessed, with an understanding of the level of certainty in the design parameters.

As per the reference list below, this chapter has been based on BS EN 1997-1:2004 and its National Annex (BSI, 2007b). The Structural Eurocodes, of which BS EN 1997-1 is part, are being revised and changes are being proposed in the nomenclature used in the transition from derived properties of the ground to the design values used in calculations. Key aspects of the proposed change terminology for BS EN 1997-1:202X (REVISED) are shown in **Figure 27.7**. The main aspect to take note of is the use of the term 'representative value' being obtained from one of two methods: statistics using the 'characteristic' term; and estimation using the 'nominal' term. The reader is advised to consult the new Eurocodes when they are published and the Guideline documents that will accompany them to obtain a most current use of terminology over and beyond **Figure 27.7**.

Acknowledgement

This chapter is a reworked version of Chapter 27 from the first edition of this book. Thanks must be paid to John Burland of Imperial College London and Stuart Pennington of Arup Australia for their work on the first version, which is a large part of this current version.

References

BSI (British Standards Institution) (2004a) BS EN 1997-1:2004: Geotechnical design – Part 1. BSI, London, UK.

BSI (2004b) NA to BS EN 1997-1:2004: UK National Annex to Eurocode 7 – Geotechnical design – Part 1. BSI, London, UK.

BSI (2005) BS EN 1990:2002+A1:2005: Eurocode – Basis of structural design. BSI, London, UK.

BSI (2007a) BS EN 1997-2:2007: Geotechnical design – Part 2. BSI, London, UK.

BSI (2007b) NA to BS EN 1997-1:2007: UK National Annex to Eurocode 7: Geotechnical design – Part 1. BSI, London, UK.

BSI (2015a) BS 8004:2015: Code of practice for foundations. BSI, London, UK.

BSI (2015b) BS 8002:2015: Code of practice for earth retaining structures. BSI, London, UK.

Chapman T and Marcetteau A (2004) Achieving economy and reliability in piled foundation design for a building project. *The Structural Engineer*, 2 June.

LDSA (2017) *Guidance Notes for the Design of Straight Shafted Bored Piles in London Clay*. London District Surveyors Association Publications, London, UK.

Morrison P, Ganesharatnam L and Barker C (2020) CFA piling in the east of London. *Proceedings of Piling 2020 Conference, London, UK*. ICE, London, UK, pp. 169–174.

Nicholson D, Che-Ming T and Penny C (1999) *The Observational Method in Ground Engineering: Principles and Applications*. CIRIA Report R185. CIRIA, London, UK.

Potts D, Dounias G and Vaughan P (1990) Finite element analysis of progressive failure of Carsington embankment. *Géotechnique* **40:** 79–101.

Further reading

Bond AJ and Harris AJ (2008) *Decoding Eurocode 7*. Taylor & Francis, London, UK.

BSI (2005) NA to BS EN 1990:2002+A1:2005: UK National Annex for Eurocode. BSI, London, UK.

BSI (2007) NA to BS EN 1997-2:2007: UK National Annex to Eurocode 7: Geotechnical design – Part 2. BSI, London, UK.

When available:

BS EN (1997-1/2/3:202xUK)
NA to BS EN 1997-1/2/3:202x
Guidelines to EN 1997-1/2/3:202X.

All chapters within Sections 1 *Context* and 2 *Fundamental principles* together provide a complete introduction to the Manual and no individual chapter should be read in isolation from the rest.

Section 3: Problematic soils and their issues

Section editor: **Ian Jefferson**

doi: 10.1680/icemge.66816.0341

Chapter 28
Introduction to Section 3

Ian Jefferson University of Birmingham, Birmingham, UK

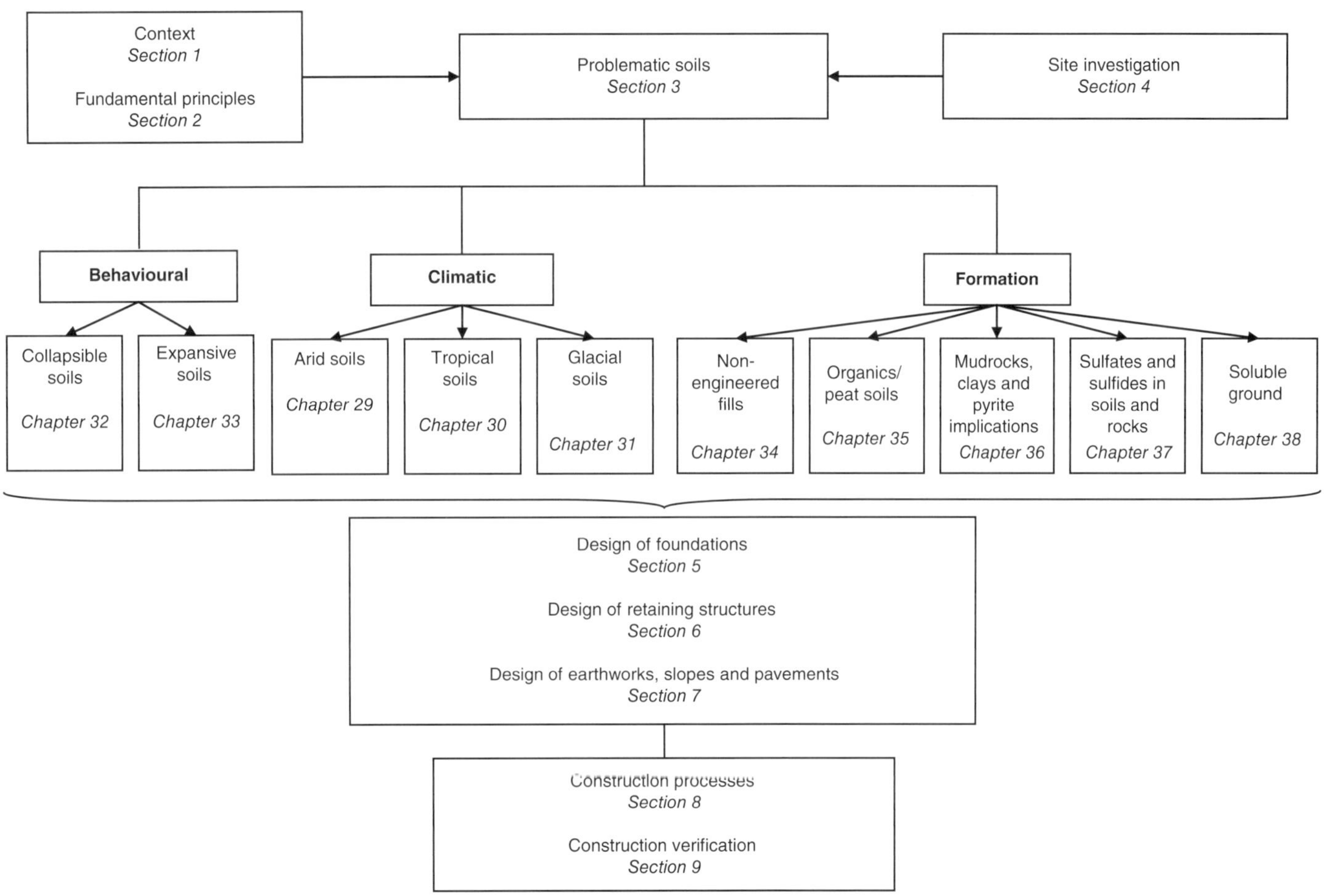

Figure 28.1 Layout of chapters in Section 3

Many soils can prove problematic to geotechnical engineering. Problematic characteristics include significant volume change, a distinct lack of strength or soils that are potentially corrosive. This creates difficulties across built and urban environments and continues to cause billions of pounds of financial loss annually across the world: costs that often exceed those associated with earthquakes, floods, hurricanes and tornadoes combined.

The degree to which a soil is problematic to engineering is a function of the

- nature of the soil itself (mineralogy, micro-fabric and geotechnical and other properties)
- geological processes that caused it to be formed (whether fluvial, glacial, aeolian or others)
- processes that are acting on it currently (weathering, erosion and human activity).

However, problematic soils are also profoundly influenced by the climatic regime in which they were developed. This creates many types of problematic soils and a number challenges when dealing with them. Nevertheless, these can be overcome by paying careful attention to

- recognition (through geology and geomorphology)
- classification and assessment (through correlations and laboratory and field testing)
- evaluation (through use of appropriate models – empirical, analytical or numerical)
- treatment by application of suitable and appropriately targeted engineering solutions.

Thus, problematic soils can be engineered effectively through observation, understanding and technological development, allowing soils that are considered problematic to be assessed in a rational manner using the approaches detailed throughout this manual. This will overcome what Leroueil (2001) considered as problematic engineering, designs, specifications or use of technologies that occur through a lack of appropriate or detailed understanding of the situation in hand. Engineers are moving to ever-better understanding of soil behaviour; however, as Vaughan (1999) suggests, many of our geotechnical materials seem unusual or 'problematic' because their behaviour does not fit the classic theories of soil mechanics. Thus, modification of basic soil mechanics is needed to incorporate problematic soils. However, as yet this has not been fully accomplished and many soils continue to cause problems globally. Further discussion on this is provided by Hirpara *et al.* (2019).

The challenges associated with problematic soils will only become more acute as the infrastructure and built environment base that society critically relies on continues to grow, necessitating ever more interaction with these soils. Add in drivers to reduce carbon impacts associated with key sustainability targets set by governments across the world, coupled to more extreme weather patterns associated with climate change, and knowledge of how to deal with problematic soils will only increase in importance in the future.

Section 3 *Problematic soils and their issues* covers the main problematic soils/ground encountered globally. These have been grouped to reflect the main process in their formation (**Figure 28.1**).

- Behavioural
  - Chapter 32 *Collapsible soils*
  - Chapter 33 *Expansive soils*
- Climatic
  - Chapter 29 *Arid soils*
  - Chapter 30 *Tropical soils*
  - Chapter 31 *Glacial soils*
- Formation
  - Chapter 34 *Non-engineered fills*
  - Chapter 35 *Organics/peat soils*
  - Chapter 36 *Mudrocks, clays and pyrite implications*
  - Chapter 37 *Sulfates and sulfides in soils and rocks*
  - Chapter 38 *Soluble ground*

These soil types or formations, which continually cause significant challenges, are associated with many of the difficulties experienced with problematic ground, across the UK and throughout the world. Moreover, these constitute the problematic soils/ground commonly encountered in geotechnical engineering. While this does not represent all possible types of problematic soils, the approaches highlighted throughout Section 3 *Problematic soils and their issues* provide a useful benchmark from which all problematic soils can be dealt with.

The ultimate aims of this section are therefore to

- highlight the importance of understanding geological and geomorphological aspects associated with problematic soils
- identify methods to assess and evaluate problematic soils and their behaviour
- present methods to treat and engineer problematic soils effectively.

This section provides a sound basis of engineering in problematic soils and thus aims to help prevent many of the costly mistakes that continue to plague engineers dealing with them.

References

Hirpara DV, Jadeja AS, Vekaria LC, *et al.* (2019) *Textbook of Problematic Soils*. Jaya Publishing House, Delhi, India.

Leroueil S (2001) No problematic soils, only engineering solutions. *Proceedings of the Problematic Soils Symposium, Nottingham, UK* (Jefferson I, Murray EJ, Faragher E and Fleming PR (eds)). Thomas Telford, London, UK, pp. 191–211.

Vaughan PR (1999) Problematic soil or problematic soil mechanics? *Proceedings of the International Symposium on Problematic Soils, Sendai, Japan* (Yanagisawa E, Moroto N and Mitachi T (eds)). Balkema, Lisse, the Netherlands, pp. 803–814.

doi: 10.1680/icemge.66816.0343

Chapter 29
Arid soils

Alexander Royal School of Engineering, University of Birmingham, Birmingham, UK

CONTENTS

Arid soils form in conditions where a water deficit (where mean annual losses through evapotranspiration exceed the mean annual volume of water entering the ground from precipitation) acts upon the soil. These soils can present several challenges to engineers due to the highly variable nature of their engineering properties, which are, in part, associated with active and ancient geomorphic processes. Arid soils can be unsaturated, cemented or uncemented, loose and poorly graded, and as a result can prove problematic to geotechnical engineering because of (but not limited to) metastable behaviour, significant shrink–swell or collapsing potentials, and/or aggressive salt environments that can weather man-made structures. Key to understanding how arid soils behave, and how these soils can be engineered, is achieved through an appreciation of the geomorphology of their formation and a suitable ground investigation. Once the ground conditions are understood, appropriate engineering assessments and interventions can be applied. Several other logistical challenges also occur with arid environments owing to the limited availability of suitable water sources for construction; through careful planning these problems can be avoided.

29.1. Introduction

Arid soils can be described as those conditioned by an arid climate; such climates occur when the potential annual evaporation exceeds annual precipitation and the result is a soil moisture deficit (Atkinson, 1994). Overall, it is estimated that one-third of the Earth's land mass currently experiences arid conditions (Warren, 1994a), yet only approximately 25% of the soils found in hyper-arid and arid regions comprises active sand dunes (**Figure 29.1**) (Warren, 1994b). Arid conditions can be found in both hot and cold locations on the Earth – that is, arid soils can comprise hot desert soils such as the Sahara and cold desert soils such as the Gobi.

In addition, movements of the tectonic plates result in soils and rock previously formed in arid conditions being found in temperate climates (such as the dune bedded sandstones (**Figure 29.2**) in the UK's Midlands), and soils not commonly associated with arid conditions being found in modern-day arid environments.

Figure 29.1 Sand dune in the Namib Desert, Namibia (Image courtesy of © R Jackson-Royal)

Arid soils can present several challenges to geotechnical engineers due to

- the highly variable nature of their engineering properties associated with active and ancient geomorphic processes
- (in hot arid conditions) the considerable diurnal temperature ranges experienced, potentially resulting in accelerated disintegration of soil and rock formations as well as man-made structures
- lithification processes, which can result in cementation of the soil particles, making excavation problematic
- problematic soils, such as expansive soils, soils exhibiting collapse potential, weak soils, chemically aggressive soils
- various problematic soils that occur in arid conditions, such as those with significant shrink–swell or collapse potentials (see Section 3 *Problematic soils and their issues* for related chapters).

Figure 29.2 Dune-bedded Permian sandstone, Shropshire, UK (Reproduced from Toghill, 2006; Airlife Publishing)

Arid conditions can mean there is a limited availability of water, potentially affecting the suitability of geotechnical processes during construction. This might necessitate the use of locally sourced water, which can contain high concentrations of dissolved salts and could affect construction activities such as the casting of concretes. Limited availability of water can create logistical problems, which must be surmounted if the project is to be completed on time, or it can raise the cost of the project.

The mechanics describing the behaviour of arid soils are not unique to these deposits; soils not currently experiencing arid conditions may still be unsaturated, cemented, prone to shrink–swell or be metastable, although the effects of these mechanisms can be pronounced in arid deposits. However, the climatic and geomorphologic processes that shape arid soils must be considered to better understand the nature and engineering behaviour of soils in arid regions. This chapter will, therefore, give the key aspects of (selected) soil mechanics and (selected) geotechnical engineering associated with arid soils along with a description of climatic and geomorphologic processes that shape arid soil conditions (and are likely to continue affecting a site after the completion of a construction project).

29.2. Arid climates

Arid climates form when more water evaporates and transpires than enters the ground from precipitation. It is this process that generates specific problematic soils occurring at specific locations on the Earth's surface; therefore this section briefly introduces

- the development of arid conditions in hot and cold regions
- the characterisation of arid regions into hyper-arid, arid and semi-arid zones.

29.2.1 Development of arid conditions

Arid conditions develop through the global circulation patterns (or macroclimate) of the atmosphere, the location of the land in relation to the sea and the prevailing winds. This is governed by several complex processes driven by differences in air temperature and pressure across the globe and the rotation of the Earth. A distribution of arid environments around the world is shown in **Figure 29.3**.

The equatorial regions of the Earth experience the greatest concentration of solar radiation, and the poles the least; this imbalance creates several convection currents within the macroclimate on either side of the equator (see Wallen and Stockholm, 1966, for further details). The associated climate and atmospheric patterns are the principal reasons why many hot arid conditions are located around the 30° line of latitude on either side of the equator (**Figure 29.3**).

Cold arid landscapes tend to form because of their location within a land mass in relation to the sea and the prevailing winds (Bell, 1998; Wallen and Stockholm, 1966) (**Figure 29.3**); the further from the sea, the more likely the water transported in the atmosphere has already precipitated, leaving the air relatively dry (see Lee and Fookes, 2005a, for more details). Topographic structures, such as mountain ranges, can also interrupt the transport of water from sea to land by inducing precipitation on the topographic structure and leaving the air flowing into the lands behind relatively dry (Bell, 1998; Wallen and Stockholm, 1966).

Cold sea currents that approach the surface and flow near to land (such as the Humboldt Current, situated off Chile's coast) can exacerbate the effect of atmospheric circulation on an arid environment by creating stable and cool air conditions at sea, making precipitation unlikely to occur (although fog

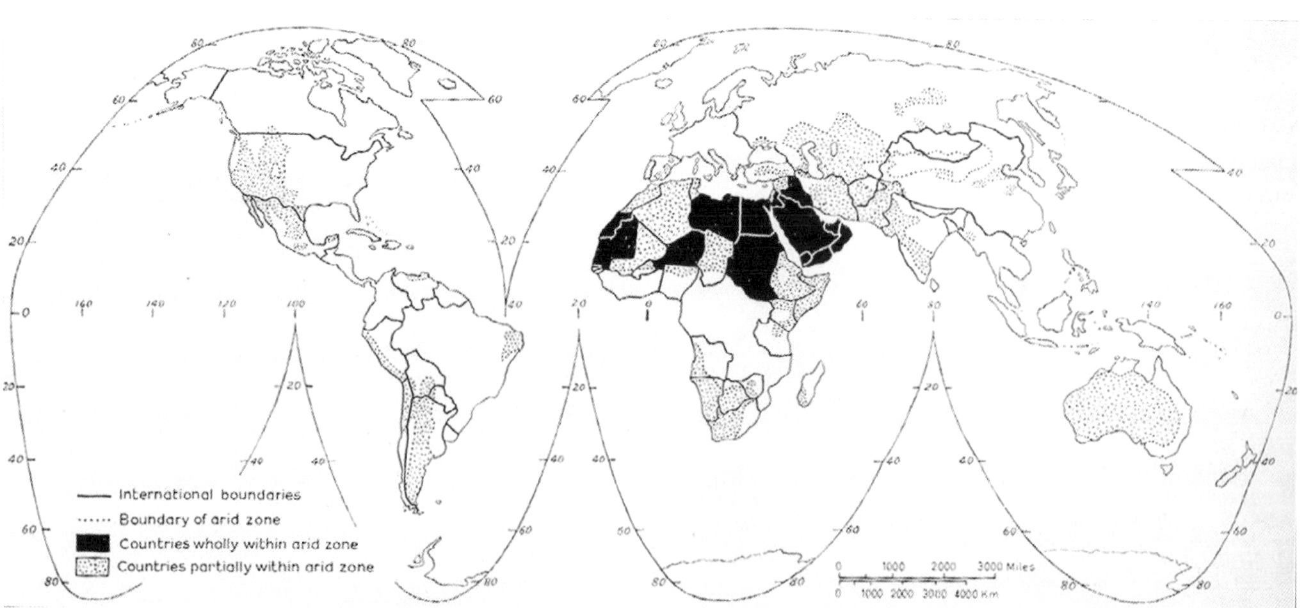

Figure 29.3 The distribution of arid environments (Reproduced from White, 1966; Methuen and Co Ltd)

and clouds can still form) (Wallen and Stockholm, 1966). The Atacama Desert, Chile, is considered one of the driest places on Earth due to the combined effects of the atmospheric circulation, the presence of a topographical structure along the coastline and the Humboldt Current (Lee and Fookes, 2005a; Wallen and Stockholm, 1966); Lee and Fookes (2005a) give data indicating that the Atacama Desert receives a mean annual precipitation of less than 0.3 mm.

29.2.2 Precipitation in arid environments

Arid environments are not necessarily under constant drought conditions, although there are locations in hyper-arid regions that may be in drought for years. However, by definition, arid environments do experience a moisture deficit. Regions can be classified as arid (or subdivided into hyper-arid and arid, depending on the classification system used), sub-humid (where the magnitude of the deficit is low) or semi-arid (transitional tracts of land between arid and sub-humid lands) – see **Figure 29.4**.

Figure 29.4 Semi-arid desert conditions in Botswana during the summer 'wet season'; note the profusion of vegetation indicating the occurrence of recent rains (Image courtesy of © R Jackson-Royal)

Lee and Fookes (2005a) give a simple relationship between mean annual precipitation and annual potential evapotranspiration (the combination of water vapour entering the atmosphere by evaporation and transpiration (Fetter, 2001)) to determine the aridity of a region (**Table 29.1**). Calculation of the annual potential evapotranspiration for a region can be determined using several approaches, although the Penman equation (not considered here) is considered suitable (for further details see Lee and Fookes, 2005b; Wallen and Stockholm, 1966). Deriving the annual potential evapotranspiration requires the quantification of several environmental parameters; and Wallen and Stockholm (1966) give an alternative method (developed by Köppen) that relates aridity to mean annual precipitation and mean annual temperature (for a winter, summer or undefined season) – see **Table 29.1**.

Arid regions can be described as having wet and dry seasons, although such an approach is perhaps more applicable to semi-arid regions than arid or hyper-arid regions. The duration of the wet season can be very short. Predicting actual precipitation patterns in hyper-arid and arid conditions can be very difficult as the annual precipitation can vary considerably, whereas semi-arid regions can experience more predictable conditions (Bell, 1998; Lee and Fookes, 2005b). Taking the mean precipitation over a reasonable time frame will result in the values quoted in the literature, although the annual precipitation will vary considerably throughout this period – for example, in some years very little precipitation, if any, might occur, while in other years far greater amounts will be recorded (Lee and Fookes, 2005b). Lee and Fookes (2005b) give further details of the distribution of arid regions, including indications of aridity, periods of drought and associated temperature and the occurrence of wet and dry seasons.

Precipitation, when it occurs, can take the form of short-lived very heavy storms (such as cloudbursts). These storms can occur as infrequently as once per decade or even once per century in certain arid environments (Hills *et al.*, 1966). However, they can be very intense events, can cause significant changes to localised topography and can damage earthworks (as illustrated in subsequent sections).

Humidity province	Characteristic vegetation[a]	P/ETP[b]	Summer[c]	Winter[c]	Undefined season[c]
Sub-humid zone	Grassland	0.50–0.75	–	–	–
Semi-arid zone	Steppe	0.20–0.50	$P \leq 20T$	$P \leq 20(T + 14)$	$P \leq 20(T + 7)$
Arid zone	Desert	<0.03	$P < 10T$	$P \leq 10(T + 14)$	$P \leq 10(T + 7)$
Hyper-arid zone	Desert	<0.03	–	–	–

P is the annual precipitation (mm), ETP is the annual potential evapotranspiration and T is the annual mean temperature (°C)
[a] Lee and Fookes (2005a);
[b] Lee and Fookes (2005b);
[c] Wallen and Stockholm (1966).

Table 29.1 Methods to determine if arid conditions apply for a given region (Data taken from Lee and Fookes (2005a, 2005b) and Wallen and Stockholm (1966))

29.3. Introduction to geomorphology of arid soils and the effect of selected geomorphic processes on the geotechnical properties of arid soils

An overview of the geomorphology of arid soils has been included in this chapter as the geotechnical properties of these soils are likely to be related to the geomorphic processes active in arid regions. Failure to understand the geomorphic processes which have acted upon, and can continue to act upon, the arid soils found on site could limit the ability to successfully engineer those soils. Therefore, this section of the chapter considers

- commonly encountered types of arid regions and the associated ground conditions
- erosion and deposition processes within arid regions (alluvial, chemical, fluvial and mechanical) and the potential effect these process have upon construction processes
- landforms that are found in arid regions and their potential for use as aggregates
- the formation of duricrusts and cemented soils and the problems these present to construction
- the formation of hammada (or hamada), dunes and other aeolian features, the conditions associated with these landforms and the geohazards these landforms present when undertaking construction
- the formation of fans, wadis and other fluvial landforms, the conditions associated with these landforms and the geohazards these landforms present when undertaking construction
- the formation of salinas, sabkhas and playas, the conditions associated with these landforms and the geohazards these landforms present when undertaking construction
- various considerations when undertaking site investigations in arid environments and potential sources of aggregates.

The geomorphic processes affecting arid environments are like those affecting other regions; arid environments are not governed by special circumstances (Hills *et al.*, 1966). However, the lack of vegetation and the aridity of the ground conditions makes aeolian erosion more prominent in arid regions than in other regions: aeolian deposits can be prone to collapse (as described in detail in Section 3 *Problematic soils and their issues*), affecting the design of structures built in such conditions. Stabilisation may be necessary before constructing on these deposits. Aeolian deposits can also be extremely mobile, potentially swamping a site and making it unfit for purpose. Arid conditions can also experience extreme changes in diurnal temperature, resulting in both increased rates of mechanical and chemical weathering of soils and man-made structures, such as concrete, and can significantly reduce the engineering life of a structure. Certain geomorphic processes might only occur infrequently, such as fluvial erosion, yet can have a pronounced effect upon conditions on site.

29.3.1 Landforms within arid environments

There are two distinct types of arid region: shields and basins (or plains) (Lee and Fookes, 2005b), with four distinct zones (from a geomorphologic viewpoint) that control the topography within an arid region: uplands, foot-slopes, plains and base-level plains (Lee and Fookes, 2005b). Each of these zones can exhibit different geotechnical problems because of their geomorphologic environments. Shields are defined as a 'major structural unit of the Earth's crust, consisting of large mass of Precambrian rocks, both metamorphic and igneous, which have remained unaffected by later mountain building' (Whitten and Brooks, 1973: p. 411). Such features are remarkably stable geological formations, having survived intact since the Precambrian eon (Precambrian rocks were formed at least 570 million years ago and can be billions of years old). Therefore, many shield deserts do not contain active volcanoes (although there are examples of recently active volcanoes in arid regions of America; Hills *et al.*, 1966) and many of the topographic features within a shield that were created by tectonic actions in a previous age have been affected by hundreds of millions of years of geomorphic processes (i.e. mountain ranges become reduced to hills by weathering and erosion). A shield may be overlain by younger rock formations, or soil, that will erode preferentially to the shield. Conversely, a basin (defined as a 'depression of large size, may be of structural or erosional origin'; Whitten and Brooks, 1973: p. 49) will act as a sink for debris eroded from higher ground by fluvial and aeolian actions and likely contain soils with a considerable range of particle sizes. A basin may have dunes.

29.3.1.1 Uplands and foot-slopes

Uplands are regions of hills, plateaus and canyons, characterised as having flat tops with scarps and ramparts (**Figure 29.5**). The topography of upland regions (with the exception of the ramparts) tends to be denuded of soil deposits by fluvial and aeolian processes, leaving barren rock or boulder fields (Lee and Fookes, 2005b). The topography of upland

Figure 29.5 Fish River Canyon, Namibia; note the flat and barren tops, steep scarps, establishment of ramparts at the base of the scarps and the development of a mesa and a butte (centre) within the canyon (Image courtesy of © R Jackson-Royal)

regions is predominantly shaped by mechanical, chemical and thermal weathering, and eroded by fluvial and aeolian action (and is also weathered by these processes), with weaker rock formations eroding preferentially to harder ones. This creates the prominent flat-topped, steeply sided features synonymous with arid environments (Monument Valley in the USA is a prime example).

Foot-slopes (**Figure 29.6**) are transitional regions between uplands and a basin. Gently dipping plateaus that form between these two regions are known as piedmonts. Eroded materials from uplands are deposited by fluvial and aeolian actions within a piedmont to form transitionary topographic structures such as alluvial fans and pediment slopes, which can exhibit a significant potential for collapse. The concept that these transitional regions are created by deposition of materials from the uplands is not a universally held view, and piedmonts are contentious within the field of geomorphology; an alternative theory suggests that they are seen as an extension of the uplands (piedmonts commonly share the same geology as the uplands) that have experienced weathering prior to degradation of the current prominent upland features (Dohrenwend and Parsons, 2009). However, from a geotechnical engineering viewpoint, piedmonts can be valuable sources of readily available aggregates and other resources for construction projects (Lee and Fookes, 2005b), as illustrated in Section 29.3.7.5.

The same geomorphic processes that created deposits on foot-slopes can also erode them, transporting the eroded material into a basin. While chemical, mechanical and thermal weathering occurs in these regions, aeolian and fluvial processes dominate the shaping of the topography in foot-slope environments.

29.3.1.2 Plains and base-level plains

The topography of plains tends to be dominated by the deposition and continued transport of previously weathered and eroded materials from uplands and foot-slopes. Aeolian and fluvial transport mechanisms dominate the changing topography of these regions – for example, they create structures such as dunes and wadis and segregate the transported particles with respect to size and density, depending on the location of the source material and the orientation of the weathering process. Base-level plains are naturally occurring depressions within a basin, and they are also affected by aeolian and fluvial actions (akin to plains), although the groundwater level can be relatively near the surface. The proximity of groundwater, the moisture deficit experienced by the soils in the vadose zone and the chemistry of the pore water can result in an accumulation of salts within the vadose zone. These highly salty conditions can accelerate the weathering of soil and rock deposits within base-level plains and can cause damage to man-made structures.

While upland features such as scarps and inselbergs (Uluru being a prime example; Hills *et al.*, 1966) can be ancient, features such as fans and dunes (**Figure 29.6**) are thought to be products of the Quaternary and are likely to be reshaped by future geomorphic processes (Lee and Fookes, 2005b). The lack of vegetation and predominantly dry surface soil can result in coarse-grained soils being readily eroded by geomorphic processes. Phreatic surfaces can be found in arid environments, although the depth of the water table can vary considerably and it can be very deep; Lee and Fookes (2005b) cite an example of a phreatic surface being 50 m below the ground surface in the Sahara, and this is echoed by Blight

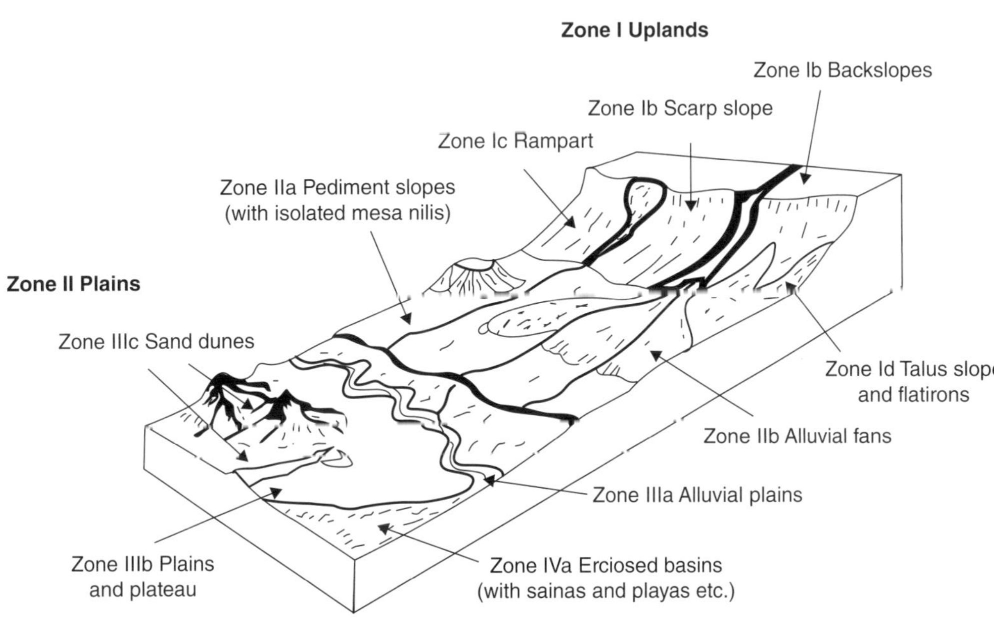

Figure 29.6 Illustration of hot desert topography (Reproduced from Lee and Fookes, 2005b; Whittles Publishing)

(1994), who cites similar groundwater levels in semi-arid conditions in southern Africa. The geomorphic processes that dominate arid environments include aeolian and fluvial erosion as well as chemical, mechanical and thermal weathering. The transitional, and potentially desiccated, nature of superficial deposits within arid landscapes, along with the bedrock being very close to the surface in upland areas, can prove to be a significant resource of aggregates for construction purposes, as illustrated in Section 29.3.7.5.

29.3.2 Development of arid soils through weathering

Weathering by aeolian, chemical, fluvial, mechanical and thermal processes occurs in arid environments. These processes produce a complex array of landforms and associated ground conditions, many of which present several significant challenges to geotechnical engineers. An understanding of this behaviour is essential. Further details and other useful references can be found in Fookes *et al.* (2005).

29.3.2.1 Mechanical weathering processes in arid environments

The process of eroding an upland is relatively slow; Lee and Fookes (2005b) cite rates for a scarp to recede as being of the order of millimetres per year. The resilience of upland formations is related to the nature of the upper layers of the topography, with rocks such as sandstone or cemented soils (known as a duricrust) forming barriers that resist the erosive processes. These layers are called capping layers and they can be slowly eroded at the scarp by mechanical weathering. If these formations are jointed then a capping layer may be undermined by block-by-block failure (Lee and Fookes, 2005b; Howard and Selby, 2009); if the formations below the capping layer are susceptible to erosion by water (if they contain smectites, gypsum, etc.) then these formations can soften when inundated by water, resulting in the collapse of the capping layer blocks above (Howard and Selby, 2009). The capping layer may also be weathered if the rock underlying the capping layer is permeable. Water seeping through this rock may weaken it; eventually voids can form within the underlying rock and ultimately lead to collapse of the weakened capping layer above (Lee and Fookes, 2005b).

Mechanically weathered material can fall from the scarp and accumulate on the ramparts, where it continues to weather, forming a debris slope (known as a talus slope). Talus (commonly known as scree) will form at or near the angle of repose; it can form metastable materials, which can slide down the slope if disturbed. Thus, talus slopes can be dangerous places for construction and may require treatment to prevent slips causing disruption.

Degradation of a scarp can result in a gradual reduction in the scarp angle (and height), producing smoother slopes, with the upland eventually being reduced to piedmont slopes (see Lee and Fookes, 2005b, for details).

Cyclic changes in climate experienced by the uplands, alternating from wet to dry, can result in the development of flatirons (wedge-shaped features, which can be of considerable size, that have detached from the rampart and slipped down the talus slope; for more information see Howard and Selby, 2009). These can be massive, or smaller-scale and encountered within a pediment or talus slope. Repeated wedge failure, instigated by weakening of internal boundaries (joints, slip surfaces, dipping rock boundaries, etc.), with a change in climate (from dry to wet), may result in rapid degradation of a scarp and result in several flatirons being created on the ramparts.

29.3.2.2 Acceleration of mechanical weathering processes with changes in temperature

The mechanical weathering of arid landscapes, and man-made structures within these arid environments, can be accelerated by changes in temperature. Hot arid environments can experience considerable changes in temperature throughout the year and, more importantly, diurnally – for example, on a site in a hot arid region during the summer the daily temperature will change from low twenties (°C) or less in the early morning to the high forties (°C) or greater in a matter of hours; therefore, the acceleration in weathering can be considerable. Rapid changes in temperature can result in the degradation of man-made structures; this is particularly true for composite structures containing materials with different thermal expansion coefficients (such as reinforced concrete), as differential expansion and contraction induce stresses within the material. The material need not be composite to experience such weathering, differential heating and cooling rates (i.e. the outer face compared to the core) can equally result in increased rates of weathering; the outer layers of the material can spall by a process known as exfoliation. Hills *et al.* (1966) suggest that this method of weathering can be very slow and that unconfined rocks, free to expand and contract, do not readily fail by exfoliation. However, rocks, or structures that are partially buried in a different medium can experience differential rates of change in temperature (the surface exposed to the ambient temperature could expand or contract at a greater rate than the buried surface), and the differential movements could result in additional stresses developing and lead to accelerated rates of weathering.

29.3.3 Degradation of arid environments and man-made structures through chemical/physical weathering involving miscible salts

The interaction between miscible salts, pore water and the ground/man-made structures in arid conditions can result in chemical/physical weathering of these structures. This can be by way of: the precipitation of salts within the joints, layers or pores, resulting in crystal growth that can exert pressures on the ground/structure (e.g. Horta, 1980; Tomlinson, 1978);

change in the form of crystal (e.g. anhydrite to gypsum) resulting in volume change, potentially causing heave (or settlement) (e.g. Alonso and Ramon, 2013; Yilmaz, 2001); dissolution of the crystals resulting in release of ions that chemically weather the ground or man-made structures, as illustrated in **Figure 29.7**; dissolution of crystals that contribute to the stiffness and strength of the soil, resulting in weakening of the soil and potential damage to structures (e.g. Horta, 1980; Kadhim, 2014; Tomlinson, 1978; van Alphen and de los Rios Romero, 1971).

Miscible salts are often found naturally occurring in soils and sedimentary rocks, either as crystals or dissolved salts within the pore water. Vertical seepage of groundwater, induced by suction within the vadose zone, can transport dissolved compounds upwards towards the ground surface. If the groundwater table is sufficiently close to the ground level, and suction due to capillarity acts on the groundwater, then the transport of water vertically upwards (see Section 29.4.6 for more information) can increase the concentration of soluble compounds within the pores of the near-surface layers of soil.

The concentration at which water can no longer support additional quantities of miscible compounds is known as the saturation concentration. The solubility of a compound is related to several parameters, including temperature and presence of other compounds (for more information, please refer to Fetter, 1999) (**Figure 29.8**). For example, gypsum solubility has been found to be a nonlinear function with temperature (Klimchouk, 1996), and dissolution rates of gypsum have been found to increase if compounds such as sodium chloride or alkali metal sulfate or nitrate compounds are also present in the pore water (James and Edworthy, 1985).

Changes in environmental conditions, including change in temperature, change in the concentration of dissolved salts, change in the mixture of dissolved salts and change in the volume of pore water, can result in the precipitation of compounds out of solution.

Diurnal changes in temperature can be sufficient to result in a reduction in the saturation concentration within the pore water, resulting in precipitation of compounds above this concentration threshold. The resulting crystal growth can apply pressure in the soil/rock fabric. Repeated diurnal (or seasonal) changes in temperature can result in cycles of precipitation and dissolution of the compounds, and this can contribute to the weathering of the natural ground conditions or man-made structures. Additionally, if the ambient temperature is sufficiently high, evapotranspiration can dry out the pores, causing the precipitation of all salts previously dissolved in the pore water. However, Goudie (1994) states that it is not necessary for pores to dry out before the precipitation of miscible compounds is an issue for naturally occurring rock, soil and man-made structures alike. Man-made structures can be particularly at risk, with concretes spalling and steel reinforcing experiencing accelerated rates of corrosion. Al-Amoudi *et al.* (1995) suggest there is also the potential for solutes to enter placed fills, such as the layers of a highway, by capillarity, and this can result in damage occurring to these structures with time.

The saturation concentration for a compound is also related to the presence of other compounds dissolved within the

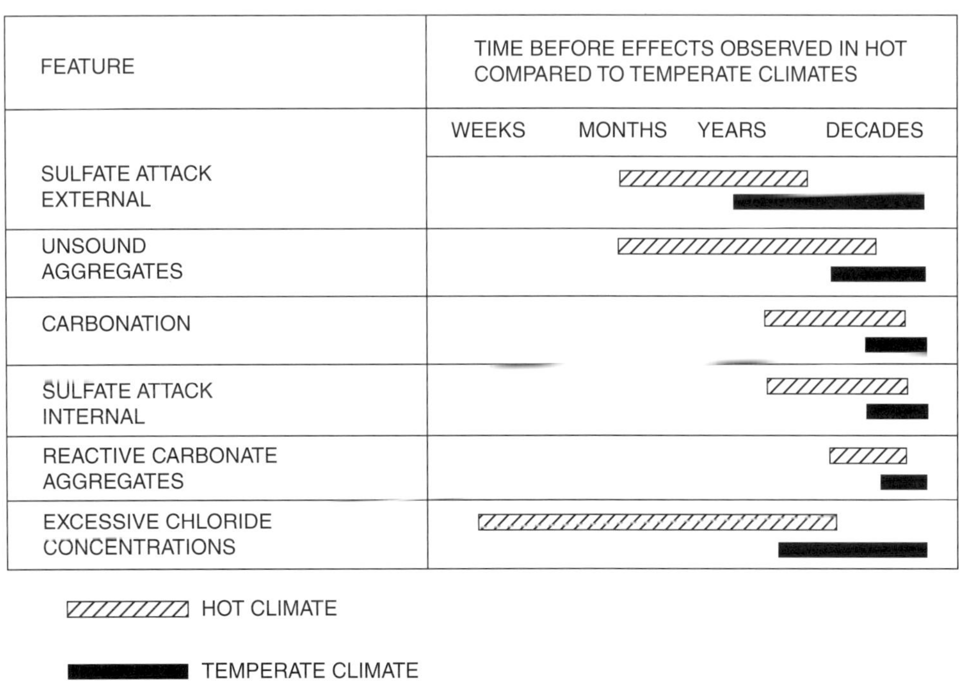

Figure 29.7 Deterioration of concrete due to environmental factors over time (Reproduced from Khan, 1982; with permission from Elsevier)

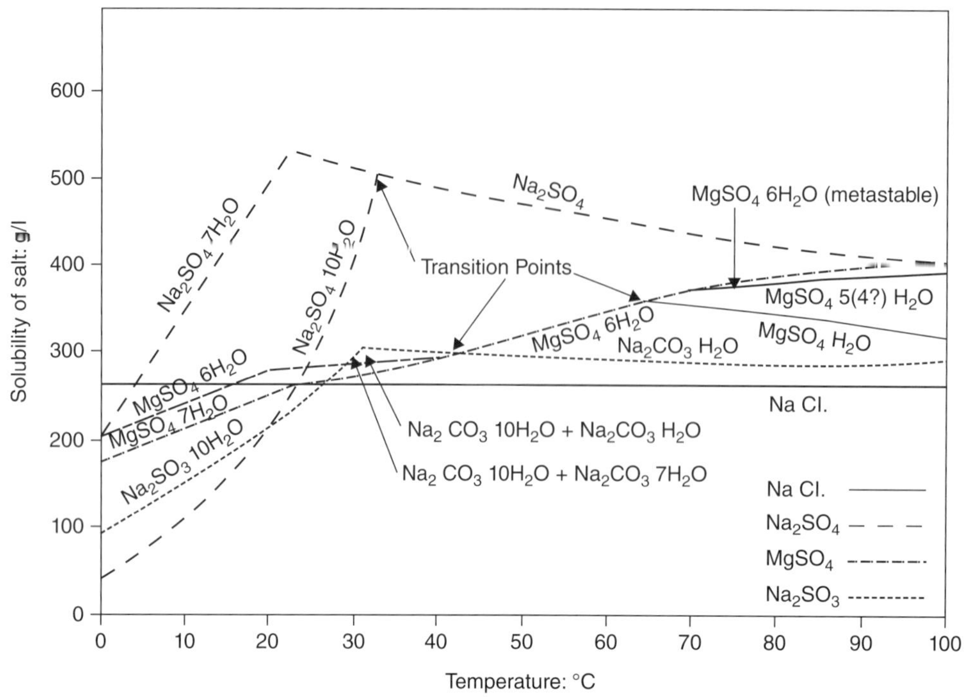

Figure 29.8 Temperature–solubility relationship for some common salts (Reproduced from Obika *et al.*, 1989; © The Geological Society)

pore water. Pore water can only undergo a finite degree of restructuring to accommodate miscible compounds; thus, if more than one compound is present the saturation concentration for each will be reduced and the concentrations of salts precipitated out of solution will be greater than predicted by summing the saturation concentrations for each compound present. This situation is colloquially known as 'salt-out'. Consideration should be given to the potential negative consequences of dissolved compounds (and the precipitation of these within the soil or structure) interacting with man-made structures or placed fills during the design stage (e.g. foundations or pavement layers). Fetter (1999) provides guidance on estimating solubility of various compounds.

The volume of a solid form of a salt is likely to be greater than that for the compound when dissolved in water, particularly if the compound hydrates on precipitation, potentially applying a pressure in the pore or fissure and causing degradation of the soil or rock structure. Bell (1998) gives crystallisation pressures for various salts: gypsum ($CaSO_4.2H_2O$) at 100 MPa, kieserite ($MgSO_4.H_2O$) at 100 MPa and halite ($NaCl$) at 200 MPa, and cites a study illustrating that the salts Na_2SO_4 and $MgSO_4$ caused a far more pronounced degradation of sandstone cubes that were exposed to solutions containing them, and that a mixture of salts within a solution can be far more destructive than when present individually. The pressures exerted by precipitation of miscible compounds can be considerable and man-made structures are also at risk of degradation by this process (Goudie, 1994).

Calcium sulfate can be encountered in more than one form (normally anhydrite, $CaSO_4$; or gypsum, $CaSO_4.2H_2O$), with the crystal form being a function of relative temperature, pressure and humidity within the pore (Klimchouk, 1996; Yilmaz, 2001); hence the crystal can transform with an associated change in volume. For example, the transformation of anhydrite into gypsum results in an increase in specific volume of approximately 60% (Yilmaz, 2001). Such a transformation can cause significant distress to ground conditions or man-made structures. Alonso and Ramon (2013) reported heave in a bridge due to such a change; pile groups (for the bridge piers) were installed within ground conditions that included claystone containing gypsum and anhydrite (anhydrite found below the base of the piles). Post-construction, parts of the bridge were observed to have heaved by up to 370 mm. This was attributed to the pile installation creating hydraulic connections between upper aquifers and the anhydritic claystone, with the anhydrite experiencing dissolution and reprecipitation as gypsum. To remediate the situation, an earthwork, 33 m deep in places, had to be constructed over the affected area in the valley to reduce the rate of heave to an acceptable level (Alonso and Ramon, 2013).

Thus, detailed investigation is necessary before undertaking construction in highly salty ground conditions to ascertain the likely effect the salts will have on the engineering life of the structure. Ground conditions with high concentrations of salts can cause physical weathering of man-made structures (due to the expansion of precipitated crystals within pore spaces, cracks, etc.) and chemical weathering (such as the accelerated corrosion of steel reinforcing) (Goudie, 1994).

29.3.4 Chemical bonding of soil creating duricrusts and cemented soils

Duricrusts (defined as 'a layer of strongly cemented material occurring in unconsolidated sediments, often found a short distance below the surface; usually formed as a result of water action; the cementing material may be calcareous, siliceous or ferruginous'; Whitten and Brooks, 1973: p. 222) are cemented soils, and often form below ground and are exposed by erosion. Duricrusts are commonly a few hundred millimetres thick (e.g. 300–500 mm), although the development of multiple crusts can result in significant duricrust thicknesses (e.g. 3–5 m) (Lee and Fookes, 2005b), and are common in arid conditions (and other environments); for example, calcrete duricrusts are estimated to cover 13% of the Earth's land mass (Dixon and McLaren, 2009).

The cementation process stabilises the soil and makes duricrusts resilient to weathering and provides strengths greater than those associated with unaffected soils. The types of duricrust commonly encountered in arid conditions include aluminous, calcareous, gypseous and siliceous duricrusts. Siliceous and aluminous duricrusts are considered to be stronger than calcareous duricrusts (Dixon and McLaren, 2009; Hills *et al.*, 1966), and these formations can prove to be very difficult to excavate. Excavation of a duricrust may necessitate drill and blast techniques or powerful plant (Lee and Fookes, 2005b). Once through the duricrust the subsequent strata may require less effort to excavate. Therefore, not only do duricrusts create soil structures resilient to erosion, but they can also create problematic conditions for construction and must not be underestimated when planning construction activities.

The formation of lightly cemented soil also occurs in arid environments. While not producing materials as resilient to erosion as the heavily cemented duricrusts, there may be an increase in stiffness and they may be self-supporting when excavated (the sides of an excavation in cemented, desiccated sands may not collapse with time if the degree of cementation is sufficient to resist movement). The soils may be cemented by precipitation of salts by the same mechanisms creating duricrusts (Dixon and McLaren, 2009).

29.3.5 Fluvial erosion, the effect of flooding and the creation of challenging engineering environments including salinas, playas and sabkhas

Many watercourses found in hot arid regions are likely to be ephemeral: they only contain water directly after a storm, and the water rapidly evaporates or permeates into the soil, leaving the watercourse dry once more. The drainage path of the water from a storm will dictate how it interacts with the surrounding environment (**Figure 29.9**); flood water will readily migrate across rocky piedmont slopes to wadi drainage channels, forming ephemeral streams; conversely, flood water entering tracts of loose sandy soil will rapidly diminish due to infiltration of the water into the soil and ephemeral streams are unlikely to form.

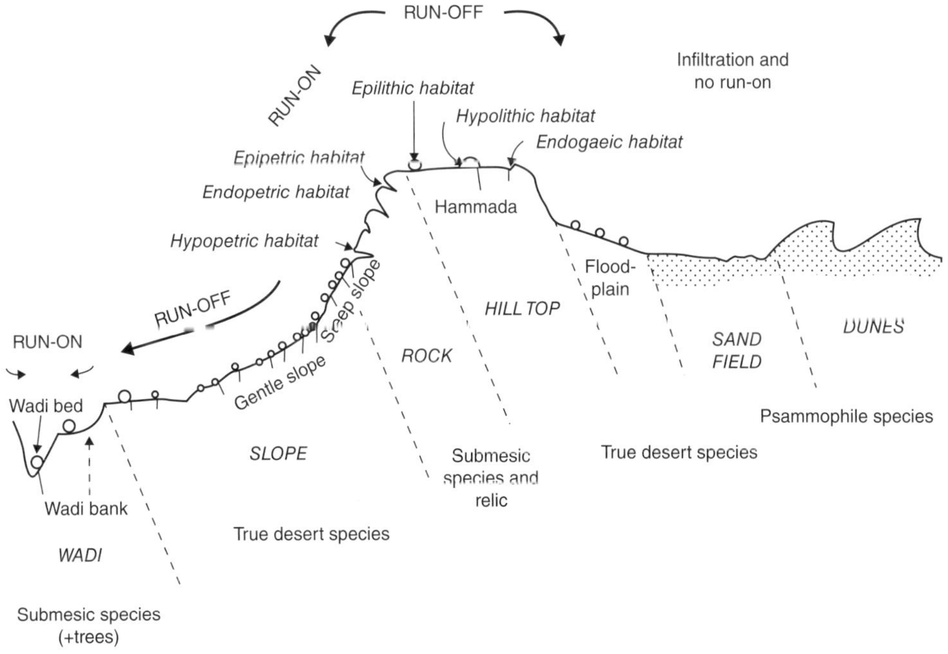

Figure 29.9 Elements of the hydrological cycle in arid lands (Thornes, 2009, reproduced from Shmida *et al.*, 1986)

Lee and Fookes (2005b) cite a study indicating that 90% of the erosion on a site over a 10-year period occurred over 5 days involving seven storms. This is because the intensity of the storms, which take place over a relatively short time frame, can result in large volumes of water migrating through previously established drainage networks (or create new networks when draining). The velocity of this water is often sufficient to transport significant quantities of sediment as it migrates through the drainage pathways, creating new topographic features as the sediment is deposited (for more information, refer to Parsons and Abrahams, 2009).

The infrequent storms encountered in arid conditions can also damage earthworks (particularly during construction when the earthworks and their drainage provisions are incomplete); discharging water can erode the surface of an earthwork, and water permeating into an earthwork can reduce suction within the soil, thus potentially reducing the strength and stiffness of the soil and inducing ground movement. The former can result in damage to the earthwork, necessitating remedial measures to reinstate the earthwork, and any ground movement can cause serviceability issues or the failure of structures on the earthwork, again necessitating remedial action. Puttock *et al.* (2011) describe a construction project in Morocco that experienced storms during the construction of an earthwork. **Figure 29.10** illustrates a drainage channel neighbouring the construction site in a flood; this occurred after a storm, and jeopardised a section of the newly compacted earthwork. There is often a lack of flood risk assessment for sites within arid environments; such an assessment should have been included during the planning stages for this project as a matter of course if potential flood risks were to be properly managed.

Figure 29.10 Ephemeral drainage channel in a flood after a storm; note the proximity of the drainage channel to the newly placed earthwork places the earthwork at risk (Reproduced from Truslove, 2010; all rights reserved)

While the risk of a flood might be significantly lower in arid environments, the potential damage arising from such an event could still be significant. This requires suitable drainage provision for the placed earthwork. If storm water collects on the surface of an earthwork until it reaches a drainage pathway and then dissipates, damage to the earthwork might occur (**Figure 29.11**), requiring remedial measures to reinstate it to a functional condition.

29.3.5.1 Sediment transport and deposition

When precipitation falls on uplands, where the upper surfaces are commonly barren or comprise boulder fields, the water can rapidly drain through a discrete network of drainage pathways, through gullies, into the plains below. The limited nature of the drainage channels and gullies eroded into uplands creates locations for the repeated deposition of eroded materials around the base of the uplands. The materials deposited within these regions can vary greatly but can prove an excellent source of aggregates for the construction industry (Lee and Fookes, 2005b).

29.3.5.2 Alluvial fans

Sediment load is a function of velocity and as this decreases the water will start to deposit sediment, resulting in poorly graded fan-shaped deposits. These fan deposits (alluvial fans) are prominent features on piedmont slopes. Repeated storms will create newly deposited alluvial fans, with the flood water flowing through preferential pathways within previously established fans; the flood water will erode sections of a previously created fan, depositing it further away from the uplands and creating distinctive 'lobe' shapes within the fan. Alluvial fans can cover significant tracts of land; Hills *et al.* (1966) reported the extent of an alluvial fan in the Atacama Desert as

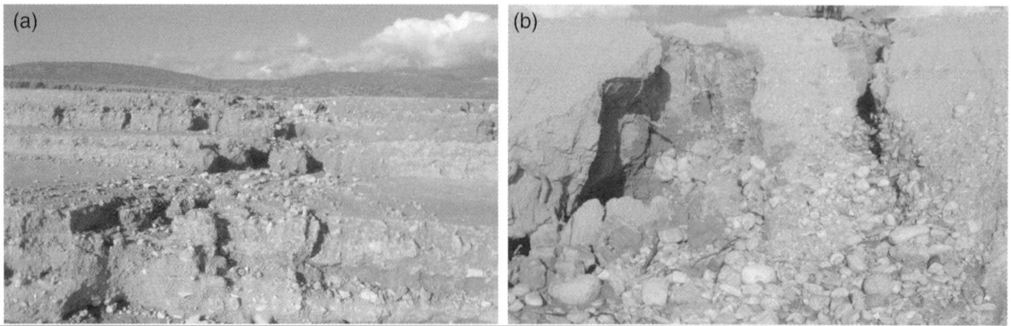

Figure 29.11 Damage to compacted earthwork ((a) Reproduced from Truslove, 2010; (b) reproduced from Puttock *et al.*, 2011; All rights reserved)

stretching 60 km from the upland source. Further details can be found in Blair and McPherson (2009).

29.3.5.3 Wadis

Wadis (valleys with intermittent streams) are ephemeral in nature and form braided streams. Whitten and Brooks (1973: p. 62) define these as 'a stream consisting of interwoven channels, constantly shifting through islands of alluvium and sandbanks; the banks are comprised of soft sediments (often self-deposited), which are easily eroded; the stream beds are wider and shallower than where meander occurs, and are more liable to flood, which in itself helps the constant shifting, widening process'. This creates new alluvium deposits along the river valley in what can seem like haphazard locations (when compared to deposition sites of meandering watercourses). Thus, subsequent patchy layers of poorly graded, loose alluvium deposits (overlying previously deposited alluvium) can be encountered when working in these conditions. Loose deposits can be eroded and redeposited in subsequent storms; thus, the topography of wadis is transitory. These conditions can also be dangerous places to work during the 'rainy' season when flash floods can occur very quickly and with little warning (Waltham, 2009).

29.3.5.4 Sabkhas, salinas and playas

Fine-grained particles are likely to settle out of suspension last, and so they can be transported considerable distances by flood water. In base-level plains the flood water can pond to form ephemeral lakes. As the lakes recede, silts and clays settle out and form a crust (an example of a crust formed as lake water evaporated can be seen in **Figure 29.12**, although this was not formed by receding flood water but due to changes in river pathways that previously fed the lake). These clay or silt crusts can accumulate with subsequent floods, forming additional layers, and will thicken the crust (Lee and Fookes, 2005b). These layers may be interspersed (or overlain) by layers of sand, deposited by aeolian action between periods of flooding, or contain evaporites (in the case of sabkhas and salinas). The surfaces of these crusts can be resilient to settlement induced by light loading or inundation of water during flooding, and Al-Amoudi and Abduljawad (1995) suggest that (in the case of sabkha deposits) this is due to the desiccated and cemented nature of the crusts.

Figure 29.12 Nxai Pan in Botswana, one of a series of salt pans created with the drying of a major lake (Makgadikgadi) due to regional changes in river pathways in southern Africa (caused by tectonic uplift; Moore *et al.*, 2012) (© A. Royal)

If the crusts form on plains adjacent to the sea, then sabkhas or salt playa can form (**Figure 29.13**). Sabkhas experience coastal flooding and are likely to have a water table close to the surface. The salt concentrations accumulated within these crusts can be extremely high (**Table 29.2**).

Salt playas can also have high salt concentrations (although not the same magnitude as sabkha) due to the precipitation of salt with the silts and clays as the flood water recedes (Lee and Fookes, 2005b). Sabkhas and salt playas may experience a degree of cementation with the precipitation of solutes from the pore water. Such precipitation can also lead to the creation of gypsum and other evaporite lenses developing within these landforms (Al-Amoudi and Abduljawad, 1995) – see also Chapters 37 *Sulfates and sulfides in soils and rocks* and 38 *Soluble ground*.

However, the soil below the crust can be metastable, it might contain minerals susceptible to dissolution (e.g. gypsum and halite) and it might be very soft and, hence, potentially susceptible to ground movement once the upper surface of the crust is compromised (Al-Amoudi and Abduljawad, 1995).

The properties of sabkha soil can also change with the season; with once trafficable surfaces becoming impassable during the 'wet season' as the cementing salts (if only partially) dissolve (Shehata and Amin, 1997). Sabkhas formed during the Quaternary (Al-Amoudi and Abduljawad, 1995), leaving deposits along what formerly was the coastline, although the sea level has fluctuated during this period and many of these sabkhas have been reworked by action of the sea to form carbonate sands and silts as well as bioclastic sands (Al-Amoudi and Abduljawad, 1995).

Care must be taken when constructing on these topographic features; the high concentration of salts within these crusts makes them problematic when coupled with the potential for flooding (Millington, 1994). Construction in such environments requires consideration of the stiffness and strength of the underlying soil (particularly for salinas), as the density of these soils can be low (Al-Amoudi, 1994). Other issues include the collapse potential and the highly salty nature of the upper layers of the soil. Buried man-made structures within these landforms (e.g. concrete foundations or steel pipes) can experience accelerated weathering due to the high salt content within the soil (Shehata and Amin, 1997). Using materials from these areas for construction processes is not recommended (see Section 29.3.7.5 for further details).

Silt or clay crusts formed by freshwater flooding are known as playas or salinas, with the difference in behaviour being

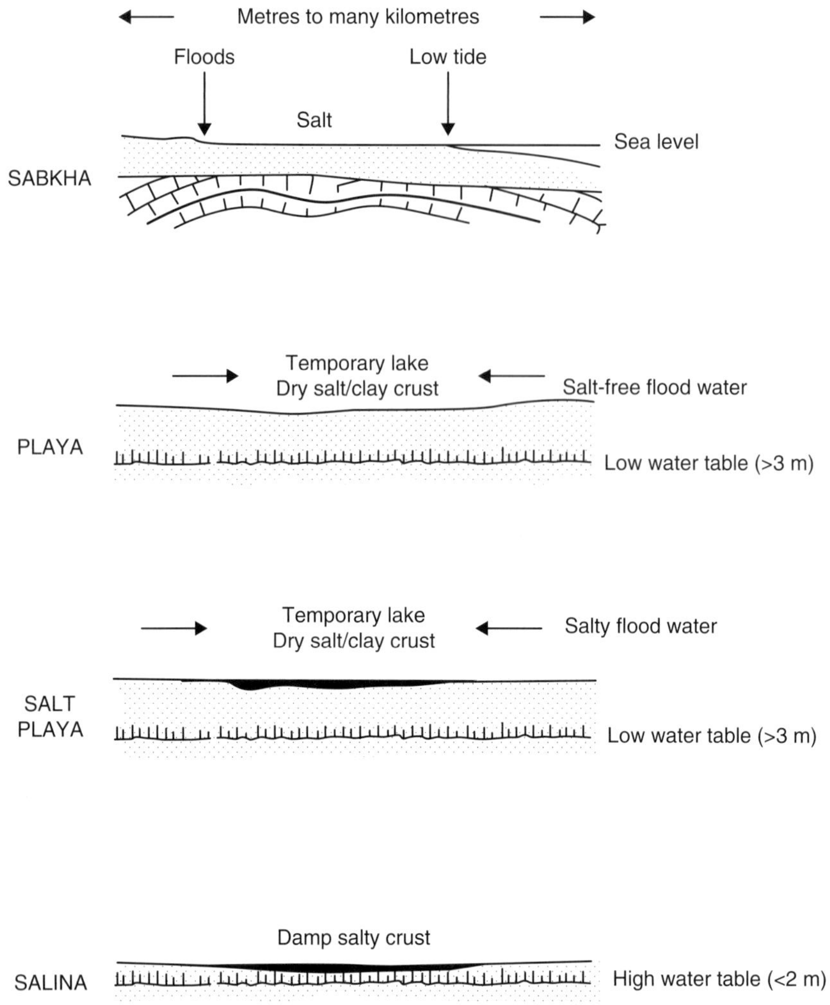

Figure 29.13 Idealised cross-sections through a sabkha, playa, salt playa and salina (Lee and Fookes, 2005b; reproduced from Fookes, 1976)

dictated by the proximity of the water table to the ground surface. In base-level plains the groundwater table can be relatively close to the surface, and suction within the vadose zone can result in a significant capillary rise (e.g. over 10 m in clays and 1–10 m in silts (Bell, 1998)). If the playa is susceptible to the movement of water by capillarity then it is called a salina and salts are transported towards the surface (Lee and Fookes, 2005b). The increase in salt concentration within the salina, akin to sabkha and salt playa, can also make construction difficult. If the movement of groundwater intersects with the ground surface, then the surface layers of the salina can be damp (**Figure 29.13**), with soft material directly below.

Ion	Open sea: mg/l	Coastal seawater, Arabian Gulf: mg/l	Sabkha water: mg/l
Ca	420	420	1250
Mg	1320	1550	4000
Na	10 700	20 650	30 000
K	380	650	1300
SO_4	2 700	3300	9950
Cl	19 300	35 000	56 600
HCO_3	75	170	150

Table 29.2 Typical compositions of seawater and sabkha water (in mg/l) (Data taken from Fookes *et al.*, 1985)

29.3.6 Aeolian erosion/deposition and the creation of challenging engineering environments including collapsible soils, hammadas and sand dunes

Arid soils are more susceptible to aeolian erosion than soils found in other regions due to climatic conditions and the resulting lack of vegetation. Aeolian erosion has two principal forms: first, the wind can transport soil particles from one location to another – the ability of the wind to transport soil particles is related to a number of factors, including wind speed and particle type (e.g. size, shape, specific gravity); and second, soil transported by the wind can be used to scour existing topographic features (akin to the sea using transported sediment to erode the base of cliff faces). The scouring power of wind should not be underestimated and has negative connotations when working on site (for more information, see Parsons and Abrahams, 2009).

The mode of aeolian transport is dependent upon the size, shape and density of the soil particles (for given wind and topographic conditions). Very fine materials (less than 20 μm) can stay in suspension for a long time; their transport can last for several days and particles can be transported thousands of kilometres (Nickling and McKenna Neuman, 2009). Larger fine-grained materials (less than 70 μm but greater than 20 μm) are too heavy to stay in suspension for a long time; transport will last a few minutes to hours, and the particles can be transported tens of kilometres (Nickling and McKenna Neuman, 2009). Coarse-grained material is generally too heavy to transport by suspension and instead can be transported by saltation or rolling. Saltation describes the process where particles (up to 500 μm) move through the air in a cycle of collection, transportation and deposition in short elliptical paths. Larger particles can roll along the surface of the ground under the force of the wind; however, there is a point where the particles become too large to be transported by aeolian action.

The segregation of particles by size, shape and density due to aeolian erosion can result in a grading of particles within a basin relative to the dominant wind direction and source of material. The loss of transportable particles by aeolian action can result in a gradual lowering of the ground surface and at the same time an accumulation of particles too large to transport by aeolian action; eventually the particles too large to be transported by the wind will cover the ground surface and prevent further erosion through aeolian action. The ground surface is then known as a hammada (or hamada) (see Stalker, 1999, for further details). These are resilient surfaces that can be beneficial to construction; the surface can be trafficable without the need for much preceding ground improvement, but once the surface is broken the affected area can once again be susceptible to aeolian erosion.

The movement of sand-sized particles – not necessarily quartz, the particles can be clay aggregations, salt crystals, non-quartz particles and so on (Warren, 1994b) – by aeolian action can lead to the creation of dunes at the windward or leeward face of a topographic feature, whether naturally occurring or man-made, or where two or more winds converge (Lee and Fookes, 2005b). The type of dune formed (see Stalker, 1999, for details of common dune types) is dependent upon several factors, including the volume of sand transported and the consistency of the direction of the wind, with wind consistency being considered of primary importance (Lancaster, 2009). The particle diameter size range commonly found in a dune is 0.1–0.7 mm, with median grain diameters of 0.2–0.4 mm (Lee and Fookes, 2005b). Thus, a dune is poorly graded (Lee and Fookes, 2005b).

The processes that create dunes can cause them to migrate across the surface of an arid region, and they are then known as active dunes. Sands accumulate on the slip face, which causes the crest of the dune to move with time and, hence, the location of the slip face. The migratory rates of dunes (see Warren, 1994b, for details of the rates of movement with height) are inversely related to the dune height and are a function of the cross-section and bulk density of the dune as well as the aeolian transport rate of the sand (Lee and Fookes, 2005b). Dunes can form across vast areas (known as dune fields or sand seas; sand seas being far larger than dune fields). For example, it is considered that about four-fifths of arid lands are barren rock or boulder fields (Bell, 1998) and that less than 25% of arid and hyper-arid regions are covered in active dunes (Warren, 1994b), yet Hills *et al.* (1966) cite an uninterrupted sand sea in the Sahara covering a surface area greater than France. In such places the dunes can achieve considerable heights; Lee and Fookes (2005b) describe extreme transverse dunes in the Sahara reaching 240 m.

The transition from arid to semi-arid conditions can result in a stabilisation of the dunes as increased precipitation levels result in increased coverage of vegetation. Dunes can also be stabilised with the cementation of the sand particles by the development of salt crystals or the weathering of the non-quartz particles within the dune (Warren, 1994b). Lancaster (2009) gives the geographic locations of dune fields and sand seas across the world and the activity of their dunes.

The aeolian erosion of arid soils (that are vulnerable to these mechanisms) is likely to occur during and after construction and this must be considered during the planning stage of any construction project. Aeolian deposits can affect structures (blocking doorways and in extreme cases burying houses; see Shehata and Amin, 1997), can affect construction processes (e.g. engine filters readily become blocked with fine materials; prevent setting out using optical devices) and can create logistical problems (**Figure 29.14**) where loose material can

Figure 29.14 A car stuck on loose, aeolian deposits covering a road. The road was perfectly trafficable two weeks before (© A. Royal)

be problematic to negotiate, particularly for large vehicles not equipped with caterpillar tracks, such as articulated lorries. Measures can be established to protect a site against the build-up of wind-transported soils, such as using porous fences as sand traps, illustrated in **Figure 29.15**, although the land required to protect a site can be considerable.

Aeolian erosion denudes landscapes of the finer-grained particles and creates transitory features such as dunes. However, particle-laden winds also act as a geomorphologic process on rock formations, scouring them (akin to commercial sandblasting, if not as intensive) and further shaping the topography of the uplands and foot-slopes; this process is slow and continuous. Laity (2009) quotes abrasion rates of the order of magnitude of 0.01–1.63 mm/year (abrasion can be far greater in places like Antarctica, where wind speeds can be considerable, although this is not considered here). A variety of different rock formations (Bell, 1998) are created depending on the type of rock, rock strength and the alignment of the layers of rock.

Superficial deposits can experience subsequent stabilisation by precipitation of salts (e.g. by unsaturated seepage of salt-laden water – see Sections 29.3.3 and 29.4.6), which can help resist subsequent erosion. In addition, the degree of stabilisation (or cementation) of what can be loose deposits can result in soils that are (apparently) relatively stiff, strong and resist deformation; making them appear ideal foundation soils for man-made structures. These apparent beneficial properties are likely to be metastable; while they might remain in a dry (or low degree of saturation) condition, wetting of these soils can result in loss of strength and rapid settlement (collapse). In addition, the stabilised deposits can exhibit brittle deformation response that can experience strain-softening post-peak strength (e.g. Haeri *et al.*, 2005; Macedo *et al.*, 2023). Metastable response of these types of soils has resulted in damage/distress being reported in: earth structures/pavements (e.g. Horta, 1980; James and Edworthy, 1985; Johnson, 2008; Kadhim, 2014; Tomlinson, 1978); earth dams and drainage/irrigation channels (e.g. Johnson, 2008; van Alphen and de los Rios Romero, 1971); structures and foundations (e.g. Akili, 2006; Sissakian and Al-Mousawi, 2007).

29.3.7 Undertaking site investigations in arid areas and classifying arid soils

Soil classification for geotechnical engineering applications should provide sufficient information so that the design and construction phases of a project are cost-effective while minimising the risks of failure. However, to ignore the nature of arid soils courts failure of the project (during or after construction) by potentially unexpected means (Smalley *et al.*, 1994). While construction projects often do not have sufficient resources to fully classify the properties of a soil found on site (in terms of mineralogy, formation, etc.), it is recommended that the classification process is extended beyond the scope of a laboratory study to determine the geotechnical parameters of the soil and to consider textural (e.g. fabric and soil structure) aspects of the soil (Rogers *et al.*, 1994).

In arid conditions, soils are often unsaturated, can contain high salt concentrations and, due to the geomorphologic process that created them, can be loose, metastable and, therefore, collapsible, or expansive (see Chapters 32 *Collapsible soils* and 33 *Expansive soils* for further details). **Figure 29.16** provides an example approach to the consideration/classification of arid soils; such approaches can be used to help identify soils that may experience changes in volume. Laboratory testing to identify physical parameters will not necessarily identify the risks posed to the foundation of a structure through chemical or physical weathering induced by high salt concentrations below the ground surface, or failure of the foundation by collapse (or expansion) of the soil skeleton upon wetting. Hence additional testing might be required. Moreover, a perfectly

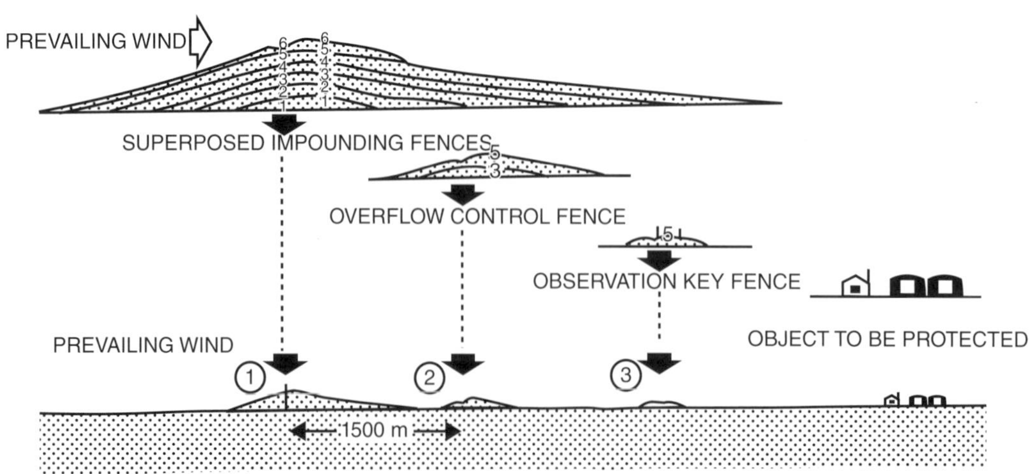

Figure 29.15 The use of porous fences as sand traps (Lee and Fookes, 2005b, reproduced from Kerr and Nigra, 1952; American Association of Petroleum Geologists)

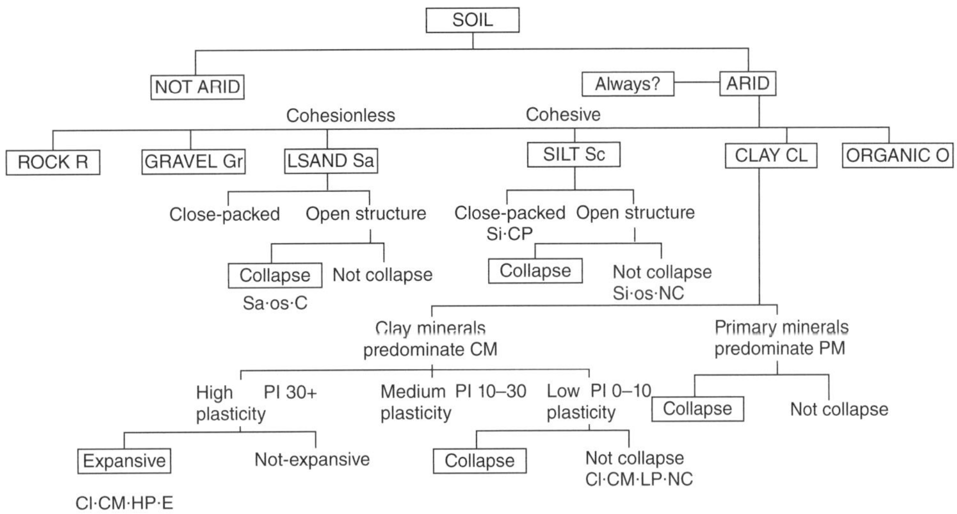

Figure 29.16 Proposed decision tree to indicate volume change problems. Si-CP, silt – close packed; Si-os-NC, silt – open structure – not collapse; Sa-os-C, sand – open structure – collapse; Sa-os-NC, sand – open structure – not collapse; Cl-CM-LP-NC, clay – clay minerals dominate – low plasticity – not collapse; Cl-CM-HP-E, clay – clay minerals dominate – high plasticity – expansive (Reproduced from Rogers *et al.*, 1994)

serviceable structure could become expensive to maintain due to the continued accumulation of wind-blown deposits.

29.3.7.1 Guidance for ground investigation in arid soils

The preceding content in this section has illustrated that the nature of soil formations encountered in arid conditions are likely to be influenced by their location within the arid environment and/or the geomorphic processes that act (or have acted) upon them (Sections 29.3.1 to 29.3.6). For example, soils found on uplands or foot-slopes can vary considerably from those in the plains or basins. This should be considered when developing a ground investigation strategy.

Furthermore, arid soils have the potential to be heterogeneous and/or metastable in nature (e.g. talus slopes or aeolian deposits such as sand dunes or loess), they can be unsaturated or dry (unless within a base level plain, where the phreatic surface can be relatively close to the surface and the soils could be saturated) and could be cemented. In addition, these soils could be vulnerable to collapse with sufficient disturbance and/or wetting. All of which could combine to make collection of undisturbed samples very challenging. Hence, the engineering parameters derived from a limited number of borehole samples or laboratory-derived data may be unreliable (or limited in nature). Therefore, any ground investigation should be underpinned by a preliminary desk study to highlight the types of soil that might be encountered and identify the types of tests and best practices required to characterise the soils on site. Section 4 of this manual provides information relating to site investigation. For more information refer directly to the relevant chapters within that section.

Due to the difficulty in obtaining sufficient quantities of undisturbed samples for a thorough laboratory investigation, there have been attempts to manufacture samples within the laboratory. While manufactured samples will not capture the complex nature of the soil structure/fabric that has formed due to the geomorphological (or biological) processes that have shaped them, these manufactured samples can be developed to exhibit similar geotechnical properties or behaviours (i.e. similar water contents, void ratios and/or dry densities; degrees of cementation; collapse potential). These samples can be manufactured in the quantities required to facilitate a systematic laboratory study to further fundamental understanding of the engineering response of these types of soil deposits; or to allow systematic investigation of proposed engineering interventions. Examples of manufactured samples, either from reconstituted natural soils or from soils sourced from suppliers, include: Assallay *et al.* (1997), Haeri *et al.* (2005), Huang and Airey (1998), Jefferson and Ahmad (2007), Lee *et al.* (2010) and Naeimi *et al.* (2022). While manufacture of problematic soils in the laboratory can help further understanding of fundamental response, this approach is not envisaged to replace the need for traditional ground investigations – it aims to augment them.

Care is needed when using standards (e.g. British or European Standards) as these may not be best suited to arid soil conditions (for tropical soil, see Chapter 30 *Tropical soils*). For example, Puttock *et al.* (2011) constructed earthworks in an arid environment and found success when combining French Standards (using *Technical Guideline on Embankment and Capping Layer Construction*; LCPC and SETRA, 1992), which they suggested were more suited to arid conditions than British Standards, with practices taken from British Standards (using end performance specifications and placed soil investigation techniques).

29.3.7.2 In situ testing

In situ testing can determine geotechnical data or classify soil properties and can complement data obtained from laboratory tests (such as characterising and index testing, physical testing including swelling and the collapse potential of reformed samples). In situ testing may disturb the soil, but in loose, unsaturated, granular deposits (or those with collapse potential) this could be less than the disturbance from extracting and transporting samples for laboratory analysis. The British and European Standards provide guidance on the suitability of various ground investigation methods in given soil conditions and describe the quality of the samples likely to be obtained (see Section 29.7.2). For example, British Standards are available for ground investigations (BS 5930:2015+A1:2020; BSI, 2020), identification and description of rock (BS EN ISO 14689:2018; BSI, 2018), testing of soils (BS 1377-1:2016; BSI, 2016) and field testing (BS 22476-14:2020; BSI, various). For more information please refer directly to the British Standards Institution website. Section 4 provides details on good ground investigation practices. For more information refer directly to that section.

An illustration of potential issues can be found in Livneh *et al.* (1998), who undertook a ground investigation for a proposed runway in arid ground conditions containing gypsum. The ground conditions included sandy soils containing clays and gravels with silt and clay lenses and gypsum-rich lenses interspersed between the sandy soils; the authors suggested that the site could be characterised as a sabkha. The gypsum formed through precipitation of soluble minerals with changes in groundwater conditions, and concern was raised over the performance of the ground if the gypsum was wetted during loading. Various in situ tests were undertaken, including dynamic cone penetrometer testing, density determination and loss of weight with wetting of the gypsum-bearing deposits. It was found that low in situ density values, coupled with the potential for dissolution of the gypsum, were the parameters that dictated the pavement design for the runway. Livneh *et al.* (1998) carried out the following laboratory tests on samples taken from the site: plasticity index testing, dry density, California bearing ratio (CBR), unconfined compressive strength (UCS) and analysis of pore water chemistry, to supplement the information gathered from in situ testing.

Various in situ tests to evaluate the collapse and swell potential are discussed in Chapters 32 *Collapsible soils* and 33 *Expansive soils*, respectively, and details of the various control, interpretation and assessment issues are given in these chapters.

29.3.7.3 Laboratory testing

Care must be taken when investigating some arid soils in the laboratory, notably those that are metastable or those that contain high concentrations of miscible salts, such as sabkhas. Laboratory tests for collapsible soils and expansive soils can be found in Chapters 32 *Collapsible soils* and 33 *Expansive soils*, respectively.

An example of an arid soil that requires careful handling in the laboratory is described in a study by Aiban *et al.* (2006). A sabkha soil was investigated and found to exhibit different behaviour depending on the water used in the laboratory tests.

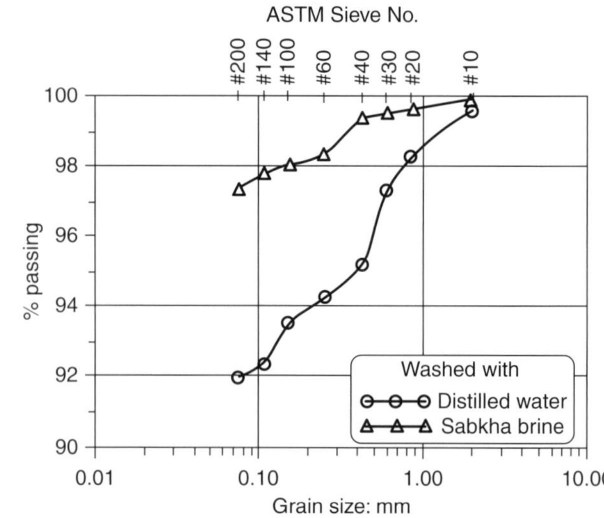

Figure 29.17 Grain-size distributions of a sabkha soil using either distilled water or sabkha brine to wash the soil (Reproduced from Aiban *et al.*, 2006; with permission from Elsevier)

Property		Value
Particle size range (passing ASTM sieve 200)		More than 50%
Liquid limit (LL)	Distilled water	42.5
	Sabkha brine	33.0
Plastic limit (PL)	Distilled water	26.3
	Sabkha brine	20.4
Plasticity index (PI)	Distilled water	16.2
	Sabkha brine	12.6

Table 29.3 Difference in plasticity index with the solution used to investigate a sabkha soil (Aiban *et al.*, 2006)

(**Figure 29.17** shows the differences in the particle size distribution and **Table 29.3** gives the plasticity index for this sabkha soil when prepared with distilled water or sabkha brine.) These differences are believed to be related to the difference in the salt concentration between the soil and the water used. Distilled water would only contain minute concentrations of the solutes found in the soil, if any; thus a proportion of the crystalline salts cementing the soil was believed to dissolve into solution, thereby changing the physical properties of the soil. Thus, it is important to consider the potential impacts that any liquid used in an investigation is likely to have on the soil and it might be prudent to attempt to simulate the liquid that will be used/encountered on site in the laboratory investigation.

29.3.7.4 Geophysical investigation of arid soils

Geophysical investigation(s) (not considered in this chapter in any detail) can potentially be successfully used in arid ground conditions, be non-disruptive, cover significant tracts of land

and augment information derived from traditional borehole or trial-pit surveys. Guidance on the use of geophysical techniques in ground investigations can be found in various publications, including the *Geological Society's Engineering Geology Special Publication 19* (McDowell *et al.*, 2002) and *Engineering Geophysics* (Bondo Medhus and Klinkby, 2022); although not written specifically for arid ground conditions, these publications introduce the geophysical techniques commonly associated with civil engineering ground investigations.

Examples of the use of geophysical techniques in arid (or semi-arid) soils include Ray and Murray's (1994) report on research into the use of airborne SAR, AVIRIS (airborne visible/infrared imaging spectrometer) and thermal imaging to investigate the nature of surface sand deposits in an arid region within the USA. They found that mobile sand deposits could be distinguished from fixed deposits using these techniques. The ability to detect mobile deposits stems from the difference in particle size distribution (active soils tend to comprise a greater proportion of particles in the saltation range and tend to be poorly graded) and the relative density of active and fixed sand deposits (Ray and Murray, 1994). Shao *et al.* (2009) investigated the use of polarimetric SAR to detect subsurface hyper-saline deposits in arid conditions and found promising results; Afshar *et al.* (2017) utilised ground penetrating radar (GPR) to identify the presence of shallow subsurface cemented soil layers; Flinchum *et al.* (2020) investigated the use of a seismic survey, correlating findings with those from other methods (such as down-hole nuclear magnetic resonance and soil sampling) to attempt to identify the position of a water table in a semi-arid environment.

The (potential) ability to detect certain problematic soils geophysically could provide an opportunity to optimise any additional investigation and hence (potentially) provide the means of better identifying geotechnical processes/interventions that will have to be undertaken before construction can begin/continue.

29.3.7.5 Material resources for construction projects in arid environments

Various landforms within uplands, foot-slopes and plains (see Section 29.3.1) are potentially sources of aggregates. **Table 29.4** indicates potential locations for sourcing aggregates and suggests potential applications within construction processes for these materials. Lee and Fookes (2005b) recommend alluvial

Feature	Aggregate type	Nature of material	Engineering properties of fills	Potential volume
Bedrock mountains	Crushed rock suitable for all types of aggregate	Angular, clean and rough texture	Depends on rock type and processing. Good fills	Very extensive
Duricrust	Road base and subbase	Often contains salts. Needs crushing and processing	May be self-cementing with time. Quite good rockfill	Often only small deposits of good quality
Upper alluvial fan deposits	Concrete and can be crushed for road base	When crushed and screened angular and clean. Otherwise sometimes dirty and often rounded	Good compaction as fill	Often very extensive but good-quality material may be found only in small deposits
Middle alluvial fan deposits	May make road base	Often high fines content	Often good compaction as fill. Good bearing	Small to extensive
Lower alluvial fan deposits	Generally not useful	High fines content. Needs processing	Poor bearing and as fill	Small to extensive
Other piedmont plain alluvium	Variable, locally good concrete aggregate	Dirty, rounded and well graded. Can contain salts	Good bearing capacity (dense sediments) and as fill. Locally poor due to clay and silt layers	Very extensive but good-quality material in small deposits
Old river deposits	Concrete	Variable	Difficult to locate in field	Often deposits patchy and thin
Dunes	Generally not good	Usually too fine and rounded	Fills of poor compaction	Locally extensive
Interdunes	Fine aggregate	Coarse to fine, angular. Needs processing	Fills of poor compaction	Very localised
Salt playas and sabkhas	Not suitable	Very salty and aggressive	Poor. Special random fill	Locally extensive
Coastal dunes	Generally not suitable	Generally too fine and rounded	Special random fill, often poor compaction	Sometimes extensive
Storm beach	Fine aggregate may be sharp or salty	Sufficiently coarse for concrete sand. Clean after processing	Random fill	Sometimes extensive
Foreshore	Generally not suitable	Fine, rounded sand	Salt contaminated, but might make random fill	Locally extensive

Table 29.4 Landforms and aggregate potential (Reproduced from Lee and Fookes, 2005b; data taken from Fookes and Higginbottom, 1980; Cooke *et al.*, 1982)

deposits (fans and wadi channels) when sourcing aggregates; it is suggested that dune deposits can be used, but with caution as the particle grading often encountered in dunes can be too fine to be suitable. It is recommended that deposits containing high concentrations of salts should be avoided (or treated to remove the salts). Lee and Fookes (2005b) also suggest that care should be taken when sourcing aggregates as cemented deposits are likely to require processing before being suitable.

The availability of water for construction projects in arid regions must be considered, as this can severely affect the processes being undertaken. Arid regions are often located in developing countries (Wallen and Stockholm, 1966), which have, potentially, limited infrastructure away from urban centres, and the climatic and geomorphologic processes affecting the site must be understood. Thus, water may have to be brought to a site, or groundwater must be extracted (if the phreatic surface is sufficiently close to the surface to make this option economically viable); groundwater can contain significant concentrations of dissolved minerals, which can have a detrimental effect on construction (casting of concrete, etc.) and these factors should be considered by a site investigation.

29.4. Aspects of the geotechnical behaviour of arid soils

When engineering on site in an arid environment it is vital to understand how the aridity will affect the work and what geomorphologic processes will (more than likely) continue to act on the site once construction is complete. This section contains a description of generalised aspects of the geotechnical behaviour of arid soils and considers

- generalised aspects of (selected) geotechnical behaviour of arid soils
- various aspects that should be considered when dealing with arid soils, including the (generalised) behaviour of unsaturated soils; the development of suction within the soil; the relationship between total stress, effective stress, pore water pressure and pore air pressure; and how the shear and volume change behaviour can be considered
- the behaviour of expansive and collapsible soils in arid regions and the effects of cemented soils, together with how such soils can be treated.

29.4.1 Generalised aspects of the geotechnical behaviour of arid soils

Arid soils, by definition, have a moisture deficit, which is likely to result in unsaturated conditions being prevalent (in the surface layers of the soil at least). This is not to suggest that groundwater cannot exist in arid locations, but it may be at a significant depth below the ground surface. The resulting suction can lead to an increase in effective stress of the unsaturated soil and can induce groundwater flow vertically upwards from the phreatic surface into the vadose zone. Both phenomena can have a marked impact on the behaviour of arid soils.

The suction in an arid soil is a function of the interaction between water and air within the pores. In simple terms, suction forms due to the creation of menisci within the pores because of surface tension between the pore water and pore air (Fleureau and Taibi, 1994) as the pore water preferentially lines the surfaces of particles before filling the pore space. However, in soils containing a distribution of pore sizes, some of the pores might fill completely with fluid (known as bulk water) while others remain unsaturated (containing meniscus water), resulting in a heterogeneous distribution of suction within the soil (Wheeler *et al*., 2003). Thus, the effective stress of the unsaturated soil is a function of the suction within the soil (see Section 29.4.2), which in turn is a function of the structure of the soil. Therefore, a change to the structure of the soil (which could occur when obtaining samples for laboratory investigation) for a constant water content can result in a different effective stress of the soil (Wheeler *et al*., 2003).

Suction within a soil may be described as matric suction (a function of the relationship between pore water and pore air pressure) or total suction (which incorporates suction due to osmotic pressure within the soil in addition to matric suction), although at high suction ($\geq$1500 kPa) total suction and matric suction are considered to be the same (Fredlund and Xing, 1994). The relationship between the matrix suction experienced within a soil and the water content is illustrated by a soil–water characterisation curve (SWCC) (Fredlund and Rahardjo, 1993a, 1993b) – **Figure 29.18**. The nature of these curves is determined by soil parameters such as the pore size distribution, the minimum radius of the pores and the stress history and the plasticity of the soil (Fredlund and Xing, 1994).

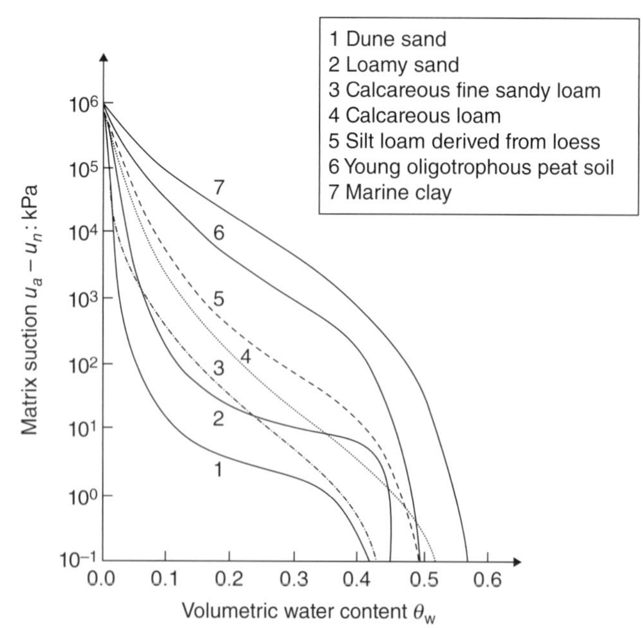

Figure 29.18 Typical soil–water characterisation curves for various soils (Mitchell and Soga, 2005; reproduced from Koorevaar *et al*., 1993, with permission from Elsevier)

In addition to suctions, arid soils can be cemented, exhibit plasticity and shrink–swell potential, be metastable and vulnerable to collapse and be vulnerable to compressibility. These factors will likely impact on the engineering response of the soil and thus need to be considered before undertaking construction on/in/with them.

Various aspects of arid soil behaviour will be introduced in turn, highlighting features pertinent to arid soils. However, only brief details are provided and further details can be found in publications such as Fredlund and Rahardjo (1993a), Fredlund *et al.* (2012), Blight (2017) and references cited therein.

29.4.2 Engineering behaviour of unsaturated, uncemented arid soils

Fredlund *et al.* (2012) presents a simple conceptual model for the vadose (or unsaturated) zone with respect to suction. In this model the vadose zone is divided into three layers: the capillary fringe (directly above the phreatic surface), the transition zone (or two-phase zone) and the residual (or dry) zone (near the ground surface). Fredlund *et al.* (2012) provide generalised ranges of degrees of saturation (S_r) for these zones: S_r 90–100% in the capillary fringe; S_r 15–90% in the transition zone; and S_r 0–15% in the dry zone. However, the boundary between zones for a given soil should be determined by interpretation of its soil–water characteristic curve, and Fredlund *et al.* (2012) provide a technique to define these boundaries (the boundary between the capillary fringe and the transition zone is known as the 'air entry point'). The suction in the capillary fringe is likely to be relatively low as the pores are dominated by water (potentially fully saturated, although the pore water pressures are negative); once the air entry point is reached the suction can increase significantly, although the largest suction would be expected within the residual zone (where the pore spaces are dominated by air) (Fredlund *et al.*, 2012).

It must be noted that the soil–water characteristic curve has two fundamental nonlinear (following a hysteresial relationship) relationships between matric suction and water contents. This is because a soil will exhibit different behaviour if it is drying (experiencing loss of water from the soil) or wetting (water re-entering the soil). Hence, the SWCC has two boundaries: the drying curve and wetting curve, with the drying curve exhibiting greater suction than the wetting curve for the same water content. If the conditions acting on the soil change (e.g. going from wetting to drying) then the soil will follow a nonlinear pathway between the two boundaries. For more information on the SWCC, see Gaspar *et al.* (2019) or Eyo *et al.* (2022).

If a soil is uncemented and has reached the air entry point, the relationship between stress and pore water pressure should be modified to consider the effect of suction in the soil on the effective stress, and, hence, the shear strength of the soil. At this point effective stress is a function of total stress, pore water pressure (u_w) and pore air pressure (u_a), which control the engineering response of an uncemented arid soil. The shear strength of uncemented, unsaturated arid soils can be approximated using Equation 29.1 (in which σ is normal stress, $(\sigma - u_a)$ is net stress, σ' is effective stress, τ_f is shear strength, ϕ' is the angle of internal shearing resistance, $(u_a - u_w)$ is matric suction, χ is a coefficient related to degree of saturation, and c' is apparent cohesion), although there are challenges associated with defining χ, which is not a constant but varies with drying/wetting response of the soil (when the soil is dry $\chi = 0$, and when the soil is fully saturated $\chi = 1$: Abed and Vermeer, 2006). Instead of combining the net stress and the matric suction into one term, which is affected by the internal angle of shearing resistance, an alternative relationship can be derived if the net stress and the matric suction are considered as separate terms (Equation 29.2, where a and b are parameters for net stress and matric suction, respectively). Substituting two angles of shearing resistance for parameters a and b, and incorporating apparent cohesion, creates an alternative definition for the effect of suction on shear strength (Equation 29.3, where ϕ^b is the angle of shearing resistance with respect to matric suction). The angle of shearing resistance with respect to matric suction is equal to the angle of shearing resistance due to frictional interactions (ϕ') at high suction but can be reduced with increasing degrees of saturation (Rassam and Williams, 1999)

$$\tau_f = c' + [(\sigma - u_a) + \chi(u_a - u_w)]\tan\phi' \quad \text{(Tian \textit{et al.}, 2023)} \qquad (29.1)$$

$$\tau \propto a(\sigma - u_a) + b(u_a - u_w) \qquad (29.2)$$

$$\tau_f = c' + (\sigma - u_a)\tan\phi' + (u_a - u_w)\tan\phi^b \quad \text{(Rassam and Williams, 1999)} \qquad (29.3)$$

The compressibility and volume change behaviours for uncemented, unsaturated soils have been studied using various models, such as the Barcelona model developed by Alonso and Gens (1994). These allow net stress, deviatoric stress, suction and specific volume (or, by extension, volumetric strains) to be assessed together for the soils (as illustrated by Abed and Vermeer, 2006; Alonso and Gens, 1994). This suggests that matric suction improves the ability of a soil to resist movement, even after the soil experiences plastic deformation. If suction within the soil dissipates, then a new stress path will develop within the soil; this stress path could result in the soil exceeding the pre-consolidation pressure, in which case the soil will yield without an increase in external load and experience plastic deformation (Alonso and Gens, 1994). The dissipation of suction results in a softening of the soil and an associated deformation. While an uncemented, unsaturated soil experiences elastic deformation with the dissipation of suction – that is, the pre-consolidation pressure has not been reached – the soil can expand when water enters its pores (Alonso and Gens, 1994). Once the pre-consolidation pressure is reached the sample will yield and experience plastic (i.e. unrecoverable) deformation and the soil will compress.

Restructuring of the soil fabric at the microscale in response to the dissipation of suction can result in disruption to the soil packing on the macroscale (Alonso and Gens, 1994). Thus, the overall effect may soften the soil and the actual pre-consolidation pressure may be less than predicted from the results of small-scale experiments. Softening may also occur if a soil experiences shearing. As the soil deforms under shear, the menisci within the pores may fail, and until the menisci reform the soil will experience a decrease in suction and as a result it will soften (Mitchell and Soga, 2005).

29.4.3 Relationship between net stress, volumetric strain and suction for collapsible and expansive arid soils

By studying the behaviour of soil samples taken from a site, Blight (1994) was able to describe the relationship between net stress, suction and volumetric strain for expansive and collapsing soils. In this section, drawing from the work of Blight (1994), a brief description of the relationship between net stress, volumetric strain and suction for collapsible and expansive soils is given. More details are given in Chapters 32 *Collapsible soils* and 33 *Expansive soils*, respectively.

Blight (1994) illustrated that an unsaturated expansive soil under low net stress will experience an increase in volumetric strain with the dissipation of suction, and the associated increase in water content, until the suction has fully dissipated (and then the soil can experience additional swelling under saturated conditions if the net stress remains low) (path A-B-C′ in **Figure 29.19**). For an expansive soil experiencing a greater magnitude of net stress, the dissipation of suction and the increase in volumetric strain may result in the expansive soil becoming unstable and, at a critical level of suction, the structure will become unable to support the load and will collapse (path D-E-F in **Figure 29.19**). Once the soil structure has collapsed, additional expansion can occur with a continued reduction in suction. The net stress experienced by the soil will dictate the degree of volumetric strain experienced by the unsaturated expansive soil. Once suction has dissipated the soil may experience swelling or shrinking with a change in stress.

Blight (1994) investigated the behaviour of a clayey silt sample using an oedometer at the natural water content of the soil and under saturated conditions. The saturated soil sample in this investigation initially experienced heave under light consolidating pressure before yielding and suffering plastic deformation (**Figure 29.20**). The unsaturated sample did not experience heave on loading and yielded at a greater pre-consolidation pressure (**Figure 29.20**). The two examples illustrate that predicting the actual deformation experienced by an unsaturated soil on site due to changes in loading can be very difficult and it is recommended that laboratory or field trials are undertaken to determine the behaviour of the soil. Blight (1994) states that arid soils can be particularly problematic due to the geomorphologic processes that formed them, as they can be prone to both expansion and collapse (echoed by Wheeler, 1994). Thus, some form of experimental investigation is considered fundamental if the behaviour of a soil is to be understood.

29.4.4 Cemented soils in arid conditions

Light cementation of soils (as opposed to the degree of cementation required to create a duricrust) is relatively common in

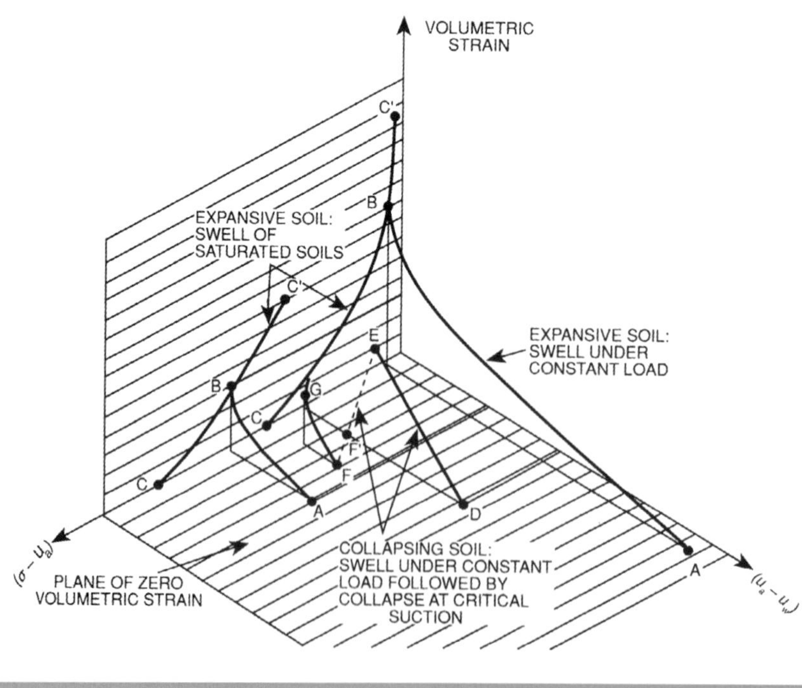

Figure 29.19 Three-dimensional stress–strain diagrams for partially saturated soil for the swell of a partially saturated soil under constant isotropic load (Reproduced from Blight, 1994)

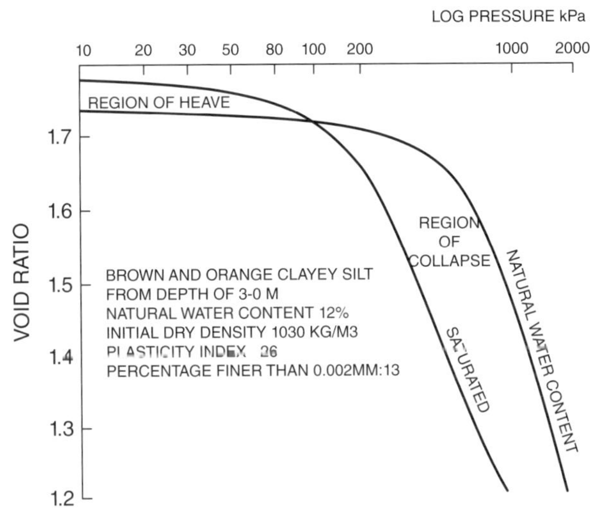

Figure 29.20 Behaviour of clayey silt at natural water content and after saturation (Reproduced from Blight, 1994)

arid environments (see Section 29.3.4) and can affect how a soil behaves with loading. Cemented soils can have increased tensile strength and can remain stable even after excavation; **Figure 29.21** shows aeolian sand deposits in a semi-arid environment during the wet season. The soil is stable after excavation due to cementation of the soil grains. Care should be taken when assuming that the walls of an excavation will remain stable due to cementation; localised loading can still cause the sides of an excavation to deform and, if of sufficient magnitude, ultimately fail. Cementation of the particles can result in an increase in the shear resistance (**Figure 29.22**), an increased 'yield' stress and, likely, increased stiffness moduli prior to yield. However, once the cemented soil yields, the critical strength of the uncemented soil and lightly cemented soils (assuming all other parameters are the same) are similar. This is because the crystals bonding the particles within the cemented soil together are relatively brittle and will likely shear at low strains (Mitchell and Soga, 2005). Once the cementation is disrupted the shear strength reduces and

Figure 29.21 Excavation with sides of stable cemented sand (aeolian deposits) during the wet season in semi-arid conditions (© A. Royal)

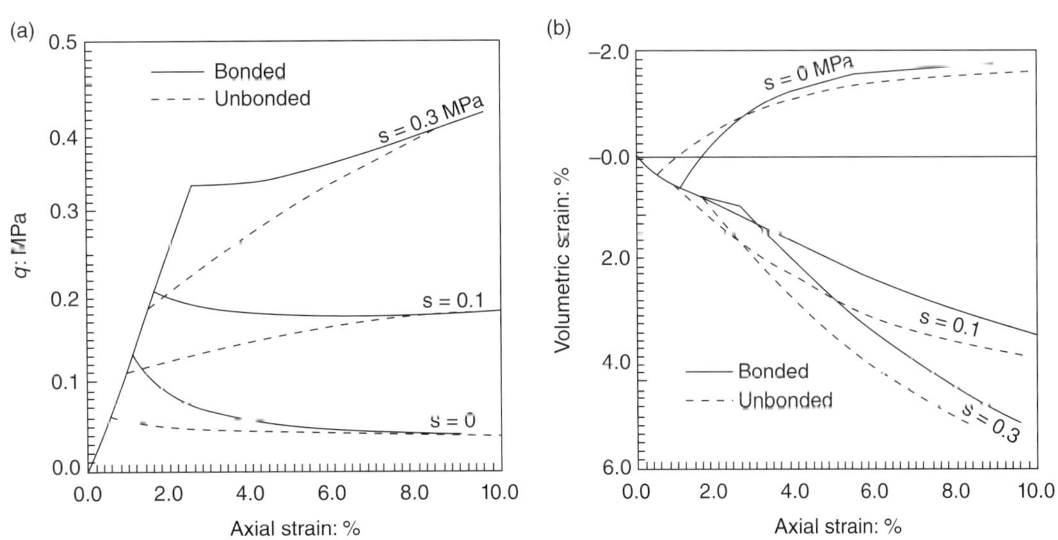

Figure 29.22 Simulated triaxial tests on unsaturated cemented and uncemented soils: (a) shear stress shear–strain relationship and (b) volumetric response (Reproduced from Alonso and Gens, 1994)

becomes a function of the behaviour of the soil particles, following the behaviour of uncemented soils.

If the degree of cementation is sufficient to create soils with relatively high stiffness and strength, and with the ability to resist deformation, these soil deposits could appear ideal for engineering applications. However, wetting of these soils can result in loss of strength and settlement (potentially due to (partial) dissipation of suction and/or (partial) dissolution of the cementing compounds). This is known as collapse settlements and has been the focus of many investigations (e.g. Al-Amoudi and Abduljawad, 1995; Alawaji, 2008; Assallay *et al.*, 1997; Fattah *et al.*, 2014; Haeri *et al.*, 2005; Hafhouf and Abbeche 2023; Macedo *et al.*, 2023). Wetting of these soils could stem from several causes, such as changes in hydrogeological conditions, changes due to construction activities or stemming from the outcomes of construction as well as (over) irrigation or from leaking utilities (Wheeler, 1994).

Experimental investigation of collapsible soils might consider deformation response (using apparatus such as triaxial, UCS, simple shear, shear box), consistency limits, chemical nature of the cementation, dry density and void ratio and collapse potential. The collapse behaviour is often considered using the oedometer, which can be undertaken using one sample (where the sample is initially investigated under low degrees of saturation before the oedometer cell is inundated with water, initiating collapse) or two samples (where one sample is compressed in a low degree of saturation state and the other saturated from the outset, with the difference in response used to estimate collapse) (e.g. Fattah *et al.*, 2022).

Al-Amoudi and Abduljawad (1995) suggested two simple equations that can be used to predict if the magnitude of the collapse in the oedometer signals problematic soil conditions on site. In Equations 29.4 and 29.5, i_{c1} and i_{c2} are the collapse potentials, Δe_c is the reduction in the void ratio on wetting, e_0 is the initial void ratio and e_1 is the void ratio at the beginning of saturation:

$$i_{c1} = \left(\frac{\Delta e_c}{1+e_1}\right) \times 100 \quad (29.4)$$

$$i_{c2} = \left(\frac{\Delta e_c}{1+e_0}\right) \times 100 \quad (29.5)$$

Equation 29.4 gives the change at the beginning of flooding and Equation 29.5 gives the collapse potential at the beginning of the consolidation test. Al-Amoudi and Abduljawad (1995) cite findings by Jennings and Knight (1975) that relate the collapse potential to the severity of the problem (see Chapter 32 *Collapsible soils* for further details).

If collapse potential is a function of (partial) dissolution of miscible salts (hence a function of the dissolution rate of the salt(s) and an equilibrium of the solid and dissolved phases of salts within the soil), then the time frame of the experimentation and the hydraulic conditions applied can have an impact on the results returned. Fattah *et al.* (2022) noted that collapse potential of naturally occurring, poorly graded sands stabilised by (predominantly) gypsum increased with increased duration of experimentation (1 day and 60 days). Al-Amoudi and Abduljawad (1995) undertook testing in the oedometer under hydrostatic and hydrodynamic conditions, noting that greater movements were observed with the sample that experienced seepage through it.

29.4.5 Sabkha soils

Sabkha soils are heterogeneous in nature, can be cemented, can be metastable and can contain high salt concentrations (Akili, 2015). In addition, these soils can be loose and highly compressible (Moon *et al.*, 2019). Much like other metastable deposits, changes in water content can induce collapse in sabkha soils. The dramatic effect of water on sabkha soil was demonstrated by Aiban *et al.* (2006), who showed how the CBR for unsoaked specimens was around 50% at the maximum dry density, compared to around 5% when soaked at the same density. This difference in behaviour was attributed to dissolution of the cementing salts.

Sabkha soils are (likely to be) metastable; once the strength of the salts cementing the particles is exceeded (or the influence of the cementing salts is reduced as dissolution occurs with changes in the pore water), then there can be rapid rearrangements of the particles within the soil, resulting in compression. This behaviour is not universally encountered in laboratory investigations of the collapse potential of metastable sabkha samples (or the magnitude of collapse potential was lower than envisaged). Al-Amoudi and Abduljawad (1995) suggested that this is a function of the volume of water introduced into the sample, with an equilibrium forming between the concentration of salts in the water and the cemented soil. Using a modified oedometer (that applied a constant head of water across the sample and allowed water to seep through the soil while under load), the test sample exhibited greater collapse potential (and achieved lower resulting void ratios) when compared to that observed in traditional oedometer testing.

Hafhouf *et al.* (2022) and Hafhouf and Abbeche (2023) considered the change in electrical conductivity, salinity, suction, shear strength, UCS and collapse potential of samples reconstituted from sabkha soil (in both studies samples were obtained from site, dried, sieved and mixed with water to reconstitute the soil, before statically compacting the samples into the desired shape with desired dry densities) when exposed to cycles of drying and rewetting. It was observed that the magnitude of these parameters would change during drying and wetting stages (e.g. suction, and hence strength, increasing when exposed to drying, reducing with rewetting), and overall (after the four cycles of drying and rewetting) there were reductions in electrical conductivity, suction and UCS, while collapse potential increased. These changes were attributed to the reduction in salts within the soil structure and the associated changes in soil structure. Such findings might indicate that consideration should be given to the long-term response of arid soils when engineering in these regions.

29.4.6 Hydraulic conductivity and groundwater flow in unsaturated soils

Darcy's Law can be applied to unsaturated soils with modification. Mitchell and Soga (2005) give a relationship for the vertical seepage velocity in unsaturated soils: Equation 29.6, where $k(S)$ is the saturation-dependent hydraulic conductivity, v_i is the vertical seepage velocity, ψ is the matric suction and $\delta z/\delta x$ a gravitational vector measured upwards in the direction of z for flow in direction i, in which $\delta\psi/\delta x$ is matric suction equivalent head.

This equation can be incorporated into a mass conservation equation to give Equation 29.7, which predicts changes in flow for changes in saturation and suction, where S is saturation, t is time, η is the porosity and $k(\psi)$ is the matric suction-dependent hydraulic conductivity derived from $k(S)$ using the soil–water characterisation curve

$$v_{\mathrm{i}} = -k(s)\left(\frac{\delta\psi}{\delta x_{\mathrm{i}}} + \frac{\delta z}{\delta x_{\mathrm{i}}}\right) \tag{29.6}$$

$$\eta = \left(\frac{\delta S}{\delta\psi}\frac{\delta\psi}{\delta t}\right) = \frac{\delta}{\delta x_{\mathrm{i}}}\left[k(\psi)\left(\frac{\delta\psi}{\delta x_{\mathrm{i}}} + \frac{\delta z}{\delta x_{\mathrm{i}}}\right)\right] \tag{29.7}$$

This solution is considered valid provided that there are no changes in the soil structure or loss of water from the system (i.e. no loss of water due to biological activity, etc.). The transitional nature of the groundwater flow in unsaturated media, with changing permeability, degree of saturation and suction with flow and time, lends itself to numerical solutions (Mitchell and Soga, 2005).

Measurement of the hydraulic conductivity in unsaturated soils can be problematic due to the dependent relationship between the hydraulic conductivity, saturation and the matric suction. The soil–water characterisation curve of a soil may be used to derive the hydraulic conductivity; an alternative approach is to use the relative permeability (see Mitchell and Soga, 2005, for further details). Equation 29.8 can be used to estimate the hydraulic conductivity k for an unsaturated soil where g is the acceleration due to gravity, k_r is the relative permeability, k_s is the saturated hydraulic conductivity, K is the intrinsic permeability and ρ_{pf} and η_{pf} are the density and dynamic viscosity of the permeating fluid, respectively. The intrinsic permeability is defined by Equation 29.9, where e is the void ratio, k_0 is a pore shape factor, T is a tortuosity factor and S_0 is the wetted surface area per unit volume of particles.

$$k = k_{\mathrm{r}}K\frac{\rho_{\mathrm{pf}}g}{\eta_{\mathrm{pf}}} = k_{\mathrm{r}}k_{\mathrm{s}} \tag{29.8}$$

$$K = \frac{1}{k_0T^2S_0^2}\left(\frac{e^3}{1+e}\right) \tag{29.9}$$

Lee and Fookes (2005b) suggest that in arid conditions the fines content can reach 30% without having a significant effect on the ability of the soil to drain.

29.5. Engineering in problematic arid soil conditions

The creation of soil deposits that can be metastable, expansive, vulnerable to collapse and/or chemically aggressive (from the viewpoint of the durability of man-made structures) within arid regions results in a number of geohazards, which must be addressed by the geotechnical engineer. This section introduces case studies of various ground conditions encountered in arid regions, including

- foundations in expansive, desiccated or collapsible soils
- foundations in chemically stabilised soil or rock formations
- compaction of placed fill
- stabilisation of sabkha soils.

Chapters 32 *Collapsible soils* and 33 *Expansive soils* describe the behaviours of collapsible soils and expansive soils, with more detailed explanations of structure–ground interactions. This section provides a brief overview of problems encountered when dealing with soil deposits in arid environments, with the aim of highlighting areas that might necessitate the attention of the geotechnical engineer during design and construction.

29.5.1 Geohazards encountered in desiccated, expansive and collapsible soils

29.5.1.1 Examples of foundation behaviour in expansive soils

A detailed review of foundations suitable for arid soils that exhibit expansive behaviour can be found in Chapter 33 *Expansive soils*. That chapter deals with assessment and the foundation or remediation options available to deal with such soils. A key aspect for foundation design in these conditions is the implication of the unsaturated state, due to the arid environment, of the expansive soil which could dictate the nature of the problem and the responses needed (e.g. swelling and softening with wetting). For example, Kropp (2010) reviewed changes in the foundation design for domestic dwellings on expansive soils in the San Francisco Bay Area. Here, the preference in the local community is for grassed or planted gardens and watering is common. It was suggested that homeowners regularly overwatered the ground to the point where the mass of water penetrating the ground exceeded that exiting the ground by evapotranspiration, and this change in localised groundwater conditions results in considerable, and potentially differential, ground movement which strip footings were ill-suited to resist.

Nusier and Alawneh (2002) report on the distress caused to a single-storey concrete building with a mat foundation (a reinforced-concrete slab) constructed on expansive clay in Jordan. The shallow foundation experienced differential ground movement as the ground under most of the house heaved, yet one corner of the building settled (due to the desiccating action of a tree); the mat was not sufficiently rigid to withstand these movements, the concrete foundation walls were not reinforced and the building suffered significant cracking. It was suggested that the soil under the building heaved due to a thermal gradient existing across the soil below the building and the soil surrounding the building directly exposed to sunlight, resulting in the migration of water from the surrounding area towards the ground below the house. Nusier and Alawneh (2002) concluded that, had the foundation been deeper, to a depth where the pressure from the building acting on the soil equalled or exceeded that of the pressure from the expanding soil, then the situation could have been avoided. Instead, the building was repaired by installing a reinforced-concrete skeleton to the outside of the building and a ring beam to the mat. This intervention cost approximately 25% of the original cost of constructing the building, highlighting the need to understand the probable behaviour of the ground before undertaking construction.

When a desiccated clayey soil experiences infiltration of water during and directly after a storm (or because of a leaking utility) the soil fabric can undergo restructuring as it swells. This phenomenon can be pronounced if the soil comprises a significant fraction of clays, or an active clay mineral such as smectite. The seasonal cycle of desiccation/wetting may also cause problems with deep foundations. Blight (1994) reports that pile tests in soil conditions in southern Africa indicate that soil shrinks away from the surface of a pile as it dries in the dry season; clearly this has significant implications when designing a deep foundation if there is a considerable reduction in the skin friction acting on a pile.

Various improvement approaches can be used to treat expansive arid soils (see Chapter 33 *Expansive soils*). For example, Rao *et al.* (2001) investigated the performance of lime-stabilised expansive-clay compacted samples exposed to repeated cycles of wetting and drying. Samples were created with lesser and greater amounts of lime than the initial consumption of lime (ICL) value of the soil and were then exposed to several wetting and drying cycles. Both the wetting and drying phases each took 48 hours to complete. It was noted that repeated cycles of wetting and drying had a detrimental effect upon the lime-stabilised soil for samples with lime contents below or slightly above the ICL. Increasing the number of wetting and drying cycles resulted in more clay detected in the stabilised samples, a higher liquid limit (although the liquid limits of the treated soils were predominantly lower than the untreated soil), a lower plastic limit (although the plastic limits of the treated soils were higher than the untreated soil) and a lower shrinkage limit (although the shrinkage limits of the treated soils were higher than the untreated soil), which Rao *et al.* (2001) attributed to the breakdown of the stabilised soil structure (disruption of particle aggregations).

A range of treatment and improvement approaches can be employed to mitigate the behaviour of expansive soils; further discussion can be found in Chapter 33 *Expansive soils.*

29.5.1.2 An example of foundation behaviour in a collapsible soil

A detailed review of how to deal with foundations in arid soils that exhibit collapsible behaviour can be found in Chapter 32 *Collapsible soils.* That chapter deals with assessment and the foundation or remediation options available to deal with such soils. A key aspect, as with expansive arid soils, is the implication of the unsaturated state, due to the arid environment, which could dictate the nature of the problem and the responses required (e.g. softening and potentially settlement with wetting).

Loess, a collapsible soil, is often considered to be a problematic soil readily collapsing after inundation by water or an increase in load. For example, during a plate loading test, Komornik (1994) applied cyclic loading to a plate in a loess soil at the natural water content and after wetting the soil. Wetting resulted in collapse, leading to large unrecoverable deformations.

The difference between the behaviour of the soil at the natural water content and the soil after wetting illustrates the necessity of designing a foundation carefully if collapse potential is considered a possibility, and also the necessity of protecting the location from possible inundation (during storms, from excessive irrigation or leaking utilities, etc.) in order to prevent movement of the foundation after completion of construction.

A range of treatment and improvement approaches can be employed to mitigate the behaviour of collapsible arid soils (e.g. dynamic compaction, cement stabilisation); further discussion can be found in Chapter 32 *Collapsible soils.*

29.5.2 Compaction of fill

In an arid environment, where water is not generally plentiful, applying too little water to the soil (or, in a hot arid environment, applying the desired amount and losing a proportion to evaporation) can result in the development of suction within the soil as it is compacted. The suction can resist the compactive effort and the net effect is a low-density fill, which may experience undesirable levels of settlement. It has been suggested that in such circumstances compacting the soil when it is dry might be more appropriate (Newill and O'Connell, 1994). Of course, dry compaction will produce fills with a high volume of air voids and measures must be taken to ensure that subsequent ingress of water is prevented or the fill may undergo restructuring of its structure, causing the fill to deform or collapse.

In large construction projects the volume of water required can be significant, which can create logistical problems when working in arid conditions. For example, Newill and O'Connell

(1994) investigated the construction of a major road in an arid environment and estimated that the water required to create one kilometre of the road could be as much as 2800 m^3 and equated to approximately 20% of the total construction cost in sourcing and transporting the water to site. This value is likely to exceed the cost of water for construction in tropical or temperate climates. It must also be considered that arid conditions are often located in developing countries (White, 1966), and the infrastructure can be, at best, extremely limited outside urban areas. Due to the potential lack of infrastructure in rural areas, the source of water may be from boreholes, and it is unlikely that any treatment will be applied to the extracted water before it is used for construction. Therefore, the extracted water can potentially contain significant concentrations of salts that will remain within the fill and can have a detrimental effect on the man-made structures placed upon it.

In addition, not all construction projects in developing countries boast the resources necessary to surmount the logistical problems of sourcing sufficient water for a project. For example, while working on a site in semi-arid conditions a sister site (approximately 100 km away) faced significant problems when it was announced that the local village borehole was at risk of running dry and a moratorium was placed on the use of water for anything other than critical functions (which did not include construction). In order to continue construction and ensure that the workforce had water for drinking and washing, 5 m^3 of water was transported by bowser daily. This placed a strain on the efficiencies of both sites, and it was difficult to justify the use of water when compacting fills when other more critical functions (drinking, washing, casting concrete, etc.) took precedence. Eventually it was decided that in order to meet the specifications (end-product specifications were being used) water would have to be used during compaction and the programmes for both sites overran.

When compacting soils in arid environments, particularly for significant geotechnical structures such as roads or embankments, the processes applied are likely to be governed by a framework or standard. Truslove (2010) and Puttock *et al.* (2011), by considering suitable earthwork specifications in arid environments, reviewed several standards used in various countries with arid lands. Truslove (2010) concluded that there were standards that permitted the use of compaction of dry fill in rural areas, whereas other standards refer to an end-product specification (minimum acceptable dry density, etc.) and do not necessarily comment on the amount of water required. Kropp *et al.* (1994) investigated a compacted fill that had suffered excessive deformation; the soil was a granular material that contained clay and had been compacted to a density greater than the minimum required in the end-product specification, although with a water content less than the optimum (up to 7% less). It was concluded that wetting had induced collapse and that, perhaps, the standard should also impose conditions on the water content of the compacted fill to try to avoid this scenario. Truslove (2010) cites studies that suggest that consideration must be given to the nature of the soil underlying the compacted fill and that care should be taken when selecting the water content of the fill as there is the potential for water to migrate from the fill into the underlying soil, which could induce structural changes within the underlying soil, jeopardising the performance of the fill above.

Sahu (2001) investigated the performance of six regionally sourced soils from locations in Botswana when mixed with fly ash with the aim of producing a range of materials that could be used as base courses in highway construction. The six soils ranged from Kalahari sand, calcrete, silts (of various plasticities) and silty sands, as well as an expansive clay soil. Various proportions of fly ash were mixed with the soils, and the mixtures were hydrated and left to cure. With the exception of the expansive clay, the inclusion of fly ash lowered the maximum dry densities. However, the inclusion of the fly ash did increase the CBR for the soils containing silts and the calcite; therefore it was suggested that this made these soils suitable for use in highway construction. The inclusion of fly ash in the Kalahari sands and the expansive clays did not have a noticeable impact on the CBR and these were deemed unsuitable for highway construction.

29.5.3 Stabilisation of sabkha soils

Al-Amoudi *et al.* (1995) cite instances where geotechnical improvement techniques such as dynamic compaction, stone columns, densification piles and the creation of drainage have all successfully been used to improve the physical properties of these notoriously weak materials. Al-Amoudi *et al.* (1995) quote unconfined strengths of around 20 kPa and SPT *N* ranges of 0–8 for these types of formations of sabkha soils, although a value of 30 kPa can be achieved at depth (Shehata and Amin, 1997).

Cement and lime have been investigated as stabilising agents for sabkha soils and found to be beneficial (e.g. Al-Amoudi *et al.*, 1995; Al-Homidy *et al.*, 2017; Danish *et al.*, 2022) (**Figure 29.23**), although there are concerns about the long-term durability of the stabilised materials (Al-Amoudi *et al.*, 1995).

Alternative ground improvement materials, such as the use of filler particles (limestone dust or marl) to reduce the void ratio of the soil and potentially increase the UCS, have been trialled with sabkha soils (Al-Amoudi *et al.*, 1995). The limestone dust (a by-product of an industry local to the study) and marl increased the maximum dry density of the sabkha soil, with the limestone dust reducing the optimum water content and the marl increasing the optimum water content. However, varying the content of limestone dust or marl did not noticeably change the UCS of the sabkha soil, suggesting that the incorporation of these materials can reduce the volume of voids of the compacted sabkha soil but not have a noticeable increase on its strength. Cement kiln dust (CKD) has also been considered (a by-product of cement production) (e.g. Al-Homidy *et al.*, 2017; Danish *et al.*, 2022), and improvements in UCS and CBR have been observed in treated samples when compared to untreated sabkha soils. However, in the studies cited the

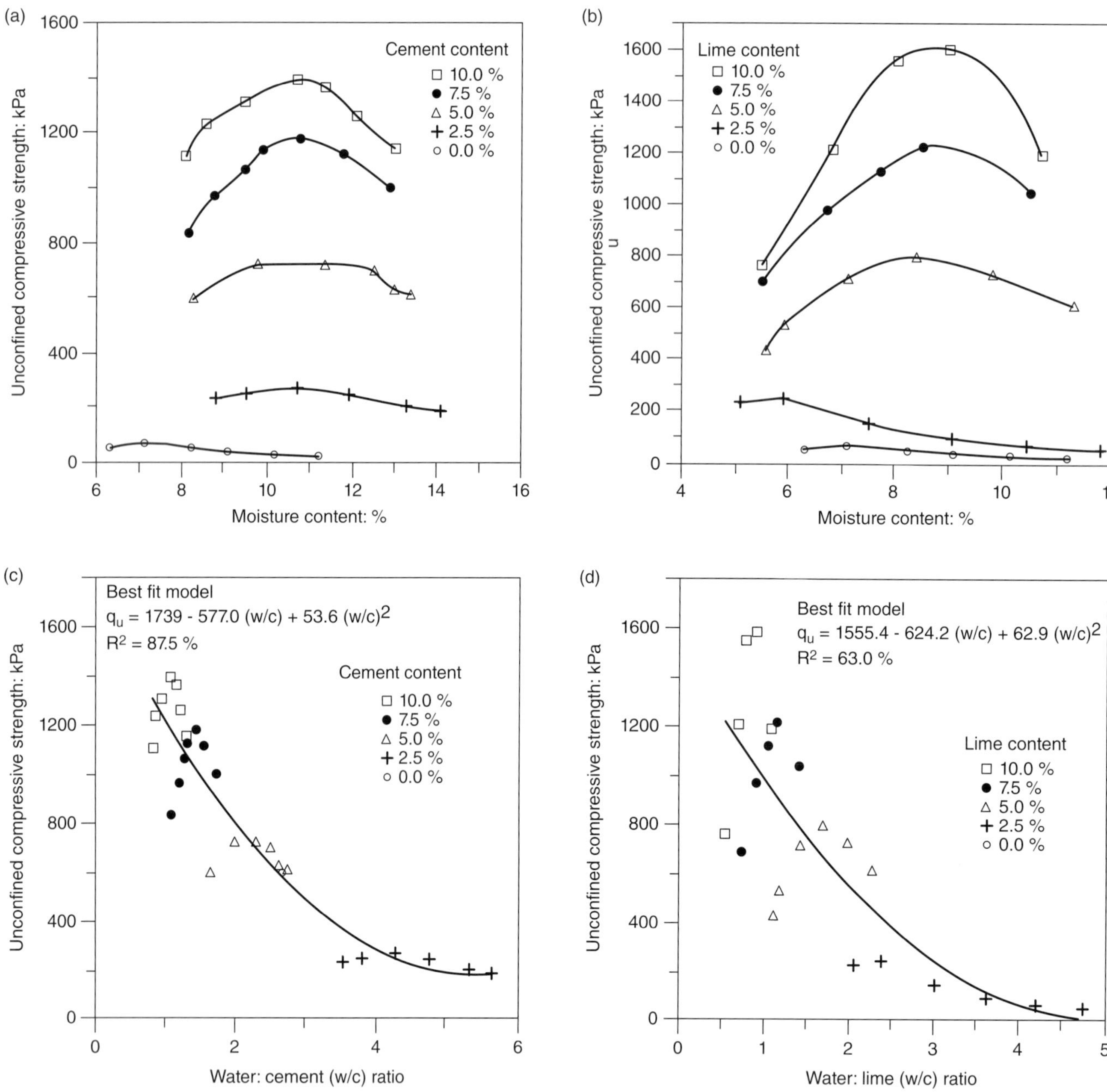

Figure 29.23 Stabilisation of a sabkha soil with cement (a) and lime (b). Illustration of the relationship between the UCS and the water/cement ratio (c) or the water/lime ratio (d) (Reproduced from Al-Amoudi *et al.*, 1995; © The Geological Society)

material replacement proportions considered have been relatively high (10–30%) and Al-Homidy *et al.* (2017) noted that these improvements would not meet given requirements; to do so required the inclusion of cement in addition to the CKD.

Damage to the base and subbase courses of highways due to movement in the underlying sabkha subgrades has resulted in investigations into the use of geotextiles between the road foundation and the subgrade (Abduljawad *et al.*, 1994; Aiban *et al.*, 2006). Static and dynamic testing of dry and wet sabkha soils underlying various geotextiles and subbase materials in a laboratory illustrated that geotextiles could increase the load required to induce deformation within the soil. **Table 29.5** illustrates the force required to induce a 30 mm deformation at the load plate on the soil for soaked and 'as moulded' conditions with and without a geotextile.

29.6. Concluding comments

Arid soils form in environments where evapotranspiration rates exceed rainfall. Arid soils cover about one-third of the

Sample code	Load at 30 mm deformation: kN	Test condition
Without geotextile		
SG0H65W	1.8	Soaked
SG0H65D	14.3	As moulded
With geotextile		
SG140H65W	2.8	Soaked
SG300H65W	6.1	Soaked
SG400H65W	6.7	Soaked
SG300H65D	23.5	As moulded

Table 29.5 Loads required to achieve a deformation of 30 mm at the load plate on a subbase overlying a sabkha subgrade, with or without a geotextile, in soaked and 'as moulded' conditions (Aiban *et al.*, 2006)

Earth's land mass and occur in hot and cold arid regions. However, soils formed under arid climates can also be found in temperate regions due to ancient geological processes.

Arid soils can present several challenges to geotechnical engineers due to

- the highly variable nature of their engineering properties associated with active and ancient geomorphic processes
- (in hot arid regions) the considerable diurnal temperature ranges experienced, potentially resulting in accelerated disintegration of soil and rock formations as well as man-made structures
- lithification processes, which can result in cementation of the soil particles, making excavation problematic
- various problematic soils, which occur in arid conditions, such as those with significant shrink–swell or collapsing potentials, or aggressive salt environments
- problematic soils such as expansive soils, soils exhibiting collapse potential, weak soils and chemically aggressive soils.

Understanding how arid soils behave, and how these soils can be engineered, is achieved through an appreciation of how geomorphology controls their formation (and could potentially continue to act on the soil after the geotechnical intervention is complete) and through investigation (e.g. laboratory or site-scale studies) to determine engineering parameters. Associated with these processes are soils that can be cemented or uncemented while likely being unsaturated. Once the geomorphologic effects are understood, the appropriate assessments (informed by suitable field and/or laboratory testing) and mitigations can be applied. A number of logistical challenges associated with water can also occur in arid environments, such as limited availability (potentially necessitating the use of groundwater that could contain high salt concentrations in construction processes, or having to bowse water from external locations to the site) or poor management (e.g. through lack of appreciation of the significant impact of flash flooding on the erosion of natural or man-made soil formations, or the impact of water containing high salt concentrations on construction processes such as the casting of concrete).

References

Abduljawad SN, Bayomy F, Al-Shaikh AM and Al-Amoudi OSB (1994) Influence of geotextiles on performance of saline sabkha soils. *Journal of Geotechnical Engineering* **120(11)**: 1939–1960.

Abed AA and Vermeer PA (2006) Numerical simulation of unsaturated soil behaviour. *Proceedings from the First Euro Mediterranean Conference on Advances in Geomaterials and Structures, Hammamet, Tunisia*, 10.13140/2.1.2836.2880

Afshar FA, Ayoubi S, Castrignanò A, Quarto R and Ardekani MRM (2017) Using ground-penetrating radar to explore the cemented soil horizon in an arid region in Iran. *Near Surface Geophysics* **15**: 103–110.

Aiban SA, Al-Ahmadi HM, Siddique ZU and Al-Amoudi OSB (2006) Effect of geotextile and cement on the performance of sabkha sub-grade. *Building and Environment* **41**: 807–820.

Akili W (2006) Salt encrusted desert flats (sabkha): problems, challenges and potential solutions. *Unsaturated Soils, Carefree, AZ*, pp. 391–402.

Akili W (2015) On foundations over salt encrusted desert flats (sabkha): properties, challenges and guidelines. *International Foundations Congress and Equipment Expo, San Antonio, TX*, pp. 21–31.

Al-Amoudi OSB (1994) Chemical stabilization of sabkha soils at high moisture contents. *Engineering Geology* **36**: 279–291.

Al-Amoudi OSB and Abduljawad SN (1995) Compressibility and collapse characteristics of arid sabkha soils. *Engineering Geology* **39**: 185–202.

Al-Amoudi OSB, Asi IM and El-Nagger ZR (1995) Stabilization of an arid, saline sabkha soil using additives. *Quarterly Journal of Engineering Geology* **28**: 369–379.

Alawaji HA (2008) Leak induced settlement of buried pipelines in collapsible soil. *ACSE Pipelines Conference, Atlanta, GA*, pp. 1–10.

Al-Homidy AA, Dahim MH and Abd El Aal AK (2017) Improvement of geotechnical properties of sabkha soil utilising cement kiln dust. *Journal of Rock Mechanics and Geotechnical Engineering* **9**: 749–760.

Alonso EE and Gens A (1994) Keynote lecture: on the mechanical behaviour of arid soils. *International Symposium on Engineering Characteristics of Arid Soils, London, UK* (Fookes PG and Parry RHG (eds)). Balkema, Rotterdam, the Netherlands, pp. 173–205.

Alonso EE and Ramon A (2013) Heave of a railway bridge induced by gypsum crystal growth: field observations. *Géotechnique* **63(9)**: 707–719.

Assallay AM, Rogers CDF and Smalley IJ (1997) Foundation and collapse of metastable particle packing and open structures in loess deposits. *Engineering Geology* **48**: 101–115.

Atkinson JH (1994) General report: classification of arid soils for engineering purposes. *International Symposium on Engineering Characteristics of Arid Soils, London, UK* (Fookes PG and Parry RHG (eds)). Balkema, Rotterdam, the Netherlands, pp. 57–64.

Bell FG (1998) *Environmental Geology: Principles and Practice*. Blackwell Sciences, Cambridge, UK.

Blair TC and McPherson JG (2009) Processes and forms of alluvial fans. In *Geomorphology of Desert Environments*, 2nd edn

(Parsons AJ and Abrahams AD (eds)). Springer, Heidelberg, Germany, pp. 413–467.

Blight GE (1994) Keynote lecture: the geotechnical behaviour of arid and semi-arid soils, Southern African experience. *International Symposium on Engineering Characteristics of Arid Soils, London, UK* (Fookes PG and Parry RHG (eds)). Balkema, Rotterdam, the Netherlands, pp. 221–235.

Blight GE (2017) *Unsaturated Soil Mechanics in Geotechnical Practice*. CRC Press, Boca Raton, FL.

Bondo Medhus A and Klinkby L (eds) (2022) *Engineering Geophysics*. CRC Press, Boca Raton, FL.

BSI (2016) BS 1377-1:2016: Methods of test for soils for civil engineering purposes – General requirements and sample preparation. BSI, London, UK.

BSI (2018) BS EN ISO 14689:2018: Geotechnical investigation and testing – Identification, description and classification of rock: Part 1 – Identification and description. BSI, London, UK.

BSI (2020) BS 5930:2015+A1:2020: Code of practice for ground investigations. BSI, London, UK.

BSI (various) BSI 22476: Geotechnical investigation and testing (parts 1-12) – Field testing. BSI, London, UK.

Cooke RU, Brunsden D, Doorkamp JC and Jones DKC (1982) *Urban Geomorphology in Drylands*. Oxford University Press, Oxford, UK.

Danish S, Chiranjeevi RR, Kiran KP and Adams J (2022) An experimental investigation on stabilisation of sabkha soils with cement and cement kiln dust (CKD) in Sultanate of Oman. *Materials Today: Proceedings* **65**: 1033–1039.

Dixon JC and McLaren SJ (2009) Duricrusts. In *Geomorphology of Desert Environments*, 2nd edn (Parsons AJ and Abrahams AD (eds)). Springer, Heidelberg, Germany, pp. 123–152.

Dohrenwend JC and Parsons AJ (2009) Pediments in arid environments. In *Geomorphology of Desert Environments*, 2nd edn (Parsons AJ and Abrahams AD (eds)). Springer, Heidelberg, Germany, pp. 377–411.

Eyo EU, Ng'ambi S and Abbey SJ (2022) An overview of soil–water characteristic curves of stabilised soils and their influential factors. *Journal of King Saud University – Engineering Science* **34**: 31–45.

Fattah MY, Al-Ani MM and Al-Lamy MTA (2014) Studying collapse potential of gypseous soil treated by grouting. *Soils and Foundations* **54(3)**: 396–404.

Fattah MY, Al-Shakarchi YJ and Al-Numani HN (2022) Effect of time history on long-term deformation of gypseous soils. *Studia Geotechnica et Mechanica* **44(3)**: 198–210.

Fetter CW (1999) *Contaminant Hydrogeology*, 2nd edn. Prentice Hall, London, UK.

Fetter CW (2001) *Applied Hydrogeology*, 4th edn. Prentice Hall, London, UK.

Fleureau JM and Taibi S (1994) Mechanical behaviour of an unsaturated loam on the oedometric path. *International Symposium on Engineering Characteristics of Arid Soils, London, UK* (Fookes PG and Parry RHG (eds)). Balkema, Rotterdam, the Netherlands, pp. 241–246.

Flinchum BA, Banks E, Hatch M, *et al.* (2020) Identifying recharge under subtle ephemeral features in a flat-lying semi-arid region using a combined geophysical approach. *Hydrology and Earth Systems Science* **24**: 4353–4368.

Fookes PG (1976) Road geotechnics in hot deserts. *Highway Engineer* **23**: 11–29.

Fookes PG and Higginbottom IE (1980) Some problems of construction aggregates in desert areas, with particular reference to the Arabian Peninsula 1: occurrence and special characteristics. *Proceedings of the Institution of Civil Engineers* **68(1)**: 39–67.

Fookes PG, French WJ and Rice SMM (1985) The influence of ground and groundwater geochemistry on construction in the Middle East. *Quarterly Journal of Engineering Geology and Hydrogeology* **18(2)**: 101–127.

Fookes PG, Lee ME and Milligan G (eds) (2005) *Geomorphology for Engineers*. Whittles Publishing, Dunbeath, UK.

Fredlund DG and Rahardjo H (1993a) *Soil Mechanics for Unsaturated Soils*. Wiley, New York, NY.

Fredlund DG and Rahardjo H (1993b) An overview of unsaturated soil behaviour. In *Unsaturated Soils: ASCE Special Publication 39*. ASCE, Reston, VA.

Fredlund DG and Xing A (1994) Equations for the soil–water characteristic curve. *Canadian Geotechnical Journal* **31(3)**: 521–532.

Fredlund DG, Rahardjo H and Fredlund MD (2012) *Unsaturated Soil Mechanics in Engineering Practice*. Wiley, New York, NY.

Gaspar TAV, Jacobsz SW, Schulz-Poblete M and Toll DG (2019) Measurement of the soil water retention curve: practical considerations. *Proceedings of the 17th African Regional Conference on Soil Mechanics and Geotechnical Engineering, Cape Town, South Africa*, pp. 227–232.

Goudie AS (1994) Keynote lecture: salt attack on buildings and other structures in arid lands. *International Symposium on Engineering Characteristics of Arid Soils, London, UK* (Fookes PG and Parry RHG (eds)). Balkema, Rotterdam, the Netherlands, pp. 15–28.

Haeri SM, Hamidi A and Tabatabaee N (2005) The effect of gypsum cementation on the mechanical behavior of gravely sands. *Geotechnical Testing Journal* **28(4)**: 380–390.

Hafhouf I and Abbeche K (2023) Impact of drying–wetting cycles on shear properties, suction and collapse of sabkha soils. *Heliyon* **9**: e13594, 10.1016/j.heliyon.2023.e13594

Hafhouf I, Bahloul O and Abbeche K (2022) Effects of drying–wetting cycles on the salinity and the mechanical behaviour of sebkha soils: a case study from Ain M'Lila, Algeria. *Catena* **212**: 106099, 10.1016.j.catena.2022.106099

Hills ES, Ollier CD and Twidale CR (1966) Geomorphology. In *Arid Lands: A Geographical Appraisal* (Hills ES (ed)). Methuen and UNESCO, London, UK and Paris, France, pp. 53–76.

Horta JDOS (1980) Calcrete, gypcrete and soil classification in Algeria. *Engineering Geology* **15(1–2)**: 15–52.

Howard AD and Selby MJ (2009) Rock slopes. In *Geomorphology of Desert Environments*, 2nd edn (Parsons AJ and Abrahams AD (eds)). Springer, Heidelberg, Germany, pp. 189–232.

Huang JT and Airey DW (1998) Properties of artificially cemented carbonate sand. *Journal of Geotechnical and Geoenvironmental Engineering* **124(6)**: 492–499.

James AN and Edworthy KJ (1985) The effects of water interactions on engineering structures. *Journal of Hydrological Sciences* **30(3)**: 395–406.

Jefferson I and Ahmad M (2007) Formation of artificial collapsible loess. *Geotechnical Special Publication 162*, 10.1061/40906(225)5.

Jennings JE and Knight K (1975) A guide to constructing on or with materials exhibiting additional settlements due to 'collapse' of grain structure. *Proceedings of the 6th Regional Conference Africa on Soil Mechanics and Foundation Engineering, Durban, South Africa*, pp. 99–105.

Johnson KS (2008) Gypsum-karst problems in constructing dams in the USA. *Environmental Geology* **53(5)**: 945–950.
Kadhim AJ (2014) Stabilization of gypseous soil by cutback asphalt for roads construction. *Journal of Engineering and Sustainable Development* **18(1)**: 46–67.
Kerr RC and Nigra JO (1952) Eolian sand control. *Bulletin of the American Association of Petroleum Geologists* **36**: 1541–1573.
Khan IH (1982) Soil studies for highway construction in arid zones. *Engineering Geology* **19**: 47–62.
Klimchouk A (1996) The dissolution and conversion of gypsum and anhydrite. *International Journal of Speleology* **25(3)**: 21–36.
Komornik A (1994) Keynote lecture: some engineering behaviour and properties of arid soils. *International Symposium on Engineering Characteristics of Arid Soils, London, UK* (Fookes PG and Parry RHG (eds)). Balkema, Rotterdam, the Netherlands, pp. 273–283.
Koorevaar P, Menelik G and Dirksen C (1993) *Elements of Soil Physics*. Elsevier, Amsterdam, the Netherlands.
Kropp A (2010) A survey of residential foundation design practice on expansive soils in the San Francisco Bay Area. *Journal of Performance of Constructed Facilities* **25(1)**: 24–30.
Kropp A, McMahon D and Houston S (1994) Field wetting tests on a collapsible soil fill. *International Symposium on Engineering Characteristics of Arid Soils, London, UK* (Fookes PG and Parry RHG (eds)). Balkema, Rotterdam, the Netherlands, pp. 343–352.
Laity JE (2009) Landforms, landscapes and processes of aeolian erosion. In *Geomorphology of Desert Environments*, 2nd edn (Parsons AJ and Abrahams AD (eds)). Springer, Heidelberg, Germany, pp. 597–628.
Lancaster N (2009) Dune morphology and dynamics. In *Geomorphology of Desert Environments*, 2nd edn (Parsons AJ and Abrahams AD (eds)). Springer, Heidelberg, Germany, pp. 557–595.
LCPC and SETRA (Laboratoire Central des Ponts et Chausées and Service d'Etudes Techniques des Routes et Autoroutes) (1992) *Réalisation des Remblais et des Couches de Forme*. Guide Technique. Fascicle 1: *Principes Générales*. Fascicle 2: *Annexes Techniques*. Laboratoire Central des Pont et Chaussées, et Service d'Etudes Techniques des Routes et Autoroutes, Paris, France.
Lee M and Fookes PG (2005a) Climate and weathering. In *Geomorphology for Engineers* (Fookes PG, Lee ME and Milligan G (eds)). Whittles Publishing, Dunbeath, UK, pp. 31–56.
Lee M and Fookes PG (2005b) Hot drylands. In *Geomorphology for Engineers* (Fookes PG, Lee ME and Milligan G (eds)). Whittles Publishing, Dunbeath, UK, pp. 419–453.
Lee MJ, Hong SJ, Choi YM and Lee W (2010) Evaluation of deformation modulus of cemented sand using CPT and DMT. *Engineering Geology* **115**: 28–35.
Livneh M, Livenh NA and Hayati G (1998) Site investigation of subsoil with gypsum lenses for runway construction in an arid zone in southern Israel. *Engineering Geology* **51**: 131–145.
Macedo J, Sotelo G, Orellana S, *et al.* (2023) Geotechnical characterisation of collapsible salty sands subjected to monotonic and cyclic loadings: a case study for areas with high seismicity. *Soils and Foundations* **63(1)**: 101252.
McDowell PW, Barker RD, Butcher AP, *et al.* (2002) *Geological Society's Engineering Geology Special Publication 19: Geophysics in Engineering Investigations*. CIRIA Report C562. CIRIA, London, UK.
Millington A (1994) Playas: new ideas on hostile environments. *International Symposium on Engineering Characteristics of Arid Soils, London, UK* (Fookes PG and Parry RHG (eds)). Balkema, Rotterdam, the Netherlands, pp. 35–40.
Mitchell JK and Soga K (2005) *Fundamentals of Soil Behaviour*, 3rd edn. Wiley, New York, NY.
Moon JS, Jung HS, Lee S and Kang ST (2019) Ground improvement using dynamic compaction in sabkha deposit. *Applied Sciences* **9**, 10.3300/app9122506.
Moore AE, Cotterill FPD and Eckardt FD (2012) The evolution and ages of Makgadikgadi palaeo-lakes: consilient evidence from Kalahari drainage evolution South Central Africa. *South African Journal of Geology* **115(3)**: 385–413.
Naeimi M, Chu J, Khosrohahi M and Zenouzi LK (2022) Soil stabilisation for dunes fixations using microbially induced calcium carbonate precipitation. *Geoderma* **429**: 1161183, 10.1016/j.geoderma.2022.116183.
Newill D and O'Connell MJO (1994) TRL research on road construction in arid areas. *International Symposium on Engineering Characteristics of Arid Soils, London, UK* (Fookes PG and Parry RHG (eds)). Balkema, Rotterdam, the Netherlands, pp. 353–360.
Nickling WG and McKenna Neuman C (2009) Aeolian sediment transport. In *Geomorphology of Desert Environments*, 2nd edn (Parsons AJ and Abrahams AD (eds)). Springer, Heidelberg, Germany, pp. 517–555.
Nusier OK and Alawneh AS (2002) Damage of reinforced concrete structure due to severe soil expansion. *Journal of Performance of Constructed Facilities* **16(1)**: 33–41.
Obika B, Freer-Hewish RJ and Fookes PG (1989) Soluble salt damage to thin bitumus road and runway surfaces. *Quarterly Journal of Engineering Geology and Hydrogeology* **22(1)**: 59–73.
Parsons AJ and Abrahams AD (eds) (2009) *Geomorphology of Desert Environments*, 2nd edn. Springer, Heidelberg, Germany.
Puttock R, Walkley S and Foster R (2011) The adaptation and use of French and UK earthworks methods. *Proceedings of the Institution of Civil Engineers, Geotechnical Engineering* **164(3)**: 160–179.
Rao SM, Reddy BVV and Muttharam M (2001) Effect of cyclic wetting and drying on the index properties of a lime-stabilised expansive soil. *Ground Improvement* **5(3)**: 107–110.
Rassam DW and Williams DJ (1999) A relationship describing the shear strength of unsaturated soils. *Canadian Geotechnical Journal* **36**: 363–368.
Ray TW and Murray BC (1994) Remote monitoring of shifting sands and vegetation cover in arid regions. *Proceedings of the Geoscience and Remote Sensing Symposium, London, UK*. IEEE, London, UK, vol. 2, pp. 1033–1035.
Rogers CDF, Dijkstra TA and Smalley IJ (1994) Keynote lecture: classification of arid soils for engineering purposes: an engineering approach. *International Symposium on Engineering Characteristics of Arid Soils, London, UK* (Fookes PG and Parry RHG (eds)), Balkema, Rotterdam, the Netherlands, pp. 99–133.
Sahu BK (2001) Improvement in California bearing ratio of various soils in Botswana by fly ash. *International Ash Utilisation Symposium, Lexington, KY*, paper 90.
Shao Y, Gong H, Xie C and Cai A (2009) Detection subsurface hyper-saline soil in Lop Nur using full-polarimetric SAR data. *Geoscience and Remote Sensing Symposium, Cape Town, South Africa*, pp. 550–553.

Shehata WM and Amin AA (1997) Geotechnical hazards associated with desert environment. *Natural Hazards* **16**: 81–95.
Shmida A, Evernari M and Noy-Meir I (1986) Hot desert ecosystems: an integrated review – ecosystems of the world. In *Hot Deserts and Arid Sublands* (Evenari M, Noy-Meir I, and Goodall DW (eds)). Elsevier, Amsterdam, the Netherlands, vol. 12B, pp. 379–388.
Sissakian VK and Al-Mousawi HA (2007) Karstification and related problems, examples from Iraq. *Iraqi Bulletin of Geology and Mining* **3(2)**: 1–12.
Smalley IJ, Dijkstra TA and Rogers CDF (1994) Classification of arid soils for engineering purposes: a pedological approach. *International Symposium on Engineering Characteristics of Arid Soils, London, UK* (Fookes PG and Parry RHG (eds)). Balkema, Rotterdam, the Netherlands, pp. 135–143.
Stalker G (ed) (1999) *The Visual Dictionary of the Earth.* Covent Garden Books, London, UK.
Thornes JB (2009) Catchment and channel hydrology. In *Geomorphology of Desert Environments*, 2nd edn (Parsons AJ and Abrahams AD (eds)). Springer, Heidelberg, Germany, pp. 303–332.
Tian W, Peiffer H, Malengier B, Liu G and Cheng L (2023) Modified equation of shear strength with respect to saturation. *Applied Sciences* **13**: 4305, 10/3390/app13074305.
Toghill P (2006) *The Geology of Great Britain: An Introduction.* Airlife Publishing, Singapore.
Tomlinson MJ (1978) Middle East highway and airfield pavements. *Quarterly Journal of Engineering Geology and Hydrogeology* **11(1)**: 65–73.
Truslove LH (2010) *Highway Construction in an Arid Environment: Towards a Suitable Earthworks Specification.* Unpublished MSc Thesis, University of Birmingham, UK.
van Alphen JG and de los Rios Romero F (1971) *Gypsiferous Soils: Notes on Their Characteristics and Management.* ILRI, Nairobi, Kenya.
Wallen CC and Stockholm SMHI (1966) Arid zone meteorology. In *Arid Lands: A Geographical Appraisal* (Hills ES (ed)). Methuen and UNESCO, London, UK and Paris, France, pp. 31–52.
Waltham AC (2009) *Foundations of Engineering Geology*, 3rd edn. Taylor & Francis, London, UK.
Warren A (1994a) General report: arid environments and description of arid soils. *International Symposium on Engineering Characteristics of Arid Soils, London, UK* (Fookes PG and Parry RHG (eds)). Balkema, Rotterdam, the Netherlands, p. 3.
Warren A (1994b) Sand dunes: highly mobile and unstable surfaces. *International Symposium on Engineering Characteristics of Arid Soils, London, UK* (Fookes PG and Parry RHG (eds)). Balkema, Rotterdam, the Netherlands, p. 4.
Wheeler SJ (1994) General report: engineering behaviour and properties of arid soils. *International Symposium on Engineering Characteristics of Arid Soils, London, UK* (Fookes PG and Parry RHG (eds)). Balkema, Rotterdam, the Netherlands, pp. 161–172.
Wheeler SJ, Sharma RJ and Buisson MSR (2003) Coupling of hydraulic hysteresis and stress–strain behaviour in unsaturated soils. *Géotechnique* **53(1)**: 41–54.
White GF (1966) The world's arid areas. In *Arid Lands: A Geographical Appraisal* (Hills ES (ed)). Methuen and UNESCO, London, UK and Paris, France, pp. 15–31.
Whitten DG and Brooks JRV (1973) *The Penguin Dictionary of Geology.* Penguin, London, UK.
Yilmaz I (2001) Gypsum/anhydrite: some engineering problems. *Bulletin of Geology and the Environment* **60(3)**: 227–230.

Further reading

Blight GE (2017) *Unsaturated Soil Mechanics in Geotechnical Practice.* CRC Press, Boca Raton, FL.
Fookes PG and Parry RHG (eds) (1994) *International Symposium on Engineering Characteristics of Arid Soils.* Balkema, Rotterdam, the Netherlands, pp. 99–133.
Fookes PG, Lee ME and Milligan G (eds) (2005) *Geomorphology for Engineers.* Whittles Publishing, Dunbeath, UK.
Fredlund DG, Rahardjo H and Fredlund MD (2012) *Unsaturated Soil Mechanics in Engineering Practice.* Wiley, New York, NY.
Parsons AJ and Abrahams AD (eds) (2009) *Geomorphology of Desert Environments*, 2nd edn. Springer, Heidelberg, Germany.

Useful websites

British Geological Survey (BGS): http://www.bgs.ac.uk
British Geotechnical Association (BGA): http://www.britishgeotech.org
British Standards Institution (BSI): https://knowledge.bsigroup.com/
US Geological Survey (USGS): http://www.usgs.gov

It is recommended this chapter is read in conjunction with:

- Chapter 7 *Geotechnical risks and their context for the whole project*
- Chapter 30 *Tropical soils*
- Chapter 32 *Collapsible soils*
- Chapter 33 *Expansive soils*
- Chapter 37 *Sulfates and sulfides in soils and rocks*
- Chapter 38 *Soluble ground*
- Chapter 40 *The ground as a hazard*
- Chapter 76 *Pavement foundation design*

All chapters in this book rely on the guidance in Sections 1 *Context* and 2 *Fundamental principles.* A sound knowledge of ground investigation is required for all geotechnical works, as set out in Section 4 *Site investigation.*

Chapter 30
Tropical soils

David G Toll Department of Engineering, Durham University, Durham, UK

doi: 10.1680/icemge.66816.0373

CONTENTS

Tropical soils are formed primarily by in situ weathering processes, and hence are residual soils. Terminology for tropical soils is confused and many classification schemes exist, based on either pedological, geochemical or engineering criteria. For classification schemes to be useful they need to include the effects of disintegration due to weathering, mineralogy (particularly the 'unusual' clay minerals that are particular to tropical soils), cementation and structure. Tropical residual soils are highly structured materials, both at macro and micro levels. The micro-structure is produced by leaching out of minerals during weathering, leaving an open structure. Tropical soils are also likely to be cemented soils due to deposition of minerals either during or after weathering. The highly structured nature of tropical soils, combined with the fact they often exist in an unsaturated state, makes them difficult to deal with as engineering materials. However, they often have good engineering properties. Nevertheless, some tropical soils can be problematic, demonstrating collapse or shrink/swell movements. Ground investigation for tropical soils poses some difficulties due to their heterogeneous nature. Sampling of tropical soils so that the original structure is maintained can be a major challenge and hence there is a strong emphasis on in situ testing methods.

30.1. Introduction

Tropical soils are formed primarily by in situ weathering processes, and hence are *residual soils*. The climatic conditions in tropical regions, with high temperatures and high levels of precipitation, lead to stronger chemical weathering of primary minerals and a greater penetration of weathering than occurs in other regions of the Earth (**Figure 30.1**).

The weathering process propagates from the Earth's surface and is controlled primarily by water movement within the joints of the parent rock. Weathering starts on the joint surfaces and progressively penetrates into the rock mass. The direction of propagation of a weathered profile will be dictated by the direction of the joints, so while there is likely to be a vertical component, weathering can also progress horizontally or at other inclinations.

Weathering is usually defined by the degree of decomposition/disintegration of the rock to form soil. This is rated as Grade I (*fresh rock*) through to Grade VI (*residual soil*), where the rock has fully decomposed to form soil. This weathering scale will be discussed in more detail below. The upper three grades – VI (*residual soil*), V (*completely weathered*) and IV (*highly weathered*) – represent a material where more than 50% of the rock has decomposed to form soil; the term *tropical residual soil* is used to describe these upper three weathering grades that are dominated by soil material (note the distinction between ***tropical*** *residual*

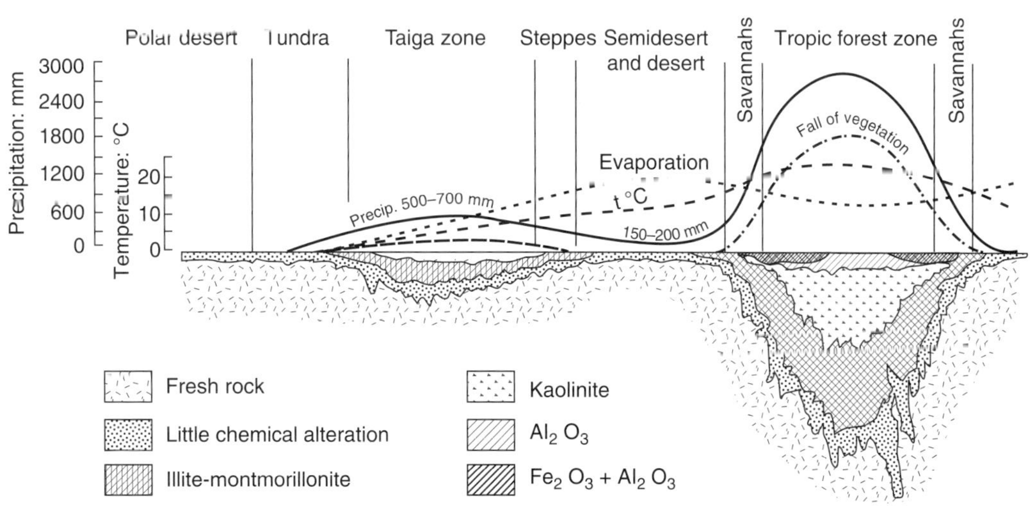

Figure 30.1 Depth of residual soil development related to climatic factors (After Strakhov, 1967)

soil, which incorporates all three weathering grades IV, V and VI, and *residual soil*, which only describes Grade VI). The term *saprolite* is used to describe completely and highly weathered material – that is, a soil which still contains some unweathered rock but does not incorporate the fully decomposed Grade VI.

The particle size distribution (grading) of tropical soils can be highly variable. Early stages of weathering are likely to produce coarser-grained soils. Large boulder- or cobble-sized fragments of unweathered material may still survive as corestones. Stronger chemical weathering will cause minerals such as feldspar and mica to decompose to form clay minerals, so the resulting residual soil may be more fine-grained. The resulting grain size distribution after weathering will depend on the parent rock. For instance, acidic igneous rocks (e.g. granite) contain a high proportion of silica that will resist weathering, so the resulting soil will contain sand-sized silica particles. **Table 30.1** shows the type of residual soil resulting from different parent rocks.

A common feature of many tropical soils is the presence of iron or aluminium oxides, often referred to as sesquioxides (Fe_2O_3 or Al_2O_3). This produces the reddish colouration common to many tropical soils. Iron and aluminium oxides released by weathering are not dissolved as in more acidic environments and remain in situ. Iron oxide is crystallised as haematite when the soil is seasonally desiccated, or as goethite in a constantly humid environment; haematite gives the soil a red colour, goethite a brown or ochreous colour. Gibbsite is the main aluminium oxide formed. During the weathering process, silica and bases (K, Na, Ca, Mg) are lost in solution or incorporated into clay minerals.

The term *laterite* has been used to describe a wide range of red soils, and the term has become almost meaningless in an engineering sense. It was used originally to describe a red clay which hardened irrecoverably on exposure to air but has subsequently been used to describe almost any soil with reddish colouration. The term *laterite* will be used here to refer to a soil with a high degree of iron cementing and the term *red tropical soil* will be used for a soil that has red colouration but without significant cementing.

The weathering process can be divided into three stages based on the mineralogical and geochemical changes (Duchaufour, 1982). These stages are based on the dominant clay minerals that are produced by weathering:

- fersiallitisation (smectite clays dominant)
- ferrugination (kaolinite and smectites)
- ferrallitisation (kaolinite and gibbsite).

As weathering proceeds (ferrugination to ferrallitisation) there is a decrease in the concentration of silica and bases and an increase in concentration of iron and aluminium oxides. The resulting soil at the final stage of weathering (ferrallitic soil) is clay-rich and also rich in sesquioxides.

The iron, in the presence of negatively charged clay particles, may be present in the reduced form as ferrous oxide (FeO). Ferrous (Fe^{2+}) iron is soluble and highly mobile, whereas ferric (Fe^{3+}) iron is virtually insoluble. If oxidising conditions become present (such as exposure to air in an excavation or by lowering of the groundwater table) the soluble ferrous iron is precipitated as ferric oxides (Fe_2O_3) such as goethite or limonite. This results in an indurated soil (i.e. a soil that becomes harder through cementation). Alternating reduction and oxidation due to fluctuating groundwater levels in the geological past leads to the development of concretionary or nodular laterites.

Indurated, rock-like *duricrusts* can be formed by cementing of tropical soils by a number of chemicals. Iron and aluminium oxides or silica released during weathering, or calcium carbonate, can be moved laterally or vertically through the soil profile by groundwater flow and may accumulate sufficiently in some horizons for crystallisation to occur, resulting in duricrust formation.

Terminology for tropical soils is confused and many classifications schemes exist, based on either pedological, geochemical or engineering criteria. The classification given by Duchaufour (1982), based on weathering stages, was adopted as the basis of a Working Party Report by the Geological Society (Anon, 1990), later published by Fookes (1997). However, it will be argued that there are other factors, such as weathering grade, degree of cementation and mass structure, that are equally important.

Tropical residual soils are often highly structured materials, both at macro and micro levels. The micro-structure is produced

Parent rock	Residual soil type	Relative susceptibility to tropical weathering
Calcareous rock (limestone, dolomite)	Gravel in clayey or silty matrix	1 (most vulnerable)
Basic igneous rock (gabbro, dolerite, basalt)	Clay (often grading into sandy clay with depth)	2
Acid crystalline rock (granite, gneiss)	Clayey sand or sandy clay (often micaceous)	3
Argillaceous sedimentary rock (mudstone, shale)	Silt or silty clay	4
Arenaceous sedimentary or metamorphic rock (sandstone, quartzite)	Sand (clayey sand in the case of residual arkose or feldspathic sandstone)	5 (least vulnerable)

Table 30.1 Types of residual soil from different parent rocks (Brink *et al.*, 1982)

Figure 30.2 Scanning electron microscope (SEM) images showing the micro-structure of a tropical residual soil from Singapore (a) well bonded with strong cementing (b) looser structure (Aung *et al.*, 2000)

by leaching out of minerals during weathering, leaving an open structure, combined with the effects of secondary cementing by minerals deposited during or after weathering has taken place. The cementing (bonding) can maintain the fabric of the soil in a metastable state – that is, such a loose structure could not exist if the bonding was not present to support it. **Figure 30.2** shows a typical example of the micro-structure of a residual soil from Singapore. The open fabric can be clearly seen, as well as variation in the degree of cementing within the soil.

Tropical soils have a reputation for being 'problematic'. This is because they do not conform to the widely used classification systems that have been developed for temperate sedimentary soils to identify likely engineering behaviour. They are also difficult to investigate, as attempts to sample or test them can destroy the cementing and structure that supports them. This chapter will identify the problems in classifying and investigating tropical soils and identify schemes and methods that are appropriate for tropical soils.

Of course, some tropical soils can be highly problematic. Residual soils can exist in a loose, structured state and can collapse on loading, or due to wetting, leading to sudden settlements. Other problematic types of tropical soil contain highly expansive smectite clays, resulting in significant heave or shrinkage as wetting or drying occurs. These aspects of behaviour of tropical soils will be considered in a range of applications: foundations, slopes and highways.

30.2. Controls on the development of tropical soils

30.2.1 Weathering processes

One of the most important factors in the engineering behaviour of tropical soils is the degree of weathering. Little (1969) first proposed a six-grade classification for tropical residual soils. This has become the well-established scheme for identifying the degree of rock weathering used in the Geological Society Working Party Report on Core Logging (Anon, 1970) and the Working Party Report on Rock Mass Description (Anon, 1977). This was the scheme used in ISRM (1978), whereby the relative percentages of 'rock' that have decomposed/disintegrated to form 'soil' are used to define the weathering grade. **Figure 30.3** shows a useful diagrammatic representation of the scheme.

British Standard BS 5930 (BSI, 2015) provides an alternative method to characterising the degree of weathering by identifying weathering grades based on three approaches, as shown in **Figure 30.4**. The Eurocode document BS EN ISO 14689 (BSI, 2018) for rock description uses a simpler scheme: *fresh*, *discoloured*, *disintegrated*, *decomposed*. As will be discussed later in this section, a full classification of tropical soils must include more than just the weathering grade and should also incorporate mineralogy, secondary cementation and structure.

As has been noted, Duchaufour (1982) identified three phases of development in the weathering process, based in changes in mineralogy and geochemistry, as opposed to the degree of decomposition/disintegration

- fersiallitisation
- ferrugination
- ferrallitisation.

As weathering proceeds (fersiallitisation → ferrugination → ferrallitisation) there are changes in mineralogical composition, in particular the formation of clay minerals. This is associated with a decrease in concentration of silica and bases (K, Na, Ca, Mg) and an increase in concentration of iron and aluminium oxides. Fersiallitic soils are dominated by 2:1 clay minerals (smectites). The main clay mineral present in ferruginous soils is kaolinite (1:1), although some smectite may be present. Ferrallitic soils are dominated by kaolinite and gibbsite (aluminium oxide). Clay minerals are alumina-silicates that are made up of sheets comprising either silica tetrahedrons (tetrahedral sheets) or alumina octahedrons (octahedral sheets). 1:1 clay minerals are made up of alternating tetrahedral and

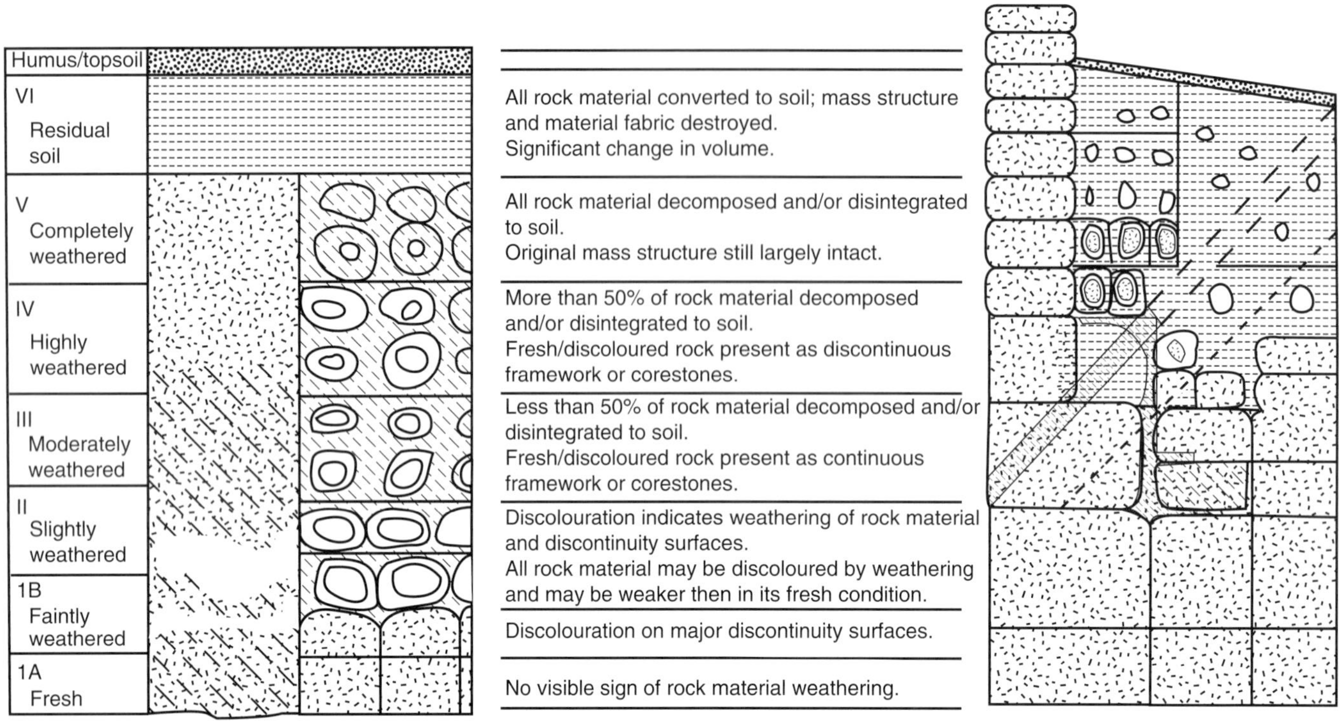

Figure 30.3 Schematic diagrams of typical weathering profiles (Anon, 1990)

octahedral sheets. The adjacent layers are closely bonded together, preventing water molecules from penetrating between the sheets. This makes them relatively low-activity minerals (i.e. low shrinkage/swelling). 2:1 clay minerals have one octahedral sheet sandwiched between two tetrahedral sheets. Adjacent layers are not held together strongly, thus allowing water molecules to penetrate between sheets, resulting in high-activity minerals (i.e. high shrinkage/swelling).

Duchaufour's stages of weathering are shown in **Table 30.2**, together with comparisons with other pedological schemes and commonly used geotechnical descriptive terms. Duchaufour's scheme was adopted by a Geological Society Working Party Report on Tropical Soils (Anon, 1990; Fookes, 1997) for its classification of tropical soils, as will be discussed in this section.

30.2.2 Parent rock

The type of residual soil will depend on the parent rock. Some typical examples of residual soil types are given in **Table 30.1**. Also shown are the relative susceptibilities to weathering in a tropical environment (Brink *et al.*, 1982).

30.2.3 Climate

The stage of weathering achieved is controlled by the climate. The effect of climate on the weathering products is shown in **Table 30.3**.

The phases of weathering defined by Duchaufour (1982) can also be related to climate. In Mediterranean or subtropical climates with a marked dry season, stage 1 (fersiallitisation) is rarely exceeded. In a dry tropical climate development stops at stage 2 (ferrugination). Only in humid equatorial climates is stage 3 (ferrallitisation) reached (**Table 30.4**).

The world distribution of the major types of tropical residual soils is shown in **Figure 30.5**. Fookes (1997) notes that these broad classes of soils extend beyond the tropics in favourable conditions. Examples are ferrallitic soils on high-rainfall subtropical continental east coasts and fersiallitic soils in west coast/Mediterranean and continental interiors in midlatitudes.

The development of duricrusts (rock-like cemented/indurated soils) is also highly dependent on climatic conditions. Ackroyd (1967) presents possible conditions under which the different stages of concretionary material develop in ferricrete (or laterite) (**Table 30.5**). The climate is categorised using the Thornthwaite moisture index (Thornthwaite and Mather, 1954). This provides a way of defining climatic conditions based on the difference between the precipitation and evapotranspiration expressed as a ratio to the potential evapotranspiration. A negative value indicates a dry environment and positive values indicate more humid environments.

However, McFarlane (1976) points out that the conditions for development of concretions are not the conditions under which the laterite minerals initially form. McFarlane identified a model of laterite formation based on a fluctuating groundwater table in an overall downward flow environment. A cycle of development may result, where the original lateritic deposit is weathered, producing iron and aluminium oxides that are mobilised and recrystallised elsewhere to form a new laterite deposit.

APPROACH 1: FACTUAL DESCRIPTION OF WEATHERING (MANDATORY)

Standard descriptions should always include comments on the degree and nature of any weathering effects at material or mass scales. This may allow subsequent classification and provide information for separating rock into zones of like character. Typical indications of weathering include:

- Changes in colour
- Changes in fracture state
- Reduction in strength
- Presence, character and extent of weathering products

These features should be described using standard terminology, quantified as appropriate, together with non-standard English descriptions as necessary to describe the results of weathering. At the mass scale the distribution and proportions of the variously weathered materials (e.g. corestones vs matrix) should be recorded.

In there a classification that can be applied unambigously?

Do NOT CLASSIFY

Rock is medium strong or stronger in fresh state

Rock is moderately weak or weaker in fresh state

Does the weathering present require a particular and local approache?

APPROACH 2: CLASSIFICATION FOR UNIFORM MATERIALS

Grade	Classifier	Typical characteristics
I	Fresh	Unchanged from original state
II	Slightly weathered	Slight discolouration, slight weakening
III	Moderately weathered	Considerably weakened, penetrative discolouration Large pieces cannot be broken by hand
IV	Highly weathered	Large pieces cannot be broken by hand. Does not readily disaggregate (slake) when dry sample immersed in water
V	Completely weathered	Considerably weakened Slakes Original texure apparent
VI	Residual soil	Soil derived by in situ weathering but retaining none of the original texure of fabric

APPROACH 4: CLASSIFICATION INCORPORATING MATERIAL AND MASS FEATURES

Class	Classifier	Typical characteristics
A	Unweathered	Original strength, colour, fracture spacing
B	Partially weathered	Slightly reduced strength, slightly closer fracture pacing, weathering penetrating in from fractures, brown oxidation
C	Distinctly weathered	Further weakened, much closer fracture spacing, grey reduction
D	Destructured	Greatly weakened, mottled, ordered lithorelics in matrix becoming weakened and disordered, bedding disturbed
E	Residual or reworked	Matrix with occasional altered random or 'apparent' lithorelics, bedding destroyed. Classed as reworked when foreign inclusions are present as a result of transporations

APPROACH 5: SPECIAL CASES

For rocks whose weathering state does not follow the other patterns indicated here, such as karst in carbonate rocks, the particular effect of arid or porcis rocks, compile or adopt a locally suitable approach.

Is a zonal classification appropriate and is there enough information available?

Use Rock Mass Classification if appropriate

APPROACH 3: CLASIFICATION FOR HETEROGENOUS MASSES

Zone	Proportions of material grades	Typical characteristics
1	100% G I – III (not necessarily all fresh rock)	Behaves as rock: apply rock mechanics principles to mass assessment and design
2	> 90% G I – III < 10% G IV – VI	Weak materials along discontinuities. Shear strength stiffness and permeability affected
3	50% to 90% G I – III 10% to 50% G IV – VI	Rock framework still locked and controls strength and stiffness; matrix controls permeability
4	30% to 50% G I – III 50% to 70% G IV – VI	Rock framework contributes to strength: matrix or weathering products control stiffness and permeability
5	< 30% G I – III > 70% G IV – VI	Weak grades will control behaviour. Corestones may be significant for investigation and construction
6	100% G IV – VI (not necessarily residual soil)	May behave as soil although relict fabric may still be significant

Figure 30.4 Description and classification of weathered rock for engineering purposes (BS 5930:2015)

Pedological classifications			Common geotechnical terminology	Colour	Mineralogy
Duchaufour	USA	FAO/UNESCO			
Vertisol (fersiallitic)	Vertisol	Vertisol	Black cotton soil	Black, brown, grey	Smectites (montmorillonite) kaolinite
Andosol (fersiallitic)	Inceptisol	Andosol	Halloysite/ allophane soil	Red, yellow, purple	Kaolinite (halloysite), allophane
Ferruginous	Alfisol	Nitosol, Alfisol, Lixisol	Red tropical soil	Red, yellow, purple	Kaolinite, hydrated iron oxide (haematite, goethite) hydrated aluminium oxide (gibbsite)
Ferrisol (transitional)	Ultisol	Ferralsol	Lateritic soil, latosol	Red, yellow, purple	Kaolinite, hydrated iron oxide (haematite, goethite), hydrated aluminium oxide (gibbsite)
Ferrallitic	Oxisol	Plinthisol	Plinthite, laterite	Red, yellow, purple	Kaolinite, hydrated iron oxide (haematite, goethite), hydrated aluminium oxide (gibbsite)

Table 30.2 Terminology for tropical soils

Reference	Rainfall: mm per annum	Clay mineral type
Pedro (1968)	<500	Montmorillonite
	500–1200/1500	Kaolinite dominant
	>1500	Gibbsite and kaolinite
Sanches Furtado (1968)	800–1000	Kaolinite and montmorillonite
	1000–1200	Kaolinite dominant
	1200–1500	Kaolinite and gibbsite

Table 30.3 Climate and weathering products (McFarlane, 1976)

30.2.4 Relief and drainage

Relief and drainage have major effects on the stages of weathering. Idealised soil profiles (catenas) are shown for three rock types in **Figure 30.6**. However, these simple catenary sequences will be modified by the drainage conditions. If the base of the slope is poorly drained, then fersiallitic soils will develop even on acid crystalline rocks. Equally, if drainage conditions are good then ferrallitic soils may develop on basic rocks.

30.2.5 Secondary cementation

Many tropical soils are bonded due to the presence of chemical cementing agents. These chemicals may develop by pedogenic processes induced by the accumulation of iron or aluminium oxides (as they are released during weathering) or the movement of leached silica, aluminium and iron oxides, gypsum or carbonates by groundwater flow. These minerals may accumulate sufficiently in some horizons for crystallisation to occur. The resulting harder varieties of pedogenic materials such as calcrete, silcrete, ferricrete or alucrete are known as duricrusts or pedocretes. The cementing agents for each type of duricrust are identified in **Table 30.6**.

Netterberg (1971) categorised calcretes into calcified material, powder calcrete, nodular calcrete and hardpan calcrete, and showed that the engineering properties were highly dependent on the type of calcrete. Charman (1988) adopted a similar classification scheme for laterite (or ferricrete) (**Table 30.7**).

Different stages of laterisation are reflected in the silica/alumina ratio (SiO_2/Al_2O_3) (Desai, 1985). Typical values are given in **Table 30.8**.

30.3. Engineering issues

30.3.1 Investigation

Ground investigation for tropical soils poses some difficulties due to the heterogeneous nature of tropical residual soils. The soil properties will be dependent on the degree of weathering, which varies within the weathered mass; there can often be ‘corestones’ of relatively unweathered rock (which can be of cobble or boulder size) contained within a matrix of more weathered material. The original rock mass structure, such as joint sets, will be represented in relict form in the tropical residual soil and can represent planes of weakness (Au, 1996; Irfan and Woods, 1988).

Phase	Soil type	Zone	Mean annual temperature: °C	Annual rainfall: m	Dry season
1	Fersiallitic	Mediterranean, subtropical	13–20	0.5–1.0	Yes
2	Ferruginous Ferrisols (transitional)	Subtropical	20–25	1.0–1.5	Sometimes
3	Ferrallitic	Tropical	>25	>1.5	No

Table 30.4 Summary of residual soil phases (Duchaufour, 1982)

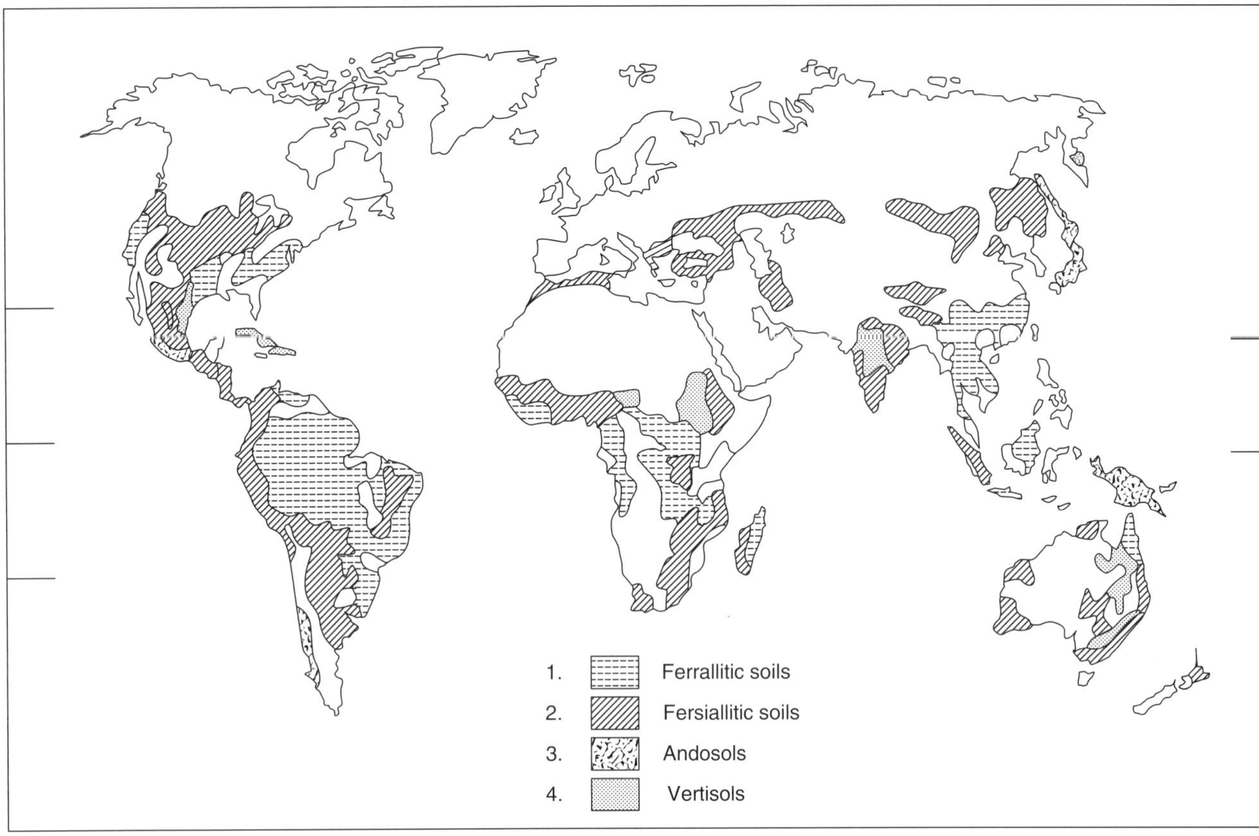

Figure 30.5 Simplified world distribution of tropical residual soils (Based on FAO World Soil Map; Fookes, 1997)

The overall degree of weathering will vary from the highest degree of weathering near the ground surface to less weathered and possibly fresh rock deeper within the ground mass. However, the changes between grades of weathering will be progressive; there are unlikely to be sharp distinctions between layers of different weathered materials, so identification of boundaries between weathering grades can be highly subjective.

In addition to this, the degree of secondary cementing can be highly variable, both in terms of the amount of the cementing minerals (e.g. iron oxides) and the strength of the cemented bonds. The 'structured' nature of tropical soils (Vaughan, 1985a) makes them particularly sensitive to disturbance during sampling and testing.

	Annual rainfall: mm		
	750–1000	1000–1500	1500–2000
Thornthwaite moisture index	−40 to −20	−20 to 0	0 to +30
Length of dry season (months)	7	6	5
Type of product	Rock laterite or cuirasse	Hard concretionary gravels	Minimum requirements for concretions to develop

Table 30.5 Possible conditions for development of concretionary laterite

Sampling of residual soils so that the original structure is maintained can be a major challenge. Driven or pushed samplers are likely to cause significant breakdown of the structure, resulting in samples that no longer represent the in situ conditions. Large block samples, trimmed by hand, may be the only way to get satisfactory samples. Details of sampling procedures are outlined in Fookes (1997).

There has been success with using rotary coring techniques, using triple tube core barrels and air foam flush, to recover good-quality samples (Phillipson and Brand, 1985; Phillipson and Chipp, 1982). Mazier core barrels (73 mm dia.) are commonly used or Treifus triple tube barrels (63 mm dia.). Plastic lining tubes should be used to protect the core on extrusion. Water drilling flush should not be used as the flushing medium can cause erosion of the core or result in changes in water content. Even compressed air flush can potentially change the suctions in samples (Richards, 1985).

Due to the difficulties in recovering high-quality, undisturbed samples there has been a strong emphasis on in situ testing to determine the engineering properties of tropical

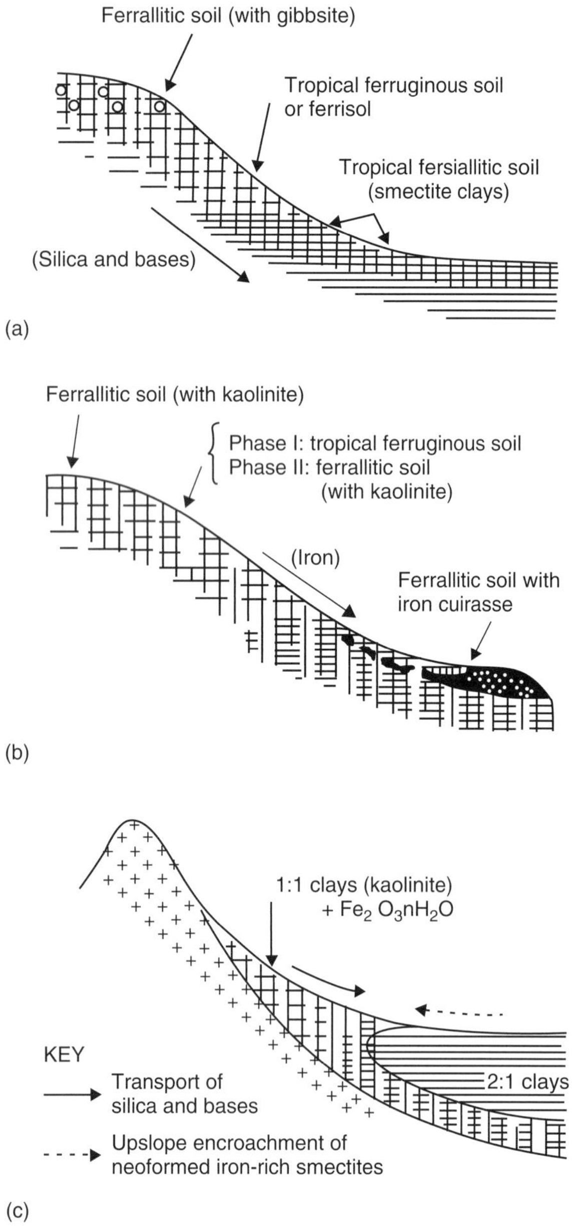

Figure 30.6 Simple soil catenas for different rock types (a) basic volcanic rocks in a humid tropical climate; (b) acid crystalline rocks in a humid tropical climate; (c) around inselbergs in the tropics (Fookes, 1997, after Duchaufour, 1982)

Duricrust	Cementing mineral
Silcrete	Silica
Calcrete	Calcium or magnesium carbonate
Gypcrete	Calcium sulfate dihydrate
Alucrete (bauxite)	Hydrated aluminium oxides
Ferricrete (laterite)	Hydrated iron oxides

Table 30.6 Duricrusts and their cementing minerals

soils. Pressuremeter and plate load tests are suitable tests for assessing in situ properties. Standard penetration testing (SPT) is also widely used. Cone penetration testing (CPT) or vane testing are unlikely to be suitable for weathered profiles containing significant amounts of rock material, as penetration of the testing device will be restricted. However, these techniques may be suitable for Grade VI residual soils.

The pressuremeter has been used for the investigation of properties of tropical soils (e.g. Schnaid and Mantaras, 2003). However, as Schnaid and Huat (2011) note, the pressuremeter response curve will be dependent on a combination of in situ horizontal stress, soil stiffness and strength parameters, and these parameters will reduce with destructuration at high shear strains. Schnaid and Huat suggest that, in residual soils, the pressuremeter should be viewed as a 'trial' boundary value problem against which a theoretical pressure–expansion curve predicted using a set of independently measured parameters can be compared to field pressuremeter tests. A good comparison between a number of observed and predicted curves can give confidence that the selected parameters used in the prediction are sensible, and can therefore be adopted in design. An example of a pressuremeter test carried out in a Brazilian residual soil reported by Schnaid and Mantaras (2003) is compared against the solution from numerical simulation in **Figure 30.7**, and shows that such good agreement can be achieved.

Plate load testing is a popular option for assessing tropical soils as it has the advantage of testing a larger volume of material, thereby giving a measure of the mass behaviour of a heterogeneous soil. Procedures for performing such tests are described by Barksdale and Blight (1997). The results can be used primarily to estimate the stiffness or compressibility of the soil. The interpretation of the results may be complicated by the structured nature of the soils and unsaturated state, as both factors will influence the initial soil stiffness and the yield stress observed. Schnaid and Huat (2011) suggest that interpretation of test data may require sophisticated numerical analysis with appropriate constitutive models, rather than the conventional interpretation methods commonly used to estimate the elastic modulus.

SPTs are widely used around the world for assessing the relative density and hence the angle of shearing resistance of soils, based on empirical correlations. It has to be recognised that such correlations have usually been established from databases of tests on sedimentary soils. They are unlikely to be appropriate for tropical residual soils as they take no account of any cementing/bonding that influences the strength of the soil. Local relationships may need to be determined for a particular tropical soil that takes account of the cementing and local variability.

Schnaid *et al.* (2004) have suggested that cementation of residual soils can be observed by considering the ratio of the elastic stiffness to ultimate strength, G_o/N_{60}, where G_o is the shear modulus at very small strains and N_{60} is the SPT value standardised to a reference value of 60% of the potential

Age	Recommended name	Characteristic	Equivalent terms in the literature
Immature (young)	Plinthite	Soil fabric containing significant amount of lateritic material. Hydrated oxides present at expense of some soil material. Unhardened, no nodules present but may be slight evidence of concretionary development	Plinthite Laterite Lateritic clay
	Nodular laterite	Distinct hard concretionary nodules present as separate particles	Lateritic gravel Ironstone Pisolitic gravel Concretionary gravel
	Honeycomb laterite	Concretions have coalesced to form a porous structure which may be filled with soil material	Vesicular laterite Pisolitic ironstone Vermicular ironstone Cellular ironstone Spaced pisolitic laterite
Mature (old)	Hardpan laterite	Indurated laterite layer, massive and tough	Ferricrete Ironstone Laterite crust Vermiform laterite Packed pisolitic laterite
	Secondary laterite	May be nodular, honeycomb or hardpan, but is the result of erosion of pre-existing layer and may display brecciated appearance	

Table 30.7 Classification of laterite (Charman, 1988)

energy of the SPT hammer. They plotted this ratio against $(N_1)_{60}$, where the N_{60} value is normalised to take account of the vertical effective stress and hence should give a closer indication of relative density of the deposit. They found that the bonded structure has a marked effect for residual soils producing values of normalised stiffness (G_o/N_{60}) that are considerably higher than those observed in uncemented materials.

30.3.2 Classification

The search for an appropriate scheme for classifying tropical soils has occupied engineering geologists, geotechnical engineers, pedologists and soil scientists for many years. A proliferation of such schemes exist: the Geological Society Working Party Report on Tropical Soils (Anon, 1990; Fookes, 1997) tabulates over 20 different schemes developed between 1951 and 1986, each with a different end use in mind. Although a number of well-developed pedological schemes exist, they are not always relevant for classifying tropical soils for engineering use. Leong and Rahardjo (1998) concluded that, in spite of the efforts to develop classification schemes by geologists, pedologists and engineers, no suitable classification system exists for the study of residual soils.

Stage of laterisation	SiO_2/Al_2O_3
Unlaterised soil	>2
Lateritic soil	1.3–2
Laterite	<1.3

Table 30.8 Typical values for silica/alumina ratio (After Desai, 1985)

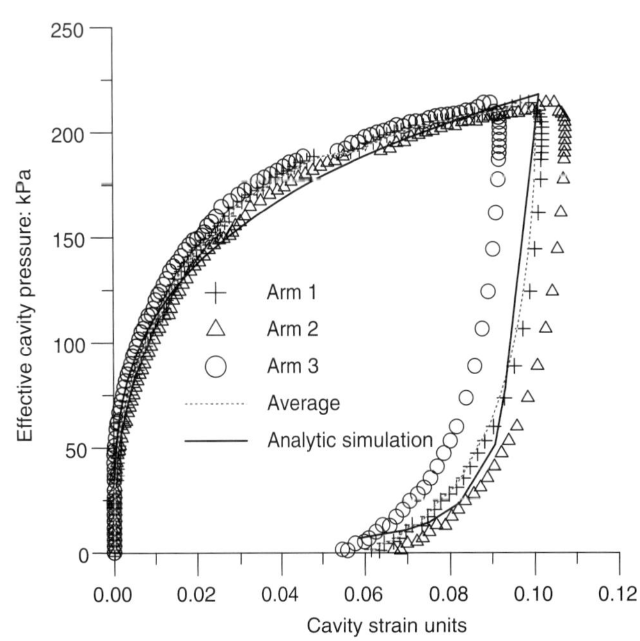

Figure 30.7 An example of a pressuremeter test carried out in saprolite from Brazil (Schnaid and Mantaras, 2003)

The Geological Society Working Party Report on Tropical Soils (Anon, 1990; Fookes, 1997) represents the most complete attempt to produce a useful classification scheme for tropical soils. The Working Party opted for a purely pedogenic classification scheme based on the work of Duchaufour (1982) (**Table 30.1**).

Another major work on residual soils (Blight, 1997) proposes a rather different classification scheme (Wesley and Irfan, 1997). They identify the factors influencing residual soil behaviour as

- physical composition (e.g. percentage of unweathered rock, particle size distribution)
- mineralogical composition
- macro-structure (layering, discontinuities, fissures, pores, etc. discernible to the naked eye)
- micro-structure (fabric, interparticle bonding or cementation, aggregation, etc.).

To take these factors into account they suggest grouping soils into three types (**Table 30.9**).

According to Wesley and Irfan, Group A (which is not strongly influenced by particular clay minerals) is typical of many tropical soil profiles. The group is further subdivided by structure components (macro-structure dominated, micro-structure dominated or soils not significantly influenced by either). They suggest that the engineering properties of Group B (which includes vertisols) will be very similar to transported soils with the same clay mineralogy. Group C, which is dominated by minerals only found in tropical soils (halloysite, allophane and aluminium and iron sesquioxides) is subdivided according to the minerals present.

A way to incorporate these different aspects is proposed by the author of this chapter, where the four factors of disintegration, mineralogy, cementation and structure (DMCS) are encoded on a six-point scale for each factor. It is suggested that the four factors are depicted as shown in **Figure 30.8** using the scales listed in **Table 30.10**. A higher number in each category indicates a more problematic material.

It has to be recognised that there will be cross-linkages between the different categories; for example, a material that is fresh rock (D = 1) but has no secondary cementing (C = 6) will not be a problematic material, even though the cementing category has a high score. Similarly, a residual soil (D = 6) that has hardpan cementing (D = 1) will not be problematic, as the cementing will overcome the decomposition due to weathering and result in a strong rock-like soil.

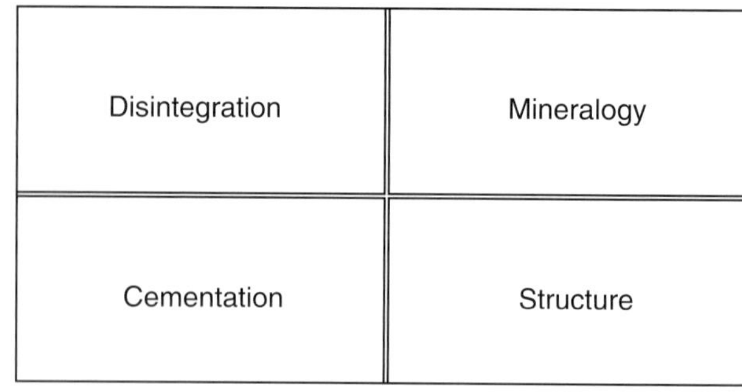

Figure 30.8 DMCS Classification scheme

Major division		Subgroup
Group A	Soils without a strong mineralogical influence	a Strong macro-structure influence b Strong micro-structure influence c Little or no structure influence
Group B	Soils with a strong mineralogical influence derived from clay minerals also commonly found in transported soils	a Montmorillonite (smectite group) b Other minerals
Group C	Soils with a strong mineralogical influence deriving from clay minerals only found in residual soils	a Allophane subgroup b Halloysite subgroup c Sesquioxide subgroup (gibbsite, goethite, haematite)

Table 30.9 Classification of residual soils (Wesley and Irfan, 1997)

Grade	Disintegration	Grade	Mineralogy[a]
6	Residual soil	6	Smectite (vertisol)
5	Completely weathered	5	Smectite/kaolin (ferruginous)
4	Highly weathered	4	Allophane/halloysite (andosol)
3	Moderately weathered	3	Kaolinite (siallitic)
2	Slightly weathered	2	Kaolinite (ferrisol)
1	Fresh/faintly weathered	1	Kaolinite/gibbsite (ferrallitic)
Grade	**Cementation[b]**	**Grade**	**Structure spacing**
6	No cementing agents present	6	Very small (<60 mm)
5	No evident cementing effect	5	Small (60–200 mm)
4	Weakly cemented	4	Medium (200–600 mm)
3	Nodular	3	Large (600 mm to 2 m)
2	Honeycomb	2	Very large (>2 m)
1	Hardpan	1	No evident macro-structure

[a]These are ranked in order of engineering behaviour rather than stages of weathering.
[b]This is a measure of secondary cementing of the weathered material, not the initial cementation of the parent rock.

Table 30.10 Proposed classification of tropical residual soils

30.3.3 Characteristics and typical engineering properties

The stages of weathering result in different clay mineralogies and these are reflected in the cation exchange capacity (CEC) of the clay fraction (Anon, 1990). Typical values are given in **Table 30.11**.

Fersiallitic soils can comprise vertisols or andosols that form at the early stage of weathering. Vertisols commonly occur in areas of impeded drainage such as on valley floors. They are often black or dark brown in colour and contain smectite clay minerals. They often exhibit excessive shrinkage and swelling properties and present major engineering problems.

Andosols are fersiallitic soils that develop from volcanic parent rocks and contain amorphous allophane or halloysite.

Stage of weathering	Cation exchange capacity: mEq/100 g
Fersiallitic soil	>25 (typically 50)
Ferruginous soil	16–25
Ferrallitic soil	<16

Table 30.11 Typical values for cation exchange capacity of the clay fraction (After Anon, 1990)

They frequently exist in an extremely loose state, and can have water contents of around 200%. Nevertheless, these high water contents are not reflected in their engineering behaviour since they generally have low compressibility and high angles of shearing resistance (Wesley, 1973, 1977).

The latter stages of weathering (ferruginous, ferrisols, ferralitic) usually result in soils with red colouration, reflecting higher iron and aluminium sesquioxide contents. They generally contain low-activity kaolinite minerals, and do not usually present major engineering problems. However, they may perform quite differently to temperate sedimentary soils and hence the application of standard classification systems can lead to difficulties.

Ferruginous soils (red tropical soils) have red colouration but do not contain high iron oxide contents. Ferrisols (lateritic soils) have higher iron contents, and contain granular nodules (pisoliths) of iron cemented material. Ferricrete (laterite cuirrasse or carapace) is an indurated deposit, heavily cemented with iron oxides, and can behave like weak rock.

The effect of secondary cementation is to improve the engineering properties. **Figure 30.9** shows the variation of void ratio, compressibility and strength properties (c' and ϕ') in a weathering profile. In the Grade VI residual soil the void ratio will be high due to leaching out of minerals and the resulting

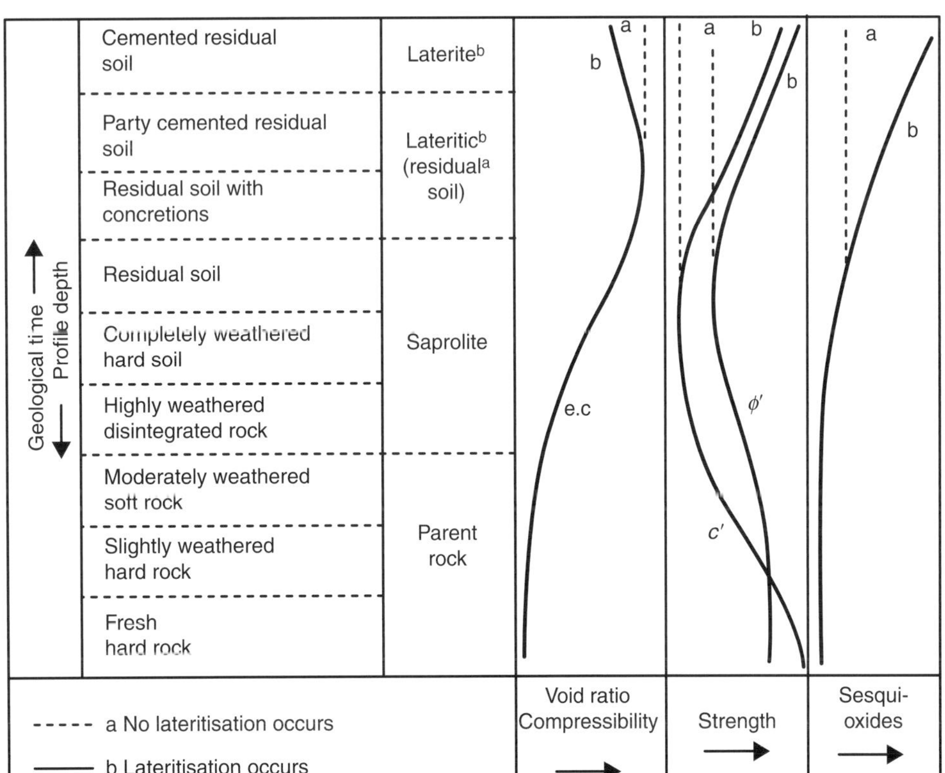

Figure 30.9 Changes in compressibility and strength in a weathering profile (Blight, 1997, adapted from Tuncer and Lohnes, 1977 and Sueoka, 1988)

structured form of the soil. This open structure results in high-compressibility and low-strength properties.

Figure 30.9 shows the effect that secondary cementing by sequioxides (laterisation) can have on the residual soil near the ground surface. This results in an upper profile with variations in the degree of cementing, from a concretionary or partly cemented layer to a fully cemented layer at the surface. The presence of the sesquioxides partially fills voids and reduces the compressibility. Secondary cementing also produces major improvements in strength properties.

The permeability of tropical residual soils is controlled to a large extent by the macro-structure provided by relict joints and so on. The more mature residual soils (Grade VI) may have lower permeability (**Table 30.12**) as the weathering will modify the macro-fabric and reduce the dominance of relict joints, as well as producing more clay minerals. However, materials that have been laterised can have higher permeability due to the open micro-structure and vesicular nature of the more cemented materials.

30.3.4 Problematic behaviour

Geotechnical classification schemes such as the Unified Soil Classification Scheme (USCS; US Army Engineer Waterways Experiment Station, 1960), which are widely used for temperate sedimentary soils, have severe limitations when applied to tropical soils. This gives tropical soils a reputation for being 'problematic' as they do not conform to these simple classification systems. However, many tropical soils, particularly the 'red' soils, can be good engineering materials and are often not problematic.

That is not to say that all tropical soils are problem-free. Some residual soils can exist in a loose state. Some ferruginous soils (red coffee soils) may exist with densities as low as 0. 6 Mg/m^3 – that is, less than the density of water. This loose structure may be sustained by an unsaturated state, where suctions give strength to clay 'bridges' supporting the coarser particles and maintaining a low density. If the soil is wetted so that the strength of the clay bridges is lost, the soil may collapse. Cementing agents may also maintain a loose metastable structure. If the soil is loaded beyond the yield strength of the bonding material, this can also lead to collapse. More information is provided in Chapter 32 *Collapsible soils.*

Zone	Relative permeability
Organic topsoils	Medium to high
Mature residual soil and/or colluvium	Low (generally medium or high in lateritic soils if pores or cavities present)
Young residual or saprolitic soil	Medium
Saprolite	High
Weathered rock	Medium to high
Sound rock	Low to medium

Table 30.12 Permeability of weathering profiles in igneous and metamorphic rocks (Deere and Patton, 1971)

A further factor to be aware of is that the particles themselves may be crushable, either due to a loss of intrinsic strength as a result of chemical attack during weathering or it may be that the coarse-grained size fraction is in fact made up of finer particles that are held together by secondary cementing or by a clay matrix. When subjected to high stresses the particles may start to crush, resulting in additional volumetric compressions (Lee and Coop, 1995).

Another major problematic type of tropical soil is vertisols, as they contain smectite clays and can be highly expansive. Information on dealing with such swelling/shrinking soils is discussed in Chapter 33 *Expansive soils.*

Other problematic aspects of tropical soils are identified below. In many cases, the behaviour is not problematic as such, but problems result from inappropriate use of classification systems that were designed for temperate soils and are not applicable to tropical soils.

30.3.4.1 Presence of 'unusual' clay minerals

Some clay minerals found in tropical soils (halloysite and allophane) exist in non-platey forms. These are unlike the common clay minerals found in temperate soils (kaolinite, illite, montmorillonite), which are generally platey in nature. The engineering behaviour of allophanous or halloysitic soils is often very different from what would be predicted by schemes like the USCS (Wesley, 1973, 1977).

Allophane is amorphous (or is poorly structured) and can hold significant amounts of water within the amorphous mineral. On drying it appears that allophane forms a more ordered structure, completely changing the nature of the soil. An allophanous clay can change in behaviour to appear like a sand after drying.

Halloysite is a member of the kaolinite family, but has a tubular habit. Water can be held within the 'tubes' which does not contribute to the engineering behaviour. Halloysite exists in two forms: hydrated halloysite and metahalloysite containing no water of crystallisation. In metahalloysite the 'tubes' may split or become partially unrolled. The transition takes place if the moisture content reduces below about 10% or the relative humidity drops below about 40% (Newill, 1961). The change is irreversible.

Because of the ability for allophane and halloysite to hold water that does not contribute to the engineering behaviour, they are often classified as troublesome soils (since they have high natural water contents and liquid limits) (Wesley, 1973). However, they generally have very good engineering properties (Wesley, 1977). They show high angles of shearing resistance (compared to kaolinite or montmorillonite). Also, since they are not platey in form, they do not show a significant reduction in angle of shearing resistance due to clay particle alignment. Therefore, the residual angles of shearing resistance are also high.

30.3.4.2 Presence of cementing agents

Many tropical soils are structured due to the presence of cementing agents that produce a physical bonding between soil particles. Schemes like the USCS are based on measurements on remoulded (or destructured) soil (Atterberg limits and particle size determination) and therefore cannot take account of soil structure. This is a severe limitation even for many temperate soils, but can be particularly limiting for tropical soils.

Under tropical weathering conditions iron and aluminium oxides are released and are not dissolved (as would occur in more acidic environments) and remain in situ. The presence of iron and aluminium oxides significantly affects the behaviour of tropical soils. Newill (1961) demonstrated that these oxides can suppress the plasticity of tropical clay soils, since on removal of the iron oxides the liquid limit was found to increase. This is the case if the oxides have an aggregating effect on the clay minerals. However, it is also possible for the oxides to contribute to plasticity, as was found by Townsend *et al.* (1971). If the oxides are present as amorphous colloids they can have a large water retention capability due to their large specific surface and will then contribute to plasticity.

30.3.4.3 Difficulties in determining Atterberg limits

Atterberg limit determinations on tropical soils are sensitive to the methods of preparation (e.g. Moh and Mazhar, 1969). Different degrees of pre-test drying (e.g. oven dried, air dried or tested from natural moisture content) can produce very significant differences in the Atterberg limits (Anon, 1990; Fookes, 1997). In addition, the amount of mixing of the soil during test preparation can also change the index properties significantly (Newill, 1961).

A comparison of the effects of different degrees of pre-test drying on the Atterberg limits is shown in **Table 30.13**. It should be noted that these changes are irreversible, and a permanent change in plasticity is produced by drying. To overcome problems relating to pre-test preparation, Charman (1988) suggests a procedure for testing the susceptibility to the method of preparation. This involves testing at different drying temperatures and different periods of mixing. If sufficient time is not available for such a detailed test programme, the best solution is to test the material without drying below the natural moisture content with a standard mixing time of 5 minutes. It is essential to report the pre-test preparation (form of drying and length of mixing/working time).

30.3.4.4 Difficulties in determining particle size distributions

Like Atterberg limits, the measurement of clay content can also be affected by pre-test drying since the drying process causes the clay particles to aggregate (Newill, 1961). These aggregations are only partially disaggregated by standard dispersion techniques, and clay fractions are often underestimated. For example, a red clay from Sasamua, Kenya showed a clay fraction of 79% when testing at natural moisture content, but this reduced to an apparent value of 47% after oven-drying (Terzaghi, 1958). Another problem is that the coarse fraction of red soils often consists of weakly cemented particles which readily break down and change grading during sieving or compaction (Gidigasu, 1972; Omotosho and Akinmusuru, 1992).

30.3.4.5 Unsaturated state

Many tropical soils exist in an unsaturated state since evapotranspiration is greater than precipitation. Water tables are often greater than 5 m deep, in many cases considerably deeper. The strength of these soils will be very dependent on moisture conditions.

Soil suction is made up of two components: *matric suction* and *osmotic suction* (also called *solute suction*). The sum is known as the *total suction*. Matric suction is due to surface tension forces at the interfaces (menisci) between the water and the gas (usually air) phases present in unsaturated soils (the surface tension effect is sometimes referred to as capillarity). Osmotic suctions are due to the presence of dissolved salts within the pore water.

In much of the soil science literature, suction is expressed in pF units (i.e. the logarithm (to base 10) of the suction expressed in centimetres of water (Schofield, 1935)). For engineering applications it is generally more convenient to use conventional stress units. The relationship to convert from pF units to kPa is given by

$$\text{Suction (kPa)} = 9.81 \times 10^{[\text{pF}-2]} \quad (30.1)$$

Location	At natural moisture content			Air dried			Oven dried (105°C)		
	LL	PL	PI	LL	PL	PI	LL	PL	PI
Costa Rica	81	29	52				56	19	37
Dominica	93	56	37	71	43	28			
Kenya (red clay)	101	70	31	77	61	16	65	47	18
Kenya (lateritic gravel)	56	26	30	46	26	20	39	25	14

Table 30.13 Effect of drying on classification tests on red soils

Suction (kPa)	Suction (pF)	Reference points	Moisture condition
1 000 000	7	Oven dry	
100 000	6		Dry
10 000	5		
1 000	4	Wilting point for plants Plastic limit	
100	3		Moist
10	2		
1	1		Wet
0.1	0	Liquid limit saturated	

Figure 30.10 The suction scale

The suction scale (showing both kPa and pF units) with some points of reference and indications of the moisture condition of a soil is shown in **Figure 30.10**.

The maximum suction that can be sustained within the pores of a soil will depend on the pore size. In a clean sandy soil, where pore sizes will be of the order of 0.1 mm or larger, the maximum suctions will be very small (typically <5 kPa). In clean silty materials, where pore sizes might be of the order of 0.01 mm, the maximum suctions are likely to be less than 100 kPa. However, in clayey soils, where pore sizes can be less than 0.001 mm, then high suctions greater than 1000 kPa can be sustained. This explains why clean sandy soils have no strength when they dry out (they lose the suction 'bonds' that hold them together) as they cannot sustain high suctions. However, clayey soils can become very strong when they dry out due to the high suctions that are maintained in the fine pores of the soil. The suctions pull the soil particles together and give the soil considerable strength in a dry state.

There are a number of different techniques for suction measurement and control. Their suitability varies according to the range of suctions operating. An indication of appropriate ranges is shown in **Figure 30.11**. It is generally necessary to use a variety of techniques in order to cover the entire suction scale.

As a soil dries out (or wets up) the suction within the soil will change. The relationship between water content and suction is known as the soil water retention curve (SWRC; also known

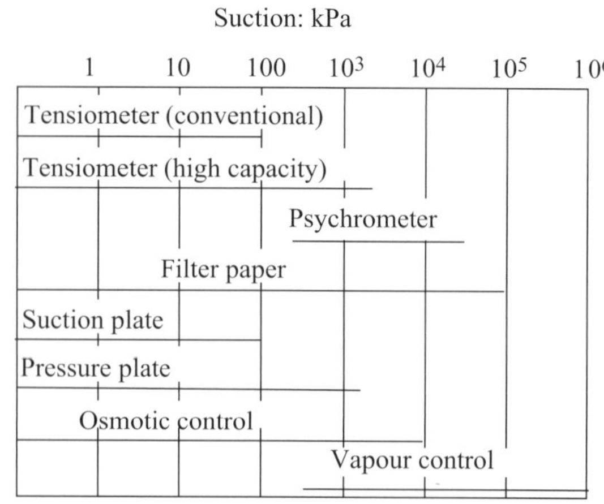

Figure 30.11 Ranges for which suction measuring/controlling devices are suitable

as the soil water characteristic curve, SWCC). Although water contents are usually defined gravimetrically (i.e. by weight) in geotechnical engineering, SWRCs are often expressed in terms of volumetric water content, θ, or degree of saturation, S_r against suction. **Figure 30.12** shows a typical SWRC.

If the soil starts from a saturated state and is then subject to drying, it will follow the *primary drying curve*. At a value of suction known as the *residual suction* (with a corresponding residual water content, θ_r) the SWRC may flatten and much smaller changes in volumetric water content result from an increase in suction. To achieve zero water content (equivalent to an oven-dried condition) requires a suction of the order of 1 GPa (pF 7) (Fredlund and Xing, 1994). On wetting from an oven-dried state the soil will follow the *primary wetting curve*. The primary drying and wetting curves define an envelope of possible states within which the soil can exist. If drying is halted partway down the primary drying curve and wetting is started, the soil will follow an intermediate *scanning curve* that is flatter than the primary wetting curve, until the primary wetting curve is reached. Therefore, different suctions can exist at a given water content depending on the pathway followed.

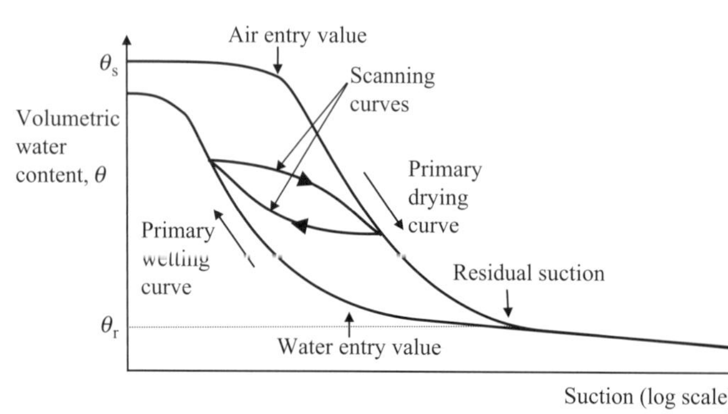

Figure 30.12 Typical soil water retention curve

The most commonly used approach to interpreting shear strength behaviour in unsaturated soils is to adopt an extended version of the traditional Mohr–Coulomb approach. This extension to unsaturated soils was put forward by Fredlund *et al.* (1978). It involves two separate angles of shearing resistance to represent the contribution to strength from the net stress (total stress referenced to the pore air pressure) and matric suction (the pore water pressure referenced to the pore air pressure), giving the shear strength equation as

$$\tau = c' + (\sigma - u_a)\tan\varphi^a + (u_a - u_w)\tan\varphi^b \qquad (30.2)$$

where τ is shear strength, c' is the effective cohesion intercept (many tropical soils demonstrate a true cohesion intercept at zero effective stress due to their bonded structure), ϕ^a is the angle of shearing resistance for changes in net stress ($\sigma - u_a$) and ϕ^b is the angle of shearing resistance for changes in matrix suction ($u_a - u_w$).

This separates the effects of net stress ($\sigma - u_a$) and suction ($u_a - u_w$) and treats them differently by having two angles of shearing resistance relating to the two components of stress. The extended Mohr–Coulomb failure surface is shown in three dimensions in **Figure 30.13**. The surface is shown by views in the net stress plane in **Figure 30.14** and in the suction plane in **Figure 30.15**.

Figure 30.14 shows that the strength envelope increases as the suction increases. This can be represented as an increase in the total cohesion, c, where

$$c = c' + (u_a - u_w)\tan\phi^b \qquad (30.3)$$

Figure 30.15 shows the increase in total cohesion, c, as the suction increases. The slope of the graph is defined by ϕ^b. The relationship between τ and ($u_a - u_w$) has been found to be non-linear by Escario and Saez (1986) and Fredlund *et al.* (1987). Below the air entry value (when the soil remains saturated) ϕ^b is equal to ϕ', but at higher suctions the value of ϕ^b reduces (**Figure 30.15**). The tangent value may fall to zero at high suctions, implying no further increase in strength at higher suctions.

A more complete model of unsaturated soil behaviour is that proposed by Alonso *et al.* (1990) and is now known as the Barcelona Basic Model. This extends the modified Cam Clay model to the unsaturated state, and provides the coupling between volumetric and deviatoric behaviour that is essential for a complete understanding of soil behaviour. It introduces the concept of a loading–collapse (LC) surface that defines yielding due either to external loading (total stresses) or wetting (loss of suction).

If the role of suction in an unsaturated soil is not recognised, then test results can be incorrectly interpreted. For instance, the compressibility of an unsaturated soil may be observed to be low at the water content at which the specimen is tested, as a result of the presence of a significant suction. However, the

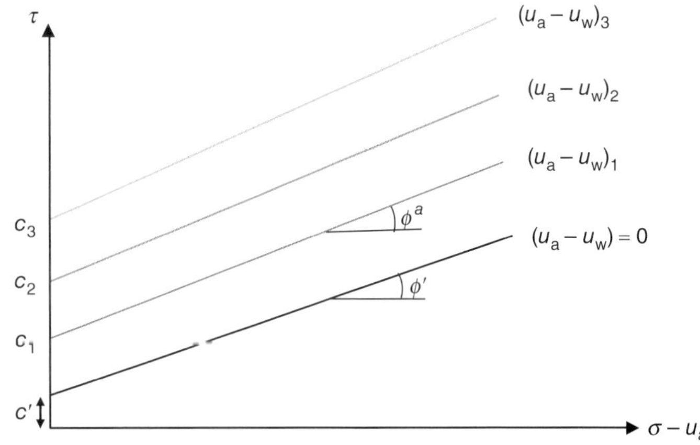

Figure 30.14 The extended Mohr–Coulomb failure envelope in net stress space

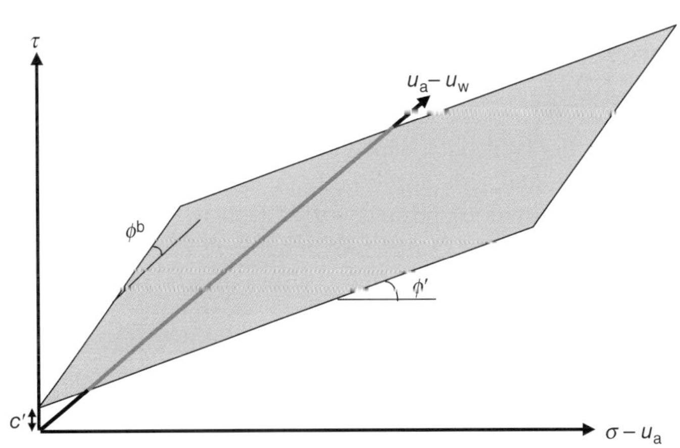

Figure 30.13 The extended Mohr–Coulomb failure envelope

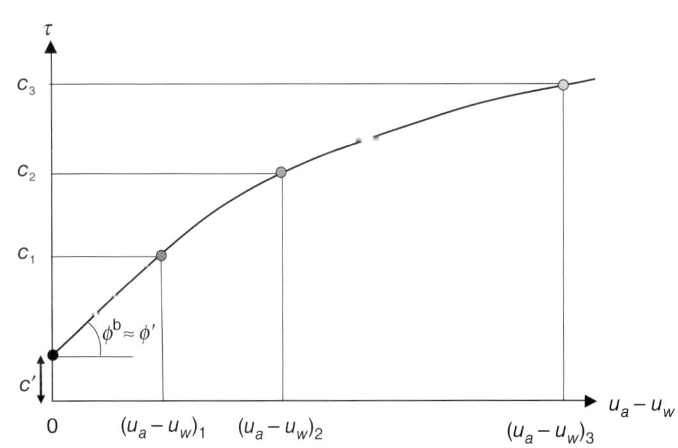

Figure 30.15 The extended Mohr–Coulomb failure envelope in matric suction space

same soil could have much higher compressibility if the soil is wetted and loses the suction. Similarly, an apparent cohesion intercept observed in strength tests may be largely due to suction rather than being a true 'cohesive' contribution to strength from bonding. Again, this component of strength will reduce (and may even be lost entirely) if the soil is wetted.

30.3.5 Foundations

Problems with shallow foundations on tropical residual soils are usually associated either with collapse problems on loose, metastable soils or shrink/swell movements on expansive vertisols.

Collapse settlements can result from overstressing the cemented micro-structure of tropical soils. The compressibility may be low if the stresses applied do not exceed the yield strength of the cementing material. However, if the foundation is loaded beyond this stress, large and rapid settlements can take place.

Collapse settlements of foundations can also occur due to wetting. Some ferruginous soils (red coffee soils) may exist with densities as low as 0.6 Mg/m^3 (i.e. less than the density of water). This loose structure is often maintained by bridges of clay minerals which support the sand- or silt-sized particles. The strength of these bridges is controlled by suction, and if the soil wets up or becomes flooded there is a rapid collapse of the loose structure, resulting in large surface settlements.

A build-up in moisture under a pad or raft foundation can occur due to cutting off evaporation and changes in the temperature regime due to construction of a covered area (e.g. a concrete foundation). However, wetting up can also be caused by simpler means, such as the construction of soakaways for buildings, or leaking services, resulting in loss of strength and failure of the foundation.

If shallow foundations are constructed on vertisols, the likely problem will be seasonal movements as water contents change beneath the foundation due to wetting and drying. Foundation heave will be observed in wet season conditions and settlements will be induced by shrinkage in dry season conditions. The zone of variation of water content will affect the edges of the foundation, while the central area of a large raft may not be affected. This results in differential movements that can be severely deleterious to the foundation and the overlying structure. Methods to deal with foundation construction on expansive soils are discussed below.

30.3.5.1 Removal and replacement

Remove expansive material and replace with non-expansive soils. Generally the expansive layer extends to depths too great to economically allow complete removal and replacement. It must then be determined what depth of excavation and fill will be necessary to prevent excessive heave.

30.3.5.2 Remoulding and compaction

The swell potential of expansive soils can be reduced by decreasing the dry density. Compaction at low densities and at water contents wet of optimum will reduce the swell potential. However, the bearing capacity of the soil at the lower density may not be adequate. Some soils have such a high potential for volume change that compaction control does not significantly reduce swell potential.

30.3.5.3 Surcharge loading

If the surcharge load applied is greater than the swelling pressure then heave can be prevented. For example, a swell pressure of 25 kPa can be controlled by 1.5 m of fill and a concrete foundation. However, swelling pressures are often too high ($\sim$400 kPa) for this to be a realistic option.

30.3.5.4 Pre-wetting

This is based on the assumption that increasing the water content will cause heave prior to construction. If the high water content is maintained, there will be no appreciable volume change to damage the structure. However, the procedure has many drawbacks. Expansive soils are normally clays with low permeability and the time required for adequate wetting may be years. Also, the increase in water content will reduce the strength of the soil and cause reductions in bearing capacity.

30.3.5.5 Moisture control by horizontal and vertical barriers

Soil expansion problems are primarily the result of fluctuations in water content. Non-uniform heave is the major cause of damage, as opposed to total heave. If changes in water content can be made to occur slowly and if the water content distribution can be made uniform, differential heave can be minimised. Moisture barriers do not prevent the heave taking place, but have the effect of slowing the rate of heave and providing a more uniform moisture distribution.

Horizontal barriers installed around a building can limit the migration of moisture into the covered area. Concrete aprons or paved areas for car parking can achieve this. The width of the barrier should be sufficient to extend the 'edge distance' – that is, the distance measured inward from the slab edge over which the soil moisture varies enough to cause soil movement (Post-Tensioning Institute, 1980).

30.3.6 Slopes

Many of the landslides which occur in the saprolitic zone of tropical residual soils are directly or indirectly controlled by relict discontinuities (Brand, 1985; Dobie, 1987; Irfan and Woods, 1988; Irfan *et al.*, 1987; Nieble *et al.*, 1985). Many types of mineral infillings and coatings may be present along relict discontinuities as a result of weathering processes, including clay minerals. Some discontinuities may be polished or slickensided as a result of internal deformation in the slopes (Irfan, 1998). These infilled or polished surfaces may have low residual angles of shearing resistance, providing a plane of weakness so that failure is constrained to occur on these relict surfaces.

Landslides are often triggered by rainfall, particularly in tropical climatic regions where rain storms can be very intense. Major landslides occur all too often, but minor landslides occur even more frequently. Although minor landslides may not lead to loss of human life, they still have economic and social impacts.

A clear linkage has been established between landslide occurrence and high rainfall in tropical regions of the world. This has been confirmed by studies in Brazil (Wolle and Hachich, 1989), Costa Rico (Sowers, 1971), Fiji (Vaughan, 1985b), Hong Kong (Au, 1993, 1998; Brand, 1984; Lumb, 1975), Japan (Yoshida *et al.*, 1991), Nigeria (Adegoke-Anthony and Agada, 1982), Papua New Guinea (Murray and Olsen, 1988), Singapore (Chatterjea, 1994; Pitts, 1985; Rahardjo *et al.*, 1998; Tan *et al.*, 1987; Toll, 2001), South Africa (van Schalkwyk and Thomas, 1991) and Thailand (Jotisankasa *et al.*, 2008).

Soil slopes in tropical regions are normally unsaturated during the dry season and the groundwater table may often be at depths of more than 10 m for most of the year. When the soil is unsaturated, suction or negative pore water pressure provides additional strength to the soil, hence stabilising the slope. This additional strength may disappear during an intense rainstorm when the soil becomes saturated and pore water pressure becomes zero.

Figure 30.16 shows rainfall data for a large number of landslides in Singapore (Toll, 2001). It shows the rainfall on the day of the landslide (*triggering* rainfall) plotted against the rainfall in the five-day period preceding it (*antecedent* rainfall). It can be seen that it is usually not a single rainstorm that produces failure; rather it is a build-up of pore water pressure over a number of days due to the antecedent rainfall followed by a storm that finally triggers the landslide event. However, there are occasions where a single storm is big enough to produce a failure even when there has been no significant antecedent rainfall in the preceding days.

It is important that when studying climate effects on slopes one does not always assume that rainfall will produce a rise in the water table level. Infiltration of rainfall at the surface can produce significant changes in pore water pressure without a change in the water table (although a perched water table may be induced at the surface) (Toll, 2006).

30.3.7 Highways

Traditionally, materials used in the construction of road bases have been clean graded aggregates, generally obtained from crushed rock. However, in tropical climatic zones good-quality rock for crushing is often unavailable because of the extensive weathering that occurs in the tropics. Even when it is available, the costs of processing and transporting the material can make this an uneconomical option. Therefore, naturally occurring materials such as lateritic gravels or calcretes are widely used for construction of roads with low traffic volumes. These natural materials generally contain a greater amount of fines, and the fines have higher plasticity than is accepted by many existing specifications for construction materials. Details of materials which have been successfully used as road base construction materials in the tropics are given by Lionjanga *et al.* (1987), Grace and Toll (1989), Gidigasu (1991), Metcalf (1991),

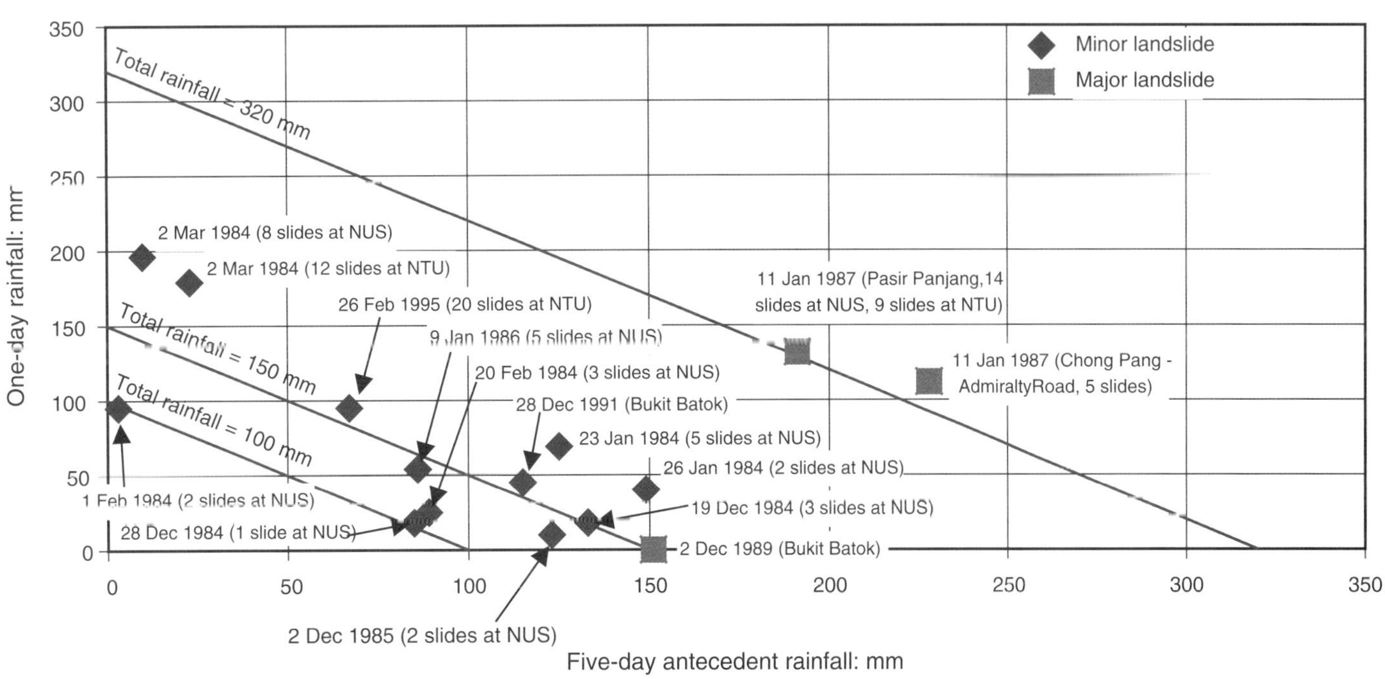

Figure 30.16 Rainfall events leading to landslides in Singapore

Netterberg (1994), Gourley and Greening (1997), Paige-Green *et al.* (2015).

Because naturally occurring gravels have greater quantities of fines, soil suction and fabric are major factors controlling their behaviour. The presence of around 10% clay in a lateritic gravel was sufficient to provide a matrix with small pore sizes which could sustain significant suctions (Toll, 1991). Yong *et al.* (1982) similarly report that a small clay fraction had an important effect in influencing the suction characteristics of a weathered granite. Therefore, the effects of suction should not be overlooked in granular materials if they contain small amounts of clay. Provided the fines are well distributed, it will be possible to develop high suction throughout the soil. The matrix of fines will then act as a binder that can hold the granular material together, thus imparting overall high strength and stiffness.

The ability to maintain soil suction, and hence maintain good performance, is highly dependent on the avoidance of wetting up of the road base material, particularly in unsurfaced roads. In cases where the water table is close to the surface, the benefits of suction cannot be relied upon and conventional specifications using good-quality aggregates must be adopted. For cases where the water table is more than 5 m below the ground surface, then reliance can be placed on suction, provided drainage measures ensure that water will not pond on the road surface. A bituminous surface is beneficial in preventing direct infiltration into the road base material.

Trial constructions in Kenya of low-volume roads surfaced with a sprayed bituminous seal coat compared lateritic gravel with conventional crushed stone for the road base (Grace and Toll, 1989). It was found that the lateritic gravel sections showed better performance. This was because failures of the bituminous surface coat allowed water to penetrate and spread within the high-permeability crushed stone base, softening the subgrade and causing large areas of cracking. On the laterite sections, a small pothole formed where the bituminous surface coat failed, but it did not spread due to the low permeability of the road base material. Any water accumulating in the pothole during a rainstorm evaporated during drying periods. Therefore, the laterite sections performed better than the 'higher-quality' construction methods.

Toll (1991) argues that a good material for a road base in a subtropical or tropical climate will have sufficient fines to allow significant suctions to develop and also produce low permeability. A small amount of clay can be beneficial in this. However, the fines content should not be so great as to suppress the dilatant tendency of the granular fraction or to significantly reduce the angle of shearing resistance. Also, any clay present should have low activity in order to restrict shrinkage and swelling. The fines should be well distributed throughout the fabric if they are to support high suctions and provide a strong binder, holding the granular fraction together. This also produces low permeability.

Property	Value
Maximum size: mm	37.5
Maximum oversize index (I_o)[a]	5%
Shrinkage product (S_p)[b]	100–365 (maximum of 240 preferable)
Grading coefficient (G_c)[c]	16–34
Soaked CBR (at 95% modified AASHTO compaction)	>15%
Treton impact value (%)	20–65

(a) I_o, the oversize index, is the percentage retained on 37.5 mm sieve.
(b) S_p = linear shrinkage × (% passing 0.425 mm sieve);
(b) G_c = (% passing 26.5 mm – % passing 2.0 mm) × (% passing 4.75 mm)/100.

Table 30.14 Recommended material specifications for unsealed rural roads (Paige-Green, 2007)

Charman (1988) provides comparisons between specifications for natural gravels for road bases used around the world and Paige-Green *et al.* (2015) provide an updated review. Paige-Green (2007) gives recommended material specifications for unsealed rural roads (**Table 30.14**) based on experience in southern Africa. This is based on distinguishing between materials that will become slippery, erode, ravel or form corrugations (Figure **30.17**).

Paige-Green suggests a minimum value of soaked CBR (at 95% modified AASHTO compaction) of 15%. This is even lower than the **minimum** value of 20% suggested by Grace (1991) for bituminous sealed roads, combined with an **average** value of soaked CBR of 40%. Gourley and Greening suggest a minimum soaked CBR of 45% (at 100% modified AASHTO compaction) for sealed roads carrying less than 0.01 million equivalent standard axles (ESA), but a higher requirement of 55-80% for more highly trafficked roads carrying 0.5 million ESAs (the lower limit for road base CBR of 55% is for a weak subgrade with CBR = 3–4% and the higher limit of 80% for a strong subgrade with CBR > 30%).

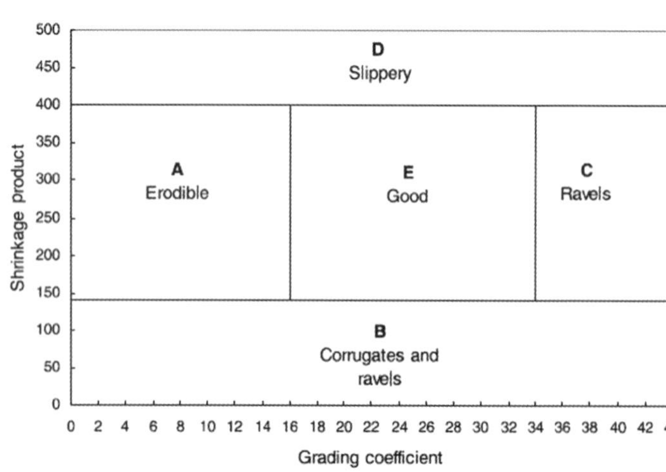

Figure 30.17 Categories of road performance for rural unsurfaced roads (Paige-Green, 2007)

30.4. Concluding remarks

Tropical soils pose many challenges for geotechnical engineers. They are highly structured at both micro and macro levels. They are often cemented due to deposition of minerals either during or after weathering has taken place. They can be highly heterogeneous. A major difficulty is that traditional classification systems that have been developed for temperate sedimentary soils cannot be used to infer likely engineering behaviours for tropical soils. Simple classification tests, such as Atterberg limits, cannot be easily determined due to the presence of unusual clay minerals, such as halloysite or allophone, or due to the effects of iron or aluminium sesquioxides.

Tropical soils are often thought of as problematic soils, largely due to the difficulty in classifying them. However, may tropical soils, particularly the 'red' soils, often have good engineering properties, such as low compressibility and high strength. The cemented structure can enhance the strength.

Nevertheless, some tropical soils do demonstrate problematic behaviour. Examples are those that exist in a loose, metastable state that can collapse when loaded, or when subject to wetting. Other problematic tropical soils are vertisols that contain active smectite clay minerals that demonstrate excessive shrink/swell behaviour when subject to drying and wetting.

References

Ackroyd LW (1967) Formation and properties of concretionary and non-concretionary soils in western Nigeria. *Proceedings of the 4th African Conference on Soil Mechanics and Foundation Engineering, Cape Town, South Africa*. Balkema, Rotterdam, the Netherlands, pp. 47–52.

Adegoke-Anthony CW and Agada OA (1982) Observed slope and road failures in some Nigerian residual soils. *Proceedings ASCE Specialty Conference Engineering and Construction in Tropical and Residual Soils, Honolulu, Hawaii*, pp. 519–538.

Alonso EE, Gens A and Josa A (1990) A constitutive model for partially saturated soils. *Géotechnique* **40(3)**: 405–430.

Anon (1970) The logging of rock cores for engineering purposes, Geological Society Engineering Group, Working Party Report. *Quarterly Journal of Engineering Geology* **3**: 1–24.

Anon (1977) The description of rock masses for engineering purposes, Geological Society Engineering Group, Working Party Report. *Quarterly Journal of Engineering Geology* **10**: 355–388.

Anon (1990) Tropical residual soils, Geological Society Engineering Group, Working Party Report. *Quarterly Journal of Engineering Geology* **23**: 1–101.

Au SWC (1993) Rainfall and slope failure in Hong Kong. *Engineering Geology* **36**: 141–147.

Au SWC (1996) The influence of joint-planes on the mass strength of Hong Kong saprolitic soils. *Quarterly Journal of Engineering Geology* **29**: 199–204.

Au SWC (1998) Rain-induced slope instability in Hong Kong. *Engineering Geology* **51(1)**: 1–36.

Aung KK, Rahardjo H, Toll DG and Leong EC (2000) Mineralogy and microfabric of unsaturated residual soil. In *Unsaturated Soils for Asia. Proceedings of the Asian Conference on Unsaturated Soils, Singapore* (Rahardjo H, Toll DG and Leong EC (eds)). Balkema, Rotterdam, the Netherlands, pp. 317–321.

Barksdale RD and Blight GE (1997) Compressibility and settlement of residual soils. In *Mechanics of Residual Soils* (Blight GE (ed)). Balkema, Rotterdam, the Netherlands, pp. 95–154.

Blight GE (ed) (1997) *Mechanics of Residual Soils*. Balkema, Rotterdam, the Netherlands.

Brand EW (1984) Landslides in Southeast Asia: a State-of-the-art report. *Proceedings 4th International Symposium. on Landslides, Toronto, ON*, vol. 1, pp. 17–59.

Brand EW (1985) Predicting the performance of residual soil slopes. *Proceedings of the 11th International Conference Soil Mechanics and Foundation Engineering, San Francisco, CA*, pp. 2541–2578.

Brink ABA, Partridge TC and Williams AAB (1982) *Soil Survey for Engineering*. Clarendon, Oxford, UK.

BSI (British Standards Institution) (2015) BS 5930: Code of practice for site investigations. BSI, London, UK.

BSI (2018) BS EN ISO 14689: Geotechnical investigation and testing – Identification and classification of rock. BSI, London, UK.

Charman JG (1988) *Laterite in Road Pavements*. CIRIA SP47. CIRIA, London, UK.

Chatterjea K (1994) Dynamics of fluvial and slope processes in the changing geomorphic environment of Singapore. *Earth Surface Processes and Landforms* **19**: 585–607.

Deere DU and Patton FD (1971) Slope stability in residual soils. *Proceedings of the 4th Panamerican Conference on Soil Mechanics and Foundation Engineering*, San Juan, Puerto Rico, pp. 87–170.

Desai MD (1985) Geotechnical aspects of residual soils of India. In *Sampling and Testing of Residual Soils* (Brandt EW and Phillipson HB (eds)). Scorpion Press, Hong Kong.

Dobie MJD (1987) Slope instability in a profile of weathered norite. *Quarterly Journal of Engineering Geology* **20**: 279–286.

Duchaufour P (1982) *Pedology, Pedogenesis and Classification*. George Allen and Unwin, London, UK.

Escario V and Saez J (1986) The shear strength of partly saturated soil. *Géotechnique* **36(3)**: 453–456.

Fookes PG (1997) *Tropical Residual Soils*. Geological Society Publishing House, Bath, UK.

Fredlund DG and Xing A (1994) Equations for the soil–water characteristic curve. *Canadian Geotechnical Journal* **31**: 521–532.

Fredlund DG, Morgenstern NR and Widger RA (1978) Shear strength of unsaturated soils. *Canadian Geotechnical Journal* **15(3)**: 313–321.

Fredlund DG, Rahardjo H and Gan JK-M (1987) Nonlinearity of strength envelope for unsaturated soils. *Proceedings of the 6th International Conference on Expansive Soils, New Delhi, India*, pp. 49–54.

Gidigasu MD (1972) Mode of formation and geotechnical characteristics of laterite materials of Ghana in relation to soil forming factors. *Engineering Geology* **6**: 79–150.

Gidigasu MD (1991) Characterisation and use of tropical gravels for pavement construction in West Africa. *Geotechnical and Geological Engineering* **9(3/4)**: 219–260.

Gourley CS and Greening PAK (1997) Use of sub-standard laterite gravels as road base materials in southern Africa. *International Symposium on Thin Pavements, Surface Treatments and Unbound Roads*, Fredericton, NB, Canada.

Grace H (1991) Investigations in Kenya and Malawi using as-dug laterite as bases for bituminous surfaced roads. *Geotechnical and Geological Engineering* **9**: 183–195.

Grace H and Toll DG (1989) The improvement of roads in developing countries to bituminous standards using naturally occurring laterites. *Proceedings of the 3rd International Conference on Unbound Aggregates in Roads, Nottingham, UK*. Butterworth, Oxford, UK, pp. 322–332.

Irfan TY (1998) Structurally controlled landslides in saprolitic soils in Hong Kong. *Geotechnical and Geological Engineering* **16**: 215–238.

Irfan TY and Woods NW (1988) The influence of relict discontinuities on slope stability in saprolitic soils. *Proceedings of the 2nd International Conference Geomechanics in Tropical Soils, Singapore*, vol. 1, pp. 267–276.

Irfan TY, Koirala NP and Tang KY (1987) A complex slope failure in a highly weathered rock mass. *Proceedings of the 6th International Congress on Rock Mechanics, Montreal, QC*, pp. 397–402.

ISRM (1978) Suggested methods for the quantitative description of discontinuities in rock masses. *International Journal of Rock Mechanics and Mining Sciences & Geomechanics Abstracts* **15(6)**: 319–368.

Jotisankasa A, Kulsawan B, Toll DG and Rahardjo H (2008) Studies of rainfall-induced landslides in Thailand and Singapore. *Proceedings of the 1st European Conference Unsaturated Soils, Durham, UK*, pp. 901–907.

Lee IK and Coop MR (1995) The intrinsic behaviour of a decomposed granite soil. *Géotechnique* **45(1)**: 117–130.

Leong EC and Rahardjo H (1998) A review of soil classification systems. In *Problematic Soils* (Yanagisawa E, Moroto N and Mitachi T (eds)). Balkema, Rotterdam, the Netherlands, pp. 493–497.

Lionjanga AV, Toole T and Greening PAK (1987) The use of calcrete in paved roads in Botswana. *Proceedings of the 9th African Regional Conference Soil Mechanics and Foundation Engineering, Lagos, Nigeria. Balkema, Rotterdam, the Netherlands*, vol. 1, pp 489–502.

Little AL (1969) The engineering classification of residual tropical soils. *Proceedings of the 7th International Conference on Soil Mechanics and Foundation Engineering, Mexico City, Mexico*, vol. 1, pp 1–10.

Lumb P (1975) Slope failures in Hong Kong. *Quarterly Journal of Engineering Geology* **8**: 31–65.

McFarlane MJ (1976) *Laterite and Landscape*. Academic Press, London, UK.

Metcalf JB (1991) Use of naturally occurring but non-standard materials in low-cost road construction. *Geotechnical and Geological Engineering* **9**: 155–165.

Moh ZC and Mazhar FM (1969) Effects of method of preparation on index properties of lateritic soils. *Proceedings of the Special Session on Engineering Properties of Lateritic Soils, 7th International Conference Soil Mechanics and Foundation Engineering, Montreal, QC*, pp. 23–35.

Murray LM and Olsen MT (1988) Colluvial slopes: a geotechnical and climatic study. *Proceedings of the 2nd International Conference on Geomechanics in Tropical Soils, Singapore*, vol. 2, pp. 573–579.

Netterberg F (1971) Calcrete in road construction, *National Institute for Road Research Bulletin 10*. South African Council for Scientific and Industrial Research, Pretoria, South Africa.

Netterberg F (1994) Low-cost local road materials in southern Africa. *Geotechnical and Geological Engineering* **12**: 35–42.

Newill D (1961) A laboratory investigation of two red clays from Kenya. *Géotechnique* **11(4)**: 302–318.

Nieble CM, Cornides AT and Fernandes AJ (1985) Regressive failures originated by relict structures in saprolites. *Proceedings of the 1st International Conference on Tropical Lateritic and Saprolitic Soils, Brasilia, Brazil*, pp. 41–48.

Omotosho PO and Akinmusuru JO (1992) Behaviour of soils (lateritic) subjected to multi-cyclic compaction. *Engineering Geology* **32**: 53–58.

Paige-Green P (2007) Improved material specifications for unsealed roads. *Quarterly Journal of Engineering Geology and Hydrogeology* **40(2)**: 175–179.

Paige-Green P, Pinard MI and Netterberg F (2015) A review of specifications for lateritic materials for low volume roads. *Transportation Geotechnics* **5**: 86–98.

Pedro G (1968) Distribution des Principaux Types d'Altération Chimique à la Surface du Globe. Présentation d'Une Esquisse Géographique. *Révue Geographie Physique et Géologie Dynamique* **10**: 457–470.

Phillipson HB and Brand EW (1985) Sampling and testing of residual soils in Hong Kong. In *Sampling and Testing of Residual Soils* (Brand EW and Phillipson HB (eds)). Scorpion Press, Hong Kong, pp. 75–82.

Phillipson HB and Chipp PN (1982) Air foam sampling of residual soils in Hong Kong. *Proceedings of the ASCE Speciality Conference on Engineering and Construction in Tropical and Residual Soils, Honolulu, HI*, pp. 339–356.

Pitts J (1985) *An investigation of slope stability on the NTI Campus, Singapore. Applied Research Project RPI/83*, Nanyang Technological Institute, Singapore.

Post-Tensioning Institute (1980) *Design and Construction of Post-Tensioned Slabs-on-Ground*. Post-Tensioning Institute, Phoenix, AZ.

Rahardjo H, Leong EC, Gasmo JM and Tang SK (1998) Assessment of rainfall effects on stability of residual soil slopes. *Proceedings of the 2nd International Conference on Unsaturated Soils, Beijing, P.R. China*, vol. 1, pp. 280–285.

Richards BG (1985) Geotechnical aspects of residual soils in Australia. In *Sampling and Testing of Residual Soils* (Brand EW and Phillipson HB (eds)). Scorpion Press, Hong Kong, pp. 23–30.

Sanches Furtado AFA (1968) Altération des Granites dans les Régions Intertropicales sous Différents Climats. *Proceedings of the 9th International Congress Soil Science, Adelaide, Australia*, vol. 4, pp. 403–409.

Schnaid F and Huat BBK (2011) Sampling and testing tropical residual soils. In *Handbook of Tropical Residual Soil Engineering* (Huat BBK and Toll DG (eds)). Taylor & Francis, London, UK, pp. 67–115.

Schnaid F and Mantaras FM (2003) Cavity expansion in cemented materials: structure degradation effects. *Géotechnique* **53(9)**: 797–807.

Schnaid F, Fahey M and Lehane B (2004) In situ test characterization of unusual geomaterials. *Proceedings of the 2nd International Conference on Geotechnical and Geophysical Site Characterization, Porto, Portugal* (Viana da Fonseca A and Mayne P (eds)). Millpress, Rotterdam, the Netherlands, vol. 1, pp. 49–74.

Schofield RK (1935) The pF of the water in soil. *Transactions of the 3rd International Congress of Soil Science, Oxford, UK.*

Sowers GF (1971) Landslides in weathered volcanics in Puerto Rico. *Proceedings of the 4th Panamerican Conference Soil Mechanics and Foundation Engineering, San Juan, Puerto Rico*, vol. 2, pp. 105–115.

Strakhov NM (1967) *The Principles of Lithogenesis, vol. 1.* Oliver and Boyd, Edinburgh, UK.

Sueoka T (1988) Identification and classification of granitic residual soils using chemical weathering index. *Proceedings of the 2nd International Conference on Geomechanics in Tropical Soils, Singapore*, vol. 1, pp. 421–428.

Tan SB, Tan SL, Lim TL and Yang KS (1987) Landslides problems and their control in Singapore. *Proceedings of the 9th Southeast Asian Geotechnical Conference, Bangkok, Thailand*, pp. 1:25–1:36.

Terzaghi K (1958) Design and performance of the Sasumua dam. *Proceedings of the Institution of Civil Engineers* **9**: 369–394.

Thornthwaite CW and Mather JR (1954) The computation of soil moisture in estimating soil tractionability from climatic data. *Climate* **7**: 397–402.

Toll DG (1991) Towards understanding the behaviour of naturally-occurring road construction materials. *Geotechnical and Geological Engineering* **9(3/4)**: 197–217.

Toll DG (2001) Rainfall-induced landslides in Singapore. *Proceedings of the Institution of Civil Engineers: Geotechnical Engineering* **149(4)**: 211–216.

Toll DG (2006) Landslides in Singapore. *Ground Engineering* **39(4)**: 35–36.

Townsend FC, Manke G and Parcher JV (1971) *The influence of sesquioxides on lateritic soil properties.* Highway Research Board, Washington, DC, Highway Research Record No. 374.

Tuncer ER and Lohnes RA (1977) An engineering classification for certain basalt-derived lateritic soils. *Engineering Geology* **2(4)**: 319–339.

US Army Engineer Waterways Experiment Station (1960) The Unified Soil Classification System. *US Army Engineer Waterways Experiment Station, Technical Memorandum No 3-357.*

van Schalkwyk A and Thomas MA (1991) Slope failures associated with the floods of September 1987 and February 1988 in Natal and Kwa-Zulu, Republic of South Africa. *Proceedings of the 3rd International Conference on Tropical and Residual Soils, Maseru, Lesotho. Balkema, Rotterdam, the Netherlands*, pp. 57–64.

Vaughan PR (1985a) Mechanical and hydraulic properties of in situ residual soils: general report. *Proceedings of the 1st International Conference on Geomechanics in Tropical Lateritic and Saprolitic Soils, Brasilia, Brazil*, vol. 3, pp. 231–263.

Vaughan PR (1985b) Pore-water pressures due to infiltration into partly saturated slopes. *Proceedings of the 1st International Conference on Geomechanics of Tropical, Lateritic and Saprolitic Soils, Brasilia, Brazil*, vol. 2, pp. 61–71.

Wesley LD (1973) Some basic engineering properties of halloysite and allophane clays in Java, Indonesia. *Géotechnique* **23(4)**: 471–494.

Wesley LD (1977) Shear strength properties of halloysite and allophane clays in Java, Indonesia. *Géotechnique* **27(2)**: 125–136.

Wesley LD and Irfan TY (1997) Classification of residual soils. In *Mechanics of Residual Soils* (Blight GE (ed)). Balkema, Rotterdam, the Netherlands, pp. 17–29.

Wolle CM and Hachich W (1989) Rain-induced landslides in southeastern Brazil. *Proceedings of the 12th International Conference on Soil Mechanics and Foundation Engineering, Rio de Janeiro, Brazil*, pp. 1639–1644.

Yong RN, Sweere GTH, Sadana ML, Moh ZC and Chiang YC (1982) Composition effect on suction of a residual soil. *Proceedings of the Conference on Engineering and Construction in Tropical and Residual Soils, Honolulu, Hawaii.* American Society of Civil Engineers, New York, NY, pp. 296–313.

Yoshida Y, Kuwano J and Kuwano R (1991) Rain-induced slope failures caused by reduction in soil strength. *Soils and Foundations* **31(4)**: 187–193.

Further reading

Blight GE and Leong EC (eds) (2012) *Mechanics of Residual Soils*, 2nd edn. CRC Press, Boca Raton, FL.

Huat BBK, Toll DG and Prasad A (eds) (2012) *Handbook of Tropical Residual Soil Engineering.* Taylor & Francis, London, UK.

It is recommended this chapter is read in conjunction with

- Chapter 7 *Geotechnical risks and their context for the whole project*
- Chapter 40 *The ground as a hazard*
- Chapter 76 *Pavement foundation design*

All chapters in this book rely on the guidance in Sections 1 *Context* and 2 *Fundamental principles.* A sound knowledge of ground investigation is required for all geotechnical works, as set out in Section 4 *Site investigation.*

doi: 10.1680/icemge.66816.0395

Chapter 31
Glacial soils

Barry Clarke University of Leeds, Leeds, UK

A large volume of geological literature exists on glacial soils, which are common throughout the world's temperate zone. In the UK they account for some 60% of all soils and globally 10%. There is little published information on geotechnical characteristics despite the number of ground investigations. Glacial soils can vary from deformed basal layers that retain many of the original features of those layers to unsorted mixtures of gravel, sands, silts and clays to laminated clays. They can be deposited through a process of pressure and shear beneath a glacier as it advances, or be deposited when the ice melts. This creates a spatially variable soil which contains features that impact on the mass behaviour and lead to wide variation in results from a single source of glacial soil, making the selection of design parameters difficult. Routine sampling may not pick up these features. Tills are deposited in such a way that their characteristics do not necessarily conform to soil mechanics theory and empirical relationships created from studies of gravitationally compacted soils. Hence developing the ground model depends on knowledge of the genetic classification, creating a regional database to enhance data from new investigations, tests on reconstituted tills and a consistent framework to evaluate the mechanical characteristics.

CONTENTS

31.1. Introduction

Glacial soils are a hazard. The modes of deposition result in spatially variable soils, which are difficult to classify and characterise. Inadequate ground investigation and the lack of understanding of the impact of depositional processes on the geotechnical behaviour leads to claims and even failures, and the lack of understanding of how structures impact glacial soils means that working with glacial soils can be a risk. Yet a significant volume of literature exists on Quaternary (glacial) geology (e.g. Benn and Evans, 2014; Bennett and Glasser, 2011; Clarke, 2017; Griffiths and Martin, 2017), and glacial soils are common throughout the world's temperate zone. In the UK they account for some 60% of the soils (**Figure 31.1**) and globally 10%. A more detailed map and GIS database related to features formed during the last glacial event in the UK (the Devensian) are provided by Clark *et al.* (2004). The majority of this literature, which is often contradictory, focuses on geological aspects of glacial till (or till). Although many ground investigations have been carried out in glacial terrains in the UK, only a fraction of the information they revealed is available to the geotechnical profession (Trenter, 1999). Hence, the issues highlighted in **Table 31.1**.

Glacial soils include those that are deposited under water (glaciolacustrine and glaciofluvial soils) resulting in consolidation processes that fit with soil mechanics theory developed for gravitationally compacted or sedimented soils. However, the deposition of the majority of glacial soils (tills) is a complex process as a result of periods of glaciation, pore water regimes and source materials leading to spatially variable, unsorted mixes of boulders, gravel, sands, silt and clay which can include lenses and layers of sand and gravel and laminated clay. Clay-sized particles found in glacial soils can be composed of clay minerals or rock flour. There is evidence that these tills do not conform to classic theories of soil mechanics (Clarke *et al.*, 1997; Hughes *et al.*, 1998), which means the application of empirical and theoretical rules based on observations of sedimented soils is inappropriate.

They can be difficult to sample and the inclusion of gravel and sand lenses can produce a range of stiffness, strength and permeability with depth. Quality sampling is usually restricted to clay matrix-dominant till which is free of cobbles. Any gravel or fissures, which is usual, may affect the sampling, and lead to a wide range of test results for a given property, making it difficult to select design parameters unless there is a clear set of rules to take into account the natural variability and the impact the fabric, composition and the sampling process have upon the derived parameters.

In conclusion, till may have been gravitationally compacted, sheared, possibly reworked and weathered. Till can contain a range of particle sizes from clays to boulders and can vary from clay matrix-dominant till that contains discrete granular particles to clast-dominated tills that contain some fines. Tills can be fissured and laminated. The implication for the construction industry is that till is a challenging material as it is difficult to predict the characteristics of a particular till. This chapter focuses on tills, although reference is made to other glacial soils to help understand the formation of tills.

31.2. Geological processes

The movement of a glacier or ice sheet across the underlying soil or rock either deforms that material creating a deformation till, or erodes it, moves it to a new location and deposits it either as a lodgement or melt-out till or as a water-lain deposit (**Figure 31.2**). These processes create a range of

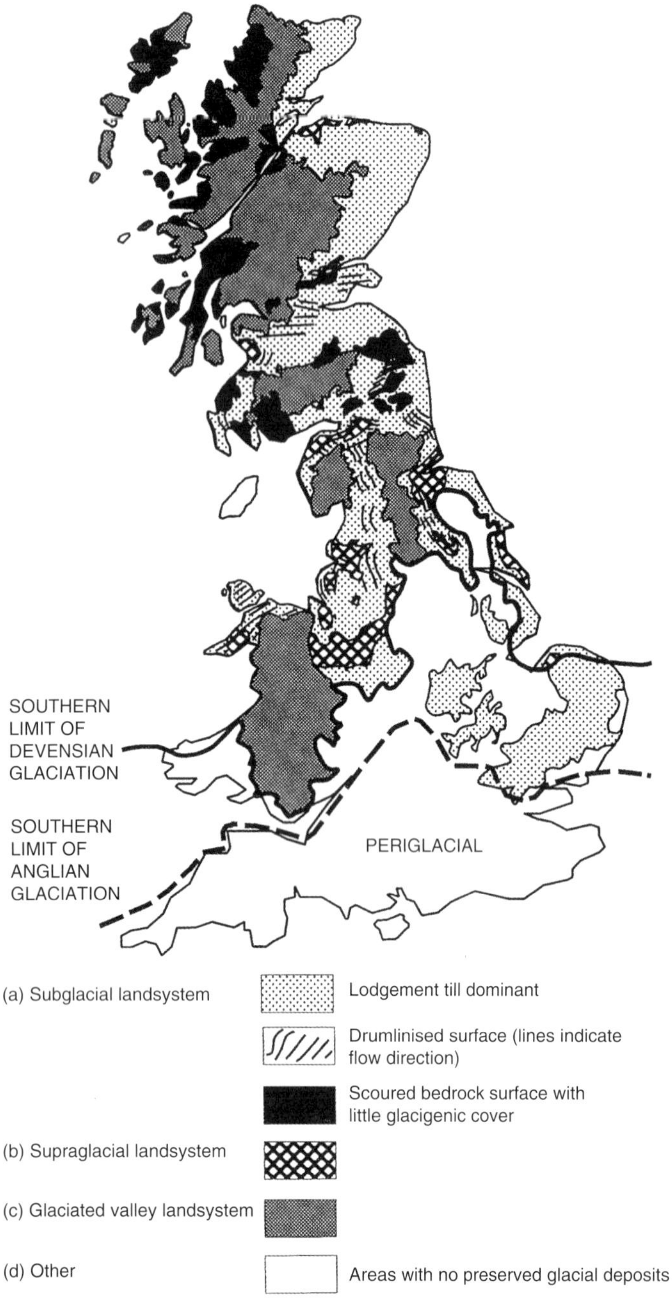

Figure 31.1 Distribution of glacial soils in the UK (Reproduced from Eyles and Dearman, 1981; with kind permission from Springer Science+Business Media)

different debris formations (see **Table 31.2**), which leads to a genetic classification based on those processes (**Table 31.3**). The erosion and transportation processes, for example, result in a breakdown of the material due to crushing, fracturing and abrasion, which may be assisted by freezing of the ground at the base of the glacier. This leads to a homogenised diamicton, which is the geological term used to describe a poorly sorted sediment. If a glacier is moving across soil and that soil is not frozen, then subglacial deformation can occur. Hence it is possible to find glaciotectonic features, such as faults and folds (**Figure 31.3**), in glacial soils. Further deformation results in the folded material breaking up, leading to pockets of original layers known as boudins within the till matrix. Glacial soils produced from ice sheets are all derived from the base of the ice sheet. Glacial soils produced from valley glaciers can include supraglacial debris that falls from the valley sides. Glacial debris is classed as subglacial (base of the glacier), supraglacial (surface of the glacier) or englacial (within the glacier). Supraglacial and englacial materials may not be modified by the movement of the ice; basal materials are modified because of abrasion and crushing. Debris can move between these three zones, depending on the physical processes that take place within the glacier as it moves and as the temperature changes. The debris (**Figure 31.4**) can be spread into place beneath a glacier (e.g. lodgement tills); deposited beneath a glacier as the glacier melts (e.g. melt-out till); deposited within a melt-water lake (e.g. laminated clays); or deposited from melt-water streams (e.g. sands and gravels). Glacial deposits are divided into those that are typically sorted or structured deposits from melt-water, and directly deposited non-sorted deposits. This simple classification can be misleading because directly deposited glacial soils can contain layers, lenses or pockets of stratified water-lain deposits if the glacier has passed over a deformable bed composed of water-lain deposits, which may or may not be a glacial deposit.

It must be noted that glaciers can move considerable distances and advance and re-advance during cold periods and retreat (or melt) during warm periods. It is estimated that the UK experienced up to 16 periods of glaciation during the Pleistocene era (approximately the last two million years) (DoE, 1994). The consequence of this is that glacial soils may have moved several times and been deposited in different forms. This contributes to the complex deposits of spatially variable glacial soils. It also implies that a simple classification of glacial soils may not be possible.

Consider a glacier moving across a rock surface (**Figure 31.5**). The rock is eroded, creating a mixture of boulders, gravels, sands and rock flour. This material may be moved a short distance and be deposited at the base of the glacier. As the glacier continues to advance, abrasion of the subglacial layer continues, leading to an unsorted mix which, when deposited, is known as lodgement till. Some of the subglacial material will migrate into the glacier. This can either be deposited as melt-out till or melt-water deposits as the glacier retreats. The next time the glacier advances it advances over till. Initially the till will deform, the amount of deformation depending on the distance moved, the pressures and the temperature profile. This till now becomes deformation till, using the definition that it is a subglacial material that has not moved very far before being deposited. Therefore, a deformation till could contain traces of previous glacial deposits (**Figures 31.3** and **31.5**). Further movement can increase the amount of abrasion and crushing, leading to a lodgement till. Therefore, it is not

Construction process	Hazard	Description
Investigation	Strength	Some tills are so strong that it is difficult to obtain undisturbed samples. The fabric of till can have a significant effect on the strength determined from routine tests. Empirical correlations developed for stiff clays may not apply to basal tills. Boreholes may not be positioned to pass through zones of weaker materials.
	Permeability	In situ permeability is often greater than that measured in the laboratory because of fissures and laminations.
	Boulders	This could lead to false identification of rock head from borehole records.
	Laminated clay	It will be difficult to identify whether this is a pocket or layer of laminated clay from borehole records.
	Sand and gravel	It will be difficult to identify whether this is a pocket or layer of sand and gravel from borehole records.
Excavation	Strength	Some glacial soils, particularly basal tills, are very dense and hence difficult to excavate; explosives and road picks have been used. Basal tills can contain pockets of more ductile soil which can influence the stability of excavations.
	Boulders	Some glacial soils may contain boulders which can vary in size up to rafts of rock. These could be an issue for excavations for piles, tunnels or diaphragm walls.
	Boulder beds	These can occur between different layers of tills or within melt-out tills and basal till. Their presence will impact on any form of excavation.
	Laminated clay	Deformation till formed from glaciolacustrine deposits can include pockets and lenses or layers of laminated clays. These may be softer than the surrounding till, leading to instability.
	Sand and gravel	Glaciolacustrine deposits and deformation till formed from glaciolacustrine deposits can include pockets and lenses or layers of sands and gravels. These could be water-bearing and, if connected to an aquifer or source of water, lead to flooding.
	Fissures	Many basal tills are fissured as a result of depositional processes. This can lead to block failure and the development of shear zones.
Design	Strength	The in situ undrained shear strength of some basal tills exceeds that measured in the laboratory. This leads to overdesigned foundations.
	Stiffness	The stiffness of some basal tills is such that settlement calculations overpredict the amount of settlement. The stiffness of some tills is such that the pore water response is lower than expected.
	Boulders	These are more of an issue for excavation but the presence of a boulder could lead to an unsafe estimate of performance of a test pile.
	Boulder beds	This could lead to differential movements of a group of end-bearing piles.
	Laminated clay	Pockets of laminated clay could lead to differential movement of a group of piles. Laminated clay beneath the base of piles could lead to excessive settlement. Laminated clay over the pile length could reduce the capacity. Laminated clay layers could trigger slope failures in the longer term.
	Sand and gravel	Pockets of sand and gravel could lead to differential movement of a group of piles.

Table 31.1 Examples of construction problems associated with glacial soils

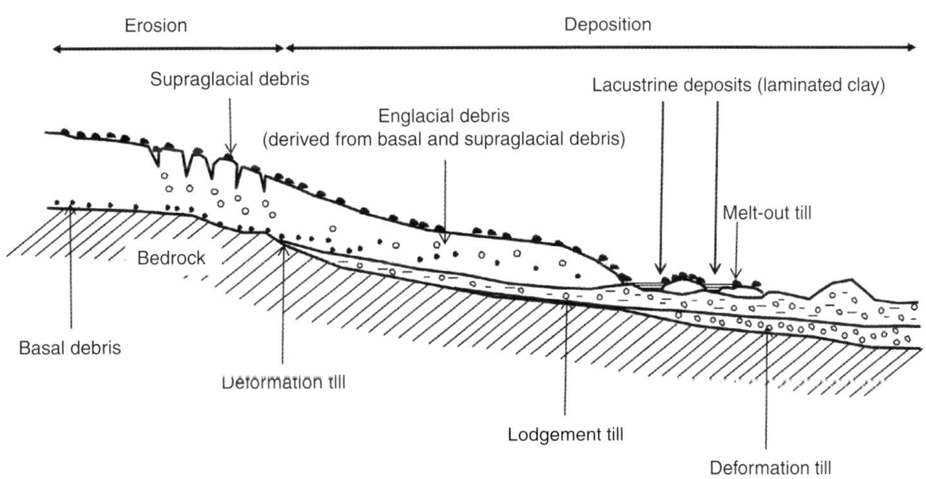

Figure 31.2 The formation of glacial soils (Reproduced from Hambrey, 1994; UCL Press)

Type of debris	Deposition environment	Type of glacial soil	Description
Supraglacial			Debris that is carried on the surface of a glacier; derived from rock falls and, possibly, from the glacier bed
Englacial			Debris that is found within a glacier either derived from supraglacial or subglacial material
Subglacial			Debris that is deposited from the base of a glacier
	Glacioterrestial		Basal deposits on land
		Melt-out till	Deposited by melting of stagnant or slowly moving debris-rich glacier ice without subsequent transport or deformation
		Flow till	Deposits that flow from the margin of a glacier
		Lodgement till	Deposited by spreading under pressure glacial debris from a sliding glacier bed
		Deformation till	Sediment which has been disaggregated and (usually) homogenised by shearing in the subglacial deformed layer
	Glaciofluvial		Sand and gravel deposited from water from a melting glacier
		Ice marginal	Adjacent to the glacier
		Pro-glacial	Deposited in front of or just beyond the outer limits of a glacier or ice sheet
		Subglacial	Deposited beneath the glacier
	Glaciolacustrine		Laminated clays found in ice-dammed lakes
		Ice contact	Deposited adjacent to the ice margin
		Distal	Deposited some distance from the glacier
	Glaciomarine		Glacial deposits found within a marine environment
		Fjord	Deposited within a fjord
		Continental	Deposited on the continental shelf
		Water-lain	Deposited by water
		Proximal	Deposited near to the glacier
		Distal	Deposited some distance from the glacier

Table 31.2 Key descriptions of glacial materials

unexpected to find lenses of water-lain deposits of laminated clays and sands and gravels within what appears to be a till. It explains why rock particles can be found many miles from their original source and still be intact and why rock particles from one source can be widely distributed. It also shows why it is difficult to be certain that an unsorted soil of glacial origin can be classified as a lodgement till.

There is a genetic classification of tills (**Figure 31.4** and **Table 31.3**) based on those produced after one cycle of advance and retreat of a glacier. This classification may not apply in practice because of the cycles of advance and retreat of glaciers leading to glacial deposits being transported and deposited several times. In addition to the glacial action, post-glacial processes can also alter the properties of glacial soils both between movements of a glacier and since the last glaciation. This adds a further complication when interpreting glacial soil profiles. For example, in northern England it is not uncommon to find three layers of glacial soil: an upper and lower red till separated by a thin layer of sand and gravel or laminated clay from a lower grey till (**Figure 31.6**). It is commonly stated that this succession was originally deposited as a grey lodgement till by a single ice sheet advance, and that the reddish colour of the upper till is solely due to post-glacial weathering (Eyles and Sladen, 1981; Lunn, 1995). There are, however, some aspects of the composition of the glacial succession that do not seem to be adequately explained by this single deposit and weathering concept. These include the red (upper) till generally having a much lower stone content (gravel, cobbles, boulders) than the grey (lower) till (Beaumont, 1968) and the frequency and extent of the sand/gravel laminated clay layers, which in some cases are well over a square kilometre in area, are very probably pro-glacial (glaciolacustrine/glaciofluvial) deposits and therefore indicate that the ice front was receding and releasing debris-laden melt-water before the area was again overlain by later ice and ice-transported deposits (Hughes *et al.*, 1998). This demonstrates the difficulty of producing a simple ground model from a complex deposition process which is not fully understood;

Criterion	Lodgement till	Melt-out till	Flow till	Deformation till
Deposition	Deposited by plastering of glacial debris from the sliding base of a moving glacier, by pressure melting and/or other mechanical processes (Hambrey, 1994).	Deposited by a slow release of glacial debris from ice, neither sliding nor deforming internally (Dreimanis, 1988).	Deposition accomplished by gravitational slope processes and may occur supraglacially, subglacially or at the ice margin (Dreimanis, 1988).	Comprises rock or unconsolidated sediment detached by the glacier from its source; primary sedimentary structures distorted or destroyed and some foreign material admixed (Elson, 1988).
Position and sequence	Lodged over older glacial sediments or on bedrock.	Usually deposited during glacial retreat.	Most commonly the uppermost glacigenic deposit.	Formed and deposited subglacially, often where the glacier moves upslope.
Basal contact	Lodgement and melt-out tills formed and deposited at glacial base. Contact with the substratum (bedrock or unconsolidated sediments) generally erosional and sharp. Glacial erosion marks and clast alignment have same orientation. Supraglacial melt-out tills may have variable basal contact.		Variable basal contact but seldom conformable over longer distances. Tills may fill shallow channels or depressions.	Variable basal contact.
Landforms	Mainly ground moraines, drumlins, flutes and other subglacial landforms.	Those ice-marginal landforms where glacier ice stagnated.	Associated with most ice-marginal landforms.	Landforms rarely diagnostic.
Thickness	Typically one to a few metres thick but may attain substantial thickness in the English lowlands; relative lateral consistency.	Single units usually a few centimetres to a few metres thick. Units may stack too much greater accumulated thickness.	Very variable. Individual flows usually a few tens of centimetres to metres thick. Units may stack to accumulated thickness of many metres.	Varies up to many metres depending upon the nature of the glacier bed.
Structure	Usually massive but may contain various consistently orientated macro- and micro-structures. Sub-horizontal jointing is common and vertical and transverse joints may also be present. Orientation of deformation structures related to stress applied by moving glacier and may be laterally consistent.	Either massive, or with faint structures partially preserved from debris stratification in basal debris-rich ice. Loss of volume with melting leads to draping of sorted sediments over large clasts.	Either massive or displaying various flow structures depending on type of flow and water content.	Primary structure may be preserved but usually deformed, especially in upper part of the sequence which may blend into other massive tills.
Grain size composition	Abrasion in traction zone during lodgement produces silt-size particles typical of lodgement tills. Most have relatively consistent grain size composition except for the basal part which may contain boulders of local glacier bed.	Winnowing of silt and clay-size particles occurs during melt-out. Some particle size variability inherited from debris bands in ice. Supraglacial melt-out tills of valley glaciers contain characteristic coarse-grained debris.	Usually diamicton with polymodal particle size distribution. Some particle size redistribution and sorting may occur during flow. Inverse or normal grading may develop.	Deformation tills derived from weak rocks contain clasts separated by minor amounts of finer matrix. Clast size reflects bedding thickness of original material.
Lithology of clasts and matrix	Lithological composition often more consistent than other tills. Composition of matrix particularly uniform. Materials of local derivation increase in abundance towards basal contact.	Supraglacial melt-out till more variable in composition with increased possibility of exotic material.	Lithological composition generally same as source material. May include incorporated glacier bed or exotic materials depending on debris source, transport and deposition.	Deformation tills generally have same lithological composition as underlying sediments. Occasional erratics present, particularly in upper part of the sequence.

Table 31.3 Characteristics of genetic till types (from Dreimanis, 1988; Elson, 1988; Hambrey, 1994; Trenter, 1999) (continued on next page)

Criterion	Lodgement till	Melt-out till	Flow till	Deformation till
Clast shapes and their surface marks	Subangular to subrounded clasts. Bullet-shaped, faceted, crushed, sheared and streaked-out clasts more common in lodgement than other tills. Lodged clasts striated parallel to direction of the lodging movement.	Variable degree of roundness but angular clasts occur where supraglacial melt-out debris is englacially or supraglacially derived.	If present, soft sediment clasts may be rounded or deformed by shear. More resistant rock clasts will retain their original shape.	Clast shape and surface marks generally inherited from original material and not diagnostic. Clasts generally transported passively and not significantly modified.
Fabric	Strong macro-fabric with clast long axes parallel to local direction of movement. Transverse orientation possible, associated with folding and shearing.	Fabric inherited from glacier transport. Melt-out process may weaken fabric, particularly micro-fabric.	Fabric may be random or strongly developed and parallel or transverse to flow direction. Fabric may vary laterally over short distances.	Preferred orientation rare and generally reflects shearing deformation.
Consolidation	Consolidation process depends on pore pressure regime, temperature profile and permeability of glacier bed. Tills can be 'lightly overconsolidated' to 'heavily overconsolidated'.	Lightly overconsolidated.	Usually normally consolidated.	Variably consolidated.
Density	Very dense often in excess of gravitationally compacted stiff clays due to combination of normal and shear stress during deposition.	Bulk density lower and more variable than lodgement till.	Density lower than in lodgement tills and typical of normally consolidated deposits.	Spatially variable densities due to presence of lower-density pockets of material and dilatency during deformation.
Strength	Very strong due to high densities. Strengths can approach those of weak rock in some cases.	Strength lower and more variable than lodgement till.	Strengths typical of normally consolidated deposits.	Spatially variable strength due to presence of lower-density pockets of material and dilatency during deformation.
Permeability	Highly impermeable if a clay matrix-dominant till.	Permeability variable due to mix of particle sizes.	Relatively permeable compared to clay matrix-dominant lodgement tills.	Spatially variable strength due to presence of lower-density pockets of material and dilatency during deformation.
Relevant references for summaries of diagnostic properties	Goldthwait (1971), Boulton (1976b), Dreimanis (1976), Boulton and Deynoux (1981), McGown and Derbyshire (1977), Eyles *et al.* (1982), Clarke and Chen (1997), Clarke *et al.* (1998, 2008)	Boulton (1976b), Dreimanis (1976), McGown and Derbyshire (1977), Lawson (1979), Boulton and Deynoux (1981), Shaw (1985)	Boulton (1976b), Lawson (1979, 1982), Boulton and Deynoux (1981), Lutenegger *et al.* (1983), Gravenor *et al.* (1984), Rappol (1985), Drewry (1986)	Elson (1988), Boulton (1979), Clarke *et al.* (2008)

Table 31.3 Continued

although a number of geological and hydrogeological domains are distinguishable based on field characteristics (e.g. see McMillan *et al.*, 2000 for further details). The consequences of a failure to understand the ground model can be dramatic, as highlighted in **Table 31.1**. The sample descriptions leading to the soil classification do not necessarily describe the macro- and micro-behaviour of the soil mass. Laminated lenses or layers may trigger failures in excavated slopes; lenses of sand and gravels can lead to local collapses in excavations and, importantly, in open hole excavations; layers of water-bearing sand and gravel can be found in excavations; boulders can impede piling and tunnelling operations; failure to correctly identify rock head can affect the design and construction of end-bearing piles and tunnels along the rock head elevation.

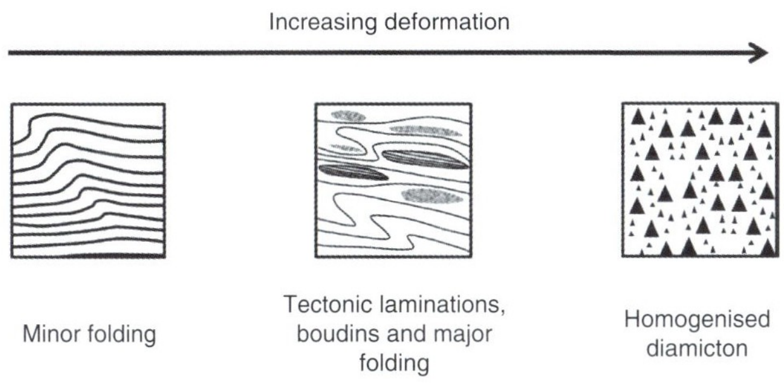

Figure 31.3 Glaciotectonic features found in deformation tills

31.3. Features of glacial soils

A genetic classification of glacial soils can be based on the depositional process (**Figure 31.4**). The source of glacial material controls the lithology of the particles and, possibly, the grain size distribution and particle morphology (e.g. Benn and Evans, 2014). The mode of transport effects the abrasion process and hence the morphology, grain size distribution,

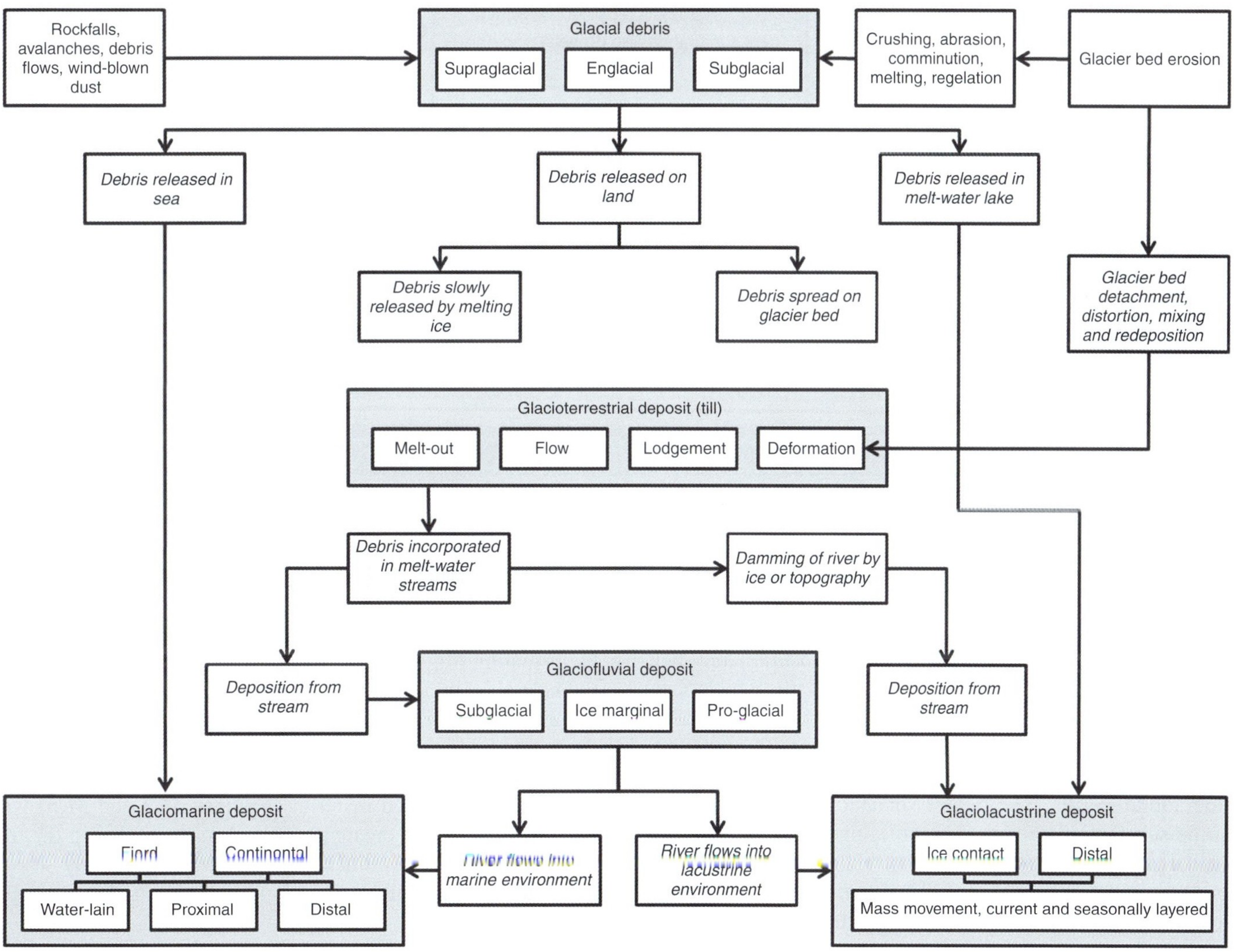

Figure 31.4 The transport and deposition processes for glacial soils leading to a genetic classification. Table 31.2 lists definitions of the key terms (Reproduced with permission from CIRIA C504, Trenter, 1999; www.ciria.org)

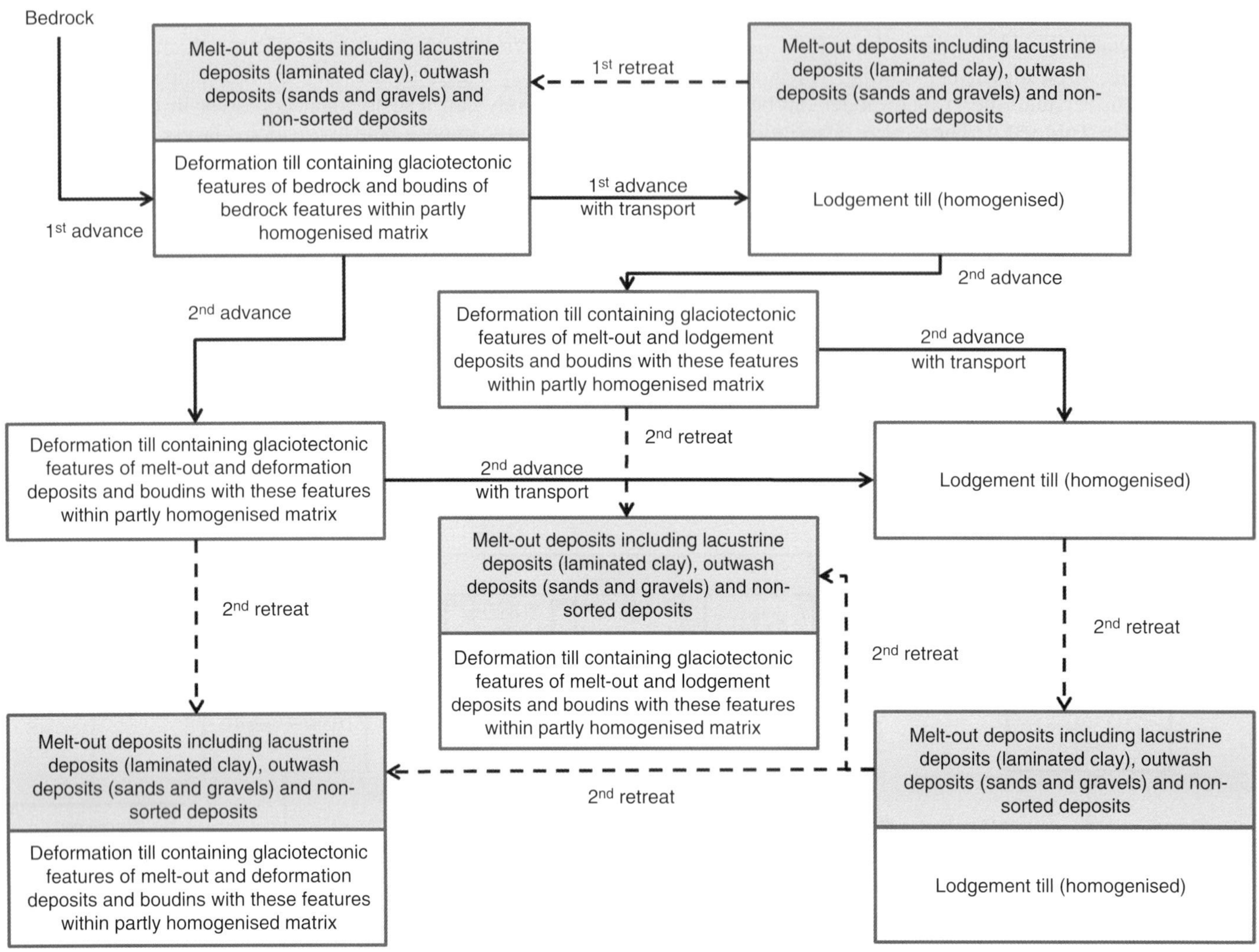

Figure 31.5 The processes at the base of a glacier showing how reworking of a glacial landscape can incorporate features of glacial soils in subsequent glacial deposits

fabric and structure. The depositional process affects the geological and geotechnical profile. Lawson (1981) defined till as a sediment deposited directly by glaciers, which is a primary deposit including lodgement till, deformation till, communition till (formed by subglacial grinding or crushing) and melt-out till. There are several types of failures of tills which occur because of their fabric (**Table 31.1**). In this case fabric refers to depositional and post-depositional processes (Trenter, 1999) which create the clasts, layers, lenses, fissures, cracks and joints found in tills. The risks associated with these features can be substantially reduced by building up a ground model of local glacial geology from other investigations and observations of excavations. Fookes (1997) describes the total ground model, which is based on the principle that you start by creating a ground model of the site from known regional and local information, and that preliminary model drives the design of the ground investigation, which is designed to confirm the model, fill in the gaps in the model and produce data to characterise the ground and produce design parameters.

31.3.1 Deformation till

Deformation till is formed of soil particles, which may or may not be glacial in origin. Ice moving across this soil will start to deform the soil, initially leading to glaciotectonic features of faults and folds, sometimes separately referred to as glaciotectonite. As the ice continues to move, these features will be further distorted. Folded material (**Figures 31.3** and **31.5**) will become detached to create pockets of material within the till. These, known as boudins, will become increasingly drawn out, forming tectonic laminations and eventually homogeneous till.

Deformation tills that have been deposited by a glacier passing over rock head will contain intact rock fragments, which can be extensive, leading to the possibility that rock head may

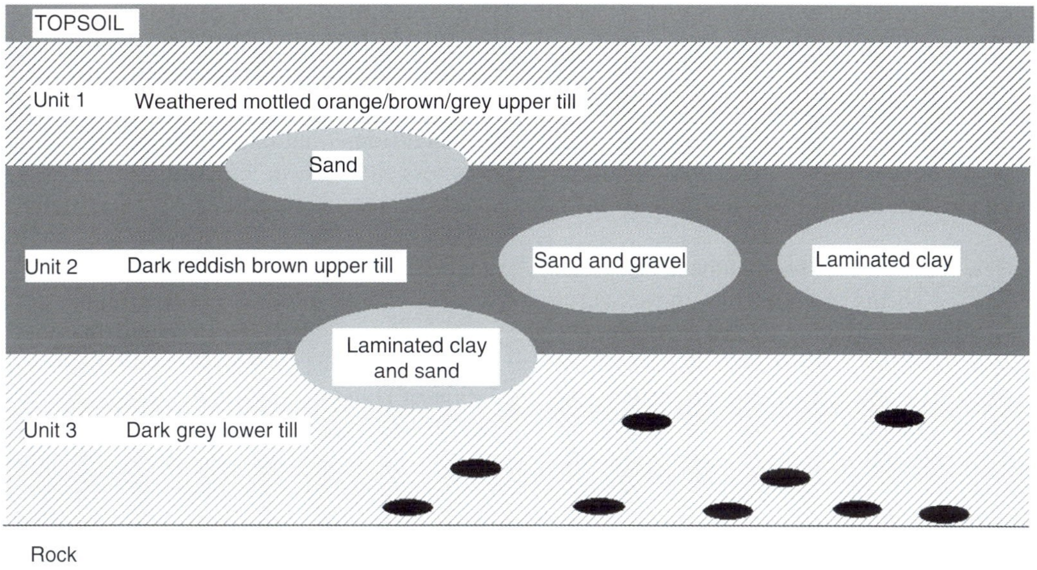

Figure 31.6 A tripartite succession found in the north of England showing three units of till separated by thin layers of laminated clays and sands and gravels (Reproduced from Clarke *et al.*, 2008)

be incorrectly identified. The lateral extent of rock encountered in a borehole is unknown, but observations from excavations indicate that the pieces of rock can vary in size from boulders to extensive rafts of rock. Therefore, it is important to prove rock head. These boulders and rafts of rock can present problems during excavation. It is very difficult to indicate how many boulders will be present from borehole records. It is prudent to assume that till contains boulders even if none are encountered in the boreholes.

Deformation tills that have been deposited by a glacier passing over glacial soils will contain remnants of those soils often within a matrix of gravel, sands, silts, clay and rock flour. These could include intact rock fragments, lodgement till, melt-water tills including laminated clays, and lenses of sands and gravels. It may be difficult to identify whether they are isolated pockets of these materials or layers of materials. It is prudent to assume that glacial soils can contain pockets and lenses of more ductile material. Laminated clay pockets could lead to zones of weakness and potential failure, especially in excavations (see **Table 31.1**). Sand and gravel lenses could be water-bearing and, if interconnected, lead to potential flooding of excavations. These features are common, suggesting many tills are deformation tills derived from melt-out, deformation and lodgement tills.

Many basal tills (or subglacial tills) are deformation tills (e.g. Bennett and Glasser, 2011). Indeed, the deformation process could be responsible for many of the hazards listed in **Table 31.1**. The fact that these tills are created from underlying soil, which may have been deposited by a previous glaciation, means that the composition of the till is the same as the underlying soil. This may make it difficult to identify the base of a deformation till. The till could contain glaciotectonic features, rafts of intact soil or rock that have not been subject to abrasion and a range of particle sizes and pockets of intact, possibly distorted, material such as laminated clay, as well as sands and gravels.

Thus, samples may not be representative of the soil mass, leading to uncertainty over the 3D ground model. There could be unidentified zones of weakness (e.g. laminated clay lenses) or misinterpreted zones of weakness. For example, failures in excavations in till have been triggered by lenses of laminated clay – especially where they dip into the excavation. There could be water-bearing layers. For example, a piling contractor took the view that the soil description referred to a clay and therefore open boreholes could be used. Unfortunately, one borehole encountered a sand layer which proved to be continuous across the site and possibly connected to a nearby river. This led to all boreholes being fully cased. Proving rock head is critical, especially when considering end-bearing piles or tunnelling operations.

31.3.2 Lodgement till

This is a subglacial or basal till, which is deposited as ice advances because the frictional resistance between the till and the bed is greater than the shear introduced by the moving ice. There are two theories of deposition: one related to effective stress at the bed of the glacier (Boulton, 1974) and the other (Hallett, 1979) in which the contact pressure between the glacial soil and bed is independent of the effective stress. In the effective-stress model, the stress is created by the weight of ice and the water pressure. The stress will depend on the temperature at the base of the glacier and the hydrological characteristics of the bed. The effective stress will increase as the permeability of the bed below a warm-based glacier increases

(**Figure 31.7**). This means abrasion will increase if the debris continues to move. However, friction between the debris and the bed will also increase, which could lead to the deposition of the debris due to lodgement. If the bed is less permeable, then the effective stress will not be as great. If the bed of the glacier is frozen, the effective and total stress will be similar. The debris can be deposited at any time, depending on the friction.

The alternative view (Hallett, 1979) is that the debris is carried by the ice. Abrasion occurs as the debris moves across the bed. The debris will be deposited as the ice slows down or melts. Both theories are correct at some time as they depend on the depth of ice, the speed of ice movement and the temperature profile (**Figure 31.7**).

Lodgement tills are generally a very dense mixture of sands, silts and clays or rock flour and could be considered a relatively uniform deposit. Often, they are a clay matrix till containing gravel, which typically has rounded edges, striated faces and is aligned with the direction of flow. The particle size distribution can be uniform or gap graded. Shear planes are evident.

Lodgement tills are difficult to sample because of their density and presence of gravel. Thick-walled sampling tubes are commonly used, which can lead to sample disturbance. For the same reasons, it is difficult to subsample these tills, hence strength tests are usually based on 100 mm dia. samples. In situ strength can exceed laboratory-measured strengths. This has affected tunnelling operations because the assumed strength was less than the in situ strength. These tills are described as clays and therefore there is an assumption that they have an undrained shear strength. Designs based on that strength are often conservative because the mobilised strength is much greater. These tills have been highly remoulded and can include shear planes, which means that their expected behaviour based on classic soil mechanics theories and the interpretation of standard laboratory tests is not observed. Atkinson (2013) suggests that the strength obtained from samples of clay reconstituted to the correct stress and water content may be more appropriate for design. Tills were not gravitationally compacted when deposited. Therefore, it is better to test samples that have been reconstituted to the in situ density. Tills are denser, stronger and stiffer in situ than predicted from routine site investigations. This does lead to conservative designs, but given the uncertainties over the 3D ground model from a site investigation this is acceptable unless use can be made of regional knowledge.

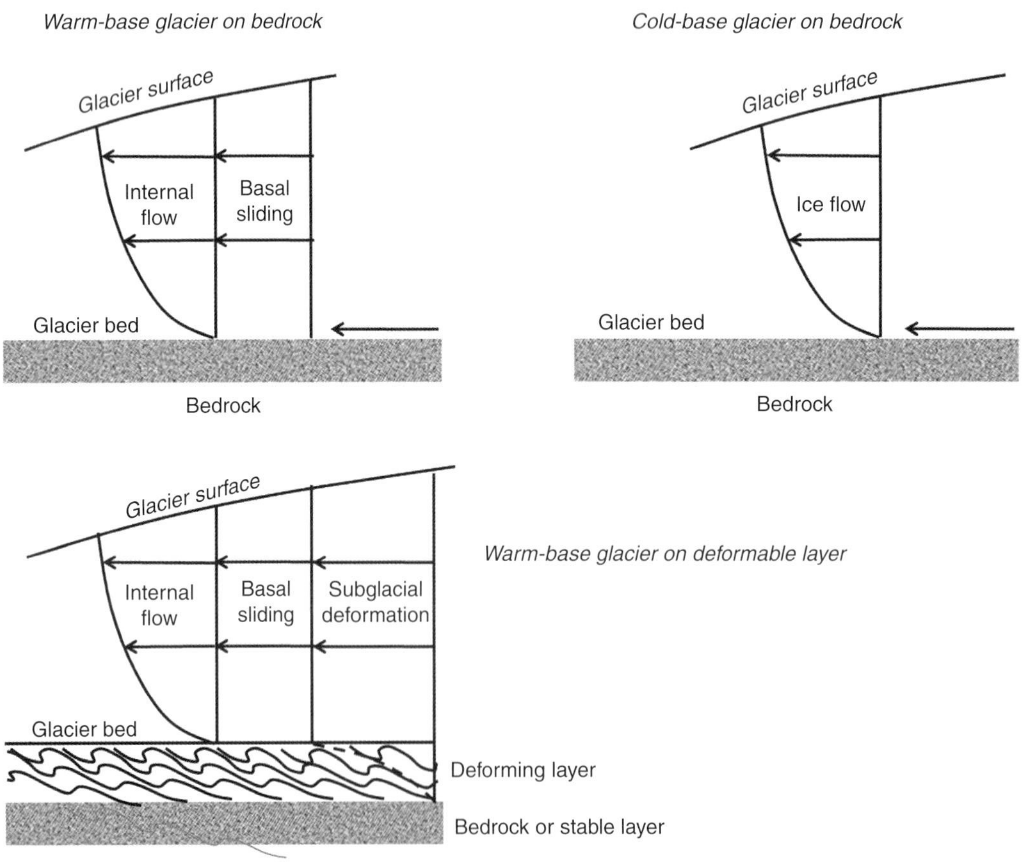

Figure 31.7 The relationship between the temperature profile and type of bed and the effect that they can have on the depositional process. The contours within the ice mass refer to the relative deformation of the ice (Reproduced from Boulton, 1976a; The Royal Society of Canada)

31.3.3 Subglacial melt-out till

This till comprises debris deposited as a glacier melts primarily due to geothermal heat. There may also be sediment flow features. A key difference between this till and lodgement till is that this till is likely to be less dense as the till was lowered into place rather than spread into place. These tills are characterised by their variability, leading to difficulty in selecting characteristic design parameters.

31.3.4 Supraglacial melt-out till

This till is formed as the surface of a glacier melts. It is typically a loose, coarse, angular deposit that has been reworked because of movements of debris associated with flow and topography. As the ice melts releasing the supraglacial and englacial debris, the debris can insulate the underlying ice leading to an irregular topography. In this case, the debris will eventually be deposited as an ablation till creating a landform of moraines (mounds or ridges) of various types, including lateral moraines formed at the margin of the ice, hummocky moraines deposited as the ice retreats and end moraines formed at the front of a glacier. If the debris layer is thin, then the ice melts uniformly, leaving a thin layer of coarse material. These layers could include outwash sediments. The total extent of supraglacial till is significantly less than that of subglacial tills because it is deposited only when the ice melts.

31.3.5 Macro features of tills

The periods of glaciation sometimes lead to layers of different tills separated by a thin layer of sands and gravels or laminated clay. The upper tills are variously described as a distinctly different lodgement till to the lower layer or a weathered version of the lower layer. They are often of a different colour, supporting the weathering hypothesis, although the alternative is that the source of the till is different.

Deformation and lodgement tills often contain shear planes (e.g. Boulton, 1970) with associated conjugate shears because of the shearing that takes place as the tills are moved. These planes can lead to local failures and block failures in excavations. Both deformation and lodgement tills are very dense and brittle. It is difficult to recreate this density in soil mechanics laboratories due to the shearing process linked to the pressure of the weight of ice. The shearing process also means that these tills are highly remoulded, implying there is little loss of strength on excavation. It is not unusual to find natural slopes in tills of 45° (Vaughan, 1994) even though the laboratory tests indicate lower strengths and hence lower slopes. The pockets and lenses within a till are likely to be less dense and more ductile.

Laminated clays, a feature of glacial lakes, are generally composed of thin alternative layers of clay and silt, creating a highly anisotropic material in terms of stiffness, strength and permeability. Layers of laminated clays are often found in basal tills. It is not uncommon for failures to occur in excavations of these basal tills if laminated clays are present.

31.4. Geotechnical classification

As with all soils the geological processes govern the geotechnical properties. Water-lain glacial soils produced at the end of the glacial period and not subject to further glacial erosion will behave in a similar manner to other water-lain soils. However, the process of deposition may create an anisotropic structure (e.g. laminated clay) or a spatially variable structure (e.g. outwash deposits primarily of coarser materials). Supraglacial tills are gravitationally deposited but not through water. Therefore, they could have similar properties, but not features, to water-lain deposits.

Subglacial or basal tills (i.e. deformation and lodgement) are either deformed or 'smeared' into place under great pressure due to the weight of ice. Basal tills are the predominant till types. For example, Eyles (1979) suggested that Icelandic glaciers produce between 1% and 7% of supraglacial till compared to basal till. This may take place under frozen or unfrozen conditions. In either case it should result in a very dense material. This density is not just a result of pressure, but of a combination of pressure and shear. Basal tills include those that are essentially deformed bed layers, which are likely to contain glaciotectonic features such as faults and folds with pre-existing features being readily identified (see **Figure 31.3**). These dense tills can contain pockets of less dense material. These pockets of material retain their density because the till was frozen or unfrozen. If unfrozen the density of the surrounding till was such that excess pore waters could not dissipate. This can also occur because the structural form of the inclusion meant that the surrounding till provided a structural seal to the pocket of material. These pockets of less dense material, often remnants of melt-out tills, can be a hazard (see **Table 31.1**).

Subglacial tills may be classed as lodgement tills based on their fabric, but in fact because of the periods of glaciation many could be considered deformation tills: that is, tills that retain much of the original features of the bed, having been transported a short distance with little abrasion. Therefore, the genetic classification may not be the best indicator of geotechnical properties. Meer *et al.* (2003) came to the conclusion that there is only one type of basal till, deformation till, since all basal tills contain deformation features.

It is possible to create a classification of glacial soils based on their geotechnical performance. This classification combines the genetic classification with observations of glacial soils to highlight the features that will impact on the interpretation of ground investigation and the behaviour of the tills during construction and in service. McGown and Derbyshire (1977) suggested that it is more appropriate to use a classification that is based on superimposing the dominant soil fraction on the genetic scheme (**Table 31.4**). The classification suggests that the dominant soil fraction and the type of till govern the relative value of the engineering properties. This qualitative assessment shows that the density and compressibility decrease as the dominant soil fraction size increases – that

Till	Dominant soil fraction	Fabric features	Relative scales (1 (low) to 9 (high)) Density	Compressibility	Permeability	Anisotropy
Lodgement till	G	MACRO: Interlayering of glaciofluvials, joints, fissures, contortions. Consistent preferred clast orientation.	4–7	1	5–6	7
	W	MESO: Fissuring. Contortion. Moderate to very high consistency of preferred orientation of clasts.	5–8	2	2–3	
	Mg		6–8	2	4–5	
	Mc	MICRO: Moderate to high degree of parallelism of fines in sympathy with clast surfaces.	6–8	3	2	
Melt-out till	G	MACRO: Occasional interlayering with glaciofluvials. Clast preferred orientation often retained from englacial state.	2–4	2–4	7–9	3–5
	W	MESO: Moderate to high preservation of preferred clast orientation from englacial state, especially in subglacial type.	2–6	3–5	4–5	
	Mg		2–6	3–6	5–8	
	Mc	MICRO: Open to moderately closed arrangements of fines with many englacial arrangements retained, especially in subglacial type.	2–7	4–7	3–4	
Flow till	G	MACRO: Interlayering with glaciofluvials common. Segregation, contortions, layering and fissuring in upper section and nose of flow.	3	2	7	7
	W	MESO: Aligned low-angle orientation of clasts conforming to flow direction rather than ice direction.	4	2–4	4	
	Mg		5	2–4	6	
	Mc	MICRO: Rather compact parallel arrangement of fines related to flow rather than direction of ice movement.	5	2–5	3	
Deformation till	G	TOTAL FABRIC: Deformed bedrock and soil (including previous glacial deposits) structures related to ice movement direction.	5–8	1–3	2–8	3–7
	W					
	Mg					
	Mc					

Table 31.4 Characteristics and geotechnical properties of glacial tills (from McGown and Derbyshire, 1977)

is, a clay matrix lodgement till is likely to be the densest and stiffest of the tills. This is not surprising given the amount of shearing and abrasion that takes place during transportation. A feature of all these tills is the fact they can be lightly or normally consolidated despite the pressures involved during deposition and the relief of those pressures due to ice melt. The link between consolidation theory and the mechanical characteristics of soils developed for gravitationally sedimented soils does not apply to tills. This means that very strong, stiff glacial soils are not necessarily heavily overconsolidated. In fact, the behaviour of till often depends on the relationship between the fabric and the geotechnical structure, with the fabric having a significant impact on the performance of the geotechnical structure (**Table 31.1**).

Consider an element of basal till. The total vertical stress on the element is equal to the pressure of the ice plus the pressure due to the overlying till. When the glacier retreats (melts) the total stress reduces due to the reduction in pressure of the ice. The pore water regime will depend on whether the basal till was frozen or unfrozen and the hydrological characteristics of the bed of the glacier. **Table 31.5** shows the possible pore water regimes. If the bed is frozen, then the pore water pressure is zero. As the glacier melts, pore water pressure will be generated due to the weight of till as the bed melts. The

Bed of glacier	Base of till	Underlying soil/rock	Pore pressure	Degree of overconsolidation
Frozen	Frozen*	Aquifer	No excess	Heavily
Frozen	Frozen*	No aquifer	Excess	Normal
Unfrozen	Frozen*	Aquifer		Heavily
Unfrozen	Frozen*	No aquifer		Normal
Unfrozen	Unfrozen	Aquifer	No excess	Heavily
Unfrozen	Unfrozen	No aquifer	Excess	Normal

*pore water pressures generated as ice melts
NB Total stress due to weight of ice for each regime

Table 31.5 Pore water pressure regimes within till during deposition total stress reduces as the weight of ice reduces. Therefore, the degree of consolidation is low. In the case of a bed that is not frozen and the bed acts as an aquifer, the degree of consolidation is high because the effective stress reduces due to the weight of ice. It is possible, therefore, for a till to be 'normally consolidated' or 'heavily overconsolidated' depending upon the pore water regime and underlying geological conditions.

The shearing, abrasion and crushing processes further complicate the stress regime. The concept of overconsolidation ratio (i.e. the ratio of the maximum past effective stress to the current effective stress) applied to gravitationally compacted sediments can be applied to tills, but since the maximum stress, related to the thickness of ice, is unknown and the deposition process introduces shear, the concept of the degree of consolidation as an indication of soil properties is not valid and therefore characteristics cannot be deduced from stress history. Indeed, density and the dominant matrix may be a better indication of likely behaviour.

31.5. Geotechnical properties

31.5.1 Classification data

It is important with glacial soils to link the geological characteristics with the geotechnical classification to ensure that the soils are correctly identified, and the appropriate properties assigned. For example, a liquid and plastic limit test on laminated clay formed of clay layers separated by silt layers will produce very different results from tests on the two layers forming the laminated clay. The test may suggest a silty clay when in fact it is a laminated clay. Indeed, in this extreme case liquid and plastic limit tests may not be appropriate. Linking empirical relationships developed for gravitationally compacted soils between liquid and plastic limit test results and mechanical properties is not advisable for tills.

Boulton and Paul (1976) showed that the plasticity indices for tills cluster around the T-line, a line parallel to the Casagrande A-line. This may be because the finer particles in tills comprise rock flour rather than clay minerals. This applies to many tills (**Figure 31.8**), which shows that tills vary from CL to CI in general. The limit tests are only carried out on the finer particles and therefore are only relevant to clay matrix-dominated tills. Therefore, one technique to identify tills is to assess the properties of the matrix. A clay matrix-dominant till is defined as a well-graded till with a fines content of 15–45% (McGown and Derbyshire, 1977). Barnes (1988) and Winter *et al.* (1998) showed that the matrix dry density reduced markedly when the gravel content exceeded 40–50%. At that point a glacial soil is considered a granular soil and has the characteristics of such soils. Gens and Hight (1979) reported that tests on reconstituted soils are unaffected by the gravel content if the gravel content is less than 15%. Thus, the grading curve is a useful indicator of characteristic behaviour.

Limit tests can be used to assess whether a till is weathered, since the weathering process increases the clay content due to the formation of clay minerals. It also increases the water content. Limit tests on weathered till still cluster around the T-line. Therefore, it would be expected to find the limit test results to migrate down the T-line with depth because the amount of weathering reduces.

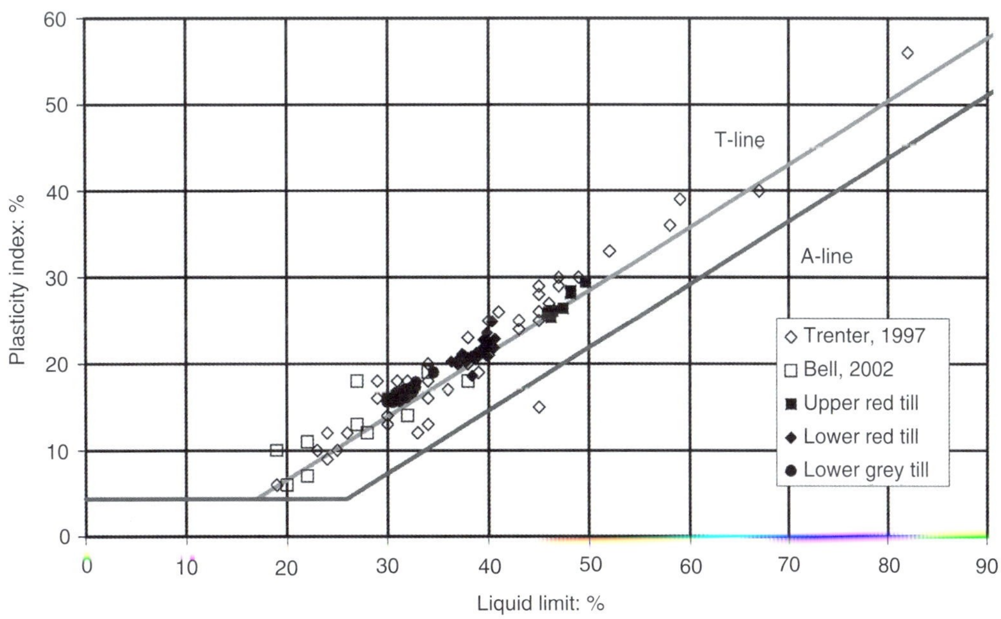

Figure 31.8 Consistency limits for UK matrix-dominated tills and clay minerals (after Trenter, 1999 and Clarke *et al.*, 2008) showing the T-line (Boulton and Paul, 1976)

31.5.2 Particle size distribution

The wearing down of particles due to abrasion and crushing during the transportation process results in a broad range of particle sizes. The percentage of fine particles increases with distance from source. A feature of lodgement tills is that the distribution of particle sizes has fractal characteristics – that is, the numbers of grains at each size when plotted on a double logarithmic plot are linearly related (**Figure 31.9**). Theoretically the slope of the line would be 2.58 (Sammis *et al.*, 1987) if the particles were produced solely by fracturing. In practice the slope is greater (Hooke and Iverson, 1995).

The particle size distribution in glacial soils can vary from pure clay to pure gravel. However, the majority of lodgement and deformation tills are clay matrix-dominant tills which sit between a silty clay (e.g. London Clay) and a sandy gravel (e.g. Thames Gravel) in terms of their grading (**Figure 31.10**). Note that boulders will not have been included in the grading curves, so the curves are not necessarily a true representation of the soil. The chart distinguishes between fine-grained soils and sands, gravels and cobbles. Fine-grained soils include clays, silts and rock flour, the rock flour being derived from the abrasive processes during deposition. Rock flour is clay and silt size but is neither clay nor silt; it is formed of particles of rock.

Clay-dominant subglacial tills are generally well graded. **Figure 31.11** shows a ternary diagram for the percentages of fines, sands and gravel (**Figure 31.11(a)**) and clay, silt and coarse particles (**Figure 31.11(b)**). Note that cobbles are removed either in the sampling process or in preparing the specimen. The gradings confirm that the tills are dominated by fine-grained particles. The averages show that the tills are distinctly different, as can be seen from **Table 31.3**.

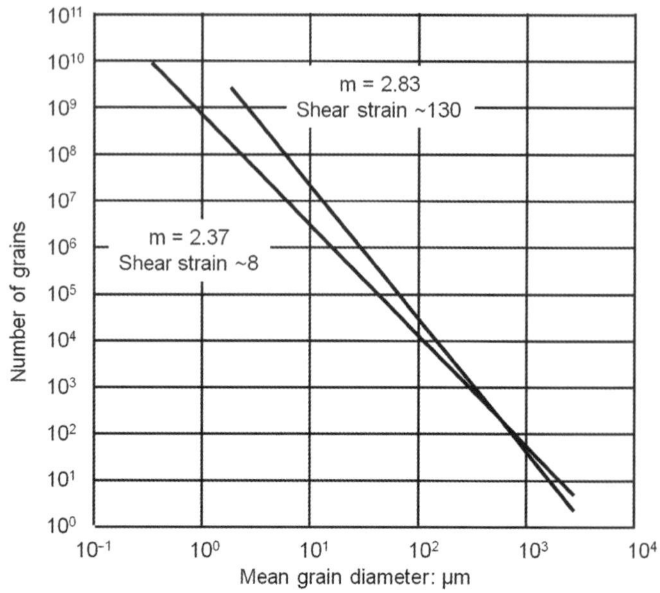

Figure 31.9 The fractal characteristic of lodgement tills (Reproduced from Iverson *et al.*, 1996; International Glaciology Society)

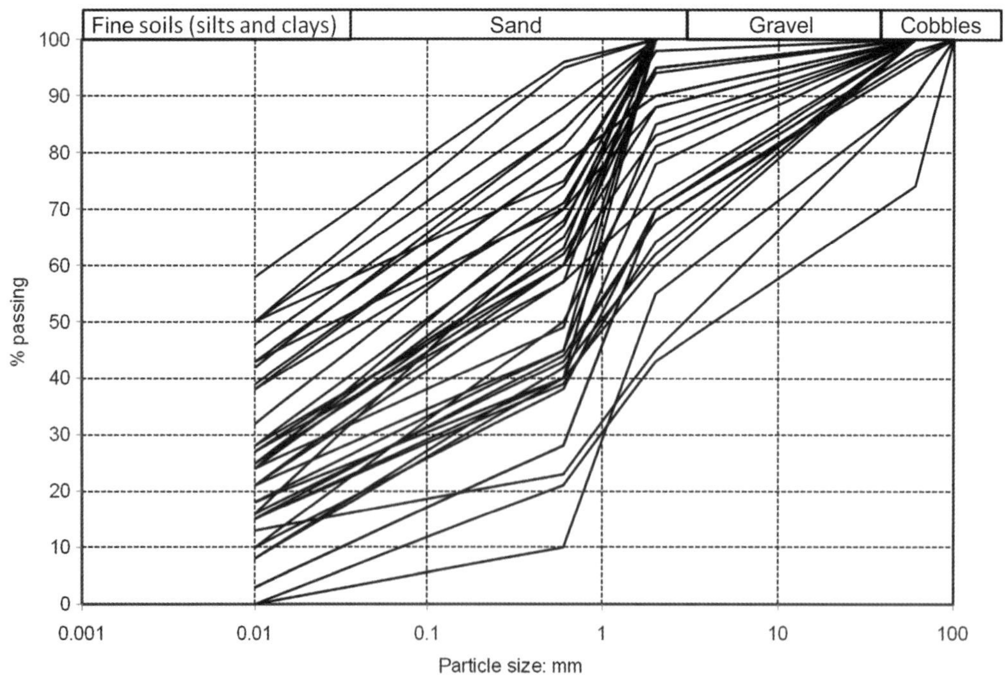

Figure 31.10 The grading of glacial tills; the data come from various sources from across the UK, including those summarised by Trenter (1999)

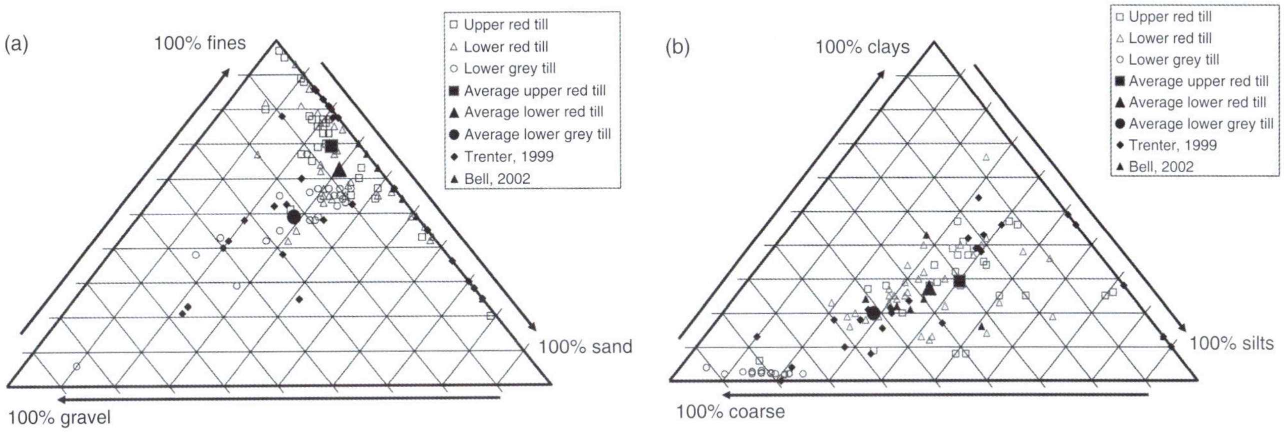

Figure 31.11 A ternary plot of a number of glacial basal tills showing (a) the relationship between fine and coarse particles and (b) between the fine particles

31.5.3 Mechanical characteristics

31.5.3.1 Undrained shear strength

Sampling clay basal tills is difficult because of the gravel content and the strength of the till. This leads to significant scatter in the results of strength and stiffness tests. It is for that reason that tests are often carried out on larger samples – in the UK, U100 samples – providing a single value of strength at the depth of the sample. It used to be practice to carry out multi-stage tests on a single 100 mm sample (Anderson, 1974). The sample is taken to failure in the first stage. The same sample is taken to failure a second and third time at different confining predetermined planes. Given the density of the till, it is likely that it will exhibit brittle behaviour during the first stage, thus creating a predetermined failure plane for the subsequent stages. The implication is that the derived shear strength from a multi-stage test does not have any meaning. However, given the difficulty of preparing subsamples from 100 mm dia. samples of basal tills because of the particle size distribution, in particular the gravel content, it is normal to test 100 mm samples. Given that there will be only one strength test from each sample, it is prudent to increase the number of samples tested for strength. Much of the sample still exists after the strength test, so it is possible to use that sample for classification tests.

The measured undrained shear strength of clay matrix-dominant tills is very dependent on the fabric and method of sampling. The density of the till means that sample distortion using driven tubes is inevitable. Gravel affect both the sampling and mode of failure. Fissures affect the mode of failure (**Figure 31.12**). McGown *et al.* (1977) suggested that a representative undrained strength can be obtained only from samples larger than those normally taken in ground investigations. This questions the value of U100 samples from ground investigations. Hence, profiles of undrained shear strength in till are likely to show much scatter (**Figure 31.13**). McGown *et al.* (1977) used the fact that the relationship between intact strength and strength of fissured soil varied with sample size, leading to the conclusion that the representative strength is typically 0.75 of the average strength measured using U100 samples of fissured till.

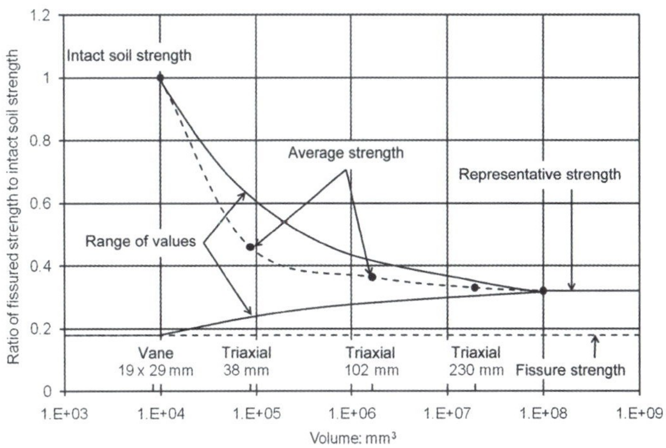

Figure 31.12 The effect of fissures on the strength of glacial till showing the ratio of fissured strength to intact strength against the size of the sample expressed in terms of volume (Reproduced from McGown *et al.*, 1977; © The Geological Society)

In some cases, because of the composition and strength, it is difficult to obtain undisturbed specimens. Therefore, an alternative method is used to indirectly measure undrained shear strength. For example, undrained strengths of some Irish tills (Farrell and Wall, 1990) are estimated from correlations between SPT blow counts and tests on remoulded specimens. A value of 6 (s_u = 6 N) for Irish tills (Orr, 1993) is commonly used, but note that the correlation was based on tests on remoulded specimens. This empirical method is commonly used for gravitationally compacted soils. Stroud (1975) presented a correlation between undrained shear strength and blow count based on plasticity index for fissured clays. This correlation used uncorrected values of SPT blow count. Ground investigations carried out in accordance with Eurocode 7 will provide corrected values.

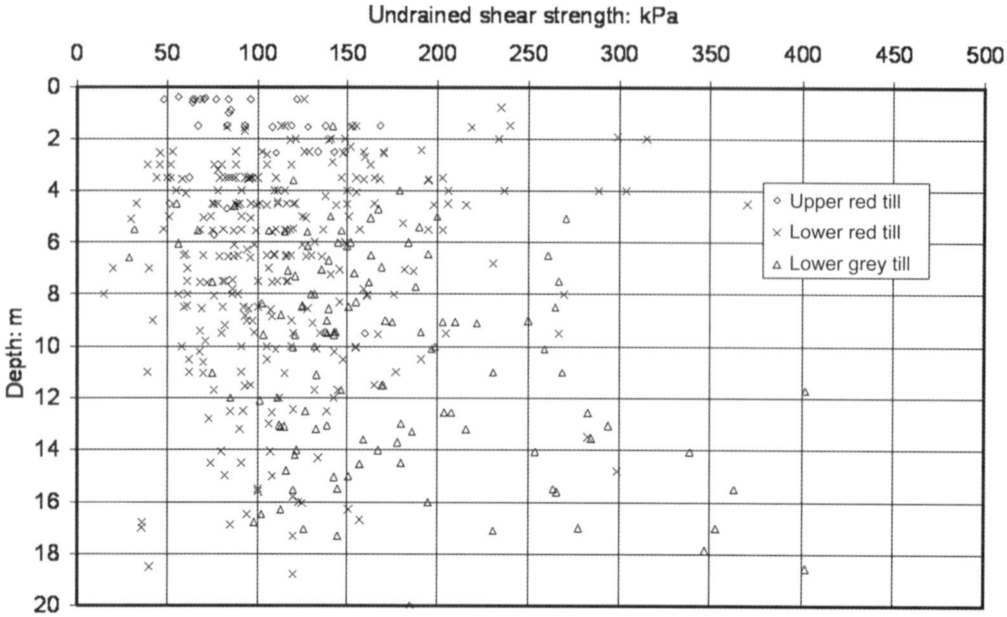

Figure 31.13 A typical profile of undrained shear strength of basal till highlighting the scatter in the data and the difficulty in selecting a characteristic value of strength (the upper and lower tills refer to the tripartite succession in northern England)

Many tills are fissured so the correlation may apply to tills, but there is evidence that shear strengths estimated from SPT results using the Stroud correlation are significantly less than the in situ strength, suggesting the correlation factor is too low. This could be because the correlation has not been developed for such dense clays or it could be because the stiffness of these tills makes them behave in a partially drained manner. It is assumed that the clay tills can be considered as undrained clays when loaded because they are considered relatively impermeable. However, they are very stiff. This means that pore pressures generated during rapid loading (as in an undrained test) may be less than expected, since some of the load will be taken by the soil skeleton as water is compressible. Further, in situ the pore pressures may dissipate more quickly than expected because the coefficient of consolidation is greater than expected for clays. The coefficient of consolidation is the ratio of the stiffness to permeability. Given that the permeability of the basal till is similar to other clays, but the stiffness can be much greater, the coefficient of consolidation is often much greater than that of sedimented clays. Thus, excess pore pressures generated could be much less than observed in sedimented clays. This may account for the apparently high undrained strengths in basal tills; in practice they are behaving in a partially drained manner.

The data used by Stroud, which covered plasticity indices between 15% and 65%, showed that the ratio between undrained shear strength and the SPT blow count varies between 7 and 4.1. **Figure 31.14** shows data from a site in northern England in which results of triaxial tests on undisturbed specimens are plotted against depth together with values of strength derived

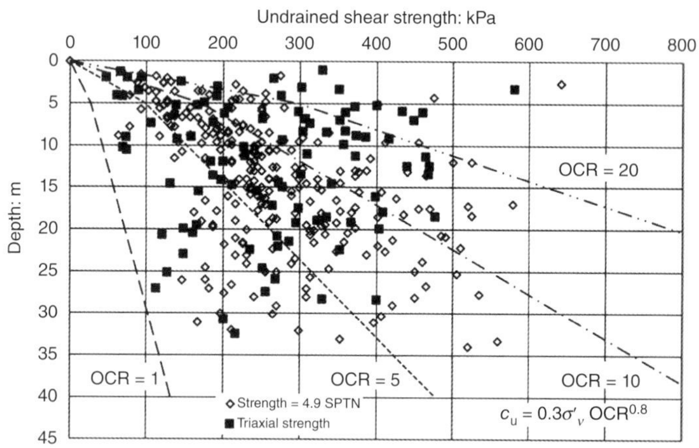

Figure 31.14 The relationship between SPT blow count and the triaxial undrained shear strength of some tills in northern England

from SPT *N* values. There is evidence from piling contractors that quoted correlations between undrained shear strength and blow count are too low. This suggests that earthworks calculations should be based on a factor of 6 since the till is remoulded during construction, whereas foundation design calculations could be based on a factor of 9, since the in situ strength is mobilised during foundation loading. Note that sometimes 9 may be considered too low, but this should be challenged unless there is evidence from the investigation that supports this view.

The consequence of this response of basal clay tills is that foundations are often overdesigned and excavation techniques inappropriate. This reinforces the need for pile tests to assess the actual performance rather than the predicted performance (see Chapter 98 *Pile capacity testing*).

31.5.3.2 Effective strength parameters

As with undrained shear strength, the fabric and composition of glacial soils influence the interpreted values of c' and ϕ'. It is important to stress that c' and ϕ' are based on the assumption that there is a linear relationship between confining pressure and deviator stress independent of the level of confining pressure and, for a given soil, independent of the variation in fabric between specimens. In tills, the latter assumption is invalid, resulting in a wide variation in values of c' and ϕ' for tills (**Table 31.8**), even if they are from the same till deposit. Increasing the density of a till increases its strength and there is an optimum fines content for a maximum strength. Thus, density is an indication of strength. Hence, when interpreting laboratory test results it is useful to disregard low values of strength from high-density samples because the fabric may have been the dominant factor in determining the strength – that is, failure may have been triggered by gravel inclusions or shear planes. This emphasises the need for quality descriptions of the samples tested and for caution when selecting characteristic strengths from a limited number of tests. The fabric may be the dominant factor that influences the mass behaviour and not the material strength.

There have been attempts to relate ϕ' with plasticity index (**Figure 31.15**), percentage fines and density (**Figure 31.16**) for clay matrix-dominant tills. Such relationships exist but they may not be appropriate for till because of the effects of fabric and sampling. Stark *et al.* (2005) proposed a relationship between ϕ' and liquid limit considering clay friction and confining pressure (**Figure 31.17**).

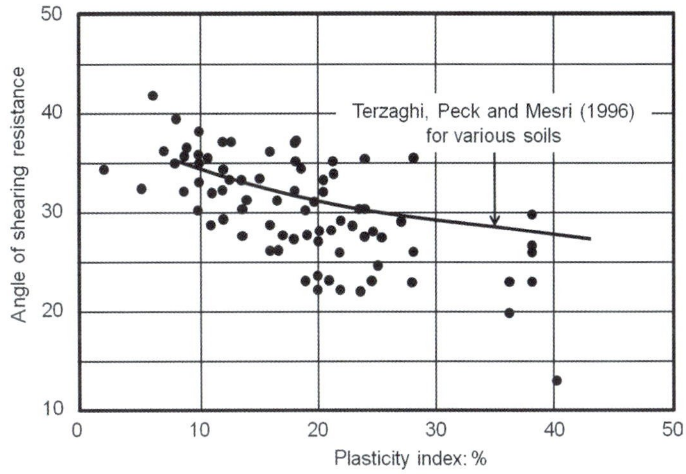

Figure 31.15 Relationship between angle of shearing resistance and plasticity index for a number of basal tills (Reproduced with permission from CIRIA C504, Trenter (1999); www.ciria.org)

The uncertainty associated with properties of glacial soils is likely to be underestimated because of their spatial variation (Grasmick *et al.*, 2020) if a global simulation is used in which the properties of a stratum are grouped according to a regional assessment. They recommended that a local simulation be

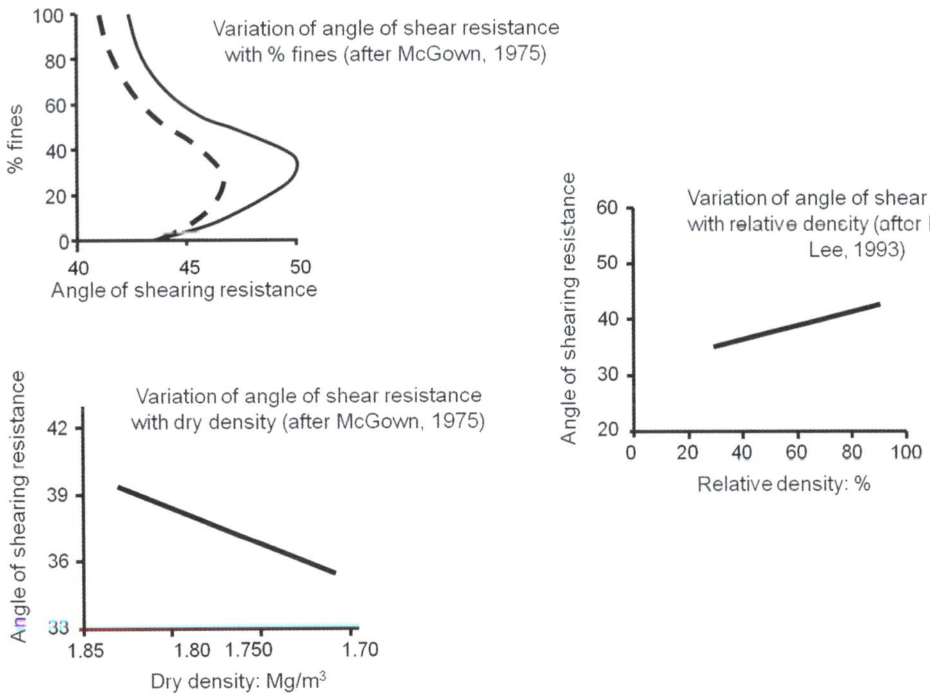

Figure 31.16 Relationship between percentage fines, density and angle of shearing resistance (Reproduced with permission from CIRIA C504, Trenter (1999); www.ciria.org)

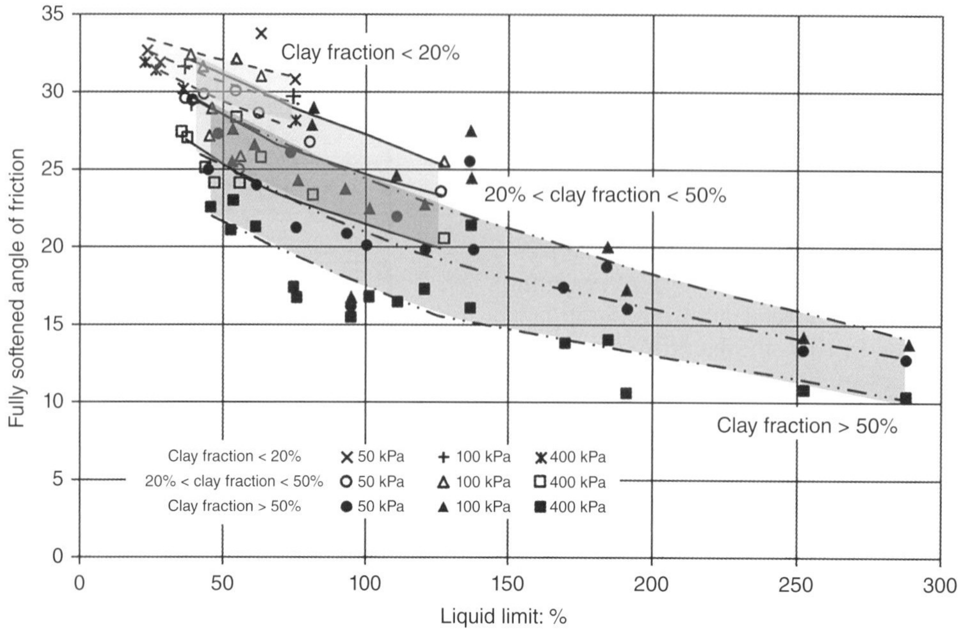

Figure 31.17 Variation in fully softened angles of friction with clay fraction and liquid limit (After Stark *et al.*, 2005)

used such that soil properties are grouped according to the physical characteristics at site level.

There are two approaches that could be used to evaluate a characteristic strength. In both cases (**Figure 31.18**) the samples must be grouped together according to their description and classification data, noting that two tills from the same site can have the same description but different liquid limits. This is because the source of the tills may be different, resulting in different particle compositions. The first approach is to plot the deviator stress and mean stress for all samples on the same graph and identify the failure envelope from these data (**Figure 31.19**). Spurious data points related to density and fabric should be given a lower weighting. This gives the characteristic value of strength provided sufficient samples are tested. Creer *et al.* (2020) used this approach to create an assessment of the strength properties of the glacial soils in Toronto. They produced evidence to show typical properties which exceed those normally recommended.

The alternative way of producing a characteristic strength, especially when there are a limited number of samples, is to test reconstituted glacial till (Burland, 1990) with all coarse material removed. In this case, the laboratory-prepared specimens are gravitationally compacted to as high a density as possible, ideally using a hydraulic press. The density will still be less than the in situ density so the test results could be considered a lower bound value. Basal tills were spread into place under vertical loads of up to 10 000 kPa; typical laboratory tests use pressures of up to 700 kPa. The scatter due to fabric is removed when testing reconstituted till (**Figure 31.20**), but the interpreted value of strength lies below the values obtained from tests on intact samples.

Note, however, in both cases the strength of the glacial till mass may be dominated by lenses or pockets of laminated clay and sand and gravel.

31.5.3.3 Stiffness

Stiffness of soils is notoriously difficult to measure, is test-dependent and depends on the quality of the soil being tested. In tills this is further compounded by the fact that the tills are spatially variable to such an extent that measurements of the stiffness of elements of tills may not be representative of the mass stiffness. The comments on the impact of fabric and sample size on strength apply to measurements of stiffness.

No matter which method is used, it is important to test a quality element of till. This means high-quality samples such as wireline and block samples from matrix-dominated till, which are only possible on major projects because of the cost. In general, U100 driven tube samples will be available. In that case it may be prudent to determine stiffness from tests on reconstituted till, although, as with strength, the value obtained will be less than the in situ value. However, as **Figure 31.21** shows, results (the shaded area) can be obtained from intact samples of matrix-dominated tills.

It is possible to measure stiffness in situ (e.g. pressuremeter, plate and cone). In these cases, the installation of the measuring device inevitably disturbs the till, leading to an underestimate of the in situ stiffness, leading in turn to the use of empirical correlations. Non-destructive geophysical methods can be used to obtain the maximum shear modulus (G_{max}) with some degree of accuracy. The shear modulus strain degradation curve obtained from quality tests on elements of till can then be normalised to G_{max} (**Figure 31.21**). These techniques

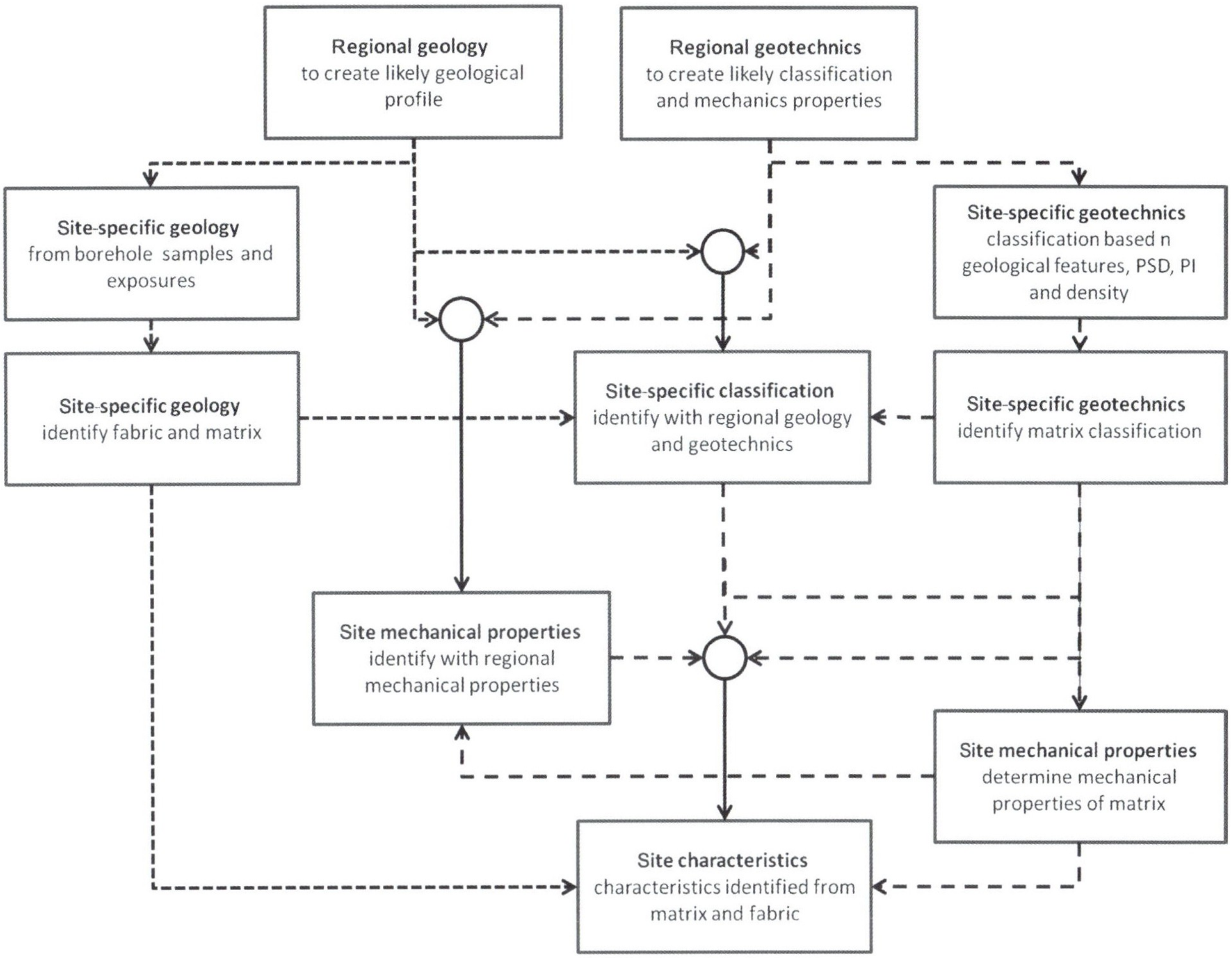

Figure 31.18 A flow chart for the assessment of soil characteristics; in particular the strength of the till

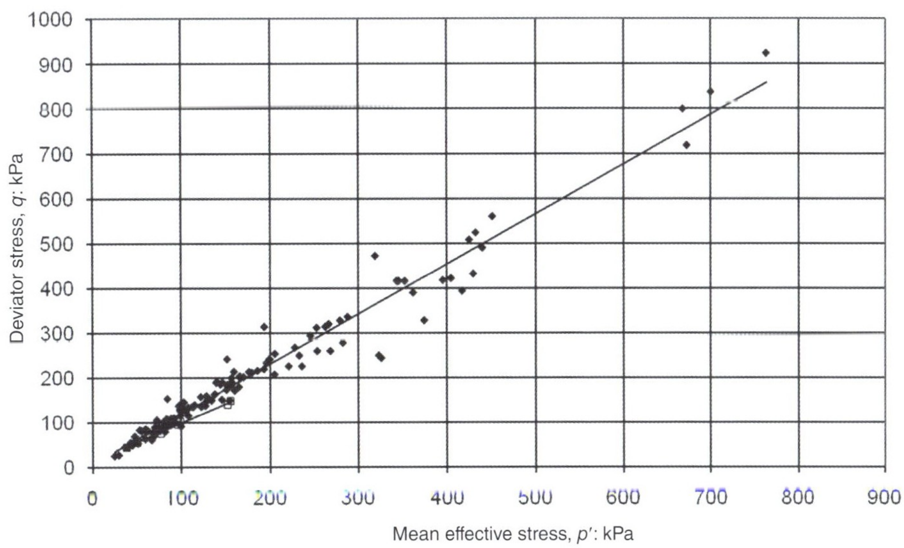

Figure 31.19 Characteristic strength derived from tests on individual specimens. The ranges of cohesion (0–27 kPa) and angles of friction (15–34°) were reduced to characteristic values of cohesion of 2.5 kPa and angle of friction of 31°; the two trend lines refer to the best fit to the data from 0 to 100 kPa (p') and 0 to 800 kPa

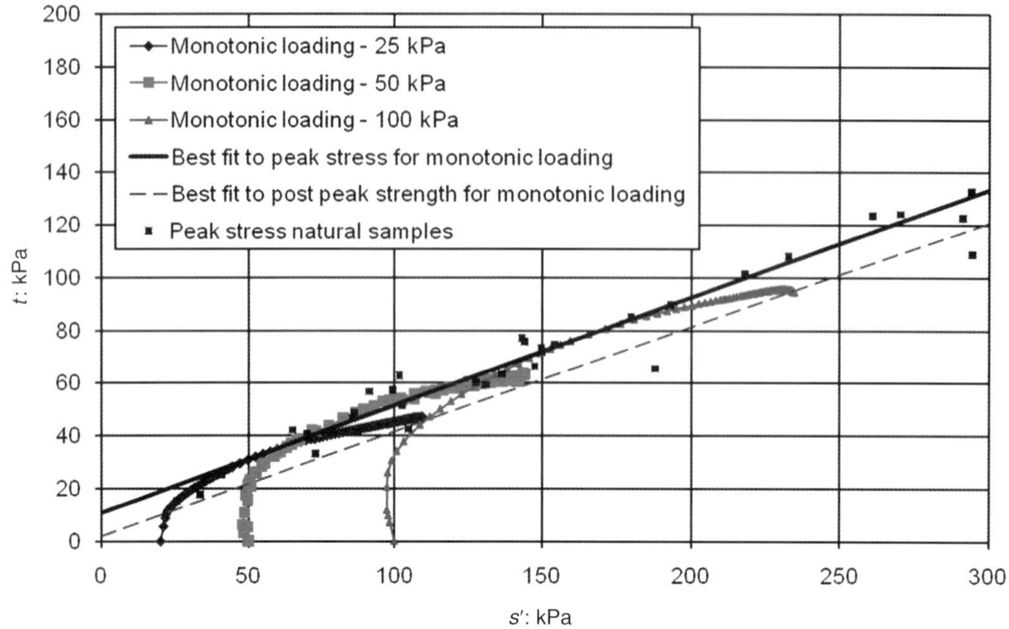

Figure 31.20 Comparison between tests on reconstituted tills and tests on intact specimens showing that the strength of the reconstituted till lies below that of the natural tills

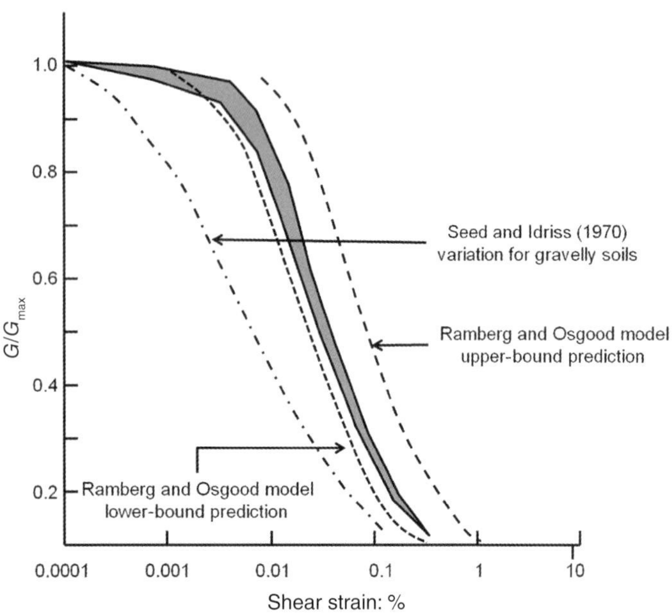

Figure 31.21 Degradation of stiffness from resonant column tests on samples of glacial till (Reproduced from Chegini and Trenter, 1996)

are expensive and may be used on large projects, but a combination of geophysical tests and tests on reconstituted samples of till may be the best combination to determine the stiffness characteristics of till (e.g. **Figure 31.22**).

31.5.4 Hydraulic conductivity

The in situ hydraulic conductivity of glacial soils will be dominated by the fabric and structure of the soil; hence, correctly describing the till and identifying its features are critical. The presence of fissures in clay tills and the laminations in glaciolacustrine deposits imply that the in situ hydraulic conductivity may be several orders of magnitude greater than that of the intact matrix. The implication is that the hydraulic conductivity measured in the laboratory will depend on sample size and orientation as well as density and grading, but in situ the fabric may be the dominant factor. Hence, as with strength, scatter in values of hydraulic conductivity is expected and a representative value may be difficult to obtain from laboratory tests. Lloyd (1983) suggested that the permeability of clay matrix-dominant tills would be between 10^{-11} and 10^{-8} m/s; and if fissured between 10^{-10} and 10^{-7} m/s. It is possible to determine the intrinsic value of hydraulic conductivity from reconstituted till (**Figure 31.23**). In practice, if the hydrological characteristics are critical, then it is best to undertake in situ borehole tests using a test pocket that is sufficiently long enough to take into account the representative fabric and features of the till. This means that it is important to establish a quality 3D ground model that adequately describes those features which could influence the mass hydrological characteristics.

Low-permeability glacial tills can contain water-bearing lenses and layers of more permeable soils, which means the mass permeability could be much greater than that of the matrix. Whether or not this is a risk depends on the transmissivity of the lenses and layers (Preene, 2021) – that is, it is important to determine the hydraulic conductivity and transmissivity of any more permeable layers encountered in matrix-dominated tills whenever excavations are being planned.

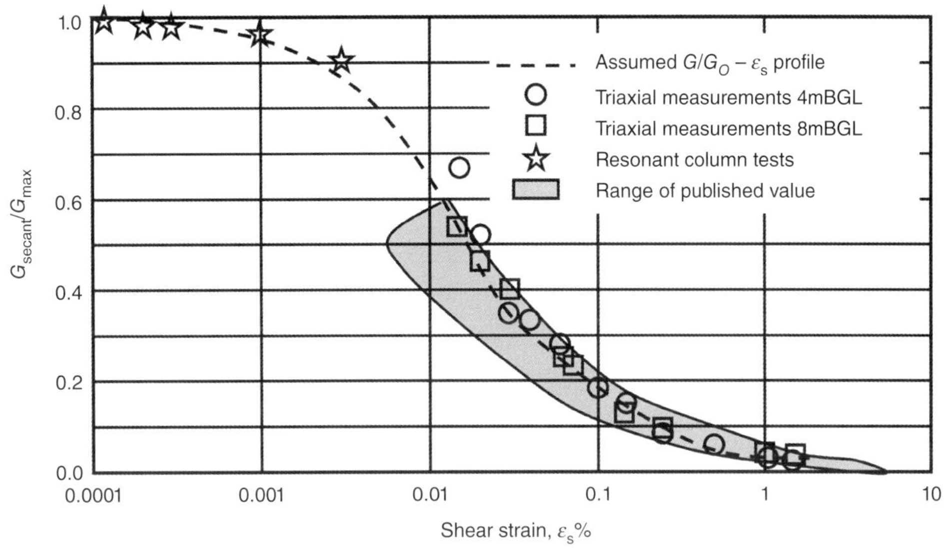

Figure 31.22 Degradation of stiffness of a lodgement till from north-east England (Reproduced from Brown *et al.*, 2006)

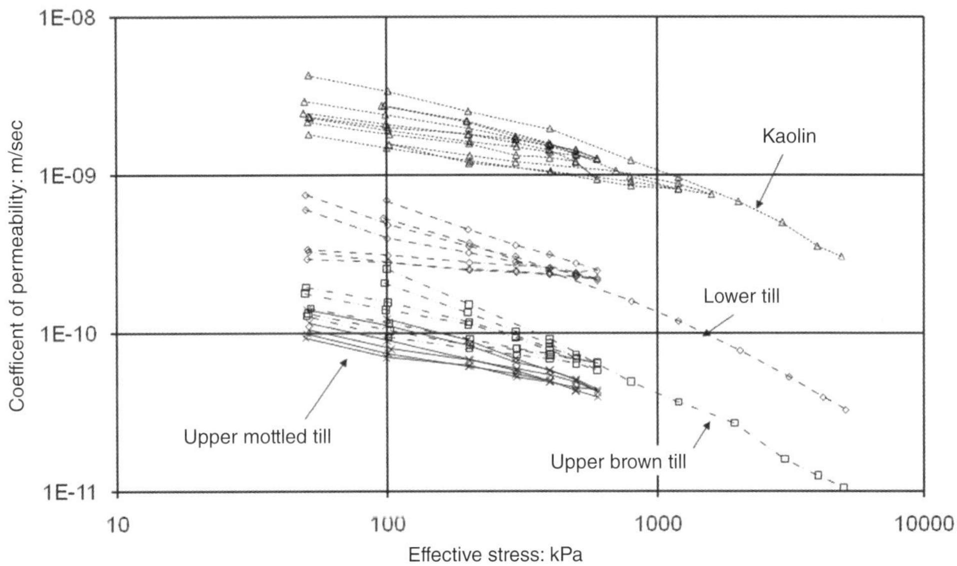

Figure 31.23 Hydraulic conductivity of glacial tills (compared to kaolin) from constant flow tests on triaxial samples (Reproduced from Clarke and Chen, 1997)

31.6. Routine investigations

Much of the development work on exploratory boreholes in UK site investigation practice took place in the south-east of England (e.g. Cooling, 1942; Cooling and Smith, 1936), where soft alluvial clays, stiff clays and sands and gravels are found. They evolved from well-boring techniques that fixed the diameter of the boreholes and sampling devices which led to the U100 sampling tube and the standard penetration test (SPT), both of which had their origins in American 'dry' sampling techniques. This then fixed the size of the test specimens.

The subject of soil mechanics was being developed at the same time as the site operations and this led to the introduction of the oedometer, triaxial and shear box tests. The testing techniques have essentially remained the same since then. While the techniques have proved successful in areas in which clays and sands are the predominant soil type, they are not necessarily the best for other areas where heterogeneous mixtures of soil particles are more commonly found. For example, it is difficult to select design parameters from standard tests on tills because of the difficulties in sampling and testing these soils using standard equipment. It may be possible to develop other techniques, but they could not be used in practice because parameters are test-dependent and confidence in design methods is based on standard tests.

Glacial soil type	Equipment/ method of sampling	Comment
Matrix-dominated (sandy clay with gravel)	U100 driven tube	Generally possible depending upon clast content: maintain frequent checks for circularity of sampler; change cutting shoes regularly and keep inside of sampler lightly oiled
Clast-dominated (clayey sandy gravel with occasional cobbles)	U100 driven tube	Generally not possible: using double tubes may make it possible to recover a (disturbed) sample sufficient for lithological description
Matrix-dominated (sandy clay with gravel)	Block	Possible: adjust block size to cope with largest expected clast dimension
Clast-dominated (clayey sandy gravel with occasional cobbles)	Block	Depends upon amount and plasticity of matrix: where sufficient to form a coherent block during excavation, the procedure may be used; adjust block size to cope with largest expected clast dimension
Matrix-dominated (sandy clay with gravel)	Rotary core double tube plus liner	Possible for strata definition but swelling on contact with flushing medium may preclude use for laboratory test purposes
Clast-dominated (clayey sandy gravel with occasional cobbles)	Rotary core double tube plus liner	Sometimes possible for strata definition but core retention may be difficult; bit wear may be high; usually unsuitable for laboratory test purposes
Matrix-dominated (sandy clay with gravel)	Wireline	Generally possible. Note that the bit is not routinely brought to the surface to permit inspection for wear, as with the conventional system
Clast-dominated (clayey sandy gravel with occasional cobbles)	Wireline	May be possible, depending on amount, nature and angularity of clasts. Note that the bit is not routinely brought to the surface to permit inspection for wear, as with the conventional system

Table 31.6 Guide to selection of sampling methods in glacial tills (After Trenter, 1999)

Table 31.6 lists the sampling methods suggested by Trenter (1999), giving an indication of the quality of the sample. The U100 driven tube is the most common method of sampling matrix-dominant tills in the UK. The quality of the sample will depend on the amount and size of gravel and cobbles present, the strength of the matrix and the presence of shear planes. A detailed description of those samples tested is required to help evaluate the significance of the results of the tests. For example, the failure of a sample in a triaxial test may be a result of the fabric and not the strength of the matrix.

31.7. Developing the ground model and design profile

Reducing the risk associated with glacial soils requires both geological and engineering knowledge, which are brought together in an engineering geological model (e.g. Baynes *et al.*, 2021; Parry and Baynes, 2014). Hence, the model is based on observations of the ground and its properties, models that anticipate the behaviour of the ground subject to change, and the changes that occur due to the engineering processes. The model depends on an understanding of the formation of glacial soils and the impact that had upon their properties. At the start of a project there are probably limited data, but as the project proceeds more data become available. Thus, the model should continuously be updated to reduce risk. However, risk remains. Therefore, it is prudent to develop geotechnical baseline reports (Essex, 2007) in which unforeseen risk is allocated. The baseline report is developed from observations (Davis, 2021) taking into account possible scenarios that may have not been observed but could exist based on an understanding of the type of ground. For example, it would be prudent to assume clay-dominant matrix tills contain lenses and layers of laminated clay and sand and gravels, are fissured and contain boulders. This assumption could be adjusted with further investigation or as construction proceeds.

In some areas (e.g. Norway and the USA) results from benchmark test sites are available (L'Heureux and Lunne, 2019). These are well-characterised and documented field test sites used to advance the state of the art in areas such as in situ testing, instrumentation, prediction of soil behaviour and foundation prototype testing. Regional databases also exist (e.g. Clarke *et al.*, 2008; Creer *et al.*, 2020).

31.7.1 Geotechnical properties

It is possible to develop a regional model which is the basis for the design of a detailed investigation and preliminary engineering designs. In that case, the model is used to provide a steer as to what to look for and for further investigation should observations not fit with the model (Fookes, 1997). This is reasonable in major urban areas where there is a significant amount of data publicly available. In the absence of a regional model a more thorough investigation is necessary, leading to site-specific correlations. While regional models are important when investigating glacial soils, it is necessary to undertake a thorough investigation because of the spatial variation in composition and properties both regionally and locally.

Figure 31.14 highlights the difficulty in assigning characteristic values to this spatially variable soil. Clarke *et al.* (2008) and Creer *et al.* (2020) have suggested that it is possible to use a regional database to create characteristic values. In that way data from a new site could be compared to the regional database and, if found to be statistically aligned with that database, then the characteristic values for the region could be used.

Box plots (**Figure 31.24**) are a visual means of assessing the normality of the data. From an engineering point of view the differences in Atterberg limits between the sites are not significant, hence the regional values have been identified. This approach can also be applied to water content and density.

Schneider (1999) suggested that the characteristic value for a new investigation is simply the mean less half the standard deviation since the regional coefficient of variation is likely to be unknown. **Table 31.7** shows that this can lead to a characteristic value with about 70% of the results exceeding that value (i.e. it is an overcautious estimate of the mean).

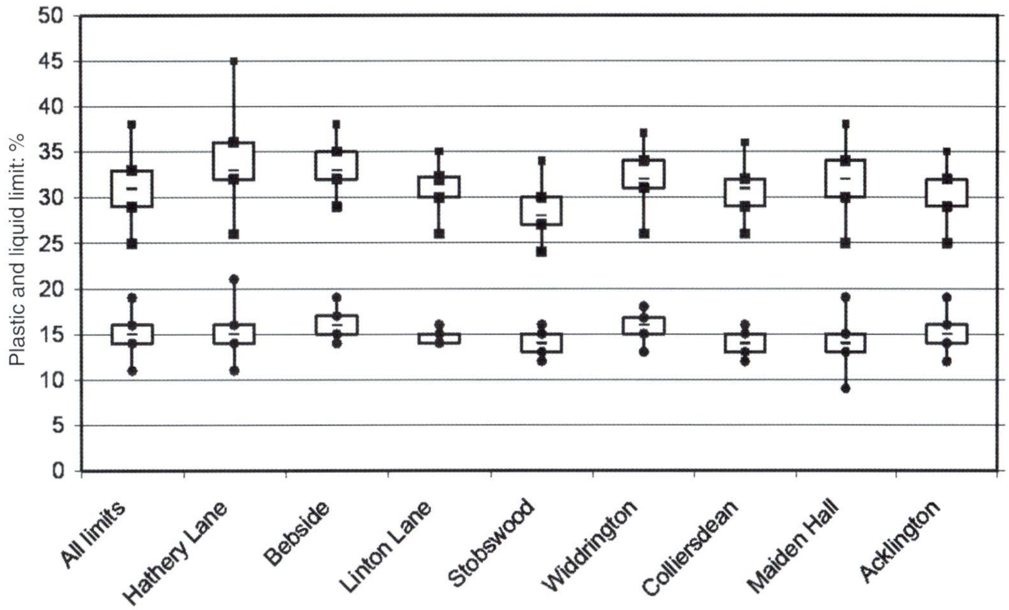

Figure 31.24 Box plots showing the regional variation of the Atterberg limits for glacial tills; the median is the data point in the middle of a box which represents the samples between the first and third quartile. The extensions refer to the limits of the data. The names refer to opencast mines in northern England

Parameter	Regional				Site-specific					
	Mean	Characteristic values			Mean	Characteristic Values				
		Schneider (low)	Cautious mean (low)	Cautious mean (high)		Schneider (low)	Cautious mean (low)	Cautious mean (high)	Bayes mean (low)	Bayes mean (high)
Till A										
γ_b: Mg/m^3	2.00	1.960	1.993	2.003	2.11	2.07	2.09	2.13	2.10	2.11
w: %	21.5	19.74	21.28	21.72	20.8	18.2	19.7	21.9	20.8	20.8
PL: %	20.3	18.98	20.17	20.53	20.1	18.3	19.2	21.0	20.1	20.1
LL: %	46.5	43.18	46.06	46.93	42.4	38.8	40.6	44.3	42.5	42.6
c_u: kPa	99	76	96	102	89	71	77	100	89	89
Till B										
γ_b: Mg/m^3	2.06	2.023	2.059	2.065	2.16	2.12	2.13	2.19	2.16	2.16
w: %	18.1	16.56	18.02	18.25	16.4	13.6	15.5	17.4	16.48	16.48
PL: %	17.6	16.39	17.51	17.69	15.6	14.3	15.0	16.3	15.7	15.7
LL: %	38.7	36.02	38.49	38.89	34.6	31.6	33.2	36.0	34.7	34.7
c_u: kPa	100	78	98	101	89	60	65	104	86	86

Table 31.7 The effect a regional database has upon the selection of characteristic values (After Clarke *et al.*, 2008)

The first step in establishing the characteristic values from a new site is to determine whether the soils are tills. This could be proven from the geological description, Atterberg limits and grading curves. If this combination indicates that the soils are likely to be typical of the region, then Bayesian statistics can be used to determine the characteristic values. An example is given in **Table 31.7**, which shows the values based on the dataset and on a priori knowledge of the regional dataset. This leads to less cautious design parameters and hence a more economical design. The Bayesian low mean or high mean is used depending on the worst condition being assessed.

31.7.2 Effective strength parameters

Consolidated undrained with pore water pressure measurements or consolidated drained triaxial tests are used to determine the effective strength of a soil. These are time-consuming especially for low-permeability soils such as glacial tills. Further, the number of tests is normally limited. Given the spatial variability in properties and fabric of glacial soils, it is not unusual to obtain a range of properties. Normally, it would not be feasible to carry out many tests in a routine investigation. Therefore, the results are unlikely to be representative of the mass. This could lead to an unsafe design especially with 'high cohesion, low angle of friction' values. For this reason, it is normal to restrict cohesion to 5 kPa.

An alternative approach is to identify the strata from classification tests and determine the best fit to all the tests for that strata. **Table 31.8** is a summary of results produced by Clarke *et al.* (2008) and Creer *et al.* (2020) from 225 and 417 triaxial tests, respectively, covering seven types of glacial soils. The

Location	Soil group	Parameter	Average	Minimum	Maximum	Standard deviation	Design value
NE England	Weathered till	Angle of friction	24.8	12.1	27.7	6.6	30.2
		Cohesion: kPa	7.2	0	23.8	7.6	0
		Bulk unit weight: kN/m^3	19.5	16.8	22.7	0.9	19.5
		Water content: %	21.0	16.7	27.3	3.2	
	Upper till	Angle of friction	27.3	15.4	33.9	3.6	28.2
		Cohesion: kPa	6.9	0	26.7	7.7	2.5
		Bulk unit weight: kN/m^3	20.0	16.9	24.7	1.0	20.0
		Water content: %	17.2	11.0	25.3	3.8	
	Lower till	Angle of friction	27.3	11.3	34.3	6.6	30.6
		Cohesion: kPa	6.8	0	17.4	6.0	0
		Bulk unit weight: kN/m^3	20.7	15.7	22.6	0.9	20.7
		Water content: %	13.6	11.0	17.3	1.9	
Toronto	Non-cohesive till	Angle of friction	40	33	46	2.9	43
		Cohesion: kPa	21.6	0	95	25.0	3.0
		Bulk unit weight: kN/m^3	22.5	17.6	25.6	1.6	22.5
		Water content: %	11.0	2.0	76.9	3.0	
	Cohesive till	Angle of friction	32	23	45	2.9	31
		Cohesion: kPa	13.2	0	78.5	16.3	9
		Bulk unit weight: kN/m^3	21.6	21.1	26.8	2.4	21.6
		Water content: %	15.2	3.0	35.0	4.2	
	Silty sand/sandy silt	Angle of friction	41	33	47	3.9	48
		Cohesion: kPa	24.0	0	200	40.2	0
		Bulk unit weight: kN/m^3	21.1	14.4	25.3	1.7	21.1
		Water content: %	18.0	2.9	35.6	4.2	
	Silty clay	Angle of friction	33	19	41	4.4	32
		Cohesion: kPa	30.8	0	125	34.8	0
		Bulk unit weight: kN/m^3	20.9	0	26.9	2.2	20.9
		Water content: %	20.3	8.0	41.8	3.6	

Table 31.8 Effective strength parameters of glacial soils derived from standard triaxial compression tests (after Clarke *et al.*, 2008, Creer *et al.*, 2020) highlighting the spatial variation in properties and suggested design values

recommended design values are based on the best fit to all the data and show an acceptable value of cohesion and an angle of friction greater than the average value.

31.7.3 Intrinsic properties

Design parameters can be selected using a deterministic model based on empirical or theoretical relationships or a stochastic model based on a statistical analysis of the data or a combination of these. The former includes relationships between the Atterberg limits and undrained shear strength of remoulded clays (e.g. Carrier and Beckman, 1984; Wroth and Wood, 1976); in situ strengths of normally consolidated clays, density and Atterberg limits (e.g. Skempton, 1957); and overconsolidated soils and Atterberg limits (e.g. Yilmaz, 2000). Given the spatial variability of tills, it is recommended that a consistent framework for the interpretation of design parameters is used which is based on a geotechnical model that encompasses a study of the deposition and post-depositional processes, soil mechanics theory and experimental data, including tests on reconstituted tills.

Burland (1990) suggested that the compressibility and strength of reconstituted clay could provide a consistent framework for the interpretation of properties of natural clay. Reconstituted clay is a natural clay mixed with water such that the water content is at about $1.25w_L$, the true liquid limit. The properties of this slurry are inherent to the material but independent of its natural state. In the case of a clay matrix-dominant till the slurry should be made from the clay matrix. This is not practical given the difficulty of separating out the fine particles. However, removing all particles exceeding 2 mm produces a soil that is essentially the matrix of the till.

This framework is shown in **Figure 31.25** in which the average shear strength for three tills is plotted against the void index I_v

$$I_v = \left(\frac{e - e^*_{100}}{C^*_c} \right) \quad (31.1)$$

where e^*_{100} is the void ratio at a vertical effective stress of 100 kPa on the intrinsic compression line and C^*_c is the slope of the intrinsic compression line (ICL). SCL is the sedimentation compression line, which is the compression line for natural or intact soils. Burland (1990) suggested that these two intrinsic properties are empirically related to the void ratio at the liquid limit, e_L

$$e^*_{100} = 0.114 + 0.581e_L \quad (31.2)$$

$$C^*_c = 0.256e_L - 0.04 \quad (31.3)$$

The ICL represents the compression curve for reconstituted clays; the intrinsic strength line, ISuL, represents the strength of reconstituted soils (Chandler, 2000). **Figure 31.25** shows that results of tests on reconstituted tills (Clarke *et al.*, 1998) fit to the ISuL, which suggests that this framework can be used to characterise the tills.

Most of the data from tests on samples of natural tills lie to the left of the ISuL. Two of the tills lie about the trend line for the regional database. The scatter in the data reflects the nature of the till, but a combination of the geological history,

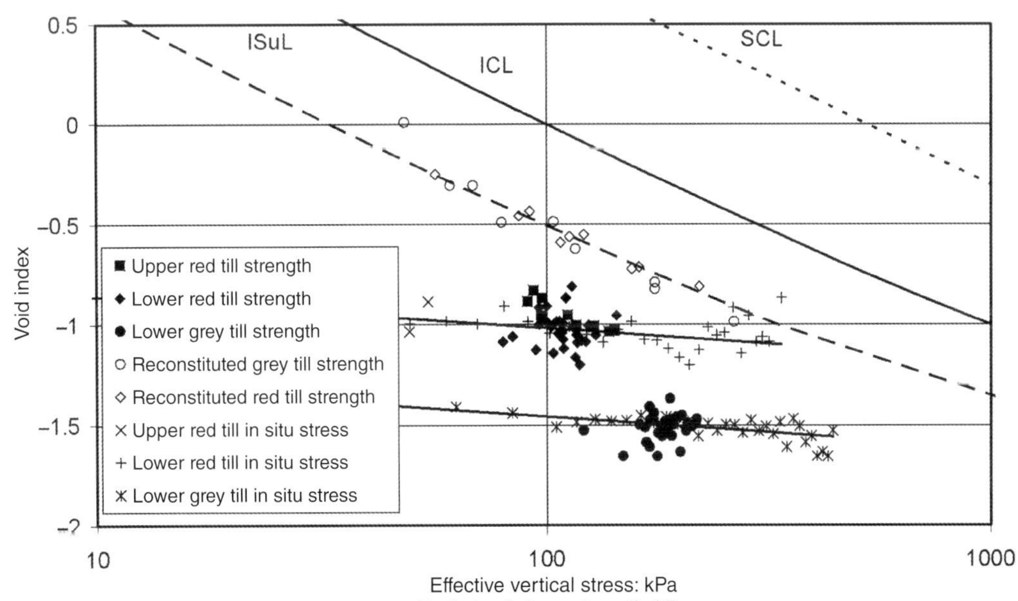

Figure 31.25 The properties of three basal tills using the intrinsic framework proposed by Burland (1990) and Chandler (2000) (Reproduced from Clarke *et al.*, 1998; © The Geological Society)

description and the classification characteristics suggests that the tills are the same and part of the regional database. The third till lies about a separate but parallel regional trend. The fact that the data points lie to the left of the ISuL suggests that they are overconsolidated. The single till on the lower trend line is there because the pre-consolidation pressure exceeded that of the other two tills. The slope of the lines is such that the average void ratio is very nearly constant, which implies that the characteristic shear strength profile could be vertical.

31.8. Earthworks

The previous sections have focused on the selection of design parameters for excavations, foundations and retaining structures. Till is also used as a construction material for earthworks and landfills. In these cases it is necessary to prove that the material can be compacted to the correct density, the compacted material is strong enough to resist failure in the design of embankments and, for landfill liners, it is ductile enough to remain relatively impermeable when subject to deformation.

It is most likely that a matrix-dominated till could be used for landfill construction including liners. The acceptable criteria are likely to be based on a 5–10% air voids line. Tills have a variable mean particle density (PD) which implies, in practice, that it is possible for site tests to plot above the 0% air voids line because of the high PD particles in the sample.

It is unlikely that the water content of the till will be altered so it is important to determine the hydraulic conductivity, strength and ductility of the till at the in situ water content. Typically, a strength of 50 kPa is required (Terraconsult, 2010) to allow clay to be compacted. The maximum hydraulic conductivity is 10^{-9} m/s. A reduction in water content increases the strength, making it more difficult to compact and possibly producing a brittle material that will fracture on straining, leading to an unacceptable increase in hydraulic conductivity. Further details of the testing regime can be

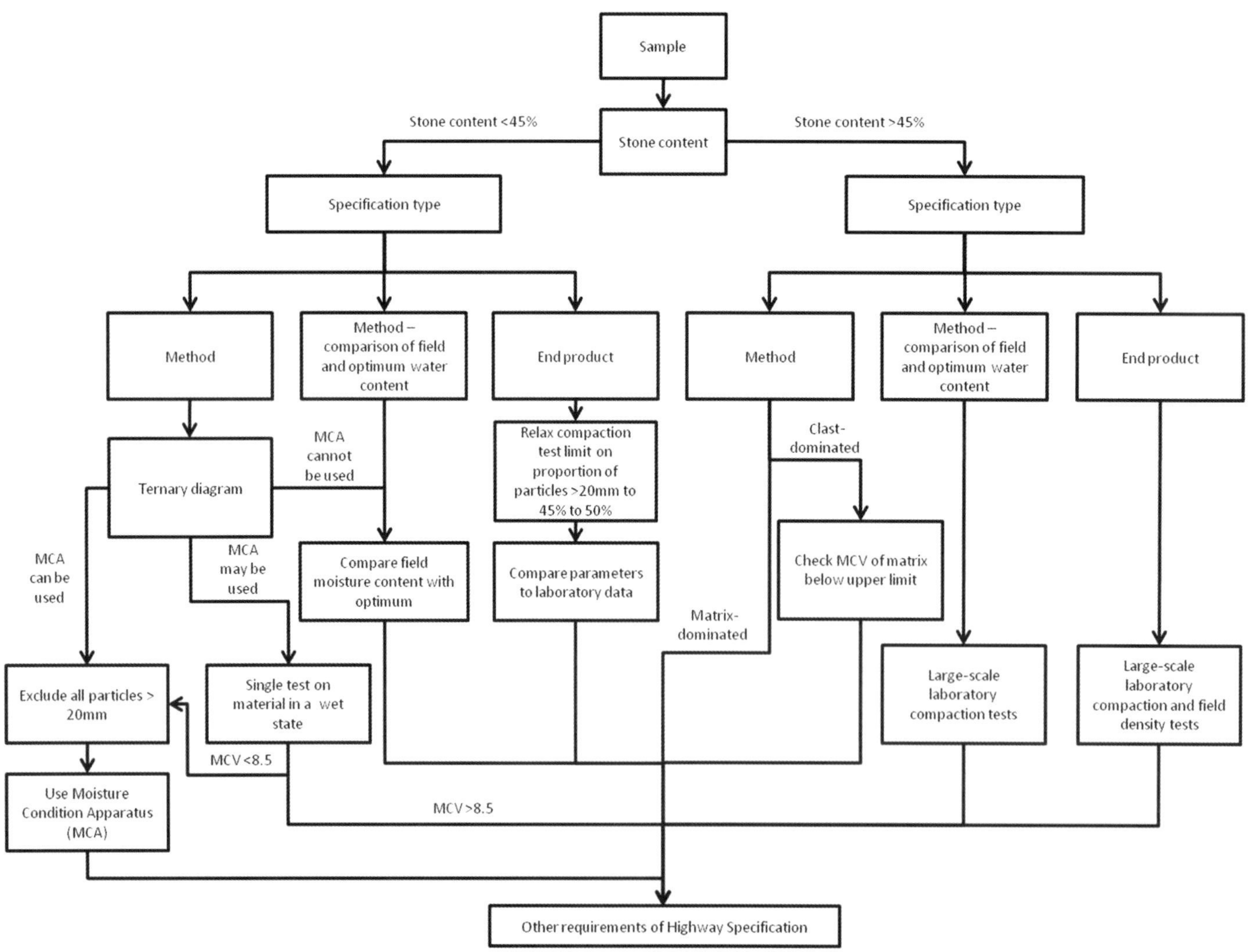

Figure 31.26 Flow chart for the acceptability of tills for earthworks (Reproduced from Winter *et al.*, 1998; © The Geological Society)

Property	'Minimum' requirement	Test
Permeability/hydraulic conductivity	Depends on Environmental Permit	BS EN ISO 17892-11:2019
Remoulded undrained shear strength	Typically ≥50 kPa or other site-specifically defined value	BS EN ISO 17892-8:2018
Plasticity index (I_p)	10% ≤ I_p ≤ 65%	BS EN ISO 17892-12:2018
Liquid limit	≤ 90%	
Percentage fines	<0.063 mm (63 μm) ≥ 20% but with a minimum clay content (particles <2 μm) of 8%	BS EN ISO 17892-4:2016
Percentage gravel	>5 mm ≤30%	
Maximum particle (stone) size	Two-thirds compacted layer thickness	Typically 125 mm but must not prejudice the liner, for instance by larger particles sticking together to form larger lumps

Table 31.9 Properties of tills for landfill liners (Data taken from Terraconsult, 2010)

found in the Environment Agency's guide to landfill design (Terraconsult, 2010).

The thickness of a compacted layer specified by the Highways Agency is 300 mm, which means that all particles more than 125 mm must be removed. Thus, it is still possible to use matrix-dominant and clast-dominated tills. However, there is a limit to the size of particles in the control tests (e.g. compaction and moisture condition value). Particles greater than 20 mm must be removed before testing. If the stone content is small (i.e. they are discrete elements in the sample) then it is possible to test the matrix (Winter *et al.*, 1998). This is the case for up to 50% stone content. Above that the stones govern the behaviour. An increase in stone content increases the maximum dry density and lowers the optimum water content, but this is not important since the stone of high density and low water content is replaced by a low-density, high water content matrix. **Figure 31.26** shows the method of selecting tills for earthworks. Note that if those earthworks include landfill liners then there are further limits, indicated in **Table 31.9**.

31.9. Concluding comments

Glacial soils are a hazard but, through a strategic approach to the ground investigation linked to an understanding of the impact the geological processes have upon the geotechnical behaviour, it is possible to mitigate against the risk. There are two extremes. The first is to base the design parameters on the properties of the dominant material; the second is to base the design on the zones of weakness. The former can lead to local failure, the latter to overdesign. Therefore, investing in the ground investigation to establish the extent of the zones of weakness and taking sufficient samples to establish the characteristic values is prudent. This should be carried out with a thorough understanding of the ground model developed from previous investigations and observations of exposures of glacial soil.

References

Anderson WF (1974) The use of multistage triaxial tests to find the undrained strength parameters of stony boulder clay. *Proceedings of the Institution of Civil Engineers* **57(2)**: 367–372.

Atkinson J (2013) Soil parameters, sampling and testing for design to Eurocode 7. *Proceedings of the Institution of Civil Engineers* **166(GE5)**: 428–430.

Barnes GE (1988) The moisture condition value and the compaction of stony clays. In *Proceedings of the Conference on Compaction Technology, London, UK*. Thomas Telford, London, UK, pp. 79–90.

Baynes FJ, Parry S and Novotný J (2021) Engineering geological models, projects and geotechnical risk. *Quarterly Journal of Engineering Geology and Hydrogeology* **54(2)**, 10.1144/qjegh2020-080.

Beaumont P (1968) A history of glacial research in Northern England from 1860 to the present day. University of Durham, Department of Geography, Occasional Paper 9. Durham, UK.

Benn EI and Evans DJA (2014) *Glaciers and Glaciation*. Edward Arnold, London, UK.

Bennett MR and Glasser NF (2011) *Glacial Geology Ice Sheets and Landforms*. Wiley, New York, NY.

Boulton GS (1970) On the deposition of subglacial and melt out tills on the margin of certain Svalbard glaciers. *Journal of Glaciology* **9**: 231–245.

Boulton GS (1974) Processes and patterns of subglacial erosion. In *Glacial Morphology* (Coates DR (ed)). State University of New York Press, New York, NY, pp. 41–87.

Boulton GS (1976a) Some relations between the genesis of tills and their geotechnical properties. In *Glacial Till: An Interdisciplinary Study* (Legget RF (ed)). Royal Society of Canada, Ottawa, ON, Canada.

Boulton GS (1976b) The development of geotechnical properties in glacial tills. In *Glacial Till: An Interdisciplinary Study* (Legget RF (ed)). Royal Society of Canada, Ottawa, ON, Canada.

Boulton GS (1979) Processes of glacier erosion of different substrata. *Journal of Glaciology* **23**: 15–38.

Boulton GS and Deynoux M (1981) Sedimentation in glacial environments and the identification of tills and tillites in ancient sedimentary sequences. *Precambrian Research* **15**: 397–420.

Boulton GS and Paul MA (1976) The influence of genetic processes on some geotechnical properties of glacial clays. *Quarterly Journal of Engineering Geology* **9**: 159–194.

Brown MJ, Hyde AFL and Anderson WF (2006) Analysis of a rapid load test on an instrumented bored pile in clay. *Géotechnique* **56(9)**: 627–638.

Burland J (1990) Thirtieth Rankine Lecture. On the compressibility and shear strength of natural clays. *Géotechnique* **40(3)**: 329–378.

Carrier WD and Beckman JF (1984) Correlations between index tests and the properties of the remoulded clays. *Géotechnique* **34(2)**: 211–228.

Chandler RJ (2000) Clay sediments in depositional basins: the geotechnical cycle (The 3rd Glossop Lecture). *Quarterly Journal of Engineering Geology and Hydrology* **33**: 5–39.

Chegini A and Trenter NA (1996) The shear strength and deformation behaviour of a glacial till. *Proceedings of the International Conference on Advances in Site Investigation Practice, London, UK* (Craig C (ed)). ICE, London, UK, pp. 851–866.

Clark C, Evans D, Khatwa A, *et al.* (2004) Map and GIS database of glacial landforms and features related to the last British Ice Sheet. *Boreas* **33**: 359–375.

Clarke BG (2017) *Engineering of Glacial Deposits.* CRC Press, Boca Raton, FL.

Clarke BG and Chen C-C (1997) Intrinsic properties of permeability. *Proceedings of the 14th International Conference on Soil Mechanics and Foundation Engineering, Hamburg, Germany.* Balkema, Rotterdam, the Netherlands.

Clarke BG, Aflaki E and Hughes DB (1997) A framework for the characterization of glacial tills. *Proceedings of the 14th International Conference on Soil Mechanics and Foundation Engineering, Hamburg, Germany.* Balkema, Rotterdam, the Netherlands, pp. 263–266.

Clarke BG, Chen C-C and Aflaki E (1998) Intrinsic compression and swelling properties of glacial till. *Quarterly Journal of Engineering Geology and Hydrology* **31**: 235–246.

Clarke BG, Hughes DB and Hashemi S (2008) Physical characteristics of sub-glacial tills. *Géotechnique* **58(1)**: 67–76.

Cooling LF (1942) Soil mechanics and site exploration. *Journal of the Institution of Civil Engineers* **18**: 37–61.

Cooling LF and Smith DB (1936) Exploration of soil conditions and sampling operations. *Proceedings of the 1st International Conference on Soil Mechanics and Foundation Engineering, Cambridge, MA*, vol. 1, p. 12.

Creer G, Olshansky D and Drevininkas A (2020) Rethinking the strength properties of soils in the Greater Toronto Area. *Proceedings GeoVirtual 2020, Resilience and Innovation* [online].

Davis JA (2021) Geotechnical baseline reports: ground models you can just make up? *Quarterly Journal of Engineering Geology and Hydrogeology* **54(2)**, 10.1144/qjegh2020-019.

DoE (1994) *Landsliding in Great Britain.* HMSO, London, UK.

Dreimanis A (1976) Tills, their origin and properties. In Glacial Till (Leggett RF (ed)). Royal Society of Canada, Special Publication 12, 11–49.

Dreimanis A (1988) Tills: their genetic terminology and classification. In *Genetic Classification of Glaciogenic Deposits* (Goldthwait RP and Marsh CL (eds)). Balkema, Rotterdam, the Netherlands, pp. 17–83.

Drewry DJ (1986) *Glacial Geological Processes.* Edward Arnold, London, UK.

Elson JA (1988) Comment on glaciotectonite, deformation till and comminution till. In *Genetic Classification of Glacigenic Deposits* (Goldthwait RP and Marsh CL (eds)). Balkema, Rotterdam, the Netherlands, pp. 85–88.

Essex R (2007) *Geotechnical Baseline Reports for Construction: Suggested Guidelines.* ASCE, Reston, VA.

Eyles N (1979) Facies of supraglacial sedimentation on Icelandic and alpine temperate glaciers. *Canadian Journal of Earth Sciences* **16**: 1341–1361.

Eyles N and Dearman W (1981) A glacial terrain map of Britain for engineering purposes. *Bulletin of the International Association of Engineering Geology* **24**: 173–184.

Eyles N and Sladen JA (1981) Stratigraphy and geotechnical properties of weathered lodgement till in Northumberland, England. *Quarterly Journal of Engineering Geology* **14(2)**: 129–141.

Eyles N, Sladen JA and Gilroy S (1982) A depositional model for stratigraphic complexes and facies superimposition in lodgement tills. *Boreas* **11**: 317–333.

Farrell E and Wall D (1990) *Soils of Dublin, Republic of Ireland Branch Seminar, Dublin, Ireland.* Geotechnical Society of Ireland and Institution of Structural Engineers, Dublin, Ireland.

Fookes PG (1997) The first Glossop Lecture: Geology for engineers: the geological model, prediction and performance. *Quarterly Journal of Engineering Geology* **30**: 293–431.

Gens A and Hight DW (1979) The laboratory measurement of design parameters for a glacial till. *Proceedings of the 7th European Conference on Soil Mechanics and Foundation Engineering, Brighton, UK*, vol. 2, pp. 57–65.

Goldthwait RP (1971) Introduction to till today. In *Till: A Symposium* (Goldthwait RP (ed)). Ohio State University Press, Columbus, OH, pp. 3–26.

Grasmick JG, Mooney MA, Trainor-Guitton WJ and Walton G (2020) Global versus local simulation of geotechnical parameters for tunneling projects. *Journal of Geotechnical and Geoenvironmental Engineering* **146(7)**, 10.1061/(ASCE)GT.1943-5606.00022

Gravenor CP, Von Brunn V and Dreimanis A (1984) Nature and classification of water lain glacigenic sediments, exemplified by Pleistocene, Late Paleozoic and Late Precambrian deposits. *Earth Science Reviews* **20**: 105–166.

Griffiths JS and Martin CJ (eds) (2017) *Engineering Geology and Geomorphology of Glaciated and Periglaciated Terrains: Engineering Group Working Party Report.* Geological Society of London, London, UK.

Hallett B (1979) A theoretical model of subglacial erosion. *Journal of Glaciology* **17**: 209–221.

Hambrey MJ (1994) *Glacial Environments.* UCL Press, London, UK.

Hooke RLB and Iverson NR (1995) Grain size distribution in deforming subglacial tills. *Geology* **23**: 57–60.

Hughes DB, Clarke BG and Money MS (1998) The glacial succession in lowland Northern England. *Quarterly Journal of Engineering Geology* **31(3)**: 211–234.

Iverson NR, Hooyer TS and Hooke RLB (1996) A laboratory study of sediment deformation, stress heterogeneity and grain size evolution. *Annals of Glaciology* **22**: 167–175.

L'Heureux JS and Lunne T (2019) Characterization and engineering properties of natural soils used for geotesting. *AIMS Geosciences* **6(1)**: 35–53.

Lawson DE (1979) *Sedimentological Analysis of the Western Terminus Region of the Matanuska Glacier, Alaska*. Cold Region Research and Engineering Laboratory, Hanover, NH, report 79-9.

Lawson DE (1981) *Sedimentalogical Characteristics and Classification of Depositional Processes and Deposits in the Glacial Environment*. Cold Region Research and Engineering Laboratory, Hanover, NH, report 81-27.

Lawson DE (1982) Mobilisation, movement and deposition of active sub aerial sediment flows, Matanuska Glacier, Alaska. *Journal of Geology* **90**: 279–300.

Lloyd JW (1983) Hydrogeological investigations in glaciated terrains. In *Glacial Geology: An Introduction for Engineers and Earth Scientists* (Eyles N (ed)). Pergamon, Oxford, UK, pp. 349–368.

Lunn AG (1995) Quaternary. In *Robson's Geology of North England* (Johnson GLA (ed)). Natural History Society of Northumbria, Newcastle, UK, pp. 297–311.

Lutenegger AJ, Kemmis TJ and Hallberg GR (1983) Origin and properties of glacial till and diamictons. In *Geological Environment and Soil Properties* (Young RN (ed)). ASCE, Reston, VA, pp. 310–331.

McGown A and Derbyshire E (1977) Genetic influences on the properties of tills. *Quarterly Journal of Engineering Geology* **10(4)**: 389–410.

McGown A, Radwan AM and Gabr AWA (1977) Laboratory testing of fissured and laminated soils. *Proceedings of the 9th International Conference on Soil Mechanics and Foundation Engineering, Tokyo, Japan*, vol. 1, pp. 205–210.

McMillan AA, Heathcote JA, Klinck BA, *et al.* (2000) Hydrogeological characterization of the onshore Quaternary sediments at Sellafield using the concept of domains. *Quarterly Journal of Engineering Geology* **33(4)**: 301–323.

Meer JJM van der, Menzies J and Rose J (2003) Subglacial till: the deforming glacier bed. *Quaternary Science Reviews* **22**: 1659–1685.

Orr TLL (1993) Probabilistic characterization of Irish till properties. In *Risk and Reliability in Ground Engineering* (Skip BO (ed)). Thomas Telford, London, UK, pp. 126–132.

Parry S and Baynes FJ (2014) Engineering geological models: an introduction. *Bulletin of Engineering Geology and the Environment* **73**: 689–706.

Preene M (2021) Conceptual modelling for the design of groundwater control systems. *Quarterly Journal of Engineering Geology and Hydrogeology* **54(2)**, 10.1144/qjegh2020-138.

Rappol M (1985) Clast fabric strength in tills and debris flows compared for different environments. *Geologie en Mijnbouw* **64**: 327–332.

Sammis C, King G and Biegal R (1987) The kinematics of gouge deformation. *Pure and Applied Geophysics* **125**: 777–812.

Schneider HR (1999). Definition and determination of characteristic soil properties. *Proceedings of the 12th International Conference on Soil Mechanics and Geotechnical Engineering, Rio De Janeiro, Brazil*, vol. 4, pp. 2271–2274.

Shaw J (1985) Subglacial and ice marginal environments. In *Glacial Sedimentary Environments* (Ashley GM, Shaw J and Smith HD (eds)). Society of Economic Paleontologists and Mineralogists, Tulsa, OK, pp. 7–84.

Skempton AW (1957) Discussion on the planning and design of the new Hong Kong airport. *Proceedings of the Institution of Civil Engineers* **7**: 305–307.

Stark TD, Choi H and McCone S (2005) Drained shear strength parameters for analysis of landslides. *Journal of Geotechnical and Geoenvironmental Engineering* **131(5)**: 575–588.

Stroud MA (1975) The standard penetration test in insensitive clays and soft rocks. *Proceedings of the European Symposium on Penetration Testing, Stockholm, Sweden*, vol. 2, pp. 367–375.

Terraconsult (2010) *Earthworks in Landfill Engineering: Design, Construction and Quality Assurance of Earthworks in Landfill Engineering*. Environment Agency, London, UK.

Trenter NA (1999) *Engineering in Glacial Tills*. CIRIA Report C504. CIRIA, London, UK.

Vaughan PR (1994). Thirty-fourth Rankine Lecture: Assumption. prediction and reality in geotechnical engineering. *Géotechnique* **44(3)**: 573–609.

Winter MG, Hólmgeirsdóttir TH and Suhardi (1998) The effect of large particles on acceptability determination for earthworks compaction. *Quarterly Journal of Engineering Geology & Hydrogeology* **31(3)**: 247–268.

Wroth CP and Wood DM (1976). The correlation and index properties with some basic engineering properties of soils. *Canadian Geotechnical Journal* **15(2)**: 137–145.

Yilmaz I (2000). Evaluation of shear strength of clayey soils by using their liquidity index. *Bulletin of Engineering Geology and the Environment* **59**: 227–229.

It is recommended this chapter is read in conjunction with

- Chapter 7 *Geotechnical risks and their context for the whole project*
- Chapter 13 *The ground profile and its genesis*
- Chapter 17 *Strength and deformation behaviour of soils*
- Chapter 40 *The ground as a hazard*

All chapters in this book rely on the guidance in Sections 1 *Context* and 2 *Fundamental principles*. A sound knowledge of ground investigation is required for all geotechnical works, as set out in Section 4 *Site investigation*.

Chapter 32
Collapsible soils

Ian Jefferson School of Engineering, University of Birmingham, Birmingham, UK
Chris DF Rogers School of Engineering, University of Birmingham, Birmingham, UK

doi: 10.1680/icemge.66816.0425

CONTENTS

Collapsible soils present significant geotechnical and structural engineering challenges the world over. They can be found in many forms – either naturally occurring or formed through human activities. However, an essential prerequisite is that an open metastable structure develops through various bonding mechanisms. Bonds can be generated through capillary forces (suctions) and/or through cementing materials such as clay or salts. Collapse occurs when net stresses (through loading or saturation) exceed the yield strength of these bonding materials. Collapse is most commonly triggered by inundation through a range of different water sources, although the impact varies, with different sources yielding different amounts of collapse. To engineer in and mitigate the effects of collapsible soils, it is essential to recognise their existence, which may not be easy, and to gather vital geologic and geomorphologic information. Collapsibility should be confirmed through direct response to wetting/loading tests using laboratory and field methods. The key challenge faced with collapsible soils is the spatial extent and the degree of wetting that will take place. Care is needed to ensure that appropriate and realistic assessments are undertaken. Ultimately, if treated using one of a suite of the possible improvement techniques available, then the potential for collapse can be eliminated effectively.

32.1. Introduction

Collapsible soils are extremely common and can be formed naturally owing to various geologic and geomorphologic processes, or be the result of human activity. These processes, although different in nature, allow the development of an open metastable structure: the essential prerequisite to the formation of a collapsible deposit (Dudley, 1970). Volume changes that occur are more sudden than those experienced through consolidation processes and typically occur in material that is non-plastic or of very low plasticity, confined and initially dry (Houston *et al.*, 2001).

Although collapsible soils are found for the most part in arid regions, arid environments tend to favour their formation. Naturally occurring collapsible soils are formed typically from debris flow (e.g. alluvial fan materials), as wind-blown sediments (e.g. loess), as cemented high-salt-content metastable soils (e.g. sabkha, see Chapter 29 *Arid soils*) and as tropical residual soil (see Chapter 30 *Tropical soils*). In addition, collapsible soils can be formed artificially through poor compaction control or where compaction is dry of optimum (e.g. non-engineered fill; see Chapter 34 *Non-engineered fills*), or as waste materials (e.g. fly ash beds; Madhyannapu *et al.*, 2006). However formed, common to almost all collapsible soils are both low densities and a relatively stiff and strong state when dry (as illustrated in **Figure 32.1**). Notable exceptions to this are post-glacial sensitive 'quick' clay soils found mainly in Canada, Alaska and Scandinavia, but these are arguably a special case, being geographically centred and in a saturated collapsible state. An additional exception is saturated slide material such as saturated sands that have been allowed to flow on slopes that have experienced liquefaction (Nieuwenhuis and de Groot, 1995).

Collapsible soils can be defined as soils in which the major structural units are initially arranged in an open metastable packing through a suite of different bonding mechanisms. If the soil is loaded beyond the yield strength of the bonding material, collapse will occur – this results in a rearrangement of particles to form a denser stable configuration (see **Figure 32.2**). Thus the collapse itself is controlled both microscopically and macroscopically. An appreciation of both these

Figure 32.1 Example of loess in its pre-collapsed state exhibiting its relatively stiff soft rock state when dry, allowing the formation of man-made caves, Slovakia

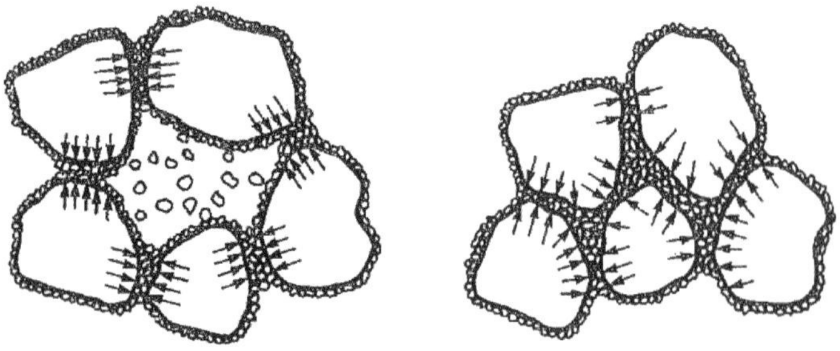

Figure 32.2 An illustration of collapse through inundation (Reproduced from Houston *et al.*, 1988; with kind permission from ASCE)

elements is essential if the true nature of collapse, and therefore its effective remediation, are to be fully understood.

Collapse is often triggered by a combination of increased stress (load) and the addition of water leading to increased degrees of saturation. However, collapse is most often triggered by an increase in water content. Li *et al.* (2016) have provided a comprehensive review of the various triggering mechanisms that produce collapse, with reference to the main ways the literature presents and deals with collapse mechanisms. There are a variety of sources of water ingress that can initiate collapse, many associated with urban environments. These include landscape irrigation, broken water or sewer pipes, run-off or poor drainage control, groundwater recharge or water content changes through capillary rise.

An example of the effects of urbanisation and associated landscaping was illustrated by a commercial building in New Mexico. After winning the city's most beautiful lawn and landscaping award, achieved through heavy watering, the building suffered US $500 000 foundation damage owing to soil collapse (Houston *et al.*, 2001). The result of infiltration with water results in a relatively sudden volume compression, and is often associated with loss of strength. This can clearly have important geotechnical consequences, including loss of serviceability resulting in expensive remediation or, on occasion, complete failure. This was illustrated by the structural failures caused by damage to foundations on collapsible soils in Egypt (Sakr *et al.*, 2008) and in the UAE (Vandanapu *et al.*, 2016), and the collapse problems caused by hydrogeological changes associated with the construction of a dam in Brazil (Vilar and Rodrigues, 2011).

Other reported problems have occurred in embankment bases (Thorel *et al.*, 2011), dam embankments (Peterson and Iverson, 1953), road embankments (Knight and Dehlen, 1963) and fill (Charles and Watts, 2001). It can be particularly problematical when collapse causes damage to historic buildings (Herle *et al.*, 2009), such as the cracking that developed in the wall of a fifteenth-century pagoda in Lanzhou, China, after the introduction of an irrigation scheme (see **Figure 32.3**).

Figure 32.3 Collapse-induced cracking in a fifteenth-century pagoda (Reproduced from Billard *et al.*, 2000; Wiley)

From this it is clear that, given the correct depositional environment, collapse has the potential to occur in any soil. In fact, given the correct stress environment, a collapsible soil may exhibit expansive behaviour (Barksdale and Blight, 1997; Derbyshire and Mellors, 1988; see also Section 32.3.3). It is therefore necessary to understand the process of collapse if problems associated with collapsible soils are to be avoided or mitigated. As collapse is for the most part triggered by increases in water content, problems are often encountered in urban and built environments and so the impact occurs where it has potential to cause the most harm. This can be particularly problematical as in many parts of the world collapsible soils exist in areas of high seismic actively (Houston *et al.*, 2003). The consequences can be catastrophic, such as the Haiyuan earthquake in which induced loess landslides in 1920 killed over 200 000 people (Derbyshire *et al.*, 2000; Zhang and Wang, 2007). Moreover, the potential for collapse will remain until full collapse has been induced either naturally through flooding and/or loading (earthquakes), or artificially by human activity. Human activity can be either accidental through poor drainage control, or deliberate through ground improvement (e.g. dynamic compaction).

32.2. Where are collapsible soils found?

Rogers (1995), Lin (1995), Bell and de Bruyn (1997), Houston *et al.* (2001, 2003) and Opukumo *et al.* (2022) discuss in detail the various forms of collapsible soils found worldwide. Collapse occurs because soil has certain inherent properties. Typical features that are found with most collapsible soils include

- an open metastable structure
- a high voids ratio and low dry density
- a high porosity
- a geologically young or recently altered deposit
- a deposit of high sensitivity
- a soil with inherent low inter-particle bond strength.

Thus, many soils can and will exhibit collapsible behaviour. This is illustrated in **Figure 32.4**, along with the various formation processes that can yield a collapsible soil.

Collapsible soils include naturally occurring soil formations, such as tropical residual soils, loess and quick clays, and anthropogenic soils such as uncompacted or poorly compacted fill. However, any compacted soil can exhibit collapse if the confining pressure is high enough (see Section 32.3.3). Further details of tropical and anthropogenic soil collapse can be found in Chapters 30 *Tropical soils* and 34 *Non-engineered fills*, respectively. Additional information on the collapse behaviour of residual soils is also provided by Barksdale and Blight (1997) and Roa and Revansiddappa (2002). Examples include decomposed gneiss (Feda, 1966), decomposed granites (Brink and Kanty, 1961) granitic gneiss (Pereira and Fredlund, 2000) and lateritic soil (Phanikumar *et al.*, 2016).

Other soils exhibiting collapse include collapsible gravels (Rollins *et al.*, 1994), granitic sands in South Africa (Jennings and Knight, 1975), collapsible sands of the southern African east coast (Rust *et al.*, 2005) and quick clays (Bell, 2000). Loess is probably the most encountered naturally occurring collapsible soil, covering some 10% of the world's landmass (Jefferson *et al.*, 2001). Many of the practices and engineering approaches used to deal with collapsible soils stem from the treatment of loess soils.

32.2.1 Wind-blown soils: collapsible loess soils

Loess consists essentially of silt-sized (typically 20–30 μm) primary quartz particles that form as a result of high energy earth-surface processes such as glacial grinding or cold climate weathering (Rogers *et al.*, 1994). These particles are transported from the source by rivers. Subsequent flooding by these rivers allows the quartz silt particles to be deposited on floodplains (Smalley *et al.*, 2007). On drying out, these particles are detached and transported by the prevailing winds until deposition leeward at distances ranging from tens to hundreds of kilometres. Cementing materials are often added after deposition or dissolved and re-precipitated at particle contacts (Houston *et al.*, 2001). This process has resulted in the almost continuous deposit draped over the landscape from the North China plain to south-east England (where it is generally referred to as 'brickearth') – see **Figure 32.5**.

It is possible to isolate five major loess regions worldwide: North America, South America, Europe (including western Russia), Central Asia and China (Smalley *et al.*, 2007). These

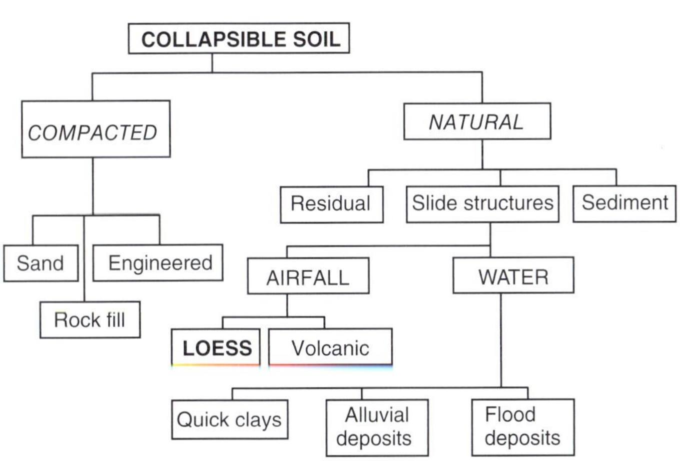

Figure 32.4 Classification of collapsible soils (Rogers, 1995)

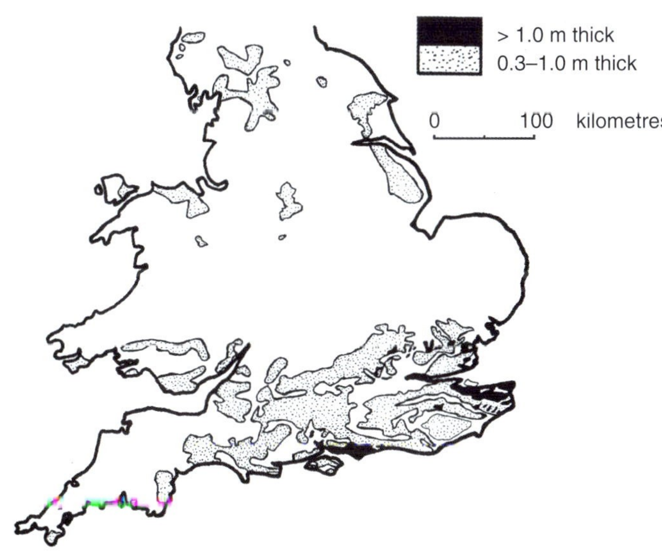

Figure 32.5 Distribution of 'brickearth' (loess) deposits in southern England and Wales (Modified from Catt, 1985; Allen & Unwin (as reproduced in Jefferson *et al.*, 2001))

loess regions often underlie highly populated areas and major infrastructure links, making them vulnerable to soil collapse. The areas of most widespread concern are concentrated in Eastern Europe, Russia and, to a growing extent, China (see Derbyshire *et al.*, 1995, 2000), although potentially serious problems of collapse exist wherever loess is found.

Figure 32.5 indicates the approximate distribution of loess/brickearth greater than 0.3 m in thickness in the UK. Significant thicknesses (greater than 1 m) are restricted to north and east Kent (see Fookes and Best, 1969; Derbyshire and Mellors, 1988), south Essex (see Northmore *et al.*, 1996) and the Sussex coastal plains. In Essex, deposits of up to 8 m have been found, although thicknesses of 4 m or so are more typical (Northmore *et al.*, 1996). However, substantial deposits can reach tens and even hundreds of metres thickness worldwide (see Smalley *et al.*, 2007 for further details).

Regional trends in loess occur in the type of deposit found around the world and these can be determined through textural and mineralogical distinctions. However, there is generally a progressive decrease in modal size with distance from its source material – suggesting that sorting of loess material occurs with wind direction during deposition (Catt, 1985).

Originally, loessic deposits would have been more extensive but they have been removed by post-depositional erosion, colluviation, deforestation/agriculture and resource-stripping activities (Catt, 1977). As a result, modern loess deposits are often only found overlying relatively permeable strata. Culshaw *et al.* (2020) provide further details of loessic and other collapsible deposits found across the UK.

32.2.2 Other collapsible deposits

As discussed in Sections 32.2 and 32.3, many soils exhibit collapsibility. Apart from those already highlighted above, a number of water-sediment deposits exhibit collapse potential. Two main groups exist: alluvial deposits and quick clays. Alluvial deposits reported to cause collapse problems include alluvial fans, alluvial floodplain deposits and mudflow deposits, with several case histories of collapse problems in alluvial fan deposits being provided by Rollins *et al.* (1994).

By comparison, saturated quick clay has become unstable owing to its post-glacial depositional environments, allowing an open structure to form by slow sedimentation under shallow marine conditions (Locat, 1995; Rogers, 1995). The resulting open fabrics are maintained by small amounts of carbonate cementation of clay minerals, with salt leaching often having occurred. Quick clays typically have liquidity indices greater than 1, with liquid limits often less than 40% (Bell, 2000). When disturbed, the particles in quick clays are remoulded into closely packed configurations. As the water content remains unchanged during the collapse process, quick clays become oversaturated and may flow as a viscous liquid. This can have devastating effects as witnessed by the 1978 landslide in Rissa, Norway, where, after the initial landslide, a series of minor slides developed which eventually covered an area of 3 300 000 m^2; or more recently by the large landslide in Leda clay in Ontario, Canada in 1993 (see Bell, 2000).

32.3. What controls collapsible behaviour?

Collapsibility occurs owing to the various geomorphologic and geologic processes as well as human activities during soil formation. The key to this is understanding the nature and processes that take place during provenance (P), transportation (T) and deposition (D) of the soil particles: it is the PTD sequence of a deposit that produces an open metastable collapsible structure of relatively high void ratio (Jefferson *et al.*, 2003b; Smalley *et al.*, 2007; Sun, 2002). The PTD approach conceptualises the processes of soil genesis, subsequent transportation and ultimate disposition, so elucidating how these influence the engineering behaviour of the final deposit.

However, some deposits have been through secondary or tertiary PTD sequences, often reducing or even eliminating collapse potential. An example to illustrate this is given in **Figure 32.6**, based on a PTD model of loess found in southern Britain, with further details discussed by Pye and Sherwin (1999) and Derbyshire and Meng (2005).

Wind-blown collapsible soils such as loess often consist of different zones of variable collapsibility, both laterally and with depth, as a result of their geomorphology. The prevailing wind direction generates zones of material laterally changing in nature from sandy loess through to clayey loess deposits (**Figure 32.7**). It should be noted that the terms clayey, silty and sandy loess are defined by particle size analysis (see Holtz and Gibbs, 1952). In turn, the degree of saturation and potential collapsibility often follows similar trends. For example,

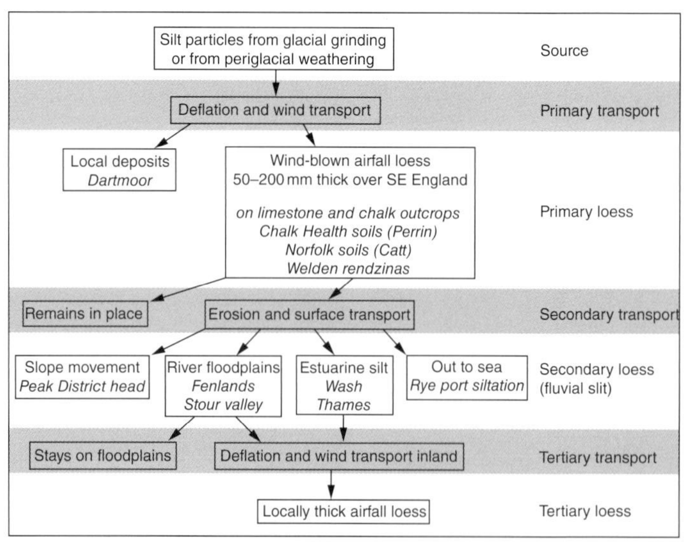

Figure 32.6 A PTD geomorphological model of loess in southern Britain. Note: Norfolk soils (Catt *et al.*, 1971); Chalk Heath soils (Perrin, 1956) (Reproduced from Jefferson *et al.*, 2003b; East Midlands Geological Society)

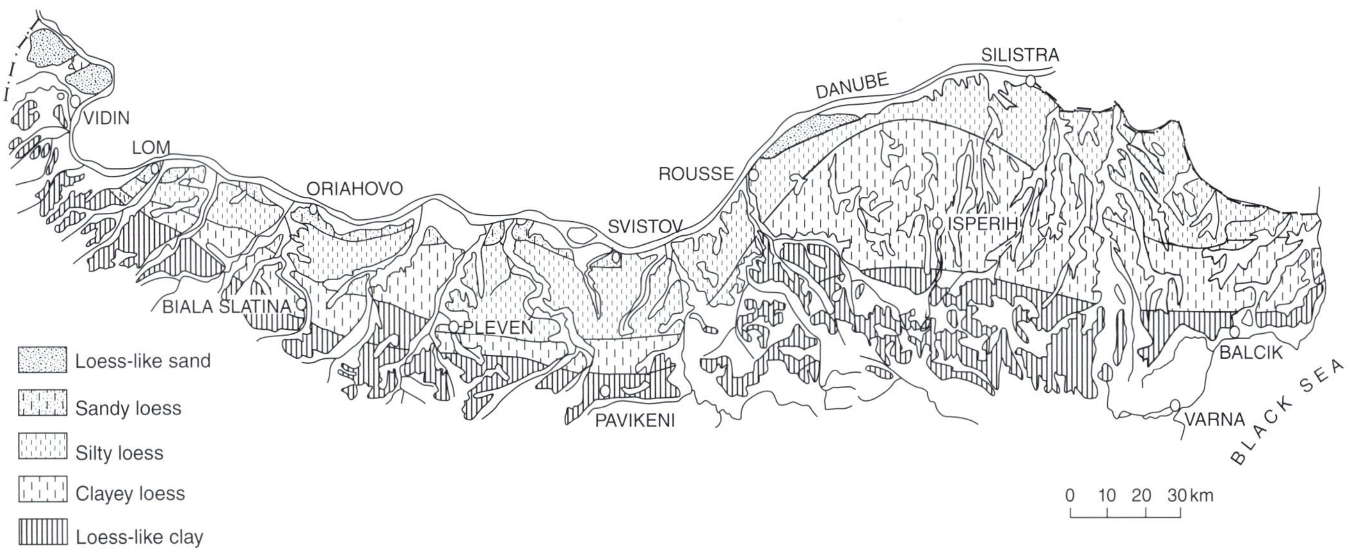

Figure 32.7 Distribution of collapsible loess soils from Danube floodplain in Bulgaria (Reproduced from Minkov, 1968; Marin Drinov Academic Publishing House/BAS Press)

the loess in Bulgaria (**Figure 32.7**) exhibits an increase in saturation as the modal size reduces (Jefferson *et al.*, 2002). The thicknesses of loess formed often follow the same pattern – with the greatest thicknesses occurring nearest to the source material.

In addition, deposits such as loess have been formed over many thousands of years and during this time climactic conditions have varied considerably. As a result, the loess soil sequence has alternating layers of loess (formed during cold periods) and clay-rich palaeosols (formed during warm periods) – see **Figure 32.8** for illustration.

The alternating nature of loess formation significantly influences the engineering behaviour and ultimately the nature of collapse, and the location (depth) where collapse occurs. This will dictate the nature of the infiltration patterns of water into the soil and as a result can yield collapse in unexpected locations. Moreover, this can influence the effectiveness of any ground improvement approach used to remove collapsibility.

As a result of the various geomorphologic processes involved in the formation of loess, these deposits typically have three zones of relative collapsibility.

- *Zone 1:* A zone at depth of collapsed material due to overburden pressure.
- *Zone 2:* A collapsible zone.
- *Zone 3:* A surface crust (which requires additional load to cause collapse). This has led countries with extensive loess deposits (including those of Eastern Europe and the former Soviet Union) to develop classification schemes related to the collapse of loaded foundations (**Figure 32.8**). Two types of collapsibility occur in such schemes
  - *Type I:* mainly loaded collapsibility (collapse deformation under overburden pressure of $\delta_n < 5$ cm)
  - *Type II:* mainly unloaded collapsibility ($\delta_n > 5$ cm, where δ_n is the collapse deformation).

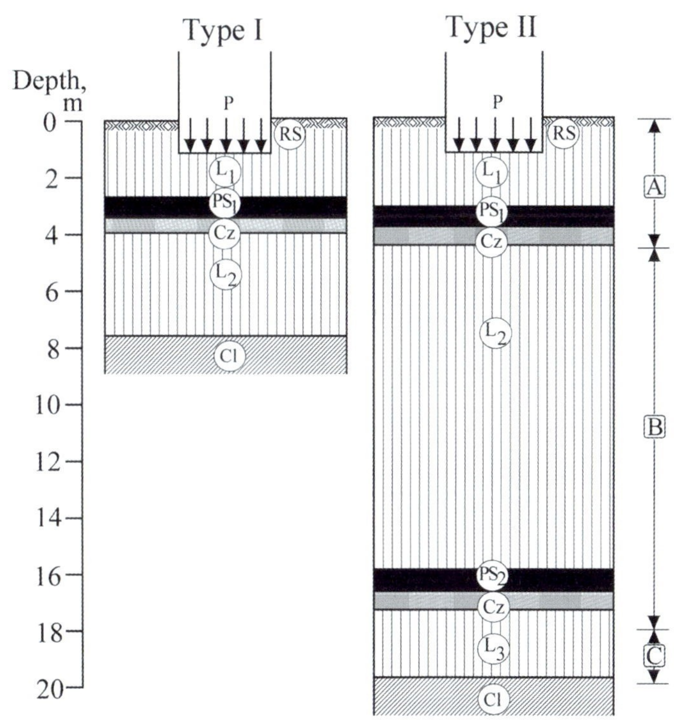

Figure 32.8 Types of collapsible loess soils (Reproduced from Jefferson *et al.*, 2005; with permission from Elsevier)

Type I loess is usually of small thickness (shown in **Figure 32.8**) and contains one or two palaeosols (PS), together with an associated carbonate zone (Cz). Collapse occurs after the foundation stress exceeds a certain critical stress, which can be determined by laboratory tests or tests in the field (see Section 32.4).

Type II loess has greater thickness – up to 50 m or more. **Figure 32.8** shows a typical case of a Danubian terrace with a deposition of loess of about 20 m. In this case, three loess horizons (L_1, L_2 and L_3) are separated by two palaeosols (PS_1 and PS_2). In the Type II loess three zones can be distinguished

- upper zone A – no unloaded collapsibility but with potential loaded collapsibility
- middle zone B – unloaded collapsibility
- lower zone C – uncollapsible (or collapsed).

Loess in zone C has previously collapsed under overburden pressure. Loess in zone A has had no unloaded collapsibility, since here the overburden pressure is small (i.e. it will not collapse under self-weight). However, it can be collapsible under additional load. In zone B, unloaded collapsibility occurs – it often contains thicker loess horizons with lower density and higher porosity, n (i.e. $n > 50\%$), and it has a higher silt content than in zone C.

Further details of the different geomorphologic processes that generate collapsibility are given in Chapters 30 *Tropical soils* and 29 *Arid soils*, and an excellent treatment of the broader subject is provided by Fookes *et al.* (2005).

32.3.1 Bonding mechanisms and fabric

For collapse to occur, an open structure with relatively large voids must exist together with a source of strength to hold soil particles in position, resisting shearing forces associated with the current stress environment. To achieve this, bonding between particles or grains of sufficient strength must occur which, when weakened by the addition of water and/or an additional load, allows particles to slide over one another – resulting in collapse. There are generally considered to be three main bonding mechanisms present in collapsible soils (Barden *et al.*, 1973; Clemence and Finbarr, 1981; Li *et al.*, 2016; Rogers, 1995)

- capillary or matric suction forces (**Figure 32.9(a)**)
- clay and silt particles at coarser particle contacts (**Figure 32.9(b)–32.9(d)**)
- cementing agents, such as carbonates or oxides (see **Figure 32.9(e)**).

The fabric of a collapsible soil takes the form of a loose skeleton built of grains (generally quartz in the case of loess) and micro-aggregations (in the case of loess assemblages of clay or clay and silty-clay particles, see **Figure 32.10**). In some soils, such as loess, carbonate diagenesis may strengthen the meniscus clay bridges between silt grains and further influence collapse potential. Recent observations by Milodowski *et al.* (2015) suggest that there are three variants of the clay bridges: (*a*) simple clay meniscus films; (*b*) clay films developed on a scaffold of an earlier meniscus of fibrous calcite; and (*c*) clay films permeated or encrusted by microcrystalline calcite and/or dolomite. Hence the strength can vary both laterally and vertically over relatively short distances. This, together with

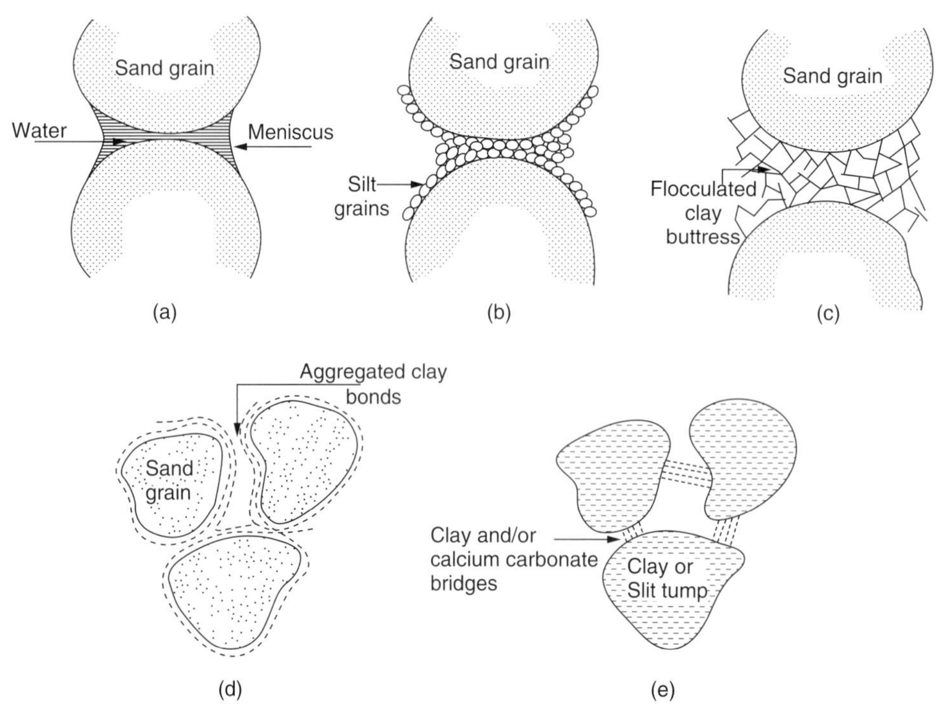

Figure 32.9 Typical bonding arrangement formed in collapsible soils (Reproduced from Popescu, 1986; with permission from Elsevier)

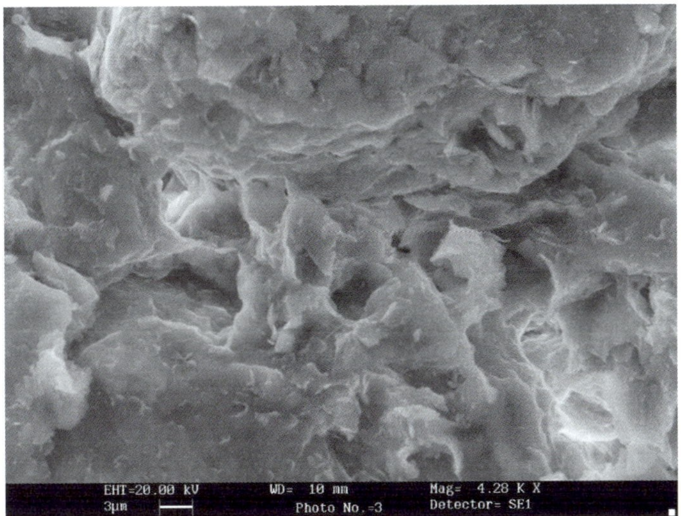

Figure 32.10 Scanning electron micrograph of bonded quartz particles in brickearth (loess) from Ockley brickworks, Sittingbourne, Kent (Reproduced from Jefferson *et al.*, 2001; all rights reserved)

the durability of the cementing material, needs to be taken into account for engineering purposes. Further discussion is provided by Li *et al.* (2016).

It should be noted that the micro-fabric of loess soils that have experienced reworking (see **Figure 32.6**) are often anisotropic and as a result exhibit much reduced collapsibility (Pye and Sherwin, 1999). In addition, as with other collapsible soils, domains of pelletised material can form and be arranged in a loose framework. Further details are given in Klukanova and Frankovska (1995), Jefferson *et al.* (2003b) and Milodowski *et al.* (2015). Derbyshire and Meng (2005) provide further discussion of fabrics associated with loess soils in China.

Other collapsible soils, such as quick clay, have essentially the same open metastable fabric, but are generated under different geomorphologic conditions. Further details are provided by Locat (1995), Bentley and Roberts (1995) and Bell (2000); see also Section 32.2.2. Further details related to other collapsible deposits are given elsewhere in this manual (see Chapters 34 *Non-engineered fills* and 30 *Tropical soils*).

It should be noted that although fabric is widely recognised as important in explaining collapse behaviour, it lacks a simple quantitative descriptor (Alonso, 1993; as cited by Pereira and Fredlund, 2000).

32.3.2 Mechanisms of collapse

Collapse in cemented soils typically involves the destruction of all three bonding types. In contrast, collapse in uncemented dry soils is solely due to the destruction of capillary forces. The strength derived from suction and cementing can be characterised in similar ways. However, on wetting suction will be lost, whereas chemical bonding is likely to be less affected by a change in suction. However, salt and clay bonds that occur at particle contacts will tend to be removed or weakened after inundation and hence collapse occurs. A detailed discussion of the key mechanisms of collapse is provided by Li *et al.* (2016).

Petrographic evidence indicates that collapse in cemented loess soils occurs in three stages after inundation (Klukanova and Frankovska, 1995; Milodowski *et al.*, 2015).

- *Stage 1:* Dispersion and disruption of clay bridges or buttresses between loosely packed silt grains, leading to initial rapid collapse of inter-ped matrix.
- *Stage 2:* Load taken up through contact between adjacent compact silt peds, which rearrange into a closer packing.
- *Stage 3:* With increased loading, progressive deformation and shearing of peds occur, resulting in further collapse as peds disaggregate and silt particles collapse into now-unsupported inter-ped areas.

Similar observations concerning the collapse process in compacted soils have been made by Pereira and Fredlund (2000), who suggest the following phases.

- *Phase 1 (pre-collapse):* High matric suction generates a metastable structure that suffers small volume deformations with decreased suction. No particle slippage occurs and structure remains intact.
- *Phase 2 (collapse):* Intermediate matric suction and a significant decrease in volume occur, altering the structure through bond breakage.
- *Phase 3 (post-collapse):* Saturation is approached and no further decrease in volume (or reductions in matric suction) will occur.

Pereira and Fredlund (2000) further observed that as the net confining pressure increases, the wetting-induced collapse becomes greater and the matric suction associated with phases 1 and 3 will be higher.

During the collapse process, the nature of pores also changes. Studies on loess have shown that most of the macropores (100–500 μm) are typically destroyed, leaving smaller intergrain and inter-aggregate pores (8–100 μm) (Osopov and Sokolov, 1995). Similar observations with respect to collapse mechanisms were made by Klukanova and Frankovska (1995) and Feda (1995), and more recently on residual soils by Roa and Revansiddappa (2002).

Cerato *et al.* (2009) observed that compacted soils with a greater number of smaller clods showed greater collapse, with collapse largely dependent on the inter-aggregate and intra-aggregate pore distribution. Moreover, when clods are in a drier than optimum water content, stronger state, higher yield stresses result and the overall soil structure is less prone to collapse. In some fill materials the parent material may also lose some strength, or aggregates within the fill may soften as its water content increases – resulting in a possible collapse (Charles and Skinner, 2001; Charles and Watts, 2001; Lawton *et al.*, 1992).

For uncemented soil, collapse is related to the destruction of capillary (matric suction) forces, with water infiltration producing wetting fronts. The volume change associated with collapse is confined to the wetted zone (Fredlund and Gan, 1995). As matric suction can be visualised as isochrones (akin to excess pore water pressures seen in consolidation) analysis can be undertaken in much the same way. Further details of this, including experimental observations, are provided by Fredlund and Gan (1995). Collapse deformations that occur owing to suction reductions have been found to depend mainly on soil density and the stress state under which collapse occurs (Sun *et al.*, 2007).

Pereira and Fredlund (2000) highlighted key features in the collapse of compacted soils.

- For any type of soil compacted dry of optimum water content, collapse can occur.
- High microforces of shear strength exist through bonding, chiefly through capillary action.
- Compressibility gradually increases and shear strength gradually decreases in collapsible soils during saturation.
- Soil collapse progresses with increasing degree of saturation. However, above a critical degree of saturation no further collapse occurs.
- Collapse is associated with localised shear failure (see further discussion in Section 32.4.5).
- During wetting-induced collapse under constant load and anisotropic oedometer conditions, horizontal stresses increase.
- For a given mean normal total stress under triaxial conditions, the magnitude of axial collapse increases and radial collapse decreases with increased stress ratio.

Fill collapse potential is thus controlled by placement conditions, water content history and stress history. Further discussion on this subject and of other poorly compacted materials is provided by Charles and Skinner (2001), Charles and Watts (2001), and elsewhere in this manual; see Chapter 34 *Non-engineered fills*.

32.3.3 Modelling approaches: collapse prediction

Most collapsible soils exist in a partially saturated state. The suctions that develop are made up of two components: matric suction and osmotic suction, the sum of which is known as total suction. Discussion of these aspects and their implications is presented in Chapter 30 *Tropical soils*, with reference to collapsible loess provided by Li *et al.* (2016) and a more detailed treatment provided by Fredlund and Rahardjo (1993) and Fredlund (2006).

This approach has important limitations and a more complete model has been developed by Alonso *et al.* (1990). Their approach extended the modified Cam Clay model to unsaturated soils and introduced the loading–collapse (LC) surface to define yielding due to either external loading (total stress) or saturation (loss of suction). Experimental evidence for this has been presented by a number of authors (e.g. Jotisankasa *et al.*, 2009). The LC model further demonstrated the stress path dependency of collapse and explains why, under lower net stress, water inundation may cause swelling and at higher net stresses collapse occurs. Therefore any modelling approaches used should treat soils as potentially expansive or collapsible in the same framework. However, it should be noted that collapse, unlike swelling, is an irreversible process.

The effects of bonding and bond yield strength are important aspects in many collapsible soils. Further details are discussed by Leroueil and Vaughan (1990), Maâtouk *et al.* (1995), Malandraki and Toll (1996) and Cuccovillo and Coop (1999).

Reviews of the constitutive models used to assess partially saturated soils are provided by D'Onza *et al.* (2011) and Sheng (2011), with discussion on their limitations provided by Zhang and Li (2011). Overall, constitutive models for partially saturated soils deal with the mechanical stress–strain and hydraulic suction–saturation relationships.

Constitutive models have allowed a number of numerical approaches to be developed. However, these often need a range of parameters for analysis and so may not be cost-effective for many projects. However, this can be mitigated to some degree using approaches developed by Nobar and Duncan (1972) and Farias *et al.* (1998) that utilise, for example, stress–strain curves for dry and saturated behaviour. A similar approach has been advocated by Charles and Skinner (2001) and Skinner (2001) when dealing with fill. In addition, discrete element methods (DEMs) are allowing micromechanical aspects of collapsible soils to be examined (see Liu and Sun, 2002, for further details).

32.4. Investigation and assessment

In order to provide an economical and efficient engineering solution, four basic steps must be undertaken when dealing with collapsible soils (after Popescu, 1986).

- *Identification:* Determine whether a collapsible soils exists.
- *Classification:* If a collapsible soils exists, how significant is it?
- *Quantification:* Assess the degree of collapse that will occur.
- *Evaluation:* Assess the design options.

However, one of the greatest problems with collapsible soils is that their existence and the extent of their collapse potential are often not recognised prior to construction (Houston *et al.*, 2001). Thus it is essential to first identify a collapsible soil and then to estimate its collapse potential, particularly (but not exclusively) on sites containing water-sensitive soils.

Engineers often mistake, or simply do not recognise the presence of, a collapsible soil. Current standards relating to soil field descriptions, used by engineers, tend to group all fine materials together (e.g. silts and clays) under a common descriptor, which does not help in this regard. While there are practical reasons for this, such groupings potentially reduce the ability of engineers to identify and assess whether a soil is collapsible. Moreover, even though a considerable database of knowledge exists globally, much of this work tends to be lost owing to use of formats and terms unfamiliar to engineers or simply suffers from language barriers (Jefferson *et al.*, 2003a).

Popescu (1986), Houston and Houston (1997) and Houston *et al.* (2001) provide excellent overviews of the key aspects associated with the identification and characterisation of collapsible soils. When characterising a collapsible soil, Houston *et al.* (2001) suggest the following stages be undertaken

- reconnaissance
- use of indirect correlations
- laboratory testing
- field testing.

These aspects are, in general, common to the investigation of expansive soils; reference to discussions in Chapter 33 *Expansive soils* would be useful.

32.4.1 Reconnaissance

Using reconnaissance to gather useful geologic and geomorphologic information can be useful in anticipating collapse by providing clues on what to look out for. The first step is to understand geologic and geomorphologic settings (see Section 32.3). For example, Lin (1995) found that there was a strong correlation between geomorphologic information and collapsibility. It may be that the underlying assumption should be that a deposit is collapsible until confirmed otherwise – as in the case of Beckwith (1995), who recommends that alluvial fans should all be assumed to be collapsible. Charles and Watts (2001) make a similar suggestion when dealing with partially saturated fill – until there is adequate evidence to the contrary. Further clues to the likelihood of collapsible soils can be gained from prior history and environmental factors.

Lin (1995) highlighted how the collapse of loess soils was influenced by age, overburden pressure and the degree of saturation and suction, illustrating the difficulties associated with determining whether or not a soil is collapsible. Popescu (1992) highlighted other aspects that need to be considered when assessing the collapsibility of soils. These include

- internal factors
  - mineralogy of particles
  - shape and distribution of particles
  - nature of inter-particle bonding/cementing
  - soil structure
  - initial dry density (which is often low)
  - initial water content
- external factors
  - availability and nature of water
  - applied pressure
  - time permitted for water percolation to occur
  - stress history
  - climate.

Thus a detailed programme of classification and quantification is necessary to fully assess collapse potential across a site.

Details of reconnaissance and its role in site investigations are dealt with elsewhere in the manual – see Chapter 45 *Geophysical exploration and remote sensing*. Further information can also be found in Fookes *et al.* (2005).

32.4.2 Indirect correlations

Various collapse coefficients related to loess have been produced, which include ever more parameters (e.g. Basma and Tuncer, 1992; Fujun *et al.*, 1998). However, these are unnecessarily complex and the traditional collapse potential, for example the one described by Gibbs and Bara (1962), is better owing to its relative simplicity.

Many criteria and correlations have been proposed in the assessment of collapse potential based on soil properties such as natural water content, void ratio or index properties (see Bell, 2000; Northmore *et al.*, 1996; Rogers *et al.*, 1994). However, these can be misleading as they are often based on remoulded and approximate soil properties; therefore, inappropriate evaluation can occur (Northmore *et al.*, 1996). Many of the correlations available have met with only moderate success owing to their weak correlations and considerable scatter (Houston *et al.*, 2001). Thus, it is more efficient and economical to use either laboratory or field tests when assessing collapse potential. This has the added advantage of providing not only identification, but also assessment data.

32.4.3 Laboratory testing

The most effective method to access collapsibility is through collapse tests. The actual collapse potential is traditionally measured using double and single oedometer tests (Jennings and Knight, 1975), which have been subsequently modified by Houston *et al.* (1988). The amount of collapse strain produced when the test specimen is flooded under a given pressure indicates a sample's susceptibility to collapse.

Figure 32.11 illustrates a typical response (Houston *et al.*, 1988), where a seating stress of 5 kPa has been used to establish an initial state. Any compression under this stress is attributed to sample disturbance. The initial compression curve A–B–C represents the response of the soil at its in situ water content. Pressure is applied until the stress on the sample is equal to (or greater than) that expected in the field. At this point, the sample is inundated and compression measured (C–D in **Figure 32.11**), after which further loading is undertaken, corresponding to line D–E. This equates to the three phases of collapse described by Pereira and Fredlund (2000).

The amount of collapse of a layer is found by multiplying the thickness of the layer by the collapse strain, using values

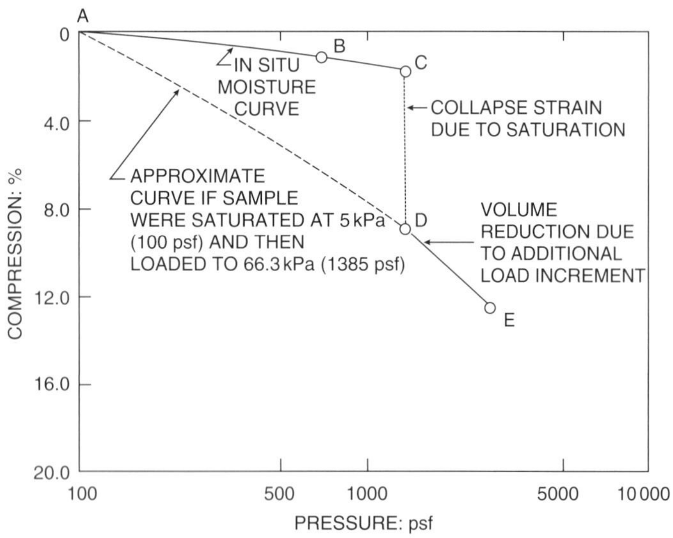

Figure 32.11 Compression curves for modified Jennings and Knight oedometer test (NB 1 pound per square foot (psf) = 0.0479 kPa) (Reproduced from Houston *et al.*, 1988; with kind permission from ASCE)

corresponding to the final stress at the midpoint of the zone in question. Further details are provided by Houston *et al.* (1988), who note that collapse strains vary both laterally and with depth, so requiring integration of strains along a vertical column of zones to estimate surface settlements.

The double oedometer test uses two near-identical specimens of soil and incremental stress increases are applied to one specimen in its natural state and to the other specimen that has been immediately saturated at a seating stress of typically 5 kPa. Based on the double oedometer test, the degree of collapsibility can be assessed and used to provide an indication of the potential severity of collapse. **Table 32.1** provides details presented by Jennings and Knight (1975), which indicate a 1% collapse can be regarded as metastable. However, this cut-off varies across the world, with values of 1.5% taken in China (Lin and Wang, 1988) and values in excess of 2% in the USA used to indicate soils susceptible to collapse (Lutenegger and Hallberg, 1988).

Collapse (%)	Severity of problem
0–1	No problem
1–5	Moderate trouble
5–10	Trouble
10–20	Severe trouble
>20	Very severe trouble

Table 32.1 Collapse percentage (defined as $\Delta e/(1 + e)$, where *e* is the void ratio) as an indication of potential severity (Data taken from Jennings and Knight, 1975)

Collapsible fill identification is often best achieved by testing samples at various water contents and dry densities over the range of stress levels expected (Charles and Skinner, 2001; Houston *et al.*, 2001).

Although only approximate, double oedometer tests do give a repeatable and reproducible qualitative indication of collapse. Often it is the collapsibility risk that is more important to assess than the actual amount of collapse that will occur. However, traditional oedometer tests suffer from sample disturbance effects and often reach saturations not commonly encountered in the field (Rust *et al.*, 2005). The extent of sampling effects has been debated, as have the relative merits of the use of block and tube sampling when testing collapsible soils (see Day, 1990; Houston and El-Ehwany, 1991; Houston *et al.*, 1988; Neely, 2010). Northmore *et al.* (1996) observed that, upon flooding in the oedometer, certain specimens of brickearth became saturated almost instantaneously with a rapid intake of water into the pore space.

Thus, at best, traditional oedometer collapse tests should be considered as index tests; for full collapse evaluation, a field trial should be conducted. When estimating collapse settlements, significant suctions may remain after wetting and soil will remain only partially collapsed (see Section 32.5.1 for further discussion). This must therefore be accounted for in any assessments.

Other test methods have been employed to characterise collapsibility; these include suction-monitored oedometers (e.g. Dineen and Burland, 1995; Jotisankasa *et al.*, 2007; Vilar and Rodrigues, 2011), Rowe cells (Blanchfield and Anderson, 2000) and triaxial collapse tests (e.g. Lawton *et al.*, 1991; Pereira and Fredlund, 2000; Rust *et al.*, 2005). Further details are provided by Fredlund and Rahardjo (1993) and Rampino *et al.* (2000). Tarantino *et al.* (2011) provide a review of techniques used for measuring and controlling suctions, while Fredlund and Houston (2009) discuss protocols for the assessment of unsaturated soil properties in geotechnical practice. Vilar and Rodrigues (2011) provide a useful example of suction measurement in the assessment of collapse. Further details are presented in Chapter 30 *Tropical soils.*

32.4.4 Field testing

Field methods have traditionally used plate loading tests (Reznik, 1991, 1995; Rollins *et al.*, 1994) and, more recently, pressuremeter tests to determine collapse potential (Schnaid *et al.*, 2004; Smith and Rollins, 1997). Francisca (2007) provides details of the use of standard penetration tests (SPTs) to evaluate the constrained modulus and collapsibility of loess in Argentina, with higher *N* values being recorded in soils of a lower collapse potential. However, care is needed to ensure uniformity of stress state in the collapse region. This is often the main disadvantage with in situ collapse tests and has led a number of researchers to develop more sensitive test methodologies.

For example, Handy (1995) devised a stepped blade method to evaluate lateral stress changes. Methods to determine

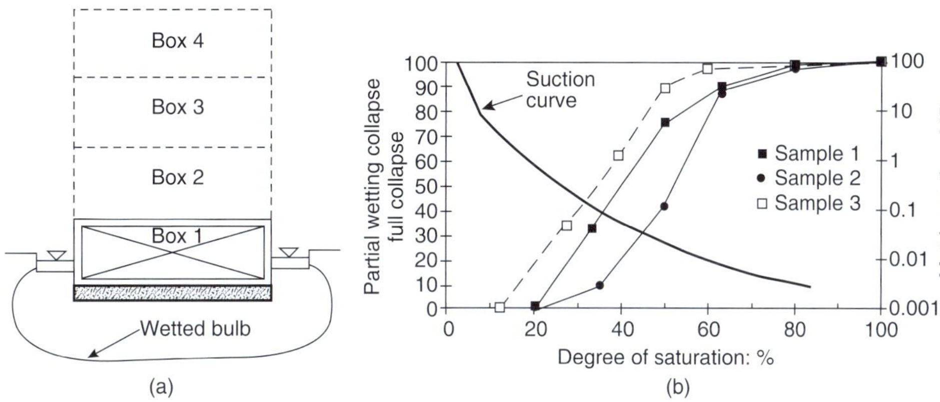

Figure 32.12 (a) Experimental layout to measure collapse of low-plasticity silt; (b) the relationship between partial wetting collapse, matric suction and degree of saturation (samples 1, 2 and 3 refer to three different silt soils of low plasticity) (Reproduced from Houston *et al.*, 2001; with kind permission from Springer Science+Business Media)

response to wetting were developed by Houston *et al.* (1995b) (the down-hole plate test) and Mahmoud *et al.* (1995) (the box plate load test). Recently, Rocha *et al.* (2023) discussed identification of collapsible soils from seismic cone tests (SCPT) using a qualitative approach. A brief but useful overview of interpretation and comparison of collapse measurement techniques, including their relative merits, is provided by Houston *et al.* (1995a). These include minimal sample disturbance, large soil volumes tested and degree of wetting likely to be similar to the prototype.

Houston *et al.* (2001) present field investigations of the collapse potential of a low-plasticity silt using an in situ test apparatus (**Figure 32.12(a)**). Boxes of concrete or steel were lowered onto a concrete pad and filled with soil, the base of the foundation was inundated with water and movements were recorded (see also Mahmoud *et al.*, 1995). The relationship between partial collapse, matric suction and degree of saturation (**Figure 32.12(b)**) highlights the importance of understanding the likely changes in water content of the soil surrounding a structure – something which Houston *et al.* (2001) suggests can be very difficult to predict accurately.

Houston *et al.* (2001) also provide a comparison between collapse predicted for a soil investigated using in situ methods (not the same tests as presented in **Figure 32.12**) and laboratory methods (**Figure 32.13**), and suggest that there are difficulties involved with the in situ approach. The load applied can be controlled, but the region affected by wetting and the degree of wetting can be very difficult to control.

Smith and Rollins (1997) investigated the use of a borehole pressuremeter to investigate the collapse potential of an arid soil. Once at the desired depth within the borehole, the pressuremeter expands radially to apply a pressure to the annulus of the borehole and measures the dry modulus of the soil (E_D). A set volume of water is discharged through the pressuremeter into the surrounding soil; the modulus during collapse (E_C) and wet modulus (E_W) post-collapse are then measured (see **Figure 32.14**). Smith and Rollins (1997) suggest that the moduli ratios E_C/E_D and E_W/E_D can be used to predict the collapsibility of a potentially collapsible soil (**Table 32.2**).

With recent improvements in technology, geophysical approaches have been advanced as a method to determine collapse potential (Evans *et al.*, 2004; Rodrigues *et al.*, 2006). The power of geophysical approaches to assess when a collapsible zone is present and its extent across a site was illustrated by Northmore *et al.* (2008) (see **Figure 32.15**).

Northmore *et al.* (2008) illustrated how depth and lateral extent of both collapsible and a non-collapsible loessic brickearth could be ascertained using a suite of different geophysical approaches, including electro-magnetic (EM31 and EM34), electrical resistivity and shear wave profiles. Further details are provided by Gunn *et al.* (2006) for shear wave measurements and Jackson *et al.* (2006) for resistivity measurements. However, it is vital that adequate calibration is undertaken through laboratory testing. For this, traditional

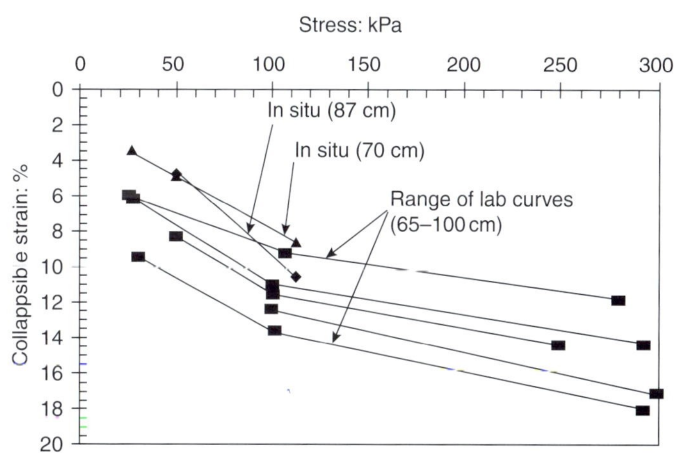

Figure 32.13 Comparison of collapsible strains for a soil investigated in situ and in the laboratory, using plate sizes of 70 cm and 87 cm in the plate bearing tests (Reproduced from Houston *et al.*, 2001; with kind permission from Springer Science+Business Media)

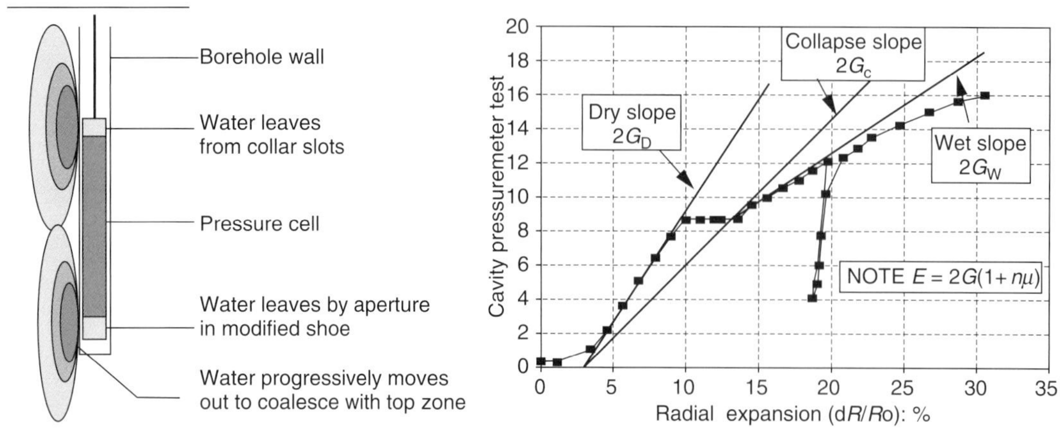

Figure 32.14 Illustration of pressuremeter (water only illustrated on left-hand side for clarity) and determination of the dry, collapsible and wet moduli, where G is shear modulus and μ is Poisson ratio (Reproduced from Smith and Rollins, 1997; ASTM)

Criteria			Collapse potential
E_C/E_D		E_W/E_D	
~1		~1	Not collapsible
<1	and	~1	Moderate collapse
~1		<1	Moderate collapse
<1		<1	Possible severe collapse

Table 32.2 Use of modulus ratios to predict collapsibility of potentially collapsible soil (Data taken from Smith and Rollins, 1997)

double oedometer and Panda probe profiling are used, where the Panda probe is a lightweight dynamic cone penetrometer which allows detailed physical soundings down to around 5 m to be taken (Langton, 1999). Once achieved, a full assessment of the presence of a potentially collapsible soil across the site can be made.

An important aspect with any field evaluation is the number of tests needed to adequately characterise the collapse potential of a loess soil. Houston *et al.* (2001) discuss statistical approaches developed to evaluate the minimum number of tests required to satisfactorily characterise a site and its collapse potential.

32.4.5 Assessment of wetting

The most challenging task for improving collapsible soils is the assessment of wetting extent and degree of potential future wetting. This is particularly true in arid or semi-arid environments where collapsible soils have not been wetted to any significant depth. However, even with collapsible deposits found in more humid environments, the deposits are often situated where significant wetting at depth has not occurred, or only part saturation has taken place, rendering the deposit still partially collapsible (Charles and Watts, 2001; Northmore *et al.*, 1996).

In general, detrimental effects of collapse occur in the zones both under foundations and within the probable wetting front (Houston *et al.*, 1988). Ponding tests may be used which can give an indication of depth and lateral extent of water migration, from which the best foundation option can be determined. El-Ehwany and Houston (1990) present results from laboratory infiltration wetting fronts and, by comparing these with observed rates, predict the depth of wetting against time. They then describe how this information can be used to predict collapse settlements, taking account of partial wetting. More recently, Yanrong *et al.* (2018) present a numerical simulation of collapse in loess subjected to rising groundwater, finding their simulation produced good agreement with both laboratory and field (in situ water immersion) tests, providing useful illustrations of how rising groundwater impacts collapse settlements.

Clearly an assessment of the extent and degree of wetting is essential to determine collapse potential and the scope and requirements for any treatment processes. Many practitioners tend to be conservative and assume the degree of wetting equates to 100%, particularly if the collapsible zone is near to the surface and does not extend too deeply (Houston and Houston, 1997). Full wetting of a collapsible soil would only be expected with rising groundwater. However, this is not often the case, and saturation only usually reaches between 35% and 60%, particularly when downward infiltration occurs. Hence the additional costs associated with such a conservative assumption may not be warranted (Houston *et al.*, 2001). El-Ehwany and Houston (1990) found that 50% saturation produces 85% of full collapse, agreeing approximately with the observations presented in **Figure 32.12(b)**. They suggest that full collapse is achieved at 65–70% saturation. Lawton *et al.* (1992) and others have shown that partial saturation will first trigger partial collapse, with full collapse occurring at saturation values as low as 60% – a figure that Bally (1988) agrees with. However, Osopov and Sokolov (1995) considered

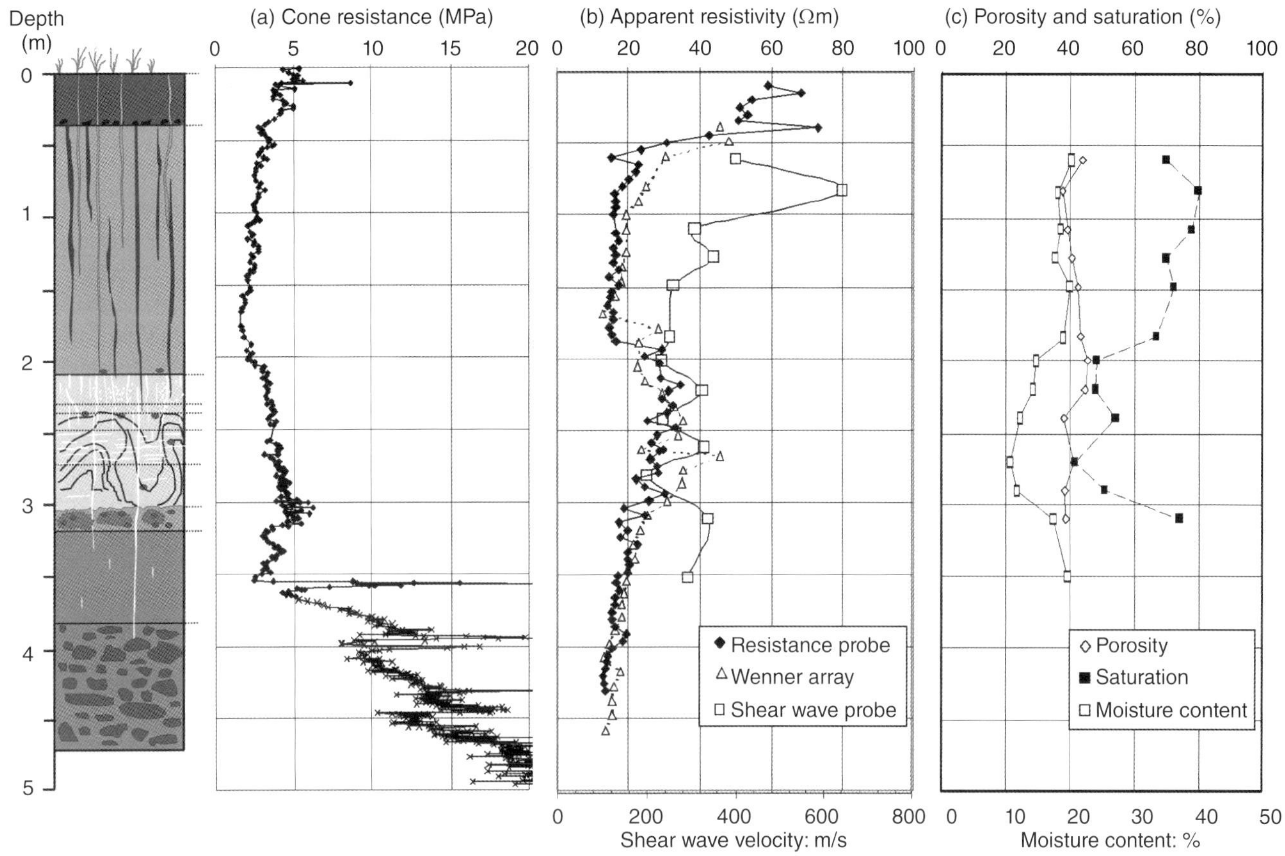

Figure 32.15 Profile through collapsible (~2–3 m) and non-collapsible (~0.5–2 m) brickearth – with collapsible deposits existing at 2–3 m, correlating to a drop in the degree of saturation (Reproduced from Northmore *et al.*, 2008; all rights reserved)

that full collapsibility would be realised only when saturation exceeded 80%.

This has implications for the prediction of collapse settlement, based on laboratory tests, which overestimate collapse strain and generally produce a greater degree of saturation than is achieved in the field; estimates put the overestimation at around 10% (El-Ehwany and Houston, 1990). However, this is not particularly large given the nature and accuracy of settlement predictions in general.

Wetting effects can be modelled using unsaturated stress state variables: net normal stress ($\sigma - u_a$) and matric suction ($u_a - u_w$) – further details can be found in Fredlund and Rahardjo (1993). The matric suction changes during wetting can be indicated by soil–water characteristic curves (SWCCs); further details are discussed in Chapter 30 *Tropical soils*. Houston *et al.* (2001) provide a range of SWCCs for collapsible loess soils from around the world. Further details of the evaluation of wetting using suction measurements have been discussed by Walsh *et al.* (1993). More recently, Al-Obaidi and Schanz (2023) examined SWCCs for three types of collapsible soils, namely a sandy gypseous, a silty loess and an artificial gypsum–sand mixture, illustrating the difference in the hysteresis variability and how this was impacted by geology, and homogeneity of the soil's fabric.

All collapsible soils will experience partial collapse under partial wetting conditions. However, the shape and position of the partial collapse curve depend on the soil type, fines content and type of bonding present. Overall, it is clear that assessment of the extent and degree of wetting is the most difficult part of collapsibility evaluation.

32.5. Key engineering issues

32.5.1 Foundation options

Four basic approaches exist when dealing with design solutions in collapsible soils (Popescu, 1992).

- Use very stiff raft foundations and a rigid superstructure to minimise the effects of differential settlements (e.g. **Figure 32.16**). This tends to be expensive and not universally successful.
- Ensure sufficient flexibility of the foundation and superstructure to accommodate ground movements without damage. This approach may be more applicable to smaller, lower-cost buildings. Alternatively, a split rigid building can be flexibly connected.
- Bypass the collapsible layer by use of piles.

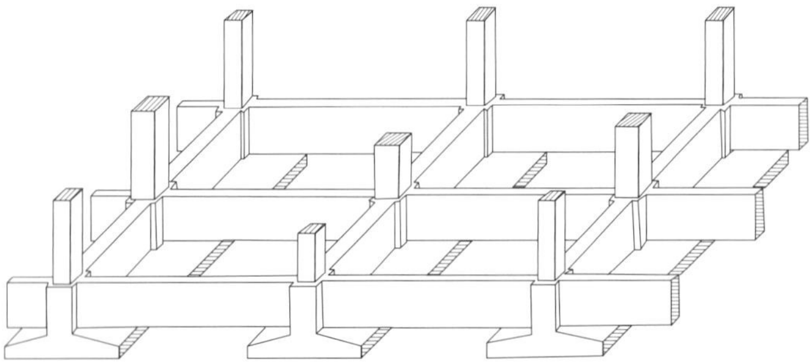

Figure 32.16 Continuous footing design for collapsible soils (Reproduced from Zeevaert, 1972; Van Nostrand Reinhold Co.)

- Control or alter ground conditions through one or more of the various improvement techniques available (see Section 32.5.5 and Chapter 25 *The role of ground improvement*).

Where the thickness of collapsible soil is relatively small, foundation recommendations are straightforward: the foundation level should be set below the collapsible soil layer. If this is not the case then some form of pre-treatment is necessary to remove collapse potential.

If the collapsible layer is deep or is of significant thickness below the surface, pile foundations are used. However, Grigoryan (1997) has reported several cases where piles used in collapsible soils have experienced loss of bearing capacity and excessive settlements immediately after inundation through negative skin friction effects. In addition, the presence of a collapsible layer may adversely affect the performance of piles during the life of a building.

Kakoli (2011) provides one of the few detailed reviews and assessments of piles used in collapsible soils, drawing mainly on the work of Grigoryan and Grigoryan (1975). They suggested that in collapsible soils, negative skin friction exists for a few hours but disappears after pile settlement. Chen *et al.* (2008) present load tests for piles in collapsible soils subjected to inundation. They measured negative skin friction across five sites in China, finding values of negative skin friction between around 20 and nearly 60 kPa. Kakoli (2011) presents numerical works (using PLAXIS) in the assessment of the effects of inundation on the performance of piles in collapsible soils, achieving results comparable to those presented by Chen *et al.* (2008). Further details of the general behaviour of piles in collapsible soils are provided by Redolfi and Mazo (1992). Noor *et al.* (2013) provide further analysis of piles in collapsible soils when subjected to inundation events with results from laboratory tests on a single pile driven into a collapsible soil comparing well with values (specifically skin friction) generated by numerical model prediction using PLAXIS.

Often less expensive foundation options can be employed: Bally (1988) and Popescu (1992) provide some examples of foundation practice in collapsible soils and Jefferson *et al.* (2005) illustrate this with a case study from Bulgaria. Other approaches that have been used successfully include the use of geogrid/geotextile reinforcement. For example, geogrid-reinforced sands over collapsible soils to improve settlement and bearing capacity characteristics (see Alawaji, 2001).

32.5.2 Transport and utility infrastructure

As with foundations, transport infrastructure (roads and railways) can experience problems due to collapse. Particular problems for roads are the non-uniform collapse and non-uniform wetting of subgrades that occur along the length of a road. These cause rough, wavy surfaces and have the potential to result in many miles of extensive damage to road structures (Houston, 1988). The potential damage to railways is more severe owing to their intolerance of longitudinal differential settlement, and the same is true for pipelines unless they are sufficiently flexible to permit longitudinal differential or lateral movements. Moreover, if water pipelines or sewers suffer fracture due to differential movement, water leakage can exacerbate the collapse phenomena. Damage can also occur to associated structural and geotechnical assets, including bridges, slopes and cuttings.

Stress applied to subgrade from highways has two components: overburden stress and imposed traffic loading (determined using common pavement analytical tools). It is likely that once wetting has occurred, short-duration loading from heavy lorries would be sufficient to cause full collapse. It should be noted that collapse strains can occur at any depth at which significant wetting occurs (see Section 32.4.5), regardless of the relative balance between overburden and imposed loads. Clearly changes in soil type along the length of a highway are important and routine non-destructive tests such as the falling weight deflectometer are useful here (Houston *et al.*, 2002).

Drainage management is of particular concern, as most collapsible soils have higher permeabilities and suctions than road foundation materials. In addition, the pavement interrupts evaporation and changes water-content regimes relative to the surrounding land. Hence significant risks of wetting collapsible soil subgrades exist; the extent and depth of wetting is controlled by the sources of water – for example, rising groundwater or surface ponding from poor drainage.

The approaches discussed in Sections 32.4 and 32.5.5 essentially apply to roads built on collapsible soils. Further details specific to roads are provided by Houston (1988) and Houston *et al.* (2002). Examples of parameters used in pavement design associated with collapsible soils are provided by Cameron and Nuntasarn (2006) and Roohnavaz *et al.* (2011). Further, Roohanavaz *et al.* (2011), when exploring the use of partially saturated loess for earthworks, discussed the benefits of repeated reworking and recompaction. More details of earthworks and impact of collapse settlements are presented by Nowak and Gilbert (2015; originally created by N.A. Trenter), who highlighted the need for special attention to manage collapse potential when working with loess.

32.5.3 Dynamic behaviour

Collapsible soils are particularly susceptible to liquefaction and dynamic settlement owing to their highly contractive nature during shearing. This can result in devastating consequences, with the Haiyuan landslide being a prime example (see Zhang and Wang, 2007). However, detection of the liquefaction and dynamic settlement potential of collapsible soils is difficult as they often have sufficient cementing when dry to prevent significant deformations during dynamic or earthquake loading (Houston *et al*., 2001). If, however, they become wetted, these bonds weaken and their liquefaction and dynamic settlement potential can significantly increase. Post-wetting behaviour is therefore of particular importance in earthquake-prone regions, especially if collapse is triggered by rising groundwater (Houston *et al*., 2003). Moreover, these soils often have insufficient fines to render them non-liquefiable if saturated. Hence compressions induced upon wetted collapse may be inadequate to mitigate liquefaction and dynamic settlement potential (Houston *et al*., 2001).

Houston *et al*. (2003) provide a brief overview of research that examines dynamic behaviour of collapsible soils. Their results show how cyclic stress ratios (CSRs) are strongly dependent on the degree of saturation (see **Figure 32.17**). Here, failure is taken to occur at CSR, causing 10% strain. CSR can be defined as the ratio of maximum shear stress (related to the cyclic shear stress amplitude of the earthquake) to vertical effective stress.

Liquefaction in loess soils is complicated by their microstructural aspects and so the process is generally less well understood than liquefaction in sands (see Hwang *et al*., 2000; Wang *et al*., 2004). However, improvement approaches that have proved successful in reducing liquefaction potential include dynamic compaction, use of compaction piles and methods that alter the soil's structure, such as grouting (Wang *et al*., 2004) – see Section 32.5.5 for more details.

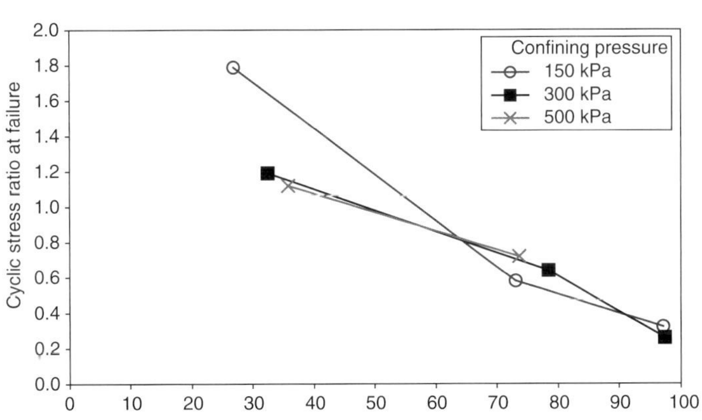

Figure 32.17 Variation of cyclic stress ratio at failure with degree of saturation (Reproduced from Houston *et al*., 2003)

Wang *et al*. (1998, 2004) provide details of lessons learned when dealing with earthquake-induced problems in loess soils through the use of dynamic triaxial tests and resonant column tests. Key earthquake-induced problems are seismically induced landslides (e.g. the 1920 Haiyuan landslide – Zhang and Wang, 2007); seismically induced subsidence and liquefaction. Similar problems can occur in other collapsible soils due to earthquakes – for instance, seismically induced landslides triggered in quick clays (Stark and Contreras, 1998).

32.5.4 Slope stability

Slope stability issues related to collapsible soils are a problem associated extensively with loess soils, although other collapsible deposits such as quick clay can exhibit significant problems – for example, the Rissa landslide (see Section 32.2.2) – with further details discussed by Bell (2000).

However, loess regions, such as the loess mantled mountainous region of Gansu in China, have suffered more than 40 000 large-scale landslides over the last century (Meng and Derbyshire, 1998). Landslides in this region are caused by seismic shocks and severe summer monsoonal rains. As a consequence of this rainfall, loess karst and sinkhole features are additional hazards.

Landslides in loess are very diverse owing to a broad range of conditions in which they occur (Derbyshire and Meng, 2005). The principal types of landslide in loess are shown in **Figure 32.18**, with detailed descriptions given in Meng and Derbyshire (1998) and Meng *et al*. (2000b). It should be noted that Tan-ta are small adjustment failures, generally less than 10 m dia. and at a depth of a few metres, occurring predominately on steep slopes in loess. After initiation of the slide, rapid disintegration takes place, often with high sliding velocities (Meng *et al*., 2000b).

The considerable experience from China had led to success with a number of control measures. These include

- landscaping (e.g. stepped slopes)
- retaining structures (e.g. retaining walls, underground drainage trenches and retaining piles).

Using these methods, most shallow and moderately deep landslides can be treated. Often a combination of approaches has proved to be best; for instance, using a combined structure of retaining walls and drainage ditches. However, terrace landslides can prove difficult to handle and often require a horizontal drainage borehole to alleviate groundwater pressures. Because of these pressures loess liquefaction, induced by slight slope displacement, can threaten whole slope stability.

Further details are provided by Meng and Derbyshire (1998) and Meng *et al*. (2000a), including details of successful treatment for landslide control. In addition, Dijkstra *et al*. (2000a) provide a detailed account of laboratory and in situ strength measurements with respect to slope stability in loess;

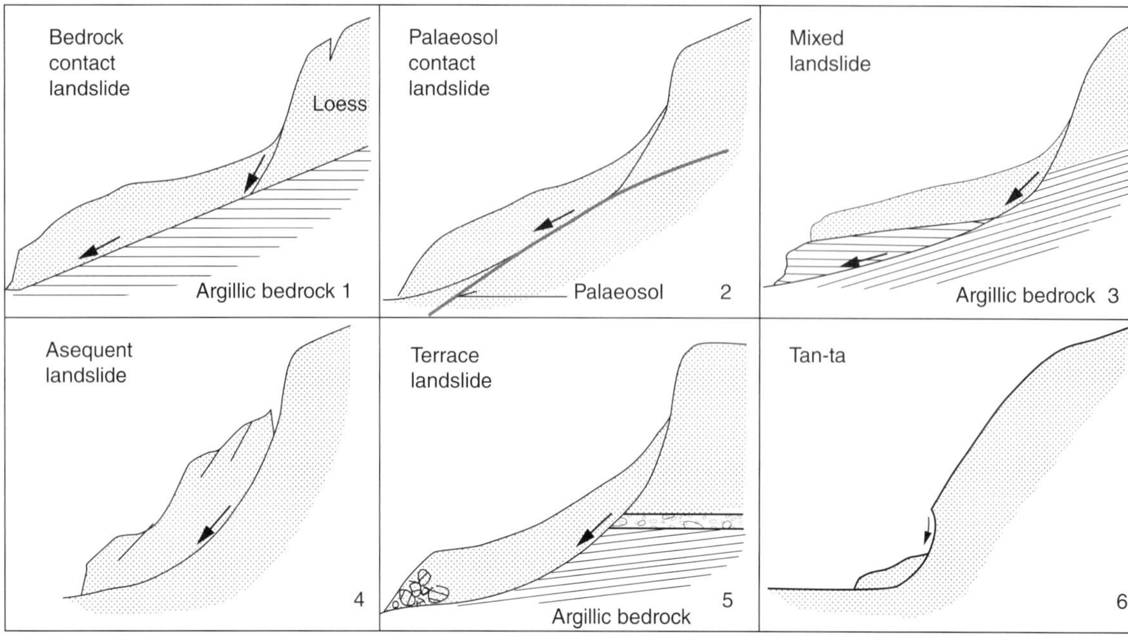

Figure 32.18 Principal types of loess landslides (Reproduced from Meng and Derbyshire, 1998; © The Geological Society of London)

details of modelling of landslides are presented by Dijkstra *et al.* (2000b).

32.5.5 Improvement and remediation

A wide variety of improvement processes exist for collapsible soils. Some of the more exotic ones have only been tried at an experimental stage. Evstatiev (1988), Houston *et al.* (2001) and Jefferson *et al.* (2005) provide excellent overviews of a range of possible treatment techniques used to improve collapsible loess soils.

Ultimately the best technique depends on several factors (after Houston *et al.*, 2001)

- when the collapsible loess soil was discovered
- how stress is to be applied to the soil
- depth and extent of the collapsible zone
- sources of wetting
- costs.

Details are provided throughout the literature, in particular the experience from Eastern Europe, especially Russia (e.g. Abelev, 1975; Evstatiev, 1995; Evstatiev *et al.*, 2002; Lutenegger, 1986; Ryzhov, 1989). Deng (1991), Wang (1991), Zhong (1991), Zhai *et al.* (1991), Fujun *et al.* (1998) and Gao *et al.* (2004), among others, allow insights into treatment techniques commonly employed in China, while researchers such as Clemence and Finbarr (1981), Rollins and Rogers (1994), Rollins and Kim (1994), Pengelly *et al.* (1997), Houston and Houston (1997), Rollins *et al.* (1998), Houston *et al.* (2001) and Rollins and Kim (2011) provide reviews of ground improvement approaches used to treat collapsible soils in North America. Rollins and Rogers (1994) provide an overview of the various advantages and limitations of a number of different possible treatment methods to reduce/eliminate collapsibility.

Table 32.3 provides an overview of techniques that can be used to treat loess ground and reduce/remove its collapse potential.

It is very common to find bands of clayey material (palaeosol) within some collapsible soils, such as loess deposits. However, it is still uncertain what role these bands play in the overall behaviour of the loess mass. It is likely that they will act as planes of weakness. What is certain, though, is that the presence of these bands will influence the effectiveness of any method used to treat a collapsible soil. These bands will affect the transmission of compaction energy and the route taken by stabilising fluids through the soil.

Post-construction treatment typically involves some form of chemical stabilisation – typically grouting – or alternatively some form of underpinning, both of which can prove expensive. The following methods have been used in cases where collapsing loess has damaged existing buildings: silicate grout injection (silicatisation), jet grouting, underpinning by root piles (pali radici), squeeze grouting (injection by 'tube à manchettes') and stabilisation by in-depth heating.

An alternative proposed by Houston *et al.* (2001) involves controlled differential wetting by separately controllable trenches built around the foundation slab. This approach has been used to tilt a structure in a controlled way. Initial trials demonstrated that, first, it was possible to re-level the foundation thereby eliminating any future collapse potential, and, second, its control was

Depth: m	Treatment method	Comments
0–1.5	Surface compaction by vibratory rollers, light tampers	Economical, but requires careful site control (e.g. limits on water content)
	Pre-wetting (inundation)	Can effectively treat thicker deposits but needs large volumes of water and time
	Vibroflotation	Needs careful site control
1.5–10	Vibrocompaction (stone columns, concrete columns, encased stone columns)	Cheaper than conventional piles but requires careful site control and assessment. If uncased, stone columns may fail with loss of lateral support on collapse
	Dynamic compaction; rapid impact compaction	Simple and easily understood but requires care with water content and vibrations produced
	Explosions	Safety issues need to be addressed
	Compaction pile	Needs careful site control
	Grouting	Flexible but may adversely affect the environment
	Ponding/inundation/pre-wetting	Difficult to control effectiveness of compression produced
	Soil mixing lime/cement	Convenient and gains strength with time. Various environmental and safety aspects, and the chemical controls on reactions need to be assessed
	Heat treatment	Expensive
	Chemical methods	Flexible; relatively expensive
>10	As for 1.5–10 m, some techniques may have a limited effect	(See above)
	Pile foundations	High bearing capacity but expensive

Table 32.3 Methods of treating collapsible loess ground (Reproduced from Jefferson *et al.*, 2005; with permission from Elsevier)

relatively straightforward by allowing site owners to control flow rates from each trench. However, as yet there are few directly relevant precedents, which limit its take-up.

Other approaches have been used in the treatment of collapsible soils, including encapsulated/encased stone columns (Al-Obaidy *et al.*, 2015; Ayadat, 2022), nanomaterials (Iranpour and Haddad, 2016) and biocementation (Cardoso *et al.*, 2023; Sun *et al.*, 2020).

Several case studies illustrate how collapsibility has been successfully treated and details of these are provided by Evans and Bell (1981), Rollins and Kim (2011) and Jefferson *et al.* (2015).

32.6. Concluding remarks

Collapsible soils are found throughout the world and are formed through various geomorphologic and geologic processes. These can be natural (through fluvial or aeolian processes) or man-made (through poor compaction). Whatever the processes involved, the key prerequisite is that an open metastable structure develops through bonding mechanisms generated through capillary forces (suctions) and/or through cementing materials such as clay or salts. Collapse occurs when net stresses (through loading or saturation) exceed the yield strength of the bonding material. Inundation is by far the most common cause of collapse and can be triggered through a range of different water sources. Different sources yield different amounts of collapse.

Thus a detailed knowledge of macroscopic and microscopic characteristics is vital to engineer these materials effectively and safely. Failure to recognise and deal with collapsible soils can have a significant impact on the built and urban environments – with catastrophic effects and potential loss of life. To engineer effectively in collapsible soils it is essential to recognise their existence, for which key geologic and geomorphologic information is vital. However, collapsibility should be confirmed through direct response to wetting/loading tests using laboratory and field methods. The key challenge with collapsible soils is to predict the extent and degree of wetting that will take place. Care is needed to ensure that appropriate and realistic assessments are undertaken, followed by treatment through a suite of the possible improvement techniques available. The collapsible potential can then be eliminated effectively.

References

Abelev MY (1975) Compacting loess soils in the USSR. *Géotechnique* **25(1)**: 79–82.

Alawaji HA (2001) Settlement and bearing capacity of geogrid-reinforced sand over collapsible soil. *Geotextiles and Geomembranes* **19**: 75–88.

Al-Obaidi QA and Schanz T (2023) Soil–water characteristic curve of unsaturated collapsible soils. *Journal of the Mechanical Behavior of Materials* **32**: 20220210, 10/1515/jmbm-2022-0210

Al-Obaidy N, Jefferson I and Gahataora G (2015) Treatment of Iraqi collapsible soil using encased stone columns. *Proceedings of 15th Asian Regional Conference on Soil Mechanics and Geotechnical Engineering, Kukuoka, Japan*, pp. 564–569.

Alonso EE, Gens A and Josa A (1990) A constitutive model for partially saturated soils. *Géotechnique* **40(3)**: 405–430.

Ayadat T (2022) Geotechnical performance of encapsulated and stabilized stone columns in a collapsible soil. *International Journal of Geomechanics* **22(6)**: 04022057.

Bally RJ (1988) Some specific problems of wetted loessial soils in civil engineering. *Engineering Geology* **25**: 303–324.

Barden L, McGown A and Collins K (1973) The collapse mechanism in partly saturated soil. *Engineering Geology* **7**: 49–60.

Barksdale RD and Blight GE (1997) Compressibility and settlement of residual soil. In *Mechanics of Residual Soils: A Guide to the Formation, Classification and Geotechnical Properties of Residual Soils, with Advice for Geotechnical Design* (Blight GE (ed)). Balkema, Rotterdam, the Netherlands, pp. 95–154.

Basma AA and Tuncer ER (1992) Evaluation and control of collapsible soils. *Journal of Geotechnical Engineering* **118(10)**: 1491–1504.

Beckwith GH (1995) Foundation design practices for collapsing soils in the western United States in unsaturated soils. *Proceedings of the 1st International Conference on Unsaturated Soils, Paris, France*, pp. 953–958.

Bell FG (2000) *Engineering Properties of Soils and Rocks*, 4th edn. Blackwell Science, Oxford, UK.

Bell FG and de Bruyn IA (1997) Sensitive, expansive, dispersive and collapsive soils. *Bulletin of the International Association of Engineering Geology* **56**: 19–38.

Bentley SP and Roberts AJ (1995) Consideration of the possible contributions of amphorous phases to the sensitivity of glaciomarine clays. *Proceedings of NATO Advance Workshop on Genesis and Properties of Collapsible Soils, Loughborough, UK*, pp. 225–245.

Billard A, Muxart T, Andrieu A and Derbyshire E (2000) Loess and water. In *Landslides in Thick Loess Terrain of North-West China* (Derbyshire E, Meng XM and Dijkstra TA (eds)). Wiley, Chichester, UK, pp. 91–130.

Blanchfield R and Anderson WF (2000) Wetting collapse in opencast coalmine backfill. *Proceedings of the ICE, Geotechnical Engineering* **143(3)**: 139–149.

Brink ABA and Kanty BA (1961) Collapsible grain structure in residual granite soils in southern Africa. *Proceedings of the 5th International Conference on Soil Mechanics and Foundation Engineering, Paris, France*, vol. 1, pp. 611–614.

Cameron DA and Nuntasarn R (2006) Pavement engineering parameters for Thai collapsible soil. *Proceedings of the 4th International Conference on Unsaturated Soils, Carefree, AZ*, pp. 1061–1072.

Cardoso R, Vieira J and Borges I (2023) On the use of biocementation to treat collapsible soils. *Engineering Geology* **313**: 106971.

Catt JA (1977) Loess and coversands. In *British Quaternary Studies, Recent Advances* (Shotton FW (ed)). Clarendon Press, Oxford, UK, pp. 221–229.

Catt JA (1985) Particle size distribution and mineralogy as indicators of pedogenic and geomorphic history: examples from soils of England and Wales. In *Geomorphology and Soils* (Richards KS, Arnett RR and Ellis S (eds)). George Allen and Unwin, London, UK, pp. 202–218.

Catt JA, Corbett WM, Hodge CA, *et al.* (1971) Loess in the soils of north Norfolk. *Journal of Soil Science* **22**: 444–452.

Cerato AB, Miller GA and Hajjat JA (2009) Influence of clod-size and structure on wetting-induced volume change of compacted soils. *Journal of Geotechnical and Geoenvironmental Engineering* **135(11)**: 1620–1628.

Charles JA and Skinner H (2001) Compressibility of foundation *fill. Proceedings of the ICE, Geotechnical Engineering* **149(3)**: 145–157.

Charles JA and Watts KS (2001) *Building on Fill: Geotechnical Aspects*, 2nd edn. CRC, London, UK.

Chen ZH, Huang XF, Qin B, Fang XW and Guo JF (2008) Negative skin friction for cast-in-place piles in thick collapsible loess. In *Unsaturated Soils: Advances in Geo-Engineering* (Toll DG, Augarde CE, Gallipoli D and Wheeler SJ (eds)). CRC Press, London, UK, pp. 979–985.

Clemence SP and Finbarr AO (1981) Design considerations for collapsible soils. *Journal of Geotechnical Engineering* **107(3)**: 305–317.

Cuccovillo T and Coop MR (1999) On the mechanics of structured sands. *Géotechnique* **49(6)**: 741–760.

Culshaw MG, Northmore KJ, Jefferson I, Assadi-Langroudi A and Bell FG (2020) Collapsible soils in the UK. In *UK Geohazards* (Giles D (ed)). Geological Society of London, London, UK, pp. 187–203.

Day RW (1990) Sample disturbance of collapsible soil. *Journal of Geotechnical Engineering* **116(1)**: 158–161.

Deng C (1991) Research of chemical stabilization effect of Lanzhou loess. In *Geotechnical Properties of Loess in China* (Lisheng Z, Zhenhua S, Hongdeng D and Shanlin L (eds)). China Architecture and Building Press, Beijing, China, pp. 93–97.

Derbyshire E and Mellors TW (1988) Geological and geotechnical characteristics of some loess and loessic soils from China and Britain. *Engineering Geology* **25**: 135–175.

Derbyshire E and Meng XM (2005) Loess. In *Geomorphology for Engineers* (Fookes PG, Lee EM and Milligan GC (eds)). Whittles, Dunbeath, UK, pp. 688–728.

Derbyshire E, Dijkstra TA and Smalley IJ (eds) (1995) *Genesis and Properties of Collapsible Soils*. Kluwer, Dordrecht, the Netherlands.

Derbyshire E, Meng XM and Dijkstra TA (eds) (2000) *Landslides in Thick Loess Terrain of North-West China*. Wiley, Chichester, UK.

Dijkstra TA, Rappange FE, van Asch TWJ, Li YJ and Li BX (2000a) Laboratory and in situ shear strength parameter of Lanzhou loess. In *Landslides in Thick Loess Terrain of North West China* (Derbyshire E, Meng XM and Dijkstra TA (eds)). Wiley, Chichester, UK, pp. 131–172.

Dijkstra TA, van Asch TWJ, Rappange FE and Meng XM (2000b) Modelling landslide hazards in loess terrain. In *Landslides in Thick Loess Terrain of North-West China* (Derbyshire E, Meng XM and Dijkstra TA (eds)). Wiley, Chichester, UK, pp. 203–242.

Dineen K and Burland JB (1995) A new approach to osmotically controlled oedometer testings. *Proceedings of the 1st International Conference on Unsaturated Soils, Paris, France*, pp. 459–465.

D'Onza F, Gallipoli D, Wheeler S, *et al.* (2011) Benchmark of constitutive models for unsaturated soils. *Géotechnique* **61(4)**: 282–302.

Dudley JH (1970) Review of collapsing soils. *Journal of Soil Mechanics and Foundation Division, ASCE* **96(3)**: 925–947.

El-Ehwany M and Houston SL (1990) Settlement and moisture movement in collapsible soils. *Journal of Geotechnical Engineering* **116(10)**: 1521–1535.

Evans GL and Bell DH (1981) Chemical stabilisation of loess, New Zealand. *Proceedings of the 10th International Conference on Soil mechanics and Foundation Engineering, Stockholm, Sweden*, vol. 3, pp. 649–658.

Evans RD, Jefferson I, Northmore KJ, Synac O and Serridge CJ (2004) Geophysical investigation and in situ treatment of collapsible soils. In *Geotechnical Engineering for Transportation Projects, Geotechnical Special Publication No. 126*. ASCE, Reston, VA, vol. 2, pp. 1848–1857.

Evstatiev D (1988) Loess improvement methods. *Engineering Geology* **25**: 135–175.

Evstatiev D (1995) Design and treatment of loess bases in Bulgaria. *Genesis and Properties of Collapsible Soils: Proceedings of NATO Advance Workshop on Genesis and Properties of Collapsible Soils, Loughborough, UK*, pp. 375–382.

Evstatiev D, Karastanev D, Angelova R and Jefferson I (2002) Improvement of collapsible loess soils from eastern Europe: lessons from Bulgaria. *Proceedings of the 4th International Conference on Ground Improvement Techniques, Kuala Lumpur, Malaysia*, pp. 331–338.

Farias MM, Assis AP and Luna SCP (1998) Some traps in modelling of collapse. *Proceedings of the International Symposium on Problematic Soils, Sendai, Japan*, pp. 309–312.

Feda J (1966) Structural stability of subsidence loess from Praha-Dejvice. *Engineering Geology* **1**: 201–219.

Feda J (1995) Mechanisms of collapse of soil structure. *Genesis and Properties of Collapsible Soils: Proceedings of NATO Advance Workshop on Genesis and Properties of Collapsible Soils, Loughborough, UK*, pp. 149–172.

Fookes PG and Best R (1969) Consolidation characteristics of some late Pleistocene periglacial metastable soils of east Kent. *Quarterly Journal of Engineering Geology* **2**: 103–128.

Fookes PG, Lee EM and Milligan GC (eds) (2005) *Geomorphology for Engineers*. Whittles, Dunbeath, UK.

Francisca FM (2007) Evaluating the constrained modulus and collapsibility of loess from standard penetration test. *International Journal of Geomechanics* **7(4)**: 307–310.

Fredlund DG (2006) Unsaturated soil mechanics in engineering practice. *Journal of Geotechnical and Geoenvironmental Engineering* **132(3)**: 286–321.

Fredlund DG and Gan JKM (1995) The collapse mechanism of a soil subjected to one-dimensional loading and wetting. *Genesis and Properties of Collapsible Soils: Proceedings of NATO Advance Workshop on Genesis and Properties of Collapsible Soils, Loughborough, UK*, pp. 173–205.

Fredlund DG and Houston SL (2009) Protocol for the assessment of unsaturated soil properties in geotechnical engineering practice. *Canadian Geotechnical Journal* **46**: 694–707.

Fredlund DG and Rahardjo H (1993) *Soil Mechanics for Unsaturated Soils*. Wiley, New York, NY.

Fujun N, Wankui N and Yuhai L (1998) Wetting-induced collapsibility of loess and its engineering treatments. *Proceedings of the International Symposium on Problematic Soils, Sendai, Japan*, vol. 1, pp. 395–399.

Gao GY, Shui WH, Wang YL and Li W (2004) Application of high energy level dynamic compaction to high-capacity oil tank foundation. *Rock and Soil Mechanics* **25(8)**: 1275–1278 (in Chinese).

Gibbs HJ and Bara JP (1962) Predicting surface subsidence from basic soil tests. In *Special Technical Publication* 332. ASTM, West Conshohocken, PA, pp. 231–247.

Grigoryan AA (1997) *Pile Foundations for Buildings and Structures in Collapsible Soils*. Balkema, Brookfield, WI.

Grigoryan AA and Grigoryan RG (1975) Experimental investigation of negative friction forces along the lateral surface of piles as soils experience slump-type settlement under their own weight. *Osnovaniya, Fundamenty i Mekhanika Gruntov* **5**: 2–5.

Gunn DA, Nelder LM, Jackson PD, *et al.* (2006) Shear wave velocity monitoring of collapsible brickearth soil. *Quarterly Journal of Engineering Geology and Hydrogeology* **39**: 173–188.

Handy RL (1995) A stress path model for collapsible loess. *Genesis and Properties of Collapsible Soils: Proceedings of NATO Advance Workshop on Genesis and Properties of Collapsible Soils, Loughborough, UK*, pp. 173–205.

Herle I, Herbstová V, Kupka M and Kolymbas D (2009) Geotechnical problems of cultural heritage due to floods. *Journal of Performance of Constructed Facilities* **24(5)**: 446–451.

Holtz WG and Gibbs HJ (1952) Consolidation and related properties of loessial soils. In *ASTM Special Technical Publication* 126. ASTM, West Conshohocken, PA, pp. 9–33.

Houston SL (1988) Pavement problems caused by collapsible soils. *Journal of Transportation Engineering* **114(6)**: 673–683.

Houston SL and El-Ehwany M (1991) Sample disturbance of cemented collapsible soils. *Journal of Geotechnical Engineering* **117(5)**: 731–752.

Houston WN and Houston SL (1997) Soil improvement in collapsing soil. *Proceedings of the 1st International Conference on Ground Improvement Techniques, Macau*, pp. 237–246.

Houston SL, Houston WN and Spadola DJ (1988) Prediction of field collapse of soils due to wetting. *Journal of Geotechnical Engineering* **114(1)**: 40–58.

Houston SL, Houston WN and Mahmoud HH (1995a) Interpretation and comparison of collapse measurement techniques. *Proceedings of NATO Advance Workshop on Genesis and Properties of Collapsible Soils, Loughborough, UK*, pp. 217–224.

Houston SL, Mahmoud HH and Houston WN (1995b) Down-hole collapse test system. *Journal of Geotechnical Engineering* **121(4)**: 341–349.

Houston SL, Houston WN, Zapata CE and Lawrence C (2001) Geotechnical engineering practice for collapsible soils. *Geotechnical and Geological Engineering* **19**: 333–355.

Houston SL, Houston WN and Lawrence CA (2002) Collapsible soil engineering in highway infrastructure development. *Journal of Transportation Engineering* **128(3)**: 295–300.

Houston SL, Elkady T and Houston WN (2003) Post-wetting static and dynamic behavior of collapsible soils. *Proceedings of the 1st International Conference on Problematic Soils, Nottingham, UK*, pp. 63–71.

Hwang H, Wang L and Yuan Z (2000) Comparison of liquefaction potential of loess in Lanzhou, China and Memphis, USA. *Soil Dynamics and Earthquake Engineering* **20**: 389–395.

Iranpour B and Haddad A (2016) The influence of nanomaterials on collapsible soil treatment. *Engineering Geology* **205**: 40–53.

Jackson PD, Northmore KJ, Entwisle DC, *et al.* (2006) Electrical resistivity monitoring of a collapsing metastable soil. *Quarterly Journal of Engineering Geology and Hydrogeology* **39**: 151–172.

Jefferson I, Tye C and Northmore KG (2001) The engineering characteristics of UK brickearth. In *Problematic Soils* (Jefferson I, Murray EJ, Faragher E and Fleming PR (eds)). Thomas Telford, London, UK, pp. 37–52.

Jefferson I, Smalley IJ, Karastanev D and Evstatiev D (2002) Comparison of the behaviour of British and Bulgarian loess. *Proceedings of the 9th Congress of the International Association of Engineering Geology and the Environment, Durban, South Africa*, pp. 253–263.

Jefferson I, Evstatiev D, Karastanev D, Mavlyanova N and Smalley IJ (2003a) Engineering geology of loess and loess-like deposits: a commentary on the Russian literature. *Engineering Geology* **68**: 333–351.

Jefferson I, Smalley IJ and Northmore KJ (2003b) Consequences of a modest loess fall over southern Britain. *Mercian Geologist* **15(4)**: 199–208.

Jefferson I, Evstatiev D, Karastanev D and Rogers CDF (2005) Treatment of metastable loess soils: lessons from Eastern Europe. In *Ground Improvement Case Histories* (Chu J and Indrartna B (eds)). Elsevier, Amsterdam, the Netherlands, pp. 727–766.

Jefferson I, Evstatiev D, Karastanev D and Rogers CDF (2015) Improvements of collapsible loess in Eastern Europe. In *Ground Improvement Case Histories* (Chu J and Indrartna B (eds)). Elsevier, Amsterdam, the Netherlands, vol. 2, pp. 215–261.

Jennings JE and Knight K (1975) A guide to construction on or with materials exhibiting additional settlement due to collapse of grain structure. *Proceedings of the 6th African Conference on Soil Mechanics and Foundation Engineering, Durban, South Africa*, pp. 99–105.

Jotisankasa A, Ridley A and Coop M (2007) Collapse behavior of compacted silty clay in suction monitored oedometer apparatus. *Journal of Geotechnical and Geoenvironmental Engineering* **133(7)**: 867–886.

Jotisankasa A, Coop MR and Ridley A (2009) The mechanical behaviour of an unsaturated compacted silty clay. *Géotechnique* **59(5)**: 415–428.

Kakoli STN (2011) *Negative Skin Friction Induced on Piles in Collapsible Soils due to Inundation*. Unpublished PhD thesis, Concordia University, Canada.

Klukanova A and Frankovska J (1995) The Slovak Carpathians loess sediments, their fabric and properties. *Proceedings of NATO Advance Workshop on Genesis and Properties of Collapsible Soils, Loughborough, UK*, pp. 129–147.

Knight K and Dehlen G (1963) Failure of road constructed on collapsing soil. *Proceedings of the 3rd Regional Conference of African Soil Mechanics and Foundation Engineering, Durban, South Africa*, vol. 1, pp. 31–34.

Langton DD (1999) The Panda lightweight penetrometer for soil investigation and monitoring material compaction. *Ground Engineering* **31(9)**: 33–37.

Lawton EC, Fragaszy RJ and Hardcastle JH (1991) Stress ratio effects on collapse of compacted clayey sand. *Journal of Geotechnical Engineering* **117(5)**: 714–730.

Lawton EC, Fragaszy RJ and Hetherington MD (1992) Review of wetting induced collapse in compacted soils. *Journal of Geotechnical Engineering* **118(9)**: 1376–1392.

Leroueil S and Vaughan PR (1990) The general and congruent effects of structure in natural soils and weal rock. *Géotechnique* **40(3)**: 467–488.

Li P, Vanapalli S and Li T (2016) Review of collapse triggering mechanism of collapsible soils due to wetting. *Journal of Rock Mechanics and Geotechnical Engineering* **8**: 256–274.

Lin Z (1995) Variation in collapsibility and strength of loess with age. *Proceedings of NATO Advance Workshop on Genesis and Properties of Collapsible Soils, Loughborough, UK*, pp. 247–265.

Lin ZG and Wang SJ (1988) Collapsibility and deformation characteristics of deep-seated loess in China. *Engineering Geology* **25**: 271–282.

Liu SH and Sun DA (2002) Simulating the collapse of un-saturated soil by DEM. *International Journal for Numerical and Analytical Methods in Geomechanics* **26**: 633–646.

Locat J (1995) On the development of microstructure in collapsible soils. *Proceedings of NATO Advance Workshop on Genesis and Properties of Collapsible Soils, Loughborough, UK*, pp. 93–128.

Lutenegger AJ (1986) Dynamic compaction of friable loess. *Journal of Geotechnical Engineering* **110(6)**: 663–667.

Lutenegger AJ and Hallberg GR (1988) Stability of loess. *Engineering Geology* **25**: 247–261.

Maâtouk A, Leroueil S and La Rochelle P (1995) Yielding and critical state of a collapsible unsaturated silty soil. *Géotechnique* **45(3)**: 465–477.

Madhyannapu RS, Madhav MR, Puppala AJ and Ghosh A (2006) Compressibility and collapsibility characteristics of sedimented fly ash beds. *Journal of Material in Civil Engineering* **20(6)**: 401–409.

Mahmoud HH, Houston WN and Houston SL (1995) Apparatus and procedure for an in-situ collapse test. *ASTM Geotechnical Testing Journal* **121(4)**: 341–349.

Malandraki V and Toll DG (1996) The definition of yield for bonded materials. *Geotechnical and Geological Engineering* **14**: 67–82.

Meng XM and Derbyshire E (1998) Landslides and their control in the Chinese Loess Plateau: models and case studies from Gansu Province, China. In *Geohazards in Engineering Geology* (Maund JG and Eddleston M (eds)). Geological Society, London, UK, pp. 141–153.

Meng XM, Derbyshire E and Dijkstra TA (2000a) Landslide amelioration and mitigation. In *Landslides in Thick Loess Terrain of North-West China* (Derbyshire E, Meng XM and Dijkstra TA (eds)). Wiley, Chichester, UK, pp. 243–262.

Meng XM, Dijkstra TA and Derbyshire E (2000b) Loess slope instability. In *Landslides in Thick Loess Terrain of North-West China* (Derbyshire E, Meng XM and Dijkstra TA (eds)). Wiley, Chichester, UK, pp. 173–202.

Milodowski AE, Northmore KJ, Kemp SJ, *et al.* (2015) The mineralogy and fabric of 'brickearths' in Kent and their relationship to engineering properties. *Bulletin of Engineering Geology and the Environment* **74**: 1187–1211.

Minkov M (1968) *The Loess of North Bulgaria: A Complex Study*. Publishing House of Bulgarian Academy of Sciences, Sofia, Bulgaria (in Bulgarian).

Neely WJ (2010) Discussion of 'Oedometer behavior of an artificial cemented highly collapsible soil' by Medero, G. M., Schnaid, F. and Gehling, W. Y. Y. *Journal of Geotechnical and Geoenvironmental Engineering* **136(5)**: 771.

Nieuwenhuis JD and de Groot MB (1995) Simulation and modelling of collapsible soils. *Proceedings of NATO Advance Workshop on Genesis and Properties of Collapsible Soils, Loughborough, UK*, pp. 345–359.

Nobar ES and Duncan JM (1972) Movements in dams due to reservoir filling. *Special Conference on Earth-Supported Structures, Lafayette, IN*, vol. 1, pp. 797–815.

Noor ST, Hanna A and Mashhour I (2013) Numerical modeling of piles in collapsible soil subjected to inundation. *International Journal of Geomechanics* **13(5)**: 514–526.

Northmore KJ, Bell FG and Culshaw MG (1996) The engineering properties and behaviour of the brickearth of south Essex. *Quarterly Journal of Engineering Geology* **29**: 147–161.

Northmore KN, Jefferson I, Jackson PD, *et al.* (2008) On-site characterisation of loessic brickearth deposits at Ospringe, Kent, UK. *Proceedings of the Institution of Civil Engineers, Geotechnical Engineering* **161**: 3–17.

Nowak P and Gilbert P (2015) *Earthworks: A Guide*, 2nd edn. ICE Publishing, London, UK.

Opukumo AW, Davie CT, Glendinning S and Oborie E (2022) A review of the identification methods and types of collapsible soils. *Journal of Engineering and Applied Science* **69**: 17, 10.1186/s44147-021-00064-2.

Osopov VI and Sokolov VN (1995) Factors and mechanism of loess collapsibility. *Proceedings of NATO Advance Workshop on Genesis and Properties of Collapsible Soils, Loughborough, UK*, pp. 49–63.

Pengelly A, Boehm D, Rector E and Welsh J (1997) Engineering experience with in situ modification of collapsible and expansive soils. In *Unsaturated Soil Engineering, ASCE Special Geotechnical Publication 68*. ASCE, New York, NY, pp. 277–298.

Pereira JHF and Fredlund DG (2000) Volume change behaviour of collapsible compacted gneiss soil. *Journal of Geotechnical and Geoenvironmental Engineering* **126(10)**: 907–916.

Perrin RMS (1956) Nature of 'Chalk Heath' soils. *Nature* **178**: 31–32.

Peterson R and Iverson NL (1953) Study of several low earth dam failures. *Proceedings of the 3rd International Conference on Soil Mechanics and Foundation Engineering, Zurich, Switzerland*, vol. 2, pp. 273–276.

Phanikumar BR, Raghav R and Bhargav K (2016) Collapse behaviour of a lateritic soil. *Geomechanics and Geoengineering* **11(2)**: 119–124.

Popescu ME (1986) A comparison between the behaviour of swelling and of collapsing soils. *Engineering Geology* **23**: 145–163.

Popescu ME (1992) Engineering problems associated with expansive and collapsible soil behaviour. *Proceedings of the 7th International Conference on Expansive Soils, Dallas, TX*, vol. 2, pp. 25–46.

Pye K and Sherwin D (1999) Loess. In *Aeolian Environments, Sediments and Landforms* (Goudie AS, Livingstone I and Stokes S (eds)). Wiley, Chichester, UK, pp. 213–238.

Rampino C, Mancuso C and Vinale F (2000) Experimental behaviour and modelling of an unsaturated compacted soil. *Canadian Geotechnical Journal* **37**: 748–763.

Redolfi ER and Mazo CO (1992) A model of pile interface in collapsible soils. *Proceedings of the 7th International Conference on Expansive Soils, Dallas, TX*, vol. 1, pp. 483–488.

Renzik YM (1991) Plate-load tests of collapsible soils. *Journal of Geotechnical Engineering* **119(3)**: 608–615.

Renzik YM (1995) Comparison of results of oedometer and plate load test performed on collapsible soils. *Proceedings of NATO Advance Workshop on Genesis and Properties of Collapsible Soils, Loughborough, UK*, pp. 383–408.

Roa SM and Revansiddappa K (2002) Collapse behaviour of residual soil. *Géotechnique* **52(4)**: 259–268.

Rocha BP, Silveira IA, Rodrigues RA, Lodi PC and Giacheti HL (2023) Identifying collapsible soils from seismic cone (SCPT): a qualitative approach. *Buildings* **13**: 830, 10.3390/buildings13030830.

Rodrigues RA, Elis VR, Prado R and De Lollo JA (2006) Laboratory tests and applied geophysical investigations in collapsible soil horizon definition. *Engineering Geology for Tomorrow's Cities, 10th International Congress, Nottingham, UK*. Geological Society, London, UK, CD-ROM.

Rogers CDF (1995) Types and distribution of collapsible soils. *Proceedings of NATO Advance Workshop on Genesis and Properties of Collapsible Soils, Loughborough, UK*, pp. 1–17.

Rogers CDF, Djikstra TA and Smalley IJ (1994) Hydro-consolidation and subsidence of loess: studies from China, Russia, North America and Europe. *Engineering Geology* **37**: 83–113.

Rollins KM and Kim JH (1994) US experience with dynamic compaction of collapsible soils. In *In-Situ Deep Soil Improvement: Geotechnical Special Publication No. 45* (Rollins KM (ed)). ASCE, New York, NY, pp. 26–43.

Rollins KM and Kim JH (2011) Dynamic compaction of collapsible soils based on US case histories. *Journal of Geotechnical and Geoenvironmental Engineering* **136(9)**: 1178–1186.

Rollins KM and Rogers GW (1994) Mitigation measures for small structures on collapsible alluvial soils. *Journal of Geotechnical Engineering* **120(9)**: 1533–1553.

Rollins KM, Rollins RL, Smith TD and Beckwith GH (1994) Identification and characterization of collapsible gravels. *Journal of Geotechnical Engineering* **120(3)**: 528–542.

Rollins KM, Jorgensen SJ and Ross TE (1998) Optimum moisture content for dynamic compaction of collapsible soils. *Journal of Geotechnical and Geoenvironmental Engineering* **124(8)**: 699–708.

Roohnavaz C, Russell EJF and Taylor HT (2011) Unsaturated loessial soils: a sustainable solution for earthworks. *Proceedings of ICE, Geotechnical Engineering* **164**: 257–276.

Rust E, Heymann G and Jones GA (2005) Collapse potential of partly saturated sandy soils from Mozal, Mozambique. *Journal of the South African Institution of Civil Engineering* **47(1)**: 8–14.

Ryzhov AM (1989) On compaction of collapsing loess soils by the energy of deep explosions. *Proceedings of the International Conference on Engineering Problems of Regional Soils, Beijing, China*, pp. 308–315.

Sakr M, Mashhour M and Hanna A (2008) Egyptian collapsible soils and their improvement. *Proceedings of GeoCongress, New Orleans, LA*, pp. 654–661.

Schnaid F, de Oliveira LAK and Gehling WYY (2004) Unsaturated constitutive surfaces from pressuremeter test. *Journal of Geotechnical and Geoenvironmental Engineering* **130(2)**: 174–185.

Sheng D (2011) Review of fundamental principles in modelling unsaturated soil behaviour. *Computers and Geotechnics* **38**: 757–776.

Skinner H (2001) Construction on fill. In *Problematic Soils* (Jefferson I, Murray EJ, Faragher E and Fleming PR (eds)). Thomas Telford, London, UK, pp. 37–52.

Smalley IJ, O'Hara-Dhand KA, Wint J, *et al.* (2007) Rivers and loess: the significance of long river transportation in the complex event-sequence approach to loess deposit formation. *Quaternary International* **198**: 7–18.

Smith TD and Rollins KM (1997) Pressuremeter testing in arid collapsible soils. *Geotechnical Testing Journal* **20(1)**: 12–16.

Stark TD and Contreras IA (1998) Fourth avenue landsliding during the 1964 Alaskan earthquake. *Journal of Geotechnical and Geoenvironmental Engineering* **124(2)**: 99–109.

Sun D, Sheng D and Xu Y (2007) Collapse behaviour of un-saturated compacted soil with different initial densities. *Canadian Geotechnical Journal* **44**: 673–686.

Sun JM (2002) Provenance of loess materials and formation of loess deposits on the Chinese loess plateau. *Earth Planet Science Letters* **203**: 845–849.

Sun X, Miao L and Chen R (2020) The application of bio-cementation for improvement in collapsibility of loess. *International Journal of Environmental Science and Technology* **18**: 2607–2618.

Tarantino A, Gallipoli D, Augarde CE, *et al.* (2011) Benchmark of experimental techniques for measuring and controlling suction. *Géotechnique* **61(4)**: 282–302.

Thorel L, Ferber V, Caicedo B and Khokhar IM (2011) Physical modelling of wetting-induced collapse in embankment base. *Géotechnique* **61(5)**: 409–420.

Vandanapu R, Omer JR and Attom MF (2016) Geotechnical case studies: emphasis on collapsible soil cases. *Proceedings of ICE, Forensic Engineering* **169(FE3)**: 103–110.

Vilar OM and Rodrigues RA (2011) Collapse behaviour of soil in a Brazilian region affected by rising water table. *Canadian Geotechnical Journal* **48**: 226–233.

Walsh KD, Houston WN and Houston SL (1993) Evaluation of in-place wetting using soil suction measurements. *Journal of Geotechnical Engineering* **119(5)**: 862–873.

Wang G (1991) Experimental research on treatment of the self-weight collapsible loess ground with the compacting method. In *Geotechnical Properties of Loess in China* (Lisheng Z, Zhenhua S, Hongdeng D and Shanlin L (eds)). China Architecture and Building Press, Beijing, China, pp. 88–92.

Wang LM, Zhang ZZ, Lin ZX and Yuan ZX (1998) The recent progress in loess dynamics and its application in prediction of earthquake-induced disasters in China. *Proceedings of the 11th European Conference on Earthquake Engineering, Paris, France*, paper 219.

Wang LM, Wang YQ, Wang J, Li L and Yuan AX (2004) The liquefaction potential of loess in China and its prevention. *Proceedings of the 13th World Conference on Earthquake Engineering, Vancouver, BC*, paper 3462.

Yanrong X, Leung CF, Jian Y and Chen W (2018) Numerical modelling of hydro-mechanical behaviour of ground settlement due to rising water table in loess. *Natural Hazards* **94**: 241–260.

Zeevaert L (1972) *Foundation Engineering for Difficult Subsoil Conditions*. Van Nostrand Reinhold, New York, NY.

Zhai L, Jiao J, Li S and Li L (1991) Dynamic compaction tests on loess in Shangan. In *Geotechnical Properties of Loess in China* (Lisheng Z, Zhenhua S, Hongdeng D and Shanlin L (eds)) China Architecture and Building Press, Beijing, China, pp. 105–110.

Zhang D and Wang G (2007) Study of the 1920 Haiyuan earthquake-induced landslides in loess (China). *Engineering Geology* **94**: 76–88.

Zhang X and Li L (2011) Limitation in constitutive modelling of unsaturated soils and solutions. *International Journal of Geomechanics* **11(3)**: 174–185.

Zhong L (1991) Ground treatment of soft loess of high moisture. In *Geotechnical Properties of Loess in China* (Lisheng Z, Zhenhua S, Hongdeng D and Shanlin L (eds)) China Architecture and Building Press, Beijing, China, pp. 98–104.

Further reading

Charles JA and Watts KS (2001) *Building on Fill: Geotechnical Aspects*, 2nd edn. CRC Press, London, UK.

Culshaw MG, Northmore KJ, Jefferson I, Assadi-Langroudi A and Bell FG (2020) Collapsible soils in the UK. In *UK Geohazards* (Giles D (ed)). Geological Society, London, UK, pp. 187–203.

Derbyshire E, Dijkstra T and Smalley IJ (eds) (1995) *Proceedings of NATO Advance Workshop on Genesis and Properties of Collapsible Soils, Loughborough, UK.*

Derbyshire E, Meng XM and Dijkstra TA (eds) (2000) *Landslides in Thick Loess Terrain of North-West China*. Wiley, Chichester, UK.

Fookes PG, Lee EM and Milligan G (eds) (2005) *Geomorphology for Engineers*. Whittles, Dunbeath, UK.

Useful websites

British Geological Survey: http://www.bgs.ac.uk

It is recommended this chapter is read in conjunction with

- Chapter 7 *Geotechnical risks and their context for the whole project*
- Chapter 14 *Soils as particulate materials*
- Chapter 29 *Arid Soils*
- Chapter 40 *The ground as a hazard*
- Chapter 58 *Building on fills*

All chapters in this book rely on the guidance in Sections 1 *Context* and 2 *Fundamental principles.* A sound knowledge of ground investigation is required for all geotechnical works, as set out in Section 4 *Site investigation*

doi: 10.1680/icemge.66816.0447

Chapter 33
Expansive soils

Lee D Jones British Geological Survey, Nottingham, UK
Ian Jefferson School of Engineering, University of Birmingham, UK

CONTENTS

Expansive soils present significant geotechnical and structural engineering challenges the world over, with costs associated with expansive behaviour estimated to run into several billion pounds annually. Expansive soils are those which experience significant volume changes associated with changes in water content. These volume changes can either be in the form of swell or shrinkage, and are sometimes known as swell–shrink soils. Key aspects that need identification when dealing with expansive soils include soil properties, suction/water conditions, temporal and spatial water content variations that may be generated, for example, by trees, and the geometry/stiffness of foundations and associated structures. Expansive soils can be found both in humid environments where expansive problems occur with soils of high plasticity index, and in arid/semi-arid soils where soils of even moderate expansiveness can cause significant damage. This chapter reviews the nature and extent of expansive soils, highlighting key engineering issues. These include methods to investigate expansive behaviour both in the field and the laboratory, and the associated empirical and analytical tools to evaluate expansive behaviour. Design options for pre- and post-construction are highlighted for both foundations and pavements, together with methods to ameliorate potentially damaging expansive behaviour.

33.1. What is an expansive soil?

Essentially, expansive soil is one that changes in volume in relation to changes in water content. The focus here is on soils that exhibit significant swell potential and, in addition, shrinkage potential. There are a number of cases where expansion can occur because of chemically induced changes (e.g. swelling of lime-treated sulfate soils – see Chapter 37 *Sulfates and sulfides in soils and rocks*). However, many soils that exhibit swelling and shrinking behaviour contain expansive clay minerals, such as smectite, that absorb water. The more of this clay a soil contains, the higher its swell potential and the more water it can absorb. As a result, these materials swell and thus increase in volume when they become wet, and shrink when they dry. The more water they absorb, the more their volume increases – for the most expansive clays expansions of 10% are not uncommon (Chen, 1988; Nelson and Miller, 1992; Nelson *et al.*, 2015). It should be noted that other soils exhibit volume change characteristics with changes in water content (e.g. collapsible soils) and these are dealt with in Chapter 32 *Collapsible soils*.

The amount by which the ground can shrink and/or swell is determined by the water content in the near-surface zone. Significant activity usually occurs to about 3 m depth, unless this zone is extended by the presence of tree roots (Biddle, 1998; Driscoll, 1983). Fine-grained clay-rich soils can absorb large quantities of water after rainfall, becoming sticky and heavy. Conversely, they can also become very hard when dry, resulting in shrinking and cracking of the ground. This hardening and softening is known as 'shrink–swell' behaviour. The effects of significant changes in water content on soils with a high shrink–swell potential can be severe for supporting structures.

Swelling and shrinkage are not fully reversible processes (Holtz and Kovacs, 1981). The process of shrinkage causes cracks which, on re-wetting, do not close perfectly and hence cause the soil to bulk out slightly, and allow enhanced access to water for the swelling process. In geological timescales, shrinkage cracks may become in-filled with sediment, thus imparting heterogeneity to the soil. When material falls into cracks the soil is unable to move back, resulting in enhanced swelling pressures.

The primary problem with expansive soils is that deformations are significantly greater than those that can be predicted using classical elastic and plastic theory. As a result, several different approaches have been developed to predict and engineer expansive soils, and these are highlighted throughout this chapter.

33.2. Why are they problematic?

Many towns, cities, transport routes and buildings are founded on clay-rich soils and rocks. The clays within these materials may be a significant hazard to engineering construction due to their ability to shrink or swell with changes in water content. Changing water content may be due to seasonal variations (often related to rainfall and the evapotranspiration of vegetation), or be brought about by local site changes such as leakage from water supply pipes or drains, changes to surface drainage and landscaping (including paving) or following the planting, removal or severe pruning of trees or hedges, as water cannot be supplied to desiccated soil as efficiently as a tree originally extracted it through its root system (Cheney, 1988). During a long dry period or drought a persistent water deficit may develop, causing the soil to dry out to a greater depth than normal and leading to long-term subsidence. This is why expansive problems are often found in arid environments (see Chapter 29 *Arid soils*). As this water deficit dissipates it is possible that long-term heave may occur.

In the UK, the effects of shrinkage and swelling were first recognised by geotechnical specialists following the dry summer of 1947, and since then the cost of damage due to the shrinking and swelling of clay soils in the UK has risen dramatically. After the drought of 1975–1976, insurance claims came to over £50 million. In 1991, after the preceding drought, claims peaked at over £500 million. Over the first part of the twenty-first century the adverse effects of shrink–swell behaviour have cost the economy an estimated £3 billion, making it the most damaging geohazard in Britain today. The Association of British Insurers (ABI) has estimated that the average cost of shrink–swell-related subsidence to the insurance industry stands at over £400 million annually (Driscoll and Crilly, 2000). More recently the ABI reported that £64 million of claims were made during July to September 2018 (BGS, 2023). In the USA the estimated damage to buildings and infrastructure exceeds $15 billion annually. The American Society of Civil Engineers (ASCE) estimates that one in four homes have some damage caused by expansive soils. In a typical year, expansive soils cause a greater financial loss to property owners than earthquakes, floods, hurricanes and tornadoes combined (Nelson and Miller, 1992).

Swelling pressures can cause heaving, or lifting, of structures, while shrinkage can cause differential settlement. Failure results when the volume changes are unevenly distributed beneath the foundation. For example, water content changes in the soil around the edge of a building can cause swelling pressure beneath the perimeter of the building, while the water content of the soil beneath the centre remains constant. This results in a failure known as 'end lift' (**Figure 33.1**). The opposite of this is 'centre lift', where swelling is focused beneath the centre of the structure or where shrinkage takes place under the edges.

Damage to foundations in expansive soils commonly results from tree growth. This occurs in two principal ways: (*a*) physical disturbance of the ground, and (*b*) shrinkage of the ground by removal of water. Physical disturbance of the ground caused by root growth is often seen as damage to pavements and broken walls. An example of vegetation-induced shrinkage causing differential settlement of building foundations is provided in **Figure 33.2**. Vegetation-induced changes to water profiles can also have a significant impact on other underground features, including utilities. Clayton *et al.* (2010), reporting monitoring data of pipes in London Clay over a two-year period, found significant ground movements (both vertical and horizontal) of the order of 3–6 mm/m length of pipe, which generated significant tensile stresses when in the vicinity of trees.

Figure 33.1 Structural damage to a house caused by 'end lift' (© Peter Kelsey & Partners)

Figure 33.2 An example of differential settlement due to the influence of trees

Such tree-induced movement has the potential to be a significant contributor to failure of old pipes located in clay soils near deciduous trees (Clayton *et al.*, 2010). Further details are discussed in Section 33.5.4.5.

33.3. Where are expansive soils found?

In the UK, infrastructure, towns and cities built on/in clay-rich soils most susceptible to shrink–swell behaviour are found mainly in the south-east of the country (**Figure 33.3**). In the south-east, many of the clay formations are too young to have been changed into stronger mudstones, leaving them still able to absorb and lose moisture. Clay rocks elsewhere in the country are older and have been hardened by processes resulting from deep burial and are less able to absorb water. Some areas (e.g. around The Wash, north-west of Peterborough – see **Figure 33.3**) are deeply buried beneath other (superficial) soils that are not susceptible to shrink–swell behaviour. However, other superficial deposits such as alluvium, peat and laminated clays can also be susceptible to soil subsidence and heave (e.g. in the Vale of York, east of Leeds – see **Figure 33.3**). Some

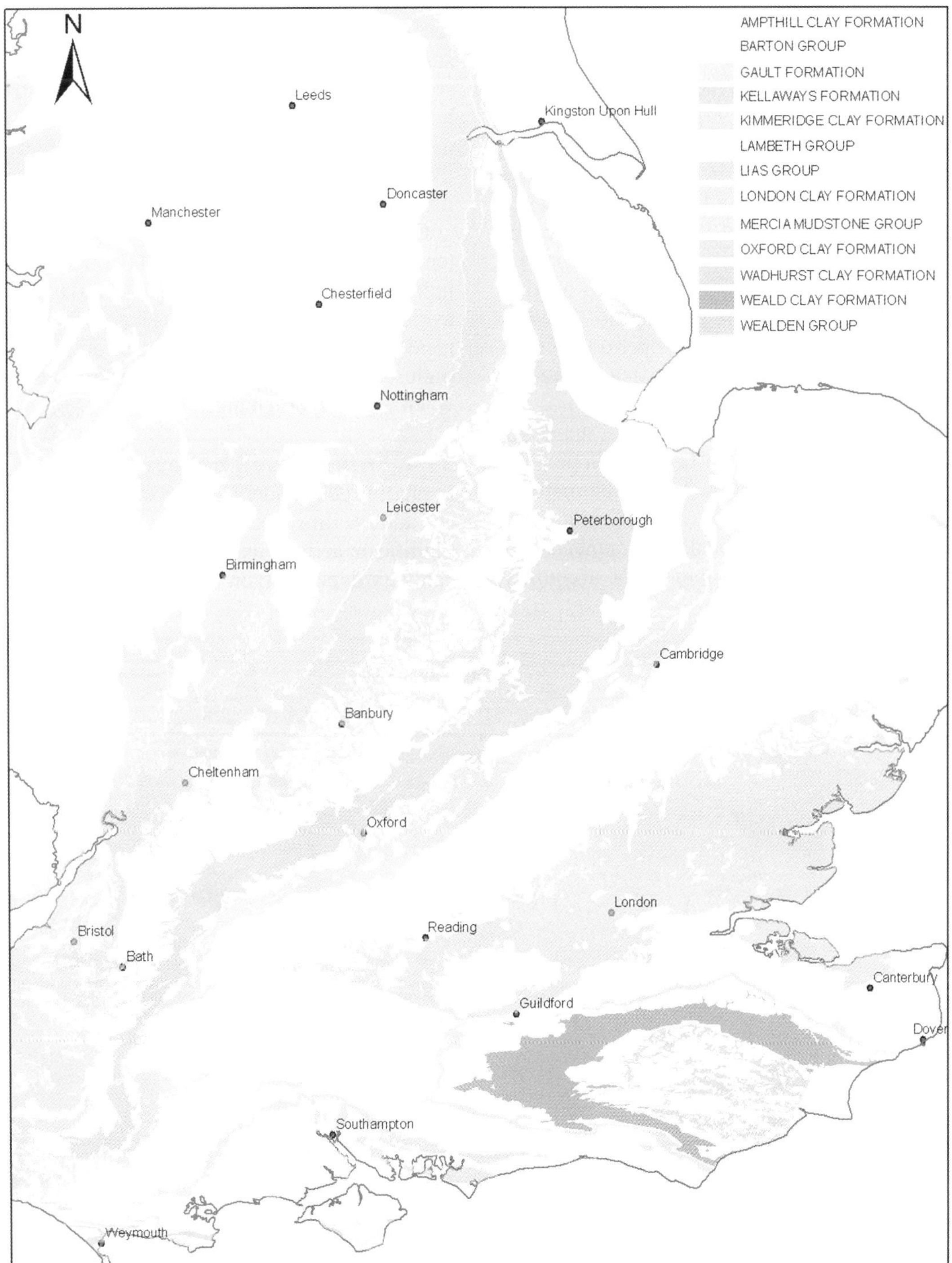

Figure 33.3 Distribution of UK clay-rich soil formations. A colour version of this figure is available online

Mesozoic and Tertiary clay soils and weak mudrocks are also susceptible to shrinkage and swelling as environmental conditions change (Harrison *et al.*, 2012).

Expansive soils are found throughout many regions of the world, particularly in arid and semi-arid regions, as well as those where wet conditions occur after prolonged periods of drought. Their distribution is dependent on geology (parent material), climate, hydrology, geomorphology and vegetation.

The literature is full of studies from all over the world, concerned with problems associated with expansive clays (e.g. Fredlund and Rahardjo, 1993; Hyndman and Hyndman, 2009, Nelson *et al.*, 2015; Simmons, 1991; Stavridakis, 2006). Expansive soils incur major construction costs around the world, with notable examples found in the USA, Australia, India and South Africa, to name but a few (**Figure 33.4**). In these countries, or significant areas of them, the evaporation rate is higher than the annual rainfall so there is usually a moisture deficiency in the soil. Subsequently, when it rains, the ground swells and so increases the potential for heave to occur. In semi-arid regions a pattern of short periods of rainfall followed by long dry periods (drought) can develop, resulting in seasonal cycles of swelling and shrinkage.

Due to the global distribution of expansive soils, many different ways to tackle the problem have been developed and these can vary considerably (Radevsky, 2001). The methods to deal with the problem of expansive soils differ in many ways and depend not only on technical developments, but also on the legal framework and regulations of a country, insurance policies and the attitudes of insurers, experience of the engineers and other specialists dealing with the problem and, importantly, the sensitivity of the owner of the property affected. In the UK in particular there is high sensitivity to relative small cracks (see Section 33.5.3). A summary of these issues is provided by Radevsky (2001) in his review of how different countries deal with expansive soil problems, and a detailed informative study from Arizona, USA has more recently been presented by Houston *et al.* (2011). The latter study demonstrated how the source of problems in expansive soils often stems from poor drainage, construction problems, homeowner activity and its adverse effects and landscaping using vegetation, or a combination of these. These aspects may cause more expansive soil problems than landscape type itself.

Overall, in humid climates, problems with expansive soils tend to be limited to those soils containing higher-plasticity index (I_p) clays. However, in arid/semi-arid climates soils that exhibit even moderate expansiveness can cause distress to residential property. This stems directly from their relatively high suction and the larger changes in water content that result when water levels change.

33.4. Shrink–swell behaviour

Excluding deep underground excavations (e.g. tunnels), shrinkage and swelling effects are restricted to the near-surface zone. Significant activity usually occurs to about 3 m depth, but this can vary depending on climatic conditions. The shrink–swell

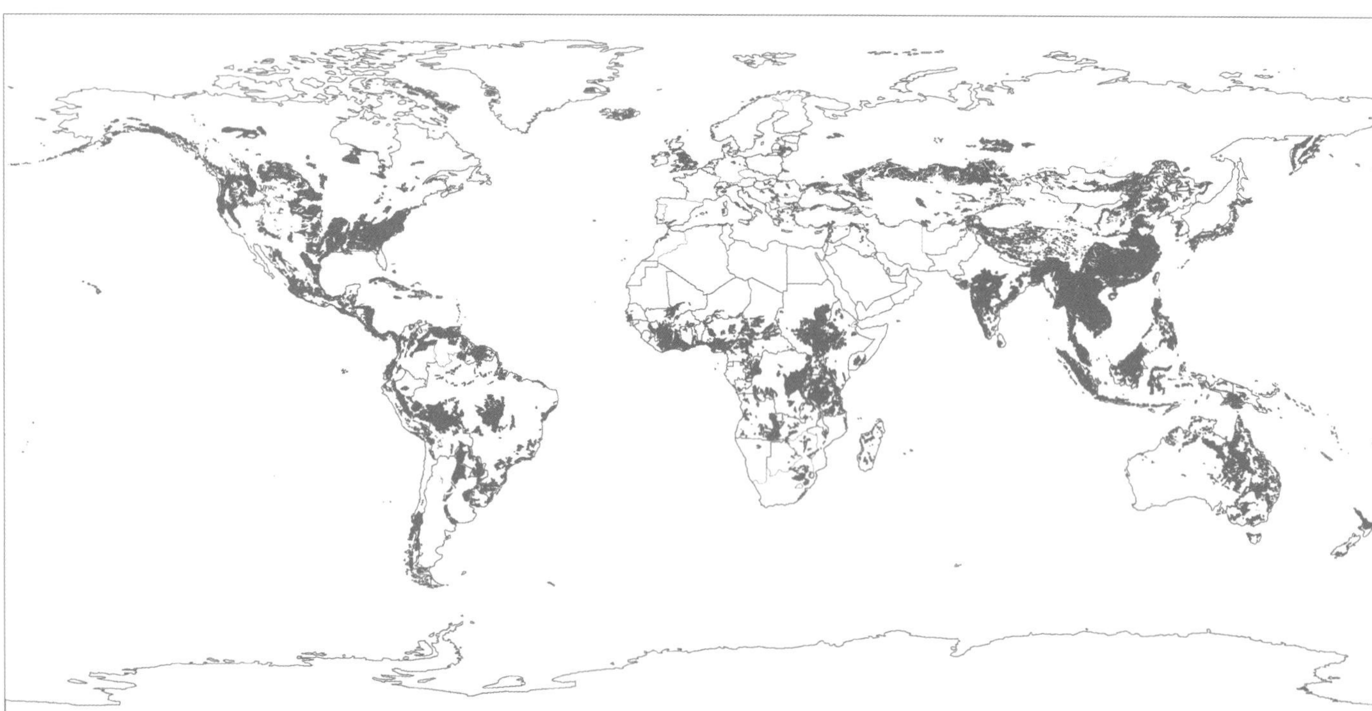

Figure 33.4 Global distribution of shrink–swell soil formations, by region (After Jones *et al.*, 2020)

potential of expansive soils is determined by its initial water content, void ratio, internal structure and vertical stresses, as well as the type and amount of clay minerals in the soil (Bell and Culshaw, 2001). These minerals determine the natural expansiveness of the soil, and include smectite, montmorillonite, nontronite, vermiculite, illite and chlorite. Generally, the more of these minerals that are present in the soil, the greater the expansive potential. However, these expansive effects may become diluted by the presence of other non-swelling minerals such as quartz and carbonate (Kemp *et al*., 2005).

The key aspect of expansive soils behaviour is the soil vulnerability of water-induced volume change. When soils with a high expansive potential are present they will usually not cause a problem as long as their water content remains relatively constant. This is largely controlled by (Houston *et al*., 2011)

- soil properties (e.g. mineralogy)
- suction and water conditions
- water content variations, both temporally and spatially
- geometry and stiffness of a structure, in particular its foundation.

In a partially saturated soil, changes in water content, or suction (increasing strength of the soil due to negative pore water pressures), significantly increase the chances of damage occurring. Changes in soil suction occur due to water movement through the soil due to evaporation, transpiration or recharge, which are often significantly influenced by interaction with trees through response to dry/wet periods of weather (Biddle, 2001). In a fully saturated soil the shrink–swell behaviour is controlled by the clay mineralogy.

33.4.1 Mineralogical aspect of expansive soils

Clay particles are very small and their shape is determined by the arrangement of the thin crystal lattice layers that they form, along with many other elements that can become incorporated into the clay mineral structure (hydrogen, sodium, calcium, magnesium, sulfur). The presence and abundance of these dissolved ions can have a large impact on the behaviour of the clay minerals. In an expansive clay, the molecular structure and arrangement of these clay crystal sheets have a particular affinity to attract and hold water molecules between the crystalline layers in a strongly bonded 'sandwich'. Because of the electrical dipole structure of water molecules they have an electrochemical attraction to the microscopic clay sheets. The mechanism by which these molecules become attached to each other is called adsorption. The clay mineral montmorillonite, part of the smectite family, can adsorb very large amounts of water molecules between its clay sheets, and therefore has a large shrink–swell potential. For further details of the mineralogy of clay minerals and their influence on the engineering properties of soils, see Mitchell and Soga (2005).

When potentially expansive soils become saturated, more water molecules are absorbed between the clay sheets, causing the bulk volume of the soil to increase, or swell. This same process weakens the inter-clay bonds and causes a reduction in the strength of the soil. When water is removed by evaporation or gravitational forces the water between the clay sheets is released, causing the overall volume of the soil to decrease, or shrink. As this occurs, features such as voids or desiccation cracks can develop.

Potentially expansive soils are initially identified by undertaking particle size analyses to determine the percentage of fine particles in a sample. Clay-sized particles are considered to be less than 2 μm (although this value varies slightly throughout the world), but the difference between clays and silts is more to do with origin and particle shape. Silt particles (generally comprising quartz particles) are products of mechanical erosion, whereas clay particles are products of chemical weathering and are characterised by their sheet structure and composition.

33.4.2 Changes to effective stress and role of suctions

Following any reduction in *total* stress, deformations will take place in the ground. A distinction can be made between (*a*) an immediate but time-dependent elastic rebound, and (*b*) swelling due to effective stress changes. In soils, as in rocks, rebound can be an important deformation process which encourages stress relief fractures and zones of secondary permeability which can localise delayed swelling. The amount of deformation depends on the undrained stiffness of the soil, which is equivalent to the modulus of elasticity for the soil, as reflected by its Young's modulus and Poisson's ratio. Subsequent swelling requires an *effective* stress decrease, and a movement of fluid into a geological formation or soil. The magnitude of strains associated with these processes depends on the drained stiffness, the extent of the stress change, the resulting water pressures in the soil or rock and the new boundary conditions. The rate of volume change depends on the compressibility, expansibility and hydraulic conductivity of the sediment and surrounding materials. In stiff homogeneous materials with a low hydraulic conductivity, several decades may be necessary to complete the process.

Accurate laboratory measurements of the controlling elastic properties at small strains in both rebound and swelling (i.e. before yield takes place) are difficult, largely because of sampling disturbance (Burland, 1989). Further discussion of these difficulties, states of stress and the other important concepts of consolidation/swelling in soils are treated in detail by many standard soil engineering texts (Atkinson, 2007; Powrie, 2018) – see also Section 2 *Fundamental principles*.

Shrinkage by evaporation is similarly accompanied by a reduction in water pressure and development of negative capillary pressures. Deformation follows the same principles of effective stress. However, Bishop *et al*. (1975) have shown by laboratory studies that the degree of saturation of unconfined dried clay samples at a given water content was less than for a similar sample consolidated in a triaxial test to the same water content – that is, there was some intake of air which affected both the modulus and strength of the soil. This process leads to a void ratio which is higher than for a clay consolidated to the same water content by simply increasing the confining load. Such a soil thus becomes inherently unstable and, if re-wetted, may collapse. Subsequent laboratory tests on partially saturated soils have shown that depending on their in situ stress conditions and fabric, some samples may also first swell and then collapse (Alonso *et al.*, 1990). The processes of shrinkage due to evaporation have also been reviewed in detail using effective stress concepts by Sridharan and Venkatappa (1971). A good overview of the core principles related to changes due to suction is provided by Nelson *et al.* (2015).

33.4.3 Seasonal variations in water content

The seasonal volumetric behaviour of a desiccated soil is complex, and this increases with the severity of the shrinkage phenomena. This is reflected by the vertical in situ suction profile, water content profile and the degree of saturation (see **Figure 33.5**).

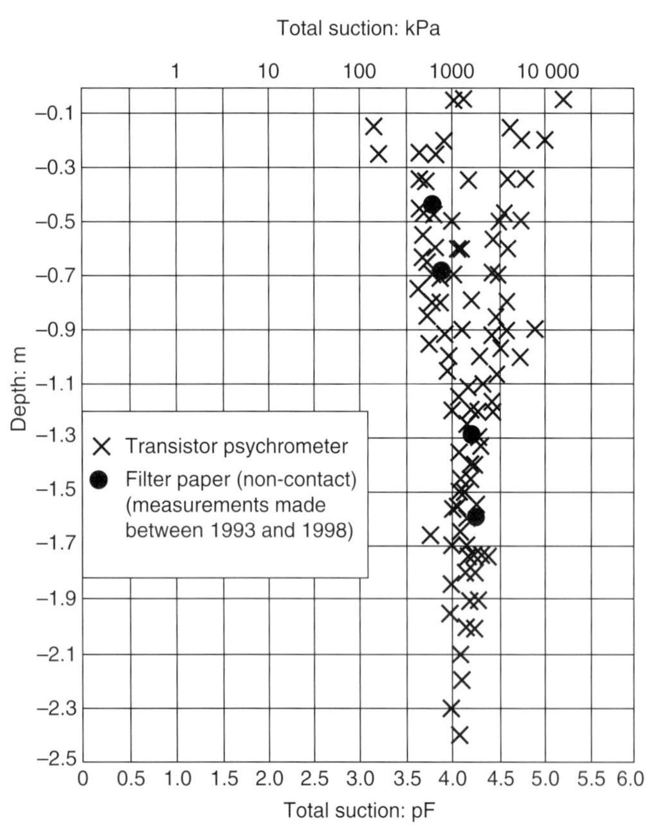

Figure 33.5 Examples of total suction profile (Reproduced from Fityus *et al.*, 2004; with kind permission from ASCE)

The relative values of suction depend on the composition of the soil, particularly its particle size and clay mineral content. The hydraulic conductivity of a soil may also vary both seasonally and over longer timescales. Secondary permeabilities can be induced through fabric changes, tension cracking and shallow shear failure during the swelling and shrinkage process, which may influence subsequent moisture movements. For example, Scott *et al.* (1986) have shown in a micro-fabric study of clay soils that compression (swelling) cracks tended to run parallel to ground contours and dip into the slope at around 60°, and could usually be distinguished from shrinkage cracks which were randomly distributed. In the London Clay soils studied, for example, they found that the ratio between shrinkage and swelling discontinuities was about 2:1. Although not discussed, it seems likely that the nature and distribution of these discontinuities will also influence bulk volumetric seasonal strains.

Expansive soil problems typically occur due to water content changes in the upper few metres, with deep-seated heave being rare (Nelson *et al.*, 2015). The water content in these upper layers is significantly influenced by climatic and environmental factors and is generally termed the *zone of seasonal fluctuations*, or *active zone*, as shown in **Figure 33.6**.

In the active zone, negative pore water pressures exist. However, if excess water is added to the surface or if evapotranspiration is eliminated, then water contents increase and heave will occur. Migration of water through the zone is also influenced by temperature, as shown in **Figure 33.6** – see Nelson *et al.* (2001) for further details. Thus it is important to determine the depth of the active zone during a site investigation. This can vary significantly with different climatic conditions – it may be 5–6 m in some countries, but typically in the UK it is 1.5–2 m (Biddle, 2001). If the drying is greater than the rehydration, then the depth of this zone will increase, with 3–4 m having been observed in some cases in London Clay (Biddle, 2001). These effects are likely to become more significant with climate change.

The term 'active zone' can have different meanings. Nelson *et al.* (2001) provide four definitions for clarity.

- *Active zone:* The zone of soil that contributes to soil expansion at any particular time.
- *Zone of seasonal moisture fluctuation:* The zone in which water content changes due to climatic changes at the ground surface.
- *Depth of wetting:* The depth that water contents have reached owing to the introduction of water from external sources.
- *Depth of potential heave:* The depth at which the overburden vertical stress equals or exceeds the swelling pressure of the soil. This is the maximum depth of the active zone.

The depth of wetting is particularly important as it is used to estimate heave by integrating the strain produced over the zone in which water contents change (Walsh *et al.*, 2009). Details of how this can be achieved and the relative merits of

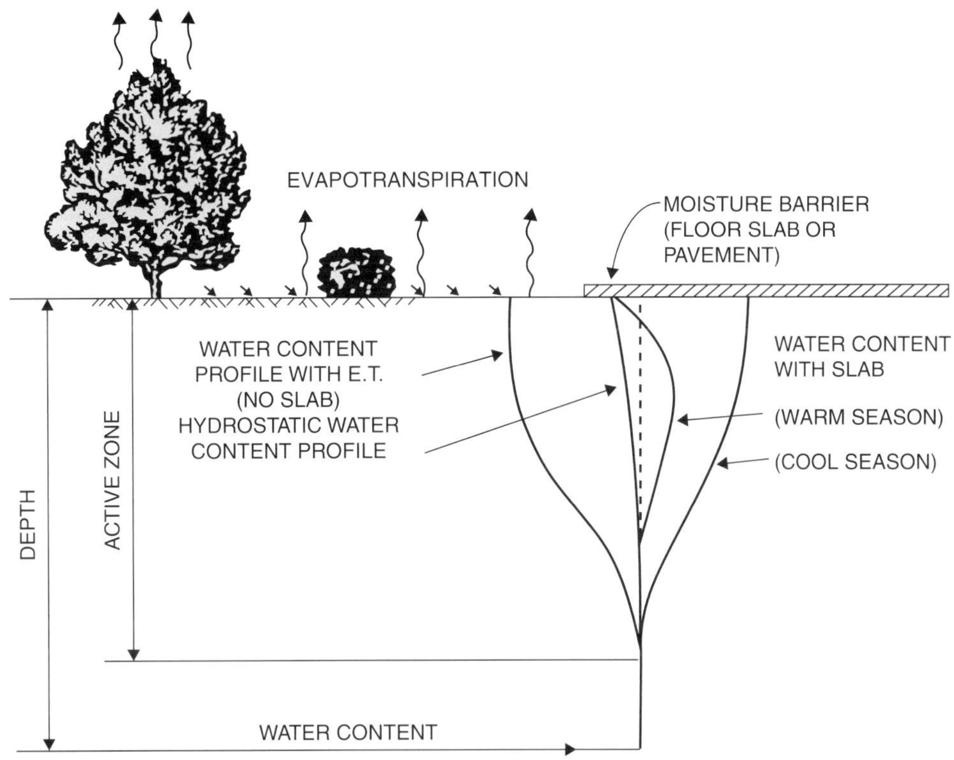

Figure 33.6 Water content profiles in the active zone (Reproduced from Nelson and Miller, 1992; Wiley)

regional and site-specific approaches are considered in detail for a post-development profile by Walsh *et al.* (2009), with further discussion presented by Nelson *et al.* (2011), Aguirre (2011), and Walsh *et al.* (2011).

33.5. Engineering issues

As has been previously stated, many towns, cities, transport routes, services and buildings are founded on expansive soils. These may be solid (bedrock) geological strata in a weathered or unweathered condition, or superficial (drift) geological strata such as glacial or alluvial material, also in a weathered or unweathered condition. These materials constitute a significant hazard to engineering construction in terms of their ability to swell or shrink, usually caused by seasonal changes in moisture content. Superimposed on these widespread climatic influences are local ones, such as tree roots and leakage from water supply pipes and drains. The swelling of shrinkable clay soils after trees have been removed can produce either very large uplifts or very large pressures (if confined), and the ground's recovery can continue over a period of many years (Cheney, 1988). It is the differential, rather than the total, movement of the foundation or superstructure that causes major structural damage. The structures most affected by expansive soils include the foundations and walls of residential and other low-rise buildings, pipelines, pylons, pavements and shallow services. Frequently, these structures only receive a cursory site investigation, if any. It is usually some time after construction that problems come to light. Damage can occur within a few months of construction, develop slowly over a period of 3–5 years or remain hidden until something happens that changes the water content of the soil.

Houston *et al.* (2011) examined the type of wetting that occurs in response to irrigation patterns. They observed that deeper wetting was common with irrigation of heavily turfed areas, and that if ponding of water occurred at the surface there was more likely to be greater distress to buildings through differential movements. Walsh *et al.* (2009) also note that when heave is deep-seated, differential movements are less significant than when the source of heave is at shallower depths.

The structures most susceptible to damage caused by expansive soils are usually lightweight in construction. Houses, pavements and shallow services are especially vulnerable because they are less able to suppress differential movements than heavier multistorey structures. For more information about design parameters and construction techniques for housing and pavements, reference should be made to

- NHBC Standards: building near trees (NHBC, 2023a)
- *Preventing Foundation Failures in New Dwellings* (NHBC, 1988)
- Planning Policy Guidance Note 14: Development on unstable land: Annex 2: subsidence and planning (DTLR, 2002)
- *BRE Digests 240–241: Low-Rise Buildings on Shrinkable Clay Soils* (BRE, 1993a)

- *BRE Digest 298: The Influence of Trees on House Foundations in Clay Soils* (BRE, 1999)
- *BRE Digest 412: The Significance of Desiccation* (BRE, 1996)
- *Criteria for Selection and Design of Residential Slabs-on-Ground* (BRAB, 1968)
- *Evaluation and Control of Expansive Soils* (TRB, 1985).

In many respects, engineering in expansive soils is still based on experience and soil characterisation, and so is often perceived as difficult and expensive (especially for lightweight structures). Engineers use local knowledge and empirically derived procedures, although considerable research has been done on expansive soils – for instance, the database on performance (Houston *et al*., 2011). However, through careful consideration of key aspects associated with expansive soils, problems and difficulties can be dealt with in a cost-effective way.

Two major factors must be identified in the characterisation of a site where a potentially expansive soil exists

- the properties of the soil (e.g. mineralogy, soil water chemistry, suction, soils fabric)
- environmental conditions that can contribute to changes in water content of the soil, such as water conditions and their variations (climate, drainage, vegetation, permeability, temperature) and stress conditions (history and in situ conditions, loading and soil profile).

Normal non-expansive site investigations are often not adequate and a more extensive examination is required to provide sufficient information. This may involve specialist test programmes, even for relatively lightweight structures (Nelson *et al*., 2015). Although there are several methods available to identify expansive soils, each with their relative merits, there are no universally reliable ones. Moreover, expansiveness has no direct measure and so it is necessary to make comparisons, measured under known conditions, to express expansive behaviour (Gourley *et al*., 1994). However, the stages of investigation needed for expansive soils follow those used for any site (see Section 4 *Site investigation* for further details).

33.5.1 Investigation and assessment

It is important to recognise the existence, and understand the potential problems, of expansive soils early on during site investigation and laboratory testing, to ensure that the correct design strategy is adopted before costly remedial measures are required. It is equally important that investigations determine the extent of the active zone.

Despite the proliferation of test methods for determining shrinkage or swelling properties, they are rarely employed in the course of routine site investigations in the UK. Further details of tests commonly employed around the world are given by Chen (1988) and Nelson *et al*. (2015). This means that few datasets are available for creating databases of the directly measured shrink–swell properties of the major clay formations, and reliance has to be placed on estimates based on index parameters, such as liquid limit, plasticity index and density (Holtz and Kovacs, 1981; Oloo *et al*., 1987; Reeve *et al*., 1980). Such empirical correlations may be based on a small dataset, using a specific test method, and at only a small number of sites. Variation of the test method would probably lead to errors in the correlation. The reason for the lack of direct shrink–swell test data is that few engineering applications have a perceived requirement for these data for design or construction.

33.5.1.1 Site investigation

A key difficulty with expansive soils is that they often exhibit significant variability from one location to another (i.e. spatial variability). These proper, adequate, site investigations in areas of potentially expansive soil are often worth the cost. Essential to the investigation of any expansive soils is a good knowledge of local geology: the use of maps provides a framework for this. These maps are particularly useful when constructing transportation networks. In some countries, such as the USA, mapping includes identification of expansive soil potential (Nelson *et al*., 2015). As with any site investigation, field observations and reconnaissance can provide valuable data on the extent and nature of expansive soils and their associated problems. Some key features may be observed locally; important observations include

- *soil characteristics*
  - spacing and width of wide or deep shrinkage cracks
  - high dry strength and low wet strength – high-plasticity soil
  - stickiness and low trafficability when wet
  - shear surfaces have glazed or shiny appearance
- *geology and topography*
  - undulating topography
  - evidence of low permeability by surface drainage and infiltration features
- *environmental conditions*
  - vegetation type
  - climate.

Sampling in expansive soils is generally done in the same way as for conventional soils, with care taken to minimise disturbances through, for example, water content changes or poor control during transportation. Further details are provided in Section 4 *Site investigation*, and an overview of practices specifically used for expansive soils in other countries is provided by Chen (1988), Nelson and Miller (1992) and, more recently, Nelson *et al*. (2015). However, the depth and frequency of sampling may need to be increased in expansive areas due to their high spatial variability.

33.5.1.2 In situ testing

A suite of different field tests can be used to evaluate expansive soils; these include

- soil suction measurements using thermocouple psychrometers, tensiometers or filter paper methods
- in situ density and moisture tests
- settlement and heave monitoring
- piezometers or observations wells
- penetration resistance
- pressuremeters and dilatometers
- geophysical methods.

Expansive soils can be tested in the field using methods that rely on empirical correlation such as the standard penetration test (SPT) or the cone penetration test (CPT) to infer soil strength parameters (Clayton *et al.*, 1995). Initial effective stresses can be estimated using a psychrometer (Fredlund and Rahardjo, 1993) or a suction probe (Gourley *et al.*, 1994), which will measure the soil suction. The undrained shear strength of the soil can be determined using a shear vane (Bjerrum, 1967). The stiffness parameters of the soil can be determined using a plate loading test (BSI, 1999), along with its strength and compressibility. Other tests include the pressuremeter and the dilatometer (ASTM, 2010), which measure strength, stiffness and compressibility parameters.

Seismic test apparatus uses the transmission of elastic waves through the ground to determine its density and elastic properties (see Chapter 45 *Geophysical exploration and remote sensing*). Electrical resistivity methods have also shown promise as a method to determine swell pressure and shrinkage of expansive soils. Resistivity was found to increase as both swell pressure and shrinkage increased (Zha *et al.*, 2006). More recently, Jones *et al.* (2009) successfully monitored tree-induced subsidence in London Clay using electrical resistivity imaging.

Monitoring should also be considered, and a number of approaches can be used which are common with non-expansive soils. Key methods are settlement and heave monitoring for volume change, and piezometers for pore water changes. Monitoring of water content profiles over several wet and dry seasons are used to establish the extent of the active zone (Nelson *et al.*, 2001). In cases where the soil is not uniform or several strata exist, a correction can be applied using the liquidity index. Nelson and Miller (1992) provide an example of this calculation.

Examples of monitoring associated with expansive soils are provided throughout the literature. Examples include Fityus *et al.* (2004), where a site near Newcastle, Australia, was instrumented and soil water and suction profiles together with ground movements were determined over the period 1993–2000. In addition, the work of the BRE at their London Clay site near Chattenden, Kent, provides details of similar monitoring regimes over several years (Crilly and Driscoll, 2000; Driscoll and Chown, 2001). Stable benchmarks are important for any monitoring in expansive soils, and design details and installation instructions are given in many papers (e.g. Chao *et al.*, 2006). Further details can be found in Sections 4 *Site investigation* and 9 *Construction verification*. For specific discussions in the context of expansive soils, see Chen (1988), Nelson and Miller (1992) and Nelson *et al.* (2015).

33.5.1.3 Laboratory testing

Considerable research work has been carried out on behalf of the oil and mining industries, especially in the USA, on the swelling behaviour of 'compact' clays and mudrocks, in particular clay shales. Swelling pressure has caused damage in tunnels (Madsen, 1979), as is the case – usually at great depths – in the mining industry. In the oil industry the swelling of shales and 'compact' clays in borehole and well linings has been a topic of interest. The laboratory test methods developed differ considerably from those applied by the civil engineering industry and tend to duplicate the particular phenomena causing problems. For example, the moisture activity index test (Huang *et al.*, 1986) duplicates changes in relative humidity in the air passing through mine tunnels, and consequent swelling of the tunnel lining. However, the confined swelling pressure test is relatively universal. As shrinkage is a near-surface phenomenon in the UK, much work has been done by the soil survey and agricultural organisations. Reeve *et al.* (1980) describe the determination of shrinkage potential for a variety of soils classified on a pedological basis.

For geotechnical purposes, a suite of different tests can be used to identify expansive soils, and include Atterberg limits, shrinkage limits, mineralogical tests such as X-ray diffraction, swell tests and suction measurements (see Nelson *et al.*, 2015 for further details). Undisturbed samples are normally used for one-dimensional response to wetting tests. However, it should be noted that when conducting swell tests in the laboratory it is important to distinguish between swelling in compacted, undisturbed and reconstituted samples, which occurs due to significant differences in their respective fabrics.

Swell–shrink tests

Swelling tests may be broadly divided into those tests attempting to measure the deformation or strain resulting from swelling, and those which attempt to measure the stress, or pressure, required to prevent deformation due to swelling. These two types are referred to here as swelling strain and swelling pressure tests, respectively. Swelling strain tests may be linear (i.e. one-dimensional, or 1D) or volumetric (i.e. three-dimensional, or 3D). Swelling pressure tests are almost always 1D and traditionally used oedometer-type testing arrangements (Fityus *et al.*, 2005). However, shrinkage tests deal solely with the measurement of shrinkage strain in either 1D or 3D.

Standards do exist for shrink–swell tests but they do not cover all of the methods in use internationally. Like many 'index'-type soils tests, some shrink–swell tests are based on practical needs and tend to be rather crude and unreliable. While measurement of water content is easily achieved with some accuracy, the measurement of the volume change of a clay soil specimen is not, particularly in the case of shrinkage. Solutions to this problem have been found by the measurement of volume change in only one dimension, or by immersion of the specimen in a non-penetrating liquid such as mercury. However, the use of mercury in this way is far from ideal. Measurement of volume change in the case of swelling, where the specimen is assumed to be saturated, is only slightly less problematic. In this case, dimensional changes are required to be made while the specimen is immersed in water. This introduces the problem of either immersed displacement transducers or sealed joints for non-immersed transducers.

Nelson *et al.* (2015) provide a detailed account of various swell and heave tests (with the oedometer being the most commonly used) which are often developed based on geographic regions with specific expansive soil problems. However, they can be considered applicable in general situations (Fityus *et al.*, 2005). These tests determine the applied stress required to prevent swelling strain when a specimen is subjected to flooding. The ability to do this is enhanced by computer control, or by at least some form of feedback control. The determination of swelling pressure should not be confused with the determination of rebound strain under consolidation stresses in the oedometer test. In the latter case, the slope of the rebound part of the familiar voids ratio to applied stress (e–log p') curve is referred to as the swelling index (C_s); that is, the rebound or decompressional equivalent of the compression index (C_c). It is common, however, for measured swell potential to be low to medium when soil units across a region have high potential; this is the result of natural soil variability (Houston *et al.*, 2011).

Mineralogical testing

In addition to the traditional approaches used, several parameters have been investigated which are either wholly or largely dependent on clay mineralogy. These are surface area (Farrar and Coleman, 1967), dielectric dispersion (Basu and Arulanandan, 1974) and disjoining pressure (Derjaguin and Churaev, 1987). The factors affecting swelling of very compact or heavily overconsolidated clays and clay shales may differ from those affecting normally consolidated or weathered clays. Physico-chemical and diagenetic bonding forces probably dominate in these materials, whereas capillary forces are negligible. It is likely that the distance between clay platelets, the ionic concentration of pore fluids and fluids used in laboratory tests relative to the clay mineral activity of such materials are the key factors in swelling. Traditional concepts of Darcian permeability and pore water pressure are thrown into doubt in these compact clays and clay shales. Diffusion may be the principal mode of fluid movement in these very low-permeability clays.

Use index tests

The volume change potential (VCP) (also known as the potential volume change, PVC) of a soil is the relative change in volume to be expected with changes in soil water content, and is reflected by shrinking and swelling of the ground; in other words, the extent to which the soil shrinks as it dries out or swells when it gets wet. More recently the BGS has developed a reliable laboratory test of undisturbed clay samples – SHRINKiT, an automated measurement system used to measure the shrinkage properties of soils safely and reliably (Hobbs *et al.*, 2014). However, despite the various test methods available for determining these two phenomena (e.g. tests 6.3–6.5 of BS 1377-2:1990 and test 4 of BS 1377-5:1990 (BSI, 1990)), they are rarely employed in the course of routine site investigations in the UK. Hence few data are available for databasing the directly measured shrink–swell properties of the major clay formations. Consequently, reliance is placed on estimates based on index parameters – namely, liquid limit, plastic limit, plasticity index, and density (Holtz and Kovacs, 1981; Oloo *et al.*, 1987; Reeve *et al.*, 1980). No consideration has been given to the saturation state of the soil and therefore to the effective stress or pore water pressures within it.

The most widely used parameter for determining the shrinkage and swelling potential of a soil is the plasticity index (I_p). Such plasticity parameters, being based on remoulded specimens, cannot precisely predict the shrink–swell behaviour of an in situ soil. However, they do follow properly laid down procedures, being performed under reproducible conditions to internationally recognised standards (Jones, 1999). A 'modified plasticity index' (I_p') is proposed in the BRE Digest 240 (BRE, 1993a) for use where the particle size data, specifically the fraction passing through a 425 μm sieve, is known or can be assumed as 100% passing (**Table 33.1**).

The modified I_p' considers the whole sample and not just the fines fraction; it therefore gives a better indication of the 'real' plasticity value of an engineering soil and eliminates discrepancies due to particle size – for example, in glacial till. This compares with a classification produced by the National House Building Council which forms the basis of the NHBC 'foundation depth' tables (**Table 33.2**), which uses the same modified I_p' approach as presented in **Table 33.1**.

The concept of 'effective plasticity index' has been described to deal with multilayered soils of different plasticity index (BRAB, 1968).

Ultimately, swelling and shrinkage *potential* may be considered to be the ultimate capability of a soil to swell and shrink, but this potential is not necessarily realised in a given moisture-change situation. These do not therefore represent the fundamental properties of a soil. However, potential may be described differently. For example, swelling potential is described by Basu and Arulanandan (1974) as 'the ability and

I_p' (%)	Volume change potential
>60	Very high
40–60	High
20–40	Medium
<20	Low

$I_p' = I_p \times (\% < 425\ \mu m)/100\%$

Table 33.1 Classification for shrink–swell clay soils (Data taken from BRE, 1993a)

I_p' (%)	Volume change potential
>40	High
20–40	Medium
10–20	Low

Table 33.2 Classification for shrink–swell clay soils (Data taken from NHBC, 2023a)

degree to which swelling is realised under given conditions'. So there is already some confusion in terminology. Oloo *et al.* (1987) differentiate between intrinsic expansiveness (swell) and heave. They define intrinsic expansiveness as that property which 'relates change in water content, and thus change in volume, to the suction change' of a clay soil. Thus a soil of high intrinsic expansiveness will exhibit a large water content or volume change compared with one of low intrinsic expansiveness for a given suction change – all else being equal. Oloo *et al.* (1987) state that no procedure has been developed to measure this property. Swell is defined as 'a measure of the volume strain, or axial strain, in a soil under a particular set of stress and suction conditions'. Heave is defined as 'the displacement of a point in the soil due to suction and stress changes interacting with the intrinsic expansiveness'. Heave is not a soil property.

Overall, there are many methods of testing for the shrinkage and swelling properties of clay soils. These methods are covered in detail by Jones (1999), where the pros and cons of each method are discussed and the reasons for the selection and rejection of methods is determined. Further evaluation of these tests is also provided by Fityus *et al.* (2005).

33.5.2 Shrink–swell predictions

Common to all geotechnical predictions of volume change is the need to define initial and final in situ stress state conditions. This requires characterisation of the stress–strain behaviour of each soil profile. Initial stress states and constitutive properties can be evaluated using a suite of approaches (highlighted by many texts, e.g. Fredlund and Rahardjo, 1993; Powrie, 2018), but it is the final stress condition that must usually be assumed. Guidelines are presented by Nelson and Miller (1992), with calculations based on knowledge of effective overburden stress (i.e. the increment of stress due to applied load and soil suction). However, each situation requires engineering judgement and consideration of environmental conditions.

Details of constitutive relationships for expansive soils have been reviewed, and a useful description of these is given by Nelson *et al.* (2015). These include unsaturated soil models dealing with matric and osmotic suctions. A detailed account of this, the theoretical basis and the associated models used to predict partially saturated soils behaviour, together with test methods used to determine key soil parameters, is provided by Fredlund and Rahardjo (1993) and Fredlund (2006).

Overall prediction methods can be grouped into three broad categories: theoretical, semi-empirical and empirical. They all rely on testing methods; particular care must be taken with empirical methods which are only valid within the bounds of the soil type, environment and engineering application for which they were developed.

Several heave predictions are available that are based on oedometer or suction tests, and Nelson *et al.* (2015) provide a detailed account of these, together with examples of associated predictions. Nelson *et al.* (2010) provide an illustration using free-field heave predictions and their use in foundation design, as well as methods for prediction heave rates.

33.5.2.1 Oedometer-based methods

Oedometer-based tests include 1D and double oedometer tests (developed by Jennings and Knight, 1957). Double oedometer tests consist of two near-identical undisturbed samples, one loaded at its natural water content and the other inundated under a small load and then loaded under saturated conditions. The use of the oedometer has distinct advantages due to familiarity among geotechnical engineers.

Tests can be conducted as free swell tests, where swelling is allowed to occur at a predetermined pressure after water is added. The swell pressure is then defined as the pressure required to recompress the swollen sample to its pre-swollen volume. These tests, however, suffer the limitation that volume change can occur and that hysteresis is incorporated into the estimation of the in situ state. An alternative approach that overcomes these problems involves inundating a sample placed in the oedometer and preventing it from swelling. The swell pressure is then the maximum applied stress required to achieve a constant volume. Typical results from these tests are shown in **Figure 33.7**, with σ_0' representing the stress when inundation has occurred and σ_s' representing the stress equated to swelling pressure.

The constant volume test may overcome the difficulties of the free swell test, but as a result it is more vulnerable to sample disturbance. To account for sample disturbance, Rao *et al.* (1988) and Fredlund and Rahardjo (1993) suggest simplifications to facilitate predictions using parameters measured by constant volume oedometer tests (pressures increase during swelling to maintain constant volume) using established techniques. This is illustrated in **Figure 33.8**.

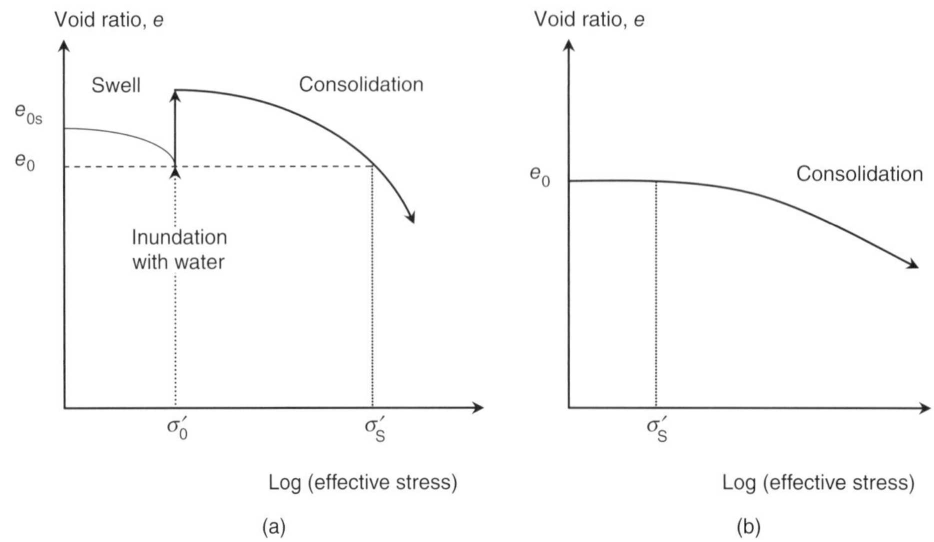

Figure 33.7 Typical oedometer swell test curves: (a) an illustration of a free swell test result; (b) an illustration of constant volume test results

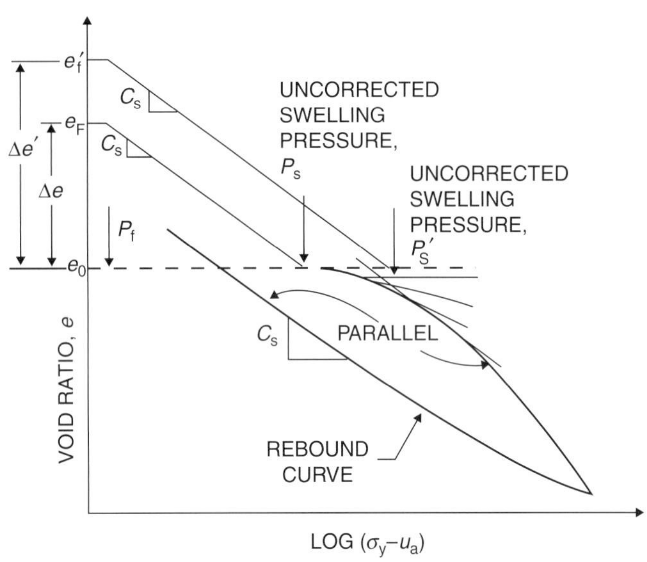

Figure 33.8 One-dimensional oedometer test results showing the effect of sampling disturbance. C_s is swell index; $(\sigma_y - u_a)$ is overburden pressure; P_f is final stress state; e_f is final void ratio, and e_f' is final void ratio corresponding to corrected swell pressure, P_s' (Reproduced from Rao *et al.*, 1988, with kind permission from ASCE)

Fityus *et al.* (2005) questioned this approach and considered that specialist apparatus not normally used in standard geotechnical engineering testing laboratories is needed to achieve meaningful results. However, not all authors agree, with Nelson *et al.* (2015) believing good-quality data and predictions can be obtained with such an approach. Moreover, several disadvantages exist, as tests where the specimen is fully wetted are conservative, as full saturation is not often reached in the field (Houston *et al.*, 2011). Thus, swell tests based on submerged samples at the level of stress of interest will overpredict heave. The effect of partial wetting may be as important as the depth to which wetting has occurred (Fredlund *et al.*, 2006).

33.5.2.2 Suction-based tests

Suction tests are used to predict soil response in much the same manner as with saturated effective stress changes. Various methods have been developed, such as the US Army Corps of Engineers Waterways Experiment Station method or the clod method, details of which (including advantages and limitations) can be found in Nelson *et al.* (2015). Fredlund and Hung (2001) have subsequently developed suction-based predictions to evaluate volume changes from both environmental and vegetation changes, and they provide useful outline example calculations.

Nelson *et al.* (2015) suggest that with careful sampling and testing it is possible to predict heave within a few centimetres. However, it is essential that the testing is conducted within the expected stress range in the field. Furthermore, experimental studies involving direct measurement of partially saturated properties are expensive and often time-consuming. Chandler *et al.* (1992) provide details of suction measurements using the filter paper method, highlighting the need for careful calibration as results can be affected by temperature fluctuations, particle entrainment in the filter paper during testing and hysteresis effects. Such approaches have a number of advantages as a means to estimate soil suction and hence suction profiles (**Figure 33.5**).

For this reason, numerical and semi-empirical methods increasingly use the soil–water characteristic curves (SWCCs) (Puppala *et al.*, 2006). The SWCCs describe the relationship between water content (either gravimetric or volumetric) and soil suction. Alternatively, they can be used to describe the relationship between the degree of saturation and soil suction. A more detailed discussion and examples of typical SWCCs are also provided in Chapter 30 *Tropical soils*.

Only a limited number of investigations have been undertaken on expansive soils, with Ng *et al.* (2000), Likos *et al.* (2003) and Miao *et al.* (2006) providing some examples of these. Puppala *et al.* (2006) detail SWCCs for treated and untreated expansive soils. Further details of this are provided by Fredlund and Rahardjo (1993), with Nelson *et al.* (2015) providing details in the context of expansive soils. However, it should be noted that suction measurements are subject to errors that can be substantial (Walsh *et al.*, 2009).

Empirical methods are still common in geotechnical engineering (Houston *et al.*, 2011). Heave is often estimated by the integration of strain over the zone in which the water content changes. However, uncertainty occurs and arises from three sources (Walsh *et al.*, 2009)

- the depth over which the wetting will occur
- the swell properties of the soil
- the initial and final suction over the depth of wetting.

Care is needed with all models used, as small changes in input parameters can lead to significant changes in an estimated soil response. The real challenge is, therefore, to understand the relationship between soil–water stress level and volume changes, coupled with a prediction of the actual depth and degree of wetting that will occur in the field. Both are related to soil properties and control of site water (Houston *et al.*, 2011).

Houston *et al.* (2011) compared predictions from a number of forensic studies from field and laboratory investigations in arid/semi-arid areas to those undertaken using numerical approaches (in this case, the simple 1D and 2D unsaturated flow model), with details of site drainage and landscape practices also considered. Comparisons were made after one year; they concluded that drainage conditions were the more important factor in the prediction of foundation problems. This study revealed that the effects of poor drainage and roof run-off ponding near a structure is the worst-case scenario. Uncontrolled drainage and water ponding near foundations led to significant suction reduction to greater depths (0.8 m was found after one year), resulting in differential soil swell and foundation movement (see **Figure 33.9**).

33.5.2.3 Numerical approaches

One-dimensional simulations also dominate numerical studies, as unsaturated flow solutions are sensitive to accurate and detailed simulation of surface flux conditions, thus requiring an extremely tight mesh and time steps (Houston *et al.*, 2011). This may result in very lengthy runtimes of several months, even for 1D assessments (Dye *et al.*, 2011). However, Xiao *et al.* (2011) demonstrated how numerical simulations could be used to assess pile–soil interactions, providing an effective way to undertake sensitivity analysis, but noted that many parameters are needed when undertaking numerical assessments.

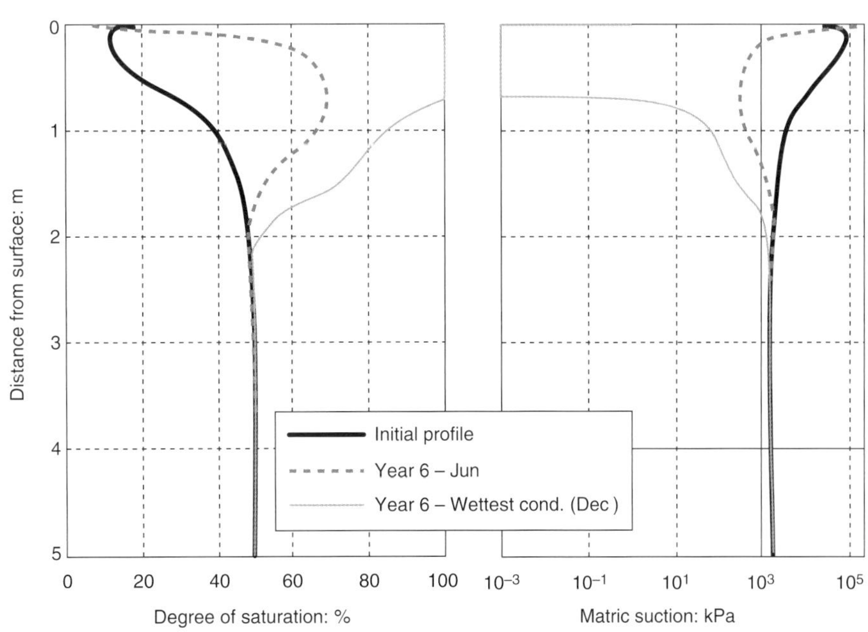

Figure 33.9 Profile for 1 year of roof run-off water ponding next to foundations after 6 years of desert landscape. Wettest and driest conditions in 1D (Reproduced from Houston *et al.*, 2011, with kind permission from ASCE)

33.5.3 Characterisation

Many attempts have been made to find a universally applicable system for the classification of shrinking and swelling in order to characterise an expansive soil. Some have even attempted to produce a unified swelling potential index using commonly used indices (e.g. Kariuki and van der Meer, 2004; Sridharan and Prakash, 2000; Yilmaz, 2006) or from specific surface areas (Yukselen-Aksoy and Kaya, 2010), but these are yet to be adopted. Examples of various schemes commonly used around the world are illustrated in **Figure 33.10**. The various schemes that have been developed lack standard definitions of swell potential, since both sample conditions and testing factors vary over a wide range of values (Nelson *et al.*, 2015).

33.5.3.1 Classification schemes

Most classification schemes give a qualitative expansion rating (e.g. high or critical). The different classification schemes can be categorised into four groups, depending on which method they employ to determine their results. These are

- free swell (see Holtz and Gibbs, 1956)
- heave potential (see Snethen *et al.*, 1977; Vijayvergiya and Sullivan, 1974)
- degree of expansiveness (see Chen, 1988; FHA, 1965)
- shrinkage potential (see Altmeyer, 1956; Holtz and Kovacs, 1981).

Since liquid limit and swelling of clays both depend on the amount of water a clay tries to imbibe, it is not surprising that they are related. Chen (1988) suggested that a relationship between the swelling potential of clay and its plasticity index can be established. While it may be true that high swelling soil will manifest high index properties, the converse is not always true.

Other classification schemes relate to expansion potential, based on the Skempton 'activity' plot (Skempton, 1953) and its development by Williams and Donaldson (1980) from Van der Merwe (1964). Details are described in Taylor and Smith (1986) with respect to various UK clay mudstone formations.

A host of schemes have been put forward for estimating shrink–swell, particularly in the USA (see Chen, 1988; Nelson *et al.*, 2015), most of which use swelling and suction

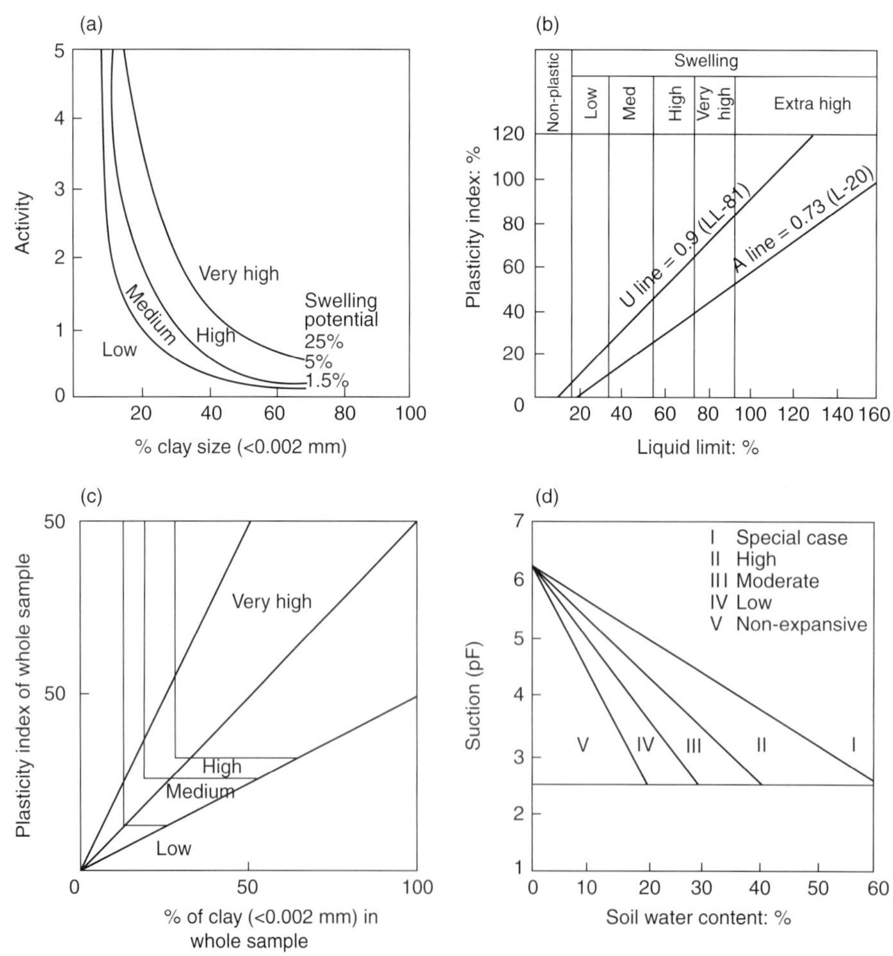

Figure 33.10 Commonly used criteria for determining swell potential from across the world (Reproduced from Yilmaz, 2006, with permission from Elsevier)

as their basis (Snethen, 1984). Sarman *et al.* (1994) concluded that swelling was not related solely to clay mineral type, but also to pore morphology. It was found that samples showing high swelling had a large pore volume combined with a high percentage of small-sized pores. The high swelling was attributed to the samples' ability to absorb and adsorb water. It was found that correlations between swelling and other parameters were unsuccessful.

With all classification schemes only indications of expansion are obtained as, in reality, field conditions vary considerably. Such ratings can be of little use unless the user is familiar with the soil type and the test conditions used to develop the ratings. Ratings themselves may be misleading and can, if used with design options outside the region where the rating was established, cause significant difficulties (Nelson *et al.*, 2015). Classifications, therefore, should only be considered to provide an indication of potential expansive problems, and further testing is needed. If such schemes are used as a basis of design, the result is either over-conservative solutions or inadequate construction (Nelson *et al.*, 2015).

33.5.3.2 UK approach

While much study has been carried out worldwide to infer swelling and shrinkage behaviour from soil index properties such as plasticity (see Section 33.5.1.1), few direct data are available in UK geotechnical databases (Hobbs *et al.*, 1998). Two schemes that are commonly used within the UK are based on the BRE and NHBC schemes.

Volume change potential has more recently been defined for overconsolidated clays in terms of a modified plasticity index term (I_p') by BRE Digest 240 (BRE, 1993a) – see **Table 33.1**. This classification aims to eliminate discrepancies due to particle size.

High-shrinkage potential soils may not behave very differently from low-shrinkage ones because environmental conditions in the UK do not allow full potential to be realised (Reeve *et al.*, 1980). The NHBC (2023a) classified VCP as shown in **Table 33.2**. This classification forms the basis of the NHBC's 'foundation depth' tables.

Since a set of soil properties will often not fit neatly into one category, the determination of shrinkage potential requires some judgement. The BRE (1993a) suggests that plasticity index and clay fraction can be used to indicate the potential of a soil to shrink or swell as shown in **Table 33.3**.

The overlap of categories reflects the fact that figures were obtained from multiple sources.

33.5.3.3 National compared to regional characteristics

A meaningful assessment of the shrink–swell potential of soil in the UK requires a considerable amount of high-quality and well-distributed spatial data of a consistent standard. The BGS's *National Geotechnical Properties Database* (Self *et al.*, 2008) contains a large body of index test data. At the time of writing the database contained data from more than 80 000 boreholes, comprising nearly 320 000 geotechnical samples, with 100 000 containing relevant plasticity data.

The BGS GeoSure *National Ground Stability Data* provides geological information about potential ground movement or subsidence, including the GeoSure shrink–swell dataset (Booth *et al.*, 2011). It should be noted that this assessment does not quantify the shrink–swell behaviour of a soil at a particular site; it indicates the potential for such a hazard to be present with regard to the behaviour of the underlying geological unit throughout its outcrop.

The VCP of a soil provides the relative change in volume to be expected with changes in soil water content. This was calculated from the I_p' values and a classification made based on the *upper quartile* value (**Table 33.4**). This is based on the BRE (1993a) scheme shown in **Table 33.1**. In this way, a VCP was assigned to each of the geological units and a map of shrink–swell potential built (**Figure 33.11**).

Looking at clays on a national scale can give a good indication of the potential problems associated with them and provide initial information regarding planning decisions. However, no two clay soils are the same in terms of their behaviour or their shrink–swell potential. Therefore, it is useful to look at a particular clay formation on a more regional basis. For illustration, the London Clay formation will be used.

The London Clay formation is of major importance in the fields of geotechnical engineering and engineering geology. This is because it has hosted a large proportion of subsurface engineering works in London over the last 150 years. It has also been the subject of internationally recognised research in soil mechanics over the last 50 years (Chandler and Apted, 1988; Skempton and DeLory, 1957; Takahashi *et al.*, 2005). London Clay is subject to shrinkage and swelling behaviour, which has resulted in a long history of foundation damage within the outcrop.

Plasticity index: %	Clay fraction < 0.002 mm	Shrinkage potential
>35	>95	Very high
22–48	60–95	High
12–32	30–60	Medium
<18	<30	Low

Table 33.3 The potential of a soil to shrink or swell (After BRE, 1993a)

Classification	I_p': %	VCP
A	<1	Non-plastic
C	1–20	Low
D	20–40	Medium
D	40–60	High
E	>60	Very high

Table 33.4 Classification of VCP

Figure 33.11 Shrink–swell potential map, based on VCP (Reproduced from Jackson, 2004; © NERC, with permission from the British Geological Survey)

Jones and Terrington (2011) follow the methodology described by Diaz Doce *et al.* (2011), using 11 366 samples across the London Clay outcrop, splitting it into four distinct areas based on geographical location, plasticity values and depth of overlying sediment. In this way, a more detailed assessment of the outcrop could be carried out, and a 3D model providing a seamless interpolation of the VCP of the London Clay was created. This model gives a visualisation of the I_p' values, allowing them to be examined at a variety of depths relative to ground level (**Figure 33.12**). This type of analysis

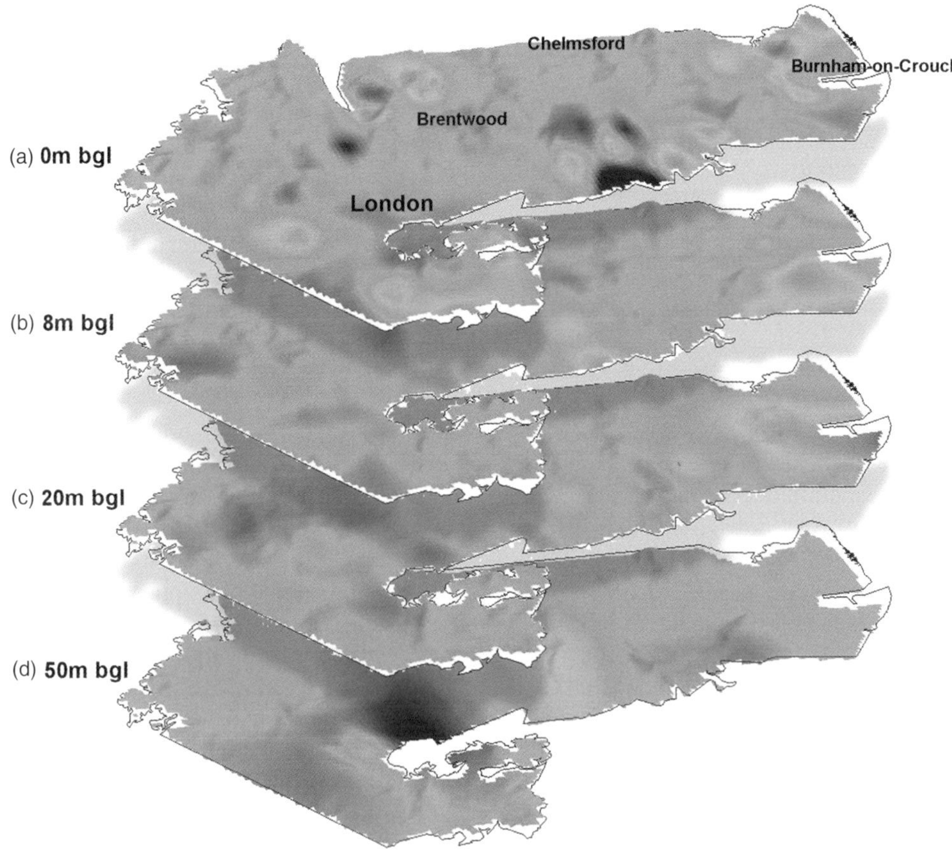

Figure 33.12 S-grid interpolations for area 3, showing surfaces at 0 m, 8 m, 20 m and 50 m bgl; blue: medium, green: high, yellow/red: very high VCP (Reproduced from Jones and Terrington, 2011; © The Geological Society of London. A colour version of this figure is available online)

indicates that 3D modelling methods have considerable potential for predicting the spatial variation of VCP within expansive clay soils, so long as they have large enough datasets.

33.5.4 Specific problems with expansive soils

The principal adverse effects of the swell–shrink process arise when either swelling pressures result in heaving (or lifting) of structures, or shrinkage leads to differential settlement. As a result, several mitigation and design options exist either in the form of specific foundation types, or through the use of a range of different ground improvement techniques. Excellent reviews of the full range of these are provided by Chen (1988), Nelson and Miller (1992) and more recently by Nelson *et al.* (2015), together with details provided by NHBC (2023a). A summary is provided in Sections 33.5.4.1–33.5.4.4, highlighting the key features associated with these options. In addition, discussion of some of the key issues faced in the UK is provided (see Section 33.5.4.5), where impact of vegetation is often the major cause of soil–structure problems in expansive soils.

33.5.4.1 Foundation options in expansive soils

A large number of factors influence foundation types and design methods (see Section 5 *Design of foundations*); these include climatic, financial and legal aspects, as well as technical issues. Importantly, swell–shrink behaviour often does not manifest itself for several months and so design alternatives must take account of this. Other issues, such as financial considerations, can place strain on this, and so early communication with all relevant stakeholders is essential. Higher initial costs are often offset many times over by a reduction in post-construction maintenance costs when dealing with expansive soils (Nelson *et al.*, 2015).

Foundation alternatives when dealing with potentially expansive soils follow three options

- use of structural alternatives (e.g. stiffened raft)
- use of ground improvement techniques
- a combination of both.

As with any foundation option, the main aim is to minimise the effects of movement, principally differential. Two strategies are used when dealing with expansive soils

- isolate structure from soil movements
- design a foundation stiff enough to resist movements.

Foundation type	Design philosophy	Advantages	Disadvantages
Pier and beam; pile and beam	Isolate structure from expansive movement by counteracting swell with anchoring to stable strata	Can be used in a wide variety of soils; reliable for soils of high swell potential	Relatively complex design and construction processes requiring specialist contractors
Raft; stiffened raft	Provides a rigid foundation to protect structure from differential settlements	Reliable for soils of moderate swell potential; no specialist equipment needed in construction	Only works for relatively simple building layout; requires full construction quality control
Modified continuous perimeter footing; deep trench fill foundations	Same as raft or stiffened raft foundation – includes stiffened perimeter beams	Simple construction with no specialist equipment needed	Ineffective in highly expansive soils or within the zone of influence of trees

Table 33.5 Foundation types used in expansive soils (Data taken from Nelson and Miller, 1992; NHBC, 2023a)

The major types of foundations used in expansive soils from around the world are pier and beam or pile and beam systems, reinforced rafts and modified continuous perimeter spread footings. These are summarised in **Table 33.5**; further details are provided by Chen (1988), Nelson and Miller (1992), Nelson *et al.* (2015) and NHBC (2023a, 2023b, 2023c), and are discussed further below. It should be noted that terminology used to describe the foundation types listed in this table vary across the world – for example, slab-on-grade is used in the USA for raft foundations.

Pier and beam/pile and beam foundations

These foundations consist of a ground beam to support structural loads, transferring the load to the piers or piles. A void is provided between the pier/pile and the ground beam to isolate the structure and prevent uplift from swelling. NHBC (2023a) provides guidance on minimum void dimensions. Floors are then constructed as floating slabs. The piers/piles are reinforced (with reinforcement taken over the whole length to avoid tensile failures) using concrete shafts with or without bell bottoms, steel piles (driven or pushed) or helical piles whose aim is to transfer loads to stable strata. Under-reamed bottoms and helical piers/piles can be effective in soils with a high swell potential, overcoming the impractical length that would otherwise be required with straight shaft piers/piles, or where there is a possibility of a loss of skin friction due to rising groundwater levels. If a stable non-expansive stratum occurs near the surface, the piers/piles can be designed as rigid anchoring members. If, however, the depth of potential swell is high, the piers/piles should be designed as elastic members in an elastic medium. **Figure 33.13** illustrates a typical pier and beam foundation from US practice. Very similar arrangements are used in the UK and are illustrated in NHBC (2023a: figures 4–6).

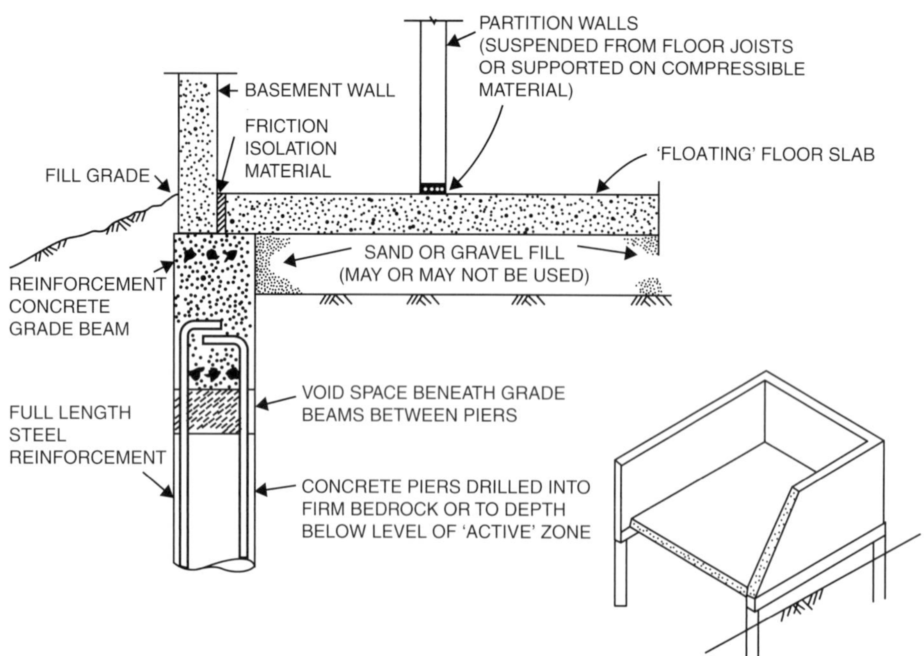

Figure 33.13 Illustration of pier and beam foundations (Reproduced from Nelson and Miller, 1992; Wiley)

Design and construction procedures for each of these systems are provided in detail (including sample design calculations) by Chen (1988), Nelson and Miller (1992) and Nelson *et al.* (2015). Additional discussion and example design calculations are provided by Nelson *et al.* (2007). It is important to ensure sufficient anchorage below the active zone. Pier/pile diameters are kept small (typically 300–450 mm). Any smaller and problems will result in poor concrete placement and associated defects (e.g. void spaces). Another problem that can occur is 'mushrooming' near the top of the pier/pile, which provides an additional area for uplift forces to act upon. To avoid this, cylindrical cardboard forms are often employed and removed after the beam is cast to prevent a means to transmit swell pressures. The size of this void space depends on the magnitude of potential swell, with 150–300 mm often being used. In the upper active zone shafts should be treated to reduce skin friction and hence minimise uplift forces. It is important that any chosen approach does not provide potential pathways to allow water to ingress to deeper layers, as this will cause deep-seated swelling.

Stiffened rafts

Stiffened slabs are either reinforced or post-tensioned systems, the latter being common in countries like the USA. Design procedures consist of determining bending moments, shear and deflections, associated with structural and swell pressure loads. The general layout used is illustrated in **Figure 33.14**, which shows examples used commonly in the USA. Similar approaches are used in the UK and are presented in NHBC (2023a, 2023b).

Designs are modelled on the soil–structure interaction at the base of the slab by considering the slab as a loaded plate or beam resting on an elastic medium. Essentially, two extremes exist – the first where a ground profile develops assuming a weightless slab, and the second where a slab of infinite stiffness is placed on the swelling soil. In reality, slabs exhibit some flexibility and so the actual heave produced by swelling soils lies somewhere between these two extremes. These modes of movement are illustrated in **Figure 33.15**.

Several design approaches have been developed, each using a range of different combinations of soil and structural design parameters. A detailed account of these is provided by Nelson *et al.* (2015), with additional discussion provided by Houston *et al.* (2011).

The primarily geotechnical information required includes size, shape and properties of the distorted soil surface that develops below the slab. These depend on a number of factors, including heave, soil stiffness, initial water content, water distribution, climate, post-construction time, loading and slab rigidity. It should be noted that the slab, through its elimination of evapotranspiration (see **Figure 33.6**), promotes the greatest increase in water content near to the centre of the slab – and hence to where long-term distortion is most severe. However, the maximum differential heave (y_m in **Figure 33.15**) has been found to vary between 33% and 100% of total maximum heave (Nelson and Miller, 1992). On occasion, edge heave can occur when the exterior of a structure experiences increases in water content before the interior.

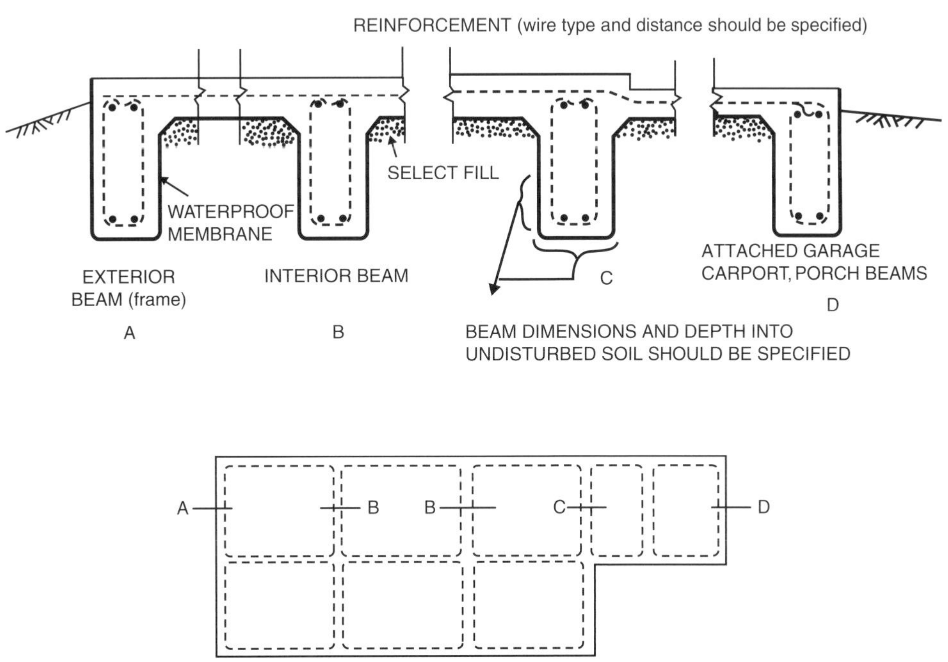

Figure 33.14 Typical detail of a stiffened raft (Reproduced from Nelson and Miller, 1992; Wiley)

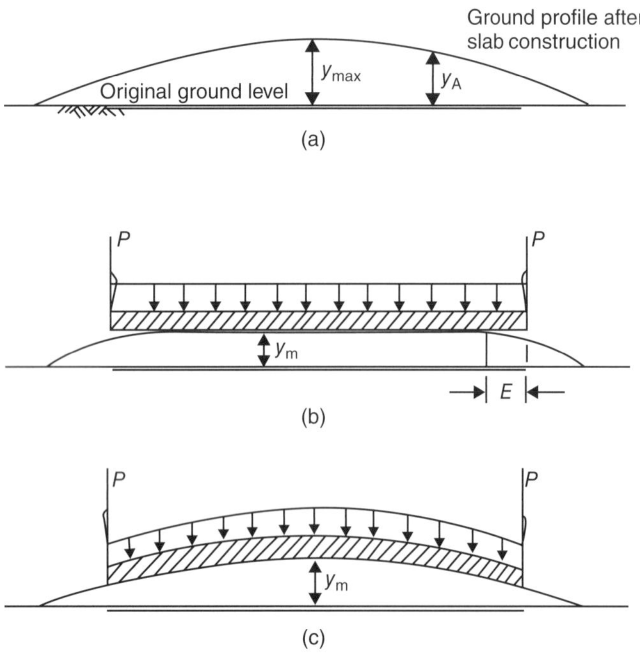

Figure 33.15 Profiles after construction for various stiffness of raft: (a) with no load applied; (b) with infinitely stiff slab; (c) with flexible slab. y_{max} = maximum heave, no foundation present – the free-field heave; y_m = maximum differential heave; E = distance from the outer edge to the point where swelling soil contacts the foundation; P = loading; y_A = height of free-field heave along the ground profile (Reproduced from Nelson and Miller, 1992; Wiley)

Modified continuous perimeter footing

Shallow footing should be avoided where expansive soils are found. However, where they are used, a number of approaches can be employed to minimise the effects of swelling/shrinkage. Modifications include

- narrowing footing width
- providing void spaces within the support beam/wall to concentrate loads at isolated points
- increasing perimeter reinforcement – taking this into the floor slab and stiffening foundations.

The use of narrow spread footing in expansive soils should be restricted to soils exhibiting 1% swell potential and very low swell pressures (Nelson and Miller, 1992; Nelson *et al.*, 2015).

NHBC (2023a) suggests that strip and trench fill foundations can be used when placed in a non-expansive layer that overlies expansive soils, provided that

- soil is consistent across the site
- the depth of non-expansive material is greater than three-quarters of the equivalent foundation depth, assuming all soil is expansive (guidance provided within NHBC, 2023a)
- the thickness of the non-expansive soil below the foundation is at least equal to the foundation width.

Case studies

Chen (1988) provides a series of case study examples of foundations and problems that arise when dealing with expansive soils, including distress caused by the following: pier/pile uplift, improper pier/pile design and construction, heaving of a pad and floor slab, heaving of a continuous floor and a rising water table. Further examples are provided by Nelson *et al.* (2015). Additional reviews of issues related to other foundation types, for example the use of post-tensioned stiffened raft foundations, are discussed by Houston *et al.* (2011). Other useful case studies are provided by Simmons (1991) and Kropp (2011). It is clear that a number of foundation failures occur and these can be summarised as

- changes in water content
  - chiefly high water tables
  - poor drainage under foundations
  - leaks due to sewer failure or poorly managed run-off
  - irrigation and garden watering
- poor construction practice
  - insufficient edge beam stiffness
  - inadequate slab thickness
  - pier length inadequate or 'mushrooming' of piers/piles resulting in uplift as swelling occurs
  - lack of reinforcement making structures intolerant to movements
  - inadequate void space
- lack of appreciation of soil profile
  - underlying geology contains inclined bedding of bedrock, causing swell to be both vertical and horizontal
  - uncontrolled fill placement
  - areas of extensive depth of expansive soil, so drilled pier and beam foundation may not be practical and a more flexible system should be used.

When assessing failure from swell–shrink behaviour it is important to isolate structural defects from foundation movement, as both can cause cracking distress in buildings (Chen, 1988). Useful reviews of geotechnical practice in relation to expansive soils have been provided by Lawson (2006) for Texas, Kropp (2011) for the San Francisco Bay Area and Houston *et al.* (2011) for Arizona. Although these are US-based, there are many lessons that geotechnical engineers can learn from these studies. Ewing (2011) provides an interesting case from Jackson, Mississippi, of a series of repairs over a 30-year period to a house (on the USA's Register of Historic Places) built on 1.5 m of non-expansive soils overlying expansive clay some 8 m thick.

33.5.4.2 Pavement and expansive soils

Pavements are particularly vulnerable to expansive soil damage, with estimates suggesting that they are associated with approximately half of the overall costs from expansive soils (Chen, 1988). Their inherent vulnerability stems from their reasonably lightweight nature, extended over a relatively large area. For example, Cameron (2006) describes problems with railways built on expansive soils where poor drainage exists, and Zheng *et al.* (2009) provide details from China of highway subgrade construction on embankments and in slopes. Damage to pavements on expansive soils comes in four major forms

- severe unevenness along significant lengths – cracks may or may not be visible (particularly important for airport runways)
- longitudinal cracking
- lateral cracking, developed from significant localised deformations
- localised pavement failure associated with disintegration of the surface.

Pavement design is essentially the same as that used for foundations. However, a number of different approaches are required as pavements cannot be isolated from the soils and it is impractical to make pavements stiff enough to avoid differential movements. Therefore it is often more economic to treat subgrade soils (see Section 33.5.4.3 for further details). Pavement designs are based on either flexible or rigid pavement systems; these procedures are discussed in Section 7 *Design of earthworks, slopes and pavements* and Chapter 76 *Pavement foundation design*. However, when dealing with expansive soils several approaches should be considered.

- Choose an alternative route and avoid expansive soil.
- Remove and replace expansive soil with a non-expansive alternative.
- Design for low strength and allow regular maintenance.
- Physically alter expansive soils through disturbance and recompaction.
- Stabilise soils through chemical additives, such as lime treatment.
- Control water content changes, although this is very difficult over the life of a pavement. Techniques include pre-wetting, membranes, deep drains and slurry injection treatment.

Nelson *et al.* (2015) provide further details on testing undertaken to mitigate expansive soil behaviour for pavement construction. Cameron (2006) has advocated the use of trees as they can be beneficial in semi-arid environments to manage poorly drained areas under railways. However, this needs careful management and may require several years to be fully effective.

33.5.4.3 Treatment of expansive soils

Essentially, treatment of expansive soils can be grouped into two categories

- soil stabilisation – remove/replace, remould and compact, pre-wet and chemical/cement stabilisation
- water content control methods – horizontal barriers (membranes, asphalt and rigid barriers), vertical barriers, electrochemical soil treatment and heat treatment.

A detailed account of the various treatment approaches is provided by Chen (1988) and Nelson *et al.* (2015), with a detailed review of stabilisation over the last 60 years provided by Petry and Little (2002). More recently, Ikeagwuani and Nwonu (2019) provided a update review of emerging trends in expansive soil stabilisation. As with any treatment approach, it is essential to undertake appropriate site investigations and evaluations (see Section 6 *Design of retaining structures* and Section 33.5.1). Special consideration should be given to the following: depth of the active zone, potential for volume change, soil chemistry, water variations within the soil, permeability, uniformity of the soils and project requirements. An overview of each of the two categories of treatments applied to expansive soils is provided below, with **Table 33.6** providing brief details of soil stabilisation approaches.

In a recent survey, Houston *et al.* (2011) found that many geotechnical and structural engineers considered chemical stabilisation approaches, such as the use of lime, as ineffective for pre-treatment of expansive soils for foundations. Beetham *et al.* (2015) state that if durability and/or long-term performance are to be optimised, treatment of medium- to high-plasticity clay soils may require adaptations to standard working practices. Preference is typically given for use of either pier/pile and beam foundations, or stiffened raft foundations. This is not true for pavements, where lime and other chemical stabilisation approaches are commonly used worldwide. The various stabilisers can be grouped into three categories (Petry and Little, 2002)

- traditional stabilisers – lime and cement
- by-product stabilisers – cement/lime kiln dust and fly ash
- non-traditional stabilisers – for example, sulfonated oils, potassium compounds, ammonium compounds and polymers.

Further details of these can be found in Petry and Little (2002). However, as with any soil treated with lime, care is needed to assess chemical as well as physical soil properties to prevent swelling from adverse chemical reactions (Petry and Little, 2002). For example, Madhyannapu *et al.* (2010) provide details of quality control when stabilising expansive subsoils using deep soil mixing, demonstrating the use of non-destructive tests based on seismic methods.

Chemical stabilisation can be used to provide a cushion immediately below foundations placed on expansive soils, such as for pavements (Ramana and Praveen, 2008). Swell mitigation has also been achieved by mixing non-swelling material such as sand (Hudyma and Avar, 2006) or granulated tyre rubber (Patil *et al.*, 2011) into expansive soils to dilute swell potential.

Improvement approach	Outline of approach	Advantages	Disadvantages
Removal and replacement	Expansive soil is removed and replaced by non-expansive fill to a depth necessary to prevent excessive heave. Depth governed by weight needed to prevent uplift and mitigate differential movement. Chen (1988) suggests a minimum of 1–1.3 m	Non-expansive fill can achieve increase bearing capacities; simple and easy to undertake; often quicker than alternatives	Preferable to use impervious fill to prevent water ingress, which can be expensive; thickness required may be impractical; failure can occur during construction due to water ingress
Remoulding and compaction	Less expansion observed for soil compacted at low densities above OWC[a] than those at high densities and below OWC (**Figure 33.16**). Standard compaction methods and control can be used to achieve target densities	Uses clay on site, eliminating cost of imported fill; can achieve a relatively impermeable fill, minimising water ingress; swell potential reduced without introducing excess water	Low-density compaction may be detrimental to bearing capacity; may not be effective for soil of high swell potential; requires close and careful quality control
Pre-wetting or ponding	Water content increased to promote heave prior to construction. Dykes or berms used to impound water in flooded area. Alternatively, trenches and vertical drains can be used to speed infiltration of water into soil	Has been used successfully when soils have sufficiently high permeabilities to allow relatively quick water ingress, e.g. with fissure clays	May require several years to achieve adequate wetting; loss of strength and failure can occur; ingress limited to a depth less than the active zone; water redistribution can occur – causing heave after construction
Chemical stabilisation	Lime (3–8% by weight) common with cements (2–6% by weight) sometimes used, and salts, fly ash and organic compounds less commonly used. Generally lime mixed into surface (~300 mm), sealed, cured and then compacted. Lime may also be injected in slurry form. Lime generally best when dealing with highly plastic clays	All fine-grained soils can be treated by chemical stabilisers; effective in reducing plasticity and swell potential of an expansive soil	Soil chemistry may be detrimental to chemical treatment; health and safety need careful consideration as chemical stabilisers carry potential risks; environmental risks may also occur – e.g. quick lime is particularly reactive; curing inhibited in colder temperatures

[a]OWC, optimum water content as determined by standard proctor test BS 1377 (BSI, 1990).

Table 33.6 Soil stabilisation approaches applied to expansive soils (Data taken from Nelson and Miller, 1992)

In some cases surcharging may be used, but this is only effective with soils of low to moderate swelling pressures. This requires enough surcharge load (see the first row in **Table 33.6**) to counteract expected swell pressures. This method is therefore only used for soil of low swell pressure and with structures that can tolerate heave. Examples include secondary highway systems, or where high foundation pressures occur. Pre-wetting – due to its uncertainties – can only be used with caution, with both Chen (1988) and Nelson and Miller (1992) indicating that it is unlikely to play an important role in the construction of foundations on expansive soils.

Fluctuations in water content are one of the primary causes of swell–shrink problems, with non-uniform heave occurring due to non-uniformity of water content, soil properties or both. Thus, if water content fluctuations can be minimised over time, then swell–shrink problems can be mitigated. Moreover, if water content changes can be slowed down and water distributions in expansive soils made uniform, then differential movements can also be reduced. In essence, this is the aim of the introduction of moisture/water barriers. These act to

- move the edge effects away from the foundation/pavement and so minimise seasonal fluctuation effects
- lengthen the time for water content changes to occur – due to longer migration paths under foundations.

Barrier techniques comprise

- horizontal barriers – using membranes, bituminous membranes or concrete
- vertical barriers – polyethylene, concrete, impervious semi-hardening slurries.

Detailed accounts of these are provided by Chen (1988) and Nelson *et al.* (2015). In addition, electrochemical soil treatment approaches are being developed that utilise electrical current to inject stabilising agents into the soils. Further details are provided by Barker *et al.* (2004) and Mosavat *et al.* (2012). As well as barrier methods, water management can be employed with restrictions applied to avoid irrigation within certain distances of the structure. However, monitoring is needed to ensure compliance with these restrictions.

33.5.4.4 Remedial options

Expansive soils cause significant damage to buildings, as discussed throughout this chapter, and so remedial action is required to repair any damage. However, it is important to establish a number of factors before embarking on a remedial plan. Key questions that should be considered are as follows (after Nelson *et al.*, 2015)

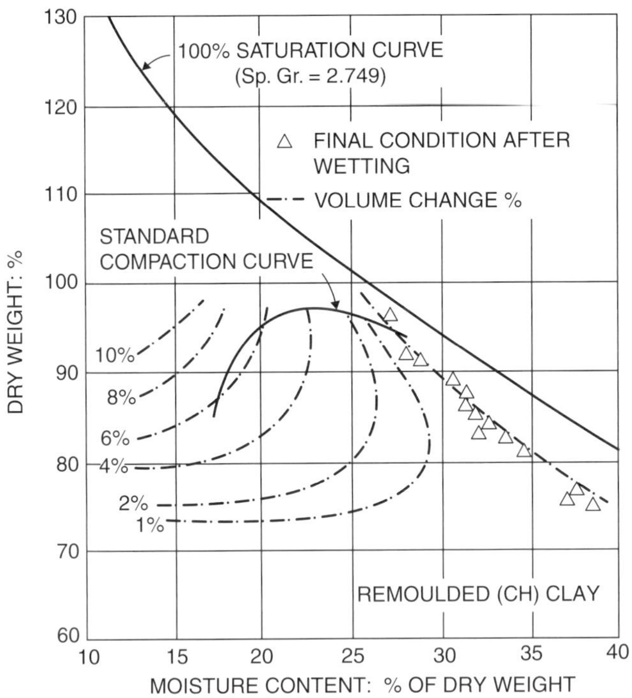

Figure 33.16 Percentage expansion for various placement conditions (see Table 33.6) (Reproduced from Holtz, 1959; all rights reserved)

- Are remedial measures needed – is damage severe enough to warrant treatment?
- Is continued movement anticipated and so would it be better to wait?
- Who will pay?
- What criteria should be selected?
- How has the damage been caused and what is its extent?
- What remedial measures are applicable?
- Are there any residual risks post-remediation?

Clearly, to select an appropriate remedial measure an adequate forensic site investigation is required. Key information required includes the cause and extent of the damage, the soil profile (as it is often difficult to determine whether settlement/heave is the cause of structural distress) and the soil's expansive potential. Other necessary information has already been discussed in Section 33.5.1. Failure to carry out an adequate site investigation can lead to false diagnoses and inappropriate remedial measures being employed. Further details are provided by Nelson and Miller (1992), Nelson *et al.* (2015) and BRE Digests 251 (1995a), 298 (1999), 361 (1991), 412 (1996) and 471 (2002).

Examples of remedial measures employed for foundations include

- repair and replace structural elements or correct improper design features
- underpin
- provide structural adjustments of additional structural supports (e.g. post-tensioning)
- stiffen foundations
- provide drainage control
- stabilise water contents of foundation soils
- install moisture barriers to control water content fluctuations.

Full underpinning of an operational structure is often impractical (and increasingly seen as unnecessary) and it is more common for underpinning work to be applied only to key parts of the foundations (Buzzi *et al.*, 2010). Moreover, localised application of underpinning to deal with differential settlements may not improve the overall performance of the foundation (Walsh and Cameron, 1997). Thus any localised treatment must be designed to take account of all factors, otherwise there is a danger of exacerbating the problems due to the inherent natural spatial variability of expansive soils. Recently, underpinning using expanded polyurethane resin has met with some success, because resin can be injected using small-diameter tubes directly where it is needed (Buzzi *et al.*, 2010). However, due to concerns about its long-term stability and the possibility that swelling in injected soils could be exacerbated if all the cracks were filled, its adoption has been slow. However, a detailed experimental study (Buzzi *et al.*, 2010) concluded that resin-injected expansive soils did not exhibit enhanced swelling as several cracks remained unfilled, providing swell relief. Problems with lateral swelling can sometimes be accommodated by cracking within the soil matrix. However, if no cracks are present, problems can occur – particularly with retaining structures. Expanded polystyrene geofoam has demonstrated some success with dealing with lateral expansion, and has been shown to reduce the subsequent impact of vertical swelling (Ikizler *et al.*, 2008).

With respect to pavements, distress can be considered as one of four possible types of damage, as highlighted in Section 33.5.4.2. Most common remedial measures are either removal and replacement, or construction of overlays. Whichever method is used, care is needed to ensure that the causes of the original distress are dealt with.

Many of the pre-construction approaches can also be used for post-construction treatments; for pavements these include moisture barriers, removal, replacement and compaction and drainage control.

33.5.4.5 Domestic dwelling and vegetation

Tree roots will grow in the direction of least resistance and where they have the best access to water, air and nutrients (Roberts, 1976). The actual pattern of root growth depends upon, among other factors, the type of tree, the depth to the water table and local ground conditions. Trees will tend to maintain a compact root system. However, when trees become very large, or where they are under stress, they can send root systems far from the trunk. There is some published guidance on 'safe planting distances' that can be used by the insurance industry to inform householders of the potential impacts of different tree species on their properties. Further details are also given in NHBC (2023a).

Paving of previously open areas of land, such as the building of patios and driveways, can cause major disruption to the soil–water system. If the paving cuts off infiltration, many trees will send their roots deeper into the ground or further from the trunk to source water. The movement of these tree roots will cause disturbance of the ground and will lead to the removal of water from a larger area around the tree. Problems occur when houses are situated within the zone of influence of a tree (**Figure 33.17**).

If an impermeable method of paving is used, it may prevent water from penetrating into the ground. This can affect the shrink–swell behaviour of the ground and also the growing patterns of nearby trees. A well-designed impermeable paving system, in good condition, may actually reduce the amount of shrink–swell activity in the ground immediately below it. Paving moderates variations in water content of the soil and thus the range of shrink–swell behaviour. However, if the paving seal is broken water can suddenly enter the system, causing swelling of the ground.

Different problems are faced when considering the distinctly separate areas of designing new-build structures and remediating existing damaged buildings. New-build guidelines for domestic dwellings recognise the need for thorough

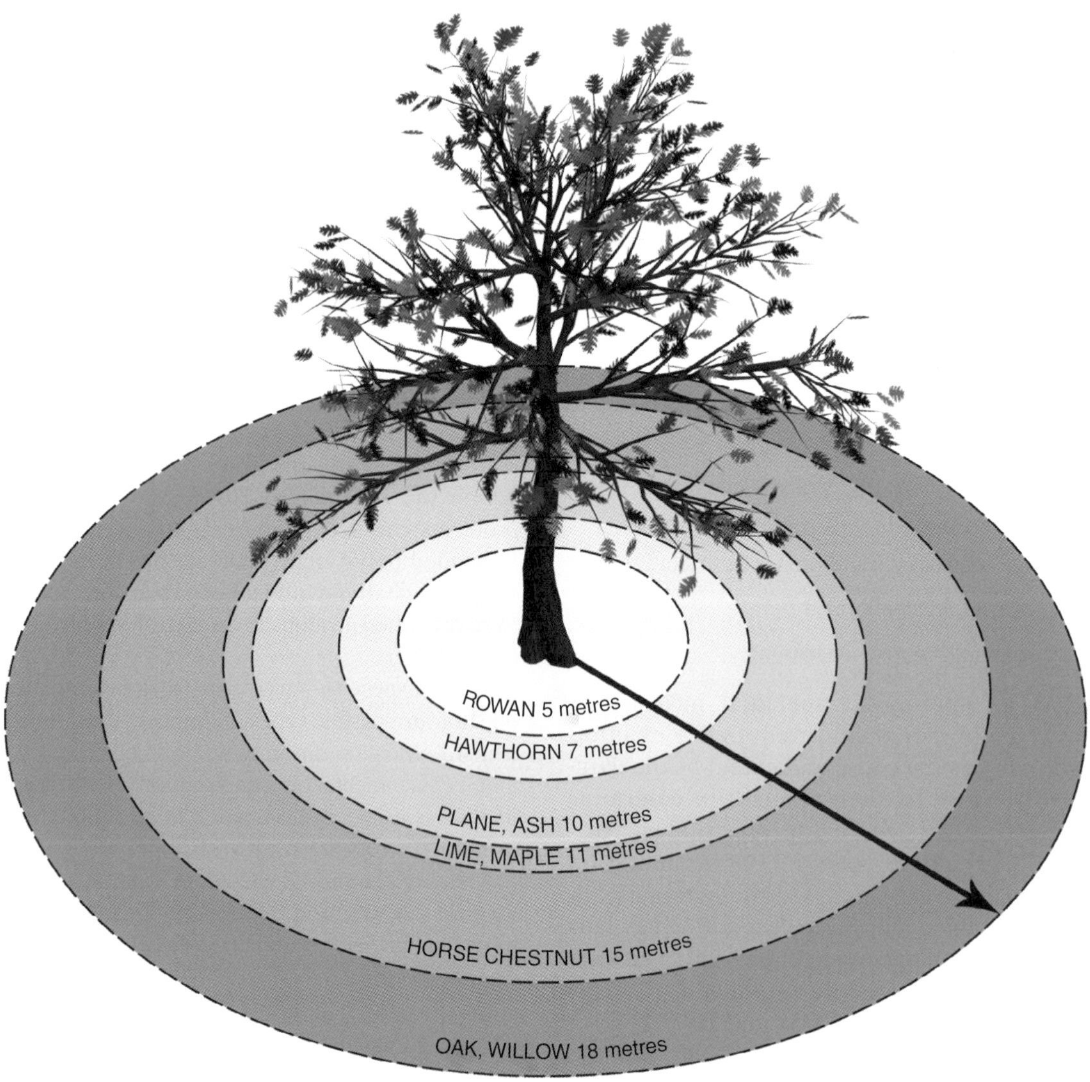

Figure 33.17 The zone of influence of some common UK trees (Reproduced from Jones *et al.*, 2006; © NERC, with permission from the British Geological Survey)

ground investigations to design systems to cope with the hazards presented by existing trees or their recent removal. Reference should be made to NHBC Standards 'Chapter 4.2 Building near trees' (NHBC, 2023a) and the *Efficient Design of Piled Foundations for Low-Rise Housing – Design Guide* (NHBC, 2010). In the case of existing dwellings, a range of reports and digests are available (e.g. BRE, 1996, 1999) and *A Good Technical Practice Guide* is provided by Driscoll and Skinner (2007).

Essentially, foundations should make allowances for trees in expansive (swell–shrink) soils and should take account of (NHBC, 2023a)

- shrinkage/heave linked to changes in water content
- soil classification
- water demand of trees (this is species-dependent)
- tree height
- climate.

In the case of existing structures, the main cause of distress results from the effects of differential settlement, where different parts of the building move by varying amounts due to variations in the properties of the underlying soil. Equal or proportionate movements across the plan area of a building, although significant in terms of vertical movement, may result in little structural damage (IStructE, 1994). However, in the UK this is rare; by far the most overwhelming cause of damage to property results from the desiccation of clay subsoil, which consequently causes differential settlements/movements, often stemming from the abstraction of water by the roots of nearby vegetation.

If vegetation is involved, it produces a characteristic seasonal pattern of foundation movement: subsidence in the summer, reaching a maximum around September, followed by upward recovery in the winter (**Figure 33.18**) If subsidence followed by recovery is occurring, there is no need to try to demonstrate shrinkable clay or desiccation. No other cause produces a similar pattern – soil drying by vegetation must be involved (unless the foundations are less than 300 mm). Furthermore, there is no need to demonstrate the full cycle as it is sufficient to confirm movement is consistent with this pattern. Monitoring upward recovery in the winter is particularly valuable in this case. Further details are given by Crilly and Driscoll (2000) and Driscoll and Chown (2001), drawn from a test site in Chattenden, Kent, set in expansive London Clay (**Figure 33.18**). In addition, both articles provide details of instrumented piles, discussing design implications.

Level monitoring can demonstrate this pattern. BRE Digest 344 (1995b) makes recommendations for the taking of measurements of the 'out-of-level' of a course of masonry or of the damp-proof course, which can be used to estimate the amount of differential settlement or heave that has already taken place. BRE Digest 386 (1993b) discusses precise levelling techniques and equipment which can monitor vertical movements with an accuracy consistently better than ± 0.5 mm. Precise levelling can be conducted easily, quickly and accurately, and so provides one of the most effective ways to distinguish between potential causes of foundation movement (Biddle, 2001).

The choice of mitigation should be proportionate to the problem and specific to the true area of the affected structure. It is important not to become distracted by extraneous but nevertheless interesting features.

Biddle (2001) suggests one of four remedial options to deal with the adverse actions of trees.

- Fell the offending tree to eliminate all future drying.
- Prune the tree to reduce drying and the amplitude of seasonal movement.
- Control the root spread to prevent drying under foundations.
- Provide supplementary watering to prevent soil from drying.

Biddle (2001) states that it is now recognised that in most situations underpinning is unnecessary and that foundations can be stabilised by appropriate tree management – usually by felling the offending tree or by carrying out heavy crown reduction. Site investigations should reflect this change and be aimed at providing the information to allow appropriate decisions on tree management. In particular

- confirmation that vegetation-related subsidence is involved
- identification of which tree(s) or shrub(s) are involved
- assessment of the risk of heave if a tree is felled or managed
- identification of the need for any other site investigations
- if the tree warrants retention, assessment of whether partial underpinning would be sufficient
- confirmation that vegetation management has been effective in stabilising the foundations
- provision of information within an acceptable timescale.

Trees are often pruned to reduce their water use and therefore their influence on the surrounding soil. However, unless the trees are thereafter subjected to a frequent and ongoing regime of management, the problems will very quickly return. While tree removal will ultimately provide an absolute solution in the majority of cases, there are situations where this is not an option (e.g. protected trees, adverse risk of heave, incomplete evidence in contentious issues and physical proximity of trees).

In the past, an obvious and often knee-jerk solution has been to provide significant and often disproportionate support to the structure through foundation strengthening schemes, incorporating various forms of underpinning. This approach is often ecologically, financially and technically incongruent with the problems faced. Alternatively, various forms of physical barriers can be used, constructed from, for example, in situ

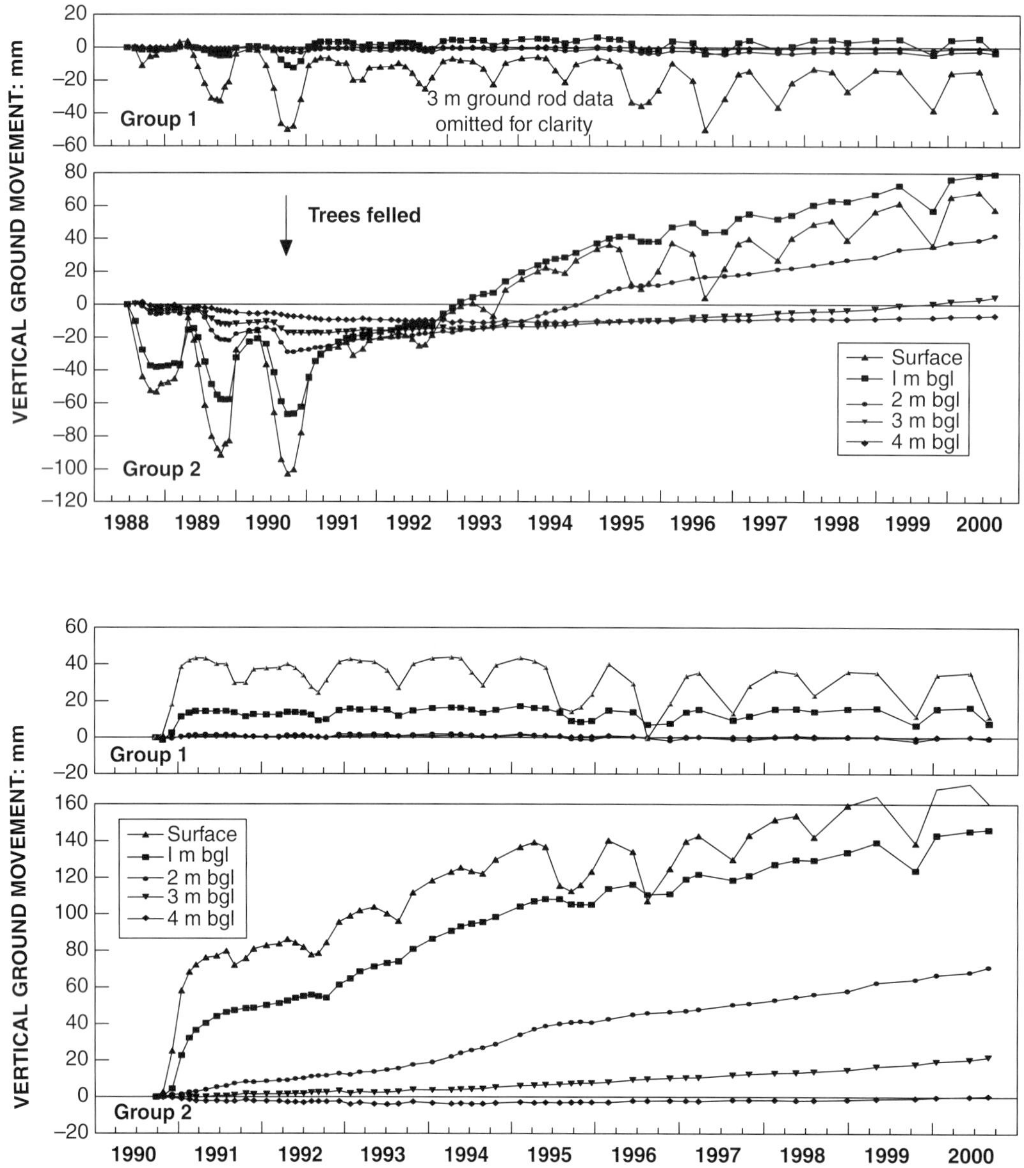

Figure 33.18 Examples of ground movements due to seasonal fluctuations at Chattenden. The upper plot shows results obtained since the first movements in June 1988. The lower plot shows an enlarged scale with results obtained since the trees were felled – group 1 is remote from trees and group 2 near to trees (Reproduced from Crilly and Driscoll, 2000 and Driscoll and Chown, 2001; all rights reserved)

concrete. However, such barriers often prove ineffective over time. Barriers are currently being developed that incorporate a bioroot barrier, which is a mechanically bonded geocomposite consisting of a copper-foil firmly embedded between two layers of geotextile. Such biobarriers are now being used specifically in arboriculture and for Japanese knotweed control where a permeable barrier is required. They act as signal barriers by diverting root growth (both biologically and physically) without making any attempt to physically restrain their progress.

Alternative remediation by supplementary watering is usually considered impractical due to the quantities required by the tree. This approach can suffer from the unavailability of water precisely when it is needed, due to prevailing drought conditions.

If a mature tree is felled, a building may incur heaving on a dry clay soil. Unfortunately, the evidence is rarely obvious; however, clues to look out for include

- the house is new – less than 20 years old
- there is expansive soil present
- the crack pattern might appear a bit odd – wider at the bottom than at the top, with no obvious cause
- cracks continue to open, even in the wet months.

Heave problems can be costly and always require thorough investigation involving soil sampling, precise levels and aerial photographs. Heave is a threat but rarely a reality where established existing properties are involved, and the structure predates the planting of the tree.

Ultimately, if the offending tree can be accurately targeted and dealt with rapidly before the next growing season, the extent of any damage and need for remedial work will be kept to a minimum (Biddle, 2001).

33.6. Conclusions

Expansive soils are one of the most significant ground-related hazards found globally, costing billions of pounds annually. They are found throughout the world – commonly in arid/semi-arid regions – where their high suctions and potential for large water content changes can cause significant volume changes. In humid regions, such as the UK, problematic expansive behaviour generally occurs in clays of high plasticity index. Either way, expansive soils have the potential to demonstrate significant volume changes in direct response to changes in water content. This can be induced through water ingress, through modification to water conditions or through the action of external influences such as trees. To understand and hence engineer expansive soils in an effective way, it is necessary to understand soil properties, suction/water conditions, water content variations (temporal and spatial) and the geometry/stiffness of foundations and associated structures. This chapter provides an overview of these features and includes methods to investigate expansive behaviour both in the field and in the laboratory, together with associated empirical and analytical tools to evaluate it. Following this design, options for pre- and post-construction are highlighted for both foundations and pavements, together with methods to ameliorate potentially damaging expansive behaviour, including dealing with the impact of trees.

References

Aguirre VE (2011) Discussion of 'Method for Evaluation of Depth of Wetting in Residential Areas' by Walsh et al. (2009). *Journal of Geotechnical and Geoenvironmental Engineering* **137(3)**: 296–299.

Alonso EE, Gens A and Josa A (1990) A constitutive model for partially saturated soils. *Géotechnique* **40**: 405–430.

Altmeyer WT (1956) Discussion following paper by Holtz and Gibbs (1956). *Transactions of the American Society of Civil Engineering* **2(1)**: 666–669.

ASTM (2010) Sections 04.08 Soil and Rock (I) and 04.09 Soil and rock (II); Building stones. In *Annual Book of Standards*. ASTM, Philadelphia, PA.

Atkinson JH (2007) *The Mechanics of Soils and Foundations*, 2nd edn. Taylor & Francis, Oxford, UK.

Barker JE, Rogers CDF, Boardman DI and Peterson J (2004) Electrokinetic stabilisation: an overview and case study. *Ground Improvement* **8(2)**: 47–58.

Basu R and Arulandan K (1974) A new approach for the identification of swell potential of soils. *Bulletin of the Association of Engineering Geologists* **11**: 315–330.

Beetham P, Dijkstra TA, Dixon N, Fleming P, Hutchinson R and Bateman J (2015) Lime stabilisation for earthworks: a UK perspective. *Proceedings of the Institution of Civil Engineers: Ground Improvement* **168(2)**: 81–95.

Bell FG and Culshaw MG (2001) Problem soils: a review from a British perspective. *Problematic Soils Symposium, Nottingham, UK* (Jefferson I, Murray EJ, Faragher E and Fleming PR (eds)), pp. 1–35.

BGS (British Geological Survey) (2023) Swelling and shrinking soil. https://www.bgs.ac.uk/geology-projects/shallow-geohazards/clay-shrink-swell (accessed 05/06/2023).

Biddle PG (1998) *Tree Roots and Foundations*. Arboriculture Research and Information Note 142/98/EXT. Forestry Commission Research Station, Farnham, UK.

Biddle PG (2001) Tree root damage to buildings. In *Expansive Clay Soils and Vegetative Influence on Shallow Foundations* (Vipulanandan C, Addison MB and Hasen M (eds)). ASCE, Reston, VA.

Bishop AW, Kumapley NK and El-Ruwayih AE (1975) The influence of pore water tension on the strength of clay. *Philosophical Transactions of the Royal Society London* **278**: 511–554.

Bjerrum L (1967) Progressive failure in slopes of overconsolidated plastic clay and clay shales. *Journal of Soil Mechanics and Foundation Division* **93**: 3–49.

Booth KA, Diaz Doce D, Harrison M and Wildman G (2011) *User Guide for the British Geological Survey GeoSure Dataset*. British Geological Survey Internal Report OR/10/066. BGS, Nottingham, UK.

BRAB (1968) *Criteria for Selection and Design of Residential Slabs-On-Ground*. Building Research Advisory Board, Federal Housing Administration, Washington, DC.

BRE (Building Research Establishment) (1991) *Why Do Buildings Crack?* BRE Digest 361. CRC, London, UK.

BRE (1993a) *Low-Rise Buildings on Shrinkable Clay Soils*. BRE Digest 240–241. CRC, London, UK.

BRE (1993b) *Monitoring Building and Ground Movement by Precise Levelling*. BRE Digest 386. CRC, London, UK.

BRE (1995a) *Assessment of Damage in Low-Rise Buildings*. BRE Digest 251. CRC, London, UK.

BRE (1995b) *Simple Measuring and Monitoring of Movement in Low-Rise Buildings: Part 2 – Settlement, Heave and Out of Plumb*. BRE Digest 344. CRC, London, UK.

BRE (1996) *Desiccation in Clay Soils*. BRE Digest 412. CRC, London, UK.

BRE (1999) *The Influence of Trees on House Foundations in Clay Soils*. BRE Digest 298. CRC, London, UK.

BRE (2002) *Low-Rise Building Foundations on Soft Ground*. BRE Digest 471. CRC, London, UK.

BSI (British Standards Institution) (1990) BS 1377: British standard methods of test for soils for civil engineering purposes. BSI, London, UK.

BSI (1999) BS 5930:1999 + 2:2010: Code of practice for site investigations. BSI, London, UK.

Burland JB (1989) Small is beautiful: the stiffness of soils at small strains. *Canadian Geotechnical Journal* **26**: 499–516.

Buzzi O, Fityus S and Sloan SW (2010) Use of expanding polyurethane resin to remediate expansive soil foundations. *Canadian Geotechnical Journal* **47**: 623–634.

Cameron DA (2006) The role of vegetation in stabilizing highly plastic clay subgrades. *Proceedings of Railway Foundations, Birmingham, UK* (Ghataora GS and Burrow MPN (eds)), pp. 165–186.

Chandler RJ and Apted JP (1988) The effect of weathering on the strength of London Clay. *Quarterly Journal of Engineering Geology and Hydrogeology* **21**: 59–68.

Chandler RJ, Crilly MS and Montgomery G (1992) A low-cost method of assessing clay desiccation for low-rise buildings. *Proceedings of the Institution of Civil Engineers: Geotechnical Engineering* **92**: 82–89.

Chao KC, Overton DD and Nelson JD (2006) Design and installation of deep benchmarks in expansive soils. *Journal of Surveying Engineering* **132(3)**: 124–131.

Chen FH (1988) *Foundations on Expansive Soils*. Elsevier, Amsterdam, UK.

Cheney JE (1988) 25 years' heave of a building constructed on clay, after tree removal. *Ground Engineering*, Jul.: pp. 13–27.

Clayton CRI, Matthews MC and Simons NE (1995) *Site Investigation*, 2nd edn. Blackwell Science, Oxford, UK.

Clayton CRI, Xu M, Whiter JT, Ham A and Rust M (2010) Stresses in cast-iron pipes due to seasonal shrink–swell of clay soils. *Proceedings of the Institution of Civil Engineers: Water Management* **163(WM3)**: 157–162.

Crilly MS and Driscoll RMC (2000) The behavior of lightly loaded piles in swelling ground and implications for their design. *Proceedings of the Institution of Civil Engineers: Geotechnical Engineering* **143**: 3–16.

Derjaguin BV and Churaev NV (1987) Structure of water in thin layers. *Langmuir* **3**: 607–612.

Diaz Doce D, Jones LD and Booth KA (2011) *Methodology: Shrink–Swell. GeoSure Version 6*. British Geological Survey Internal Report IR/10/093. BGS, Nottingham, UK.

Driscoll R (1983) The influence of vegetation on the swelling and shrinking of clay soils in Britain. *Géotechnique* **33**: 93–105.

Driscoll RMC and Chown R (2001) Shrinking and swelling of clays. *Problematic Soils Symposium, Nottingham, UK* (Jefferson I, Murray EJ, Faragher E and Fleming PR (eds)), pp. 53–66.

Driscoll R and Crilly M (2000) *Subsidence Damage to Domestic Buildings. Lessons Learned and Questions Asked*. BRE Press, Watford, UK.

Driscoll RMC and Skinner H (2007) *Subsidence Damage to Domestic Building: A Good Technical Practice Guide*. BRE Press, Watford, UK.

DTLR (2002) *Planning Policy Guidance Note 14: Development on Unstable Land: Annex 2: Subsidence and Planning*. DTLR, London, UK.

Dye HB, Houston SL and Welfert BD (2011) Influence of unsaturated soil properties uncertainty on moisture flow modeling. *Geotechnical and Geological Engineering* **29**: 161–169.

Ewing RC (2011) Foundation repairs due to expansive soils: Eudora Welty House, Jackson, Mississippi. *Journal of Performance of Constructed Facilities* **25(1)**: 50–55.

Farrar DM and Coleman JD (1967) The correlation of surface area with other properties of nineteen British clay soils. *Journal of Soil Science* **18**: 118–124.

FHA (1965) *Land Development with Controlled Earthwork*. Land Planning Bulletin 3, Data Sheet 79G – Handbook 4140.3. FHA, Washington, DC.

Fityus SG, Smith DW and Allman MA (2004) Expansive soil test site near Newcastle. *Journal of Geotechnical and Geoenvironmental Engineering* **130(7)**: 686–695.

Fityus SG, Cameron DA and Walsh PF (2005) The shrink swell test. *Geotechnical Testing Journal* **28(1)**: 1–10.

Fredlund DG (2006) Unsaturated soil mechanics in engineering practice. *Journal of Geotechnical and Geoenvironmental Engineering* **132(3)**: 286–321.

Fredlund DG and Hung VQ (2001) Prediction of volume change in an expansive soil as a result of vegetation and environmental changes. In *Expansive Clay Soils and Vegetative Influence on Shallow Foundations* (Vipulanandan C, Addison MB and Hasen M (eds)). ASCE, Reston, VA, pp. 24–43.

Fredlund DG and Rahardjo H (1993) *Soil Mechanics for Unsaturated Soils*. Wiley, New York, NY.

Fredlund MD, Stianson JR, Fredlund DG, Vu H and Thode RC (2006) Numerical modeling of slab-on-grade foundation. *Proceedings of UNSAT'06, Reston, VA*, pp. 2121–2132.

Gourley CS, Newill D and Schreiner HD (1994) Expansive soils: TRL's research strategy. In *Engineering Characteristics of Arid Soils* (Fookes PG and Parry RHG (eds)). Balkema, Rotterdam, the Netherlands, pp. 247–260.

Harrison AM, Plim JFM, Harrison M, Jones LD and Culshaw MG (2012) The relationship between shrink–swell occurrence and climate in south-east England. *Proceedings of the Geologists Association* **123(4)**: 556–575.

Hobbs PRN, Hallam JR, Forster A, *et al.* (1998) *Engineering Geology of British Rocks and Soils: Mercia Mudstone*. Technical Report WN/98/4. BGS, Nottingham, UK.

Hobbs PRN, Jones LD, Kirkham MP, *et al.* (2014) A new apparatus for determining the shrinkage limit of clay soils. *Géotechnique* **64(3)**: 195–203.

Holtz WG (1959) Expansive clay: properties and problems. *Quarterly of the Colorado School of Mines* **54(4)**: 89–125.

Holtz WG and Gibbs HJ (1956) Engineering properties of expansive clays. *Transactions of the American Society of Civil Engineers* **121**: 641–663.

Holtz RD and Kovacs WD (1981) *An Introduction to Geotechnical Engineering*. Prentice Hall, Hoboken, NJ.

Houston SL, Dye HB, Zapata CE, Walsh KD and Houston WN (2011) Study of expansive soils and residential foundations on expansive soils in Arizona. *Journal of Performance of Constructed Facilities* **25(1)**: 31–44.

Huang SL, Aughenbaugh NB and Rockaway JD (1986) Swelling pressure studies of shales. *International Journal of Rock Mechanics, Mining, Science and Geomechanics* **23**: 371–377.

Hudyma NB and Avar B (2006) Changes in swell behavior of expansive soils from dilution with sand. *Environmental Engineering Geoscience* **12(2)**: 137–145.

Hyndman D and Hyndman D (2009) *Natural Hazards and Disasters*. Brooks/Cole, Cengage Learning, Boston, MA.

Ikeagwuani CC and Nwonu DC (2019) Emerging trends in expansive soil stabilisation: a review. *Journal of Rock Mechanics and Geotechnical Engineering* **11**: 423–440.

Ikizler SB, Aytekin M and Nas E (2008) Laboratory study of expanded polystyrene (EPS) geoform used with expansive soils. *Geotextiles and Geomembranes* **26**: 189–195.

IStructE (1994) *Subsidence of Low Rise Buildings*. Thomas Telford, London, UK.

Jackson I (2004) *Britain Beneath Our Feet. British Geological Survey Occasional Publication 4*. BGS, Nottingham, UK.

Jennings JEB and Knight K (1957) The prediction of total heave from the double oedeometer test. *Transactions of the South African Institution of Civil Engineering* **7**: 285–291.

Jones GM, Cassidy NJ, Thomas PA, Plante S and Pringle JK (2009) Imaging and monitoring tree-induced subsidence using electrical resistivity imaging. *Near Surface Geophysics* **7(3)**: 191–206.

Jones LD (1999) *A Shrink/Swell Classification for UK Clay Soils*. Unpublished B.Eng. thesis, Nottingham Trent University, Nottingham, UK.

Jones LD and Terrington R (2011) Modelling volume change potential in the London Clay. *Quarterly Journal of Engineering Geology* **44**: 1–15.

Jones LD, Venus J and Gibson AD (2006) *Trees and Foundation Damage*. British Geological Survey Commissioned Report CR/06/225. BGS, Nottingham, UK.

Jones LD, Jefferson I and Banks V (2020) Swelling and shrinking soils. In *UK Geohazards* (Giles D (ed)). Geological Society, London, UK, pp. 223–242.

Kariuki PC and van der Meer F (2004) A unified swelling potential index for expansive soils. *Engineering Geology* **72**: 1–8.

Kemp SJ, Merriman RJ and Bouch JE (2005) Clay mineral reaction progress: the maturity and burial history of the Lias Group of England and Wales. *Clay Minerals* **40**: 43–61.

Kropp A (2011) Survey of residential foundation design practice on expansive soils in the San Francisco Bay Area. *Journal of Performance of Constructed Facilities* **25(1)**: 24–30.

Lawson WD (2006) A survey of geotechnical practice for expansive soils in Texas. *Proceedings of Unsaturated Soils, Carefree, AZ*, pp. 304–314.

Likos WJ, Olsen HW, Krosley L and Lu N (2003) Measured and estimated suction indices for swelling potential classification. *Journal of Geotechnical and Geoenvironmental Engineering* **129(7)**: 665–668.

Madhyannapu RS, Puppala AJ, Nazarian S and Yuan D (2010) Quality assessment and quality control of deep soil mixing construction for stabilizing expansive subsoils. *Journal of Geotechnical and Geoenvironmental Engineering* **136(1)**: 119–128.

Madsen FT (1979) Determination of the swelling pressure of claystones and marlstones using mineralogical data. *Proceedings of the 4th ISRM Conference*, vol. 1, pp. 237–241.

Miao L, Jing F and Houston SL (2006) Soil–water characteristic curve of remoulded expansive soil. *Proceedings of Unsaturated Soils, Carefree, AZ*, pp. 997–1004.

Mitchell JK and Soga K (2005) *Fundamentals of Soil Behavior*, 3rd edn. Wiley, New York, NY.

Mosavat N, Erwin Oh E and Chai G (2012) A review of electrokinetic treatment technique for improving the engineering characteristics of low permeable problematic soils. *GEOMATE Journal* **2(4)**: 266–272.

Nelson JD and Miller DJ (1992) *Expansive Soils: Problems and Practice in Foundation and Pavement Engineering*. Wiley, New York, NY.

Nelson JD, Overton DD and Durkee DB (2001) Depth of wetting and the active zone. In *Expansive Clay Soils and Vegetative Influence on Shallow Foundations* (Vipulanandan C, Addison MB and Hasen M (eds)). ASCE, Reston, VA, pp. 95–109.

Nelson JD, Chao KC and Overton DD (2007) Design of pier foundations on expansive soils. *Proceedings of the 3rd Asian Conference on Unsaturated Soils, Beijing, China*. Science Press, Beijing, China, pp. 97–108.

Nelson JD, Overton DD and Chao K (2010) An empirical method for predicting foundation heave rate in expansive soil. *Proceedings of GeoShanghai, Shanghai, China*, pp. 190–196.

Nelson JD, Chao KC and Overton DD (2011) Discussion of 'Method for evaluation of depth of wetting in residential areas' by Walsh et al. (2009). *Journal of Geotechnical and Geoenvironmental Engineering* **137(3)**: 293–296.

Nelson JD, Chao KC, Overton DD and Nelson EJ (2015) *Foundation Engineering for Expansive Soils*. Wiley, New York, NY.

Ng CWW, Wang B, Gong BW and Bao CG (2000) Preliminary study on soil–water characteristics of two expansive soils. In *Unsaturated Soils for Asia* (Rahardjo H, Toll DG and Leong EC (eds)). Balkema, Rotterdam, the Netherlands, pp. 347–356.

NHBC (1988) *Registered House-Builder's Foundations Manual: Preventing Foundation Failures in New Buildings*. NHBC, London, UK.

NHBC (2010) *Efficient Design of Piled Foundations for Low-Rise Housing: Design Guide*. NHBC, London, UK.

NHBC (2023a) 'Standards 4.2: Building near trees'. NHBC, London, UK.

NHBC (2023b) 'Standards 4.4: Raft, pile, pier and beam foundations'. NHBC, London, UK.

NHBC (2023c) 'Standards 4.3: Strip and trench fill foundations'. NHBC, London, UK.

Oloo S, Schreiner HD and Burland JB (1987) Identification and classification of expansive soils. *6th International Conference on Expansive Soils, New Delhi, India*, pp. 23–29.

Patil U, Valdes JR and Evans MT (2011) Swell mitigation with granulated tire rubber. *Journal of Materials in Civil Engineering* **25(5)**: 721–727.

Petry TM and Little DN (2002) Review of stabilization of clays and expansive soils in pavement and lightly loaded structures: history, practice and future. *Journal of Materials in Civil Engineering* **14(6)**: 447–460.

Powrie W (2018) *Soil Mechanics Concepts and Applications*, 3rd edn. CRC Press, Boca Raton, FL.

Puppala AJ, Punthutaecha K and Vanapalli SK (2006) Soil–water characteristic curves of stabilized expansive soil. *Journal of Geotechnical and Geoenvironmental Engineering* **132(6)**: 736–751.

Radevsky R (2001) Expansive clay problems: how are they dealt with outside the US? In *Expansive Clay Soils and Vegetative Influence on Shallow Foundations* (Vipulanandan C, Addison MB and Hasen M (eds)). ASCE, Reston, VA, pp. 172–191.

Ramana MV and Praveen GV (2008) Use of chemically stabilized soil as cushion material below light weight structures founded on expansive soils. *Journal of Materials in Civil Engineering* **20(5)**: 392–400.

Rao RR, Rahardjo H and Fredlund DG (1988) Closed-form heave solutions for expansive soils. *Journal of Geotechnical Engineering* **114(5)**: 573–588.

Reeve MJ, Hall DGM and Bullock P (1980) The effect of soil composition and environmental factors on the shrinkage of some clayey British soils. *Journal of Soil Science* **31**: 429–442.

Roberts J (1976) A study of root distribution and growth in a *Pinus Sylvestris L.* (Scots pine) plantation in East Anglia. *Plant and Soil* **44**: 607–621.

Sarman R, Shakoor A and Palmer DF (1994) A multiple regression approach to predict swelling in mudrocks. *Bulletin of the Association of Engineering Geology* **31**: 107–121.

Scott GJT, Webster R and Nortcliff S (1986) An analysis of crack pattern in clay soil: its density and orientation. *Journal of Soil Science* **37**: 653–668.

Self S, Entwisle D and Northmore K (2008) *The Structure and Operation of the BGS National Geotechnical Properties Database.* British Geological Internal Report IR/08/000. BGS, Nottingham, UK.

Simmons KB (1991) Limitations of residential structures on expansive soils. *Journal of Performance of Constructed Facilities* **5(4)**: 258–270.

Skempton AW (1953) The colloidal activity of clays. *Proceedings of the 3rd International Conference on Soil Mechanics, Zurich, Switzerland*, vol. 1, pp. 57–61.

Skempton AW and DeLory FA (1957) Stability of natural slopes in London Clay. *Proceedings of the 4th International Conference on Soil Mechanics, London, UK*, pp. 378–381.

Snethen DR (1984) Evaluation of expedient methods for identification and classification of potentially expansive soils. *Proceedings of the 5th International Conference on Expansive Soils, Adelaide, Australia*, pp. 22–26.

Snethen DR, Johnson LD and Patrick DM (1977) *An Evaluation of Expedient Methodology for Identification of Potentially Expansive Soils.* Report FHWA-RD-77–94. US Army Engineer Waterways Experiment Station, Vicksburg, MS.

Sridharan A and Prakash K (2000) Classification procedures for expansive soils. *Proceedings of the Institution of Civil Engineers: Geotechnical Engineering* **143**: 235–240.

Sridharan A and Venkatappa RG (1971) Mechanisms controlling compressibility of clays. *Journal of Soil Mechanics and Foundations* **97(6)**: 940–945.

Stavridakis EI (2006) Assessment of anisotropic behaviour of swelling soils on ground and construction work. In *Expansive Soils: Recent Advances in Characterization and Treatment* (Al-Rawas AA and Goosen MFA (eds)). Taylor & Francis, London, UK.

Takahashi A, Jardine RJ and Fung DWH (2005) Swelling effects on mechanical behaviour of natural London Clay. *Proceedings of the 16th International Conference on Soil Mechanics, Osaka, Japan*, pp. 443–446.

Taylor RK and Smith TJ (1986) The engineering geology of clay minerals: swelling, shrinking and mudrock breakdown. *Clay Minerals* **21**: 235–260.

TRB (1985) *Evaluation and Control of Expansive Soils.* Transportation Research Board, London, UK.

Van der Merwe DH (1964) The prediction of heave from the plasticity index and percentage clay fraction of soils. *Transaction of the South African Institution of Civil Engineers* **6**: 103–107.

Vijayvergiya VN and Sullivan RA (1974) Simple technique for identifying heave potential. *Bulletin of the Association of Engineering Geology* **11**: 277–292.

Walsh KD and Cameron DA (1997) *The Design of Residential Slabs and Footings.* Standards Australia, Sydney, Australia.

Walsh KD, Colby CA, Houston WN and Houston SL (2009) Method for evaluation of depth of wetting in residential areas. *Journal of Geotechnical and Geoenvironmental Engineering* **135(2)**: 169–176.

Walsh KD, Colby CA, Houston WN and Houston SL (2011) Closure to discussion of 'Method for evaluation of depth of wetting in residential areas' by Walsh *et al.* (2009). *Journal of Geotechnical and Geoenvironmental Engineering* **137(3)**: 299–309.

Williams AAB and Donaldson G (1980) Building on expansive soils in South Africa. *Proceedings of the 4th International Conference on Expansive Soils,* Denver, CO, vol. 2, pp. 834–844.

Xiao HB, Zhang CS, Wang YH and Fan ZH (2011) Pile interaction in expansive soil foundation: analytical solution and numerical simulation. *International Journal of Geomechanics* **11(3)**: 159–166.

Yilmaz I (2006) Indirect estimation of the swelling percent and a new classification of soils depending on liquid limit and cation exchange capacity. *Engineering Geology* **85**: 295–301.

Yukselen-Aksoy Y and Kaya A (2010) Predicting soil swelling behaviour from specific surface area. *Proceedings of the Institution of Civil Engineers: Geotechnical Engineering* **163(GE4)**: 229–238.

Zha FS, Liu SY and Du YJ (2006) Evaluation of swell–shrink properties of compacted expansive soils using electrical resistivity methods. In *Unsaturated Soil, Seepage, and Environmental Geotechnics.* ASCE, Reston, VA, pp. 143–151.

Zheng JL, Zhang R and Yang HP (2009) Highway subgrade construction in expansive soil areas. *Journal of Materials in Civil Engineering* **21(4)**: 154–162.

Further reading

Al-Rawas AA and Goosen MFA (eds) (2006) *Expansive Soils: Recent Advances on Characterization and Treatment.* Taylor & Francis, London, UK.

BRE (1996) *The Significance of Desiccation.* BRE Digest 412. CRC, London, UK.

BRE (1999) *The Influence of Trees on House Foundations in Clay Soils.* BRE Digest 298. CRC, London, UK.

Chen FH (1988) *Foundations on Expansive Soils.* Elsevier, Amsterdam, the Netherlands.

Fredlund DG and Rahardjo H (1993) *Soil Mechanics for Unsaturated Soils.* Wiley, New York, NY.

Ikeagwuani CC and Nwonu DC (2019) Emerging trends in expansive soil stabilisation: a review. *Journal of Rock Mechanics and Geotechnical Engineering*, **11**: 423–440.

Nelson JD, Chao KC, Overton DD and Nelson EJ (2015) *Foundation Engineering for Expansive Soils.* Wiley, New York, NY.

Nelson JD and Miller DJ (1992) *Expansive Soils: Problems and Practice in Foundation and Pavement Engineering.* Wiley, New York, NY.

NHBC (2023) 'Standards 4: Foundations'. NHBC, London, UK.

Vipulanandan C, Addison MB and Hasen M (eds) (2001) *Expansive Clay Soils and Vegetative Influence on Shallow Foundations.* ASCE, Reston, VA.

Useful websites

Association of British Insurers: http://www.abi.org.uk
British Geological Survey (BGS): http://www.bgs.ac.uk
International Society of Arboriculture, UK and Ireland Chapter: http://www.isa-arboriculture.org
Royal Institution of Chartered Surveyors: http://www.rics.org
Subsidence Claims Advisory Bureau: http://www.subsidencebureau.com
The Clay Research Group, UK: http://www.theclayresearchgroup.org
The Subsidence Forum: http://www.subsidenceforum.org
US Geological Survey (USGS): http://www.usgs.gov

It is recommended this chapter is read in conjunction with

- Chapter 7 *Geotechnical risks and their context for the whole project*
- Chapter 29 *Arid soils*
- Chapter 40 *The ground as a hazard*
- Chapter 76 *Pavement foundation design*

All chapters in this book rely on the guidance in Sections 1 *Context* and 2 *Fundamental principles.* A sound knowledge of ground investigation is required for all geotechnical works, as set out in Section 4 *Site investigation.*

doi: 10.1680/icemge.66816.0479

CONTENTS

Chapter 34
Non-engineered fills

Hilary D Skinner COWI UK, London, UK
Fred G Bell British Geological Survey, Keyworth, UK
Martin G Culshaw University of Birmingham, Birmingham, UK, and British Geological Survey, Keyworth, UK

The term 'fill' (or 'made ground') is used to describe material that has been deposited by human processes. The material could be natural, or have been altered artificially prior to deposition. Fill is either engineered or non-engineered depending on whether any specific treatment during deposition takes place. As a result, a number of engineering challenges may exist and will require different remediation approaches to improve and mitigate problematical behaviour.

Non-engineered fill may consist of domestic waste, building waste, slag, mining and quarry waste, industrial waste and soil waste. However, methodologies for the classification and description of fills are not fully developed or standardised nationally or internationally. This affects the way in which fills are presented on geological maps, although standardisation in the UK is becoming established.

Non-engineered fills may settle variably, have poor bearing capacity and may suffer significant movements due to causes other than the imposed loading. Other problems associated with some wastes include contamination, spontaneous combustion and the emission of gas. The extent to which non-engineered fill will be suitable as a foundation material depends largely on its age, composition, uniformity, properties and the method by which the material was placed.

34.1. Introduction

The term 'fill' refers to material that has been deposited artificially as, for instance, in an excavation or as ground made by human activity. Collectively, these deposits, together with unfilled excavations, are sometimes termed 'artificial ground' with reference to their geographical distribution and method of placement, rather than their composition (e.g. Ford *et al.*, 2006). Non-engineered fill, as opposed to engineered fill, is material that has not been subject to some degree of controlled placement to ensure that the geotechnical properties of the finally placed material comply with a predetermined design. It is dumped, with little control, in deep lifts so that it is typically poorly compacted, and so is likely to be in a loose state with varying geotechnical properties, both laterally and with depth. Terminology, in practice, can be confusing, as Norbury (2010) has pointed out, in that engineered fill is often referred to as 'fill', while non-engineered fill is referred to as 'made ground'. This usage does not coincide with that of the British Geological Survey (BGS) in mapping artificial ground (see Section 34.3.1).

Due to the shortage of prime land for development in urban areas, artificial ground, of necessity, is being – and increasingly will be – used for building purposes (Charles, 2008). However, non-engineered fill is derived from a wide variety of materials, including domestic refuse, ashes, slag, clinker, building waste, chemical waste, mining and quarry waste, as well as different types of soils. These materials, by their nature, exhibit significant physical and chemical variability. Thus, the extent to which an existing non-engineered fill may be suitable as a foundation material depends largely on its composition and uniformity; its engineering behaviour is influenced by its method of placement and subsequent history. In the past, the control exercised in placing many fills was frequently insufficient to ensure appropriate engineering properties (hence the term 'non-engineered'). Consequently, non-engineered fills may have variable settlement characteristics (**Figure 34.1**) or poor bearing capacity, and may be compositionally variable over short distances and suffer significant movements due to causes other than the imposed loading. Other problems can include contamination, spontaneous combustion and the emission of gas, depending on the nature of the fill material. In addition, time-related changes in volume and composition are likely to take place or may be triggered by any stress or environmental changes.

This chapter first discusses the generic characteristics of non-engineered fills that cause their problematic properties and their often unpredictable engineering behaviour, which can yield significant problems for geotechnical engineers. An overview of how non-engineered fills are classified (as 'artificial ground'), mapped and described is followed by a discussion of the nature, properties and behaviours of a range of non-engineered fills. Chapter 58 *Building on fills* gives specific details of building and foundation aspects for fill materials.

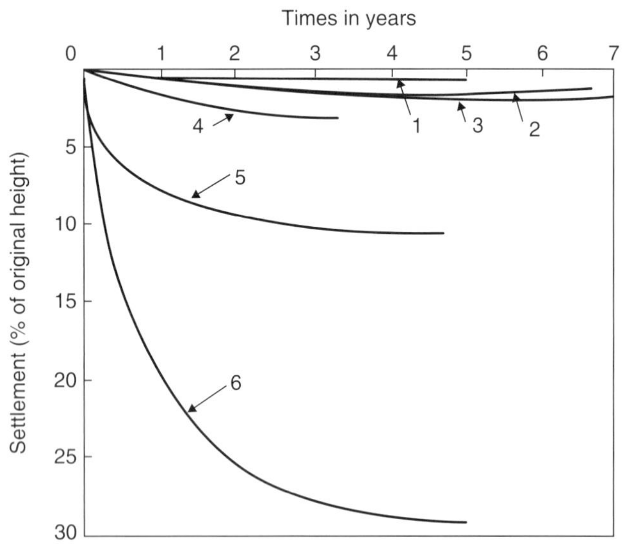

Figure 34.1 Observations of the settlement of various types of fill due to consolidation under its own weight: (1) well-graded sand, well compacted; (2) rockfill, medium state of compaction; (3) clay or chalk, lightly compacted; (4) sand, uncompacted; (5) clay, uncompacted; (6) mixed refuse, well compacted (Reproduced from Meyerhof, 1951; Institution of Structural Engineers)

34.2. Problematic characteristics

Fills exhibit a range of engineering properties quite as wide as that of natural soils. The behaviour of natural soils and rocks is strongly influenced by their structure and fabric. Although soil structure arises from many different causes, its effect on behaviour is similar. As strength increases, the stress domain over which a soil exhibits stiffer behaviour is enlarged. In fills formed from natural soils, such structure will have been largely destroyed during excavation and placement, with little opportunity for it to be reinstated. Consequently, fill behaviour may be inferior to that of apparently similar natural soils.

Variability cannot be overemphasised. Whether between areas of infilled ground and natural ground, or between different areas of fill, variability can result in differential settlement or structural loading that is unacceptable. All analyses of fill must address the likely variations in properties and behaviour. Where fill density and history can be estimated, tests on representative, recompacted samples may give an indication of fill behaviour. The engineering behaviour of the material should be identified within the required context: for example, load-carrying capacity, volume changes in response to load or other factors, or strength. In addition, the potential for the generation of deleterious substances should be identified.

It is helpful to compare and contrast the properties of fills with the properties of natural soil, with the following factors in mind that will affect their engineering behaviour.

- *Nature of the material:* A fill may be composed of the same type of material as a natural soil (such as clay, sand or rock) or it could be composed of a range of other materials (such as wastes from chemical and industrial processes). The nature of the material will indicate the underlying best-case strength and stiffness behaviour and will affect the potential for any change in density, whether by mechanical, hydrological, biological or chemical causes.
- *Method of deposition:* A fill may have been deposited in a manner that is quite similar to the deposition of some natural soils (e.g. sedimentation under water) or it may have been deposited in a quite different manner (e.g. compaction). The method of placement and any subsequent activities will affect the density, water content and variability. Generally, like natural materials, a higher density at a lower moisture content is likely to demonstrate better engineering behaviour, although care should be taken where cementation or other 'metastable' conditions may occur.
- *Age:* Many fills have been placed quite recently and the deposits are much younger than natural soils. Often, younger deposits are still undergoing movements arising from their placement or material type. In addition, changes in composition may be anticipated; for some fills that may mean poorer behaviour should be anticipated – for example, the ash content of more recent domestic refuse is lower, and the accompanying behaviour poorer, than older deposits.

See the further reading section, and Chapter 58 *Building on fills*, for more information.

34.3. Classification, mapping and description of artificial ground

34.3.1 Classification

Geotechnical engineers have typically divided 'fills' into those that are engineered and those that are not. However, when carrying out a site investigation on existing artificial ground it is not usually easy to make this distinction. Traditionally, geological surveyors have ignored artificial ground when creating a geological map. Engineering geologists, too, have paid little attention to the mapping of these deposits. For example, a report by the Geological Society Engineering Group Working Party on engineering geological mapping (Knill *et al.*, 1970) only recognised 'made ground', 'spoil tip (made ground above natural surface level)' and 'backfilled opencast site or excavation'. However, the increased demand for land in urban areas has resulted in pressure to redevelop brownfield sites, including artificial ground. In turn, this has encouraged mapping geologists and engineering geologists to include these deposits on their maps. However, the simple mapping of artificial ground without systematically distinguishing between different categories does not meet the needs of developers and geotechnical engineers for more detailed information. In the last decade, a simple classification of artificial deposits has been used on geological maps in the UK. Ford *et al.* (2006) have extended the classification of Rosenbaum *et al.* (2003) and identified five principal classes of artificial ground (**Figure 34.2**).

1. *Made ground* is where material has been placed upon the pre-existing ground surface. Most engineered fill would fall into this class.

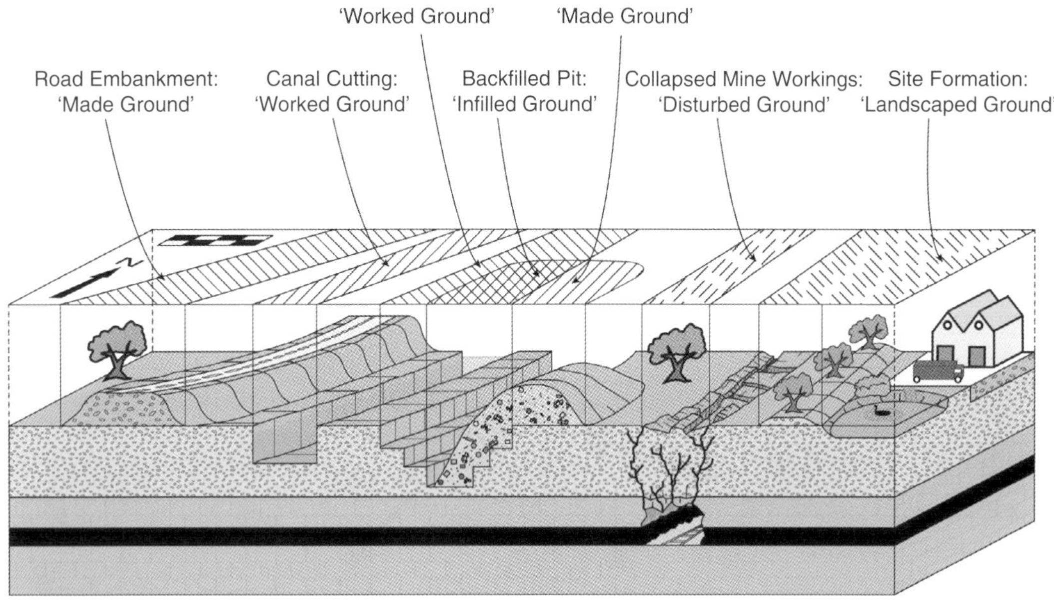

Figure 34.2 The main types of artificial ground and the way in which they are depicted on geological maps in the UK (Reproduced from Price *et al.*, 2011; The Royal Society)

2. *Worked ground* is where there have been excavations into the pre-existing ground surface.
3. *Infilled ground* is a combination of one or more phases of excavation involving the extraction of material (worked ground) and one or more phases of deposition of material on the excavated surface or made ground (Price *et al.*, 2011).
4. *Disturbed ground* is where there has been surface or near-surface disturbance so that ill-defined excavations, areas of subsidence, spoil and other irregular features are associated in a complex way with each other; areas around mines often fall into this category.
5. *Landscaped ground* is where the original surface has been extensively remodelled, but where it is impractical or impossible to separately delineate areas of worked and made ground.

For each of the classes 1, 2 and 4 a three-tier hierarchy (class, type, unit) provides a flexible and logical subdivision into an increasingly descriptive set of lithostratigraphical categories for mapping purposes. The structure allows the user to select an appropriate level of detail according to available requirements, while providing a structure that is readily amenable to digital recording. Examples of the classification hierarchy are shown in **Figure 34.3**.

The typical usage is intended to be intuitive. The unit level gives the highest level of detail, allowing specific types of artificial ground to be recorded. This level of detail is suitable for mapping at 1:10 000 scale, which is the scale of most geological maps in the UK. Should a confident unit-level classification be impossible due to insufficient data or time, then a classification at the type level can be considered. The range of type-level categories provides a non-specific set of parent classes that retain a moderate level of detail. Class level corresponds to the level of detail currently available on 1:50 000 scale geological maps published in the UK in the last 20 years or so. This is a very general classification level and specific observations and significant information may be lost through generalisation.

For landscaped ground, the classification is not subdivided beyond type level. This is intended to avoid duplication of information that is available from other UK sources (e.g. Ordnance Survey topographical maps and the National Land Use Database, which provides a comprehensive, consistent and up-to-date record of previously developed land in England that may be available for development).

34.3.2 Mapping

The mapping of artificial ground utilises the same principles as for superficial deposits and, to a lesser extent, bedrock. Mapping considers *morphology* (the shape of the ground), *origin* and *genesis* (the 'process' that has formed the deposit), *age* (important where several generations of deposits are present) and *lithology* (the composition of the materials themselves). This can be illustrated by comparing the way in which a railway embankment (class – made ground; type – embankment; unit – railway embankment) and an esker (an embankment-like natural deposit formed as a glacier retreats) are mapped. The two have similar morphologies, although an esker may well be more sinuous. Their genesis is clearly different: eskers are formed as glaciers retreat while railway embankments are

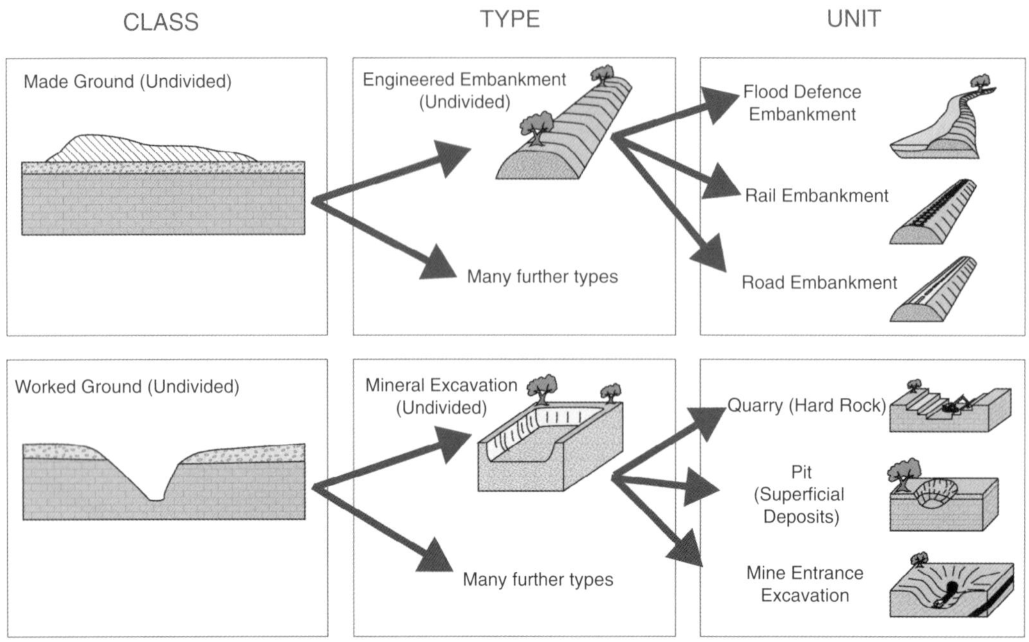

Figure 34.3 Examples of the artificial ground classification hierarchy (Reproduced from Price *et al*., 2011; The Royal Society)

formed during the construction of railway systems. Eskers have ages of thousands of years, while railway embankments are less than 200 years old. Eskers are composed largely of sand and gravel, while railway embankments can be composed of a variety of materials, both natural and artificial, which have experienced varying degrees of engineering (usually compaction).

34.3.3 Description

The Geological Society Engineering Group Working Party reports on core logging and on engineering geological mapping (Geological Society Engineering Group Working Party, 1972; Knill *et al*., 1970) do not refer to how fill materials should be described. However, the International Association of Engineering Geology's Commission on Engineering Geological Mapping has produced a brief guide to the mass characteristics of non-engineered fill that should be described (IAEG Commission, 1981)

- mode of origin, including method of placement
- presence of larger objects
- voids
- chemical and organic matter
- toxic materials, including gases
- age.

In terms of non-engineered fills, Norbury (2010) has suggested a more detailed approach. First, a sample of non-engineered fill material is divided into very coarse, gravel and fine fractions. The proportion of each is determined and how the material is likely to behave in different situations (e.g. in the sides of an excavation or in a spoil heap) is assessed. The material is then described in terms of its composition, identifying

(*a*) natural soil materials

(*b*) artificial materials such as

- asphalt
- concrete
- masonry, blockwork and brickwork
- glass, metal, plastics
- wood, paper
- burned products such as slag, clinker and ash, including evidence by strong colours
- large objects and obstructions, which may be critical to piling operations
- hollow objects, compressible and collapsible objects or voids, which could be important for ground stability
- chemical wastes and dangerous or hazardous materials
- decomposable materials, with a note of the degree of decomposition.

The proportions of each are also determined.

Norbury (2010) suggested that further information should be provided where possible, covering

- any dating for papers or other artefacts
- signs of current or recent heat or combustion

- smells or odours
- layers and their inclination as a guide to the method of placement
- origin of materials, if identifiable.

34.4. Types of non-engineered fill

Non-engineered fills generally are associated with the backfilling of opencast mineral extraction sites, disposal of waste and with derelict land. Although a modern waste disposal site is engineered so that its construction should minimise the production of leachate and its infiltration into the ground, as well as dealing with any landfill gas produced, it is not necessarily designed to support the subsequent erection of buildings.

34.4.1 Filled opencast sites

34.4.1.1 Working of opencast sites

Stratified mineral deposits that occur at or near the surface, such as coal or sedimentary iron ore in relatively flat terrain, are generally mined by strip or opencast mining. A deposit will either be horizontal or gently dipping. In Britain, opencast coal mining involves the exploitation of shallow seams at sites normally from 10 to 800 ha in area. This usually involves creating a box-cut (an initial, trench-like excavation) to reveal the coal. The opencast working of coal involves its excavation to maximum depths of around 100 m below the surface with stripping ratios (the ratio of the amount of overburden or non-coal material to the amount of coal recovered) of up to 25:1. Hence, opencast mining has been a major producer of deep fills, the geotechnical properties of which are of notable importance if, after restoration, they are to be considered for building development. In this context the major factor according to Charles (2008) is the possibility of long-term settlement of the backfill.

An opencast area will be worked on a broad front, with face lengths of 3–5 km not being uncommon. Once the coal has been extracted, the box-cut is filled with overburden from the next cut. Draglines may be used for overburden (primarily sandstone and mudrock) removal, along with face shovels, dump trucks and scrapers. The topsoil is removed separately by scraper and placed in dumps around the site for use in subsequent restoration. Similarly, the subsoil is stripped and stored temporarily in separate dumps about the site, again for restoration purposes. Face shovels are used to excavate the initial box-cut and for subsequent forward reduction of overburden.

The material from the box-cut is placed above ground in a suitable position in relation to filling the excavation so that later re-handling is minimised. Draglines excavate the lower seams in a progressive strip cut behind the face shovels. Drilling and blasting is carried out where necessary. Because of the short lifespan of spoil heaps they can often be maintained at their natural angle of repose.

Restoration can begin before a site is closed; indeed, this usually is the more convenient method. Hence, worked-out areas behind the excavation front are filled with rock waste. This means that the final contours can be designed with less spoil movement than if the two operations were undertaken separately. Furthermore, more soil for spreading can be conserved when restoration and coal working are carried out simultaneously. Because of high stripping ratios (often 15:1 to 25:1) there is usually enough spoil to more or less fill an excavation. Once the spoil has been regraded it is covered with topsoil. The restored land is generally used for agriculture or forestry, but it can be used for country parks, golf courses, housing or light industrial estates with suitable engineering.

34.4.1.2 Properties and engineering behaviour

The properties of some backfills of opencast mines were given by Watts and Charles (2015) and have been summarised by Charles (2008); see **Table 34.1**. However, many fills are heterogeneous and may contain large boulders of mudstone, siltstone or sandstone, so the geotechnical properties quoted should be regarded as indicative values at best. Hence, field measurements

(a) Coarse backfills

Location	Fill type	Silt and clay: %	Dry density (ρ d): Mg/m^3	Particle density (ρ s): Mg/m^3	Moisture content (w): %	Porosity (n): %	Air voids (V_a): %
Horsley	Mudstone and sandstone	10	1.70	2.54	7	33	21
Blindwells	Mudstone and sandstone	20	1.56	2.45	7	38	23
Tamworth	Clay with shale fragments	45	1.78	2.62	9	32	16

(b) Clay backfills

Location	Moisture content (w): %		Liquid limit (w_L): %	Undrained shear strength (w_u): kPa
	Mean	Range		
Ilkeston	19	12–25	41	150
Corby	18	7–28	28	100

Table 34.1 Some properties of opencast mining backfills (Reproduced from Charles, 2008; all rights reserved)

taken at opencast sites are of special value in understanding the geotechnical behaviour of a backfill. Instrumentation and monitoring of a fill is carried out once filling is finished (see Chapter 58 *Building on fills* for further details).

The water table at many opencast sites is lowered by pumping to provide dry working conditions in the pit. If, when worked out, a site is to be restored for agricultural use or for forestry, then this takes place without control of compaction. However, if a site is to be built over then settlement is likely to be a problem without proper compaction. In particular, significant settlements of opencast backfill can occur when the partially saturated material becomes saturated by rising groundwater after pumping has ceased. For example, Charles *et al.* (1993) referred to an opencast site in Northumberland where the waste was backfilled without any systematic compaction. They recorded that when the water table rose some 0.33 m into the fill, settlement occurred where the fill was 63 m in depth.

Settlement due to wetting collapse is more significant than that due to the self-weight of the backfill, which can give rise to difficulties in settlement prediction (Blanchfield and Anderson, 2000). Therefore, if a site is to be built on after restoration, then ideally the waste should be compacted properly. However, the subsequent development of a site will frequently be some time after an opencast area has been worked out and backfilled. As a consequence, such areas require investigation prior to any building development to determine the moisture condition of the backfill material and to determine how much it has been compacted, if at all (Occupational Health and Safety Information Service, 1983). Consequently, Charles (1993) recommended that if an opencast site is to be built over but the backfill was not properly compacted, then collapse compression could be realised prior to development by inundation. Indeed, Charles and Skinner (2001) maintained that the susceptibility to collapse compression on inundation often represents a notable hazard when construction is to take place on fill. Most types of partially saturated fill, which are in a sufficiently loose or dry state, are susceptible to collapse compression over a wide range of applied loads when first inundated. Inundation can result from the downward percolation of water infiltrating from the surface or can be due to rising groundwater levels after pumping has ceased. Collapse compression can cause serious damage to any buildings placed upon such fills. However, in many instances flooding is difficult to carry out in a controlled and effective way. Alternatively, the ground could be preloaded or subjected to deep compaction by either vibrocompaction or dynamic compaction, or the structures could be designed to accommodate any subsequent ground movements. However, placing vibro stone columns in opencast backfill, according to Watts and Charles (2015), might provide a means of allowing surface water into the fill and thereby cause collapse compression.

Alternatively, where backfilled opencast coal workings exceed 30 m in depth, because greater settlements may occur, Kilkenny (1968) recommended that the minimum time that should elapse before development takes place should be 12 years after restoration is complete. He noted that the settlement of opencast backfill appeared to be complete within 5–10 years after the operation. For example, comprehensive observations of the opencast restored area at Chibburn, Northumberland, which was 23–38 m in depth, revealed that the ultimate settlement amounted to approximately 1.2% of the fill thickness and that some 50% of the settlement was complete after 2 years and 75% within 5 years. In shallow opencast fills – that is, up to 20 m in depth – settlements of up to 75 mm have been observed. Greater settlement may occur if there is movement of groundwater through a permeable fill, including a rising water table as noted above, as a result of the breakdown of point-to-point contacts in the shale or sandstone components of the fill. In such circumstances settlement may continue over a greater length of time than noted by Kilkenny.

Restoration of other opencast mineral workings can be dealt with in a similar way to those for coal. Another mineral that has been extracted on a large scale by opencast mining was sedimentary iron ore; for example, this was worked near Scunthorpe, Lincolnshire, and near Corby in Northamptonshire. In the latter area Penman and Godwin (1975) noted that the maximum rates of settlement occurred immediately after the construction of two-storey semi-detached houses on an old opencast site that had been backfilled. Settlement decreased to small rates after about 4 years. They suggested that two of the causes of settlement in this fill were creep, which is proportional to log time, and partial inundation. Similar conclusions were reached by Sowers *et al.* (1965). The houses at Corby were constructed 12 years after the fill was placed. The amount of damage that they suffered was relatively small and was attributable to differential settlement; it was not related to the type of foundation structure used. These were either strip footings or reinforced-concrete rafts with edge beams.

34.4.2 Domestic waste disposal or sanitary landfill

With increasing industrialisation, technical development and economic growth, the quantity of waste produced by society has increased immensely. In addition, in developed countries the nature and composition of waste has evolved over the decades, reflecting industrial and domestic practices. For example, in Britain domestic waste has changed significantly since the 1950s, from largely ashes with little putrescible content of relatively high density to low-density, highly putrescible waste. Although domestic waste is disposed of in a number of ways, quantitatively the most important method is placement in a landfill. In England in 2003–2004, for example, of approximately 25.2 million tonnes of domestic solid waste produced, about 72% was disposed of in landfills (Parliamentary Office of Science and Technology, 2005). However, by 2007–2008 the amount recycled had increased to 35% (Office of National Statistics, 2014). Landfill is controlled through the

European Landfill Directive (Council of the European Union, 1999), implemented in the UK through the Landfill (England and Wales) Regulations 2002 and the Landfill (Scotland) Regulations 2003. Since the introduction of these regulations, the amount of waste sent to landfill has reduced significantly. Increasingly, waste materials from different sectors (e.g. construction or household) are recycled. However, geotechnical engineers commonly encounter historical landfill sites which, depending on their time of operation, will exhibit significantly different physical and chemical characteristics. Moreover, landfills are active sites with a number of time-dependent changes occurring throughout their lives. The Environment Agency has published a series of guides covering waste acceptance; monitoring of landfill leachate, groundwater and surface water; landfill engineering; landfill gas; and landfill permitting and surrender.

34.4.2.1 Composition

Waste disposal or sanitary landfills are usually very mixed in composition (**Table 34.2**) and suffer from continuing organic decomposition and physico-chemical breakdown. They can consist of a heterogeneous collection of almost anything: waste food, garden rubbish, paper, plastic, glass, rubber, cloth, ashes, building waste, metals and so on. In future, recycling is likely to change the character of waste. Nonetheless, matter exists in the gaseous, liquid and solid states in landfills and all landfills comprise a delicate and shifting balance between the three states. Any assessment of the state of a landfill and its environment must take into consideration the substances present in it, and their mobility now and in the future. Much of the material of which a landfill is composed is capable of reacting with water to give a liquid rich in organic matter, mineral salts and bacteria, commonly known as leachate. Leachate is formed when rainfall infiltrates into a landfill and dissolves the soluble fraction of the waste, and from soluble products formed as a result of chemical and biochemical processes occurring within the decaying waste. Methane and carbon dioxide are the primary gases produced in landfills, together with traces of organic gases, carbon monoxide and hydrogen sulfide. The concentration of methane can range from 20% to 65% and carbon dioxide from 15% to 40%, while other gases typically make up <1% (Wilson *et al.*, 2007). In very general terms, gas production is likely to have reduced to a very low level after about 50–60 years (Nastev, 1998). Gases are produced in the breakdown process and accumulations of gas in pockets in fills have led to explosions – for example, following the escape of methane at Loscoe in Derbyshire (Williams and Aitkenhead, 1991). Some materials such as ashes and industrial waste may contain sulfate and other products that are potentially deleterious to concrete.

Material	Characteristics as fill
Garbage: food, waste	Wet, ferments and decays readily, compressible, weak
Paper, cloth	Dry to damp, decays and burns, compressible
Garden refuse	Damp, ferments, decays and burns, compressible
Plastic	Dry, decay resistant, may burn, compressible
Hollow metal (e.g. drums)	Dry, corrodible and crushable
Massive metal	Dry, slightly corrodible, rigid
Rubber (e.g. tyres)	Dry, resilient, burns, decay resistant, compressible
Glass	Dry, decay resistant, crushable and compressible
Demolition timber	Dry, decays and burns, crushable
Building rubble	Damp, decay resistant, crushable and erodible
Ashes, clinker and chemical wastes	Damp, compressible, active chemically and partially soluble

Table 34.2 Municipal waste materials incorporated in fills

34.4.2.2 Landfill design

The design of a landfill site is influenced by the physical and biochemical properties of the waste and the need to control leachate production. Post-closure settlement of sites is taken into account at a very early stage in the design of a landfill. Indeed, modern landfill disposal facilities require detailed investigations to ensure that appropriate design and safety precautions are undertaken. Furthermore, legislation generally requires that those responsible for waste disposal facilities guarantee that sites are suitably contained so as to prevent harm to the environment (for further details see the Environment Agency's website). Geological and hydrogeological conditions must be taken into account when selecting a landfill site (Proske *et al.*, 2005). However, the selection also involves economic and social factors. It should be noted, also, that the methods used to contain landfill have changed considerably. It was not until the late 1970s and early 1980s that landfill liners were used; prior to this leachate and gas control was through dilute-and-disperse techniques. Since then, landfill liner and capping construction have become ever more sophisticated, with double-lining systems now being used, combining clay (or a suitable substitute using bentonite-enriched sands or clays, or engineered spoil material, such as colliery spoil) and plastic HDPE (high-density polyethylene) liners. Included in a lining system there will also be a series of leachate and gas collection layers and protection media, so liners are complex engineered inclusions within a landfill matrix. Great care is needed to avoid compromising the lining system during investigation or ground treatment (for example, through piling), otherwise pollution linkages will be generated allowing possible migration to a number of sensitive receptors. **Figure 34.4** shows the early stages of construction of a cell within a landfill site.

It is vital that monitoring continues during and after the construction of a landfill, particularly with regard to leachate

Figure 34.4 Early stages in the construction of a landfill cell in north-east England. The liner and a high-permeability drainage layer are being placed (© Paul Nathanail, with permission)

Figure 34.5 Cellular structure of landfill near Kansas City, USA

and gas emissions. Boreholes through the base of a leachate-containing landfill to check what is happening should only be used if precautions are taken to avoid leakage. Geophysical methods are being used increasingly for this purpose (e.g. Kuras *et al.*, 2006).

At the present time there is no standard rule as to how waste should be dumped or compacted. Nonetheless, compaction is important since it reduces settlement and hydraulic conductivity, while increasing shear strength and bearing capacity. Moreover, the smaller the quantity of air trapped within landfill waste, the lower the potential for spontaneous combustion. Modern well-designed landfills usually possess a cellular structure, as well as a lining and a cover – that is, the waste is contained within a series of cells formed of clay (see **Figure 34.5**). However, certain leachates have been troublesome for some clay liners. Those consisting of organic solvents and those containing high levels of dissolved salts can give rise to cracking and the development of pipes in clay. Clays can be treated with special polymers to reduce their sensitivity to potential contaminants. Uneven settlement of a landfill can cause fractures within the cover, which allow water to percolate into the fill. If conventional compaction techniques are used they do not always achieve effective results, especially with non-uniform waste. An additional complication is the time dependency of waste degradation, resulting in further settlement and the potential for cracking of capping layers.

34.4.2.3 Geotechnical properties

Waste materials in landfills have very variable properties, with dry densities varying from 0.16 to 0.35 Mg/m^3 when tipped; after compaction the density may exceed 0.60 Mg/m^3. The moisture content ranges from 10% to 50%, the average specific gravity of the solids varies from 1.7 to 2.5, and the low bearing capacities are 19–34 kPa. Baling or shredding the waste materials before deposition improves their in situ properties but at greatly increased cost. Further details of typical engineering properties of landfill are given by Dixon *et al.* (1998).

34.4.2.4 Decomposition

Initially, the decomposition of organic waste is aerobic. Once decomposition starts the oxygen in the waste is rapidly exhausted and so the decomposition becomes anaerobic. There are basically two processes for the anaerobic decomposition of organic waste. Initially, complex organic materials are broken down into simpler organic substances, which are typified by various acids and alcohols. Any nitrogen present in the original organic material tends to be converted into ammonium ions that are readily soluble and may give rise to significant quantities of ammonia in the leachate. The reducing environment converts oxidised ions such as those in ferric salts to the ferrous state. Ferrous salts are more soluble and, therefore, iron is leached from the landfill. The sulfate in the landfill may be reduced biochemically to sulfide. Although this may lead to the production of small quantities of hydrogen sulfide, the sulfide tends to remain in the landfill as highly insoluble metal sulfides. In a young landfill the dissolved salt content may exceed 10 000 mg/l, with relatively high concentrations of sodium, calcium, chloride, sulfate and iron, whereas as the landfill ages the concentration of such inorganic materials usually decreases. Also, suspended particles may be present in the leachate due to the washout of fine material from the landfill.

The second stage of anaerobic decomposition involves the formation of methane. In other words, methanogenic bacteria use the end products of the first stage of anaerobic decomposition to produce methane and carbon dioxide. Methanogenic bacteria prefer neutral conditions, so excessive acid formation in the first stage will inhibit their activity.

As a landfill ages, because much of the readily biodegradable material has been broken down, the organic content of the leachate decreases. The acids associated with older landfills are not readily biodegradable. Consequently, the biochemical oxygen demand (BOD) in the leachate changes with time, increasing to a peak as the microbial activity increases. This peak is reached between 6 months and 2.5 years after tipping.

Methane production can constitute a dangerous hazard because methane is combustible and in certain concentrations in air is explosive (5–15% by volume in air), as well as asphyxiating. In many instances landfill gas is able to disperse safely into the atmosphere from the surface of a landfill. However, when a landfill is completely covered with a soil capping of low permeability to limit leachate generation, the potential for the gas to migrate along unknown pathways increases and there are cases of hazards arising from methane migration. Furthermore, there are unfortunately cases on record of explosions occurring in buildings due to the ignition of accumulated methane derived from landfills near to, or on, which they were built (Williams and Aitkenhead, 1991).

The source of gas should be identified so that remedial action can be taken. Generally, the connection between a source of methane and the location where it is detected can be verified by detecting a component of the gas that is specific to the source or by establishing the existence of a migration pathway from the source to the location where the gas is detected. Identification of the source can involve a drilling programme and analysis of the gas recovered. For instance, in a case referred to by Raybould and Anderson (1987), distinction had to be made between household gas and gas from sewers, old coal mines and a landfill before remedial action could be taken to eliminate the gas hazard affecting several houses. Proper closure of a landfill site can require gas management by passive venting, power-operated venting or the use of an impermeable barrier. Measures to prevent migration of the gas include impermeable barriers (clay, bentonite, geomembranes or cement) and gas venting. Venting either wastes the gas into the atmosphere or facilitates its collection for utilisation. Where atmospheric venting is insufficient to control the discharge of gas, an active or forced venting system is used. Further details on the investigation of gas hazard in landfills and of remedial measures are given by Wilson *et al.* (2007) and references therein.

Other hazardous situations can occur due to microbial activity in a landfill, which generates heat. Hence, the temperature in a landfill rises during biodegradation to between 24°C and 45°C, although temperatures of up to 70°C have been recorded. In some cases landfills have caught fire, increasing the risk still further, although the risk is relatively low; Roche (1996) reported fire problems in around 1 in 600 landfill sites in England and Wales compared with surface water pollution, groundwater pollution and gas migration, each affecting about 1 in 60 sites. A fire at the Lean Quarry landfill site near Liskeard in south-east Cornwall in June 2010 took over 12 hours to extinguish. A fire in January 2011 at the Kerdiffstown landfill near Naas in Ireland burned for several weeks.

34.4.2.5 Settlement

McDougall *et al.* (2004) noted that settlements associated with landfills are likely to be large and irregular. According to Sowers (1973) the mechanisms responsible for the settlement of waste disposal fills include mechanical distortion, bending, crushing and reorientation of materials, which cause a reduction in the void ratio; ravelling (that is, the transfer of fines into the voids); physico-chemical and biochemical changes such as corrosion, combustion and fermentation; and the interactions of these various mechanisms. The initial mechanical settlement of waste disposal fills is rapid and is due to a reduction in the initial void ratio. It takes place with no build-up of pore water pressure. This primary compression takes place within a few days of further material being placed on top (Powrie *et al.*, 1998). Settlement continues due to a combination of secondary compression (that is, material disturbance) and physico-chemical and biochemical action; Sowers showed that the settlement log–time relationship is more or less linear. The rate of settlement produced by ravelling and combustion is erratic. Hence, the ultimate settlement is related to the initial void ratio and the environmental conditions favourable to deterioration, decay, ravelling and combustion.

The prediction of the settlement at a landfill site is a vital element for effective design, particularly in relation to post-closure settlement during site redevelopment. However, a determination of the amount and rate of settlement of a landfill is not a simple task. Secondary settlement of 15–50% of the original fill thickness has been suggested (Powrie *et al.*, 1998) while Ling *et al.* (1998) gave figures of 30–40% for overall settlement. The heterogeneous nature of the materials involved, the fact that different materials may have been disposed of in different places and at different times, means that traditional soil mechanics for settlement prediction is generally unsatisfactory. Not only would settlement, especially large differential settlement, adversely affect any future construction on such a site, but it may damage the cover system or lining – for example, geomembrane linings may be torn. Accordingly, Ling *et al.* (1998) reviewed the various empirical methods of estimating the settlement of landfills and then proposed a further method for evaluating the amount and rate of settlement. Watts and Charles (1999) also discussed the settlement characteristics of landfills and suggested various ground improvement techniques that could be used; these are discussed in Chapter 84 *Ground improvement*.

34.4.2.6 Stabilisation

The stabilisation of soft landfills can be realised by mixing with soil, fly ash, incinerator residue, lime or cement. When a minimum unconfined compressive strength of 24 kPa occurs due, for example, to irregularities in mixing at the site, a geosynthetic material (e.g. a geoweb) should be used to span such areas and should be securely anchored in a trench at the

(a)

(b)

Figure 34.6 (a) Spoil tips (one of which has been 'landscaped') at International Colliery, Blaengarw, Garw Valley, South Wales (© NERC, with permission from the British Geological Survey); (b) spoil tip at the abandoned Greenside lead mine tip in the Glenridding Valley, Cumbria, England (© Paul Nathanail, with permission)

perimeter of the site. A geomembrane should be placed over the geoweb to provide a secondary seal protecting against the infiltration of surface liquids. Clay soil above the geomembrane acts as a seal but it must be as thin as possible so as not to overload the geoweb and geomembrane.

34.4.3 Coarse colliery discard

Spoil heaps associated with coal mines are ugly blemishes on the landscape and have a blighting effect on the environment (**Figure 34.6**). In the past these had the potential for catastrophic failure with devastating consequences (e.g. the Aberfan failure in October 1966). Spoil heaps consist of coarse discard – that is, run-of-mine material that reflects the various rock types (mudrock, siltstone and sandstone) that are extracted during mining operations. Coarse discard also contains varying amounts of coaly material that has not been separated by the preparation process. Obviously, the characteristics of coarse colliery discard differ according to the nature of the spoil. The method of tipping also influences the character of a coarse discard. In addition, some spoil heaps, particularly those with relatively high coal contents, may be burned, or still be burning, and this affects their mineralogical composition.

34.4.3.1 Composition and properties

In Britain, illites and mixed layer clays are the principal components of unburned spoil in English and Welsh tips (Taylor, 1975). Although kaolinite is a common constituent in Northumberland and Durham, it averages only 10.5% of the discard of other areas. Quartz exceeds the organic carbon or coal content, but the latter is significant in that it acts as the major diluent – that is, the clay mineral content tends to decrease as the carbon content increases. Sulfates, feldspars, calcite and other materials average less than 2%.

The chemical composition of spoil material reflects the mineralogical composition. Free silica may be present in concentrations up to 80% and above, and combined silica in the form of clay minerals may range up to 60%. Concentrations of aluminium oxide may be between a few per cent and 40% or so. Calcium, magnesium, iron, sodium, potassium and titanium oxides may be present in concentrations of a few per cent. Lower amounts of manganese and phosphorus also may be present, with copper, nickel, lead and zinc in trace amounts. The sulfur content of fresh spoils is often less than 1% and occurs as organic sulfur in coal, and in pyrite. Pyrite is an unstable mineral, breaking down quickly under the influence of weathering (see also Chapter 36 *Mudrocks, clays and pyrite implications*). The primary oxidation products of pyrite are ferrous and ferric sulfates and sulfuric acid. A chemical attack by sulfate on concrete can lead to its deterioration. The oxidation of pyrite within spoil heap waste is governed by the access of air, which, in turn, depends upon the particle size distribution, water saturation and the degree of compaction. However, any highly acidic oxidation products that may form will be neutralised by alkaline materials in the waste material. Charles and Watts (2001) noted that pyrite in some shales has led to expansion through slow oxidation.

The moisture content of spoil increases with increasing content of fines. It also is influenced by the permeability of the material, the topography and climatic conditions. Generally, it falls within the range 5–15% (**Table 34.3**).

The range of specific gravity values depends on the relative proportions of coal, shale, mudstone and sandstone in the waste and tends to vary between 1.7 and 2.7. The proportion of coal is of particular importance: the higher the coal content, the lower the specific gravity. The bulk density of the material in spoil heaps shows a wide variation, with most material falling within the range 1.5–2.5 Mg/m^3. Low densities are mainly a function of low specific gravity. The bulk density tends to increase with increasing clay content.

The argillaceous content influences the grading of the spoil, although most spoil material is essentially granular

	Yorkshire Main	Brancepath	Wharncliffe
Moisture content: %	8.0–13.6	5.3–11.9	6–13
Bulk density: Mg/m³	1.67–2.19	1.27–1.88	1.58–2.21
Dry density: Mg/m³	1.51–1.94	1.06–1.68	1.39–1.91
Specific gravity	2.04–2.63	1.81–2.54	2.16–2.61
Plastic limit: %	16–25	Non-plastic to 35	14–21
Liquid limit: %	23–44	23–42	25–46
Permeability: m/s	$(1.42–9.78) \times 10^{-6}$	?	?
Size <0.002 mm: %	0–17	Most material of sand-sized range	2–20
Size >2.0 mm: %	30–57		38–67
Friction angle, ϕ	31.5–35.0°	27.5–39.5°	29–37°
Cohesion intercept, c: kPa	19–21	4–39	16–40

Table 34.3 Examples of soil properties of coarse discard (Data taken from Bell, 1996)

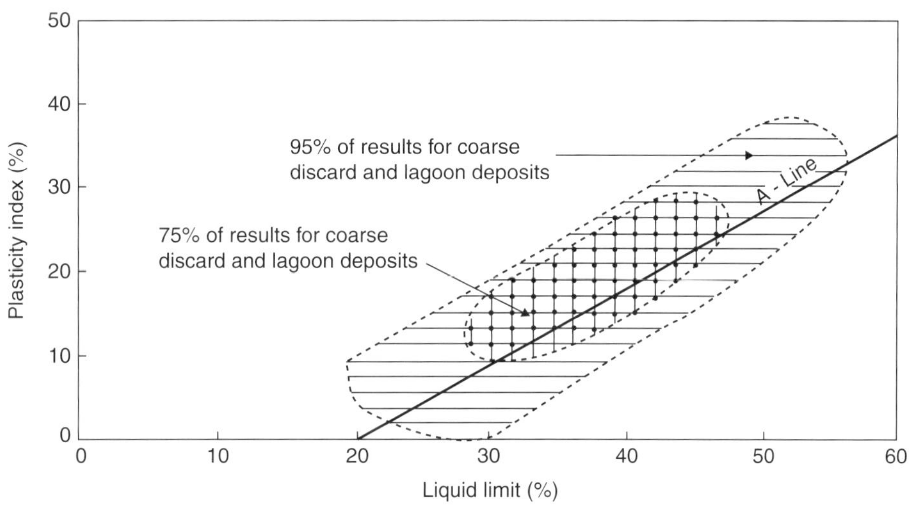

Figure 34.7 General range of plasticity characteristics of the coarse discard found in spoil heaps in the UK (Reproduced from Thomson, 1973; National Coal Board)

(**Table 34.3**). In fact, as far as the particle size distribution of a coarse discard is concerned there is a wide variation; often most material may fall within the sand range, but significant proportions in the gravel and cobble ranges may also be present. Indeed, at placement a coarse discard very often consists mainly of gravel to cobble-sized particles, but subsequent breakdown on weathering reduces the particle size. Once buried within a spoil heap, particles of a coarse discard undergo little further reduction in size. Hence, older and surface samples of spoil contain a higher proportion of fines than those obtained from depth.

The liquid and plastic limits can provide a rough guide to the engineering characteristics of a soil. However, for a coarse discard they are only representative of that fraction passing the 425 μm BS sieve, which frequently is less than 40% of the sample concerned. Nevertheless, the results of consistency tests indicate a low to medium plasticity, while in certain instances spoil has proved to be virtually non-plastic (**Figure 34.7**). Plasticity increases with increasing clay content.

34.4.3.2 Weathering

The most significant change in the character of a coarse colliery discard due to weathering is the reduction of particle size. The extent to which breakdown occurs depends upon the type of parent material involved and the effects of air, water and handling between mining and placing on the spoil heap. After a few months of weathering the debris from sandstones and siltstones usually consists of larger than cobble-sized particles. After that, the degradation to component grains takes place at a slow rate. However, mudstones, shales and seatearth exhibit rapid disintegration to gravel-sized particles. Although a coarse discard may reach its final level of degradation within a matter of months, with the degradation of many mudstones and shales taking place within days, once it is buried within a spoil heap it suffers little change. When spoil material is burned it becomes much more stable with respect to weathering.

Taylor and Smith (1986) stated that mudrock breakdown resulted from either mechanical (air breakage) or physico-chemical mechanisms. Seedsman (1993) divided physico-chemical swelling into crystalline and osmotic swelling. The former involves an increase in the volume of the clay mineral by adsorption of water at its surface. The latter involves the addition of water into the interlayer space of the clay mineral in response to chemical conditions and a consequent increase in volume. Taylor (1988) maintained that the disintegration of British Coal Measures mudrocks (which form a significant part of colliery waste) takes place as a consequence of air breakage after a sufficient number of cycles of wetting and drying.

In other words, if mudrocks undergo desiccation, then air is drawn into the outer pores and capillaries as high suction pressure develops. Then, on saturation, the entrapped air is pressurised as water is drawn into the rock by capillarity. Such slaking, therefore, causes the fabric of the rock to be stressed. The size of the pores is more important than the volume of the pores for the development of capillary pressure. Capillary pressure is proportional to surface tension. Previously, however, Badger *et al.* (1956) had contended that air breakage occurred only in Coal Measures shales that were mechanically weak, while the presence of dispersed colloidal material appeared to be a general cause of disintegration. They also found that the variation in the disintegration of different shales in water was usually not connected with the total amount of clay colloid or the variation in the types of clay mineral present. Rather, it was controlled by the type of exchangeable cations attached to the clay particles and the accessibility of the latter to attack by water, which, in turn, depended on the porosity of the shale. Air breakage could assist this process by presenting new surfaces of shale to water. Subsequently, Dick and Shakoor (1992) maintained that the influence of clay minerals on slaking diminishes in those mudrocks that contain less than 50% of clay-sized particles and that micro-fractures are the dominant lithological characteristic controlling durability. In such cases, slaking is initiated along micro-fractures. Moon and Beattie (1995), in a study of Coal Measures mudrocks from Waikato, New Zealand, found that the proportion of clay mineral present in the mudrock had the greatest influence on durability. Further details of mudrocks are given in Chapter 36 *Mudrocks, clays and pyrite implications*.

Intraparticle swelling was considered by Taylor and Spears (1970) to have a major effect on the durability of mudrocks from Coal Measures in Britain that contained significant amounts of expandable mixed layer clay minerals. Moreover, they suggested that those mudrocks that exhibited the greater breakdown were found to contain a higher exchangeable sodium ion content in the mixed layer clay. In fact, the presence of expandable mixed layer clay is thought to be one of the factors promoting the breakdown of argillaceous rocks, particularly if Na^+ is a prominent interlayer cation. Present evidence nevertheless suggests that in England colliery spoils with high exchangeable Na^+ levels are uncommon. Sodium may be replaced by Ca^{2+} and Mg^{2+} cations originating from sulfates and carbonates in the waste, but dilution by more inert rock types is possibly the most usual reason for the generally low levels of sodium. Moreover, after going through the preparation process and after exposure as a new surface layer of a spoil heap prior to burial, the disintegration of the mudrock and shale material will have been largely achieved. However, it appears that the low shear strength of certain fine-grained English spoils is associated with comparatively high exchangeable Na^+ levels.

34.4.3.3 Spontaneous combustion

Spontaneous combustion of carbonaceous material, frequently aggravated by the oxidation of pyrite, is the most common cause of burning spoil. It can be regarded as an atmospheric oxidation (exothermic) process in which self-heating occurs. Coal and carbonaceous materials may be oxidised in the presence of air at ordinary temperatures, below their ignition point. Generally, lower-rank coals (that is, coals with a lower carbon content) are more reactive and accordingly more susceptible to self-heating than coals of higher rank. Spoil heaps that have been subject to spontaneous combustion are made up of a mixture of burned, partially burned and unburned material. Therefore, the characteristics of the spoil are more variable.

The oxidation of pyrite at ambient temperature in moist air leads to the formation of ferric and ferrous sulfate, and sulfuric acid (see Chapter 37 *Sulfates and sulfides in soils and rocks*). This reaction also is exothermic. When present in sufficient amounts, and especially when finely divided, pyrite associated with 'coaly' material increases the likelihood of spontaneous combustion (Michalski *et al.*, 1990). When heated, the oxidation of pyrite and organic sulfur in the coal generates sulfur dioxide. If there is not enough air for complete oxidation, then hydrogen sulfide is formed, which is a highly toxic gas.

The moisture content and grading of spoil are also important factors in spontaneous combustion. At relatively low temperatures an increase in free moisture increases the rate of spontaneous heating. Oxidation generally takes place very slowly at ambient temperatures, but as the temperature rises oxidation increases rapidly. In material of large size the movement of air can cause heat to be dissipated, while in fine material the air remains stagnant, which means that burning ceases when the oxygen is consumed. Accordingly, ideal conditions for spontaneous combustion exist when the grading is intermediate between these two extremes and hotspots may develop under such conditions. These hotspots may have temperatures around 600°C or occasionally up to 900°C (Bell, 1996). Furthermore, the rate of oxidation generally increases as the specific surface of the particles increases.

Spontaneous combustion may give rise to subsurface cavities in spoil heaps, the roofs of which may be incapable of supporting a person. Burned ashes may also cover zones that are red hot to appreciable depths. Thomson (1973) recommended that during restoration of a spoil heap a probe or crane with a drop-weight could be used to prove areas of doubtful safety that are burning or underlain by cavities. Badly fissured areas should be avoided and workers should wear lifelines if they walk over areas not proved safe. Any area that is suspected of having cavities should be excavated by a dragline or drag scraper rather than allowing plant to move over suspect ground. An example of an exposed cavity in an Indian spoil tip is shown in **Figure 34.8**.

Figure 34.8 Cavity exposed in a burning coal spoil tip in India (© Laurance Donnelly, with permission)

When steam comes in contact with red-hot carbonaceous material, water gas (carbon monoxide and hydrogen) is formed, and when the latter is mixed with air, over a wide range of concentrations, it is potentially explosive. If a cloud of coal dust forms near burning spoil when reworking a heap, then it can ignite and explode. Damping with a spray may prove useful in the latter case.

Noxious gases are emitted from burning spoil. These include carbon monoxide, carbon dioxide, sulfur dioxide and, less frequently, hydrogen sulfide. Each may be dangerous if inhaled at sufficiently high concentrations, which may be present at fires on spoil heaps (**Table 34.4**). The rate of production of these gases may be accelerated by disturbing burning spoil through excavating or reshaping. Carbon monoxide is the most dangerous gas since it cannot be detected by taste, smell or irritation, and may be present in potentially lethal concentrations. By contrast, sulfur gases are readily detectable and are usually not present in high concentrations. Even so, when diluted they may still cause distress to someone with a respiratory ailment. Nonetheless, sulfur gases are mainly a nuisance rather than a threat to life. In certain situations a gas-monitoring

Gas	Concentration by volume in air: ppm	Effect
Carbon monoxide	100	Threshold limit value (TLV) under which it is believed nearly all workers may be repeatedly exposed day after day without adverse effect
	200	Headache after about 7 hours if resting or after 2 hours if working
	400	Headache and discomfort, with possibility of collapse, after 2 hours at rest or 45 minutes' exertion
	1 200	Palpitations after 30 minutes at rest or 10 minutes' exertion
	2 000	Unconsciousness after 30 minutes at rest or 10 minutes' exertion
Carbon dioxide	5 000	TLV. Lung ventilation slightly increased
	50 000	Breathing is laboured
	90 000	Depression of breathing begins
Hydrogen sulfide	10	TLV
	100	Irritation to eyes and throat: headache
	200	Maximum concentration tolerable for 1 hour
	1 000	Immediate unconsciousness
Sulfur dioxide	1–5	Can be detected by taste at the lower level and by smell at the upper level
	5	TLV. Onset of irritation to the nose and throat
	20	Irritation to the eyes
	400	Immediately dangerous to life

Notes
1. Some gases have a synergistic effect – that is, they augment the effects of others and cause a lowering of the concentration at which the symptoms shown in the table occur. Further, a gas which is not itself toxic may increase the toxicity of a toxic gas, for example by increasing the rate of respiration; strenuous work will have a similar effect.
2. Of the gases listed, carbon monoxide is the only one likely to prove a danger to life, as it is the commonest. The others become intolerably unpleasant at concentrations far below the danger level.

Table 34.4 Effects of noxious gases (Reproduced from Thomson, 1973; National Coal Board)

(a)

(b)

Figure 34.9 (a) Burning spoil in an old coal tip, Wharncliffe, near Barnsley; (b) burning coal spoil tip, India (Courtesy of International Mining Consultants Limited)

programme may be required. Where danger areas are identified, personnel should wear breathing apparatus.

Burning spoil material may represent a notable problem when reclaiming old tips (**Figure 34.9**). Spontaneous combustion of coal in colliery spoil can be averted if the coal occurs in an oxygen-deficient atmosphere that is humid enough with excess moisture to dissipate any heating that develops. Cook (1990), for example, described shrouding a burning spoil heap with a cover of compacted discard to smother the existing burning material and prevent further spontaneous combustion. Thomson (1973) also recommended blanketing and compaction, as well as digging out, trenching, injection with non-combustible material and water, and water spraying as methods by which spontaneous combustion in spoil material may be controlled. Bell (1996) described the injection of a curtain wall of pulverised fuel ash (PFA), extending to the original ground level, as a means of dealing with hotspots.

34.4.3.4 Restoration

The configuration of a spoil heap depends upon the type of equipment used in its construction and the sequence of tipping the waste. The shape, aspect and height of a spoil heap affect the intensity of exposure, the amount of surface erosion that occurs, the moisture content in its surface layers and its stability. Although the mineralogical composition of coarse discards from different mines varies, pyrite frequently occurs in shales and coaly material is present in spoil heaps. When pyrite weathers it promotes acidic conditions in the weathered material. Such conditions do not aid the growth of vegetation. Indeed, some spoils may contain elements that are toxic to plant life. Furthermore, the uppermost slopes of a spoil heap are frequently devoid of near-surface moisture. To support vegetation a spoil heap should have a stable surface in which roots can become established, and must be non-toxic and contain an adequate and available supply of nutrients. Hence, spoil heaps are often barren of vegetation.

The restoration of a spoil heap will require large-scale earth-moving. The restored land may be used for agriculture, for public amenities such as golf courses or for buildings. Since restoration invariably involves spreading the waste over a larger area, this may mean that additional land beyond the site boundary has to be purchased. Where a spoil heap is very close to a disused colliery, spoil may be spread over the latter area. This involves the burial or removal of derelict colliery buildings and sometimes old mineshafts or shallow old workings may have to be treated (Johnson and James, 1990). Watercourses may have to be diverted, as may services, notably roads.

The landscaping of spoil heaps is frequently carried out to allow them to be used for agriculture or forestry. This type of restoration is generally less critical than when structures are to be erected on the site, since bearing capacities are not as important and steeper surface gradients are acceptable. Most spoil heaps offer no special handling problems other than the cost of regrading and possibly the provision of adjacent land, so that the gradients on the existing site can be reduced by the transfer of spoil to the adjacent land. However, as noted, some spoil heaps present problems due to spontaneous combustion.

The surface treatments of spoil heaps vary according to the chemical and physical nature of the spoil, and the climatic conditions. The preparation of the surface of a spoil heap also depends upon the use to which it is to be put subsequently. Where it is intended to sow grass or plant trees, the surface layer should not be compacted. Drainage plays an important part in the restoration of a spoil heap and during landscaping there should be erosion control. If the spoil is acidic, then it can be neutralised by liming. The chemical composition influences the choice of fertiliser used and, in some instances, spoil can be seeded without the addition of topsoil, if suitably fertilised. More commonly topsoil is added prior to seeding or planting.

34.4.3.5 Acid mine drainage

Acid mine drainage (AMD) or acid rock drainage (ARD) may be associated with a coarse colliery discard since the coaly material and mudrock are likely to contain sulfide minerals, notably pyrite (Geldenhuis and Bell, 1998). Such drainage results from the natural oxidation of sulfide minerals in the waste when it is exposed to air and water. It is a consequence of the oxidation of sulfur in the mineral to a higher state of oxidation and, if aqueous iron is present and unstable, the precipitation of ferric iron with hydroxide occurs. AMD can also be associated with opencast workings as well as underground workings, tailings ponds or mineral stockpiles (Brodie *et al.*, 1989). If AMD is not controlled it can pose a threat to the environment since acid generation can lead to elevated levels of heavy metals and sulfate in surface water or groundwater, which will obviously have a detrimental effect on water quality (see Coal Authority, 2015). Acid generation tends to occur in the surface layers of spoil heaps where air and water come into contact with sulfide minerals. Methods for remediating AMD have been reviewed by Johnson and Hallberg (2005). Broadly, there are three main types. Active methods require the pumping of the mine water and treatment with chemical reagents such as hydrogen peroxide or caustic soda. Passive methods use natural systems such as reed beds to improve water quality and do not use pumping; the water from the mine is allowed to flow under gravity. Hybrid active and passive schemes may utilise pumping and some chemical treatment at the start of the process before water flows into reed bed systems (Coal Authority, 2017). Around 50 abandoned coal mines have been treated in Britain, mostly using passive systems. For the Wheal Jane tin mine in Cornwall, Whitehead and Prior (2005) described a passive treatment system developed as part of a research project. Three separate systems were installed, each with aerobic reed beds, anaerobic cells and rock filters. The three systems differed in the method used to raise the pH of the mine water entering them. The Coal Authority website briefly describes the treatment approaches used at a number of former coal mining sites.

34.4.4 Other types of fill

34.4.4.1 Pulverised fuel ash

Pulverised fuel ash (PFA) is waste that is produced by coal-fired power stations and is defined as the solid material extracted by electrostatic or mechanical means from the furnace flue gases (Winter and Clarke, 2002). Charles and Watts (2001) pointed out that most of the PFA is metal oxides along with some trace constituents and pozzolanic materials. About 80% of the ash from a coal-fired power station is PFA and the remaining 20% is furnace-bottom ash. After extraction, the PFA is referred to as fresh PFA and can be conditioned with an optimum amount of water to produce conditioned ash. Conditioning PFA allows it to be transported more easily and reduces the problem of dust. When PFA is stored as conditioned ash it is called stockpiled ash.

When PFA is disposed of as waste it is usually mixed with water so that it can be transported as slurry to a lagoon, for example. It can be disposed of in this manner in old clay pits. This ash can be subsequently retrieved as stockpiled ash. The water in the lagoon can be decanted when the ash has settled out of the slurry. Developments on PFA lagoons are rare, but some are mentioned by Charles and Watts (2001). One is the Metro Centre at Gateshead and another is housing in Peterborough. They also referred to an oil tank being constructed on a 5 m thickness of lagoon PFA in Denmark.

According to Yang *et al.* (1993), the properties of fresh PFA are governed by the type and chemical composition of the coal used, the source of the coal, the degree of pulverisation, the efficiency of combustion, the variation in firing conditions and the extraction process. Consequently, the character of PFA does vary. In fact, it is not only variable between sources but also on a day-to-day basis from a single source. Nonetheless, the particles of PFA are primarily of coarse silt to fine-sand size and are more or less spherical in shape. Their specific gravity ranges from around 1.90 to 2.72, depending on their source, and they are non-plastic. Ash may exhibit 'cohesive' properties, although Yang *et al.* (1993) noted that such properties might be attributable to the suction forces that develop during compaction of the ash. Hence, they may dissipate with time or when inundated, so leading to loss of strength. Some ashes are self-hardening because of the presence of cementitious material such as free lime or calcium sulfate compounds. In addition to Yang *et al.* (1993), reviews of the engineering properties of PFA can be found in Sutherland *et al.* (1968), Gray and Kin (1972) and Leonards and Bailey (1982). PFA has a number of other civil engineering applications – for example, as a cement replacement or additive.

Most PFA is used for land reclamation projects or for general and structural fills (e.g. embankments, foundation fills or fills behind retaining walls). PFA has frequently been used for the construction of road embankments. For instance, Ekins *et al.* (1993) mentioned the use of lightweight PFA in the construction of the Western Approach Road, Southampton. PFA is also used as fill behind the abutments of road bridges. The use of PFA as a structural fill for the foundations of a precipitator at a power-generating station in Indianapolis, Indiana, was described by Leonards and Bailey (1982). In addition, a notable quantity of PFA is used in the manufacture of cement as some types of PFA are pozzolanic. The pozzolanic activity is attributable to the reaction of the fine glass (mullite) particles with free lime in the presence of water.

Conditioned PFA can be used as high-quality selected fill, either as fill to structures and reinforced earth or for stabilisation with cement to form capping. Enough water must be added to the PFA so that it is suitable for compaction. Winter and Clarke (2002) indicated that the optimum moisture content of conditioned ash is around 25%, but that it may be as high as 35% or more. They further noted that the optimum moisture content of some conditioned ash from Scotland may be as low as 18%, whereas that of lagoon ash may be as high as 28%. They quoted maximum dry densities varying from 1.0 to 1.65 Mg/m^3.

34.4.4.2 Industrial waste

There are many types of industrial waste, such as chemical wastes and wastes from ore smelting. Not only may industrial fills be very heterogeneous and poorly compacted, but they may be chemically problematic in that the ground may be aggressive. Industrial wastes can take a number of forms, such as smelter slag, various types of residues and sludge. Furthermore, industrial wastes are commonly associated with derelict sites. Such sites represent a wasted resource and are normally located within urban areas so that their rehabilitation is highly desirable. Mabey (1991) reviewed the position of derelict land in Britain, noting that in 1988 there were some 40 500 ha of derelict land. Of the land attributable to industrial dereliction, 94% was considered to need remediation. More recently there has been great emphasis on the reclamation of derelict land, but national statistics show that there may be nearly 80 000 ha in this condition in Britain. According to the National Land Use Database, the amount of derelict land in England has remained fairly constant in recent years, falling from about 66 000 ha in 2002 to nearly 62 000 ha in 2009. In Scotland there were just over 8000 ha of derelict land in 2010 (Anon, 2011).

Unfortunately, many industrial wastes have been contaminated to a greater or lesser extent (Bell *et al.*, 2000). Contaminated land may also emit gases or may be a fire hazard. Details of such hazards should be determined during a site investigation (BSI, 2011). Site hazards result in constraints on the freedom of action, necessitate stringent safety requirements, may involve time-consuming and costly working procedures and affect the type of development. For instance, Leach and Goodyear (1991) noted that constraints might mean that the development plan has to be changed so that the more sensitive land uses are located in areas of reduced hazard. Alternatively, where notable hazards exist, then a change to a less sensitive end use may be advisable. There are a number of general references that provide further information on how to deal with and remediate hazards from contaminated land (e.g. Bardos *et al.*, 2007; Nathanail and Bardos, 2004). Bunce and Braithwaite (2001) described the remediation of an 80 ha site at Pride Park, Derby. The land had been used for gravel extraction, waste disposal, railway works and a gas and coke plant. The site was heavily contaminated and for the worst areas a bentonite containment wall was constructed to prevent contaminant migration. Water collecting within the wall was pumped out using 18 abstraction wells, treated and finally discharged into the River Derwent. About 75% of the site was usable as a result of the remedial work.

Some industrial wastes, notably those that are non-contaminated, can be used as bulk fill for land reclamation schemes. Indeed, derelict sites often require varying amounts of filling, levelling and regrading. Fill should be obtained, if possible, from the site. Furthermore, some industrial wastes can be used as raw materials; for example, blast furnace slag can be used for road aggregate.

34.4.4.3 Building waste

Building waste is associated with demolition work and mainly includes concrete and brick rubble. In addition, timber, steel reinforcement, glass, plaster, stone, slate and tiles are likely to be included in building waste to a greater or lesser extent so that it is essentially heterogeneous in composition. Some of these materials are unsuitable for use as fill. For example, Reid and Buchanan (1987) mentioned that the derelict Queen's Dock in Glasgow had been filled with 1.3 million m^3 of demolition rubble from old buildings in the vicinity, degradable material such as timber having been removed. Similarly, Hartley (1991) referred to the use of old foundation material that was crushed and used as fill at a site in Middlesbrough. Building wastes are usually poorly compacted and may contain gypsum products. The latter could be a source of sulfate, which could attack concrete. Again building wastes, like industrial waste, may be contaminated and therefore appropriate assessment will be needed (see discussion above). However, it is increasingly the case that building waste following demolition is separated, crushed and graded (where appropriate) and recycled.

34.5. Conclusions

The reclamation of derelict land is an important aspect of government policy. More recently there has been an emphasis on siting new building developments on brownfield rather than greenfield sites and many of these brownfield sites are covered in substantial depths of fill. Thus, the subject of building on fill has acquired considerable prominence in recent years. The geotechnical problems in achieving safe and economic developments on filled ground are substantial.

Therefore, it is important to understand both the distribution of different types of fill and their likely behaviour. It has been shown that the engineering behaviour of a fill is strongly influenced by its composition, method of deposition and subsequent stress history. These factors control the key aspects in successfully reusing areas underlain by non-engineered fill. It is important, therefore, to assess the variability of a fill (in terms of composition and physical, mechanical and chemical properties), and to assess how the behaviour of a fill may change over time. In particular, a fill may be subject to significant volume change on loading. This can be caused by a number of factors, including the original loading as the fill was deposited (often over wide areas), changes in water content influenced by variable permeability, swelling and shrinkage of clay fills, collapse compression, biodegradation and chemical reactions.

Past (largely) industrial land use indicates the likely nature of non-engineered fills encountered. The principal types of fill include filled opencast sites (widely distributed in coalfield areas), domestic waste and sanitary landfill (widely distributed in former quarries and other excavations in urban, peri-urban and rural areas), colliery discard (found close to former coal mines), PFA (both as a waste and used for land reclamation), industrial wastes (which can be both physically and chemically variable) and building wastes (often found in central urban areas).

Chapter 54 *Single piles* describes the requirements for building on fills in further detail, with reference to the investigation and prediction of behaviour.

Acknowledgements

This chapter is published with the permission of the Executive Director of the British Geological Survey (NERC).

References

Anon (2011) Scottish vacant and derelict land survey 2010. *Statistical Bulletin, Planning Series, PLG/2011/1.* Scottish Government, Edinburgh, UK.

Badger CW, Cummings CD and Whitmore RI (1956) The disintegration of shale. *Journal of the Institute of Fuel* **29**: 417–423.

Bardos P, Nathanail J and Nathanail P (2007) *Contaminated Land Management Ready Reference.* EPP Publications/Land Quality Press, London, UK.

Bell FG (1996) Dereliction: colliery spoil heaps and their rehabilitation. *Environmental and Engineering Geoscience* **2**: 85–96.

Bell FG, Genske DD, Hytiris N and Lindsay P (2000) A survey of contaminated ground with illustrative case histories. *Land Degradation and Development* **11**: 419–437.

Blanchfield R and Anderson WF (2000) Wetting collapse in opencast coal mine backfill. *Proceedings of the Institution of Civil Engineers, Geotechnical Engineering* **143**: 139–149.

Brodie MJ, Broughton IM and Robertson A (1989) A conceptual rock classification system for waste management and a laboratory method for ARD prediction from rock piles. In *Draft Technical Guide (British Columbia Acid Mine Drainage Task Force)*, vol. 1. British Columbia Acid Mine Drainage Task Force, Vancouver, BC, Canada, pp. 130–135.

BSI (British Standards Institution) (2011) BS 10175:2011: Investigation of potentially contaminated sites: code of practice. BSI, London, UK.

Bunce D and Braithwaite P (2001) Reclamation of contaminated land with specific reference to Pride Park, Derby. *Proceedings of the Symposium on Problematic Soils, Nottingham, UK* (Jefferson I, Murray EJ, Faragher E and Fleming PR (eds)). Thomas Telford, London, UK, pp. 121–127.

Charles JA (1993) Engineered fills: space, time and water. *Proceedings of a Conference on Engineered Fills, Newcastle Upon Tyne, UK* (Clarke BG, Jones CJFP and Moffat AIB (eds)). Thomas Telford, London, UK, pp. 42–65.

Charles JA (2008) The engineering behaviour of fill materials: the use, misuse and disuse of case histories. *Géotechnique* **58**: 541–570.

Charles JA and Skinner HD (2001) Compressibility of foundation fills. *Proceedings of the Institution of Civil Engineers, Geotechnical Engineering* **149**: 145–157.

Charles JA and Watts KS (2001) *Building on Fill: Geotechnical Aspects*, 2nd edn. BRE Press, Watford, UK.

Charles JA, Burford D and Hughes DB (1993) Settlement of opencast backfill at Horsley 1973–1992. *Proceedings of a Conference on Engineered Fills, Newcastle Upon Tyne, UK* (Clarke BG, Jones CJFP and Moffat AIB (eds)). Thomas Telford, London, UK, pp. 429–440.

Coal Authority (2015) Appendix 3: treating mine water. http://coal.decc.gov.uk/en/coal/cms/environment/about_m_water/about_m_water.aspx (accessed 17/05/2023).

Coal Authority (2017) Coal mine water treatment. https://www.gov.uk/government/collections/coal-mine-water-treatment (accessed 21/05/2023).

Cook BJ (1990) Coal discard-rehabilitation of a burning heap. In *Reclamation, Treatment and Utilization of Coal Mining Wastes* (Rainbow AKM (ed)). Balkema, Rotterdam, the Netherlands, pp. 223–230.

Council of the European Union (1999) Directive 1999/31/EC on the landfilling of wastes. *Official Journal of the European Communities* L182.

Dick JC and Shakoor A (1992) Lithological controls of mudrock durability. *Quarterly Journal of Engineering Geology* **25**: 31–46.

Dixon N, Murray EJ and Jones DRV (eds) (1998) *Geotechnical Engineering of Landfill: Proceedings of a Symposium.* Thomas Telford, London, UK.

Ekins JDK, Cater R and Hounsham AD (1993) Highway embankments: their design, construction and performance. *Proceedings of a Conference on Engineered Fills, Newcastle Upon Tyne, UK* (Clarke BG, Jones CJFP and Moffat AIB (eds)). Thomas Telford, London, UK, pp. 1–17.

Ford J, Kessler H, Cooper AH, Price SJ and Humpage AJ (2006) *An Enhanced Classification for Artificial Ground.* British Geological Survey Internal Report IR/04/038. British Geological Survey, Nottingham, UK.

Geldenhuis S and Bell FG (1998) Acid mine drainage at a coal mine in the eastern Transvaal, South Africa. *Environmental Geology* **34**: 234–242.

Geological Society Engineering Group Working Party (1972). The preparation of maps and plans in terms of engineering geology. *Quarterly Journal of Engineering Geology* **5**: 293–382.

Gray DH and Kin YK (1972) Engineering properties of compacted fly ash. *Proceedings of the American Society of Civil Engineers, Journal of the Soil Mechanics and Foundation Engineering Division* **98**: 361–380.

Hartley D (1991) The use of derelict land: thinking the unthinkable. In *Land Reclamation, an End to Dereliction* (Davies MCR (ed)). Elsevier, London, UK, pp. 65–74.

IAEG Commission (1981) Rock and soil description and classification for engineering geological mapping. *Bulletin of the International Association of Engineering Geology* **24**: 235–274.

Johnson AC and James EJ (1990) Granville colliery land reclamation/coal recovery scheme. In *Reclamation and Treatment of Coal Mining Wastes* (Rainbow AKM (ed)). Balkema, Rotterdam, the Netherlands, pp. 193–202.

Johnson DB and Hallberg KB (2005) Acid mine drainage remediation options: a review. *Science of the Total Environment* **338**: 1–14.

Kilkenny WM (1968) *A Study of the Settlement of Restored Opencast Coal Sites and Their Suitability for Development.* Department of Civil Engineering, University of Newcastle upon Tyne, Newcastle upon Tyne, UK.

Knill JL, Cratchley CR, Early KR, *et al.* (1970) The logging of rock cores for engineering purposes: report by the Geological Society Engineering Group Working Party. *Quarterly Journal of Engineering Geology* **3**: 1–24.

Kuras O, Ogilvy RD, Pritchard J, *et al.* (2006) Monitoring leachate levels in landfill sites using automated time-lapse electrical resistivity tomography (ALERT). *Proceedings of the 12th Annual Meeting of the European Association of Geologists and Engineers on Near Surface Geophysics, Helsinki, Finland.*

Leach BA and Goodyear HK (1991) *Building on Derelict Land*. CIRIA SP78. CIRIA, London, UK.

Leonards GA and Bailey B (1982) Pulverized coal ash as structural fill. *Proceedings of the American Society of Civil Engineers, Journal of the Geotechnical Engineering Division* **108**: 517–531.

Ling HI, Leshchinsky D, Mohri Y and Kawabata T (1998) Estimation of municipal solid waste settlement. *Proceedings of the American Society of Civil Engineers, Journal of the Geotechnical and Geoenvironmental Engineering Division* **124**: 21–28.

Mabey R (1991) Derelict land: recent developments and current issues. In *Land Reclamation, an End to Dereliction* (Davies MCR (ed)). Elsevier, London, UK, pp. 3–39.

McDougall J, Pyrah I, Yuen STS, *et al.* (2004) Decomposition and settlement in land-fill and other soil-like materials. *Géotechnique* **54**: 605–610.

Meyerhof GG (1951) Building on fill with special reference to settlement of a large factory. *Structural Engineer* **29**: 46–57.

Michalski SR, Winschel IJ and Gray RE (1990) Fires in abandoned mines. *Bulletin of the Association of Engineering Geologists* **27**: 479–495.

Moon VG and Beattie AG (1995) Textural and microstructural influences on the durability of Waikato Coal Measures mudrocks. *Quarterly Journal of Engineering Geology* **28**: 303–312.

Nastev M (1998) *Modeling Landfill Gas Generation and Migration in Sanitary Landfills and Geological Formations*. Unpublished PhD Thesis, Laval University, Quebec City, QC, Canada.

Nathanail CP and Bardos RP (2004) *Reclamation of Contaminated Land*. Wiley, New York, NY.

Norbury D (2010) *Soil and Rock Description in Engineering Practice*. Whittles Publishing, Caithness, UK.

Occupational Health and Safety Information Service (1983) *Fill. Part 2: Site Investigation, Ground Improvement and Foundation Design*. Digest 275. Building Research Establishment, Watford, UK.

Office of National Statistics (2014) UK statistics on waste. https://www.gov.uk/government/statistics/uk-waste-data (accessed 21/05/2023).

Parliamentary Office of Science and Technology (2005) Recycling household waste. *Post note 252*. POST, London, UK.

Penman ADM and Godwin EW (1975) Settlement of experimental houses on land left by opencast mining at Corby. In *Settlement of Structures (British Geotechnical Society)*. Pentech Press, London, UK, pp. 53–61.

Powrie W, Richards DJ and Beaven RP (1998) Compression of waste and its implications for practice. In *Geotechnical Engineering of Landfill* (Dixon N, Murray EJ and Jones DRV (eds)). Thomas Telford, London, UK, pp. 3–18.

Price SJ, Ford JR, Cooper AH and Neal C (2011) Humans as major geological and geomorphological agents in the Anthropocene: the significance of artificial ground. *Philosophical Transactions of the Royal Society A* **369**: 1056–1084.

Proske H, Vlcko J, Rosenbaum MS, Culshaw M and Marker B (2005) Special purpose mapping for waste disposal sites. Report of IAEG Commission 1: Engineering Geological Maps. *Bulletin of Engineering Geology and the Environment* **64**: 1–54.

Raybould JG and Anderson JG (1987) Migration of landfill gas and its control by grouting: a case history. *Quarterly Journal of Engineering Geology* **20**: 78–83.

Reid WM and Buchanan NW (1987) The Scottish Exhibition Centre. *Proceedings of a Conference on Building on Marginal and Derelict Land*, Glasgow, UK. Thomas Telford, London, UK, pp. 435–448.

Roche D (1996) Landfill failure survey: a technical note. In *Engineering Geology of Waste Disposal* (Bentley SP (ed)). Geological Society Engineering, London, UK, pp. 379–380.

Rosenbaum MS, McMillan AA, Powell JH, *et al.* (2003) Classification of artificial (man-made) ground. *Engineering Geology* **69**: 399–409.

Seedsman RW (1993) Characterizing clay shales. In *Comprehensive Rock Engineering*, vol. 3 (Hudson JA (ed)). Pergamon Press, Oxford, UK, pp. 151–165.

Sowers GF (1973) Settlement of waste disposal fills. In *Proceedings of the Eighth International Conference on Soil Mechanics and Foundation Engineering, Moscow, Russia*, vol. 2, pp. 207–212.

Sowers GF, Williams RC and Wallace TS (1965) Compressibility of broken rock and settlement of rock fills. In *Proceedings of the Sixth International Conference on Soil Mechanics and Foundation Engineering, Montreal, Qc*. University of Toronto Press, Toronto, ON, Canada, vol. 2, pp. 561–565.

Sutherland HB, Finlay TW and Cram IA (1968) Engineering and related properties of pulverized fuel ash. *Journal of the Institution of Highway Engineers* **15**: 1–9.

Taylor RK (1975) English and Welsh colliery spoil heaps: mineralogical and mechanical relationships. *Engineering Geology* **7**: 39–52.

Taylor RK (1988) Coal Measures mudrock composition and weathering processes. *Quarterly Journal of Engineering Geology* **21**: 85–99.

Taylor RK and Smith TJ (1986) The engineering geology of clay minerals: swelling, shrinking and mudrock breakdown. *Clay Minerals* **21**: 235–260.

Taylor RK and Spears DA (1970) The breakdown of British Coal Measures rocks. *International Journal of Rock Mechanics and Mining Science* **7**: 481–501.

Thomson GM (1973) *Spoil Heaps and Lagoons: Technical Handbook*. National Coal Board, London, UK.

Watts KS and Charles JA (1999) Settlement characteristics of landfill wastes. *Proceedings of the Institution of Civil Engineers, Geotechnical Engineering* **137**: 225–233.

Watts KS and Charles JA (2015) *Building on Fill: Geotechnical Aspects*, 3rd edn. IHS BRE Press, Watford, UK.

Whitehead P and Prior H (2005) Bioremediation of acid mines drainage: an introduction to the Wheal Jane Mine wetlands project. *Science of the Total Environment* **338**: 1–2, 15–21.

Williams GM and Aitkenhead N (1991) Lessons from Loscoe: the uncontrolled migration of landfill gas. *Quarterly Journal of Engineering Geology* **24**: 191–208.

Wilson S, Oliver S, Mallett H, Hutchings H and Card G (2007) *Assessing Risks Posed by Hazardous Ground Gases to Buildings (Revised)*. CIRIA Report C665. CIRIA, London, UK.

Winter MG and Clarke BG (2002) Improved use of pulverised fuel ash as general fill. *Proceedings of the Institution of Civil Engineers, Geotechnical Engineering* **155**: 133–141.

Yang Y, Clarke BG and Jones CJFP (1993) A classification of pulverized fuel ash as an engineered fill. *Proceedings of a Conference on Engineered Fills, Newcastle Upon Tyne, UK* (Clarke BG, Jones CJFP and Moffat AIB (eds)). Thomas Telford, London, UK, pp. 367–378.

Further reading

Bell FG and Donnelly LJ (2006) *Mining and its Impact on the Environment*. Taylor & Francis, Abingdon, UK.

Bentley SP (ed) (1996) *Engineering Geology of Waste Disposal*. Geological Society Engineering, London, UK.

Coventry S, Woolveridge C and Hillier S (1999) *The Reclaimed and Recycled Construction Materials Handbook. Part 21 – Pulverised Fuel Ash (PFA)*. CIRIA Report C513. CIRIA, London: UK.

Crawford JF and Smith PG (1985) *Landfill Technology*. Butterworths, London, UK.

Highways Agency (2007). Treatment of fill and capping materials using either lime or cement or both. In *Design Manual for Roads and Bridges, Vol. 4 (Geotechnics and Drainage)*. Highways Agency manual HA 74/07. The Stationery Office, Norwich, UK, Section 1 (Earthworks), Part 6.

Ingoldby HC and Parsons AW (1977) *The Classification of Chalk for Use as a Fill Material*. Report LR806. Transport Research Laboratory, Crowthorne, UK.

Kwan JCT, *et al.* (1997) *Ground Engineering Spoil: Good Management Practice*. CIRIA Report R179. CIRIA, London, UK.

Leach BA and Goodger HK (1991) *Building on Derelict Land*. CIRIA SP78. CIRIA, London, UK.

National Coal Board (1973) *Spoil Heaps and Lagoons*. National Coal Board, London, UK.

Norbury D (2010) *Soil and Rock Description in Engineering Practice*. Whittles Publishing, Caithness, UK.

Owies IS and Khera RP (1990) *Geotechnology of Waste Management*. Butterworths, London, UK.

Sarsby RW (ed) (1995) *Waste Disposal by Landfill*. Balkema, Rotterdam, the Netherlands.

Sherwood PT (1994) *A Review of the Use of Waste Materials and By-products in Road Construction*. Report 358. Transport Research Laboratory, Crowthorne, UK.

Syms P (2004) *Previously Developed Land: Industrial Activities and Contamination*, 2nd edn. Blackwell Publishing, Oxford, UK.

Winter M and Clarke BG (2001) *Specification of Pulverised Fuel Ash for Use as General Fill*. Report 519. Transport Research Laboratory, Crowthorne, UK.

Useful websites

BRE Group (formerly Building Research Establishment): http://www.bre.co.uk

British Geological Survey (BGS): http://www.bgs.ac.uk

Coal Authority: http://www.coal.decc.gov.uk

Construction Industry Research and Information Association (CIRIA): http://www.ciria.org

Environment Agency: http://www.environment-agency.gov.uk

National Land Use Database: http://www.homesandcommunities.co.uk/ourwork/national-land-use-database

Northern Ireland Environment Agency: http://www.ni-environment.gov.uk

Ordnance Survey: http://www.ordnancesurvey.co.uk

Scottish Environment Protection Agency: http://www.sepa.org.uk

Transport Research Laboratory (TRL): http://www.trl.co.uk

It is recommended this chapter is read in conjunction with

- Chapter 7 *Geotechnical risks and their context for the whole project*
- Chapter 58 *Building on fills*
- Chapter 69 *Earthworks design principles*
- Chapter 75 *Earthworks material specification, compaction and control*

All chapters in this book rely on the guidance in Sections 1 *Context* and 2 *Fundamental principles*. A sound knowledge of ground investigation is required for all geotechnical works, as set out in Section 4 *Site investigation.*

doi: 10.1680/icemge.66816.0499

Chapter 35
Organics/peat soils

Eric R Farrell Geotechnical Consultant, Dublin, Ireland

Peat and organic soils cover about 6% of the landmass of Great Britain and Ireland as a whole, with Ireland and Scotland having percentages of 17% and 10%; extensive deposits are to be found in many other countries worldwide, such as Canada, Russia, Finland and Indonesia, to mention a few. Their engineering behaviour depends on their morphology, which must be reflected in the description and classification of soil and in turn must be included in the interpretation of geotechnical parameters. Water content is the most important index test and can be used to estimate some of the engineering properties. Other important indices are the loss on ignition and the degree of decomposition as indicated by the von Post scale. Secondary compression dominates when estimating settlements of peat and it is generally regarded that this occurs during the primary consolidation (hydrodynamic) phase as well as after dissipation of the excess pore pressure. Sophisticated soil models have been developed which reflect this continuous secondary compression; however, at present these are not readily used for routine design. Surcharging of peat is an effective method of reducing the long-term creep movements under embankments. The permeability of peat decreases significantly with compression and this affects the consolidation behaviour in the field.

The laboratory-determined effective stress parameters of fibrous peats depend on the type of test and the orientation of the sample. The fibres also affect the in situ test methods used to determine the undrained shear strength, particularly the in situ vane which give c_u values which are dependent on the size of the vane used. Laboratory tests also recorded relatively high $c_u/\sigma_{v'}$ ratios.

Critical design issues include the design and construction approach for roads, structures, slopes, canals and dykes founded in or with peats and organic soils.

CONTENTS

35.1. Introduction

Peats and organic soils would generally be considered to be 'problematic' in a geotechnical context because of their high compressibility, creep behaviour, low bulk density and generally low undrained shear strength. The high compressibility and creep behaviour have given rise to large settlements under structures and to undulating roads that become a hazard to traffic. Regionally, for example, changes in the hydrogeological regime have resulted in settlements of several metres and large bog slides have wiped out villages, with many fatalities, in Ireland. Historically, dykes and canals have been constructed on or with peat, which leaves a legacy of high-risk structures within our society. The engineering behaviour of peat and organic soils is therefore an important aspect of geotechnical engineering, both in the design of structures on peat and organic soils and in the assessment of existing structures on these soils.

The percentage of land cover by peat is about 33.5% in Finland, 18% in Canada, 17% in the Republic of Ireland and Sweden, 13.7% in Indonesia, 12.4% in Northern Ireland, 10.4% in Scotland and 6.3% across the United Kingdom as a whole (Hobbs, 1986). Large tracts of peatlands are also to be found in Russia and the USA. A variety of terms in the English language are used to describe the areas within which peats and organic soils are formed, such as swamp, bogland, moor, fen and muskeg, but *mire* is the internationally accepted term to cover all of these ecosystems (Gore, 1983). This chapter mainly discusses the particular properties of peats; however, there is obviously a transition in the behaviour from highly organic soils, which would have many similar properties to peats, to soils with low organic content, which would have similar behaviour to mineral soils.

35.2. Nature of peats and organic soils

Peat and organic soils covers a wide range of soils with different morphologies and characteristics that impact on their engineering behaviour. The organic content can come from plants or animals, but most of this chapter will relate to the organics from plant debris. Peat itself is formed from the accumulation of the remains of dead vegetation in various stages of decomposition, and this accumulation can build up where there is a reservoir of water to promote growth and to preserve the remains. However, there are a variety of conditions where this can occur. For example, raised bogs obtain their nutrients from rainwater, which results in a particular type of vegetation growth, and are made up almost entirely of organic material, whereas other bogs are fed from surface and groundwater flow which inevitably includes some inorganic particles that are generally referred to as mineral particles in this context.

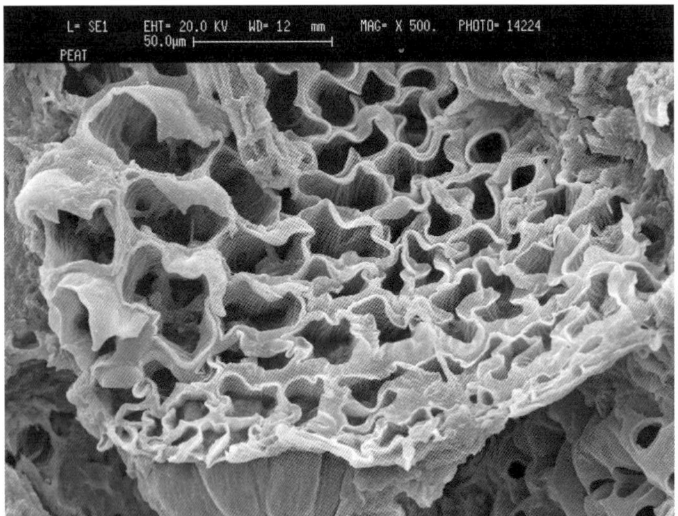

Figure 35.1 Electron microscope image of Balltdermot peat (Reproduced with permission from Hebib and Farrell, 2003; © Canadian Science Publishing or its licensors)

The remains of plants that make up peats and organic soils have an open cellular structure which give rise to their high compressibility and high moisture contents. The open structure of a peat fibre can be appreciated from the electron microscope image of a typical stem, shown in **Figure 35.1**. The cell structure can thus deform when transferring interparticle forces. Water is held in three forms (*a*) free water in large cavities; (*b*) capillary water in the narrower cavities; and (*c*) water bound physically, colloidally and osmotically (MacFarlane and Radforth, 1964). There is also the reinforcing effect of the fibres, which can significantly affect the ability of peats to resist load. These differences from clays and sands should, in theory, make many of the traditional methods of analysis inapplicable. However, many of the correlations between various properties and behavioural parameters are little different from those of clay soils, except in regard to its strength (Hobbs, 1986).

The decomposition of the plant remains can have a significant effect on the characteristics of a peat, and this is normally termed the degree of humification in geotechnical engineering. Plant remains are broken down by soil microflora, bacteria and fungi; in non-acidic soils earthworms can contribute to the breakdown (Hobbs, 1986). The initial process is aerobic, where the plant matter is broken down to give off gas and water. Immersion in water dramatically reduces the availability of oxygen, thus reducing not only the aerobic microflora but encouraging the anaerobic species with different and less rapid metabolic activity. The partially decayed vegetation matter can thus accumulate as peat. The upper active layer with undecayed fibrous plants is called the acrotelm (generally 100–600 mm thick), with the lower layer being called the catotelm. Initially the decomposition of the cellulose structure effects the plant leaves, but this process can develop to the stems and roots of the plants, eventually to completed decomposition into an amorphous-granular material consisting of mainly gelatinous organic acids that have a sponge-like structure (Landva and Pheeney, 1980). However, the complete breakdown of the plant remains is rare in the UK and Ireland. The rate of decay generally increases with an increase in temperature and with pH. This process of plant degradation is not uniform and therefore can exacerbate the natural non-uniformity of peatlands. The nature of peat can also be affected by historic fires, anthropogenic factors, previous landslides and other external factors.

Mires can be broadly separated into fen and bog, with bog being subdivided into raised bog (over fen bog) and blanket bog. Raised bogs are formed over what was originally a fen bog and includes a transitional stage, which must be considered when making an engineering assessment. The separation between fen and bog lies in the origin and chemistry of their water supplies. Moore and Bellamy (1974) and Hobbs (1986) illustrate the different formation environments in which mires are formed.

Fens are predominantly fed by groundwater or flowing water sources, and generally are neutral to alkaline, although there are also acidic fens. The alkaline environment increases the rate of decay and hence can result in a greater degree of humification than in raised bogs. Fens form in a variety of wetlands, from water-filled depressions to river banks to lakes shores, and can support a variety of vegetation growth, depending on their water sources and genesis. Floating bogs can form by vegetation growth from the side of a lake, leaving clear water between the base of the peat and the lake bottom.

Bogs are formed in ombrotropic (rain-fed) conditions and are acidic. Raised bogs start off by organic growth on or from the banks of lakes (i.e. fen peat) that expands to encompass the lake, at which stage the landscape would commonly be referred to as a fen (Hobbs, 1986). These fen peats would frequently be formed over the original very soft lake bed deposits, where these exist, which can be the critical feature in the selection of appropriate design/construction options on these deposits. Where the conditions are appropriate there is a transitional stage where the upward growth is such that the plants come to rely more on the rainfall for nutrients rather than on the natural groundwater, and eventually a mounded mass of organic material is formed which obtains its entire nutrient from the rainwater. At this stage it is called a raised bog. Sphagnum moss and cotton grass tussocks are common forms of vegetation.

Blanket bogs are a form of raised bog with rainwater as the principal source of nutrients, but which form directly on the original soil rather than on a lake. These typically form on the sides of mountains in the colder and temperate regions of Earth, where there is an excess of precipitation over evaporation. Given that these are generally on sloping ground, the nature of the original ground on which it is formed and the hydrogeological conditions can be important considerations when assessing the slope stability of the mass, together with the characteristics of the peat itself. Features of some of these blanket bogs are the presence of subterranean pipes or channels which have developed over time and also, in some locations, the presence of a hardpan layer in mineral soils close to the base of the peat which was formed by the accumulation of leached minerals from the peat. Donnelly (2008) documents the presence of soil pipes of up to 1.5 m dia. on parts of the Pennine Moors and of pseudo-sinkholes caused by their collapse.

Layers of marl can be associated with peat deposits in calcareous regions. These layers form due to photo-reduction of dissolved carbon dioxide by submerged green plants.

Peats have also been deposited in estuarine and coastal environments and are sometimes buried under mineral soils due to change in the depositional environment. For example, Kidson and Heyworth (1976) have shown that large areas of Somerset, known as the Levels, are underlain by considerable thickness of Quaternary sands, gravels, clays and peats, which were formed within the contemporary tidal range, the sedimentation being influenced by rise in sea level over the last 9000 years.

35.3. Characterisation of peats and organic soils

35.3.1 Classifications systems

There are a number of classification systems for peat deposits in existence, most of which have a botanical basis suitable for agriculture and ecology, but these are generally too elaborate for engineering applications. An example of a comprehensive classification system is given by Clymo (1983). Engineering classification systems have been proposed by von Post (1922), Radforth (1969) in Canada, Landva *et al.* (1983) and Hobbs (1986, 1987), among others. The von Post system, which was developed for horticultural, agricultural and forestry use, includes a simple 'squeeze' test as a means of assessing the degree of decomposition of peats, which has been incorporated into many of the other classification systems, albeit with some modification.

The important characteristics to be identified in any engineering classification system are the type of peat, the structure, the degree of decomposition and the amount of mineral particles present. The in situ colour description can be useful; however, a characteristic of organic soils is the rapid change in colour on exposure to air due to oxidation, which can make colour descriptions from recovered samples problematical.

The Eurocodes system references two standards which cover the identification and classification of soils, namely EN ISO 14688-1:2017 (Identification and description) and EN ISO 14688-2:2017 (Principles for a classification). EN ISO 14688-1:2017 sets out a summary of the identification of soils rich in organic matter – see **Table 35.1**. BS5930:2015 (BSI, 2015) has description categories of fibrous, pseudo-fibrous and amorphous peat. Gygttja is a particular Scandinavian term which is applied to organic bottom deposits in lakes or the sea. It also states that the colour should be described as it can indicate the organic content, whereby grey would be associated with a slightly organic soil, dark grey with an organic soil and black with a very organic soil. The degree of decomposition of wet peat in EN ISO 14688-1:2017 (BSI, 2017a) is assessed on the basis of a simple squeeze test – see **Table 35.2**, which is a simplification of the von Post system where the degree of decomposition (H_n) is measured on a scale of 1–10, a modification of which has been incorporated into the ASTM standards – see **Table 35.3**. The von Post system is commonly used where a detailed description of peats is required.

Term	Description
Fibrous peat	Fibrous structure, easily recognisable plant structure, retains some strength
Pseudo-fibrous peat	Mixture of fibres and amorphous paste
Amorphous peat	No visible plant structure, mushy consistency
Topsoil or humus	Plant remains, living organisms and their excretions together with inorganic constituents
Gyttja	Sediment formed in nutrient-rich water and mainly consisting of more or less decomposed remains of plants and animals (detritus)
DY	Sediment formed in nutrient-poor water and mainly consisting of precipitated colloidal humic substances (dy matter)

Table 35.1 Identification and description of organic soil (Data taken from EN ISO 14688-1:2017)

Term	Decomposition	Remains	Squeeze
Fibrous	None	Clearly recognisable	Only water; no solids
Pseudo-fibrous	Moderate	Recognisable	Turbid water; <50% solids
Amorphous	Full	Not recognisable	Paste; >50% solids

Table 35.2 Degree of decomposition of wet peat as determined by squeezing (Data taken from EN ISO 14688-1:2017)

EN ISO 14688-2:2017 (BSI, 2017b) (Principles for a classification) classifies a soil with more than 20% organics as high-organic soil – see **Table 35.4**. However, there is no particular definition of when an organic soil becomes a peat or when a clayey peat becomes a peaty clay or of how the organic content is determined. BS 5930:2015 gives a classification system for soils where the organic content is the secondary constituent but does not classify a peat. ASTM classifies a peat as having less than 25% ash content, which is equivalent to an organic content in excess of 75%.

35.3.2 Index tests

The water content (w) is the most important property when dealing with peat and organic soils as it is a simple test to which many of the engineering properties can be related. Consequently,

Degree of humification	Nature of material extruded on squeezing	Nature of plant structure in residue
H1	Clear, colourless water; no organic solids squeezed out	Unaltered, fibrous, undecomposed
H2	Yellowish water; no organic solids squeezed out	Almost unaltered, fibrous
H3	Brown, turbid water; no organic solids squeezed out	Easily identifiable
H4	Dark brown, turbid water; no organic solids squeezed out	Visibly altered but identifiable
H5	Turbid water and some organic solids squeezed out	Recognisable but vague, difficult to identify
H6	Turbid water; half of sample squeezed out	Indistinct, pasty
H7	Very turbid water, half of sample squeezed out	Faintly recognisable; few remains identifiable, mostly amorphous
H8	Thick and pasty; two-thirds of sample squeezed out	Very indistinct
H9	No free water; nearly all of sample squeezed out	No identifiable remains
H10	No free water; all of sample squeezed out	Completely amorphous

Table 35.3 Degree of humification or decomposition, ASTM D5715-00 (ASTM, 2006) (Based on von Post, 1922)

Soil	Organic content ($\leq$2 mm) % dry mass
Low organic	2–6
Medium organic	6–20
High organic	>20

Table 35.4 Classification of soils with particle sizes $\leq$2 mm with organic content (Data taken from EN ISO 14688-2:2017)

w can be used to quantify variations in the peat and organic soil deposits. There is evidence that there may be charring of certain types of peat when the water content is determined at the normal 105–110°C, and MacFarlane and Allen (1964) suggest that a temperature of 85°C may be appropriate if combustion of material or oxidation of the peat is to be avoided. However, Hobbs (1986) considers that the normal 105°C is acceptable for engineering purposes and this would be consistent with the findings of O'Kelly (2005) and with general practice.

The organic content is a very useful index test and is normally taken as being equal to the loss on ignition (N%). This index test quantifies the amount of organics present and also has the very useful relationship with the average specific gravity of the particles ($G_{s_{ave}}$) given as Equation 35.1 (Skempton and Petley, 1970). This equation is based on the specific gravity of the cellulose being about 1.4 and of the mineral particles being about 2.7. This specific gravity can be used to estimate the bulk density using Equation 35.2; however, the degree of saturation S_r can be less than 100% in some peats due to its gas content

$$\frac{1}{G_{s_{ave}}} = \frac{N}{1.4} + \frac{(1-N)}{2.7} \tag{35.1}$$

$$\rho = \frac{G_{s_{ave}} \rho_w (1+w)}{1 + \frac{wG_{s_{ave}}}{S_r}} \tag{35.2}$$

The loss on ignition (N) is determined on a specimen initially dried at 50°C and fired in a muffle furnace at 440°C ± 25°C as recommended in BS 1377-3:2018 (BSI, 2018). Clay minerals present lose fixed water at temperatures in excess of 450°C, causing an error in the ignition loss. The ASTM uses the ash content (D) rather than the organic content, where $D = 1 - N$, and includes two tests for its determination, one for a firing temperature of 440°C (initially dried at 105°C) until it is completed ashed, and the other for 750°C. Some researchers have found that fresh plant roots are not removed at temperatures below 900°C. Hobbs (1987) summarises the correction factors to be applied to the loss on ignition value determined at temperatures above 450°C to account for the water loss from minerals. BS 1377-3:2018 also includes a chemical titration test (the Walkley and Black method) which is generally only used for soils with low organic content.

The liquid limit (w_L), which can be a useful indicator to the morphology of the peat and its degree of decomposition, can generally be carried out on peats with a degree of decomposition greater than H_3 on the von Post scale (von Post scale given in **Table 35.3**). The structure of the peat/organic soil generally has to be broken down to a suitable size using a mechanical device, thus not only breaking down the structure but also some of the decomposed material (Hobbs, 1986). Hanrahan *et al.* (1967) found the w_L value to be sensitive to the chemistry of the water, therefore water from its natural source should be used. Bog peats (w_L = 800–1500%) typically have higher w_L than fen peats (w_L = 200–600%). Hobbs (1986) showed, by observing the decomposition of a cauliflower, that both the liquid limit and the water content decreased with the degree of decomposition, and considered that the water content (natural) was equal to about $1.38w_L$ and $0.875w_L$ for bog and fen peat, respectively. The plastic limit can only be carried out on peats with some clay particles; consequently it is generally of little interest to bog peats. The addition of organic content to a mineral soil can alter its liquid and plastic limit significantly, not only because it introduces an additional material but also because the organic particles, which are negatively charged, may be strongly adsorbed on the mineral surfaces, with the effect that the properties of both materials are altered (Mitchell and Soga, 2005). It also means that the classification of an organic soil based on organic content alone would not be expected to capture all the significant geotechnical features of a soil.

The bulk density is also a useful index and there is generally a good agreement between measured bulk densities and those determined from Equation 35.2, using $G_{s_{ave}}$ estimated from loss on ignition tests. Hobbs (1986) plotted the bulk density measured on five UK mires and showed that the degree of saturation/gas content was the main influence on bulk density when the water content was in excess of about 600%. About half the samples had bulk densities of less than 1 Mg/m^3 when the water content was greater than 500%, indicating that the peat would be buoyant in water. For water contents in excess of about 200%, the volume of gas as a percentage of the total volume is approximately $1 - S_r$, where S_r is the degree of saturation. The average gas volume indicated by the results of Hobbs (1986) is about 7.5%.

Although not normally considered an index test, the pH can be a useful indicator of the nature of peats and organic soils (Hobbs, 1986). Metabolic activity is influenced by the temperature and acidity; the higher the temperature and pH (lower acidity) the faster the decomposition. There is a considerable degree of overlap, but bog peats have a pH in the range 4.5 to 3.3, and transitional peats from 6 to 4. The pH of a fen peat will depend on the water source; most tend to have pH greater than 5, with the acid fens being less than 5.

35.3.3 Sampling methods

It can be difficult to obtain 'undisturbed' samples of highly organic soils, particularly fibrous peats. Simple sampling methods that can be used to get continuous or almost continuous disturbed samples, such as gouge augers, can be used to determine a water content/depth profile and thus to identify important stratigraphic variations. Sharp-edged piston samples have been used successfully; however, the samples generally compress and it is important to record the percentage recovered in order to assess the degree of compression due to sampling. Undisturbed samples can be cut from the in situ material, but this becomes impractical with depth. Special samplers, such as the Sherbrooke sampler (Terzaghi *et al.*, 1996), which essentially allows a block sample to be cut from the base of a borehole using a system of blades, can be used on special projects. Oxidation generally changes the colour of recovered samples when exposed to the atmosphere and biodegradation can occur within the samples. The colour changes are typically limited to the exposed surfaces; however, microbial processes can lead to gas formation which may contribute to excess pore water pressures (Kellner *et al.*, 2005). The effect of such microbial process on the long-term settlements were investigated by Fox *et al.* (1999) by comparing the long-term settlements of an oedometer sample which had been sterilised using radiation to kill bacteria and fungi with untreated samples. This work suggests that biodegradation may be an issue in long-term creep tests, but additional research is required before this can be confirmed.

35.4. Compressibility of peats and organic soils

35.4.1 Introduction

The large compressions that can occur in virgin peats are clearly illustrated in **Figure 35.2**, which presents the vertical strain plotted against the logarithm of the effective stress,

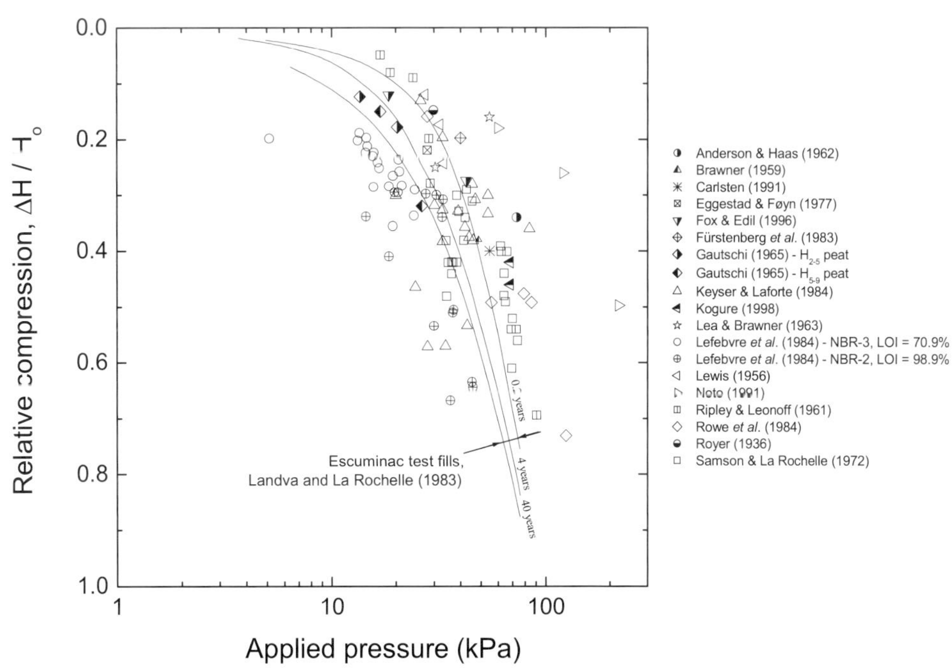

Figure 35.2 Relative compression plotted against applied pressure (Landva and La Rochelle, 1983; updated by O'Loughlin, 2007) (Reproduced from O'Loughlin, 2007)

as recorded in a number of case histories (Landva and La Rochelle, 1983; additional cases by O'Loughlin, 2007), although some shear deformation and secondary compression is included in some of these data. These results show that a peat strata can reduce in thickness by more than 50% under a relatively modest increase in effective stress of the order of 50 kPa. Furthermore, secondary compression can be the dominant process, and not 'a rather tiresome appendage' to the consolidation process (hydrodynamic) as is generally assumed (Hobbs, 1986). The large strains will necessarily distort the structure of peats and organic soils and hence the permeability and compressibility characteristics change as the soil compresses. Given the genesis of raised bogs with vegetation growth patterns, the properties are not uniform in either the vertical or horizontal direction. There are therefore many complexities in peats compared with mineral soils which must be considered when selecting an appropriate model to estimate settlements.

The term 'primary' and 'secondary' compression can lead to confusion and to controversy (den Haan, 1996), and many researchers consider that it is better to consider 'primary consolidation' as a hydrodynamic phase in the settlement process which is controlled by the dissipation of excess pore water pressures and that secondary compression is a continuous process that occurs during both the hydrodynamic phase and continues when the excess pore pressures have essentially dissipated. This introduces some interesting concepts – for example, that secondary compression movements should be estimated from the start of loading rather than from the end of primary consolidation (EOP), as is assumed in the EOP approach, which also affects the interpretation of 'pre-consolidation' or critical pressure.

35.4.2 Magnitude of settlement (constrained deformation)

35.4.2.1 Different philosophies

Secondary compression can be a dominant process in organic soils and there are two fundamentally different schools of thought on the way it should be modelled: one which considers that it begins at the end of primary consolidation (EOP method); and the other which considers that secondary compression occurs both during and after the primary consolidation phase, the primary consolidation being considered as the hydrodynamic phase. The significance of these two approaches is illustrated in **Figure 35.3**. The EOP assumption simplifies the estimation of settlements in that the primary settlements can be estimated in the traditional fashion and the secondary compression included as a separate calculation. The EOP approach would not include the additional settlement that would arise from secondary compression that occurs during the primary phase for a thick layer. The standard 24-hour consolidation test on peats would include some secondary compression, which may reduce the error in practice (Hobbs, 1986).

The EOP approach is frequently used in practice as the calculation models for the continuous approach to date are complex and the additional complexity is generally not justified when other uncertainties are considered, except for very large projects.

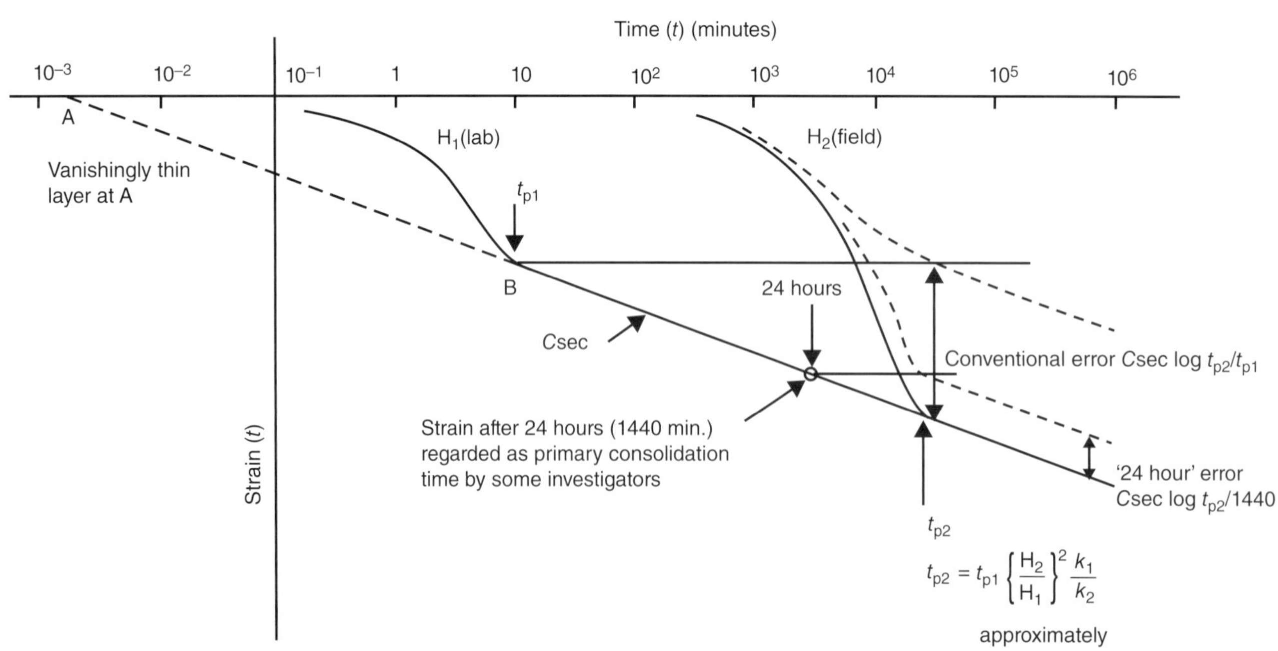

Figure 35.3 Comparison between EOP and continuous assumptions (Reproduced from Hobbs, 1986; © The Geological Society)

35.4.2.2 EOP method

The advantage of the EOP approach is that the primary settlement can be estimated from an e-log σ_v' plot in the normal fashion, but necessarily requires some assumption/knowledge of a 'pre-consolidation/yield' pressure. Field and laboratory tests have shown that virgin peats do have an 'apparent' yield point, possibly due to capillary suction forces when in the acrotelm or due to the weight of snow (Hobbs, 1986; Mesri, 2007). This may be estimated from the e-log σ' curve in the normal Casagrande (1936) construction, by using Janbu's method (1963), or from relationships such as that given by Mesri (2007).

For those soils which have an essentially bilinear e-log σ' relationship, which may not be true for highly organic soils and peats as will be discussed later in this section, the compression and swelling indices, C_c and C_s, can be used to determine the change in void ratio due to a change in effective stress. Hobbs (1986) developed useful relationships for UK peats of $C_c = 0.0065w$ for bog and $C_c = 0.008w$ for fen peat. Mesri *et al.* (1997) suggest that $C_c = 0.01w$. Hobbs (1986) showed that there was a relationship between C_c (from 24-hour incremental load consolidation tests) and initial void ratio, which had a limiting value for peats of $C_c \approx 0.45e_o$. Given that $e_o \gg 1$ for peats, this implies that $C_c/(1 + e_o) \approx 0.45$ is a limiting value. It also means that for peats the settlement is virtually independent of the void ratio or water content.

It is generally more useful to consider the settlements in terms of strain resulting from an effective stress increase from σ_{vo}' to $\sigma_{vo}' + \Delta\sigma_{vo}'$ which, for increases in vertical effective stress from below to above the pre-consolidation pressure, is

$$\varepsilon = \frac{\Delta e}{1+e_o} = \frac{C_s}{1+e_o}\log\left(\frac{\sigma_c'}{\sigma_{vo}'}\right) + \frac{C_c}{1+e_o}\log\left(\frac{\sigma_{vo}' + \Delta\sigma_v'}{\sigma_c}\right) \quad (35.3)$$

$$\varepsilon = \frac{\Delta e}{1+e_o} = \text{RR}\log\left(\frac{\sigma_c'}{\sigma_{vo}'}\right) + \text{CR}\log\left(\frac{\sigma_{vo}' + \Delta\sigma_v'}{\sigma_c'}\right) \quad (35.4)$$

where CR is the compression ratio and RR is the recompression ratio.

The use of the compression and recompression ratios, $\text{CR} = C_c/(1 + e_o)$ and $\text{RR} = C_s/(1 + e_o)$ respectively, requires an assumption regarding the initial void ratio and is therefore not a fundamental soil parameter. However, it can be useful in practice to give an indication of the variation in compressibility of normally or lightly consolidated soils. The CR values determined from 24-hour standard incremental consolidation tests are essentially independent of water contents for liquid limits in excess of about 200% (Farrell and O'Donnell, 2007), which supports the contention of limiting values of this parameter (**Figure 35.4**). As stated previously, these 24-hour results inevitable include some secondary consolidation.

Laboratory tests on highly compressible peat samples have shown that C_c is not constant at large strains – see **Figure 35.5**. This nonlinearity can be considered in the EOP method by using the actual curve e-log σ' (Mesri, 2007) or alternatively

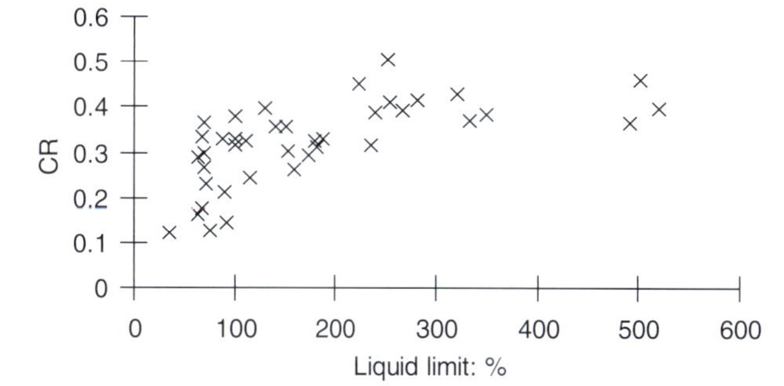

Figure 35.4 Variation of CR with w_L (Reproduced from Farrell and O'Donnell, 2007)

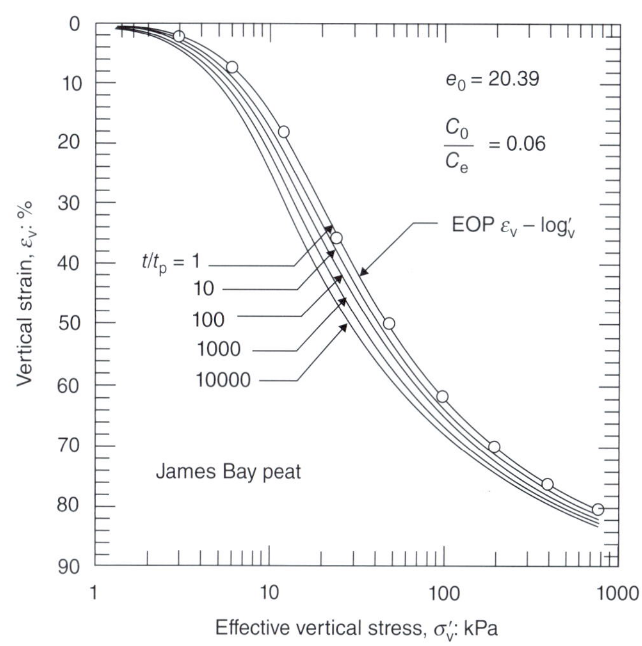

Figure 35.5 Vertical strain plotted against σ_v' (Reproduced from Mesri, 2007; with permission from ASCE)

the plot can normally be linearised by using natural strain e^H in place of linear strain e^c, where $e^H = -\ln(1 - e^c)$.

The void ratio and strain due to soil secondary compression under a constant effective stress generally increases linearly with log time; however, there are two parameters used to express this linearity, namely C_{sec}, which is the strain $\{\Delta e/(1 + e_o)\}$ per log cycle of time, and C_α, which is Δe per log cycle of time. The strain is normally referenced to the initial height rather than from the height at the end of primary consolidation. There is some evidence from laboratory tests that these coefficients may increase in time, called tertiary creep (Dhowain and Edil, 1980). The values of C_{sec} or C_α are generally considered to be independent of the applied stress for those stresses above the pre-consolidation/yield stress and for those soils which can be modelled by a constant C_c. There is a stress dependency to C_{sec} or C_α for stresses around the pre-consolidation/yield stress, and for those highly organic soils which cannot be modelled by a constant C_c. Mesri and Godlewski (1977) consider that similar processes are involved in C_c and C_α and by defining $C_c = \Delta e/\log\sigma$ (which can be used above and below the pre-consolidation/yield stress), have shown that the relationship C_α/C_c is relatively constant. Mesri (2007) gives $C_\alpha/C_c = 0.06 \pm 0.01$ for a fibrous peat. Some researchers consider that the laboratory-determined values of C_{sec} or C_α tend to underestimate the field creep by a factor of the order of 2 (Hobbs, 1986), whereas others consider that the values from field observations and laboratory tests are comparable (Mesri, 2007). The change in strain at constant σ_v' can be estimated from the relationship $\Delta\varepsilon = C_{sec} \log \frac{t}{t_p}$ where t is the time from the initial loading and t_p is the time to the end of primary consolidation. Thus, the strain at different time increments can be drawn.

35.4.2.3 More sophisticated time–settlement models

The EOP method would be considered to be an empirical model for time-dependency behaviour of soil (Liingaard *et al.*, 2004). Some of the complexities of modelling creep in organic soils can be appreciated in the extension of the 1D empirical model concept developed by den Haan (1996), called the abc model. The abc model incorporates the continuous creep, as well as other features such as the natural strain (e^H) and varying permeability with void ratio, which will be discussed below. The volume changes in the abc model are expressed in terms of specific volume ($\upsilon = 1 + e$) rather than in terms of void ratio, and it assumes that the creep strain rate is uniquely defined by present stress and strain. The normally consolidated stress–strain relationship is $\ln v = \ln v_1 - b\ln\sigma'$ where v_1 is a reference specific volume at an effective stress of unity. Creep is defined by lines of constant creep rate $\left(\frac{d\varepsilon_c^H}{dt}\right)$, called isotaches, which have a slope b on a $\ln\upsilon$–$\ln\sigma$ plot, as shown in **Figure 35.6**. The parameters used in this model are $b = \frac{\delta \ln \nu}{\delta \ln \sigma_v'}$; $c = \frac{\delta \ln \nu}{\delta \ln t}$, which are analogous to C_c and C_α respectively, and 'a', which is equivalent to a reload compression index and is defined as $a = \frac{\delta\varepsilon^H}{\delta \ln \sigma_v}$, where the superscript H refers to natural strains which are attributed to Hencky. The model also introduces the parameter called the intrinsic time (τ). Creep at constant effective stress, ε^H is assumed to be linear when plotted against the logarithm of intrinsic time τ. The use of intrinsic time is necessary because of the complications that arise with a logarithmic scale and may be considered as the time which would be necessary to achieve the present volume if the present stress had been applied immediately to the soil in its freshly sedimented state. The intrinsic time of the baseline is given the symbol τ_0. The isotaches are given by

$$-\ln\frac{\nu}{\nu_1} = b\ln\sigma_v' + c\ln\frac{\tau}{\tau_0} \tag{35.5}$$

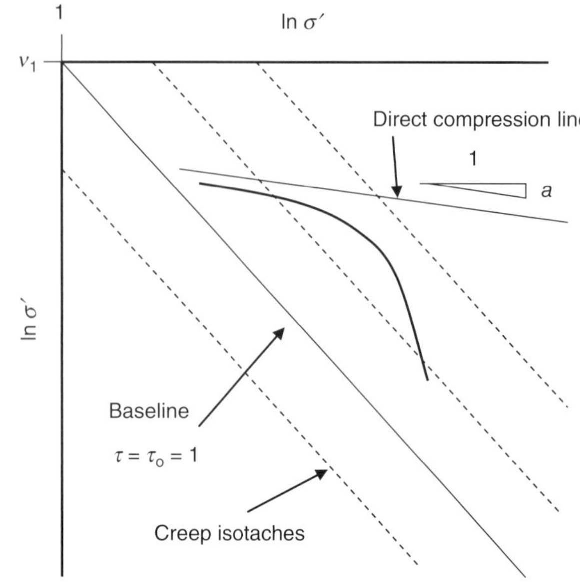

Figure 35.6 Stress plotted against creep for the abc model

The total strain rate $\left(\frac{d\varepsilon^H}{dt}\right)$ is the sum of direct strain rate $\frac{d\varepsilon_d^H}{dt} = a\frac{d\ln\sigma_v}{dt}$ and the creep strain rate $\frac{d\varepsilon_c^H}{dt}$ which is related to intrinsic time by $\frac{d\varepsilon_c^H}{dt} = \frac{c}{\tau}$. Thus, the total strain rate is

$$\frac{d\varepsilon^H}{dt} = \frac{d\varepsilon_d^H}{dt} + \frac{d\varepsilon_c^H}{dt} = \frac{a}{\sigma_v}\frac{d\sigma_v}{dt} + \frac{c}{\tau} \tag{35.6}$$

More sophisticated 1D and 3D methods have been developed which take into account other factors related to creep. Liingaard *et al.* (2004) present a comprehensive review of the different methods which have been developed to model the time-dependent behaviour of soils, which fall into the categories of (*a*) empirical, of which the EOP is an example; (*b*) rheological; and (*c*) general stress–strain–time models.

Berry and Poskitt (1972) have used rheological models for peats with some success, but these have not been adopted for general design.

General–strain–time models are available in computer codes which have different degrees of complexity and sophistication. One model, the Soft Soil Creep (SSC) model, which is formulated mainly within the framework of visco-elasto-plasticity, has the advantage of using input parameters which are related to familiar consolidation parameters – for example, C_c, C_s and C_α, among others. The background to this model is given in the Plaxis manual (Plaxis, 2020). Tan (2008) has reported the successful use of this model in a case study of highway construction on peat, as have Long *et al.* (2022) for trial bund embankments on peat. However, as with all sophisticated programs, careful calibration of the model parameters is required.

35.4.3 Surcharging

Secondary compression settlements, which can give rise to significant serviceability issues for roads and structures founded on peats, can be reduced by surcharging, particularly as it can benefit from the high initial permeability of fibrous peats. Care must be taken, however, that the increased load does not give rise to instability. The effectiveness of this method depends on the surcharge ratio $R'_s = \dfrac{\sigma'_{vs} - \sigma'_{vf}}{\sigma'_{vf}}$, where σ'_{vf} is the post-construction effective vertical stress and σ'_{vs} is the effective vertical stress reached during surcharging (Hanrahan, 1953; Mesri, 2007). The design criterion for a surcharge is usually to build out sufficient settlement during construction which would include both the primary and secondary settlement to be expected over the design life of the structure if a surcharge were not applied. The benefits of a surcharge are illustrated on **Figure 35.7**, where Curve A is the estimated settlement/time using, for example, Equation 35.4, with an embankment at the design height with no surcharge, with the total estimated settlement after the design period (t_{design}) to be X. A surcharge is selected to give the settlement X within the construction period as shown on **Figure 35.7**. The soil is allowed to consolidate under a surcharge $\Delta\sigma$, there is a rebound on removal, computed using $C_s \approx 0.1C_c$, long-term rebound (Samson and La Rochelle, 1972), then essentially no creep settlement until the creep line for that effective load is reached.

The longer the maximum load is maintained the greater the delay in recurrence of the secondary compression. In both the field and laboratory, the rebound shows a marked increase with surcharge ratios exceeding about 3 (Hobbs, 1986). Some of the secondary consolidation/creep models discussed in Section 35.4.2.3 can assist in the assessment of the effectiveness of surcharge.

35.4.4 Rate of settlement

The general experience is that the field permeability of fibrous peat is significantly greater than that recorded in the laboratory consolidation tests (Hobbs, 1986). This is considered to be due to the limited size of the samples tested in the laboratory,

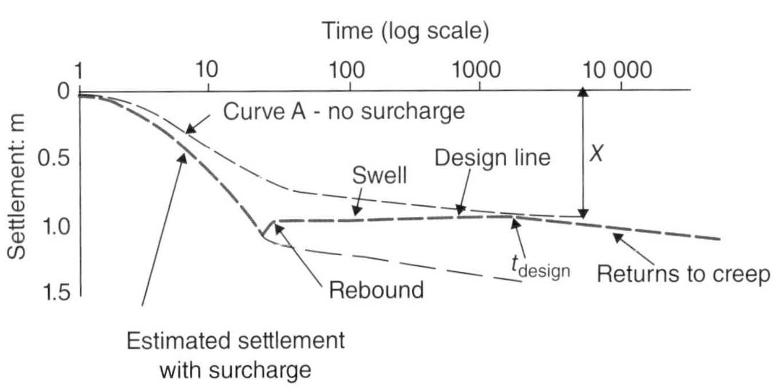

Figure 35.7 Simplified method of assessing effectiveness of surcharging (t_p = time to primary consolidation)

to gas and to the natural heterogeneity of peat to be expected from the plant growth patterns. The relatively high permeability of the acrotelm and the anisotropy of the peat deposits are also considered to contribute to increased field consolidation rates. The reported comparison between the time for primary consolidation of laboratory samples (t_s) and field consolidation times (t_f), and the respective sample height (H_s) and field thickness (H_f) using Equation 35.7 give i values from 0 to 2 for peats. Thus $i = 0$ implies that the time for primary consolidation is independent of the thickness of the peat, whereas if the peat complied with Terzaghi's theory of consolidation, $i = 2$

$$\frac{t_f}{t_s} = \left\{\frac{H_f}{H_s}\right\}^i \tag{35.7}$$

Hobbs (1986) considers that the time for primary consolidation for peats can only be determined from field trials, which would require the installation of piezometers. Generally researchers have found that the time for primary consolidation in the laboratory, inferred in the normal way using the semi-log plot or Taylor's method, was approximately consistent with the end of the dissipation of excess pore pressures (Lefebvre *et al.*, 1984; Mesri, 2007). O'Loughlin (2007), however, found that Taylor's method, and to a lesser extent Casagrande's method, consistently underestimated the time for primary consolidation.

The variation in permeability with effective stress would be expected to vary significantly with the type of peat and organic soil which depends on the macro- and micro-structure of the peat as well as on its degree of decomposition. It is common to express the change in permeability with respect to void ratio as $C_k = \Delta e/\Delta \log k_v$ (where Δe is the change in void ratio and k_v is the permeability in the vertical direction). Mesri (2007) considers that $C_k = 0.25e_o$ for fibrous peat, where e_o is the in situ void ratio. Hanrahan (1954) considered the change in permeability to be linear on a log e-log k plot and derived $k = k_{vo}\left\{\dfrac{e}{e_o}\right\}^n$, where k_o and e_o are the in situ permeability and void ratio, respectively. Hobbs (1986) interpreted from Hanrahan (1954) $k_{vo} = 4 \times 10^{-6}$ m/s, $e_o = 12$, $n = 11.03$ for a fibrous peat, where k_{vo} is the initial coefficient of permeability in the vertical direction. Hogan *et al.* (2006) reported k values of 2×10^{-2} m/s near the surface of a fen peat, which reduced to 10^{-5} to 10^{-6} m/s at 2–3 m depth in virgin peat. Hanrahan (1954) reported a decrease in k_v from an initial value of 4×10^{-6} m/s to 2×10^{-8} m/s under a pressure of 55 kPa for 2 days, which decreased further to 8×10^{-11} m/s after 7 months. Generally the horizontal permeability has been found to be significantly greater than the vertical by a factor of 3–10 (Mesri, 2007). The in situ horizontal permeability of peat can be estimated in the field by carefully executed variable head tests; however, the reduction in this value with the decrease in void ratio that results from an increase in effective stress would be an important consideration in the interpretation of the test results. Back-analysis of preloaded sections of field trials to failure carried out in the Netherlands on a fibrous peat by Zwanenburg and Jardine (2015) indicated a relatively short initial consolidation time that gradually extended as loading levels increased.

35.5. Shear strength of peats and organic soils

35.5.1 Introduction

Generally the shear strength of peats and organic soils is determined in the traditional manner with c_u for undrained loading and c', φ' for drained conditions, without special consideration being given to the fibre content, high compressibility or relatively high permeability and gas content of some of the fibrous peats. The high compressibility and fibre content affect the behaviour of fibrous peats in the laboratory and field tests and is poorly considered in current methods used to interpret these tests. In laboratory tests, for example, the high compressibility of peat samples results in necking and dimensional non-uniformities of samples during the consolidation phase of triaxial tests; also, the fibre effects mean that drained triaxial tests result in an essentially one-dimensional behaviour with the result that the soil cannot be taken to 'failure' as defined by a Mohr–Coulomb criteria. The fibres also affect the interpretation of effective stress parameters in undrained triaxial tests as the low Poisson's ratio of the fibre content of peat means that the pore water pressures rapidly build up to equal the cell pressure, thus making the effective lateral stress equal to zero. Determining the appropriate field strength also has its difficulties as the fibre content affects field vane tests and typically the recorded c_u decreases with an increase in the size of the vane (Landva, 1980).

In practice, the effect of fibres is eliminated in ring shear tests and these give the lower values of φ' for conservative estimates. Recently more attention is being given to direct simple shear (DSS) tests for determining undrained shear strength parameters (Boylan *et al.*, 2008; Farrell *et al.*, 1999; Zwanenburg and Jardine, 2015). The tensile strength of the upper zone of a fibrous peat may be relevant to some design situations and the measurement of this is given by O'Kelly (2017), which also includes a comprehensive review of shear strength testing of peat in general.

35.5.2 Effective stress parameters

Peats and organic soils are predominantly frictional materials with values of φ' which are very much dependent on the method of test. For example, φ'_{peak} values of the order 50–60° have typically been recorded in consolidated undrained triaxial tests on fibrous peats, compared with values of the order of 32–40° for direct shear tests and for ring shear tests on the same material. The accuracy of the interpretation of φ' on fibrous peats from undrained triaxial tests is affected by the low lateral effective stresses recorded at failure in these tests, which frequently are essentially zero. Similarly, it is generally not possible to bring a sample of fibrous peat to failure in a drained triaxial test due to the continual compression of the fibres which can be compared to the compression of a mattress. DSS tests have been used to estimate the effective stress parameters; however, it is necessary to make an assumption regarding the effective horizontal stress at failure. Farrell *et al.* (1999) considered, on the basis of a finite element analysis of the simple shear test, that φ' could be estimated from $\varphi' = \sin^{-1}\left(\frac{\tau_f}{\sigma'_{vf}}\right)$, where τ_f and σ'_{vf} are the shear stress and vertical effective stress at failure. Tests by Yamaguchi *et al.* (1985a, 1985b) recorded φ' of 35° on triaxial compression tests on horizontal samples compared with 51–55° on vertical samples, with even higher values of 62° being recorded in triaxial extension on vertical samples, which illustrates the inherent anisotropy. It is considered that ring shear tests, and possibly direct shear tests, give the intrinsic φ' of the constituents and that the higher values recorded in undrained compression and extension tests on vertical samples include the effects of the fibre structure. The relevance of these higher values must therefore be related to the actual failure mechanism being considered. There is little information on the effect of the possible presence of gas on the effective stress parameters. The results of a series of undrained and drained compression and extension triaxial tests on a fibrous peat reported by Farrell and Hebib (1998) is shown in **Figure 35.8**. Note that the drained tests, which were taken to 30% strain, did not reach the failure line and the φ' in compression of 55° was considerably greater than the φ' of 39° and 18° measured in the extension tests at initial consolidation pressures of 40 kPa and 60 kPa, respectively. The direct shear and ring shear tests on this peat recorded φ' of 38°.

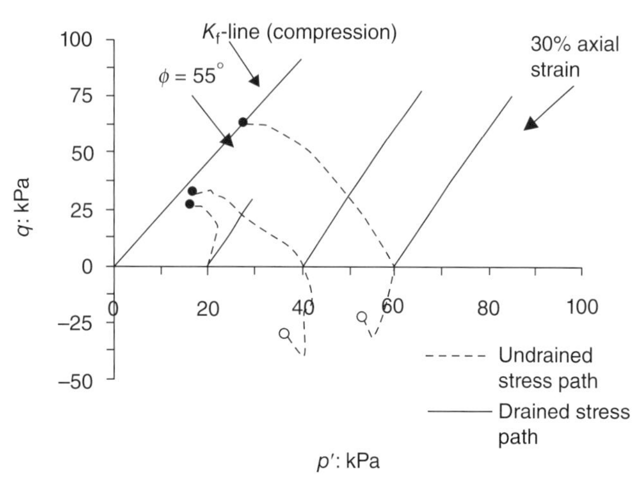

Figure 35.8 Triaxial tests on a fibrous peat (Reproduced from Farrell and Hebib, 1998)

The effective stress parameters for the organic soil which do not have a fibre structure can be obtained using the normal methods.

The at-rest pressure (K_o), which is normally related to φ' by Jaky's equation $K_o = 1 - \sin\varphi'$, has generally been measured at between 0.3 and 0.35 on organic soils with water contents in the 330–850% range, which would suggest φ' values of the order of 40–44° (Mesri, 2007). Edil and Wang (2000), on the basis of laboratory tests, considered that K_o was of the order of 0.49 for amorphous peats and 0.33 for fibrous peats.

35.5.3 Undrained shear strength

The undrained shear strength of organic soils (c_u) can be determined using the traditional methods of consolidated triaxial tests, in situ vanes, CPTu and so on; however, the results of these tests are affected by the fibres in the more fibrous peats. Furthermore, the undrained shear strength of some peats is very low, of the order of 2–4 kPa, which is at the limit of the accuracy of normal testing arrangements. The gas content of peats can also affect their in situ behaviour, and this is frequently eliminated in laboratory tests by using a backpressure. This gas content can be lost when sampling the fibrous peat unless special measures are taken, as is evident from the length of samples recovered compared to the length pushed into the ground. There is a move towards testing larger triaxial samples of fibrous peat to ensure that the fibre length is captured in the tests and also to use modified end caps to limit the effects of end restraint (Murano and Jommi, 2021; Zwanenburg and Van, 2015).

Laboratory tests have shown that peats have a high c_u/σ'_{vc} ratio in compression of about 0.5–0.6, which compares with ratios of about 0.22–0.3 for inorganic soils (Mesri, 2007). Similar high ratios have been recorded in extension. The DSS test is receiving more attention in recent years as the failure mechanism is considered to be relevant to peat slides. Ratios of c_{uDSS}/σ'_v of 0.45 (Porbaha *et al.*, 2000), 0.46 (Foott and Ladd, 1981) and 0.5 (Farrell and Hebib, 1998) were recorded on fibrous peats, considerably higher than the ratio of about 0.22–0.3 expected for inorganic clays.

The field measurement of the in situ undrained shear strength of peats is problematic due to the influence of fibres on the results and also because of the low strengths measured. Long and Boylan (2012) report that temperature variations can give rise to significant inaccuracies in CPTu probes, which often result in the interpretation of negative strength values, and advocate the use of full-flow penetrometers such as the T-bar and ball penetrometer for peats, which test a greater volume of the peat than the standard CPT probe (Boylan and Long, 2006). The methods of determining the undrained shear strength of fibrous peats by field tests is an area that requires further research.

In the absence of an alternative field test, the vane (FVT) is frequently used to determine a c_u of peat despite its significant limitations which have been discussed by Landva (1980). For example, the vane tended to pull the fibres along a surface outside the vane itself; there was compression of the peat in front of the vane; and a gap frequently opened at the back of the vane – conditions far removed from a cylindrical shearing surface assumed when interpreting the results. Furthermore, the interpreted c_u values are generally found to decrease with the size of the vane, as is illustrated on a comparison of c_u values measured in a fibrous blanket bog in **Figure 35.9**. These results are consistent with those reported by Landva (1980), where the c_u interpreted from a 55 × 110 mm vane were about twice that of a 75 × 150 mm vane and the interpreted undrained strength of this fibrous peat was as low as 2 kPa. Notwithstanding these limitations, the in situ vane is frequently used to assess the in situ strength of fibrous peats, as well as the increase in strength with consolidation, with the 55 mm dia. vane becoming the preferred size (Noto, 1991) and the test being considered to be an index test only. Mesri (2007) tentatively suggests that c_{uvane} be reduced by 0.5 to get the operational c_u; however, Noto (1991) estimated that the ratio of c_{ufield}/c_{uvane}, with c_{ufield} values being back-calculated from failure experiments on a testing embankment, to vary from 0.38 to 1.24, with an average of 0.74.

Zwanenburg and Jardine (2015) present the findings of relatively rapid full-scale field trials to failure at Uitdam, the Netherlands, on a fibrous peat (75% < N < 92%;

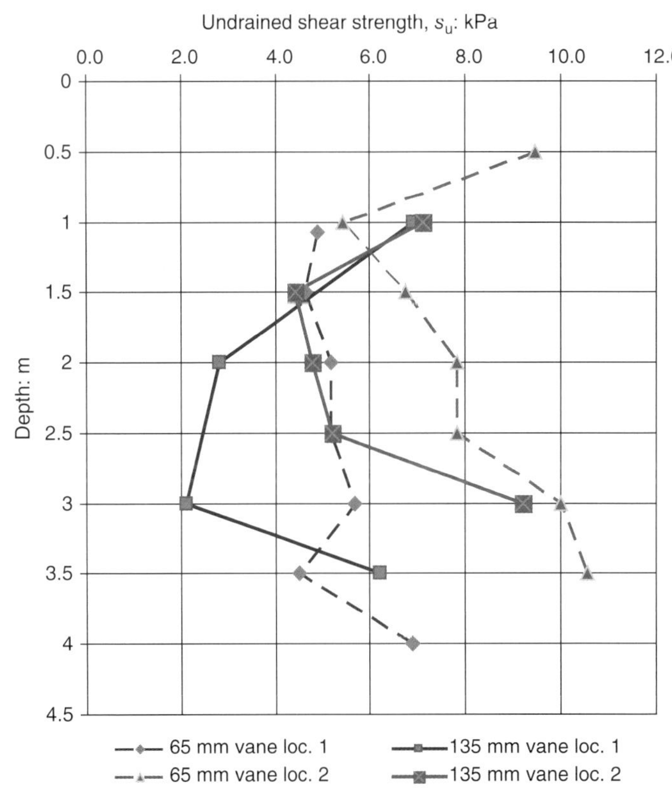

Figure 35.9 Comparison of s_u measured with 65 and 135 mm dia. vanes in a fibrous peat

645% < w < 1240%) which was tested with a comprehensive series of laboratory and field tests. Representative operational c_u ranges were determined from simple Tresca-based numerical back-analyses. These analyses tended to confirm the reducing factor of 0.5 for c_{uvane}, and interpreted cone factors for CPTu and ball penetrometers. The analyses of field test results highlighted the substantial variations around the mean c_u profile trends of ± 30, 20 and 40% for the CPTu, ball and FVT devices, respectively. A vertical yield stress profile was estimated from 24-hour staged loading oedometers tests using the approach of Becker *et al.* (1987). The analyses of the laboratory and field data appear to show that it is possible to predict the operational shear strengths by applying a Shansep approach based on oedometer yield stress and normally consolidated c_u/σ'_v ratios developed from either triaxial or DSS tests, interpreted as outlined in their paper. The c_u/σ'_v ratios are significantly higher than those found for clays, which is consistent with the results of laboratory tests.

35.6. Critical design issues in peats and organic soils

35.6.1 Roads

Most of the design/construction options for construction on soft inorganic soils can be used for roads on peats and organic soils, but these must be modified to consider the high compressibility/large deformations, the significant secondary compression, the low density of peats and the local geology within which the peat was formed. Munro (1991) gives a useful overview of some of the construction methods adopted by road authorities in Northern Europe.

The excavate/replace option is widely used for major roads on relatively thin layers of peats, generally up to about 3–4 m in thickness, as it avoids the uncertainties of long secondary compression movement which may arise even with surcharging. However, it becomes uneconomical for significant depths of peats due to cost and the environmental considerations of disposal of the excavated material. Apart from the cost, there is increasing consideration being given to constructing on peat to limit the environment effects arising from its removal. Additionally, the requirement for the replacement material to extend out from the footprint of the road to avoid surface cracks in the road pavement due to lateral consolidation (**Figure 35.10**) can also significantly affect costs and land-take requirements. Lateral failure can occur if the granular replacement material is placed against a vertical or near-vertical peat face due to the low density of the peat compared with a granular material. The replacement of upper peat layers may make it difficult to install vertical drains should these be required to speed up consolidation in underlying soft clay layers.

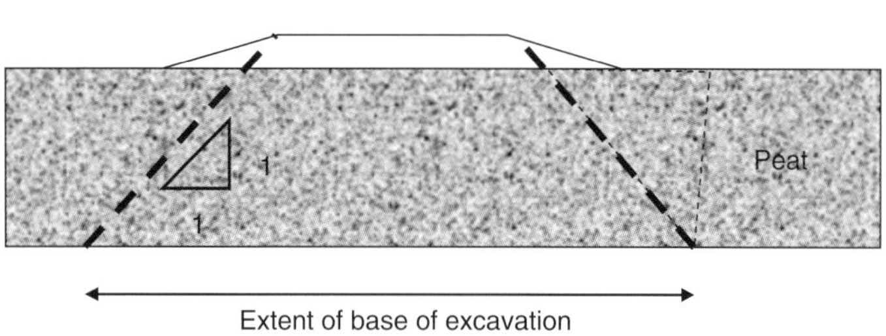

Figure 35.10 Extent of excavate/replace required

Historically, there is a legacy of old roads that have been constructed directly on top of the peats that currently serve in the minor road networks in the UK and Ireland. These roads, which generally started as gravel roads but were subsequently paved, develop an undulating profile under traffic and require regular maintenance and rehabilitation. The increased deformation under modern traffic also gives rise to transverse cracking which can be mitigated to some extent by including geosynthetics to reinforce the bituminous layers or sometimes the subgrade. Raised bogs frequently start their life over the very soft lake clay deposits; consequently the practice of placing additional surface material to periodically regulate the profile has resulted in past failures of the entire road due to shear failure through the soft clays. This problem can be avoided, and the secondary compression settlements reduced, by the use of lightweight fill in road improvement works.

'Floating' roads are a common method of forming access and low-traffic roads on peats. Traditionally these were constructed using tree trunks as a base, but current practice is to use geosynthetics as basal reinforcement to a gravel layer where trees are not available. Such roads, which can carry heavy loads in fibrous peats due to the reinforcing effects of the fibres, are frequently used in forestry and windfarm sites in mountain areas; consequently their effect on the overall stability must be assessed in any design. Scottish Natural Heritage (2010) give some guidance on the design and construction of floating roads on peat.

Lightweight fill, together with surcharge, is an acceptable construction option on peats and organic soils. Large-scale use of these materials tends to be limited by their costs. Lightweight fill is very efficient when used in combination with surcharge as the effectiveness of surcharge depends on the surcharge ratio, as discussed in Section 35.4.3. When using lightweight fill the design must consider the risk of the road floating under flood conditions.

Stage construction with surcharging is the most common method for constructing on organic soils, and is sometimes used for peats. The terms 'preload' and 'surcharge' are considered to differ in that a preload is taken up to the design load whereas a surcharge is taken beyond the design load, and in the case of peats and organic soils is designed to reduce secondary compression settlements. The principles of the design of surcharges are discussed in Section 35.4.3. Thus, if the expected primary settlements and secondary settlements under a design load are estimated to be X, then the design load and a surcharge are left in place until this amount of settlement has occurred. Secondary compression settlements will occur after that design period, but at a lesser rate than would

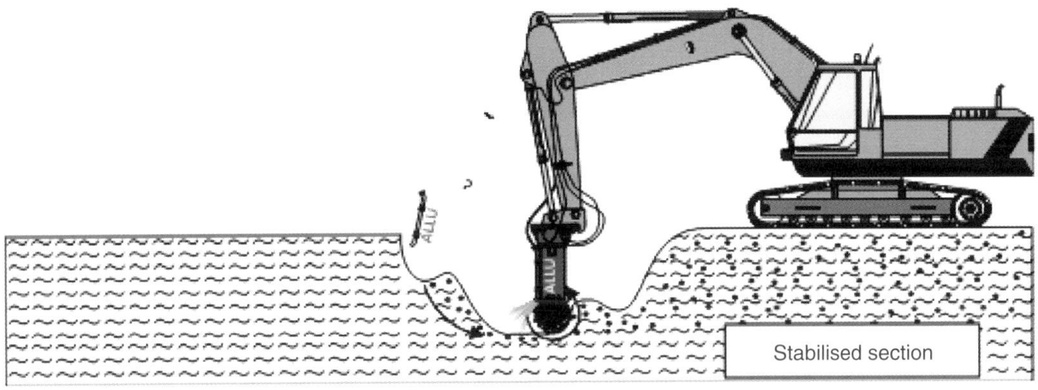

Figure 35.11 Mass stabilisation (Reproduced from BRE, 2002)

arise at the end of primary compression. In some situations the settlements can exceed the thickness of fill placed, and it is necessary to apply more fill or to use lightweight fill for the initial loadings. The reduction in the effective stress due to submergence of the fill must be considered when estimating the settlements. Where the surcharge is not left in place for a sufficient time, care must be taken to ensure that there are no residual excess pore water pressures within low-permeability layers after its removal.

The applicability of vertical drains to speed up the consolidation process depends on the type of peat or organic soil. Their usefulness for low embankments on fibrous peat which have high initial permeability is questionable, particularly as the drains themselves may be effected by the high compressions; however, the permeability decreases significantly at high effective stresses, thus the drains may be required where higher effective stresses are applied. General experience is that sand drains can act to reinforce peat, thus increasing its stability and reducing vibrations under load. Vacuum consolidation has benefits in speeding up consolidation but requires further research for use with the highly organic peats.

Stabilisation of peats and organic soils by mixing in a binder is a common practice in Sweden, Finland and some other countries, but has not been widely used in the UK or Ireland to date. The binder can be mixed in using an attachment to an excavator as shown in **Figure 35.11** to form a stabilised mass, or alternatively a mixing tool can be used to form stabilised columns, as shown in **Figure 35.12**, or a combination of both. Two types of mixing methods can be used: the dry method where the binder is injected pneumatically; and the wet process where the binders are mixed in a slurry form before being injected from the mixing tool. The design and construction aspects using this method are set out in the manual *Design Guide: Soft Soil Stabilisation* (BRE, 2002) and some experiences with Irish raised bogs are given by Hebib and Farrell (2003). The efficiency of this process is very dependent on the mineral content of the peat/organic soil and on the humic acid in the peat, the process being more cost-effective in organic clays than in the high-organic peats. Stabilisation of the

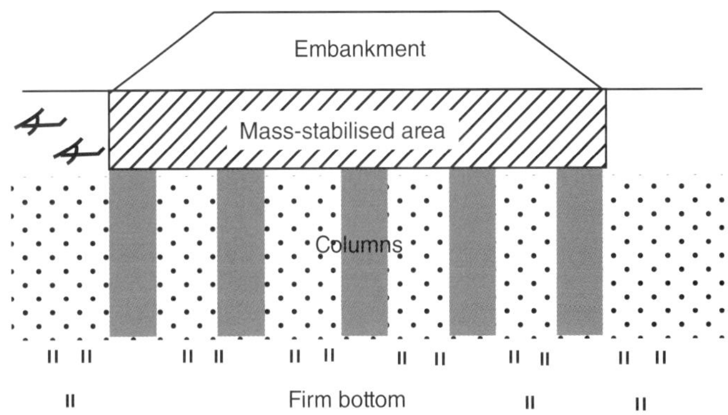

Figure 35.12 Column and mass stabilisation (Reproduced from BRE, 2002)

high-organic peats in Ireland has not proven to be cost-effective to date as about 150–200 kg of cement per cubic metre of peat is required to achieve the required engineering properties (Hebib and Farrell, 2003).

The efficiency of stone columns will depend on the organic content of the soil due to the inevitable lateral consolidation and creep of the soils, and also due to the low strength of the peats. The actual feasibility of constructing in situ concrete columns in very soft ground could limit the use of that method.

Piled supported basal reinforced embankments have been constructed on very compressible and very soft peats (Orsmond, 2008). Issues that must be considered include the lateral stability of the entire embankment, the difficulties in maintaining the position of piles, the type of pile and the construction platform to access the locations. Basally reinforced granular layers have generally provided a suitable platform for light piling rigs, even on very soft peats. Driven piles are commonly used. The feasibility of using cast-in-place or CFA piles will depend on the ability to get access for the rigs and also on the ability of the very soft material to support liquid concrete without radial failure. Although rarely reported in professional journals for legal reasons, failures have occurred in basally reinforced piled embankments due to stockpiling of material adjacent to the structure, which give rise to failure of the piles due to the lateral loadings. This failure mechanism also has implications for long-term development of land adjacent to such structures.

35.6.2 Structures

The design of structures on peats and organic soil follows the normal procedure for building on soft soils, taking into account the high compressibility of peats. Structures are normally piled, although surcharging (see Section 35.6.1) has been used in some situations with a rafted structure. It is normally necessary to construct a gravel-on-geogrid working platform to allow access for an appropriate piling rig and this platform will inevitably settle away from any piled structure, leaving a gap in the long term and also giving rise to downdrag forces on the piles. The risk of buckling of piles is assessed in the normal manner.

35.6.3 Slopes

Bog bursts and peat slides are major geohazards, particularly in Scotland and Ireland, which can be caused by environmental or anthropological factors. Major slides have occurred on slopes as shallow as 3° and have involved mass of up to 100 000 tonnes of material covering about 0.5 km in width and over 1.5 km in length. Villages have been destroyed in the past, with one particularly unfortunate incident in Castlegarde, Co Limerick, Ireland in 1708 in which 21 people died. Recent research for the development of a landslide database has shown a proliferation of historic landslides in upland peat areas of Ireland. The Geological Survey of Ireland produced a booklet on landslides (Creighton, 2006), including bogslides, and the Scottish Executive (2006) have published a document on peat landslide hazard and risk assessment. Given that a peat bog sitting on the side of a hill is made up of over 90% water, it is not surprising that slope stability is a major issue.

Bog failures are generally thin relative to the length of material involved and therefore resemble a planar-type slide, as is discussed in Chapter 23: *Slope stability*. Typically the upper layers of bogs are fibrous and the reinforcing effect of these fibres prevent a local failure and thus contribute to the formation of a large mass movement with a failure surface at or close to the base of the slide. The determination of the disturbing force ($W \sin\alpha$, where α is the angle of the base to the horizontal) is relatively straightforward; however, the estimation of the resisting force depends on whether the failure is considered in terms of effective or total stress. With reference to Section 23.3.2, and using the terms defined in that chapter, the resistance (R) is

$$R = (W\cos\alpha - u_\mathrm{A})\tan\varphi' + c'_\mathrm{A} \qquad (35.8)$$

Given the high water tables that are normally present in these bogs, and the fact that the bulk density of peat can be close to or less than that of water, the $W\cos\alpha - u_\mathrm{A}$ term can be negligible and the resistance is then c'_A. Considering the slope in terms of total stress, the disturbing force is the same as for the effective stress analysis, namely $W\sin\alpha$, and the resistance is $c_\mathrm{u}A$, thus the difference in the analyses is whether c' or c_u should be used.

Some blanket bogs are known to have ‘pipes’ at their base through which water flows during heavy rainfalls (Donnelly, 2008). Water entry to the base of these bogs, with a consequential build-up of water pressures, is also facilitated by tension cracks forming in the peat during periods of dry weather. Given that the bulk density of peat is very close or even less than that of water, the effective stress can be very small or negligible and the resisting force therefore depends on the apparent cohesion, if this is present. The failure surface of a well-documented peat slide in Pollatomish, Co May (Long and Jennings, 2005), which is shown in **Figure 35.13**, was at the interface of the peat and an underlying sand layer following a period of heavy rain, confirming that slide was caused by a loss in effective stress. The risk of such failures is increased by a prolonged period of dry weather which can dry out the peat, thus reducing the bulk density, and also give rise to opening of tension cracks which facilitate the access of water. The build-up of water pressures at the base of the peat can also be exacerbated by the presence of an iron pan layer which is sometimes noted at the base of blanket bogs.

When considering the stability of peat layer of thickness D against planar failure in terms of total stress, with a c_u assigned to the peat, the disturbing force per unit basal area is $\gamma D\sin\alpha$ and the resisting force is a function of c_u, assuming there are no external loads and that the tension capacity of the peat is negligible compared with other forces. The ratio of resisting to disturbing force is then $c_\mathrm{u}/\gamma D\sin\alpha$. This ratio therefore decreases with increased thickness, increased slope angle and increased bulk density. Some of the bog failures have occurred

Figure 35.13 Peat slide at Pollatomish, 2003

Figure 35.14 Edenderry canal failure, Ireland, 1989 (Reproduced from Piggott *et al.*, 1992)

on very flat slopes (as low as 3°, but frequently on slopes of about 4–7°). Using this total stress approach, back-analysis of some of the bog slides would indicate c_u values at failure of the order of 2 kPa, which is very low and difficult to measure in the field. Farrell (2010) proposed a mechanism to explain how a major bog slide could progress downhill if a local failure is initiated at the higher elevations if the material at the base of the slide is sensitive.

The stability of bogs can be affected by forestry drainage ditches and other anthropogenic factors.

35.6.4 Retaining structures, canals and dams

The special issues arising from the use of peats and organic soil in relation to retaining structures, canals and dams are their low density, high compressibility, high creep rate and generally low strength. The water-retaining structures require particular attention because small changes in the hydrogeology can result in very low effective stresses. Major failures have occurred in the past, such as the Edenderry canal failure of 1989 (Piggott *et al.*, 1992), which is shown in **Figure 35.14**, which resulted in the horizontal movement of a 225 m by 105 m block of peat a distance of about 60 m.

35.7. Conclusions

Peats and organic soils, which are to be found in many countries and are very extensive in some, present particular challenges to the geotechnical engineer because of their high compressibility, creep, low density and generally very low shear strength. Despite the presence of the cellular structure of organic remains, many of the correlations between the various properties and their behavioural patterns are little different from those of a clay, other than their shear strength.

Various processes can be involved in the formation of organic soils and the nature of the process can be important when interpreting the relevant geotechnical parameter and soil stratification appropriate to a particular design/construction situation. Classification systems have been developed to assist in the characterisation of the materials, the von Post (1922) or a variant being the system most generally adopted. The water content is the most important index test as it is a simple parameter to which many engineering properties can be related; however, the organic content, determined by the loss on ignition test, and the pH are also useful, as is the liquid limit, which can generally be carried out on peats with a degree of decomposition greater than H_3 on the von Post scale.

The recovery of 'undisturbed' samples of highly organic peats, particularly fibrous peats, is difficult due to the compression of the material during sampling even with sharp-edged piston samplers. Furthermore, the samples may undergo oxidation when exposed, which can lead to decomposition and gas generation, although the effect of this is generally not considered significant except possibly for long-term testing.

The methods of estimating settlements and rates of settlements used for clays are also used for organic soils, although the effect of secondary consolidation/creep on these estimates increases with the organic content. Because of the large settlements involved in the highly organic soils, which can be greater than the original thickness of fill placed, and the natural variability within these deposits, simple behavioural models are usually used which consider that the secondary consolidation/creep starts at the end of the primary phase. Most sophisticated models which consider that secondary consolidation occurs continuously are more difficult to apply but are being incorporated into some software programs. Surcharging of the peat is an effective method of reducing secondary compression/creep.

The undrained shear strength of organic soils and peats is normally determined using in situ vanes, cone penetration tests and laboratory triaxial or DSS tests, although the most appropriate tests for a particular failure mechanism in fibrous peats requires careful consideration. The very low undrained shear strength of some peats, sometimes of the order of 2 kPa, is below the accuracy of many of the current measurement methods. Similarly, the effective stress parameters determined in triaxial tests can be affected by fibres and by the method of test.

The unique properties of peats and organic soils have to be considered when designing roads and foundations or when assessing the stability of slopes in these soils. Some of the methods that have been adopted have been presented within this chapter.

References

Anderson KO and Haas RC (1962) Construction over muskeg on the Red Deer Bypass. *Proceedings of the 8th Muskeg Research Conference. Ottawa, ON*, pp. 31–41.

ASTM (2006) D5715-00: Standard test method for estimating the degree of huminification of peat and other organic soils (visual/manual method). ASTM, West Conshohocken, PA.

Becker DE, Crooks JHA, Been K and Jefferies MG (1987) Work as a criterion for determining in situ and yield stress in clays. *Canadian Geotechnical Journal* **24(4)**: 549–564.
Berry PL and Poskitt TJ (1972) The consolidation of peat. *Géotechnique* **22**: 27–52.
Boylan N and Long M (2006) Characterisation of peat using full flow penetrometers. *Proceedings of the 4th International Conference on Soft Soil Engineering, Vancouver, Canada*. Taylor and Francis, London, UK, pp. 403–414.
Boylan N, Jennings P and Long M (2008) Peat slope failure in Ireland. *Quarterly Journal of Engineering Geology and Hydrogeology* **41**: 93–108.
Brawner CO (1959) The muskeg problem in British Columbia highway construction. *Proceedings of the 4th Muskeg Research Conferences. Ottawa, ON*, pp. 45–53.
BRE (2002) *Design Guide: Soft Soil Stabilisation*. BRE Press, Watford, UK.
BSI (British Standards Institution) (2015) BS 5930: Code of practice for site investigations. BSI, London, UK.
BSI (2017a) EN ISO 14688-1:2017: Geotechnical investigation and testing – Identification and classification of soil – Part 1: Identification and description. BSI, London, UK.
BSI (2017b) EN ISO 14688-2:2017: Geotechnical investigation and testing – Identification and classification of soil – Part 2: Principles of classification. BSI, London, UK.
BSI (2018) BS 1377-3:2018 Methods of tests for soils for civil engineering purposes – Chemical and electro-chemical tests. BSI, London, UK.
Carlsten P (1991) Embankment on soft organic soil using the preloading technique. *Proceedings of the 10th European Conference on Soil Mechanics and Foundation Engineering*. Florence, pp. 359–362.
Casagrande A (1936) The determination of the preconsolidation load and its practical significance. *Proceedings of the 1st International Conference on Soil Mechanics and Foundation Engineering, Cambridge, MA*, vol. 3, pp. 60–64.
Clymo RS (1983) Peat. In *Ecosystems of the world 4A: Mires – Swamp, Bog, Fen and Moor*. Elsevier, Oxford, UK, pp. 159–224.
Creighton R (2006) *Landslides in Ireland: A Report of the Irish Landslides Working Group*. Geological Survey of Ireland, Blackrock, Ireland.
den Haan EJ (1996) A compression model for non-brittle soft clays and peat. *Géotechnique* **46**: 1–16.
Dhowain AW and Edil TB (1980) Consolidation behavior of peat. *Geotechnical Testing Journal* **3(3)**: 105–114.
Donnelly LJ (2008) Subsidence and associated ground movements on the Pennines, northern England, *Quarterly Journal of Engineering and Geology* **41**: 315–332.
Edil TB and Wang X (2000) Shear strength and K_0 of peats and organic soils. In *Geotechnics of High Water Content Materials*. ASTM, West Conshohocken, PA, pp. 209–225.
Eggestad A and Føyn T (1977) Settlement observations beneath a test fill on peat. Väg-ogvatten-byggaren. *Journal of Swedish Society of Civil Engineers* **23(8–9)**: 44–45.
Farrell ER (2010) Lessons learnt: problems to solve. *Proceedings of the BCRI Conference, Cork, Ireland*, pp. 181–198.
Farrell ER and Hebib S (1998) The determination of the geotechnical properties of organic soils. In *Problematic Soils* (Yanagisawa M (ed)). Balkema, Rotterdam, the Netherlands, pp. 33–36.
Farrell ER and O'Donnell C (2007) Comparison of predicted and observed performance of some embankments on soft ground in Ireland. *Soft Ground Engineering, Portlaosie, Ireland*, paper 3.1.
Farrell ER, Jonker SK, Knibbelerb AGM and Brinkgreve RBJ (1999) The use of direct simple shear test for the design of a motorway on peat. *XII European Conference on Soil Mechanics and Geotechnical Engineering, Lisbon, Portugal*, pp. 1027–1033.
Foott R and Ladd CC (1981) Undrained settlement of plastic and organic clays. *Journal of the Geotechnical Engineering Division* **107(8)**: 1079–1094.
Fox PJ and Edil TB (1996) Effects of stress and temperature on secondary compression of peat. *Canadian Geotechnical Journal* **33**: 405–415.
Fox PJ, Roy-Chowdhury N and Edil TB (1999) Secondary compression of peat with or without surcharging, *Journal of Geotechnical and Geoenvironmental Engineering* **125(2)**: 160–162. Discussion.
Fürstenberg A, Lechowicz Z, Szymanski A and Wolski W (1983) Effectiveness of vertical drains in organic soils. *Proceedings of the 8th European Conference on Soil Mechanics and Foundation Engineering*. Helsinki, pp. 611–616.
Gautschi MA (1965) Peat as foundation soil [Torf als baugrund], Research summary report, Norwegian Geotechnical Institute, Oslo.
Gore AJP (1983) Introduction. In *Ecosystems of the World 4A. Mires – Swamp, Bog, Fen and Moor*. Elsevier, Oxford, UK, pp. 1–34.
Hanrahan ET (1953) The mechanical properties of peat with special reference to road construction. *Transactions of the Institution of Civil Engineers of Ireland* **78**: 179–215.
Hanrahan ET (1954) An investigation into some physical properties of peat. *Géotechnique* **4**: 108–123.
Hanrahan ET, Dunne JM and Sodha VG (1967) Shear strength of peat. *Proceedings of the Geotechnical Conference, Oslo, Norway*, vol. 1, pp. 193–198.
Hebib S and Farrell ER (2003) Some experiences on the stabilization of Irish peats. *Canadian Geotechnical Journal* **40(1)**: 107–120.
Hobbs NB (1986) Mire morphology and the properties and behaviour of some British and foreign peats. *Quarterly Journal of Engineering Geology* **19**: 7–80.
Hobbs NB (1987) A note on the classification of peat. *Géotechnique* **37(3)**: 405–407.
Hogan JM, van der Kamp G, Barbour SL and Schmidt R (2006) Field methods for measuring hydraulic properties of peat deposits. *Hydrological Processes* **20**: 3635–3649.
Janbu N (1963) Soil compressibility as determined by oedometer and triaxial tests. *Proceedings of the European Conference of Soil Mechanics and Foundation Engineering, Wiesbaden, Germany*, pp. 19–25.
Kellner E, Waddington JM and Price JS (2005) Dynamics of biogenic gas bubbles in peat: potential effects on water storage and peat deformation. *Water Resource Research* **41**: W08417.
Keyser JH and Laforte MA (1984) Construction and performance of pavement over muskegs. *Transportation Research Record* **978**: 68–80.
Kidson C and Heyworth A (1976) The quaternary deposits of the Somerset Levels. *Quarterly Journal of Engineering and Geology* **9**: 217–235.
Kogure K (1998) Consolidation and settlement of peat under loading. *Proceedings of International Symposium on Problematic Soils. Japan*, pp. 14–29.

Landva AO (1980) Vane testing in peat. *Canadian Geotechnical Journal* **17**: 1–19.

Landva AO and Pheeney PE (1980) Peat fabric and structure. *Canadian Geotechnical Journal* **17**: 416–435.

Landva AO and La Rochelle P (1983) Compressibility and shear characteristics of Radforth peats. In *Testing of Peat and Organic Soils* (Jarrett PM (ed)). ASTM, West Conshohocken, PA, pp. 157–191.

Landva AO, Korpijaakko EO and Pheeney PE (1983) Geotechnical classification of peats and organic soils. In *Testing of Peat and Organic Soils* (Jarrett PM (ed)). ASTM, West Conshohocken, PA, pp. 37–54.

Lea ND and Brawner CO (1963) Highway design and construction over peat deposits in Lower British Columbia. *Highway Research Record* **7**: 1–31.

Lefebvre G, Langlois P, Lupien C and Lavallee J (1984) Laboratory testing on in situ behavior of peat as embankment foundation. *Canadian Geotechnical Journal* **21**: 322–337.

Lewis WA (1956) The settlement of the approach embankments to a new road bridge at Lockford, West Suffolk. *Géotechnique* **3**: 106–114.

Liingaard M, Augustesen A and Lade PV (2004) Characterization of models for time-dependent behaviour of soils. *International Journal of Geomechanics* **4(3)**: 157–177.

Long M and Boylan N (2012) In-situ testing of peat: a review and update on recent developments. *Geotechnical Engineerings Journal of SEADS & AGSSEA* **43(4)**: 41–55.

Long M and Jennings P (2005) Analysis of the peat slide at Pollatomish, County Mayo, Ireland. *Journal of Landslides* **3**: 51–61.

Long M, Paniagua P, Grimstad G *et al.* (2022) Engineering properties of Norwegian peat for calculation of settlements. *Engineering Geology*, Oct.: p. 308.

MacFarlane IC and Allen CM (1964) An examination of some index test procedures for peat. *Proceedings of the 9th Muskeg Research Quebec City, ON*, pp. 171–183.

MacFarlane IC and Radforth NW (1964) A study of the physical behaviour of peat derivatives under compression. *Proceedings of the 10th Muskeg Research Conference. Ottawa, ON*, pp. 159–164.

Mesri G (2007) Engineering properties of fibrous peats. *Journal of Geotechnical and Geoenvironmental Engineering, ASCE* **133**: 850–866.

Mesri G and Godlewski PM (1977) Time– and stress–compressibility interrelationship. *Journal of the Geotechnical Engineering Division* **103**: 417–430.

Mesri G, Stark TD, Ajlouni MA and Chen CS (1997) Secondary compression of peat with or without surcharging. *Journal of Geotechnical and Geoenvironmental Engineering* **123**: 411–421.

Mitchell JK and Soga K (2005) *Fundamentals of Soil Behavior*, 3rd edn. Wiley, New York, NY.

Moore PD and Bellamy DJ (1974) *Peatlands*. Springer, New York, NY.

Munro RSP (1991) *Road Construction Over Peat*. Department of Roads and Transport, Highland Regional Council, Inverness, UK.

Murano S and Jommi C (2021) Determination of the shear strength of peat from standard undrained triaxial tests: correcting for the effects of end restraint. *Géotechnique* **71(1)**: 76–87.

Noto S (1991) *Peat Engineering Handbook*. Civil Engineering Research Institute, Hokkaido Development Agency, Hokkaido, Japan.

O'Kelly B (2005) Method to compare water content values determined on the basis of different over drying temperatures. *Géotechnique* **55(4)**: 329–332.

O'Kelly B (2017) Measurement, interpretation and recommended use of laboratory strength properties of fibrous peat. *Geotechnical Research* **4(GR3)**: 136–171.

O'Loughlin CD (2007) Simple and sophisticated methods for predicting settlement of embankments constructed on peat. *Proceedings of the Conference on Soft Ground Engineering, Portlaoise, Ireland*, paper 1.4.

Orsmond W (2008) A1N1 Flurrybog piled embankment design, construction and monitoring. *Proceedings of the 4th European Geosynthetic Conference, Edinburgh, UK*, paper 290.

Piggott PT, Hanrahan ET and Somers N (1992) Major canal construction in peat. *Proceedings of the Institution of Civil Engineers: Water, Maritime and Energy* **96**: 141–152.

Plaxis (2020) *Plaxis Connect Material Models*. Version 20. Plaxis B.V., Delft, the Netherlands.

Porbaha A, Hanazawa H and Kishida T (2000) Analysis of a failed embankment on peaty ground. *Slope Stability* 2000, *Proceedings of the Geo-Denver 2000 Geo-Institute Soft Ground Technology Conference*, Reston, VA, pp. 281–293.

Radforth NW (1969) Classification of muskeg. In *Muskeg Engineering Handbook* (MacFarlane IC (ed)). University of Toronto Press, Toronto, ON, Canada.

Ripley CF and Leonoff CE (1961) Embankment settlement behaviour on deep peat. *Proceedings of the 7th Muskeg Research Conference. Ottawa, ON*, p. 71.

Rowe RK, MacLean MD and Barsvary AK (1984) The observed behaviour of a geotextile-reinforced embankment constructed on peat. *Canadian Geotechnical Journal* **21**: 563–576.

Royer JA (1936) Experimental road for heavy traffic on very compressible soil (peat-bog-ground). *Proceedings of the International Conference on Soil Mechanics and Foundation Engineering*, pp. 235–238.

Samson L and La Rochelle P (1972) Design and performance of an expressway constructed over peat by preloading. *Canadian Geotechnical Journal* **22**: 308–312.

Scottish Executive (2006) *Peat Landslide Hazard and Risk Assessment*. Scottish Executive, Edinburgh, UK.

Scottish Natural Heritage (2010) *Floating Roads on Peat*. SNH, Inverness, UK.

Skempton AW and Petley DJ (1970) Ignition loss and other properties of peats and clays from Avonmouth, King's Lynn and Cranberry Moss. *Géotechnique* **20**: 343–356.

Tan Y (2008) Finite element analysis of highway construction in peat bog. *Canadian Geotechnical Journal* **45**: 147–160.

Terzaghi K, Peck RB and Mesri G (1996) *Soil Mechanics in Engineering Practice*, 3rd edn. Wiley, New York, NY.

von Post L (1922) Sveriges Geologiska Undersoknings torvinventering och nogra av dess hittils vunna resultat [SGU peat inventory and some preliminary results]. *Svenska Mosskulturforeningens Tidskrift, Jonkoping, Sweden* **36**: 1–37.

Yamaguchi H, Ohira Y, Kogue K and Mori S (1985a) Undrained shear characteristics of normally consolidated peat under triaxial compression and extension tests. *Japanese Society of Soil Mechanics and Foundation Engineering* **25(3)**: 1–18.

Yamaguchi H, Ohira Y, Kogue K and Mori S (1985b) Deformation and strength properties of peat. *Proceedings of the 11th*

International Conference on Soil Mechanics and Foundation Engineering, vol. 4, pp. 2461–2464.
Zwanenburg C and Jardine RJ (2015) Laboratory, in situ and full-scale load tests to assess flood embankment stability on peat. *Géotechnique* **65(4)**: 309–326.
Zwanenburg C and Van MA (2015) Comparison between conventional and large scale triaxial compression tests on peat. *Proceedings of the 6th International Symposium on Deformation Characteristics of Geomaterials, Buenos Sires, Argentina* (Rinaldi VA, Zeballos ME and Clariá JJ (eds)). ISO Press, Amsterdam, the Netherlands, pp. 543–550.

Further reading

Bell FG (2000) *Engineering Properties of Soils and Rocks*, 4th edn. Blackwell Science, Oxford, UK.

It is recommended this chapter is read in conjunction with

- Chapter 7 *Geotechnical risks and their context for the whole project*
- Chapter 40 *The ground as a hazard*
- Chapter 48 *Geo-environmental testing*
- Chapter 49 *Sampling and laboratory testing*

All chapters in this book rely on the guidance in Sections 1 *Context* and 2 *Fundamental principles.* A sound knowledge of ground investigation is required for all geotechnical works, as set out in Section 4 *Site investigation.*

doi: 10.1680/icemge.66816.0517

Chapter 36
Mudrocks, clays and pyrite implications

Mourice A Czerewko AECOM Ltd, Chesterfield, UK
John C Cripps University of Sheffield, Sheffield, UK

CONTENTS

Mudrock is the general name given to a large variety of fine-grained, clay-rich sedimentary rocks. As these materials occur frequently in the sedimentary rock sequence they commonly form the natural ground on construction sites and they may also be encountered as made ground, fill or construction materials. Depending on compositional and structural factors, they can display a range of engineering behaviours – some of them problematic, including low strength, low durability and susceptibility to volume changes. In addition, it can be difficult to assess their mass properties. This chapter reviews the present state of knowledge concerning the performance of mudrocks in engineering situations. The intention is to provide the reader with a framework for understanding the geotechnical properties of mudrocks, especially with regard to effects and changes brought about by construction activities.

36.1. Introduction

The term mudrock is used here to denote a diverse group of fine-grained, clay-rich sedimentary rocks. Since such materials occur widely in the geological sequence, they are very frequently encountered in civil and environmental engineering projects. They are important because they can be the cause of a range of problems both during and after construction. A generally accepted geological definition of mudrocks that is suitable for engineering purposes is proposed by Stow (1981) as follows: mudrocks have a dominant grain size of <63 μm comprising more than 50% of the constituents; and more than 50% of the mineral constituents are composed of siliciclastic mineral fragments (including clay minerals, quartz and feldspar). Other fine-grained sediments with different compositions are designated accordingly as carbonate mudrocks, silica mudrocks and so on. As explained in Section 36.1.1, there are discrepancies in terminology resulting in disagreement among different standards.

Mudrocks typically range from soft soil-like clay deposits to very stiff soils and medium strong rocks. Due to the low strength and high compressibility of the former types, coupled with shrink–swell behaviour and, in the case of harder types, rapid breakdown in response to changes in moisture content and weathering processes, mudrocks present various challenges in civil engineering design and construction work. The controls on the performance of mudrocks and several forms of potentially problematic behaviour in different engineering situations are illustrated in **Figure 36.1**. Problematic behaviour often arises because of the combined effects of natural weathering processes and the impact of engineering works, but it is also strongly influenced by the composition and structure of the material.

Although **Figure 36.1** is an extensive list of potentially problematic characteristics, mudrocks do have redeeming features. For example, their low permeability makes them ideal for the construction of low-permeability barriers and, as most types also possess high cation exchange capacity, they can capture potentially harmful dissolved contaminant species (e.g. heavy metals) from groundwater. Mudrocks are also used as fills and in the construction of earthworks. Because most mudrocks have low strength they are relatively easy to excavate, but nonetheless most types perform satisfactorily as a bearing medium for moderately to highly loaded, low-sensitivity foundations.

Unfortunately, the reasons listed under 'Mode of Formation' in **Figure 36.1** can hamper attempts to sample, investigate and ultimately predict problematic forms of behaviour in mudrocks. However, an understanding of the composition and mode of formation of these materials, changes brought about during burial and global tectonic changes, through to more recent changes caused by geomorphological processes, the selection of suitable methods of investigation and the correct interpretation of test results can all assist with this. Attempts are made to avoid excessive jargon but, where necessary, the reader should refer to the glossary at the end of this chapter for definitions of the relevant terms.

36.1.1 Definition and classification of mudrocks

The purpose of this section is to define the materials being described in this chapter. Besides the different uses of the terminology by geologists and engineers, and by workers from different countries, there is unfortunately a lack of agreement in the literature (including British and Eurocode standards) as to what constitutes mudrock.

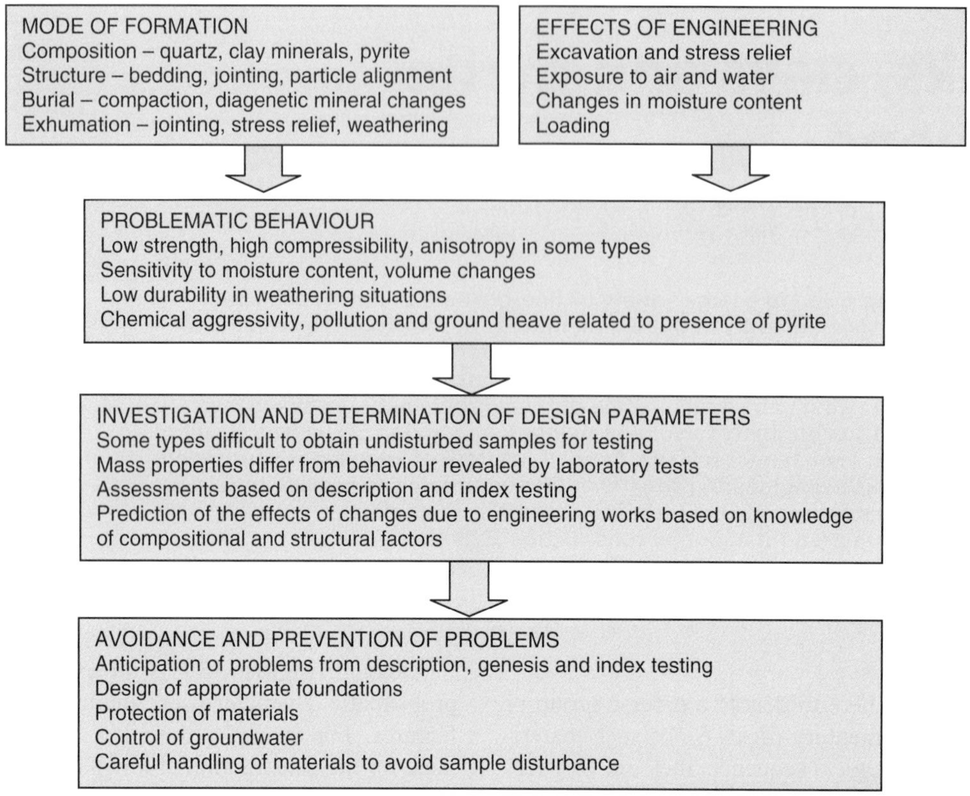

Figure 36.1 Interactions between mode of formation, effect of engineering and engineering problems in mudrocks

Figure 36.2(a) shows the classification for mudrocks given in the current Eurocode 7 guidance document. BS EN ISO 14689-1 (BSI, 2018c) provides very little support on the naming of mudrocks. It clearly states that the guidance offered for naming rocks is to 'provide the engineer with limited geological knowledge means of assigning a rock a name, which may not be strictly correct geologically'. With regards to mudrocks, the definition comes under siliceous sedimentary rock types which are classified based on grain size of the constituents, with siltstone defined as 2–63 μm grain size and material with a grain size of <2 μm defined as claystone and mudstone, which is unclear. The current version of the national standard, BS 5930 (BSI, 2020), as reproduced in **Figure 36.2(b)**, only defines siltstone and mudstone with similarly inadequate explanation. BS 5930 provides guidance only for naming rocks and suggests that geological training is required for the satisfactory identification of rocks. Mudrocks which are defined as argillaceous of siliceous nature and based on grain size description are named as siltstone for material with 0.063–0.002 mm grain size, further described as 'mostly silt', and mudstone for material with <0.002 mm grain size. Mention is made that combinations of rock names are possible, including carbonaceous mudstone, which occurs widely in the UK, but no further definition is provided. Unfortunately, siltstones, mudstones or claystones all predominantly comprise both silt- and clay-sized particles, but the relative proportions are not stated and therefore they are not clearly explained.

The distinctions between 'soil' and 'rock' are described in the current editions of BS EN ISO 14688-1 (BSI, 2018a) and -2 (BSI, 2018b) (for soils), BS EN ISO 14689-1 (for rocks), and BS 5930 (for soils and rocks) based on compressive strength, where soils are defined as materials with an undrained shear strength lower than 300 kPa (uniaxial compressive strength (UCS) of 600 kPa), and rocks are materials which are stronger than this. As also shown later in **Table 36.6**, this corresponds with a material that can be broken by hand pressure.

The term 'shale' is no longer used in Eurocode 7 and UK national standards, but this continues to be a widely used term, and may be retained in schemes elsewhere, but it should not be used for generically naming or referring to mudrocks. In its correct definition it was used to denote a fissile mudstone; in other words, one that will split into thin sheets (see Section 36.2.4.1). The term 'fissile mudstone' is preferred over the term shale, except where it is used as part of a formation name (e.g. Bowland Shale). This is because, when applied correctly, in some mudrocks fissility only becomes manifested as a result of weathering action or following removal of confinement, so the same material could be called 'shale' or 'mudstone', depending on its weathering grade or the elapsed time between sampling and examination. But otherwise it has historically also been applied for describing non-fissile blocky mudstones and therefore becomes ambiguous. The differences between the standards and classification schemes are subtle and difficult to

GENETIC GROUP		SEDIMENTARY				
		CLASTIC SEDIMENTARY				
Usual structures		BEDDED				
Composition		Grains of rock, quartz, feldspars and clay minerals			At least 50% of grains are of carbonate	
Predominant grain size (mm)	0.063	Argillaceous or lutaceous	MUDSTONE SHALE: fissile mudstone	SILTSTONE: 50% fine-grained particles	MARLSTONE	
	0.002			CLAYSTONE: 50% very fine-grained particles		

(a)

Grain size (mm)		Bedded rocks (mostly sedimentary)				
Grains size boundaries approximate		Grain size description			At least 50% of grains are of carbonate	
	0.063	ARGILLACEOUS		SILTSTONE Mostly silt	Calcareous mudstone	
	0.002			MUDSTONE Mostly clay		

(b)

Figure 36.2 Classification of mudrocks according to (a) BS EN ISO 14689:2003 and (b) BS 5930+A2:2010

recognise and apply, and with this lack of agreement between standards it is vital to specify the scheme being used for classification. Further discussion of this issue is presented below and in Section 36.1.2, which includes details of Grainger's (1984) comprehensive classification of mudrocks.

BS EN ISO 14688-2 specification for the classification of soils defines clays as possessing an average grain size of less than 0.002 mm, and advice is offered in Section 36.3.2.2 of this chapter concerning the identification and description of such soils using simple physical tests. BS 5930 also provides guidance on the identification and classification of soils, including fine-grained ones such as clays based on simple hand tests. It accordingly indicates that the predominant grain size of clays is less than 0.002 mm and suggests that classification should be based on consideration of plasticity rather than grain size alone. In view of the average strength criterion mentioned above, most overconsolidated clays fall within the definition of clay soils.

Care must be exercised over the use of the term 'clay', which is used in three different ways

- an average particle size of <2 μm
- a type of mineral, such as illite, kaolinite, smectite, chlorite
- a type of soil, including overconsolidated clays, that contains clay minerals and other constituents.

It needs to be understood that whereas clay minerals tend to be very small, they commonly form larger crystals – for example, aggregations or flocs may exceed 10–20 μm in size. Although 2 μm is used in European classifications (including British ones) as the upper limit of clay size, some American standards regard 4 or 5 μm as the upper limit. A potential source of confusion arises from the names of some geological formations. For example, although Oxford Clay and similar deposits contain much clay, they also include other lithologies, such as fine-grained sandstones, argillaceous limestones and limestone.

As Reeves *et al.* (2006) explain, although particle size is used as the main characteristic to classify mudrocks, it can be difficult to accurately determine the grain size distribution as most of the grains are too small to discern with the naked eye or even with optical microscopy. Furthermore, disaggregation can result in breakage of grains rather than breakage of the bonds holding them together. For instance, Davis (1967) noted that in Mercia Mudstone clay mineral grains are cemented together in clusters, so the measured clay-size fraction is less than the actual amount of clay present. Features that can be used to assist with assessment of the grain size and identification are listed in Section 36.3.2.

Various authors, including Spears (1980), have suggested that mineralogical data, rather than size, can be used as a guide to the sizes of particles present. However, this approach entails the use of relatively expensive X-ray diffraction or scanning electron microscope (SEM) methods, which therefore hampers its use in everyday investigations.

36.1.2 Compaction and cemented mudrocks

Whether the material behaves in a rock-like or soil-like fashion is highly relevant in many applications. Mead (1936) suggested the use of the terms

- compaction mudrock – formed just by compaction due to burial
- cemented mudrock – the grains are attached together by a significant amount of natural cement (often iron oxide, quartz or carbonates).

As further explained in Section 36.2.2, in the former the grains are not bonded together and the materials display soil-like behaviour, whereas intergranular cementation and recrystallisation result in rock-like behaviour in the latter. Mead used the terms compaction and cementation shales, but for the reasons stated in Section 36.1.1 it is preferable to refer to them as mudrocks. Morgenstern and Eigenbrod (1974), Wood and Deo (1975) and Hopkins and Dean (1984) suggest tests that can be used to distinguish between these types of mudrock.

The scheme shown in **Figure 36.3**, in which mudrocks are classified in terms of grain size, UCS (σ_c) and second-cycle slake durability index (Id_2) (see Section 36.3.2.3), was devised by Grainger (1984) to provide a classification that reflects the engineering performance of mudrocks

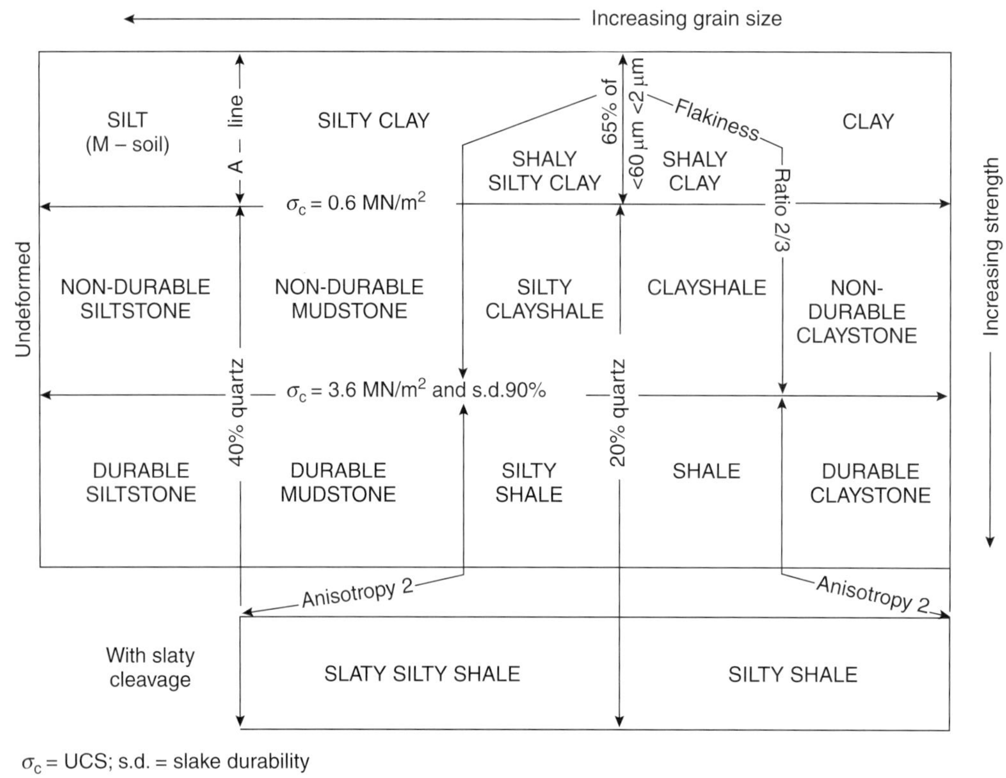

Figure 36.3 Grainger's (1984) classification of mudrocks

- soil (clay): $\sigma_c < 0.60$ MPa
- non-durable mudrock: $0.60 < \sigma_c < 3.6$ MPa; $Id_2 < 90\%$
- durable mudrock: $3.6 < \sigma_c < 100$ MPa; $Id_2 > 90\%$
- metamudrock: $\sigma_c > 100$ MPa (metamudrock is a mudrock affected by low-grade metamorphism).

As mentioned previously, a UCS of 0.60 MPa is equivalent to an undrained shear strength of 300 kPa, where measurement is made perpendicular to any fissility or cleavage in the rock (see Section 36.2.4.1). The quartz content distinguishes between claystone, mudstone and siltstone as

- claystone: <20% quartz
- mudstone: 20–40% quartz (*in Grainger's (1984) paper he also refers to shale and silty shale)
- siltstone: >40% quartz

where the latter figure conforms with the suggestion made by Spears (1980). Grainger's (1984) classification also separates fissile from non-fissile material, where rocks with a flakiness ratio (i.e. shortest dimension of a fragment divided by the intermediate dimension) greater than two-thirds are regarded as fissile, while those with a ratio less than one-third are massive. For durable and slaty mudrocks the anisotropy may be determined using point load test or cone indenter test to determine the compressive strengths parallel and perpendicular to the predominant weakness surfaces (bedding or foliation), with a proposed anisotropy ratio >2 distinguishing fissile mudstones. As noted in Section 36.1.1, to avoid the possibility of the rock name changing as the rock becomes weathered, it is preferable to avoid the term shale and refer to fissile mudstone or claystone instead. Argillite is an indurated clay-rich rock with a random arrangement of particles, whereas in slate the rock possesses cleavage due to a preferred orientation of grains.

36.1.3 Distribution of mudrocks in the UK

As **Figure 36.4** shows, mudrocks occur widely in the UK. Generally speaking, younger, less indurated varieties underlie large areas of the south and east of England, including the London area. Thus, a few important urban areas and infrastructure routes lie on the outcrop of mudrock formations. Older, more indurated mudrocks occur in the Triassic, Permian and Carboniferous rocks of the Midlands, west and north England, south and north Wales and the central valley area of Scotland. These regions coincide with large urban areas and transport links.

Older terrain, ranging from Precambrian to Devonian in age, outcrops in the western and northern parts of the UK. Whereas most mudrocks in these areas have been metamorphosed into slates and schists, there are isolated outcrops of less indurated

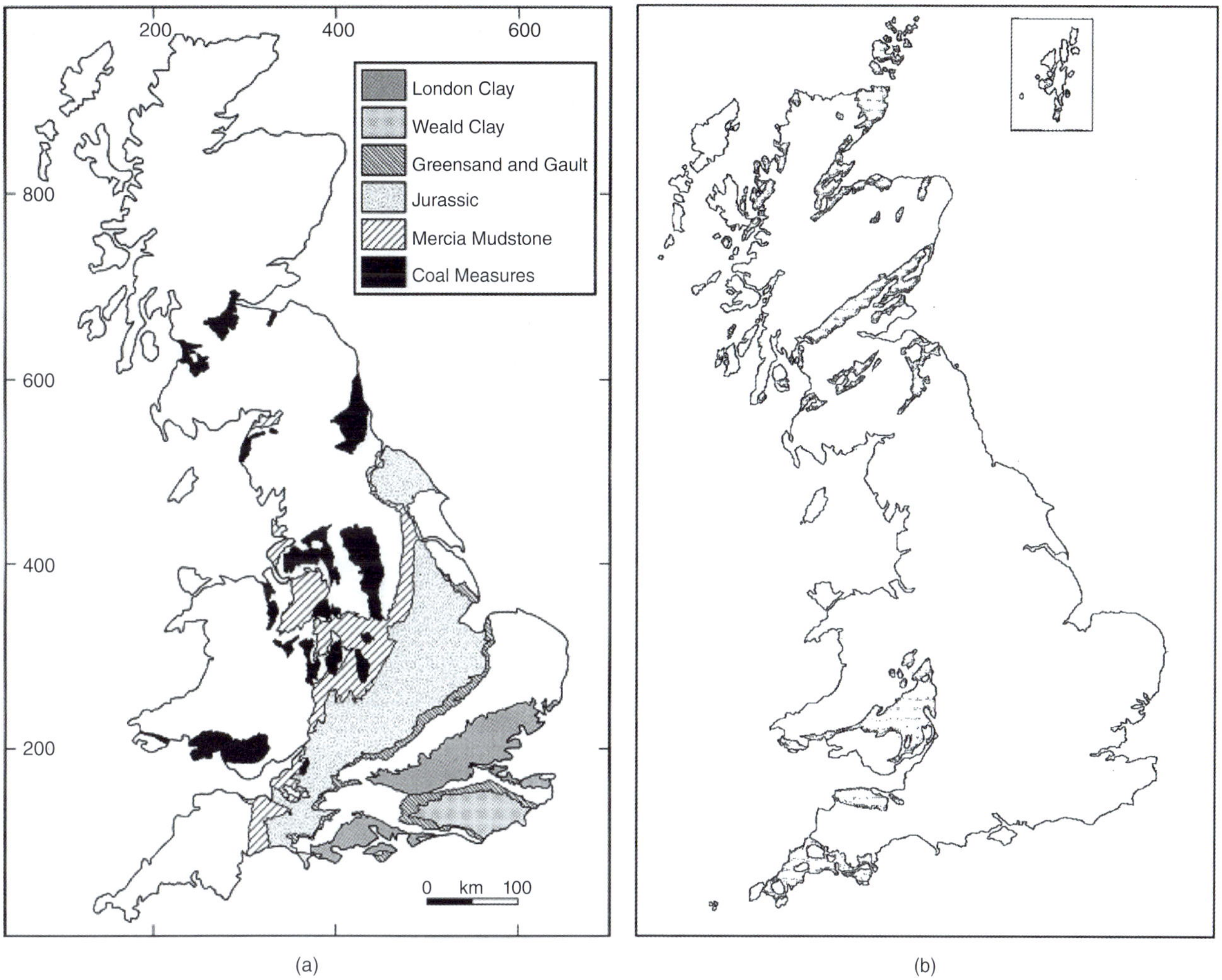

Figure 36.4 Distribution of main mudrock formations in the UK: (a) Carboniferous to Eocene; (b) Pre-Cambrian to Devonian mudrocks (Reproduced from Reeves *et al.*, 2006; © The Geological Society of London)

rocks. Although less populated, the presence of mudrocks may still affect engineering design and construction in these areas.

36.2. Controls on mudrock behaviour

By definition, mudrocks consist of deposited mineral grains such as quartz and clay minerals, with an average size of less than 0.06 mm. In addition to clay minerals and fine-grained quartz, minor amounts of feldspar, calcite and mica together with organic matter and, as explained in the following sections, various diagenetic minerals, may be present. Although the properties of mudrocks are very much affected by the nature and the relative amounts of the mineral constituents, the density and structure are also important.

36.2.1 Constituents of mudrocks

As clay minerals tend to exert a dominant control over the physical properties of mudrocks, knowledge of these minerals assists in understanding mudrock behaviour. Clay minerals have a sheet-like structure typified by mica which, as shown in **Figure 36.5**, is composed of alternating tetrahedral layers of silica (Si_2O_4) and alumina ($Al_2(OH)_6$), accompanied by intermediate layers of exchangeable cations. Within this basic framework a complex mineralogy exists, based on the number of silica–alumina sheets and their arrangement. Ionic substitution within the basic sheet framework by ions such as Mg^{2+}; Fe^{2+} also influences the properties of the mineral, as well as its cation exchange capacity and shrink–swell behaviour.

The formation of clay minerals by the chemical weathering of silicate minerals such as feldspar is illustrated in **Figure 36.6** (see also Section 36.2.3). Whether mudrocks originate as in situ residual or transported deposits depends mainly on the climatic conditions prevailing in the area. Residual deposits, usually dominated by kaolinite, commonly occur in hot and humid weathering environments (see Lum, 1965). On the other hand, in temperate climates the products of weathering are usually dominated by illite.

The type of rock being weathered also influences the clay mineral assemblage; for instance, kaolinite is derived from

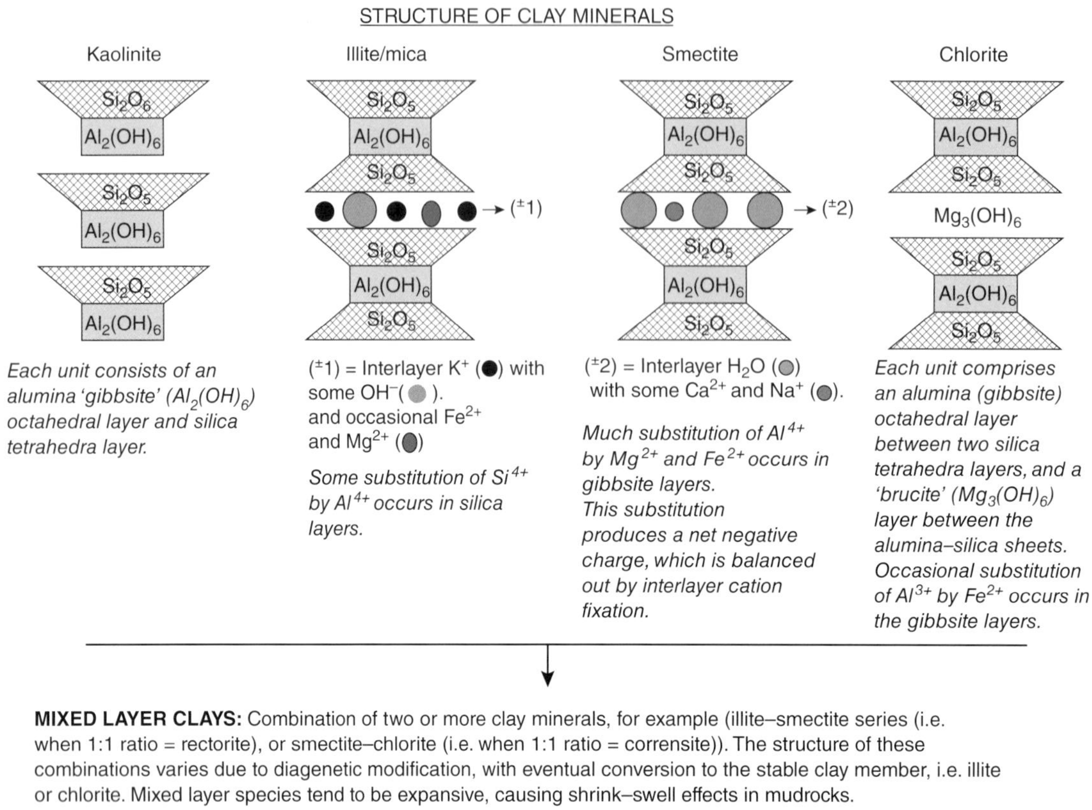

Figure 36.5 Platy structure of typical clay minerals showing potential for swelling by ion exchange

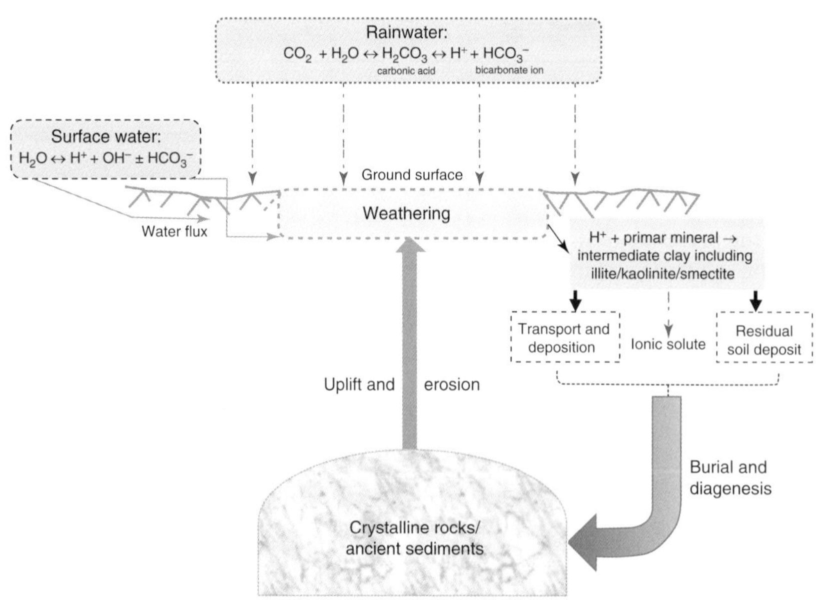

Figure 36.6 Weathering of existing rocks to produce clay minerals

granite degradation, whereas illite results from the weathering of pre-existing mudrocks and mica-rich rocks. Smectite forms in volcanic terrains and low-silica parent rock. As explained below, some clay minerals referred to as authigenic minerals are precipitated from pore solutions. Chlorite may be present in most mudrocks, particularly those derived from weathering of metamorphic materials such as argillite, phyllite or schist. Several non-clay minerals, such as quartz, various carbonates and pyrite, together with organic matter, are commonly present.

Because of the atomic structure of clay minerals, they typically occur as relatively small flat or platy particles where, as illustrated in **Table 36.1**, the average size, specific surface area and activity contrast dramatically with those of equidimensional minerals like quartz and feldspar. The relatively high surface area of swelling clays (such as smectite) is further increased as the interior of the crystal and the exchangeable ions may be accessed by the pore fluid. Reeves *et al.* (2006) explain that minerals such as illite and smectite and, to a much smaller extent, kaolinite, have a net negative charge on their composite layers due to the substitution of high-valency cations by lower-valency ones. Such charges are balanced by the interlayer cations and, when suspended in water, each clay particle is surrounded by a diffuse double layer of hydrated cations. Due to broken ionic bonds at the edges of clay particles, the edges become negatively or positively charged depending on whether conditions are alkaline or acidic, respectively. The ions adsorbed onto clay particles may be exchanged by other ions in the solution, resulting in changes in the thickness of the double layer and of the mineral, giving rise to cation exchange capacity in clays. The exchange of cations into and out of the double layer occurs in response to imbalances in the concentration of cations in the double layer and the surrounding solution.

The relatively large surface area of clay minerals means that they adsorb large amounts of water onto their surfaces, and the shared water between particles imparts plasticity to clays. The activity, which is a function of plasticity and percentage of particles less than 2 μm in size, is an indication of the sensitivity of the material to water. A high activity indicates a more plastic behaviour and greater shrink–swell potential in response to changes in moisture content.

The relatively small size of clay particles also means that they are attracted to each other by van der Waals forces, which operate over very short distances, and in addition they are also subject to electrostatic attraction and repulsion forces which act over longer distances. The strength of the latter forces depends on the ionic strength and composition of the pore water surrounding the particles as well as the distance between them. Thus, particles in solution are subject to forces of attraction and repulsion depending on the distance to neighbouring particles and the chemistry of the solution. Generally speaking, in conditions of high ionic strength, in other words saline conditions, the diffuse double layer is suppressed and attraction forces tend to dominate – which causes the particles to form flocculated groups with many face-to-edge contacts. This so-called card-house structure gives rise to a relatively low-density, high-water-content clay deposit. In low ionic strength conditions the repulsive forces dominate and the clay particles are deposited in a dispersed state which has lower porosity. High amounts of organic matter also increase the tendency for clay minerals to form dispersed structures.

The non-clay constituents of mudrocks include quartz, feldspar, micas, carbonates, heavy minerals, organic matter, pyrite and iron minerals. Quartz and finely crystalline (cryptocrystalline) forms of silica, such as flint or chert, are usually present as detrital (or deposited) grains. Mudrocks are liable to contain iron oxides and hydroxides as well as the iron sulfide mineral pyrite which, as explained in Section 36.2.4.3, is a common and very important, although minor, constituent.

36.2.2 Deposition, burial and diagenesis

The environment of deposition exerts a strong control – not just over the type of material deposited, but also over the presence and spacing of bedding surfaces and other sedimentary structures. As explained in Section 36.2.1, clay particles may be deposited as dispersed individual particles in fairly dense arrangements, with the platy particles aligned approximately horizontally or in flocculated aggregations of clay particles, with an open card-house or honeycomb-type structure in which the particles are in a random arrangement. Such deposits possess relatively high porosity – in the region of 70–90% – and high water content.

	Montmorillonite	Illite	Kaolinite	Chlorite	Allophane	Quartz
Basal diameter (μm)	0.0–2	0.1–2	0.1–4	0.1–2		50–5 000
Basal layer thickness (Å)	10	10	7.2	14		
Particle thickness (Å)	10–100	50–300	500	100–1 000		10^5–10^7
Specific surface area (m²/g)	700–800	80–120	5–20	80		10^{-4}–10^{-6}
Cation exchange capacity (meq/100g)	80–100	15–40	3–15	20–40	40–70	
Area per charge (Å²)	100	50	25	50	120	
Liquid limit (%)	100–900	60–120	30–110	44–47	200–250	
Plastic limit (%)	50–100	35–60	25–40	36–40	130–140	
Shrinkage limit (%)	8.5–15	15–17	25–29			
Activity	1.5–7	0.5–1	0.5	0.5–1.2		

Table 36.1 Specific surface area and activity of quartz and certain clay minerals (Data taken from Mitchell and Soga, 2005 and Hillel, 1980)

Burial of the sediment beneath younger sediments removes access to air and increases both the vertical stress and temperature. An associated reduction in porosity and moisture content is accompanied by the formation of diagenetic minerals, changes to clay minerals and the precipitation of cements, as illustrated in **Figure 36.7**. The exclusion of air can result in sulfate-reducing bacteria releasing oxygen from seawater-derived sulfate (SO_4^{2-}) by reaction with detrital iron in the sediment. The result of this process is the formation of pyrite 'framboids' in the upper levels of the sediment. As **Figure 36.8(a)** indicates, framboids comprise clusters (measuring about 1–10 μm across) of minute pyrite grains. In more oxygenated conditions nodules or amorphous masses of calcite or iron carbonate-rich (siderite) claystone may be formed (**Figure 36.9**).

Burial of muddy sediment, comprising a saturated mixture of silt and clay particles, beneath subsequent deposits brings about progressive reductions in porosity and moisture content, as well as changes (diagenesis) to some minerals and production of new authigenic minerals (**Figures 36.10** and **36.11**). These diagenetic effects increase with burial depth and duration of burial, until depths of 10–12 km are reached, where there is a transition to low-grade metamorphism.

The results of studies into the effects of burial of mud consisting of silt and clay slurry to different depths are summarised in **Figure 36.7** and the related structural changes to mudrocks are illustrated in **Figures 36.11** and **36.12**. Czerewko and Cripps (2006b) explain that during this process clay minerals are squeezed and reorientated to fill the voids between the clastic grains. Any surviving clay flocs are destroyed, and porosity, water and volatiles are lost. In clay-rich deposits or clay layers in laminated deposits (**Figure 36.12(c)**), this results in the reorientation of clay particles towards the horizontal (see **Figure 36.12(b)**), which will tend to render the material anisotropic – so it may be weaker in one direction. The tendency to develop a preferred orientation in deposits comprising clay mixed with silt particles is hindered because the clay particles become packed into the spaces between silt grains (**Figures 36.12(a)** and **36.12(e)**).

Compaction mudrocks (see Section 36.1.2) are liable to contain at least some swelling clay minerals such as smectite and mixed layer illite–smectite. Kaolinite and vermiculite may also be present, and pyrite will typically occur as framboids or very small crystals (see **Figure 36.8(a)**). The compaction experienced by the deposit will have increased its density, with a commensurate reduction in moisture content, but the mineralogical changes will have been relatively minor.

Burial of about 1 km is required to bring about significant changes to clay minerals and pyrite. Besides further reductions in moisture content and increases in density, smectite and smectite–illite mixed layer clays are progressively converted to illite and, ultimately, by metamorphic processes, to

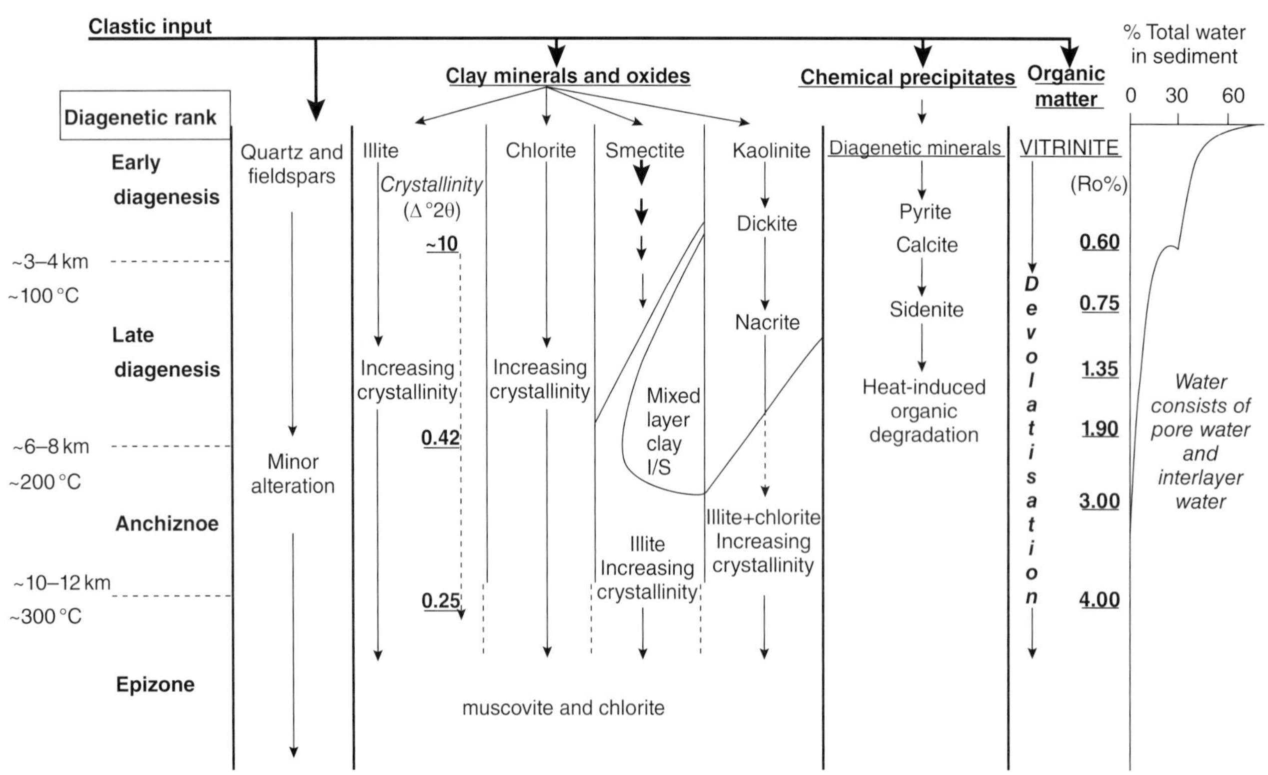

Figure 36.7 Changes in mudrock mineralogy and moisture content associated with burial mixed layer clay; I/S = illite/smectite (Reproduced from Czerewko and Cripps, 2006b)

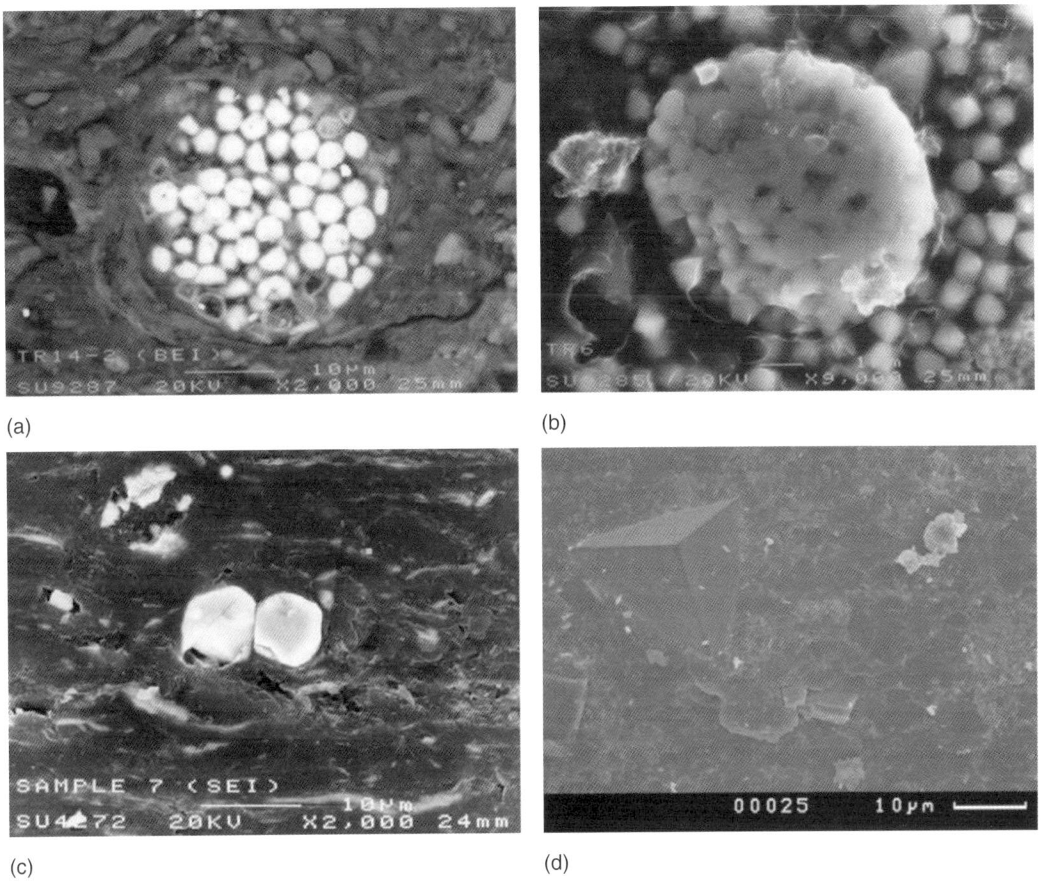

Figure 36.8 Diagenetic changes in pyrite morphology associated with burial and recrystallisation, from poorly indurated compaction mudrock in (a) to well-indurated mudrock in (d). (a) Pyrite framboid comprising a collection of discrete pyrite crystallites (Lias Clay). (b) Partly cemented framboid with partial recrystallisation of discrete crystallites (Oxford Clay). (c) Pyritohedral pyrite crystals resulting from cementation of former framboids (Upper Carboniferous mudrock). (d) Pyrite cube from a well-indurated mudrock (Lower Carboniferous)

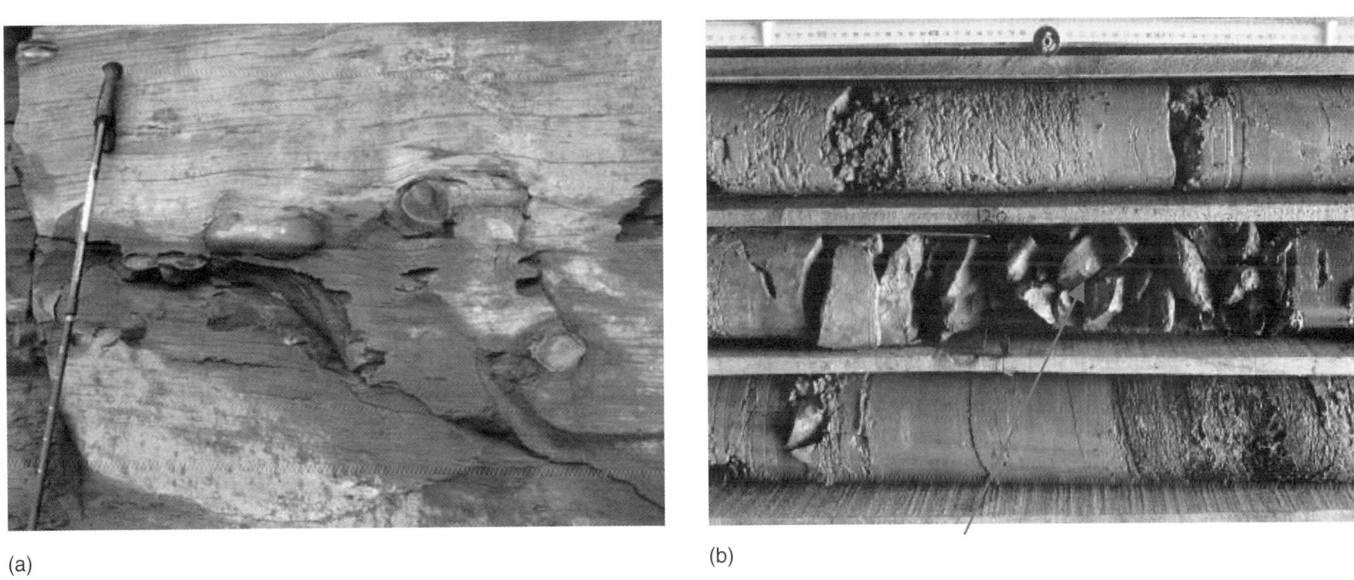

Figure 36.9 (a) Nodules of pyritic claystone in Jurassic mudstone from north-east England (note wrap-round distortion of bedding due to differential compaction of sediment); (b) fragmentation of core due to presence of strong claystone nodules in Carboniferous Coal Measures mudstone from the Midlands, England. The arrow is pointing to a claystone nodule

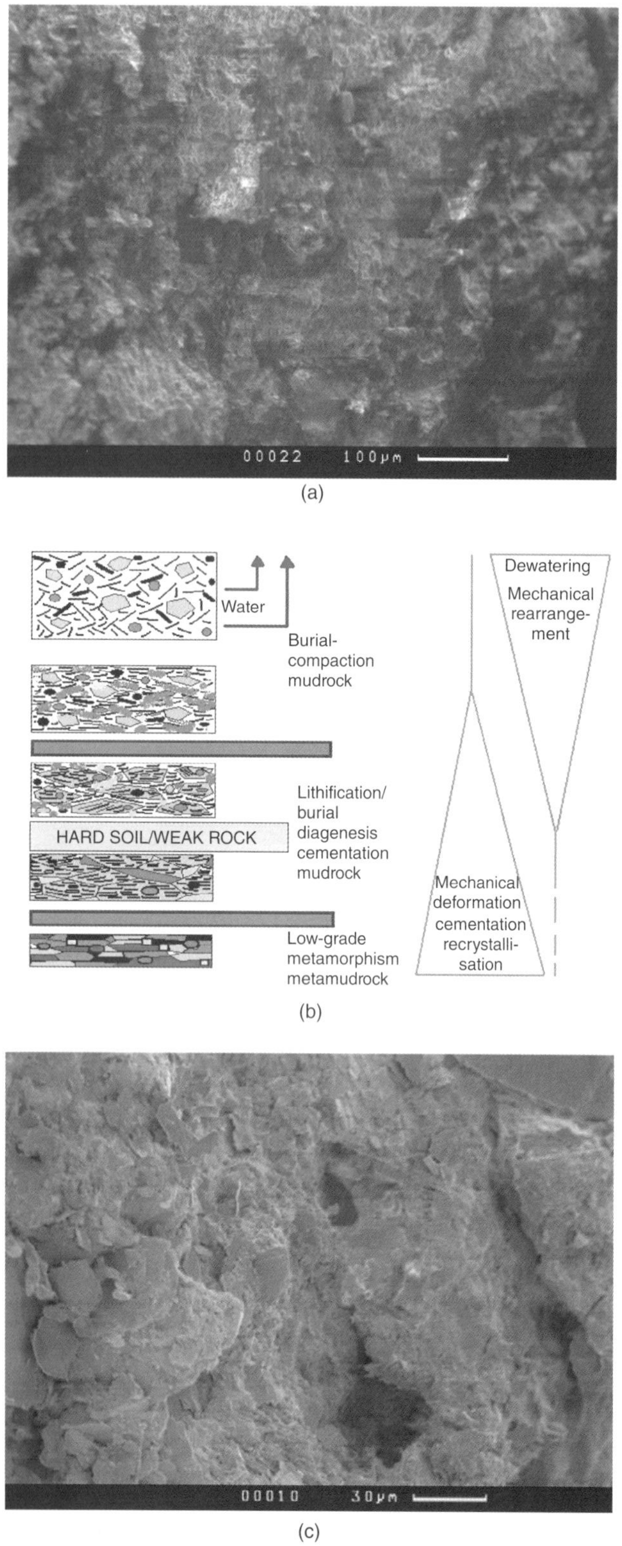

Figure 36.10 Effects of burial on mudrock textures: (a) open 'loose' texture of a poorly indurated mudrock as seen in a sample of London Clay from south-east England; (b) changes occurring in conversion of a mud into a mudrock; (c) dense 'tight' texture with encapsulation of silt-sized quartz grains by clay minerals seen in indurated Carboniferous Namurian mudstone

muscovite or sericite. This process involves the release of ions and water from the interlayer sites of these minerals, which may participate in the precipitation of authigenic clay minerals and cements in the pore space. If the clastic components are sufficient in quantity then, at a certain degree of compaction, a skeletal framework is formed within the deposit (**Figure 36.12(e)**). Where no skeletal clastic framework develops, as in clay-rich muds, then (as illustrated in **Figure 36.12(b)**) a deposit with an approximately horizontal preferred orientation of clay particles is the likely outcome and a fissile mudstone or claystone (see Section 36.2.4.1) is produced. Weakly cemented mudrocks may contain some swelling clay minerals with illite of enhanced crystal size, together with a partial skeletal framework. Chlorite may also form during deep diagenesis and vermiculite may be converted during burial diagenesis through mixed layer corrensite to chlorite at depth.

Burial to 2–3 km usually results in a porosity reduction of 3–5% in strongly cemented mudrocks with elimination of most, if not all, of the low-stability clay minerals. A series of discrete bonded crystal domains is produced which, with a continued increase in burial depth and temperature, increase in size. They eventually coalesce to form a continuous framework of cemented grains such that the material progresses to become a cemented mudrock (**Figure 36.12(f)**). Ultimately, at burial depths of approximately 5 km, all the grains may become cemented. Not only is the total pore volume in the deposit decreased, but a reduction in pore-throat diameter, which controls the movement of water, also occurs. This (as explained in Section 36.3.2.3), together with the removal of swelling clay minerals and increased size of pyrite crystals (see **Figures 36.8(a)** to **36.8(d)**), considerably enhances the durability of the material. In this process, any micro-fractures and other structural discontinuities in the material are healed – which augments its resistance to breakdown. As illustrated in **Figure 36.8**, the progressive recrystallisation of pyrite results in the conversion of framboids to fine-grained and larger crystals to the extent that eventually visible cubic crystals are produced.

Ancient residual deposits that have been indurated by compaction and cementation may form mudrock formations like those derived from transported clay-rich deposits. However, some residual deposits will appear to display rock-like behaviour because of high pore suction pressures caused by desiccation. Suction pressures that hold the particles in such deposits are destroyed by even slight wetting of the material, and with further wetting the material very rapidly reverts to soft clay.

36.2.3 Uplift, unloading and weathering

The changes in void ratio or moisture content, shear strength and effective horizontal stress that accompany burial of muddy sediment comprising a saturated mixture of silt and clay particles, followed by reduction of overburden, are illustrated by Cripps and Taylor (1981) in **Figure 36.13**. The effects of burial on mineralogy and micro-structure are described in Section 36.2.2 and are illustrated in **Figures 36.7** and **36.10**.

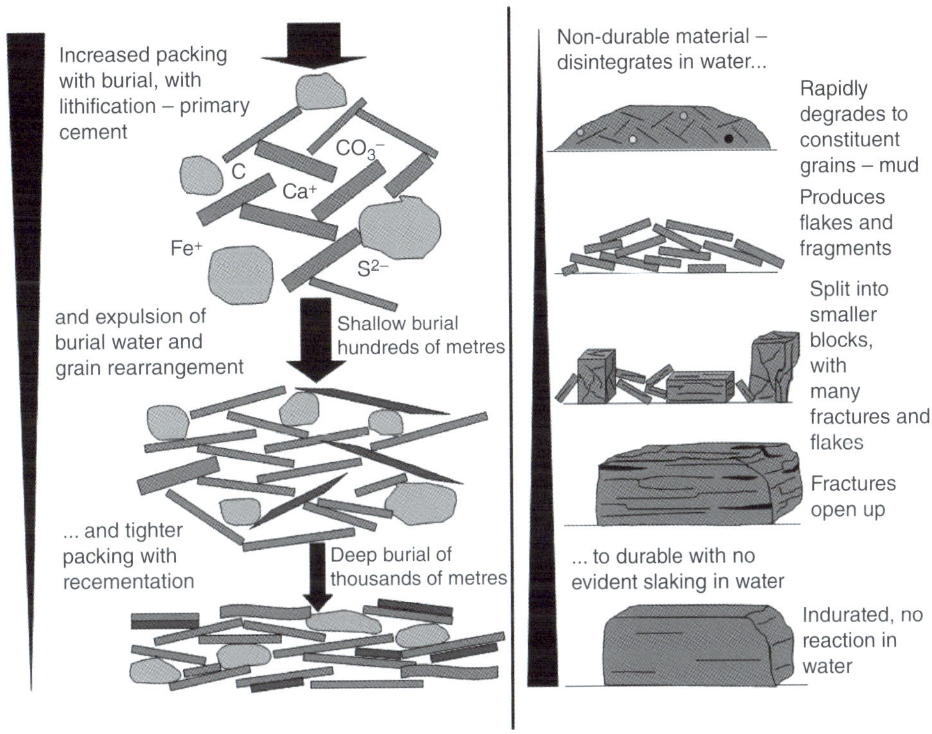

Figure 36.11 Effects of burial on the structure and behaviour of mudrocks

As explained in Chapter 17 *Strength and deformation behaviour of soils*, unloading results in the deposit becoming overconsolidated; in other words, it is denser and stiffer than it was for the same value of vertical effective stress during burial. Weathering processes entailing both physical breakdown and chemical alteration result in the fragmentation of the material, together with progressive dissolution of cements and other mineralogical changes. As **Figure 36.13** shows, these processes eventually result in the loss of the effects of diagenetic compaction and cementation, and the material is returned to a condition similar to the state when it was deposited.

One of the physical processes of weathering caused by the removal of overburden entails the vertical expansion of the material and the development of stress relief joints that lie roughly parallel to the ground surface. Due to lateral restraint, the horizontal stress induced by the original overburden pressure is not reduced in a commensurate manner, resulting in high horizontal stress. This is manifested by high K_o values in near-surface overconsolidated clay formations, as shown in **Figure 36.13**, as well as by the formation of systems of inclined fractures. These discontinuities facilitate access for water and gases that participate in both physical and chemical weathering processes. Weathering processes cause progressive opening of discontinuities as well as the development of new ones. Softening and expansion of the material occurs as a result of the uptake of water and the weakening of bonds. The loss of strength of the material near to the ground surface means that there is a rapid drop in the K_o value at shallow depths.

Whereas in temperate climates physical weathering processes predominate, in tropical zones the bedrock is typically mantled with tens of metres of residual soil formed by chemical alteration of the underlying rock (see Fletcher, 2016). For example, in Hong Kong the granitic and volcanic areas are mantled by up to 60 m of clay-rich residual soils (see Fletcher, 2016; Hencher, 2012; Lum, 1965). The principal weathering effects will be concentrated mainly along discontinuities, and a progressive increase in the proportion of weathered products is accompanied by a decrease in the size and number of corestones towards the surface.

Chemical weathering processes are illustrated in **Figure 36.6**; however, as explained in Section 36.2.4.3, microbial activities that participate in the weathering of pyrite are also very important. Other biological processes, such as the growth of plant roots, can have both physical and chemical impacts.

The impact of unloading, and the style and rate of changes to the material in response to weathering action are strongly influenced by structure and composition, themselves products of the original material and diagenetic changes. Generally speaking, in compaction mudrocks and overconsolidated clays the effects of burial are rapidly lost. However, in cemented mudrocks degradation may occur over a protracted period. The release of stored elastic strain energy leads to the progressive development of fractures with little initial impact on the intact mudrock between them. Subsequently the inter-particle bonding and cementation will be weakened and further fractures will form. Ultimately, the effects of burial and diagenetic

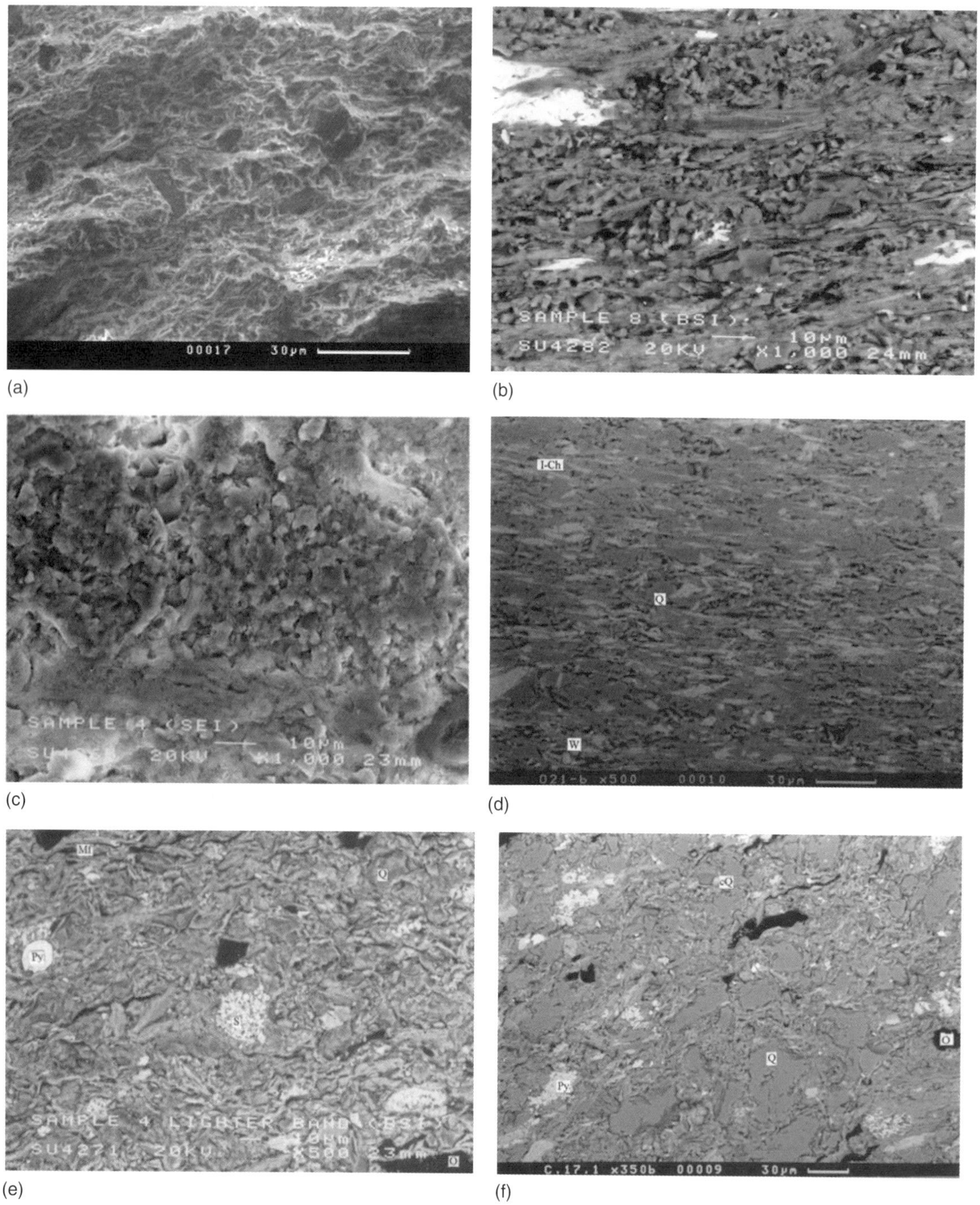

Figure 36.12 Scanning electron microscope (secondary and backscatter) images illustrating the effects of compaction and recrystallisation on the textures of mudrocks. (a) Random orientation of clay particles with wrapping around silt-size quartz grains in Jurassic Lias Clay. (b) Preferred alignment of grains in fissile mudrock from the Carboniferous Upper Coal Measures. (c) Thin lamination of silt-grade quartz in Carboniferous Lower Coal Measures laminated mudrock. (d) Evidence of recrystallisation and cementation in Ordovician argillite. (e) Randomly orientated clay minerals in tightly compacted well-indurated massive mudrock from north-east England. (f) Cemented mudrock from Scotland. Bright areas: pyrite, medium-bright areas: calcite cement; grey: clay; black: organic matter

changes are lost as the material becomes disaggregated and returns to a normally consolidated condition.

The fragmentation process in indurated mudrocks is strongly influenced by their degree of micro-fracturing (see **Figure 36.14**). In indurated mudrocks, diagenetic cements may provide a rigid framework that encapsulates strained mineral grains and crystals (see **Figure 36.12(e)**). The removal of constraint due to overburden and confinement puts this framework and diagenetic bonding under stress and with weathering-induced removal of the cement and weakening of inter-particle bonds,

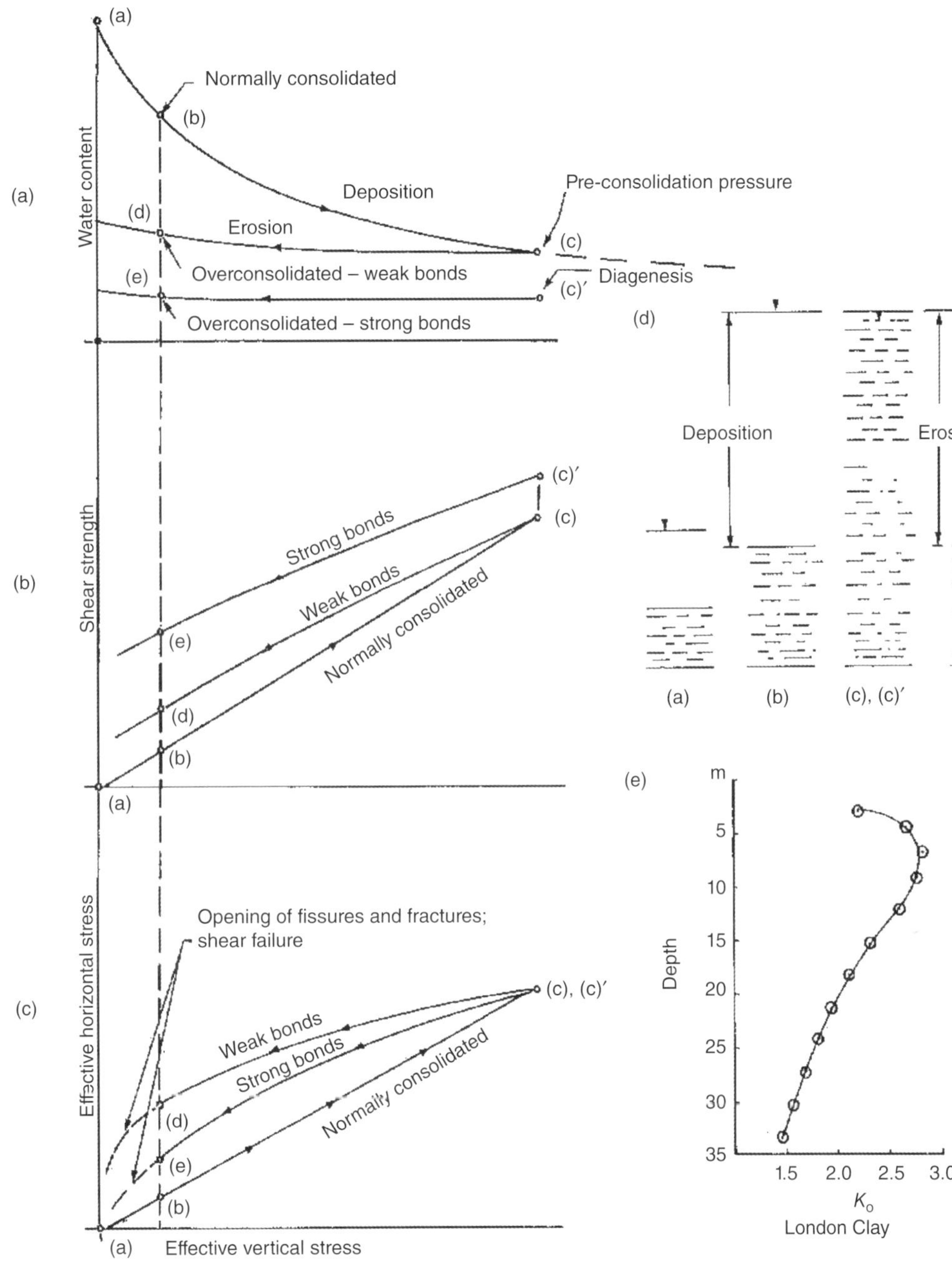

Figure 36.13 Changes in water content, shear strength and horizontal stress due to loading and unloading of sediments (Reproduced from Cripps and Taylor, 1981; © The Geological Society of London)

the material is apt to slake into sand- or gravel-sized pieces (see **Figure 36.15**). In some rare cases, such as that shown in **Figure 36.16**, rock bursts occur with an explosive release of energy.

36.2.4 Genetic features of mudrocks

36.2.4.1 Laminations and fissility

Laminations comprise discrete bedded units less than 20 mm thick and, as **Figure 36.12(c)** illustrates, the units may comprise horizons of differing grain size, or they may be successive units of the same grain size. Mudrocks that possess a laminated structure usually display anisotropy such that they break preferentially along the laminae. The fact that a mudrock can be split into thin laminae defines it as fissile. In some clay-rich mudrocks fissility is enhanced by a preferred alignment of platy clay particles. As explained in Section 36.1.1, fissile mudrocks are sometimes called shales, but because this feature may become apparent only after a degree of weathering action, it is better not to use the term in this way.

Figure 36.14 Effect of rock micro-structure on the morphology of the weathered product. (a) Platy or 'fissile' (Ordovician, West Wales). (b) Flaky (Carboniferous Coal Measures, South Yorkshire); (c) blocky (Carboniferous Coal Measures, South Yorkshire)

36.2.4.2 Nodules and concretions

Nodules or concretions of various shapes, types and sizes are commonly found in mudrocks. They consist of harder, more durable material and are formed by diagenetic processes (see Section 36.2.2) in which localised precipitation of minerals has cemented the sediment. Examples are shown in **Figure 36.9**. As explained in Section 36.3.1, nodules may sometimes be formed in soil horizons or weathered mudrocks.

36.2.4.3 Pyrite in mudrocks

Pyrite is an iron sulfide mineral which, although constituting only a few per cent of the composition of mudrocks and overconsolidated clays, can give rise to several types of problems, as further detailed in Sections 36.4.2 and 36.4.6. It is particularly associated with dark-coloured organic-rich mudrocks. As Czerewko *et al.* (2003a) explain, pyrite is not the only potentially troublesome sulfide mineral, but it is the most commonly occurring one. Pyrite commonly occurs as framboids, which are spherical clusters typically less than 10–20 μm across (see **Figure 36.8(a)**). Due to their great collective surface area, pyrite framboids are very reactive in oxygenated and humid weathering environments. **Table 36.2** lists many of the UK sediments that contain pyrite. It occurs in mudrocks of all ages but it is particularly common in marine overconsolidated clays and mudrocks of Jurassic, Cretaceous and Cenozoic ages which occupy about 25% of the land surface of southern and eastern England, and small areas of Wales and Scotland.

(a)

(b)

Figure 36.15 Examples of slaking on exposure of Carboniferous Coal Measures mudrocks. (a) Indurated mudrock slaked to gravel size domains after 2 months' exposure. Carboniferous Coal Measures mudrocks from north Derbyshire. (b) Rapid slope deterioration due to slaking in Coal Measures mudrock, Sheffield

(a)

(b)

Figure 36.16 Effects of strain energy release in mudrocks: (a) failure of cut slope in Mercia Mudstone, East Midlands, UK; (b) failure of roof in mine – Coal Measures, north England

Diagenetic processes (referred to in Section 36.2.2) progressively convert open-textured framboids into octahedral or cubic crystal forms up to several millimetres across. At each stage of the process the pyrite becomes more resistant to oxidation. Exhumation of mudrock formations allows them access to air and water, so the pyrite may become oxidised by the process illustrated in **Box 36.1**.

Exposure to oxygen and water facilitates slow chemical oxidation of pyrite according to reaction 1. This reaction is self-limiting at $pH < 4$. However, ferric iron – which is produced by the activities of sulfate-reducing bacteria – acts as an oxidant to pyrite. The oxidation reaction is exothermic, and the conditions become more acidic due to the release of H^- ions. As the bacteria are at their most active in warm ($\sim 30°C$) acid conditions ($pH \sim 3$), the reaction proceeds 3–100 times faster than for purely chemical oxidation.

If there is insufficient water present for the products of pyrite oxidation to remain in solution, the reaction becomes arrested with the formation of hydrated iron sulfate. In fact, the process can have a desiccating effect on clays (see Cripps and Edwards, 1997). Although iron sulfate is stable under acidic conditions, flowing water removes the soluble minerals

Stratigraphic unit	Pyrite content (% FeS_2)	Carbonate mineral content (%)	Carbonate mineralogy
Eocene			
London Clay – blue grey fissured silty clay	0.6–4	0.4–5.2	Calcite, dolomite
Basal London Clay – sandy clay	2.5–41	N/A	Calcite
Palaeogene			
Lambeth Group – laminated and shelly clays	0.4–19.2	0–18	Calcite
Lambeth Group – mottled clays	<0.2	0	Absent
Cretaceous			
Weald Clay	0.6	1–6	Calcite
Gault Clay	0.7–1.0	0.7–33	Calcite
Jurassic			
Ancholme Clay	0.2–4	2.7–4.3	Calcite
Kimmeridge Clay	0.4–4	8–36	Calcite
Oxford Clay	3–15	4.5–19	Calcite
Whitby Mudstone	3–17	1–16	Calcite, siderite
Lias Clay	1–8	2–49	Calcite
Carboniferous			
Coal Measures – claystone	0.1–6.9	2.2–3.2	Siderite, calcite
Coal Measures – mudstone	1.2–8.2	0–9	Calcite, siderite
Coal Measures – siltstone	0–2.3	0–37	Siderite
Limestone Rock	0.2–10	N/A	Calcite
Culm Measures	2.4	N/A	N/A
Namurian Mudstone	0–6	0–0.5	Calcite
Devonian			
SW England – mudrock – argilite	0–4.3	0–12	Calcite, dolomite, siderite
Silurian			
Wales and Southern Uplands, Scotland – metamudrock and argilite	0.1–4.2	0–2.2	Calcite, siderite
Ordovician			
Wales – mudrock – slate	0.4–7.1	0–2.7	Calcite, siderite
Cambrian			
North Wales – slate	0.5–5.4	0–0.8	Dolomite
N/A: not applicable			

Table 36.2 Pyrite content range and significant carbonate mineralogy of selected UK mudrock formations (After Czerewko *et al.*, 2003a)

and sulfuric acid in solution, and produces highly aggressive acid pore waters as are typified by the highly polluting acid mine discharges (AMD) from flooded mines (Banks, 1997) and mine waste spoil tips.

With abundant water in the system, the acidity is moderated to pH > 4 and iron oxyhydroxide [$Fe(OH)_3$], which is a distinctive highly insoluble orange-brown residue known as ochre, is usually produced. The sulfuric acid generated in the oxidation process is biotoxic and highly corrosive to metals, concrete and rocks. As a strong leachant, its effect is to release potentially harmful elements, including heavy metals, into the environment.

If calcium carbonate is present, it will react with the acid to produce gypsum ($CaSO_4.2H_2O$) or selenite, as **Figure 36.17** shows. Once selenite crystallises it reluctantly redissolves, thus acting as a nucleus for further precipitation of calcium sulfate. The conditions remain acidic (pH < 2.5) if the amount of calcium carbonate is low – so acid attack on illite and albite feldspar produces jarosite ($KFe_3(SO_4)_2(OH)_6$) and alunite ($KAl_3(SO_4)_2(OH)_6$), respectively. On the other hand, a high calcium carbonate content (see **Table 36.2**) renders the environment less hospitable to microbes, so the rate of reaction decreases.

36.3. Engineering properties and performance

36.3.1 Physical properties and characteristics of mudrocks

Mudrocks display a wide range of physical properties and characteristics, depending mainly on composition and structure, which depend on the origins and history of the materials (see Sections 36.2.2 and 36.2.3). In terms of engineering performance, mudrocks can be divided into three main types, depending upon the extent of intergranular cementation and

Box 36.1 Process of the oxidation of pyrite

1. Oxidation of pyrite by molecular oxygen produces ferrous iron in a purely chemical reaction. Under saturated conditions iron remains in solution:

$$2FeS_{2(s)}(\text{pyrite}) + 2H_2O + 7O_2 \rightarrow 2Fe^{2+}(\text{ferrous iron}) + 4SO_4^{2-} + 4H^+_{(aq)}.$$

2. Oxidation of ferrous iron by molecular oxygen to ferric iron which is a strong oxidising agent:

$$4Fe^{2+}\ (\text{ferous iron}) + 4H^+_{(aq)} + O_2 \rightarrow 4Fe^{3+}\ (\text{ferric iron}) + 2H_2O.$$

3. Further oxidation of pyrite by ferric iron which acts as an electron acceptor producing ferrous iron, sulfate anion and protons:

$$FeS_2\ (\text{pyrite}) + 14Fe^{3+}\ (\text{ferric iron}) + 8H_2O \rightarrow 15Fe^{2+}\ (\text{ferrous iron}) + 2SO_4^{2-} + 16H^+_{(aq)}.$$

4. If calcium carbonate is present the acid will be buffered and gypsum and carbon dioxide will be produced. The reactions require sufficient water for the reaction products to be in solution and gypsum will be precipitated if the water is evaporated or consumed so that the solution becomes saturated in that compound

$$CaCO_3(s) + 2H^+_{(aq)} + SO_4^{2-}{}_{(aq)} + H_2O_{(l)} \rightarrow CaSO_4 \cdot 2H_2O_{(s)} + CO_{2(g)}.$$

recrystallisation. Although the burial history rather than age exerts a strong control over the state of induration, a general trend of increasing induration and strength with age occurs in UK mudrocks (see Taylor, 1984) as

- overconsolidated clays: Cenozoic, Cretaceous and most Jurassic clays/mudstones
- weaker mudrocks: Lias (Lower Jurassic), Mercia Mudstone (Triassic), Permian and some Coal Measures (Carboniferous) mudstones
- stronger mudrocks: Carboniferous mudstones, Devonian and older mudrocks/metamudstones.

In some cases well-cemented mudrocks may occur in younger sequences. Conversely, weathering action and/or fracturing can result in weaker materials being present in older formations.

Table 36.3 presents strength and also compressibility and plasticity data typical of UK unweathered mudrocks of different types. The performance of these materials is discussed below. Further data on other formations are given by Reeves *et al.* (2006).

Published values of geotechnical parameters such as those in **Table 36.3** are site-specific and should be used solely for guidance for preliminary designs and to assist with appropriate scoping of investigation and testing programmes. This is because the values obtained from geomechanical tests depend on the sampling, test conditions and other factors. Because weathering action has an extremely variable impact on mudrocks, only very general guidance about the likely effects of weathering can be given. The remarks given below are intended only to demonstrate very broad differences in the ways the parameters change for different lithologies.

36.3.1.1 Compaction mudrocks and overconsolidated clays

Only relatively weak inter-particle bonds are formed during the burial of compaction mudrocks, so in many engineering situations they have little impact and effective stress has a larger influence on inter-particle friction forces. In the UK, as shown in **Figure 36.4**, several important overconsolidated clay formations of the Jurassic period and younger occur widely in the east and south of England. The older formations, such as the Lower Jurassic Lias Clay and siltier parts of others, such as Oxford Clay and London Clay, display properties which place them on the boundary between 'very high strength or very stiff' soils and 'extremely weak or very weak' rocks as defined by BS EN ISO 14689-1 and BS 5930.

Clay within about 3 m of the ground surface will typically be in partly saturated (partly desiccated) condition and will tend to be stiffer than the material it overlies. Wetting reduces the effective stress and causes rapid softening of desiccated clay, which may then break down into individual grains or partly bonded domains. In compaction mudrocks the effects of overconsolidation are rapidly destroyed by weathering

(a) (b) (c)

Figure 36.17 Effects of gypsum precipitation in mudrocks. (a) Bowland Shale formation from Mam Tor, Derbyshire, UK showing expansive gypsum crystallisation along bedding discontinuities. (b) Random gypsum crystal growth in Lias Clay destroying the structure of less indurated rock and causing volume increase. (c) Gypsum 'selenite-star' and induced micro-fracturing in Bowland Shale formation from Mam Tor, Derbyshire, UK producing silt-size fragments

Palaeogene – London Clay: Overconsolidated clay – compaction mudrock with soil-like characteristics. When fresh, tends to be very stiff to hard, fissured blue grey locally calcareous pyritic silty clay

Id_3	I_j'	Durability	Strength c_u: kPa	Compressibility m_v: m²/MN	MC: %	LL: %	PL: %	PI: %	Void ratio, *e*
0–12	8	ND	70–280	0.01–0.002	19–28	50–106	22–35	40–65	0.6–0.8

Jurassic – Lias Clay: Overconsolidated clay – compaction and weakly cemented mudrock with soil-like characteristics. When fresh, tends to be very weak to weak, laminated grey locally calcareous silty mudstone

Id_3	I_j'	Durability	Strength UCS: MPa	Strength c_u: kPa	Compressibility m_v: m²/MN	*E*: MPa	MC: %	LL: %	PI: %	Void ratio, *e*
2–8	6–8	ND	0.38–45	80–1200	0.002–0.004	22–35	16–22	30–63	8–55	0.5–0.7

Triassic – Mercia Mudstone: Overconsolidated clay to mudrock – compaction and cementation mudrock with soil-like and weak rock-like characteristics. Is weak when fresh, may be medium strong when cemented; red-brown locally calcareous or dolomitic mudstone and silty mudstone

Id_3	I_j'	Durability	Strength UCS: MPa	Strength c_u: kPa	Compressibility m_v: m²/MN	*E*: MPa	MC: %	LL: %	PI: %	Void ratio, *e*
19–81	2–7	D-ND	0.55–36	80–2800	0.008–0.03	100–1200	16–22	19–35	8–35	8–55

Carboniferous – Coal Measures: Mudstone: Coal Measures mudrocks include claystone, mudstone and siltstone; only mudstone is considered here. Tends to be mudrocks with variable properties depending on silt content and degree of cementation, but generally classed as a cementation mudrock with rock-like characteristics. When fresh is weak to strong, locally interlaminated to thinly bedded grey mudstone

Id_3	I_j'	Durability	Strength UCS: MPa	*E*: GPa	Is (50): MPa	MC: %	MC_{ABS}: %	E_V: %	E_A: %	LL^c: %
10–95	2–7	ED-ND	3.4–50.1[a] 33–128[b]	5–50	0.21–3.69[a] 0.6–7.2[b]	1.8–5.7[a] 1.4–2.9[b]	2.3–10.3	0.28–3.12	0.6–7.8	27–33

Lower Palaeozoic – Ordovician and Cambrian: Metamudstone to slate: tend to be cementation deposits of moderately high strength depending on silt content and induration. Tend to be of variable properties. When fresh, are medium strong to very strong, occasionally interlaminated fissile to cleaved grey, purple, green-grey mudstone, metamudrock, to slate.

Id_3	I_j'	Durability	Strength	*E*: GPa	Is (50): MPa	MC: %	MC_{ABS}: %	E_V: %	E_A: %	UCS: MPa
98–99	1–3	ED	66–272	≥31	2.1–15.8	0.11–1.03	0.1–2.2	0.005–0.17	0.02–0.43	

[a]Mudrocks from England and Scotland; [b]Mudrocks from South Wales; [c]Material prepared by mechanical means from fresh mudrock; e: void ratio; *E*: Deformation modulus, analogous to Young's modulus but behaviour not strictly elastic, and varies with confining pressure; E_A: axial swell; ED: Extremely durable; D: Durable; ND: Non-durable; E_V: volumetric swell; Is (50): point load test strength; LL: liquid limit:%; MC: moisture content:%; MC_{ABS}: moisture absorption:%; PI: plasticity index = LL – PL:%; PL: plastic limit; Id_3: 3-cycle slake durability index value; Ij': Jar slake test index; UCS: uniaxial compressive strength

Table 36.3 Examples of mudrock properties from a number of prominent UK stratigraphic units showing typical range of values encountered in fresh unweathered material (Data taken from Cripps and Taylor, 1981; Reeves *et al.*, 2006; Czerewko and Cripps, 2006b)

processes and breakdown is likely within an engineering timescale and, in some cases, a construction timescale. The relative strength of inter-particle bonding in mudrocks (which will provide an indicative rate of deterioration) may be assessed from the results of durability tests such as the static jar slake test (Czerewko and Cripps, 2001) or the dynamic slake durability test (Franklin and Chandra, 1972).

Higher values of strength and density with lower moisture content are found for siltier and more deeply buried clays; in some cases they are due to the presence of inter-particle bonds and some cementation of particular horizons. Higher values of plasticity would indicate the presence of smectite and mixed layer clays, which would indicate lower diagenetic rank. Weathering is typically responsible for reducing the undrained shear strength to about half of the unweathered value, whereas weathering processes result in an increase in moisture content of up to 50% and liquid limit may increase by about 10%. In parallel, density is typically decreased by about 5% and compressibility increased by about one order of magnitude.

Some formations contain systems of fissures such that the mass strength will be less than the strength of small laboratory samples that are recovered from between fissures and remain independent of their influence when subjected to mechanical laboratory testing. Some display anisotropy due to the presence of laminations and/or the preferred alignment of platy clay minerals. For reasons explained in Section 36.2.3, the horizontal in situ stress can be 3–4 times greater than the vertical stress in the upper 10 m or so of the formation. In the UK, weathering effects generally extend to depths of 7–8 m, and in the upper few metres deformation relieves the horizontal stress. On the other hand, joint and fissure development, which are the result of overburden reduction, extends to tens of metres below the surface.

The glacial conditions that occurred during the Pleistocene had very important consequences on the behaviour of mudrocks throughout the UK. In compaction mudrocks and overconsolidated clays, localised zones of disturbance to near-surface strata were formed and solifluction deposits mantled most escarpments and outcrops on sloping ground. The solifluction deposits contain shear surfaces that mobilise shear strength near to the residual value, so they can undergo shearing if subjected to relatively low values of shear stress. Weeks (1969) and Symonds and Booth (1971) describe hillsides that became unstable due to changes in groundwater conditions, loading or the removal of support.

36.3.1.2 Weakly cemented mudrocks

The physical and mineralogical changes that characterise cemented mudrocks also enhance durability by restricting the access of air and water to any reactive (swelling) clay minerals. However, because distorted mineral particles are restrained by intergranular bonds and cements, the reduction in confining stress and weathering effects can result in time-dependent strain energy release. Such materials tend to undergo rapid slaking as the volume changes that accompany alternating changes in moisture content result in the breakage of bonds and inter-particle cements.

One of the initial consequences of overburden reduction is the development of systems of sub-horizontal and inclined joints that facilitate air and water penetration into the rock mass. Shrink–swell effects due to changes of moisture content also result in the development of joints and micro-fractures along bedding discontinuities and grain boundaries (see **Figure 36.14**). Generally speaking, these processes result in degradation to sand- to fine gravel-sized grains. Further breakdown of the material to individual mineral grains then usually requires breakage of the inter-particle bonds by a combination of fatigue failure – caused by repeated volume changes and dissolution of cements (see **Figure 36.15**).

Weakly cemented mudrocks may well be weak rocks in terms of their UCS, but they tend to undergo rapid slaking when they are exposed in excavations and slopes, as seen in **Figure 36.14(b)**.

The changes in parameters due to weathering action are greater in cemented mudrocks than in compaction mudrocks. The former are initially more indurated but the product materials are similar, notwithstanding that degradation is more rapid for compaction mudrocks. The moisture content of highly weathered mudstone is typically 30–35%, whereas bulk density will be 1.95–2.05 Mg/m^3, undrained shear strength 70–100 kPa, UCS 10–500 kPa and coefficient of volume compressibility 0.1–0.24 m^2/MN. Higher values of moisture content, plasticity and compressibility, and lower values of strength and density, are liable to be found in less deeply buried, poorly indurated clay-rich deposits, or in more indurated materials containing swelling or mixed layer clays.

Disturbance during the sampling process and sensitivity to changes in stress and moisture conditions render many weakly cemented mudrocks challenging to test and classify, which can lead to the underestimation of the in situ behaviour. Soil classification tests, such as particle size analysis and Atterberg limits which require destructuring of material, do not usually provide an accurate indication of behaviour even on a small scale. For instance, many published values of high activity in Mercia Mudstone are probably the result of the underestimation of the <2 mm fraction rather than an indication that they contain high values of swelling clay minerals.

36.3.1.3 Strongly cemented mudrocks

In strongly cemented mudrocks, a skeletal framework consisting of bonded clay grains and a continuous framework of cemented grains are present; any low-strength micro-discontinuities will have been eliminated. In addition, stored strain energy due to the presence of distorted mineral grains will have been gradually eradicated by recrystallisation of minerals and any remaining swelling clay minerals will have been replaced by more stable forms. In practice, like weakly cemented mudrocks, ultimate breakdown to constituent particles is a two-stage process in which the rock mass is initially broken down into sand- or gravel-sized fragments which may then degrade into mineral grains. In strongly cemented mudrocks the fragments are usually larger and take longer to break down.

The liquid limit of disaggregated strongly cemented mudrock usually ranges between 35% and 45%, which reflects a predominance of illite in the clay mineral fraction. The extent of degradation due to weathering is greater than for other mudrocks. Typical values of parameters for weathered cemented mudrocks are as follows: moisture content 15–30%, liquid limit 35–45%, bulk density 1.9–2.0 Mg/m^3, undrained shear strength 15–70 kPa.

The impact of weathering on well-bonded mudrocks is particularly apparent in triaxial strength tests on fresh and weathered samples. In the results of tests by Taylor (1988) shown in **Figure 36.18**, a curved Mohr envelope typical of Carboniferous and other fresh cemented mudrocks was obtained. There is apparent cohesion at low values of confining pressure, but at higher confining stress the breakage of grains occurs. The removal of the bonding effects by weathering action results in a linear envelope with a lower friction angle typical of unbonded 'unconsolidated' clays. Shearing results in a further reduction in strength to the residual value because of the orientation of clay minerals on shear surfaces.

36.3.2 Investigation, description and assessment

In this section, the techniques of investigation are reviewed, particularly regarding some of the problems that may arise with mudrocks. Features of mudrocks that help with the anticipation and avoidance of adverse forms of behaviour are also discussed.

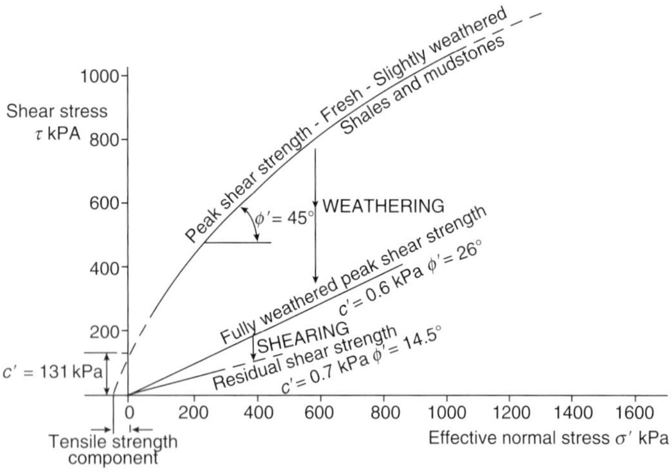

Figure 36.18 Mohr envelopes for bonded and weathered Carboniferous mudrocks (Reproduced from Taylor, 1988; © The Geological Society of London)

As explained in Section 36.1.1, BS 5930 provides guidance on the procedures for carrying out ground investigations for civil engineering works in the UK. It is argued that the money spent on carrying out ground investigations is saved by avoiding additional costs due to changes and delays in the construction programme. Savings in materials and construction costs may be possible because a more refined design can be adopted.

Laboratory classification tests for soils are described in BS 1377-2 (BSI, 2022). These documents specify minimum standards for commonly occurring engineering situations. So far as mudrocks are concerned, there are a number of deficiencies, and due to uncertainty of meaning some of the terminology and methods need to be clarified.

The density of the material is an extremely useful parameter for assessing its state of induration and likely performance. As mudrocks have broadly similar compositions it may also be indicative of mineral composition and void volume.

36.3.2.1 Soil and rock

Deciding whether to base designs on soil mechanics or rock mechanics principles is not a straightforward issue for some mudrocks, and assuming a worse case would not normally provide the best engineering solution. Soil mechanics deals with material as assemblages of discrete particles or a continuum. Design is usually based on limit state modelling of a plastic material of low shear strength in which changes in moisture content bring about significant changes in plasticity and volume. On the other hand, the rock mechanics approach considers an intact material of significant rigidity where the mass properties are controlled by the nature and geometry of dislocations present in the rock mass. The engineering assessment of mudrocks and clays needs to consider the most appropriate design approach, given the material and the effects on the material of the imposed stresses and the environmental conditions at different stages of the project.

At a certain stage of induration, the network of cemented grains will resist deformation such that failure of the material requires brittle fracture of the framework. Such behaviour results in fragmentation if the confining stress is low, whereas the same material can deform in a plastic manner in high confining stress conditions. As mentioned in Section 36.1.2, this makes the material difficult to classify and, as these changes are progressive, the effects may vary within the formation.

Norbury (2010) points out that the boundary between soil and rock is taken as an undrained shear strength of 300 kPa, which is equivalent to a UCS of 0.6 MPa, as specified by BS 5930. He recommends the following field tests to distinguish between soils and rocks. With rocks

- fingernail does not penetrate the surface
- fragments behave in a brittle manner in the hands (i.e. breakage rather than bending)
- material does not immediately soften or disintegrate on contact with water.

Changes with time must also be considered, as some mudrocks degrade rapidly. A small change in weathering grade, which might occur during or after the construction period, can have a significant impact on the appropriateness of a construction process or design. It can be very difficult to decide whether it is appropriate to regard the material as an engineering soil or as a rock; sometimes it is necessary to apply both approaches and then optimise the design so that both forms of behaviour are accommodated. Further relevant discussion is presented in the next section.

36.3.2.2 Description

In view of the high possibility of sample disturbance influencing the results of laboratory tests, it is particularly important that descriptions of mudrocks are accurate and clear. Both the material and the structures in the rock or soil mass should be recognised and correctly characterised. In practice, an accurate interpretation of laboratory test results depends on descriptions of the different materials and of their spatial relationships with each other. It is usually possible to obtain good-quality undisturbed samples of either strongly cemented mudrocks, which survive the sampling procedures and preparation for testing, or compaction mudrocks, because of their low strength and plasticity.

The aim of description is to record unambiguously essential features of the material so that (*a*) similar material can be recognised elsewhere, and (*b*) accurate prediction of the engineering behaviour is possible. Norbury (2010) provides highly useful advice on the description of mudrocks and other geomaterials.

Unfortunately, the job of logging cores and exposures is regarded by many as a menial task to be assigned to junior, less experienced staff. This can result in a weak link in the transfer of information about the ground conditions to designers and construction engineers. It may also be wasteful given the high investment of time, effort and money expended on ground investigations, and the additional costs and delays that inappropriate design can cause. Therefore, description of mudrocks should be undertaken in a well-informed and focused manner with thorough attention to detail. Distinguishing between soil and rock is an essential part of description as there are different approaches to describing these materials. However, as discussed in Section 36.3.2.1, mudrocks can span the range from rock- to soil-like behaviour and some rock or soil masses contain both materials. It is often necessary to describe the weathered parts of a rock mass using the terminology for soils, whereas some soils contain hard inclusions that need to be described in the manner of rocks. The proportions and relationships between the different materials should also be recorded.

Although it is preferable to use a standard approach to description, according to a specific standard, it is also necessary to ensure that all information that will enable a material to be recognised (and its behaviour to be anticipated) has been recorded. The standard schemes given in BS EN ISO 14688-1, BS EN ISO 14689-1 and BS 5930 should be regarded as the minimum amount of information to be logged. The recording of additional information relating to factors controlling the behaviour of mudrocks facilitates more reliable assessments of their performance in engineering situations. This approach is supported by Franklin and Dusseault (1989), Dick and Shakoor (1992) and Hawkins and Pinches (1992), who emphasise that a thorough geological description is the essential first stage of making an engineering assessment of mudrocks, with further classification based on the results of suitable index tests accompanied by strength and slake durability tests, as discussed by Cripps and Czerewko (2017).

Although both BS EN ISO 14689-1 and BS 5930 contain deficiencies, they provide good bases for descriptions, and only the aspects requiring additional details are described below. As discussed in Section 36.1.1, a particular inadequacy of the standards is the choice of rock names; the scheme given in **Table 36.4**, which groups the materials into classes that relate to their engineering performance, will usually prove more appropriate.

Using **Table 36.5**, the distinction between soil and rock material is based upon the state of induration determined by the tests described by Norbury (2010), as shown in Section 36.3.2.1. Rocks do not readily disaggregate when a dry sample is immersed in water and it is not possible to indent rock with a fingernail. The grain size, in the silt and clay portions, is judged according to the surface texture and appearance of the material. In the case of non-indurated materials, the dilatancy, dry strength and plasticity tests (see BS 5930 and BS EN ISO 14688-1) should be used. The distinction between indurated mudstone and metamudrock is also based on strength, where the latter is typically medium strong or stronger (UCS $>$ 25 MPa).

According to BS EN ISO 14688-1, BS 5930 and BS EN ISO 14689-1, information that needs to be recorded in rock and soil material should include the following aspects, although (as noted above) this should be regarded as the minimum and attention should be given to the details listed in **Table 36.5**.

- strength: very soft to very stiff for soils; extremely weak to extremely strong for rocks (**Table 36.6**)
- structure: lamination and bedding and so on; see **Table 36.7**, and in particular, the footnotes
- colour: describe in terms of depth, chroma and hue, preferably with reference to a Munsell chart, and record mixing such as mottling or banding
- fabric: spatial arrangement of solid grains and associated voids, such as graded or granular
- texture: shape and size of constituent grains, including groundmass, inclusions and clasts and so on
- rock or soil name: in capitals, such as MUDSTONE or SILTSTONE (see **Table 36.4**)
- other relevant information: includes minor constituents such as fossil debris, calcareous and ferruginous nodules, disseminated or crystalline pyrite or iron oxide
- geological formation: where known, such as LOWER LIAS CLAY or LOWER COAL MEASURES.

The second part of the description deals with the structure of the rock or soil mass, and includes the following characteristics.

- Weathering: description is required of observed material changes including colour changes, differences in strength and fracture state, as well as the presence and character of any weathered products, as is mandatory in BS EN ISO 14689-1, and recommended in BS 5930. Where possible, a weathering classification scheme should be used but only if it can be applied unambiguously to denote the degree and style of weathering (see **Table 36.8(a)** and **36.8(b)**).
- Discontinuities: information should be recorded about the distribution and features of discontinuities including the type, orientation, spacing, geometrical form, shape, rugosity (surface roughness) and evidence of alteration of rock adjacent to the discontinuities, as well as the nature and thickness of any infill material.

The level of detail it is possible to record depends upon the scale of exposure; for example, it may not be possible to comment on the planarity of joints in borehole cores. **Table 36.5** provides guidance about the terms that should be used in the description of mudrocks. The strength categories in **Table 36.6** are based on measured in situ or laboratory shear strength values, whereas the indicative strength assessed using the manual ('tactile' hand) tests for material description should use the consistency terms 'very soft' to 'very stiff'. Details of manual tests are given in BS EN ISO 14688-1, BS 5930 and Norbury (2010), and classifications based on these tests are presented in BS EN ISO 14688-2. For rock descriptions, guidance is provided in BS EN ISO 14689-1, but as some of these do not give reliable results for mudrocks the tests detailed in **Table 36.6** are usually more helpful.

		Average grain size < 63 μm (<50% of grains less than 63 μm in size)			
Mudrock type		% < 2 μm	Characteristics	Material type	
Non-indurated	S_u < 300 kPa σ_c < 0.6 MPa Can be broken by hand Rapid breakdown in water	66–100	Smooth, soapy feel with fingers Plastic[a] Some particles visible with hand lens	Clay	
		33–65	Slightly gritty feel with fingers Slightly plastic[a] Particles visible with hand lens	Silty clay Clayey silt	
		0–32	Gritty to touch with fingers. Non-plastic[a] Some particles visible with naked eye	Silt	
Indurated	σ_c 0.6–50 MPa Retains form in water, but surface softens Immersion in water causes slaking	66–100	Smooth non-gritty feel with blade/fingers Some particles visible with hand lens Surface softens with water and water becomes cloudy	Claystone	
		33–65	Slightly gritty feel with blade/fingers Particles visible with hand lens Surface softens with water and water becomes slightly cloudy	Mudstone	
		0–32	Gritty to touch with blade/fingers Some particles visible with naked eye Does not soften with water Sparkly appearance	Siltstone	
				Massive	Foliated
Metamudrock – very low grade	σ_c > 25 MPa[b]	66–100	Partly to wholly recrystallised Grains visible with hand lens	Argilite	Slate
		33–65	Some grains visible to naked eye Slate: flat platy particles may be broken from edges with blade		
		0–32	Partly to wholly recrystallised Grains visible with hand lens Grains visible to naked eye	Quartz argilite	Siltstone-quarzite
Metamudrock – low grade	σ_c > 25 MPa[b]	66–100	Recrystallised Some grains visible to naked eye	Argilite-hornfels	Phyllite
		33–65	Phyllite: silky sheen and crenulated cleavage		
		0–32	Recrystallised Grains visible to naked eye Quartzite: lustrous appearance	Hornfels-quartzite	Siltstone-quarzite

[a] Tests to assess plastic or non-plastic behaviour in terms of dilatancy, plasticity and dry strength are described by BS EN ISO 14688-1:2002.
[b] Strength values for unweathered (fresh) material.

Table 36.4 Identification and classification of mudrocks (Based on data from Potter *et al.*, 1980; Attewell, 1997)

Although BS EN ISO 14688-1 and BS EN ISO 14689-1 provide schemes for classifying the spacing of structural features, these lack subdivisions within the thinly and thickly laminated categories that are useful for assessing the potential style and rate of weathering degradation.

The colour of rocks and soils depends on whether the surface is wet or dry. Generally speaking, the material should be described in the condition it was in when it was taken from the ground – but this should be stated. Care must be taken to ensure that the surface of the sample is clean of coating

Attribute	Descriptive adjectives
Induration	Enables decision on description as soil or rock. If resistant to slaking in water and hard – rock; if susceptible to slaking in water, deformable and 'earthy' consistency – soil. Strength depends on moisture state; dry sediment is stronger than wet, and rock strength varies with moisture content; sampling may impair strength.
Strength	Strength is designated based on the degree of induration. For soil, use field consistency values based on manual assessment, e.g. stiff; when shear strength measurements are made, use strength terms, e.g. high strength. For rock, definition based principally on manual field assessment using geological hammer and knife, may be confirmed with UCS measurement: indurated mudrocks range from extremely weak to medium strong; metamudrocks are stronger depending on weathering.
Structure	Standard terms – see **Table 36.7**. Include description of lithology and textural interrelationship, as complex features may be present with structured strata such as 'thin beds of cross-bedded' mudstone.
Colour	Use Munsell colour chart for consistency. Important for correlation; likely environment of formation and indication of likely behaviour of material, i.e. red colour – likely formation under oxidising continental environment. Most important to mudrocks is relationship between colour on the Fe^{3+}/Fe^{2+} ratio. A decrease in this ratio gives an increase in colour from red → green → grey (more Fe^{2+} indicates the presence of pyrite). Organic carbon controls colour: <0.2–0.3%C = light grey to olive grey; 0.3–0.5%C = mid-grey; >0.5%C = dark grey to black.
Accessory minerals	Calcareous (slightly to very) – based on level of effervescence when assessed with HCl. May also be carbonaceous, dolomitic, ferruginous, glauconitic, gypsiferous, pyritic, micaceous, sideritic, phosphatic, etc.
Rock name	See **Table 36.4.**
Additional information	Presence of fossils – record type (generic such as bivalve and retain for identification), abundance, condition, orientation. Inclusions – nodules (with mineral type and details); gravel, sand, silt partings or pockets, etc.
State of weathering	Alteration seen as distinct discolouration, significant strength reduction to discontinuities and presence of lithorelicts (note orientation). Standard Eurocode 7 approach too limited for mudrocks, use specific schemes where available as guides, i.e. Lias and London Clay, Mercia Mudstone.
Fractures	Use EC7 standard terms and procedures. For rock supplement with details such as nature of fragmentation, e.g. conchoidal, hackly, brittle, splintery, slabby, fissile.

Full engineering descriptions of the intrinsic condition and mass properties using descriptive adjectives provided in EC7 (BS EN ISO 14688-1:2002; BS EN ISO 14688-2:2004; BS EN ISO 14689-1:2003, supplemented with additional information to assist assessment.
This list offers limited guidance for key features particular to mudrocks.

Table 36.5 Guide to the description of mudrock features

materials which may be different colours from the bulk of the material. Samples should be broken open to facilitate a description of both the interior and exterior. In addition, if there are any changes in the appearance of broken surfaces some hours later, they should be recorded.

Texture of soils and rocks refers to the relationships between the constituent grains, including the variation in size. Mudrocks tend to have a fine granular texture, implying that the main constituents are detrital grains that are cemented together. The particle shapes and crystallinity should be recorded if possible, but this is often impossible to see even using a hand lens. Details of the composition should also be recorded – for example, the presence of organic material (by visual assessment or smell) or carbonate (as indicated by effervescence when hydrochloric acid is applied), using the categories slightly, highly and so on, according to the criteria given in BS EN ISO 14688-1, BS EN ISO 14689-1 and BS 5930.

In some cases it is best to describe the material rather than attempting to determine an accurate geological name. The rock or soil type should always be written in capitals, which implies that the rest of the description should be presented in sentence case.

The presence of constituents that are not implied by the rock name should be specified to aid recognition of the material. The description should also include details of observable features not already covered. Examples would include the presence of fossils, nodules, inclusions and so on. Where it is known, the geological formation name should be specified.

The presence of pyrite is a possible adverse feature of mudrocks that should be considered in the description of the materials. Characteristics of mudrocks that make the presence of pyrite more likely are

- dark-coloured (grey or dark grey), organic-rich mudrock or clay
- finely disseminated gold-coloured crystals, greenish grey hue or orange-brown and brown staining
- sulfurous odour when hit or scratched
- clear or white-grey crystals on surfaces which can be scratched and broken with a fingernail, or a white surface coating when the material is dried (liable to be gypsum/selenite).

Recording the effects of weathering is a highly important aspect of the description of a rock or soil mass. The appearance of the material should be recorded, particularly any changes in colour, strength or fracture spacing that are a consequence of weathering processes. This is difficult where the material has not been viewed in its fresh state for comparison. If it is possible and can be applied unambiguously, the condition of

Term	Strength	Description
Strong mudrock	σ_c 50–100 MPa	Can only be scratched by knife or pick end of a geological hammer, and can only be broken with more than one firm hammer blow.
Medium strong mudrock	σ_c 25–50 MPa	Can be deeply scored by a knife or pick end of a geological hammer, and a thin slab can be broken by heavy hand pressure. Specimen is readily fractured with a single firm blow of geological hammer or split with a knife blade. Cannot be peeled with a pocket knife.
Weak mudrock	σ_c 5–25 MPa	Small gravel-sized fragment can be deformed with heavy finger pressure, shallow indentations readily made by firm blow with point of geological hammer. Can be peeled by a pocket knife with difficulty.
Very weak mudrock	σ_c 1–5 MPa	Crumbles under firm blows with point of geological hammer, can be peeled by a pocket knife.
Extremely weak mudrock	σ_c 0.6–1 MPa	Can be indented by thumbnail.
Extremely high strength clay	s_u 300–600	Field description will generally be as a 'very stiff clay'. Crumbles, does not remould, can be indented by thumbnail.
Very high strength clay	s_u 150–300	Determine by testing – field description as 'very stiff clay'.
High strength clay	s_u 75–150	
Medium strength clay	s_u 40–75	Field description will generally be as a 'stiff clay'. Crumbles, breaks, remoulds to lump.
Low strength clay	s_u 20–40	Field description will generally be as a 'firm clay'. Cannot be moulded, rolls to thread.
Very low strength clay	s_u 10–20	Field description will generally be as a 'soft clay'. Moulds by light finger pressure.
Extremely low strength clay	s_u <10	Field description will generally be as a 'very soft clay'. Extrudes between fingers.

σ_c uniaxial compressive strength.
s_u undrained shear strength; $\sigma_c = 2 \times s_u$. Extremely high strength clay soils and extremely weak mudrocks have similar characteristics – choice of term depends on the geological setting and engineering situation.
Field consistency descriptions for clays and strength tests descriptions should be used separately and treated as mutually exclusive.

Table 36.6 Strength criteria for mudrocks

Thickness	Stratification		Parting	Composition
200 mm – 60 mm	Thin	Bedding	Slabby	Clay and organic content increase (downward); Sand, silt and carbonate content increase (upward)
60 mm – 20 mm	Very thin	Bedding	Slabby	
20 mm – 6 mm	Thick	Lamination	Flaggy	
6 mm – 1 mm	Medium	Lamination	Platy	
1 mm – 0.5 mm	Thin	Lamination	Fissile	
< 0.5 mm	Very thin	Lamination	Papery	

Discontinuity spacing	
2000 mm – 600 mm	Wide
600 mm – 200 mm	Medium
200 mm – 60 mm	Close
60 mm – 20 mm	Very close
< 20 mm	Extremely close

Mudrocks that occur without laminations or bedded stratification are described as massive.
Textural features should be used as *descriptive adjectives* when describing mudrocks and not for naming rocks similar to generic schemes used for other rock types, as this may be confusing and ambiguous; for example 'shale' can mean a fissile, or laminated, or massive mudstone or claystone; in preference, use the adjective as a prefix such as fissile claystone, laminated mudstone.
Stratification is a structural feature concerned with the interrelationship of textural features and lithology caused by differences of grain size, fabric and composition; partings is the tenacity of a mudrock to split along lamination, bedding or mineral orientation, which is greatly enhanced by weathering; whereas discontinuities are mechanical breaks or fractures with low to no tensile strength.

Table 36.7 Description of structural features of mudrocks (stratification, partings and discontinuities) (Data taken from BS EN ISO 14689-1:2003; Potter *et al.*, 1980)

the rock or soil mass should be classified. Although weathering begins along discontinuities in mudrocks, there also tends to be an impact on the intervening blocks, whereas in stronger rocks there is usually little effect on these so that corestones are formed. Both BS 5930 and BS EN ISO 14689-1 include the same classification scheme as Approach 4, which is suggested as being suitable for mudrocks and material that is 'moderately weak or weaker' in a fresh state. But in BS EN ISO 14689-1, Approach 4 is included in Appendix 2 as informative, and a preferred mandatory classification for rock masses is presented in the main text; this comprises six distinct grades that classify the changes in terms of progression from rock to rock-and-soil and finally soil. This approach may be difficult to apply in all situations, particularly in lithological sequences comprising mixed successions of limestones and sandstones with mudrocks, and it is suggested that for mudrocks the schemes presented in **Table 36.8(a)** and **Table 36.8(b)** are used to produce a more informative classification of weathering condition, where the former provides a scheme for non-indurated mudrocks and clays and the latter provides one for indurated mudrocks and metamudrocks.

The features that need to be recorded in the description of structures should include, where possible, the origin of discontinuities. Where this is not discernible, the term 'fracture' should be used. In some mudrocks, especially weakly cemented types, it can be difficult to distinguish between in situ and induced discontinuities, where the latter have been formed during or after removal from the ground. Stress relief, change in moisture content and physical disturbance may all result in such fracturing. Usually such fractures have a fresher and cleaner appearance than in situ ones, but this is not always the case.

Figure 36.14 demonstrates how the characteristics of the weathered material can be used to deduce the internal structure of the material. Platy fragments will be produced in a rock possessing tectonic cleavage or a laminated structure; flaky and blocky fragments are produced by a uniform distribution of the minerals in the rock. The bedding is thicker in the latter case.

36.3.2.3 Durability

Durability is a measure of the rate that a rock will degrade when exposed to weathering action. Compared with most other types of rocks, mudrocks degrade rapidly. Hence, significant changes to geotechnical properties occur on timescales of significance to engineering designs, while for some low-durability mudrocks changes may occur within the construction period. In the process, a seemingly competent rock material becomes transformed into clay or silty clay, whereas a rock mass may well become degraded into sand- and gravel-sized fragments with a serious loss of strength and increase in compressibility. It should also be remembered that degradation processes may

Classifier	Nature	Typical characteristic	
Unweathered	Fresh	No visible sign of weathering. Typically grey colour, clean fissured clay or mudrock.	Pyrite
	Faintly weathered	Trace of brown surface discolouration on discontinuities otherwise intact fissured clay or mudrock.	
Partially Weathered	Slightly weathered	Fissures more closely spaced, slight strength reduction evident as surface softening some bedding disturbance. Discontinuity surfaces with pervasive brown discolouration and slight penetration typically 1–2 mm. Rare selenite crystals on discontinuity surfaces.	Increasing oxidation
	Moderately weathered	Fissures more closely spaced, slight strength reduction evident as surface softening, some bedding disturbance. Discontinuity surfaces with pervasive brown discolouration and slight penetration typically few mm to cm with centre of fissure bound block still grey. Occasional selenite crystals.	Selenite
Distinctly Weathered	Highly weathered	Fissure spacing much closer with lensoidal fissuring – development of matrix bound regular angular block and platy orientated 'lithorelicts' and matrix as few mm to cm of claysilt on discontinuities, with occasional polished fissures or gleying. Material is generally brown to grey brown with completed discolouration – occasional fissure bound blocks with trace of grey core. Selenite crystals common in matrix and blocks.	Oxidation complete increase in reduction producing gleying
Destructured	Completely weathered	Greatly weakened material with numerous horizontally aligned ordered lithorelicts in clay-silt matrix. Bedding disturbed. Generally brown and grey brown, may have light grey mottling. Fissure blocks and lithorelicts brown throughout. Selenite crystals common.	
		Greatly weakened material comprising numerous randomly orientated lithorelicts in a clay-silt matrix. Lithorelicts are completely discoloured to brown or grey brown and softened. Gypsum crystals common. Bedding heavily disturbed.	
Reworked	Completely reworked or residual soil	Unbound sediment with occasional randomly orientated predominantly sub-angular to sub-rounded lithorelicts and occasional foreign material. No original bedding or structure remaining. Generally mottled brown, orange-brown and light grey. Rare selenite crystals may be present but generally absent.	Surface water infiltration – leaching

Table 36.8(a) Weathering classification and description for non-indurated (compaction mudrocks) or overconsolidated clays

Classifier	Nature	Typical characteristics
Unweathered	Fresh	No visible signs of weathering or alteration. Primary colours of dark grey, grey, purple, red-brown, green grey.
	Faintly weathered	Colour of original material is changed – discolouration tends to be confined to major discontinuity surfaces – the colour change and extent of discolouration should be described. Generally change to light brown, orange-brown, light brown and red-brown discolouration.
Partially weathered	Slightly weathered	Colour of original material is changed – discolouration tends to be confined to discontinuity surfaces with pervasive penetrative discolouration. Discolouration should be described and any evidence of material weakening along discontinuity should be described and quantified where possible.
	Moderately weathered	Considerable weakening with alteration and comminuition to matrix of clay and silt penetrating in from discontinuities accompanied by penetrative discolouration. Rock framework intact with <50% of material decomposed. Rock framework generally locked and controls strength and stiffness. Matrix affects permeability.
Distinctly weathered	Highly weathered	Generally >50% of rock material decomposed. The rock framework may contribute to strength but weathered decomposed material forming the matrix controls stiffness and permeability – consider structural design based on soil mechanics principles.
Destructured	Completely weathered	Rock material largely decomposed and considerably weakened. The weak 'soil-like' material controls behaviour. Lithorelicts may be present with random orientation. Hard corestones/concretions may be present and affect construction – consider structural design based on soil mechanics principles.
Reworked – residual	Residual/reworked soil	Unbound soil material comprising of clay and silt with occasional or rare randomly orientated soil lithorelicts and foreign material. No original bedding, structure or fabric preserved. Properties of soil, moderate compressibility, low permeability, low to moderate soil strength.

Table 36.8(b) Weathering classification and description for cementation or rock-like mudrocks

also affect the results of laboratory tests on samples and that if this occurs tests may return misleading results.

The main process of degradation of mudrocks is slaking, which usually causes breakage of the mass into sand- and gravel-sized fragments and occurs due to alternating changes in water content. Because volume changes to clay minerals are greater for swelling clay minerals, mudrocks containing smectite or mixed layer smectite–illite are particularly vulnerable to this type of degradation. Pore water suctions, produced as water is drawn out of narrow pores during drying cycles, also cause shrinkage. This effect is greatest in fine-grained rocks, especially where the pore size has been reduced through a certain degree of compaction and cementation. Taylor (1988) attributed the breakdown of mudrocks in colliery spoils to the process of air breakage – resulting from high pore air pressures produced as water enters narrow capillary pores during the saturation of mudrocks. The shrinkage and swelling of the rock, associated with changes in moisture content, stress the intergranular bonds and partly cemented skeletal framework possessed by weakly cemented mudrocks such that breakage predominantly occurs along existing structural weaknesses within the material. This causes the rock mass to break down into sand- or gravel-sized pieces, as illustrated in **Figure 36.15**.

Durability tests are designed to provide both qualitative and quantitative assessments of the susceptibility of the rock to degradation processes. Ideally, to provide accurate guidance about the performance of a material at different stages of degradation. The test used should simulate the dominant degradation processes to which the material will be subjected in the particular engineering environment. Unfortunately, only the slake durability test described by Franklin and Chandra (1972) and ISRM (1979) has been standardised. The test entails subjecting samples to both wetting and drying action and mechanical disturbance. The degree of degradation is assessed by the proportion of particles of less than 2 mm produced by the degradation process expressed as a percentage of the original dry mass of the sample. In the standard test, the number of cycles of drying and slaking (n) equals 2, where $Id_2 > 90\%$ distinguishes non-indurated from indurated mudrocks. However, Taylor (1988) suggested that the third cycle durability value $Id_3 > 60\%$ is a more reliable indication of this boundary; this was later corroborated by Czerewko and Cripps (2001). Two of the drawbacks of the method are that: (*a*) the degradation to particles larger than 2 mm is not directly assessed and (*b*) there is a lack of sensitivity in compaction mudrocks and some weakly cemented mudrocks, which make it preferable to carry out static slaking tests. Furthermore, complete drying of the material in the test does not simulate the more subtle changes in moisture content that occur in natural weathering environments. Drawback (*a*) can be overcome by carrying out plasticity testing on the <2 mm fraction. Difficulty (*b*) can also be overcome by carrying out plasticity testing on the <2 mm fraction and/or by particle size analyses of the drum-retained fragments. Regrettably these important aspects of durability assessment are frequently omitted.

The static jar slake test procedure was advocated by Lutton (1977) and further developed by Czerewko in 1999 (see Czerewko and Cripps, 2001); it more accurately simulates degradation that does not entail mechanical disturbance.

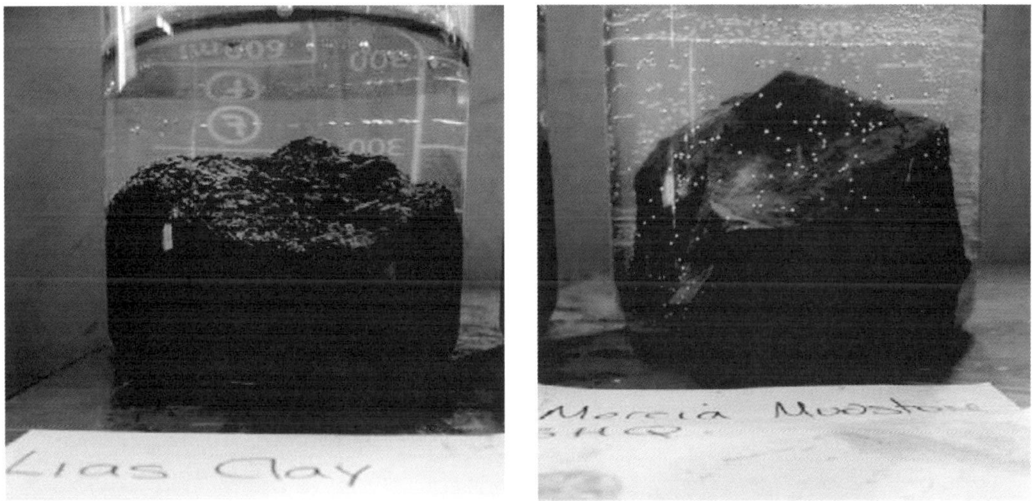

(a) (b)

Figure 36.19 Behaviour of non-durable and durable mudrocks to water immersion in jar slake tests. (a) Lias Clay from East Midlands, England: rapid slaking after 3-hour immersion to sand- to gravel-sized material. (b) Durable Basal Mercia Mudstone from East Midlands, England showing no slaking after 24 hours of immersion in water

Following immersion of an oven-dried sample in water, the degree of fracture development and of slaking are monitored over a 24-hour period. **Figure 36.19** shows the results of jar slake tests on two mudrocks of contrasting durability. Experience with a wide range of UK mudrocks and using strength (UCS), static jar slake (I_j') and dynamic slake durability (Id_3) test procedures suggests that UCS = 5 MN/m^2, $Id_3 > 60\%$ and $I_j' < 6$ effectively distinguish bonded (cemented) and non-bonded (compaction) mudrocks from one other, as detailed in **Table 36.9** and further discussed by Cripps and Czerewko (2017).

The likelihood of rapid degradation of mudrocks and the size of the breakdown products may also be ascertained by observing samples or outcrops that have been subjected to short periods of exposure (see **Figures 36.14** and **36.15**). For example, **Figure 36.20** shows the complete degradation of Lias mudstone over a period of 41 days.

Slake durability – Id_s: %	Jar slake – I_j'	Compressive strength – UCS: MPa	Classification
<60	6–8	<5 (≤ very weak)	Non-durable mudrock
60–80	3–6	5–100 (weak to strong)	Durable mudrock
80–97	1–3		Extremely durable mudrock
>97	1–2	>100 (≥ very strong)	Metamudrock

Table 36.9 Durability classification of mudrocks for engineering purposes

36.3.2.4 Investigation techniques

Ascertaining the special distribution and engineering properties of different rock units for the design of engineering works and to facilitate safe and economic construction procedures usually entails the excavation of trial pits and/or the drilling of boreholes. These direct methods enable the different materials to be identified and for samples to be taken for subsequent laboratory tests. Other techniques, such as surface geological mapping and geophysical surveys, may also provide relevant data about the ground. Although some mudrock units are exposed at the surface, due to the low durability of most mudrocks if exposures exist the material exposed in outcrops will be in a degraded condition. This means that most investigations in mudrock terrain will require the drilling of boreholes and the excavation of trial pits. Boreholes may be drilled to facilitate the deployment of geophysical logging tools and in situ testing devices, but more usually their primary purpose is to facilitate identification of the rock types present and the taking of samples.

Sample disturbance is a significant problem with borehole investigations of mudrocks, especially where they are weakly cemented as these are prone to becoming distorted by the processes of taking or extracting samples. Low-durability mudrocks may be reduced to mud consisting of silt and clay slurry and be lost in the drilling fluid, so they are entirely lost. Besides the possibility of degradation and increased fragmentation, mudrocks also tend to be vulnerable to volume changes in response to removal of overburden pressure and changes in moisture content. The use of wireline coring, triple-tube core barrels or plastic core-lining tubes in double-tube core barrels greatly enhances the amount of core recovered and the quality

Figure 36.20 Example of the field assessment of material weathering due to pyrite oxidation, allowing consideration for design assumptions and beneficial construction sequencing. (a) Lias Clay material proposed for use as earthwork fill was exposed to atmospheric conditions to observe for signs of deleterious reaction of pyrite. Pyrite: 2% content, with calcite observed as shell debris and nodules. (b) Presence of framboidal pyrite confirmed by SEM analysis. (c) Complete breakdown of sample shown in (a) in a 41-day period with oxidation of majority of pyrite and production of selenite crystals and significant softening, shown in (d)

of cores. Other methods for obtaining undisturbed samples of mudrocks, which are potentially suitable for strength and compressibility testing, are outlined in **Table 36.10**.

It has been common practice in the UK to begin drilling investigation boreholes using cable percussive boring techniques to advance through and recover samples of material with a soil-like consistency and then swap over to rotary coring once progress with cable percussive techniques becomes impeded. Unfortunately, this process involves a change of drilling equipment and recovery techniques that often result in loss of or excessive disturbance to recovered material that is unrepresentative of the ground conditions. The recent development of remotely controlled, compact, tracked hydraulic drilling machines that are equipped with a combination of percussive and rotary drill heads such as that illustrated in **Figure 36.21** has improved the situation by enabling a smooth switch over between techniques as the nature of the ground changes. This equipment has been found to be particularly beneficial for the investigation of weathered mudrock sequences at shallow depth. The multipurpose rigs are capable of drilling using conventional rotary and wireline coring, rotary augering techniques as well as the carrying out of dynamic, standard

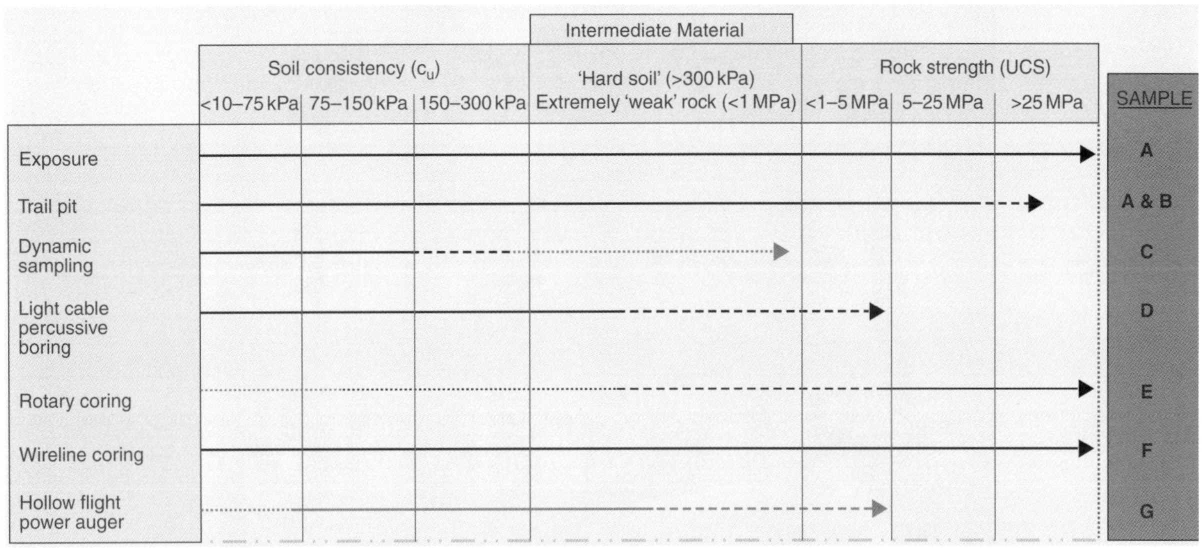

A = In situ examination of exposure allows accurate assessment of mass properties and recovery of block samples – limited by age and size of exposure.
B = Trial pits allow identification, description and sampling of disturbed excavated material to depth of approximately 4–6 m; support required for inspection and logging of mass properties and weathering and recovery of Class 1 block samples.
C = Dynamic samples allow recovery of continuous ground profile; material retained in rigid plastic liner; material damage by percussive driving method; material Class 2–3.
D = Standard UK procedure; percussive recovery of U100 as Class 2–3 sample; weak rock recovered as fragments for confirmation; allows use of hydraulic thin-walled push sampler with recovery of Class 1–2 material, limited to approx 100–120 kPa strength material. Gravel inclusions or nodules liable to cause damage to samples and sample tubes.
E = Hard soil and weak rock recovery, improved by use of triple-tube retractor barrels (at high cost); use of large-diameter (>100 mm) double/triple-tube core barrels to aid appropriate recovery of weak material; use of conventional diameter (76–92 mm) core barrels for stronger competent mudrock; choice of tungsten carbide bits for weak material, and diamond impregnated and diamond surface set bits for harder rock; consider cost; sample preservation is paramount.
F = Excellent recovery of all material type as Class 1; may lose borehole progress with bit wipe-out; very expensive and requires heavy equipment.
G = Uncommon in UK due to cost and requirement of heavy equipment; drive samples Class 2–3; unsuitable for continuous sample recovery.
Solid line: Suitable technique
Dashes: Less suitable technique
Broken line: Unsuitable

Table 36.10 Outline of sampling options for mudrocks and their limitations

penetration (SPT; see BSI, 2011) and thin-walled undisturbed (UT) sampling (BSI, 2007).

As explained in BS EN ISO 14688-1 and BS EN ISO 22475-1 (BSI, 2006), physical characteristics may be determined using laboratory 'mechanical' test procedures, the undrained shear strength and effective stress shear strength parameters may be determined by triaxial testing, and consolidation and swelling characteristics (including C_c – compression index; C_s – swelling index; m_V – coefficient of volume compressibility; and c_V – coefficient of vertical consolidation) can be determined by oedometer testing. The relevant test procedures are given in BS 1377 and are described by Head (1994, 1998). Guidance given by ISRM (Ulusay, 2014) is provided for strength and deformability tests on rocks.

Rock masses may be classified using geomechanical rock quality classification systems (see Barton *et al.*, 1974; Bieniawski, 1976) which are based on intact UCS values and mass discontinuity characteristics, usually from detailed descriptions of exposed rock or good-quality borehole cores. Additionally, the point load strength (PLS) may be determined on cores or samples from trial pits or exposures. But care should be exercised as the point load test is less reliable on weak rock that may spall and in which the conical test platens can embed themselves, recording unrepresentative test results. Bell (1992) states that the point load test is limited to rock with a UCS exceeding 25 MPa (i.e. point load index above 1 MPa). The ASTM test procedure (ASTM D5731, 2007) indicates that the test procedure is suitable for 'hard' rocks over 15 MPa in compressive strength. But it is likely that point load testing will be the only viable means of strength determination in weakly cemented or friable mudrocks and therefore an adequate number of samples should be tested to derive a representative strength.

The slake durability test can be used to determine the rate and extent of breakdown sustained by these rocks. Gamble's (1971) slake durability–plasticity scheme is the most suitable classification for compaction mudrocks; in low-durability mudrocks the slake durability–strength schemes of Grainger (1984) and Taylor (1988) are more suitable; finally, the strength–swell coefficient approach of Olivier (1979) is best for medium strong to strong mudrocks. Triaxial testing of rock to determine shear strength (c' and φ') and elastic deformation (Young's modulus, E; Poisson's ratio, ν) parameters necessitate the preparation of

Figure 36.21 Examples of modern compact multipurpose drill rigs fitted with both rotary drill head and percussive configurations; (b) includes a dedicated SPT tool

intact samples which will not usually be possible in all but the strongest and most intact mudrocks.

An alternative approach is to estimate strength and deformation parameters from comprehensive descriptions and use of index testing. Various schemes are provided by Hoek and Brown (1997), Hudson and Harrison (1997) and Wyllie and Mah (2004). In situ pressuremeter tests avoid problems with obtaining samples, but disturbance of borehole walls can still give unrepresentative results. However, examination of excavations and, if possible, boreholes using borehole cameras (optical or acoustic) may well be beneficial for obtaining a more accurate assessment of the rock mass condition. It is often difficult to recover mudrocks in an intact condition from boreholes, in which case in situ inspection of the drill hole using an optical televiewer (OPTV) provides a means of assessing the rock structure and lithological variations. The devices facilitate the recording of dip and azimuth of structural features. These direct observations may prove particularly useful for confirming positions of key horizons, such as confined water-bearing strata or horizons affected by dissolution if incomplete core recovery would result in uncertainty in the recorded core log, as is demonstrated in **Figure 36.22**.

Due to the difficulties with recovering good-quality undisturbed samples of mudrocks from shallow boreholes, the preferred method of assessment is through inspection and sampling of the ground exposed in trial excavations. Such excavations can be rapidly dug to depths of up to 4.5 m in favourable conditions, although due to the possibility of collapse logging will be limited to examination from the surface with inspection of excavation arisings, which is unlikely to enable a sufficiently detailed examination of the rock mass to be carried out. Safe entry into trial excavations may be facilitated using hydraulic shoring systems such as that shown in **Figure 36.23**. Obtaining accurate descriptions and good-quality samples will usually amply justify the additional cost of the equipment. On larger earthworks projects the creation of large benched trial excavations such as that illustrated in **Figure 36.23** that provide a favourable means of recording the rock mass conditions and taking samples should be programmed into the construction sequence.

36.3.2.5 Assessment of pyrite

As explained in Section 36.2.4.3, pyrite is a commonly occurring (but minor) constituent of many mudrocks that usually has a negative impact in engineering applications. As **Figure 36.8** illustrates, the mineral occurs in a number of different forms, depending upon the state of induration of the rock (see Section 36.2.2) and other factors. It can be widely and erratically distributed through the deposit, which makes it difficult to detect and accurately quantify, particularly when relying on samples recovered from targeted boreholes that do not facilitate proper assessment of the lateral variability. Only in the crystalline form is it easy to recognise pyrite in hand specimens; furthermore, in this form it often occurs as gold-coloured cubes that are sufficiently hard to scratch normal glass and that produce a slightly sulfurous odour when struck. Czerewko *et al.* (2003b)

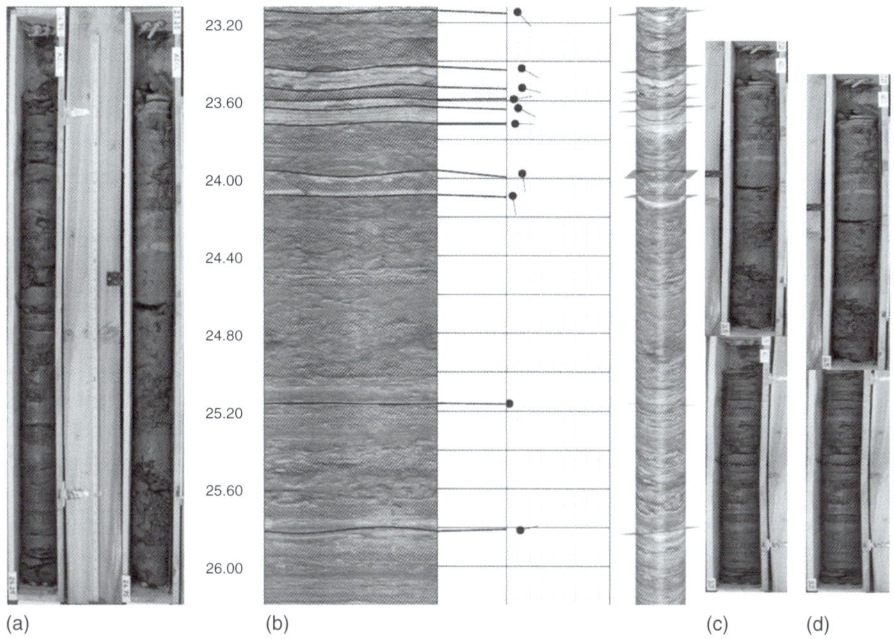

Figure 36.22 Example of an optical televiewer (OPTV) complimenting core recovery from fragile Mercia Mudstone strata. (a) Core recovered using wireline coring showing some core loss; (b) OPTV image from the same section records variable low-angle laminations and depth sequences of the interbedded lithologies; (c) core loss would result in uncertainty of the depths of the skerry (siltstone) beds; (d) correctly positioned siltstone beds assigned with reference to the OPTV records as seen in (b)

Figure 36.23 Examples of in situ inspection of mudrock sequences through trial excavations: (a) unsupported trial pit assessed from ground surface and by examination of excavated material showing change from weathered to fresh sequence, with limited recording of details and lithological variation; (b) and (c) trial pits with supported sides allowing safe entry to undertake detailed in situ inspection, sampling and in situ testing; (d) a wide trial excavation with benched sides enabling safe entry for detailed mapping of the rock mass, sampling, in situ testing and 3D geometrical assessment of structural features as seen in (e)

explain that other sulfur minerals may also cause problems in engineering situations and guidance concerning the identification and evaluation of the various forms of sulfur is provided. In compaction mudrocks, pyrite usually occurs in an indiscernible, framboidal or fine-grained dark-coloured form (see **Figure 36.8(a)**). The framboids and small crystals may be a dark greenish grey colour with slightly sparkling appearance and may, when rubbed or hit, produce a slightly sulfurous odour, but in this condition certain identification requires the use of optical or electron microscopy or X-ray diffraction analysis (see Reeves *et al*., 2006). These methods will facilitate a definite identification but only provide an estimate of the quantity present. The accurate determination of concentration of pyrite in mudrocks requires the use of chemical analysis methods which are discussed in detail in TRL447 (Reid *et al*., 2005). These test methods are also described in the current BS 1377-3 (BSI, 2021). TRL447 (Reid *et al*., 2005) test 3 is an analysis method specific to sulfide minerals; however, the procedure requires a high degree of technical proficiency and is not suitable for routine testing but it is included for reference as Method 7.11 in BS 1377-3. Usually the amount of pyrite is instead estimated from the difference between the total sulfur and the amount of acid-soluble sulfur present (BS 1377-3, Methods 7.9 and 7.10). If sulfur is present in other substances, for example organic material, or acid-insoluble minerals such as barites ($BaSO_4$) are present, this difference method will overestimate the quantity of pyrite present. The carbonate content should also be determined (BS 1377-3) as this provides an indication of the potential for buffering the acid produced by pyrite oxidation (see Section 36.2.4.3). In addition, to determine the potential for sulfate attack of concrete and other construction materials, the water-soluble sulfur should be ascertained.

The assessment of aggregates requires chemical testing in accordance with BS EN 1744-1 (BSI, 2013a), which comprises procedures that were originally employed for the assessment of industrial products and by-products, including cements and slags. The production of these materials involves high temperature that destroys disulfide minerals such as pyrite, for which the test procedures included in BS EN 1744-1 are therefore unsuitable. However, although the standard states that the Clause 11.1 test is suitable for analysing materials that contain reduced minerals such as pyrite, the procedure is liable to seriously underestimate the amount of pyrite present. Therefore, it is unsuitable for testing geological materials and should be used solely for Eurocode product conformance testing under such standards as BS EN 13242 (BSI, 2013b) and then the results should not be relied on for any design purpose. BS EN 1744-1 also includes a procedure for determination of 'acid-soluble sulfides' in Clause 13. This procedure should not be used for determination of pyrite, for which it is unsuitable. The procedure using hydrochloric acid digestion is only effective for monosulfide mineral species such as pyrrhotite, and is ineffectual for digestion of disulfide minerals such as pyrite. Further detailed reviews are presented by Reid *et al*. (2017) and Czerewko *et al*. (2016).

BRE Special Digest 1 (2005) deals with the determination of pyrite with respect to chemical attack of buried concrete and TRL447 (Reid *et al*., 2005) with the performance of fill associated with highway structures. These publications also make recommendations concerning the taking and storage of samples required for these assessments, which are also discussed in the revised BS 1377-3. Relevant information on the storage of samples is also provided by Hawkins and Pinches (1986) and Czerewko *et al*. (2003c). In particular, precautions are required to ensure that the chemical and mineralogical determinations accurately measure the condition of pyrite in the ground. Contamination of samples with groundwater or drilling water can also cause problems.

Table 36.2 shows UK mudrock formations known to contain pyrite. This list does not include all pyrite-bearing formations and inclusion in the list does not imply the material will always be problematic. In addition, the development of difficulties due to pyrite oxidation depends on diverse environmental factors.

Care must be exercised over application of the guidance, as forms of sulfur other than pyrite may be present in some rocks. Also, as the guidance is intended for the design of commonly occurring applications of concrete and fill materials, the ground aggressivity classification values obtained do not necessarily apply in other situations – for example, in open excavations or in fill placed to support floor slabs and shallow foundations. Several cases affecting various applications in ground engineering are summarised by Cripps *et al*. (2019) and useful references for the potential implications of pyrite-bearing mudrocks in engineering applications are provided.

36.4. Engineering considerations

Problems in engineering are often exacerbated by being unexpected. It is argued here that by taking account of (*a*) the controls exerted by the composition and structure of mudrocks, and (*b*) the impact of the environmental changes resulting from the engineering works, unexpected forms of behaviour can be avoided. As explained in Sections 36.2.2 and 36.2.3, mineralogy, compaction, cementation and structures are linked and depend on the original sedimentary processes, burial diagenesis and changes due to exhumation and weathering. This section includes descriptions of commonly occurring situations in which problems are liable to arise unless suitable countermeasures are taken.

36.4.1 Foundations

Indurated strongly cemented mudrocks in a structurally sound condition present few problems in most engineering situations. Allowable foundation loads are usually derived on an empirical basis such as that described by Lambe and Whitman (1979). It is advisable to construct foundations immediately after excavation to minimise deterioration of the rock mass by exposure to the atmosphere. However, where closely spaced discontinuities or relatively weak rocks are present, the bearing capacity of shallow foundations is liable to be inadequate. It is common practice to use Young's modulus of elasticity as the deformation modulus, notwithstanding the possibility of non-recoverable strain and a nonlinear stress–stain relationship. Anisotropy, displayed by many mudrocks, may cause appreciable differential settlement which would necessitate the use of a strong reinforced-concrete raft or provision for articulation in the structure.

In cemented mudrocks settlements will usually occur immediately when loads are imposed during construction, with insignificant post-construction settlement. However, consideration needs to be given to the possibility of weaker, more compressible layers being present within the ground below the structure. The impacts of discontinuities and weaker zones need to be assessed by undertaking loading or pressuremeter tests. If the magnitude and duration of long-term settlements are considered unacceptable, options such as preloading or deep foundations may be considered. It is also essential that the weathering grade is correctly assessed and considered in designs. This includes estimation of the extent and effect of weathering, and the proportions of weathered and unweathered materials so that the foundation may be located at an appropriate level in the ground.

Where ground conditions are inappropriate for shallow foundations, piled foundations may be used. Bored piling allows more control over the installation process and causes less disturbance to the surrounding ground, although the presence of softened material at the base of the hole may result in low end-bearing resistance. On the other hand, driving piles can result in shattering of the ground and changes in groundwater conditions. This is a particular problem with cemented mudrocks. A typical design of end-bearing piles would provide for a socket some 3–4 pile diameters deep to be formed in sound rock and reassurance of sufficient load-bearing capacity would be confirmed by carrying out representative pile load–settlement tests.

Due to the difficulties of taking and testing samples of mudrocks, end-bearing piles and deep foundations are often designed on the basis of the results of dynamic penetration, plate bearing and pressuremeter tests. Although the standard penetration test (SPT) is a useful means of identifying the variability of ground conditions, the parameters are not normally used for design purposes. However, in lower-durability mudrocks in situ plate bearing and cone penetrometer tests usually provide reliable data for design. In stronger mudrocks penetrometer testing may be used to provide the required design parameters.

The end-bearing capacity of piles is usually based on the UCS and discontinuity spacing of the bearing strata, where the former is determined by laboratory index tests.

36.4.2 Ground volume changes

As noted in Section 36.2.4.3, the oxidation of pyrite produces sulfate-rich acidic pore waters. Neutralisation of the acid by reaction with calcite, concrete and clays can result in the expansive formation of gypsum and other minerals. Precipitation may occur some distance away from areas of oxidation, particularly in areas of least stress such as bedding planes, joints or construction interfaces. Nixon (1978) describes structural damage to buildings in north-east England where expansion of pyritic fill was the cause. Several cases involving mudstone bedrock in eastern Canada are described by Penner *et al.* (1970), and Hawkins and Pinches (1987) give details of a similar problem in Cardiff in South Wales.

Significant problems with damage to houses in Ireland have been blamed on the expansion of pyrite-bearing fill placed beneath ground-supported concrete floor slabs. Theoretical heave pressures are given in **Table 36.11** and are more than sufficient to be responsible for the damage. Heave may be more than 100–150 mm, depending on the loading. Hoover and Lehmann (2009) calculate the potential heave based on the pyrite content of the ground, but the result is dependent on many assumptions.

In compaction mudrocks and overconsolidated clays, pyrite tends to occur as finely disseminated framboids which readily oxidise and produce randomly distributed gypsum crystals, as shown in **Figure 36.17(b)**. On the other hand, in cemented and indurated mudrocks (see **Figure 36.17(c)**), the gypsum will tend to grow as bladed crystals in a stratified manner along discontinuities such as bedding planes and joints, tending to push them apart, which results in breakage and expansion of fragments. In fills, expansive crystal growth may also occur in the pore spaces between particles.

The crystallisation of selenite gypsum may occur at some distance from the site of oxidation due to solute transportation. It is very difficult to specify a lower limit for the amount of pyrite acceptable in a fill. Experience in Canada (Comité Technique Québécois, 2001) suggests that problems are unlikely to be encountered if less than 0.5% pyrite is present, but that there is a high risk for values above 1%; whether higher amounts cause problems depends on the circumstances. Also, it is uncertain how differences in building construction and climatic conditions between Canada and the UK would influence this value.

The chemistry of the ground adjacent to concrete is specified by BRE Special Digest 1 (2005), which provides concrete design suitable for ground sulfate classes, and requirements for additional protective measures are specified for more

	Concentration of solution c/c_s		
Temperature: °C	2	4	10
5	29.2	58.5	97.1
10	29.7	59.7	98.7
20	30.8	61.8	102.2

Table 36.11 Theoretical pressures (MPa) due to the precipitation of gypsum (Data taken from Winkler and Singer, 1972)

aggressive ground conditions. Requirements for construction fills are given by the current European Standard BS EN ISO 13242 and National Standards Agency Ireland (NSAI, 2014) document SR21:2004. The total sulfur contents of aggregates is declared in conformance with BS EN ISO 13242, but no thresholds are suggested other than if pyrrhotite, the unstable form of iron sulfide, is present in the aggregate, in which case a maximum total sulfur content of 0.4% as 'S' applies. But SR21 states that the total sulfur value should be less than 0.1% (equivalent to 0.2% pyrite). BRE Digest DG522 (Longworth, 2011) recommends that the oxidisable sulfide (OS) content should not exceed 0.8% SO_4, equivalent to 0.5% pyrite for the ground adjacent to high-quality concrete. For highway fills, TRL 447 (Reid *et al.*, 2005) proposes limits of oxidisable sulfur of less than 0.3% sulfate (0.2% pyrite), and this report warns against the use of material in which pyrite is in the framboidal form. Hawkins and Pinches (1992) recommend caution as regards ground heave if equivalent pyrite as sulfate is above 0.5% (0.3% pyrite).

As explained in Section 36.4.5, when considering ground improvement involving the use of cementitious additives such as lime or cement, consideration needs to be given to the possibility of sulfates causing expansion, with a disruptive increase in volume, as described by Britpave (2019). Sulfates may be derived from pyrite during weathering and it is common for oxidation of sulfide minerals to take place rapidly during the treatment process. Further discussion of stabilised soils in highways construction is given in Section 36.4.5.

Assessment of pyrite-bearing mudstones and overconsolidated clay deposits such as Oxford Clay, Lias Clay and London Clay and superficial deposits derived from these should consider that they will have a variable distribution of pyrite and oxidation products such as selenite gypsum and other associated constituents, both laterally and with depth within the weathered zone. This is likely to have a bearing for various ground engineering activities situated at shallow depths, including earthworks, foundations, routeways and ground-bearing concrete and steel structures, where the weathering zone may extend variably between 2–3 m to 7–10 m, dependent on geological and geomorphological situation. Depth variations are influenced by former groundwater levels that may result in horizons containing concentrations of selenite gypsum at various depths. Contemporary and former root zones that may extend to depths of 3–5 m below ground level at the time of formation may also influence the sulfate mineral distribution derived through pyrite oxidation. These and other factors can result in a complex distribution of significant features associated with pyrite weathering, including concentrations of selenite crystals along interface shear zones often located between intact and weathered material; pockets of concentrated sulfate and iron oxide minerals; and concentrations or focused masses of disseminated saccharoidal selenite along root tracks of root zones, as demonstrated in **Figure 36.24**. Further details that may assist with ground assessment are discussed in BRE SD1 (2005), and cases are presented by Czerewko and Cross (2015).

Whether a particular pyrite-bearing mudrock will cause heave problems depends on many factors, including the nature of the pyrite, the presence of other minerals in the mudrock, the environmental conditions and the design of the structure. For example, in well-engineered earthworks the presence of a relatively impermeable capping layer can limit the extent of pyrite oxidation, so reactions become self-limiting. An alternative strategy for avoiding problems in earthwork fills is the provision of free-draining granular separation layers that control the migration of pore fluids through the material (see Czerewko *et al.*, 2011).

Figure 36.17(a) illustrates the breakdown of indurated mudrock (Bowland Shale) brought about by the forceful precipitation of selenite gypsum; in **Figure 36.17(b)** the structural disruption to less indurated Lias Clay is illustrated. Cripps and Edwards (1997) suggest that desiccation due to the removal of water and the action of acid on swelling clays can cause shrinkage and breakdown of mudrocks.

36.4.3 Degradation of mudrock

The failure in 1984 of the Carsington Dam in Derbyshire, UK, during the latter stages of construction highlighted a few issues related to the presence of pyrite in the mudrocks used for constructing the embankment (Reid and Cripps, 2019). Pyrite oxidation in the mudstone fill may have compromised the effectiveness of the limestone drainage layers contained within it. Pye and Miller (1990) and Anderson and Cripps (1993) suggest that an overly optimistic view had been taken of the shear strength parameters used in the design, especially in view of the rapidity with which the mudstone degraded during construction operations. Further consideration of the engineering problems due to rapid degradation of mudrocks is given in Section 36.4.4.

Breakdown processes in mudrocks in colliery spoils were extensively investigated by Taylor (1984), and a summary of this research is given in Taylor (1988). He notes that, depending on the character of the rock, more particularly with the presence of structural weaknesses, slaking occurs due to wetting and drying under atmospheric conditions, and the volume changes due to swelling clays, if present. Czerewko and Cripps (2001) provide guidance on the assessment of durability and the performance of several UK mudrock formations of different ages and states of induration.

Although the use of specific index properties has been found to be useful for predicting the behaviour of mudrocks in particular situations, applying the classification schemes to other formations or different applications has proved less successful. For example, although Gamble's (1971) scheme – based on second-cycle slake durability and Atterberg limits of the fines – improved on previous classification schemes, it was found to be inappropriate by Olivier (1979) when applied to tunnelling projects in South Africa. Olivier's scheme was

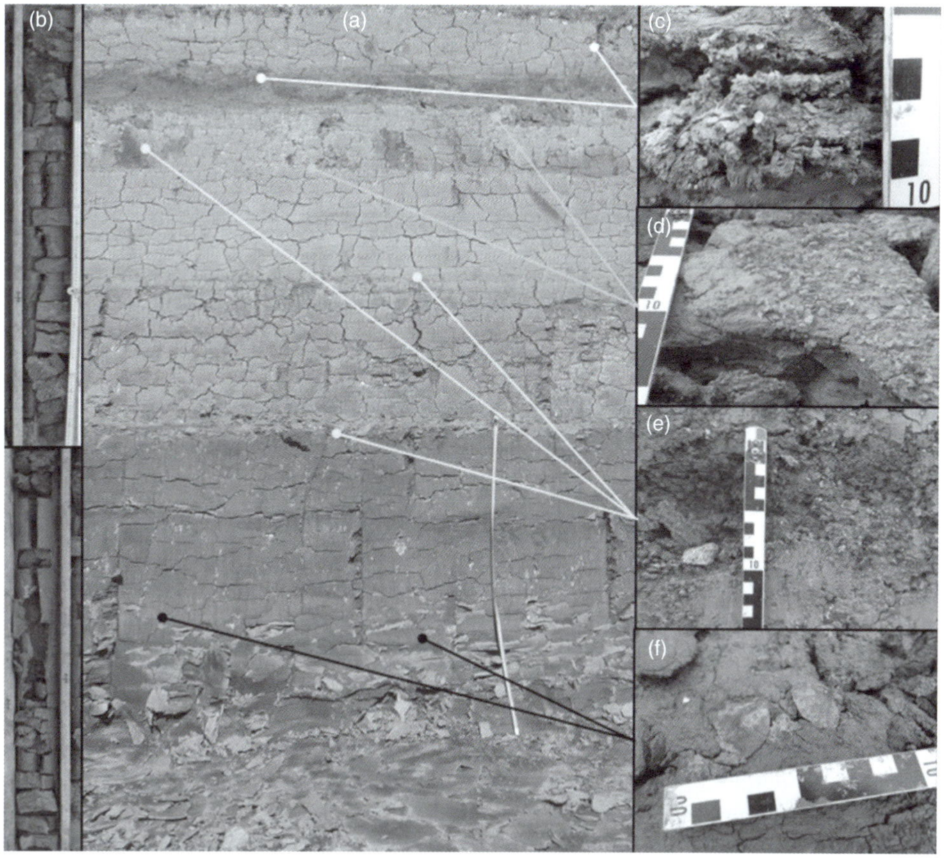

Figure 36.24 Detailed ground assessment of weathered Oxford Clay in support of earthworks design. (a) Variable distribution of pyrite weathering constituents seen in mass excavation, details of the nature and distribution of features observed through close examination are shown on the right. (b) Borehole core from series of targeted boreholes showing a lack of visibility of features seen in excavation (a). (c) Concentrated saccharoidal selenite along root zones at shallow depth. (d) Concentration of small crystals at former groundwater horizons and formerly saturated gleyed ground. (e) Detail of nodules of concentrated selenite crystals in ferruginous cement marked former groundwater interfaces. (f) Base of weathering where interface is marked by low-angled shears with abundant large selenite crystals. The colour change seen in the profile is not reproduced and shows change from an orange-brown colour at the top, through light brown to dark blue grey of unweathered material at the base

based on UCS and uniaxial free-swell testing. However, Varley (1990) reported that excessive slurrying was not predicted by Olivier's scheme in tunnels in UK Carboniferous Coal Measures mudrock strata. Problems with the operation of tunnelling machines due to the breakdown of Mercia Mudstone in Leicester, UK, were attributed by Atkinson *et al.* (2003) to the presence of swelling clay minerals in the rock.

36.4.4 Excavations and slopes

Most types of mudrocks are relatively easy to excavate using earth excavation machinery. The methods of excavation are usually determined on the basis of the seismic velocity measurements or on an assessment of UCS and fracture spacing.

The excavation of tunnels, shafts and other underground cavities requires assessment of the slaking properties of the mudrock. This is to allow determination of the stand-up time for the excavations and of the handling and disposal methods for the spoil. Low silt–high clay content compaction mudrocks may be sufficiently strong to remain stable during the excavation process, but may then undergo rapid deterioration before support from the lining is fully operable.

In excavations, mudrocks may undergo lateral displacement as a result of stored strain energy being released. Some low-durability cemented or compaction mudrocks, especially those containing swelling clay minerals, undergo significant swelling when the overburden and lateral restraint is reduced, resulting in uplift and lateral displacements. Rigid support systems can become overloaded and, on reloading, excessive settlement may occur. This process may be exacerbated by the presence of water and tends to be more rapid in overconsolidated clay deposits than in more durable cemented mudrocks. Heave effects may often be avoided by minimising the time of exposure – for example, by immediately covering exposed mudrock with sprayed concrete. Although covering does not prevent uplift due to pyrite oxidation and gypsum precipitation, preventing water entry does reduce clay mineral swelling and slaking.

In sequences of rocks in which mudrocks are interbedded with more durable sandstones or limestones, more rapid erosion of the mudrocks may leave overlying more competent beds in an unsupported condition. Although common in surface exposures, this effect may also lead to instability in underground openings. Behaviour of this type may be predicted by examination of the material with associated laboratory testing to assess the relative erosion rates and strengths. The stability is highly dependent on the attitudes of the beds and the rock-mass jointing. Solutions to problems of this type include measures such as removal and control bolting of potentially unstable blocks, sprayed concrete or masonry work. The latter methods protect the mudrocks from weathering and erosion as well as regulating run-off and groundwater. Eyre (1973) provides details of these and similar measures required to ensure stability of steep rock faces in sequences of cemented mudstones and limestones in south-west England.

Many slopes in mudrock formations in the south of England are mantled by considerable thicknesses of solifluction deposits that contain shear surfaces which mobilise residual strength. Relatively small increases in loading, the removal of support or increases in pore water pressures can lead to instability (see Symonds and Booth, 1971; Weeks, 1969). Some indurated mudrock formations in other parts of the country are also mantled by solifluction or 'head' colluvium deposits, which are liable to become unstable during engineering works.

36.4.5 Highway construction

In highway construction problems with mudrocks containing pyrite can be reduced by leaving the weathered material in place for as long as possible, or by covering freshly exposed ground with a layer of compacted clay or a geomembrane barrier. An alternative strategy is to expose the mudstone to weathering for a suitable period of time so that pyrite oxidation and the reaction products are generated before construction takes place. A potential drawback of this procedure is the delay to the construction programme; furthermore, the sulfate now present may participate in chemical attack and cause heave due to the dissolution and reprecipitation of sulfate minerals if entry of surface or groundwater is not adequately managed.

Compaction of mudrock fills is best performed using a sheepsfoot roller, which thoroughly mixes the material and facilitates a more effective reduction of voids, while avoiding the formation of smooth, potentially low-strength surfaces within a layered fill. However, the final proof rolling with a smooth roller will form a smooth, low-permeability surface, which will tend to shed water. Both hot and dry *and* wet weather can cause excessive break-up and/or softening of the fill during construction, which may render the material unsuitable for use.

The weight of free-draining uniformly graded granular capping layers placed over freshly exposed pyritic mudrocks in cuttings can act as a surcharge which inhibits expansion. If expansion does occur, such layers can also act as a compressive layer, and thus reduce uplift. Such an interface has the further effect of forming a barrier to the capillary rise of sulfate-rich water that could otherwise cause chemical attack or heave problems. However, it is important that a separator layer prevents migration of degraded mudrock into the granular fill.

Improvement in the workability and bearing characteristics of clay soils for road and building construction involves soil stabilisation with one of the following: (*a*) lime, slaked lime ($Ca(OH)_2$) and cement; (*b*) granulated ground blast furnace slag (GGBS) and cement; or (*c*) pulverised fuel ash (PFA) and cement. Such methods are cheaper and more sustainable than the use of rock aggregates. Pozzolanic effects produce a cementitious product known as hydraulically bound material (HBM) or cement-bound material (CBM). Documented cases in the UK of problems due to expansive sulfate minerals formation include the M40 near Banbury on Lias Clay and the A10 Wadesmill bypass in Hertfordshire. Snedker and Temporal (1990) explain that in the case of the M40, heave of 150 mm occurred in a 250 mm thick stabilised layer over a few months. It would appear that the treatment of the ground when it is in a dry condition makes heave more likely to occur. Thus, timing the treatment and allowing time for reactions to occur, for example over a winter, reduce the risk of problems with expansion. Road and foundation construction may be further proofed against heave problems by the provision of appropriate drainage to maintain the groundwater level and capillary fringe below the stabilised horizon. Well-designed flexible drainage is also beneficial in the case of construction of earthworks involving the use of pyritic mudrocks. Definitive guidance for the assessment of ground suitability for stabilisation is given by BP62 (Britpave, 2019), which also provides detailed discussion and recommendations.

Potential problems were detected on recent and ongoing highway earthwork construction projects involving Jurassic Ancholme Clay and Lower Lias Clay (Czerewko and Cross, 2015; Czerewko *et al.*, 2011) in the East Midlands of England. Investigations provided data on the rates of pyrite oxidation and selenite formation in these pyritic, calcite-bearing mudrocks. Prior to the use of the mudrock excavated on site for use as fill it was subjected to a three-week period of exposure oxidation and selenite formation, therefore mitigating post-construction heave.

36.4.6 Aggressive ground conditions

As explained in Section 36.2.4.3, the weathering of pyrite gives rise to acidic solutions rich in sulfates that engage in chemical reactions that are detrimental to concrete, steel and other materials (see BRE SD1, 2005; Czerewko and Cripps, 2006a). For many decades it has been standard practice to determine the sulfate content and aggressive nature of soils and fills for the design of cement and concrete in buried structures, thus sulfate attack on buried concrete has been avoided. However, the discovery of serious deterioration of the concrete foundations of 30-year-old bridge structures on the M5 motorway in south-west England revealed serious deficiencies in the assessment methodology used at that time. The problem was attributed to thaumasite sulfate attack on the concrete, which may be much more extensive than previously thought, as there has probably been a lack of reporting of cases of deterioration and, where it has been reported, thaumasite has not been correctly identified.

BRE SD1 (2005) provides comprehensive guidance on procedures to predict, prevent and mitigate against deterioration of concrete due to the presence of aggressive ground conditions. Various preventive and protective measures are possible, including modified mix designs, the use of bitumen coatings to concrete and the addition of a capillary break to prevent contact with aggressive solutions.

36.5. Conclusions

As mudrocks occur extensively in the geological sequence and are geographically widely distributed, they are frequently encountered in engineering situations. Certain types can be troublesome, in that it is difficult to predict their behaviour during construction and to determine reliable values of geotechnical parameters for design. It has been explained that mudrocks are the product of the burial and subsequent exhumation of muddy sediments comprising a saturated mixture of silt and clay particles, which have undergone change due to these processes. The resulting materials are a function mainly of the composition and structure of the original sediment, the depth of burial and local geothermal gradient. It has been suggested that understanding the process of formation of these materials can help with predicting how they will perform in different engineering situations. It is convenient to consider mudrocks under the following headings in order of increasing degree of induration

- compaction mudrocks and overconsolidated clays
- weakly cemented mudrocks
- strongly cemented mudrocks.

Compaction mudrocks are weak and compressible materials. They will degrade rapidly at the ground surface and in most engineering situations they behave as soils. Most types contain swelling clays and they will therefore undergo relatively large volume changes in response to changes in moisture content. They may be anisotropic if clay-rich and usually contain pyrite in its more reactive framboidal form.

Weakly cemented mudrocks are stronger and less compressible, ranging across the boundary from strong soils to weak rocks. They are difficult to characterise, and in some situations will behave as rocks but in other situations as soils. Due to stored strain energy, the removal of overburden, and the presence of a partly bonded structure and a skeletal framework, they undergo time-dependent degradation in response to weathering. Cyclical changes in moisture content are liable to lead to slaking in which the rock degrades into sand- and gravel-sized fragments that are quite durable. Pyrite may be present in its framboidal form.

Strongly cemented mudrocks are stronger and less compressible. They are liable to range downwards into weak rocks, but in most engineering applications they will behave as rocks. Weathering action can result in them consisting of a mixture of weaker and stronger materials. Pyrite is liable to be in a more stable crystalline form, although polyframboidal pyrite may be present.

Guidance has been given concerning the investigation, classification and description of mudrocks. The fact that many mudrock formations also contain pyrite gives rise to various problems that the generation of chemically aggressive solutions and volume changes produce. Finally, a few potential difficulties with engineering works involving mudrocks, together with possible responses, were presented.

References

Anderson WF and Cripps JC (1993) The effects of acid leaching on the shear strength of Namurian Shale. In *The Engineering Geology of Weak Rock* (Coulthard JM, Cripps JC, Culshaw MG, *et al.* (eds)). Balkema, Rotterdam, the Netherlands, pp. 159–168.

ASTM (2007) D5731-07: Standard test method for determination of the point load strength index of rock and application to rock strength classifications. ASTM, West Conshohocken, PA.

Atkinson JH, Fookes PG, Miglio BF and Pettifer GS (2003) Destructuring and disaggregation of Mercia Mudstone during full-face tunnelling. *Quarterly Journal of Engineering Geology and Hydrogeology* **36**: 293–303.

Attewell PB (1997) Tunnelling and site investigation. In *Geotechnical Engineering of Hard Soils–Soft Rocks* (Anagnostopoulos A, Schlosser F, Kalteziotis N, *et al.* (eds)). Balkema, Rotterdam, the Netherlands, pp. 1767–1790.

Banks D (1997) Hydrochemistry of Millstone Grit and Coal Measures groundwaters, South Yorkshire and North Derbyshire, UK. *Quarterly Journal of Engineering Geology* **30**: 237–256.

Barton NR, Lien R and Lunde J (1974) Engineering classification of rock masses for the design of tunnel support. *Rock Mechanics* **6(4)**: 189–239.

Bell FG (1992) *Engineering Properties of Soils and Rocks*. Butterworth Heinemann, London, UK.

Bieniawski ZT (1976) Rock mass classification in rock engineering. *Symposium Proceedings Exploration for Rock Engineering, Cape Town, South Africa* (Bieniawski ZT (ed)). Balkema, Rotterdam, the Netherlands, vol. 1, pp. 97–106.

BRE (Building Research Establishment) (2005) *Concrete in Aggressive Ground. Part 1: Assessing the Aggressive Chemical Environment*, 3rd edn. BRE SD1. BRE, Watford, UK.

Britpave (2019) *Stabilisation of Sulfate-Bearing Soils. Guidelines for Best Practice*. British Cementitious Paving Association, Wokingham, UK.

BSI (British Standards Institution) (2006) BS EN ISO 22475-1:2006: Geotechnical investigation and testing – Sampling methods and groundwater measurements – Part 1: Technical principles for execution. BSI, London, UK.

BSI (2007) BS EN 1997-2:2007: Eurocode 7 – Geotechnical design – Part 2: Ground investigation and testing. BSI, London, UK.

BSI (2011) BS EN ISO 22476-3:2005+A1:2011: Geotechnical investigation and testing – Field testing – Part 3: Standard penetration test. BSI, London, UK.

BSI (2013a) BS EN 1744-1:2009+A1:2012: Tests for chemical properties of aggregates – Part 1: Chemical analysis. BSI, London, UK.

BSI (2013b) BS EN ISO 13242:2013: Aggregates for unbound and hydraulic bound materials for use in civil engineering work and road construction. BSI, London, UK.
BSI (2018a) BS EN ISO 14688-1:2018: Geotechnical investigation and testing – Identification and classification of soils. Part 1: Identification and description. BSI, London, UK.
BSI (2018b) BS EN ISO 14688-2:2018: Geotechnical investigation and testing – Identification and classification of soils. Part 2: Principles for a classification. BSI, London, UK.
BSI (2018c) BS EN ISO 14689:2018: Geotechnical investigation and testing – Identification, description and classification of rocks. BSI, London, UK.
BSI (2020) BS 5930:2015+A1:2020: Code of practice for site investigations. BSI, London, UK.
BSI (2021) BS 1377-3:2018+A1:2021: Methods of test for soils for civil engineering purposes – Part 3: Chemical and electrochemical tests. BSI, London, UK.
BSI (2022) BS 1377-2:2022: Methods of test for soils for civil engineering purposes – Part 2: Classification tests and determination of geotechnical properties. BSI, London, UK.
Comité Technique Québécois (2001) *Appraisal Procedure for Existing Residential Buildings*. CTQ-M200. Comité Technique Québécois d'étude des Problèmes de Gonflement Associés a la Pyrite, Quebec, ON, Canada.
Cripps JC and Czerewko MA (2017) The influence of diagenetic and mineralogical factors on the breakdown and geotechnical properties of mudrocks. In *Geomechanical and Petrophysical Properties of Mudrocks* (Rutter EH, Mecklenburgh J and Taylor KG (eds)). Geological Society, London, UK, pp. 271–293.
Cripps JC and Edwards RL (1997) Some geotechnical problems associated with pyrite-bearing rocks. *Proceedings of the International Conference on the Implications of Ground Chemistry/Microbiology for Construction, Bristol, UK* (Hawkins AB (ed)). Balkema, Rotterdam, the Netherlands, pp. 77–87.
Cripps JC and Taylor RK (1981) The engineering properties of mudrocks. *Quarterly Journal of Engineering Geology* **14**: 325–346.
Cripps JC, Reid JM, Czerewko MA and Longworth I (2019) Tackling problems in civil engineering caused by the presence of pyrite. *Quarterly Journal of Engineering Geology and Hydrogeology* **52**: 481–500.
Czerewko MA and Cripps JC (2001) Assessing the durability of mudrocks using the modified jar slake index test. *Quarterly Journal of Engineering Geology and Hydrogeology* **34**: 153–163.
Czerewko MA and Cripps JC (2006a) Sulfate and sulfide minerals in the UK and their implications for the built environment. *Proceedings of the 10th IAEG International Congress, Engineering Geology for Tomorrow's Cities, Nottingham, UK.* Geological Society, London, UK, paper 121.
Czerewko MA and Cripps JC (2006b) The implications of diagenetic history and weathering on the engineering behaviour of mudrocks. *Proceedings of the 10th IAEG International Congress, Engineering Geology for Tomorrow's Cities, Nottingham, UK.* Geological Society, London, UK, paper 118.
Czerewko MA and Cross SA (2015) The benefits of a granular interface over pyritic subgrade. *Geotechnical Research* **2(3)**: 97–122.
Czerewko MA, Cripps JC, Reid JM and Duffell CG (2003a) Sulfur species in geological materials: sources and quantification. *Cement and Concrete Composites* **25**: 657–671.
Czerewko MA, Cripps JC, Duffell CG and Reid JM (2003b) The distribution and evaluation of sulfur species in geological materials and man-made fills. *Cement and Concrete Composites* **25**: 1025–1034.
Czerewko MA, Cripps JC, Reid JM and Duffell CG (2003c) The effects of storage conditions on the sulfur speciation in geological material. *Quarterly Journal of Engineering Geology and Hydrogeology* **36**: 331–342.
Czerewko MA, Cross SA, Dumelow PG and Saadvandi A (2011) Assessment of pyritic Lower Lias mudrocks for earthworks. *Proceedings of the Institution of Civil Engineers Geotechnical Engineering* **164**: 59–77.
Czerewko MA, Longworth I, Reid JM and Cripps JC (2016) Standardized terminology and test methods for sulphur mineral phases for the assessment of construction materials and aggressive ground. *Quarterly Journal of Engineering Geology and Hydrogeology* **49**: 245–265.
Davis AG (1967) On the mineralogy and phase equilibrium of Keuper Marl. *Quarterly Journal of Engineering Geology* **1**: 25–46.
Dick JC and Shakoor A (1992) Lithological controls of mudrock durability. *Quarterly Journal of Engineering Geology* **25**: 31–46.
Eyre WA (1973) The revetment of rock slopes in the Clevedon Hills for the M5 motorway. *Quarterly Journal of Engineering Geology* **6**: 223–229.
Fletcher CJN (2016) *Geology for Ground Engineering Projects.* CRC Press, London, UK.
Franklin JA and Chandra A (1972) The slake durability test. *International Journal of Rock Mechanics and Mining Sciences* **9**: 325–341.
Franklin JA and Dusseault MB (1989) *Rock Engineering.* McGraw-Hill, New York, NY.
Gamble JC (1971) *Durability–Plasticity Classification of Shales and Other Argillaceous Rocks.* Unpublished PhD thesis, University of Illinois, Urbana, IL.
Grainger P (1984) The classification of mudrocks for engineering purposes. *Quarterly Journal of Engineering Geology* **17**: 381–387.
Hawkins AB and Pinches GM (1986) Timing and correct chemical testing of soils/weak rocks. In *Engineering Group of the Geological Society, Special Publication 2*. Geological Society, London, UK, pp. 59–66.
Hawkins AB and Pinches GM (1987) Sulfate analysis on black mudstones. *Géotechnique* **37**: 191–196.
Hawkins AB and Pinches GM (1992) Engineering description of mudrocks. *Quarterly Journal of Engineering Geology* **25**: 17–30.
Head KH (1994) *Manual of Soil Laboratory Testing, volume 2*, 2nd edn. Wiley, Chichester, UK.
Head KH (1998) *Manual of Soil Laboratory Testing, volume 3*, 2nd edn. Wiley, Chichester, UK.
Hencher S (2012) *Practical Engineering Geology.* Spon Press, London, UK.
Hillel D (1980) *Fundamentals of Soil Physics.* Academic Press, New York, NY.
Hoek E and Brown ET (1997) Practical estimates or rock mass strength. *International Journal of Rock Mechanics and Mining Sciences & Geomechanical Abstracts* **34(8)**: 1165–1186.
Hoover SE and Lehmann D (2009) The expansive effects of concentrated pyritic zones within the Devonian Marcellus Shale formation of North America. *Quarterly Journal of Engineering Geology and Hydrogeology* **42**: 157–164.

Hopkins TC and Dean RC (1984) Identification of shales. *Geotechnical Testing Journal ASTM* **7**: 10–18.

Hudson JA and Harrison JP (1997) *Engineering Rock Mechanics: An Introduction to the Principles*. Pergamon, Oxford, UK.

ISRM (1979) Suggested methods for determining water content, porosity, density, absorption and related properties, and swelling, and slake–durability index properties. *International Journal of Rock Mechanics and Mining Science and Geomechanical Abstracts* **16**: 141–156.

Lambe TW and Whitman RV (1979) *Soil Mechanics, SI Version*. Wiley, New York, NY.

Longworth I (2011) Hardcore for supporting ground floors of buildings. Part 1: Selecting and specifying materials. BRE DG522. HIS BRE Press, Watford, UK.

Lum P (1965) The residual soils of Hong Kong. *Géotechnique* **15**: 180–194.

Lutton RJ (1977) *Design and Construction of Compacted Shale Embankments, Vol. 3. Slaking Indexes for Design*. US Army Engineer Waterways Experiment Station, Vicksburg, MI.

Mead WJ (1936) Engineering geology of dam sites. *Transactions of the 2nd International Congress on Large Dams, Washington, DC*, vol. 4, pp. 183–198.

Mitchell JK and Soga K (2005) *Fundamentals of Soil Behavior*, 3rd edn. Wiley, New York, NY.

Morgenstern NR and Eigenbrod KD (1974) Classification of argillaceous soils and rocks. *Journal of the Geotechnical Engineering Division, ASCE* **100**: 1137–1156.

Nixon PJ (1978) Floor heave in buildings due to use of pyritic shales as fill materials. *Chemistry and Industry* **4**: 160–164.

Norbury N (2010) *Soil and Rock Description in Engineering Practice*. Whittles Publishing, Dunbeath, UK.

NSAI (2014) SR21:2014: Guidance on the use of I.S. EN 13242:2002+A1:2007. Aggregates for unbound and hydraulic bound materials for use in civil engineering work and road construction. NSAI, Dublin, Ireland.

Olivier HJ (1979) Some aspects of the influence of mineralogy and moisture redistribution on the weathering behaviour of mudrocks. *Proceedings of the 4th International Conference on Rock Mechanics, Montreux, Switzerland*, vol. 3, pp. 467–475.

Penner E, Gillot JE and Eden WJ (1970) Investigations to heave in billings shale by mineralogical and biogeochemical methods. *Canadian Geotechnical Journal* **7**: 333–338.

Potter PE, Maynard JB and Pryor WA (1980) *Sedimentology of Shale*. Springer, New York, NY.

Pye EK and Miller JA (1990) Chemical and biochemical weathering of pyritic mudrocks in a shale embankment. *Quarterly Journal of Engineering Geology* **23**: 365–381.

Reeves GM, Sims I and Cripps JC (eds) (2006) *Clay Materials Used in Construction*. Geological Society, London, UK.

Reid JM and Cripps JC (2019) Geochemical lessons from Carsington Dam failure of 1984 and reconstruction. *Quarterly Journal of Engineering Geology and Hydrogeology* **52**. 414–424.

Reid JM, Czerewko MA and Cripps JC (2005) *Sulfate Specification for Structural Backfills*. TRL447. Transport Research Laboratory, Crowthorne, UK.

Reid JM, Czerewko MA, Longworth I and Cripps JC (2017) Sulfur compounds in soils and rocks: avoiding pitfalls in terminology, test procedures and ground assessment. *Ground Engineering*, Jan./Feb.

Snedker EA and Temporal J (1990) M40 Motorway Banbury IV contract: lime stabilisation. *Highways and Transportation*, Dec.: pp. 7–8.

Spears DA (1980) Towards a classification of shales. *Journal of the Geological Society, London* **137**: 125–129.

Stow DAV (1981) Fine-grained sediments: terminology. *Quarterly Journal of Engineering Geology* **14**: 243–244.

Symonds IF and Booth AI (1971) *Investigation of the Stability of Earthworks Construction on the Original Line of the Sevenoaks Bypass, Kent*. Report 393. Transport and Road Research Laboratory, Crowthorne, UK.

Taylor RK (1984) *Composition and Engineering Properties of British Colliery Discards*. National Coal Board, London, UK.

Taylor RK (1988) Coal Measures mudrocks: composition, classification and weathering processes. *Quarterly Journal of Engineering Geology* **21**: 85–99.

Ulusay R (ed) (2014) *The ISRM Suggested Methods for Rock Characterisation, Testing and Monitoring: 2007–2014*. Springer, Cham, Switzerland.

Varley PM (1990) Susceptibility of Coal Measures mudstones to slurrying during tunnelling. *Quarterly Journal of Engineering Geology* **23**: 147–160.

Weeks AG (1969) The stability of natural slopes in S.E. England as affected by periglacial activity. *Quarterly Journal of Engineering Geology* **2**: 49–61.

Winkler EM and Singer PC (1972) Crystallization pressure of salts in stone and concrete. *GSA Bulletin* **83(11)**: 3509–3514.

Wood LE and Deo P (1975) A suggested method of classifying shales for embankments. *Bulletin of the Association of Engineering Geologists* **12**: 39–55.

Wyllie DC and Mah CW (2004) *Rock Slope Engineering: Civil and Mining*, 4th edn. Spon Press, Oxford, UK.

Further reading

Bell FG (1992) *Engineering in Rock Masses*. Butterworth Heinemann, London, UK.

Cripps JC and Taylor RK (1986) Engineering characteristics of British over-consolidated clays and mudrocks: I – Tertiary deposits. *Engineering Geology* **22**: 349–376.

Cripps JC and Taylor RK (1987) Engineering characteristics of British over-consolidated clays and mudrocks: II – Mesozoic deposits. *Engineering Geology* **23**: 213–253.

Cripps JC, Hawkins AB and Reid JM (1993) Engineering problems with pyritic mudrocks. *Geoscientist* **3(2)**: 16–19.

Fookes PG (ed) (1997) *Tropical Residual Soils*. Geological Society, London, UK.

Greensmith JT (1989) *Petrology of the Sedimentary Rocks*. Unwin Hyman, London, UK.

Head KH (1992) *Manual of Soil Laboratory Testing, volume 1*, 2nd edn. Wiley, Chichester, UK.

Hencher S (2016) *Practical Rock Mechanics*. CRC Press, London, UK.

IAEG (1979) Classification of rocks and soils for engineering geological mapping. Part 1: Rock and soil materials. *Bulletin of the International Association of Engineering Geology* **19**: 364–371.

Jaeger JC and Cook NGW (1979) *Fundamentals of Rock Mechanics*. Chapman & Hall, London, UK

Jeans CV (1989) Clay diagenesis in sandstones and shales: an introduction. *Clay Minerals* **24**: 127–136.

Longworth I (2004) Assessment of sulfate-bearing ground for soil stabilisation for built development: technical note. *Ground Engineering*, May.

Newman A (1998) Pyrite oxidation and museum collections: a review of theory and conservation treatments. *The Geological Curator* **6**(**10**): 363–370.

Olivier HJ (1980) A new engineering-geological rock durability classification. *Engineering Geology* **14**: 255–279.

Sasaki M, Tsunekawa M, Ohtsuka T and Konno H (1998) The role of sulfur-oxidising bacteria *Thiobacillus thio-oxidans* in pyrite weathering. *Colloids and Surfaces A: Physiochem and Engineering Aspects* **133**: 269–278.

Shaw DB and Weaver CE (1965) The mineralogical composition of shales. *Journal of Sedimentary Petrology* **35**: 213–222.

Shaw HF (1981) Mineralogy and petrology of the argillaceous sedimentary rocks of the U.K. *Quarterly Journal of Engineering Geology* **14**, 277–290.

Terzaghi K and Peck RB (1967) *Soil Mechanics in Engineering Practice*, 2nd edn. Wiley, NY.

Thaumasite Expert Group (1999) *The Thaumasite Form of Sulfate Attack: Risks, Diagnosis, Remedial Works and Guidance on New Construction*. Department of the Environment, Transport and the Regions, London, UK.

Tucker ME (1991) *Sedimentary Petrology: An Introduction to the Origin of Sedimentary Rocks*, 2nd edn. Blackwell Science, Oxford, UK.

Waltham AC (1994) *Foundations of Engineering Geology*. Blackie, Glasgow, UK.

Weaver CE (1989) *Clays, Muds, and Shales*. Developments in Sedimentology Vol. 44. Elsevier, Amsterdam, the Netherlands.

It is recommended this chapter is read in conjunction with

- Chapter 7 *Geotechnical risks and their context for the whole project*
- Chapter 33 *Expansive soils*
- Chapter 37 *Sulfates and sulfides in soils and rocks*
- Chapter 40 *The ground as a hazard*

All chapters in this book rely on the guidance in Sections 1 *Context* and 2 *Fundamental principles*. A sound knowledge of ground investigation is required for all geotechnical works, as set out in Section 4 *Site investigation*.

Glossary

The list below contains definitions for terms used in this chapter. Words presented in italics are themselves defined elsewhere in the list. Further definitions are given in *Dictionary of Earth Sciences*, edited by M. Allaby, Oxford University Press, 2008.

Air breakage	Process of *slaking* controlled by air pressures produced by the capillary water in the pore spaces of fine-grained, partly saturated rocks.
Anisotropy	Term describing the physical property which depends on the direction relative to a defined axis, such as *bedding* or *foliation*.
Argillaceous	Sediments or sedimentary rocks that are composed predominantly of *clay* and *silt*-sized material, comprising significant quantities of *clay minerals*.
Argillite	Compact, strong, fine-grained, *clay*-rich sedimentary rock which has been strongly *lithified* by deep burial or *low-grade metamorphism*, but lacking *lamination*, *fissility* or *cleavage* structures.
Authigenic	*Mineral* that has formed in situ in deposits during progressive *diagenesis*, resulting from changes in burial and temperature.
Burial diagenesis	*Diagenesis* that occurs due to burial of sediments beneath later sediments.
Cation	An atom (or group of atoms) which is deficient in one or more electrons to produce a net positive charge, e.g. Al^{3+}.
Cation exchange capacity	The maximum quantity of *cations* that may be sorbed onto the surfaces of *mineral* particles at a given pH value.
Cement	Mineral material precipitated in the pore space and inter-particle chemical bonds that hold grains together.
Cementation	Process by which the grains of geomaterials become attached together by the precipitation of mineral matter between the grains and the formation of inter-particle chemical bonds.
Clast	A fragment of rock or *mineral* grain that has been derived by erosional processes from older rock.
Clastic	*Sediment* or *sedimentary* rock that is composed of (broken) fragments of other rocks and minerals derived by *weathering* and *erosional* processes.
Clay grade	An average grain size of less than 0.002 mm. Although clay grade material usually contains significant amounts of *clay* minerals, other material such as quartz, organic matter, sulfides and carbonates may also be present.
Clay mineral	A group of *minerals* of aluminosilicate composition and a phyllosilicate crystal structure similar to that of mica – that is, a sheet layer structure with strong inter- and intrasheet bonding, but weak interlayer bonding. Commonly occurring clay minerals include illite, kaolinite and smectite (montmorillonite).
Claystone	An *argillaceous* sedimentary rock of *clay*-size particles (grade) mainly comprising *clay* minerals.
Colliery spoil	Waste rock material produced in the course of coal mining or coal processing. Usually contains a high proportion of mudrocks.
Compaction	In geology, the physical processes that convert loose unconsolidated sediment into a dense coherent rock, but without significant change in mineralogy. In geotechnics, the process of expelling air and moisture from a fill material by application of a dynamic load such as rolling or vibration.

Consolidation	In geotechnics, the reduction in porosity due to the application of a constant *pressure*.
Detrital	Rock or mineral grains derived from pre-existing parent rock by process of *weathering* and/or *erosion*, transported as detritus and deposited as sediments.
Diagenesis	The chemical, physical and biological changes due to burial of sediment, including *consolidation, cementation*, growth of *authigenic minerals* and increases in grain size. Excludes changes due to *weathering* or *metamorphism*.
Diffuse double layer	Part of the ion exchange process common in *clay* minerals caused when clay particles are suspended in water and interlayer *cations* on the surface of the clay particles go into solution, producing a negatively charged *clay* surface surrounded by a diffuse double layer of hydrated cations.
Elastic modulus	Amount of *stress* to produce a unit amount *strain*, where removal of the stress results in recovery to the original dimensions.
Fissility	The ability of a rock to be broken along closely spaced parallel planes, resembling the pages of a book.
Flocc	An aggregation of *clay* particles bound to each other by electrostatic attraction present in a neutral electrolyte fluid.
Flocculation	The process of aggregation of *clay* particles into larger groups, either through chemical or biological means.
Foliation	The planar arrangement of textural features or platy particles in a rock such as schistosity in *metamorphic* rocks.
Framboid	A microscopic spherical aggregate of pyrite grains in a form resembling a raspberry in *argillaceous* deposits.
Induration	The natural process of hardening of sediments into rock by pressure, heat or *cementation*.
K_o value	Earth pressure at rest, which is the ratio between horizontal and vertical effective stresses in the ground.
Laminae	Distinct layers of sediment, often with different grain size or colour and typically less than 10 mm (an upper boundary of 20 mm is used in the engineering description of rocks and soils).
Liquid limit	Minimum *moisture content* for wet soil to behave as a liquid under specified conditions.
Low-grade metamorphism	Alteration by the action of heat and/or *pressure* of existing rocks, under conditions in which recrystallisation and new mineral growth in the solid state is just possible.
Lutaceous	A term used to describe a sedimentary rock formed from mud (clay- and/or silt-sized particles).
Moisture content	Mass of water contained in a soil expressed as a percentage of the mass of dry solids.
Mudrocks	A rock formed by the *compaction* or *induration* of mud with a modal grain size in the mud grade (<0.063 mm). A generic group name for mudstone, siltstone and claystone, and *overconsolidated* clay.
Mudstone	A fine-grained sedimentary rock that is composed of *silt* and *clay*-sized particles.
Normally consolidated	A condition in which the present *overburden stress* being borne by a deposit is the maximum it has experienced since deposition.
Overconsolidated	A condition in which the *overburden stress* acting on a deposit has been greater in the past than it is at present; thus the deposit has been subject to a cycle or cycles of loading and unloading giving it greater density, *strength*, *durability* and lower compressibility compared to a similar normally *consolidated* deposit.
Periglacial	Climatic conditions experienced in the vicinity of retreating ice sheets or glaciers, entailing freeze–thaw action.
Petrology	The study of rocks through their mineralogy, geochemistry, textures, field relationships and their formation.
Phyllite	An argillaceous rock that has been altered by *low-grade metamorphism*; of coarser grain size and less perfectly cleaved than a slate, but finer grained and better cleaved than a schist.
Phyllosilicates	Sheet or layer silicate minerals composed of tetrahedral and octahedral sheets joined together to form 1:1 or 2:1 layers. In between the layers there may be interlayers of hydrated cations and water, which may compensate excess negative charges arising from the substitution of intra- and intersheet cations by other cations of lower valency.
Plastic limit	Minimum moisture content for soil to behave in a plastic manner – in which it is possible to mould the material by hand pressure and it will retain its shape.
Pressure	Force per unit area.
Schist	A type of 'medium-grained' *metamorphic* rock characterised by a distinct crenulated particle alignment (foliation) known as schistosity.
Shale	A *fissile mudrock*. (As fissility is a feature that can develop in mudrocks as a consequence of weathering action, it is preferable to refer to such materials as fissile mudrocks.)
Shear strength	The minimum *shear stress* that causes a specimen to fail under conditions of zero *normal* stress. In a saturated clay, this is half the uniaxial compressive strength.
Siltstone	A fine-grained sedimentary rock consisting of particles of average size between 0.002 and 0.063 mm.
Slaking	A degradation process in geomaterials caused by alternate cycles of wetting and drying.
Solifluction	Process by which layers of soil move downslope due to freeze–thaw action under *periglacial* conditions.
Uniaxial/unconfined compressive strength	Compressive *pressure* to bring about failure of a specimen where loading is applied along only one axis and the specimen is an otherwise unconfined condition.
Weathering	A series of physical, chemical and biological processes by which a material undergoes degradation that brings it into equilibrium with its present environment.

doi: 10.1680/icemge.66816.0559

Chapter 37
Sulfates and sulfides in soils and rocks

J Murray Reid retired, formerly TRL Limited, Wokingham, UK
Mourice A Czerewko AECOM Ltd, Chesterfield, UK
John C Cripps University of Sheffield, Sheffield, UK
Ian Longworth retired, formerly BRE, Garston, UK

CONTENTS

Sulfur is widespread in rocks and soils and occurs in two principal forms; in an oxidised state as sulfate, principally gypsum ($Ca.SO_4.2H_2O$); and in a reduced state as sulfide, principally pyrite (FeS_2). Other less stable and possibly more problematic forms occur in some situations. The oxidation of pyrite and other sulfides can occur as a result of geotechnical activities when soils and rocks are disturbed and exposed to air and water. This can lead to the generation of high concentrations of sulfate and low pH in percolating water, which can have a number of adverse consequences, including chemical attack on construction materials such as concrete and steel, precipitation of gypsum leading to heave of foundations and floor slabs, and pollution of watercourses with precipitation of hydrous iron oxides (ochre). Pyrite oxidation can also be triggered in ground stabilisation involving mixing susceptible clay materials or pyrite-bearing aggregates with lime or cement; the resulting high pH values lead to formation of expansive sulfates (ettringite and thaumasite) that cause heave of the stabilised material. Test methods are available to identify and characterise problem materials but must be used with care because of the variability of sulfur compounds. Geotechnical engineers need to recognise the types of material that could lead to problems in civil engineering works and understand how these problems can be avoided or mitigated.

37.1. Introduction

Although most UK soils and rocks encountered in civil engineering works are chemically inert, the possibility of problems due to adverse chemical changes cannot be dismissed. Sulfur compounds underlie several serious problematic impacts in engineering. Sulfur is widespread in rocks and soils and occurs in two principal forms; in an oxidised state as sulfate, principally gypsum ($CaSO_4.2H_2O$) and in a reduced state as sulfide, principally pyrite (FeS_2). Other forms occur in special circumstances, such as gaseous hydrogen sulfide (H_2S) in landfill sites. The problems that affect engineering works are largely due to sulfate in solution and to changes from one sulfur compound to another. Excavation can allow air and water to reach sulfide minerals, leading to oxidation to sulfates and the generation of acidic conditions. This can result in attacks on concrete, steel and other construction materials; heave of fill materials and overlying pavements, floor slabs and foundations; failure of engineered slopes and cuttings; blocking of drains and filters; and pollution of watercourses. Treatment of soils and rocks containing sulfates and sulfides with lime or cement can lead to the formation of expansive sulfate minerals that cause heave, resulting in disruption to overlying construction materials.

A feature common to many of the problems is that they do not occur immediately, but months or years after construction, although sometimes the problems manifest during construction. The problems can be challenging and expensive to remediate, as they may necessitate removal of the affected ground or fill and extensive repairs to, or removal of, the overlying structures. Therefore, it is important for geotechnical engineers to be able to recognise potentially susceptible materials, avoid inappropriate construction methods or take mitigating measures. It is possible to achieve satisfactory engineering solutions in many situations provided the potential problems are foreseen and appropriate countermeasures adopted.

This chapter describes the potential situations where problems can arise, how to predict them and how to provide engineering solutions. It is based on climatic conditions and materials found in the UK, but the principles will be applicable to other countries with similar conditions.

The problems due to sulfur compounds mainly relate to damage to engineering structures and pollution of the environment. These are serious enough, but sulfur compounds can also lead to death, although thankfully very rarely. **Figure 37.1** shows the memorial to four workers who were killed due to carbon dioxide (CO_2) poisoning in a manhole during the construction of Carsington Dam, Derbyshire, UK in 1982. The dam was being constructed of local highly weathered mudstone fill, with limestone drainage layers. The fill was acidic, with high levels of sulfate, which attacked the limestone so producing gypsum and carbon dioxide. The latter

Figure 37.1 Memorial to workers killed by CO_2 poisoning, Carsington Dam, Derbyshire

travelled through the drainage system to the manhole where the workers entered (Reid and Cripps, 2019). There are also circumstances in which poisonous hydrogen sulfide (HS_2) gas can be generated. The behaviour of sulfur compounds is, thus, more than just an academic or engineering issue; it can be a matter of life and death, and something that geotechnical engineers need to include in their assessments.

37.2. Sulfur compounds in soils and rocks

As indicated in **Table 37.1**, sulfur can occur in a variety of forms, and these can change from one to another under certain conditions. The most common forms are sulfides, which form under reducing conditions, and sulfates, which form in oxidising conditions. Pyrite (FeS_2) is the most common sulfide and is a common minor constituent of many igneous, metamorphic and sedimentary rocks, and the soils derived from them. Sulfide minerals generally have very low solubility and are not in themselves a hazard to construction materials. However, weathering of sulfides gives rise to soluble sulfates (SO_4^{2-}) that are liable to undergo chemical reactions with natural and manufactured construction materials. Under natural conditions, weathering of sulfides normally leads to the production of gypsum ($CaSO_4.2H_2O$), which has limited solubility under neutral pH conditions, and this is the normal form of sulfate in near-surface soils.

Sulfates also occur naturally in evaporite sequences as gypsum, anhydrite ($CaSO_4$), epsomite ($MgSO_4.7H_2O$) and other salts. Barytes ($BaSO_4$) occurs as a minor constituent in a wide range of rocks. Barytes is extremely insoluble, but anhydrite, epsomite and other salts are highly soluble, such that in the UK they have generally been leached out of the near-surface zone where most civil engineering works are carried out and are only likely to be encountered in unweathered rocks and soils at depth. They can cause serious problems to civil engineering works in these situations, including corrosion of construction materials and expansion, especially where chlorides and other salts are also present.

Pyrite is immediately recognisable when present in large crystals as 'fool's gold' (**Figure 37.2**). It is bright brassy yellow, heavy (specific gravity 4.9), often occurring as cubes and octahedra. However, it can also be present in very fine-grained forms that are invisible to the naked eye (**Figure 37.3**); these forms are potentially much more problematic because the

Mineral	Formula	Description
Anhydrite	$CaSO_4$	White crystals, slightly soluble in water at neutral pH. Found in evaporite strata at depth.
Barytes	$BaSO_4$	White crystals, heavy (SG 4.4), insoluble. Common minor constituent in a wide range of rocks.
Celestine	$SrSO_4$	Rare mineral, insoluble, heavy (SG 3.9–4.0), occasionally found in evaporite sequences, argillaceous limestones and dolomite.
Epsomite	$MgSO_4.7H_2O$	Common in evaporite sequences. Highly soluble, only found at depth.
Gypsum	$CaSO_4.2H_2O$	White crystals, slightly soluble at neutral pH, soluble in acid.
Jarosite	$KFe_3(OH)_6(SO_4)_2$	Weathering product of pyrite, often found as a yellow-green powder on fissure and joint surfaces.
Marcasite	FeS_2	Similar to pyrite but has different crystal structure. Found as nodules in chalk and limestone.
Mirabilite (Glauber's salt)	$Na_2SO_4.10H_2O$	Occurs in evaporite strata. Highly soluble, only found at depth.
Pyrite	FeS_2	The most common form of sulfur, ranges from large yellow cubes to fine-grained crystals invisible to the eye. Heavy (SG 4.9), insoluble but fine-grained varieties liable to oxidation.
Pyrrhotite	FeS	Bronze to dark brown crystals or powder, heavy (SG 4.6), soluble in acid with H_2S evolved (rotten egg smell). Found in association with pyrite and marcasite.
Organic S	S	Common in peat and organic soils.
Reaction products of sulfates with cement and aggregates		
Ettringite	$3CaO.Al_2O_3.3CaSO_4.31H_2O$	Soft white fibrous crystals forming in cracks and causing concrete to break up.
Thaumasite	$CaSiO_3.CaCO_3.CaSO_4.15H_2O$	Soft white crystals formed on the outside surface of concrete and working inwards, turning the cement to a soft, mushy paste.

Table 37.1 Sulfur minerals in UK rocks and soils

Figure 37.2 A visible well-formed cube of pyrite in Cambrian slate (rule markings 1 mm apart) (Reid *et al.*, 2017)

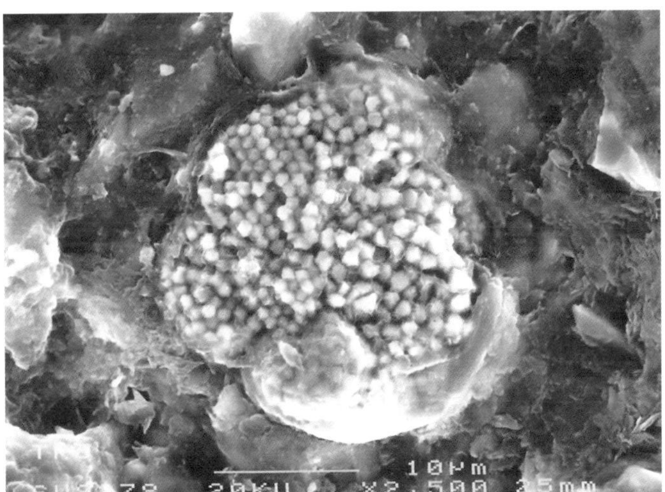

Figure 37.3 Microscopic framboidal pyrite (Reid *et al.*, 2017)

Figure 37.4 Precipitate of gypsum on the surface of Ancholme Clay, A63 Melton Bypass

crystals will react much more rapidly when exposed to air and water than the large crystals in **Figure 37.2**. Gypsum can occur in a wide variety of forms, from large translucent crystals in undisturbed strata to fine efflorescent white powdery crystals growing on exposed surfaces of reactive clays or mudstones. **Figure 37.4** shows white crystals of gypsum left behind when a pool of rainwater evaporated in a road cutting in Ancholme Clay, a Jurassic mudstone rich in sulfides and sulfates. Such fine white powders often form on the surface of fill materials that have a high sulfate or sulfide content and are a sure sign of possible trouble. They are often accompanied by orange-brown deposits of hydrous iron oxides (ochre), which are another product of pyrite oxidation (see Equation 37.1). These deposits can block drains and blanket the beds of watercourses, leading to the death of aquatic life. It is important for geotechnical engineers to be able to recognise these tell-tale signs; they are very effective indicators of potential trouble.

Further details of sulfur minerals can be found in Czerewko *et al.* (2016) and standard mineralogical textbooks such as Gribble (2005).

37.3. Weathering of pyrite

37.3.1 Weathering reactions

The main weathering reaction under natural conditions is the oxidation of pyrite and other sulfide species to sulfates. The top 2–3 m of natural strata in the UK often contains very low concentrations of sulfates and sulfides, as they have been removed by infiltrating rainfall, but there may be accumulations of sulfate nodules at the base of this leached zone. Where these occur at former high groundwater levels or tree root zones, they may cause a hazard for foundations. At greater depths sulfide is the dominant form except in evaporite sequences. The transition from sulfate to sulfide corresponds to the depth of surface weathering, often marked by a change of colour from black or dark grey deposits at depth to yellow, brown or mottled materials in the weathered zone. Profiles like this are particularly common in overconsolidated clays and mudstones, which are widespread in the UK (see Chapter 36 *Mudrocks, clays and pyrite implications*). Complex profiles can occur in places where the groundwater level has varied in the past, or where the level of sulfide minerals in the strata varies over short vertical or horizontal distances, such as in thinly interbedded mudstones and limestones (Hawkins, 1998; St. John, 2020). This can lead to large variations in sulfate and sulfide concentrations over short distances.

Civil engineering works can accelerate weathering when materials are excavated and placed as fill open to percolating rainwater and air. Free-draining embankment fill is a highly oxidising environment in which acidic conditions can quickly develop, leading to greatly increased solubility of sulfates and potential attack on construction materials. Oxidation of sulfides is often greatly accelerated by microbiological activity (Pye and Miller, 1990).

The main chemical reactions in the oxidation of pyrite are summarised below. While the details of the reactions can be complex and dependent on the specific situation, the overall result is the conversion of pyrite to iron hydroxides with the production of sulfuric acid and generation of pH values as low as 2 or 3

$$4FeS_2 + 15O_2 + 14H_2O \rightarrow 4Fe(OH)_3 + 8H_2SO_4 \qquad (37.1)$$

If carbonate minerals such as calcite are present, the acid will react with them to produce gypsum and carbon dioxide

$$CaCO_3 + H_2SO_4 + H_2O \rightarrow CaCO_4{\cdot}2H_2O + CO_2 \qquad (37.2)$$

The precipitation of gypsum involves a large net increase in volume and can lead to heave of the strata involved. Even if void space is present that would accommodate the gypsum, crystallisation on bedding planes and in joints in fill particles and at the points of contact of particles will cause overall expansion of the material. Because the gypsum formed expands the discontinuity without totally filling it, the overall expansion of the material can be much greater than expected from consideration of the quantities of pyrite, calcite and gypsum involved (Matheson and Jones, 2015; McCabe *et al.*, 2015).

If carbonate minerals are not present, the acid may attack clay minerals, producing minerals such as jarosite (**Table 37.1**). The drainage water from pyrite-bearing fills can be highly acidic, with pH values in the range 2–4, and very high concentrations of dissolved sulfates and metals such as iron, manganese and aluminium. When these waters encounter the atmosphere, the iron is rapidly oxidised and precipitates as ochre, causing pollution of watercourses and blockage of filters, pipes and drains (**Figure 37.5**).

Figure 37.5 Drainage outflow from mudstone embankment dam showing precipitation of ochre

37.3.2 Rate of reactions

A key factor in whether pyrite or other sulfides will cause problems in civil engineering works is the rate at which it will be oxidised to sulfate in any given situation. This depends on several variables

- mineral grain size
- form of mineral present
- access to air and water
- microbiological activity
- total amount of sulfide present.

37.3.2.1 Mineral grain size

One of the most important influences on the rate of reaction of any mineral is the grain size. Reactivity is related to the specific surface area of individual particles, which is related to the inverse square of the grain size. Large crystals visible to the naked eye will be much slower to react than very fine-grained crystals at a scale of a few microns, which leads to the paradox that pyrite is generally troublesome when it cannot be seen; it might be dubbed 'the invisible menace'. Microscopic clusters of pyrite crystals known as 'framboids' are particularly reactive. These can be identified by electron microscopy, as in **Figure 37.3**, or by reflected light optical microscopy, but these are not tools generally used in routine ground investigations and assessments (see Section 37.5.8).

37.3.2.2 Form of mineral

The reactivity of sulfides is also linked to the form or degree of crystallinity of the minerals. In general, the older the mineral, the more likely it is to have a more ordered structure, and the slower it is likely to react. This may explain why sulfides in granites and slates (**Figure 37.2**) tend to be less reactive when exposed to the atmosphere than those in mudstones and overconsolidated clays. Framboids in recently deposited alluvial sands and silts (**Figure 37.6**) are liable to be highly reactive. Pyrite in strata that have been subject to weathering before being disturbed for civil engineering works are likely to be more reactive than in fresh strata, regardless of their age, and older rocks that have been weathered may be susceptible to rapid oxidation when exposed in engineering works. This aspect is harder to quantify than grain size, but it is important to bear in mind when assessing how likely a particular sulfide will be to undergo oxidation reactions.

Problems may arise in alluvial or estuarine sediments, where iron sulfides form in reducing conditions owing to the decay of organic matter. They generally only occur in situations such as backfill to sheet piles in marine works just below mean water level (Tiller, 1997). These iron sulfides are particularly corrosive and cause major problems for steel sheet piles.

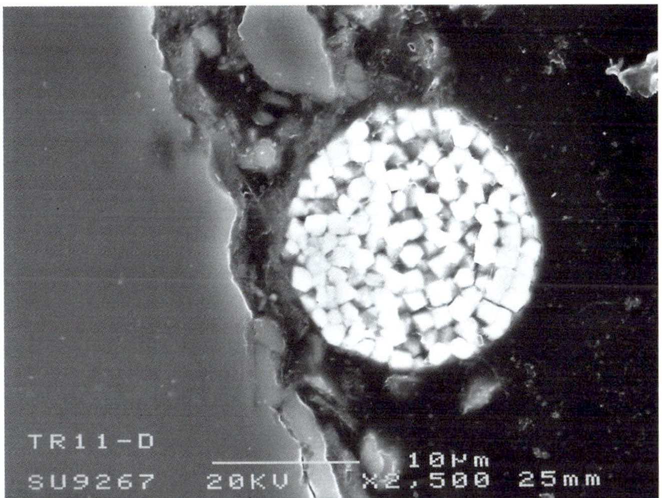

Figure 37.6 Framboidal pyrite in alluvial sand

37.3.2.3 Access to air and water

Both air and water, in sufficient quantities, are required for sulfide oxidation (Equation 37.1). However, unless the products of reaction are removed and a fresh supply of air and water provided, equilibrium will be reached and no further reaction will occur. If the strata are below the water table, very little reaction will occur because the supply of air is limited, even if water is abundant. Above the water table, a steady percolation of aerated water through the ground will drive the reactions, remove the products through drainage and potentially cause problems elsewhere, such as precipitation of gypsum where calcareous materials are encountered. The flow of water can also move materials upwards by capillary flow through fill below floor slabs, leading to expansive formation of gypsum, heave in the fill and disruption to the floor slabs. The movement of solutions from one location and chemical attack and precipitation of expansive minerals in another allows the reactions to continue. Pyritic oxidation is especially liable to occur in embankments constructed out of compacted pyrite-bearing soil or rock fills (Davies and Reid, 1997; Pye and Miller, 1990).

37.3.2.4 Microbiological activity

The oxidation of pyrite occurs in several stages, which are discussed by numerous authors (e.g. Reid *et al.*, 2005), and in Chapter 36 *Mudrocks, clays and pyrite implications.* Many of these stages are catalysed by bacteria that derive energy and nutrients from the reactions and greatly accelerate pyrite oxidation reactions (Pye and Miller, 1990). They can exist and even thrive in the most extreme environments, including the very acidic conditions generated by oxidation of sulfides. The exothermic nature of these oxidation reactions also assists bacterial activity by raising the ambient temperature.

37.3.2.5 Total amount of sulfide present

It might be expected that the total quantity of sulfide would be the main factor in determining the extent of the reaction, but grain size, mineral form and exposure to air and water are more important. Thus, some materials with relatively high sulfide contents may not be as troublesome for engineering works where the sulfide is in the form of large crystals, whereas other materials with small quantities of highly active sulfides can cause very significant problems. In the case of fine-grained and framboidal pyrite, the sulfide concentration must be almost at the limit of detection to ensure trouble-free performance. This causes difficulties for geotechnical engineers, as reference to limiting values may give a conservative view of the likelihood of sulfide oxidation and its potential consequences. Moreover, while the total sulfur content of soils and rocks can be determined relatively reliably and cheaply, determination of the total sulfide content is more difficult (see Section 37.5.6). Even if this can be determined accurately, an assessment must then be made as to whether the sulfide is likely to be oxidised and, if so, at what rate and what effect this will have on the works or the environment. This may be evaluated by simple block exposure observations of fresh recovered pyritic material in natural weathering tests (Czerewko *et al.*, 2011).

Only by taking all the circumstances of each case into account and being aware of the likely geochemical behaviour of the materials can predictions be successfully made. It may then be possible to avoid or mitigate deleterious consequences.

37.4. Occurrence of sulfur compounds

Lists of geological formations in the UK known to contain pyrite in sufficient quantities to cause problems are given in BRE Special Digest 1 (BRE, 2005), HA Advice Note 74 (Highways Agency *et al.*, 2007) and Czerewko *et al.* (2016). Some of the materials that are particularly likely to cause problems are discussed below; however, it is important that every situation is evaluated on its merits.

37.4.1 Mudstones and overconsolidated clays

Dark-coloured marine mudstones and overconsolidated clays are particularly likely to contain significant concentrations of pyrite in fresh strata and a mixture of pyrite and gypsum in weathered strata. Potential problems due to sulfates and sulfides should always be considered when engineering works in these strata are proposed. Among the key strata in this group are:

- mudstones, particularly those of Carboniferous age
- Mercia Mudstone (formerly Keuper Marl – sulfates only)
- Penarth – Lilstock and Westbury Formations
- Lower Lias Clay
- Kimmeridge Clay (the Ancholme Clay in **Figure 37.4** is a lateral equivalent of the Kimmeridge Clay)
- Oxford Clay
- Weald Clay

- Gault Clay
- London Clay
- glacial till or other drift deposits derived from any of the above.

These strata are also likely to pose other problems for engineering works, including seasonal shrink–swell behaviour, long-term shear strength reduction in slopes and excavations and the presence of organic matter. These aspects and the distribution of these formations in the UK are discussed in Chapter 36 *Mudrocks, clays and pyrite implications*.

37.4.2 Alluvial deposits

Problems due to sulfur compounds have been reported in alluvial sands and gravels (Reid *et al*., 2005; Sandover and Norbury, 1993). In these cases, the sulfide is often in the highly reactive framboidal form (**Figure 37.6**). As these deposits often have little or no carbonate content, highly acidic conditions can develop if they are excavated and used as fill, even if the sulfide and sulfate content are quite low. This means that deleterious reactions with other construction materials may rapidly occur.

37.4.3 Colliery spoil and burnt oil shale

Colliery spoil consists of strata excavated during mining for coal and includes rock from immediately above and below the coal seams, fragments of coal and strata from the driving of underground roadways and shafts. The spoil can be highly variable in composition. The dominant material tends to be mudstone, often containing significant amounts of sulfides and sulfates, particularly in marine mudstones, which often form the roofs of coal seams. The spoil may also contain significant amounts of coal, which also has high sulfide and sulfate content. The carbonate content can be highly variable, with significant concentrations of calcite in some marine mudstones, whereas the mudstones and sandstones immediately below coal seams generally have a low calcite content. Fresh colliery spoil generally has a pH in the range 7.5–8.5, but on exposure to the atmosphere the pH can drop to as low as 2–3.

The potentially corrosive properties of colliery spoil have long been recognised (West and O'Reilly, 1986) and it is not permitted as backfill to concrete and structural metallic elements by the UK Specification for Highway Works (Highways England *et al*., 2016a).

Colliery spoil that has been placed loosely in tips has often suffered ignition, resulting in combustion of the coal particles and changes to the chemical and physical properties of the material. Most noticeable in this 'burnt shale' is a change in colour from dark grey to red, accompanied by the oxidation of pyrite and an increase in the particle strength. Although it can contain high concentrations of sulfates, it is permitted in a range of applications by the Specification for Highway Work (Highways England *et al*., 2016a).

Spent oil shale, which is the residue from the extraction of oil from oil shale of Carboniferous age in the Lothian area of central Scotland, is similar in many ways to burnt colliery spoil. Heating the shales in retorts has resulted in the production of red shale material with variable quantities of sulfate but generally low sulfide content.

37.4.4 Recycled and manufactured aggregates and secondary materials

With the rise of the sustainability agenda, there has been increasing pressure to reuse and recycle materials rather than disposing of them in landfills. The proportion of the UK aggregate market taken by recycled and manufactured aggregates is just under 30% (Mineral Products Association, 2021). The vast bulk of this consists of recycled aggregate derived from construction and demolition waste and recycled concrete aggregate, derived exclusively from concrete. Small amounts of manufactured aggregates such as blast furnace and steel slag are sometimes used as unbound aggregates and in ground (comminuted) form as binders and activators for hydraulically bound materials (HBM) and replacements for cement in concrete. Other secondary materials that are sometimes used in civil engineering works include pulverised fuel ash (PFA) and furnace-bottom ash (FBA) from coal-fired power stations, and incinerator-bottom ash aggregate (IBAA) from energy-from-waste (EfW) plants.

Recycled, manufactured and secondary materials can contain high concentrations of sulfur compounds, particularly where the source of the material is variable over time, such as recycled aggregates and IBAA. The sulfur will generally be in the form of sulfate, although blast furnace slag can contain reactive sulfides. Some materials, such as IBAA, may also contain other potentially reactive constituents such as aluminium, which can react with sulfate to form highly expansive minerals such as ettringite when the materials are mixed with cement. Unweathered slags and ashes may also contain unhydrated lime and magnesium oxides, which can result in expansive reactions when exposed to air and water, similar to the expansive reactions caused by sulfur compounds. The production of manufactured and secondary aggregates should include a period of weathering to allow these oxides to hydrate.

All aggregates used in construction applications should be produced under a system of factory production control in accordance with the relevant standards and specifications, which include appropriate tests and limiting values for sulfur compounds.

37.5. Sampling and testing for sulfur compounds

37.5.1 Standards and specifications

The appropriate standard for chemical tests at the site investigation stage is BS 1377-3:2018+A1:2021 (BSI, 2021). This part of BS 1377 describes test methods for chemical substances in samples of soil and extremely weak and very weak rocks, as defined in BS 5930 (BSI, 2020), and groundwater. The test methods in BS 1377-3 are suitable for use with BRE Special Digest 1 (BRE, 2005) and BRE Digest 522 Parts I and II (Longworth, 2011a, 2011b).

Ground investigation and testing in the UK is now covered by BS EN 1997-2:2007 (BSI, 2007b) and the corresponding National Annex, NA to BS EN 1997-2:2009 (BSI, 2009). These documents state that chemical testing of soils should continue to be carried out in accordance with BS 1377-3 except for pH value determination (acidity and alkalinity), where Annex N.5 of BS EN 1997-2 may be used. Further information on how to use the various tests in BS 1377-3 is given in NA3.15 of the National Annex.

The standard for chemical testing of aggregates is BS EN 1744-1:2009+A1:2012 (BSI, 2012). This standard is stipulated by the UK Specification for Highway Works, Series 500 (drainage), 600 (earthworks) and 800 (unbound and hydraulically bound subbase).

The test methods used for different forms of sulfur in BS 1377-3:2018+A1:2021 and BS EN 1744-1:2009+A1:2012 are broadly similar but differ in some important particulars, with the potential for serious underestimation of the amount of pyrite present if the wrong methods are used. The differences between test methods and their potential effects are discussed by Czerewko *et al.* (2016), Reid *et al.* (2017) and Cripps *et al.* (2019).

37.5.2 Sampling and storage

To assess the potential impact of sulfur compounds on civil engineering works, it is necessary to take representative samples and store them in a way that preserves the compounds in the form they were in when the sample was taken. Concentrations of sulfur compounds and the nature of the minerals can vary rapidly over short distances horizontally and vertically in many strata, particularly in the weathering zone where there may be sudden and dramatic changes from sulfate to sulfide. Samples that are taken at set depth intervals may miss much of this complexity, potentially leading to underestimation of the risk posed to the civil engineering works. Sampling strategies to take account of this variability are given for pyritic ground by Czerewko and Cross (2015) and St. John (2020).

Because of the variability, sufficient samples should be taken to capture the likely range of values, and characteristic values for design should be based on the highest values. BRE SD1 (2005) recommends using the highest sulfate value if fewer than five tests are available for the material being considered, the mean of the highest two tests if 5–9 samples are available and the mean of the highest 20% if 10 or more results are available.

Both soil strata and groundwater should be sampled for chemical testing during the site investigation. Samples of groundwater are particularly valuable, as this is the medium in which sulfur compounds can cause most damage to materials or the environment, but samples must be taken and processed in accordance with best practice guidelines. Basic parameters should be determined at the time of sampling, in particular temperature, pH value, conductivity and redox potential/dissolved O_2. These will aid interpretation of the laboratory results.

37.5.3 Water-soluble sulfate (WS)

The most common test is the sulfate content of water samples or the water-soluble sulfate content (WS) of soil or rock samples. This is the form of sulfate that attacks construction materials and instigates other chemical reactions in fill materials and waters; it represents the immediate danger posed by the material or water. Water-soluble sulfate is used as an index parameter for assessing the potential for harm. It does not represent the total amount of available sulfate in the sample, as that depends on the solubility of the sulfate minerals that are present.

For soil or rock samples, sulfate is brought into solution with a 2:1 extraction in de-ionised water for 24 hours, after which the solution is filtered and the sulfate concentration determined by the methods in BS 1377-3 or BS EN 1744-1, respectively.

The results should be expressed as mg/l SO_4, as this is the form used by most guidance documents when setting limiting values for specific applications and is the form recommended by BS 1377-3:2018+A1:2021. However, various formats are used by different standards. The BS EN 1744-1 test result is reported as % SO_3. Sulfate occurs as SO_4, so this is the most useful way to express the results. Concentrations in SO_3 can be converted to SO_4 in the same units by multiplying by 1.2, and % SO_3 can be converted to g/l SO_3 by multiplying by 5 because the extraction ratio is fixed at 2:1. Concentrations in % SO_3 can therefore be converted directly to mg/l SO_4 by multiplying by 6000.

When determining the water-soluble sulfate content of a soil or rock sample, it is useful to measure the pH value of the 2:1 water extract at the same time, as recommended in BS 1377-3:2018+A1:2021. The pH value of soil may be determined in accordance with BS 1377-3, but there is no test method for the pH value of aggregates, so it is important to measure pH during the water-soluble sulfate test.

The water-soluble sulfate content is assigned the symbol WS in guidance documents such as BRE SD 1 (2005) and in the UK Specification for Highway Works (Highways England *et al.*, 2016a). The water-soluble sulfate test will extract readily soluble sulfates such as epsomite and mirabilite (**Table 37.1**). Gypsum is sparingly soluble in water at neutral pH, so if this is the only sulfate compound present WS will not exceed about 1500 mg/l SO_4. If the pH is acidic, however, much higher values of WS can be obtained even if gypsum is the only sulfate present. Determination of the pH value, thus, helps to interpret the WS value and assess the threat posed by the material or water.

37.5.4 Acid-soluble sulfate (AS)

While the WS test determines the soluble sulfate content of a soil or rock in contact with de-ionised water, the sulfate content may be significantly higher than this if the sulfate is in the form of sparingly soluble minerals such as gypsum. The acid-soluble sulfate (AS) represents the 'sulfate reservoir' that can be brought into solution by water over time. Test methods for this are given in BS 1377-3:2018+A1:2021 (Clause 7.9) and BS EN 1744-1 (Clause 12). The tests involve grinding a sample to pass a 212 μm sieve before extracting it with an excess of 10% hydrochloric acid. The sulfate content is determined from the extract gravimetrically by precipitation of barium sulfate (BS EN 1744-1) or instrumentally by ICP-OES or similar methods (BS 1377-3:2018+A1:2021). The gravimetric method is also permitted under BS 1377-3:2018, although an instrumental test is preferred. The BS EN 1744-1 test result is reported as % SO_3. The BS 1377-3:2018+A1:2021 result is reported as % SO_4. The SO_3 values can be converted to SO_4 by multiplying by 1.2.

The acid-soluble sulfate content is assigned the symbol AS in guidance documents such as BRE SD 1 (2005) and the Specification for Highway Works (Highways England *et al.*, 2016a). It is used directly in the Specification for Highway Works as a limiting value for some applications such as aggregates for use in HBM (Series 800, table 8.14). In this case, the limiting value is expressed as % SO_3, so if AS has been determined by the BS 1377-3:2018 methods, the results must be divided by 1.2 to convert them to % SO_3.

The AS content will include all potentially reactive sulfate minerals, but unreactive sulfates such as barytes are not mobilised by the extraction. AS is thus not a measure of total sulfate content and the results of AS sulfate tests should not be referred to as total sulfate.

37.5.5 Total sulfur (TS)

The total sulfur content (TS) of a soil or rock includes all the sulfur present as sulfate, sulfide and other forms such as organic sulfur.

The test method for total sulfur in BS 1377-3:2018+A1:2021 is rapid high-temperature combustion. A chemical method for TS content is given in BS EN 1744-1 as the reference method (Clause 11.1); in this method, sulfides are converted to sulfates by treatment with hydrochloric acid and hydrogen peroxide, which is a powerful oxidising agent, and the sulfate content is determined gravimetrically as before. The results are given as % S. There are considerable doubts over how effective this chemical treatment is in bringing all the disulfide forms of sulfur into solution (Czerewko *et al.*, 2016) and it is not recommended. The high-temperature combustion method is also permitted in BS EN 1744-1 as an alternative method (Clause 11.2) and this method should be used for determination of the total sulfur content of aggregates.

The total sulfur content is assigned the symbol TS in guidance documents such as BRE SD1 (2005) and the Specification for Highway Works (Highways England *et al.*, 2016a). It is rarely used directly as a limiting value, but is used to calculate two parameters that are used as limiting values.

The total potential sulfate content (TPS) is the total amount of sulfate (as % SO_4) that could be generated if all the sulfur compounds in the sample were converted to sulfate. It gives a worst-case figure for the sulfate threat from a particular material. It is calculated as

$$\mathrm{TPS}(\%\ \mathrm{SO_4}) = 3 \times \mathrm{TS}(\%\ \mathrm{S}) \tag{37.3}$$

The oxidisable sulfides content (OS) is the sulfate (as % SO_4) that could be produced if all the sulfur present as sulfide in the sample is oxidised to sulfate. It is lower than the TPS and is usually calculated indirectly from TS and AS as

$$\mathrm{OS}(\%\ \mathrm{SO_4}) = [3 \times \mathrm{TS}(\%\ \mathrm{S})] - \mathrm{AS}(\%\ \mathrm{SO_4}) \tag{37.4}$$

These three tests – WS, AS and TS – are the main ones that are routinely performed on soils, weak rocks and aggregates. They have the advantage of being relatively rapid and cheap, thus allowing large numbers of samples to be tested without incurring excessive cost or delay.

37.5.6 Total reduced sulfur (TRS)

The content of reduced sulfur compounds can be estimated from the difference between TS and AS (see Section 37.5.5); however, this figure will also include other inert forms of sulfur such as organic sulfur or recalcitrant forms of sulfur such as barytes, and will be an overestimate of the sulfide content. It will also be of limited accuracy, particularly at low values, as there are significant error margins associated with the measurement of both TS and AS. A direct test for the reduced sulfur content would improve the accuracy of the estimates. Reid *et al.* (2005) developed a chemical method for total reduced sulfur (TRS) which is included in BS 1377-3:2018+A1:2021 as Clause 7.11. It is a specialist chemical test that should only be carried out by skilled personnel; unlike the tests for WS, AS and TS it is not automated and may be expensive to perform. It is useful in specific situations to identify the nature of sulfur compounds present in a material in more detail, or in forensic studies. The results are reported as % S.

A test for sulfides is given in BS EN 1744-1 (Clause 13); however, this test seriously underestimates the total sulfide content (Czerewko *et al.*, 2016) and is not recommended. It is, however, suitable for the determination of monosulfides (Section 37.5.7).

Oxidisable sulfides (OS) can be determined directly from TRS as

$$\mathrm{OS}(\%\ \mathrm{SO_4}) = 3 \times \mathrm{TRS}(\%\ \mathrm{S}) \tag{37.5}$$

37.5.7 Monosulfide sulfur (MS)

Monosulfides are highly reactive and can be important in certain situations, such as where the mineral pyrrhotite (FeS) is present, in estuarine and organic sediments from marshes, bogs and peat, and in some manufactured aggregates such as blast furnace slag. In the determination of monosulfides the sample is treated with hydrochloric acid, which converts it to gaseous hydrogen sulfide (H_2S). This is then trapped by precipitation as copper sulfide and the quantity determined by the reduction in concentration of copper in the trapping solution. This test is usually combined with the determination of AS but can be carried out on its own if required. The monosulfide content (MS) is reported as % S. The test is included in BS 1377-3:2018+A1:2021 as Clause 7.12. The test for acid-soluble sulfides in BS EN 1744-1 (Clause 13) is very similar and measures highly reactive monosulfides. Like the TRS test, it involves hazardous chemicals and should be carried out by suitably skilled and experienced personnel.

37.5.8 Other tests

Techniques such as X-ray diffraction (XRD) can be used to test for the presence of pyrite, gypsum and other minerals on a qualitative and semi-quantitative basis in soils and rocks. This technique also gives information about other minerals, such as carbonates and clay minerals. X-ray fluorescence (XRF) analysis of soils or rocks gives the overall chemical composition, including an estimate of the TS content, but is less accurate than direct chemical determination. Transmitted and reflected light microscopy and scanning electron microscopy (SEM) can be used to investigate the crystal form and grain size of fine-grained sulfur compounds, as in **Figure 37.6**. SEM is particularly useful as it has capacity for high magnification and 3D imaging, as well as having EDX capability that facilitates the chemical analysis of constituents. SEM can be very helpful in the identification of framboidal pyrite and detecting whether it has undergone any oxidation. It can also assist with understanding the reactivity of compounds in a particular situation; however, these techniques are expensive and may not be suitable for routine testing purposes.

37.6. Problems and solutions

37.6.1 Attack on buried concrete

It has been known for over 80 years that sulfates and acids in the ground can attack concrete, resulting in cracking, expansion and softening (**Figure 37.7**). Guidance on how to assess the risk and design concrete to withstand attack has been issued by the BRE in a series of digests and guidance notes dating back to 1939. The most recent version is BRE SD1, published in 2005. This sets out detailed guidance for all parties involved in the design and construction of concrete structures, including the geotechnical specialist who is responsible for carrying out the site investigation and assessing the design sulfate class (DS class) and the aggressive chemical environment for concrete class (ACEC class) for the structure and passing this information to the designer of the building or structure.

Figure 37.7 Deterioration of reinforced-concrete footing buried in colliery spoil due to chemical attack (Reid *et al.*, 2017)

Five DS and ACEC classes are defined, ranging from low concentrations of sulfates and sulfides and neutral pH values for DS-1 and AC-1 to very high values of sulfates and sulfides and highly acidic pH values for DS-5 and AC-5. The ACEC classes are labelled AC-1 rather than ACEC-1.

The limiting values for the DS and AC classes are for design purposes rather than pass/fail criteria. However, limiting values for backfill to concrete, cement-bound materials, other cementitious materials or stabilised capping forming part of the permanent works are given in the UK Specification for Highway Works (Highways England *et al.*, 2016a), Clause 601.16. These limiting values are based on the top of the DS-2 class from BRE SD1 and roughly correspond to the solubility of gypsum at neutral pH values. The limiting values are

- WS content not exceeding 1500 mg/l SO_4 (BS EN 1744-1, Clause 10)
- TS content not exceeding 1% for aggregates not including air-cooled blast furnace slag or 2% for air-cooled blast furnace slag (BS EN 1744-1, Clause 11).

In addition, the material shall conform to at least one of the following options

- when described in accordance with BS EN 932-3 and BS EN 13242 Annex A (BSI, 2007a), limestone, chalk, dolomite, blast furnace slag, steel slag or crushed concrete are predominant
- the sulfide content of the material determined in accordance with BS EN 1744-1 Clause 13 is less than 0.5% (as SO_4).

The second set of requirements are designed to prevent materials containing oxidisable sulfides, which are not detected by the WS test, from being used as backfill, as this could lead to high concentrations of sulfate and acidic leachate attacking the concrete (Highways England *et al.*, 2016b). However, the BS EN 1744-1 test will not detect disulfides such as pyrite, which is insoluble in the acid specified (Cripps *et al.*, 2019; Czerewko *et al.*, 2016), so the BS 1377-1:2018+A1:2021 Clause 7.11 test should be used instead. The BS EN 1744-1 test only detects monosulfides such as pyrrhotite that are likely to be present in very low concentrations in natural materials or monosulfide forms that may be present in blast furnace slag (Cripps *et al.*, 2019; Czerewko and Cross, 2015; Hawkins, 2014; Reid and Cripps, 2019; Reid *et al.*, 2005). The list of materials exempt from sulfide requirements is based on them having adequate acid-neutralising capacity to absorb any acid likely to be generated by oxidation of a small amount of sulfide-bearing material in the fill.

The limiting values for backfill are designed to ensure that materials placed directly against the concrete will not cause sulfate or acid attack; this is different to the BRE SD1 approach, which is based on the existing ground conditions. Backfill to structures can either be generated from arisings on site or imported to site. Hence, it is necessary to set pass/fail limits for this application.

Note: The text above is based on the content of the SHW Series 600 at the time of writing (May 2023). The SHW is updated from time to time and readers should always check the current requirements, which can be found at Series 0600 – Earthworks on the Standards for Highways website.

Structural concrete may be exposed to fill materials above ground or percolation of water through overlying embankment fill. When assessing the DS and AC classes, the composition of any fill material that impacts on the concrete should be considered. An embankment is a highly oxidising environment, so if materials containing active sulfates and sulfides are used, the implications for concrete structures need to be assessed. The limiting values for backfill in the Specification for Highway Works normally only apply to material up to a distance of 500 mm from the structure, so it is possible for sulfate-rich drainage that could attack concrete or other construction materials to be generated from materials used outside this zone.

Problems can arise with bulk fill materials such as colliery spoil that frequently contain high concentrations of sulfates and sulfides. This was the case in two highway improvement schemes in Yorkshire and Humberside: the A64 at Tadcaster and the A63 Melton Bypass. Both schemes required the import of large volumes of bulk fill for embankments, and colliery spoil was the most readily available material, but chemical testing showed it had high concentrations of both sulfates and sulfides. While it was suitable for use as bulk fill, it would not have been appropriate in locations where it could shed drainage onto structural concrete. The problem was solved by using local limestone to form a wedge rising over the colliery spoil in a series of benches away from the structure. In this way the colliery spoil was always below the limestone backfill and could not drain into it. This enabled large volumes of colliery spoil to be used as bulk fill, while preserving primary aggregates for higher value uses such as structural backfill.

37.6.2 Attack on buried metallic structural elements

Metallic elements forming part of the permanent works are particularly susceptible to corrosion from sulfates, sulfides and acids; it is very important to ensure that they are not exposed to aggressive drainage from fill materials. Applications where structural metallic elements are used in civil engineering include galvanised steel and stainless steel reinforcement in anchored earth and reinforced soil, and corrugated steel buried culverts (**Figure 37.8**). Table B.1 of 'BS EN 14475:2006: Execution of special geotechnical works – Reinforced fill' (BSI, 2006), and tables 6/1 and 6/3 of the Specification for Highway Works (Highways England *et al.*, 2016a) specify allowable limits for the types, and chemical properties, of acceptable materials. Guidance is also given in the *UK Design Manual for Roads and Bridges* in 'CD375 Revision 1: Design of corrugated steel buried structures' (Highways England *et al.*, 2020). BS 8006 (BSI, 2016) refers to Table B.1 of BS EN 14475 for electrochemical properties of fills used with metallic reinforcement.

It has been known for many years that some materials, such as colliery spoil and argillaceous rocks in general, are liable to cause corrosion of structural metallic elements (West and O'Reilly, 1986; Winter *et al.*, 2002), and the Specification for Highway Works (Highways England *et al.*, 2016a) prohibits these materials as backfill to reinforced soil, anchored earth or corrugated steel buried structures. However, there have been problems with other materials such as alluvial sands and gravels (Reid *et al.*, 2005) in which acid generated by pyrite

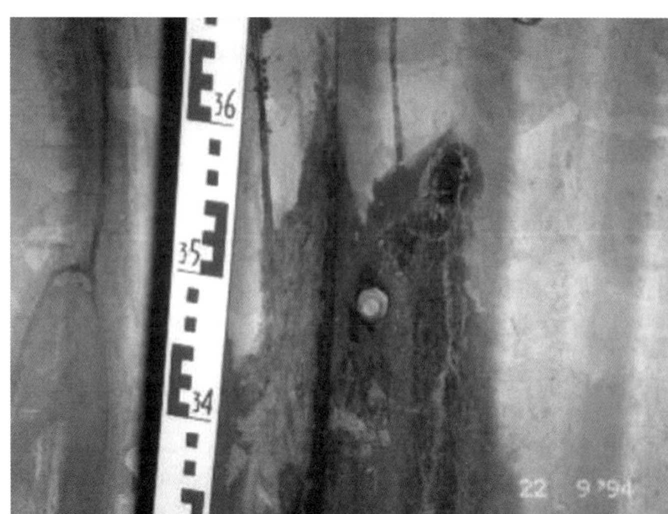

Figure 37.8 Corrosion of galvanised steel culvert on the A50 trunk road in Staffordshire

oxidation caused corrosion of corrugated steel buried culverts in a highway scheme in England (**Figure 37.8**). The culverts and the backfill material had to be replaced before the road could be opened, causing considerable delay and expense.

The limiting values for backfill are shown in **Table 37.2**, which is an extract from table 6/3 and Clause 616.17 in the UK Specification for Highway Works (Highways England *et al.*, 2016a). The same concerns expressed in Section 37.6.1 regarding the use of the BS EN 1744-1 Clause 13 test as a measure of oxidisable sulfides also apply to its use in this application. Limiting values are also given in table 6/3 for chloride content, organic content, resistivity, redox potential and microbial activity index.

It should be noted that the limiting values in table 6/3 of the Specification for Highway Works and **Table 37.2** do not apply to backfill to ancillary metallic structures such as manholes and drain covers; they only apply to structural metallic elements forming part of the permanent works. Where the metallic elements are encased in concrete the limiting values for backfill to concrete should be used (see Section 37.6.1).

The text above is based on the content of the SHW Series 600 at the time of writing (May 2023). The SHW is updated from time to time and readers should always check the current requirements, which can be found at Series 0600 – Earthworks on the Standards for Highways website.

An investigation by Reid and Avery (2009) demonstrated that although some aggregates have OS values exceeding the limits of **Table 37.2**, they would be suitable for use as backfill to structural metallic elements. Problems are more likely if new sources or site-won aggregates are used, rather than established, well-investigated materials.

37.6.3 Attack on buried steel sheet piles

Steel sheet piles are widely used in civil engineering and are susceptible to corrosion from acid, sulfates and sulfides. In oxidising situations (e.g. above the water table in granular soils) corrosion will result from the mobilisation of sulfates and acidity. However, where oxygen is scarce or absent, a different set of reactions may occur as sulfates may be reduced to sulfides, leading in some cases to generation of highly corrosive monosulfides of iron or gaseous hydrogen sulfide. These reactions can occur in organic silty sediments that occur frequently in estuarine areas, which are often locations where extensive sheet piling is employed in harbour works. The high salinity associated with such environments, plus the diurnal change of water level due to tides, leads to conditions particularly well suited to the formation of reactive monosulfides. Tiller (1997) describes the mechanisms of corrosion specific to these environments together with various methods for combating them, such as use of cathodic protection.

37.6.4 Degradation of bulk fill materials: geotechnical and environmental aspects

The consequences of pyrite oxidation in bulk fill materials in terms of potential attacks on concrete and structural metallic elements have already been described in Sections 37.6.1 and 37.6.2. However, chemical reactions can also have significant effects on the geotechnical properties of the fill and leaching of polluted drainage waters into nearby watercourses.

The materials most susceptible to degradation are weak rocks such as mudstones and argillaceous limestones, and materials such as colliery spoil, which contain large quantities of these components. These materials are widely used as bulk fill for highway and other embankments and for embankment dams. Embankment dams are expected to have a service life of 200 years or more, so any potential long-term deterioration in their geotechnical properties is important, as this could lead to a decrease in the factor of safety and loss of serviceability. Embankment dams are generally constructed using the materials available within the footprint of the proposed reservoir, as it is seldom practicable or economic in the UK to import the large volumes of fill material required. It is, therefore, particularly important to understand any potential degradation of the fill materials and design the embankment to cope with it.

37.6.4.1 Embankment dams

Geochemical issues were investigated for two major embankment dams in the UK during the 1980s and 1990s. The dams were Carsington in Derbyshire and Roadford in Devon, which were constructed using locally available weathered

Reinforcing element material	Properties of fill pH value Min	Max	Max. water-soluble sulfate content (WS) as mg/l SO_4	Max. oxidisable sulfides (OS) content as % SO_4	Total sulfur (TS) content as % S	Total sulfur (TS) content as % S – air-cooled blast furnace slag
Galvanised steel	5	10	300	0.06	1	2
Stainless steel	5	10	600	0.12	1	2

WS, TS and OS shall be determined in accordance with the methods described in BS EN 1744-1 Clauses 10, 11 and 13.
The limiting values for OS shall not apply if the material is predominantly limestone, chalk, dolomite, blast furnace slag, steel slag or crushed concrete when described in accordance with BS EN 932-3 or BS EN 13242 Annex A.
When determining WS, TS or sulfide content, at least five samples of each material shall be tested. The mean of the highest two values shall be used for comparison with the limiting values. This shall also apply if 6–9 results are available. If 10 or more results are available, the mean of the highest 20% of the results shall be used for comparison with the limiting values.

Table 37.2 Limiting values of sulfur compounds for backfill to reinforcing elements (Extract from table 6/3 in UK Specification for Highway Works; Highways England *et al.*, 2016a)

Carboniferous mudstones containing varying quantities of sulfates and sulfides. An overview of the geochemical issues and the lessons learned at Carsington is given by Reid and Cripps (2019), who also outline the situation at Roadford. Among the geochemical issues identified were

- loss of shear strength of the fill materials due to leaching by sulfuric acid
- chemical attack on drainage materials by sulfuric acid
- clogging of drains by precipitation of gypsum or ochre (**Figure 37.9**)
- increased compressibility of the fill
- attacks on concrete and metal
- pollution of downstream watercourses by acidic drainage waters with high concentrations of metals (**Figure 37.5**)
- potential for contamination of the impounded water by acid, sulfates and metals.

The results of the investigations and the records of measures taken during construction are also described by Chalmers *et al.* (1993) for Carsington and Wilson and Evans (1990) and Davies and Reid (1997) for Roadford.

For both Carsington and Roadford Dams, important information on the chemical reactions was obtained from observations of trial embankments constructed during the ground investigations. These were primarily designed to give information about the compaction behaviour of the various fill materials, but also yielded much information about the nature and scale of the chemical reactions. Observations of precipitates of ochre and gypsum in drainage pools at the toes of the embankments and the formation of white powdery gypsum on the surface of the embankments a few weeks after construction gave a very visible indication of the chemical reactions that occurred when the materials were excavated and exposed to air and water. Chemical analyses of the drainage waters gave an indication of the scale of the problem and enabled design decisions to be made to deal with it. Much of this might have been missed had it not been for the sharp eyes and enquiring minds of the inspectors and engineers supervising these trial embankments. It is important for engineering geologists and geotechnical engineers to recognise the signs of chemical reactions and understand their significance.

Figure 37.9 Limestone drainage layer in mudstone fill, Carsington dam, showing formation of gypsum and iron hydroxide minerals in the drainage layer

37.6.4.2 Highway embankments

Problems of accelerated weathering have also occurred in highway schemes, as the following two case studies illustrate. The first concerns the A63 Melton to Ferriby improvement. The scheme included the construction of a grade-separated junction and associated link roads over a 1.5 km section of the A63 dual carriageway west of Hull, built between 2004 and 2007.

The site is partially located over a prominent north–south trending ridge of pyrite-bearing Jurassic age Ancholme Group mudstone. The presence of pyrite in the Ancholme Clay was confirmed by chemical testing recording total sulfur contents of up to 2.6% S.

The rapidity of pyrite oxidation following exposure and subsequent deterioration of the mudrock following a period of heavy rain had not been anticipated (Czerewko and Cross, 2015). **Figure 37.4** shows a formerly flat-rolled surface that, due to drying in the sun, developed polygonal desiccation cracks and extensive white efflorescence.

A 600 mm flexible inert granular capping to the subgrade and benched side slopes were constructed to protect against any long-term effects from ongoing pyrite oxidation processes. The construction programme was managed to avoid exposure of prepared surfaces to bouts of inclement weather. The section of road was opened in 2006 and has performed satisfactorily since.

The second case study concerns the A46 in Nottinghamshire. Between March 2009 and the summer of 2012, a 28 km dual carriageway section of the A46 Newark to Widmerpool trunk road was constructed to improve road links between the East and West Midlands. A 7 km section of this route was located over pyritic Lower Lias Mudstone deposits of Jurassic age (Czerewko *et al.*, 2011).

Following the preliminary phases of ground investigation, the pavement was designed in accordance with the Highways Agency's Interim Advice Note 73/06 (Highways Agency, 2009). It was initially concluded that a capping layer would not be required. However, concerns arose over the potential for instability across the pyrite-bearing Liassic mudrock section near Cotgrave, Nottinghamshire, where gently undulating ground with low-lying rounded summits showed signs of slope movement and groundwater discharge. Ground investigations recorded weak, dark grey, calcareous mudstone, becoming stiff brown calcareous clay where weathered. The fabric of the weathered clay was observed to be closely fissured and friable, and the mudstone was closely jointed, with development of extensive shallow slip planes which had accommodated progressive gypsum crystal growth. The clay and mudstone beds contained calcareous nodules and fossil debris with occasional pyrite nodules, cubic pyrite crystal clusters and pyritised fossil fragments, together with occasional thin limestone beds. Discrete gypsum crystals and clusters up to 80 mm in size were recovered from moist joint surfaces in trial pits.

Czerewko *et al.* (2011) describe detailed laboratory studies and field weathering trials that were carried out to assess the implications to embankment construction and the stability of cut slopes. Field weathering trials showed that rapid pyrite oxidation and mudrock degradation (**Figure 37.10**) presented a high likelihood for ground heave problems.

Construction proceeded with pre-conditioning of the clay by placing it in loose stockpiles prior to compaction. The loose stockpiling encouraged pyrite oxidation, thereby mitigating much of the heave potential. Water ingress was managed with a smooth-rolled finish and construction interface. The construction sequence was managed to optimise material quality and a final cover system was engineered to inhibit water entry and limit pavement deterioration. A 300 mm thick protective layer of inert granitic aggregate was placed over the subgrade to facilitate construction traffic movements. By limiting capillary rise at the interface, this prevented both deterioration of the earthwork and the formation of gypsum crystals at the road interface. This layer was left in place in the final road construction as it would also function as a flexible strain mitigation barrier. To date there have been no signs of deterioration and the pavement remains in a satisfactory condition.

Figure 37.10 Rapid degradation and growth of gypsum crystals in pyritic mudstone following exposure (Lower Lias Clay on A46 trunk road) (Reid *et al.*, 2017)

These case studies illustrate how earthworks can be successfully constructed using pyritic materials provided the potential problems are identified and accommodated in the design.

37.6.5 Heave of stabilised soils

The treatment of soils with binders such as lime, PFA, slag or cement is often carried out to improve the geotechnical properties of the materials. This may simply involve drying out wet soils to make them suitable for use as bulk fill materials, but also includes creating stronger layers in situ for use as capping or as an unbound subbase. This topic is covered in more detail in the *Design Manual for Roads and Bridges* HA74/07, 'Treatment of fill and capping materials using either lime or cement or both' (Highways Agency *et al.*, 2007) and in the Britpave guide, *Stabilisation of Sulfate Bearing Soils – Guidelines for Best Practice* (Britpave, 2019).

Reactions between the binders and the soil result in the formation of cementitious compounds in a strongly alkaline environment at pH values higher than 12. At these high pH values, sulfates in the soil react with the cementitious compounds to form a highly expansive calcium sulfo-aluminate hydrate called ettringite ($3CaO.Al_2O_3.3CaSO_4.31H_2O$) and gypsum; these minerals also cause expansive attacks on concrete (Section 37.6.1). This is compounded because high pH conditions accelerate the oxidation of pyrite, thus generating even more sulfate. The reactions often do not occur immediately, but several months later, possibly due to seasonal ingress of water, by which time the resulting heave of the stabilised layers cause disruption to the pavement or building constructed there. Usually, the remediation is expensive and disruptive to carry out, given the frequent need to replace the expansive material.

Guidance on how to assess materials for potential swelling is given in HA 74/07 (Highways Agency *et al.*, 2007) for bulk fill and capping applications and Britpave BP/51 (Britpave, 2019). Rather than setting specific limiting values, engineers are required to carry out trial mixtures and measure the expansion caused by soaking them for 28 days. Materials that are particularly susceptible to these reactions include overconsolidated clays, superficial deposits derived from them and colliery spoil (see Section 37.4).

Where soil is to be stabilised in situ to form a hydraulically bound subbase layer, a limit of 0.25% as the SO_4 content for the total potential sulfate (TPS) is set in the Specification for Highway Works (Highways Agency *et al.*, 2020, Clause 840.4). This very low limiting value will exclude many commonly used construction materials.

Unlike use as bulk fill (Section 37.6.4), where many potential problems can be overcome, there are few options available for stabilisation if the material to be treated contains significant quantities of sulfates or sulfides. There is some evidence that the use of granulated ground blast furnace slag reduces swelling compared to cement or lime (Higgins *et al.*, 2002). Great care should be taken when assessing the potential for stabilisation of clay soils to ensure that they will not be susceptible to swelling.

Further examples of the problems that can arise when soils containing pyrite or gypsum are stabilised are given by Cripps *et al.* (2019). A case study involving the A1(M) in Cambridgeshire is described by Czerewko and Cross (2015).

Traditionally road construction inadvertently avoided the possibility of these problems by using a granular capping layer beneath the pavement, which would arrest the upward migration of moisture and absorb stresses associated with precipitation of pyrite oxidation products. Revision to the guidance and changes to construction practice in the last few decades with emphasis on increasing the use of site-won materials has driven this change. However, as this case study and those for the A63 and A46 in Section 37.6.4.2 illustrate, where the subgrade consists of weak, weathered, pyritic mudrocks or overconsolidated clays, a granular capping layer has provided a successful solution (Czerewko and Cross, 2015).

Recycled, manufactured and secondary aggregates are sometimes used in higher strength HBMs such as stabilised subbase. These materials can sometimes contain significant quantities of sulfates and sulfides, which can react with cement to produce ettringite and thaumasite. This can be exacerbated where the materials also contain aluminium, which is commonly found in IBAA. As with stabilised soils, these reactions do not take place immediately but cause progressive deterioration of the HBM over time. Where it is proposed to use such materials, their chemistry and potential short- and long-term reactions with cement should be investigated to prevent such deterioration from occurring.

37.6.6 Heave of floor slabs

Heave of floor slabs can occur due to precipitation of gypsum in the underlying fill material (**Figure 37.11**). This can occur where the underlying material contains high sulfate or sulfide levels and there is a high water table. The building prevents surface infiltration and therefore drying occurs in the underlying fill. If the interior of the building is heated, groundwater is drawn into the fill material below the floor slab by capillary action. The water evaporates, leading to crystallisation of gypsum in the void spaces in the fill material. Ultimately, the void spaces become filled and further crystallisation leads to expansion of the fill and heave of the overlying floor slab. Crystallisation can also occur within and between particles before the void space is totally filled (Matheson and Jones, 2015), which generates very high swelling pressures. Besides heave and disruption of the fill constituents, ground-bearing elements and the floor slab may be uplifted (Czerewko and Cripps, 2022).

This type of problem is particularly common where buildings are founded on mudrocks or overconsolidated clays that

Figure 37.11 Cracking of the concrete floor slab of a factory caused by 80 mm sulfate-induced ground heave (Reid *et al.*, 2017)

contain significant concentrations of sulfides and sulfates. The disturbance due to construction works allows air to oxidise some of the sulfides, the products of which may then be transported by capillary action to the fill immediately below the floor slab, which is uplifted due to the expansive formation of gypsum. Examples involving both fills and bedrock from the UK have been reported by Hawkins and Pinches (1987) and Wilson (1987).

While many incidents have been due to construction directly on sulfate- and sulfide-rich strata with a high water table, similar problems can also occur if fill/hardcore materials for use below floor slabs contain significant quantities of sulfates, sulfides or other expansive minerals. This is illustrated by the problems with numerous houses and other buildings in the Dublin area in the 2000s, where pyritic mudrock was widely used as fill below floor slabs (Czerewko and Cripps, 2022; Hawkins, 2014; Tuohy *et al.*, 2012). Investigations revealed that locally sourced calcareous mudstone containing pyrite had been used as fill beneath the cast in situ concrete floor slabs of these structures. Affected structures typically suffered domed uplift of floors, sometimes with extensive distortion and diagonal cracking of walls, particularly at the corners of doorway and window openings (Czerewko and Cripps, 2022; Hawkins and Stevens, 2014). The expansion is progressive, such that damage to the structure occurs some years after construction and continues after any damage has been repaired. A Government of Ireland expert panel (Tuohy *et al.*, 2012) noted that 30,000–70,000 homes were likely to be affected by heave, with the only practical solution being the highly disruptive and expensive procedure of complete removal and replacement of the pyritic fill with non-expansive material.

Guidance on the identification of fills likely to cause expansion is given by Matheson and Quigley (2016), whereas guidance on the selection of fill for supporting cast concrete floor slabs is provided by BRE Digest 522 (Longworth,

2011a, 2011b), in which methods of analysis developed by Reid *et al.* (2005) are specified for the evaluation of materials. These test methods have now been incorporated in BS 1377-3:2018+A1:2021.

37.7. Discussion: lessons learned

Reid and Cripps (2019) discuss the geochemical lessons learned from the reconstruction of Carsington Dam, Derbyshire, UK. The paper covers the period from the investigation of the failure of the original embankment in 1984 and includes the site investigation and design for the reconstruction of the embankment, the construction works and subsequent monitoring of the performance of the new embankment, and covers many of the issues raised in the previous sections. The lessons learned are intended for application to all civil engineering projects where geochemistry is an issue.

Lesson 1: Geochemistry matters

As illustrated by **Figure 37.1**, geochemistry can be a matter of life and death. Failure to understand the geochemistry during the construction of the original Carsington Dam led to four deaths. Geochemistry is not often a case of life and death, but it can cause major problems during and after construction, leading to delays, increased costs and reputational damage for those found to be responsible. One of the saddest features is that the same problems tend to occur again and again, indicating that lessons have not been learned (Czerewko and Cripps, 2022). Perhaps the most obvious recent example was the use of low-quality calcareous mudstone containing pyrite as fill below floor slabs in numerous housing developments in Dublin and the resulting problems of heave and cracking in the buildings, as discussed in Section 37.6.6. Consideration of the geochemistry at an early stage can avoid these problems.

Lesson 2: If you don't know what is happening, ask someone who does

Geochemical reactions can be confusing even for expert geologists and geochemists, so it is not surprising that civil engineers are often confused by them. Although understanding the reactions may involve the use of experts and detailed geochemical investigations and testing, it may simply involve having a discussion with someone in the organisation or project team who has experience of the materials and knows the potential problems. This can be problematic if teams are fragmented, working in different offices or on their own from home, and not familiar with the local materials involved on a project.

Lesson 3: Use your eyes; geochemical reactions often leave visible traces

Site staff undertaking site investigations or construction works should be aware of the potential signs of chemically active strata. These include: white efflorescence on dried-up puddles, spoil heaps or trial embankments (**Figure 37.4**); iron staining and ochreous precipitates in pools at the base of spoil heaps or embankments or in drainage waters (**Figure 37.5**); crystals of gypsum in strata exposed in trial pits and excavations (**Figure 37.10**); and extensive brown, yellow or green staining on joint and fissure surfaces in near-surface strata due to formation of alteration products of pyrite such as iron hydroxides, jarosite and alunite that are often accompanied by a sharp, acrid, vitriolic smell (**Table 37.1**). These observations should be followed up with targeted testing of waters and the relevant materials to confirm what reactions are taking place.

Trial embankments are particularly useful in this regard, as are demonstration sections of HBMs if they can be left intact for a reasonable length of time to allow any reactions to take place. Even leaving the spoil heaps from trial pits or recovered block samples exposed for a few weeks will reveal much about their susceptibility to both chemical and physical weathering.

Lesson 4: Integrate geochemistry into the design process

Many geochemical problems arise because the reactions were not anticipated at the design stage and the sampling and testing for geochemical issues was not sufficiently detailed to pick up potential problems. Where the geochemical risks are properly understood, as in the reconstruction of Carsington Dam, they can be allowed for in the design and the construction completed successfully (Reid and Cripps, 2019).

Lesson 5: Correct storage of samples is essential

Sulfide and sulfate minerals in soils and weak rocks can potentially be unstable when they are removed from the ground and exposed to different conditions of temperature, exposure to air and water. In aerated and moist conditions unstable sulfides can rapidly oxidise to sulfate, whereas in anaerobic conditions sulfates can be reduced to sulfides (Hawkins and Wilson, 1990). There is no point in taking samples and carrying out chemical tests if the results are misleading because the samples were not stored correctly. Guidance on storage of samples is given in BS 1377-3:2018+A1:2021.

Lesson 6: Understand the test methods and use the ones that are appropriate for the situation

Test methods for chemical species can be complex, but if the wrong ones are used the results can be misleading. Slight changes in the methods can have a significant effect on the results and choosing the wrong method can give misleading results (Cripps *et al.*, 2019; Czerewko *et al.*, 2016; Reid *et al.*, 2017).

Lesson 7: Monitor outcomes whenever possible

Monitoring for geochemical reactions may involve regular chemical analyses of drainage waters, as at Carsington and Roadford Dams (Reid *et al.*, 2005), or visual inspection for signs of chemical reactions or heave. Routine monitoring gives reassurance that conditions remain within the bounds anticipated at the design stage. It also enables any problems that do occur to be picked up at an early stage and correctly diagnosed. Compared with geotechnical monitoring, geochemical monitoring is relatively straightforward and inexpensive and can be increased or decreased over time in response to observed conditions. Unfortunately, adherence to standards and recommended limits on the composition of materials does not guarantee that problems will not arise, as standards apply to an assumed set of circumstances that may not apply in all situations.

37.8. Conclusions

This chapter has indicated how sulfur compounds can be a hazard in geotechnical engineering works, potentially leading to corrosion of concrete and steel, clogging of drains, loss of shear strength, heave of pavements and floor slabs and pollution of watercourses. These hazards can be avoided if they are recognised at an early stage and appropriate measures taken, either to adopt mitigating measures or by avoiding the use of unsuitable materials. Guidance is available to deal with most of the situations discussed; the main concern is for engineers to be aware of the potential dangers and to ensure that they are assessed properly during the ground investigation and design stages of the project. This does not just involve carrying out laboratory tests and comparing the results against various sets of limiting values, but understanding the materials and how they will behave when used in the proposed engineering works.

References

BRE (Building Research Establishment) (2005) *Concrete in Aggressive Ground*. BRE Special Digest 1. BRE, Watford, UK.

Britpave (2019) *Guidelines for Stabilisation of Sulfate Bearing Soils*. Britpave Guide BP/51. Britpave, Wokingham, UK.

BSI (British Standards Institution) (2006) BS EN 14475:2006: Execution of special geotechnical works – Reinforced fill. BSI, London, UK.

BSI (2007a) BS EN 13242+A1:2007: Aggregates for unbound and hydraulically bound materials for use in civil engineering work and road construction. BSI, London, UK.

BSI (2007b) BS EN 1997-2:2007: Eurocode 7 – Geotechnical design – Part 2: Ground investigation and testing. BSI, London, UK.

BSI (2009) NA to BS EN 1997-2:2009: UK National Annex to Eurocode 7 – Geotechnical Design – Part 2: Ground investigation and testing. BSI, London, UK.

BSI (2012) BS EN 1744-1:2009+A1:2012: Tests for chemical properties of aggregates – Part 1: Chemical analysis. BSI, London, UK.

BSI (2016) BS 8006-1:2010+A1:2016: Code of practice for strengthened/reinforced soils and other fills. BSI, London, UK.

BSI (2020) BS 5930:2015+A1:2020: Code of practice for ground investigations. BSI, London, UK.

BSI (2021) BS 1377-3:2018+A1:2021: Part 3: Chemical and electrochemical tests. BSI, London, UK.

Chalmers RWC, Vaughan PR and Coats DJ (1993) Reconstructed Carsington Dam: design and performance. *Proceedings of the Institution of Civil Engineers: Water, Maritime and Energy* **101**: 1–16.

Cripps JC, Reid JM, Czerewko MA and Longworth I (2019) Tackling problems in civil engineering caused by the presence of pyrite. *Quarterly Journal of Engineering Geology and Hydrogeology* **52**: 481–500, 10.1144/qjegh2019-024.

Czerewko MA and Cripps JC (2022) Investigation of destructive ground heave attributed to pyritic fill affecting new-build properties in the Dublin area of Ireland. *Engineering Geology* **299**: 106534, 10.1016/j.enggeo.2022.106534.

Czerewko MA and Cross SA (2015) The benefits of a granular interface over pyritic subgrade. *Proceedings of the Institution of Civil Engineers: Geotechnical Research* **2(3)**: 97–122, 10.1680/jgere.15.00003.

Czerewko MA, Cross SA, Dumelow PG and Saadvandi A (2011) Assessment of pyritic Lower Lias mudrocks for earthworks. *Proceedings of the Institution of Civil Engineers: Geotechnical Engineering* **164**: 59–77, 10.1680/geng.2011.164.2.59.

Czerewko, MA, Longworth I, Reid JM and Cripps JC (2016) Standardized terminology and test methods for sulphur mineral phases for the assessment of construction materials and aggressive ground. *Quarterly Journal of Engineering Geology and Hydrogeology* **49(3)**: 245–265, 10.1144/qjegh2016-010.

Davies SE and Reid JM (1997) Roadford dam: geochemical aspects of construction of a low grade rockfill embankment. *Proceedings of the International Conference on the Implications of Ground Chemistry/Microbiology for Construction, Bristol, UK* (Hawkins AB (ed)). Balkema, Rotterdam, the Netherlands, pp. 111–131.

Gribble CD (2005) *Rutley's Elements of Mineralogy*, 27th edn. CBS Publishers and Distributors, New Delhi, India.

Hawkins AB (1998) Engineering significance of ground sulfates. In *Geotechnical Site Characterisation* (Robertson PK and Mayne PW (eds)). Balkema, Rotterdam, the Netherlands, pp. 685–692.

Hawkins AB (2014) Engineering implications of the oxidation of pyrite: an overview, with particular reference to Ireland. In *Implications of Pyrite Oxidation for Engineering Works* (Hawkins AB (ed)). Springer, Cham, Switzerland, pp. 1–98.

Hawkins AB and Pinches GM (1987) Cause and significance of heave at Llandough Hospital, Cardiff: a case history of ground floor heave due to pyrite growth. *Quarterly Journal of Engineering Geology* **20**: 41–58, 10.1144/GSL.QJEG.1987.020.01.05

Hawkins AB and Stevens M (2014) Problems with the use of pyritiferous fill at Ballymun youth facility, Dublin. In *Implications of Pyrite Oxidation for Engineering Works* (Hawkins AB (ed)). Springer, Cham, Switzerland, pp. 133–166.

Hawkins AB and Wilson SLS (1990) Sulphate increase in laboratory prepared samples. *Quarterly Journal of Engineering Geology and Hydrogeology* **23**: 383–385.

Higgins DD, Thomas D and Kinuthia J (2002) Pyrite oxidation, expansion of stabilised clay and the effect of GGBS. *Fourth European Symposium on the Performance of Bituminous and Hydraulic Materials in Pavements*, Nottingham, UK, pp. 161–168.

Highways Agency (2009) *Interim Advice Note 73/06: Design Guidance for Road Pavement Foundations (Draft HD25)*. The Stationery Office, London, UK.

Highways Agency, Transport Scotland, Welsh Assembly Government and the Department for Regional Development Northern Ireland (2007) Treatment of fill and capping materials using either lime or cement or both. HA 74/07. In *Design Manual for Roads and Bridges*. The Stationery Office, London, UK.

Highways England, Transport Scotland, Welsh Government and the Department for Infrastructure Northern Ireland (2016a) *Specification for Highway Works*. The Stationery Office, London, UK.

Highways England, Transport Scotland, Welsh Government and the Department for Infrastructure Northern Ireland (2016b) *Notes for Guidance for the Specification for Highway Works*. The Stationery Office, London, UK.

Highways England, Transport Scotland, Welsh Government and the Department for Infrastructure Northern Ireland (2020) CD 375 Revision 1: Design of corrugated steel buried structures. In *Design Manual for Roads and Bridges*. The Stationery Office, London, UK.

Longworth I (2011a) *Hardcore for Supporting Ground Floors of Buildings: Part 1 – Selecting and Specifying Materials*. BRE Digest DG522 Part 1. BRE, Watford, UK.

Longworth I (2011b) *Hardcore for Supporting Ground Floors of Buildings: Part 2 – Placing Hardcore and the Legacy of Problem Materials*. BRE Digest DG522 Part 2. BRE, Watford, UK.

Matheson GD and Jones GLI (2015) The habit and form of gypsum crystals in Irish mudstone aggregate affected by pyrite-induced swelling. *Quarterly Journal of Engineering Geology and Hydrogeology* **48**: 167–172, 10.1144/qjegh2014-106.

Matheson GD and Quigley P (2016) Evaluating pyrite-induced swelling in Dublin mudrocks. *Quarterly Journal of Engineering Geology and Hydrogeology* **49**: 47–66, 10.1144/qjegh2014-103.

McCabe BA, McKeon ÉP, Virbukiene RJ, Mannion PJ and O'Connell AM (2015) Pyritiferous mudstone-siltstone: expansion rate measurement and prediction. *Quarterly Journal of Engineering Geology and Hydrogeology* **48(1)**: 41–54, 10.1144/qjegh2013-067.

Mineral Products Association (2021) *Profile of the UK Mineral Products Industry 2020*. Mineral Products Association, London, UK.

Pye EK and Miller JA (1990) Chemical and biochemical weathering of pyritic mudrocks in a shale embankment. *Quarterly Journal of Engineering Geology and Hydrogeology* **23**: 365–381, 10.1144/GSL.QJEG.1990.023.04.11.

Reid JM and Avery K (2009) Measuring the potential reactivity of sulfides in aggregates. Final project report for MIST Project MA/7/G/5/001.

Reid JM and Cripps JC (2019) Geochemical lessons from Carsington Dam failure of 1984 and reconstruction. *Quarterly Journal of Engineering Geology and Hydrogeology* **52(4)**: 414–424, 10.1144/qjegh2018-184.

Reid JM, Czerewko MA and Cripps JC (2005) *Sulfate Specification for Structural Backfills*. TRL447. TRL, Wokingham, UK.

Reid JM, Czerewko MA, Longworth I and Cripps JC (2017) Sulfur compounds in soils and rocks: avoiding pitfalls in terminology, test procedures and ground assessment. *Ground Engineering*, Jan./Feb.: pp. 30–36.

Sandover BR and Norbury DR (1993) Technical note: on the occurrence of abnormal acidity in granular soils. *Quarterly Journal of Engineering Geology* **26(2)**: 149–153, 10.1144/GSL.QJEG.1993.026.02.06.

St. John TW (2020) Geotechnical characterization of sulfur species in UK Jurassic mudrocks. *Quarterly Journal of Engineering Geology and Hydrogeology* **53(4)**: 598–608, 10.1144/qjegh2019-148.

Tiller AK (1997) An overview of the factors responsible for the degradation of materials exposed to the marine/estuarine environment. *Proceedings of the International Conference on the Implications of Ground Chemistry/Microbiology for Construction, Bristol, UK* (Hawkins AB (ed)). Balkema, Rotterdam, the Netherlands, pp. 337–353.

Tuohy B, Carroll N and Edger M (2012) *Report of the Pyrite Panel*. https://www.pyriteboard.ie/Pyrite/media/Pyrite/Updated/Report-of-Pyrite-Panel-June-2012.pdf (accessed 09/06/2023).

West G and O'Reilly MP (1986) *An Evaluation of Unburnt Colliery Shale as Fill for Reinforced Earth Structures*. TRL Research Report RR97. TRL, Wokingham, UK.

Wilson AC and Evans JD (1990) The use of low grade rockfill at Roadford Dam: the embankment dam. *Proceedings of the 6th Conference of the British Dam Society, Nottingham, UK*. Thomas Telford, London, UK, pp. 21–27.

Wilson EJ (1987) Pyritic shale heave in the Lower Lias at Barry, Glamorgan. *Quarterly Journal of Engineering Geology* **20**: 251–253, 10.1144/GSL.QJEG.1987.020.03.07.

Winter MG, Butler AM, Brady KC and Stewart WA (2002) Investigation of corroded stainless steel reinforcing elements in spent oil shale backfill. *Proceedings of the Institution of Civil Engineers: Geotechnical Engineering* **155**: 35–46, 10.1680/geng.2002.155.1.35.

Further reading

Czerewko MA and Cripps JC (2015) Investigation of destructive ground heave affecting new-build properties constructed using pyrite bearing fill in the Dublin area of Ireland. *Proceedings of the XVI European Conference on Soil Mechanics and Geotechnical Engineering, Edinburgh, UK*. Thomas Telford, London, UK, pp. 2361–2366.

Czerewko MA and Cripps JC (2023) Implications of sulfur mineralogy and consequences of pyrite oxidation for ground engineering. *Journal of the Geological Society*, 10.1144/jgs2022-101.

Hawkins AB (2014) *Implications of Pyrite Oxidation for Engineering Works*. Springer, Cham, Switzerland.

Useful websites

Standards for Highways: https://www.standardsforhighways.co.uk

It is recommended this chapter is read in conjunction with

- Chapter 7 *Geotechnical risks and their context for the whole project*
- Chapter 36 *Mudrocks, clays and pyrite implications*
- Chapter 40 *The ground as a hazard*
- Chapter 48 *Geo-environmental testing*
- Chapter 49 *Sampling and laboratory testing*

All chapters in this book rely on the guidance in Sections 1 *Context* and 2 *Fundamental principles*. A sound knowledge of ground investigation is required for all geotechnical works, as set out in Section 4 *Site investigation*.

Chapter 38
Soluble ground

Clive Edmonds Engineering geologist, Subsolve Ltd and Stantec (UK) Ltd, Reading, UK
Tony Waltham Engineering geologist, Nottingham, UK

Principal soluble rock types include limestone, chalk, gypsum and salt, which can dissolve in natural water, which over various time spans – depending upon solubility – can create cavities in the ground. These cavities, which are characteristic features of karst terrains, vary enormously in scale, density and infilling (air/water/soil/rock). They can be the cause of ground failure and surface subsidence which can pose risks to people, structures and infrastructure. The dominant karst geohazard is the development of new sinkholes within soil profiles overlying soluble rocks, where the rock remains stable but either soil is washed downwards into open cavity features, or an existing metastable cavity feature infill is destabilised, usually by the action of inundation. Most sinkholes are induced by water leakage into the ground or by loading/vibration associated with engineering activity. Careful controls are required to prevent triggering instability. Karstic geohazards pose particular problems to the safe construction of foundations and surface water drainage disposal. The nature of karst weathering affecting ground conditions can be difficult to assess, but the broad characteristics of karst are now well known and should be appreciated in practice. The availability of karst databases and case histories can greatly assist with the initial assessment of soluble ground geohazards at sites.

doi: 10.1680/icemge.66816.0577

CONTENTS

38.1. Introduction

Naturally occurring soluble rocks include limestones (including dolomite and chalk), gypsum (also anhydrite) and salt (halite), in order of decreasing abundance and increasing solubility. Exposure of these soluble rocks to rainwater, stream water and groundwater commonly leads to the development of subsurface cavities, together with associated risks of collapse and ground subsidence. It is common for soluble rock terrains to be mostly dry at the surface because the passage of water has been diverted underground through solution-widened joints, fissures, bedding planes and caves. This leads to the creation of a distinctive suite of landforms (e.g. dry valley features, limestone pavements) known as karst. Karst terrains pose varying degrees of geohazard ranging from slow ground subsidence to catastrophic collapses and destructive sinkhole development.

38.2. Soluble ground and karst

Soluble rocks present a variety of geohazards where they lie within a few tens of metres of the ground surface (and sometimes also at greater depths). Direct dissolution of the rock is only rapid enough to undermine a built structure within its lifetime (assuming a 50–100-year design life) in the case of salt, and to a lesser extent gypsum. Notable loss of ground by salt/gypsum dissolution would take place only where it is exposed to a significant flow of aggressive water. For strong limestones and weaker limestones, such as chalk, dissolution by comparison is much slower, taking many thousands of years to impact on a structure founded upon it. The hazards for limestone and gypsum are created by open fissures and cavities that are dissolved out of the bedrock over geological timescales. The chalk surface can be similarly impacted by dissolution penetrating to depth along joint intersections. The main hazards relate to soil-filled features within solution-widened fissures penetrating the buried rockhead. Viewed in plan these can be linear in nature or be formed at point locations, often in clusters. In chalk such features are usually referred to as solution pipes. These karst features commonly form very difficult ground for founding structures because of the subsidence risk posed. The soil-filled karst features are prone to destabilisation by water inundation, piping and erosion processes that can trigger a sinkhole to occur in timescales ranging from hours to days. However, by comparison, collapses of rock over open caves are rare.

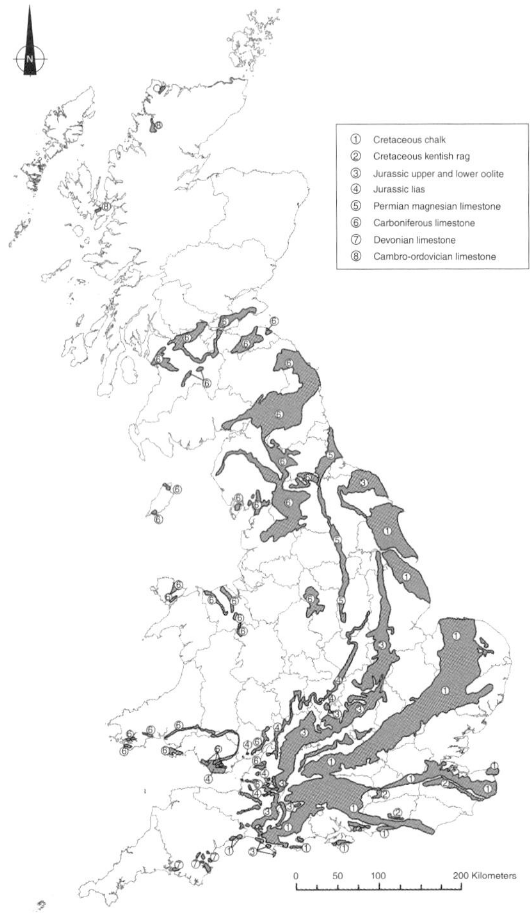

Figure 38.1 The schematic distribution of limestones in Britain that are prone to karst geohazards and where recognisable karst landscapes are only widely developed on the cavernous limestones and to some degree upon the chalk outcrop

38.3. Distribution of soluble rocks

Soluble rocks are widely distributed in nearly all sedimentary sequences around the world, with limestones and varieties of carbonate rocks being the most widespread, although the scale of karst development varies. The geology of Britain exhibits a range of limestones that pose a variety of scales and types of geohazard (**Figure 38.1**). It is notable in Britain that only some of the limestones are old, strong and cavernous, and have widespread karst geohazards; however, their impacts upon construction are limited because spatially they tend to occur in more upland areas away from population centres, where development is restricted by landscape protection designation (often associated with National Parks and Areas of Outstanding Natural Beauty). The other weaker limestones rarely cause major difficulties for construction, except for chalk. Chalk is a special case, where not only does rock strength and weathering present engineering challenges (Lawrence *et al.*, 2018; Lord *et al.*, 2002), but also karstic dissolution features and ground subsidence (Edmonds, 2008, 2018) can have particular impacts upon the design, construction and longevity of urban development and infrastructure. Worldwide, limestone karst causes the most extensive engineering difficulties in huge swathes of southern China, in large parts of the eastern USA and in the Dinaric karst that mainly extends from northeast Italy through Slovenia, Croatia, Bosnia and Herzegovina, Montenegro and Albania.

Gypsum is far less widespread than limestone. In Britain (**Figure 38.2**) limited outcrops occur in the Midlands and Northern England within Permian and Triassic strata, their presence being under-represented when comparisons are made worldwide. The American Midwest and Ukraine are the two large regions most impacted by gypsum karst. Salt is of even more restricted extent, both in Britain (**Figure 38.2**) and worldwide, but its geohazard impacts upon development are well known in Britain and America, although it is the Middle East of Southwest Asia that has the largest areas directly underlain by salt.

38.4. Influences on the geohazard scale of limestone karst

Dissolution causes the most conspicuous karst features in pure, strong limestones (unconfined compressive strength, UCS $>$ 70 MPa) where extensive fractures can be enlarged to create fissures and caves between blocks of intact strong rock. Weaker and softer limestones, such as England's Cotswold oolites, are more porous, so have more diffuse groundwater flows through micro-fissures. Chalk also has diffuse flows, but much of its groundwater is transmitted through open and solution-widened fissures, and minor caves can occur (Edmonds, 2008; Lord *et al.*, 2002; Maurice *et al.*, 2021). On a broad scale, the age of the limestone is irrelevant: Carboniferous limestones in England, Jurassic limestones in France and Tertiary limestones in Southeast Asia have identical engineering properties. Dolomites and dolomitic limestones are rather less prone to dissolution, so generally have a reduced scale of karst features. Larger cavities develop where streams of aggressive water enter the limestone from adjacent or overlying impermeable rock outcrops, so there is a tendency for increased karst development along the geological boundaries. Beyond that, rock structure and lithology influence cavity development, which is mainly at large fractures and chemically favourable inception horizons, and the guiding features can be recognised in most mapped cavities. However, the spatial distribution and exact positions of caves cannot be predicted with certainty, given the hugely variable structure of natural ground conditions.

Although defining the exact positions of buried karstic features without a surface expression usually requires some form of ground investigation, it has been found by research for chalk in England, at least, that it is possible to broadly zone the landscape into areas of higher and lower potential for the occurrence of karstic features. Study of spatial patterns of dissolution features has identified various causal factors that enable semi-quantitative predictions to be made concerning the likely occurrence of karstic features (Edmonds, 2001a, 2001b). The results of geohazard modelling and mapping can

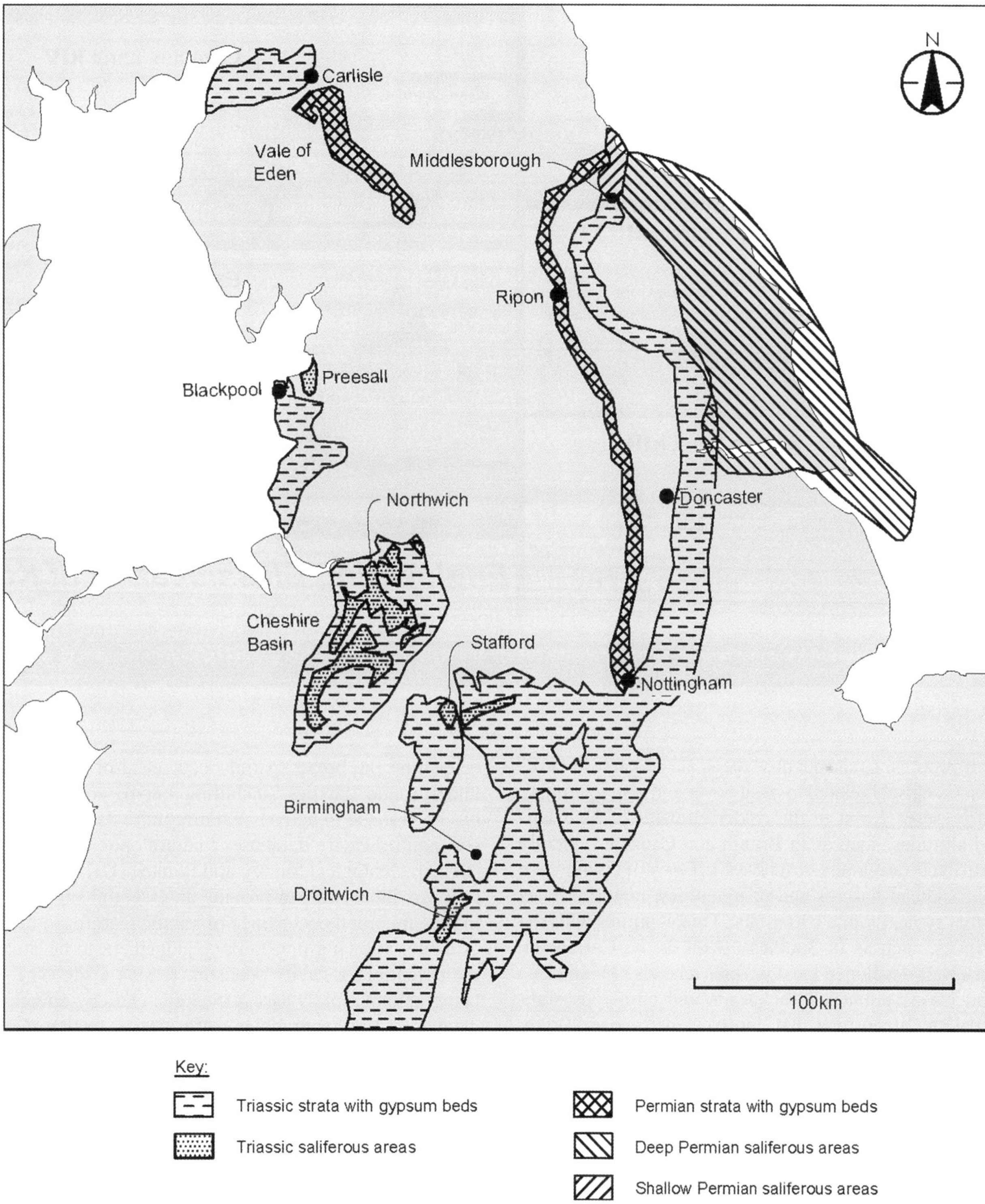

Figure 38.2 The schematic spatial distribution of gypsum- and salt-bearing strata in the Midlands and Northern England areas of Britain (Based on Cooper, 2020a, 2020b)

be used to guide the design of ground investigation layouts and intensity of investigation effort.

The scale of karst development varies considerably worldwide, and the broad classification of karst ground conditions (classes) is usefully illustrated by Waltham and Fookes (2003; **Figure 38.3**). Limestone dissolution in water is dependent on carbon dioxide to create soluble bicarbonate ions, and most dissolved carbon dioxide is derived from biogenic sources

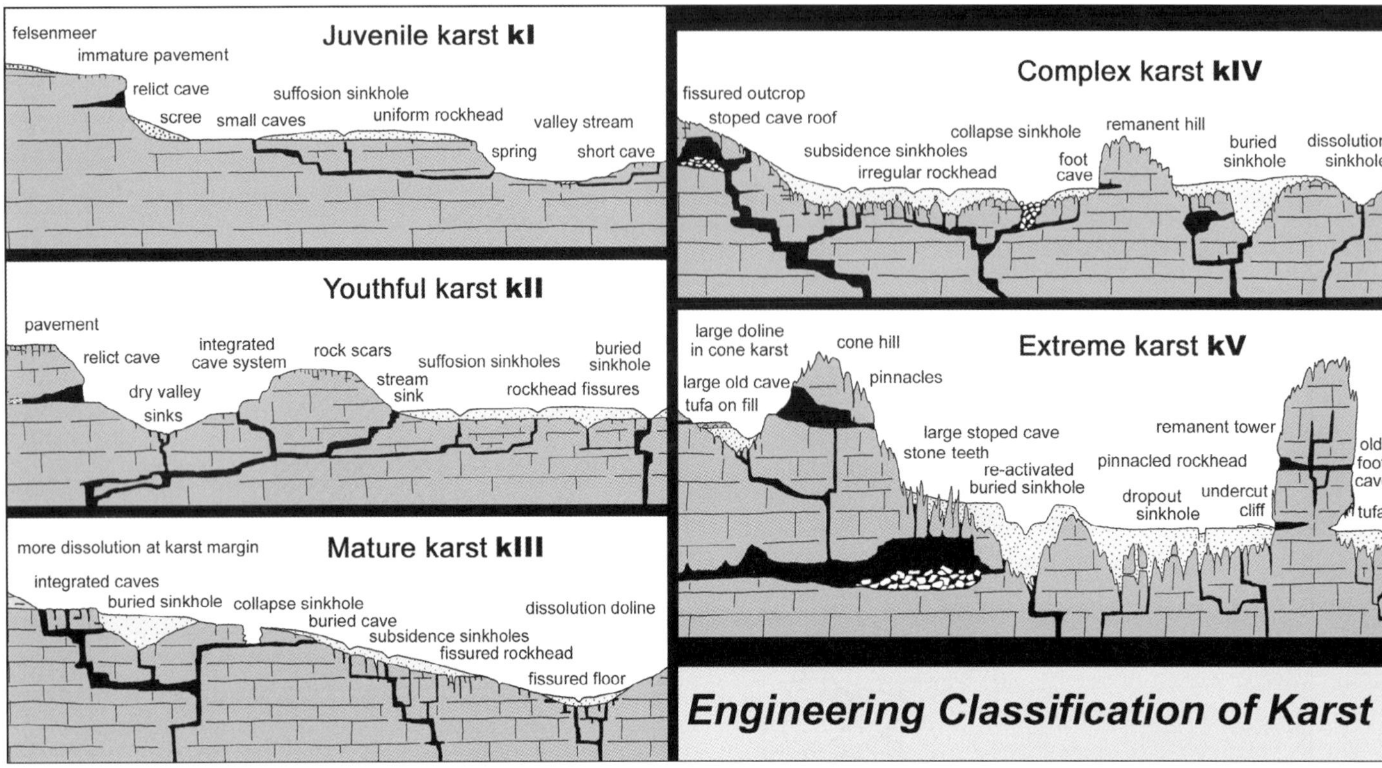

Figure 38.3 The five classes of karst that broadly demonstrate the variety and scale of landforms and ground conditions relevant to engineering (Modified from Waltham and Fookes, 2003)

within soil profiles. Consequently, the scale of karst features (karst class) is closely related to plant cover and also to climate (past and present). Karst in the colder climates of high latitudes and altitudes, such as in Britain and Canada, typically shows restricted development (classes kII or kIII), with cave passages, rockhead fissures and relatively few new sinkholes forming that typically are <10 m dia. This contrasts with the humid tropics, such as in Southeast Asia or the Caribbean, where very well-developed karst terrains (classes kIV and kV) have giant caves, pinnacled rockheads and numerous sinkholes, all with dimensions that approach or exceed 100 m. Limestones in hot and dry environments, such as in Australia and the Middle East, have modern karst development restricted by low rainfall, but commonly have isolated large caves and other relict karst landforms from prior wetter climates during the Quaternary.

Assessment of karst for engineering purposes should take into account not only the limestone lithology and the karst class, but should also consider typical cave size, frequency of new sinkholes and rockhead relief (Waltham and Fookes, 2003). Sometimes, however, development site sizes and the budget availability may restrict what can be practically and cost-effectively achieved. In Britain, following the publication of a national study (Applied Geology Ltd, 1993) to determine the broad spatial occurrence of naturally formed underground cavities (including karst), considerable effort has been made to actively capture more data to create extensive natural cavity databases that are operated commercially (British Geological Survey and Stantec (UK) Ltd). Reference to these databases can provide an essential snapshot regarding the recorded occurrence of karstic features in the vicinity of a site, allowing preliminary judgement to be made about the potential for cavity occurrence upon a site. The value of being able to rapidly assess the karst class by referring to the databases in a geographical information system (GIS) environment can be appreciated with reference to **Figure 38.4**, particularly as the metadata for each cavity data point can be interrogated.

Although the natural environments dictate the overall scale of karst development, it is important to recognise that individual events of ground subsidence or collapse are commonly generated by man's activities that bring a sudden change to equilibrium situations. Engineering works that either remove or introduce water, or add an imposed load, all happen at faster rates than normal geological evolution. In a natural and undisturbed karst, ground movements and collapses do occur, but typically as isolated events perhaps separated by hundreds or thousands of years. However, ground movement events may

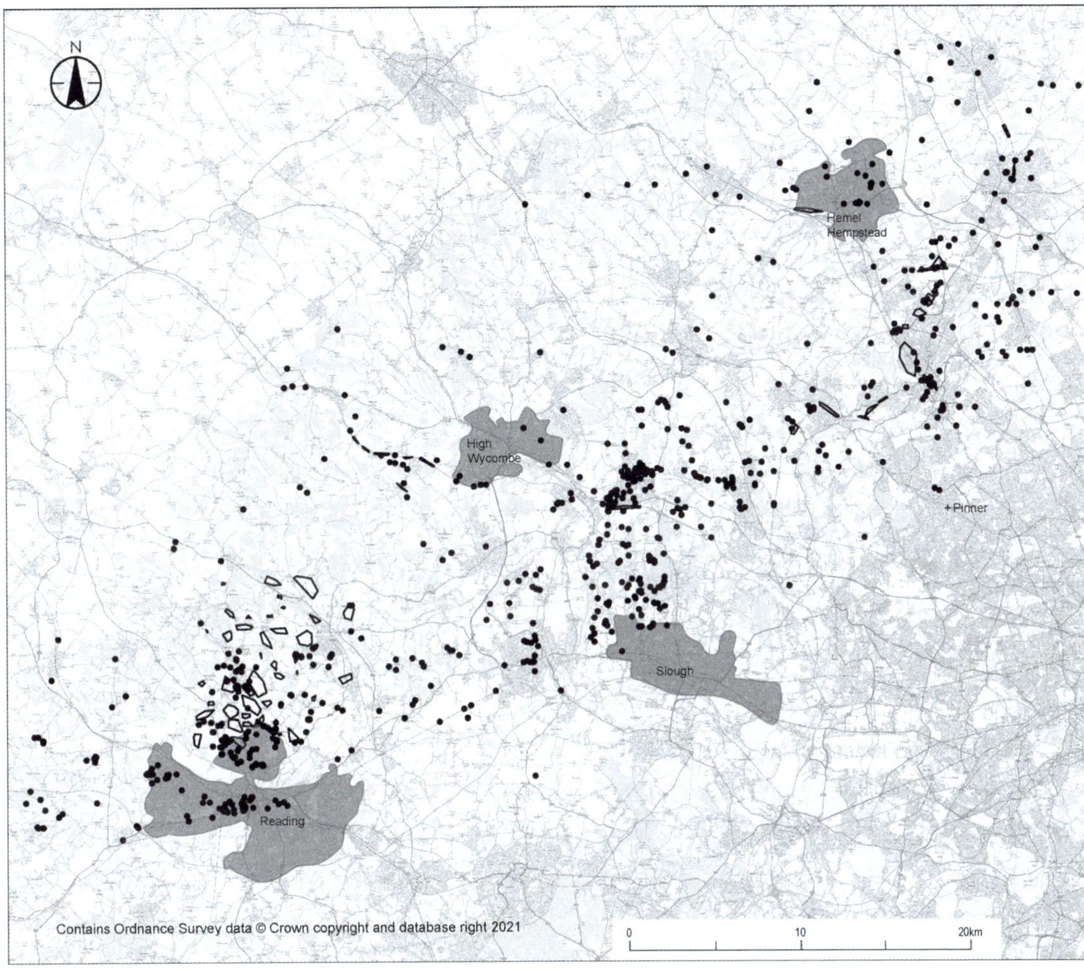

Figure 38.4 Map extract of recorded karstic features (shown as dots and polygons) for the south-west portion of the Chiltern Hills, southern England (Courtesy of Stantec (UK) Ltd)

be triggered prematurely by inappropriate engineering activity. Prevention of ground movement hazards requires appropriate precautions to avoid undue disturbance of the existing environment; these include thorough control of the drainage to avoid inundation settlement and soil loss, and the avoidance of excessive loads imposed on unstable rock or metastable karstic cavity infillings (Waltham, 2016).

38.5. Sinkhole geohazards upon limestones

The term sinkhole, meaning a closed depression, dominates American and engineering literature, although was originally known as a doline. Sinkholes can vary in size, ranging from 1 to 100 m in diameter and depth, and are the diagnostic landforms of karst terrains (Waltham *et al.*, 2005). The term implies that the ground has sunk, and water can sink into them too. Where visible streams (permanent or seasonal) sink into them, they are often referred to as swallow holes. It is important to recognise the various types of sinkholes (**Figure 38.5**).

The main karst geohazard is the formation of subsidence sinkholes in the soil or non-soluble rock cover over fissured and cavernous limestone rock (**Figure 38.6**), which can pose a geohazard to surface development and engineering works. Not only are these the most abundant, but new subsidence sinkholes can form by rapid movement of the soil cover and, therefore, have the greatest impact on engineered structures. The suffosion and dropout types differ due to the properties of their soil cover and, therefore, in their mode of failure – either the slow ravelling in a sandy soil or an arch failure into an undermined void in a more fine-grained soil. A soil-filled buried sinkhole constitutes an extreme form of rockhead relief, the soil infill often being soft and weak, and sometimes developing a shallow compaction sinkhole in the surface above. Collapse sinkholes form where the limestone cave roof collapses (**Figure 38.7**) and are found in most karst terrains, but the chances of a new failure are so low at any site that they mostly present a negligible risk to engineering. Large collapse sinkholes tend to have expanded in sequences of failure

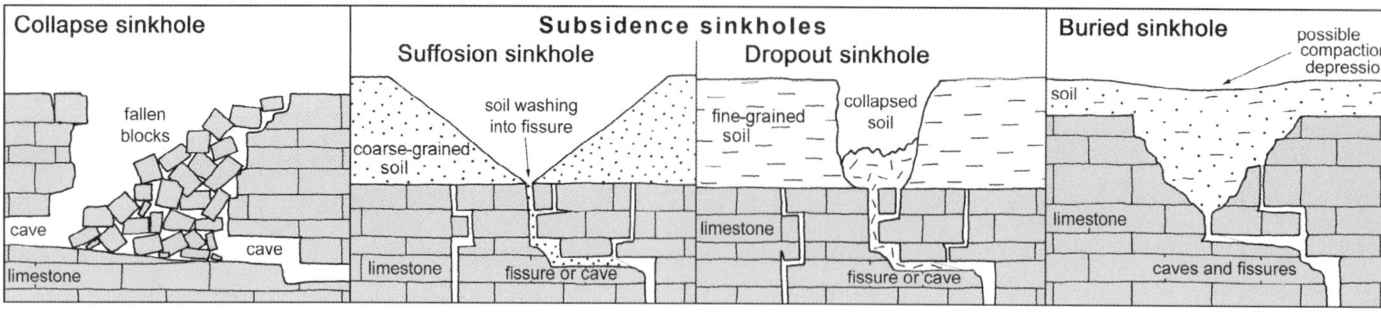

Figure 38.5 The main types of sinkholes developed in or above soluble rocks in karst terrains (Modified from Waltham and Fookes, 2003)

Figure 38.6 Example of a subsidence sinkhole with ravelling sides in Millstone Grit over Carboniferous-aged limestone strata in South Wales

events maturing through time, hence the risk of new massively destructive ground collapses is reduced. Sinkholes may also develop purely by rock dissolution, but these are long-term erosional features comparable to valleys in non-karst terrains, and have little engineering significance except that the ground beneath is likely to be more cavernous than adjacent ground.

Weaker and softer limestones such as chalk also develop sinkholes that pose a significant engineering hazard to surface development, and while they are similar to those described above for harder limestones, they have their own particular characteristics. The subsidence sinkholes usually form in the non-soluble geological cover deposits (Palaeogene and/or Quaternary in age) above the karstic chalk surface where underdrainage through geological time has produced solution pipes within the karstic chalk surface (**Figure 38.8**). Such dissolution features can be typically up to 10 m dia. and up to 20 m deep. Exceptional-sized features can be much greater in diameter and up to 50 m deep. The karstic features are effectively a hybrid version of the subsidence sinkholes and buried sinkholes described by Waltham and Fookes (2003) in **Figure 38.5**. Long-term studies of recorded recently formed sinkhole occurrence upon English chalk (Edmonds, 2018) show that the size category 'up to 3 m depth and diameter' accounts for about 78% of sinkholes, and the next size category 'up to 5 m depth and diameter' accounts for about 95% of sinkholes. Just a few sinkholes were larger in diameter (up to 8 m) but shallower, mostly not exceeding 1 m depth. The latter examples were more akin to the shallow compaction sinkholes referred to above. An earlier study (Edmonds, 2008) suggested a multiphase formative process that would explain how ancient sinkholes with much larger dimensions

Figure 38.7 Example of a collapse sinkhole in limestone, Yucatan, Mexico (Yokdzonot cenote)

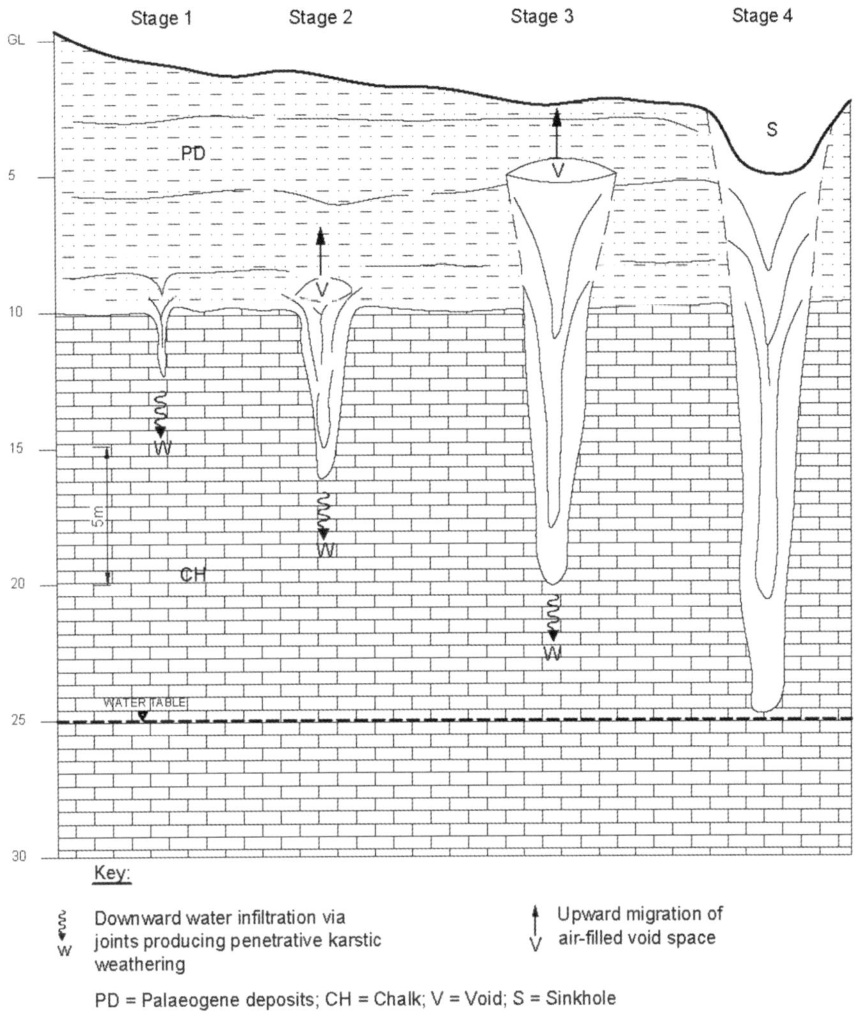

Figure 38.8 Schematic explanation of sinkhole development above maturing solution pipes in chalk karst terrains

can be formed (for example, Culpeppers Dish in Dorset which is about 21 m deep and 86 m dia.). The same study also identified a number of ground subsidence events in areas underlain by chalk considered most likely to be collapse sinkholes as described above upon the stronger limestones, together with a variety of locations where caves have been discovered in chalk which supports the potential for collapse sinkholes to occur, albeit relatively rarely. Sinkholes formed entirely by chalk dissolution are mostly absent; the few possible examples known are more likely created by glacial and/or periglacial weathering activity.

The development of new subsidence sinkholes within the soil or non-soluble rock cover over fissured limestone bedrock can happen rapidly, impacting structures within their design life and sometimes even during their construction (**Figures 38.9** and **38.10**). Any unconsolidated soil lying over karst bedrock is prone to loss by downward migration, known as suffosion or ravelling, into underlying bedrock voids. Clean sand flows with ease and can, therefore, cause a slow lowering of the surface until the stable profile of a suffosion sinkhole is reached, with a throat at a bedrock opening (**Figure 38.5**). A soil with a degree of clay content has the capability to develop an arch above the rockhead surface as the soil below collapses downwards and gets washed through underlying fissures. In time this can create a developing soil void beneath an unstable soil arch, with no surface indication of the potential for failure until the thinned soil arch fails, instantly producing a dropout sinkhole (**Figure 38.5**). In reality, most soils have some degree of apparent cohesion, created in part by negative pore water pressures, so they can develop soil voids and produce surface failures that vary between instantaneous and progressive. Following failure, the profiles of most dropout sinkholes degrade into those of suffosion sinkholes either within a few days of the wall slumping or over longer periods. Thicker soil sequences over limestone (>10 m thickness) tend to reduce infiltration and the number of new sinkholes formed.

Figure 38.9 Example of recently formed subsidence sinkhole in glacial sand and gravel deposits over Cretaceous-aged chalk strata. Ground movement and structural damage was triggered by water leakage from a sewer which resulted in loss of the corner house foundation, a sudden drop in the adjacent patio level and tilting of the garden shed

Figure 38.10 Collapse of a road on a bridge approach undermined by a new subsidence sinkhole in alluvial soils in Pennsylvania; this was one of many ground failures that developed within the zone of the decline of the water table around a deep quarry that was kept dry by continuous pumping

However, this is not an absolute limit, and ground failures have been recorded where the cavernous rock lies more than 100 m below the surface level. Messerklinger (2014) recorded formation of sinkholes in the Lar Valley of north-west Iran, where the thickness of alluvial deposits above limestone bedrock is about 200–300 m.

In karst terrains developed upon weaker and softer limestones such as chalk the formation of sinkholes has occurred at locations where the cover deposit thickness (Palaeogene- and/or Quaternary-age deposits) can be within the range of up to 25 m or more.

38.6. Cave geohazards upon limestones

While the primary concern for ground stability is associated with the formation of sinkholes as discussed above, there is also a secondary concern that is associated with the possible occurrence of unseen underground caves. In nearly all karst terrains there is only a small statistical chance of a building being threatened by a sufficiently large cave that is directly beneath and possesses a critically thin rock cover. Many structures have been inadvertently located over large caves in tropical terrains, but collapse has not been instigated by the imposed loads, which are commonly very modest compared to the self-loads of the rock. Guidelines exist for the thickness of roof required in limestones of various qualities in order to safely span caves of various sizes when loads are applied to them (Waltham and Lu, 2007). A cave generally has to be significantly wider than its cover thickness to be a stability threat (**Figure 38.11**). This means that, except in tropical karsts where large caves are typical, only a few metres of sound rock cover are generally required to provide integrity and render any deeper cave irrelevant to structural loading. Cases of rock failure in karst, as opposed to soil failure over karst, are generally rare. The potential requires detailed ground investigation and analysis, otherwise an inadequate ground investigation can lead to problems, delaying projects and adding costs as remedial measures may be required.

An example of this approach was taken to check the stability of a location overlying Clearwell Cave within the Forest of Dean, underlain by karstic Carboniferous-age limestone, south of Coleford. The karstic cave, enhanced historically by mining of iron ore and ochre, crossed below a main road and land access point. Underground laser surveying of the void space in 3D was carried out together with a detailed surface topographical survey and a geophysical survey (microgravity) to build a detailed ground model to understand the size and distribution of voids below the road at the surface. Finite element modelling, using the survey data and engineering parameters for the limestone, found that the mine/cave roof was 'marginally stable to unstable', indicating that more detailed investigation leading to remedial strengthening measures were likely to be needed. **Figure 38.12** shows a failing support pillar observed underground.

The main geohazard from open caves in karst is to end-bearing piles with high point loads, and particularly to any heavily loaded bored piles or caissons. An example of encountering this kind of hazard occurred below deeply piled piers being installed to support a new viaduct along the line of HS1 (formerly Channel Tunnel Rail Link), where it crossed the Aveley Marshes, Essex (Edmonds, 2008). Deep below the shallow groundwater table level, during pile installation, loss of concrete and unexpected settlement of groups of large-diameter piles occurred at depth in the chalk, which suggested collapse of clusters of cave-like voids up to at least 1 m high below the pile bases.

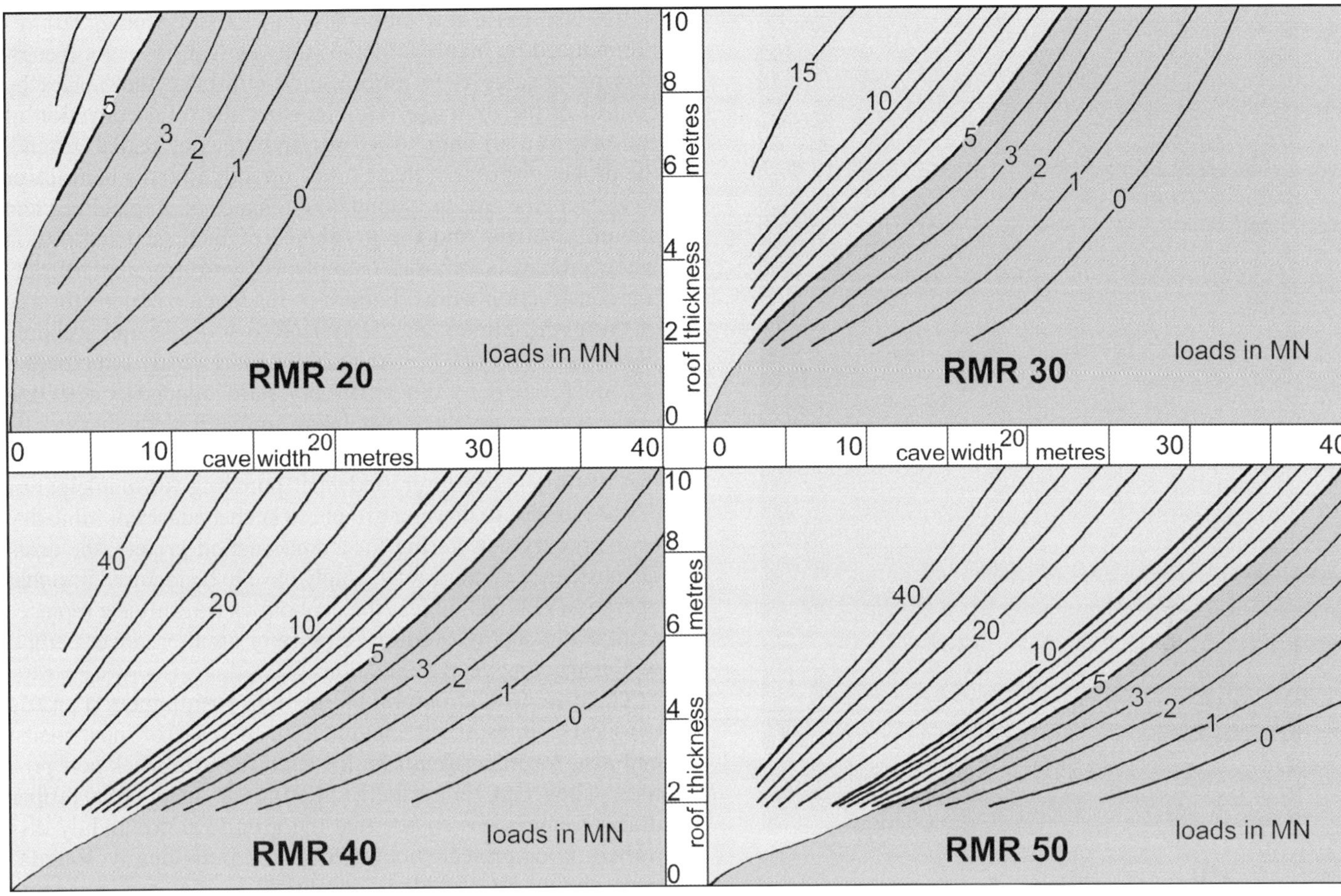

Figure 38.11 Nomograms that relate failure loads to cave width and roof thickness in ground of various rock mass ratings; the grey shaded areas represent situations that should be regarded as unsafe for footings with imposed loads of 1 MN that lie directly over the caves (Modified from Waltham and Lu, 2007)

38.7. Drainage-induced sinkhole hazards on limestones

Percolating water causes soil saturation and flow, collapse of metastable soil arches, washing through of fines and erosion, which can result in downwashing of soil (suffosion) into a fissured limestone surface. A similar inundation settlement process can occur within metastable solution pipe infilling deposits associated with karstic chalk surfaces, resulting in induced sinkhole formation (Edmonds, 1988). It might be expected, therefore, that many new sinkholes will develop during or soon after major rainstorms. However, the locations of new sinkholes above a fissured limestone surface are not predictable, as they will be located over unseen open fissures in the rockhead that are mantled by soil until the sinkhole develops (**Figure 38.13**). In a similar way, where solution pipes are developed within the chalk rockhead without any surface expression it is not possible to identify which solution pipe feature will be triggered to produce subsidence and which will retain its equilibrium stability at a particular moment in time.

It should also be noted that climate change effects can potentially influence triggering of sinkhole instability. Extremes of climate, wetness or dryness can have potential unforeseen impacts. During longer periods of hotter and drier weather there is a tendency for soils to lose moisture and become desiccated, promoting tension cracking, especially in more cohesive soils prone to shrinkage with a clay mineralogy sensitive to volume change. Such clay soils are common in areas of south-east England, often overlying karstic chalk areas. Deep tension cracking, for example, can lead potentially to rapid infiltration rainfall into the ground should a sudden storm occur, which could have a destabilising effect upon sinkhole-prone ground conditions.

It is significant that subsidence sinkholes are most likely to develop where and when there is a change to an existing equilibrium in rainfall infiltration, groundwater flow or leakage of water from a broken utility. Effects can also be enhanced by concentrated run-off from a built structure, infiltration from a soakaway or from a drainage storage pond. Equally, the change may be the increased drawdown of surface water

Figure 38.12 Failing support pillar within the Old Ham Mine linked to Clearwell Cave, observed during underground surveying works to evaluate ground stability

Figure 38.13 Numerous small subsidence sinkholes in a thin soil of glacial till over limestone in the Yorkshire Dales; there is no pattern to the sinkholes, as each has developed over an open fissure in the buried limestone, comparable to the exposed fissures in the bare patches of the limestone pavement

due to a water table that has declined within an aquifer that is over-pumped. More than 90% of new sinkhole appearances are related to the disturbance of the drainage equilibrium by civil engineering activities (Edmonds, 1988; Newton, 1987; Waltham, 2016; Waltham *et al.*, 2005).

Because most new sinkholes and karstic ground failures are induced by man's activities, the key feature to good engineering practice is to minimise or eliminate the hazard by control of the drainage. This includes the total collection of run-off from all built structures and areas of sealed ground, the proper disposal of that run-off directly into the bedrock or away from the site, the sound maintenance of all pipelines and stream channels and the avoidance of increased infiltration when ground is exposed or replaced with granular fill during construction works. Failure or inadequate actions for any of these will almost inevitably lead to soil loss and sinkhole development, which will require far more costly remediation at some future date. Where sinkholes are induced by a decline of the water table, there may be no simple remedy that can be applied within the confines of a site or construction project. If that decline is due to groundwater pumping for municipal or private supply, to dewater ground to keep a quarry or mine dry or temporary dewatering for a construction project, the costs of subsidence damage may simply have to be factored into that operation's budget, unless the economics determine a replacement water supply, a quarry closure or an alternative method of constructing deep foundations.

The placing of structural foundations within the soil profile over karst limestone is inevitable in the case of most roads, and also for many small built structures over thick soil profiles. They can be perfectly appropriate where the proper drainage measures ensure that the ground is not unduly disturbed. Each project should be treated individually; foundations should be provided appropriate to that particular site (Sowers, 1996; Waltham *et al.*, 2005). A host of alternatives may involve engineered soils, geogrids, rafts and mattresses, partial excavations, replacement soils, load transfer to bedrock pinnacles and reinforced foundations; all of these measures should be designed to minimise soil ravelling and to bridge any small voids that may develop subsequently (useful examples are provided by Lei and Liang, 2005; Vandevelde and Schmitt, 1988).

Grouting can be appropriate, but the sealing of all fissures within a karst limestone is likely to prove very expensive because huge quantities of fluid grout can be lost into large but unseen open voids. Compaction grouting, within the soil just above the rockhead, can effectively prevent the downward loss of soil, which is densified in the same process, and this commonly yields better results (Henry, 1987; Stapleton *et al.*, 1995).

Where an existing structure exhibits settlement damage following a history of stability, the key to remediation is to determine and rectify the feature that caused the new movement. In many cases this will be caused by a leaking water utility or pipeline failure, and repairs may be all that is required to recover the integrity of the structure. Pressure grouting within the soil may be appropriate if action has not been taken before settlement damage becomes too severe. New open sinkholes commonly require repair, and this should entail backfilling with selected sizes of materials so that the sinkhole is choked to prevent further soil loss, but it may still be required to drain surface water safely into bedrock fissures (Waltham *et al.*, 2005).

38.8. Ground investigation approach to limestones

Gauging the scale and extent of karst development at a site underlain by limestone, whether the limestone occurs at outcrop or subcrop at depth below overlying soil cover or rock strata cover (may be tens of metres thickness), at the outset of a project is a challenge. This can be made particularly difficult if there is no surface expression of the underlying karst such as sinkholes. Establishing an initial understanding of the intensity of local karst development is essential for broadly evaluating the key parameters of sinkhole frequency, cave size and rockhead relief that may be anticipated. Interpretation of these factors usually benefits from wider experience in karst terrains. However, in Britain at least, the availability of commercial digital natural cavity databases (British Geological Survey and Stantec UK Ltd) helps to provide information about the recorded spatial occurrence of karstic features across the country, reducing the need to research specialist publications and data sources from the outset. The information can also be combined readily with ground investigation data within 3D geological models and GIS to aid interpretation of karstic ground conditions. In addition, other complementary information can be obtained and used to inform the karst class identification, such as carrying out a site walkover survey incorporating geohazard mapping using geomorphological techniques, viewing and interpreting aerial photograph and remote sensing imagery (the infrared spectrum is useful for detection; Edmonds *et al.*, 1987), as well as using satellite technology like InSAR (interferometric synthetic aperture radar) to detect relative ground deformation movement patterns through time.

The morphological complexity of karst fissures and cavities means that any ground assessment based on borehole logs will inevitably be oversimplified; this applies to both unseen caves and rockhead profiles (**Figure 38.14**). For this reason, depending upon site conditions and access, it is often best to carry out a geophysical survey in advance of intrusive investigations to characterise the karstic rock interface and detect whether voids might be present (Edmonds, 2020). The success of such surveys will often depend upon the depth and size of the natural cavity, its infill and whether its characteristics provide sufficient contrast within the surrounding rock mass to be detected. However, while it cannot be guaranteed to detect every cavity, usually a number of cavity features (anomalies) will be found which provide a focus for follow-on ground investigation works and direct exposure for recognition of the

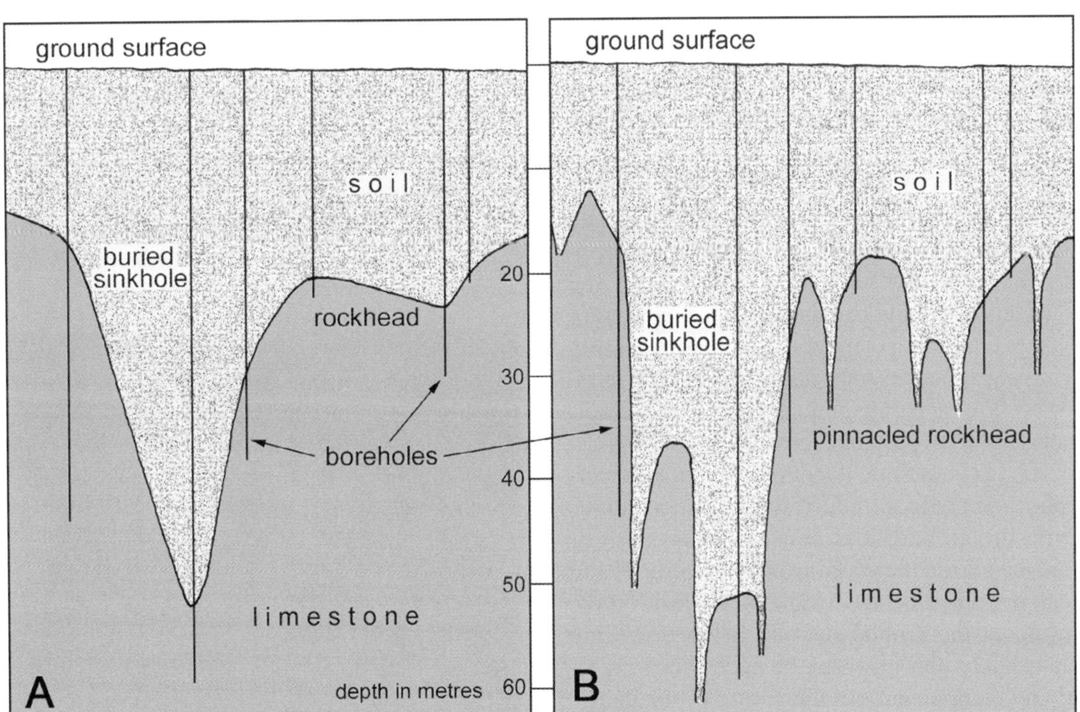

Figure 38.14 Two interpretations of the rockhead profile investigated beneath the site of a new tower block in Kuala Lumpur, Malaysia: (a) the basic interpretation from the borehole data; (b) a more likely interpretation based on an understanding of the pinnacled rockheads that are well known in the area

detailed engineering characteristics. Experience has shown that techniques like ground conductivity (EM), electrical resistivity tomography (ERT), microgravity and seismic surveys can be useful for detecting karstic features either because of the weakness or voiding within features or because of the sudden level variability of the soluble rockhead surface. For best results the geotechnical engineer should discuss the range, types and scale of possible natural cavity targets in advance with the geophysicist to optimise the survey approach and to consider the re-interpretation of the survey results if considered economically beneficial after the ground investigation results are known for critical target locations identified by the survey.

Once an anomaly or target is detected by the geophysical survey then direct investigation by trial pits, boreholes and/or probing, as appropriate, can be carried out to better define (ground truth) the scale and nature of the karstic feature. The karst class assessment may indicate that karstic hazards have the potential to be present at depth across an entire site area. For such sites, the detection of all karstic features is likely to be prohibitively costly and so it may be more pragmatic and cost-effective to instead concentrate the investigative effort upon the footprint of the structure that is intended for construction, for example. It is usually the case that a grid of exploratory holes at an appropriate spacing (perhaps 3 m, 5 m or more spacing) is undertaken and a 3D interpretation of the results completed. Isolated boreholes to test whether karstic features are present are rarely useful and cannot confirm the extent of the karst hazard alone even if a feature is encountered. There is no simple answer to how many boreholes should be drilled to assess a karst terrain, the number required being as many as needed to provide confidence to the engineer that the likely ground conditions are revealed and understood to a level that is adequate and appropriate for the risks involved in the particular project, taking account of the sensitivity of the structure to ground movement and damage. Ground investigation data in a digital format (e.g. AGS format) can be readily incorporated into geological modelling software (e.g. Leapfrog) and used with GIS datasets to aid understanding of the geohazard posed to the development. The potential for sudden variability in karst weathering always needs to be appreciated since modelling will never be fully accurate and cannot entirely eradicate risk and uncertainty in limestone terrains.

When rotary probe hole drilling (whether open hole or coring), it is also important not to introduce significant quantities of water flush into the ground; it is best to use air flush or mist flush to avoid triggering the washing through of fines and destabilising weak or voided infills to karstic cavities. If there are uncertainties about the ground stability below a drilling rig position, then a safety platform may be necessary to safeguard personnel and equipment until the nature of the underlying ground conditions is better defined. It is also important to ensure that boreholes are sealed after drilling to prevent unwanted infiltration of water into the ground and possible triggering of instability.

Interpretation of limestone karst investigation needs to be considered in a 3D context to understand the degree of variability present in the rockhead, both vertically and laterally. Dissolution-widened fissures may be widely, irregularly or closely spaced, leaving blocks or pinnacles of bedrock between them. A pinnacled rockhead has a forest of narrow rock pinnacles between networks of wide fissures (**Figure 38.15**). Individual blocks or pinnacles may be loose due to dissolution penetrating along sub-horizontal fractures or bedding planes, and are only held in place by the surrounding soil. The completely loose blocks within the soil are known as floaters. The depths of fissures, and the heights of pinnacles, are commonly some metres in any karst, and may be many tens of metres in tropical karsts of classes kIV and kV. It is, therefore, not abnormal to find depths to the rockhead varying by tens of metres at adjacent investigation boreholes. For weaker limestones such as chalk containing solution pipes that penetrate down into the rockhead surface, the pipes can be tens of metres deep, producing highly variable borehole profiles.

For chalk it is also significant to interpret the weathering profile; if the rockhead possesses a widespread, relatively thin highly weathered zone of chalk then the effect of penetrative karst weathering is usually minimal, but if the highly weathered zone persists to depth then it may be indicating close proximity to a lateral transitional zone adjacent to a deeply penetrating solution pipe (**Figure 38.16**) or other complex natural cavity feature perhaps containing a karstic weathered 'raft' of chalk. Large-scale variations in the buried limestone rockhead profile, perhaps 100 m or more wide and deep, may be completely filled with weak non-soluble materials derived from weathering or other younger geological deposits. The soft soil fills can give rise to compaction sinkholes and the limestone bedrock interface at depth may be highly fissured and irregular, thereby providing poor-quality foundation conditions and presenting a challenge to economic foundation design.

Figure 38.15 Pinnacled rockhead in mature limestone karst (of class kIV) exposed on a construction site in southern China; some pinnacles have already been broken down (with sledgehammers) to create a solid and almost level footing for a new hotel; the grey top of the pinnacle on the right originally projected above the soil level, as do some undisturbed pinnacles in the background

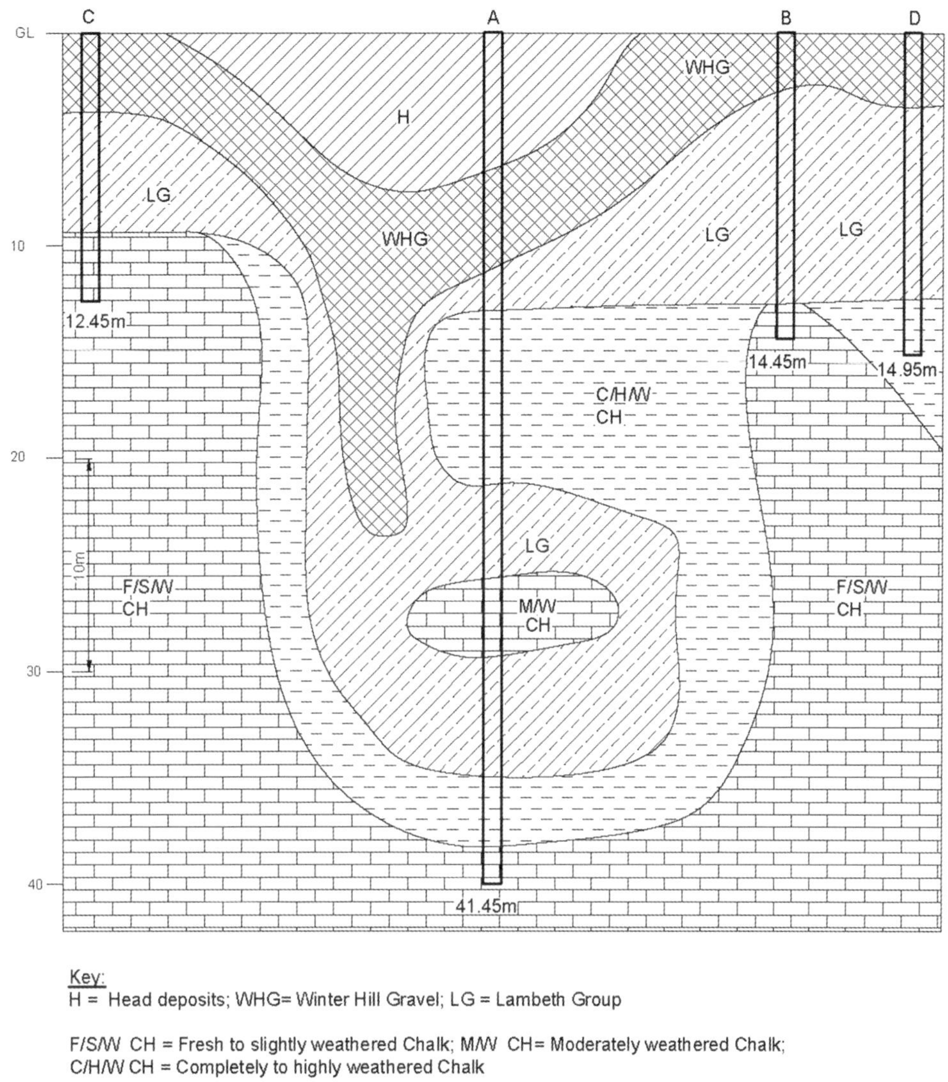

Figure 38.16 Schematic section showing lateral and vertical effects of chalk rockhead weathering in relation to karst development based on experienced interpretation of borehole profiles

Should open air-filled or water-filled voids be encountered within a limestone rockhead by boreholes, then it is important to check the size and interconnected nature of the void. Voids can be viewed directly by down-hole cameras with light sources to illuminate the void to identify its size, orientation and nature. If the void appears extensive and its position relative to the development footprint at the surface is uncertain, then it is possible to carry out a down-hole laser survey to obtain a measured 3D image of the void and to relate its position by accurate survey with the surface so that its significance and hazard potential can be further evaluated. For open voids the critical stability factor is the void width in relation to the thickness of the limestone roof, as shown in **Table 38.1**.

38.9. Engineering works on limestone bedrock

Foundations carried through to the bedrock eliminate the major dangers of soil movement and sinkhole development within the overburden, but karst ground buried beneath soil cover is very variable and not easily assessed. Karstic limestone rockhead poses two major difficulties associated with (*a*) an uneven, fissured or pinnacled rockhead, and (*b*) the threat of open caves present at varying depths. These conditions pose problems to the economic and safe design of foundations.

Most deeply fissured and cavernous karst limestones are strong materials, so structural loads may be carried safely on pinnacles that have been proven to be of adequate width and are not disconnected from the underlying bedrock. In such cases reinforced foundations can bear on the pinnacles and span the intervening soil-filled fissures and buried sinkholes. Driven piles with any significant end bearing on rockheads of strong limestone (UCS > 70 MPa) can suffer from a range of installation difficulties (**Figure 38.17**). Providing a safe seating for piles on steeply sloping bedrock surfaces is difficult to achieve, so it is normally necessary to drill sockets into the

Rock	Imposed load: kPa	Karst class	Cave width, likely maximum: m	Safe roof thickness: m
Strong karstic limestone	2000	kI–kIII	5	3
		kIV	5–10	5
		kV	>10	7
Weak limestone and chalk	750		5	5
Gypsum	500		5	5

Table 38.1 Safe roof thicknesses for various cave situations, and, therefore, a guideline to the depths to be proven by probing prior to construction (Data taken from Waltham *et al.*, 2005)

rock, which can also be difficult through a steeply inclined interface. Rockhead voids can be infilled with grout to create a more stable base for foundations, although additional stability might be needed in the form of dowels. Pile installation needs to be carefully instrumented and monitored in case of the pile losing integrity (e.g. loss of concrete during pile installation) or loss of vertical alignment. There may also be concerns that the basal tips of the piles might be underlain by karstic voids, in which case exploratory test drill probes can be undertaken to check the stability of the underlying bedrock.

In chalk terrains, while the ground profile generally lacks a sharp contrast between soft soil and strong rock, the zone of weathered chalk around and below dissolution pipes penetrating the rockhead presents challenges for installation of piled foundations. Piles stopped within the weathered chalk can be subject to pile tip failure adjacent to weak and steeply dipping dissolution pipe margins and so foundations need to be extended to depths where the chalk is structured, and its strength is sufficiently high to provide adequate bearing capacity (i.e. below the depth of the karstic weathered zone). Dissolution pipes (buried sinkholes) penetrating the chalk are frequently the sites of surface subsidence whenever soil movement is reactivated due to modified drainage or leaking pipelines (Edmonds, 2008) and for this reason the pipe infills do not provide reliable bearing for foundations. The weak infills can be spanned by a raft or a grillage of ground beams

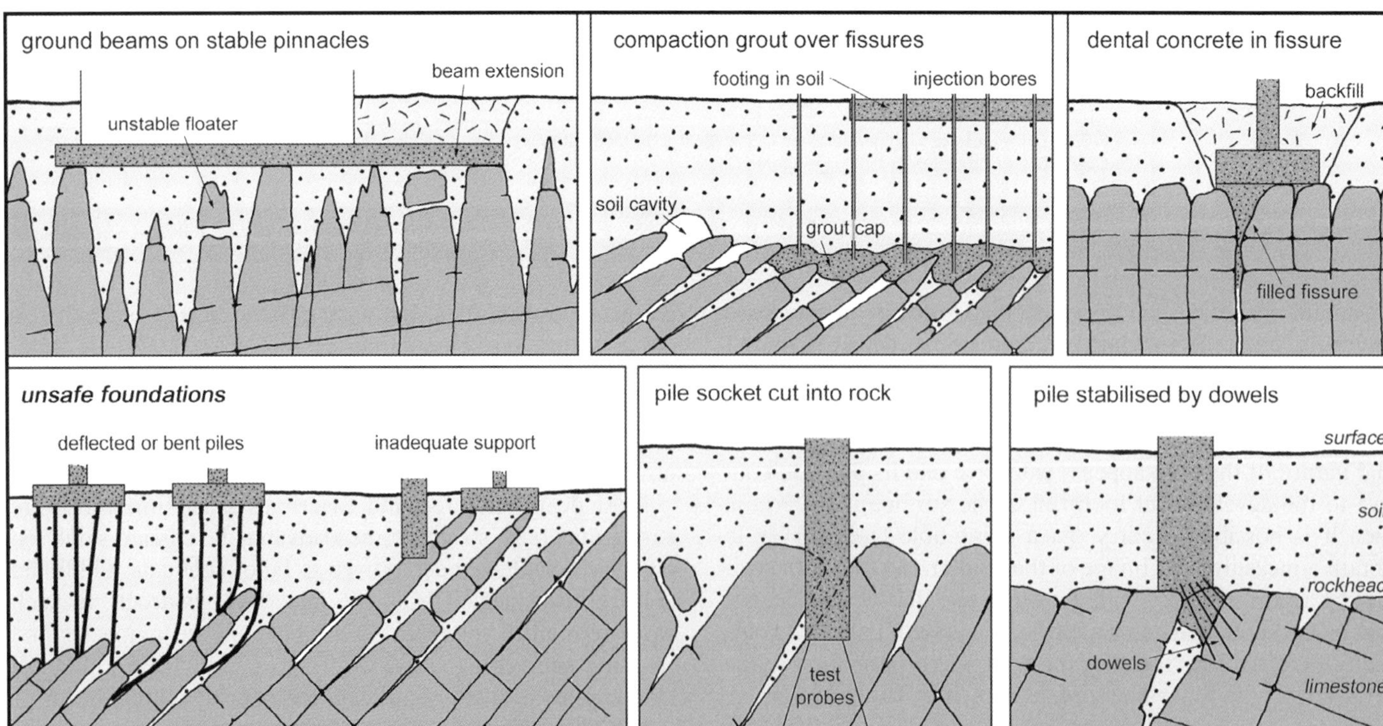

Figure 38.17 Various examples of good and bad foundations on soil-covered limestone karst (Waltham *et al.*, 2005, with due credit to the ideas and experience of the late George Sowers)

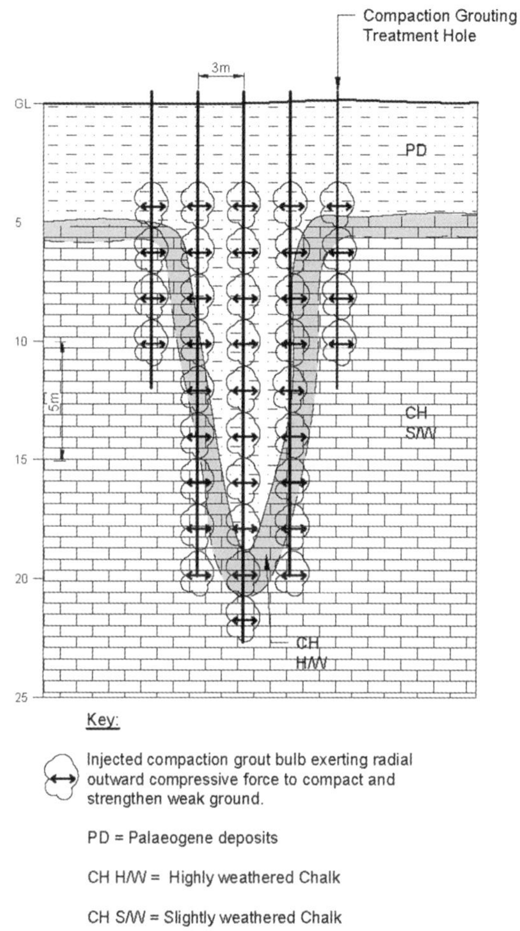

Figure 38.18 Example of using compaction grouting to stabilise the weak/loose infill of a dissolution pipe ahead of shallow foundation installation

provided the dimensions of the dissolution pipe can be defined by close-centred investigation probes or exposed by excavation. Usually foundation spans of up to 5 m across features can be economic but become considerably less economic for larger spans, in which case it is common for the metastable infills to dissolution pipes to be stabilised by ground treatment techniques like compaction grouting (**Figure 38.18**). Ascending grout bulb injection producing radial compaction of the ground through end-cased holes over a close-centred grid matching a building footprint has a good track record for overcoming instability problems associated with chalk karst. Shallow foundations can then be installed directly upon the upper surface of treatment or upon an engineered fill platform above, raised to the level required.

Dam foundation and reservoir impoundment attract a whole series of problems when carried out on, or partially on, karstic limestone. Most limestones are strong enough to bear the loads imposed, but the problems of leakage can be massive and very complex, and are beyond the scope of this chapter. Both foundation problems and hydrological situations are comprehensively reviewed by Milanovic (2004), based on extensive experience in the Dinaric karst. The Kalecik Dam in Turkey provides an accessible case history of karst leakage and its remediation (Turkmen, 2003). In addition, Messerklinger (2014) also refers to subsidence problems associated with the construction and impounding of the Lar Valley Dam and reservoir in Iran.

38.10. Geohazards on gypsum terrains

Gypsum karsts are similar to those on limestone, except that there are no climatic constraints on the extent and distribution of gypsum dissolution. As gypsum is mechanically weaker and more rapidly dissolved than limestone, the karst terrains lack the range of extreme landforms found upon limestone. The most widespread geohazard on gypsum is the development of new sinkholes within the soil and/or rock cover, and in this respect the conditions, the processes, the extents and their assessment are comparable to those in a limestone karst (Johnson and Neal, 2003). Geohazards on gypsum terrains have been widely reported from Europe (Great Britain, Spain, France, Germany, Lithuania, Ukraine, Turkey), the Middle East (sabkha areas), Russia, China and the USA.

Unlike limestone, gypsum is much more rapidly dissolved in natural water, such that rock removal by dissolution in timescales of decades can become a factor within the design lifetime of an engineered structure. As with limestones, though, the triggering and sudden development of sinkholes over preexisting dissolution features is a concern for site stability and such processes are more rapid than the timescale for dissolution. As for limestones, drainage control is a key factor of surface stability in gypsum karst, since disposal of large flows of water into gypsiferous bedrock can lead to enhanced dissolution activity, rapidly creating new cavities and modifying the existing drainage. In Britain gypsum is associated with Permian and Triassic strata as described by Cooper (2020a), who has particularly studied the spatial occurrence of subsidence problems in and around Ripon, North Yorkshire. Significant thicknesses of gypsum occur within two geological horizons underlying Ripon, known as the Edlington Formation, with ~35–40 m of gypsum, and the higher Roxby Formation, with ~10 m of gypsum. A schematic section through local ground conditions based on ground investigation work and ground treatment work is presented in **Figure 38.19** and is further discussed below. It appears that random caves and cavities are present that can cause surface instability as collapse zones (breccia pipes) which rise to the surface to form sinkholes (**Figure 38.20**).

The scale of development of caves (**Figure 38.21**) and cavities (bedrock voids) in gypsum is generally smaller than for limestones, as might be expected because of the comparative weakness of gypsum, but even so collapse sinkholes of up to 30 m across and 20 m deep have been recorded, although most are smaller than this. In the Ripon urban area new sinkholes causing structural damage tend to be recorded at intervals of every few years. It appears that larger collapse sinkholes

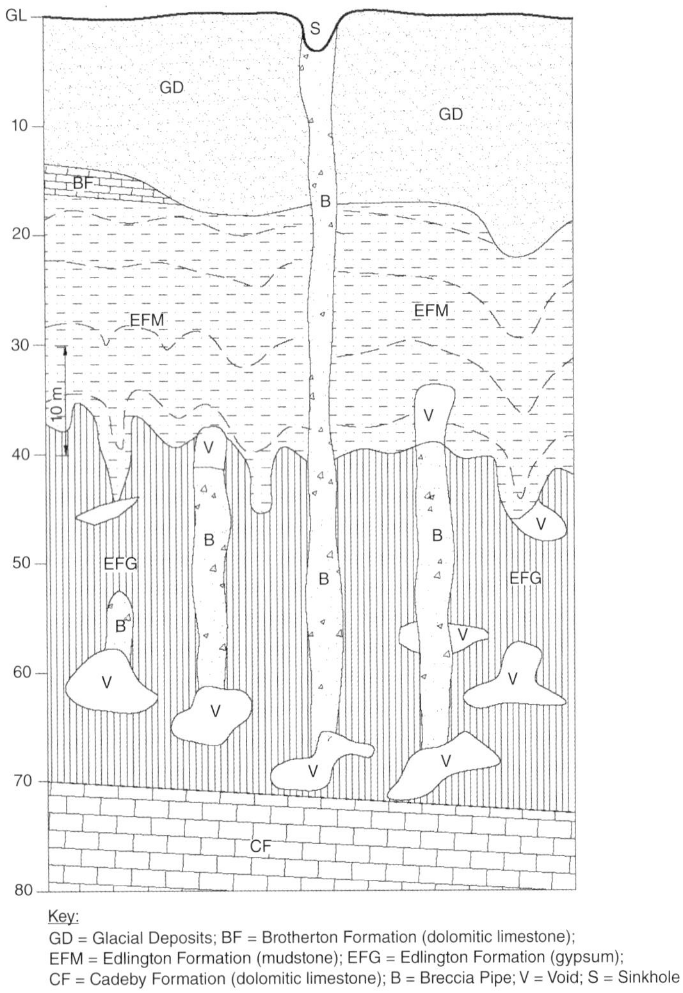

Figure 38.19 Schematic illustration of gypsum karst development below a development at Ripon (Based on ground investigation and ground treatment drilling data, and with reference to Cooper, 2020a)

Figure 38.20 Example of reactivation of a collapse sinkhole (1.5 m dia., 0.5 m deep) at a proposed development site in Ripon. Original collapse measured 5 × 7 m and contained standing water at 1.6 m depth below ground level before backfilling with gravel

Figure 38.21 Cavity migration by progressive roof failure over a cave passage in massive gypsum at Pinega, Russia; flakes that are peeling away from the roof deform and sag before they break off

(> 100 m across), as found in some gypsum karsts, have probably developed as a result of multiple ground failures that have laterally expanded the area of collapsed ground over time, rather than being caused by a single event. This field observation suggests that following collapse of non-soluble fill materials down into a void within gypsum forming an upward-migrating breccia pipe, the focus of dissolution moves radially outwards around the collapse zone.

Another feature of gypsum to be aware of is that it is formed as hydrated calcium sulfate ($CaSO_4 \cdot 2H_2O$) but as it is buried deeper it dehydrates to become anhydrite ($CaSO_4$). The dehydration usually takes place at depths of 450 m or more (Klimchouk and Andrejchuk, 1996), depending upon geothermal gradient, circulating groundwater fluids and overlying strata conditions. In certain circumstances, however, the process can be reversed, resulting in ~60% increase in volume as anhydrite reverts to gypsum. This process induces high pressures in the ground, leading to deformation, and is often accompanied by uplift which can be damaging to surface development. For example, Fleuchaus and Blum (2017) reported ground surface uplift of 58 cm and lateral movement of 43 cm during an 8-year period, resulting in extensive damage to buildings at Staufen, Germany, where the reversion of anhydrite to gypsum was facilitated by the installation of open loop ground source heat pumps.

38.11. Ground investigation approach to gypsum

In principle the ground investigation approach for gypsum terrains can be similar to that of limestone karst areas. It is best to start with a desk study that includes definition of the geological profile and recorded depth(s) at which gypsum is expected to be present to understand the ground model, together with historical research about evidence of past instability and the scale of ground movement known. Background information concerning the scale of karst development, hydrogeology and geomorphology of the area of interest is also important. As discussed above, in Britain the existence of digital natural cavity databases can provide significant preliminary information for a site of interest ahead of the ground investigation stage.

While a selection of boreholes will assist in the general confirmation of the geological profile and provide some indication of the depth of karst, it will always be insufficient to determine the detailed ground conditions below proposed new-build structures. The geological model for the site may be improved by the use of geophysical surveying as mentioned above for limestone karst and making reference to airborne/satellite imagery or InSAR, but ultimately it is usually the case that detailed grid pattern drilling or probing will be required. Spacing and depth of holes will need to be a combination of financial and risk management strategy while addressing the technical needs to mitigate the hazards posed. Therefore, a holistic approach is best, and the geotechnical interpretation undertaken by an experienced engineer with appropriate karst knowledge utilising GIS techniques and 3D geological modelling.

There are also additional considerations when working in gypsum terrains, such as not wanting to introduce groundwater flows to enhance circulation and dissolution at depth (especially if there is a risk of artesian water pressures at depth) and using drilling fluid mixes that are benign to gypsum. On completion of exploratory holes it is also essential to grout-seal the holes using a sulfate-resistant grout mix so that the integrity of the ground conditions is maintained. A useful summary of the ground investigation approach in gypsum karst terrains with particular reference to Britain is provided by Cooper (2020a), who advocates quality ground investigation and hazard avoidance where possible.

An example of a ground investigation strategy for a new building in Ripon in an area underlain by gypsum karst comprised the following approach. An initial stage of geophysical surveying (electro-magnetic conductivity and electrical resistivity surveying) was carried out which revealed a series of potential karstic ground anomalies below the development footprint. Preliminary shallow ground investigation boreholes helped provide some confirmation of the ground profile but did not particularly define the nature of the underlying karst. To further clarify the ground profile and to avoid drilling penetration of the deeper-level gypsum karst, the ground conditions were initially characterised by undertaking a grid of cone penetration testing holes across the building footprint. They were used to check the strength consistency and thickness of the glacial deposits above the Edlington Formation rockhead and to look at variations in the depth and nature of the weathered rock surface. This approach provided a reasonable preliminary model of the broad karstic rockhead surface variations and identified weak zones of upward-migrating breccia pipes anticipated to overlie voids at depth (**Figure 38.19**). To maximise the value of the data all geotechnical information was input to a 3D geological modelling software package and subjected to experienced interpretation and evaluation by a team familiar with karstic ground conditions.

38.12. Engineering works over gypsum bedrock

Gypsum bedrock terrains present significant challenges to the provision of constructing safe building developments. Cooper (2020a) discusses approaches taken to mitigate against ground instability in the Ripon area to overcome the engineering impacts of gypsum karst. Such approaches include the installation of reinforced raft foundations suitably designed to span loss of ground support should subsidence take place. Grillages of ground beams with extended cruciform corners offer another design approach to increase void spanning and cantilevering. Problems can also affect connecting utility services, so flexible service pipes with swan-neck connections can be useful. Another concern is safeguarding against rupture of water-carrying services, and so strategies such as placement of soil-reinforcing geogrid below pipes and lining trenches with waterproof membranes can help to prevent problems.

For a new large building development in Ripon the foundations strategy approach has been to stabilise a bridge of ground in situ using compaction grouting techniques (using sulfate-resistant grout mixes) to stabilise the weathered, foundered Edlington rockhead and overlying glacial deposits. This approach has created treated stable ground spanning the main karstic zone at depth and reduced the potential for upward migration of voids. At the surface shallow reinforced raft foundations have been installed that are capable of spanning voids up to 5 m dia. Additionally, a protective monitoring and warning system has been installed comprising three depth level extensometer nests at several positions below the building footprint to warn of any future upward migration of voids breaking through the grouted zone, plus pairs of strain gauges arranged orthogonally have been installed at the same locations within the raft foundation to check for stress/strain build-up should ground movement occur. By this approach, grouting of the main voided karstic gypsum zone at depth has been avoided to prevent displacement of groundwater flows leading to enhancement of dissolution at depth while providing substantial subsidence movement protection at a higher level in the ground. The monitoring system will be active for the duration of the building design life, and should future significant ground movement be detected then remedial ground treatment (e.g. angled grout treatment holes) can be undertaken to mitigate the hazard if necessary.

Just as for limestones, linear structures like roads, railways and pipelines may need special protection where these structures cross hazardous karstic zones underlain by gypsum. Such measures might include enhanced reinforced foundations (e.g. reinforced soil embankments), which may need to account for redundancy loss in the support systems. Additionally, it is important not to change the groundwater percolation regime to concentrate water flows into the ground that might lead to enhanced dissolution and subsidence problems in the future. Hydraulic structures such as dams, reservoirs and canals can lead to diversion and concentration of water flows, which can cause enhanced karstic dissolution problems and leakage failure of structures. The water flows can also wash out infills within existing dissolution features as well. Detailed ground investigation, hazard evaluation and ground modelling are required to identify and mitigate arising problems. Abstraction for water supply can also cause accelerated dissolution, and potential for ground subsidence, by drawing in flows of unsaturated water. A recently revealed example of this was noted at a new development site in Ripon where, during ground stabilisation treatment by grouting, there was a significant increase in grout takes in the ground surrounding the position of an old abandoned well.

The extensive list of references in Cooper (2020a) provides a useful source of published papers relevant to a wide range of gypsum engineering problems in many parts of the world.

38.13. Geohazards in salt terrains

Salt (also known as rock salt or halite) is an evaporite material found widely within sedimentary sequences deposited in arid environments. Salt is rapidly dissolved in rainwater, and so its occurrence and geohazard are restricted to three environments: modern desert salars, salt domes in semi-arid terrains and buried rock sequences elsewhere.

Salars, salt pans, playas and continental sabkhas are areas of newly deposited evaporite salt, commonly with other minerals in thicker sequences. Groundwater is normally saturated, thereby restricting dissolution. Karstic cavities some metres across, which lie at shallow depths and may be revealed by smaller collapse openings at the surface, are generally restricted to marginal areas where the salt has been reached by aggressive surface run-off or groundwater input from adjacent hills. As active salt areas are largely prone to seasonal inundation, construction on them is very limited, but concealed cavities close to their margins can be hazardous to roads and driving routes that cross them.

Salt diapirs, or salt domes, rise from stratiform salt at depths of some kilometres. The salt is generally lost to rainfall at an outcrop, so that only caps of less soluble gypsum, anhydrite and clay survive at outcrop. Actively rising salt diapirs form significant salt mountains, with or without salt glaciers flowing from them, but there are few beyond the many in the semi-arid Zagros Mountains of southern Iran (Talbot and Aftabi, 2004). These constitute mobile and cavernous terrains that are unsuitable for any development (Bosak *et al.*, 1999). Their thick mantles of residual soil are pitted by closely packed subsidence sinkholes that are actively collapsing and ravelling into dissolution cavities within the underlying salt (**Figure 38.22**). Cave chambers are up to 15 m across, with block failure and roof migration continuing at rates that are orders of magnitude faster than in limestone caves, so that bedrock collapses impact the ground surface as frequent events. Only when a diapir's rise virtually ceases does surface lowering dominate, and then the residual mantle can become thick enough to prevent most rainfall reaching the salt, allowing its surface to approach stability. By that stage, the salt subcrop lies beneath a lowland terrain with chaotic sinkhole topography, but it is stable enough to be crossed by highways with minimal subsidence problems.

Figure 38.22 Collapsing ground on a salt dome in southern Iran; the person is standing at the edge of an active sinkhole in about 4 m of residual soil that is being undermined by rapid dissolution of the heavily eroded and cavernous salt beneath

In areas subject to significant rainfall and groundwater infiltration salt deposits do not appear as surface outcrops, only surviving at depth beneath a mantle of dissolution residue (or breccia) formed mostly of foundered non-soluble strata extending to depth below a soil capping. This form of buried salt occurrence is discussed below.

38.14. Subsidence over buried salt

In Britain, the Cheshire Plain in north-west England comprises a glacial drift 10–50 m thick overlying dissolution breccia that is around 50 m thick and consists of collapsed blocks of the mudstone that was originally interbedded with salt (Waltham, 1989). Below the dissolution residue, thick salt beds remain in situ, creating wide subcrops due to their low angles of dip. Ground stability is ensured where any voids within the salt are filled with saturated brine, but subsidence takes place over any sites where the input of freshwater allows renewed dissolution of the salt. This is most extensive along linear subsidences (typically 1–10 m deep, 100–400 m wide and 1–5 km long) that develop over 'brine streams' of concentrated groundwater flow through the permeable dissolution breccia at the rockhead. These linear subsidence features are often referred to as 'flashes'.

Historically, brine was pumped from these zones on an industrial scale, thereby drawing freshwater into the breccia and producing destructive ground collapses and large sinkholes, especially where the brine was also removed from old flooded mine workings, causing the support pillars to dissolve as freshwater was drawn into the mines. Eventually, due to legal cases against extraction companies and because of the destructive nature of brine removal by pumping, the activity was stopped in many places by the early 1970s. The activity, although reduced in scale, continued in some places such as Northwich in Cheshire, where ground movement and surface subsidence damage continued to occur (Branston and Styles, 2003) until operations ceased in about 2006. A useful summary account of the various salt fields in England and Northern Ireland and their history of extraction is set out in Cooper (2020b).

It is also known that salt beds with no outcrop can create major surface subsidence problems where water and brine are able to flow through them. Some events are due to poorly managed brining operations, but the USA has many cases of large collapse sinkholes that developed after nearby wells, drilled for freshwater or petroleum through salt beds, were left uncased or poorly cased, thereby allowing brine outflow into aquifers that previously had no connection (Johnson and Neal, 2003).

Continuing flows from brine springs indicate that dissolution of buried salt is happening as a natural process in an undisturbed natural environment. The result of this over time is ground subsidence, albeit on a modest scale, that is uncontrollable. In former areas where wild brine pumping took place (e.g. Droitwich, Worcestershire), despite the salt extraction process being stopped some 50 years ago, there is still evidence of ongoing ground subsidence movement. It is debatable whether the movement continues to relate to the establishment of a new quasi-equilibrium stability status following historical brine extraction which massively disturbed the hydrogeological conditions, or whether it now relates to entirely newly established natural dissolution (post brine extraction) or is even a hybrid state. In recent times this situation resulted in the rejection of plans to construct a residential housing development in the vicinity of the Stoke Works (former saltworks north-east of Droitwich). At the location, detailed historical ground-level surveying combined with interpretation of InSAR data demonstrated the continuing active nature of ground subsidence and the suspected upward, overturning movement of the buried bedrock margin immediately adjacent to the main subsidence zone. This case study, worked upon by the author, illustrates the uncertainty and blighting effects of such ground conditions.

An appropriate engineering response may be the construction of houses on rafts, which prevent structural damage and are prone only to tilting; these may then be jacked back to horizontal, in the manner that was common on a large scale in past times of active brining in Cheshire, although current movements are generally too small to warrant such an operation. Natural subsidence, generally on a small scale but locally with sinkhole development, is widespread over salt beds in the USA (Johnson, 2005), and can extend to the rare formation of large collapse sinkholes such as at the McCauley Sinks in remote country in Arizona (Johnson and Neal, 2003).

Suitable and economically viable mitigation approaches to subsidence-prone salt terrains is challenging and difficult. In the past (e.g. in Cheshire) buildings have been constructed upon rafts in areas where significant ground movement took place such that where the buildings tilted they could be jacked back to the horizontal. Requirements for adjusting building levels at intervals, difficulties with prevention of damage for connecting utility services, concerns over safety and obtaining insurance, among other issues, largely rule out this mitigation option in present times. Although in Britain the historical concerns for major ground subsidence linked to past salt and uncontrolled brine extraction have ceased, it is still evident from the case study above that ground stability concerns continue. This has acted as a considerable disincentive for new development and the economic prosperity of historical places built over salt-bearing areas. The costs to carry out ground stabilisation in such areas to mitigate the blight is considerable. For example, a former salt-mined area below the northern edge of Northwich where there were concerns for continued dissolution and collapse of the mines affecting existing development and preventing investment for new development relied upon the grant of some £32 million by central government during the early 2000s to carry out extensive grout infilling remedial works to stabilise the ground up to 90 m below the surface across a 32 ha area. Further details are presented by Brooks *et al.* (2007) and Cooper (2020b). On this basis it appears unlikely that normalised investment and development of urban areas severely afflicted by past salt extraction and continuing dissolution effects is possible without state assistance.

38.15. Karst geohazards on sabkha

Geohazards within the sabkha environment of arid coastal plains have grown in importance with the huge increase of construction activity on the lands fringing the Arabian Gulf. Coastal sabkhas are supratidal flats up to about 10 km wide, formed of evaporites and carbonates typically with a thin cover of aeolian sand (Warren, 2006). Continental sabkhas in inland basins are dominated by halite, with the attendant hazards due to dissolution discussed above. Coastal sabkha lithologies are normally only a few metres thick, but can accumulate to significant thicknesses along the margins of subsiding basins. When subsequently uplifted, the carbonate and sulfate materials are subject to erosion and the consequent development of more conventional karst geohazards.

The special concern for sabkha relates to the modern sediments that immediately underlie the coastal plains. These include relict and buried sabkha horizons within sequences dominated by aeolian sands that extend some kilometres inland from the active sabkhas. Sediments include carbonate sands, algal mats and dolomitised facies, or are dominated by clastic quartz with variable amounts of carbonate. Typically these are poorly consolidated to depths of 5–10 m with sandy materials that have weak cements of anhydrite, gypsum and calcite, which are prone to loss by dissolution when the groundwater regime is changed. Denser, competent and more lithified sediments persist at greater depths, where any halite in the coastal sabkhas may have been lost by dissolution during diagenesis.

The primary source of karst geohazard on sabkha is due to gypsum that occurs at shallow depths within the poorly lithified sequence. This is deposited both within the capillary zone and beneath the shallow water table. Beds of almost pure gypsum are generally no more than 1 m thick; they contain nodules of anhydrite and have an open texture with small cavities (vugs) and high primary permeability. Halite is locally present, but is not in massive beds; carbonates are present, but only in some of the sand grains. Groundwater in a sabkha is dominated by brines leaking upward from buried aquifers, and these are normally saturated with respect to gypsum. Consequently, evaporation causes interstitial precipitation of gypsum, which converts easily to anhydrite with the characteristic chicken-wire texture, producing ground with locally reduced bearing capacity.

In the undisturbed sabkha environment where solute precipitation from saturated waters dominates and very low rainfall occurs to disturb the equilibrium, then conditions are unfavourable for dissolution to take place. Despite this, sinkhole development is widespread across the Gulf coastal regions. It appears that new sinkholes are mostly no more than 5 m across and 2 m deep, usually appearing at the perimeter of construction sites; therefore damage to structures is rare and they tend to be infilled and forgotten about. It appears that the new sinkholes are formed in relation to engineering activities, including leaking pipelines, uncontrolled drainage disposal, site dewatering and drilling with water flush that is not sulfate-saturated. It is thought that the sinkholes have formed in response to rapid dissolution of sabkha gypsum horizons at multiple levels within the Quaternary sequences, mostly within a few metres of the surface where stable groundwater conditions have been displaced by the input of unsaturated water relating to the engineering works. Suffosion of the unconsolidated cover sands then produced the sinkholes (**Figure 38.23**).

A second subsidence mechanism occurs where the new input of unsaturated water dissolves the sulfate or halite cement of the sabkha sands, allowing localised compaction and displacement of the loose material. There are many undocumented 'collapse settlements' in Saudi Arabia, including one event at a large, steel-framed desalination plant near Jubail that involved a number of column bases, founded on pads 1 m below the surface level, each sinking by about 50 mm (Sabtan, 2005). It is believed that this resulted from leakage of old, corroded and fractured pipelines, which had been observed for some time before the implications were appreciated. The structure was remediated with mini-piles, 15 m long, driven into stable ground. Such ground conditions imply that the collapse potential is high, but they do not show instantaneous compaction in the style of hydrocollapse, because time is required for the dissolution of the natural cement.

Figure 38.23 A small new sinkhole, already partly backfilled, which developed within a construction site on the coastal sabkha of the Arabian Gulf, after engineering works appeared to have induced dissolution of the gypsum that lies within either the bedrock sequence or the Quaternary cover (Photo courtesy of Laurance Donnelly, Halcrow)

Since the dissolution of sabkha soils and their subjacent gypsum sequences is largely or entirely at shallow depths, the situation can be mitigated by installing deep pile foundations to support structures. This is normal practice for large built structures in the Gulf region, hence they have not been affected by the sinkholes in adjacent ground triggered by the construction works. However, the situation does create problems for roads and infrastructure that lack deep foundations, and where there is failure to control all water movement. While it appears that direct rainfall can infiltrate the ground without disturbing the chemical equilibrium, any points of localised run-off from large built structures and areas of hard standing may displace saturated groundwater and permit new dissolution activity after rare rainfall events. For this reason the use of soakaways and infiltration basins that can introduce freshwater into the ground are not advisable. Mitigation actions for voids include induced compaction, sinkhole grouting and spread footings.

Finally, it should be noted that ground investigation works may reveal little or no evidence of karstic features since the majority of problems develop either post-investigation (ground disturbed by input of drilling flush) or as a result of the site construction activities. Therefore it is important to recognise the potential at the outset of a project where halite and sulfate lithologies are present in the ground conditions, and to take suitable precautions, especially to control surface water infiltration drainage at such sites.

38.16. Concluding remarks

Karst ground conditions are probably the most variable that a civil engineer may encounter, and are therefore particularly challenging where engineering solutions to their extreme and unpredictable variability have to be risk-managed within a reasonable budget. Managing client expectations to foresee potential impacts and options is important. Major errors, unwarranted expense and possible catastrophic ground movements occur where the special nature of karst ground conditions is not recognised at an early stage within a project. Those with no experience or understanding of karst can make the most elementary of errors. All too often, karst is only considered on a site after the first new sinkhole has appeared amid the construction works. Yet if karst is recognised early, if it is given due consideration in the design strategy and if all site drainage is properly controlled, then the sinkhole geohazard may be mitigated to acceptable levels. For building construction over karstic ground conditions, stable support to foundations remains a challenge and may need to be addressed by ground investigations appropriate to the scale of the building footprint to improve confidence in mitigating the subterranean geohazard posed. A key strategy is to recognise the nature and scale of karst that may be present, to determine the natural cavity types that will be the targets for ground investigation and to utilise appropriate techniques to characterise and understand the karst geohazard ground model.

Acknowledgements

Thanks are due to Tony Waltham for the many reused portions of original text and illustrations in this new edition of this chapter. Thanks are also due for previous contributions made by Laurance Donnelly (Halcrow Group) and Andrew Farrant (British Geological Survey) concerning the sabkha environment. Additional thanks are due to discussions with fellow geohazard practitioners and preparation of new figures by David Cotton (Stantec UK Ltd).

References

Applied Geology Limited (1993) *Review of Instability Due to Natural Cavities in Great Britain: Regional Reports (Volume 1), Technical Reports (Volume 2) and Summary Report*. Applied Geology Limited, Leamington Spa, UK.

Bosak P, Bruthans J, Filippi M, Svoboda T and Smid J (1999) Karst and caves in salt diapirs, S E Zagros Mountains, Iran. *Acta Carsologica* **28**: 41–75.

Branston MW and Styles P (2003) The application of time-lapse microgravity for the investigation and monitoring of subsidence at Northwich, Cheshire. *Quarterly Journal of Engineering Geology and Hydrogeology* **36**: 231–244.

Brooks TG, O'Riordan NJ, Bird JF, Stirling R and Billington D (2007) Stabilisation of abandoned salt mines in North West England. *Proceedings of Engineering Geology for Tomorrow's Cities, Nottingham, UK*, paper 781.

Cooper AH (2020a) Geohazards caused by gypsum and anhydrite in the UK: including dissolution, subsidence, sinkholes and heave. In *Geological Hazards in the UK: Their Occurrence, Monitoring and Mitigation, Engineering Group Working Party Report* (Giles DP and Griffiths JS (eds)). Geological Society, London, UK, pp. 403–423.

Cooper AH (2020b) Geological hazards from salt mining, brine extraction and natural salt dissolution in the UK. In *Geological Hazards in the UK: Their Occurrence, Monitoring and Mitigation, Engineering Group Working Party Report* (Giles DP and Griffiths JS (eds)). Geological Society, London, UK, pp. 369–387.

Edmonds CN (1988) Induced subsurface movements associated with the presence of natural and artificial underground openings in areas underlain by Cretaceous Chalk. In *Engineering Geology of Underground Movements* (Bell FG, Culshaw MG, Cripps JC and Lovell MA (eds)). Geological Society, London, UK, pp. 205–214.

Edmonds CN (2001a) Predicting natural cavities in chalk. In *Land Surface Evaluation for Engineering Practice* (Griffiths JS (ed)). Geological Society, London, UK, pp. 29–38.

Edmonds CN (2001b) Subsidence hazard in Berkshire in areas underlain by chalk karst. In *Land Surface Evaluation for Engineering Practice* (Griffiths JS (ed)). Geological Society, London, UK, pp. 97–106.

Edmonds CN (2008) Karst and mining geohazards with particular reference to the chalk outcrop, England. *Quarterly Journal of Engineering Geology and Hydrogeology* **41**: 261–278.

Edmonds CN (2018) Review of collapse events on chalk since 2000 and the opportunity for improved engineering practice. *Engineering in Chalk: Proceedings of the Chalk 2018 Conference, London, UK* (Lawrence JA, Preene M, Lawrence UL and Buckley R (eds)). ICE Publishing, London, UK, pp. 53–73.

Edmonds CN (2020) Dissolution: carbonates. In *Geological Hazards in the UK: Their Occurrence, Monitoring and Mitigation, Engineering Group Working Party Report* (Giles DP and Griffiths JS (eds)). Geological Society, London, UK, pp. 389–401.

Edmonds CN, Kennie TJM and Rosenbaum MS (1987) The application of airborne remote sensing to the detection of solution features in limestone. In *Planning in Engineering Geology* (Culshaw MG, Bell FG, Cripps JC and O'Hara M (eds)). Geological Society, London, UK, pp. 125–131.

Fleuchaus P and Blum P (2017) Damage event analysis of vertical ground source heat pump systems in Germany. *Geothermal Energy* **5**: 1–15.

Henry JF (1987) The application of compaction grouting to karstic limestone problems. In *Karst Hydrogeology: Engineering and Environmental Applications* (Beck BF and Wilson WL (eds)). Balkema, Rotterdam, the Netherlands, pp. 447–450.

Johnson KS (2005) Subsidence hazards due to evaporite dissolution in the United States. *Environmental Geology* **48**: 395–409.

Johnson KS and Neal JT (2003) *Evaporite Karst and Engineering/Environmental Problems in the United States*. Oklahoma Geological Survey, Oklahoma City, OK.

Klimchouk A and Andrejchuk V (1996) Sulphate rocks as an arena for karst development. *International Journal of Speleology* **25**: 9–20.

Lawrence JA, Preene M, Lawrence UL and Buckley R (eds) (2018) *Engineering in Chalk. Proceedings of the Chalk 2018 Conference, London, UK*. ICE Publishing, London, UK.

Lei M and Liang J (2005) Karst collapse prevention along Shui-Nan Highway, China. In *Sinkholes and Subsidence: Karst and Cavernous Rocks in Engineering and Construction* (Waltham T, Bell F and Culshaw M (eds)). Springer, Berlin, Germany, pp. 293–298.

Lord JA, Clayton CRI and Mortimore RN (2002) *Engineering in Chalk*. CIRIA Report C574. CIRIA, London, UK.
Maurice L, Farrant AR, Mathewson E and Atkinson T (2021) Karst hydrogeology of the Chalk and implications for groundwater protection. In *The Chalk Aquifers of Northern Europe* (Farrell RP, Massei N, Foley AE, Howlett PR and West LJ (eds)). Geological Society, London, UK.
Messerklinger S (2014) Formation mechanism of large subsidence sinkholes in the Lar Valley in Iran. *Quarterly Journal of Engineering Geology and Hydrogeology* **47**: 237–250.
Milanovic PT (2004) *Water Resources Engineering in Karst*. CRC Press, Boca Raton, FL.
Newton JG (1987) *Development of Sinkholes Resulting from Man's Activities in the Eastern United States*. US Geological Survey, Reston, VA.
Sabtan AA (2005) Performance of a steel structure on Ar-Rayyas Sabkha soils. *Geotechnical and Geological Engineering* **23**: 157–174.
Sowers GF (1996) *Building on Sinkholes: Design and Construction of Foundations in Karst Terrain*. ASCE Press, New York, NY.
Stapleton DC, Corso D and Blakita P (1995) A case history of compaction grouting to improve soft soils over karstic limestone. In *Karst Geohazards* (Beck BF (ed)). Balkema, Rotterdam, the Netherlands, pp. 383–387.
Talbot CJ and Aftabi P (2004) Geology and models of salt extrusion at Qum Kuh, central Iran. *Journal of the Geological Society London* **161**: 321–334.
Turkmen S (2003) Treatment of the seepage problems at the Kalecik Dam, Turkey. *Engineering Geology* **68**: 159–169.
Vandevelde GT and Schmitt NG (1988) Geotechnical exploration and site preparation techniques for a large mall in karst terrain. In *Geotechnical Aspects of Karst Terrains: Exploration, Foundation Design and Performance, and Remedial Measures* (Sitar N (ed)). ASCE, New York, NY, pp. 86–96.
Waltham AC (1989) *Ground Subsidence*. Blackie, Glasgow, UK.
Waltham AC and Fookes PG (2003) Engineering classification of karst ground conditions. *Quarterly Journal of Engineering Geology and Hydrogeology* **36**: 101–118.
Waltham T (2016) Control the drainage: the gospel accorded to sinkholes. *Quarterly Journal of Engineering Geology and Hydrogeology* **49**: 5–20.
Waltham T and Lu Z (2007) Natural and anthropogenic rock collapse over open caves. *Geological Society Special Publications* **279**: 13–21.
Waltham T, Bell F and Culshaw M (2005) *Sinkholes and Subsidence: Karst and Cavernous Rocks in Engineering and Construction*. Springer, Berlin, Germany.
Warren JK (2006) *Evaporites: Sediments, Resources, Hydrocarbons*. Springer, Berlin, Germany.

Further reading

Beck BF (ed) (1984) *Sinkholes: Their Geology, Engineering and Environmental Impact – Proceedings of the First Multidisciplinary Conference on Sinkholes*. Balkema, Rotterdam, the Netherlands.
Edmonds CN (2018) Five decades of settlement and subsidence. *Quarterly Journal of Engineering Geology and Hydrogeology* **51**: 403–416.
Ford DC and Williams PW (2007) *Karst Hydrogeology and Geomorphology*. Wiley, New York, NY.
Gutierrez F, Johnson KS and Cooper AH (2008) Evaporite karst processes, landforms and environmental problems. *Environmental Geology* **53(5)**: 935–1098 [Special Issue].
Jennings JN (1985) *Karst Geomorphology*. Blackwell, Oxford, UK.
Neal JT and Johnson KS (2003) A compound breccia pipe in evaporite karst: McCauley Sinks, Arizona. *Oklahoma Geological Survey Circular* **109**: 305–314.
Parise M and Gunn J (eds) (2007) *Natural and Anthropogenic Hazards in Karst Areas: Recognition, Analysis and Mitigation*. Geological Society, London, UK.
Waltham AC, Vandenven G and Ek CM (1986) Site investigations on cavernous limestone for the Remouchamps viaduct, Belgium. *Ground Engineering* **19(8)**: 16–18.
Waltham T (2005) Karst terrains. In *Geomorphology for Engineers* (Fookes PG, Lee EM and Milligan G (eds)). Whittles, Dunbeath, UK, pp. 318–342.
Waltham T (2008) Sinkhole hazard case histories in karst terrains. *Quarterly Journal of Engineering Geology and Hydrogeology* **41**: 291–300.
Younger PL, Lamont-Black J and Gandy CJ (2005) Risk of subsidence due to evaporite solution. *Environmental Geology*, **48(3)**: 285–409 [Special Issue].
Yuhr L, Alexander EC and Beck B (eds) (2007) *Sinkholes and the Engineering and Environmental Impacts of Karst*. ASCE, New York, NY.

It is recommended this chapter is read in conjunction with

- Chapter 7 *Geotechnical risks and their context for the whole project*
- Chapter 40 *The ground as a hazard*

All chapters in this book rely on the guidance in Sections 1 *Context* and 2 *Fundamental principles*. A sound knowledge of ground investigation is required for all geotechnical works, as set out in Section 4 *Site investigation*.

Section 4: Site investigation

Section editor: **John A Davis**

doi: 10.1680/icemge.66816.0601

Chapter 39
Introduction to Section 4

John A Davis Geotechnical Consulting Group, London, UK

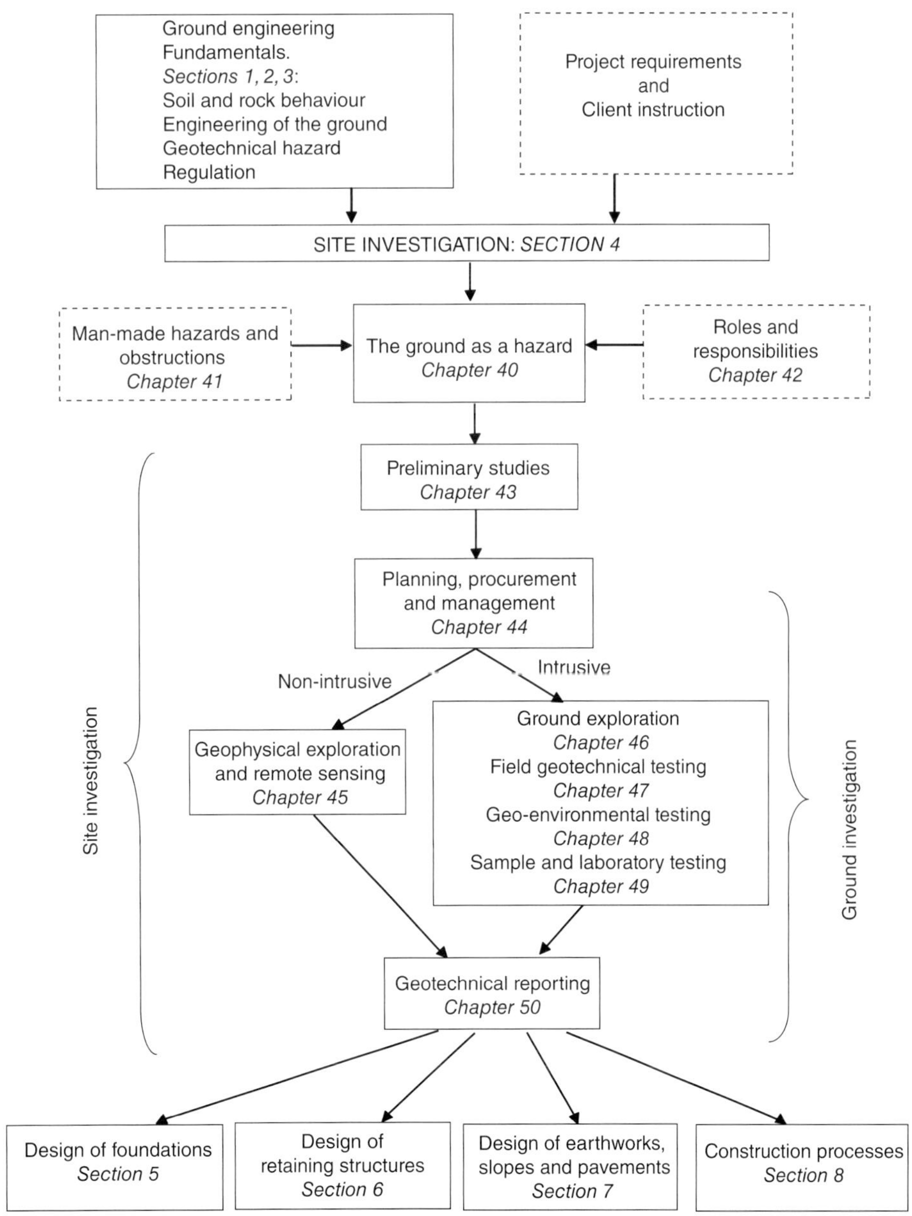

Figure 39.1 The structure of Section 4 and its relationship to other sections

Site investigation provides a bridge to link the requirements of a specific project with its successful design and execution. Knowledge of the fundamental principles of material behaviour, ground hazards and construction risk are necessary at the outset, and these are discussed in the preceding sections.

In preparing this section, the authors have focused on providing general guidance on principles, without overly delving into detail. For example, general classifications systems for soils and rock are provided in standard texts such as BS EN ISO 14688 (BSI, 2002, 2004) and 14689 (BSI, 2003), and reference to other detailed texts may be required for material-specific classification systems. The emphasis here is on the importance of thought and knowledge to site investigation practice, and the section begins with an appeal for its application.

An understanding of land processes, topography, geology and the influence of human activity on the environment are essential. As discussed in Chapters 40 *The ground as a hazard*, 41 *Man-made hazards and obstructions* and 48 *Geo-environmental testing*, practitioners should have a rudimentary knowledge of potential ground hazards, including natural hazards, man-made hazards and contamination. The legal framework and the roles and responsibilities of all parties should be clearly defined and understood, and this subject is outlined in Chapter 42 *Roles and responsibilities*.

As shown in **Figure 39.1**, the term 'site investigation' describes a process that encompasses a desk study and geotechnical reporting. The term 'ground investigation' refers to the physical activity of carrying out intrusive or non-intrusive investigations: laboratory and field testing are a part of this process. Desk studies are described in Chapter 43 *Preliminary studies* and planning and procurement in Chapter 44 *Planning, procurement and management*.

Chapters 45 *Geophysical exploration and remote sensing* to 49 *Sampling and laboratory testing* deal with the mechanics of ground investigation: geophysics, intrusive exploration, geotechnical and environmental field testing, sampling and laboratory testing. The product of the site investigation process is, of course, the geotechnical report, which, as discussed in Chapter 50 *Geotechnical reporting*, must be clear and unambiguous. The report should be as easily understood and as informative years after writing as it should be immediately after writing. To this end, it is important that any standards or classification systems used are clearly defined; not only should there be a description of what was done, there should also be an explanation of what could not be achieved and why, together with highlighted areas of uncertainty or potential error.

All too often, it is the failure of this important process that has led to compromised safety and economic loss. It is hoped that the discussion within this section will be of interest to readers and help practitioners cross the bridge between aspiration and the successful execution of projects.

Key to all of the above is the ground model concept. This concept is woven through the content of this section. An initial conceptual model of the ground should be developed in the early desk study stages. The conceptual model sets out what the geological and developmental history of the site says about what might be present in the ground. It does not rely on procured investigation data.

As historic and/or procured data is added to the model in the later desk study and investigation stages it becomes an observational model. This observational model (typically recorded in factual and interpretative reports) is then simplified into an analytical model for design and possibly also into a commercial ground model in the form of a Geotechnical Baseline Report (GBR, see Chapter 50 *Geotechnical reporting*). Since ground investigations cannot successfully characterise the entirety of the ground, the observational model always contains uncertainty. Uncertainty equals design and construction risk.

In a design process, this uncertainty tends to be addressed by cautious simplification in analytical models (e.g. by design standards that require selection of cautious characteristic values). This approach may deal with design risk but does not always deal with construction that encounters 'unforeseen' conditions.

The mitigation for any identified design or construction risk arising from this uncertainty should be based on feedback loops to the conceptual model. These should consider: the plausible geological and site history features not identified by the observational model, the inbuilt limitations in many ground investigations; any investigation access constraints; and the potential consequences of difficult encounters with the ground during construction.

A new version of Eurocode 7 is due to be published in 2024. The drafts of the new edition (published for comment in 2022) contain new requirements relating to the reporting of ground models. In 2022 the International Association for Engineering Geology and the Environment (IAEG) published comprehensive guidance on the application of ground models in projects.

References

BSI (British Standards Institution) (2002) BS EN ISO 14688-1:2002: Geotechnical investigation and testing – Identification and classification of soil. Identification and description. BSI, London, UK.

BSI (2003) BS EN ISO 14689-1: 2003: Geotechnical investigation and testing – Identification and classification of rock. Identification and description. BSI, London, UK.

BSI (2004) BS EN ISO 14688-2:2004: Geotechnical investigation and testing – Identification and classification of soil. Principles for a classification. BSI, London, UK.

IAEG (International Association for Engineering Geology and the Environment) (2022) *Guidelines for the Development and Application of Engineering Geological Models on Projects*. IAEG, London, UK. Commission 25 Publication No. 1.

doi: 10.1680/icemge.66816.0603

CONTENTS

Chapter 40
The ground as a hazard
An appeal for an intelligent view of the ground

Jackie A Skipper Geotechnical Consulting Group, London, UK

How humans interact with hazardous ground conditions governs the resultant ground risk. Potentially hazardous ground conditions frequently attract engineering development because of important aesthetic or economic considerations. Many are successfully undertaken, but success requires sufficient investment, sound ground knowledge and an appropriate design. Yet in the majority of projects, the ground is the least well understood material, and this can lead to expensive problems being discovered during the construction phase.

This chapter encourages the reader to become actively knowledgeable about the ground of their project by first discussing the main types of geological hazard which may be encountered on a UK engineering site or project. Second, sources of ground information (many of which are now freely available online) are discussed, along with a system by which these can be interrogated to increase knowledge of ground conditions and design an effective site investigation. Finally, the use of ground models in the communication of these conditions has been shown to be extremely important in the success of engineering projects, ensuring that the engineer avoids becoming a victim of potentially hazardous ground.

40.1. Introduction

Humans often see the ground as hazardous. However, the earth is simply influenced by its own chemistry and physics: gravity, water, heat or lack of it, plants and animals and the sun. On the other hand, humans often *make* the ground hazardous (or rather, it is their behaviour that increases risk) by where they choose to live and work. For example, crumbly cliffs may be potentially hazardous, but they do not represent a risk until humans decide they like the view from the steep crumbly cliffs overlooking the sea, and make their home there. The success of humans as a species has often propelled us to develop, live and work in areas which were previously uninhabited (e.g. steep mountains, bogs), often for good economic or social reasons. Humans, however, have remarkably short memories of risks related to hazardous ground events and frequently emotionally weight an advantage (e.g. a lovely view, a perfect location) over a disadvantage (unstable ground). Fortunately, because of modern advances in civil engineering, most environments can be engineered if enough money, a good understanding of the ground and an appropriate design are involved. Ultimately, ground hazards should only become risks if they are unforeseen and unmitigated (for detail on ground profiles, see Chapter 13 *The ground profile and its genesis*). This introduction is therefore an encouragement to understand the ground better before attempting to engineer it.

Geotechnical engineering is one of the most challenging of engineering specialities because the ground is far more diverse and variable than any man-made material. Considering how important the ground is in the success of most construction projects (amazingly, it frequently keeps buildings upright, sometimes for many hundreds or thousands of years), it is one of the poorest understood of the materials utilised on a regular basis. On the majority of projects, for example, there is a far better sampling and testing regime and an infinitely better understanding of the behaviour of concrete and grout mixes on site than there is of the ground – the ultimate material. The exception is, of course, when a contractor wants to make a claim for unexpected ground conditions.

On some projects it is clear that people end up as victims of the ground rather than understanding and mastering it. 'Understanding the ground' is, however, not the same thing as doing a 'preliminary investigation' (see Chapter 43 *Preliminary studies*) and a 'design investigation' (see Chapter 44 *Planning, procurement and management*). Although part of those processes, understanding is the process by which ground information is initially combined into a Conceptual Ground Model, later augmented with ground investigation data as the Observational Ground Model, communicated to all the relevant parties in the project, and if possible added to throughout the project (Baynes *et al.*, 2021).

Why bother understanding the ground better? Why not, as an engineer once asked, just drill a couple of boreholes, test what comes up and base your design on that? My response then was to use an analogy: if the ground is represented by a wedding cake at the bottom of a black plastic bag, and the borehole is your hand reaching into the black plastic bag, completely blind, to grab a piece, you may end up basing your design on a sugar rosebud. Boreholes represent a tiny fraction of the ground and, moreover, the ground between boreholes seldom consists of strata in perfectly horizontal beds. Hence the ground in *between* the boreholes will not necessarily be the same as that fraction that is recovered in the boreholes.

Many clients (and therefore the companies attempting to get work from them) attempt to save money by having a 'cheap' ground investigation, and this is one of the most common practices in which ground data are likely to be lost. As Chapman and Walsh demonstrate in Chapter 7 *Geotechnical risks and their context for the whole project*, saving money with a 'cheap' site investigation is practically the worst investment in the construction industry over the long term – hence the inverted commas. However, despite improvements in standards of site investigation over the past few years (see Chapters 42 *Roles and responsibilities* to 48 *Geo-environmental testing*), there will always be economic downturns. 'Cheap' site investigations will always be sought and may not be avoidable, but can miss important variability in the ground. This is not only because fewer boreholes, maybe of a cheaper type, will give you less data, but because lower-paid staff are frequently less well trained in what they are logging, sampling and testing. They may have cold and poorly lit sheds in which to work, leading to disenchanted workers with poor colour/texture perception – and even with a full Eurocode 7-approved site investigation it is possible for important information about the ground to slip through the cracks between the specification and the Ground Investigation Report. So it is important to squeeze the most out of every piece of available information as soon as possible when starting on a project.

In geotechnical engineering it may be possible to compensate for less adequate site investigations by adopting a conservative design – by taking fewer risks, or (as many contractors chose to do) taking the basic minimal information available and taking the risk that the ground will be fine. However, those extremes have been shown repeatedly (Chapman and Marcetteau, 2004) not to be the smartest and most cost-effective solutions. Understanding the ground to the best of our ability, and designing with the ground in mind, makes the best economic sense (see also Chapter 2 *Foundations and other geotechnical elements in context – their role*).

40.2. Ground hazards in the UK

Most of the common types of hazardous ground in the UK are dealt with in this volume and the relevant sections in this publication are linked here. These ground hazards can generally be grouped into three categories: *geological hazards*, *geomorphological and topographical hazards* and *anthropomorphic hazards*, and are outlined below.

40.2.1 Geological hazards

This category includes the results of events that occur naturally on Earth but which are often considered unreasonable by humans. Examples include dramatic geohazards such as volcanic activity – resulting in debris or mudflows – and earthquakes – resulting in (for example) the possibility of ground movements or tsunamis. Although common in many parts of the world, the UK is generally considered low risk for these hazards since it is a long way from tectonically active centres. However, there is regular minor earthquake activity in the UK – particularly occasional earthquakes in certain areas – and have in the not too distant past (e.g. the 1755 Lisbon earthquake) been affected by tsunamis due to earthquakes or underwater landslide failures. Seismic hazards do therefore need to be considered for long-life or sensitive projects (such as large tunnels and nuclear power stations) and for some sensitive coastal projects. The British Geological Survey (BGS) provides an abundance of free historical and recent information and data on this subject (BGS, 2023a).

Volcanoes are even less common in the UK – the last ones to erupt on the UK mainland were probably active about 50–60 million years ago. However, bear in mind that *many* of our soils and rocks still contain layers of volcanic ash laid down or reworked within them, and these ash deposits frequently contain swelling clays which may cause heave or settlement, precipitate or exacerbate landslips, slow down tunnelling operations and impede soil handling and reuse (see Section 40.2.2). If in doubt, the clay minerals should be analysed (see NHBC, 2022). The BGS website contains very useful information, historical data and risk maps for these types of geohazards (BGS, 2023b).

The second part of this category includes ground types which are harder, weaker, softer, looser, have more and bigger holes or caves in them, are more variable, less stable, more aggressive or contain water or gas at higher pressures than desirable for this particular project. These types of ground are covered in detail in Section 3 *Problematic soils and their issues*, while folded or faulted rocks are dealt with in Chapter 18 *Rock behaviour*.

In an ideal engineering world, all geological deposits would be laterally extensive, of uniform and predictable thickness and homogeneous in nature. Unfortunately, as previously discussed, the ground is usually variable. While it is possible to be exceptionally lucky and encounter tens of metres of thickness of well-behaved, unweathered, unfaulted, horizontally bedded strata across an entire project site, it is actually not the norm. The reason for this is the variability of processes which contribute to the deposition of soils and rocks, and which alter them afterwards (again see Chapter 13 *The ground profile and its genesis*). Generally the most uniform, homogeneous and laterally extensive rocks are marine continental shelf sediments, which can be deposited as essentially the same type of soil or rock over hundreds of thousands of kilometres; but do not be lulled into a false sense of security. Even amazingly thick, fully marine deposits sediments like Chalk, Carboniferous Limestone and London Clay vary vertically in strength, texture and permeability due to changes in water depth when they were deposited. The nearer to the shore deposition occured, the more variable the sediments get (think Carboniferous Coal Measures, which were deposited in swamp to delta environments, or the Mercia Mudstone Group, which varied from desert to floodplain in origin). In addition, worldwide fluctuations in sea level and the extremes of climate variation over geological time can end up with a stack of potentially very variable sediments indeed – and this variability increases further with the inclusion of the deposition or intrusion of igneous rocks or the processes of folding, faulting and metamorphosis. However, it is important to remember that variability is still only a potential risk *if it is not anticipated.* A preliminary understanding of the environment of deposition can therefore be extremely helpful, and enable an early idea of ground risk.

40.2.2 Geomorphological and topographical hazards

This category includes all the hazards which are primarily concerned with how gravity and erosion have affected the ground, and how water interacts with the ground near or above its surface. Landslips and landslides, mudslides, coastal erosion and flood risk are all within this category. Tectonism (the action of the Earth's tectonic plates moving in relation to each other, resulting in, among other things, mountain-building events) has a lot to answer for. Topographically higher ground is (in geological terms) just waiting for gravity to act on it, to be weathered and fall down, to be talus until it falls further and gets washed into a river, where eventually it breaks down into sand and gets deposited in the sea as sediment. It is important to recognise once more that humans love to build in high places, near the sea and everywhere in between – but that hilly areas do fall down with time, and that coastal and low-lying areas often flood. With the climate warming these hazards will increase due to extremes in temperature, rainfall and storminess.

40.2.3 Anthropomorphic hazards

In the last few thousand years, humans have had an enormous impact on the surface and subsurface of the Earth. Just the last 200 years of human activity have changed the planet more than in all the previous several thousand years put together. Our skills as humans have, in this short time, enabled us to use fossil fuels and minerals to drive transport and industry (often leaving large holes in the ground where they were extracted from) and our skills in civil engineering have enabled us to master many formerly forbidding environments and locations.

An awareness of how an area has been changed by human activity over time may reveal a wide range of human activities, many of which will have implications for particular projects. For example, in the area of east London where this author used to live there was, in reverse chronological order

- a boggy area, previously part of the Holocene Thames floodplain
- a Bronze Age trackway development (early human development, archaeology showing historic faulting)
- a former coal gas production plant (probable contamination)
- Joseph Bazalgette's Northern Sewage Outfall (Victorian tunnels and obstructions)
- past and present industry associated with the London Docks (archaeology, pollution and obstructions).

Superimposed on these layers and issues were

- an abundance of munitions dropped in the area during the Second World War of 1939–1945 (possible unexploded ordnance)
- the relatively new Dockland Light Railway and its tunnels, and normal everyday services
- the newly constructed Thames Tideway Project.

There are many companies who will, for a reasonable fee, help steer a way through this potential multiplicity of man-made hazards, provide site-specific environmental risk information and historical mapping for a given site area – but always be aware that your own basic observations or knowledge may pick up additional human activities that have eluded these reports.

40.3. Predicting what the ground may have in store

Personal computers provide access to more information about the ground of a potential site faster than ever before. The following are a selection of available resources (software and website locations are constantly changing, and the sites listed may change capability or web address with time).

- Google Earth can provide aerial imagery and an idea about geomorphology which previously required long waits or searches, and possibly special viewing equipment. Not only are site records available in seconds, but also the approximate elevation and distances between interesting points. The *layers* options can provide additional information, including local photographs, administrative boundaries, locations of transport infrastructure and much more. It is possible to exaggerate topography and 'fly over' your area to see it from all angles.
- Using Street View within Google Earth (or in Google Maps) can provide a close-up of topography and buildings in an area.
- Historical maps of the area (e.g. https://maps.nls.uk/) can show previous land use, industry, quarries, infrastructure and more.
- Use the BGS GeoIndex (onshore or offshore version as appropriate) and BGS Map Portal to investigate the geology of your site area. Links to both are shown in the References.
- UK flood-risk maps available from GOV.uk provide maps by postcode of flood risk from a variety of sources.
- Google Scholar is an academic search engine that is always worth a look. Enter your site or nearest town name and see if any relevant papers are suggested.
- Standard search engines can be useful to search for your site and local area name together with a few hazard types (e.g. 'landslip' 'sinkholes' 'collapse') to see if anything emerges.

Armed with these resources and a geological map of your site area (see Section 40.4), you can potentially have in minutes a detailed history of the previous uses of the site and of the presence of any industries that may have left adverse ground conditions.

To help concentrate the mind amid all this information, try using the *topography–water–anything odd* (TWA) system.

Topography Running the cursor over the aerial view of the site and the surrounding area in Google Earth will give you an idea of variations in topography. Is the terrain precipitous? Very low-lying? Does the topography make sense and, if not, why is it like this? If you look at the historical (and geological) map as well, is there an obvious reason?

Water Where is the water in this area? Is it in rivers, streams, lakes, marshes, an estuary, the sea? Does it make sense where the water is, or is there poor drainage due to impermeable strata? Is the site in an obvious flood-risk area (see the flood-risk maps)? Is there no water at all in this area? Why? Could it have gone underground into soluble strata?

Anything odd? Consider the following.

- Is there anything that does not seem quite right, or any colours or shapes on the ground that do not make immediate sense in terms of natural or human activity?
- Are there any areas where buildings 'should be' but are not? Could flooding or landslip be responsible for this?
- Road names can be useful indicators of geological or former engineering hazards. Are there road or lane names near your site which include such words as *watermeadow*, *flood*, *spring*, *cave*, *swallow* or *swallowhole* (possible flood or solution hazards), *brick*, *kiln*, *mine* or *quarry* (former mining or quarrying), *undercliff*, *zigzag*, *slip* (possible ground instability problems)? The names *cave* or *dene* can also indicate natural or man-made cavities.
- Using Google Earth, Street View (or Google Maps satellite view) can be another useful source of information, picking up uneven ground surfaces, major cracks in walls and repeatedly re-made roads.

40.4. Geological maps

Geological maps (for example the invaluable BGS GeoIndex, and the BGS Maps Portal) may, to the average engineer, be slightly confusing. They use bizarre colours and odd symbols. However, they do contain vast amounts of useful information if you have a little patience. If you want help, ask a geologist or look online for how to read geological maps.

A note of caution: engineers often think UK geological maps are 'gospel'. They are not – they are an interpretation produced by clever geologists who do a lot of fieldwork, incorporate as many good-quality boreholes as possible and then extrapolate their findings. Without X-ray vision, their geological maps, while a very good best guess of what the ground consists of, can occasionally be wrong. Another important point to mention here is that just because a geological formation is called Kimmeridge Clay, Lias Clay or Gault Clay and so on, it does not mean that the formation is only clay. Likewise, formation names with 'sand' at the end (e.g. Lower Greensand, Arden Sandstone) rarely consist entirely of sand. Soils and rocks of any age or name usually contain naturally occurring harder layers or weaker layers, or indeed sandier layers – be prepared and read the geological map carefully. Again, the BGS has very useful tools on their website such as the Lexicon of Named Rock Units (also useful for looking up soils – 'rock' means rock *or* soil in this instance). The BGS onshore borehole historic database (BGS GeoIndex) is now largely free, and in combination with the Lexicon will probably tell you far more than you ever thought you needed to know about the strata in the area of your project, and this information will act as the basis for a desk study and contribute to a refined ground model based on a well-designed site investigation.

40.5. Conclusions

Having a better understanding of the ground is a huge leap forward and will help enormously in planning and designing for a project, but what appears to make the greatest difference is communicating this understanding to others (Skipper, 2008). This author first experienced this principle in 2003 when working on the Dublin Port Tunnel, where communication of the ground model to *all* levels of staff (from site investigation staff to designers, supervisors and site workers) made a huge difference to their cooperation, communication and feedback. This improved level of understanding allowed the use of the observational method in what was complex and challenging geology in a very sensitive location (Long *et al.*, 2003). Since then this author has been involved in the use of this principle in a wide range of projects from the small to the very large and has seen it result in a wide range of improvements, from better specified site investigations to improved ground descriptions and interpretations, in turn leading to optimised design and better risk management. To conclude, an intelligent understanding of the ground makes the best technical and economic sense, and communicating this understanding with and to others maximises this value to engineering projects.

References

Baynes FJ, Parry S and Novotný J (2021) Engineering geological models, projects and geotechnical risk. *Quarterly Journal of Engineering Geology and Hydrogeology* **54(2)**, 10.1144/qjegh2020-080

BGS (British Geological Survey) (2023a) Earthquakes around the British Isles in the last 60 days. https://earthquakes.bgs.ac.uk/earthquakes/home.html (accessed 21/03/2023).

BGS (2023b) Swelling and shrinking soils. https://www.bgs.ac.uk/geology-projects/shallow-geohazards/clay-shrink-swell (accessed 21/03/2023).

Chapman T and Marcetteau A (2004) Achieving economy and reliability in piled foundation design for a building project. *The Structural Engineer*, 2 June: pp. 32–37.

Long M, Menkiti CO, Kovacevic N, *et al.* (2003) An observational approach to the design of steep sided excavations in Dublin

Glacial Till. *Proceedings of the Underground Construction, London, UK*, pp. 443–454.

NHBC (National House Building Council) (2022) 4.2 Building near trees. NHBC, Milton Keynes, UK.

Skipper JA (2008). Project specific geological training: a new tool for geotechnical risk remediation? *Proceedings of Euroengeo 2008, International Association of Engineering Geology Conference, Madrid, Spain.*

Further reading

Bryant E (2004) *Natural Hazards.* Cambridge University Press, Cambridge, UK.

Griffiths JS (2002) *Mapping in Engineering Geology.* Geological Society, London, UK.

Mitchell C and Mitchell P (2007) *Landform and Terrain, the Physical Geography of Landscape.* Brailsford, Birmingham, UK.

Useful websites

BGS Geoindex: https://mapapps2.bgs.ac.uk/geoindex/home.html
BGS Lexicon of Named Rock Units: www.bgs.ac.uk/lexicon
BGS map viewers: https://www.bgs.ac.uk/geological-data/map-viewers
Google Earth: www.google.com/earth/index.html
National Library of Scotland (covers the UK) https://maps.nls.uk/
UK Government flood-risk service: https://check-long-term-flood-risk.service.gov.uk/postcode

It is recommended this chapter is read in conjunction with

- Chapter 7 *Geotechnical risks and their context for the whole project*
- Chapter 8 *Health and safety in geotechnical engineering*
- Chapter 13 *The ground profile and its genesis*

All chapters in this book rely on the guidance in Sections 1 *Context* and 2 *Fundamental principles.* A sound knowledge of ground investigation is required for all geotechnical works, as set out in Section 4 *Site investigation.*

Chapter 41
Man-made hazards and obstructions

John A Davis Geotechnical Consulting Group, London, UK
Clive Edmonds Subsolve Ltd and Stantec UK Ltd, Reading, UK

doi: 10.1680/icemge.66816.0609

CONTENTS

Man-made hazards in the ground take many forms. The principal forms discussed here are voids, contamination, unexploded weapons, obstructions and buried services. Many different geological materials have been mined in the UK over thousands of years. These include materials used for fuel (coal and oil shale), a wide variety of metal ores, non-metals and stone itself. Voids left by mining activity occur in a wide range of geological settings and both the mining methods and the nature of the voids themselves are similarly varied. References are made to methods of void detection and treatment. Man-made contamination in the ground is common in previously developed sites, especially those with an industrial history. A legacy of war brings the risk of encountering unexploded munitions in the ground. Processes for minimising the risks associated with contamination and unexploded weapons are discussed in this chapter. Man-made obstructions in the form of buried structures or parts of structures are commonly encountered in ground investigations and ground works. These can be particularly hazardous if they contain voids. Underground services are an almost ever-present hazard for ground investigations. Risk minimisation techniques are also discussed for these classes of hazard.

41.1. Introduction

This chapter briefly describes man-made hazards and obstructions that might delay or compromise the implementation of ground investigations. These man-made hazards and obstructions may also affect subsequent design or construction, creating obstructions to penetration of foundations, hard areas or variability. This is discussed in only a limited way here.

In the context of ground investigations a hazard is taken here to be something man-made within the ground that has the capacity to cause harm in an occupational health and safety sense. Above-ground operational health and safety hazards are not discussed.

An obstruction is taken here to be something man-made within the ground that has the capacity to delay or prevent progress of a ground investigation. Some examples of pre-ground-investigation bureaucratic processes that can cause delay if not addressed are also given.

The possible occurrence of both the physical and bureaucratic issues raised here should be addressed in the desk study, and the scope and form of any subsequent ground investigation should be adjusted accordingly. See Chapter 43 *Preliminary studies* for more detail on the scope and content of desk studies.

41.2. Mining

Great Britain has a long and rich history of mineral extraction using mining techniques, dating back to Roman and even Neolithic times. The desire to obtain minerals of special value and use has driven the development of mining activity through time. As a result, significant areas of land contain mined ground and a legacy of ground instability.

Historical mine workings present a significant subsidence hazard to existing and new-build construction, as well as a risk of personal injury or death. From a health and safety perspective it is therefore necessary to identify where the mine workings are, the nature and extent of the void space and the degree of risk posed to the surface.

For coal mining areas the investigation and treatment of workings and mine entries needs to be notified and agreed in advance with the Coal Authority. Outside of coalfields it is usual to agree such matters with the local council, who may wish to involve the Mineral Valuer and the Health and Safety Executive. Awareness of the legal framework, such as the Mines and Quarries Act 1954, is important as well.

41.2.1 Mineral types

From a historical mining perspective it is convenient to think of mineral types in terms of coal and related minerals (e.g. fire clay, gannister, iron ore, oil shale) extracted within the many coalfield areas of Great Britain (**Figure 41.1**) and those minerals that were mined outside of coalfield areas.

Minerals that were extracted in non-coalfield areas include metal ores (e.g. iron, lead, zinc, copper, silver, gold, tin); non-metals (e.g. fluorspar, calcite, barytes); evaporites (e.g. gypsum, salt); and stone (e.g. slate, sandstone, limestone, chalk). For reference, the recorded spatial distribution of non-coal mining is also shown in **Figure 41.1**.

Currently mining is restricted to salt, potash, gypsum and others (silica sand, limestone, barytes, fluorspar, slate and haematite). Coal mining has virtually ceased. Further details can

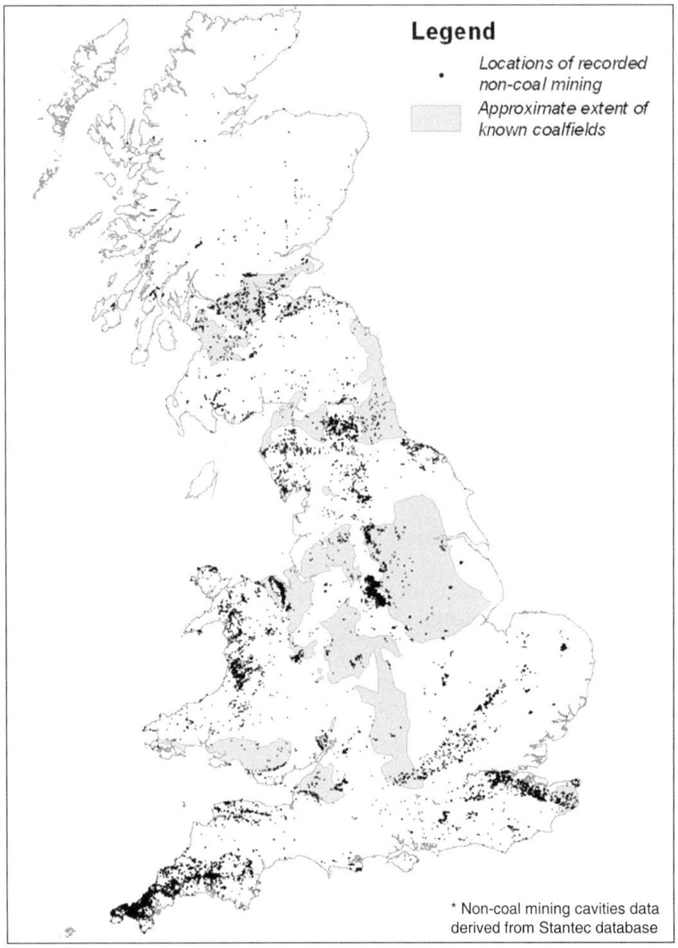

Figure 41.1 Spatial overview of mining in Great Britain

be obtained from publications such as the *United Kingdom Minerals Yearbook* (Bide *et al.*, 2019).

41.2.2 Methods of mining

Methods of mining have changed with time and technology, but also reflect differing geological settings.

A mineral at outcrop would be extracted by quarrying. However, as the overburden rock/soil thickness that needs to be removed from above the mineral layer increases, extraction by adits driven through the quarry face becomes more efficient for continuing the extraction underground without the need to remove the overburden (see **Figure 41.2**). For minerals that occur at shallow level in layers that do not dip steeply, the next simplest style of mining is the 'bellpit'. A narrow, unsupported shaft was dug down to the mineral layer and the extraction continued outwards from the shaft base radially to form a 'bell shape' (**Figure 41.2**). Other regional variants were also developed, for example in the chalk of southern and eastern England deneholes, chalkwells and chalkangles were used (Edmonds *et al.*, 1990).

Later, more advanced forms of mining, referred to as 'pillar-and-stall workings', were developed. The workings, accessed through shafts and adits, comprised a series of tunnels (stalls) excavated outwards from the entry point, linked by cross-tunnels to create a grid of intervening pillars to support the mine roof (**Figure 41.2**). For coal extraction the pattern of tunnels and pillars that were created, and the names referring to them, varied regionally (Littlejohn, 1979).

For many minerals (e.g. limestone, chalk and gypsum) this form of mining continued as the preferred method of working. However, for coal, the mining style was further developed into the 'longwall' method. This technique involved coal being completely extracted along a 'longwall' face, progressing forwards below large areas. Access to the working face and removal of the mined coal was achieved through connecting perpendicular roadways. Propping was used to support the working face. In the areas between the roadways the roof was allowed to progressively fail and subside as the working face moved forwards. The original mines were hand-dug, but in time the process became mechanised (Healy and Head, 1984, now superseded by Parry and Chiverrell, 2019).

Other methods of mining were developed to suit the particular characteristics and geological setting of the mineral. Where a mineral exists in the form of an irregular ore body or steeply dipping vein it is common to follow the depositional pattern extending through the host rock. In limestones the ore or vein may be accessed through naturally occurring karstic cavities. Larger ore bodies or veins can be mined upwards (stoping). These forms of past mining are typical in areas like the Pennines, North and South Wales, Peak District, Lake District, Mendip Hills and Cornwall.

The soluble nature of salt has been used to extract it by brine pumping (solution mining), particularly in parts of Cheshire (Bell, 1975), Lancashire, Somerset, Staffordshire and Worcestershire (Cooper, 2020). A useful regional summary of mineral exploitation is presented in Parry and Chiverrell (2019), with additional geohazard insights provided by Giles and Griffiths (2020).

41.2.3 Mining history and information sources

Many of the mining techniques described above date from the Middle Ages and earlier; however, mining activities notably intensified from the late 1600s, through the 1700s and 1800s, continuing into the 1900s. It is evident that mining activities may have started and ceased in an area before publication of the first edition of the Ordnance Survey (OS) maps (typically dating from the 1870s to 1880s). Consequently, caution should be applied when making use of historical OS maps alone to research the history of a site in a mining area.

Information on coal and non-coal mining is available from a wide variety of sources, often not mainstream or easily accessible. Specialist knowledge may be required to fully interpret and understand the data. However, since the early

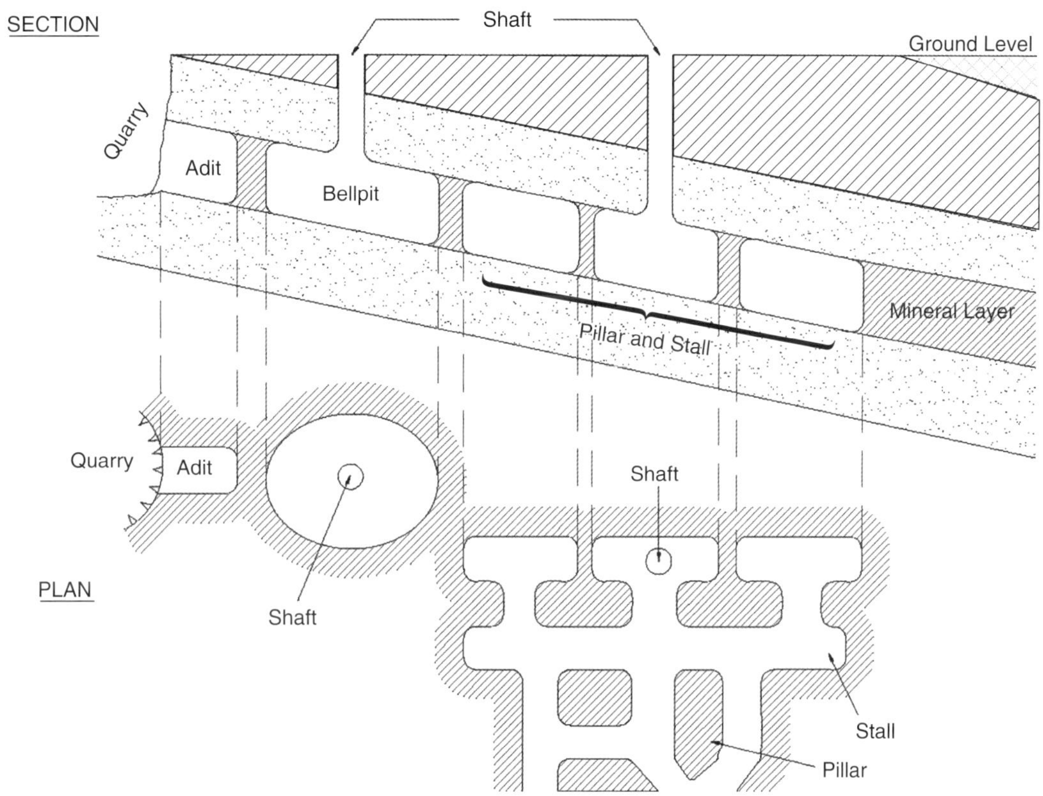

Figure 41.2 Illustration of mineral extraction techniques used as the mineral layer deepens from outcrop

1990s a number of national and regional studies have been published (e.g. Arup Geotechnics, 1990), making information more accessible – now further updated by Parry and Chiverrell (2019). The recorded mining data have been captured digitally and made available by the British Geological Survey and Stantec UK Ltd. For coalfield areas, a large digital mining archive is held by the Coal Authority, mostly dating from 1872 onwards when it became a legal requirement to produce an accurate plan of the underground mine workings. Other smaller collections of data are held by various councils, the British Geological Survey online as well as caving and mining-history organisations. Obtaining relevant mining information from these sources is essential for a desk study to check for the potential for current and past mining activities.

41.2.4 Ground instability hazard

Collapses over abandoned mines pose a ground instability hazard. Common subsidence triggers include leaking water-carrying services and soakaways, changes in water table level, heavy rainfall, disturbance by excavation and surface loading.

The most common form of instability relates to the collapse of old shaft entries, often left poorly sealed and backfilled. Large ground movements can result if a shaft has deteriorated, and failing ground can move both vertically down and laterally into open, connected void space at depth (see **Figure 41.3**). Subsidence may also occur over shallow workings and adit entrances, but again depends on the failure mechanism and the extent of remnant interconnected mine void space at depth.

Figure 41.3 An example of the results of collapse of old shaft entries (Courtesy of Peter Brett Associates LLP)

Large-scale mining of a succession of coal seams can produce widespread subsidence troughs, which are problematic for properties and infrastructure at the surface (Parry and Chiverrell, 2019). Other forms of longer-term ground movement over coal mining areas can be associated with fault lines and major discontinuities where stress relief may be concentrated (e.g. Donnelly *et al.*, 2008).

A range of ground instability problems caused by abandoned mines associated with a variety of other mineral types is presented in the Technical Review Case Study Reports prepared by Arup Geotechnics (1990) as part of a national review study. More recent case studies and reviews are presented by Edmonds (2018), Parry and Chiverrell (2019) and Giles and Griffiths (2020).

Discussion of how local authorities and developers should take account of ground instability relating to mining, in the context of planning and development, is set out in the National Planning Policy Framework (NPPF) (Ministry of Housing, Communities and Local Government, 2021) Clauses 183 and 184, which supersede the previous more detailed published planning guidance (PPG 14).

41.2.5 Investigating mine workings

A wide variety of direct and indirect investigation techniques exist for detecting mine workings. For investigations of coal mine workings, advance permission must be obtained from the Coal Authority.

Indirect techniques include the use of aerial photography and satellite imagery to look for shafts, adits or associated surface excavations, spoil heaps and infrastructure. Light detection and ranging (LiDAR) techniques are also useful for finding surface features associated with former mining sites, particularly since the pulsed laser light is capable of penetrating surface vegetation. The viewing of historical images as well as utilising different parts of the electro-magnetic spectrum can be useful. There are many texts describing the types of analogue/digital data available, together with methods of image manipulation and interpretation (e.g. Lillesand *et al.*, 2015). In addition, increasing use is being made of interferometric synthetic aperture radar (InSAR) to view possible evidence of ground movement over former mining areas.

Other forms of indirect techniques comprise geophysical surveying. In a mining context the techniques need to be sensitive to the detection of open voids, partially to fully collapsed voids, water-filled voids, broken or collapsing rock/soil sequences, made ground backfills, brick or metal shaft linings and so on. Depth to the mine void and the surface environmental setting of the survey both play a part in the final choice of survey – if a survey is even feasible. There are many texts describing the theory and practice of geophysical surveying (e.g. Reynolds, 2011).

Again, there is a range of direct investigation techniques that can be used to detect mine workings. Considerations for the choice of method depend on a number of factors, such as geology, depth of mineral, style of mining, whether intact or collapsed, access and health and safety risks. The potential for hazardous gas or polluted water escapes also needs to be evaluated. In the case of soluble strata or collapse situations, the choice of drilling flush is important. Inaccessible open voids may be viewed directly using down-hole closed circuit television (CCTV) cameras, which allow the style of mining and condition of the workings to be evaluated. In circumstances where there is a need to measure the size and orientation of open workings, down-hole laser survey techniques can prove useful (see **Figure 41.4**). Where access is possible and safe, direct visual inspection can then be undertaken (see **Figure 41.5**). The various investigation techniques are described in detail elsewhere in this manual.

41.2.6 Mine working investigation case history

In order to illustrate how a combination of indirect and direct investigation techniques can be used effectively to detect

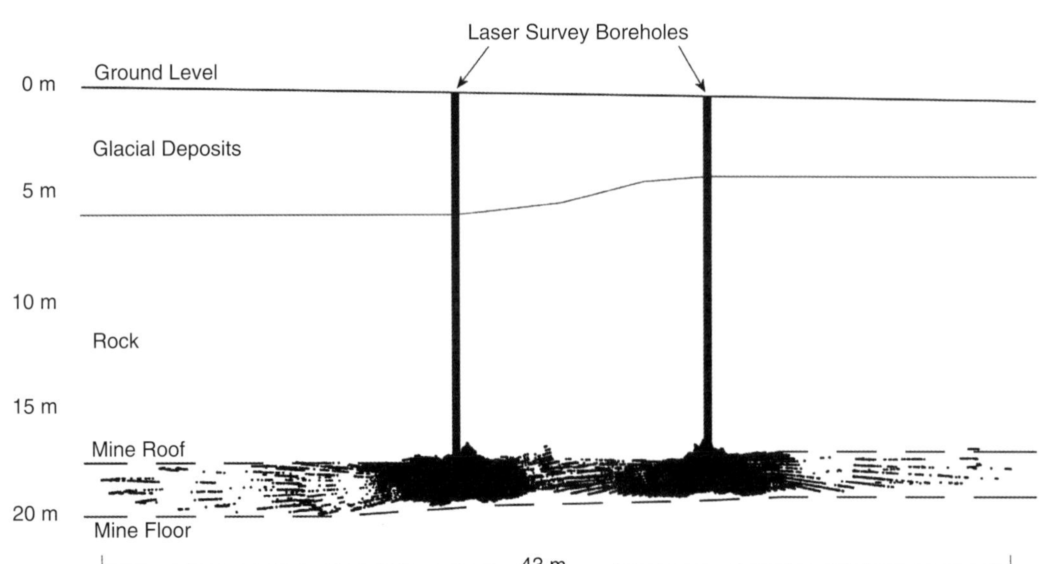

Figure 41.4 Results of down-hole laser survey techniques

Figure 41.5 Direct visualisation of mine workings (Courtesy of Peter Brett Associates LLP)

Figure 41.6 A major ground subsidence, 4 January 2000 (Courtesy of Peter Brett Associates LLP)

historical mine workings, the following account briefly describes the investigation approach at Field Road, Coley, Reading (SU710728 area).

A major ground subsidence occurred on 4 January 2000, causing the collapse and loss of the fronts of two houses, their gardens, the pavement and adjacent road. The collapse measured about 9 m across and was several metres deep (**Figure 41.6**).

After initial infilling of the collapse, a surrounding ring of close-centred dynamic probes was undertaken to establish the likely cause of the collapse. The results revealed a series of open voids that were interpreted to be old chalk mine workings, having a mine roof level at about 7.5–11 m below ground level (bgl) and a mine floor level at about 13–14 m bgl. A typical probing profile is shown in **Figure 41.7**. The ring of probes was extended outwards in a grid pattern that revealed yet more mine workings forming an irregular pillar-and-stall style of working. A smaller number of light cable percussion boreholes was positioned strategically to provide sampled geological profiles to confirm the interpretations being made for the

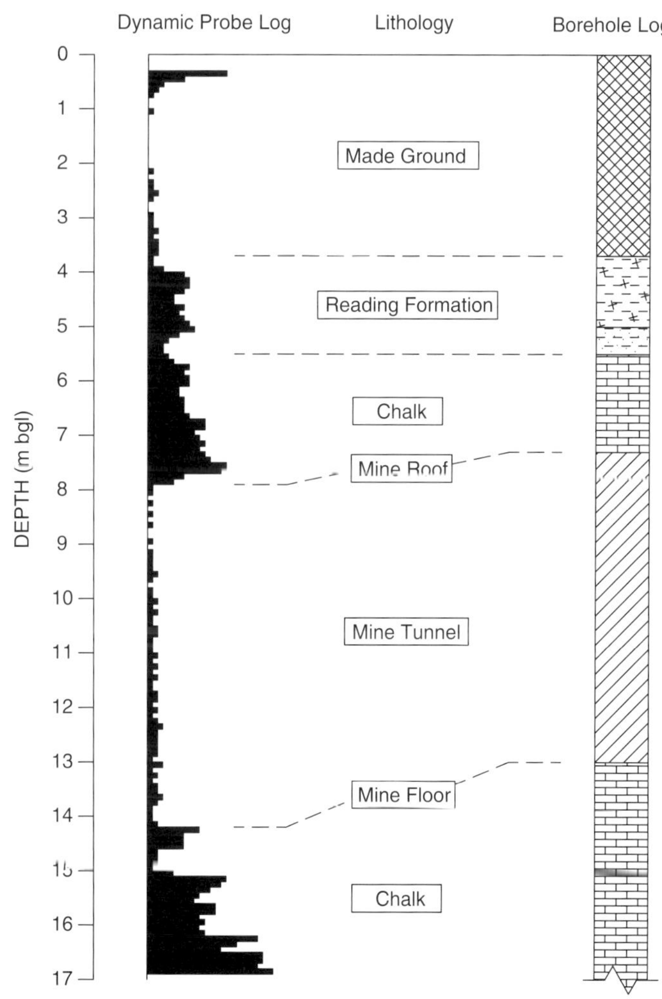

Figure 41.7 A typical probing profile

probing profiles. A good geological correlation was achieved (**Figure 41.7**).

To check how far the mine workings might extend, a microgravity geophysical survey was conducted initially as a trial along the road, south of the collapse. The trial was successful in detecting the pattern of mine workings as a series of Bouguer anomalies, and so the survey was extended further along the road in a southerly direction and outwards into the adjacent garden areas. The survey interpretation was then checked by carrying out additional dynamic probes. **Figure 41.8** shows an example

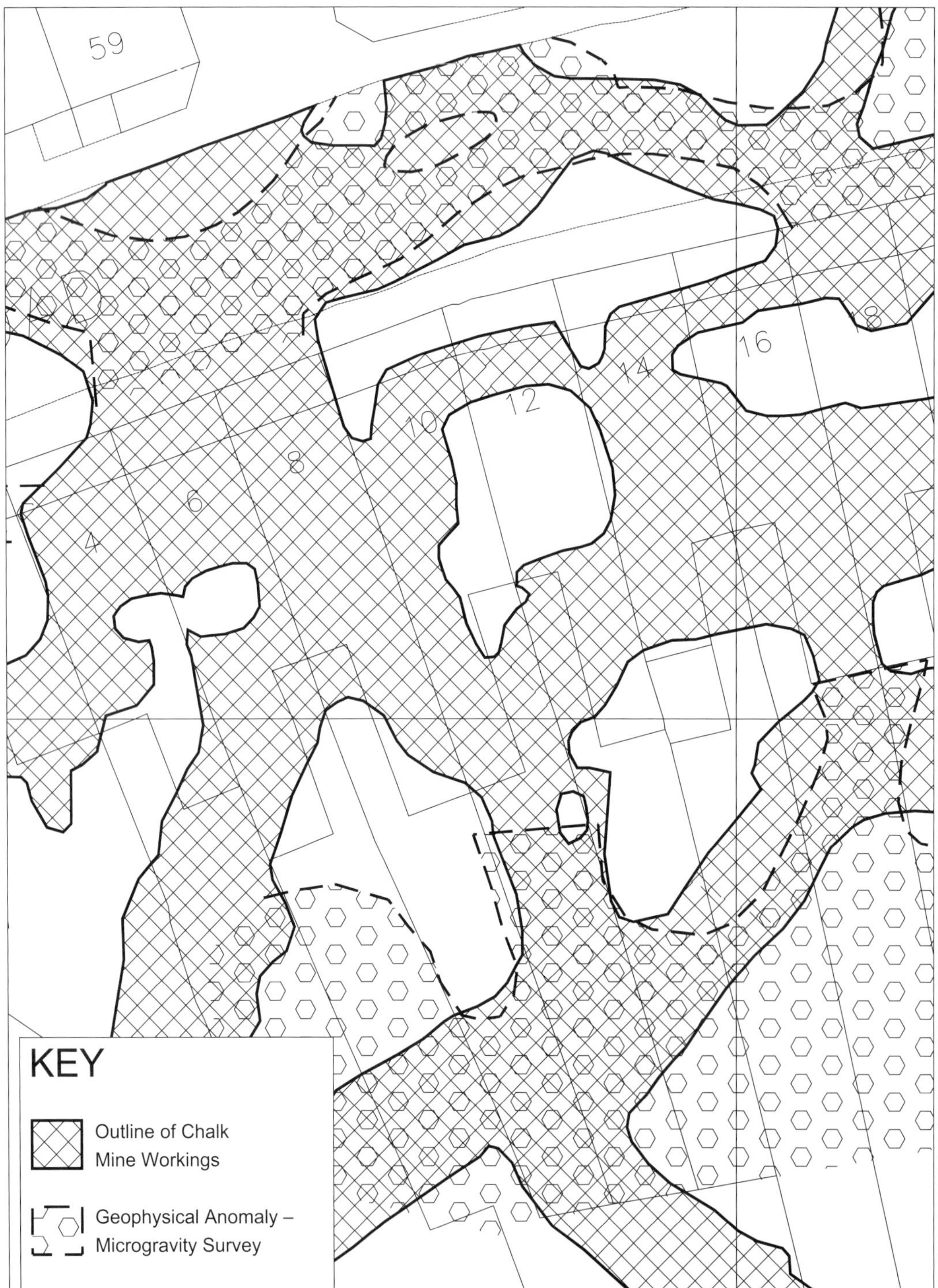

Figure 41.8 An example of the microgravity survey results compared with the actual mapped mine workings detected by dynamic probing

of the microgravity survey results compared with the actual mapped mine workings detected by dynamic probing. The geophysical survey results demonstrated that the microgravity technique provides a good indication of where mine workings are to be found, so useful in a reconnaissance sense, although locating the exact mine tunnel positions was less precise.

When faced with the task of mapping out the pattern and extent of larger-scale pillar-and-stall mine workings, there is a tendency to use a grid pattern of exploratory holes. If the mine workings are in an advanced state of collapse, or have collapsed, then this approach may be unavoidable. However, where there are sections of open, intact mine workings it is possible to reduce the number of exploratory holes being bored. At Field Road, for example, if a probe encountered a void the access to the void was enlarged using a window sampler borehole and plastic casing inserted into the hole to the roof level of the void. A small-diameter borehole CCTV camera was then introduced by way of the casing, and a visual survey of the mine void was undertaken (**Figure 41.9**). The direction of the mine tunnel was inferred from the survey, and the next probes were positioned accordingly, some 5–10 m away to intersect the tunnel further along, and the survey process was repeated again. This offered a cost-effective way of mapping the mine tunnels present, greatly reducing the numbers of probes required compared with adopting a conventional grid-pattern approach. Where dynamic probes cannot penetrate to sufficient depth to map mine workings, or angled holes beneath a building are required, rotary drilled boreholes can be used instead (**Figure 41.10**).

41.2.7 Treatment of mine workings

Depending upon the nature and extent of the mine workings defined by the ground investigation, there are a variety of remedial mine treatment options. It is often the case that historical mine workings will comprise a mix of open voids, collapsed ground and further disturbed/broken ground over and around the mine workings.

Sometimes access to the mine workings may be possible from ground level, providing a means to infill and stabilise the workings using grouting techniques such as progressive infilling behind shuttering (**Figure 41.11**).

Figure 41.9 Visual survey of a mine void using a small-diameter borehole CCTV camera

Figure 41.10 An example of the use of rotary drilled boreholes (Courtesy of Peter Brett Associates LLP)

Figure 41.11 Infill and stabilisation of mine workings using grouting techniques (Courtesy of Peter Brett Associates LLP)

Often it is unsafe to enter the mine workings, in which case the infilling works have to be undertaken from the surface, involving grout injection by way of boreholes. Open voids can be stabilised using bulk infilling methods (e.g. pumped as grout or another filler such as sand or pulverised fuel ash washed in using water). Broken rock sequences overlying collapsed workings can be strengthened using pressure grouting techniques (high-slump grouts) to seek out and infill dispersed voids or within unconsolidated sequences using compaction grouting techniques (low-slump grout) to compact the loose material.

Where shaft entries have collapsed they may be infilled with a granular material, which is then solidified with a cementitious grout as necessary to counter settlement and loss at the base with time. The shaft top is then capped with a reinforced slab, whose depth below surface and size depends upon the nature of the near-surface ground conditions, shaft diameter and construction (Parry and Chiverrell, 2019).

Often raft foundations are utilised in areas of past mining to protect the structure from subsidence damage. However, if there is still a risk of voids rising to the surface, consideration needs to be given to the design span and cantilever of the raft compared to the likely size of the crown hole. In situations where the near-surface ground conditions are of poor quality but the risk of subsidence is deemed to be low, then piled foundations can be considered.

For shallow workings it might be feasible to pile down through the mine floor level, end bearing into the stable ground below. Such piles need to be sleeved through the ground above the mine floor level to resist negative skin friction as the ground settles/collapses and to free-stand through voids if necessary.

In mined areas consideration also needs to be given to protecting services and infrastructure against subsidence damage. Some additional protection can be given by placing high-tensile-strength geogrid within the road pavement construction and below service runs.

It is also very important to carefully consider how surface water drainage is to be collected and discharged at sites that may be underlain by mine workings. Current preferences are to encourage the use of sustainable drainage methods (SuDS), but if this results in discharge of water into the ground above mine workings then it might trigger ground instability. The design approach needs to be carefully risk assessed so that the potential for instability and how best to mitigate it is fully understood.

The various ground treatment techniques and foundation types are described in detail elsewhere in this manual.

41.3. Contamination

Contamination of the ground by pollutants can be both a hazard and an obstruction. The nature of any possible contamination should be addressed in the desk study and carried through to construction design and management (CDM) risk assessments for site activities by the investigation designer, regardless of the status of the investigation in CDM terms. The ICE Ground Investigation Specification (Site Investigation Steering Group, 2022) has a simple but somewhat dated site health and safety classification scheme for contaminants. This is useful in a contractual sense and should be completed, but it does not replace the need for a comprehensive designer's risk assessment. See Chapter 48 *Geo-environmental testing* for further discussion of the philosophy, sampling and testing of potentially contaminated ground.

If it is not addressed directly, contamination can also be a significant bureaucratic hazard in the UK. Where ground investigations are for projects involved in the Local Authority Planning process then third parties (the Local Authority and the Environment Agency (EA), or SEPA in Scotland) have a significant interest in the content of ground investigations and can control their content by the use of planning conditions – see Planning Policy Statement 23 Annex 2, which is available online (see the Useful Websites section at the end of the chapter).

Failure to consider the needs of the Local Authority or the EA/SEPA or consult with them when designing a ground investigation could result in further investigation being required by the planning conditions. This is a significant matter. Such conditions would typically require an additional investigation to be completed and assessed before development can begin. The text of typical conditions is given in Appendix 2B of Planning Policy Statement 23 Annex 2. An example is given below; it is clear that failure to discharge this condition could lead to a great deal of delay. This type of condition would apply to any type of site, even those where the apparent risk of 'contamination' appears to be very low. What the Local Authority seeks is a demonstration that the risk of contamination is negligible, or measures to deal with any confirmed source–pathway–target linkages.

> The development hereby permitted shall not begin until a scheme to deal with contamination of land and/or groundwater has been submitted and approved by the LPA and until the measures approved in that scheme have been fully implemented. The scheme shall include all of the following measures unless the LPA dispenses with any such requirement specifically and in writing:
>
> 1. A desk-top study carried out by a competent person to identify and evaluate all potential sources and impacts of land and/or groundwater contamination relevant to the site. The requirements of the LPA shall be fully established before the desk-top study is commenced and it shall conform to any such requirements. Two full copies of the desk-top study and a non-technical summary shall be submitted to the LPA without delay upon completion.
> 2. A site investigation shall be carried out by a competent person to fully and effectively characterise the nature and extent of any land and/or groundwater contamination and its implications. The site investigation shall not be commenced until:

(i) a desk-top study has been completed satisfying the requirements of paragraph (1) above;
(ii) The requirements of the LPA for site investigations have been fully established; and
(iii) The extent and methodology have been agreed in writing with the LPA. Two full copies of a report on the completed site investigation shall be submitted to the LPA without delay on completion.

3. A written method statement for the remediation of land and/or groundwater contamination affecting the site shall be agreed in writing with the LPA prior to commencement and all requirements shall be implemented and completed to the satisfaction of the LPA by a competent person. No deviation shall be made from this scheme without the express written agreement of the LPA. Two full copies of a full completion report confirming the objectives, methods, results and conclusions of all remediation works shall be submitted to the LPA.

Appendix 2B, Planning Policy Statement 23 Annex 2

Ground investigations have the ability to promote the transport of contaminants. Great care needs to be given to the design of ground investigations where mobile contaminants exist adjacent to aquifers or surface waters. A careless approach can result in prosecution. See Chapter 48 *Geo-environmental testing* for more detail.

41.4. Archaeology

The possibility of archaeological remains being present on a site should be addressed in the desk study, preferably by approaching the Local Authority Planning Department in the UK. Where archaeological remains are likely to be found the Local Authority is likely to seek to influence the content and form of the ground investigation. This may have a significant impact on the duration, content and conduct of the investigation, and early consultation is strongly recommended. Specific archaeological investigations may be required, either as part of the ground investigation or as a separate exercise. See Chapter 43 *Preliminary studies* for more detail on desk studies and archaeology.

The archaeological content of combined archaeological and geotechnical investigations could range from a watching brief for trial pits and boreholes, where an archaeologist examines the near-surface deposits as the geotechnical investigation passes through them, to specific boreholes and trial pits. Where archaeology-specific boreholes are required they usually involve continuous sampling of near-surface deposits. The destructive examination of these continuous samples, often off site, usually precludes the use of these parts of the borehole for any geotechnical use. Finding archaeological remains, especially human bones or burial grounds, can cause significant delay to a ground investigation. Where bones are found that are suspected to be human the police should be called immediately. In these cases the police are likely to initially consider the site to be a crime scene. This will cause delay; however, the consequences of not contacting the police could be significantly more onerous. Testing for specific bacteria and viruses may be required, depending on the context of the find.

Archaeology can present significant restrictions to design and construction works. In a number of instances the archaeology must be left in place, and either bridged over or minimally disturbed by piling. See the English Heritage (2019) guidance note on piling and archaeology. Construction may be subject to delays in a similar fashion to the site investigation.

41.5. Ordnance and unexploded ordnance (UXO)

There are obvious hazards and potential delays associated with inadvertently finding buried explosives in a ground investigation. Guidance on assessing such risks and on mitigation techniques is given in CIRIA Report C681 (Stone *et al.*, 2009). Higher-risk areas are typically those in urban and industrial areas of major cities that, because of their historic situation, are such that the entry points of unexploded bombs might have gone unnoticed. Examples would be intertidal areas or areas of already demolished or damaged buildings. Desk study information that can be used to assess UXO risks includes

- historical maps (comparison of pre- and post-war maps)
- wartime bomb damage maps
- wartime records of 'abandoned' unexploded bombs.

Detailed guidance is given in CIRIA Reports C784 (Cooper and Cooke, 2016) and C785 (Bowman *et al.*, 2019).

Unexploded bombs can move significant distances laterally below ground at the time of impact. They can also move significant distances after impact where they have fallen into large tidal rivers. There are even known instances of unexploded bombs being unknowingly picked up, placed in fill and then dug up again several years later.

Backfilled explosion craters also raise the possibility of otherwise unexpected ground conditions and the selection of inappropriate ground investigation techniques. These issues should be considered at the desk study stage. Where backfilled craters are expected, low-cost simple techniques that can be used at relatively close centres can be used to identify variations in backfill thickness. These would include trial pits in general and window sampling for pits filled with the local natural soil if that is generally suitable for this technique. Buried craters in urban areas are more likely to be backfilled with coarse demolition rubble. Craters in more rural settings, such as airfields, are more likely to be backfilled with the soil displaced by the impact and explosion. Geophysical techniques may also be of use in this scenario. The choice of technique will depend on the nature of the backfill compared to the natural soil.

Where there is considered to be a higher than usual risk of encountering UXO in a borehole, the usual mitigation choices are to move the borehole, to abandon the borehole or to employ surface or more usually down-hole magnetometer techniques to assess the presence of a bomb before the borehole reaches it. This activity requires the use of stainless steel casing and, as such, must be pre-planned.

As the site investigation is the first intrusion onto a site, the requirements of the investigation are often more onerous than during later construction. However, probing or further magnetometer surveys may be required during piling, and strict procedures followed when employing bulk excavation.

41.6. Buried obstructions and structures

Made ground can contain man-made obstructions in the form of cobbles and boulders of natural and man-made materials, either placed as fill or as a result of in situ demolition. Typically, any large clasts like these are going to be stronger materials (rock, concrete, brick, steel and timber). These are likely to cause some degree of difficulty for any drilling or probing technique and possibly for trial pitting using smaller machines. The use of larger machines may be constrained on site by space limitations and off site by difficulties in accessing the site with low loaders. The desk study can indicate to a modest degree the likelihood of encountering these types of obstructions.

There are many different types of buried structure that a ground investigation might experience as an obstruction. Examples include foundations, old floors, basements, tunnels, culverts, machine chambers and membranes. These structures can be solid or can contain space filled with other solid material, liquids or gases, possibly under pressure. Consideration of these matters is a basic part of a desk study, and identification of significant obstructions would normally be a defined aim of ground investigations. In addition to probing or other intrusive methods of locating these, various geophysical techniques can be used (see Chapter 45 *Geophysical exploration and remote sensing*).

When considering this issue during a desk study it is useful to classify buried structures in one of two ways: those in use and those that are disused. Buried structures that are in use typically have owners who are keen to protect them. Their advice should be sought. This advice may significantly constrain the form and content of the investigation – for example, restrictions on drilling close to tunnels. Examples of such structures would be major sewers, major pipelines, communications ducts or tunnels, railway and road tunnels, buried storage tanks, culverts and penstocks.

Disused and 'ownerless' underground structures are rare and are necessarily harder to identify. The types of structure that might be encountered are all those listed above plus air-raid shelters and other historical military and civil defence infrastructure.

Since many people are very interested in these types of underground spaces the internet is a useful source of data. In the UK, Subterranea Britannica can be a good source of information and further links, particularly for disused military and civil defence infrastructure.

The desk study and walkover (Chapter 43 *Preliminary studies*) present the best opportunity to identify disused buried structures before ground investigation begins. This issue should be actively considered in each and every desk study. The ground investigation has the potential to significantly damage a buried structure and in so doing create hazards by instigating collapse or by releasing liquids or gases. The desk study should address the likelihood of encountering buried structures and identify strategies for dealing with that occurrence. Buried structures can also be affected by ground investigations even though the ground investigations do not directly encounter the structure. Examples of this possibility include excavations leading to settlement damage, excavations damaging tree roots and pumping of groundwater leading to settlement of nearby structures.

Where ground investigations have finding buried structures as a specific aim, then the choice of technique will depend on safety considerations. Questions about whether the structure can afford to be damaged, and consideration of the consequences of any damage, need to be addressed before a technique can be selected. Machine-dug trial pitting is often used for shallow robust structures. Shallow structures which must not be damaged will often be investigated using hand-dug pits with suitable support. If the structures contain voids, non-intrusive geophysical techniques such as ground probing radar may be useful (see Chapter 45 *Geophysical exploration and remote sensing*).

41.7. Services

Buried services constitute one of the greatest health and safety hazards in ground investigation. If encountered and damaged they can also be the source of significant delay and expense and can cause pollution of the ground. All ground investigations, especially those on brownfield and urban sites, should incorporate some or all of

- risk assessment
- searches with utility providers
- surveys to locate and trace services
- hand-dug starter pits.

Contractual responsibilities for these activities should be clear, and sufficient time and money should be allowed for their completion. See Chapters 42 *Roles and responsibilities* and 44 *Planning, procurement and management* for more information on roles and responsibilities.

Since searches with utility providers can take several weeks to complete, this activity should be started as early as possible in the desk study process. Clients and designers have responsibilities in this area, particularly under CDM Regulations (see Chapter 42 *Roles and responsibilities* for more detail). Consideration should be given to carrying out pre-investigation service traces. The cost of this can be mitigated if this work is combined with topographic surveys in cases where the project requires such surveys anyway. Ideally such surveys will involve lifting manhole covers and inducing signals along services that can be reliably traced at the surface. This information can then be rechecked on site at the time of the ground investigation using portable CAT scanning equipment. This equipment, however, is not effective with all services and great care should be used when relying on it in the absence of other information or surveys.

Disused and derelict sites may contain both live and dead services. Dead or disused services can still be hazardous and should be treated with the same caution as live services. Disused gas pipes can still contain explosive gas, fuel pipes can contain fuel and so on. Older high voltage (HV) cables often contain asbestos paper wraps and can contain toxic oil-filled jackets. It is often useful to consider the former and current uses of a site. Was HV required and where? How were structures heated? Was fuel (gas/oil) distributed around the site and how?

References

Arup Geotechnics (1990) *Review of Mining Instability in Great Britain*. Contract No. PECD 7/1/271 for the Department of the Environment. Arup Geotechnics, Newcastle upon Tyne, UK.

Bell FG (1975) Salt and subsidence in Cheshire, England. *Engineering Geology* **9**: 237–247.

Bide T, Brown TJ, Idoine NE, and Mankelow JM (2019) *United Kingdom Minerals Yearbook*. British Geological Survey, Keyworth, UK.

Bowman R, Davies P and Baptie P (2019) *Unexploded Ordnance (UXO) Risk Management Guide for Land-Based Projects*. CIRIA Report C785D. CIRIA, London, UK.

Cooper AH (2020) Geological hazards from salt mining, brine extraction and natural salt dissolution in the UK. In *Geological Hazards in the UK: Their Occurrence, Monitoring and Mitigation*. Geological Society Engineering, London, UK, pp. 367–387.

Cooper N and Cooke S (2016) *Assessment and Management of Unexploded Ordnance (UXO) Risk in the Marine Environment*. CIRIA Report C784D. CIRIA, London, UK.

Donnelly LJ, Culshaw MG and Bell FG (2008) Longwall mining-induced fault reactivation and delayed subsidence ground movement in British coalfields. *Quarterly Journal of Engineering Geology and Hydrogeology* **41**: 301–314.

Edmonds CN (2018) Five decades of settlement and subsidence. *Quarterly Journal of Engineering Geology and Hydrogeology* **51**: 403–416.

Edmonds CN, Green CP and Higginbottom IE (1990) Review of underground mines in the English chalk: form, origin, distribution and engineering significance. *Chalk: Proceedings of the International Chalk Symposium, Brighton, UK*. Thomas Telford, London, UK.

English Heritage (2019) Piling and archaeology: guidance and good practice. https://historicengland.org.uk/images-books/publications/piling-and-archaeology (accessed 21/03/2023).

Giles DP and Griffiths JS (2020) *Geological Hazards in the UK: Their Occurrence, Monitoring and Mitigation*. Geological Society Engineering, London, UK.

Healy PR and Head JM (1984) *Construction over Abandoned Mine Workings*. CIRIA Special Publication 32. CIRIA, London, UK.

Lillesand TM, Kiefer RW and Chipman JW (2015) *Remote Sensing and Image Interpretation*, 7th edn. Wiley, New York, NY.

Littlejohn GS (1979) Surface stability in areas underlain by old coal workings. *Ground Engineering* **12(3)**: 22–30.

Ministry of Housing, Communities and Local Government (2021) *National Planning Policy Framework*. Ministry of Housing, Communities and Local Government, London, UK.

Parry D and Chiverrell C (eds) (2019) *Abandoned Mine Workings Manual*. CIRIA Report C758D. CIRIA, London, UK.

Reynolds JM (2011) *Introduction to Applied and Environmental Geophysics*, 2nd edn. Wiley, New York, NY.

Site Investigation Steering Group (2022). *Specification for Ground Investigation*, 3rd edn. Thomas Telford, London, UK.

Stone K, Murray A, Cooke S, Foran J and Gooderham L (2009) *Unexploded Ordnance (UXO): A Guide for the Construction Industry*. CIRIA Report C681. CIRIA, London, UK.

Legislation

HMG (Her Majesty's Government) (1954) Mines and Quarries Act 1954 (Great Britain). The Stationery Office, London, UK.

HMG (2015) The Construction (Design and Management) Regulations 2015. The Stationery Office, London, UK. Statutory Instrument 2015 No. 51.

Further reading

Atkinson B (1988) *Mining Sites in Cornwall and South West Devon*. Dyllansow, Redruth, UK.

Atkinson B (1994) *Mining Sites in Cornwall*, vol. 2. Dyllansow, Redruth, UK.

Bell FG and Donnelly LJ (2006) *Mining and its Impact on the Environment*. Spon Press, London, UK.

British Geological Survey (2008) *Directory of Mines and Quarries*, 8th edn. BGS, Keyworth, UK.

Burgess P (2006) *East Surrey Underground* (self-published).

Culshaw MG and Waltham AC (1987) Natural and artificial cavities as ground engineering hazards. *Quarterly Journal of Engineering Geology* **20**: 139–150.

Department of the Environment (1983) *Limestone Mines in the West Midlands: The Legacy of Mines Long Abandoned*. DoE, London, UK.

Edmonds CN (2008) Karst and mining geohazards with particular reference to the Chalk outcrop, England. *Quarterly Journal of Engineering Geology and Hydrogeology*, **41**: 261–278.

Ford TD and Rieuwerts JH (2000) *Lead Mining in the Peak District*. Landmark Publishing, London, UK.

Howard Humphreys and Partners Ltd (1993) *Subsidence in Norwich*. Contract No. PECD7/1/362 for the Department of the Environment. HMSO, London, UK.

Lord JA, Clayton CRI and Mortimore RN (2002) *Engineering in Chalk*. CIRIA Report C574. CIRIA, London, UK.

McAleenan C and Oloke D (2010) *ICE Manual of Health and Safety in Construction*. Thomas Telford, London, UK.

National Coal Board (1975) *Subsidence Engineers Handbook*. NCB Mining Department, London, UK.

National Coal Board (1982) *The Treatment of Disused Mine Shafts and Adits*. NCB Mining Department, London, UK.

Price L (1984) *Bath Freestone Workings*. Resurgence Press, Bath, UK.

Richards AJ (2007) *Gazeteer of Slate Quarrying in Wales*, rev. edn. Llygad Gwalch, Pwllheli, UK.

Site Investigation Steering Group (1993) *Site Investigation in Construction* (4 parts). Thomas Telford, London, UK.

Tonks E (1990) *The Ironstone Quarries of the Midlands. Part IV: The Wellingborough Area*. Runpast Publishing, Cheltenham, UK.

Tonks E (1991) *The Ironstone Quarries of the Midlands. Part V: The Kettering Area*. Runpast Publishing, Cheltenham, UK.

Tonks E (1991) *The Ironstone Quarries of the Midlands. Part VIII: South Lincolnshire*. Runpast Publishing, Cheltenham, UK.

Tuffs P (2003) *Catalogue of Cleveland Ironstone Mines* (self-published).

Tyler I (2006) *The Lakes & Cumbria Mines Guide* (self-published).

Useful websites

British Geological Society: www.bgs.ac.uk

Coal Authority: www.coal.gov.uk

Cornish Mining World Heritage: www.cornish-mining.org.uk

Environment Agency: www.environment-agency.gov.uk

History of Bathstone Quarrying: www.choghole.co.uk/Main%20page.htm

Kent Underground Research Group: www.kurg.org.uk

Mine-Explorer, the Home of UK Disused Mine Exploration: www.mine-explorer.co.uk

Planning Policy Statement 23: Planning and Pollution Control Annex 2: www.communities.gov.uk/publications/planningandbuilding/pps23annex2

Scottish Environment Protection Agency (SEPA): www.sepa.org.uk

Slate Industry of North and Mid Wales: www.penmorfa.com/Slate

Subterranea Britannica (a good source of information and further links, particularly for disused military and civil defence infrastructure): www.subbrit.org.uk

UK Minerals Yearbook: www.bgs.ac.uk/downloads/browse.cfm?sec=12&cat=132

It is recommended this chapter is read in conjunction with

- Chapter 7 *Geotechnical risks and their context for the whole project*
- Chapter 8 *Health and safety in geotechnical engineering*
- Chapter 48 *Geo-environmental testing*

All chapters in this book rely on the guidance in Sections 1 *Context* and 2 *Fundamental principles.* A sound knowledge of ground investigation is required for all geotechnical works, as set out in Section 4 *Site investigation.*

doi: 10.1680/icemge.66816.0621

Chapter 42
Roles and responsibilities

Phil Smith Geotechnical Consulting Group, London, UK
Jim Cook Buro Happold Ltd, London, UK

This chapter provides guidance on the roles and responsibilities of the participants involved in site investigations. The guidance is based upon key documents, which have been published by various organisations involved in this industry. These documents are referenced and it is recommended that they be read. The chapter focuses on the non-technical issues regarding site investigation work. Comments are given on the Institution of Engineers/AGS Specification for Ground Investigation and on the obligations under the Construction (Design and Management) Regulations 2015, corporate manslaughter and health and safety. The section on conditions of engagement provides commentary on tendering requirements and submissions. The timing of when ground investigations ought to be carried out is discussed in relation to the design stages recommended by the Royal Institute of British Architects. Commercial issues relating to consultants and ground investigations are discussed with comments on the traditional methods of procurement, the one-stop shop approach and the situation when the consultant carries out the ground investigation. Underground services are also mentioned and the chapter concludes with guidance on commercial issues that need to be considered, such as insurance.

CONTENTS

42.1. Introduction to site investigation guides

It is important to understand the elements of work that make up a site investigation, and which people are going to do what and their individual responsibilities. It is usual that the work elements of a site investigation are undertaken by separate organisations. There are usually two entities – a consultant and a specialist contractor – both of whom work on behalf of the client (the employer).

The client, aside from their normal obligations, is usually responsible for obtaining all the appropriate planning and building regulation approvals and possession of the site prior to the start of any site inspection or intrusive investigation work.

It is incumbent upon the organisation (usually the consultant) that is producing the site investigation tender documentation to provide all available information, including client-supplied information, or advise where it can be obtained or viewed.

The objective and scope of a site investigation should reflect the requirements of the intended structure or building. It is important that the briefing section of the tender document provides all relevant information and any particular requirements or potential hazards in undertaking an intrusive investigation (e.g. boreholes or trial pits constructed close to or near to existing structures that may be above or below ground).

The method to be adopted for carrying out a site investigation should be appropriate to the ground conditions at the site and the ground engineering testing and sampling requirements. These methods are usually specified by the consultant, who would approve the specialist contractor's method statement.

Cottington and Akenhead (1984) provide useful guidance on the contractual responsibilities of the parties involved in carrying out a site investigation.

Figure 42.1 shows the sequence of activities for a site investigation. Notice that it is an iterative process.

First, a desk study and an initial ground model is produced. This is usually undertaken by a consultant. This is followed by fieldwork, which is usually called the ground investigation. This is followed by laboratory testing and by factual reporting. Updates to the ground model should be considered every time new data are acquired. The ground investigation can be intrusive or non-intrusive (or a combination of both) and it may be carried out in several phases. This element of the investigation is usually undertaken by a specialist contractor. The factual report provided by the specialist contractor is generally reviewed by the consultant and the interpretative report and recommendations are produced by the consultant.

There are variations on the above, generally dependent upon the size of the project, the client's requirements and so on. See Chapter 50 *Geotechnical reporting* for further discussion regarding the terms 'factual report' and 'interpretative report', and how these terms have in part been superseded.

A basic responsibility for the consultant, who has been appointed to arrange and procure the investigation on behalf of the client, is to ensure the work is undertaken in accordance with the appropriate specification, standards and conditions of engagement. All work should be carried out in accordance with what is known as 'best practice'. The Institution of Civil Engineers (ICE) has produced a specification and conditions of contract document and the Association of Geotechnical and Geoenvironmental Specialists (AGS) provides very good

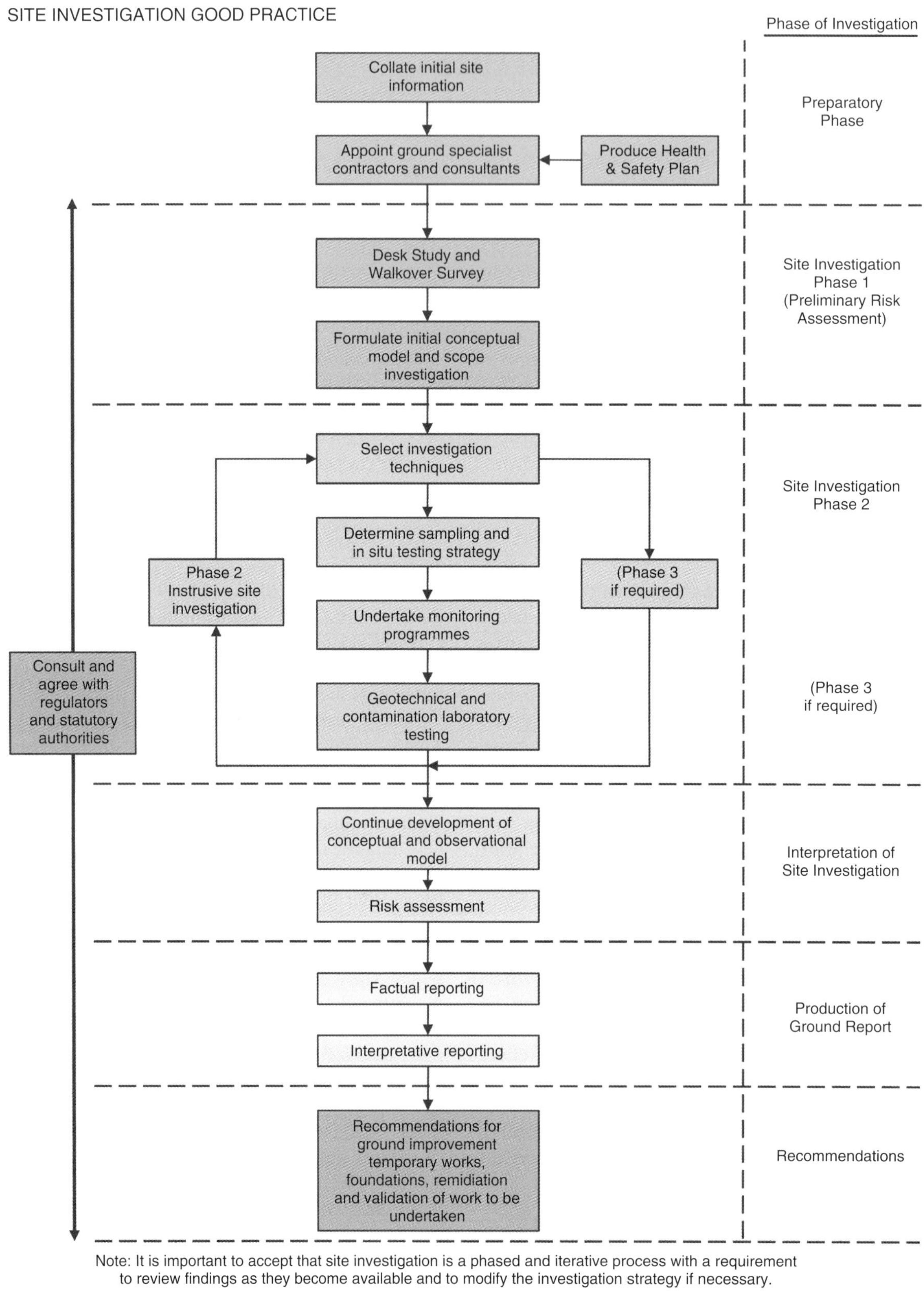

Figure 42.1 Site investigation activities (Reproduced from AGS, 2006)

guides on these matters (details are given in Further reading). Details on codes and standards and their relevance can be found in Chapter 10 *Codes and standards and their relevance*.

ICE produced a series of documents, the *Site Investigation in Construction Series* (ICE, 1993), under the guidance of the ICE Site Investigation Steering Group (SISG), a body which includes various learned societies and trade associations. These documents were supported and endorsed by many major organisations, such as the Association of Consultancy and Engineering (ACE), the British Geotechnical Association (BGA), ICE, the Department for Transport (DfT), the Department of the Environment, Transport and the Regions (DETR), the Construction Industry Research and Information Association (CIRIA) and the National House Building Council (NHBC), to name but a few. The above documents quickly became accepted as the industry norm, but since their publication the methods and techniques for undertaking a ground investigation have changed and geo-environmental aspects have come more to the fore in view of the increase in brownfield development. The original series has been superseded by a second edition, including most notably *Effective Site Investigation* (ICE, 2013) and a second edition specification; these in turn are being updated, with the third edition of the *UK Specification for Ground Investigation* (AGS and ICE, 2022) currently available, and the update to *Effective Site Investigation* currently in progress and anticipated to be published in 2024.

42.2. CDM Regulations (2015)

Site investigation is an investigation into the ground conditions at a particular site, which could be a greenfield site, where no development has previously been carried out, or, quite frequently in the UK, a brownfield site, where the land is to be redeveloped. Any work carried out on site, be it a simple walkover survey as part of a desk study or a full intrusive investigation, involves varying degrees of risk (see Chapter 43 *Preliminary studies*).

The Construction (Design and Management) (CDM) Regulations 2015 (HMG, 2015) apply to all construction projects, including site investigation works. Section 2.1 of the regulations details their application and gives the responsibilities required; these are discussed below.

42.2.1 Application of the CDM Regulations (2015)

The application and obligations of the client regarding ground investigations are fully described in AGS's *Clients' Guide to CDM 2015: Compliance for the Geotechnical Projects* (AGS, 2015). Also recommended is the HSE *Guidance on Regulations, L153* (HSE, 2015). The following section is mainly based upon these publications, which should be read, with supporting comments from the author.

It is important to note that under the CDM Regulations the client has the ultimate responsibility even if the investigation is procured or planned by a third party. In the case of a ground investigation any intrusive fieldwork, such as boreholes or trial pits, is considered as construction work.

The CDM Regulations are applicable to all ground investigations that involve any form of intrusive work in the ground. A key issue concerning the CDM Regulations is whether a ground investigation is notifiable to the Health and Safety Executive (HSE). A ground investigation is notifiable if it forms part of a project where the construction period (including the investigation) is likely to comprise 500 person-days or more, or the construction work is likely to last 30 project working days or more and have more than 20 workers working simultaneously at any point in the project. If the ground investigation is included within an existing or intended project, it will be notifiable as part of that particular project. It is the client's duty to submit a notice, in writing, to the relevant enforcing authority.

For a ground investigation where no specific development is planned at the time of the investigation and the work is wholly non-intrusive, the CDM Regulations 2015 probably do not apply. However, if site visits or walkover surveys are to be undertaken, due consideration must still be given to health and safety issues for personnel undertaking this type of work – that is, site visit risk assessments and method statements will be required anyway.

For site investigations that are to be carried out in potentially high-risk environments (e.g. over water, on or adjacent to railways and highways or underground), specialist training is likely to be required within a more rigorous health and safety regime.

It should be noted that there have been changes to the defined roles and responsibilities under the latest CDM 2015 Regulations compared to the previous versions of the regulations which applied when the first edition of this manual was published.

Under CDM, all parties have a number of general duties; while the specific guidance documents referred to will provide more detailed information, these general duties broadly require all parties to cooperate with each other and any other party that may be affected/affect the works (such as neighbouring sites), to provide clear information and instruction, and to report dangerous conditions.

42.2.1.1 Client's obligations

The client cannot abdicate their responsibilities but can delegate particular tasks to others, including third parties. It should be noted that specific conditions apply to domestic clients, who are defined as people who have construction work carried out on their own home or the home of a family member that is not done as part of a business, whether for profit or not (see HSE (2015) for more details).

The client's key responsibility is to make suitable arrangements for managing a project. This includes ensuring adequate allocation of time and other resources to ensure that the works can be completed safely. Specific duties include ensuring

- notification of the project is given, if relevant
- suitable arrangements for managing the project with respect to health safety and welfare are in place
- suitability qualified designers and contractors are hired
- that the principal designer and principal contractor are appointed as soon as possible, and that lines of communication between all parties are clear
- all relevant parties (including designers and contractors) receive appropriate pre-construction information as soon as it becomes available
- a construction phase plan is prepared before construction begins
- all parties (including the client) have sufficient resources at all stages of the project
- the project timeline is appropriate for each stage of the project
- the principal designer prepares a health and safety plan for the project
- adequate welfare facilities are in place on site before work begins.

Note that the designer has an obligation to make the client aware of their duties with regard to the CDM Regulations.

42.2.1.2 Designer's obligations

Where intrusive work is undertaken as part of a ground investigation, the CDM Regulations apply even if the project is not notifiable. The design of the ground investigation may often be undertaken in advance of the main project design, and potentially by a party other than the principal designer. It is important to note that design of a ground investigation is design, as defined by CDM, and therefore whoever designs the ground investigation, whether a structural engineer, specialist geotechnical consultant, contractor or someone else is performing a design and thus has the duties of a designer.

The designer of the ground investigation will rarely be the principal designer unless they are also responsible for the main construction works.

The designer's duties include ensuring that the client is aware of their own duties and responsibilities, and the general duties that apply to all parties under the CDM Regulations.

They also have a key role in ensuring that the design incorporates the general principles of prevention. It is of particular relevance in relation to intrusive investigation works that the designer ensures adequate pre-construction information is made available to the investigation contractor. Historically, there has been a tendency to place responsibility on the contractor for checking for subsurface utilities, services and other hazards. While the contractor retains responsibility for satisfying themselves that work is safe to proceed, the designer must take reasonable practical steps to obtain this information, and allow for time and cost as necessary to do so (for more information on this issue, see HSE, 2014).

42.2.1.3 Contractor's obligations

Just as the designer of the ground investigation is a designer under the CDM Regulations, so the contractor carrying out the works is a contractor, with defined duties.

Since intrusive works are often undertaken ahead of any other works, the investigation contractor is likely to be the principal contractor under the regulations; while there is not always a principal designer role, there is always a principal contractor.

As with all other parties, the contractor must meet the general duties. The contractor must also produce the construction phase plan for their works. The principal contractor has further duties since they control the site. They must ensure that the site is secure and that all personnel accessing the site have adequate induction. They have duties to provide welfare facilities and to ensure workers on the site are properly consulted with respect to health and safety. The principal contractor must also ensure cooperation between all contractors working on the site.

Note that the principal contractor role can be transferred to other contractors; thus the site investigation contractor may hold this role while on site prior to any construction activity, but might revert to having only a 'contractor' role if construction works begin and the general works contractor takes over as the principal contractor.

42.3. Corporate manslaughter

Corporate manslaughter is on the UK statute books through the Corporate Manslaughter and Corporate Homicide Act 2007 (HMG, 2007). This is a landmark in law and came into force on 6 April 2008. Under this act, companies can be found guilty of corporate manslaughter (or corporate homicide in Scotland, as there is no such thing as 'manslaughter' under Scottish law) as a result of serious management failures resulting in a death or deaths as a direct result of a gross breach of their duty of care. The act clarifies the criminal liabilities of companies, including large organisations, where serious failures in the management of health and safety result in a fatality.

An organisation is guilty of corporate manslaughter if the way in which its activities are managed or organised causes a death and amounts to a gross breach of a relevant duty of care to the deceased. Companies and organisations should keep their health and safety management systems under review – in particular the way in which their activities are managed and organised by senior management.

Sadly the first conviction under this act occurred in February 2011, when a geotechnical company was deemed to be responsible for the death of a young geologist who was killed while working inside an unsupported excavation. Concerning this very important piece of legislation, the reader should refer to *A Guide to the Corporate Manslaughter and*

Corporate Homicide Act 2007 published by the Ministry of Justice (2007) and the HSE and Crown Prosecution Service websites given in the references.

42.4. Health and safety

The British Drilling Association (BDA) document *Guidance for Safe Intrusive Activities on Contaminated or Potentially Contaminated Land* (BDA, 2008) provides health and safety information and recommendations on best practice for ground investigations.

This document describes the rationale and the need for new guidance for ground investigations, mainly due to an increase in statutory and regulatory legislation, changes to working practices and an increase in brownfield site development.

This guide is comprehensive: it has 21 chapters ranging from appropriate legislation, including four acts and 11 regulations, followed by chapters on competence, training and qualifications, managing health and safety, risk assessment and so on.

Further information regarding health and safety can be found in Chapter 8 *Health and safety in geotechnical engineering*.

42.5. Conditions of engagement

A ground investigation specialist can be engaged using bespoke terms, which are drawn up specifically for a particular project, or using the terms and conditions of contract produced by public bodies such as the Highways Agency.

ICE previously produced the *ICE Conditions of Contract for Ground Investigation*, the second edition of which was published in November 2003 together with a set of guidance notes (ICE and CECA, 2003). The second edition was specifically written to focus on the contractual requirements of ground investigation and in particular on fieldwork, including laboratory testing and factual reporting.

However, ICE now formally endorses NEC3 contracts, and transferred its interest in the ICE contracts range to the ACE and the Civil Engineering Contractors Association (CECA).

Whatever conditions of contract are used, it is important to ensure all parties understand their roles and responsibilities.

Typically, the appointment of a ground investigation contractor is carried out through a tendering process, with the designer producing the scope of the investigation works. The tender documents need to include a specification, preferably in accordance with the AGS/ICE (2022) published *UK Specification for Ground Investigation* (3rd edition), which incorporates a bill of quantities, and which includes specific definitions of contractor staff roles within the investigation. The tender documents also need to include all relevant pre-construction information, in accordance with CDM 2015.

For smaller ground investigation projects it is not uncommon for a simple exchange of letters, between either an architect or a structural engineer acting on behalf of the client, or perhaps even directly by the client, and a specialist site investigation contractor, to form the basis of the contract.

In this situation it is important that the specialist site investigation contractor has the appropriate level of resources, competence and experience to undertake all the work required, including a desk study, designing a specification, fieldwork, laboratory testing and producing an interpretative ground engineering report. When work is instructed in this 'informal' manner, it is critical to ensure that all parties understand their duties under the CDM Regulations, and that full pre-construction information is provided as applicable. This form of 'informal' instruction is often used where the intrusive works are tightly constrained by the budget for the works, and consideration needs to be given to how the scope of works can be changed to reflect the actual nature of the ground as found during the course of the works. This leads to consideration of the role of the investigation supervisor.

42.5.1 The investigation supervisor and contractor staff roles

Generally, ground investigations within the UK should be expected to be undertaken in accordance with the UK *Specification for Ground Investigation* (AGS and ICE, 2022). The current third edition of this document sets down roles and responsibilities for a variety of the contractor's staff, which provides for a significant increase in clarity with regard to what services are being provided and how staff are accounted for in relation to payment.

The *Specification for Ground Investigation* also identifies the role of the investigation supervisor (this being a role introduced in the second edition of the specification and continuing into the third edition). The investigation supervisor occupies a role that was previously generally referred to as 'the engineer'; drawing directly from the text of the specification, the 'Investigation Supervisor means the named individual having responsibility to see that the technical objectives and quality of the investigation are met within the programme and cost constraints. The Investigation Supervisor shall act in a professional and independent manner in order to achieve the technical objectives.' Thus, the investigation supervisor needs to have a clear understanding of the purposes of the investigation, and sufficient authority to be able to instruct variations to the scope of works to address the precise ground conditions actually encountered.

42.6. When should a ground investigation be carried out?

The intrusive part of a site investigation can be carried out during a single site visit or over several phases of fieldwork. It is usual that at least a desk study is undertaken for a project that is going to be submitted for planning approval. The desk study is often referred to as the phase 1 site investigation.

Subject to project requirements, an initial ground investigation may be undertaken focusing on geo-environmental concerns more than on geotechnical issues.

The phase 1 site investigation work should include geotechnical as well as geo-environmental aspects since it is generally more prudent and economic to have a combined phase 1 site investigation focusing on both aspects. However, ground models produced for planning authorities that relate to contaminated land issues should not be merged with ground models for consideration of design or construction matters (explicit reporting of ground models is highly likely to be mandatory in revisions to BS EN 1997-1 and 1997-2, due to be published in 2024). The two types of ground model will often have a common basis in data, but their focus is entirely different.

Ground investigations generally should be carried out on a phased basis and need to take into account the design process. There are several design phases in building design; these generally comprise

- feasibility and concept design
- scheme design
- design development
- final (detailed) design
- production information and tender
- construction to practical completion
- after practical completion.

RIBA phasing has design stages 0–7 (replacing the earlier A–L stages) given in the RIBA Plan of Work 2020. The ground engineering input required in the RIBA stages is shown in **Figure 42.2**.

From **Figure 42.2** it can be seen that the site investigation spans the early stages of design for RIBA (stages 0–2). A phase 1 ground investigation would typically be undertaken during stages 0 or 1 during the site appraisal and the phase 2 detailed ground investigation would be completed ideally before stage 3 begins.

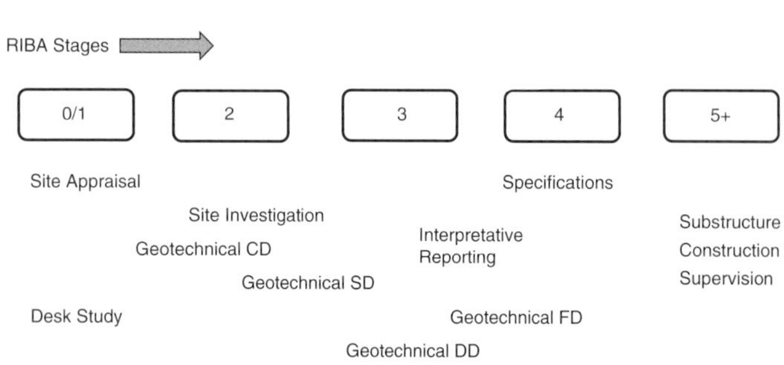

Figure 42.2 Typical ground engineering input during RIBA stages; CD concept design, SD scheme design, DD detailed design and FD final design (Courtesy of Buro Happold Ltd)

42.7. Consultants and ground investigations

42.7.1 Background

It is becoming more commonplace that consulting engineers are asked to provide a 'one-stop shop' and to incorporate the provision of ground investigation and laboratory services within the scope of their design contract with the client or the client's agent.

Consequently, in this instance, the consultant is required to employ a ground investigation contractor directly or to perform the ground investigation using their own staff with their own or hired equipment. The consultant, therefore, becomes the employer of the specialist ground investigation contractor and is acting in the role of a contractor (rather than just a consultant) to their client or their client's agent. For some consultants acting as a contractor is an entirely new situation.

AGS has produced guidance (AGS, 2022), concerning this change of role. The following comments are not an exhaustive description of how to deal with all the issues associated with the changes that consulting engineers must now adopt; instead they are intended to raise awareness of some of the issues that will be faced by consulting engineers so that further guidance, as appropriate, can be obtained.

One of the key issues here is that the consultant bears the financial risk of the cost of the specialist ground investigation contractor.

Traditionally for site investigation work, the consultant has no direct contractual arrangement with the ground investigation contractor.

In circumstances where the consultant procures the ground investigation services on behalf of the ultimate client under the contractual scope of supply, there will be obligations to the ground investigation and laboratory contractor that will need to be properly dealt with in the contract with the client. The consultant will need to be remunerated for the additional effort and business risks.

Generally the most successful ground investigations have been performed within a contractual relationship that readily allows flexibility in work scope due to variations in anticipated ground conditions.

There has also been a tendency for clients or their agents to request 'lump sum' ground investigations (which include laboratory testing). It is likely that a lump sum form of contract that constrains the ground investigation work scope, with resulting inflexibility, potentially places the party holding the risk of the lump sum in a conflict between commercial and professional obligations.

See also Chapter 44 *Planning, procurement and management*.

42.7.2 Traditional approach

The traditional procurement route of engagement is where the client engages a consultant engineer who undertakes the procurement of a ground investigation contractor on behalf of the client to undertake a ground investigation.

The risks to the client (e.g. an increase in overall contract costs) and the responsibilities of the client (e.g. prompt payment) are generally clearly understood. The duties of 'the engineer' (e.g. certification of invoicing) and the responsibilities of the contractor (e.g. carrying out the work to the agreed programme) are also normally clearly defined.

The commercial arrangements for the client in this approach mean that there has to be a separate contract for the consultant to deal with the requirements of the ground investigation and there has to be another contract for the ground investigation contractor.

This two-contract approach may seem cumbersome and this has probably resulted in the adoption of the single appointment, the one-stop shop.

In the traditional approach, the agreement between the client and the consultant, and that between the client and the ground investigation contractor may be of the same form, or differ, depending on the preference of the parties involved.

42.7.3 One-stop shop approach

The one-stop shop approach is a non-traditional arrangement that allows the client to have a single agreement with the consultant to deal with the requirements of the ground investigation and the ground investigation contractor.

This arrangement transfers a certain degree of the client's commercial risk to the consultant, who would usually lead the arrangement. These issues include risks such as responsibility for payment of the ground investigation contractor, whose input into the project is usually substantially more than the fees attributable to the consultant's input. The risk of default by the client and funding claims for extra work can be substantial.

The additional responsibilities that the consultant takes on need careful consideration and may be considered as being unfair. There are also possible concerns from the ground investigation contractor's point of view.

Where a consultant is appointed to procure the ground investigation, they may have multiple roles under the CDM Regulations, including designer, investigation supervisor and possibly also contractor (potentially principal contractor). It is important to ensure that the duties of each role are clearly and definitely completed, and that the lack of a third party expecting delivery of information does not allow duties under the CDM Regulations to be neglected.

A variant of the one-stop shop approach is where the client directly appoints the ground investigation contractor, who fully scopes and completes the investigation works (as discussed in Section 42.5); the output of the investigation may then go to a design consultant or to a design-and-build contractor.

Additional areas of concern for the consultant include the provision of necessary and suitable insurance, as standard insurance arrangements typically held by a consultant may not be adequate. The consultant would be well advised to ensure project cover is in place (typically through a project contractor's all-risks policy).

42.7.4 Other insurance

In addition to the contractor's all-risks insurance mentioned above, there are three important types of insurance that should be considered by both consultants and ground investigation contractors: professional indemnity (PI) insurance, public liability (PL) insurance and employer's liability insurance.

PI insurance is usually carried by consultants or ground investigation contractors to cover liabilities arising from their interpretation, engineering, design and recommendations for a project. For geo-environmental aspects the level of cover offered by insurance companies can be much lower than those offered for geotechnical aspects due to the perceived higher risk of dealing with contaminated ground.

PL insurance is usually taken out by companies to cover the potential affects to third parties caused in carrying out their business activities. This includes bodily harm or death to members of the public or damage to property.

Regarding PI and PL insurance, organisations should consider all potential risks in order to decide upon the level of insurance needed and the amount of insurance cover deemed appropriate. It is important to note that the level of insurance cover offered in a contract does not necessarily limit the degree of exposure (total liability) to the sum insured unless specifically agreed.

The Employer's Liability (Compulsory Insurance) Act 1969 (HMG, 1969) sets out a minimum level of employer's liability insurance cover, which is currently £5 000 000. This insurance cover is needed as the employer is responsible for the welfare of employees when they are at work, either in the office or on site.

Further information should be obtained from the various insurance companies and brokers, who offer advice and guidance. The Insurance Institute of London provides publications on the above and other types of insurance.

42.8. Underground services and utilities

The likelihood of there being underground services and utilities is extremely high for sites in developed areas and less so for more greenfield sites, although it should be noted that even in greenfield locations there can be significant subsurface assets (oil pipes, railway tunnels, etc.). It is also worth mentioning that there are above-ground services and utilities that should be considered and these are not dealt with here. In undertaking any form of intrusive ground investigation there is a risk of encountering services and utilities.

In general, the client's engineer (the investigation designer, in accordance with the CDM) provides the location plans, showing the existing services and utilities for the project, to the ground investigation contractor. These may be provided to the engineer by the service or utility-owner company. The ground investigation contractor would then arrange for a non-intrusive survey (cable and pipe locating) to be carried out to confirm the location on the ground and subsequently excavate a test pit to confirm the existence of the services and utilities if appropriate, and/or undertake hand-dug/vacuum extraction inspection pits to confirm the absence of services at the location of each borehole or trial pit.

The question of responsibility is one which needs to be carefully considered. The HSE document HSG47 *Avoiding Danger from Underground Services* (HSE, 2014) provides guidance and also outlines dangers and risk reduction methods (see also BSI, 2022).

HSG47 focuses specifically on health and safety, but also points out that by considering this there will be a reduced risk of damage to services and property. This guide is targeted at all parties that could be involved, including owners, clients, designers, planning supervisors, contractors, operators and employees of the aforementioned.

The dangers of damaging underground services and safe systems of work are mentioned together with a flow diagram. There are sections on training and supervision, planning, design and also where the CDM applies, includes duties and responsibilities. This document is comprehensive and also provides appendices on legislation and suggested advice for site personnel.

42.9. Contamination

The risk and likely extent of contamination of a site is normally determined during the first phase of a geo-environmental site investigation, the phase 1 site investigation. This study would usually be carried out early in the project in order to meet the requirements of obtaining outline planning approval.

Following the phase 1 investigation, if deemed necessary because of the identification of potential contaminants, or simply due to a requirement to provide greater evidence that the site is free from contamination, a more detailed geo-environmental investigation will possibly be required together with a geotechnical site investigation. From the results of the desk studies, site work and laboratory chemical tests, a geo-environmental risk assessment will be made of the site and remediation options recommended if necessary.

In the event that elevated levels of previously unidentified contaminants are discovered on site during construction, the responsibility for dealing with these will be subject to the contractual arrangements between the contractor, consultant and client. The general principle is 'polluter pays'; however, this is subject to commercial arrangements between the original owners and the purchaser, and the nature of any land-use changes for the new development.

AGS collaborated with the Environmental Industries Commission (EIC) and ACE to produce *The Terms upon which Contaminated Land Consultants are Employed* (AGS *et al.*, 2007). This document is available from the AGS website. This guide was developed to support the commercial arrangements for studying contaminated land and more specifically to highlight the common contractual issues that should be considered by consultants. It provides guidance on key contractual issues such as the limitation of liability, collateral warranties, strict obligations and fitness for purpose, insurance clauses, indemnities, incorporation of standard forms, assignment and duration of risk.

Note

This chapter is generally based upon literature published by AGS and others, which has been referenced and acknowledged. These documents are generally readily available for download on the organisations' websites. At the time of writing some of the documents and guidance notes referenced are subject to continual update and change. There have been recent changes in legislation and there are likely to be more in the future.

Disclaimer

Although every reasonable effort has been made to check the information and validity of the guidance provided in this chapter and the AGS documents referred to, neither the author, any individuals, members of Buro Happold, members of the Working Group, nor the AGS provide any warranty or guarantee as to the accuracy of the contents and do not accept any liability or responsibility whatsoever for any inaccuracy, misstatement or misrepresentation contained herein or for any loss, damage, expense, cost, claim or the like arising directly or indirectly from any reliance upon it, howsoever arising.

References

AGS (Association of Geotechnical and Geoenvironmental Specialists) (2006) *AGS Guidelines for Good Practice in Site Investigation (Issue 2)*. AGS, Beckenham, UK (believed to have been withdrawn).

AGS (2015) *Clients' Guide to CDM 2015: Compliance for Geotechnical Projects*. AGS, Beckenham, UK.

AGS (2022) *Loss Prevention Guidance*. AGS, Beckenham, UK.

AGS and ICE (2022) *UK Specification for Ground Investigation*, 3rd edn. ICE, London, UK.

AGS, EIC and ACE (2007) *The Terms upon which Contaminated Land Consultants are Employed*. AGS, Beckenham, UK.

BDA (British Drilling Association) (2008) *Guidance for Safe Intrusive Activities on Contaminated or Potentially Contaminated Land*. BDA, Upper Boddington, UK.

BSI (British Standards Institution) (2022) PAS128: Underground utility detection. BSI, London, UK.

Cottington J and Akenhead R (1984) *Site Investigation and the Law*. Thomas Telford, London, UK.

HMG (Her Majesty's Government) (1969) Employers' Liability (Compulsory Insurance) Act 1969. Elizabeth II. Chapter 57. The Stationery Office, London, UK.

HMG (2007) Corporate Manslaughter and Corporate Homicide Act 2007. Elizabeth II. Chapter 19. The Stationery Office, London, UK.

HMG (2015) Construction (Design and Management) Regulations, 2015. The Stationery Office, London, UK, Statutory Instrument 2015 No. 51.

HSE (Health and Safety Executive) (2014) *Avoiding Danger from Underground Services*, HSG47, 3rd edn. HSE, London, UK.

HSE (2015) *Managing Health and Safety in Construction. Construction (Design and Management) Regulations 2015. Guidance on Regulations* L153. HSE, London, UK.

ICE (Institution of Civil Engineers) (1993) *Site Investigation in Construction Series.* Thomas Telford, London, UK (superseded).

ICE Site Investigation Steering Group (2013) *Site Investigation in Construction: Effective Site Investigation.* Thomas Telford, London, UK.

ICE and CECA (Civil Engineering Contractors Association) (2003) *ICE Conditions of Contract for Ground Investigation*, 2nd edn. Thomas Telford, London, UK.

Ministry of Justice (2007) *A Guide to the Corporate Manslaughter and Corporate Homicide Act 2007.* The Stationery Office, London, UK.

Further reading

AGS (2006) *Client's Guide to Professional Indemnity Insurance.* AGS, Beckenham, UK.

AGS (2023) *Client's Guide for Ground Engineering and Project Timelines.* AGS, Beckenham, UK.

Clayton CRI (2001) *Managing Geotechnical Risk.* Thomas Telford, London, UK.

Clayton CRI, Matthews MC and Simons NE (1995) *Site Investigation*, 2nd edn. Blackwell Science, London, UK.

HMG (1990) Environmental Protection Act 1990. HMSO, London, UK.

HMG (1990) Environmental Protection Act 1990 Part 2A. HMSO, London, UK.

McAleenan C and Oloke D (eds) (2010) *ICE Manual of Health and Safety in Construction.* Thomas Telford, London, UK.

Useful websites

AGS Client Guides: http://www.ags.org.uk/publications/

Association for Consultancy and Engineering (ACE): http://www.acenet.co.uk

Association of Geotechnical & Geo-environmental Specialists (AGS): http://www.ags.org.uk

British Geotechnical Association (BGA): http://www.britishgeotech.org

CIRIA (Construction Industry Research and Information Association): http://www.ciria.org

Civil Engineering Contractors Association (CECA): http://www.ceca.co.uk

Corporate manslaughter, CPS (England/Wales) guidance: http://www.cps.gov.uk/legal/a_to_c/corporate_manslaughter

Corporate manslaughter, HSE: http://www.hse.gov.uk/corpmanslaughter

Institution of Engineers (ICE): http://www.ice.org.uk

Insurance Institute of London: http://www.iilondon.co.uk/home

National Housing Building Council (NHBC): http://www.nhbc.co.uk

NEC Contracts: http://www.neccontract.com

Royal Institute of British Architects (RIBA): http://www.architecture.com

UK Government (legislation): http://www.legislation.gov.uk

It is recommended this chapter is read in conjunction with

- Chapter 8 *Health and safety in geotechnical engineering*
- Chapter 96 *Technical supervision of site works*

All chapters in this book rely on the guidance in Sections 1 *Context* and 2 *Fundamental principles.* A sound knowledge of ground investigation is required for all geotechnical works, as set out in Section 4 *Site investigation.*

Chapter 43
Preliminary studies

Phil Smith Geotechnical Consulting Group, London, UK

Preliminary geotechnical studies are a key element in managing ground risk in construction. They take the form of a desk study in which relevant available documentary information about the ground at a site is identified, obtained and reviewed. A walkover site survey adds significant value. The preliminary study should be completed before scoping and specifying a ground investigation to help focus the ground investigation and to maximise its value and relevance.

doi: 10.1680/icemge.66816.0631

CONTENTS

43.1. Scope of this guidance

No guidance manual could supply the diversity of knowledge, skills and experience necessary to generate comprehensive preliminary geotechnical studies appropriate for every possible project, at every possible site. This section of the manual, therefore, sets out to: (*a*) outline the types and sources of information that should typically be reviewed when writing a preliminary geotechnical study; and (*b*) encourage the writers of preliminary studies to regard their reports as the all-important first step in the process of managing ground-related risks in a construction project. Focus is given to preliminary studies for sites in the UK, but much of the guidance is applicable to sites internationally.

43.2. Why do a preliminary geotechnical study?

Preliminary studies, often called desk studies, are an efficient and cost-effective way of gathering and scrutinising available information at the earliest stages of a project with the aim of understanding the ground and ground-related hazards at a site. Within the construction industry, project delays and cost overruns can frequently be linked to encountering 'unforeseen ground conditions' when construction starts on site (Chapter 7 *Geotechnical risks and their context for the whole project*). A well-planned, well executed preliminary study can assist greatly in reducing – although it can never wholly eliminate – the likelihood of meeting unexpected problems in the ground during construction.

Undertaking a preliminary study is a requirement of all relevant codes in the UK

- BS 5930: Section 1, Clause 7 and Section 2 (BSI, 2020)
- Eurocode 7 BS EN 1997-1, Clauses 3.1 and 3.2.2 (BSI, 2004)
- National House Building Council Standards Chapter 4.1 (NHBC, 2022).

At the time of writing, the second generation of Eurocode is under development and is expected to be published in 2024. It is anticipated that the revised document will include significantly more comprehensive guidance on the contents of the desk study.

The preliminary study should be viewed as the essential first link in the geotechnical project risk management chain (**Figure 43.1**). The chain progresses through each stage of the geotechnical engineering of a construction project (Chapter 42 *Roles and responsibilities*).

- The preliminary study identifies likely ground-related hazards.
- The ground investigation explores the hazards and their associated risk (see **Box 43.1**).
- The geotechnical design reassesses the ground risks and mitigates the associated hazards to appropriate levels.

Box 43.1 Hazards and risks are not the same

A **hazard** is a source of potential harm. The **risk** associated with a hazard is the likelihood of that harm occurring, factored by its impact if it were to occur. For example, flooding is a hazard. The risk would depend on how prone an area is to flooding, and the degree of damage and loss that would be caused by a flood event.

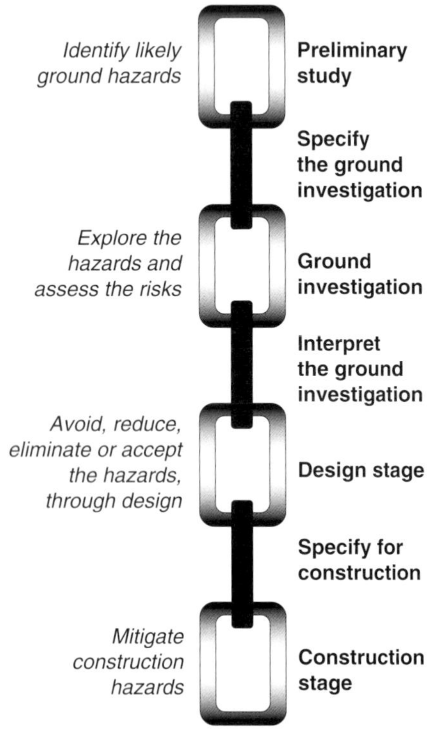

Figure 43.1 The geotechnical risk management chain

- Any remaining (known) ground-related hazards that are not mitigated at the design stage can, in liaison with the project management team, be communicated to the construction team through specifications and other contract documents, to help ensure that these issues do not come as a surprise on site during construction.

A key concept here is the concept of the ground model. The different types of ground model (conceptual, observational, analytical and commercial) are described in the IAEG (2022) report *Guidelines for the Development and Application of Engineering Geological Models on Projects*. This is a critical reference for all authors of preliminary studies and is freely available as a download. The preliminary study is a critical component in establishing the conceptual and early-stage observational models. The conceptual model is the one that sets out 'what might be there' in the ground and thus determines the initial design and investigation scopes.

43.3. What goes into a preliminary geotechnical study?

The factual core of a preliminary geotechnical study typically encompasses the following factors.

- *Site details* include the location, address, grid reference, boundaries, topography, present use, proposed use, access routes, land ownership, neighbouring land usage and more.
- *Site history* includes an interpretative review of historical maps, ideally extending back to rural fields; aerial photographs and satellite imagery, including LiDAR data; prior site usage; archaeological potential; listed buildings and scheduled monuments; watercourses; tunnels; mine (including brine extraction) workings (Chapter 41 *Man-made hazards and obstructions*); possible unstable ground, whether natural or due to human activities; underground infrastructure; flood records; buried obstructions; old foundations; underground services and utilities; possible sources of contamination from current or previous site activities and initial assessment of these; historical changes in topography; presence of trees and more.
- *Site walkover* includes observations of the site and its vicinity.
- *Site geology* includes an interpretative review of all available geological maps and memoirs, including old map series, well records and borehole logs, data from previous ground investigations at the site or in its vicinity, an assessment of the likely site stratigraphy, groundwater regime and hydrogeology and seismicity if relevant.

There are also a variety of project-specific and site-specific variations on these core themes.

Guidance on the archaeological assessment of sites is published in the *Standard and Guidance for Historic Environment Desk-Based Assessment* by the Chartered Institute for Archaeologists (2020), and is not repeated here. In relation to site contamination, a preliminary study is termed a Phase 1 study, and a ground investigation is a Phase 2 study. These must follow prescribed legislative requirements. Guidance on the requirements for Phase 1 studies is presented in 'BS 10175: Investigation of potentially contaminated sites' (BSI, 2017); the Environment Agency's 'Land contamination risk management' guidance (Environment Agency, 2020); and in Strange *et al.*'s (2017) *Contaminated Land Guidance: The Route to Sustainable Economic Solutions*. Chapter 48 *Geo-environmental testing* addresses Phase 2 studies.

The preliminary geotechnical study report should also cover the following interpretative elements, each of which demands some knowledge of the proposed development, as well as knowledge of the site itself.

- *Recommendations for the ground investigation:* outlining the scale and scope of the proposed ground investigation, and highlighting any special site-specific or project-specific aspects identified in the desk study that should be addressed in the ground investigation.
- *Details of ground-related site constraints:* cataloguing the identified site-specific factors that may affect ground investigation works, foundation options and foundation construction.
- *List of ground-related hazards:* describing and prioritising the identified ground hazards (see **Box 43.2** for some examples), and outlining how these could be investigated further and how they could be mitigated. Ground hazards can usefully be treated in three categories: topographic, geological and man-made. The format of the listing should give priority and emphasis to the site-specific and project-specific geohazards that are particular to the site: these should not be lost or overwhelmed by a long list of generic issues that could apply at any site.

Box 43.2 A partial list of hazards: some ideas to consider...

Topography
slope stability issues
site access
possible GI drill rig locations
potential for flooding

Groundwater
need to dewater? effects?
rising groundwater
tidal variations
potential for aquifer contamination
discharge of pumped groundwater

Voids/soft spots
mining
karsts
backfilled quarries
backfilled basements
old wells

Trees
present on site? – walkover
recently removed? – archive photos
clay soils: shrink/swell potential
species, height and maturity
removal or retention

Buried features
operational and disused tunnels
services and utilities
foundations, basements, tanks
UXO (unexploded ordnance)

Heritage issues
listed buildings
archaeological planning constraints
need for a watching brief during GI
impact on programme of finds
Scheduled Ancient Monuments

Neighbours
proximity to existing structures
sensitivity of existing structures
ground movements at excavations
noise and vibration
need for party wall agreement?

The inclusion of a list of ground-related hazards in a preliminary geotechnical study benefits both the client and the project team because it

- helps convert the preliminary study findings into risk management information that is of direct use from the earliest stages of a project, by identifying ground-related factors that may affect costs and the programme
- translates the preliminary study findings into a practical format that is readily understood by non-geotechnical colleagues and clients
- acknowledges the importance of risk-based methods in contemporary project planning, and integrates the preliminary geotechnical study within that process from the outset of the project
- brings considerable value to a project, yet makes use of material that the geotechnical team has already gathered and should already have thought about.

43.4. Who should write a preliminary geotechnical study?

Preliminary geotechnical studies should be written by suitably experienced geotechnical engineers or engineering geologists. One can speculate that if an engineer and a geologist were set to work in isolation of each other on desk studies about the same site, they would probably generate very differently focused reports. In practice, for all sites, a combination of both technical perspectives is invaluable: each specialist can gain a deeper insight of a site by liaising and collaborating with the other. The balance in the input needed from each specialist will depend largely on the site's location and geology: a pipeline in a remote mountainous region would typically benefit most from an engineering geologist, whereas a building in a city on an alluvial plain would probably benefit most from a geotechnical engineer.

Appropriate knowledge and experience are essential in order to understand fully the significance of the information gathered. The responsibility for a preliminary study, therefore, should not be assigned to the most junior member of the team, although, if properly supervised by an experienced senior colleague, preliminary studies are valuable training grounds for junior engineers and engineering geologists.

Beyond knowledge and experience, useful characteristics for writers of preliminary studies include a high level of observant curiosity, a persistent and creative attitude to seeking data, and a methodical and thorough approach to reviewing data. These attributes are needed because a preliminary study often must draw together disparate – and sometimes incomplete – data to form as complete and insightful a picture as possible of the ground at the site of interest. That cannot be achieved in a routine or formulaic way: it requires inquisitiveness underpinned by a systematic approach to obtaining and assembling relevant information.

43.5. Who should read a preliminary study report?

The report should be circulated to the client and the design team, including the architectural team. If a preliminary study has been undertaken thoroughly and effectively, the resulting report should

- be essential reading for whoever is tasked with scoping and specifying the ground investigation

- probably contain some items of useful information that were not previously known to the rest of the design team, and which may well influence their thinking
- summarise the currently available ground information in one convenient volume, including ground-related hazards
- engage and interest the client, especially about the history of their property.

The contents, or a summary of the contents, of a preliminary geotechnical study should usually be made available to the construction team (the contractor) alongside the ground investigation factual reports. Any concerns about releasing an interpretative study to a third party can be dealt with using a suitably phrased third-party caveat in the text of the report. In terms of controlling ground risk in a project, it is beneficial that the construction team are as fully informed as possible about a site's history and the anticipated ground conditions at the site. They can then assess the likelihood and potential impact of these factors on their site works.

43.6. How to get started: sources of information in the UK

In the UK, the following are useful starting points for gathering data from which to develop a preliminary geotechnical study. This list is not exhaustive, and new sources of information – especially online sources – become available each year.

- Geological maps and cross-sections published by the British Geological Survey (BGS) at 1:10 000 and 1:50 000 metric scale. Both drift and solid maps should be reviewed, along with the relevant geological memoir. Older maps can also be highly informative, despite changes in the names of geological formations. Additional information can be gained from mineral extraction maps.
- The GeoRecords service at the National Geoscience Data Centre can be used to obtain old well records and borehole logs from the archive held at the BGS. The archive can be searched using an online map.
- Ordnance Survey (OS) maps dating from the mid-nineteenth century to the present can assist greatly in tracking historical changes at a site. Small-scale OS maps (1:1250 and 1:2500 at metric scale) show local detail, including street names (see **Box 43.3**) and the outlines of individual buildings. OS maps include topographic data, expressed as spot heights and contours, relative to sea level (see **Box 43.4**).
- Historical maps were produced by a variety of map makers. Some maps were semi-pictorial views, but others were more rigorously cartographic, notably those of Roque, Cary, Horwood and Stanford. Copies of old maps are available from online suppliers, and some have been published as books.
- From the late 1830s, church tithe maps showed land parcel boundaries and building outlines, but the maps were of variable quality. Most local history centres hold the tithe maps for their area.
- Maps made for fire-insurance purposes by Charles E. Goad Ltd from the mid-1880s to 1970 show the commercial districts of many UK towns and cities. At a scale of 1:480, they show a building-by-building mapping, with details such as industrial use, construction materials, number of storeys and basements. The Map Library at the British Library has an extensive collection of UK Goad fire maps. Scanned Goad maps are also available commercially from online map suppliers.
- The Coal Authority Mining Reports Office (the Coal Authority) can be used to obtain details of documented mining operations, and mining and brine-extraction ground stability reports (see Chapter 41 *Man-made hazards and obstructions*). Its website includes a gazetteer indicating the areas of Great Britain where mining-related ground stability is likely to be an issue.
- Topographic data can be obtained from several sources, including the OS.
- Documentary searches of underground services and utilities plans can be commissioned from commercial search firms (see **Box 43.5**). As well as shallow utility services, the search should include transportation and utility tunnels.

Box 43.3 Place names

Place names can provide valuable information and insight into the history of a site, its topography and even its hydrogeology. An unusual place name should always trigger an enquiry into the origin of the name, and an assessment of its implications for the proposed works at the site. For example, a street called Flask Walk in Hampstead is close to Well Lane and Well Walk: all three derived their names from the iron-rich mineral waters that were sourced and bottled there from the seventeenth century onwards. Many older place names, especially Anglo-Saxon names, provide information about the natural landscape. Some examples of the many Celtic, Saxon and Viking root words used within place names in the UK include

- allt: the side of a hill
- beck, bourne, burn: a stream
- glan: the bank of a river
- gill: a valley
- hamps: a stream that is dry in summer
- ings: a meadow or marsh
- keld, kelda: a spring
- mere: a pool or lake
- moss: a marsh, bog or swamp
- pant: a hollow
- slack: a stream within a valley.

Box 43.4 Liverpool and Newlyn

When using early OS maps or old well logs dating from before the 1920s, be aware that the UK height datum system changed during the 1920s. From 1844 the height elevation datum was at Victoria Dock in Liverpool. After 1921, the UK national height datum system was based on tidal measurements at Newlyn in Cornwall. The difference in heights between the Liverpool and Newlyn systems is not large, but it is just enough to cause confusion unless it is taken into account. The height conversion factor between the two systems differs by differing amounts across the country. The OS has an online conversion calculator.

Box 43.5 Utility searches by commercial agencies

In the UK, several commercial organisations offer 'one-stop shops' for obtaining, collating and reporting documentary searches of utility plans of statutory service undertakers (gas, water, electricity, telecoms, etc.). A request for a documentary utility search should state clearly the footprint and the depth of the proposed development. The searches can take several weeks to complete, as they rely on receiving responses from third parties. Be aware that the documentary search results will show only the approximate location of the buried utilities – not exact positions – and they will not show buried utilities within a private property boundary or within buildings.

Some of these companies provide a comprehensive service collating environmental risk information, including historic and present land use, geological maps, UXO risk assessment and so on, as well as utilities information. However, there are also free online services such as 'Linesearchbeforeudig', and it is strongly recommended that such dedicated services should always be consulted prior to any breaking of the ground.

All utilities searches should be undertaken with reference to PAS128 (ICE and BSI, 2022).

- The possible presence of disused historic tunnels should not be overlooked (for example, the Post Office railway in London or Williamson's tunnels in Liverpool). Identifying such infrastructure often relies upon local knowledge, or access to it.
- Public libraries usually hold a collection of local history books, including out-of-print publications that may not be readily available elsewhere.
- Certain below-ground infrastructure such as emergency bunkers and strategic telecoms tunnels are, by their nature, not widely advertised. Some disused facilities have been reported by enthusiastic amateurs (such as at Subterranea Britannica). For operational facilities, enquiries should be made to the relevant bodies to ascertain where any 'hidden' infrastructure is in relation to the proposed development.
- The archives at the Institution of Civil Engineers are a useful resource for obtaining information and records of historic engineering structures (Chrimes, 2006).
- Details of grade-listed buildings and Scheduled Ancient Monuments in England are available from English Heritage and its searchable online archive of images of grade-listed structures.
- Historic aerial photography of England from the 1940s onwards, including aerial photographs and wartime images showing bomb damage, can be obtained from the National Monuments Record (NMR) Centre (Historic England). The NMR archive in Swindon can be visited in person. Alternatively, a significant amount of data held by the NMR is available online. Detailed interpretation of aerial photography is a specialist skill, as is the photogrammetric analysis of stereo-pairs of images to quantify dimensions and levels from aerial imagery.
- Comparable services in Wales and Scotland are provided by, respectively, the Royal Commission on the Ancient and Historical Monuments of Wales and the Royal Commission on the Ancient and Historical Monuments of Scotland, with the Scottish records now available from Historic Environment Scotland.
- Information about wartime bomb damage and unexploded ordnance (UXO) can be obtained from several sources. For London, hand-coloured OS maps were used to record the degree of bomb damage; these are available in book format (Ward, 2015). CIRIA Report C681 (Stone *et al.*, 2009) sets down the procedure for undertaking UXO risk assessment for a site and also includes a list of the reported 'abandoned bombs' within the London area. In assessing the potential for UXO risk, it is important to be aware of the existence of wartime decoy sites in rural areas (as listed in Dobson, 2000); there is often little or no trace of these sites today, but they have raised UXO risk. There are presently numerous specialist companies providing UXO risk assessments. The likelihood of UXO being present at a site is partly – but only partly – a function of the density of bombing that occurred in that district. A readily accessible summary of bombing densities across the UK is available from Zetica UXO.

43.7. Using the internet

The internet is a powerful – but not all-powerful – tool. Internet searches can offer a quick and low-cost way to find information about a site. Maps, satellite imagery (e.g. Google Maps), reports, journal papers and images can often be found online and downloaded almost immediately. For much of the UK, LiDAR data are now publicly available online; comparison of the different historical datasets provides a quick and easy check on what, if any, changes have occurred at a site over time, particularly in relation to possible ground surface changes.

At its best, the internet can provide, or lead to, useful information that would otherwise be difficult, time-consuming and costly to obtain. At its worst, however, the internet provides low-grade, unverified and sometimes erroneous information.

When using the internet as part of a preliminary study, be conscious of the following.

- Data available on the internet are often incomplete, undated and unreferenced – and, hence, potentially unreliable. Online data should be checked independently where possible. Be aware that some websites simply recycle material from other online sites, although the material may have been in error in the original source.
- The successful exploitation of internet search engines relies on selecting appropriate and well-chosen keywords, and then refining and refocusing the search as it progresses.
- Material pre-dating the internet may not be readily revealed by an internet search. Traditional desk study research methods – and some legwork – may still be needed. For example, local history centres and archives often list their catalogues online, but a visit in person to the collection may be necessary to see the material. Be aware that an online search will only reveal documents that have been catalogued using the same keywords used in a search. Discussion with a knowledgeable curator may uncover other relevant material, which may not have been revealed by a simple keyword search.

43.8. The site walkover survey

If possible, a site walkover survey should always be undertaken. Ideally, the walkover should be done about two-thirds of the way through the period assigned for the preliminary study, by which time enough background documentary information will have been gathered to inform what is seen on site, and sufficient time will still be available to follow up any unexpected observations and to integrate the walkover results into the overall site interpretation.

The planning of a site walkover visit should include due consideration of safety, and an appropriate risk and mitigation assessment should be carried out beforehand. Potential hazards are site-dependent, and may range from aggressive livestock in rural areas to unfriendly locals in urban areas. In most cases, at least two people should undertake a walkover visit together to avoid the hazards of lone working.

What to look for during a walkover survey will depend on the context: walkover surveys at a hill farm and at a city street would be substantially different. An experienced practitioner will have a mental checklist of basic items to look for (such as utility service covers; repairs in roads and pavements; drains and ditches; marshy ground; uneven ground; cracks in structures; geomorphology; rock outcrops; slopes and slope instability; evidence of cut and fill; species of trees; types of buildings; presence of basements; possible sources of contaminants; access for ground investigation drill rigs; likely locations for boreholes; overhead constraints) plus a long list of project-specific and site-specific items.

On site, all relevant features should be noted and photographed. Photocopies of site maps or satellite images can be annotated with observations and comments. For large or rural sites, a hand-held GPS to record coordinates of features can be helpful. Photography is an essential tool. Plenty of photographs should be taken: not all need to be used in the report, but a photograph can often help answer an unforeseen query when reviewing data back at the office. The preliminary study report should include a photographic key plan showing the location and direction of each photograph reproduced in the report.

The walkover survey should not be limited to the site itself, but should also look beyond the site boundaries. If access onto neighbouring land is not possible, then a visual assessment can usually be made across the boundary line.

It can be useful to talk to local people, as they usually know their area and its history. However, it is important to verify anecdotal recollections independently, and to indicate clearly any anecdotal information that is included in the report. When undertaking a walkover survey, especially when talking to locals, a degree of circumspection may be appropriate. Before undertaking a walkover visit, one should check how much of the proposed project is public knowledge, and whether the project is considered sensitive either locally or commercially, particularly if planning consent has not yet been granted.

43.9. Writing the report

When writing a preliminary report, much inspiration can be gained by looking at good examples of previous reports for other sites, but these should never be relied upon wholly. Each site is unique. Factors that may not have been relevant at another site, and which were, therefore, disregarded in its report, may be critical at the site of interest.

As with all technical report writing, the preliminary study report should be written clearly and carefully. Useful guidance on writing geotechnical reports is presented in two guides produced by the Association of Geotechnical & Geoenvironmental Specialists (AGS)

- *Guide to Good Practice in Writing Ground Reports*
- *AGS Guide to Ground Investigation Reports.*

Reference to the AGS *Client Guide: The Purpose and Use of a Ground Condition 'Desk Study'* is also recommended.

The AGS guides emphasise the importance of taking care when writing a technical report because, for example, a negligent and damaging error could potentially lead to claims against the organisation that issued the report.

The structure of the report will be partly dictated by the material that has been reviewed. The list of items in Section 43.3 can be used as a basis for the overall report structure, and as headings for sections and subsections.

As in any professional technical report, it is important to distinguish between factual and interpretative elements. The report should include a clear statement of what sources of information have been reviewed in preparing the report. Equally, if any key items have *not* been reviewed (perhaps a document that is known to exist but has not been received in time, or has been withheld by a third party), this should be stated in the report as it affects the scope and completeness of the report.

The source of all material used in the report should be clearly stated, even if the material is not copyrighted (**Box 43.6**). The aim should be to provide readers with enough information to enable them to find the same material again. As a minimum, this should include the source and date of the material. By doing this, a signal is sent to the reader about the provenance and, hence, the likely reliability of the information upon which the preliminary study has been based.

Box 43.6 Copyright

Care is needed when using material, especially images, taken from copyrighted sources. Preliminary studies are produced for clients, and are, thus, commercial reports. Under UK copyright law, the consent of the copyright owner should be sought if an extract of their work, or an adaptation of their work, is to be included within a commercial report. Also, the owner of the copyrighted material should be clearly stated in the report. For fuller details of copyright issues and UK copyright law, check the UK Intellectual Property Office's website.

If reproducing OS maps, these should be clearly labelled as being Crown copyright material. The OS operates a licensing system for businesses wanting to reproduce OS maps in professional reports.

43.10. Summary

The preliminary geotechnical report is the essential first link in the chain of geotechnical project risk management (see **Figure 43.1**). Site information, ground models, geohazards, uncertainties and ground-related risks identified in the preliminary study will form the basis for the next stages of the project, and may potentially influence the direction of the project and the engineering solutions adopted.

A well-planned, well-executed preliminary study that is followed through with a suitable ground investigation can help significantly in reducing the likelihood of unexpectedly encountering problematic ground conditions during construction, although it cannot reduce that likelihood to zero.

References

BSI (British Standards Institution) (2004) BS EN 1997-1:2004: Eurocode 7: Geotechnical design – General rules. BSI, London, UK.

BSI (2017) BS 10175+A2:2017: Investigation of potentially contaminated sites – Code of practice. BSI, London, UK.

BSI (2020) BS 5930+A1:2020: Code of practice for site investigations. BSI, London, UK.

Chartered Institute for Archaeologists (2020). *Standard and Guidance for Historic Environment Desk-Based Assessment*. Chartered Institute for Archaeologists, Reading, UK.

Chrimes M (2006) Historical research: a guide for civil engineers. *Proceedings of ICE Civil Engineering* **159(1)**: 42–47.

Dobson C (2000) *Fields of Deception: Britain's Bombing Decoys of WWII*. Methuen and English Heritage, London, UK.

Environment Agency (2020) Land contamination risk management. www.gov.uk/government/publications/land-contamination-risk-management-lcrm (accessed 21/03/2023).

IAEG (International Association for Engineering Geology and the Environment) (2022) *Guidelines for the Development and Application of Engineering Geological Models on Projects*. IAEG, London, UK. Commission 25 Publication No. 1.

ICE and BSI (2022) PAS128:2022: Underground utility detection, verification and location – Specification. BSI, London, UK.

NHBC (2022). *NHBC Standards 2022*. NHBC, Milton Keynes, UK.

Stone K, Murray A, Cooke S, Foran J and Gooderham L (2009) *Unexploded Ordnance (UXO): A Guide for the Construction Industry*. CIRIA Report C681. CIRIA, London, UK.

Strange J, Langdon N and Large A (2017) *Contaminated Land Guidance: The Route to Sustainable Economic Solutions*, 3rd edn. Thomas Telford, London, UK.

Ward L (2015). *The London County Council Bomb Damage Maps, 1939–1945*. Thames and Hudson, London, UK.

Useful websites

Association of Geotechnical and Geoenvironmental Specialists (AGS): http://www.ags.org.uk
British Geological Survey (BGS): http://www.bgs.ac.uk
British Library: http://www.bl.uk
British Standards Institution (BSI): http://www.knowledge.bsigroup.com
The Coal Authority: www.gov.uk/government/organisations/the-coal-authority
English Heritage: http://www.english-heritage.org.uk
English Heritage online archive of grade-listed structures: http://www.historicengland.org.uk/images-books/archive/history/images-of-england
Institution of Civil Engineers (ICE): http://www.ice.org.uk
LiDAR data:
England: https://environment.data.gov.uk/DefraDataDownload/?Mode=survey
Wales: https://nrw.maps.arcgis.com/apps/webappviewer/index.html?id=3f8007ca0b8847948ca47de0a27319e2
Scotland: https://remotesensingdata.gov.scot
National House Building Council (NHBC): http://www.nhbc.co.uk
Ordnance Survey (OS): http://www.ordnancesurvey.co.uk and http://www.ordnancesurvey.co.uk/gps/legacy-control-information/liverpool-to-newlyn
Royal Commission on the Ancient and Historical Monuments of Scotland: http://www.historicenvironment.scot and http://www.scotlandsplaces.gov.uk/digital-volumes/rcahms-archives
Royal Commission on the Ancient and Historical Monuments of Wales: http://www.rcahmw.gov.uk
Subterranea Britannia: http://www.subbrit.org.uk
Summary of bombing densities: http://www.zeticauxo.com/downloads-and-resources/risk-maps
UK Intellectual Property Office: http://www.gov.uk/government/organisations/intellectual-property-office

It is recommended this chapter is read in conjunction with

- Chapter 7 *Geotechnical risks and their context for the whole project*

All chapters in this book rely on the guidance in Sections 1 *Context* and 2 *Fundamental principles*. A sound knowledge of ground investigation is required for all geotechnical works, as set out in Section 4 *Site investigation*.

doi: 10.1680/icemge.66816.0639

Chapter 44
Planning, procurement and management

Tim Chapman Arup, London, UK
Alister Harwood Arup, London, UK
Marianne Walsh Arup, London, UK

CONTENTS

Coherent planning is required for every ground investigation to ensure that it is carried out simply and cost-effectively and produces all the requisite information. Poorly planned ground investigations cause significant problems in terms of damage to neighbourly relations and damage to buried utilities, the latter exposing the site operators to severe safety risks. A well-planned two-phase campaign can offer advantages for longer and more complex investigations, as information discovered in the first phase can be further investigated in the second. A clear form of contract and specifications reduces the risk of disputes. Regular review of the information recovered is the ultimate success of a ground investigation campaign. A well-planned campaign is essential to anticipate future likely changes in the scheme. It saves construction cost and programme by revealing important site features at an early stage which can then be planned for.

44.1. Overview

The success of a ground investigation depends on the extent of prior planning, the clarity of the procurement processes used and the care with which it is managed. These factors become even more important for larger ground investigations. The trend towards frameworks and large infrastructure projects means that some ground investigations are now becoming very large indeed.

A ground investigation can be considered to have failed in the following cases.

- The process does not gather all the data intended, or a party subsequently realises that more or other data should have been gathered at the same time, or the ground investigation reveals aspects of the site which should have been considered before. Note that ground investigations can never discover everything about a site, and the possibility of surprises should never be discounted. The discovery of legitimate new information in the investigation should be regarded as a success rather than a failure.
- The process results in a significant misunderstanding or dispute involving the client, the contractor, a landowner, an authority or some other body. These aspects should have been considered beforehand and the contract documents should have set out clear rules to avoid such an occurrence.

The standard guidance documents for ground investigation are those of the *Site Investigation in Construction* series, produced by the Site Investigation Steering Group, introduced in Chapter 42 *Roles and responsibilities*. This chapter addresses the geotechnical aspects of a ground investigation for design and construction and does not address the implications of ground and groundwater contamination in detail.

One of the main purposes of a ground investigation is to aid in the development of a ground model. An initial conceptual model of the ground should be developed in the early desk study stages, which sets out the geological and developmental history of the site and attempts to identify what *might be* present in the ground.

As data are obtained from the ground investigation and added to the model, it becomes an observational model. This observational model (typically recorded in factual and interpretative reports) is then simplified into an analytical model for design and possibly also into a commercial ground model in the form of a geotechnical baseline report (GBR, see Chapter 50 *Geotechnical reporting*). Since ground investigations cannot successfully characterise the entirety of the ground, the observational model always contains uncertainty. Uncertainty equals design and construction risk. In a design process this uncertainty tends to be addressed by cautious simplification in analytical models (e.g. through design standards that require selection of cautious characteristic values). This approach may deal with design risk but does not always deal with construction that encounters 'unforeseen' conditions.

The mitigation for any identified design or construction risk arising from this uncertainty should be based on feedback loops to the conceptual model. These should consider: the plausible geological and site history features not identified by the observational model, the inbuilt limitations in many ground investigations, any investigation access constraints and the potential consequences of difficult encounters with the ground during construction.

A new version of Eurocode 7 is due to be published in 2024. The drafts of the new edition (published for comment in 2022) contain new requirements relating to the reporting of ground models.

44.1.1 Key messages in this chapter

- Ground investigation is done most efficiently if properly planned in advance.
- Planning a ground investigation is firmly based on logic, but balancing the many conflicting constraints and priorities sometimes makes it akin to an art form. The planner must not lose sight of the main objective, which is to gather sufficient data for the design of the planned development and to minimise ground risk. The various constraints imposed on the ground investigation must not compromise the final information to any unacceptable level.
- Safety is a key objective. There may in some circumstances be some conflict between the safety of the site investigation operatives and the need to gather information to minimise the risk to the workforce on the main development. In those circumstances, careful consideration will be needed. Any residual risk in the site investigation must be fully communicated to all those going on site and every effort must be made to mitigate the remaining risks to tolerable levels.
- Ground investigations often take longer than many people would wish. Completion of the site works is not the end of the story. Sufficient time should be allowed in the overall project programme for timely completion of the ground investigation before the information is needed for design.
- Choosing the right people to undertake the various geotechnical activities – design and investigatory – depends on the relative strengths of the various project participants. But it is important to facilitate as far as possible a coherent geotechnical design process – disjointed steps in that process are a major source of project risk.
- As for any construction activity, a clear specification and contract with equitable terms should reduce the risk of disputes.
- Locating unforeseen conditions in a ground investigation should be seen as a success and not a failure, even if the ground investigation is pushed over budget.

44.2. Planning the ground investigation

44.2.1 Developing a coherent campaign

44.2.1.1 Meeting the needs of clients, designers and contractors

Planning for a ground investigation can only take place after an understanding has been gained of the likely geology (see Chapter 40 *The ground as a hazard*) and the desk study has identified the key uncertainties and hazards on the site (see Chapter 43 *Preliminary studies*). Rudolf Glossop said in his 1968 Rankine lecture: 'If you do not know what you should be looking for in a site investigation, you are not likely to find much of value' (Glossop, 1968).

Planning of a ground investigation should target the particular development scheme proposed and it is important that all aspects of the scheme and its implications are known, for instance

- a six-storey building over a two-level basement; perimeter retaining wall needed on north side; foundations may be piled to 30 m depth, large surface car park to west.

Thus, if a tall tower is added or if the building location changes, further supplementary ground investigation will be needed. The effects of the location of buried utilities, highways works and the necessary traffic management on the investigation also need to be considered early, as these issues often require lengthy negotiation and agreement (see Chapters 41 *Man-made hazards and obstructions* and 42 *Roles and responsibilities*). Leaving these aspects until too late, for instance delegating them to the contractor, who cannot start the negotiations until appointed, can lead to delay and extra costs.

The development of a coherent campaign of ground investigation requires considerable planning, particularly where third-party property owners believe they will be affected by the site works. Site occupants and immediate neighbours will be directly affected. Generally the ground investigation is the first physical sign of the new development that they experience, and so any opposition or concern to the development may be focused on the ground investigation fieldwork campaign. Thus special care at this stage is needed, as a carelessly executed ground investigation that causes much inconvenience can sour neighbourly relations for some time and complicate the main construction process.

Often, ground investigations are carried out while existing buildings on the site remain fully occupied and operational. This is because the site is only usually vacated in preparation for demolition; few clients will commit to demolishing while they can continue to receive rent, and will delay evicting occupants until they have let the contract to build the new structure or building. The demolition cannot be let until the structural design has been completed, which relies on the completed ground investigation. The ground investigation therefore usually has to fit around occupants and operational business. This can be a significant cause of conflict unless the planning of the site investigation ensures that occupants are not unduly disturbed, operation of the facilities is not unduly impaired and safety of the normal site occupants is not compromised. These aspects are discussed in more detail in Section 44.2.3.

Often formal permissions or licences may be needed in order to undertake the investigation, especially when access is required across neighbouring land, or where parts of the investigation have to be undertaken off the site, for instance on the public highway. The programme for the investigation needs to allow time to make all agreements, obtain all necessary licences and follow any restrictions legitimately placed on the operations by relevant parties.

Ground investigations next to live railways tend to be the most tightly regulated because of the severe consequences if a rig overturns or a slope is destabilised next to the operational railway. The person planning the investigation needs to be aware of all specific rail-related regulations and procedures that may affect the work, including training of site workers and appointment of railway look-outs. Ground investigations next to water should receive similar attention in planning.

The Party Wall Act etc. 1996 often governs the interface with neighbours when any form of construction work is being undertaken. A very thorough explanatory leaflet is available from the Department for Communities and Local Government, published in May 2016.

Section 6 of this Act applies where a landowner wants to excavate, or excavate and construct foundations for a new building or structure, within

- 3 m of a neighbouring owner's building or structure, where that work will go deeper than the neighbour's foundations or
- 6 m of a neighbouring owner's building or structure, where that work will cut a line drawn downwards at 45° from the bottom of the neighbour's foundations (**Figure 44.1**).

The limit for excavations is sometimes taken to mean excavations for ground investigations, although it is not thought that that was the intention of the Act. Unfortunately, excavations for ground investigation are not explicitly excluded under the Act, so the Act can sometimes be deemed to apply even to small boreholes or limited trial pits. Where the Party Wall Act may apply, it is therefore prudent to obtain the requisite permissions, especially if neighbourly relations are not good.

44.2.2 Considering all aspects of the site and all users for the data

Ideally the ground investigation is planned to address several needs at the same time, for instance

- carrying out geotechnical design of foundations and substructures for the new development, including all significant ancillary works such as retaining walls and car parks
- producing information for the design of temporary construction works, such as excavation slope stability or dewatering during construction, as otherwise tendering contractors will have no other source for the data and will make conservative assumptions that will push up the construction price
- gaining knowledge of current foundations for existing buildings in order that any necessary structural changes can be designed to allow the new development to proceed
- understanding the state of contamination of ground or groundwater or establishing levels of deleterious or harmful substances in the ground
- determining underground water flows under or around the site
- collecting data on existing buried services, sometimes just to ensure that they are avoided by the ground investigation but often to ensure their locations are known for main construction so that they can be avoided, diverted or protected

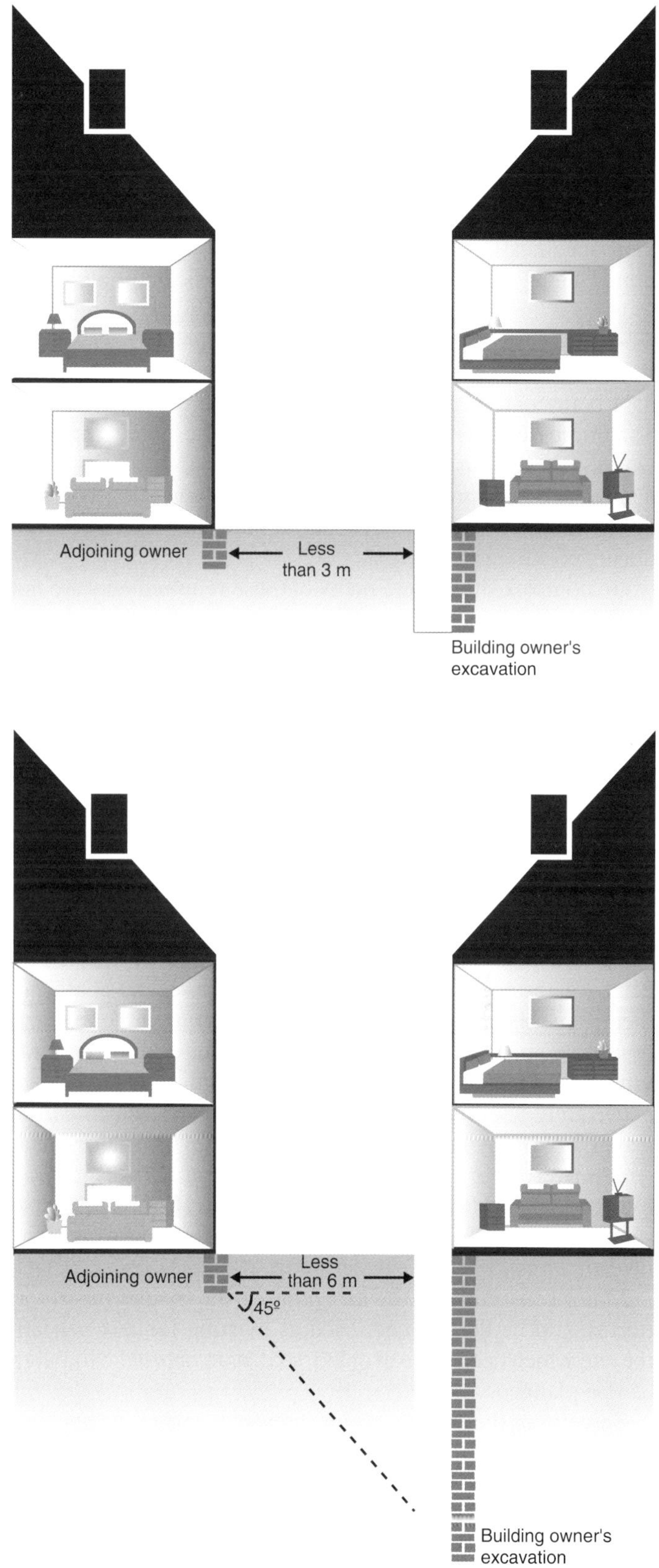

Figure 44.1 Application of Party Wall Act (Reproduced from DCLG, 2004) © Crown Copyright

- examining the ground for any evidence of surviving archaeology and coordinating the campaign with any necessary archaeological excavations.

In such cases the site operations need to be organised and phased so that all relevant consultants have the opportunity for input. Their input is needed in planning to integrate all requirements, and during the site works to satisfy them that their specifications are being followed and the right information is being obtained in a holistic way, efficiently and minimising effects on people affected by the work. Sometimes these operations are carried out in a disjointed way, with little coordination – such an approach usually leads to fewer data being obtained at greater cost and increased disruption. Sometimes aspects are neglected, such as those for temporary works – these omissions can prove exceptionally costly, either due to lack of appropriate information for tendering contractors or failure to identify key features until they are discovered during construction.

On complex sites where many of these requirements are not straightforward, one person should be responsible for ensuring that the campaign is executed coherently. That person may be an engineer's representative for the employer or a well-briefed agent for the contractor, depending on which approach is most appropriate.

A two-phase ground investigation campaign is recognised as being a valuable way to investigate the ground, particularly on sites with greater ground uncertainty, because it allows the second phase to build on the results of the first one, to refine what is found and to respond to new information. Generally the first phase would be a sparse but deep investigation, to detect the major geological formations and identify those that need greatest focus in the second phase. The second phase can then refine the results of the first phase, filling in details and providing appropriate design parameters for each major stratum. This approach is most valuable when there is significant uncertainty about the geology beneath the site, which can at least be partially clarified in the first phase. On major infrastructure projects there may be several phases of investigation. Eurocode 7, Part 2 (BS EN 1997-2:2007) (BSI, 2007) also supports a two-phase approach for larger and more complex investigations.

Major constraints to the execution of the ground investigation should be determined, such as existing tunnels beneath the site which need to be avoided, so that an optimal campaign can be planned.

Sometimes the site constraints prevent a coherent campaign being planned to properly investigate the hazards. In that case, two options are possible.

- Defer the troublesome activities until a later time when they become possible, for instance by waiting for a window between the site being vacated and the start of demolition. This means that important information will not be gathered until relatively late in the design process. This may not matter much provided the first phase of investigation gathers enough information for design, but perhaps omits information relevant to construction planning; however, it is not appropriate to delay the collection of information that may lead to a misunderstanding of the ground conditions in design.
- Accept the increased risks and make the development less dependent on them – for example, adopt a more conservative design basis which can accommodate greater uncertainty. This may have a higher direct cost, but may be expedient in terms of a quicker programme and hence be more economical in terms of overall costs to the client.

An example of the first approach would be on a site where the variation in level of a particular stratum interface is in doubt. The first phase would confirm the general level and the design parameters. Proving the level of an interface sufficiently for construction (e.g. choice of temporary casing length for piling) need only be done immediately prior to letting the construction contract, so that accurate prices can be obtained from the tendering contractors.

On some sites, and particularly those smaller sites where environmental concerns are greater, sometimes an initial geotechnical investigation is merged into a combined 'phase 1' geotechnical and geo-environmental investigation, as explained in Chapter 42 *Roles and responsibilities*. Such investigations sometimes fail to collect important geotechnical information such as adequate depth of investigation, proper soil descriptions, water level measurements and strength determinations. Failing to collect these data at an opportune time is a missed opportunity that increases project risk – every effort should be made to ensure that a phase 1 environmental investigation also provides meaningful geotechnical information.

Providing easy access to all the data produced can be a challenge, particularly for large ground investigations. Hence consideration of how the data will be managed is important. For larger investigations, incorporation of the results into a geographic information system (GIS) can pay dividends in the subsequent design phases and provide a valuable tool for the development of the analytical ground model.

44.2.3 Timing, programme and space: balancing the constraints of site and project

Knowledge is more useful the earlier it can be obtained. However, a ground investigation usually has to be delayed until

- the scheme it is to address has crystallised into what is likely to be built, at least in broad terms – doing it too early may mean a supplementary ground investigation is required later
- the client is sufficiently committed to the scheme to invest significant resources in it – although ground investigations are a very small proportion of the total project cost, they are often the first construction-related activity on site and so represent the client's first large payment to a contractor; hence this hurdle is sometimes a useful barometer for a client's commitment to a project

- the start of ground investigations does not alarm current site occupants or neighbours, who in some circumstances may be opposed to the development objectives and plans.

A way to reduce these effects and to allow earlier collection of ground investigation data is to use a two-phase approach to the ground investigation, as introduced in Section 44.2.2.

- The first phase is to check key hazards, confirm any variation in the stratigraphy across the site and obtain outline design parameters.
- The second phase is to provide more complete coverage and to confirm the applicability of design parameters.

The two-phase approach is especially valuable for sites where the ground conditions are not well known at the outset, as it permits some understanding to be gained by the ground investigation designer between collection of the data used to formulate the basic ground model and filling in the detailed information needed for each layer.

An early ground investigation will normally take place while the site is still occupied, so sensitivity is needed to avoid causing excessive disruption to occupants, users and neighbours. The main forms of disruption are as follows.

- Occupation of space – in addition to the space needed at each exploration point, the ground investigation contractor will need space for stores, spare drilling equipment (e.g. extra lengths of casing), waste skips, offices and parking.
- Restriction of access or impairment of site operations – some of the exploration positions may impede main access routes for vehicles and pedestrians or block loading bays.
- Noise and vibration – the exploration is often the first major construction activity for the new development, and so noise and vibration could be a cause of concern to neighbours and residents. On very restricted sites, boreholes or trial pits will inevitably encounter buried reinforced concrete or masonry, which will have to be broken out by pneumatic tools. Limiting their use to particular times can do much to improve neighbourly relations but will increase ground investigation cost and programme. BS 5228: Part 2 (Code of practice for noise and vibration control on construction and open sites – vibration) (BSI, 2009) and CIRIA TN142, *Ground-Borne Vibrations Arising from Piling* (Head and Jardine, 1992), although primarily intended for piling operations, provide useful guidance on controlling noise and vibration and appropriate limits for ground investigation sites.
- Disruption to services – potentially, the investigation can lead to buried services being damaged or cut, which will affect the site occupants and may cause financial losses due to business interruption. Ideally, services will have been identified and deactivated when planned works are in close proximity to them.
- Impairment of safety by extra vehicular movements or lack of care.
- Level of reinstatement – whether tarmac is fully reinstated; whether standpipes are made flush with the ground or raised and so on. These aspects are best agreed in advance so that the contractor's cost can allow for the level that will be required, and to prevent higher costs being sought at a late stage.

A ground investigation can be modified to use less space, but cost and programme will consequently have to increase, and the technical output may be compromised. Examples include the following.

- A normal cable percussion ('shell and auger') boring rig needs a working area of at least 7 × 3 m and direct vehicular access. A man-handleable portable rig can be used in confined areas, but is restricted in borehole diameter, depth and numbers of 'strings' of casing that can be inserted.

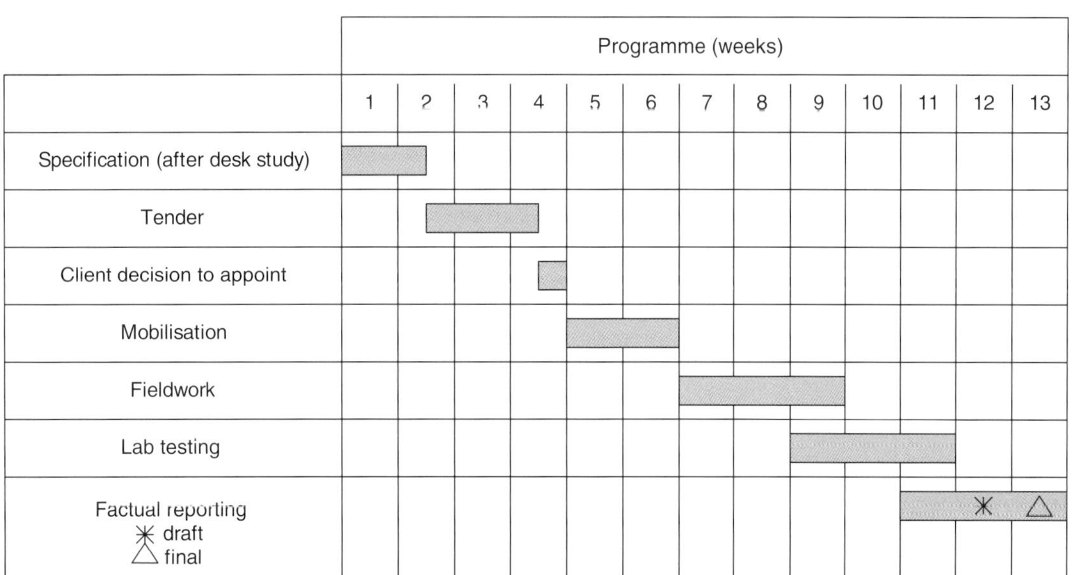

Figure 44.2 Typical overall ground investigation programme

- Machine-dug trial pits may need a space of 3 m square; hand-dug trial pits can be done in a smaller area, but need to be shored and are significantly slower with increased concerns for safety. They may still need to break through buried reinforced concrete, so may not be quieter, and if slower may prolong the agony of neighbours.

A typical programme for the procurement and execution of a ground investigation is given in **Figure 44.2**. The programme is often significantly longer than clients expect, so it is worthwhile explaining the rationale for each of the programme elements in order that an instruction to proceed is given appropriately early and results can be available when needed.

44.2.4 Parameters and information needed

44.2.4.1 Information and requirements of different projects

The parameters collected in a ground investigation depend on the nature of the site and what one plans to build on it. A particularly hazardous site or a complex development will obviously need a more elaborate analysis to address the key risks in the most economical way. The more detailed the analysis, the more parameters usually need to be determined for the results of the analysis to be meaningful. There needs to be a degree of iteration around the *scheme–analysis–parameters–investigation* loop to ensure that the ground investigation is carefully targeted. A geotechnical adviser with experience across all these disciplines is most suitable to plan the ground investigation, draw together the disparate strands and not produce an overall investigation that responds disproportionately to the most vocal demands.

The need for ground investigation depends on what is intended to be built – and in turn, in extreme circumstances, what can be built depends on the results of the ground investigation. Thus the ground investigation needs to correctly anticipate the likely form of foundations. This cycle is illustrated graphically in **Figure 44.3**.

As a minimum, the ground investigation should determine

- stratigraphy and its variation across the site
- groundwater regime on and around the site; pressure variation with depth and time; directions of flow across the site
- characteristics of principal strata in terms of their likely behaviour
- strength, stiffness, in situ stresses and permeability for key strata.

A further aspect needing investigation is the near-surface ground. This is to detect

- obstructions and buried services – see Chapter 41 *Man-made hazards and obstructions*
- extent and levels of neighbouring foundations
- weaker ground that could present a hazard for piling platforms (see BRE, 2004) or other relevant temporary needs
- tree roots in plastic clays, which could lead to desiccation shrinkage and movement of shallow foundations.

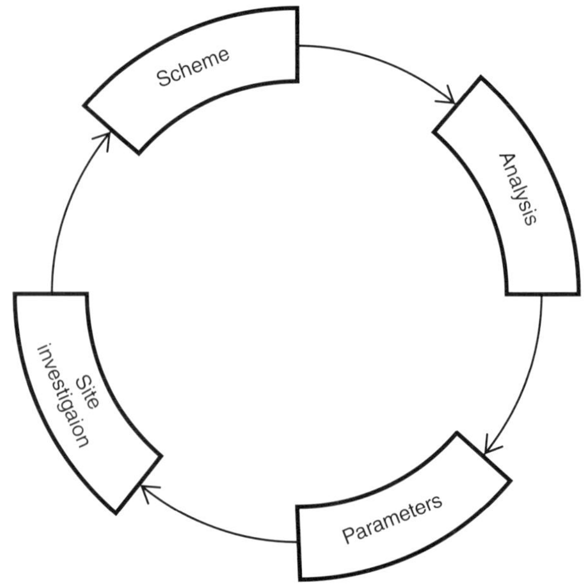

Figure 44.3 Loop connecting scheme with geotechnical analysis, soil parameters and site investigation

The ground investigation should appreciate the possibility of real geological variability and not just expect to validate the planner's preconceptions. The sorts of variability are explained in more detail in Chapter 40 *The ground as a hazard*, but could include the presence of a significant sand channel in an estuarine deposit such as the Lambeth Group, or heavily reworked sand or clay such as the very deep post-glacial scour features that occur in London Clay sporadically but often are associated with the confluence of old rivers.

It is important at the outset that the designer of the ground investigation is clear about the information that will be needed in the later analyses. It is seldom possible to repeat elements of the ground investigation without considerable disruption, expense and embarrassment.

The parameters that need to be gathered depend on the analyses required to complete the design. They must address all the key design issues in order to allow the most economical methods of construction and least project risk, as shown for piled foundations in **Table 44.1**.

44.2.5 Examples of scope issues

The following points need to be considered in determining the scope of a ground investigation.

- If piles are expected, boreholes will generally need to go deeper.
- If foundations on clay are next to trees, the extent of clay desiccation will need to be assessed.
- If the main construction works are likely to need extensive dewatering, the permeability for each of the strata will need to be estimated.

Element being designed	Analysis method	Parameters that need to be measured	Method of measurement
Pile design	Hand calculations	Stratigraphy, water pressures, ground strength	SPT/CPT tests in boreholes; lab tests on samples recovered from boreholes
Pile constructability		Borehole seepages, borehole collapses during boring	Index tests on borehole samples to identify changes in material
Concrete durability	BRE SD1	Sulfate content	Lab tests on borehole samples

Table 44.1 Relationship between pile design and the need for information from the ground investigation

Clearly, on any site, the ground investigation for a low-rise housing development will be very different from that for a tall office block over a deep basement. Therefore ground investigation reports will always be scheme-specific rather than site-specific. While an old geotechnical report should provide very useful information about a site, its scheme-specific recommendations should be used with caution for subsequent and adjacent developments.

A ground investigation scope is always a balance between more comprehensive coverage and a limited budget. On any project funds will be tight, particularly for up-front activities. On the other hand, it is not cost-effective to have to repeat site operations because parts of the ground investigation were omitted. It is therefore important that the geotechnical consultant should consult widely with the other team members, prior to completing the scope to ensure the following.

- Include all requirements, not just those of the structural engineers needing foundations, as explained at the start of this chapter. Others may include the site-wide infrastructure engineer planning roads and car parks, the public health engineer who might be considering SUDS (sustainable urban drainage system), the services engineer who may wish for trial pits to explore existing service routes, and the architect who might be concerned about zones of ground with greater archaeological potential.
- Consider the effects of the whole development on the ground. Thus the car parking requirements may lead to a need for tiered retaining walls; the most cost-effective way to build a basement may be in open cut, for which an increased knowledge is needed of the site's hydrogeology and strata permeabilities.

When planning the ground investigation, a number of choices have to be considered such as the likelihood of

- shallow or deep foundations
- retaining walls or open cut
- alternative access road routes
- finding contamination or archaeology.

44.2.6 Appropriate testing

44.2.6.1 Matching the testing schedule to the information requirements

The AGS guide *The Selection of Geotechnical Soil Laboratory Testing* (AGS, 2018) addresses this subject in excellent detail. Soil testing fulfils two essential aims.

- It confirms quantitatively the variation in nature of each stratum, spatially and with depth – this allows particular strata to be more simply considered together for engineering design, or indicates where a more complex sequence of strata needs to be considered (geologists tend towards more detail, whereas civil engineers tend towards simplification in terms of considering the engineering properties of different geological strata).
- It provides data for design. Engineering design of foundations and substructures requires a quantitative basis in terms of strength and stiffness for the subsequent analyses.

When planning tests it is often worthwhile considering these two aspects separately. A simple investigation, such as following the NHBC rules for house foundation design, will be principally concerned with the general characterisation of the soils beneath the site. A major building with a deep basement will need to more accurately quantify strength and stiffness for each layer, and may also need other parameters such as permeability. Before planning the test strategy it is useful to have clarity on the division into likely key strata.

The relative reliance on field and laboratory tests varies between different countries according to local practice. A balance has to be struck between

- making measurements crudely on the actual ground at the correct density and in situ stresses and in relatively undisturbed state
- taking discrete samples back to a laboratory where tests can be made under more controlled conditions, albeit on samples that have been disturbed during fieldwork and subsequent transportation and examination.

Attempts have been made to improve the level of technology applied in field tests (e.g. pressuremeters) and to make laboratory tests more representative (e.g. stress path triaxial tests). In fact, for most typical design cases the usefulness of all tests depends on their ability to simply apply previously observed behaviour of real foundations in the field to the new situation.

A simple classification of all common test types used on soils in the UK is given in **Table 44.2**.

The range of possible contamination tests is very wide and is addressed in Chapter 48 *Geo-environmental testing.*

For larger structures where more complex analyses are planned, the data gathered in the ground investigation need to be geared towards those analyses – inputting less accurate data into complex analyses can result in meaningless output. Therefore some redundancy is needed to allow a more refined

	Field tests	Laboratory tests
Characterisation	SPT cone	Grading (sieve and pipette) Atterberg limits mineralogy
Strength	Hand vane pocket penetrometer cone Pressuremeter indirectly from SPT	Triaxial shear box
Stiffness	SPT pressuremeter	Triaxial with local strain measurement
Permeability	Pumping test with pumping well and observation wells (standpipes) Rising/falling head tests	Oedometer test Permeameter

Table 44.2 Range of common soil tests

choice of parameters, so sometimes parameters should be deduced from several test types, both field and laboratory.

An advantage of field tests over laboratory tests is that their results are available relatively quickly. A drawback of advanced laboratory tests – particularly tests on clay either in the triaxial cell or shear box at a drained rate of shearing – is that the results can take a considerable length of time to become available. The results may arrive after important design decisions for the whole project have already been taken. The geotechnical designer should not wait for these parameters as the basis for conservative analyses in order to retain the confidence of the rest of the design team and to be able to influence the design correctly.

Laboratory and field techniques for investigating the ground are described in more detail in Chapters 47 *Field geotechnical testing* to 49 *Sampling and laboratory testing.*

44.2.7 Frequency and depths of fieldwork

44.2.7.1 Choosing the right techniques and equipment

The amount of investigation that is appropriate depends on the consequences of missing possible variations. Basing a design on just one exploration point suggests confidence that all stratum boundaries are reliably flat over the whole site and soil properties do not vary. In previously unexplored conditions, three boreholes are needed to define the angle of dip of key stratum boundary planes, assuming no folding or faults in the planes.

The key aim of any investigation is to sensibly investigate the strata and interface levels that are important for design. Thus a presumption towards a building founded on piles might indicate a much more complex investigation than one where only shallow foundations need be considered. The implications of these ground investigation planning decisions should be explained to the rest of the design team, so that factors which might invalidate them can be explored.

If parameter uncertainty is the only variable, a poor-quality or less extensive investigation can be compensated for by a more conservative design, albeit at a likely higher overall direct cost to the project. For instance, a conservative assessment of skin friction in pile design can obviate the need for more boreholes and more testing. However, reducing the extent of the site investigation may cause one to miss a significant change in ground conditions that would invalidate the choice of pile type. The project will then suffer serious cost and programme penalties while the team designs a more appropriate pile type and the contractor waits with expensive equipment standing idle.

BS 5930:2015 + A1:2020 (BSI, 2020) recommends the spacing between exploration points.

- A relatively close spacing between points of exploration (e.g. 10–30 m) is often appropriate for structures.
- For structures small in plan area, exploration should be made at a minimum of three points unless other reliable information is available in the immediate vicinity.
- Where a structure consists of a number of adjacent units, one exploration point per unit may suffice.
- Certain engineering works, such as dams, tunnels and major excavations, are particularly sensitive to geological conditions, and the spacing and location of exploration points should be related more closely to the detailed geology of the area than is usual for other works.

Eurocode 7, Part 2 (BS EN 1997-2:2007; BSI, 2007), Clause 2.4.1.3, advises:

> The locations of investigation points and the depths of the investigations shall be selected on the basis of the preliminary investigations as a function of the geological conditions, the dimensions of the structure and the engineering problems involved.

It goes on to require:

> The depth of investigations shall be extended to all strata that will affect the project or are affected by the construction. For dams, weirs and excavations below groundwater level, and where dewatering work is involved, the depth of investigation shall also be selected as a function of the hydrogeological conditions. Slopes and steps in the terrain shall be explored to depths below any potential slip surface.

Appendix B3 to BS EN 1997-2 provides useful guidance on investigation point spacings for a range of situations, including a spacing of 15–40 m for 'high-rise and industrial structures', less than 60 m for 'large area structures' and 20–200 m for long linear structures such as roads, railways, retaining walls, pipelines or tunnels. It should be noted that these spacings do not necessarily indicate frequent deep and expensive boreholes – just enough exploration points to detect the more critical changes in ground conditions. For pile design in London Clay the critical variable with the likely widest variation is the surface level of the London Clay: it can sometimes vary by several metres due to the routes of old rivers, or even by several tens of metres due to scour hollows. This means that more boreholes need to reach and prove the surface of the London Clay than boreholes that go right through the whole stratum and prove its base. Similarly, made ground is much more variable and may change over very short distances. Its surface thus needs to be proven more frequently, and its inherent variability needs to be recognised and allowed for in design. Adopting this philosophy should result in the most cost-effective extent of fieldwork that is economical but still effective in reducing the key risks.

Where the extent of the development is uncertain, a balance has to be struck between the right level of fieldwork for the most extensive development and the savings that can be made by planning for a more modest development, with the possibility of a return visit later. The choice may need to be discussed with the client to agree what is most appropriate – the client may be prepared to live with the risk that more work will be required later to realise a potential saving. Particularly where the difference in cost is relatively small and the project programme is tight, the client may not be able to live with the risk of missing critical information. Sometimes a compromise is possible where sufficient data are gathered to enable scheme design for all options, but there are only enough data for complete design of the less extensive option. It also depends on the procurement route chosen for the main contract and the mechanism for transferring risk.

As stated earlier, the ground investigation should provide the data for the likely temporary works (so far as they can be predicted at the time). There is unlikely to be time later in the temporary works design for fresh information to be obtained, so the project's overall risks (and probably costs) will be higher if this is not done at an early stage.

44.2.8 Clear campaign aims

Once a campaign of ground investigation has been planned, it is important there is a clear strategy for its execution. For larger and more complex investigations it is good practice to set out the strategy on a site plan and on an idealised geological section. As each hole is completed, its contribution to the overall campaign can be tracked. This is especially important when there are a number of aspects of the investigation which need to be resolved spatially and with depth. The strategy, if properly executed, should ensure that sufficient results are obtained without breaching the budget. Features that often need a formal strategy at the start of the campaign include

- stratigraphy – proving the surface of each important layer to get a good spread spatially
- water pressures – variation with depth, spatially and in each stratum
- geotechnical test results – variation in each stratum with depth; detecting spatial variations
- contamination test results – variation of different chemical determinands in each area of suspected contamination.

An example is shown in **Figure 44.4**. Without clearly defined aims at the outset, a ground investigation can quickly collapse into chaos, as a huge flow of unrelated data can overwhelm the recipients.

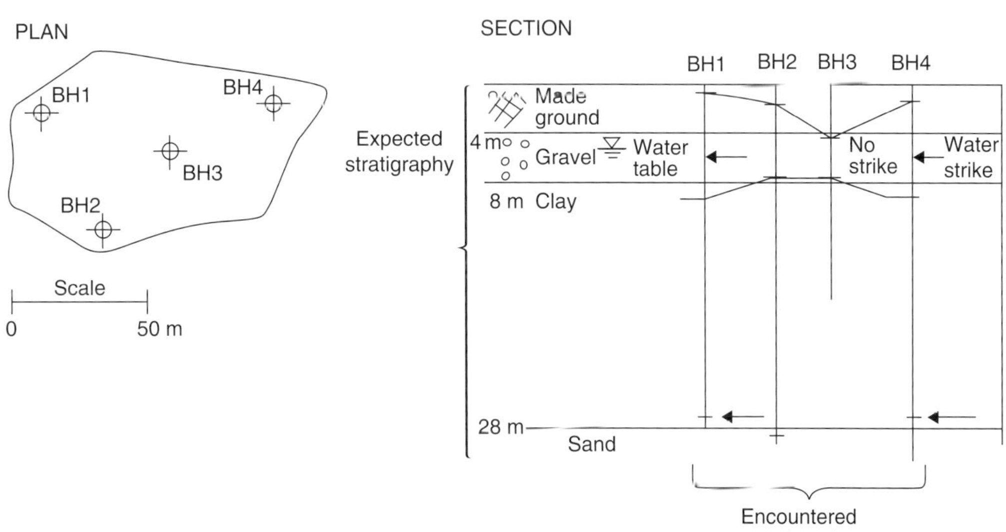

Figure 44.4 Example showing a comparison of expected and encountered stratigraphy to assist with managing a ground investigation

44.3. Procuring the site investigation

44.3.1 Typical costs

The site investigation operations of desk study and ground investigation are some of the best-value operations possible in terms of reducing risks on construction projects. The old adage says, 'you pay for a ground investigation whether you have one or not'. A poorly conceived or planned ground investigation often costs more in the end than a good one, because the cost of project change typically increases as the programme progresses.

Typical costs for ground investigations are shown in **Table 44.3**. Although these data are old, the values for buildings correspond to current experience.

Current costs of ground investigation for buildings are typically in the range 0.1–0.2% of the building cost. For small domestic works, investigation costs can often be higher – perhaps 0.5–1.0% of the works cost.

Thus, for a £30 million structure, the cost of the ground investigation would typically be £30 000 to £60 000. Higher rates apply either where the ground hazards are greater or where the proposed foundation methods may make higher demands on the ground than normal. The lower proportion normally applies when the site investigation is being carried out for a 'park' of several adjacent buildings, and efficiencies are possible, or where the structural form chosen is particularly expensive, but the foundations use simpler techniques.

For major infrastructure projects, current costs of the ground investigation phase prior to construction are in the range 0.1–0.15% of the project value. This does not include the cost of any initial phases of investigation carried out at preliminary design phases.

The ground investigation must target the specific site hazards, so arbitrary limits on its cost based on a fixed sum or typical proportions are likely to lead to a greater exposure to risk for the project. As explained in Chapters 7 *Geotechnical risks and their context for the whole project* and 40 *The ground as a hazard*, ground risks are some of the most extreme faced by most projects and economising at the expense of investigating hazards tends to result in increased costs later, so it is a very false economy.

A ground investigation is essentially an exploratory process, so its out-turn costs may vary enormously depending on what is found. It must be recognised by all parties at the outset that a ground investigation rig being delayed by obstructions is much cheaper and less programme-disruptive than, for example, the piling rig being delayed by the same obstructions.

Laboratory testing is a significant element of the ground investigation budget, so the extent of testing should not be underestimated when compiling the budget.

A rough rule of thumb is to allow £100–200 per linear metre of borehole drilled, including most types of field and laboratory testing. Thus, an investigation for a £300 million highways project involving 120 boreholes with an average depth of 20 m would be expected to cost approximately £360 000.

A ground investigation cost overrun as a result of unexpected ground conditions being encountered may be regarded as a success, because potential problems have been discovered early and measures can be taken to guard against them during the main construction works. Close, proactive control of the investigation by the geotechnical engineer and a responsive approach from the ground investigation contractor should allow the correct parameters to be determined within the overall investigation budget.

It is prudent to allow a sensible contingency which can be expended during the ground investigation in response to the evolving understanding of the ground. Normally, the contingency is derived partially from a generous estimate for rates which are more likely to be exceeded, plus an extra 10% to allow for unexpected conditions. A full schedule of rates covering all foreseeable tests should be obtained with the ground investigation tender.

If very unusual conditions are revealed there may be a need for the project team to re-evaluate and agree a new ground investigation budget. The contractual form should allow for the investigation to be varied according to the findings.

Type of work	Percentage of capital cost of works	Percentage of earthworks and foundation cost
Earth dams	0.89–3.30	1.14–5.20
Embankments	0.12–0.19	0.16–0.20
Docks	0.23–0.50	0.42–1.67
Bridges	0.12–0.50	0.26–1.30
Buildings	0.05–0.22	0.50–2.00
Roads	0.20–1.55	(1.60)?–5.67
Railways	0.60–2.00	3.5
Overall mean	0.7	1.5

Table 44.3 Typical site investigation costs (Data taken from Rowe, 1972)

44.3.2 Selecting consultants and contractors

Buying a ground investigation is like any other purchase – the buyer wants to make sure they get what they want, when they want it, and expects that there will be no nasty surprises on price or programme. The analogy of buying any commodity can be helpful: the buyer wants to be protected against things they do not expect. The more careful buyer will read all the small print and might negotiate to get what they want. But once the contract is signed, they are committed to their particular purchase.

A first step is to make sure that a proper brief is prepared. As few clients are expert in buying ground investigations, professional assistance is needed. The success of the purchase will depend on the clarity of thought that has gone into defining what will be required, and the clarity of specification in explaining that to the supplier. The next two sections explain how to choose the necessary professional assistance. The following sections explain how to describe the work needed in a contractual framework that provides protection to the client.

For a large building project, the **architect** may take on the role of *principal technical adviser* recommended in Part 2 of the *Site Investigation in Construction* series (Site Investigation Steering Group, 2012): *Planning, Procurement and Quality Management* (see Chapter 42 *Roles and responsibilities*). Normally, though, for a building project, it is usually the **structural engineer** who recognises the need for geotechnical support and who advises when it should be commissioned. Most development teams rely on their structural engineer to recommend the approach for geotechnical investigation and design. Their advice is likely to rely on the following factors.

- Geotechnical abilities of the structural engineering firm – especially whether it employs sufficiently skilled geotechnical professionals with experience in assessing ground variability and hazards.
- Attitude to geotechnical risk of the structural engineer – some tend to delegate as much responsibility as possible to other organisations in the belief that their own exposure to the geotechnical risk will correspondingly reduce. This is obviously not true if the choice leads to a greater overall risk to the project from disjointed design. If overall project risk increases, then all will be affected by the consequences.
- Special skills demanded by the particular development, for example whether special investigation techniques or advanced geotechnical analysis will be required.

A structural consultant who also operates as a ground investigation contractor may offer a complete 'one-stop shop' service providing integrated structural and geotechnical design and geotechnical investigation capabilities.

To some extent it is up to the project promoter to decide whether or not to seek the advice of the structural engineer also for geotechnical matters. This will depend on the confidence they have developed in the structural engineer to control what are potentially the highest project risks. A key factor for assessing the competence of the structural engineer for allocating ground risk is to understand the geotechnical ability of the company. If they directly employ or have ready access to experienced geotechnical advice, then they are more likely to engage in the reduction of ground risks.

The project roles outlined at the start of this chapter offer a useful means for assessing the geotechnical competence of the organisations proposed. The *geotechnical adviser* has a key role; previous successful experience and track record on similar projects are good indicators of the suitability of the individual. Before appointing a geotechnical adviser, it is worth challenging their

- experience with projects of similar structural complexity and demands on the ground
- familiarity with the expected ground conditions and special hazards faced
- understanding of the full investigation, design and construction process – beware people who offer a scope of investigation without understanding the likely extent of basement and types of foundations
- views of previous clients for previous projects: Did they perform their duties well? Were they proactive in spotting problems and reducing risks?
- ability to add value to the project and call on extra resources if needed.

An enthusiastic geotechnical adviser should offer a strategy for completing all the geotechnical work for the project, with project-specific ideas on how to reduce risks and improve buildability at each stage. The new UK Register of Ground Engineering Professionals (RoGEP) provides an appropriate qualification route.

The *Site Investigation in Construction* series documents (Site Investigation Steering Group, 2012) make sound recommendations for the employment of a geotechnical adviser to oversee a coherent geotechnical design process. They suggest that the need for this geotechnical adviser should be suggested to the client by the principal technical adviser, who for many projects will be the architect.

Figure 44.5 shows the typical decision-making process for ground investigations proposed by the Site Investigation Steering Group.

Some geotechnical advisers work for companies that are pure consultancies – that is, they need ground investigation contractors to be separately employed – while some work for companies that also operate equipment and laboratories. There are pros and cons to both alternatives, but generally the combined approach offers advantages for smaller and simpler commissions and ground investigations, while the separated approach is better for larger and more complex investigations, particularly where the client is suspicious of an integrated contractor proposing extra ground investigation purely to increase its own revenue and profit.

44.3.3 Defining the scope of work

The scope of work can be defined either by the consulting engineers providing geotechnical services for the project or by the ground investigation contractor. The former route permits a competitive tender for the work to be carried out fairly – it is unfair (and bad practice) to invite tendering contractors to compete on the extent of work, as that rewards those who are prepared to suggest the least work. This apparent saving will mean that hazards are investigated less thoroughly; contractors tendering for the main construction work will face greater uncertainty and risk, for which they will price, or which will not become clear until unexpected conditions develop during construction.

The scope of ground investigation depends on the scope of the planned construction.

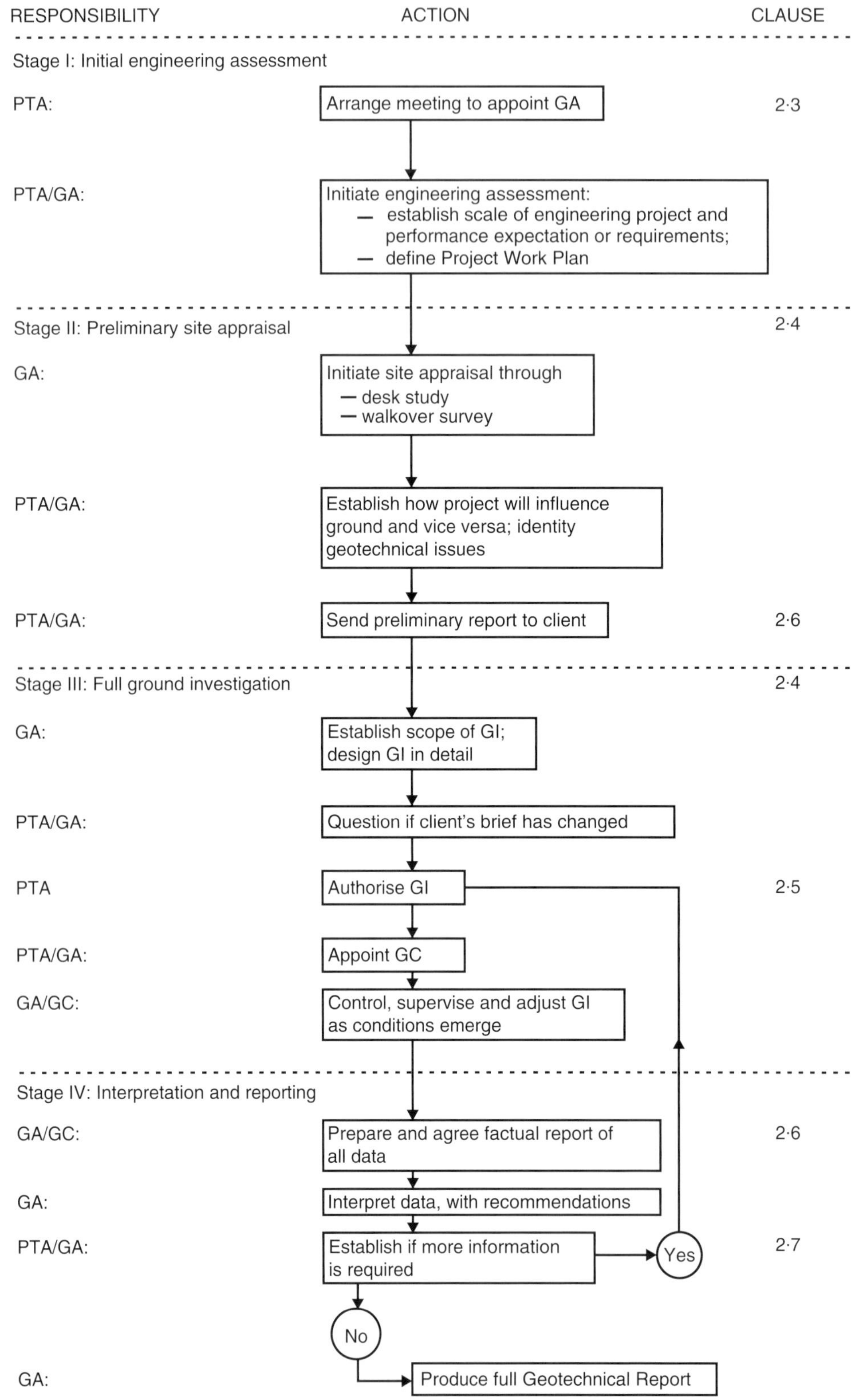

Figure 44.5 Decision-making process of site investigation (Reproduced from Site Investigation Steering Group (2012) Part 2: *Planning, Procurement and Quality Management*)

When planning the scope of work needed in a ground investigation, beware of artificial restrictions on its extent – normally imposed by an overly cost-conscious client who does not understand the increased risks the restrictions might lead to. If limitations are placed on the scope, those limitations should be spelled out in the report: either budgetary or on extent caused by lack of access. This allows

- others to understand risks identified but not followed up, so that the design can address that additional uncertainty
- responsibility for lack of coverage to be properly apportioned.

44.3.4 Ground investigation contract documents

A clear set of contract documents is needed for any ground investigation to describe the task and to regulate how it is carried out. These are introduced in Chapter 42 *Roles and responsibilities*. In broad terms, it is helpful to separate these concepts into *specification* and *contract*. The specification should provide the technical requirements and scope, while the contract documents should regulate how the work is carried out in order to avoid disputes. The two parts are distinct and input should clearly be for one purpose or the other, and not mixed. The bill of quantities will reflect both.

44.3.4.1 Specification

The purpose of a technical specification is to clearly define the work that is to be carried out, in terms of both extent and quality. A clear specification offers the following advantages.

- Prior to tender it allows the design team to discuss and reach agreement on what work needs to be done.
- During tender it unambiguously informs each tendering contractor what should be allowed for in the price.
- During the works it provides a basis for discussion between the client team and contractor, should any disagreement occur.
- After the works are complete it allows fair payment to be made.

Phrases like 'to the satisfaction of the engineer' are unhelpful in specifications, because the contractor can never know how easy the engineer will be to satisfy. It rewards more lax contractors who might assume a lower standard over more responsible or experienced contractors who might make a more realistic or generous allowance. It is seldom necessary to use such phrases, as the engineer can normally replace them with the criteria that they would wish to invoke.

A clear technical specification describes:

- the background to the work – its purpose (i.e. what will be built); the expected ground conditions; the current usage of the site and so on, including access and necessary constraints
- the work that is needed
- the standards to which the work should be carried out, referring as much as possible to national standards.

Part 3 of the *Site Investigation in Construction* series (Site Investigation Steering Group, 2012 (third edition due for publication in 2023)), *Specification for Ground Investigation* ('The Yellow Book'), provides a very well-formatted national specification for the UK. National specifications are better than client or engineer specifications because:

- contractors should automatically carry out work to the required standard
- contractors should be aware of the normal requirements, which facilitates easy tendering with less risk of omissions
- all work in the UK should comply with the minimum standards, so when a completed ground investigation report is given to a new consultant, the consultant should find it easier to accept its findings.

The Yellow Book includes a general specification which addresses most of the issues needed for most ground investigations. It also includes draft bills of quantities. It allows the specifier freedom to fully describe the scheme-specific details in a series of schedules.

- Schedule 1 gives the background to the project and describes in words what is needed, together with key constraints, such as access.
- Schedule 2 is a table that lists for each exploration hole what is needed. It can give options – such as borehole depth, which can either be given in absolute terms or can require the ground investigation contractor to continue to a certain penetration into a stratum.
- Schedule 3 is a list of facilities needed by the site supervisor – normally the engineer's representative under ICE forms of contract.
- Schedule 4 is a list of any amendments that the specifier might wish to make to the standard specification.
- Schedule 5 is a list of additions to the specification, either where extra requirements are being imposed or where new areas of work are required which are not covered by the standard specification.

44.3.4.2 Forms of contract

For all forms of construction a pre-agreed contract is necessary for the share of risk between client and contractor to be clear. This is also true for ground investigation contracts to govern what happens if things go wrong. Some common disputes in ground investigations are payment (like all contracts!) and who is responsible if buried services are damaged during the fieldwork.

The conditions of contract must address

- what the contractor is required to do
- delegated decision-making powers
- start, programme and completion dates
- alterations and change processes

- delay procedures
- measurement protocols
- certification requirements
- insurances, ownership of the works and safety
- method of payment (lump sum or re-measurement to bills of quantities), including retentions (e.g. to protect the client, who derives little benefit from the fieldwork until the contractor's factual report is delivered; or in case of inadequate reinstatement) and bonds
- repair of defective work
- termination clause
- resolution of disputes.

Many different forms of contract are available. What the investigation will find is uncertain at the outset, so the contract should allow the supervisor to change the order and scope of the works, without too cumbersome a control and approval process.

ICE forms of contract have been withdrawn and been replaced with the *Infrastructure Conditions of Contract*, 2011 (current at the time of writing) – a standard suite of contract documents for infrastructure works jointly sponsored by the Association for Consultancy and Engineering (ACE) and the Civil Engineering Contractors Association (CECA). Within this set of contracts is a Ground Investigation Version which is intended for use by an employer contracting with a contractor to carry out the ground investigation, but not to design it nor interpret the results. If other forms of contract are used it is important that they are made so as to allow a swift reaction to unexpected findings. Other forms of contract, such as JCT forms, tend not to work well as they are intended to limit flexibility on site, which can be counterproductive. Ground investigation contractors may add a risk allowance to cover unfamiliar contract conditions.

Ideally, for a ground investigation, a resident engineer from the designer will be on site to witness key elements of work and to understand what has been uncovered. If only part-time coverage is provided, the contractor must have clear instructions of how to proceed if unexpected conditions evolve, and a fair way of paying the contractor so that the individual takes the initiative to get the best results for the whole project. The key is to involve the designer, geotechnical adviser and ground investigation contractor in a collaborative approach to solving problems and obtaining the appropriate parameters for the project.

The complete contract documents should encompass

- conditions of contract, including the form of agreement
- technical specification
- priced bills of quantities
- accompanying drawings, possibly including site location plan, borehole location plans, service routes and historical or geological figures from the desk study.

44.3.4.3 Bill of quantities

The bill of quantities (or sometimes schedule of rates with provisional quantities, to emphasise the potential for variation) allows a tender price to be agreed. The method of measurement should also be defined, as otherwise differences in measurement understanding may lead to disputes. Beware using provisional sums or 'rate only' items, as in the former case no cost is agreed, and in the latter case the contractor's tender total is not penalised by the use of high rates.

44.3.5 Risk sharing

In most contractual relationships the client tries to pass risk to the contractor. In ground investigations, because of the fundamental uncertainty about what might be found, there is greater risk sharing and the client must accept the possibility that more work may be required. A ground investigation that identifies serious features meriting greater attention, resulting in cost and programme overruns, should be seen as a great success and not a failure. The consequences of those features remaining undiscovered until construction would be far worse than identifying them at an early stage.

Where a main contractor is on site and suitable plant is available, the main contractor may be in a better position to negotiate access for the ground investigation with landowners and provide temporary access tracks for investigation plant. Responsibility for, and thus risk on, access could then be removed from the ground investigation contractor's contract, leading to a more competitive price.

44.3.6 Tender evaluation and award

It is important that all tenders are fairly evaluated on the same basis. Therefore, careful consideration is needed to ensure that each restriction or exclusion is evaluated. While the tenders should be evaluated for the understanding demonstrated of the site and tasks, it is too late to regret seeking tenders from a less competent contractor who happens to submit the lowest price. If there were doubts about the contractor's competence, they should not have been invited to submit a tender.

44.4. Managing the site investigation

44.4.1 Safety management

All construction sites are potentially dangerous places. This section highlights some essential features of safety on ground investigations, but should not be taken as a definitive guide. Statutory guidance changes with time, and users must ensure that they comply with the latest regulations – in particular the Construction (Design and Maintenance) Regulations (2015), also known as the 'CDM Regs' – see Chapter 42 *Roles and responsibilities*. The following are some typical hazards in site investigations.

- Trial-pit collapses, often due to inadequate shoring. This topic is addressed by the *Excavations* section of the HSE Construction Safety Topics and the document 'Excavations: What you need to know as a busy builder CIS64' (HSE, 2019). Regulation 12 of the Construction (Health, Safety and Welfare) Regulations 1996 requires that 'All practicable steps shall be taken, where necessary to prevent danger to any person, to ensure that any new or existing excavation or any part of such excavation which may be in a temporary state of weakness or instability due to the carrying out of construction work (including other excavation work) does not collapse accidentally.' This used to be set at a 1.2 m safe depth before shoring was required, but the current regulations are more general.

- Injuries in cable percussion boring caused by use of heavy weights – particularly to fingers.
- Hitting buried services – this is a potential cause for major injury to site workers, dispute and disruption to neighbours and businesses. The hazard is so great that time should always be made to obtain plans from all owners of buried services and to carry out site scanning before work starts.
- Investigation of potentially contaminated sites.

Establishing the location of buried services is critical prior to digging holes on a site. The various steps that should be taken are as follows.

- Contact all the likely owners of buried services to receive contemporary plans – these are often inaccurate and seldom show domestic connections. Sometimes unusual services can be found, such as strategic oil pipelines. Ask around. In the UK contact an organisation like Groundwise.com for the normal range of buried utilities or Linesearch.org for pipelines. For sites where the owner has a maintenance department, talk to the people who have worked on the site for a long time.
- Lift manhole and drain covers on the site to locate likely service routes. Lines of replaced concrete or tarmac are often associated with service laying.
- Use a scanner to try to detect pipes and cables – note they seldom find plastic pipes. This can be done in conjunction with representatives from the companies owning the services on site for the best results.
- Depending on the likelihood of buried services remaining undiscovered, hand-dig 'starter pits' at each location to reduce the chances of damage being caused. Vacuum excavation techniques are also available to ensure the safety of site workers.

Sometimes major services are a significant constraint on ground investigation operations – both buried utilities and overhead cables. See Chapter 41 *Man-made hazards and obstructions*.

The Control of Substances Hazardous to Health Regulations 2002 (COSHH) require employers to control exposure to hazardous substances which could put people's health at risk.

In buildings dating even to the mid-1980s, the presence of asbestos in lagging to pipes, walls and ceilings should always be suspected. If the presence of asbestos is suspected, no intrusive work should be carried out until its absence has been verified by sight of the building's asbestos register.

The person planning the ground investigation must not locate boreholes and trial pits in locations which could be more hazardous if other safer locations are feasible.

- Boreholes under overhead power lines pose a significant hazard for those erecting and using the boring rig.
- Trial pits next to a cliff pose a significant threat to the operator of the excavator.
- Machine-excavated trial pits over live buried services pose a threat to all in the vicinity.

However, it may be that after consideration of the aims of the ground investigation, no other locations offer such benefits, in which case the designer must try to mitigate the risks as far as possible, for instance by following the guidance for digging near services, and by only using air blowing or vacuum excavation or hand-digging techniques until all the services have been found.

Particular care is needed when planning investigations inside buildings. Where it is necessary to bore from ground level through a basement, a number of checks are needed to make sure that

- no cables, pipes or ducts will be damaged at any level by the breaking out or boring works
- the suspended slab is capable of supporting the weight of the boring rig, especially when a hole has been broken in the floor.

44.4.2 Supervision

A strong case can be made for the ground investigation contractor to provide full-time technical supervision on site. This person will be involved in maintaining technical quality, ordering the works, reacting to anomalies, logging pits, borehole samples and drill runs, and selecting suitable samples for later laboratory testing. Doing these activities rapidly is often essential for samples for subsequent chemical analysis. However, it must be noted that there is a cost associated with provision of such a person, and standard fieldwork rates are seldom adequate to enable the ground investigation contractor to provide that person without some payment. Hence it is helpful if the specification and bill of quantities define the expected basis.

Additional audits should be carried out, normally by the designer or geotechnical adviser, to ensure that supervision is being carried out properly and that the specification is being reliably followed. Such visits should apply both to siteworks and to the testing in laboratories. Often laboratory checks are neglected as it is difficult to arrange the timing so that the particular samples from the site can be witnessed.

It is also of great value if the geotechnical designer can see the actual ground conditions on site, and so additional visits for this purpose should be made. Visits coinciding with the opening of trial pits prove most valuable as the stability of excavations and the inflow of water can be assessed.

44.4.3 Quality management and quality control

Quality management should be undertaken by a registered quality system to ISO 9001 by both consultants and ground investigation contractors. Quality control is vital to ensure that the tasks are carried out diligently and to a reliable set of procedures. Such quality procedures are very important for the simple repetitive testing carried out in soil testing laboratories, where small changes in operator procedure can have a profound effect on the results.

44.4.4 Data management

A ground investigation produces considerable quantities of data, and these are usually best managed in electronic form. The ground investigation contract should specify that all data are to be produced in AGS format. This simple, standardised format allows the data to be imported into all spreadsheet packages and specialist borehole management software. From here, the data can be imported into GIS and 3D virtual construction models to aid integration with the design model.

Current products under development include robust site datalogger tablets. These will enable the ground investigation operatives to input data direct from the borehole or trial pit and transmit them in real time to the geotechnical adviser. Other useful innovations include GPS tagging of site photographs and barcoding or QR coding samples.

44.4.5 Fulfilling vital campaign aims

The ground investigation should have had clear aims at the outset. Those are sometimes changed as the work takes place, but they should remain visible and coherent. Hence, at key stages it is worth reviewing whether those aims have been met, before it becomes too late. Such reviews can be tricky as at such key stages there are usually many more data and clarity can be difficult. It is worth formally reviewing the aims a week before the ground investigation contractor leaves site, after which time any further fieldwork will become extremely expensive and disruptive, and at the time of scheduling the last tranche of laboratory tests, after which time there is a severe risk of delaying the main deliverable from the ground investigation – the contractor's factual report.

44.4.6 Effects on site occupants and neighbours

The works for the ground investigation are usually the first manifestation of the new project on site. For the reasons outlined in Section 44.2.3, occupants and neighbours may not welcome the start of the ground investigation. Therefore, it needs to be started sensitively, and with good supervision to intervene between anxious residents and the site operatives. It is often a good idea to agree a simple communications plan to handle queries, complaints and correspondence in a consistent and professional way.

Every reasonable effort should be made to reduce the level of disruption. If not, the local authority environmental health officer (EHO) in UK locations may become involved and impose severe restrictions on working hours. Where periods of noisy work are inevitable it is best to inform stakeholders – including the EHO – beforehand, and then keep any promises made.

The level of reinstatement needs to be carefully considered. On the one hand, it can be seen as wasted expenditure, as the main development will obliterate it. On the other hand, a degraded site environment will make it less attractive to users, may render parts of the site unusable and could make it unsafe. Issues to consider include

- reinstatement of slabs, hard standings, roadways and grassed areas
- future protection of standpipe and piezometer stopcock covers.

44.4.7 Reinstatement

It is important that the site after completion of the fieldwork is left in a condition acceptable to the site occupiers/owners. This means

- paved areas should be reinstated as needed
- grassed areas should be rehabilitated, including wheel marks for access
- structures should be repaired, and at least made safe – it is not unusual in investigations in buildings for boring to be made from ground floor level through a basement: in such cases the holes should be securely fixed and may need to be made weatherproof to prevent water leaking into the basement, creating a further safety hazard
- all the old spoil should be cleared away; some contractors have even been known to abandon equipment at the site.

In addition, the investigation should not leave the subsoil in a condition that could cause problems in the future.

- Boreholes should be grouted up on completion. Ungrouted boreholes can leave a connection between different aquifers, which can spread pollution and can also be a source of hazard for piling and tunnelling work.
- Trial pits should be well compacted. Loosely compacted trial pits later become zones of unexpectedly weak ground. This can be a hazard for shallow foundations and for temporary piling platforms. For this reason, for housing developments it is preferable to position pits away from the footprint of the houses where known.

44.4.8 Contract administration and completion

As in all contracts, standard resident engineer (RE) records should be kept so that payment items can be fairly evaluated. The RE may be required to resolve misunderstandings with neighbours, particularly when the contractor has no professional staff on site. Payments need to be processed according to the contract, with certificates provided as necessary.

The RE records should be kept, as they may be instructive in understanding any anomalies detected on the site subsequently – for instance, a length of rotary cored drill hole over which no core was recovered may have been thought to have been lost because of poor drilling but it may later transpire to be due to cavities in the ground.

References

ACE (Association of Consultants and Engineers) and CECA (Civil Engineering Contractors Association) (2011) *Infrastructure Conditions of Contract*, 1st edn. ACE, London, UK.

AGS (Association of Geotechnical and Geoenvironmental Specialists) (2018) *Guide to the Selection of Geotechnical Soil Laboratory Testing*. AGS, Beckenham, UK.

BRE (Building Research Establishment) (2004) *Working Platforms for Tracked Plant. Report 470*. BRE, Bracknell, UK.

BSI (2007) BS EN 1997-2:2007: Eurocode 7 – Geotechnical design: Ground investigation and testing. BSI, London, UK.

BSI (2009) BS 5228-2:2009: Code of practice for noise and vibration control on construction and open sites: vibration. BSI, London, UK.

BSI (2020) BS 5930:2015 + A1:2020: Code of practice for site investigations. BSI, London, UK.

DCLG (Department for Communities and Local Government) (2004) *The Party Wall etc. Act 1996: Explanatory Booklet*. DCLG, London, UK.

Glossop R (1968) The rise of geotechnology and its influence on engineering practice. Eighth Rankine Lecture, Institution of Civil Engineers. *Géotechnique* **18**: 105–150.

Head JM and Jardine FM (1992) *Ground-Borne Vibrations Arising from Piling*. CIRIA Report TN142. CIRIA, London, UK.

HSE (Health and Safety Executive) (2019) Excavations: what you need to know as a busy builder. Construction Information Sheet 64, rev. 2. HSE, London, UK. https://www.hse.gov.uk/pubns/cis64.pdf (accessed 26/05/2023).

Rowe PW (1972) The relevance of soil fabric to site investigation practice. Twelfth Rankine Lecture, Institution of Civil Engineers. *Géotechnique* **22**: 195–300.

Site Investigation Steering Group (2012) *Site Investigation in Construction*. Thomas Telford, London, UK.

Further reading

HSE (2002) *Control of Substances Hazardous to Health*, 5th edn. HSE, London, UK.

McAleenan C and Oloke D (2010) *ICE Manual of Health and Safety in Construction*. Thomas Telford, London, UK.

Site Investigation Steering Group (2011) *UK Specification for Ground Investigation*, 2nd edn. ICE Publishing, London, UK.

Site Investigation Steering Group (in press) *Effective Site Investigation*, 2nd edn. ICE Publishing, London, UK.

Site Investigation Steering Group (in press) *Guidance for Safe Investigation of Contaminated Land*, 2nd edn. ICE Publishing, London, UK.

It is recommended this chapter is read in conjunction with:

- Chapter 8 *Health and safety in geotechnical engineering*

All chapters in this book rely on the guidance in Sections 1 *Context* and 2 *Fundamental principles.* A sound knowledge of ground investigation is required for all geotechnical works, as set out in Section 4 *Site investigation*.

doi: 10.1680/icemge.66816.0657

Chapter 45
Geophysical exploration and remote sensing

John M Reynolds Reynolds Geo-Solutions Ltd, Mold, UK

This chapter provides a brief overview of the range of geophysical methods (surface and down-hole) and remote sensing available for use in engineering investigations. The techniques described are well proven and, when used in a properly designed site investigation, can add significantly to the robustness and cost-effectiveness of such a survey, especially when used in combination with intrusive methods.

The chapter explains the role of geophysics and describes how best to procure geophysical surveys. Current best practice is to engage the services of an independent engineering geophysical adviser to design the works, supervise the fieldwork and undertake the detailed analysis, modelling and integrated interpretative reporting. Guidance is provided as to which geophysical techniques are best suited to which applications.

The most commonly used geophysical techniques are described, with examples of outputs available. These include microgravity and magnetometry; electrical methods (resistivity tomography, induced polarisation, self-polarisation); electromagnetic methods (including ground penetrating radar); seismic methods (refraction, reflection, surface waves and ground stiffness); over-water surveys (hydrographic and sub-bottom profiling); drone-deployment; and borehole logging techniques.

A brief overview of remote sensing image analysis is also provided for optical (Landsat, SPOT, ASTER), synthetic aperture radar (SAR, InSAR), LIDAR and thermal infrared (TIR) techniques.

CONTENTS

45.1. Introduction

Investigating the ground remotely can be achieved with a range of techniques, from satellites orbiting the Earth using optical imaging, through to radio waves being generated from a hand-held sensor on a concrete slab inside a building to locate steel reinforcement bars. Sensors can be deployed from different platforms to produce information for regional investigations over thousands of square kilometres to ultra-high-resolution testing on a decimetre-scale.

With advances in technology over the last few years there is also a blurring of the boundaries between the modes of deployment of techniques. For example, LiDAR (Light Detection And Ranging) using laser measurements was originally made from fixed-wing aircraft to produce high-resolution digital elevation models of the ground surface. Now the same technique is also available from ground stations looking laterally at rock cliffs to monitor slope stability and from boats to image marine infrastructure, for example. Individual geophysical sensors have become smaller and more sensitive, and capture data faster with modern electronic and computer technology. Multiple sensors can be mounted on a wide range of platforms: hand-held, towed devices; booms and stingers on aircraft; 'birds' suspended from helicopters; quadrocopter drones, single sensors, horizontal and vertical gradiometers and multi-channel systems with many sensors mounted in a range of configurations.

Many geophysical techniques have been transformed over the last few years by faster and more efficient data acquisition, so that much more information can be captured in much less time, making it possible to increase the spatial resolution by at least an order of magnitude. Sensors have become more sensitive, so permitting the detection of ever smaller targets. There has also been massive improvement in geophysical data processing capability, and in the availability of increasingly sophisticated software. It has become increasingly common and practical to integrate geophysical, geotechnical and borehole data analytically into 3D geological ground models.

This chapter provides a very brief overview of the range of geophysical and remote sensing methods that are available. Every modern engineer should be aware of these techniques, which have all become well established and are thoroughly proven. The chapter is divided into an overview of geophysical methods (including over land and water, airborne and down boreholes) and a brief description of key remote sensing techniques.

45.2. The role of geophysics

The primary role of geophysics is to reduce risk. By undertaking a professionally designed and executed survey it should be possible to

- *reduce* health and safety risks – avoiding buried utilities, locating voids and so on
- *reduce* professional indemnity risks – not missing key features on a site

- *reduce* over-engineering of structures to cover uncertainty in ground investigations
- *reduce* project delays and associated costs through 'unforeseen ground conditions'.

And it should be possible to

- *increase* the technical reliability of site investigation information
- *increase* the cost-effectiveness of the site investigation through targeting intrusive tests on anomalous areas
- *increase* the technical robustness of ground models through integrating multiple datasets and types.

The primary benefits of geophysical surveying (Reynolds, 1996) include

- rapid areal coverage (hectares per day)
- fine spatial resolution (to $\ll$1 m)
- volumetric sampling rather than spot measurements
- non-invasive and environmentally benign nature
- time-lapse measurements
- quantitative rather than qualitative data.

One of the primary benefits of geophysical investigations is that they are environmentally benign by being largely non-intrusive. Most methods involve a technician moving a sensor either across the ground or just above its surface, or placing metal electrodes or spiked geophones into only the top few centimetres of the ground surface. The measurements are made without disturbing the subsurface, which makes some methods ideal for time-lapse surveys, where profiles are re-surveyed after a period of time to determine any changes in the subsurface environment. In contrast, a borehole might provide a pathway for fluid flow, thereby affecting the local hydrology; removal of the borehole casing does not restore the situation to a pre-borehole status.

Geophysical methods are able to sample at a fine spatial interval, often of the order of every 10 cm, or in the case of ground penetrating radar (GPR) at less than every 1 cm. By covering whole swathes of a site these techniques provide almost complete coverage of the available ground. At the most basic level, such surveying provides 2D reconnaissance by which anomalous areas can be identified. This on its own is only of qualitative use. It might indicate relative anomalies that might show high and low values of apparent conductivity, but not what might be causing such values to occur. This is where the intrusive testing is so valuable. As will be demonstrated later, intrusive investigations, where deep enough and appropriately detailed, can complement geophysical techniques and permit the translation of variations in a geophysical parameter to specific ground conditions (Reynolds, 2004). By correlating the material types with the apparent conductivity anomalies, the spatial extent of these materials might be indicated.

Where geophysical techniques are used to provide information about both horizontal and vertical variability along a profile, and where the analysis provides the means of determining depths to subsurface targets, detailed borehole logs can be invaluable in providing ground truth data. Borehole logs can define layer thicknesses and material types that can be used to constrain geophysical modelling, so reducing a degree of freedom in the interpretation process. It is important to emphasise that geophysical methods do not necessarily reduce the number of intrusive tests required but permit the positioning of such tests in places where they are needed. This makes the combination of methods far more effective than intrusive testing alone and can reduce overall costs significantly.

It is crucial, however, that geophysical surveys are designed to meet a particular technical objective, such as to map the boundary of a former closed landfill. This dictates the spatial sampling required, the likely techniques to be used, what kind of analysis is needed to yield the necessary information, and how the survey should be undertaken. A common mistake in procuring geophysical investigations is to choose a method based on its familiarity to the specifier or on the general availability of equipment, rather than selecting the most appropriate method(s) to meet the survey objectives. Data acquisition is a means to an end, not an end in itself. Many unsuccessful geophysical surveys undertaken have been a waste of money because they have been incorrectly specified, poorly designed or undertaken with inadequate field practices (sometimes using inappropriate techniques), with inadequate position fixing and/or even misusing the equipment. Furthermore, there are many examples of cases where perfectly adequate field data have been wrongly interpreted, as little thought has been given at the design and specification stage to the analytical methods required once the data have been collected. It is fundamental to the integrity of each geophysical survey that thought is given first to the design and execution of the analysis, data processing and interpretation of the geophysical results, and only second to the specification of the data acquisition. For many modern-day surveys, a specialist independent engineering geophysical adviser (EGA) should be appointed to the project team, along with a geotechnical adviser, as recommended by the guidelines on engineering geophysics (McDowell *et al.*, 2002). Best practice is for the EGA to design the overall geophysical investigation (from data acquisition through to integration of results and interpretation) and also to supervise the site works and undertake the detailed interpretation of the final results. Data acquisition can be undertaken by suitably experienced geophysical contractors. Further advice on the use of geophysics in engineering has been provided by Schoer (1999), Darracott and McCann (1986), Anon (1988), Reynolds (1996, 2011) and McCann *et al.* (1997) and, for archaeological investigations, David *et al.* (2008), Schmidt *et al.* (2016) and El-Qady and Metwaly (2019).

Given the range of geophysical methods generally available, it is possible to categorise the applications to which they are most suited or not appropriate at all. An extremely general guide to the possible usefulness of the various methods is given in **Table 45.1**. However, in many cases it is advisable to use more than one method to reduce possible ambiguity in the interpretation and to respond to different physico-chemical characteristics within the same environment. For example, in the investigation of a former landfill, electrical resistivity tomography is an ideal method to sense the conductive waste and possible leachate plumes migrating away through the base and/or sides. The resistivity results would respond predominantly to the electrical properties of the fluids present rather than to the lithological boundaries, for which the seismic refraction method might be better suited. Consequently, these two techniques are both applicable and are complementary. Similarly, in environmental investigations over brownfield sites, a common pairing of techniques is electro-magnetic mapping with magnetometry. Coincidence of anomalies can help to increase confidence in the interpretation. The information given in **Table 45.1** should not be used prescriptively for survey design or specification purposes. To increase the benefit of any geophysical survey, the design and specification should be prepared by an experienced and suitably qualified EGA. A guide to the physical properties of materials and the corresponding geophysical techniques sensitive to those properties is given in **Table 45.2**.

45.3. Surface geophysics

There are four groups of geophysical techniques in common usage, and each of these is described briefly in turn. While the emphasis is on the use of these methods over the ground (or water) surface, some techniques can also be readily deployed in airborne and/or down-hole modes. Most commonly, 1D

Geophysical methods	Depth to bedrock	Stratigraphy	Lithology	Fractured zones	Fault displacement	Buried channels	Natural cavity detection	Mine workings/ adits/shafts	Groundwater exploration	Landfill investigations	Leachate plumes and migration	Brownfield site mapping	UXO detection	Buried artefacts/ USTs
Potential field methods														
Gravity	1	0	0	0	2	2	4	4	1	3	0	1	0	1
Magnetics	0	0	0	0	2	1	0	2	0	3	0	4	4	3
Electrical methods														
Resistivity – sounding	4	3	3	2	2	3	2	2	4	4	4	0	0	1
Resistivity – tomography	3	2	2	4	3	3	3	3	4	4	4	2	0	3
Induced polarisation	2	2	3	1	1	2	0	0	3	3	2	2	0	0
Self-potential	0	0	0	2	2	1	1	1	4	4	3	2	0	0
Electro-magnetic methods														
FEM	2	2	2	4	2	3	4	4	4	4	4	4	4	3
TEM	2	2	2	3	2	3	1	1	3	3	3	1	1	1
VLF	0	0	0	1	1	1	2	2	3	3	0	1	0	0
Ground penetrating radar	2	3	1	2	3	2	3	3	2	2	0	2	3	4
Seismic methods														
Refraction	4	3	2	3	4	4	1	1	2	4	0	1	0	1
Surface wave profiling	3	4	3	4	3	3	2	2	2	4	0	2	0	1
MASW	4	3	2	3	4	4	1	1	2	4	0	1	0	1
Reflection – land	2	2	2	1	2	1	2	2	2	2	0	0	0	1
Reflection – over water	4	4	2	2	4	4	0	0	0	0	0	0	0	2

Key: 0 = Not considered applicable or untried; 1 = limited use; 2 = used, or could be used, but not the best approach, or has limitations; 3 = excellent potential but not fully developed; 4 = generally considered an excellent approach, and techniques well developed.
Applicability of techniques is dependent upon a range of parameters, including physical/chemical contrast with host material, target dimensions, scale of survey and intended depth penetration, site suitability, site dimensions and access, surface topography and material type, local noise. It is always best to select the method(s) on a site-specific basis and with respect to the nature of the targets being sought. It is emphasised that this table is a very general guide only and actual suitability of a technique to a site may vary from the values indicated above. Always seek advice from a professional geophysicist.
FEM, frequency-domain electro-magnetic; MASW, multi-channel analysis of surface waves; TEM, time-domain electro-magnetic; UST, underground storage tanks; UXO, unexploded ordnance; VLF, very low frequency.

Table 45.1 Suitability of geophysical methods for a range of shallow engineering applications (Data taken from McCann *et al.*, 1997)

Geophysical methods	Dependent physical property	Hydrocarbon exploration	Regional geological studies	Mineral exploration/ development	Engineering site investigations	Environmental investigations	Hydrogeological investigations	Detection of subsurface cavities	Mapping of leachate and contamination plumes	Location of buried metal objects	Archaeogeophysics	Forensic geophysics
Potential field methods												
Gravity	Density	P	P	s	s	s	s	s	!	!	s	!
Magnetics	Susceptibility	P	P	P	s	P	!	m	!	P	P	!
Electrical methods												
Resistivity	Resistivity	m	m	P	P	P	P	P	P	s	P	m
Induced polarisation	Resistivity; capacitance	m	m	P	m	s	s	m	m	m	m	m
Self-potential	Potential difference	!	!	P	m	s	P	m	m	m	!	!
Electro-magnetic methods												
FEM	Conductivity; inductance	s	P	P	P	P	P	P	P	P	P	m
TEM	Conductivity; inductance	P	P	P	m	s	s	!	m	s	!	!
VLF	Conductivity; inductance	m	m	P	m	m	s	s	s	m	m	!
Ground penetrating radar	Dielectric permittivity; conductivity	!	!	m	P	P	P	P	s	P	P	P
Seismic methods												
Refraction	Elastic modulus; density	P	P	m	P	m	s	s	!	!	!	!
MASW/ground stiffness	Elastic modulus; density	!	!	!	P	!	s	m	!	!	!	!
Reflection	Elastic modulus; density	P	P	m	s	!	s	m	!	!	!	!

Key: P = Primary method; s = secondary method; m = may be used but not necessarily the best approach or has not been developed for this application; (!) = unsuitable.
FEM, frequency-domain electro-magnetic; MASW, multi-channel analysis of surface waves; TEM, time-domain electro-magnetic; VLF, very low frequency.

Table 45.2 Common geophysical methods, their dependent physical properties and their generalised main applications (Data taken from Reynolds, 2011)

soundings and 2D profiles are acquired. Data from closely spaced 2D profiles can be incorporated into a 3D volume. For specialist applications it is possible to acquire true 3D datasets by deploying a grid of detectors, all of which sense signals from a common source (3D resistivity and seismic reflection surveys). Re-surveying profiles/grids after a significant period of time permits the acquisition of time-lapse surveys, by which it is possible to determine changes as a function of time. It is beyond the scope of this chapter to explain the principles of operation in any detail; readers are directed to more comprehensive sources of information (Milsom and Eriksen, 2011; Reynolds, 2011).

45.4. Potential field methods

45.4.1 Gravity

The gravity method is sensitive to variations in the density of the ground, and gravimeters are designed to detect changes in the Earth's gravity field. This is measured by determining the changes in length of a spring resulting from the force exerted on a mass suspended on the spring. Modern microgravity meters are capable of making measurements to an accuracy of better than 5 μGal with a daily productivity rate of 50–80 stations per day, with typical anomaly amplitudes of several tens of μGal in engineering investigations (such as in void detection) (1 μGal = 1 cm/s^2). It is essential to be able to survey each measurement point to an accuracy of better than ±0.1 m horizontally and ±0.02 m vertically to achieve the best results. Given their heightened sensitivity, modern gravimeters pick up noise and vibrations, including distant earthquakes, which can degrade the data quality and slow the survey rate. The instrument provides values of gravity measured over discrete time periods, typically 60 s at each measurement 'station'. The values are averaged either to give a mean figure over that time period or until a designated standard deviation has been achieved. It is essential also to establish a base station at which values of gravity are measured at intervals typically no greater than every 2 hours, depending upon the type and characteristics of the instrument being used. These base station values are used to derive the diurnal drift of the instrument, which has to be compensated during the data correction process.

Gravity values (g_{obs}) are influenced by earth tides, the position on the Earth's surface with respect to its shape (oblate spheroid) and surface elevation, terrain, isostasy (in the case of larger-scale surveys), physical movement if the gravimeter is being used from a moving platform (ship or aircraft) and the physical presence of the ground itself. These corrections are required to reduce the values of g_{obs} to what is known as the Bouguer anomaly (Δg_B). This may need to have the regional gravity gradient removed and also the values corrected for the presence of buildings and walls, so that the data to be interpreted are corrected for all above-ground influences. The resultant anomaly is known as the residual Bouguer anomaly, which should indicate only the variations in the density of the materials beneath the gravimeter at a given station.

The residual Bouguer anomaly data can be plotted as 2D contour maps showing the variation in values across a site. Low values represent areas of density deficiency that might be attributable to the presence of voids or poorly consolidated backfill; high values may represent dense infill or higher-density bedrock. Commonly, data along extracted profiles can be modelled to produce 2D sections indicating the dimensions of a target, its depth of burial and its density contrast with the surrounding material (indicating, for instance, if a void is air-filled, partially filled with loose material) (**Figure 45.1**).

45.4.2 Magnetometry

Modern caesium vapour magnetometers measure the total intensity of the Earth's magnetic field, which varies from around 22 000 nT (nanotesla) at the equator to 67 000 nT at the geomagnetic poles (Alken *et al.*, 2021). If magnetisable objects are present, they distort the Earth's field in characteristic ways. Magnetometers are capable of measuring to 0.01 nT sensitivity with a sampling rate of up to ten measurements per second, with common anomaly amplitudes ranging from a few tens of nanotesla to many hundreds. The ground coverage achievable is dependent upon the speed at which the instrument is transported along a profile (transect). At walking speed this equates to about one measurement every 10–20 cm. Position fixing with a differential global positioning system (dGPS) can be achieved and recorded simultaneously with the magnetic field intensity values. It is possible to use magnetometers either singly or with multiple sensors to measure horizontal and/or vertical magnetic gradients.

Recorded data can be displayed as 2D colour-contoured maps of the magnetic field intensity and horizontal or vertical magnetic gradient (nT/m), and as discrete profiles. Interpretation is significantly more complicated than for microgravity, as magnetic measurements are a function of both the induced magnetisation of the target given a magnetic susceptibility, and any inherent remanent (permanent) magnetisation. Where induced magnetisation is the predominant cause, the anomaly shape and size are also dependent upon the orientation of the target, its dimensions and its depth of burial. Magnetic profiles can be modelled, and field values compared with synthetic values derived for a user-selected model, which is changed until the difference between the observed and

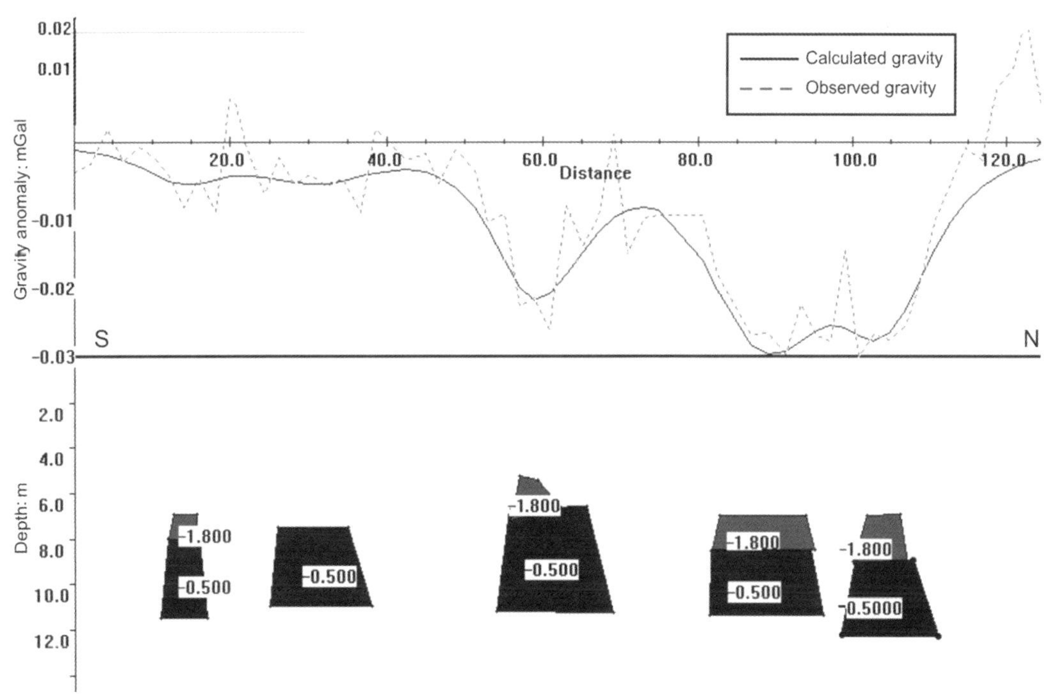

Figure 45.1 Bouguer anomaly microgravity profile showing observed and modelled values over mine galleries in chalk. The negative values refer to the density contrast with the host rock, with −1.8 Mg/m^3 indicating an air-filled void and −0.5 Mg/m^3 a partially backfilled void. These results were confirmed by subsequent probe drilling

synthetic data falls within a predetermined standard deviation or RMS error (**Figure 45.2**).

Common applications of magnetometry in engineering investigations are to locate ferrous metal objects (underground storage tanks (USTs), iron pipes, steel drums), but it can also be used to map magnetically susceptible material such as ash, clinker and bricks, and to delineate zones with different materials within landfills. A growing use of magnetic methods is in the detection of unexploded ordnance (UXO), such as ferrous metal bomb casings and mortar shells. However, the technique will not find UXO devices specifically designed to avoid detection, such as modern plastic landmines.

45.5. Electrical methods

45.5.1 Electrical resistivity

Electrical resistivity techniques detect changes in the electrical potential between a pair of electrodes generated by a separate pair of electrodes through which a low-frequency alternating current is injected. By increasing the separation between electrodes, greater depth penetration can be achieved. The technique effectively measures the resistance for a given electrode geometry to produce an *apparent* resistivity (ρ_a). This value in itself is a form of volumetric average and has no direct physical significance. There are two principal modes of use: (*a*) vertical electrical sounding (VES), and (*b*) electrical resistivity tomography (ERT).

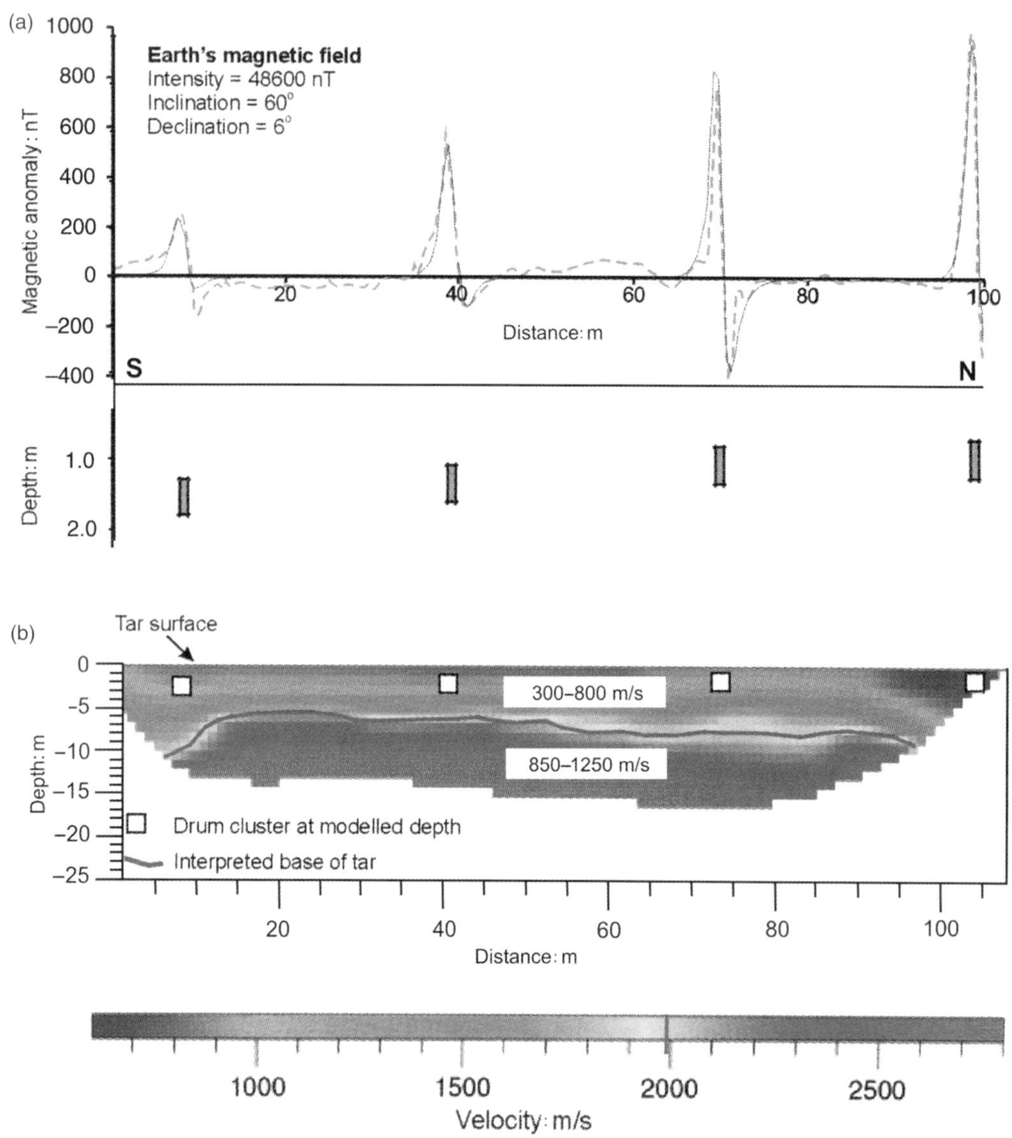

Figure 45.2 (a) Profile of magnetic field intensity showing observed results (solid line) and those generated by modelling (dashed line) the depth of burial of steel drum targets in an acid tar lagoon. (b) Small squares indicate the position of the steel drums as derived from the magnetic modelling, superimposed on the results of seismic refraction velocity modelling, indicating their positions within viscous tar above a solid base (Reproduced from Reynolds, 2002, European Association of Geoscientists and Engineers)

VES is achieved by measuring ρ_a using an expanding array of electrodes to produce a one-dimensional vertical profile (sounding) of ρ_a as a function of electrode separation. VES curves can be analysed (using a process called 'inversion') with commercially available software (e.g. IX1D, Interpex Ltd.) to produce an interpretation comprising a series of horizontal layers whose thickness and *true* resistivity are derived by the inversion process. The *true* resistivity is characteristic of the type of material present.

The ERT method uses a series of electrodes, commonly 72 or more, connected by a multi-core cable to a resistivity meter by which a 2D panel of apparent resistivity values is constructed. The resistivity data can be inverted using specialist software (e.g. RES2DINV, Geotomo Software) to produce 2D sections showing modelled values of true resistivity along the profile as a function of depth (**Figure 45.3**). The software also permits topographic corrections to be applied.

ERT is used more frequently than VES nowadays, although the latter is extremely useful in providing better vertical resolution of a layered structure; soundings are, however, sensitive to 3D structures, making the 1D inversions unreliable under these circumstances. Software is available to produce 2D and 3D models of resistivity so as to be able to test the detectability of particular structures and targets.

Resistivity techniques are used particularly where fluids in the pore spaces within materials have contrasting resistivity values, such as in mapping leachate plumes and imaging landfills. They are also useful for investigating natural and man-made voids and mine workings, for hydrogeological investigations, for imaging through landslides and in geohazard studies.

Two other electrical techniques are also used in geoengineering and environmental applications, namely *induced polarisation* (IP) and *self (spontaneous) potential* (SP).

45.5.2 Induced polarisation

IP profiles are acquired in a similar manner to ERT sections, but in addition to measuring the apparent resistivity, values of chargeability – which are a measure of the amount of polarisation observed – are also acquired. IP is used in base metal, geothermal and groundwater exploration, but also increasingly in environmental investigations.

45.5.3 Self (spontaneous) potential

The SP method uses two non-polarisable electrodes to measure the electrical potential between them generated by the flow of ion-rich water. The method is used in landfill studies and especially to detect leaks through embankments and natural dams, and also in hydrogeological investigations.

45.6. Electro-magnetic (EM) methods

45.6.1 EM methods: non-GPR

There are a wide range of EM techniques available, of which the commonest and most widely used (and abused) is GPR. In this section, non-GPR techniques are described. There are two main types of EM equipment: (*a*) frequency-domain EM (FEM) and (*b*) time-domain EM (TEM). FEM uses a pair of coils at a fixed separation (the dipole length); one coil is used to generate an EM field that propagates into the ground. A subsurface conductor responds to the primary EM field by setting up eddy currents within itself that in turn generate a secondary field that travels back to the surface and is measured by the secondary receiver coil. The ratio of the primary and secondary fields is a measure of the *apparent* conductivity of the volume of material being sampled. Conductivity is the inverse of resistivity; apparent conductivity is a function of the conductance of the ground and the geometry of the

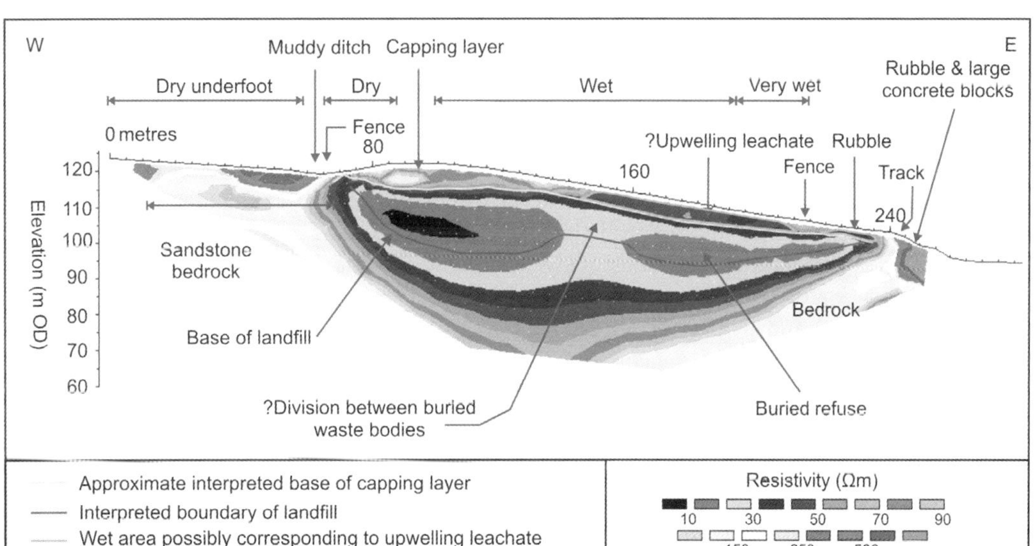

Figure 45.3 Electrical resistivity tomography section across a closed landfill

Figure 45.4 Geonics EM ground conductivity meters: (a) an EM31 with GPS antenna, and (b) an EM34 (white coils)

dipole arrangement, and in itself is not physically diagnostic of material properties. The most popular EM equipment is made by Geonics Ltd. They produce a family of EM ground conductivity meters, ranging from the smallest, the EM38 (nominal depth of penetration to 1.5 m) that is used predominantly for archaeological and agricultural surveys, through the EM31 (nominal depth of penetration to 6 m) with two coils in a rigid boom (**Figure 45.4(a)**) to the EM34, which has three sets of dipole separations (10 m, 20 m and 40 m) with separate coils (**Figure 45.4(b)**), giving nominal depths of penetration to 15 m, 30 m and 60 m respectively. Other manufacturers (e.g. Geophex Ltd) have developed instruments (e.g. GEM2) that comprise a fixed dipole separation with two coils in a rigid boom but achieve a graduated depth of penetration by using different frequencies, with the highest frequencies being used for the shallowest investigations and increasing depth of penetration with lower frequencies. Typically, these instruments have 5–8 different frequencies available. All of these instruments measure two components of the EM field, the in-phase and quadrature components. The latter, which is 90° out of phase, is a measure of the apparent conductivity, while the in-phase component gives an indication as to the presence of metalliferous material (ferrous and non-ferrous). Data can be recorded simultaneously with position-fixing data from connected dGPS systems, permitting very rapid ground coverage and display of apparent conductivity and in-phase component at each measurement point. All of these systems are capable of recording up to ten values per second. These displays show the relative variation in apparent conductivity across a site and are very useful at the reconnaissance stage of an investigation.

FEM surveys are extremely effective where the targets are conductive, such as in landfill investigations, leachate plumes, brownfield investigations, searches for metal culverts, detection of live electric cables and so on.

TEM systems work on a different principle from FEM systems. A coil of wire is energised by passing a current through it. This generates an EM field that propagates into the ground. The current is then switched off and a receiver coil is used to measure the residual response of the decaying secondary field. The rate of decay is affected by the conductivity of materials present. Measurements of the secondary field are made as a function of time following shutting off the current. Coil sizes are very large where great depth of penetration is required (many hundreds of metres). Special shallow-sensing devices known as very early time-domain EM (VETEM) instruments are used where higher resolution is required in the near-surface, typically the top few tens of metres. TEM systems are used to produce a 1D vertical sounding at each location. Profiles are produced by undertaking a series of soundings along a transect and then gridding the results to form a 2D section. TEM soundings can be inverted like 1D VES profiles to produce an interpretation with layer thicknesses and *true* conductivities that are physically diagnostic of the material present. The results from adjacent 1D soundings can be correlated and gridded to form a 2D true conductivity–depth model. Geonics Ltd. also make geophysical metal detectors based on the TEM principle (EM61-MK2 and EM63). Two small flat-lying coils are mounted vertically above each other, and the time decay measured. These instruments have a shallow depth of penetration (<5.5 m) and high lateral resolution and are used effectively and extensively for the detection of UXOs and USTs.

A relatively new development, from 2018, is the portable TEM Loupe system (Street and Duncan, 2020). It is designed as a two-person Slingram configuration with a 10 m intercoil separation that enables mapping of conductivity variations from the near-surface to around 20–30 m, depending on the ambient ground electrical properties. Unlike other TEM systems, it is designed to be used in continuous operation while the operators are walking.

45.6.2 EM methods: GPR

GPR works on the basis of transmitting radio waves with a known centre frequency from an antenna and measuring the time it takes for signals to be reflected back to a receiving antenna. By measuring the two-way travel time (TWTT) and by knowing (or assuming) the speed at which the radio waves travel through the material present, it is possible to measure the depth to a target. In many respects the technique is conceptually similar to the seismic method, where sound waves are used.

GPR systems have been developed for specific uses. Broadly, these fall into very low-frequency systems (frequencies <10 MHz) for deep investigations (⩾100 m), particularly of polar ice sheets and glaciers; those for geological applications (25–500 MHz); those for engineering investigations (500 MHz to 1.5 GHz); and those for non-destructive testing (NDT) investigations (frequencies >900 MHz to 2.5 GHz), where the required depth of penetration is of the order of decimetres but the resolution needed is down to around 1 cm or less. Special GPR systems have been developed for utility mapping and to be used by non-specialist technicians. GPR antennae can be deployed by dragging an antenna system across the ground, towing it behind a vehicle or mounting it on a boat or aircraft. As a general rule: high frequency, fine vertical resolution; low frequency, coarse vertical resolution. GPR techniques are not effective in highly conductive environments due to the high attenuation of the radar signal – that is, they should not be used in saline or conductive clay-rich environments.

GPR surveying works on the principle that reflections will occur where there is a contrast across an interface between the relative dielectric permittivities of the materials present. Consequently, the reflection characteristics can be as important in interpretation as the TWTT values. The biggest uncertainty in interpreting GPR data is the variability in radio wave velocities when translating time-domain radargrams (the raw data images) to depth images. GPR will not achieve any significant depth of penetration in areas with high conductivity, such as in saline environments or over seawater. It is, however, capable of imaging through freshwater. In the UK, typical depths of penetration are of the order of a few metres; in resistive environments imaging can be achieved to more than 10 m using a low-frequency system.

The GPR method is used in myriads of applications, including: geological mapping; environmental, engineering and archaeological investigations; surveys of highways and airfields; bridge decks; ballast surveys for railways; and in mapping animal burrows (Nichol *et al.*, 2003). It can be used to measure the water table and its characteristics, and groundwater and contaminant movement (DNAPLs and LNAPLs – dense and light non-aqueous phase liquids, respectively). Hydrocarbons floating on the water table can be imaged in the right conditions.

Raw radar data can be processed much in the same way as seismic data, with a wide range of processing techniques available to filter, amplify, enhance and migrate the data to aid interpretation (**Figure 45.5**). As with seismic data processing, it is full of potential pitfalls and opportunities to make mistakes for the unwary, and GPR data processing and interpretation should be undertaken only by skilled and experienced radar processors and interpreters.

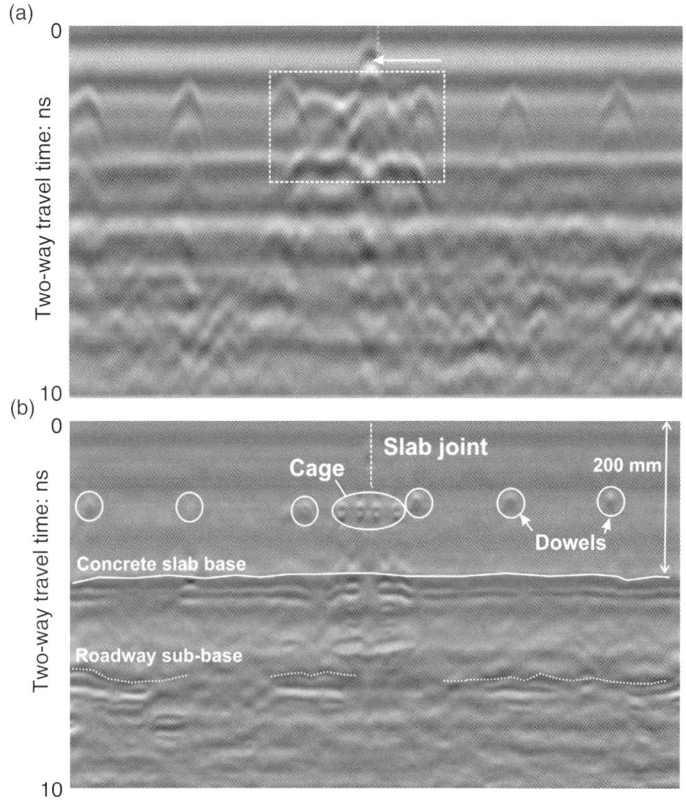

Figure 45.5 Example of (a) raw and (b) migrated GPR data acquired across the junction of two adjacent reinforced-concrete slabs within a road pavement (Reproduced from Reynolds, 2011, John Wiley & Sons, Inc)

45.7. Seismic methods

Seismic methods use the generation and propagation of elastic body waves through the ground and the measurement of their respective travel times. A wide variety of sources exist by which the signals are generated, such as a sledgehammer strike onto a base plate, blank shotgun shells fired vertically into the ground (e.g. 'Buffalo' gun), accelerated weight drops and explosives, to name but four. The signals are detected by geophones (on land) or hydrophones (in water). The type of source is chosen based on anticipated ground conditions and target depths. As with the GPR technique, there is a trade-off between vertical resolution and depth of exploration.

There are three broad styles of seismic surveys, ranging from seismic refraction and reflection through to the use of surface (Rayleigh) waves. Each of these will be described briefly in turn.

45.7.1 Seismic refraction

A seismic refraction survey comprises an array of geophones and an energy source, such as a hammer strike onto a ground plate ('shot'). Each shot generates seismic waves that travel to a boundary across which it is assumed the seismic velocity increases; the energy is refracted along the interface and re-radiated back towards the surface at the critical angle (i_c) of refraction, where $\sin i_c = V_1/V_2$ where V_1 and V_2 are the seismic velocities above and within the refractor, respectively, and $V_1 < V_2$. By measuring the time taken from the shot to when the signals arrive at each of the geophones, a time–distance graph can be constructed, from which it is possible to compute the seismic velocities of the surface, and deeper, layers causing the refraction. After careful processing, seismic velocity images can be produced showing 2D sections along a profile that indicate the lateral and vertical variations in sound wave velocity (**Figure 45.6**). Depths to refractor interfaces can also be derived.

Refraction surveys are used to determine the depth to bedrock, to locate disturbed zones indicating faulting, to measure the depth of landfills and backfilled quarries, for foundation surveys for large dam construction, to locate major solution features and for ground investigations for other major engineering structures (tunnels, roads, embankments, etc.). For shallow engineering investigations, depths of investigation are typically less than 20 m.

45.7.2 Reflection seismology

The seismic reflection method is used to measure the two-way travel time (TWTT) taken for a seismic wave to travel from a source vertically down into the ground and back to the surface. It provides information about the geometry of subsurface structures, and on the physical properties of the materials present. These techniques are highly developed in the hydrocarbon industry and there is an increasing transfer of such technology to the high-resolution shallow engineering and environmental sector.

The reflection method uses the same types of sources as refraction surveys, and depths of investigation range from at least 20 m (at the shallowest) down to several hundred metres. There are many ways to process the recorded data, using a wide range of commercially available software. Typically, the raw data are edited, filtered, amplified, processed to remove the effects of the geophone array geometry, and subjected to the analysis of seismic velocities to produce a 2D section which is typically what is then interpreted. Further data processing is possible to enhance the data to aid interpretation and to translate time sections to depth sections. Details of the type of data processing available have been described extensively by Yilmaz (2001, 2015). Seismic reflection surveying is commonly used in engineering studies for deep bridge foundations, and road and tunnel route surveys to map geological structures.

45.7.3 Surface waves

There are a number of different techniques that use surface shear waves or Rayleigh waves to determine physical properties of subsurface materials. For the purposes of this chapter, two common methods are discussed: (*a*) ground stiffness sounding, and (*b*) multi-channel analysis of surface waves (MASW).

45.7.3.1 Ground stiffness sounding

The majority of strain levels within the ground at working loads are typically less than 0.1%. It is therefore important that stiffness values to be used in geotechnical calculations should also be measured at these small-strain levels. One way to measure the very small strain stiffness, G_{max}, is to use ground stiffness 'sounding' or continuous surface wave seismics (CSW), which is a non-invasive analogue of the cone penetration test (CPT) or standard penetration test (SPT) (Butcher and Powell, 1996). A vertical 1D sounding comprises the determination of the variation of G_{max} as a function of depth to provide the required information on the geomechanical properties of the subsurface layers (i.e. hard and brittle, soft and plastic, etc.). The technique works on the basis of using a multi-frequency vibrating source, typically weighing 70 kg (**Figure 45.7(a)**), which can be tuned to a given frequency (f) in the range from 1 Hz to several hundred hertz. At the chosen frequency, the seismic vibrator generates Rayleigh waves through the surface of the ground that are detected by a short array of 2–6 geophones. Nominal depths of penetration are equivalent to one-third of the wavelength, which is controlled by the frequency of the source wave. By measuring the phase difference and wavelength across the geophone array for a given frequency it

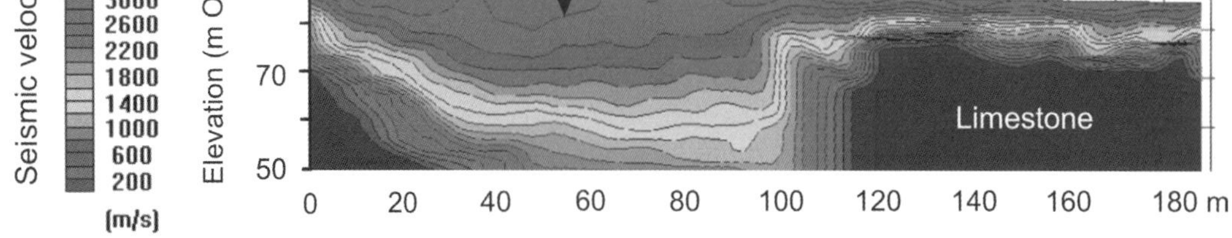

Figure 45.6 Seismic refraction velocity section across a closed landfill in a former quarry in limestone

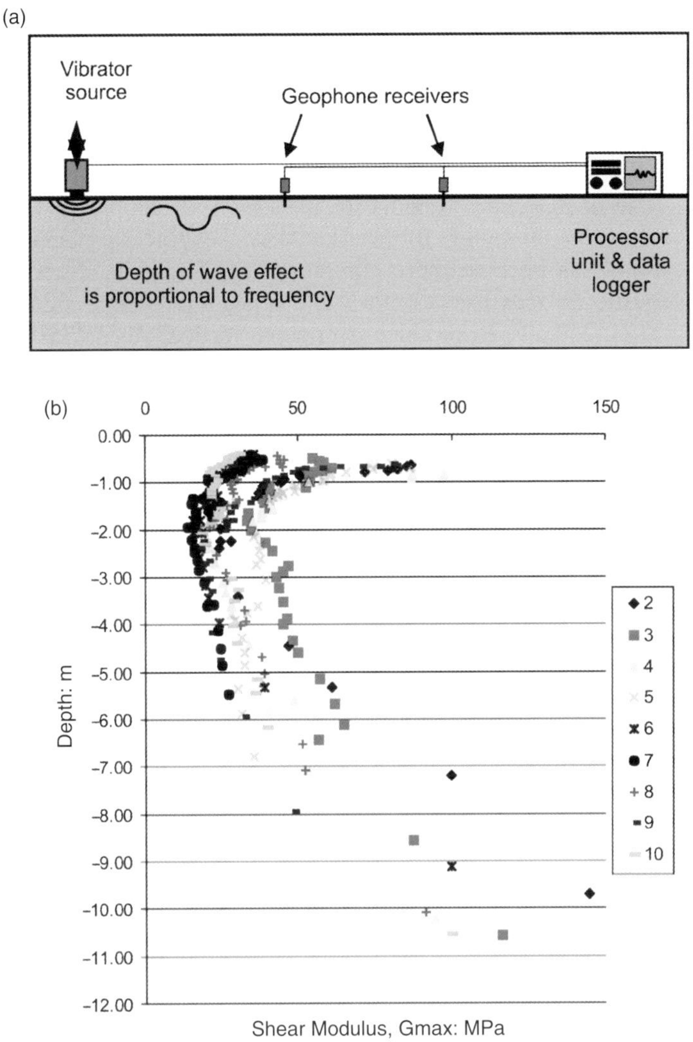

Figure 45.7 (a) Schematic showing the general layout of equipment to derive a ground stiffness vertical profile using a Rayleigh wave generator. (b) Example 1D ground stiffness profiles through landfill material

is possible to derive the ground stiffness at a range of depths. A 1D ground stiffness sounding is achieved by taking measurements at discrete frequencies from high to low frequencies, giving the mechanical strength of the materials in terms of the resistance to shearing forces. Rayleigh wave velocity (V_r) is typically 5% lower than that of shear waves (V_s), so the derived velocities are adjusted accordingly (Butcher and Powell, 1996), as G_{max} is proportional to V_s^2. Ground stiffness soundings can reach to depths of 20 m or more (**Figure 45.7(b)**), depending upon the materials present, and can reveal detailed variations in value up to several hundred megapascal (MPa).

The MASW method uses the same layout as for seismic refraction. A shot record is produced and processed, and the shear wave velocity is determined as a function of depth below the given geophone array. The results of each sounding are juxtaposed to form a 2D section displaying the variation in shear wave velocity as a function of depth and position along the array (**Figure 45.8**). The method is used to map zones of weak ground, fractured rock and voiding.

45.7.3.2 Over-water hydrographic and seismic surveys

For engineering investigations over water, such as for bridge foundations, mapping palaeo-channels, offshore wind farms and so on, hydrographic and seismic surveys are commonly used. The principle of these surveys is similar to those used for hydrocarbon exploration, but they are suitably scaled down for the intended applications. Over-water hydrographic and seismic surveys can be undertaken in lakes and artificial lagoons, rivers, estuaries, docks and canals, in water depths as shallow as 0.5 m.

Side-scan sonar – This is a water-borne technique where high-frequency (100–600 kHz) pulses of sound are beamed from a towed transducer through the water column vertically below the sensor and to either side as the vessel progresses. Reflections are produced from objects within the water column below the sensor and from the seabed surface, and their locations and acoustic characteristics are recorded and displayed on a sonograph, which is effectively an image

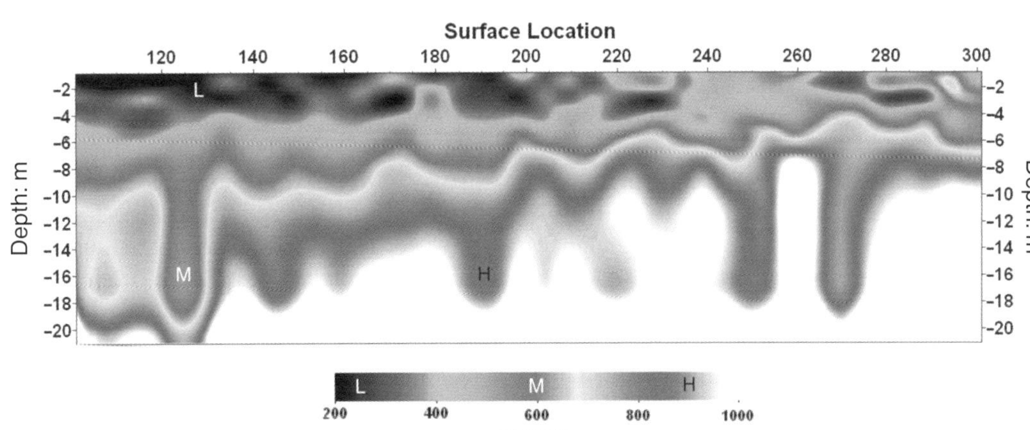

Figure 45.8 An MASW shear wave velocity section across an area with areas of weak ground

of insonified features. Side-scan sonar images are processed to form scaled mosaic pictures of the seafloor (analogous to aerial photographs of the ground surface), from which seabed features (sand waves and ripples, gravel and clay patches, iceberg and trawl scours, boulders, wrecks, anchor chains, etc.) can be identified. Individual features, such as a boulder as small as 0.3 m high, can be imaged.

Swath bathymetry – Traditional echo sounding provides an estimate of water depth vertically below the acoustic transducer. Bathymetric maps are produced by contouring echo sounding values across an area. However, far greater spatial accuracy is achieved by using multi-beam echo sounding (MBES; swath bathymetry), which uses an oblique-facing transducer to insonify a swathe of ground at right angles to the direction in which the transducer is being moved. This permits the determination of values of water depth for pixels of 0.5 m size, with vertical resolution of 0.1 m across a swathe width typically four times the water depth. Systems can also produce a back-scatter image analogous to **side-scan sonar** that produces an acoustic image of the seafloor or lakebed. By geo-referencing each pixel, 2D side-scan sonar images can be draped over the swath bathymetry results to produce 3D acoustic images of the seafloor or lakebed.

To investigate below the seafloor or lakebed, it is necessary to deploy a high-resolution **sub-bottom profiling** system. Seismic sources typically consist of a single transducer system with very high-resolution (pinger), or a boomer plate source, which is used with a string of hydrophones in a single oil-filled buoyant flexible tube (streamer). Vertical discrimination is of the order of 30 cm. Where greater depth penetration is required (at the expense of vertical resolution) higher-energy sources such as a sparker (which generates a sound pulse by an electrical discharge in salt water), air gun or water gun can be used. The acoustic source generates sound waves that travel in all directions. The waves that penetrate into the subsurface are reflected back to a string of hydrophones towed close to the seismic source to produce a seismic image. At a basic level of interpretation, such as for dredge studies, several reflections are picked to produce simple line interpretations showing the elevations of selected interfaces along each survey profile. However, if the data are converted to an internationally recognised data format standard called SEG-Y, it is possible to undertake much more sophisticated data processing and analysis, such as minimising the effects of multiple reflections, and filtering to enhance the primary reflections. It is also possible to analyse the data in oil-industry-standard 2D and 3D interpretation software, in which it is possible to display borehole and CPT data. Lithological boundaries can be identified and projected onto the seismic data to facilitate the selection of the corresponding reflection events, which can be gridded to produce 3D seismic surfaces and 3D geological ground models (Reynolds *et al.*, 2017a). While the levels of post-processing and analysis are much higher than for the most basic single-channel interpretation, the resulting interpretations have a greater level of confidence and technical robustness. Such high levels of analysis are justified for major engineering investigations, such as for large-bore tunnels (e.g. Crossrail and Thames Tideway projects) in London and for offshore wind farm investigations prior to finite element engineering design for pile foundations (Reynolds *et al.*, 2017b).

Use of 3D chirp systems for ultra-high-resolution (UHR) over-water surveying for shallow-water engineering applications has become more commonplace. These fixed array acoustic devices use a known and repeatable source signature (1–24 kHz) to produce decimetre vertical resolution of the subsurface (Bull *et al.*, 2005). The design incorporates four source transducers (1.5–13 kHz). The receiver array consists of 240 hydrophones in 60 groups with group-centres separated by 0.25 m in both horizontal directions. The entire source-receiver configuration is hard-mounted into a towable unit 2.0 m long and 2.5 m wide, with four positioning antennae at its corners. The system can also be hard-mounted onto the bow of a shallow-draft survey vessel.

For use in more open water with greater water depths (e.g. more than 25 m), for surveys over larger areas (10–50 km^2), multi-streamer surveys with high-frequency acoustic sources can be used, such as the P-cable multi-streamer configuration that was developed in 2003 (Planke *et al.*, 2009). Typically, 18–24 hydrophone streamers 50–100 m long are deployed from a cross-cable (the P-cable) with multiple high-frequency (>1 kHz) acoustic sources (e.g. sparkers or boomers) towed aft of the survey vessel but ahead of the streamer array. The acquired seismic data processing workflows are similar to those used in conventional 3D streamer processing. These systems are being used more commonly in surveys for the energy transition (carbon capture and storage and offshore wind farm surveys) (MacGregor *et al.*, 2022).

45.7.3.3 Integration with other datasets

Geophysical data are commonly interpreted without reference to other information – that is, an unconstrained interpretation (satisfies statistical error limits, appears physically plausible and is internally consistent). Many geophysical models can have an almost infinite number of equally valid solutions. Where complementary geophysical techniques have been used over the same ground, separate interpretations can be produced for each method and compared with each other to achieve consistency. For example, it might be possible to use seismic refraction results to yield estimates of depth to bedrock. A corresponding resistivity section might then show that, for example, conductive leachate is migrating into and through this otherwise resistive bedrock, as the resistivity sections would respond to the pore fluid component rather than the solid matrix on which the seismic method depends. Although it is possible to jointly process complementary geophysical datasets to yield a single interpretation, this is non-standard currently.

A more comprehensive approach would be to integrate other forms of information, such as historical data from a

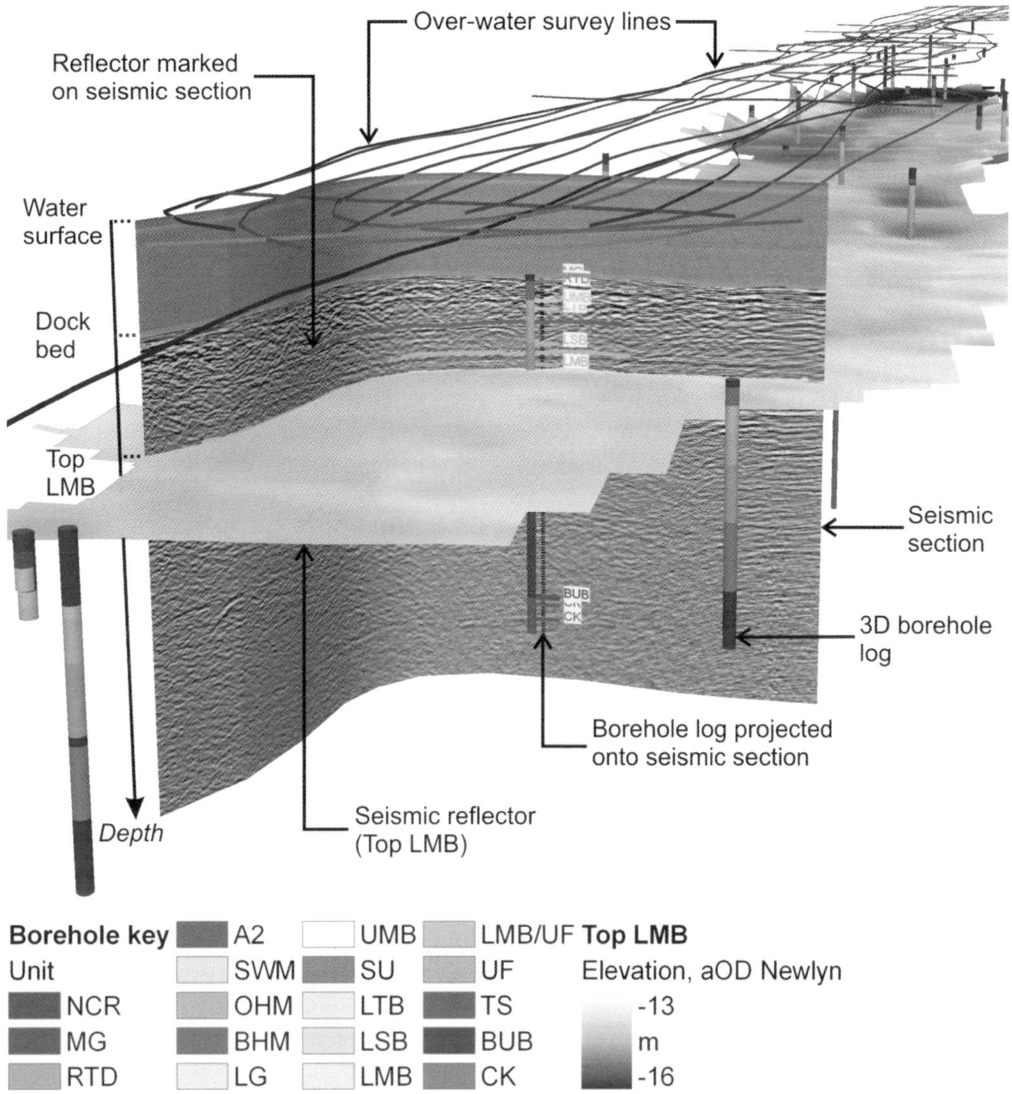

Figure 45.9 Boomer seismic reflection data, boomer survey track plots and borehole data integrated within a 3D interpretation package

desk study, aerial photographs, remote sensing images, airborne or down-hole geophysical results and so on, so that a more holistic interpretation can be produced. This can be achieved where each dataset is geo-referenced to a common coordinate system; where appropriate, the information can be managed through geographical information system (GIS) software. Some dedicated geophysical software also permits this approach, such as Oasis Montaj (Geosoft) and Surfer (Golden Software). It is possible to integrate geophysical software with GIS, which also permits the incorporation of borehole data in digital (AGS) format (**Figure 45.9**).

45.7.3.4 3D visualisation

If 2D geophysical data can be incorporated into a 3D interpretation package, it is possible to produce 3D interpolations and surfaces that can then be viewed from a 3D perspective. Surfaces can be shaded and illuminated from different directions to aid the visualisation of the 3D model. Fly-throughs of the model can also be generated.

45.7.3.5 UXO investigations

A sad legacy of war and other forms of military conflict is the presence of UXO, much of which in the UK was delivered through aerial bombardment during the Second World War. However, UXOs are also found on military bombing and firing ranges. Consequently, there is an ongoing requirement to develop techniques that will successfully and unambiguously detect remnant ordnance so that it can be rendered harmless and cleared. In 2009 CIRIA published the first comprehensive guidance for the construction industry on unexploded ordnance (Stone *et al.*, 2009). Despite the fact that geophysical techniques form one of the most significant methods of remote detection of UXOs, this publication barely mentions the techniques, or more worryingly, the limitations of such methods.

Primarily, magnetic, electro-magnetic (FEM and TEM) and GPR techniques are used for UXO detection. Commonly, instruments are deployed using a single operator, who may carry a number of magnetic sensors, for example walking across a site; or instruments can be deployed as multiple sensor arrays from a variety of platforms to investigate a swathe of ground in a single pass. For coverage of large areas, such as bombing ranges, helicopter-mounted multiple sensor arrays of magnetometers or TEM coils can be used, where the helicopter flies with a ground clearance as low as 1 m (**Figure 45.10**). Where the presence of UXOs is suspected in sediments beneath water, suitable submersible sensors can be deployed, such as marine magnetometers or gradiometers, or TEM systems (such as the Geonics EM61-SS).

Magnetometers can also be used with a CPT rig, where the sensor is pushed into the ground or inserted into a predrilled borehole (with non-magnetic casing), the hole being constructed in 1–1.5 m stages and the sensor lowered into each hole extension incrementally. Both the borehole and CPT approach require the sensor to be able to detect the presence of ferrous UXO within 1–1.5 m ahead of the hole.

Figure 45.10 Helicopter-borne systems for UXO detection: (a) FEM and (b) magnetic multisensor gradiometer systems (Courtesy of Batelle Inc., USA)

Some software packages are available that provide specific tools to help pick potential UXO targets. However, such programs do not guarantee that all possible UXOs will be identified, and a manual interpretation by a geophysicist experienced in UXO detection should also be undertaken to help reduce uncertainties in the interpretation. Neither is there an increased guarantee of higher quality of work if a military search team is involved. In all cases where UXO risk is to be assessed, clients must spend time determining the qualifications and experience of the teams to be contracted to investigate a site. The 2009 CIRIA guidance is useful in providing information on how to do this.

45.8. Borehole geophysics

An important adjunct to surface ground investigations is the deployment of geophysical sensors down boreholes. These can be as single sensors, or down-hole tools in which a number of different sensors are installed. There are two main forms of deployment: (*a*) single *down-hole techniques* (borehole logging), where the output is a vertical log of a measured parameter as a function of depth down the hole; and (*b*) *cross-hole tomography*, where a pair of normally vertical boreholes is used with some form of geophysical source being placed at various depths in one hole and a series of detectors being located in the other hole. Signals are transmitted between the holes to produce a 2D panel of data between the boreholes. Borehole tomography is commonly undertaken using seismic, electrical or radar methods. The separation between boreholes should not normally be more than ten times the minimum dimension of the target being sought – that is, boreholes should be no more than 10 m apart to image a target with a 1 m dia. Cross-hole tomography uses techniques that are analogous to their surface equivalents and will not be discussed further here.

There is a wide range of borehole logging tools available (**Table 45.3**) for a large number of different applications. As with surface geophysical techniques, tools should be selected depending upon the most diagnostic physical properties of the targets being sought. Borehole logging can have a very high vertical resolution, but may have a very limited penetration beyond the wall of the borehole. However, inter-borehole correlation can also be used to define geological structure (**Figure 45.11**). Geophysical CPT sensors (soil moisture probe, seismic cone) and traditional down-hole tools should also be considered in survey design.

Specialist borehole logging contractors are available through whom advice can be sought about the most appropriate logging tools to be used for a given purpose. Before constructing access boreholes, the size of the down-hole tools, and whether they can work in cased or uncased boreholes or in air- or fluid-filled boreholes, should be determined.

Borehole geophysical methods	Depth to bedrock	Stratigraphy	Lithology	Fractured zones	Fault displacement	Buried channels	Natural cavity detection	Mine workings/ adits/shafts	Groundwater exploration	Landfill investigations	Leachate plumes and migration	Brownfield site mapping	UXO detection	Buried artefacts/ USTs
Borehole logging techniques generally	4	4	4	3	4	3	2	2	4	1	2	2	2	0
Cross-hole seismic	0	2	3	4	4	3	3	3	0	0	0	0	0	0
Cross-hole GPR	0	3	3	4	4	3	4	3	0	0	0	0	0	0
Borehole magnetometer	0	0	0	0	0	0	0	0	0	0	0	0	3	1

	Bed boundaries	Bed thickness	Bed type	Porosity	Density	Permeable zones	Borehole fluid quality	Formation fluid quality	Fluid movement	Direction of dip	Shale/sand indication	Fractures/joints	Casing	Diameter	Type of hole
Self-potential	•	•	•					•			•				O,W
Long and short normal, and lateral resistivity	•	•	•	•				•				•	•	•	O,W
Natural gamma	•	•	•								•	•			A
Gamma gamma	•	•	•	•	•	•						•			A
Spectral gamma	•	•	•												A
Neutron	•	•	•	•		•									A
Fluid conductivity						•	•		•			•			L,O,W
Fluid temperature						•			•			•			L,O,W
Flowmeter						•			•			•			L,O,W
Dipmeter	•									•		•	•		O,W
Sonic (velocity)	•	•	•	•								•			L,O,W
Calliper	•	•				•					•	•	•	•	A
Borehole televiewer	•	•				•						•			O,W

Key: 0 = Not considered applicable or untried; 1 = limited use; 2 = used, or could be used, but not best approach or has limitations; 3 = excellent potential but not fully developed; 4 = generally considered an excellent approach, techniques well developed. Types of borehole construction: L = lined; O = open hole; W = water-/mud-filled; A = all types including air-filled.
Note: See the notes to **Table 45.1**.

Table 45.3 Suitability of borehole geophysical methods for a range of shallow engineering applications

45.9. Remote sensing

Remote sensing represents a wide spectrum of techniques relating to optical, infrared and radar imaging, predominantly from orbiting satellites. However, there is an increasing range of additional platforms from which observations can be made, such as aircraft (fixed-wing and helicopters for aerial photography, LiDAR, thermal infrared (TIR), hyperspectral imaging), blimps (unpowered, unmanned inflatable platforms), drones and related autonomous unmanned vehicles (AUVs – airborne and submersible) and elevated aerial platforms and masts from surface vehicles, such as for near-surface TIR imaging. Hyperspectral imaging uses between 100 and 200 spectral bands of relatively narrow bandwidths (5–10 nm), in contrast to multi-spectral datasets, which usually comprise about 5–10 bands of relatively large bandwidths (70–400 nm). For the purposes of this chapter, discussions are restricted to satellite imaging, LiDAR and TIR, and synthetic aperture radar (SAR) methods.

45.9.1 Optical remote sensing

Optical remote sensing has revolutionised the geosciences in being able to image large areas of the Earth's surface with

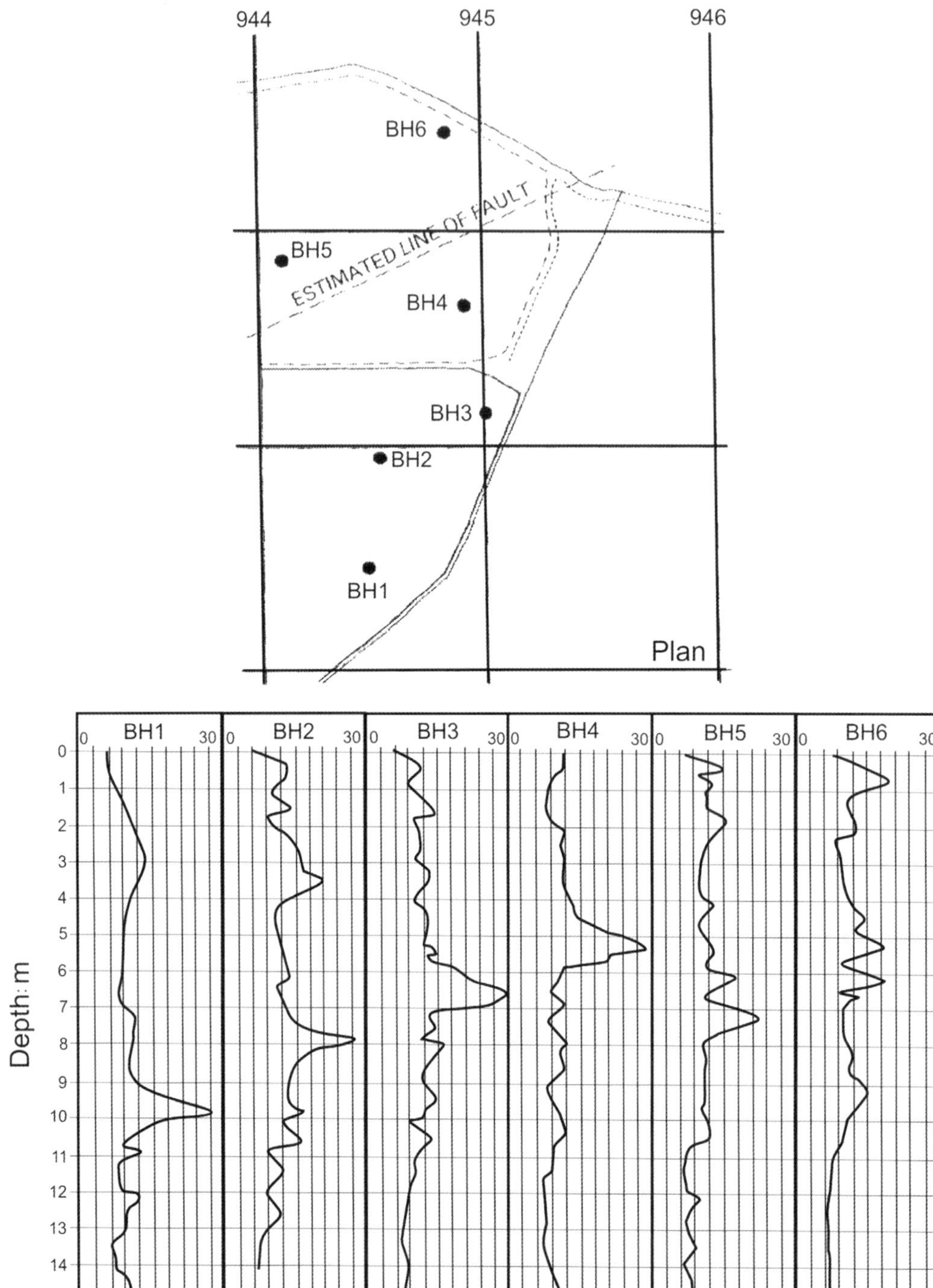

Figure 45.11 Correlation of natural gamma logs between closely spaced boreholes in an engineering site investigation (Reproduced from Reynolds, 2011, John Wiley & Sons, Inc.; modified from Cripps and McCann, 2000, Elsevier)

increasing levels of accuracy and ground resolution. Not only are 'pictures' produced, but also multi-spectral datasets that can be analysed in ever more sophisticated ways to determine a number of properties of the Earth's surface. The choice of image type is a function of the ground area to be covered, the required ground resolution, and the purpose of obtaining the imagery (with the largest image sizes having the coarsest ground resolution). For example, different applications might require panchromatic (black and white only), four-colour (red, green, blue and near-infrared (NIR)) or more spectral bands. The level of processing of the image data is also dictated by the application – that is, whether to use the image for illustrative purposes (lowest level of processing) through to full ortho-rectification, which is the technique by which distortions are corrected in image geometry resulting from the combined effect of variations in terrain elevation and non-vertical angles from the satellite to each point in the image at the moment of acquisition. Ortho-rectification allows the image to be incorporated into a GIS to maintain full spatial referencing in a designated coordinate system.

Widespread coverage of the Earth's surface was originally provided by Landsat satellites, the first of which was launched on 23 July 1972. Subsequent missions have provided improved sensor capabilities, with Landsat 5 being launched on 1 March 1984 with both multi-spectral (resolution 57 × 79 m) and thematic mapper (resolution 30 m) sensors on board; the thermal image scene has a resolution of 120 m. Landsat 7, launched on 15 April 1999, has eight spectral bands (three visible, two NIR, one TIR and one panchromatic (15 m resolution)), with 30 m ground resolution, plus a thermal band with 120 m ground resolution. Landsat 5, 7, 8, and 9 provide scenes with an area of 170 km by 183 km. Landsat 8 and 9 were launched in 2013 and 2021, respectively, with operational land manager (OLI) and thermal infrared sensors (TIRS), to give continuity of imaging through to beyond 2030. Since October 2009 all Landsat imagery has been freely available.

The ASTER system comprises three separate sensors – visible and near-infrared (VNIR; 15 m ground resolution), short-wave infrared (SWIR; 30 m) and TIR (90 m), with a scene size of 60 × 60 km. The system images to latitudes of ±83° of the equator. In 2009, an ASTER Global Digital Elevation Model (DEM) was made available at 30 m resolution. Geo-rectified images can be draped over this DEM to provide 3D perspectives (**Figure 45.12**).

Higher-resolution multi-spectral and panchromatic images are available from a number of satellite platforms, such as SPOT, IKONOS, FORMOSAT-2, KOMPSAT-2 and Quickbird, with increasing levels of ground resolution. The latest SPOT imagery (SPOT5) is available in scenes 60 × 60 km, with ground resolutions of 2.5 m, 5 m, 10 m and 20 m in panchromatic and colour, with an additional option of colour at 20 m ground resolution. Stereo imaging is also possible; from this, DEMs can be derived, although for best results users should provide accurate ground control points within the area of coverage of the scenes to calibrate the DEM. IKONOS has been available since September 1999 with panchromatic and four-colour images at 1 m and 4 m ground resolution, respectively, and a revisit period of three days. FORMOSAT-2 has 2 m and 8 m ground resolution in panchromatic and four-colour mode, respectively, with a daily revisit capability, making it ideally suited for rapid imaging after natural disasters. KOMPSAT-2 has 1 m and 4 m ground resolution in panchromatic and colour modes with a 28-day revisit period and scene size of 15 × 15 km. The satellite with the finest ground resolution is Quickbird, which was launched in October 2001. It achieves resolutions of 0.61 m and 2.44 m in panchromatic and four-colour modes, respectively, over a scene size of 16.5 × 16.5 km. The EROS satellite produces panchromatic-only images over 13.5 × 13.5 km with a 1.8 m resolution, although hyper-sampling can refine this to 1 m over an area of 9.5 × 9.5 km.

Figure 45.12 A colour ASTER image pan-sharpened and draped over a digital terrain model for a scene in Tibet (shown as grey scale)

Interpretation of images can range from photo-interpretation techniques through to software-controlled terrain analysis. Time-separated images can be used to measure rates of change by tracking the lateral movement of surface objects or through comparison of changes in surface emissivity.

45.9.2 Synthetic aperture radar

While optical satellite imagery can be restricted by daylight and visibility limited by cloud cover, radar imaging can be used at any time and through cloud cover. A number of radar satellite imaging systems are available, such as interferometric SAR (InSAR) and TerraSAR-X. The latter provides three ranges of ground resolution: *Spotlight*, ≤1 m over an image 10 km wide and 5 km long; *Stripmap*, ≤3 m, 30 km wide by 50 km long; and *Scansar*, ≤18 m, 100 km long by 150 km wide.

Radar sensing can be used not only to provide images of the ground covered by extensive cloud cover, but also can be used to derive changes in surface elevation arising from earthquakes or hydrocarbon exploitation – for instance, using interferometric methods. Rates of lateral movement of the ground surface can be measured through rapid repeat imagery, where flow rates as small as 2 cm per day can be resolved over Himalayan debris-covered glaciers. InSAR research into detection and monitoring of landslides is becoming far more sensitive and meaningful in the way the data are being analysed, such as in landslide deformation tracking (van Natijne *et al.*, 2022).

A major application of airborne SAR has been in determining biomass, such as in monitoring of forests to determine tree health (Quegan, 2022). Critical to this is the wavelength used in the various SAR wavelength bands: X-band (3 cm),

L-band (24 cm), P-band (0.7 m) and VHF (3 m). When imaging an Austrian pine tree, for example, the different bands would 'see': details of the pine vegetation; twigs, branches and tree trunk; branches and tree trunk; and tree trunk only, respectively. Presently, P-band SAR is only available for airborne deployment. The first satellite-based system is due for launch in 2024. For (hydro-)geological and engineering applications, the advent of spaceborne P-band SAR could be transformative. In areas such as dry hot deserts, L-band SAR can penetrate 1–2 m but can result in blurred images from the superficial sand. However, P-band SAR can penetrate down to 5 m in dry sand and provides much clearer images of the shallow subsurface. Once spaceborne P-band SAR imagery starts to be acquired, data should be available for larger areas at a much lower cost than for the equivalent airborne P-band SAR surveys.

45.9.3 LiDAR

LiDAR has now become an established, very high-resolution, remote mapping tool that is predominantly deployed from aircraft but is increasingly becoming available for ground-based lateral scanning of near-vertical faces. As the survey aircraft flies over a designated survey area, a wide-aperture laser beam scans across the line of flight, measuring the transit time from up to four returns from the reflective surfaces beneath, each return generating a data point within a point cloud. The transit time is converted to distance from the sensor, and, knowing the exact height and aspect of the aircraft, the elevation and position of the reflective surface can be determined; thus, a digital surface model (DSM) can be created. The first return may be associated with the tree canopy and vegetation, while the last return is normally associated with the ground; from this, a digital terrain model (DTM) can be derived. Intermediate returns may provide information from within the tree canopy (giving information about the species of trees, for example). Using specialist software, it is possible to remove the effects of trees and buildings from a DSM to generate the DTM. This can provide detailed topographic information otherwise obscured by vegetation. LiDAR has been used highly effectively to map landslides and archaeological remains concealed beneath dense tree canopies. Ground resolution is of the order of 0.25 m and vertical topography determined to within 0.1 m, subject to survey control.

Similar technology is now available for terrestrial lateral LiDAR scanning by which detailed point clouds can be obtained within a range of ~1.2 km of steep surfaces, such as rock cliffs or man-made superstructures. Point clouds can be used to generate 3D images and also to map changes in surface elevations over time, such as for pseudo-real-time monitoring of slope stability. It is now also possible to deploy LiDAR scanners from boats; images can be combined seamlessly to produce a continuous image from the seafloor and the submerged part of marine infrastructure using high-resolution multi-beam swathe bathymetry to the exposed superstructure using boat-based LiDAR.

LiDAR images can be used to generate topographic maps, DTMs and a wide range of topographic products, as well as providing the base onto which aerial photographic or satellite images can be draped to produce 3D isometric images and animated 'fly-throughs'. As the data produced are geo-rectified, they can be integrated into GIS databases and merged with other spatially rectified datasets in common coordinate sets. For example, LiDAR data can be integrated with airborne geophysical datasets, and airborne geomagnetic data can be presented in colour-contoured format overlain on panchromatic satellite imagery draped over a LiDAR-derived DTM.

45.9.4 Thermal infrared

TIR imaging can be achieved using sensors deployed on satellites and airborne or surface-mounted platforms, such as telescoping masts. The principle is based on being able to convert radiated or reflected heat into real-time pictures or images and depends on there being a detectable surface thermal contrast present, such as from a hot lava flow, exothermic methanogenesis within a landfill, an underground fire within a tyre dump or man-made infrastructure (pipes, buildings, electric cables, etc.). For engineering applications, ground-based or aerial surveys are undertaken. Subsurface heat sources may be indicated at the ground surface through thermal conduction. Timing of surveys should be planned to enhance the thermal contrasts, such as just after dawn. While thermal imaging can be undertaken at night, it is more often undertaken in daylight in conjunction with optical imaging (photography/video) to aid interpretation.

References

Alken P, Thébault E, Beggan CD, *et al.* (2021) International Geomagnetic Reference Field: the thirteenth generation. *Earth, Planets and Space* **73**: 49, 10.1186/s40623-020-01288-x.

Anon (1988) Engineering geophysics: report of the Engineering Group Working Party. *Quarterly Journal of Engineering Geology* **21**: 207–271.

Bull JM, Gutowski M, Dix JK, *et al.* (2005) Design of a 3D chirp sub-bottom imaging system. *Marine Geophysical Researches* **26(2–4)**: 157–169, 10.1007/s11001-005-3715-8.

Butcher AP and Powell JJM (1996) Practical considerations for field geophysical techniques used to assess ground stiffness. In *Advances in Site Investigation Practice* (Craig C (ed)). Thomas Telford, London, UK, pp. 701–714.

Cripps AC and McCann DM (2000) The use of the natural gamma log in engineering geological investigations. *Engineering Geology* **55**: 313–324.

Darracott BW and McCann DM (1986) Planning engineering geophysical surveys. In *Site Investigation: Assessing BS5930* (Hawkins AB (ed)). Geological Society, London, UK, pp. 85–90.

David A, Linford N and Linford P (2008) *Geophysical Survey in Archaeological Field Evaluation*, 2nd edn. English Heritage, Portsmouth, UK.

El-Qady G and Metwaly M (2019) *Archaeogeophysics*. Springer, New York, NY.

MacGregor L, Scott L, Cooper R and Nicholls J (2022) Ultra-high resolution seismic: applications of P-cable in the energy transition. *First Break* **40(11)**: 7–70.

McCann DM, Culshaw MG and Fenning PJ (1997) Setting the standard for geophysical surveys in site investigation. In *Modern Geophysics in Engineering Geology* (McCann DM, Eddleston M, Fenning PJ and Reeves GM (eds)). Geological Society, London, UK, pp. 3–14.

McDowell PW and Working Group Members (2002) *Geophysics in Engineering Investigations*. CIRIA and the Geological Society, London, UK.

Milsom J and Eriksen A (2011) *Field Geophysics*, 4th edn. Wiley, Chichester, UK.

Nichol D, Lenham JW and Reynolds JM (2003) Application of ground penetrating radar to investigate the effects of badger setts on slope stability at St Asaph Bypass, North Wales. *Quarterly Journal of Engineering Geology and Hydrogeology* **36(2)**: 143–153.

Planke S, Eriksen FN, Berndt C, Mienert J and Masson D (2009) P-cable high-resolution seismic. *Oceanography* **22(1)**: 85.

Quegan S (2022) Weighing forests from space. *Astronomy & Geophysics* **63(6)**: 24–27.

Reynolds JM (1996) Some basic guidelines for the procurement and interpretation of geophysical surveys in environmental investigations. *Proceedings of the Fourth International Conference on Construction on Polluted and Marginal Land*, London, UK, pp. 57–64.

Reynolds JM (2002) The role of geophysics in the investigation of an acid tar lagoon, North Wales UK: Llwyneinion. *First Break* **20(10)**: 630–636.

Reynolds JM (2004) Environmental geophysics investigations in urban areas. *First Break* **22(9)**: 63–69.

Reynolds JM (2011) *An Introduction to Applied and Environmental Geophysics*, 2nd edn. Wiley, Chichester, UK.

Reynolds JM, Catt LML, Fernandes AB and Knight P (2017a) Development of 3D ground models for offshore wind farms: Rhiannon OWF, Irish Sea. *Offshore Site Investigation and Geotechnics: Smarter Solutions for Future Offshore Developments. Proceedings of the 8th International Conference, London, UK*. Society for Underwater Technology, London, UK, vol. 2, pp. 1307–1314.

Reynolds JM, Catt LML, Salaün G, *et al.* (2017b) Integration of geophysical, geological, and geotechnical data for offshore wind farms: East Anglia ONE OWF, southern North Sea, a case history. *Offshore Site Investigation and Geotechnics: Smarter Solutions for Future Offshore Developments. Proceedings of the 8th International Conference, London, UK*. Society for Underwater Technology, London, UK, vol. 2, pp. 1291–1298.

Schmidt A, Linford P, Linford N, *et al.* (2016) *EAC Guidelines for the Use of Geophysics in Archaeology*. Europae Archaeologiae Consilium (European Archaeological Council), Brussels, Belgium.

Schoer B (1999) Briefing: geophysics for civil Engineers: an introduction. *Briefing sheet*, ICE, London, UK.

Stone K, Murray A, Cooke S, Foran J and Gooderham L (2009) *Unexploded Ordnance (UXO): A Guide for the Construction Industry*. CIRIA, London, UK.

Street G and Duncan A (2020) Trials using the Loupe TEM system for contaminated sites. *Fast Times* **25(2)**: 120–124.

van Natijne AL, Bogaard TA, van Leijen FJ, Hanssen RF and Lindenbergh RC (2022) World-wide InSAR sensitivity index for landslide deformation tracking. *International Journal of Applied Earth Observations and Geoinformation* **111**, 10.1016/j.jag.2022.102829.

Yilmaz O (2001) *Seismic Data Analysis: Processing, Inversion, and Interpretation of Seismic Data*. Society of Exploration Geophysicists, Tulsa, OK.

Yilmaz O (2015) *Engineering Seismology with Applications to Geotechnical Engineering*. Society of Exploration Geophysicists, Tulsa, OK.

Useful websites

ASTER: http://www.asterweb.jpl.nasa.gov
Landsat: http://www.landsat.usgs.gov
SAR data: https://data.world/datasets/insar.
Satellite images – SPOT: http://www.spotimage.com
Satellite images – other: http://www.infoterra.co.uk
Satellite images (Sentinel) and SAR data: https://scihub.copernicus.eu/dhus/#/home

It is recommended this chapter is read in conjunction with

- Chapter 13 *The ground profile and its genesis*
- Chapter 24 *Dynamic and seismic loading of soils*

All chapters in this book rely on the guidance in Sections 1 *Context* and 2 *Fundamental principles*. A sound knowledge of ground investigation is required for all geotechnical works, as set out in Section 4 *Site investigation*.

doi: 10.1680/icemge.66816.0677

Chapter 46
Ground exploration

John A Davis Geotechnical Consulting Group, London, UK

This chapter describes the main intrusive techniques available for ground exploration. These techniques are some of the tools by which the ground profile and soil behaviour elements of the geotechnical triangle are established.

CONTENTS

46.1. Introduction

Ground exploration is defined as the collection of data from intrusive ground investigation techniques. Non-intrusive techniques such as geophysics or surface loading are not discussed here.

Ground exploration includes techniques to expose the ground or collect samples for visual inspection or testing, intrusive testing of the ground without collection of samples and any subsequent installation of monitoring instrumentation in the ground. Describing and testing of recovered samples and any interpretation of recovered in situ or monitoring data is covered in other chapters (Chapters 49 *Sampling and laboratory testing* and 50 *Geotechnical reporting*).

The major ground exploration activities are divided into two types: those that involve intrusive means to characterise and/or sample the ground, and those involving subsequent monitoring of the ground.

When planning a ground investigation it is important to choose a suite of ground exploration techniques that are appropriate to the likely ground conditions and that also provide the data the project requires. Guidance on this is given in Chapters 43 *Preliminary studies*, 47 *Field geotechnical testing*, 48 *Geo-environmental testing* and 49 *Sampling and laboratory testing.*

It is important to understand that each technique provides both opportunities and constraints for data collection. Data requirements range from soil or rock description, through quantifying design parameters (by way of laboratory and in situ testing), to provision of monitoring opportunities. Other chapters provide further guidance on these issues. An alternative source of detailed advice on the applicability of the various ground investigation techniques is a specialist ground investigation contractor.

46.2. Techniques

This chapter describes the nature of the main techniques and outlines the pros and cons for

- different types of soils and rocks
- sampling, in situ testing
- monitoring.

Techniques are divided into three classes: excavation techniques, probing techniques and drilling techniques. The list of techniques discussed is not exhaustive.

46.3. Excavation techniques

46.3.1 Trial pits

Trial pits can be hand-dug or machine-dug. Hand-dug trial pits are typically used as starter pits to identify and avoid services or to uncover details of existing shallow foundations. They are normally limited to relatively easy-to-excavate soils and to very shallow depths. The need to provide shoring and the inevitable slow rate of progress mean deeper hand-dug pits are not often specified.

Suitable ground conditions and depths for machine-dug trial pits are determined by the capacity of the excavator. All natural soils are suitable, as are some weaker weathered and fractured rocks. Care must be taken in very soft soils and sands and gravels below the water table, as excavations of these materials are particularly liable to collapse. In these cases it is not usually possible to recover the required information from below the water table.

Trial pits are excellent for displaying the fabric of the ground and for obtaining large disturbed samples, such as those required for some compaction tests. Generally trial pits are only used to obtain disturbed samples. Shallow (<1 m) trial pits are also often used to carry out plate bearing tests or in situ California bearing ratio (CBR) tests. Care should be taken to site trial pits in areas where the backfilled pit will not cause problems for any subsequent construction. It is possible to install standpipes for monitoring gas and groundwater in trial pits as they are backfilled.

46.3.2 Trial trenches

These are similar to trial pits but are long compared to their width. They are useful for finding features at shallow depths that might have a limited extent in plan, such as old foundations, or systematic changes in rockhead level, such as the edges of buried cliffs.

46.3.3 Trial excavations

Trial excavations are used to investigate the feasibility of much larger construction excavations. A trial excavation is defined here as an excavation that contains the excavation plant.

46.4. Probing techniques

Probing techniques are those that recover information about the ground but do not recover a sample. The principal types are

- dynamic probing
- static cone penetration testing.

In dynamic probing, a series of rods with a standard cone-shaped shoe are driven rapidly into the ground by repeated blows of a standard-weight hammer falling through a repeated fixed distance. A penetration resistance in the form of blow count per unit length is recorded and logged, typically blows per 100 mm. Different standards exist for hammer weight and hammer drop length. More information on the various types can be found in British Standards or Eurocodes. The technique is broadly analogous to continuous standard penetration testing (SPT) (see below). However, it is important to realise the results cannot normally be used to safely generate equivalent SPT 'N' values. The technique is useful for rapid and inexpensive determination of variations in penetration resistance, such as where soft spots are thought to exist and their extent needs to be determined. The equipment used is typically light and portable and can be used in spaces with restricted access, such as basements. The technique is unlikely to achieve significant penetration in stiff clays or dense granular deposits.

In static cone penetration testing (CPT), a series of rods are driven hydraulically and continuously into the ground, typically from a purpose-built heavy vehicle which provides a reaction to the hydraulic thrust. The lowest rod (the 'cone') can contain various sensors used to measure properties of the ground it passes through. The most commonly used sensors measure the resistance the ground provides to the end of the cone as it is pushed into the ground, and the frictional resistance the ground provides to the sides of the cone. A continuous and instantaneous readout of these values is provided. These parameters can be interpreted to give an indication of the soil type and to provide various strength and stiffness parameters. These values of cone resistance and cone friction can also be used directly in some design calculations. One of the great advantages of CPT is its ability to identify very thin clay or granular layers in other strata. These can have significant impacts on geotechnical processes and are easily missed by many other ground exploration techniques.

Other types of cone can be used to give measurements of pore pressure and seismic velocity. Other varieties of cone also exist. This is discussed in greater detail in Section 46.5.3.3. Cone penetration trucks can also be used to recover very small-diameter continuous disturbed samples, although not at the same time as or exactly in the same location as any other CPT as described above.

The technique is applicable to most soils; achievable penetration depths depend on the strength of the soil and the reaction provided by the equipment the cone is mounted on. Typically, significant penetration is limited to clays that have a consistency less than stiff, and to granular soils whose relative density is less than medium dense. Penetration will be greater in finer soils compared to coarse soils. Penetration in very coarse soils (gravels and coarser) will be limited regardless of relative density. Penetration is not limited by the presence of groundwater. In soft or loose soils, great depths (around 40 m) can be achieved very rapidly, and in these conditions a single truck-mounted cone can achieve several hundred metres of testing in a day. A typical cone 'truck' is a highway-legal road vehicle weighing approximately 18 tonnes. Lighter vehicles with large low-pressure tyres are available for very soft sites. Cone penetration equipment can also easily be mounted on pontoons or jack-up rigs for offshore work and can be skid-mounted for work in restricted access sites, although in this case consideration needs to be given to provision of a reaction by the use of props, anchors or kentledge.

Neither dynamic probing nor static CPT provides for the installation of instrumentation or post-fieldwork monitoring.

46.5. Drilling techniques

46.5.1 Window sampling

A window sampler is a tube with an open slot or window down one side which is driven rapidly into the ground using high-speed percussion. A windowless sampler lacks this slot. Window or windowless samplers are typically 1–3 m long and less than 100 mm dia. and may or may not contain a rigid plastic liner. The logging of the soil can be done through the window. 'Windowless' samplers are also available. In suitable self-supporting ground they can achieve depths of 7 m or more by repeated extraction and reinsertion of progressively smaller sampler tubes using rods to extend the depth. They are only suitable for obtaining disturbed samples. It is sometimes possible to install standpipes for monitoring gas and groundwater using window sampling equipment. The principal advantages of window sampling are speed and cost-effectiveness. The equipment is small and portable and can easily be used in areas with restricted access. It is often used alongside other techniques as an inexpensive way of providing a widespread and large number of sampling points for contaminated land testing.

Window sampling of groundwater is difficult. Recovery in soils where the largest clasts, such as gravel, are similar to or larger than the internal diameter of the tube is not easy. Recovery of samples in granular soils below the water table is very difficult. High soil strengths and relative densities also restrict the depths achievable with this technique. Another drawback is that the equipment is noisy for its size.

46.5.2 Auger drilling

In geotechnical investigation (as opposed to mineral prospecting) auger drilling normally uses hollow-stem continuous fight augers. The hollow stem typically has a removable plug at the lower end and the auger is 'corkscrewed' into the ground. The material retained on the auger flights on extraction gives an approximate profile of the strata encountered, and the hollow stem and the plug can be removed from the surface to allow the use of driven samplers in the hollow stem. The technique is fast and can cope with most soil types, although sands and coarser soils generally inhibit the use of the hollow stem. Anything other than relatively shallow depths requires a powerful rig. The technique is not commonly used in geotechnical investigations and tends to be restricted to sampling for contamination testing.

46.5.3 Cable percussion boring and rotary drilling

46.5.3.1 Casing

Casing is used in both cable percussion and rotary boreholes to support the borehole walls and to seal off inflows of groundwater. It is typically formed of lengths of robust steel tubing with threaded ends allowing simple connection. A cutting shoe or bit is always present at the bottom of the lowest section of casing. Casing is not always required. Typically the base of the casing during drilling is always slightly higher than the base of the borehole. Casing is typically installed using the principal method of the rig involved – percussion for cable percussion rigs and rotary drilling techniques on rotary rigs.

46.5.3.2 Cable percussion boring

Cable percussion boring is also commonly known as shell and auger boring, and is the mainstay of the UK geotechnical investigation industry. It is used routinely in most materials from soft organic soils up to very weak rocks. The term 'shell and auger' is misleading as augers are almost never used in this context in the UK.

A cable percussion rig consists of a winch with a friction clutch or brake that winds a wire rope to and fro over a pulley at the apex of two connected A-frames. The winch is used to repeatedly lift and drop a variety of steel boring tools into the ground to create the borehole and to remove the displaced soils.

A 'claycutter' is, unsurprisingly, used to advance the borehole in clay soils. It has a similar form to a window sampler in that it is a steel cylinder, open at the bottom end, where it is fitted with a cutting shoe and with two open slots in the side. When this relatively heavy tool is dropped onto the base of the borehole, clay is forced up inside it. The cutter is then raised to the surface and the clay is manually knocked out through the shoe by inserting a metal bar through the open slots.

A shell is used to recover sands and gravels. This is a steel cylinder, open at the bottom end, where it is fitted internally with a hinged circular plate which forms a non-return valve or 'clack'. The sands and gravels are often recovered by surging the shell up and down just above the base of the hole. This draws water into the hole and loosens the soils, allowing them to pass above the clack and be retained by it. Water is often added to the borehole to facilitate this process. One of the disadvantages of this technique is that the samples it produces tend to under-represent the finer-grained fractions of the sands and silt that may actually be present in the ground.

A variety of chisels are also available for breaking through hard layers.

The recovered materials are removed frequently (typically at least every 0.5 m) by dropping or knocking them from the shell or claycutter onto a board adjacent to the top of the hole. Some or all of this material is collected as both small ('jar') and bulk ('bag') samples for logging by the engineer or geologist either at the rig site or later elsewhere. These samples can also be used for laboratory index testing.

Two other important tools are routinely used: the SPT hammer, in many materials, and the U100 sampler, in clay soils.

46.5.3.3 The standard penetration test

The SPT is a commonly used in situ index test. It can be carried out successfully in almost all soils and in some weak rocks. The SPT equipment consists of a trip hammer which drops a standard-sized weight (65.5 kg) through a standard distance (760 mm). The hammer drives either

- a solid rod of a standard length (610 mm) and diameter (2 ft or 50.8 mm) with a standard-shaped cone on the lower end or
- an open-ended split sample tube of a standard length and diameter with a standard-shaped cutting shoe or 'spoon' on the lower end.

The solid rod is used in rocks and soils with significant gravel or larger content, and the split spoon sampler in finer-grained soils. The samples obtained in this way are taken as small disturbed samples.

The test is carried out by repeatedly dropping the hammer and recording the penetration for each of six successive penetrations of 75 mm. The lowest four sets of 75 mm blow counts are totalled to give the SPT 'N' value. Various corrections can be made to this N value to account for the effects of variations in grain size, overburden pressure and hammer calibration. N values have been correlated with many soil and rock parameters. For further information on this, see Section 46.6. Where resistance to penetration is very high, equivalent N values are often extrapolated from tests with less than the normally required penetration.

The test is not always very successful in very weak soils, where the tool can sink under its own weight or penetrate several hundred millimetres with a single blow. The test is also unreliable in soils with many larger clasts (gravel and larger), where the driving resistance of individual clasts dominates the results. The test is very useful in some very weak rocks (such as high-porosity sandstones), where it is sometimes the only source of quantitative information. The test is now considered to be an unreliable indicator of geotechnical properties in chalk. The test can also be carried out in rotary bored holes.

46.5.3.4 The U100 sampler

The U100 sample is a low-quality undisturbed sample that is obtained by driving an open metal cylindrical tube into the soil. The tube is 450 mm long and has an internal diameter of 100 mm. The lower edge of the U100 sample tube sits within a steel cutting shoe containing a sprung core catcher, and its upper edge is connected to a driving head and rod. A sliding hammer connected to the rig winch moves over this rod and is repeatedly lifted and dropped, driving the sample tube into the ground. The material retained in the cutting shoe is usually sampled separately as a small disturbed sample. The open ends of the U100 tube are normally sealed with wax and covered with plastic end caps. The samples obtained in the U100 sample tube are commonly referred to as 'undisturbed'. This is not strictly true, and if high-quality samples are required for advanced triaxial testing then thin-walled push samples or samples taken from a rotary core should be used. U100 sampling is used in all types of clay soils. It is often not very successful in very soft or strong clay soils or clay soils with a high stone content. The number of blows used to fill the sample tube cannot be used in any way to correlate with SPT N values.

Eurocode 7 classifies a U100 sampler as a thick-walled sampler (see BSI, 2006: Table 3). This effectively disallows its use as a means of obtaining samples for quantitative design (a requirement for most EC7-compliant design). Modified thinner-walled U100 sampling tubes, known as UT100 samplers, are now commonly available and the use of these overcomes this EC7 sample classification problem. These sampling tubes are inevitably less robust than their thicker-walled predecessors, and this can cause problems in some soils traditionally sampled with U100s, such as tills and weaker clay-based rocks. In most cases, EC7-compliant sampling in these soils would be better achieved by using rotary coring methods.

46.5.3.5 Thin-walled and piston samplers

This type of sampler is typically used to obtain high-quality undisturbed samples in soft clays where other forms of sampling would be likely to cause excessive sample disturbance. They vary in diameter but are typically 100 mm or larger. Lengths also vary. Thin-walled samplers feature a non-return valve at the top of the sampler to allow escape of water or air during sampling. Thin-walled samplers are typically pushed continuously from the surface using a hydraulic jack or using the rig winch in conjunction with pulleys. Piston samplers are a type of thin-walled sampler. The sample tube is pushed by the drill rods into the soil over a close-fitting piston head which can be held in a fixed position by surface-operated clamps. Both types can be used with rotary rigs and cable percussion rigs. Piston samplers for use in stiff clay soils have also been developed.

46.5.3.6 Rotary drilling

There are many different types of rotary drilling technique used in ground exploration. One of the main distinctions is between cored and non-cored techniques. Within the cored category there are further distinctions between conventional, wireline and sonic coring. Choice of a flushing medium is also an important consideration. The sections below give an outline of these techniques. This, however, is a complex subject and it is recommended that these choices be discussed with a prospective ground investigation contractor at an early stage of the investigation procurement process.

46.5.3.7 Flush

Most rotary drilling systems involve the use of some sort of flush (air or liquids; the liquids include muds, foams and mists). The flush is circulated by pumps down through the drill bit and back up to the surface through the annulus between the boundary of the hole and the drill 'string' (the sequence of bit core barrel and drill rod or cable in the borehole). One of the main purposes of the flush is to remove cuttings away from the bit and transport them up the hole to the surface. Another significant benefit of flush is in resisting any sub-artesian groundwater pressures encountered during drilling.

46.5.3.8 'Open hole' rotary drilling

Rotary drilling that does not involve the collection of a core is often called 'open hole' drilling. In open hole drilling the bit occupies the full cross-sectional area of the borehole as it is advanced, and all the material below the bit is fragmented and passed up the borehole as cuttings by the flush. This is a very rapid drilling technique. Typical applications are those that require high metreages to be drilled but do not require the recovery of cores. Examples are searching for voids, such as those found in abandoned mines, and the rapid creation of boreholes for the installation of instrumentation where the ground conditions are well known or established. Open hole techniques are also used within short lengths of borehole to pass through strata where coring techniques might struggle to progress.

The only way to log rotary open hole boreholes is by examination of 'cuttings' in the flush returns. In soils, these are likely to be minimal. In rocks, examination of the small rock chips returned in the flush gives a broad view of the nature of the strata being drilled. This examination must be done *at the rig at the time of drilling*. Voids such as coal mine workings are often identified by a loss of flush. Unworked coal seams are usually identified by the distinctive arrival of black coal chips in the cuttings.

Where boreholes are being entirely drilled using open hole techniques, a system commonly known as ODEX is used. ODEX is a trade name that has come to be associated with equipment that allows the borehole casing to be advanced at the same time as the bit. Such systems often utilise 'down the hole hammers' (DTH) to increase the rate of penetration.

46.5.3.9 Rotary coring

There are three main types of rotary cored drilling: conventional, wireline and sonic. Conventional and wireline coring share many characteristics and will be discussed first. Coring of soils and rocks involves the recovery of a cylindrical 'stick' of the ground using a rotating 'core bit' and 'core barrel'. These combine to create a cylindrical tube with a cutting surface around the lower edge. Typically in geotechnical investigations most barrels are 'double-tubed'. In these systems the inner barrel is decoupled from the outer rotating barrel that carries the cutting edge to minimise disturbance to the core. This system usually also incorporates a rigid 'core liner' within the inner barrel. This liner is a plastic tube in which the core is collected and retained as it is placed in the core box at the end of the core run.

Each run has its own fresh length of liner. Spring systems are used to retain the core within the barrel as it is returned to the surface at the end of the core run. Core barrels come in a variety of lengths and diameters. Typical geotechnical investigation core diameters are between 70 and 100 mm. The wider diameters tend to maximise the likelihood of good core recovery. Core barrels are typically less than 3 m long. Core 'run' lengths (the length of borehole drilled in a single attempt) are typically restricted to less than the barrel length.

The difference between conventional and wireline systems is speed of operation in deeper (say greater than 30 m) boreholes. In conventional rotary coring the barrel is delivered to the bottom of the hole using coupled rods; these provide a rigid means of delivering the rotary motion and the flush to the bit. If the ground conditions require the hole to be cased then a typical sequence of operations would be as follows.

- Attach the barrel to the rotary head on the rig and lower it into the existing hole until the head is just above ground level.
- Clamp the top of the barrel, remove the head and pick up a rod, attach the rod to the top of the core barrel, unclamp it and lower the rod and barrel until the head is just above ground level.
- Repeat until the bit is at the base of the hole.
- Turn on the flush pumps, core a length of the ground (the 'run'), then repeat the sequence in reverse to remove and empty the barrel.
- Add a length of casing to the top of the casing string and drill this down to just above the base of the new hole created by the barrel.
- Repeat to the specified base of the hole.

In a wireline system a length of casing is added at the top of the borehole and the barrel is lowered into the casing using a cable. At the base of the hole the barrel locks into the lowest section of casing and the casing is used to deliver the rotary motion and the flush to the bit. When the run is complete the barrel is unlocked from the casing and withdrawn using the cable; this is much faster than the constant connection and disconnection of rods in the conventional method.

Sampling of rock core for testing is straightforward for most competent rocks: the samples can be taken directly from the core liner or core box. Samples of rocks that are likely to dry out prior to strength testing (most rocks) should be protected from moisture content change by immediately wiping away excess drilling fluids, wrapping in clingfilm and waxing. It is even more important that similar treatment be given to any core samples of overconsolidated clays required for anything other than simple index testing, as soon as possible after the core is removed from the barrel.

SPT is possible in rotary cored or open holes, although the process can be time-consuming compared to normal rates of drilling. Incorporation of SPT can be one of the few reliable sources of information in some weak rocks, particularly weak porous sandstones, which tend to be destroyed by the coring process and are often recovered as sand. Thin-walled push samplers, used for obtaining high-quality samples of clays, can also be used in suitable soils by most rotary rigs.

46.5.3.10 Sonic coring

Sonic coring is an entirely different arrangement. It is a rapid coring technique capable of achieving great depths and continuous complete recovery through almost any material. The technique utilises a high-frequency vibration either with or without rotation to advance what is typically a long single-tube core barrel (rigid liners can be used in some circumstances). No flush is used. As with the techniques discussed above, the drilling technique can be adopted to provide opportunities for conventional coring, SPT and piston or U100 sampling. The technique is not often used to obtain samples for laboratory testing, but is used in areas where poor recovery in conventional coring has been experienced and for the installation of geotechnical instrumentation during construction. A number of constraints limit the use of the technique for obtaining samples for laboratory testing.

- The high-frequency vibrations produce high sample temperatures and densify some soils; this means that any solid core samples recovered by this technique in soils or weak rocks should be considered disturbed from a laboratory testing point of view. The vibrations and temperatures can make even simple laboratory index test results unreliable in many soils.
- The vibrations can produce rock cores with a low solid core content compared to conventional rotary coring in the same materials.
- The use of single barrels and the lack of flush (to balance groundwater pressures) mean that significant quantities of information can be lost when coring strata that contain groundwater under artesian and sub-artesian pressures.
- The cores are typically removed from the barrel by the use of vibration, increasing the potential for disturbance.

46.6. In situ testing in boreholes

This topic is discussed in detail elsewhere. The notes below relate only to some of the practical issues involved in utilising these techniques.

Dynamic probing, static CPT and SPT are all forms of in situ testing and are all discussed above.

46.6.1 Permeability testing

Open boreholes, partially and fully cased boreholes, standpipes and standpipe piezometers are all potentially suitable for carrying out variable head permeability tests. Inflow tests using single or double packers (lengths of hollow rod surrounded by a rubber membrane that can be inflated to seal off a section of open borehole) can be carried out in stable open rotary cored boreholes.

46.6.2 Vane testing

The borehole vane test is directly analogous to the laboratory shear vane test. A large cruciform vane is lowered into the ground below the base of the borehole (typically into soft clays) and rotated, with the torque required to rotate the vane being recorded; the data can be converted into a value of undrained shear strength. The procedure is typically carried out using a cable percussion rig.

46.6.3 Plate bearing tests

It is possible to carry out plate bearing tests in boreholes. There are, however, many practical difficulties involved with obtaining a clean borehole base and continuous contact between the base of the hole and the inserted plate. Such tests are rarely carried out.

46.6.4 Pressuremeters and dilatometers

Pressuremeters and dilatometers are devices which use expanding membranes to measure strength stiffness and in situ stresses in the ground. Most of these instruments are utilised by lowering them into a test section of borehole that has been rotary cored and that is of a diameter only slightly larger than the instrument. Tight tolerances between the instrument and hole are needed to minimise the amount of expansion the instrument undergoes before contacting the borehole wall. As a result, materials that are prone to collapse or swell if the borehole is left unsupported can be difficult to test in this way. These problems can be overcome by the use of a 'self-boring' pressuremeter. Strata that tend to produce rough, irregular borehole walls can puncture the instrument's membranes. The pressure capacity of the instrument needs to be broadly matched to the strengths and stiffnesses that are to be measured.

46.7. Monitoring installations

46.7.1 Groundwater

Measurement of groundwater levels and pressures is a key objective of almost all geotechnical investigations. Successful design of groundwater instrumentation requires some knowledge of basic hydrogeology, in particular the distinctions between unconfined and confined aquifers and the difference between groundwater 'levels' and groundwater pressures (see Chapters 15 *Groundwater profiles and effective stresses*, 16 *Groundwater flow* and 80 *Groundwater control*).

There are two main types of instrumentation for measuring groundwater: standpipes and piezometers. There are also several different types of piezometer, which are described below. A common feature of all types is an area within the instrument which allows groundwater to enter the instrument or which responds to groundwater pressure: this area is placed in the borehole, where groundwater information is required (the response zone).

All piezometers are installed in boreholes in a similar way. The borehole is backfilled, usually with bentonite cement grout, to the base of the response zone. A permeable material, the nature of which varies with the type of instrument, is then backfilled around the instrument throughout the length of the response zone. Another bentonite cement grout seal is added above the response zone. The location of this upper seal in relation to the groundwater level varies according to the type of instrument.

46.7.1.1 Standpipes

Standpipes measure groundwater levels in unconfined aquifers (i.e. in relatively permeable strata). They typically consist of a tube with a diameter between 19 and 50 mm, with a slotted section that is placed within the response zone. In a standpipe the response zone and the slotted section extend above the unconfined groundwater level. The groundwater level measured in the tubing directly replicates the groundwater level in the ground. Groundwater level is measured using a dipmeter. This is a tape measure containing wires connected to a battery and a buzzer at the surface. At the other end of the tape is a sensor. When the sensor breaks the surface of the water in the tube an electrical circuit is made and the buzzer sounds. The depth to groundwater can be read off the tape.

Measurements of ground gas composition are also a routine part of ground investigations. Standpipes of 50 mm dia. are typically used for this purpose. They are installed with continuous slotted sections above the water table and an upper bentonite seal close to the ground surface. The top of the tube is fitted with a bung containing a gas tap to which a gas analyser and flow meter can be attached.

46.7.1.2 Piezometers

Piezometers measure groundwater levels or pore pressures in specific zones within the borehole. Zones are targeted for measurement on the basis of the aims of the investigation and the observations made during drilling. There are several types of piezometer.

46.7.1.3 Standpipe piezometers

Standpipe piezometers are the commonest and simplest form of piezometer. They typically consist of 19 or 24 mm tubing that is fitted with a porous 'element' of known permeability (typically 10^{-4} m/s) at its base; this element varies in length but is typically <300 mm in UK ground investigation practice. The porous element is installed in a response zone filled with sand. This sand filter is isolated in the borehole by bentonite seals above and below. The porous element allows water in the ground to enter the piezometer tubing. The height that the water reaches in the tubing reflects the pore pressure present in the response zone. If correctly located, the bentonite seals either side of the response zone isolate the pore pressure in the response zone from any other different pressures that may be present in the ground.

Practical considerations involved in achieving seals above and below response zones mean that it is unusual to install more than two standpipe piezometers in a borehole.

Standpipe piezometers can be used in all types of ground. However, in low-permeability materials the height of water in the tubing can take a very long time to come to equilibrium with the pore pressures in the ground. Standpipe piezometers can be used to carry out variable head permeability tests. Water level in the tubing is measured using a dipmeter as described above.

46.7.1.4 Other types of piezometers

In a pneumatic piezometer, twin pneumatic tubes run from the piezometer tip to the surface. The top of the piezometer contains a diaphragm which separates water pressure in the ground and gas pressure fed down one of the tubes from the surface. Gas pressure is provided by bottled nitrogen. Water pressure readings are obtained by adjusting the surface gas pressure with the water pressure. When they are equalised, gas is forced into the second tube and this is detected at the surface and the pressure recorded. These systems typically form part of long-term datalogged monitoring systems where a single datalogger and gas control system can be set up to log several boreholes. A pneumatic piezometer can be used to measure short-term negative pore pressures.

In a vibrating wire piezometer a tensioned wire in the piezometer tip is made to vibrate using an electro-magnetic field. The tension in the wire varies according to external water pressure on a diaphragm in the tip. This varies the frequency at which the wire vibrates. The variation in this frequency can be measured and calibrated in units of water pressure. This system is suitable for use with automatic dataloggers or with hand-held readout units. The principal advantage of vibrating wire piezometers is that they come to equilibrium very rapidly in low-permeability soils. This allows them to be installed directly in cement bentonite grouts rather than in conventional sand filters. Vibrating wire piezometers cannot reliably measure negative pore pressures and can become unreliable for measuring positive pressure if they are exposed to negative pressures. Great care is needed in their installation.

Pumping wells in ground investigations are installed with an associated array of monitoring piezometers when there is a need to carry out a pumping test. This test provides information on permeability, transmissivity and storage within an aquifer and is required when a project involves dewatering. The key differences between a pumping well and a piezometer are usually greater diameter (to permit submersible pump access) and the presence of a well screen. A well screen is normally used to line the borehole in the region where the water is to be pumped from, while the rest of the hole is normally lined with impervious casing. The screen's purpose is to permit passage of water from the ground to the hole while at the same time supporting the sides of the hole and controlling the quantity of material drawn into the hole by the flowing water. Successful screen design can be complex and consideration should be given to seeking expert advice before carrying out installation and testing.

46.7.2 Movement and load monitoring

46.7.2.1 Monitoring horizontal movement

Inclinometers are used to measure and monitor horizontal movement within the ground. This need typically arises where there are slope instability concerns or where excavations and retaining structures are being monitored. An inclinometer installation in a borehole takes the form of a circular tube installed to the full depth of the borehole, where the bottom of the borehole is well below the expected zones of movement. The annulus between the borehole and tubing is filled with grout and the lower end of the tubing is closed so the inside of the tubing remains full of water and/or air. The inside of the tubing contains four grooves or 'keyways' at 90° intervals. The lengths of tubing are connected using external sleeves in such a way as to ensure that the four keyways are continuous throughout the borehole and that each keyway is in the same orientation.

As the ground moves, the inclinometer tubing bends with it. The degree and location of the distortion of the tube is measured using an inclinometer. This is a metal 'torpedo' containing two pairs of wheels on sprung arms at either end of the torpedo. Both pairs of wheels are in the same plane. The wheels sit in a pair of the keyways in the tubing and the sprung arms hold the torpedo centrally in the tube. The torpedo contains a series of accelerometers orientated in the plane of the wheels and also at 90° to this. These accelerometers measure information on the inclination of the torpedo and these inclinations can be used to calculate horizontal displacement of the inclinometer tubing. In use the torpedo is lowered to the base of the tubing using its data cable. It is then pulled up 0.5 m at a time, with a reading taken every 0.5 m, and this information is usually recorded on a datalogger. The whole exercise is repeated with the torpedo turned through 90° and inserted in the remaining pair of keyways. The resulting data are processed using specialised software to give a profile of horizontal movement against depth. As this is repeated through a programme of regular monitoring a picture of displacement against depth against time is built up. Inclinometers that are used horizontally are also available. These are occasionally used to check settlement profiles in earthwork embankments.

Inclinometers that use strings of sensors that remain 'in place' in the borehole ('in place inclinometers') are available. These can provide near-continuous data for lateral displacement of the inclinometer tubing.

46.7.2.2 Monitoring vertical movement

Extensometers are used to measure vertical movement within the ground. The vertical movement can be heave or settlement. Typically a borehole installation involves the placing of a central access tube similar to inclinometer tubing to below the zone of likely movement. A series of magnetic 'spiders' are slid over this tubing at intervals during the grouting of the borehole/tubing annulus and grouted into place. As the ground moves up or down the 'spider' legs are held by the grout and ground and move up or down relative to the tube. The location of the spiders relative to the tubing is measured using a probe lowered down the central access tube.

Deep datum installations are also used to monitor the movement of the ground surface relative to a fixed point in the ground below any zones of movement. This is typically required where settlement or heave might be generated over a wide area and a reliable survey datum might otherwise have to be far from the site. These installations involve anchoring the base of a stainless steel or fibreglass bar into the ground at depth and isolating the remainder of the bar from the ground and borehole grout by some means, typically using an inner casing or sleeve. The top of the bar at ground level is then used as a datum for conventional topographic surveying techniques.

Monitoring of loads caused by soil or rock fill is often a part of earthworks operations rather than ground investigations. Plate-like pressure cells, typically vibrating wire pressure cells, are installed at the points of interest and wired to remote dataloggers for long-term monitoring. Such cells can be installed vertically behind retaining structures to verify active and passive pressure design assumptions.

46.8. Other considerations

Many of the issues identified in the desk study and walkover survey can have a considerable impact on the conduct of a ground investigation; these include

- access for rigs
- working space for rigs
- location of underground services
- water and power supplies
- underground and overhead obstructions
- soft ground in relation to stability of investigation plant
- many drilling rigs find operating on sloping ground difficult
- railway work has many unique bureaucratic requirements and access issues
- over-water and intertidal work often requires special plant to enable access.

46.9. Standards

The UK standard that applies to execution of ground investigations is currently BS EN 1997-2, also known as Eurocode 7, Part 2 (BSI, 2007). At the time of writing, this standard (and Part 1 on geotechnical design) is being revised, with publication expected in 2024. The proposed new structure of the revised Eurocode 7 will contain three parts rather than two. BS EN 1997-2 will become Part 3 of the new standard and a new Part 2 will deal with ground properties. 'BS 5930:2015 Code of practice for ground investigations (+A1:2020)' (BSI, 2015) is still current.

There are a large number of 'attachments' to Eurocode 7, Part 2. These are listed below. At the time of writing many of these had not been published and some parts of older British Standards still applied. This situation is complex and regular checks should be made with the BSI as to the current status of these documents.

46.9.1 Ground investigation and testing: sampling methods and groundwater measurement

BS EN ISO 22475-1 Part 1: Technical principles for execution

BS EN ISO 22475-2 Part 2: Qualification criteria for enterprises and personnel

BS EN ISO 22475-3 Part 3: Conformity assessments of enterprises and personnel by third parties

46.9.2 Ground investigation and testing: field testing

BS EN ISO 22476-1 Part 1: Electrical cone and piezocone penetration tests

BS EN ISO 22476-2 Part 2: Dynamic probing

BS EN ISO 22476-3 Part 3: Standard penetration test

BS EN ISO 22476-4 Part 4: Ménard pressuremeter test

BS EN ISO 22476-5 Part 5: Flexible dilatometer test

BS EN ISO 22476-6 Part 6: Self boring pressuremeter test

BS EN ISO 22476-7 Part 7: Borehole jacking test

BS EN ISO 22476-8 Part 8: Full displacement pressuremeter test

BS EN ISO 22476-9 Part 9: Field vane test

BS EN ISO 22476-10 Part 10: Weight sounding test

BS EN ISO 22476-11 Part 11: Flat dilatometer test

BS EN ISO 22476-12 Part 12: Mechanical cone penetration test

BS EN ISO 22476-13 Part 13: Plate loading test

BS EN ISO 22476-14 Part 14: Borehole dynamic probing

BS EN ISO 22476-15 Part 15: Measuring while drilling

46.9.3 Ground investigation and testing: geohydraulic tests

BS EN ISO 22282-1 Part 1: General rules

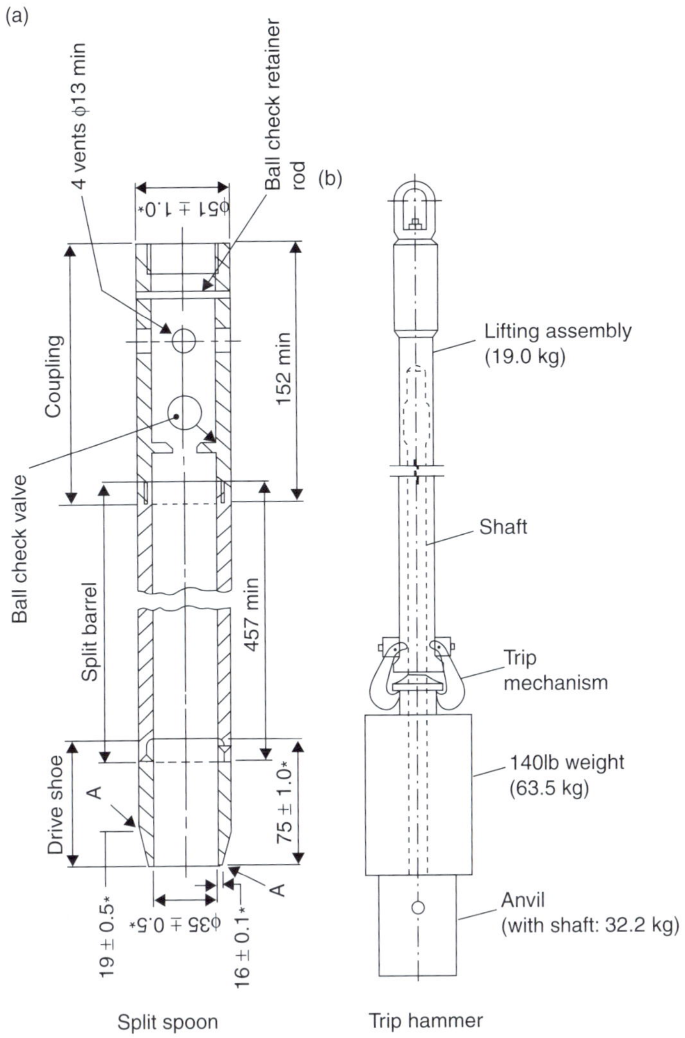

Figure 47.1 Standard penetration test (SPT): split spoon sampler (left) and automatic trip hammer (right). Corners marked 'A' may be 'slightly' rounded. (Left reproduced with permission from BS EN ISO 22476-3:2005 + A1:2011; BSI, 2005b (although dimensions in BSI (2005b) are now slightly different); Right reproduced with permission from CIRIA R143; Clayton, 1995)

bring practices into closer alignment). The SPT *N* value is the number of blows required to achieve a penetration of 300 mm, after an initial seating drive of 150 mm; corrections can then be applied to obtain a 'corrected' *N* value for a standard hammer energy and (in coarse-grained soil) overburden pressure (see below).

Despite widespread criticism, the SPT continues to be popular worldwide because

- it uses simple equipment
- it can be used in almost any ground conditions (including clays, sands, coarse granular soil and fractured weak rock)
- it provides a sample, allowing basic visual identification of the ground in which it is conducted.

It can be used, through empirical correlations, to estimate many of the strength and stiffness parameters needed for design.

Apart from the ground conditions in which the test is made, the result of the SPT is influenced by three main groups of factors

- the drilling or boring technique used to reach the test depth
- the SPT equipment used for the test
- the test procedure.

The most significant of these factors are described below (the wider adoption of the EN ISO standard, if adhered to, should improve the situation; however, these factors need to be considered when using historical information). For a more complete review see Clayton (1995) or Clayton *et al.* (1995).

In coarse-grained soil, *borehole disturbance* can be very large. Experience has shown that UK practice, using a 'shell' and light percussion boring can reduce the SPT *N* value by a factor of up to 5. Hollow-stem auger drilling, when used badly, is thought to be capable of loosening the soil so that *N* is reduced by a factor of about 3. The SPT *N* value can also be increased through the compaction of granular soil ahead of the drilling tools. Control of soil loosening can be achieved by

- ensuring that the borehole is full of water at all times, so that piping does not occur as a result of the upward inflow of groundwater
- using a small-diameter borehole, so that the depth of borehole disturbance (typically about 1.5–3 times the borehole diameter) is restricted, and preferably less than the 150 mm seating drive (BS ISO EN 22746-3 requires that the borehole diameter is always reported)
- avoiding the use of a casing (e.g. by using drilling mud to support the bottom of the borehole), thus reducing the concentration of groundwater inflow
- reducing the vigour with which the driller advances the borehole – for example, by restricting the rate at which the hollow-stem auger plug is pulled from the centre of the auger, immediately prior to the SPT.

In most clays where SPT is appropriate it is considered that the borehole disturbance will not be as severe as in silts and sands. The effects may be large when testing in soft and sensitive soil, but the SPT should be considered inappropriate in such materials due to the potential for disturbance and also as it will yield too low an *N* value to be of great use in design.

SPT *equipment* is known to have a number of important effects, the two most significant of which are due to

- the hammer design and use
- the split spoon design.

A wide variety of hammer designs are used across the world and most can be used in a variety of ways. They will deliver different amounts of energy per blow, and can be more or less consistent, blow-by-blow, depending upon the design. Since the SPT *N* value is inversely proportional to the energy delivered per blow, this means that it will vary according to hammer type and use. Therefore, the energy imparted by the SPT hammer must be measured and reported. BS EN ISO 22476-3 (BSI, 2005b) requires that all SPT hammers are calibrated regularly. BS EN ISO 22476-3 (BSI, 2005b) suggests that SPT *N* values should be corrected to 60% of the free-fall energy, N_{60}, and this is now established in British practice through BS EN ISO 22476-3 (BSI, 2005b). It has been estimated (Clayton, 1995) that hammer energy variations around the world can lead to variations in the *N* value of –25% to +35%.

In the UK it is routine practice to use an automatic trip hammer with a relatively large anvil. Measurements suggest that this will typically deliver about 70% of the free-fall energy, producing an *N* value of the order of 15% lower than that of the 60% standard energy. Reading *et al.* (2010) show quite clearly the wide variations in energy imparted by 'standard' hammers in the UK, in some cases simply due to lack of maintenance but also related to equipment type.

When very short rod lengths are used between the hammer and the split spoon there will be some reduction in energy transfer. BS EN ISO 22476-3 (BSI, 2005b) suggests that in sands a correction to the *N* value should be considered.

The design of the split spoon can also have an effect on penetration resistance, if varied from the standard. The user should be wary of two particular variations. In a few countries (including the UK) the practice has been to substitute a 60° solid cone for the split spoon cutting shoe when testing in gravel. Evidence from the UK suggests that in sands this can approximately double the measured penetration resistance. The use of a solid cone should be avoided if possible, and if used should be clearly reported as an SPT(C) test in accordance with BS EN ISO 22476-3 (BSI, 2005b).

Another potential problem occurs where, as in some countries, the split spoon is fitted with a liner to allow easy sample collection. The omission of a liner from a spoon designed to take one has been found to cause a decrease in the *N* value of about 10–30%.

The *test procedure* is fairly uniform in the UK, where blow counts are typically recorded by the rig foreman for six 75 mm sections of drive (the first two are for the seating and the remaining four are the test drive) – an amendment to BS EN ISO 22476-3 (BSI, 2005b) originally stipulated 150 mm drive lengths but now allows for 75 mm drives, as per UK practice. However, procedures can vary when testing either in loose or weak soil or in dense soil and weak rock. In the former case, penetration under the self-weight of the rods and hammer historically may or may not have been recorded, but now, under BS EN ISO 22476-3, it has to be reported. In the latter case the full 450 mm of penetration may not be achieved, and in this case it is now stated in BS EN ISO 22476-3 (BSI, 2005b) that it is acceptable to report the number of blows and the penetration achieved; a clear note identifying this variation is required in the report.

As with other in situ penetration tests, the SPT is useful for profiling, for classification and for parameter determination. When used for profiling ideally more tests should be carried out than is common in UK practice, say at 1 m centres. As with the dynamic penetration test (described below), the SPT split spoon could be driven continuously, but this is not normal, as friction and adhesion on the drive rods tend to increase penetration resistance at depth. Classification is possible through the use of a sample (albeit of very low quality in clays, and unable to provide representative gradings in coarse granular soils) combined with the measured penetration resistance. This is based upon the relative density in granular soils, the consistency in fine-grained soils and the strength in weak rocks (see Clayton, 1995, for example).

The SPT is a small-diameter dynamic, rapid test to failure. The principal soil properties affecting the SPT *N* value are therefore

- effective angle of friction (granular soils)
- relative density (granular soils)
- effective stress level (granular soils)
- grain size (coarse granular soils and silty granular soils)
- undrained shear strength (cohesive soils)
- cementing (weak rocks, granular soils)
- jointing (weak rocks).

Because interpretation depends upon the soil type, this must be known. Fortunately, as noted above, in most ground conditions a small sample will be retrieved from the split spoon.

The effective stress level needs to be taken into account when interpreting tests carried out in granular soil. For example, an SPT carried out from the original ground level would be expected to be higher (and sometimes very considerably higher) than one carried out in a general excavation (below the original ground level) during construction, as a result of the reduction in the vertical effective stress that has occurred. Modern practice is to correct *N* as measured in granular soils to its equivalent value under a vertical effective stress of 100 kPa (1 ton/ft^2), which is termed 'N_1'. *N* corrected for both hammer energy and for effective overburden pressure is termed '$(N_1)_{60}$'.

Soil grain size is an important consideration when interpreting SPT data. The standard correlations and interpretations used in granular soil are for sands. Evidence suggests (Tokimatsu, 1988) that once the mean particle size (D_{50}) exceeds about 0.5 mm, penetration resistance climbs rapidly, approximately doubling at D_{50} = 10 mm. Correlations obtained for sands cannot be assumed to hold true for gravels, and larger-diameter penetrometers have therefore been used in investigations of, for example, the liquefaction potential of gravels. Coarse-grained materials are not the only issue – the fines content is thought to contribute to lower penetration resistance (due to the development of pore pressures during driving) in silty sands.

In clays the penetration resistance seems to be primarily controlled by the undrained shear strength. Unless there is a very significant drilling disturbance (which may well be a factor in soft and sensitive soils), this will not change significantly during drilling. Therefore the SPT *N* value is not corrected for overburden pressure in clays. N_{60} is used in design. Care must always be taken when looking at correlations or design procedures as to the form of *N* value (corrected or not, and, if corrected, specifically what corrections were used; different forms of correction can be found in the literature for the same thing – see BS EN 1997-2). Guidance on interpretation both for geotechnical parameters (see **Table 47.1**) and direct design methods can be found very widely in the literature (e.g. Clayton, 1995; see also the conferences listed in the further reading section) and in Annex F of BS EN 1997-2 (BSI, 2007b), with the associated caveats in the UK National Annex (BSI, 2009a), but care must always be taken as to their applicability for the ground under investigation.

47.2.2 Dynamic probing

Because it does not require a borehole, dynamic probing provides a simple, rapid and relatively inexpensive way of profiling a site. It is a dynamic, intermittent test. It is suitable for small depths of exploratory work and to fill in or extrapolate data from more expensive and sophisticated tests, using site-specific correlations. The equipment is generally light and compact and is ideal for work on sites with restricted access. A dynamic probing test (DPT) consists of driving a solid steel cone vertically into the ground using an anvil and extension rods, with successive blows of a free-fall hammer. Lightweight hammers may be raised by hand, but in most cases a motorised device, incorporating an automatic latch and release mechanism, is used. **Figure 47.2** shows a motorised dynamic probing rig.

The number of blows required to drive the cone a fixed distance (10 or 20 cm, depending on the test specification) is recorded. The cone may be fixed to the extension rods, or loose-fitting, in which case it will be lost when the rods are withdrawn at the end of the test (sacrificial cone). Driving is halted every metre to add a further extension rod; at this point the torque required to rotate the rods in the ground is measured to assess the ground friction acting on them, which also has the effect of relieving friction for subsequent tests. The torque value can be used to correct the blow count to compensate for friction.

A variety of equipment has been developed and standardised over the years, with varying sizes of cones and hammer weights. Dynamic probing is now standardised in the UK through BS EN ISO 22476-2 (BSI, 2005a), which lists five configurations of equipment with varying cone size and drop weight (DPSH-A is not used in the UK). These configurations are generally the same as those that have been common in the UK and other countries (based on ISSMFE, 1989) for many years, except DPM, which has now changed to using a larger cone (the configuration for DPM is now that of DPM15, given in earlier versions of BS 5930 (BSI, 1999)). The range gives increasing specific work per blow from DPL through to DPSH. (A down-hole system used in some parts of Europe with slightly different configurations has been standardised in BS EN ISO 22476-14 (BSI, 2020); it uses an N_{30}.).

Figure 47.2 Motorised dynamic probing rig

The different equipment configurations detailed in BS EN ISO 22476-2 (BSI, 2005a) are reproduced in **Table 47.2**. Standard ranges of blows are given as 3–50 for DPL, DPM and DPH, and 5–100 for DPSH; it is recommended that if the minimum is not reached then the results cannot be used for quantitative purposes.

Figure 47.3 shows the acceptable shapes of the fixed and sacrificial cones. In the UK many companies use sacrificial cones of the same shape as the fixed cone. In general, DPL, DPM and DPH have the same hammer drop height but different hammer weights, and DPH and DPM use the same extension rod mass. The DPSH-B test is designed to closely resemble the SPT, as can be seen in **Table 47.2**, but differs from the other tests because it counts the number of blows per 20 cm penetration rather than 10 cm. BS EN ISO 22476-2

Factor	Test specification DPL	DPM	DPH	DPSH-A	DPSH-B
Hammer mass: kg	10 ± 0.1	30 ± 0.3	50 ± 0.5	63.5 ± 0.5	63.5 ± 0.5
Height of fall: m	0.5 ± 0.01	0.5 ± 0.01	0.5 ± 0.01	0.5 ± 0.01	0.75 ± 0.02
Mass of anvil and guide rod (max): kg	6	18	18	18	30
Rebound (max): %	50	50	50	50	50
Rod length: m	$1 \pm 0.1\%$	$1 \pm 0.1\%$	$1 \pm 0.1\%$	$1 \pm 0.1\%$	$1 \pm 0.1\%$
Mass of rod (max): kg	3	6	6	8	8
Rod eccentricity (max): mm	0.2	0.2	0.2	0.2	0.2
Rod OD (d): mm	22 ± 0.2	32 ± 0.2	32 ± 0.2	32 ± 0.3	32 ± 0.3
Rod ID: mm	6 ± 0.2	9 ± 0.2	9 ± 0.2	9 ± 0.2	9 ± 0.2
Cone apex angle	90°	90°	90°	90°	90°
Cone area (nominal) (A): cm²	10	15	15	16	20
Cone diameter new (D): mm	35.7 ± 0.3	43.7 ± 0.3	43.7 ± 0.3	45.0 ± 0.3	50.0 ± 1.0
Cone diameter worn (min): mm	34	42	42	43	49
Mantle length: mm	35.7 ± 1	43.7 ± 1	43.7 ± 1	90 ± 2	50.5 ± 2
Number of blows per *x* cm penetration (*Nx*)	N10 :10	N10 :10	N10 :10	N20 :20	N20 :20
Standard range of blows	3–50	3–50	3–50	5–100	5–100
Specific work per blow (Mgh/A): kJ/m	50	98	167	194	238

Table 47.2 Details of dynamic probing test specifications (Data taken from BS EN ISO 22476-2:2005; BSI, 2005a)

(BSI, 2005a) recommends that the imparted energy should be measured for each new device, in a similar way to the SPT. In addition it recommends the actual energy E_{meas} transmitted to the drive rods is found from calibration when the test results are to be used for quantitative evaluation purposes (i.e. parameter determination).

47.2.2.1 Results and analysis of data

The probing results are recorded as blows for 10 cm (N_{10}) or 20 cm (N_{20}) penetration and should fall within the standard range of values stated in BS EN ISO 22476-2 (BSI, 2005a). Note that the values will be specific to the configuration used and the annexes of the standard suggest that the use of an extra subscript, N_{10L} or N_{10M}, for example, would avoid any confusion when presenting results. With BS EN ISO 22476-2 (BSI, 2005a) it is now a requirement to measure the torque on the rods at least every metre, and this has to be recorded and reported. The effect of rod friction on the $N_{10/20}$ values can be significant and now it is a requirement in the standard that this shall be considered, but there is no mention of how this should be done. Examples of the effects are given in the annexes to the standard.

The N_{10} values can be converted into unit cone resistance (r_d) or dynamic cone resistance (q_d), and this can allow different configurations of equipment to be brought to a common basis; both parameters can be derived from pile-driving formulae; see BS EN 22476-2 Annex E, thus

$$q_d = \frac{M}{M + M'} r_d \tag{47.1}$$

$$r_d = \frac{Mgh}{Ae} \tag{47.2}$$

where r_d and q_d are resistance values in Pa, M is the mass of the hammer in kilograms, g is the acceleration due to gravity in m/s², h is the height of fall of the hammer in metres, A is the projected area of the cone in m², e is the average penetration in metres per blow ($0.1/N_{10}$ from DPL, DPM15, DPM and DPH, and $0.2/N_{20}$ from DPSH), and M' is the total mass of the extension rods, the anvil and the guiding rods in kilograms.

The value of r_d represents the work done in penetrating the ground and q_d modifies the r_d value to take account of the inertia of the driving rods, anvil and hammer during impact. The inclusion of the hammer weight, height of fall and cone size permits a comparison of the different configurations by way of r_d, but the inclusion of the size and mass of the extension rods in the calculations of q_d better normalises the results from the different equipment configurations. **Figure 47.4** shows typical data for a stiff clay site when three equipment configurations (DPL, DPM15 and DPH) were used. The results show how the different profiles for N_{10} are largely the same when plotted as r_d and q_d.

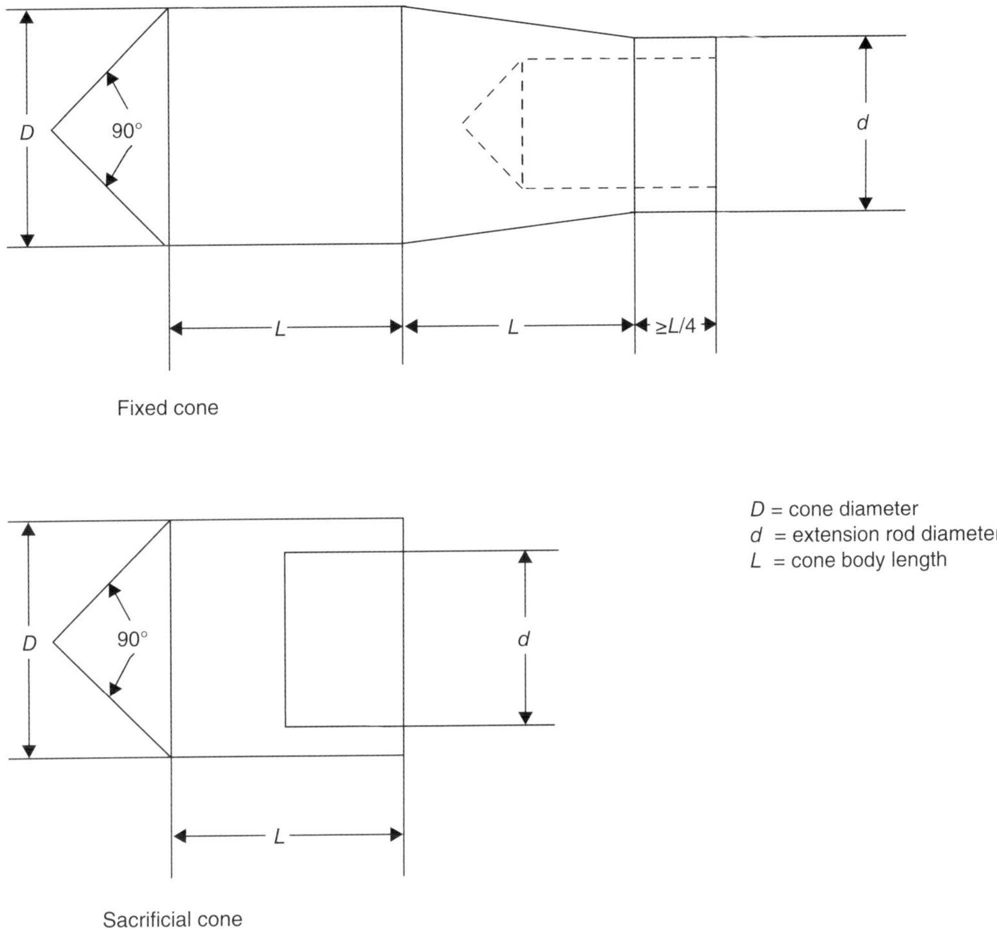

Figure 47.3 Acceptable shapes for fixed and sacrificial DPT cones

47.2.2.2 Rod friction

Friction on the rods will affect the recorded values of N_{10} or N_{20}. Experience has shown that torque readings in excess of 200 Nm generally mean that the driving rods have been forced off line and that further driving would permanently bend them, probably resulting in the driving rods no longer complying with BS EN ISO 22476-2 (BSI, 2005a). To avoid this, tests should be terminated when a torque reading reaches 120 Nm.

BS EN ISO 22476-2 (BSI, 2005a) provides for the introduction of drilling mud (bentonite slurry) to reduce rod friction. A specially perforated rod carries the cone and the drilling mud is introduced down the hollow rods and out through the perforations. For stiff clay sites, the mud can be poured into the rods to be sucked out or forced out by gravity into the annulus behind the probe cone. The level of bentonite should be topped up after the addition of each extension rod. For soft clay, a pressure just above the mean in situ stress needs to be applied to force the mud into the space behind the probe cone.

In its most common use, dynamic probing is used to look at variations within a profile and to map their extent laterally; however, in many countries considerable experience has been built up to convert the results into other parameters (see **Table 47.1**), particularly in coarse-grained deposits. For example, in the Annexes to BS EN ISO 22476-2 (BSI, 2005a) there are examples of the various factors that can affect DPT results and in BS EN 1997-2 Annex G (BSI, 2009a) some examples are given of interpretation of DPT results in coarse-grained materials. Other examples can be found in the literature (see the further reading section). It is vital to ensure that any correlations used have been validated for the local soil types and DPT configuration; in the future with better calibration and energy measurements it may be possible to transfer correlations between configurations. Little has been published on their interpretation in clays, but Butcher *et al.* (1995) give examples of the potential for interpretation in clays to determine geotechnical parameters. BS EN ISO 22476-2 (BSI, 2005a) does point out that because of hammer fall energy losses it is recommended to know by calibration the actual energy E_{meas} transmitted to the drive rods when this test is used for quantitative evaluation purposes.

47.2.3 The cone penetration test

Among the many in situ test devices available, the electric static cone penetrometer (CPT) and the piezocone (CPTU)

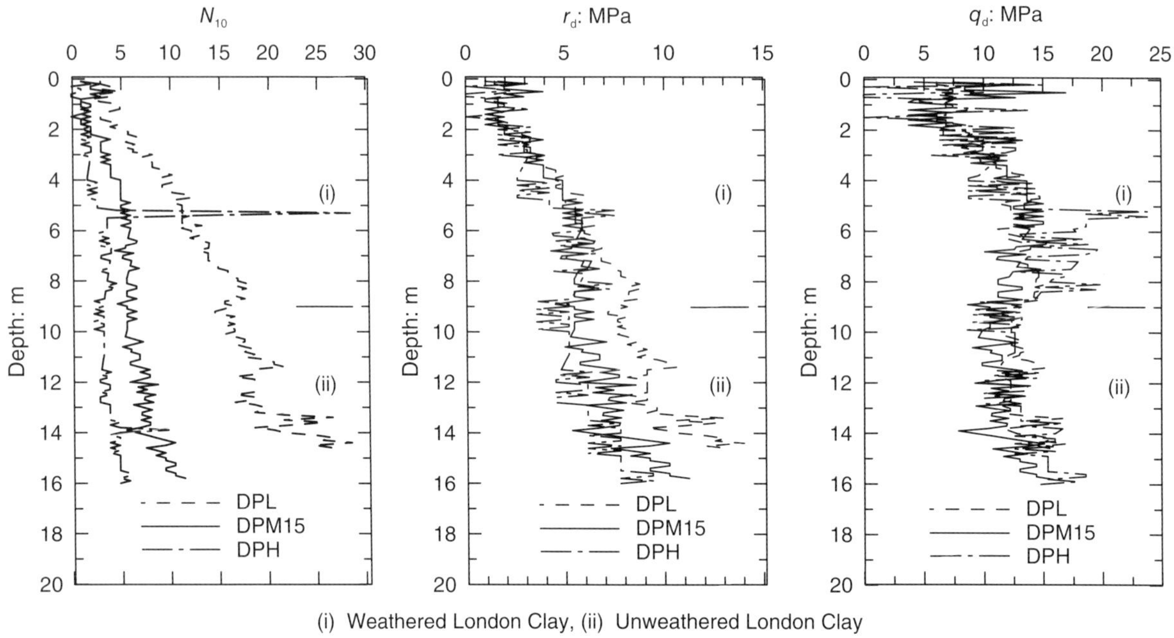

Figure 47.4 Results (N_{10}, r_d and q_d values) for three configurations of dynamic probing (DPT) for a stiff clay site

are among the best. CPT is a quasi-static test that is near-continuous; it is generally undertaken without a borehole but can be used with boreholes to advance past obstructions or to increase the depths achieved. Tests are rapid, can generate detailed ground profiles and can classify the ground, and the results can be used to calculate a wide range of geotechnical parameters that accurately reflect ground properties. The test is covered by BS EN ISO 22476-1 (BSI, 2023) and BS 1377-9 and BS EN 1997-2 (BSI, 2007b). In some countries older-style cones may be encountered that use remote mechanical systems for force measurement (CPTM) – these are covered by BS EN ISO 22476-12 (BSI, 2009b).

47.2.3.1 Test equipment

The diameter of the standard 60° cone is 35.7 mm (giving a cross-sectional area of 10 cm^2) and the area of the friction sleeve is 150 cm^2. The cone resistance and sleeve friction are normally measured using strain-gauged load cells. **Figure 47.5** illustrates the main components and terminology of cone penetrometers. The CPT and CPTU systems generally include an inclinometer to warn if the cone is going off vertical, which can easily happen if hard obstructions are encountered.

Although a cone penetrometer with a 10 cm^2 base area cone and an apex angle of 60° is specified in BS EN 22476-1 (BSI, 2023) as the reference test (being the same size as the original mechanical cone), other sizes of cone exist. Apart from the 10 cm^2 device, the 15 cm^2 cone is the only one routinely found in land-based practice, but the use of 2 and 5 cm^2 devices can be found in the offshore environment (along with the larger sizes).

The standards detail all necessary requirements with regard to shape, dimension and area, and their tolerances, as well as operational procedures. Provided the equipment dimensions are kept within these tolerances, the likely errors in measured parameters, simply from wear, will be kept to less than a few per cent.

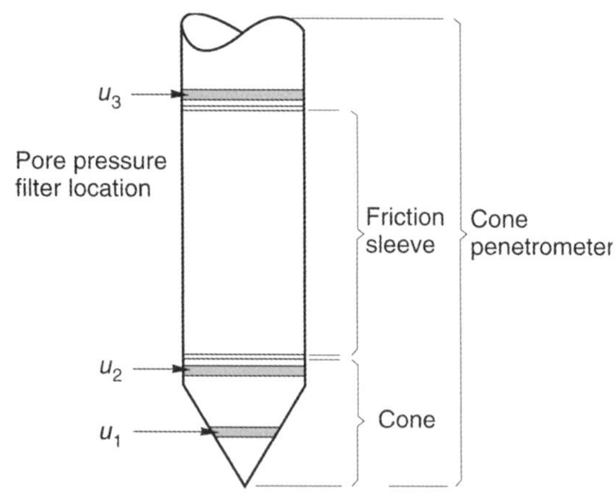

Figure 47.5 Main components of the CPT cone, also showing the positions for pore pressure measurement

The addition of a pore water pressure measuring system within the CPT (CPTU) has greatly enhanced its capability for both profiling and parameter determination. **Figure 47.5** shows that the pore pressure may be measured at several locations: 'u_1' (on the cone), 'u_2' (behind the cone) and 'u_3' (above the friction sleeve). Generally, it is recommended that the pore pressure is measured just behind the cone shoulder (at the 'u_2' position) for the following reasons

- good protection from damage, relative to 'u_1'

- relatively easy saturation
- gives good profiling detail
- generally good dissipation data
- correct location to determine q_t (see corrections below).

However, the other locations may be useful in certain conditions (see Lunne *et al.*, 1997).

The pore pressure measurement system should be designed in such a way that it is easy to saturate and to keep saturated. This is because successful CPTU profiling depends very much on good saturation of the pore pressure measurement system, particularly in clays.

In order to attain satisfactory accuracy, it is essential that the instrument is calibrated regularly and over load ranges appropriate for the ground being tested (BS EN 22476-1; BSI, 2023) goes into great detail on this topic). On every project, and at regular intervals on a large project, the engineer should check that the calibration information being used is current and appropriate. It is also important to take, report and review sensor base readings before and after each full penetration – this is a requirement in BS EN ISO 22476-1 (BSI, 2023). Calibration enables the assessment of the stability of the sensors and any change due to damage, especially of the load cells. In both cases significant changes in the base readings for the load cells can be converted into measured load and then into an assessment of the potential inaccuracy in the results. The BS EN ISO 22476-1 (BSI, 2023) changes slightly from earlier versions and introduces the idea of cone penetrometer class based on laboratory calibration (**Table 47.3** here and table 2 in the standard) and these set requirements for the accuracy of the equipment as a function of test categories (table 3 in the standard) which are related to the perceived accuracy of measurements of cone parameters related to soil type and likely end use of the results – for example, profiling or property determination (**Table 47.4** here and table A.1 in the standard). They require that the CPTU is used when geotechnical parameters are to be determined from the results. These classes are a critical item in the specification of appropriate tests, and in their interpretation.

47.2.3.2 Test method

The cone, on the end of a string of rods, is pushed into the ground at a constant rate and near-continuous or intermittent measurements are made of the resistance to penetration of the conical tip. Measurements are also made of the resistance to penetration of the friction sleeve, above the cone. When using a piezocone, pore pressure measurements are also made.

The cone is normally pushed into the ground at a rate of 20 ± 5 mm/s. A wide range of systems have been developed over the years to carry out the pushing operation, both on land

Cone penetrometer class	Measurand	Allowable maximum measurement uncertainty[a]	Ambient temperature stability[b]	Transient temperature stability[c]	Bending influence[d]
		$U_{qc:class}$ $U_{fs:class}$ $U_{u:class}$	Δa_{qc} or Δa_{qcc} Δa_{fs}, or Δa_{fsc} Δa_u or Δa_{uc}	Δt_{qc} or Δt_{qcc} Δt_{fs}, or Δt_{fsc} Δt or Δt_{uc}	Δb_{qc} Δb_{fs} Δb_u
0	Cone resistance	15 kPa or 0.5%	0.5 kPa/°C	2 kPa/°C	0.3 kPa/N
	Sleeve friction	5 kPa or 1%	0.1 kPa/°C	0.5 kPa/°C	0.1 kPa/N
	Pore pressure	3 kPa or 0.5%	0.1 kPa/°C	0.5 kPa/°C	0.05 kPa/N
1	Cone resistance	35 kPa or 1%	2 kPa/°C	10 kPa/°C	0.3 kPa/N
	Sleeve friction	5 kPa or 1%	0.1 kPa/°C	0.5 kPa/°C	0.1 kPa/N
	Pore pressure	10 kPa or 0.5%	0.1 kPa/°C	0.5 kPa/°C	0.05 kPa/N
2	Cone resistance	100 kPa or 2%	10 kPa/°C	50 kPa/°C	1 kPa/N
	Sleeve friction	15 kPa or 2%	0.5 kPa/°C	2.5 kPa/°C	0.5 kPa/N
	Pore pressure[e]	25 kPa or 1%	1 kPa/°C	2.5 kPa/°C	0.1 kPa/N
3	Cone resistance	200 kPa or 5%	10 kPa/°C	100 kPa/°C	2 kPa/N
	Sleeve friction	25 kPa or 10%	1 kPa/°C	5 kPa/°C	1 kPa/N
	Pore pressure[e]	50 kPa or 5%	1 kPa/°C	5 kPa/°C	0.2 kPa/N

(a) The maximum allowable uncertainty of the measured parameter is the larger value of the two quoted. The relative uncertainty applies to the measured value and not the measuring range.
(b) The values of ambient temperature stability represent the maximum allowable for each class.
(c) The values of transient temperature stability represent the maximum allowable for each class.
(d) The values of bending influence represent the maximum allowable for each class.
(e) Pore pressure applies only to CPTU.

Table 47.3 Classification of cone penetrometers under laboratory conditions from EN ISO 22476-1 (Data taken from BS EN ISO 22476-1; BSI, 2023)

and at sea. On land, pushing usually consists of hydraulic jacking against reaction systems specially built for the purpose; but sometimes the pushdown of an anchored drill rig is used. Hydraulic jack stroke lengths of 1–1.2 m are typically used, but shorter lengths may be adopted when working in limited headroom. Offshore systems exist that can either operate continuously down a drill string or by pushing from a jacking unit on the seabed.

During penetration, measurements are taken from all sensors at a fixed frequency so as to give discrete readings for forces, pore pressures and inclination, as appropriate (i.e. for the accuracy or application classes). Typically the minimum frequency is once every second or 2 cm of penetration; faster rates can be used to increase the detail of the ground profile (see Lunne *et al.*, 1997). In this way near-continuous profiles of the measured parameters are generated. A typical mixed profile is shown in **Figure 47.6**.

At any point during a test, penetration can be stopped and the CPTU left stationary, with measurements of pore pressure taken as a function of time (many standards also require cone resistance and sleeve friction to be recorded); this is referred to as a pore pressure 'dissipation test'. These tests can be used to determine the coefficient of consolidation. When penetration pore pressures are left to dissipate it is possible to estimate the in situ static pore pressure–depth relationship.

47.2.3.3 Cone parameters

The measured total force, Q_C, acting on the cone end, divided by the projected area of the cone, A_C, gives the measured cone resistance, q_c. The total measured force acting on the friction sleeve, F_S, divided by its surface area, A_S, gives the measured sleeve friction, f_s. In CPTU the pore pressure is measured as discussed above. The friction ratio, R_f, is obtained by dividing the sleeve friction by the cone resistance and is expressed as a percentage.

Pore pressure measurements can be used to correct measured results for geometrical effects (see below). However, a pore pressure measurement may not always be achievable – for example, in unsaturated soils and above the groundwater table.

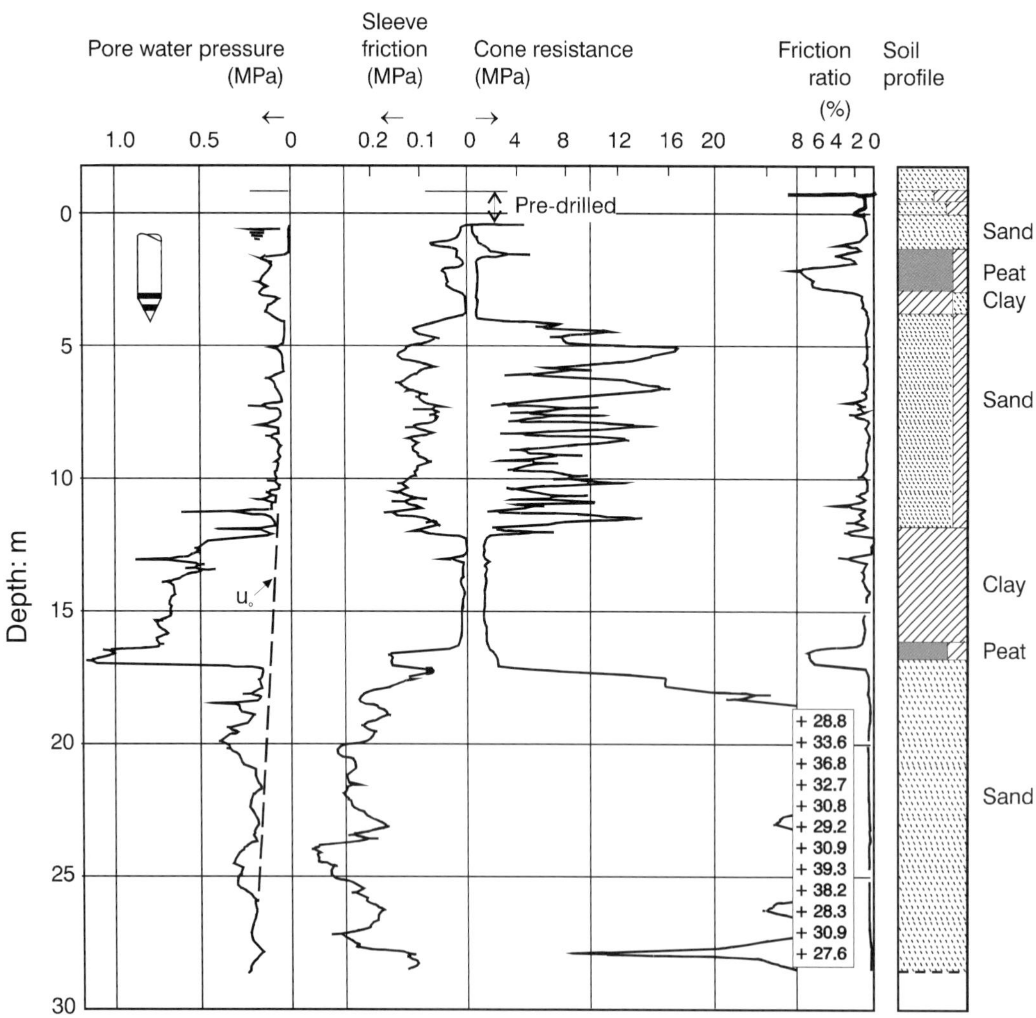

Figure 47.6 Example profile of measured and derived parameters from a CPTU profile

47.2.3.4 Effects of penetration rate, pore pressure and verticality

The results of a test are to some extent dependent on how fast the penetrometer is inserted into the ground. The speed of penetration should be 20 ± 5 mm/s according to the IRTP (ISSMGE, 1999) and BS EN ISO 22476-2 (BSI, 2005a). The rate of penetration, especially when using the piezocone in fine-grained soil, affects pore pressure generation and there may in addition be viscosity effects on q_c and f_s. The penetration rate may also influence creep and particle crushing. Typically a tenfold increase in the rate causes a 10–20% increase in the measured cone resistance in stiff clays and 5–10% in soft clays (e.g. Powell and Quarterman, 1988). If the standard is followed, the rate should not be a problem in clays and sands; however, in intermediate soils the effect of rate changes may be more significant for the measured parameters. The rate may change most noticeably when the pushing equipment is reaching capacity and may slow down.

As has been noted, the most common onshore insertion method is to push the cone penetrometer into the soil using a rig with hydraulic jacks, with a stroke length of about 1 m. At the end of each stroke, the pushing stops and the jacks return to their start position. This is referred to as 'intermittent testing'. In some soils this causes 'unrepresentative readings' every time the test is stopped and a new rod added. These effects can be even more pronounced after pauses for dissipation tests when, because of the drop in pore water pressure and resultant consolidation, local increases in both q_c and f_s can be observed; there may also be a delay in the pick-up of the pore pressure response. This behaviour could be misinterpreted as a local ground feature if not anticipated.

The measured cone resistance is affected by pore water pressures acting downwards, on the top of the cone (**Figure 47.7**). The magnitude of the effect will vary from cone to cone depending on the geometry and will result in significantly different (40%) measurements of q_c in the same soft fine-grained soil. This variability of results may be avoided by using the CPTU since the pore water pressure effects may be corrected for by using

$$q_t = q_c + u_2(1 - a) \tag{47.3}$$

where q_t is the corrected cone resistance, q_c is the measured cone resistance, u_2 is the pore water pressure measured just behind the cone and a is the area ratio (the area affected by the pore water pressure ranges from 0.35 to 0.95, typically 0.7–0.85; see Lunne *et al.*, 1997).

This correction is a requirement in BS EN ISO 22476-1 (BSI, 2023) for some 'accuracy' or 'application' classes and when the results are to be used to interpret geotechnical parameters. Friction sleeves are also affected by pore water pressure effects but correction is more difficult because

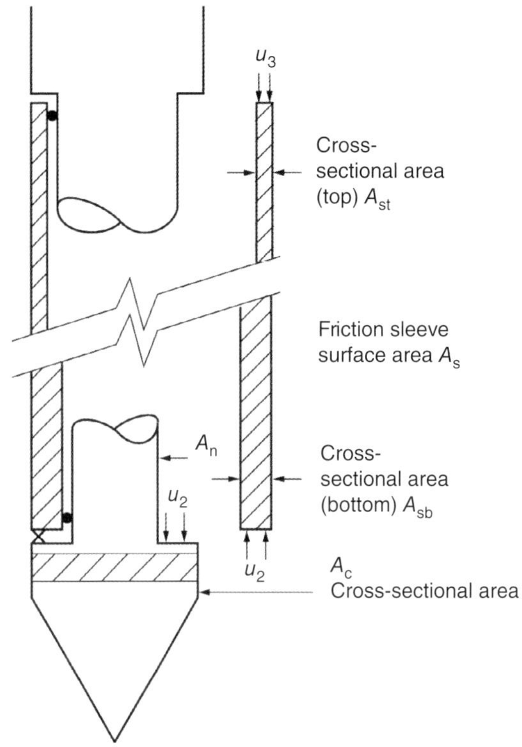

Figure 47.7 Effect of pore pressure on cone resistance

it requires measurement of both u_2 and u_3, as shown in **Figure 47.7**. The effects can be reduced significantly if the end areas A_{st} and A_{sb} are equal.

Another correction that should be considered is for errors in depth measurement due to deviation of the cone from the vertical. Most electric cone penetrometers contain simple slope sensors to measure the non-verticality of penetration. This is useful in avoiding damage to equipment due to sudden deflection. Where penetrations are deep, inclinations in excess of 45° are not uncommon, especially in stratified soil, and can result in wrongly identifying the depth of changes of the strata. This correction is a requirement in BS EN ISO 22476-1 (BSI, 2023) for some test categories.

The CPT and CPTU test probably have the widest range of ground parameters that can be derived from their results, mostly by correlation (see **Table 47.1**). The applicability of any correlation used must be ensured if confidence in the resulting parameters is to be achieved. Deriving 'site-specific' correlations is often useful. Results from CPT and CPTU tests can also be used directly in design (e.g. pile capacity, settlement predictions, for assessment of liquefaction potential or the effect of ground improvement). A number of sources can be used – see Lunne *et al.* (1997), BS EN 1997-2 Annex D (BSI, 2007b), as well as various technical conferences involving CPTs, some of which are listed in the further reading section.

47.3. Loading and shear tests

Loading and shear tests are of most benefit when representative samples cannot be taken – for example, in made ground, very soft and sensitive clays, granular or stony soils or fractured rocks. Some (e.g. the vane test) are relatively easy and low-cost. Others (e.g. large-diameter plate tests and the more sophisticated pressuremeter tests) are expensive, can be time-consuming and are not used during routine investigations.

47.3.1 The vane test

The vane shear test is, and has been, used in the UK to determine the undrained shear strength of soft soils (although it is used in stiffer materials in some countries). It is standardised in BS EN ISO 22476-9 (BSI, 2020), which now supersedes BS 1377-9 (BSI, 2007a) and BS EN 1997-2 (BSI, 2007b) and elsewhere by ASTM (2008b). The test most commonly used consists of pushing a cruciform vane mounted on a solid rod into the soil and rotating it from the ground surface while measuring rotation and torque. In different countries, vane tests may be carried out either by pushing the device from ground level, or from the base of a borehole, or both. The use of a borehole brings with it the risk of strength reduction as a result of borehole drilling disturbance, and this must be recognised and dealt with. As a result of developments in the offshore industry, where measurement at the surface would be unrealistic, equipment is now available for both land and offshore work where the torque and rotation are measured electronically (down-hole) close to the vane head.

In its conventional form (**Figure 47.8**) the field vane has four rectangular blades and has a height-to-diameter ratio (*H*/*D*) of 2. In the UK the dimensions of field vanes are controlled by BS 1377 (BSI, 2007a) (**Table 47.4**) and BS EN ISO 22476-9 (BSI, 2020). BS EN ISO 22476-9 provides guidance on suitable dimensions based on experience.

BS EN ISO 22476-9 (BSI, 2020) recognises several different configurations of equipment and also assigns a level of accuracy

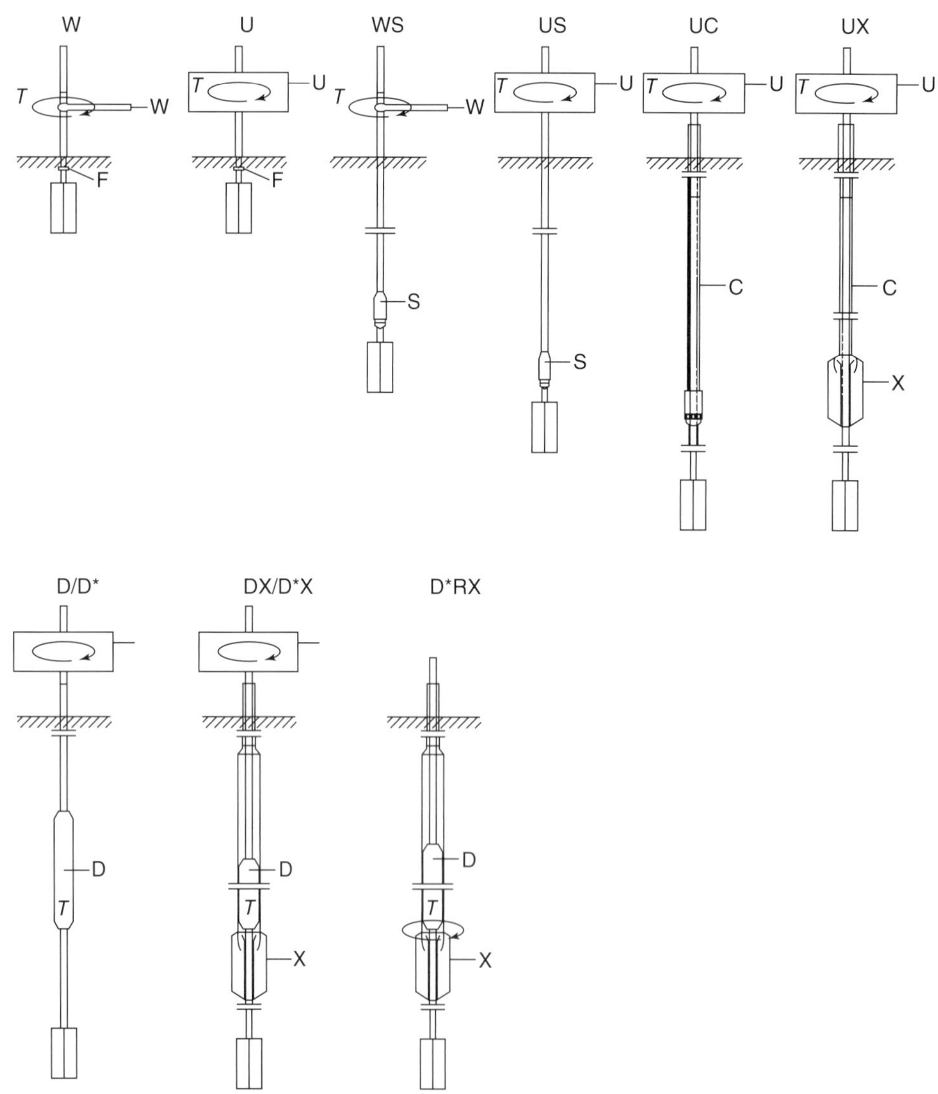

Figure 47.8 Field vane testing equipment (fig 3 of BS EN ISO 22476-9)

<table>
<tr><td rowspan="2">Application</td><td rowspan="2">Confidence level</td><td colspan="4">Cone penetrometer class</td></tr>
<tr><td>0</td><td>1</td><td>2</td><td>3</td></tr>
<tr><td rowspan="3">Characterisations of geotechnical proprieties of soil depoits with $q_{c;max} \leq 1$ MPa</td><td>High</td><td>A</td><td></td><td></td><td></td></tr>
<tr><td>Medium</td><td colspan="2">B</td><td></td><td></td></tr>
<tr><td>Low</td><td colspan="3">C</td><td></td></tr>
<tr><td rowspan="3">Characterisations of geotechnical proprieties of soil depoits with 1 MPa < $q_{c;max} \leq 3$ MPa</td><td>High</td><td colspan="2">B</td><td></td><td></td></tr>
<tr><td>Medium</td><td colspan="3">C</td><td></td></tr>
<tr><td>Low</td><td colspan="4">D</td></tr>
<tr><td rowspan="2">Characterisations of geotechnical proprieties of soil depoits with $q_{c;max} \leq 3$ MPa</td><td>High</td><td rowspan="2">Not recommended</td><td colspan="2">B and C</td><td></td></tr>
<tr><td>Medium</td><td colspan="3">D</td></tr>
<tr><td colspan="6">Note: A,B,C and D are the test categories according to Table 3.</td></tr>
</table>

Table 47.4 Confidence levels of measurements for the characterisation of geotechnical properties depending on cone type and test category (table A.1 in BS EN 22476-1)

of measurements to these which take into account potential errors or inaccuracies due to the equipment type and measuring system.

By implication, BS 1377 considered that the field vane would not be suitable for testing soils with undrained strengths greater than about 75 kPa. However, BS EN ISO 22476-9 (BSI, 2020) now recognises that the vane test can be used in stiffer soils – see **Table 47.5**.

The test is generally not suitable for fibrous peats, sands or gravels, or in clays containing laminations of silt or sand, or stones.

Four types of vane are in use around the world. In the first, the vane is pushed unprotected from the bottom of a borehole or from the ground surface. In the second, a vane housing is used to protect the vane during penetration, and the vane is then pushed ahead of the bottom of the vane housing before the test is started. In the third, the vane rods are sleeved to minimise friction between the ground and the rods during the test. Finally, some vanes incorporate a swivel just above the blades, which allows about 90° of rod rotation before the vane is engaged. This simple device allows the measurement of rod friction (which can be significant when the vane is pushed from ground level) as an integral part of the test.

In all tests it is important that the vane is pushed ahead of any disturbance caused either by the vane housing or any boring operations. BS EN ISO 22476-9 specifies that the vane should be pushed five vane housing diameters ahead of the vane housing before testing, and that when a borehole is used to get down to the test depth the vane should be advanced at least five borehole diameters ahead of the bottom of the borehole. During this process it is important that vane rotation follows the relevant standard.

Once the vane has been pushed to the required test depth, the test procedure is as follows.

In its simplest form attach a torque wrench, or preferably a purpose-built geared drive unit, to the top of the vane rods. Down-hole systems are motor-driven and turn the rods at a steady but continuous rate. This should be fast enough to ensure the test is undrained but slow enough to ensure readings can be taken and potential rate effects are not significant. A rate of rotation of 6–12° min is typically used. The standard suggests failure should occur in about 2 minutes. During this phase record the relationship between rod rotation and measured torque by taking readings of both at intervals of 15–30 s.

- Once maximum torque is clearly achieved, or in the case of BS EN ISO 22476-9 a rotation of 100°, or a set number of degrees, is reached.
- If residual strength is required then the vane is then rotated rapidly through a minimum of 10 revolutions. The shearing is then restarted at the previous slow rate, to determine the remoulded strength of the soil.

A torque–rotation curve is obtained from the measurements made during the test stages, but the reliability of this will depend on the accuracy of the rotation measurements (or assumed rotation); BS EN ISO 22476-9 gives the requirements for the accuracy of these measurements.

BS EN ISO 22476-9 should be consulted before specifying and undertaking vane tests.

The vane test is routinely used only to obtain the 'undisturbed' peak undrained shear strength (based on the maximum or peak torque), and the remoulded undrained shear strength (using the torque in the retest) of soft soils, and thereby to give an assessment of a soil's sensitivity. BS EN ISO 22476-9 (BSI, 2020) also defines a 'residual' undrained shear strength derived from the torque at 180° rotation in the initial phase.

Test Configuration	Recommended maximum insertion length	Confidence level		
		1 c_{fr} < 100 kPa	c_{fv} > 100 kPa	$c_{fv\text{-}f}$
D*RX DX/D*X D/D*	>30 m	High	Medium	Not applicable
UX UC	30 m	High	Medium	Not applicable
US	15 m	Medium	Low	Not applicable
U	0.5 m	High	Medium	Low
WS W	3 m 0.5 m	Medium	Medium	Low

Table 47.5 Application of field vane test configurations (table 4 of BS EN ISO 22476-9)

The test results should be reported as 'vane' shear strengths. BS EN 1997-2 (BSI, 2007b) and BS EN ISO 22476-9 (BSI, 2020) use c_{fv} for the 'undisturbed' shear strength, c_{pv} for a post-peak value and c_{rv} for the remoulded value. In BS EN ISO 22476-9 (BSI, 2020) the sensitivity is c_{fv}/c_{rv}. In interpretation it is assumed, among other things, that the penetration of the vane causes negligible disturbance, that no drainage occurs during shear, and that the soil fails on a cylindrical shear surface, whose diameter is equal to the width of the vane blades.

The results of a vane shear test may be influenced by many factors, namely

- type of soil, especially when a permeable fabric exists or stones are present
- strength anisotropy
- disturbance due to insertion of the vane
- rate of rotation (strain rate)
- time lapse between insertion of the vane and the beginning of the test
- progressive failure of the soil around the vane.

It is necessary to carry out the vane test rapidly, in an attempt to ensure that the shear surface remains reasonably undrained. The presence of sand or silt lenses or laminations within the test section will certainly make this assumption invalid, but (again in common with many in situ tests) it is not normally possible to know what type of material is about to be tested. Free-draining materials should not be tested. In higher-plasticity clays, the shear strength is likely to be increased by viscous effects, and there is evidence (Parry and McLeod, 1967) that unreasonably high values can be obtained.

The presence of stones or fibrous peat will mean that the assumption of a cylindrical shear surface with a diameter equal to the vane blade width is invalid. Again, the shear strength will tend to be overestimated.

Perhaps the most serious problem can result from the disturbance induced in the ground by the insertion of the vane blades. La Rochelle *et al.* (1973) reported that thicker vane blades result in lower undrained shear strength values because of greater soil disturbance and also because of the induced increase in pore water pressure in the soil surrounding the vane. In a typical test, a torque is applied shortly after insertion, and this pressure does not have time to dissipate. Hence, the time interval between the moment of vane intrusion and the time of failure is also of importance in influencing measured strength. BS 1377-9 (BSI, 2007a) specifies that the area ratio (the volume of soil displaced, divided by the volume of soil within the assumed cylindrical shear surface), which is given by

$$A_r = \frac{\left[8t(D-d) + \pi d^2\right]}{\pi d^2} \tag{47.4}$$

where t is the vane blade thickness, D is the vane diameter and d is the diameter of the vane rod, below any sleeve and including any enlargements due to welded joints, shall not exceed 12%. The disturbance caused by vane insertion is a matter for concern, and engineers should ensure that the dimensions of the vanes they use are recorded.

A torque–rotation curve is generally obtained from the measurements of rotation and torque, but the reliability of this will depend on the accuracy of the rotation measurements (or assumed rotation); BS EN ISO 22476-9 (BSI, 2020) gives the requirements for the accuracy of these measurements.

It is usual to correct 'vane' undrained shear strengths before using them in design, and several examples are given as guidance in BS EN 1997-2 Annex I (BSI, 2007b). The applicability of these corrections must always be checked for the ground conditions under investigation.

47.3.2 Plate testing

Techniques for carrying out the plate loading test have been described in ASTM D1194-72 (ASTM, 1972), BS EN ISO 22476-13 (BSI, in preparation), BS 5930 (BSI, 2015) and BS 1377-9 (BSI, 2007a). BS EN ISO 22476-9 (BSI, 2020) has replaced BS 1377-9. In the most common form of the test a relatively rigid metal plate (minimum diameter 300 mm) is bedded onto the soil to be tested, either using sand/cement mortar or plaster of Paris. A load is applied to the plate in successive increments of about one-fifth of the design loading, and held until the rate of settlement reduces to an acceptable level (**Figure 47.9**).

BS 1377-9 (BSI, 2007a) suggests that 'the load should preferably be maintained at each increment until the penetration of the plate has ceased. Tests on cohesive (fine-grained) soils should be continued at least until all the primary consolidation is complete, judged according to the settlement versus log time plot.' Clayton *et al.* (1995) suggest that each load be held until the rate of settlement is less than 0.004 mm/mm, measured for a period of at least 60 minutes, and that load increments are applied either until

- shear failure of the soil occurs or, more commonly
- the plate pressure reaches two or three times the design bearing pressure proposed for the full-scale foundation.

The load is usually applied to the plate using a calibrated hydraulic load cell and a hydraulic jack. The hydraulic jack may either bear against beams supporting concrete blocks or (preferably) a reaction may be provided by tension piles or ground anchors installed on each side of the load position. For lighter loads, a reaction against a suitable piece of plant is also possible.

At each load increment a note is made of the load on the plate and dial gauge readings are made on a 'square of the integer' basis (i.e. 1, 4, 9, 16, 25 mm, etc.) after load application. This will ensure sufficient readings in the early stages of each load application when movement occurs most rapidly.

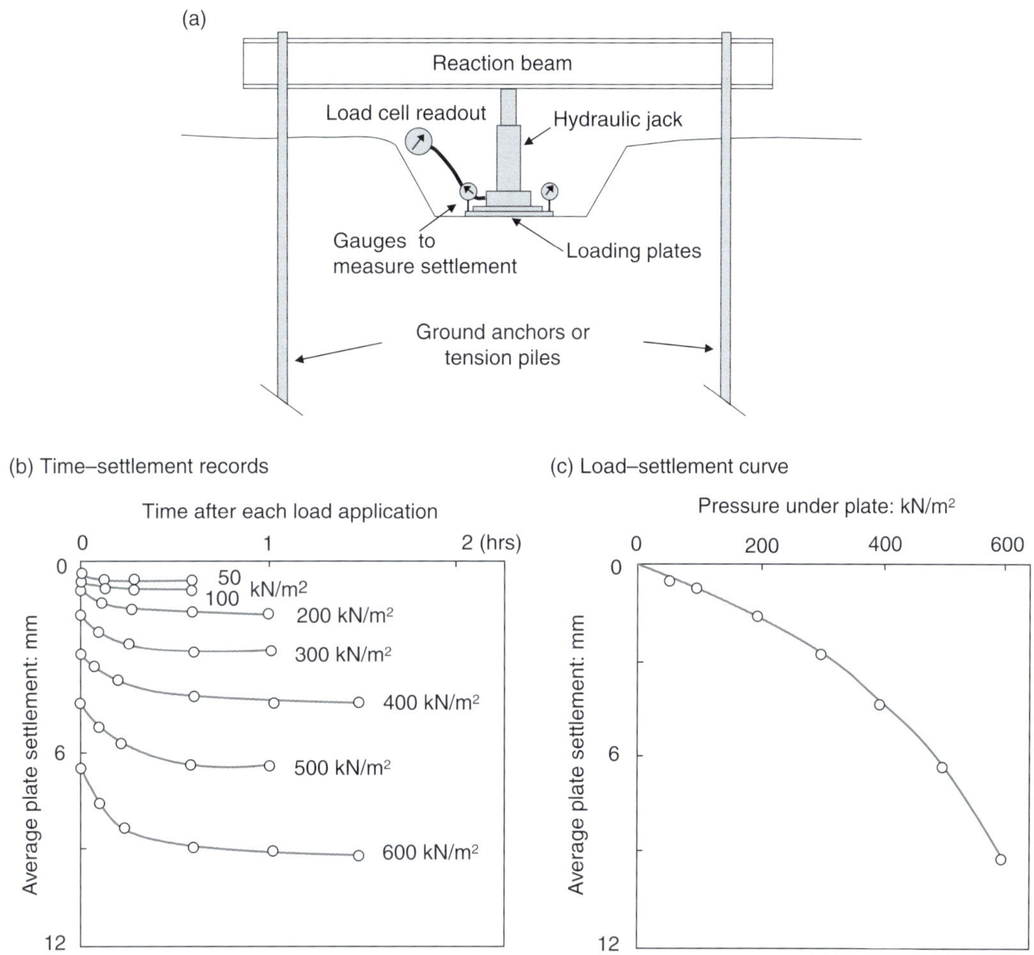

Figure 47.9 Plate load test equipment and results. (b,c reproduced from Clayton *et al.*, 1995; Wiley-Blackwell)

The results of these measurements are normally plotted in two forms: a time–settlement curve for each applied load and a load–settlement curve for the entire test.

The number of tests required depends on both the soil variability and the consequences of poor data for geotechnical design. Tests should not normally be carried out in groups of less than three, and in order to allow assessments of variability any plate testing should be carried out at the end of a site investigation, or as part of a supplementary investigation.

The size and location of plate tests should be assessed on the basis of in situ testing and visual examination of the soil or rock to be investigated. As a general rule of thumb, the plate diameter should never be less than either six times the maximum soil particle size or six times the maximum intact rock block size. The use of this rule of thumb ensures that enough discontinuities or inter-particle contacts exist in the stressed zone to give representative results, but it does not aid in extrapolating results when tests are carried out only at the proposed foundation level.

It is quite clear from the literature that the extrapolation of settlement from small plates to large loaded areas on granular soils is rather unreliable, and therefore the plate loading test on granular material should be regarded as giving a modulus of compressibility value for the soil immediately beneath the test location. Elastic stress distributions indicate that the soil will only be significantly stressed to a depth below the plate of about 1.0–1.5 times the width of a square or circular loaded area.

Plate tests on rocks appear to present a rather more attractive proposition, partly because reliable methods of predicting the settlement on rocks are almost non-existent, and also because the sorts of structure for which good estimates of the settlement on rock are required will normally justify the high expenditure necessary. Most civil engineering structures will be founded in the upper, more weathered rock zones. Ward *et al.* (1968) and Hobbs (1975) have shown that in these zones it is the compressibility of the discontinuities, and not the intact rock, which controls the compressibility of the rock in the mass. The compressibility of joints and bedding planes can be assessed visually, based on experience, but actual values under working loads can only be obtained satisfactorily from a (sufficiently large) in situ loading test.

A logical approach to the problem is to use a plate of sufficient size to determine a modulus value, and either to carry out tests at different levels, or to correlate tests on different materials with their weathering grades. However, the cost of large borehole plate tests is normally prohibitive. For plate tests intended to give elastic moduli values for soils or rocks, BS EN 1997-2 Annex K.2 (BSI, 2007b) uses the equation for a uniformly loaded rigid plate on a semi-infinite elastic isotropic solid

$$E = \frac{\pi q B}{4} \frac{(1-\nu^2)}{\rho} \quad (47.5)$$

where E is the elastic modulus, q is the applied pressure between the plate and the soil, B is the plate width, ρ is the settlement under the applied pressure q and ν is Poisson's ratio.

For granular soils and soft rocks, Poisson's ratio will normally be between 0.1 and 0.3, and so the term $(1 - \nu^2)$ has a relatively small effect. Where plate tests are carried out in the stressed zone of a proposed foundation the value of q can be taken as the vertical foundation stress to be applied at the level of the plate test, or alternatively, a safety margin can be incorporated by taking q to be 50% (for example) higher than the estimated applied stress.

Where plate tests are intended to give values for the shear strength or bearing capacity in fine-grained soils, the load is not applied in stages. The plate is pushed downwards to give a constant rate of penetration, and the undrained shear strength is deduced from the ultimate bearing capacity; see BS EN 1997-2 Annex F (BSI, 2007b).

Two enhancements have sometimes been used in conjunction with larger, more expensive, plate tests, such as are used for major investigations (reactor foundations or underground caverns) in weak rocks.

- Multi-point borehole extensometers may be placed under the plate in order to allow the determination of strain levels at various distances away from the loading (e.g. see Barla *et al.*, 1993; Marsland and Eason, 1973). Stress changes at the measuring points must be determined from elastic theory, even though this may be rather unreliable.
- An oil-filled pad (similar to a flat jack) may be placed between the plate and the rock in an attempt to remove the concentration of stresses at the plate edge due to its rigidity. When this is done, the estimates of stress change are improved, but an interpretation of surface movements should be made on the basis of the settlement of a fully flexible loaded area on an elastic half-space.

Smaller plate loading tests are routinely carried out downhole in some countries during investigations that rely upon correlation with a visual description. This combination has proved particularly valuable above the water table in hard, gravelly or unsaturated and saprolitic soils, all of which can be very difficult to sample and test. The plate test is carried out in a large-diameter hole, which is formed using an auger-piling rig. Another adaptation of the plate test is the 'skip test'. Here, a heavy-duty waste disposal skip is used to simulate the relatively low levels of loading produced, for example, by low-rise housing. This type of test is now the subject of a standard: BS 1377-9 (BSI, 2007a). Settlements are measured using levelling.

47.3.3 Marchetti dilatometer test (DMT)

The Marchetti dilatometer test (DMT) (BS EN ISO 22476-11; BSI, 2017), BS EN 1997-2 (BSI, 2007b), ASTM D6635 (ASTM, 2007) is a versatile tool for in situ ground testing. It is relatively quick and robust, can generate profiles of information and the results can be interpreted for a wide range of geotechnical parameters or used directly for design. Despitc this, it is relatively rarely used in the UK, and for that reason will only be discussed briefly.

The blade (**Figure 47.10**) is pushed vertically into the ground using rods. It is 250 mm long, 94 mm wide and 14 mm thick, with a tip angle of 16°. It has a flat, circular steel membrane mounted flush on one side. The blade is normally jacked into the ground using available field equipment such as the rigs normally used for the cone penetration test (CPT), or drill rigs. The general layout of the dilatometer test is shown in

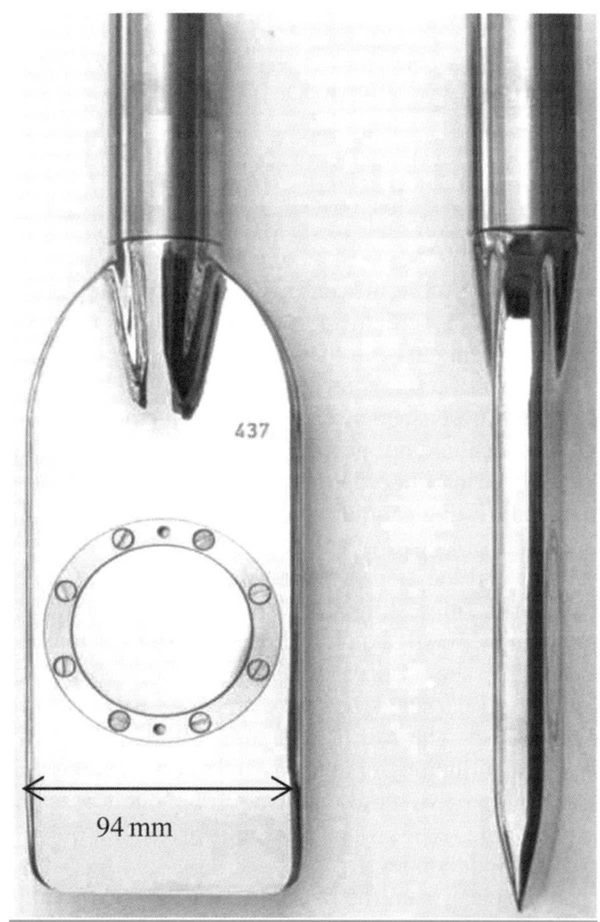

Figure 47.10 Marchetti dilatometer blade

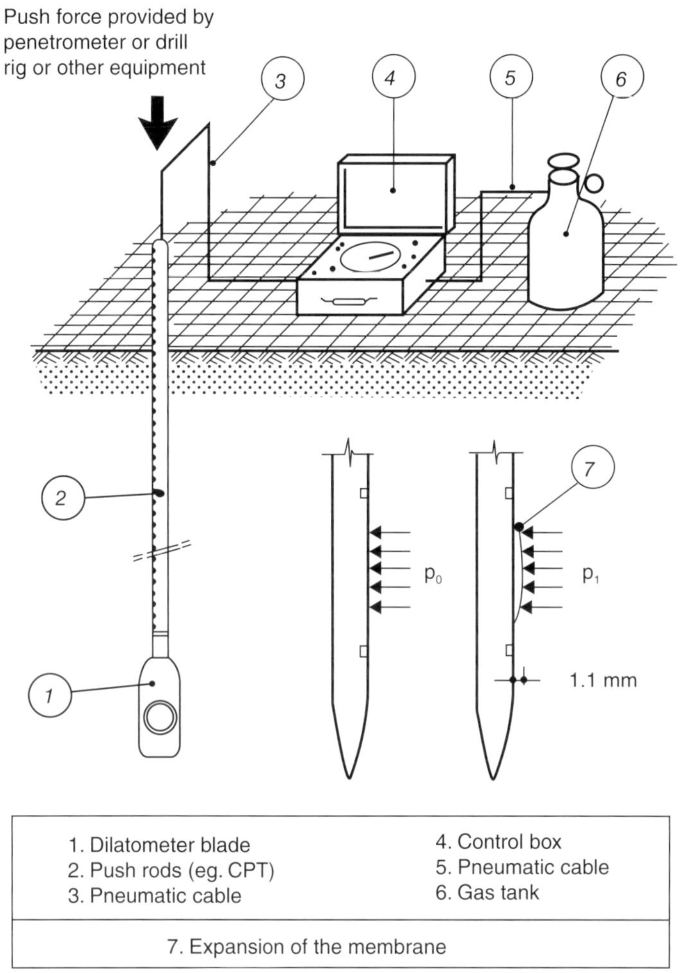

Figure 47.11 Layout of test equipment for DMT testing

Figure 47.11. At regular intervals (typically every 0.2 m) penetration is halted and a test is performed by inflating the membrane by gas pressure.

The blade is connected to a control unit at the ground surface by a tube that transmits both the gas pressure and an electrical signal, running through the push rods. After penetrating to a test depth, the operator inflates the membrane using the control unit, and takes, in about 1 minute, two or three readings.

- The pressure required to just move the membrane against the soil; this 'lift-off' pressure, denoted the A pressure, is identified by the cessation of an audible signal.
- The pressure required to move the centre of the membrane 1.1 mm into the soil; this B pressure is identified by the audible signal restarting.
- Optionally, a third 'closing pressure', denoted the C pressure, can be taken by slowly deflating the membrane soon after B is reached until the membrane returns to the 'lift-off' position; this pressure is identified by the audible signal starting again after deflation.

The pressure readings A, B and C are corrected by calibration to take into account the membrane stiffness. They are then converted into pressures termed p_0, p_1 and p_2. Using p_0, p_1 and p_2, the in situ pore pressure and the vertical stress, three 'intermediate dilatometer parameters' are derived, namely

- the material index, I_D
- the horizontal stress index, K_D
- the dilatometer modulus, E_D.

These 'dilatometer' parameters form the basis of an interpretation through a wide range of correlations to geotechnical properties such as soil type, shear strength, overconsolidation ratio, stiffness, density and more. **Figure 47.12** shows a typical profile of results in a layered sand and clay deposit. The DMT is suitable for use in sands, silts and clays, where the grains are small compared to the membrane diameter (60 mm), with a very wide range of strengths, from extremely soft clay to a hard soil or soft rock. It is not suitable for gravels, although the blade is robust enough to pass through gravel layers of no more than about 0.5 m thickness. The DMT has also been quoted as being used for a variety of direct applications such as the settlement of shallow foundations in clays and sands, the axial capacity of piles, the lateral behaviour of piles, compaction control, the liquefaction of sands and the detection of slip surfaces.

Further information can be found in Marchetti *et al.* (2001) and experience in UK soils is described in Powell and Uglow (1988). More information is available in the in situ test conferences listed in the further reading section. As with the interpretation of any in situ test, it is important to ensure that the source of the correlations used is fully appreciated and applicable to the ground conditions under investigation.

47.3.4 Pressuremeters and dilatometers

Although a number of different designs and test methods exist, all pressuremeters consist of a cylindrical cell, which is installed vertically and applies uniform radial pressure to the ground through a flexible membrane, connected by cabling and tubing to a control and measurement system at the ground surface. Higher-pressure devices of this type, designed for use in hard soils or rocks, are sometimes referred to as 'dilatometers'. The aim of a pressuremeter test is to obtain the stiffness, and in weaker materials the strength, of the ground, by measuring the relationship between the applied radial pressure and the resulting deformation. BS 5930 (BSI, 2015) and BS EN 1997-2 (BSI, 2007b) give some guidance.

Three principal types of pressuremeter are in use.

- *Borehole pressuremeter*: A borehole is formed using any conventional type of drilling rig capable of producing a smooth-sided test cavity. The pressuremeter has a slightly smaller outer diameter than the diameter of the hole, and can therefore be lowered to the test position before being inflated. Borehole pressuremeters

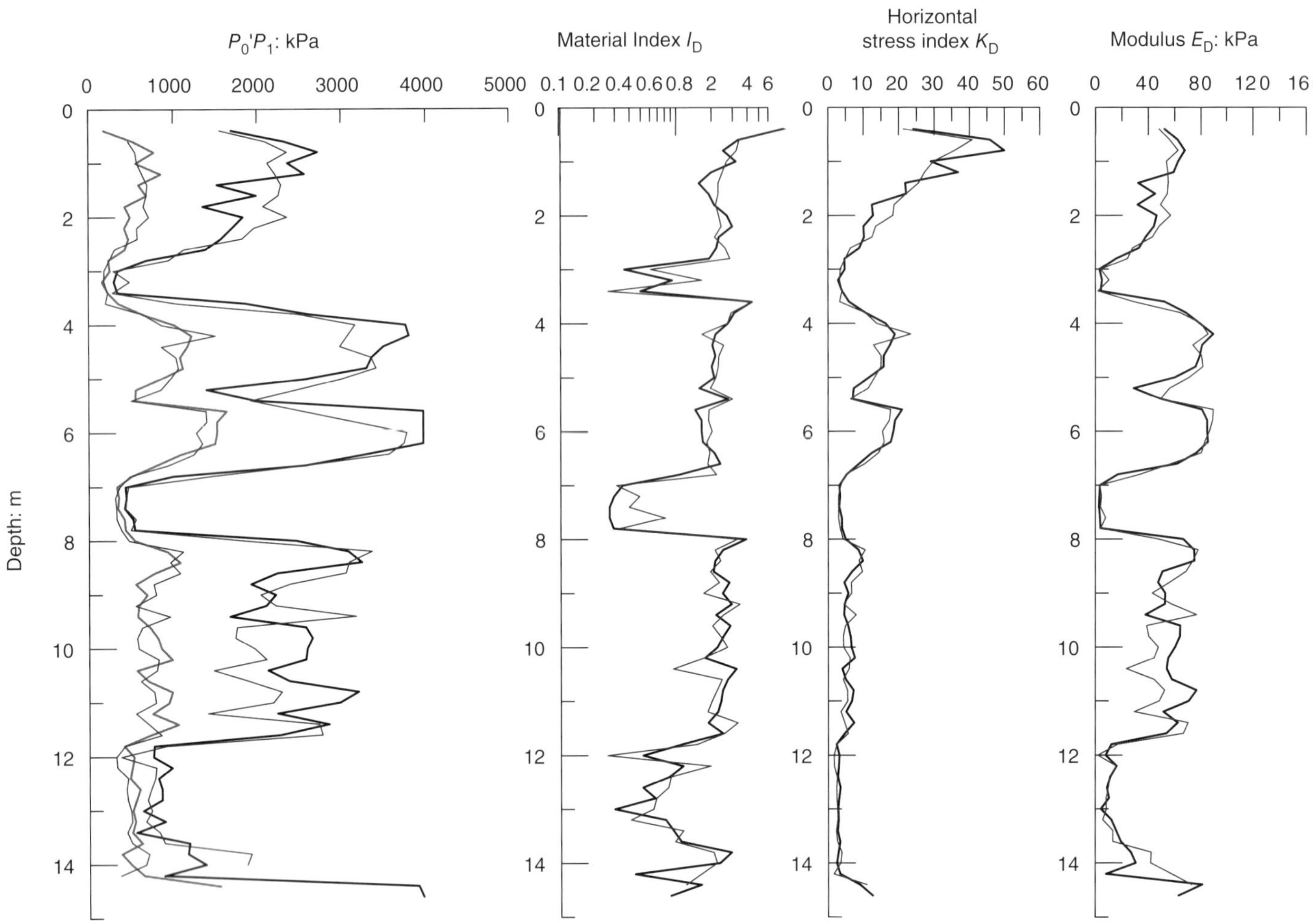

Figure 47.12 Typical profile of DMT results in layered sand and clay

have been widely used around the world, but they are not in common use in the UK. A typical test arrangement for a borehole pressuremeter test is shown in **Figure 47.13** and is covered by BS EN ISO 22476-5 (BSI, 2023). The most commonly used version of this type worldwide is the Ménard pressuremeter (MPM), covered by BS EN ISO 22476-4 (BSI, 2021).

- *Self-boring pressuremeter (SBP):* Borehole disturbance can have a very great effect on the soil properties determined from in situ testing, as already noted in the case of SPT. An SBP incorporates an internal cutting mechanism at its base; the probe is pushed hydraulically from the surface, while the cutter is rotated and supplied with flush fluid. The soil cuttings are flushed to the ground surface through the hollow centre of the probe, as the pressuremeter advances. SBPs have been widely used in the UK for high-value investigations in stiff clays and are covered by BS EN ISO 22476-6 (BSI, 2018a).
- *Full-displacement pressuremeter (FDP):* Displacement pressuremeters are pushed into place and have, to date, been used only rarely in conventional, onshore site investigations, but are seeing increased interest. The push-in pressuremeter (PIP) was primarily intended for offshore investigations, where it was used with wireline drilling equipment. The cone-pressuremeter (CPM) is a full-displacement device mounted above a CPT and is seeing increased interest. FDPs are covered by BS EN ISO 22476-8 (BSI, 2018b).

Because of the wide variety of this type of device, and their sophistication, a detailed description of pressuremeters is beyond the scope of this chapter (see Clarke, 1995 for more details).

A pressuremeter is inserted to a predetermined depth and then expanded. Expansion is either pressure-controlled (most common) or strain/expansion-controlled. Pressure-controlled tests expand the membrane using increments of pressure each of which is held for a short period while the expansion is recorded. Pressure increments continue until either the maximum system pressure or the maximum expansion of the membrane has been reached. In expansion-controlled tests the rate of radial expansion is controlled and these tests therefore generally require more sophisticated systems (SBPs are generally of this type as are many of the full-displacement systems). At the completion of a test, a plot of applied pressure against radial expansion can be generated, examples of which are shown in **Figure 47.14**.

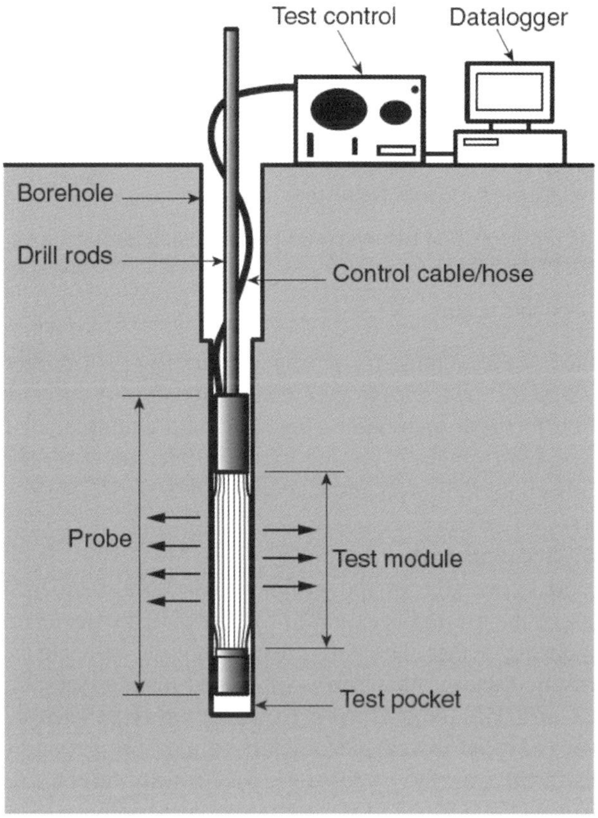

Figure 47.13 Borehole pressuremeter equipment

The three types of device generate very different relationships between the pressure and radial strain, although these are essentially different parts of the same pressure–expansion curve.

The principal differences between the three classes of pressuremeter described above lie in the stresses applied to the probe at the start of the test. Borehole pressuremeters start from a horizontal total stress level close to or equal to zero. SBPs should start the test at approximately the horizontal total stress level in the ground before insertion. Displacement pressuremeters (because they push soil aside during installation) start with a horizontal total stress, which is usually much greater than originally existed in the ground. The increase in horizontal total stress applied during a test is intended to take the soil to failure, although in rock this may not be achievable because of limited system pressure. Unload/reload iterations can be performed during any of the tests.

The interpretation of pressuremeter test data may be semi-empirical (typically the case for borehole pressuremeters and Ménard tests) or based on first principles. (Because the boundary conditions are well defined, a more rigorous theoretical analysis than other in situ tests can be performed, based on the expansion of a cylindrical cavity. There are many methods for interpreting these tests and they can even derive full stress–strain curves of the material from the test results.) But essentially, all types of pressuremeter give data that is normally used to calculate, or estimate

- the in situ horizontal total stress

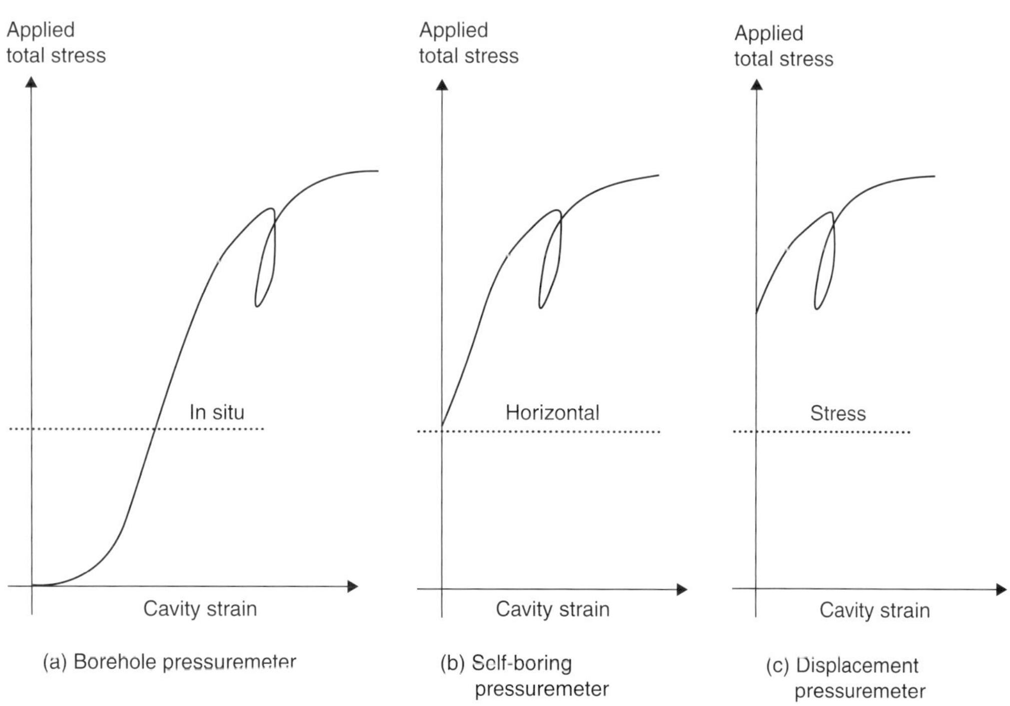

Figure 47.14 Applied pressure/radial expansion curves for three types of pressuremeter (Reproduced from Clayton *et al.*, 1995; Wiley-Blackwell)

- the stiffness, normally the horizontal Young's modulus, or shear modulus
- the strength (where maximum system pressure and volumetric strain permit)
- the drained parameters in sands (although more difficult than testing in clays).

All three types of device are useful and should be considered with regards to cost, reliability and consistency of the data generated.

SBP tests require significant experience if they are to be used successfully. Potentially they can give the best results but are expensive. MPM and pre-bored tests are simpler but require 'consistent' drilling and installation. Softening can occur around the borehole walls. FDP tests can generally be installed with 'repeatable' disturbance but they need further development for some areas of interpretation.

In principle, SBPs should provide the best estimate of the in situ horizontal stress and shear modulus, provided (as should be the case) that installation disturbance is controlled. Borehole pressuremeters and dilatometers are the most rugged of the devices, but drilling disturbance and bedding between the cell and the borehole wall make the interpretation of in situ stress rather more uncertain, and this can lead to significant underestimates of stiffness. A displacement pressuremeter imposes very significant strains on the soil, and may destructure it; however, the disturbance is generally 'repeatable'. Experience has shown that, when installed well, in clays the strength derived from all three types is similar and, provided the soil is not cemented, very similar results can be obtained for the stiffness from unload/reload iterations. The pre-bored devices cannot measure as small a strain level as the other devices. When interpretation methods exist, pre-bored and displacement pressuremeters can give reliable indications of the in situ horizontal stress (see **Table 47.1**).

The Ménard pressuremeter test generally relies on determining parameters specific to the test method – see BS EN ISO 22476-4 (BSI, 2021), and these are then used in 'rules' during design – see BS EN 1997-2 Annex E (BSI, 2007b). As with all these tests, the 'rules' should be validated for the soil types under investigation.

More information on the interpretation of pressuremeter tests both for geotechnical parameters and their use in direct design can be found in the further reading section.

47.3.5 Field geophysics

Field measurement of the seismic shear wave velocity can be an effective way to determine the shear modulus of made ground, natural ground and fractured rocks, since the relationship between the shear wave velocity and the very small strain shear modulus is a simple one

$$G_o = \rho V_s^2 \tag{47.6}$$

where G_o is the shear modulus at very small strain levels, V_s is the measured shear wave velocity and ρ is the bulk density of the ground.

Suitable methods for determining the required values include

- continuous surface wave testing
- down-hole testing (which could be, in suitable ground, a version of CPT or DMT)
- cross-hole testing.

Surface wave testing is discussed briefly in Chapter 45 *Geophysical exploration and remote sensing*. A brief introduction to these techniques, together with examples of their use and interpretation, can be found in Clayton *et al.* (1995), McDowell *et al.* (2002) and Clayton (2011).

47.4. Groundwater testing

A good knowledge of the groundwater regime and permeability of the ground is essential for any realistic geotechnical engineering design. The pore pressure regime on a site can be determined using other forms of in situ testing (see Section 47.2.3 on CPT, for example). However, on most sites it will be necessary to install piezometers to measure groundwater levels in the soil. On rocky sites, packer tests can be used to determine the permeability and its spatial variability.

The permeability of a granular, layered, anisotropic or fissured soil is unlikely to be obtained with sufficient accuracy from laboratory tests on specimens from normal-diameter boreholes, and therefore in situ permeability tests are in widespread use. They can be carried out in soils or rocks, in open boreholes, using piezometers, or in sections of drill hole sealed by inflatable packers. Two common types of test are

- rising and falling head tests
- packer or Lugeon tests.

47.4.1 Rising and falling head permeability tests

The rising or falling head test is generally used in relatively permeable soils, such as sands and gravels – see BS 5930 (BSI, 2015) and BS EN ISO 22282-2 (BSI, 2012a). It is usually carried out in a cased borehole or a simple piezometer set in a sealed borehole section, such as a Casagrande piezometer. Where the groundwater level is below the bottom of the borehole or piezometer, the test is started by adding water from a bowser. Where the groundwater level is above the base of the borehole, it is preferable to start the test by bailing water out, since this will reduce the effect of any smear caused by boring. Water level measurements are then taken at suitable time intervals until the water level returns to equilibrium (see **Figure 47.15**). The data are interpreted using Hvorslev's basic time lag method (**Figure 47.15(c)**). This requires that the equilibrium water level is known, so that H_0, the driving head at the start of the test, can be calculated. If

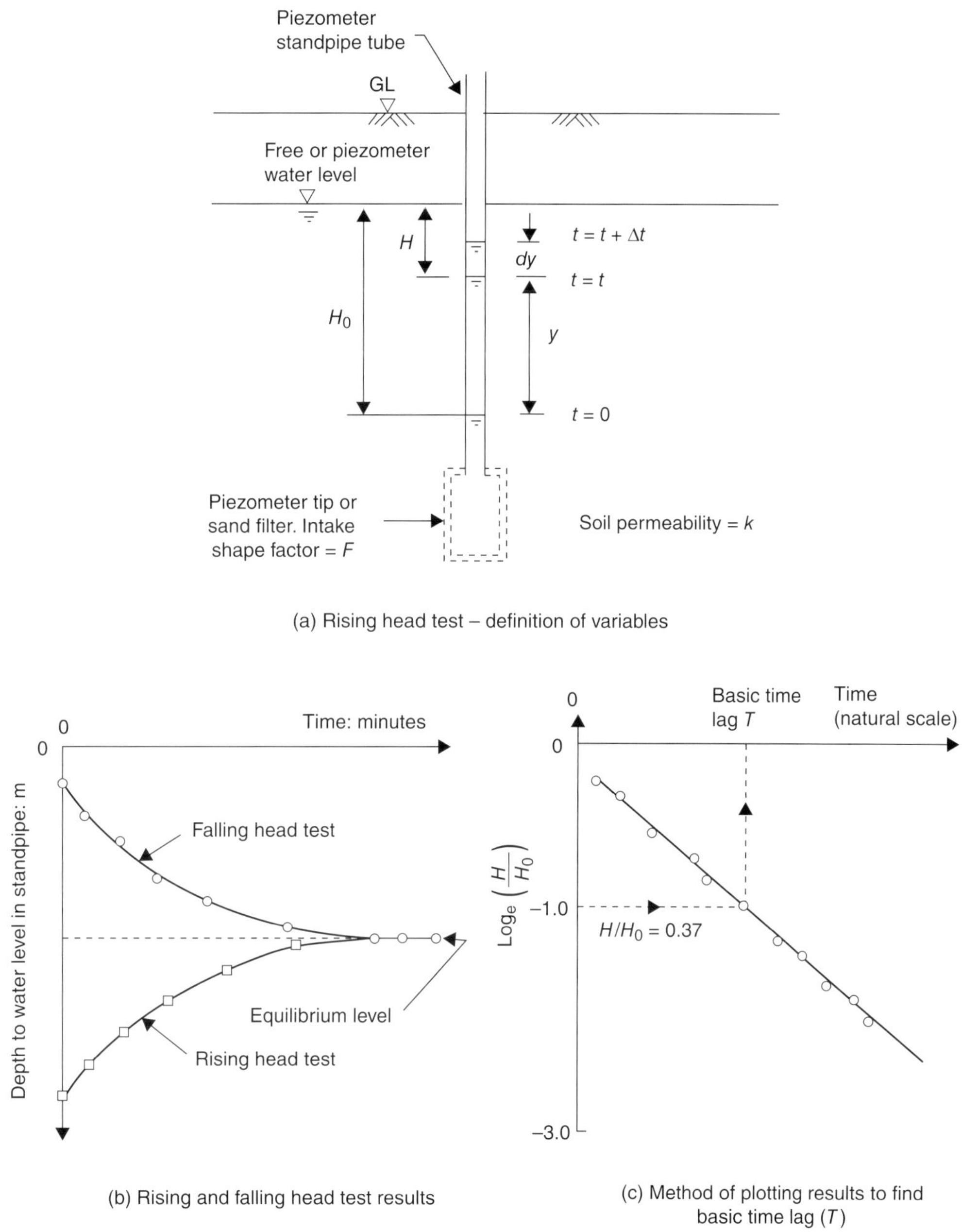

Figure 47.15 Rising and falling head permeability tests (Reproduced from Clayton *et al.*, 1995; Wiley-Blackwell)

the equilibrium value is unknown it can be found by trial and error, since for the correct value a plot of $\log_e (H/H_0)$ against (linear) time will yield a straight line. When the time equals the basic lag then

$$T = \frac{A}{Fk} \text{ so that } k = \frac{A}{FT} \tag{47.7}$$

where k is the permeability of the soil, T is the basic time lag and F is the shape factor of the test section (normally a pocket of sand).

For a cylindrical piezometer or standpipe sand pocket, or a cased borehole, of length L and diameter D

$$F = \frac{2\pi L}{\log\left[\frac{L}{D} + \sqrt{\left(1 + \left(\frac{L}{D}\right)^2\right)}\right]}. \tag{47.8}$$

If $\log_e (H/H_0)$ is plotted as a function of time, the value of the basic time lag can be found from the straight line at $\log_e (H/H_0) = -1.0$.

The test interpretation assumes that the soil will not swell or consolidate, that the geometry of the test pocket is known, that it contains material that is very much more permeable than the in situ soil under test, that there is no smear on the soil wall and that other test errors, such as those due to air in the soil or pipes, do not occur.

The time necessary to conduct a rising or falling head test increases dramatically as soil permeability reduces. Test times are prohibitively long in clayey silts and silty clays. At the same time, the assumption that there is no contribution from swelling or consolidation to the measured rate of flow becomes unreasonable. Laboratory testing can be used, providing that undisturbed samples can be taken, and these are large enough to contain representative fabric (e.g. laminations, varves and fissures).

47.4.2 Packer tests

Packer tests, also sometimes known as Lugeon tests, are used to determine the permeability of fractured rock masses – see BS 5930 (BSI, 2015) and BS EN ISO 22282-3 and BS EN ISO 22282-6 (BSI, 2012b, 2012c). A test may be carried out in the base of a drill hole using a single inflatable packer to seal off the test section, or after the hole is complete testing may be carried out at a variety of depths using a double packer to seal the top and bottom of the test section.

The construction of a packer is critical if leakage is to be avoided, and the greater the length of a packer the more effective the test. A test is carried out by lowering the packer or packers to the required depth and inflating them using gas pressure supplied from a nitrogen bottle. When expanded, the length of each packer should be at least five times the borehole diameter. The test section is often about 3 m long. Packers are supported on drill rods, which are also used to supply water under pressure to the test section. At the top of the borehole the rods are connected by a water swivel to a system for measuring pressure and volume take, and a water pump.

Testing is carried out in constant pressure stages, which cycle up to a maximum pressure and then down again. A maximum pressure should be specified to ensure that hydraulic fracture is avoided. A test is normally carried out using stages such as 1/3, 2/3, 1, 2/3 and 1/3 of the maximum allowable gauge pressure. At each stage, the pressure is held constant and the volume measured over two or more periods of 5 min. Permeability is calculated from the length and diameter of the test section, the measured volume of flow and the net dynamic head applied to the test section. Alternatively, for deeper test sections, the results can be quoted in terms of 'lugeons', where one lugeon is the water take measured in litres per metre of test section per minute at a pressure of 10 kg/cm^2 (=1000 kN/m^2).

References

ASTM (1972) D1194-72: Standard test method for bearing capacity of soil for static load and spread footings. ASTM, Philadelphia, PA (withdrawn 2003).

ASTM (2007) D6635-01: Standard test method for performing the flat plate dilatometer. ASTM, Philadelphia, PA.

ASTM (2008a). D1586-8a: Standard test method for standard penetration test (SPT) and split-barrel sampling of soils. ASTM, Philadelphia, PA (revision A).

ASTM (2008b). ASTM D2573-08: Standard test method for field vane shear test in cohesive soil. ASTM, Philadelphia, PA.

Barla G, Sharp JC and Rabagliati U (1993) Stress strength and deformability assessment for the design of large storage caverns in a weak Eocene chalk. *Proceedings of the 3rd Conference on 'Meccanica e Ingegneri dell Rocce', Torino, Italy*, pp. 22-1–22-21.

BSI (1999) BS5930:1999: Code of practice for site investigations. BSI, London, UK (now amended).

BSI (2005a) BS EN ISO 22476-2:2005 + A1:2011: Geotechnical investigation and testing – Field testing: Part 2 – Dynamic probing. BSI, London, UK.

BSI (2005b) BS EN ISO 22476-3:2005 + A1:2011: Geotechnical investigation and testing – Field testing: Part 3 – Standard penetration test. BSI, London, UK.

BSI (2007a) BS 1377-9:2007: Methods of test for soils for civil engineering purposes: Part 9 – In-situ tests. Incorporating amendments Nos. 1 and 2. BSI, London, UK. Under review.

BSI (2007b) BS EN 1997-2:2007: Eurocode 7: Geotechnical design – Part 2: Ground investigation and testing. BSI, London, UK.

BSI (2009a) UK NA to BS EN 1997-2:2007: UK National Annex to Eurocode 7: Geotechnical design – Ground investigation and testing. BSI, London, UK.

BSI (2009b) BS EN ISO 22476-12:2009: Geotechnical investigation and testing – Part 12: Mechanical cone penetration test (CPTM). BSI, London, UK.

BSI (2012a) BS EN ISO 22282-2:2012: Geotechnical investigation and testing: Geohydraulic tests – Part 2: Water permeability test in borehole without packer. BSI, London, UK.

BSI (2012b) BS EN ISO 22282-3:2012: Geotechnical investigation and testing: Geohydraulic tests – Part 3: Water pressure test in rock. BSI, London, UK.

BSI (2012c) BS EN ISO 22282-6:2012: Geotechnical investigation and testing. Geohydraulic tests – Part 6: Water permeability tests in a borehole with packer. BSI, London, UK.

BSI (2015) BS 5930:2015 + A1:2020: Code of practice for site investigations. Incorporating amendments Nos. 1 and 2. BSI, London, UK.

BSI (2017) BS EN ISO 22476-11:2017: Geotechnical investigation and testing. Field testing – Part 11: Flat dilatometer test. BSI, London, UK.

BSI (2018a) BS EN ISO 22476-6:2018: Geotechnical investigation and testing – Field testing: Part 6 – Self boring pressuremeter test. BSI, London, UK.

BSI (2018b) BS EN ISO 22476-8:2018: Geotechnical investigation and testing – Field testing: Part 8 – Full displacement pressuremeter test. BSI, London, UK.

BSI (2020) BS EN ISO 22476-9:2020: Geotechnical investigation and testing: Part 9 – Field vane test. BSI, London, UK.

BSI (2021) BS EN ISO 22476-4:2021: Geotechnical investigation and testing. Field testing: Part 4 – Ménard pressuremeter test. BSI, London, UK.

BSI (2023) BS EN ISO 22476-1:2023: Geotechnical investigation and testing: Field testing – Part 1: Electrical cone and piezocone penetration tests. BSI, London, UK.

BSI (in preparation) BS EN ISO 22476-13: Geotechnical investigation and testing – Part 1: Plate loading test. BSI, London, UK.

Butcher AP, McElmeel K and Powell JJM (1995) Dynamic probing and its use in clay soils. *Proceedings of the International Conference on Advances in Site Investigation Practice, London, UK*. ICE, London, UK, pp. 383–395.

Clarke B (1995) *Pressuremeters in Geotechnical Design*. Blackie, London, UK.

Clayton CRI (1995) *The Standard Penetration Test (SPT): Methods and Use*. CIRIA Report 143. CIRIA, London, UK.

Clayton CRI (2011) Stiffness at small strain: research and practice. *Géotechnique* **61(1)**: 5–37.

Clayton CRI, Matthews MC and Simons NE (1995) *Site Investigation*, 2nd edn. Blackwell Science, Oxford, UK.

Hobbs NB (1975) Factors affecting the prediction of settlement of structures on rock: with particular reference to the Chalk and Trias – general report and state-of-the-art review for Session 4. *Proceedings of the Conference on Settlement of Structures, Cambridge, UK*. Pentech Press, London, UK, pp. 579–610.

ISSMFE (1989) *Report of Technical Committee on Penetration Testing of Soils TC 16 With Reference Test Procedures*. Swedish Geotechnical Institute, Linköping, Sweden.

ISSMGE (1999) International reference test procedure for the cone penetration test (CPT) and the cone penetration test with pore pressure (CPTU). Report of the ISSMGE Technical Committee 16 on Ground Property Characterisation from In-Situ Testing. *Proceedings of the 12th European Conference on Soil Mechanics and Geotechnical Engineering, Amsterdam, the Netherlands*, vol. 3, pp. 2195–2222.

La Rochelle P, Roy M and Tavernas F (1973) Field measurements of cohesion in Champlain Clays. *Proceedings of the 8th International Conference on Soil Mechanics and Foundation Engineering*, Moscow, Russia, vol. 1.1, pp. 229–236.

Lunne T, Robertson PK and Powell JJM (1997) *Cone Penetration Testing in Geotechnical Practice*. Spon Press, London, UK.

Marchetti S, Monaco P, Totani G and Calabrese M (2001) *The Flat Dilatometer Test (DMT) in Soil Investigations*. ISSMGE TC16 Report. ISSMGE, London, UK.

Marsland A and Eason BJ (1973) Measurements of the displacements in the ground below loaded plates in deep boreholes. *Proceedings of the BGS Symposium on Field Instrumentation in Geotechnical Engineering*, London, UK, vol. 1, pp. 304–317.

McDowell PW, Barker RD, Butcher AP, *et al.* (2002) *Geophysics in Engineering Investigations*. Geological Society, London, UK.

Parry RHG and McLeod JH (1967) Investigation of slip failure in flood levee at Launceston, Tasmania. *Proceedings of Australia–New Zealand Conference on Soil Mechanics and Foundation Engineering*, Auckland, Australia, pp. 294–300.

Powell JJM and Quarterman RST (1988) The interpretation of cone penetration tests in clays, with particular reference to rate effects. *Proceedings of the International Symposium on Penetration Testing, Orlando, FL*, vol. 2, pp. 903–910.

Powell JJM and Uglow IM (1988) The interpretation of the Marchetti dilatometer test in UK clays. *Proceedings of the Conference on Penetration Testing in the UK*, Birmingham, UK, pp. 269–273.

Reading P, Lovell J, Spires K and Powell JJM (2010) The implications of measurement of energy ratio (Er) for the standard penetration test. *Ground Engineering*, May: pp. 28–31.

Tokimatsu K (1988) Penetration testing for dynamic problems. *Proceedings of the 1st International Symposium on Penetration Testing (ISOPT), Orlando, FL*, pp. 117–136.

Ward WJ, Burland JB and Gallois RW (1968) Geotechnical assessment of a site at Mundford, Norfolk, for a large proton accelerator. *Géotechnique* **18**: 399–431.

Further reading

Briaud J-L (1992) *The Pressuremeter*. Balkema, Rotterdam, the Netherlands.

Clarke BG (1995) *Pressuremeters in Geotechnical Design*. Blackie, London, UK.

Clayton CRI (1995) *The Standard Penetration Test (SPT): Methods and Use*. CIRIA Report R143. CIRIA, London, UK.

Clayton CRI, Matthews MC and Simons NE (1995) *Site Investigation: A Handbook for Engineers*. Blackwell Science, Oxford, UK.

CPT'95 (1995) *Proceedings of the International Symposium on Cone Penetration Testing, Linköping, Sweden*.

CPT'10 (2010) *Proceedings of the Second International Symposium on Cone Penetration Testing, Huntington Beach, CA*.

DMT2 (2006) *Proceedings of the Second International Flat Dilatometer Conference, Washington, DC* (Failmezger RA and Anderson JB (eds)).

ISC-1 (1998) *Geotechnical and Geophysical Site Characterization: Proceedings of the Second International Conference on Site Characterization, Atlanta, GA (*Robertson PK and Mayne PW (eds)). Balkema, Rotterdam, the Netherlands.

ISC-2 (2004) *Geotechnical and Geophysical Site Characterization: Proceedings of the Second International Conference on Site Characterization, Porto, Portugal* (Viana da Fonseca A and Mayne PW (eds)).

ISOPT 1 (1998) *Proceedings of the First International Symposium on Penetration Testing (ISOPT), Orlando, FL*. Balkema, Rotterdam, the Netherlands.

ISP 3 (1990) 'Pressuremeters': *Proceedings of the Third International Symposium on Pressuremeters (ISP 3), London, UK*. Thomas Telford, London, UK.

ISP 4 (1995) *The Pressuremeter and its New Avenues: Proceedings of the Fourth International Symposium on Pressuremeters (ISP 4), Sherbrooke, QC*.

ISP 5 (2005) '50yrs of Pressuremeters': *Proceedings of the Fifth International Symposium on Pressuremeters (ISP-5), Paris, France*.

Lunne T, Robertson PK and Powell JJM (1997) *Cone Penetration Testing in Geotechnical Practice*. Taylor & Francis, London, UK.

Mair RJ and Wood CM (1987) *In-situ Pressuremeter Testing: Methods Testing and Interpretation*. Butterworths, London, UK.

Marchetti S, Monaco P, Totani G and Calabrese M (2001) *The Flat Dilatometer Test (DMT) in Soil Investigations.* ISSMGE TC16 Report. ISSMGE, London, UK.
Meigh AC (1987) *Cone Penetration Testing: Methods and Interpretation.* Butterworths, London, UK.
Powell JJM and Uglow IM (1988) Marchetti dilatometer testing in UK soils. *Proceedings of the 1st International Symposium on Penetration Testing (ISOPT), Orlando, FL*, vol. 1, pp. 555–562.
Schnaid F (2009) *In Situ Testing in Geomechanics: The Main Tests.* Spon Press, Oxford, UK.
Tomlinson MJ (1980) *Foundation Design and Construction.* Pitman Publishing, London, UK.

It is recommended this chapter is read in conjunction with

- Chapter 13 *The ground profile and its genesis*
- Chapter 17 *Strength and deformation behaviour of soils*
- Chapter 18 *Rock behaviour*

All chapters in this book rely on the guidance in Sections 1 *Context* and 2 *Fundamental principles.* A sound knowledge of ground investigation is required for all geotechnical works, as set out in Section 4 *Site investigation.*

doi: 10.1680/icemge.66816.0711

Chapter 48
Geo-environmental testing

Jo Strange Card Geotechnics Ltd, Godalming, UK
Fraser Chamley Arup, London, UK (ex Card Geotechnics Ltd)
Nick Langdon Card Geotechnics Ltd, Aldershot, UK
Cathy Lee (nee Swords) Card Geotechnics Ltd, Aldershot, UK

CONTENTS

Under current UK legislation and planning regulations the assessment of a site has to address potential contamination and associated environmental risks and waste management issues. Typically this will involve a qualitative and/or quantitative risk assessment which requires collection and interpretation of appropriate geo-environmental data and relevant environmental surveys. Data are required relating to contamination sources, including the chemical composition of soils, groundwater and soil gas, in addition to potential receptors which may be impacted. The quality of this data depends on the use of good practice in the selection, collection, handling, storage and analysis of samples. The value of the risk assessment further depends on the processing and interpretation of the data. In the context of a ground investigation, there are significant benefits to combining the collection of geo-environmental data and geotechnical requirements, provided it is done in an appropriate and targeted manner.

48.1. Introduction

Geo-environmental testing commonly forms part of ground investigations tasked with providing information to support both geotechnical design and contamination issues, such as risk assessment and remediation design. The current framework guidance for investigating and assessing potential risks from contamination are set out in the Environment Agency's Land Contamination Risk Management (LCRM) guidance (EA, 2020) and DEFRA's EPA Part 2A Guidance (DEFRA, 2012). Additional guidance for undertaking investigation of potentially contaminated sites is provided in BS 10175 (BSI, 2017), and general guidance on ground investigation is provided in BS 5930 (BSI, 2020a) and Eurocode 7 (BSI, 2007).

Understanding of the ground conditions and contaminant linkages is paramount to the success of an investigation, and geo-environmental testing plays a major part in obtaining data to develop that understanding. The need for geo-environmental testing can be driven by various factors and the scope is usually determined from the initial desk study supplemented by site observation. The desk study should identify where samples are required for testing based on the identified potential chemical hazards in the ground. Obtaining, transporting, handling, preparing and testing of samples must be addressed when planning and undertaking an investigation.

Typically, geo-environmental testing may be undertaken to enable qualitative and quantitative risk assessment of contaminative linkages based on current UK guidance (LCRM). For soil-based contamination this generally involves screening recorded concentrations with assessment criteria, such as SGVs and GACs, derived using the EA's CLEA model (EA, 2009), which models estimated contaminant exposure in relation to HCVs (see the glossary at the end of this chapter for the abbreviations used throughout). Testing may also be required to undertake modelling to derive remediation targets or for classification of material for waste disposal. The suite of analysis applied to samples should be bespoke to the site conditions and the purpose of the testing, albeit likely to be based around generic suites of contaminants to assist the laboratory and minimise costs. However, the validity and value of geo-environmental testing will be questionable if it is not done with full understanding of the reasons behind it, the objectives in undertaking such testing and the factors that influence the quality of data and their assessment.

48.2. Philosophy

The philosophy behind geo-environmental testing is distinctly different from that found in conventional geotechnical testing. First, testing is done with a view to some comparison with criteria set out by national regulators and specialist technical industry groups and not for the purposes of pure investigation of a particular quality, such as shear strength or particle size. Second, the accuracy to which these determinations are made depends on equipment, sample preparation and the laboratory in a way that even the most leading-edge geotechnical testing is not. The testing of blank samples and the routine re-evaluation of results from commercial analytical laboratories is not a process normally seen in geotechnical testing. Third, some criteria set by the regulators, particularly with regard to water standards, cannot be achieved routinely by commercial laboratories. Finally, the testing is regularly specified to prove the absence of a particular contaminant. This is akin to undertaking a plastic limit test on a sand to prove it is a non-plastic material.

Added to this subtle difference in philosophy are the myriad changes in guidance from regulators and the steady improvement in the ability to measure very small concentrations of any particular contaminant. When this is set against a material being tested, usually made ground or a dynamically changing fluid source, which are thoroughly non-homogeneous and potentially unrepeatable, the treating of test results in isolation as 'absolute' is nonsensical.

Once results are obtained they are routinely processed statistically to quantify the risk to receptors – typically, human health, structures, buried concrete and services, controlled waters, ecology – and provide waste classifications based on the soil as a whole rather than assessing individual samples. Once these risks are established, a remediation strategy can be adopted to mitigate or address the risks. Typically, there is a requirement for verification testing against criteria set out in a RIP. The nearest analogy is earthworks control testing, but with tens of separate variables – rather more than are usually tracked in an earthworks contract.

The consequence is that sampling and testing protocols and the impacts of variations need to be clearly understood.

48.3. Sampling

Given the variability within soils sampled for geo-environmental purposes, sampling is not an exact science. However, the aim of sampling is to obtain *appropriate* samples for testing: 'targeted' to assess what is likely to be the worst contamination, or 'representative' for waste classification or verification.

Targeted sampling will be based on background information from the desk study (see Chapter 43 *Preliminary studies*) plus observations of the specific ground conditions including visual and olfactory evidence. Representative sampling may be undertaken at regular intervals within an exploratory hole or soil mass (**Figure 48.1**).

Figure 48.1 Sampling from trial pit

Composite samples are rarely used for geo-environmental purposes, unless to test bulk, basically homogeneous, materials. The inherent human factor involved in sampling decisions cannot be eliminated, but good sampling practice can minimise the scope for error.

48.3.1 Protocols

48.3.1.1 Soils

There are many and varied ways of getting soil samples from the ground. Generally, methods are variations on those techniques used to obtain the samples used for geotechnical purposes, as discussed in Chapter 46 *Ground exploration*. However, these variations may be critical to the validity of the samples for the required testing or to the protection of the environment. Some typical examples of good practice are identified below.

- Drilling of boreholes in certain geological conditions may result in the transfer of contaminants to deeper levels by introducing a pollutant migration pathway. In this case, clean drilling techniques may be required in contaminated soils using dual string casing with appropriate seals.
- It is often necessary to add water to cable percussion boreholes to aid drilling. The quantities of water should be kept to a minimum and only clean water used.
- Lubrication of the drilling tools should be kept to a minimum during drilling and only vegetable oils should be used, if absolutely necessary, and their use recorded.
- In the case of rotary boreholes, drilling fluids need careful consideration and recording.
- During window sampling, plastic core liners can also be used to limit the potential for cross-contamination (**Figure 48.2**).

Figure 48.2 Sampling from open window sampler

- All samples must be taken using clean equipment, including trowels, spades and gloves; where necessary, power washers and detergents (e.g. Decon 90) should be used. Equipment cleaning must be done between holes and samples, if necessary, to minimise cross-contamination.

Typically, geo-environmental samples are small disturbed samples, but collection into a plastic bag is not appropriate. Samples should be compacted into a container to minimise headspace and sufficient sample volume taken to enable the required testing. Typically a 1 kg plastic tub, 250 g amber glass jar and 60 g amber glass jar with PTFE liner are ample, but larger samples may be needed for some testing suites and requirements for individual laboratories vary. Specific sampling methods may also be required for testing for VOCs and PFOS/PFOAs. Detailed guidance is available in 'BS 10176:2020 Taking soil samples for determination of volatile organic compounds' (BSI, 2020b).

Once a material has been identified for sampling, a full description is required, ideally to the current BS 5930, and full records kept of the process. Photographs are also useful for later review. In addition, records are needed of anything that may require the laboratory to take precautions, such as broken glass, possible asbestos or potential biological contamination (details may be found on the Health Protection Agency website). Such samples should be double-bagged, sealed and clearly labelled with reference to the potential hazard.

48.3.1.2 Water

Groundwater (and gas, see Section 48.3.1.4) monitoring wells may be installed in boreholes or window sampler holes, but installations within trial pits should be avoided. Correct borehole installation is critical for monitoring purposes and construction details will be determined by site conditions. Relevant issues are listed below.

- The well size must be appropriate for the required sampling method.
- The response zone should be placed such that it monitors groundwater in the correct position in the aquifer. This will take account of the potential for free product (i.e. hydrocarbon in liquid or non-aqueous phase) and/or tidal/seasonal fluctuations. In the case of free product, the well screen should extend above the water table such that free product may be captured.
- Where more than one aquifer is present (e.g. perched waterbody and a deeper aquifer) the response zone must not cross both water bodies. Additionally, a seal (typically bentonite or similar) must be placed between the two water bodies to limit the potential for cross-contamination.
- Well construction materials should not interfere with the prevailing groundwater. PVC casing must be threaded and no glue or other adhesive used to seal the joints, and the filter material should comprise washed chemically inert granular material.
- Beware of using wells for combined gas and groundwater monitoring, without careful consideration and appropriate design.

On completion of the installation, a full and accurate written log of the well design should be produced and saved for future reference. The well should be developed to settle the gravel pack, to bail out any non-representative water and to cleanse the strata around the borehole. This involves bailing or pumping the well until the recovered water is free from suspended solids. This may require removal of ten or more borehole volumes. The well should ideally be left for 1 week following development before purging and sampling.

If required, gas monitoring including vapour or gas sampling should be completed prior to water sampling. A record of site conditions should be made, and the well purged to achieve representative water conditions for sampling. Typically at least three times the well volume should be purged prior to sampling. Alternatively, measurements of geochemical parameters such as conductivity, pH, temperature, dissolved oxygen and ORP can be continuously monitored during purging until these parameters stabilise. It is good practice to record these parameters during purging and sampling.

Groundwater samples can be taken using a bailer (stainless steel or Teflon) or by using various pumping techniques, depending on conditions and the contaminants of concern. Samples should be taken with minimal disturbance and aeration of the sample to avoid chemical changes in the water. All equipment should be cleaned thoroughly between sampling points and dedicated bailers (and string) or tubing should be used for each well.

Sample containers appropriate for the type of testing, as provided by a laboratory, and of sufficient volume, should be used. Containers for VOC analysis should be filled first and all containers should be filled until a positive meniscus forms on the top of the container to avoid air/headspace in the sample.

48.3.1.3 Surface water

When taking surface water samples, the following points should be considered.

- Samples of the water surface should be avoided unless sampling for floating free product.
- Sample containers must be submerged below the surface and solid material such as plant debris should be removed.
- Avoid taking samples from the bank of the river or stream.
- Samples should be taken downstream of the potential source in the direction of the flow of water.
- Sample the flowing parts of the watercourse where possible, avoiding areas of stagnant or standing water.

48.3.1.4 Gas and vapour

Gas wells should be installed in accordance with BS 8485:2015+A1:2019 (BSI, 2019) and BS 8576:2013, and the final design should be based on the ground conditions encountered. Note that

- it is vital that an adequate bentonite seal is placed around the top of the borehole and should generally be a minimum thickness of 0.5 m

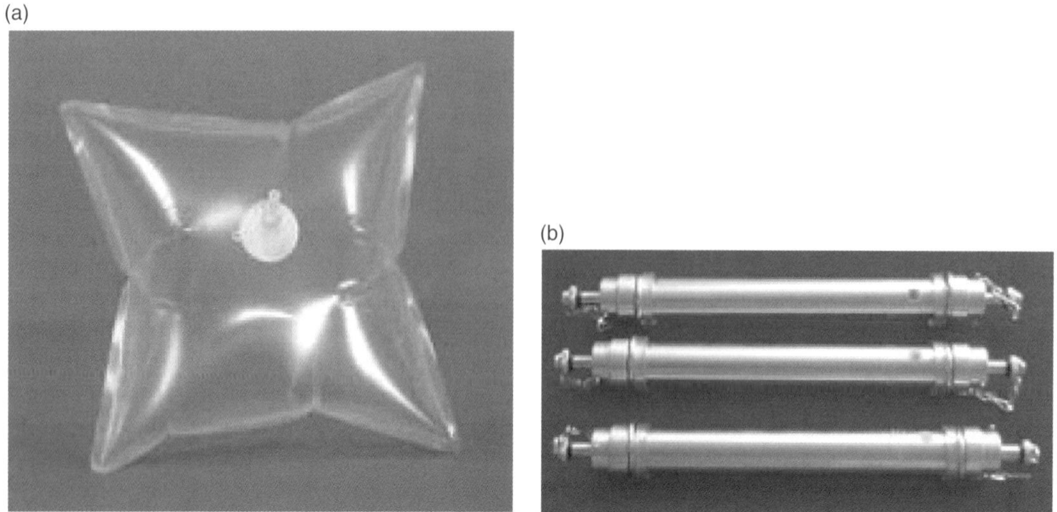

Figure 48.3 (a) Tedlar bag and (b) Gresham tubes

- pipework comprising threaded joints should be used – glue or adhesive should not be used to seal pipework.

Typically, bulk gas is measured in situ (see Section 48.4.2.3), but for laboratory analysis, especially if trace gases or organic vapours are of interest, a sample would be required. There are numerous sampling options with different advantages and disadvantages; commonly used techniques include vacuum containers or pumped steel sample containers, adsorption tubes or Tedlar bags (**Figure 48.3**). There are also specialist samplers (e.g. Vaporpin) to obtain vapour samples from below concrete slabs.

There are a number of possible occurrences in gas-monitoring wells which must be borne in mind during sampling as they can impact the results obtained. First, gas can layer in wells, as methane is lighter than air and carbon dioxide is heavier than air. In addition, during construction the soil gases within the well reflect those of atmospheric air. Over time, carbon dioxide and methane, and others, will diffuse into the well. In high-permeability soils, equilibrium can be reached fairly quickly. Wells in low-permeability soils may take longer to equilibrate. This should be taken into account when planning the monitoring programme. In addition, depending on the site model, environmental conditions such as atmospheric pressure, changes and rate of change, wind conditions and direction and recent precipitation events and groundwater level changes can impact on the gas regime and should be taken into account when monitoring or sampling gases and vapours.

48.3.2 Handling and transport

All sample containers should be labelled accordingly with indelible marker, on the sides rather than the lids. Sample containers should be placed immediately in a cool box containing ice packs (and preferably maintained below 4°C) and dispatched to the laboratory as soon as possible. All samples should be carefully packed to prevent damage during transportation.

Each sample will be recorded on a 'chain of custody' documentation, which is required to accompany the samples when transported. The chain of custody form is one of the most important documents in verifying quality assurance during environmental sampling and is often the key weakness in legal cases. This documents the transport and arrival dates of samples at the laboratory and the condition in which they were received. Each laboratory has a standard version.

All details requested on the form should be filled in and a copy retained for the project file for tracking purposes. Once samples have been received at the laboratory, a receipt should be issued, along with confirmation of the testing schedule, as appropriate.

48.3.3 Storage

Sample holding times can significantly affect the quality of results obtained and certain tests have limited holding times (e.g. those for BOD). Analyses performed outside of the recommended holding time will usually be considered 'deviating', invalid or not representative. Laboratories will advise on holding times for specific analyses. If appropriate, sample containers with preservatives may be used.

48.4. Testing methods

48.4.1 Purpose

Geo-environmental testing not only allows quantification of pollutants in various media, but also enables an understanding of the geochemical environment in the soil and groundwater to predict the fate and transport of pollutants. It is important that the scope for geo-environmental testing is considered in relation to the findings of the desk study and with due regard for future requirements or more detailed assessments.

It is absolutely critical that the testing method produces results which are relevant for the assessment required, and that limits of detection are below the appropriate assessment criteria. Inappropriate testing is a waste of time and money and can lead to, at best, an over-conservative assessment and, at worse, wrong interpretation.

48.4.2 In situ testing

In situ testing is primarily used as a support tool in geo-environmental assessments to complement laboratory testing. The benefit of in situ testing is that it can provide a large volume of data in a cost-effective manner. Real-time results also enable 'on the spot' decisions to be made. However, in situ testing is not a replacement for laboratory testing, due to issues associated with accuracy, detection limits and reliability. During the COVID-19 pandemic of 2020–2022, the EA issued guidance on the use of RMTs within the LCRM (Stage 1).

48.4.2.1 Soils

In situ testing of soils is typically used to fill data gaps in conventional laboratory testing and is also a useful method to delineate areas of contamination for remediation or verification purposes. There are various in situ testing methods currently available, but generally they fall into three main categories, as discussed below.

Hand-held field testing kits

Portable hand-held field testing kits are available for a wide range of contaminants such as hydrocarbons and heavy metals (**Figure 48.4**). All have limitations which need consideration before use. Since many hand-held testing kits are sensitive to various parameters, have a range of limitations and may involve the use of dangerous or controlled substances (e.g. solvents, X-ray technology), they must only be operated by appropriately trained individuals.

Mobile and site laboratory units

Mobile and site laboratories are a useful resource, particularly on large-scale projects, where accurate real-time data are needed. Such laboratories analyse samples in a similar way to that undertaken in the base laboratory and are operated by a suitably qualified laboratory chemist. A broad range of contaminants can be measured, depending on the specification of the mobile laboratory. Chemical test results from mobile and site laboratory units can be UKAS accredited. However, such laboratories, where available, are generally only economic on larger projects or projects with strict programme constraints (**Figure 48.5**). Detailed pre-use discussions with the regulatory bodies are advisable to ensure that the testing methods have been agreed up-front and that any concerns are addressed.

Modified probes

In situ probes can provide real-time data in a fast, efficient and economical way to characterise large areas of contamination in soils and groundwater. They also enable decisions to be made regarding targeting of site investigation or remediation and can provide data to give a rapid estimate of the volume of contamination within the subsurface.

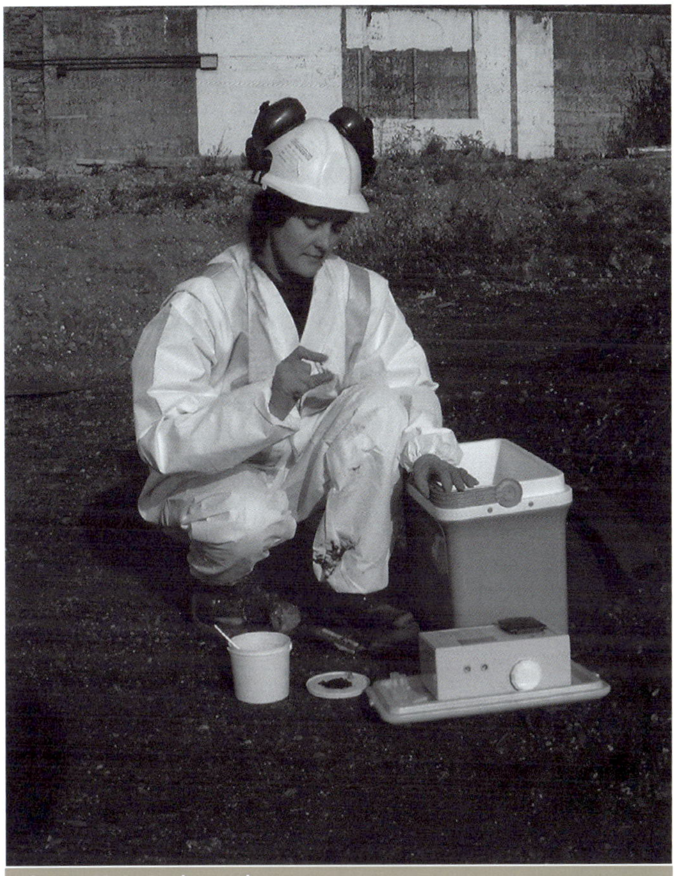

Figure 48.4 In situ testing

Cone penetration testing, typically used for geotechnical applications, has been modified to detect an array of contaminants in the ground. Techniques have been developed which use fluorescence to identify petroleum hydrocarbons by way of a modified cone driven using standard CPT equipment in both soils and groundwater. Free-phase hydrocarbon product can be detected as well as residual concentrations in soil, and it can also provide an indication of the type of petroleum

Figure 48.5 Mobile laboratory

hydrocarbon present (e.g. diesel, petrol, oil). Volatile organic contaminants such as chlorinated hydrocarbons and BTEX compounds can also be detected using MIPs.

48.4.2.2 Waters

In situ testing of waters is also routinely undertaken for various physico-chemical parameters to provide information on the geochemical conditions in the waterbody. These parameters are used to assess whether natural degradation of a contaminant may occur and thereby support the remediation decision-making process. The in situ geochemical parameters that are usually measured in groundwaters and surface waters are summarised in **Table 48.1**.

These parameters can be measured in the laboratory. However, in situ testing of these parameters is preferred since some parameters, such as dissolved oxygen and pH, are very sensitive to exposure to atmospheric air and changes in pressure – which can result from bringing water samples from depth to ground surface. For these reasons, taking a representative reading from groundwater at depth overcomes various challenges.

Hand-held meters come equipped with probes on long cables that can be lowered into a groundwater monitoring well. These meters can provide an accurate in situ reading, although they are generally expensive.

Hand-held probes require a water sample to be brought to the surface and decanted into a vessel from which the measurement is taken. These are relatively inexpensive and provide indicative information on aquifer conditions, but the disturbance caused during sampling reduces confidence in the results obtained, which should not be relied upon for decision-making.

Current best practice combines the use of low-flow sampling techniques with the use of a flow-through cell. The low-flow sampling results in minimal disturbance and aeration of the aquifer during sampling, and the flow-through cell minimises contact with atmospheric air as readings are taken.

48.4.2.3 Gas

Typically bulk gas measurements are taken using in situ measurement techniques in boreholes with specialist equipment capable of measuring gas concentrations and flow. In terms of assessing gas risks for construction purposes, monitoring for concentrations alone – without the corresponding flow reading – is of low value.

Prior to monitoring, the monitor should be run in clean atmospheric air to purge residual gas and warm up the sensor. Flow measurements should be taken first, then measurement of gas components at regular intervals until steady-state conditions are reached. Records should be kept of atmospheric pressure and air temperature. On completion, the groundwater level should be recorded.

Certain conditions can affect the validity of monitoring results. Wind blowing across the vent outlet of the flow meter can seriously affect the recorded flows, and layering of gas within the borehole should be allowed for, with sufficient time spent to observe changes in gas proportions as gas is extracted from the full depth of the well.

Gas measurements are not limited to boreholes, and where an understanding of direct surface emissions is necessary, flux box testing may be of use. This places a sealed container on the ground and emitted gases are trapped within the container for subsequent measurement. A portable FID can also be used to measure trace surface gases at multiple locations.

Parameter	Description	Reason for measurement
pH	Measure of acidity or basicity of the waterbody	Can affect microbial activity, solubility of contaminants, speciation of metals and can interfere with reduction/oxidation processes
ORP	Describes the oxidation state of the waterbody	To determine if conditions in the waterbody are likely to be oxidising or reducing, either of which can impact upon the degradation of contaminants. For example, the natural degradation of some chlorinated hydrocarbons can be enhanced by reducing conditions. For other contaminants (such as benzene), reducing conditions would inhibit natural degradation
Temperature	Temperature of the waterbody	Can affect microbial activity and solubility of contaminants
Electrical conductivity	This measures the ability of the water to conduct an electric current	This can indicate if the water is freshwater, brackish or saline
Total dissolved solids	This is a measure of the concentration of dissolved ions in the water	This is essentially measured by converting measurements for electrical conductivity using a conversion factor. This helps to determine if the water is freshwater, brackish or saline
Dissolved oxygen	This a measure of the amount of oxygen dissolved in the waterbody	Provides an indication of whether conditions in the waterbody are aerobic or anaerobic
Turbidity	This is an optical measurement of clarity ('cloudiness')	Measures the presence of suspended matter, but cannot identify the substance(s). (Note: turbidity can be measured in various units, FTU, NTU, etc. which are *not* interchangeable)

Table 48.1 Summary of typical geochemical parameters measured in water bodies

'Spot testing' of ground gas should be scheduled where possible to obtain 'worst-case' data over a suitable period of time and frequency, targeting environmental/atmospheric conditions for that objective.

In sensitive settings where detailed information on changes in concentrations within the gas may be critical (e.g. for landfill monitoring or suspected gas ingress into residential properties), continuous monitoring using automatic datalogger equipment, such as a GasClam, may be valuable. These provide regular measurements over prolonged periods to allow for assessment of changes in the gas concentrations with atmospheric/environmental conditions to increase the confidence that the worst-case conditions have been captured. In situ information using in-line techniques can also be useful for detailed assessments of gas movements, but is usually only required for detailed assessment or landfill gas monitoring.

48.4.3 Laboratory testing

Laboratory testing gives an accurate measurement of concentrations of specific chemicals or substances in a sample, such as soil, water, leachate or gas. Typically only a small percentage of the sampled material is tested and the portion used for each test is randomly selected. Therefore the validity of the testing is dependent on good sampling practice, combined with the selection of suitable methods. It should also be noted that most samples for chemical testing require subdivision and preparation in the laboratory, either by crushing, drying, solvent extraction or leaching. Therefore, all laboratory testing involves a delay between samples arriving and the testing procedure beginning. Typically, in the UK, most testing can be completed within 5–8 days. A more rapid turnaround, where possible, usually incurs additional costs. Some tests, such as leachability, require a set time period that cannot be reduced.

48.4.3.1 Methodologies

The choice of analytical method is important. Some screening tests are quick, inexpensive and provide the required data, and may be more appropriate than a slower, expensive and more complex, accurate test. The method selected should reflect the future purpose of the data and be able to achieve a LOD below the relevant assessment criteria. If the LOD is too high, all samples may have to be assumed to exceed permitted concentrations, leading to false positives and over-conservative risk assessments. Conversely, too stringent low LODs can result in unnecessary expense. It is outside the scope of this manual to discuss all of the various methods available. All laboratories can provide a list of the sample methodologies they offer, the quality accreditations associated with each method and the LODs. These should be checked for any new laboratory used.

It should be noted that the LOD is also a function of the quality procedures and statistical manipulation of the data to comply with quality accreditations. The LOD does not necessarily reflect the resolution of the testing machine, although it can mean that the laboratory cannot test to a sufficient resolution, especially where scientific advances refine the assessment criteria and the laboratories have to adapt testing methods as a result. The reason for LODs in excess of assessment criteria should be checked with the laboratory before test scheduling if they are critical to the assessment.

It is recommended for consistency that the same laboratory is used for the whole of a project, but should this not be possible, the testing methods should be kept consistent between laboratories so that data are comparable.

48.4.3.2 Suites

For practical purposes, geo-environmental testing is usually broken down into standard suites, such that similar substances are tested together (e.g. metals, PCBs, hydrocarbons, pesticides, VOCs/SVOCs). Standard suites can also be added together to produce generic suites, as may be offered by a laboratory, or bespoke suites for particular customers or projects. Agreeing to or on a suite of tests is generally much more cost-effective than specifying individual tests on each sample.

However, the appropriateness of suites needs to be checked for each project and, where necessary, parameters added or subtracted from generic suites, based on the contaminants of concern identified from the Conceptual Site Model. Typically, a site without any known contaminative history would be tested for a basic generic suite including metals (e.g. As, Cd, Cr, Cu, Ni, Zn, Pb, Hg, Se, V), basic inorganics (e.g. pH and sulfate), basic organics (e.g. TPH, PAH and phenols) and asbestos. Other contaminants, such as cyanides, solvents, pesticides, herbicides, PCBs, PFOS/PFAS would be added, depending on site-specific or surrounding conditions. For construction projects where material is to be disposed of off-site, a generic plus targeted suite of testing will be necessary for all made ground, and WAC testing may also be required. In the case of unusual substances, the specialist analytical chemists employed by the laboratories can often provide advice on the type and suitability of tests available.

For some testing the methodology must comply with specifications set out in guidance documents and regulations. For example, waste acceptability testing requires testing to a specific method.

48.4.3.3 Accreditation

A UKAS-accredited laboratory should be used for all geo-environmental testing requirements. Additionally, MCERTS-accredited testing procedures should be used, where available, for all testing which will be used for legal or regulatory purposes. However, it is noted that only soils are covered by MCERTS (EA, 2000b) at present and certain tests or determinands may not yet be accredited. This will vary between laboratories.

48.4.4 Complementary testing

In addition to chemical testing, geo-environmental assessment and detailed groundwater quantitative risk assessment in particular require some basic geotechnical information. Typically such information relates to porosity, permeability, organic content and moisture content.

48.5. Data processing

48.5.1 Assessment criteria

48.5.1.1 Human health

Current best practice for the assessment of geo-environmental data for human health risk is laid out in the EA LCRM guidance. It is necessary to understand the basis and limitations of how the chemical data can be statistically interpreted and guidance is available from CL:AIRE (2020). The latest approach suggests the use of statistics to determine a two-way confidence interval from the available dataset and then compare this to a critical concentration, usually a GAC. These GAC are generally calculated values derived in the CLEA model from toxicological information and a selection of assumptions on soil parameters and land usage to estimate an acceptable soil concentration for each contaminant. These GAC will vary with the proposed land-use parameters and it is important to understand the land usage and associated exposure model which needs to be assessed. The EA SGVs consider three main land uses; allotment, residential and commercial. At present there is not a set of statutory assessment criteria and only a limited number of the SGVs remain in use alongside a set of criteria set out for the Part 2A regime referred to as the C4SLs. Various sets of commercially available criteria have also been derived by industry groups or companies, such as the Atkins AtRisk SSVs, LQM/CIEH S4ULs (2015) and SoBRA Groundwater Vapour GAC, which have been derived using the CLEA model.

Additional land uses have also been added to the original three considered in the SGVs. Individual consultancies are likely to have in-house sets of generic criteria for standard contaminants. GAC are generally derived to be protective of land uses without detailed assessment. They are conservative values based on reasonable worst-case conditions for standard receptors and exposure pathways. Therefore, they may not be appropriate for sites which differ substantially from the assumed conditions and may be overly conservative. Detailed assessment for individual sites can be undertaken using site-specific assessment criteria derived for the site using the CLEA model and site-specific input parameters.

48.5.1.2 Waste classification

The scope and, in some cases, the methods of environmental testing required for waste classification are similar, but different from that required for environmental risk assessment. The drivers for testing are the Hazardous Waste Directive and the Landfill Directive, which are translated into law in England and Wales by way of the Hazardous Waste (England and Wales) Regulations (HMG, 2005) and the Landfill (England and Wales) Regulations (HMG, 2002), respectively. (Scotland has separate waste management regulations.)

The Hazardous Waste Directive (EU, 1991) defines hazardous wastes as those possessing one or more of 15 identified hazardous properties. In the UK the hazard properties of compounds are set out in accordance with the GB Classification, Labelling and Packaging Regulations (HMG, 2008), under which the HSE maintains a database of the codes of individual compounds referred to as the GB MCL (HSE, 2023). A similar system is also maintained under the European Waste Catalogue, which the UK system is currently based on. Some wastes may be either hazardous or non-hazardous and are known as ‘mirror entries’. All other wastes are ‘absolute entries’ classified as either hazardous or non-hazardous waste, but this category includes a further subdivision of inert waste. Guidance on the waste classification process and codes is presented within WM3 (EA, 2021).

These classifications correlate with the three categories of landfill (i.e. landfills licensed to accept hazardous, non-hazardous or inert waste) (EU, 1999). Within the hazardous waste category, some wastes are also unacceptable for disposal to landfill if they exceed certain limits. In addition, under the Landfill Regulations, certain wastes are automatically prohibited from disposal to landfill (e.g. liquid, tyre, clinical, flammable, explosive, corrosive, oxidising and chemical waste). The process of waste classification is summarised in **Figure 48.6**.

Once a waste has been identified and confirmed as acceptable for disposal, the process continues with assessing if it is hazardous. Once waste has been assessed as either hazardous or non-hazardous, it may require testing for acceptability for landfill. ‘Absolute entry’ substances only need testing for acceptability to landfill. Material that is non-hazardous and not inert does not need acceptability testing. However, in most geo-environmental/geotechnical contexts, the waste will be a natural soil or made ground, which is a mirror entry, European Waste Catalogue code 17 05 03* or 17 05 04 – soils and stones. The asterisk (*) denotes hazardous waste containing dangerous substances. The waste will require testing to determine its component substances and to demonstrate whether it is hazardous or not. This process is based on assessing the waste materials against limit values for various chemicals, determined using the hazard properties and hazard statement codes allotted to various substances in the waste as indicated in the current classification, labelling and packaging regulations of chemicals.

WAC testing is a set suite of soil and leachate tests, the latter being carried out on leachates prepared in a specified process (typically BS EN 12457-2; BSI, 2002).

It is crucial that testing of the soil for total contaminant concentrations is carried out before or alongside WAC testing. WAC testing alone is generally meaningless as the test is mostly a leachability test and assessment limits relate specifically to inert or hazardous waste categories. The waste category cannot be determined without data on total soil concentrations (see **Figure 48.6**).

WAC testing can be carried out as a full suite where the waste type is unknown and data are needed for expediency,

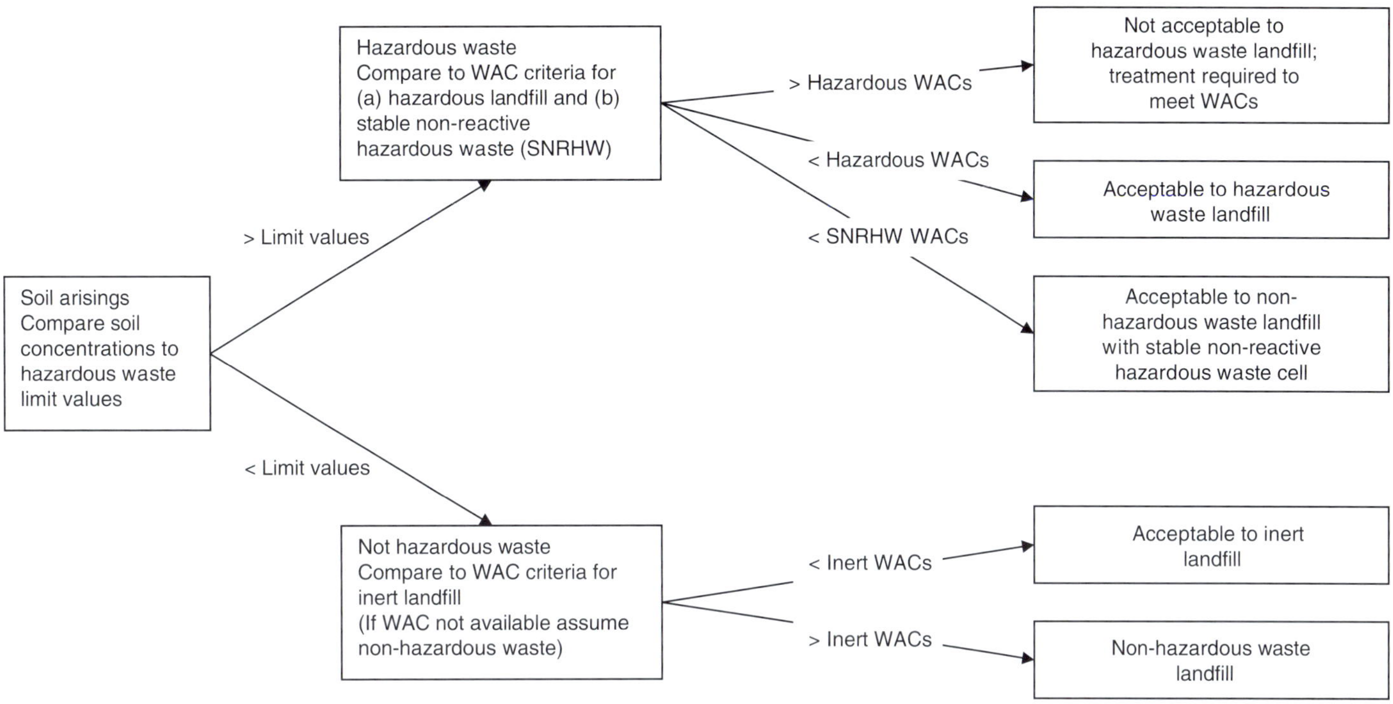

Figure 48.6 Flow chart for waste classification

or limited to a reduced suite specific to inert or hazardous wastes, if known. WAC testing is not needed for non-hazardous waste.

48.5.1.3 Waters

Water test results are generally assessed against a number of criteria depending on the aquatic sensitivity of the site. For a site which sits within a groundwater protection zone (HMG, 1998), or which otherwise provides a potable water supply (HMG, 2018), standards relevant to drinking water apply. Typically, these are UK standards but other bodies, such as the WHO, can also be used. For groundwater or surface water which does not impact on a potable supply, environmental quality standards are more appropriate, but not necessarily less stringent (e.g. WHO, 2006).

Unlike with soil assessment, there is no compulsory statistical analysis required. Where there is soil contamination which may pose a risk to groundwater, it can be assessed initially by comparing leachate test results against the relevant water quality standards. For more detailed quantitative assessment there are two UK models in common use and advocated by the EA. The remedial targets methodology (EA, 2006) is a deterministic groundwater contaminant fate and transport model which estimates the maximum allowable soil and/or groundwater concentration to limit contamination of controlled waters to a stated maximum. ConSim (Golder Associates, 2003) is a probabilistic contaminant fate and transport model which calculates the probability of exceeding a groundwater concentration from a known source.

48.5.1.4 Building materials

Some chemical testing is carried out to assess the suitability of soils for underground construction, such as pipes, concrete and membranes.

For protecting concrete in the ground, specific testing for pH and sulfates is required, with assessment in accordance with the identified groundwater and soil conditions and associated flow charts and tables published in BRE Special Digest 1 (2005). The results of testing for pH, sulfate and occasionally magnesium in soil and groundwater are used to determine the design sulfate (DS) class and the aggressive chemical environment for concrete (ACEC) site classification.

Some contaminants, such as phenols, acidic hydrocarbons and ammonia in the soil or groundwater in contact with structural concrete, can also damage concrete. Specialist advice on the design of concrete to resist such conditions should be sought.

Water supply pipes are particularly sensitive to contamination in the ground. As a rule, plastic pipes will not be permitted within contaminated soil due to the potential for damage to the pipes and/or possible tainting of the water supply. UK Water Industry Research Ltd (UKWIR) has published specific guidance on investigation and analysis to support selection of pipe materials (UKWIR, 2010). In addition, each water company may have specific requirements for testing and assessment of soils into which water supply pipes are to be laid. These can be quite onerous, and the water company should be contacted if plastic pipes are proposed to ensure that conditions are acceptable.

In situations where a barrier membrane may be installed on the ground, or be in contact with ground-borne liquids or gases, the durability and resistance of such membranes must be checked against the chemicals present in the ground. This information may be available on datasheets or from the manufacturer. For unusual chemicals, or low-specification membranes, this information may not be available.

Ground gas and vapour protection measures may be utilised to prevent ingress to buildings. Ground gas protection should be designed in accordance with BS 8485. Guidance on the use of vapour barriers is given in CIRIA Report C748 (CIRIA, 2014) and generally for gas ground risk in CIRIA C665 (CIRIA, 2007).

48.5.1.5 Verification

Verification testing is carried out to confirm whether remediation works have achieved the target concentrations. These are generally a set of values agreed with the regulator (i.e. the local authority or the EA (in England, SEPA in Scotland and the NIEA in Northern Ireland)).

Soil concentrations at or below those values are acceptable and concentrations above are usually unacceptable. Typically these values are based on the result of quantitative risk modelling. The use of SGVs and water quality standards may not be appropriate criteria for remediation and may be overly stringent. Verification data for soil is relatively limited in availability and is compared directly to the target concentrations. Statistical analysis of such data is rare, but is occasionally used if a limited number of tests out of a large batch marginally exceed limits or if such analysis is appropriate given site conditions and usage. However, the use of such an approach should be agreed in advance with the regulator.

Groundwater validation data are generally collected over a period of time, typically ranging from 3 months to 5 years, to assess trends as well as direct comparison of concentrations against targets.

48.5.2 Rogue data

All analytical results should be reviewed by a suitably qualified person to check for potential errors or unusual results; this would include

- looking for transcription errors or missing data
- considering if the data are within the expected or valid ranges for that contaminant or the site (based on previous monitoring rounds)
- understanding the site conditions (e.g. potential for contamination, soil type) – do the results match the observed contamination such as free product?
- looking at potential conflict between results (e.g. low PAH concentrations but high aromatic TPH concentrations, or the presence of ammonia and high dissolved oxygen)
- assessing other field measurements which may explain spurious results (e.g. high suspended solid content).

Identified rogue data should be checked with the laboratory and, if necessary, samples re-tested, or re-sampling undertaken to confirm the results. The majority of anomalies will be readily explained by re-testing or by the application of common sense; others will actually appear to be correct, only because of the absence of an explanation to the contrary. Contamination concentrations can, however, radically vary across a relatively small area, such as found in a 'hotspot'. In some circumstances, such as with volatile material, concentrations will also vary significantly with time and exposure.

48.6. Quality assurance

48.6.1 Site model

It is usually a requirement for a Conceptual Site Model to be developed for an individual site as part of a planning condition in the UK. The source, pathway and receptor for contaminants have to be set out, and one of the most simple processes for any QA checking is to compare the Conceptual Site Model with the type and amount of testing done in any investigation. For example, the identification of the potential for a contaminant plume of a specific hydrocarbon which is not accompanied by testing for the specific range of hydrocarbons across a site in a number of places instantly places the quality of the investigation into doubt.

48.6.2 Sampling strategy

The distribution patterns of sampling positions are much loved by statisticians and academics alike. Unfortunately they generally assume a homogeneous media in which the contaminant lies – and a client with a very large testing budget. Both are rare. The locating of previous contaminative processes from historical maps and plans becomes crucial and the numbers of tests becomes a matter of judgement, not statistical random sampling. That said, it is easier to identify the wrong strategy rather than to specify a generic and therefore inadequate ideal. The following should act as a guide.

- Do not take samples only from shallow depth when fluid contaminants such as diesel might have spread downwards into natural horizons. Many environmental investigations sample only the top 500 mm of a site.
- Do not only test the man-made soils; test the underlying natural soils to establish if contamination has migrated and impacted other soils.
- Do not only focus testing on potential 'hotspots' such as around buried tanks or process building, but look to establish the potential for 'background' or 'unaffected' levels.
- Do not ignore the groundwater; remediation of this can be many times more expensive than dealing with soils.
- Do not ignore soil gas as a remediation issue in itself and an indicator of other contamination.
- Samples can be too few in number, particularly on small sites, and can overwhelm the interpretation on larger sites without a clear assessment of the causes of distribution.

Sampling strategies should be flexible and adapted to the conditions encountered during investigation which may alter the expected CSM.

Guidance on the selection of sampling approaches and considerations of limitations is given in CLR 4 (Department for the Environment, 1994).

48.6.3 Laboratory

Laboratories are commercial organisations and it should be expected that despite accreditation and various levels of QA procedure their performance and reliability will vary significantly. It is important to consider that procedures are insufficient and that human and mechanical errors will occur at times in the best of laboratories.

As a principle, a single laboratory should handle all the testing from one site, from investigation to remediation. Indeed, a known and planned throughput can see the benefits of economies of scale.

On large projects it is recommended that the laboratory accreditation is checked; new labs could be visited to inspect and discuss testing requirements, timescales and alternative options of analysis. In certain circumstances it might also prove necessary for a back-up laboratory to be identified and compatibility of analytical techniques and sample preparation to be thoroughly checked.

Laboratories should be able to provide appropriate accreditation across the widest range of testing suites.

A principal safeguard is the 'believability' check by an experienced geo-environmental engineer, who should be able to critically appraise the results received and identify rogue data.

However, on certain projects it may be necessary to undertake quality control sampling to provide a measurement of data errors and variability in sampling. There are a number of ways of taking quality control samples, including duplicate samples, blank samples and standard/spiked samples. The types of quality control samples and the types of errors they detect are summarised in **Table 48.2**. A summary of the advantages and disadvantages of the three quality control sample types are presented in **Table 48.3**.

48.6.4 Checking and review

The number of data points and the statistical manipulation of them can obscure the random error from the checking and review process and it is important that there is a specific independent data verification check done before conclusions are drawn from the processed raw data. Ideally this is done as part of the initial calculation process by staff of similar experience. More senior staff tasked with the review of the processed data can only realistically select random samples and follow these through set processes. As with all checking operations, the process is as valuable as the experience of the individual carrying out the check.

It should be expected that the final review should be signed off by an appropriately professionally qualified individual holding chartered status as an environmentalist or similar relevant profession.

QC sample location	Duplicate QC sample	Blank QC samples	Standard/spiked QC samples	Errors or variability detected
Waterbody	Sampling duplicate (i.e. repeat entire sampling procedure)	Not possible	Not possible	Total of: purging/short-term natural variability plus errors below
Sampling equipment	Equipment duplicate (i.e. repeat use of sampling equipment)	Equipment field blank[a]	Equipment field standard/spike[a]	Total of: sampling equipment/some short-term natural variability (in the case of duplicate), plus errors below
Prior to treatment (e.g. filtering/ preservation)	Pre-treatment duplicate (split sample prior to treatment, then treat both samples identically)	Pre-treatment field blank	Pre-treatment field standard/spike	Total of: field treatment (filtering/ preservatives), plus errors below
Prior to bottling	Post-treatment duplicate	Post-treatment field blank	Post-treatment field standard/spike	Total of: ambient conditions plus errors below
Prior to transport	Not possible	Trip blank	Trip standard/spike	Total of: handling and transport plus errors below
Sample bottles	Not possible	Bottle bank (i.e. place de-ionised water in bottle and submit for analysis)	Bottle standard/spike (i.e. place standard or spiked sample in bottle and submit for analysis)	Total of: bottle material and preparation plus errors below
Delivery to laboratory	Laboratory duplicate	Laboratory blank	Laboratory standard/spike	Total laboratory errors[a]
Type of errors detected	Random only	Random and systematic gains only	Random and systematic gains and losses	

[a]Only possible if equipment is removable. For dedicated sampling equipment, this QC sample becomes less important. NB: this table only relates to the sampling process. Further QC samples should be prepared in the laboratory in order to detect errors during the laboratory handling and analytical process.

Table 48.2 Summary of QC sample types (Data taken from Environment Agency (2000a) © Crown Copyright)

QC sample type	Advantages	Disadvantages
Duplicate	• Sampling process itself can be duplicated (sampling duplicate), providing information on errors in the entire sampling/analysis process • Relatively easily performed • Can be applied for all determinands	• Only detects random errors; systematic errors are not detected
Blank	• Easily performed • Can be applied to all determinands • Detects some random and systematic errors	• Cannot be applied to initial sampling • Only detects gains in determinands; losses are not detected
Standard/spike	• Detects all random and systematic errors from point of QC sampling	• Requires laboratory-prepared standard solution. Can be more difficult to perform • Each sample usually applies to only one determinand

Table 48.3 Comparison of QC sample types

References

BRE (2005) Concrete in Aggressive Ground. *Special Digest 1*. BRE, Watford, UK.

BSI (British Standards Institution) (2002) BS EN 12457-2:2002: Characterisation of waste. Leaching. Compliance test for leaching of granular waste materials and sludges. One stage batch test at a liquid to solid ratio of 10 l/kg for materials with particle size below 4mm (without or with size reduction) BSI, London, UK.

BSI (2007) BS EN 1997-2:2007: Eurocode 7: Geotechnical design – Ground investigation and testing. BSI, London, UK.

BSI (2017) BS 10175:2011+A2:2017: Investigation of potentially contaminated sites – Code of practice. BSI, London, UK.

BSI (2019) BS 8485:2015+A1:2019: Code of practice for the design of protective measures for methane and carbon dioxide ground gases for new buildings. BSI, London, UK.

BSI (2020a) BS 5930:2015+A1:2020: Code of practice for site investigations. BSI, London, UK.

BSI (2020b) BS 10176:2020: Taking soil samples for determination of volatile organic compounds. BSI, London, UK.

CIRIA (2007) *Assessing Risks Posed by Hazardous Ground Gases to Buildings*. CIRIA Report 665. CIRIA, London, UK.

CIRIA (2014) *Guidance on the Use of Plastic Membranes as VOC Vapour Barriers*. CIRIA Report C748. CIRIA, London, UK.

CL:AIRE (2020) *Professional Guidance: Comparing Soil Contamination Data with a Critical Concentration*. Version 2. CL:AIRE, London, UK.

DEFRA (2012) Environmental Protection Act 1990: Part 2A. Contaminated Land Statutory Guidance. DEFRA, London, UK.

Department for the Environment (1994) *Sampling Strategies for Contaminated Land: Contaminated Land Research Report 4 (CLR 4)*. DoE, London, UK.

EA (Environment Agency) (2000a) *Guidance on Monitoring of Landfill Leachate, Groundwater and Surface Water*. Environment Agency, London, UK.

EA (2000b). *MCERTS (Monitoring Certification Scheme)*. Environment Agency, London, UK.

EA (2006) *Remedial Targets Methodology: Hydrogeological Risk Assessment for Land Contamination*. Environment Agency, London, UK.

EA (2009) *Contaminated Land Exposure and Assessment (CLEA) Version 1.05, Human Health Toxicological Assessment of Contaminants in Soils C050021/SR2 and Updated Technical Background to the CLEA Model SC050021/SR3*. Environment Agency, London, UK.

EA (2020) Land Contamination Risk Management (LCRM). https://www.gov.uk/government/publications/land-contamination-risk-management-lcrm (accessed 13/04/2023).

EA (2021) *Waste Classification: Guidance on the Classification and Assessment of Waste*. Technical Guidance WM3. Environment Agency, London, UK.

EU (1991) Council Directive 91/689/EEC of 12 December 1991 on hazardous waste. *Official Journal of the European Communities* **L377**.

EU (1999) Council Directive 1999/31/EC of 26 April 1999 on the landfill of waste. *Official Journal of the European Communities* **L182**.

Golder Associates (UK) Ltd (2003) ConSim version 2.0. Golder Associates, Doncaster, UK.

HMG (Her Majesty's Government) (1998) Groundwater Regulations. The Stationery Office, London, UK, Statutory Instrument 1998 No. 2746.

HMG (2002) Landfill (England and Wales) Regulations. The Stationery Office, London, UK.

HMG (2005) The Hazardous Waste (England and Wales) Regulations. The Stationery Office, London, UK, Statutory Instrument 2005 No. 894.

HMG (2008) Regulation (EC) No 1272/2008: Classification, Labelling and Packaging of Substances and Mixtures (CLP Regulations). The Stationery Office, London, UK.

HMG (2018) The Water Supply (Water Quality) Regulations 2016 (England) (with 2018 Amendments Consolidated). The Stationery Office, London, UK.

HSE (Health and Safety Executive) (2023) The GB CLP Regulations. https://www.hse.gov.uk/chemical-classification/legal/clp-regulation.htm (accessed 11/05/2023).

LQM/CIEH (2015) *The LQM/CIEH S4Uls for Human Health Risk Assessment*. Land Quality Press, Nottingham, UK.

UKWIR (2010) *Guidance for the Selection of Water Supply Pipes to be Used in Brownfield Sites*. UKWIR, London, UK, Report 10/WM/03/21 (revised edition).

WHO (2006) *Guidelines for Drinking Water Quality, Vol. 1*, 3rd edn, incorporating 1st and 2nd addenda. WHO, Geneva, Switzerland.

Further reading

CIRIA (2002) *Brownfield: Managing the Development of Previously Developed Land.* A Client's Guide. CIRIA Report C578. CIRIA, London, UK.

DCLG (2004) Planning and Pollution Control. Planning Policy Statement 23. DCLG, London, UK.

DEFRA (2006) Environmental Protection Act 1990; Part 2A. Contaminated Land. Circular 01/2006. DEFRA, London, UK.

NHBC/CIEH (2008) *Guidance for the Safe Development of Housing on Land Affected by Contamination.* NHBC, Milton Keynes, UK.

Strange J and Langdon N (eds) (2008) *Contaminated Land: Investigation, Assessment and Remediation*, 2nd edn. Thomas Telford, London, UK.

Useful websites

CABERNET (Concerted Action on Brownfield and Economic Regeneration Network): http://www.cabernet.org.uk/index.asp?c=1124

CL:AIRE (Contaminated Land: Applications in Real Environments): http://www.claire.co.uk

ContamLinks, a portal for quality information about the assessment, management and remediation of contaminated land: http://www.contamlinks.co.uk/index.htm

Environment Agency: http://www.environment-agency.gov.uk

Environment Agency publications: http://publications.environment-agency.gov.uk

EUGRIS, portal for soil and water management in Europe: http://www. eugris.info/index.asp

Health Protection Agency: http://www.hpa.org.uk

NICOLE (Network for Industrially Contaminated Land in Europe): http://www.nicole.org/index.asp

Portal for Contaminated Land Information in the UK: http://www.contaminated-land.org/index.html

It is recommended this chapter is read in conjunction with

- Chapter 8 *Health and safety in geotechnical engineering*

All chapters in this book rely on the guidance in Sections 1 *Context* and 2 *Fundamental principles*. A sound knowledge of ground investigation is required for all geotechnical works, as set out in Section 4 *Site investigation*.

Glossary

BOD	Biological oxygen demand
BTEX	Chemical compounds benzene, toluene, ethylbenzene and xylene. BTEX are constituents of petrol
C4SLs	Category 4 screening levels
CLEA	Contaminated land exposure assessment
CLR	Contaminated land report
Contaminant linkage	Interaction between a contaminant source, transfer pathway and vulnerable receptor. The linkage only exists if all three are in place
CPT	Cone penetration test
CSM	Conceptual site model
DEFRA	Department for Environment, Health and Rural Affairs
EA	UK Environment Agency
EPA Part 2A	Environmental Protection Act 1990 Part 2A – Contaminated Land.
FID	Flame ionisation detector
Free product	Hydrocarbons in a liquid phase (i.e. 'free' or non-aqueous)
GAC	Generic assessment criteria – an assessment value derived by non-government agency using a contaminant fate and transport model – usually CLEA in the UK
Hazardous waste	Waste which, because of its quantity, concentration or characteristics, may be hazardous to human health or the environment when improperly treated, stored, transported or disposed of
HCV	Health criteria value – typically in the form of an index dose (ID) or tolerable daily intake (TDI), this is the estimate of the daily intake of a contaminant that can be experienced over a lifetime without appreciable health risk (or cancer risk in the case of an ID)
Inert waste	Waste which is neither chemically nor biologically reactive and will not decompose
LCRM	Land Contamination Risk Management
LOD	Limit of detection – the lowest value to which a test result is reported
MCERTS	Monitoring Certification Scheme introduced by the Environment Agency
MIP	Membrane interface probe
Non-hazardous waste	Waste which is neither hazardous nor inert
ORP	Oxidation reduction potential – measure of how likely oxidation or reducing chemical reactions are to occur, also referred to as redox potential
PAH	Polycyclic aromatic hydrocarbon
PCB	Polychlorinated biphenyl
PFOA	Perfluorooctanoic acid
PFOS	Perfluorooctanesulfonic acid
PTFE	Polytetrafluoroethylene
Purging	Process of removing stale water from a borehole to ensure representative sampling

QA	Quality assurance
RIP	Remediation implementation plan
RMT	Rapid measurement techniques
SGV	Soil guideline value – an assessment value derived by the Environment Agency using a CLEA model
SVOC	Semi-volatile organic compounds
TPH	Total petroleum hydrocarbons
UKAS	United Kingdom Accreditation Scheme
VOC	Volatile organic compound – a human health assessment value
WAC	Waste acceptability criteria
WHO	World Health Organization
WM3	Waste Classification Technical Guidance (WM3)

Chapter 49
Sampling and laboratory testing

Chris S Russell Russell Geotechnical Innovations Limited, Weybridge, UK

The development of a good ground model relies upon a successful ground investigation with appropriate sampling techniques and laboratory testing. The involvement of all stakeholders in the process, open communication and supervision are essential for high-quality parameter determination for the foundation or temporary works design. Various sampling techniques and their validity for parameter determination are discussed in a process-based fashion along with an insight into the various laboratory tests available. The list of laboratory tests is not exhaustive but includes many of the common test types (and some specialist considerations) associated with modern design requirements from low-rise buildings to advanced construction. The reader is given the main principles of sampling and sample disturbance with reference to the effects on laboratory testing and parameter determination, which may be used as a basis for investigations worldwide.

doi: 10.1680/icemge.66816.0725

CONTENTS

49.1. Introduction

From previous chapters and sections the engineer will have created and planned the site investigation and have a list of requirements needed to satisfy the construction programme from the initial desk study phase through to the finished construction (see Chapter 4 *The geotechnical triangle*). These plans will have been laid out in a 'best practice' manner but should allow for some deviation, especially when dealing with the 'unknowns of nature' in what lies beneath the surface. Plans that are too rigid may lead to problems later when dealing with the running of the project and certainly may cause problems in the ground investigation phase. The budget for ground investigations is always only a few per cent of the overall project cost, but it is here that major savings can be made in the overall design phase if it is given thought and care is taken.

The design of the ground investigation will lead on from the desk study, which will have identified the expected stratum and ground conditions. The ground investigation will verify these conditions and identify any deviations which may require further attention or categorisation, leading to a ground model suitable for the engineering or design purposes intended. Laboratory tests are routinely used to calibrate ground models; however, in situ conditions (sometimes dominated by discontinuities or complex horizons and geological structures) may require more expensive and time-consuming field tests for full categorisation (see Chapter 47 *Field geotechnical testing*). This sounds simple on paper but is in fact crucial for the design and construction stage. Failings here could have disastrous consequences. If in doubt seek advice. It should be noted that this chapter focuses on the physical properties of the ground and so does not address testing for chemical properties or ground contamination (refer to Chapter 48 *Geo-environmental testing*). For further information with regard to the task/project in hand, you may wish to refer to (among others) Chapters 13 *The ground profile and its genesis* and 43 *Preliminary studies* to 46 *Ground exploration*, plus Section 3 *Problematic soils and their issues*.

49.2. Construction design requirements for sampling and testing

The categorisation of the site should be comprehensive and provide the best possible parameters for the foundation design, whether it be for simple load-bearing calculations for a strip footing through to finite element analysis, which is often required for more complex or fragile construction. The required parameters drive the testing schedules for laboratory testing in order to gain accurate knowledge of the physical (and chemical – refer to Chapter 48 *Geo-environmental testing*) properties of the site (the ground model) to indicate uniformity/non-uniformity of the ground (both laterally and with depth). It is this coverage which is required for any form of foundation design or physical modelling. Soil tests should be identified to provide the parameters required for design. Basic categorisation can be carried out on site through trial pitting, but drilling and sampling will be required for depth profiles to be identified. Each site or contract should be treated as unique and specifications must be reviewed for each contract in light of this. The complexity of the construction and the nature of the ground play a most important role in the parameters required. Interestingly the process of desk study followed by ground investigation and subsequent laboratory testing has required much thought for the layout and sequence of this chapter. It is the output in parameters required for the design that will dictate the sample types that need to be taken during the initial ground investigation phase, or a two-phase ground investigation may be required if the construction is complex or fragile. In this respect the order of description in this chapter has had to be reversed as the parameters for design will dictate the test types and therefore the sample quality, which will control the sampling methods and preparation of the samples for the tests. In other words, think about what you need before you try to achieve it. The cost savings are in the planning of this, and it can be expensive (as well as commercially embarrassing) to get this wrong. The parameters (and some associated tests) you may require are summarised below.

49.3. The parameters and associated test types

In their simplest form, the results of laboratory testing may be used to categorise our area of interest and identify uniformity or non-uniformity of ground characteristics, both laterally and vertically, while more advanced parameters are based around the stability or reaction of the soil with regard to the loading (or unloading) of the construction, both in the short and long term. These values are related to soil strength and stiffness (possibly anisotropy), compressibility and permeability. For the short-term requirements and very low-permeability materials undrained scenarios may be more relevant, while for the long-term requirements or high-permeability materials fully drained conditions would be more relevant. These properties all vary with different stress states and again the ground investigation should be targeted to gain knowledge of these so that laboratory testing can be representative of the in situ conditions and be used with confidence in the foundation design stage. Test standards are listed in the Appendix to this chapter for most common laboratory tests and some history/background of these tests and their basic methodology can be found in the *Manual of Soil Laboratory Testing* (Head, 2014).

49.4. Index tests

These are the most common and routine of all site characterisation techniques and are used as a profiling tool both vertically and laterally to build the ground model suitable for the project. They may be carried out using both disturbed and undisturbed material and are relatively cheap and quick to perform. Index tests may be used to support 'expected' basic ground behaviour and to identify where more expensive tests (or increased parameter interest) may be required for design. Such routine tests generally comprise moisture content determinations, particle size distribution and Atterberg limits. When combined these provide very useful profiling tools which can also be used to calibrate ground models and verify the results of any further laboratory tests. They provide information that can be combined with drilling logs to identify and corroborate the height of the water table, variation of soil type and the expected soil behaviour, so giving clarity to the ground model.

49.4.1 Moisture content

This is the simplest and cheapest of the soil tests to carry out in laboratories and can be scheduled on both undisturbed and disturbed samples. The test consists of a small sample of soil being weighed before and after drying to determine the ratio of solid particles to water. Interestingly, it is the moisture content of soils or weak rocks which often dictates or certainly can dominate their engineering behaviour. The 'engineering' of moisture content only improves the workability or placement of some materials (say, for compaction in a landfill liner or dam core), but can cause catastrophic failure through loss of shear strength/cohesion if calculated or carried out incorrectly. Moisture content profiles of depth and lateral distance can be used to indicate zones of differing soil properties. Clays with 'high' moisture content are often soft (or softened) compared to their drier counterparts (with identical mineral composition). They are also more likely to compress and collapse. The effect of desiccation and the possibility of ground heave can be identified through moisture content profiling in the vicinity of nearby vegetation or drawdown situations. Higher localised moisture content can even be used to identify the location of broken drains and water pipes, which may be the cause of undermined foundations (among other engineering problems).

'Natural' moisture content determinations require the material being tested to be totally sealed in its natural state. This may appear obvious, but many people get it wrong. The sample should be taken from the ground at whatever depth without the influence of outside water/drill fluid/evaporation and should be immediately sealed in a fully airtight fashion. If the bag is not sealed immediately it will change its moisture content and be unrepresentative (if it is raining, water may enter the sample; if it is warm and sunny, water may evaporate from the sample). Cohesive materials recovered from rotary cores or where drill flushes have been in contact with the material should be subsampled away from the periphery of the material (in contact with the drill flush). The natural moisture content in these material types will be preserved in the centre of the core for a while due to their low permeabilities. Conversely, it is almost impossible to gain a natural moisture content for many non-cohesive materials (especially sands and gravels) as the high permeability of such materials prevents retention of the water during sample extraction from the ground.

Thought should also be given to the materials scheduled for these tests, especially if they are likely to contain hydrated minerals such as gypsum. In such instances the oven-drying temperature used in the laboratory determination should be below the level at which such minerals 'dehydrate' or release water from their crystal matrix (if present). If the oven temperature is above 110°C (hence 105–110°C for standard BS 1377 determinations), volatile fluids other than water will be evaporated, leading to erroneous test values. For materials containing (or suspected to contain) gypsum the oven temperature should not be more than 80°C. The drying stage is complete when successive weighings are within 0.1% at four-hour intervals (BS 1377; BSI, 2022). Saline pore waters can also lead to incorrect determinations and in such instances other tests may be more appropriate.

49.4.2 Atterberg limits

These are used to classify fine-grained soils and commonly identify two of the original seven limits defined by Albert Atterberg. The limits are based upon the moisture content of the soil and can be obtained for both undisturbed and disturbed samples. The plastic limit (PL) is the moisture content at which the soil changes from a semi-solid to a plastic state; the liquid limit (LL) is the moisture content at which the soil changes from a plastic to a viscous state.

LL determinations are carried out either by measuring the penetration of a calibrated cone into a known volume of fully mixed material at four increasing moisture contents (four-point cone method) or by 'bumping' material in a calibrated Casagrande system using a grooving tool, again at four different moisture contents. There are alternatives to the four-point systems described here, but they are not ideal. The PL is determined by the point at which soil can be 'rolled' in a calibrated way to form a thread of 3 mm dia. which shears both transversely and longitudinally. The PL test may yield variable results due to differing operators and levels of experience. This is a basic test but not a simple one to carry out!

The difference between the PL and the LL is known as the plasticity index (PI). The relationship allows approximate determinations of compressibility, permeability and strength and is therefore very useful for soil classification. The derived PI can also be used to determine the amount of clay present. High PI values indicate significant clay content, while low PI values indicate the dominance of silt particles. A PI of zero indicates the absence of both clay and silt and is termed 'non-plastic'. Generally the higher the PI value the greater the soil's potential to change volume. High PI values would signify a large volume change when wetted and large shrinkage when dried and so on. This general rule, however, does not take into account the presence of particles larger than 425 μm (removed by sieving before the test begins) and so the modified PI (I_p') is often more appropriate, but only for overconsolidated clays. The calculation for the modified PI was founded by the Building Research Establishment (BRE) and is

$$\text{Modified PI}\,(I_p') = \text{PI} \times (\% < 425\,\mu\text{m}/100\%) \qquad (49.1)$$

The National House Building Council (NHBC) has used this same calculation but has modified the percentages that identify high-, medium- and low-volume change/shrinkage potentials. Values should be recorded as percentages as the general terminology of high, medium and low I_p' values are slightly different between the BRE and NHBC references. It can be seen very quickly that such tests can yield very good information and confidence in the material properties of fine-grained materials, especially when profiled.

49.4.3 Particle size distribution (PSD) analysis

This test can be carried out on both undisturbed and disturbed samples. A PSD determination is the mass of particles within designated size ranges expressed as a percentage of the complete sample mass. The range of sample sizes split the soil into its component groups, ranging from clay to silt, followed by sand and gravel upwards in size (through to cobbles and boulders). For a complete analysis, two distinct test types are performed. For particles larger than 63 μm (for British Standard and BS EN ISO tests; 75 μm for ASTM standards) the material is graded by passing through sieves of decreasing sizes. The definitive method for these 'coarse' grains is by wet sieving, while a quantitative test may be carried out by dry sieving (for soils containing insignificant quantities of silt and clay). For wet sieving the particles smaller than the smallest sieve size are washed from the material and retained for the second part of the test. Particles smaller than the smallest test sieve are then graded by settling from suspension (with time) in either the hydrometer test or by pipette methods.

From visual description of soils estimations are made of the percentages of the various sediment sizes that make up the sample. The PSD determination scientifically derives the exact percentages of each soil size fraction within the sample. It is therefore possible that a visual description may be slightly different to that recorded from a PSD analysis, but they can be used to calibrate each other and expose inaccuracies in logging and drilling records. The PSD of a soil will also indicate the permeability and, possibly, compressibility characteristics to be expected from other tests, which may allow test specifications to be designed for the material types in question.

Beware: it is not unknown for the sampling of some materials to be carried out badly, especially when retrieving granular material from depth. It is very easy to wash out the fines in the drill fluid or allow the sample tube to drain out water (carrying away fines in suspension), which will lead to an inaccurate PSD analysis. The author has even witnessed junior lab technicians pour the coarse material from the sample container/bag to be tested and leave the fine sediment in the bottom to be discarded.

Both of these instances would lead to a very inaccurate PSD analysis and an erroneous judgement of material properties.

Many other index tests exist and may be used in combination or with the main three tests listed above in order to complement soil characterisation. It is most important to remember that any test result is only as good as the representative sample taken and delivered to the laboratory and should be representative of the stratum from which it was taken. Coarse material should be taken in large quantities in order to be representative (see BS 1377-2:2022 and BS EN ISO 17892-12:2018; BSI, 2018 for required sample sizes for PSD and other analyses).

49.4.4 Compaction-related tests

These are a series of tests which identify the relationship of density (often with changing moisture content) with a known compactive force. These tests are mainly carried out on disturbed material or material to be classified for engineered fill (which by their nature are 'disturbed'). Compaction itself is a process where the density of the soil is increased by packing the soil particles closer together and so reducing the volume of air (without significantly changing the moisture content). The addition or reduction of moisture content for each test stage simply alters the strength characteristics of the soil and its 'compactibility'. These tests are common for field design of engineered fills as they will provide optimum moisture contents for the fill material with regard to compactive effort available (see Chapter 75 *Earthworks material specification, compaction and control*). Common forms of these tests for varying engineering uses and soil types are

- determination of dry density/moisture content relationship (2.5 kg rammer)
- determination of dry density/moisture content relationship (4.5 kg rammer)
- determination of dry density/moisture content relationship (vibrating hammer)
- determination of maximum and minimum dry density
- determination of moisture condition value (MCV)
- determination of California bearing ratio (CBR)
- determination of chalk crushing value (CCV).

Please refer to the relevant current standards for full descriptions of the above methods with regard to the soil types to be tested. Of the tests listed above, it is only the CBR which provides an empirical strength criterion (CBR value), and is often associated with pavement construction (see Chapter 76 *Pavement foundation design*).

49.5. Strength

Strength is defined as the limiting shear stress that a material can sustain as it suffers large shear strains (Atkinson, 2007). The response of the soil in both strength and stiffness are related to the 'state' of the soil. This state is related to the density of the material and stress level, which is linked to pore pressure, and therefore 'effective stress'. It is interesting that so many parameters and stress states are closely linked, but all are ultimately controlled by the principles of effective stress. 'All measurable effects of a change in stress, such as compression, distortion, and change of shearing resistance, are due exclusively to changes of effective stress' (Atkinson, 2007). The following is a list of laboratory tests, including triaxial and direct shear types (among others). Triaxial tests are a family of tests whose subtleties are controlled by varying boundary, drainage and loading conditions, but 'appear' to use similar equipment.

49.5.1 Triaxial test types

For basic boundary, drainage and loading conditions, see **Figure 49.1**.

- *Unconfined compressive strength* (UCS). This is a total stress test (no pore pressure measurement) and is carried out without radial confinement. It may also be termed 'unconfirmed' compressive strength, and unfortunately has the same acronym associated with the uniaxial compressive strength (UCS test) carried out on rocks. Although the applied stresses are all the same in these tests, the standards, methods and test equipment used for soils and their rock equivalents are distinctly different.
- *Unconsolidated undrained triaxial test* (UU). Again this is a total stress test as pore water pressure is not measured. For this test a radial (confining) pressure (σ_3) is applied to the sample and is of a magnitude which relates to the depth of sample origin. Shearing rates are standardised for these tests and reference should be made to the relevant regional standards for clarification. This test should not be confused with a UUP test (see below), which often goes under the same name.
- *Unconsolidated undrained triaxial test with pore pressure measurement* (UUP). An effective stress test which gives the undrained shear strength for the material. Be aware that the effective stress measured for such tests may not be representative of the mean effective stress of the material in situ due to the effects of sample disturbance. For more advanced testing the pore pressure may be measured both at the base and mid-height of the sample.
- *Isotropically consolidated undrained triaxial test* (CIU). As above, but the sample is isotropically consolidated to a mean effective stress relevant to the in situ depth of the sample or a particular stress/depth condition which is to be modelled.
- *Isotropically consolidated drained triaxial test* (CID). This is for the 'drained' or long-term condition of the above test. For clay materials this may be a test of very long duration due to the very low permeability associated with such particle sizes and mineralogies, but for sands and free-draining materials it is usual to carry out drained triaxial tests rather than undrained shearing because the short- and long-term conditions should approximate (due to the high permeabilities). Undrained conditions are unusual for sands (unless in a fully confined state) within the ground unless one is trying to model a very specific ground or construction condition. Undrained shear tests in non-cohesive

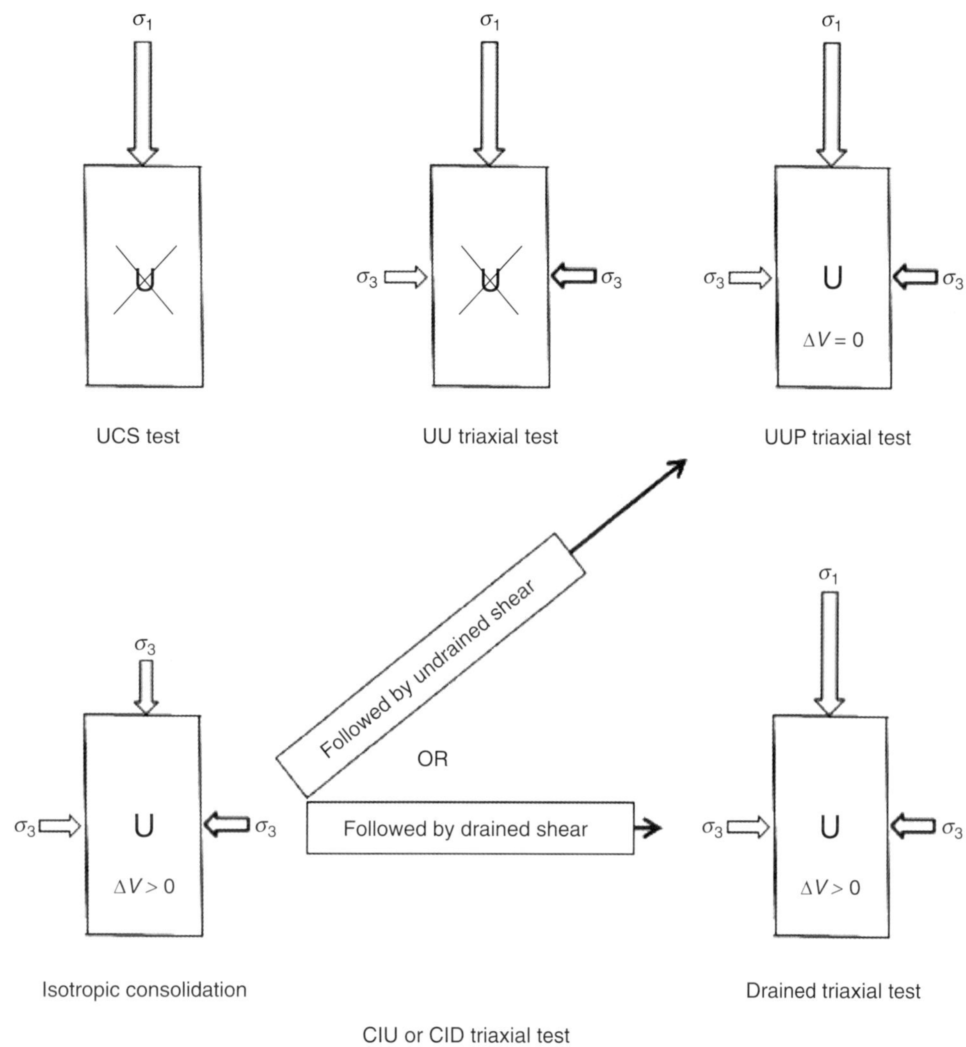

Figure 49.1 Triaxial test: boundary and loading conditions

materials cause immediate dilation of the material which is often unrepresentative of field conditions.

- *Anisotropically consolidated triaxial tests* (CAUC, CAUE, CADC, CADE – see Section 49.6).

Reference should be made to Chapter 17 *Strength and deformation behaviour of soils* for the interpretation of different strength parameters and 'ideal' triaxial tests. All of the triaxial tests listed above require undisturbed samples to be taken. Information and guidance is given later in this chapter as to the sampling types and methodologies, which should be planned from the outset of the ground investigation. It can also be seen from the list of tests above that rates of testing are mentioned. For effective stress tests pore water equalisation is required to measure strength and stiffness correctly (see Section 49.6).

Figure 49.2 shows a typical triaxial loading frame with a modern cell pressure and backpressure control system on the left and the transducer logging system on the right. Historically, pressures were applied manually, and the measuring instruments were read manually by either reading dial gauges or writing down the specific outputs from digital readout units. Modern technology has advanced triaxial testing so that stress control and instantaneous measurements of transducers are becoming increasingly the norm. Load measurement can also be carried out in several different ways in laboratories, and **Figure 49.3** shows the types of load measurement commonly available.

The original external load-measuring devices commonly used were in the form of a load ring whose calibrated deflection could be read manually by an operator. These were superseded by load rings with integral digital readouts which can be manually read or their output logged by computers and read remotely. Electronic load measurement devices have become common now, but as with all previous versions the drawback of all of these is that they are generally not waterproof and therefore need to be mounted externally to the cell. This leads

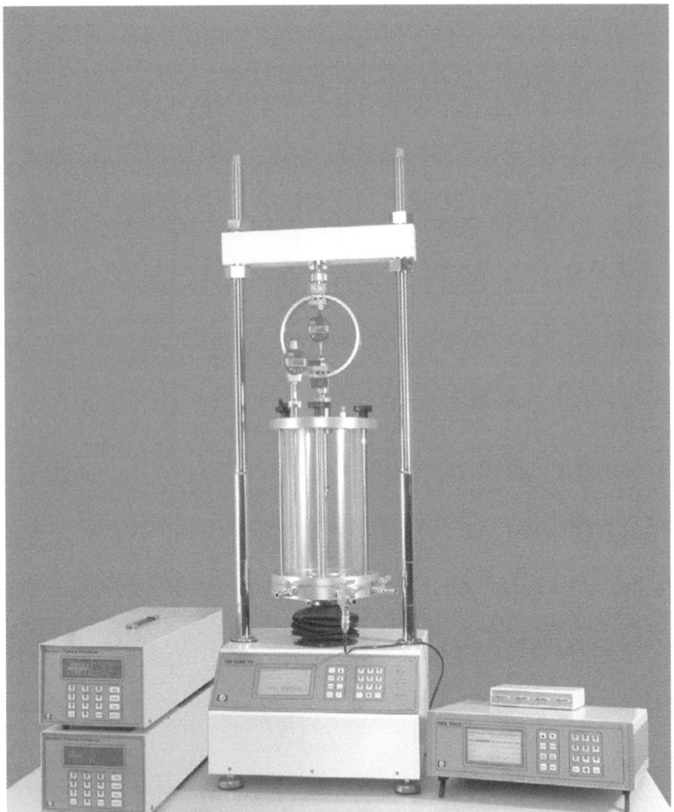

Figure 49.2 Triaxial test equipment (Courtesy of VJ Tech Ltd)

to the load transducer measuring the friction of the ram as it passes into the cell, leading to errors in load measurement. Calibrations of ram friction can be made to minimise this error; however, slight non-concentric loading can lead to the ram 'sticking' and giving 'false' load readings. It is possible to utilise rotating bushes where the ram passes into the cell, but these are not without problems themselves. The ultimate load-measuring device is the submersible load cell which is mounted on the end of the ram and remains in direct contact with the sample within the cell while any loading occurs. In essence, any load which is applied to the sample will also be seen by the submersible load cell without any external effects and so is the ultimate load-measuring instrument. The cost of such devices is often prohibitive in many institutions and so external load measurement is most likely to be the norm.

It should be noted that testing of organic soils, especially peat, can produce rather 'unexpected' test results compared to those associated with sands and clays. This is due to the type, fabric, percentage and orientation of organic material which may be present in such samples. Undrained tests do not take into account the high compressibility of such materials and would yield low shear strengths, while drained shearing stages will display very high strains and again possibly unrealistic shear strengths due to the complex nature of the material and the boundary conditions which exist in triaxial samples. These tests are certainly possible, but care should be taken in the design and expectations of shear strength tests on such materials. More will be said about such materials later.

Rock strength is dominated by its mineralogy and cementation, along with the presence and orientation of discontinuities. Reference should be made to Chapter 18 *Rock behaviour* for further understanding and categorisation of the material. Commonly strength tests are carried out as uniaxial compressive strength tests (unconfined), but may also be carried out as confined or even effective stress tests using specialist high pressure/stress equipment. Reference is made to the main standards for such common test types in the Appendix to this chapter. It should be noted that rock tests require very different equipment for preparation and testing than soil tests; however, there is a grey area where a soil might be classified as a weak rock and vice versa. In this instance, experience will prevail over which tests and equipment to adopt. As with all testing, these will require specialised personnel with significant skill and experience. The 'art' of these tests is in

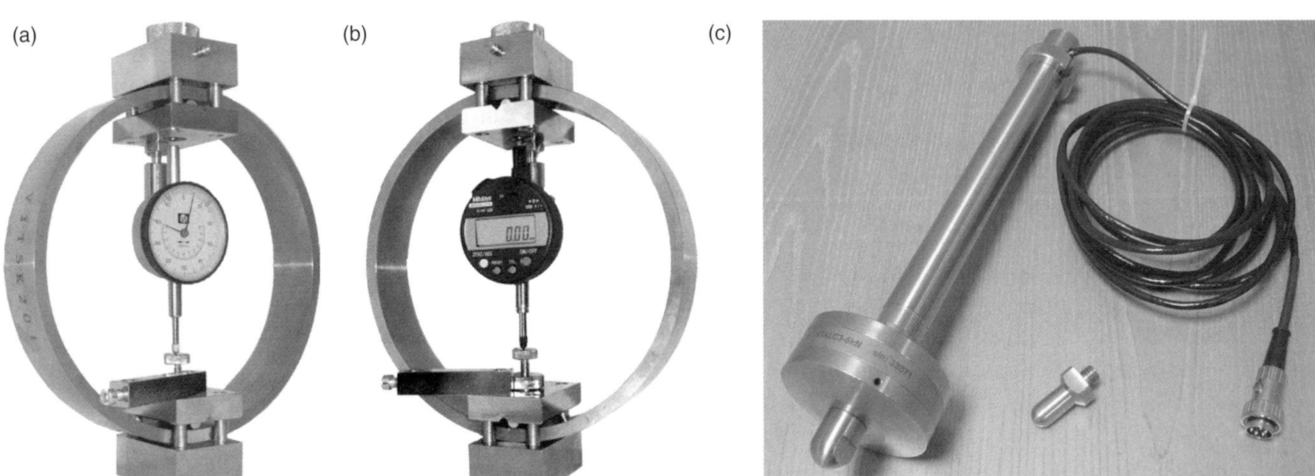

Figure 49.3 Triaxial load-measuring equipment: (a) load ring with manual dial gauge; (b) load ring with digital dial gauge; (c) submersible electronic load cell (Courtesy of VJ Tech Ltd)

the sample preparation and the specific equipment used (along with sample orientation with regard to discontinuities and preferred fabric). Check with your nominated testing laboratory that they have the correct equipment and expertise to carry out such tests.

The cutting and facing of the sample if carried out incorrectly can reduce the strength of the test specimen by up to two-thirds by the introduction of point loads and non-parallel faces. Cutting equipment should have very thin diamond blades and work by the rock core being moved (while rigidly supported) across the cutting blade, not vice versa as in concrete cutting equipment. All cutting marks must then be removed by facing. This involves polishing the sample surface until both ends are completely flat and parallel (tolerances can be found within the standards listed in the Appendix at the end of this chapter). The samples must then be mounted on specially hardened platens (which are calibrated for flatness) which have a diameter either the same or no greater than 2 mm larger than the specimen diameter, one of which will have a spherical seat and be placed on the top of the specimen. The loading surfaces of the compression machine itself will be rigid, parallel and unable to rotate. Please appreciate that rock testing is a highly specialist form of material testing and there are few laboratories which can carry out this form of testing correctly.

49.5.2 Direct shear tests

Alternative strength tests such as the shearbox test are also common where phi angles and cohesion intercepts are required for design purposes. Such tests can be carried out on both undisturbed and disturbed samples (depending upon the desired engineering use for the material). The historical test is the shearbox, which can yield values for phi and cohesion (see **Figure 49.4** for the test mechanism).

Originally designed for testing sand, the equipment is now used commonly on non-cohesive and cohesive materials alike. The equipment comes in a range of sizes which provide testing for a range of particles from fine through to coarse. The shearbox consists of a 'hollow box' which is split horizontally and into which a sample is placed. The sample is consolidated to a desired normal stress and then sheared horizontally. The bottom half of the box is displaced, while the top half of the box reacts against a load-measuring device measuring the resistance to shear. These tests are usually carried out as a set of three tests, where the first test is consolidated at half the calculated effective stress for the material, the second is at the calculated effective stress and the third is at double the calculated effective stress. The values from these tests should be used in total stress calculations only (although there is some argument). The only point at which the effective stress of the material is known (during the test) is at the end of the consolidation stage. As pore pressures are not measured throughout the following shear stage it is not possible to verify the true effective stress of the material at failure and certainly along the plane of shear.

Consolidated peak strengths are obtainable from this equipment along with residual strengths.

Residual strengths in clays in particular are often in error from this equipment as it is not possible to form a perfectly flat shear plane. Residual values are reached only at very high strains compared to peak characteristics, and the shearbox has limited travel. Some attempts to overcome this are in the test standards by reversing the direction of shearing until a consistent 'apparent' residual state is measured. For non-cohesive materials and silts this method may work, but due to the 'platey' nature of the clay particles the perfect (flat and polished) shear plane can only develop by continued shearing in a singular direction (no reversals as per the standard shear box). This problem was overcome with the invention of the ringshear apparatus (E. Bromhead). The normal loading and relative displacements induced during the test can be seen in **Figure 49.5**, while a commercially produced ringshear apparatus is shown in **Figure 49.6**.

49.5.3 Ringshear test

Rather than linear movement of a block of soil, the ringshear rotates constantly and so the linear displacement (in a single direction) is limitless (or at least until all the soil has been 'squeezed' from the test annulus). Note that two opposing load-measuring devices are used on this equipment. This is necessary as the cell rotates and it is actually torsion which is measured. The use of two load measurements (being of matched stiffness) balance the top cap under rotation and are designed to prevent friction being measured from the central locating pin for the top cap.

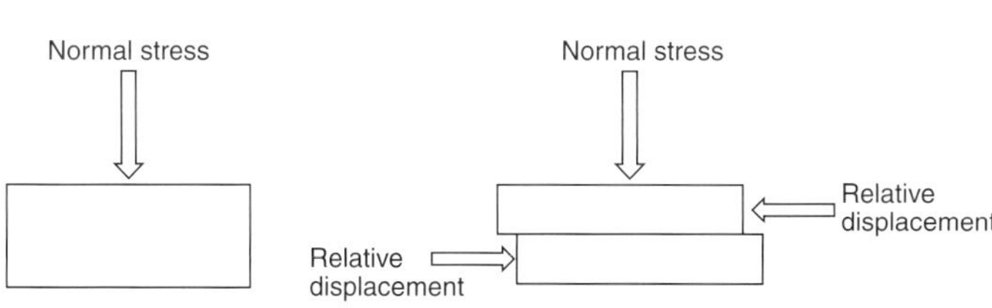

Figure 49.4 Loading and shear displacement in shearbox test

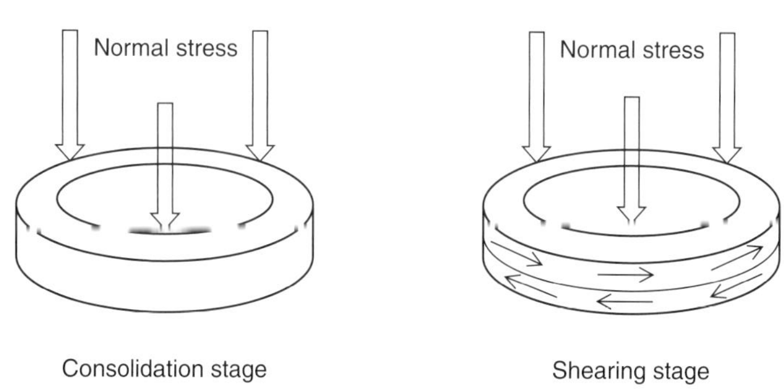

Figure 49.5 Loading and shear displacement in ringshear test

Figure 49.6 Ringshear apparatus (Courtesy of VJ Tech Ltd)

If residual values are needed (say for slope stability calculations) in clays, this is the test to perform. Peak strengths are not available from the 'Bromhead' or 'small' ringshear due to the fact that the sample must be remoulded as part of the test preparation. The material needs to be pressed into a tight annulus and benefits from having any structure or bonding destroyed. This allows for accelerated reorientation of the clay particles (shortening test times) and more repeatable test results through the homogeneity of the sample. This equipment is suitable for pure clays only. Coarse particles may roll along the shear surface during the test and will destroy it by reorientating the clay particles. This will lead to higher (non-repeatable) residual values being obtained, which may be disastrous where the true residual angle is actually lower still. For clays containing coarser material and non-cohesive samples the engineer should revert to the standard shearbox test (with its known and well-documented limitations).

49.6. Stiffness

There are many technical definitions of stiffness, but they all relate to the gradient of the line of stress plotted against strain and it is most important to consider that soils display a nonlinear stress–strain response (outside the highest levels of research). It should be remembered that the strength of the soil dictates its ultimate load-bearing capacity with large strains, while stiffness identifies the compressibility (or strains) in the material at working loads (**Figure 49.7**).

In general these parameters are not derived from the routine tests listed previously and require the highest-quality undisturbed samples, specialist capabilities, instrumentation (**Figure 49.8**) and high levels of knowledge and experience. For additional information on stiffness of rock material, reference should be made to Chapter 18 *Rock behaviour*. It is possible to fully instrument the following triaxial tests for the determinations of Young's modulus: UUP, CIU and CID. Shearing of samples may take place in either compression or

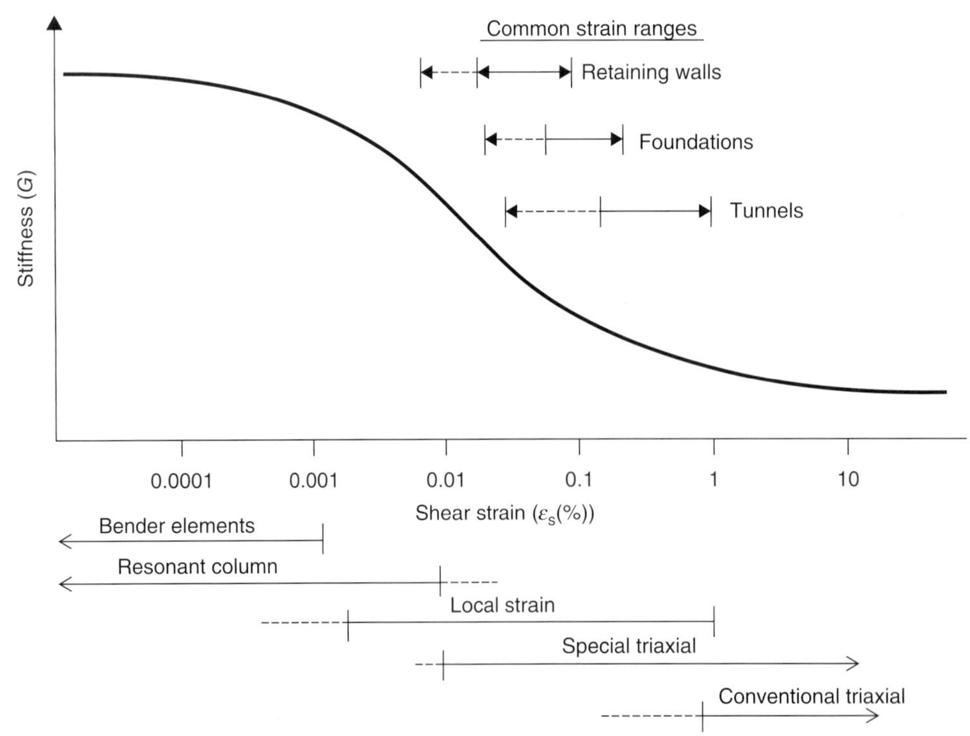

Figure 49.7 Idealised variation of stiffness with strain

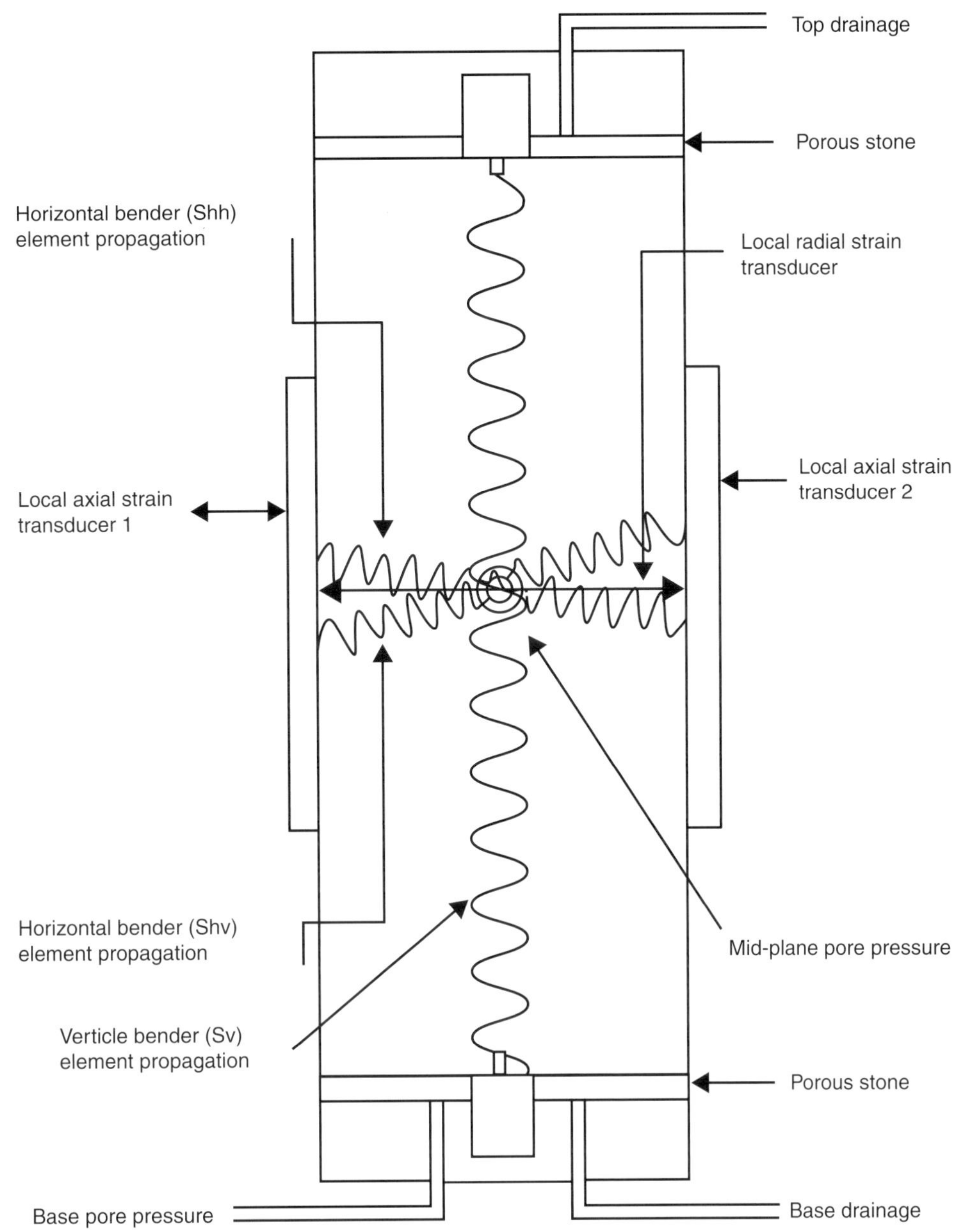

Figure 49.8 Typical advanced triaxial instrumentation for CAUC, CAUE, CADC and CADE test types

extension and so a suffix may be added to the above abbreviations in the form of a C (compression) or E (extension).

The highest level of triaxial effective stress test is known as a stress path test or anisotropically consolidated undrained (or drained) triaxial test (CAU or CAD) and again may have the C or E suffix depending upon the final shearing direction. High-resolution transducers are attached directly to the sample to measure axial strains for the determination of Young's modulus, and when combined with a radial strain transducer can be used to determine Poisson's ratio, shear modulus (G) for undrained shearing and bulk modulus (K) for drained shearing. In addition, it is possible to measure G_{max} using bender elements which measure sample stiffness beyond the resolution of local small-strain instrumentation (see **Figure 49.8** for typical advanced instrumentation).

These tests are designed to take the specimen through its recent stress history in order to minimise any effects of sample disturbance and return the specimen to its true in situ mean effective stresses or to a particular stress level required for modelling.

Figure 49.9 shows an advanced triaxial test with local axial and radial strain instrumentation, base/mid-plane pore water pressure measurement along with measurement of G_{max} using bender elements in all three possible directions. This specialist

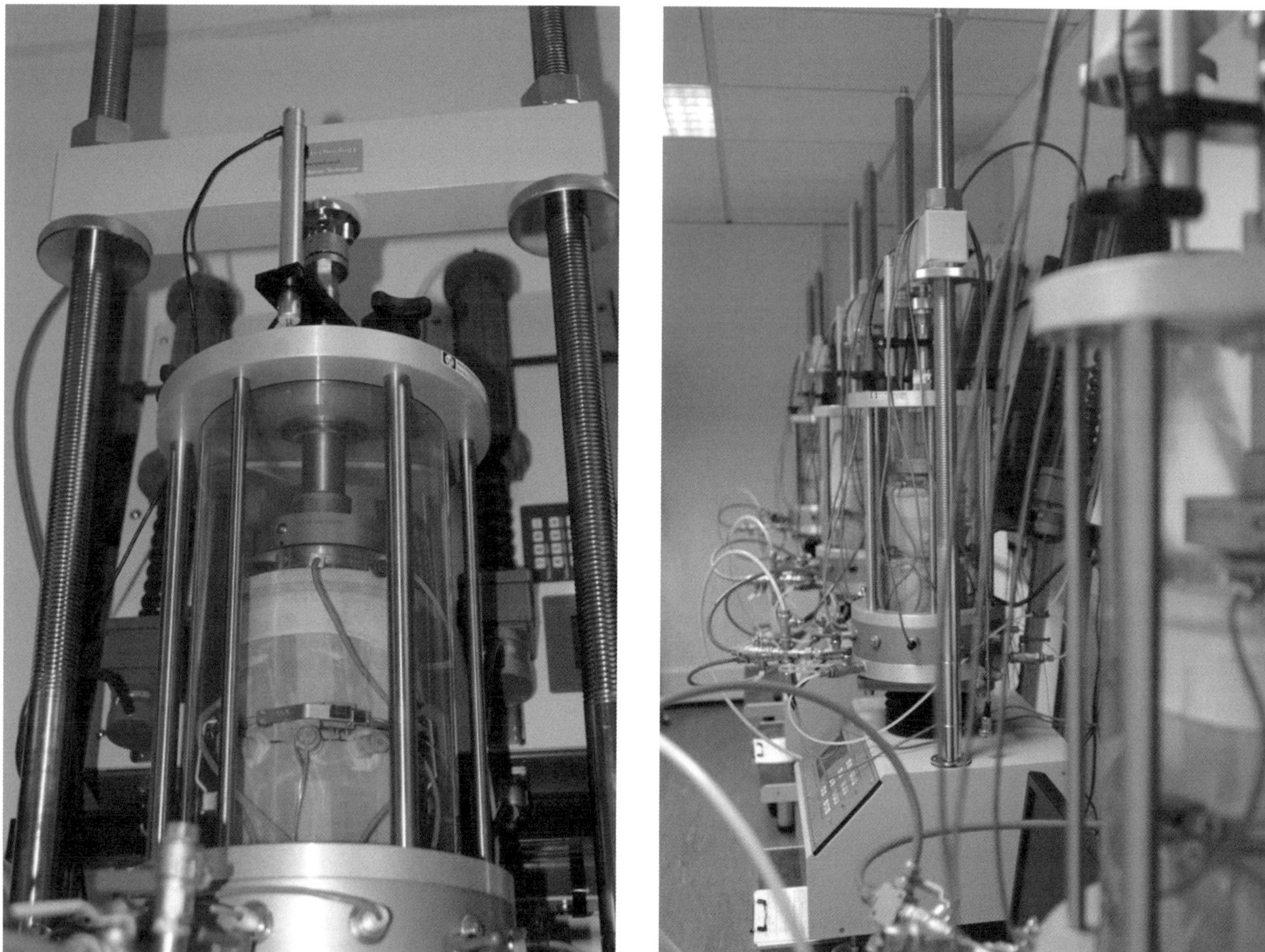

Figure 49.9 Triaxial stress (CAUC) tests with advanced instrumentation (Courtesy of Russell Geotechnical Innovation)

instrumentation is used not only for the measurement of very small strain stiffness, but the utilisation of both base and mid-plane pore water pressure measurement allows verification of effective stress measurement.

The use of local axial and radial strain transducers allows the measurement of small strains directly on the sample and minimises the boundary effects (bedding and localised stress distributions) caused by the soil being in contact with the (much stiffer) end platens. Bender elements measure stiffness at even lower strains than local strain instrumentation and provide a completely non-destructive measurement of stiffness. They can be mounted in three orientations on a triaxial specimen and provide a very good indication of general sample condition and/or possible anisotropy.

During normal triaxial testing the pore water pressure is traditionally measured only at the base of the sample throughout the test. As the sample is in contact with far more rigid (usually metal) base pedestal and top cap, the areas at the specimen ends suffer boundary effects caused by the contact with the far stiffer material. For the calculation of shearing rates (time to failure) as per BS 1377:2022 (or, alternatively, BS EN ISO 17892, among others) one can calculate that pore pressure dissipation is 95% only at the time of failure of the sample. The use of mid-plane pore pressure allows a second reference for pore pressure measurement and not only is unaffected by the metal pedestal (as it is in the central region of the sample), but can be used to display full pore water pressure equalisation throughout the sample (and not just at one end). This is ideal for drained shearing stages where the base and mid-plane pressures should remain the same for shearing at the correct rate. If the mid-plane pressure begins to deviate (increase) from the base measurement, then excess pore pressures are being generated due to the sample being sheared too fast. For undrained shearing the pore pressures should react in unison and in the same direction (whether in compression or extension), again providing evidence that the correct shearing rates have been used. Interestingly, deviation between the base and mid-plane may occur later during sample rupture as pore pressures may be generated/dissipated in different ways depending upon the inclination and orientation of any shear planes which may form in the specimen (and continued shear along rupture surfaces).

For rocks one can measure stiffness by carrying out a more advanced version of the uniaxial compressive strength test which also has additional high-resolution instrumentation that measures axial and radial strains. Because rock specimens fail at strains (usually an order of magnitude) lower than those of soils, specialist strain transducers (unsuitable for soil testing due to their very small range) are bonded axially and radially to the central third of the rock specimen. These are logged throughout the loading test and used to measure Young's modulus and Poisson's ratio.

Again, high technical competence is required for these tests. An ideal example is that the bonding agent used to bond the strain transducers to the sample must be able to not only prevent the gauges from creeping on the sample by bonding the gauge fully and remaining so throughout the test, but be of a lower stiffness than the sample itself so that they measure the natural strains evolving in the sample under load rather than the artificial strains caused by a stiffer bonding agent which may have locally filled the voids within the sample (causing a localised 'stiffer' response).

49.7. Compressibility

This is a term often used in soil mechanics and largely describes the relationship between stress and strain. It is the stiffness of the ground which determines the strains and displacements with changing stress, and so by combining the stress level and stiffness of the material its compressibility can be determined. The following are laboratory tests associated with these parameters.

- *Oedometer consolidation test.* Useful for Cv and Mv. The coefficient of consolidation (Cv) has the units M^2/year, which is the 'scaled up' time (for consolidation) from the laboratory test to the full-scale field material being modelled. The coefficient of volume compressibility (Mv) has the units M^2/kN and is the slope of the porosity against the applied effective stress curve resulting from the test. Since porosity and void ratio are related quantities, it also follows that Mv can be calculated from the void ratio against effective stress curve. It should always be remembered that the value of Mv is dependent on the stress level applied.
- *Hydraulic cell consolidation test.* This is the full effective stress version of the above and is able to impose various loading conditions and drainage paths on the specimen. In addition, the hydraulic cell can be used to measure the permeability of the sample at each consolidation stress along the drainage path used for the test.

Typical test equipment for these two types of test can be seen in **Figure 49.10** and a simplified line drawing of the loading and boundary conditions is given in **Figure 49.11**.

Reference should be made to **Figure 49.12** for an idealised consolidation curve. The first stage (primary consolidation) is the result of reorientation and repacking of the soil particles with the expulsion of water from the voids. Secondary consolidation is the actual compression of the soil particles themselves with a further expulsion of water (mainly from the particles as they compress). It is most important to identify the presence of organic materials, especially peats, and understand their

(b)

(a)

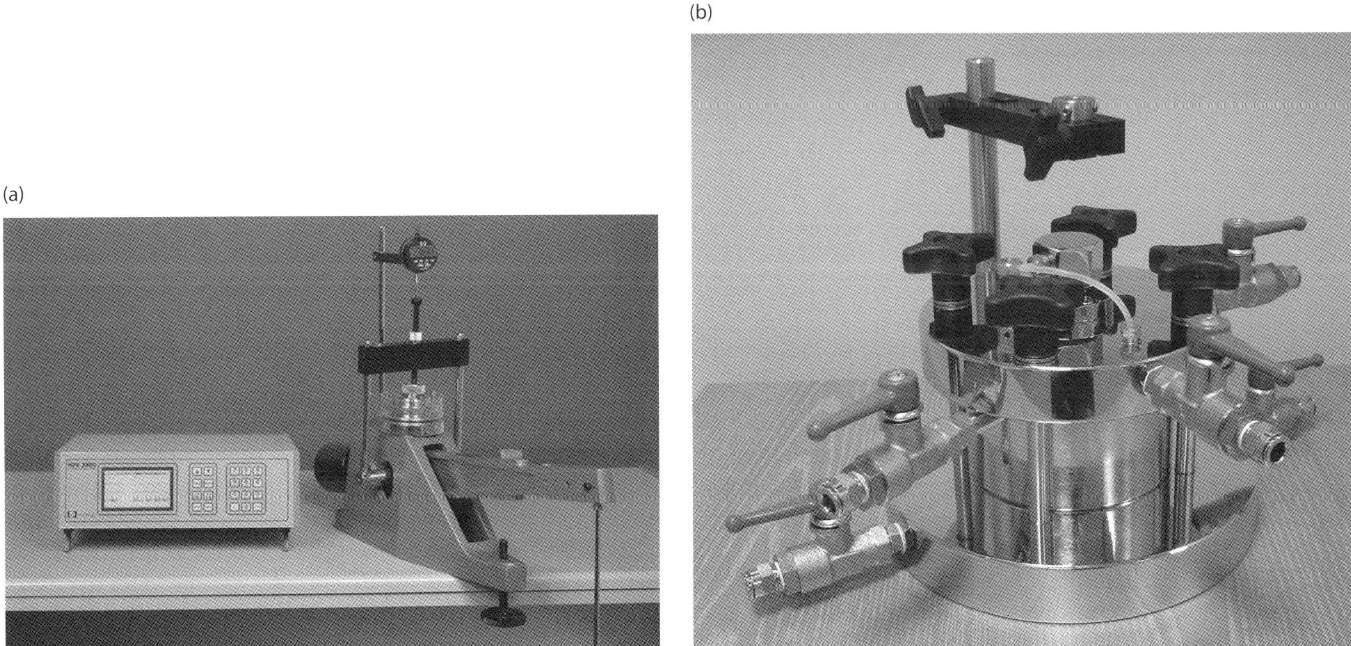

Figure 49.10 Oedometer (a) and hydraulic cell (b) (Courtesy of VJ Tech Ltd)

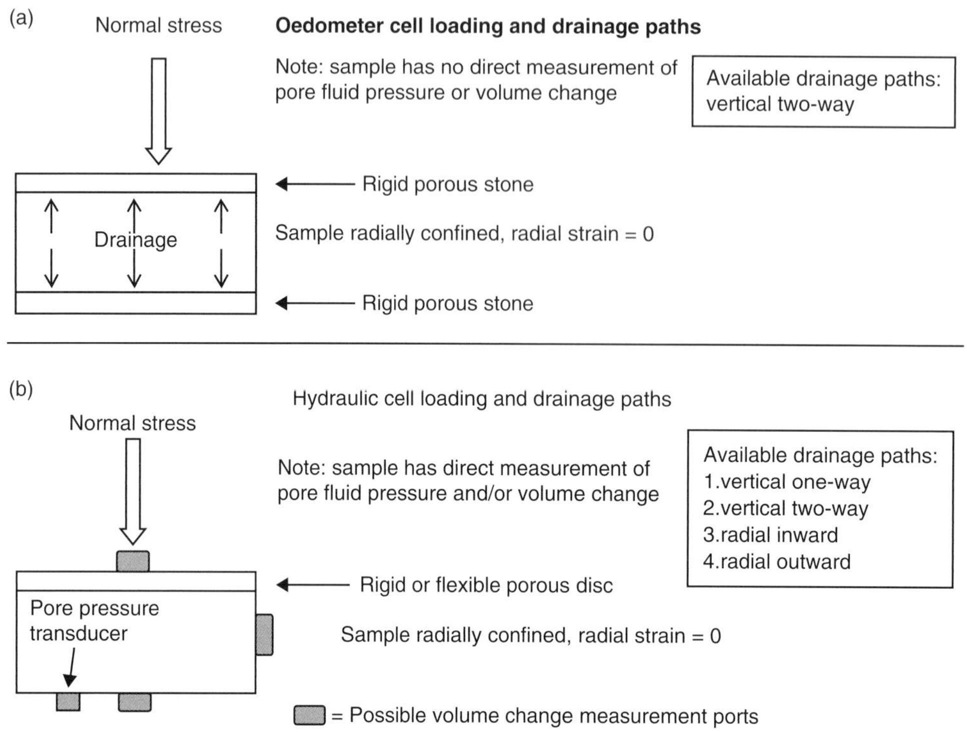

Figure 49.11 Oedometer (a) and hydraulic cell (b) loading and drainage paths

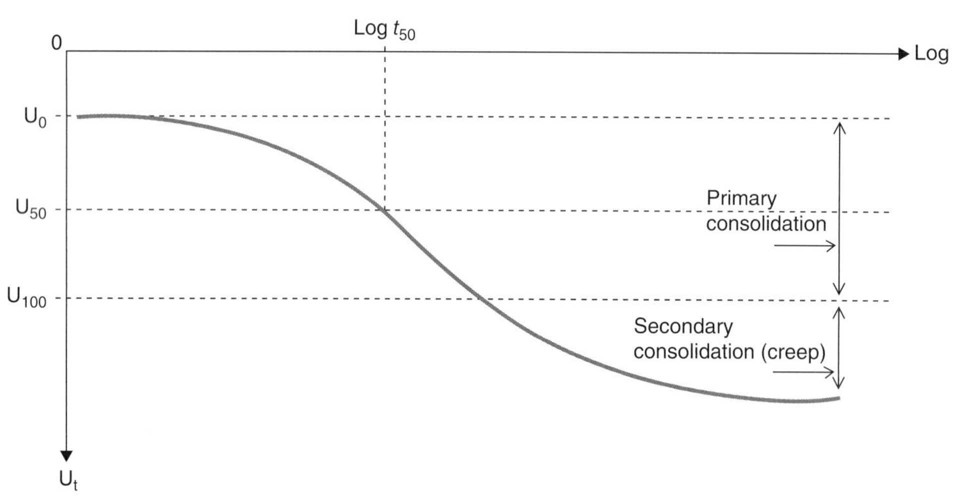

Figure 49.12 Log time consolidation curve

compressibility characteristics as settlements may be orders of magnitude higher than those associated with non-organic soils. Both the oedometer and hydraulic cell can also be used for secondary consolidation or 'creep' monitoring. Secondary consolidation is the continued compression of a soil after primary compression is complete and is caused by viscous behaviour of the soil grain–water system or the physical compression of organic matter. For quartz sands creep (secondary compression) is expected to be negligible as the sand grains are generally non-compressible. Other sand types (such as calcareous or carbonate sands) may show different behaviour or 'sudden collapse' if the normal stress exceeds the ultimate strength of the grains (and their asparites), leading to failure of the grains themselves. For peats and organic soils the soil 'solids' are themselves highly compressible and so a two-phase consolidation process is often seen with organic materials.

Secondary consolidation also occurs in clays and can play an important role especially when dealing with very soft clays.

When scheduling tests with the requirement for secondary consolidation parameters, always bear in mind that these tests may take days, sometimes weeks, per loading stage and have not only time but cost implications to match over the 'standard' tests where the interest is in primary consolidation only. It should also be noted that due to the fabric/structure of peaty materials they may show highly anisotropic behaviour especially with regard to drainage directions due to the orientation of the long axis of the vegetative material causing preferred drainage paths. In instances where these preferred paths have been identified, it may well be preferable to schedule the use of the hydraulic cell and schedule a suitable drainage path/direction.

For rock materials compressibility from the intact strength of the rock material is often assessed along with its discontinuity spacing and aperture (rock quality designation – RQD). It is possible to test rock samples for compressibility in high-stress equipment; however, remember that you are only testing the intact material. In situ the bulk strength and compressibility of rock materials are usually dominated by their discontinuity spacing/orientation, aperture spacing/orientation and aperture contact areas. Other means (preferably field-based) should be used to assess these engineering characteristics.

It can be seen from the parameters gained from such tests that the soil structure and fabric (which controls drainage paths) are the controlling factors (material properties) that are being measured. Undisturbed samples of known orientation are the basic prerequisite for such tests. Remoulded samples may be used if the construction requires design parameters to be used for such 'engineered' materials. Attention to sample quality and test preparation is paramount, especially as such small samples are tested and then the results scaled up to the field model. Small errors magnify!

49.8. Permeability

For design purposes or interest in seepage problems drained and undrained conditions or short- and long-term behaviour of the ground need to be considered. This is dominated by the particle size, orientation and packing of the soil grains and whether they are cemented (in the case of hard pans or calcareous zones), cohesive or non-cohesive. Careful thought should be given to the sampling of such materials as the size, orientation and packing of the particles can lead to strong anisotropy in situ, leading to high variations in permeability with flow direction. Sampling and the correct orientation of samples selected for laboratory testing is of paramount importance if true ground conditions are to be modelled representatively. The following are laboratory tests associated with the determination of permeability.

- *Constant head permeameter.* For non-cohesive materials. The sample is radially confined, therefore lateral strain is zero.
- *Constant head permeability determination in a triaxial cell.* Normally for cohesive materials and where desired mean effective stresses are required. The sample is isotropically consolidated and a pressure differential applied (driving head) to the separate top and base drainage lines causing flow to occur.
- *Permeability determination in a hydraulic cell.* This equipment, mentioned previously, can be used to measure permeability along different vertical or horizontal drainage paths and can be carried out as part of a test which also gives consolidation parameters. For the possible flow directions available, see **Figure 49.10**.

For rock samples, intact specimens can be tested (usually constant head), but it should be remembered that the permeability of the in situ material may be dominated by the presence of discontinuities and their aperture/orientation, infill, spacing and persistence.

49.9. Non-standard and dynamic tests

When discussing the broad range of basic (and some more advanced) parameters that can be obtained from the test types outlined previously, it would also be prudent to write a little about the more 'advanced' types of test that are available in a few specialist laboratories. These tests are certainly not routine and require the use of highly specialist equipment and highly experienced staff.

Due to modern construction requirements and the need for design with dynamic loadings (such as wind turbine monopiles), geotechnical engineers are increasingly asked for the dynamic parameters more associated with those for foundation design in earthquake regions. Other tests are not necessarily 'dynamic' but are geared more towards the research end of testing. As with the anisotropic triaxial tests and their advanced instrumentation, these test types are often more appropriate to advanced numerical analysis designs and studies. They are certainly not routine and often come with a price tag to match; however, you do get what you pay for (as long as open and clear communication prevails throughout).

The following list is not exhaustive but is intended to highlight some of the more common research-level tests available and the apparatus associated with them.

49.9.1 Cyclic triaxial test

This is a 'hybrid' triaxial testing frame which is built to 'cycle' the soil/weak rock sample by either stress or strain control around a mean level at rates commonly around 0.1–1 Hz, with data capture of the transducers at many times a second. Cyclic strength depends upon many factors, including density, confining pressure, applied cyclic shear stress, stress history, grain structure, age of soil deposit, specimen preparation procedure and the frequency, uniformity and shape of the cyclic waveform (ASTM D5311; ASTM, 2013). In addition, it should be noted that non-uniform stress conditions are imposed by the specimen end platens, which may cause a redistribution of void ratio within the specimen during the test. These tests are often carried out on non-cohesive soils and, since such materials are unable to withstand tension, the maximum cyclic shear stress that can be applied to the specimen is equal to one-half of the initial total axial pressure (ASTM D5311). Care should obviously be taken in the design of such tests, and thought given to the fact that uneven pore pressure distributions throughout the sample may result, depending upon the permeability of the soil and the rate at which it is cycled. Young's modulus and soil damping properties can also be evaluated for specialist design using this test type (ASTM D3999; ASTM, 2020).

49.9.2 Simple shear

This test can be carried out either as a monotonic test or as a dynamic cyclic test. The shear strength is measured under constant volume conditions that are equivalent to undrained conditions for a saturated specimen; hence, the test is applicable to field conditions where soils have fully consolidated under one set of stresses, and are then subjected to changes in stress without time for further drainage to occur (**Figure 49.13**). The constant volume (undrained strength) is a function of stress conditions (plane strain) and the principal stresses continuously rotate due to the application of shear stress. This simple shear stress condition occurs in many field situations, including zones below a long embankment and around axially loaded piles (ASTM D6528; ASTM, 2017).

49.9.3 Resonant column

Figure 49.14 shows a resonant column with the cell top removed. The top part of the equipment is used to induce torsional movement to the top of the sample by an electro-magnetic drive system which can run at a range of frequencies to determine the resonant frequency of a sample.

Such tests and test equipment are used to evaluate the shear moduli and damping characteristics of soil at very small strain amplitudes. Although there are two distinctly different equipment types for these tests, both apply torsion/rotation to the top of the sample in order to find the resonant frequency of the material at a controlled stress. These test methods are non-destructive if the strain amplitudes are less than 10^{-4} radians and many measurements may be made on the same sample and with various states of ambient stress (ASTM D4015; ASTM, 2021).

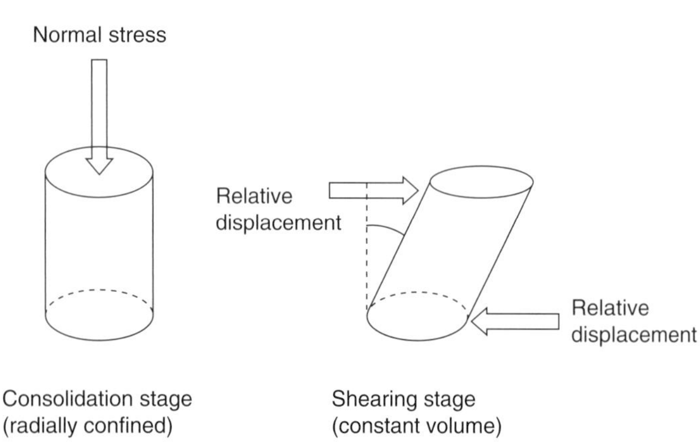

Figure 49.13 Simple shear mechanism

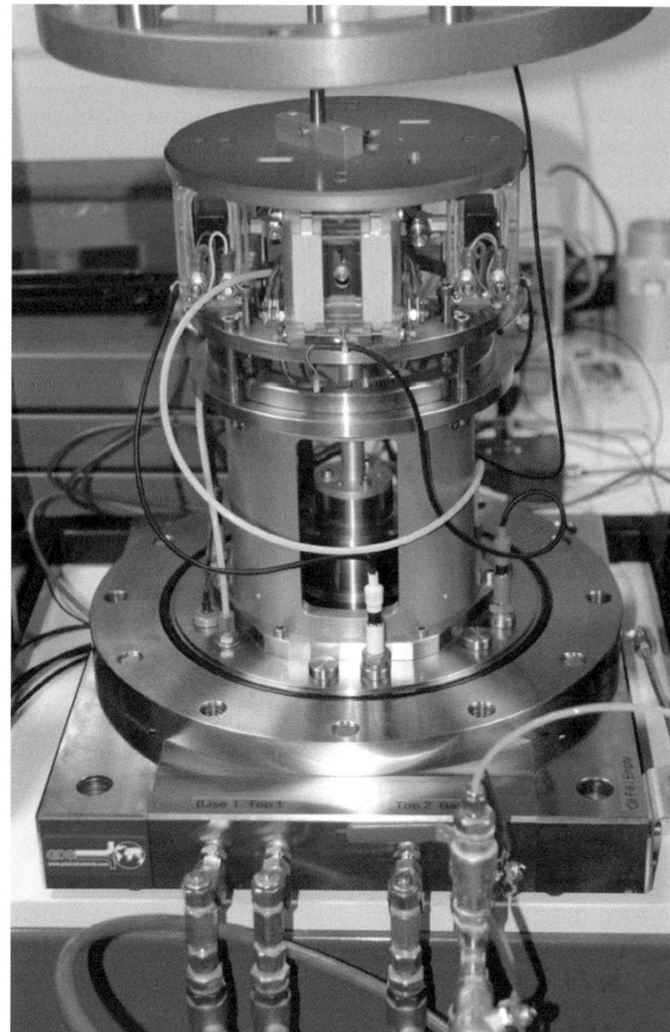

Figure 49.14 Resonant column apparatus (Courtesy of Russell Geotechnical Innovations)

49.9.4 Hollow cylinder test

Certainly the rarest of commercial tests, this equipment allows a rotational displacement to be imposed on a 'hollow' cylindrical specimen where independent control can be maintained for all three principal stresses (unlike triaxial tests which can only independently control two of the three principle stresses (where $\sigma_2 = \sigma_3$)). For this reason studies can be made of the intermediate principal stress (σ_2), sample anisotropy and the effects of principal stress rotation. These are 'research-level' tests and the parameters derived are usually only used for the most advanced numerical analysis. Tests are available for both soil and rock and the hollow cylinder is most useful in the definition and determination of anisotropic material properties.

Simpler but equally non-routine tests can be used to model specific behaviour of soils and materials. The shearbox can be used to model the frictional behaviour which occurs at the interface between soil and a geotextile/steel/concrete surface. A ringshear can be used to model the soil behaviour at an interface between soil and a steel pile (Jardine *et al.*, 2005). In addition, tests can be carried out to determine the dispersion or erodibility of soils and rocks when abraded or exposed to persistent high moisture content or flowing water. The tests listed in this section are certainly non-exhaustive and complete volumes could be written about laboratory testing and parameter determination. A good general series of volumes to read is the *Manual of Soil Laboratory Testing* by Head (2014).

49.10. Test certificates and results

Test results are issued as certificates which identify the parameter requirements, data and graphs in accordance with the test methods and standards used. There should be no problem requesting further detail from the laboratories involved, to understand or verify any particular test conditions or methods used. It is here that open communication is important, and always remember that you are the paying customer. This line of communication should also allow the laboratory to freely communicate any observations or potential sample problems which may lead to unexpected results. It is this knowledge sharing that will improve the resultant parameter and overall design quality.

To supplement the ground model and design requirements careful planning at all stages of the ground investigation process is required – hence why the natural process order of this chapter has been reversed somewhat. This process of investigation from sampling techniques through to sample storage, transport and laboratory testing requires careful planning and supervision. Any loss in integrity of the sample material properties at any stage in this sequence could have catastrophic effects on the parameters supplied from the testing house. These parameters are largely interlinked, especially moisture content and the sample physical integrity, which dominate the effective stress characteristics. Geotechnical engineering is one of the only sciences which starts with a natural material being removed from a stable environment (the ground) and taken through a series of potentially damaging processes, through water addition/loss, exposure to atmosphere, physical handling (including jarring and vibration), temperature cycles and, finally, to a 'stable' environment where the material is tested for specific parameters which are to be representative of the ground from whence it came. How is this done? This is the difference between a good site investigation (and all of the processes involved therein) and a not so good site investigation.

Fundamental in this way of thinking is the preservation of the sample moisture content. The determination of moisture content in the laboratory is only accurate if the sample retains the same moisture content that it had at that time in the ground. Bulk density, strength, effective stress and stiffness, to name but a few, are all controlled by moisture content. If this is allowed to change between sampling (or altered by the sampling method) and the final test then the design will be based on erroneous values.

To realise this from the planning stage will help to identify possible problems and build them into the specification, and drive the ground investigation in the correct way from the start. If all stakeholders are involved in the initial stages and the project expectations and responsibilities are clear, then the majority of these 'integrity loss' components can be minimised. The sample tested in the laboratory is only as good as the sample received there and this also assumes that the correct test was scheduled and that the laboratory was proficient in that particular method.

49.11. Sampling methods

This part of the project will have been conceived at an early stage as the ground investigation contractor will need the relevant equipment and correct experience for the project awarded. They will need to extract the samples from the ground and to deliver them to the laboratory for testing in the best possible condition. A chain of custody is formed and the 'smooth' operation of this will depend upon the sharing of information and open communication which will allow some flexibility to be built in for 'on-the-job' improvement. In an ideal world the laboratory will test a soil or rock which is in the same condition (and is therefore entirely representative) of the material in situ. This chain begins with the excavation of a trial pit or the drilling of a hole. A sample is then taken in various ways (to be explained later in more detail) and sealed in order to maintain its integrity. This sample is then either stored or immediately transported to a laboratory for testing, where again it may be stored (in a queue) while awaiting testing. Sampling for chemical and contamination testing is dealt with in Chapter 48 *Geo-environmental testing*, which should be referenced as necessary. This chapter deals specifically with sampling of the ground for the physical testing required for design parameters. There are many sampling methods available globally, but they may be categorised simply as bulk samples, block samples, tube samples and rotary cored samples. Each category has associated levels of disturbance and some indication of these is given along with the basic requirements for sample preservation. The following list is certainly not exhaustive but should indicate the main principles for 'good practice'. The reader must also adhere to the provisions of Eurocode 7 or other prevalent standards, depending upon the geographical location of the investigation or agreed project requirements. Eurocode 7 is very prescriptive in terms of the sample types that may be used for various types of test, as are many of the other standards generally used worldwide.

49.12. Bulk samples

This constitutes probably the simplest but the most disturbed sample type. Samples are often hand- or machine-excavated from a trial pit or spoil heap and placed in bags for logging purposes or index tests only. These samples should be of sufficient size to be representative of the horizon of interest, uncontaminated by material from other horizons and of sufficient quantity for the testing required.

For bulk samples the material should be sealed in a bag with as much air evacuated as possible in order to prevent the sample from 'sweating' or the production of mould during storage. No samples, even if fully sealed, should be left in direct sunlight (or near heat sources) as this will not only cause the sample to 'sweat', but will also cause non-uniform heating leading to expansion/contraction of any fissures or textural fabric or aid the growth of mould, fungus or microbial organisms which again may alter the material properties. Any sample should be kept at a constant temperature and away from any localised heat sources. In the UK such temperatures should be no more than 20°C and no less than 5°C (under which the sample may begin to freeze). This is a range of temperature in which the sample may be kept, but it should not be cycled more than 3°C over a mean temperature if at all possible.

49.13. Block samples

Block samples are undisturbed hand-dug blocks which are usually some 0.5 m square or diameter and at least 0.3 m in depth. They are then sealed with an impermeable barrier followed by a rigid supporting container constructed around them. This is followed by careful paring from the substrata and removal for total sealing and support for storage and transport. Bearing in mind the preservation of the sample which is required, site- and environment-specific methods would be required for taking block samples and preserving the in situ characteristics of the medium sampled. Such samples are limited by access and depth of interest as space is required for personnel to cut the block safely and extract it from the horizon of study.

49.14. Tube samples

Tube samples are taken in a variety of ways depending upon equipment available and access.

The most common tube sampler used in the UK is the UT100 (for class 1 samples – BS EN 22475-1; BSI, 2021) and is often used in conjunction with light percussion drilling techniques (tripod-type rigs), where the tubes are driven into the ground. Developed in 2009, this sampler replaces the original U100, which is lowered to class 2 in BS EN 22475-1). One can easily visualise the disturbance the material may undergo by having a tube 'hammered' into it. This is part of the reason why U100 tube samples were deemed unsuitable for undisturbed testing parameters (but fine for index tests only and logging purposes). The replacement (the UT100) sampler is an aluminium tube (with square threads) which has a redesigned cutting shoe with a lower area ratio. These tubes were deemed as providing 'undisturbed' samples, allowing for effective stress triaxial tests and so on to be carried out.

Interestingly, as time has passed a distinct problem has arisen. Where the thread stops at the top of the tube, the metal is very thin and has a tendency to break when pulling the sample tube out of the bottom of the borehole. This has led to many drillers using steel UT100 tubes as an alternative. Unfortunately, as soil is moist the sampled material has a very strong tendency to react with the steel tube if not extruded quickly. The result is that the soil sample corrodes the inside of the tube and almost 'cold welds' itself to it. On extrusion (if it is actually possible to extrude it) the sample exterior is entirely altered and the pore fluid chemistry has changed, leading to the sample being highly disturbed (and unusable for almost all forms of testing in the opinion of the author). This issue should really be addressed! Tubes may also be 'pushed' into the substrate using piston-type (fixed piston) equipment which 'jacks' the tube into the bottom of the borehole. This method is superior, and when used in conjunction with thin-walled sample tubes provides undisturbed samples of acceptable quality. For thin-wall push samples, make sure the end of the tube has been sharpened, has no burrs and is straight and true (at least before it is used). After sampling, check that the tube has remained straight and true. If the tube has buckled or deformed during the sampling process then the soil within will have deformed (strained) as well, rendering it unsuitable for high-quality parameter determination. For tube sampling the basic stages are outlined in **Figure 49.15** (Hight, 2000). From this the stresses to which the sample is exposed can be determined. Disturbance may be caused at any stage in the process, but probably the most destructive is when the sample tube penetrates the bottom of the borehole (Hight, 2000).

For sampling to take place, the tube (and cutting shoe for the U100/UT100) is required to be pushed into the ground. Due to this additional material (sampler) being introduced into the natural soil the bulk density of the material into which they are inserted changes while the sampler is intruded into the bottom of the borehole. With this in mind, in theory the sampler used should have as thin a wall and as sharp a cutting tip as possible in order to prevent local ground densification and undue strains or fabric disturbance. The sampler should

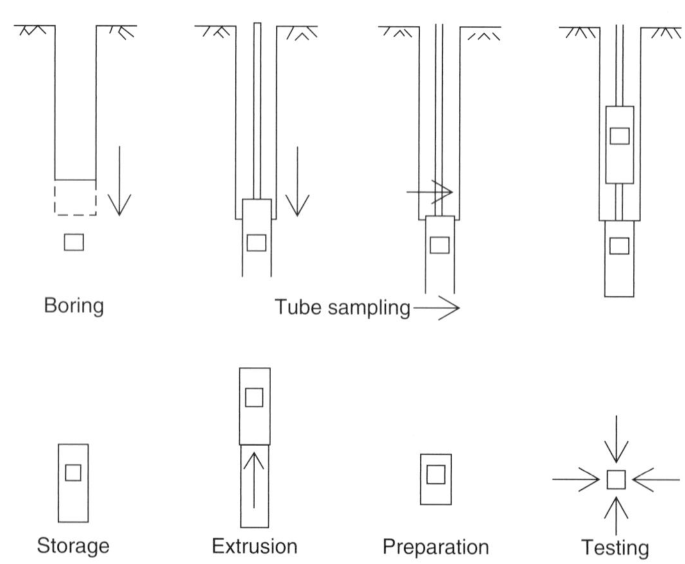

Figure 49.15 U100 sampling stages (Reproduced from Hight, 2000)

be pushed in smoothly and without side movement or jarring (not percussive techniques). From this it can be understood that the area ratio (%) of a sampler plays a huge role in the disturbance of the sample taken. The area ratio is calculated as the volume of soil displaced by the sampler as a proportion of the sample volume (calculated by measuring the internal and external diameter of the cutting shoe or edge). In theory, the lower the value the less disturbed the material within. Clayton and Siddique (1999) examined the different effects of tube geometries and used examples from the four main sample tube geometries used in the UK at that time plus a fifth experimental design. The geometries of these tubes are shown in **Figure 49.16**.

Sampler 1 is the geometry of the cutting shoe used on standard 'metal' U100 sample tubes and has an area ratio of 27%. Inside clearance is obtained by a step out from the cutting shoe where it screws onto the sample tube above it. A U100 sampler is defined as a 'thick-walled open sampler' and at best will produce a class 2 sample as described in BS EN 22475-1. The UT100 replacement has an area ratio of approximately 15%.

Sampler 2 is an upgraded version of sampler 1. It has a very similar area ratio, but with an inner step which is replaced with a slight taper and the cutting edge tapers that have been reduced (made sharper).

Sampler 3 is the version of the UK cutting tube used for U100 samplers with plastic liners, which are no longer validated but have been shown here as a warning. The area ratio is considerable at 48% and again a small step inside produces the inside clearance between the shoe and the liner. This liner produces class 3 quality samples (BS EN 22475-1)

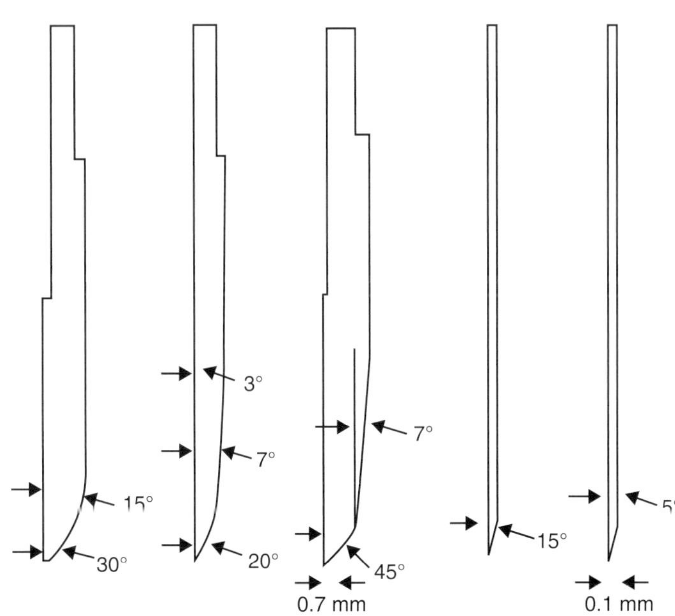

Figure 49.16 Tube sampler geometries (Reproduced from Clayton and Siddique, 1999)

Sampler 4 is the 'thin-wall' push sampler that has been widely used in many circumstances to produce quite high-quality samples. The original tube was described by Harrison (1991) and consists of a tube (normally stainless steel) with a 15° taper on the cutting edge.

Sampler 5 is an experimental sampler (Hight, 2000) which is similar to sampler 4 but is sharper (5° taper) and has a 0.1 mm flat at the cutting tip.

This is not intended to be an academic publication, but it is prudent to give some background to the sampler types with regard to sampling and disturbance and link them to the parameters required from the samples. From research testing and some very high grade commercial tests it is known that on natural stiff clays, such as London Clay, failure occurs at axial strains in the region of 0.75–2.0%. Clayton and Siddique (1999) studied the sampler geometries and made strain predictions along the centre line of the sample as it would be taken using the different geometries (**Figure 49.17**).

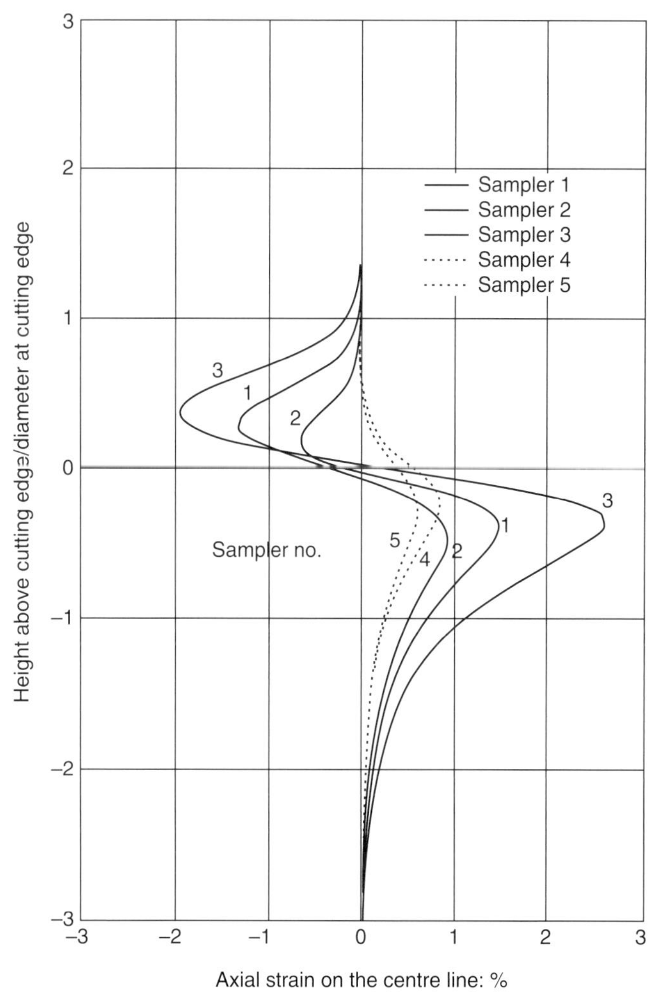

Figure 49.17 Sampler predicted axial strains (Reproduced from Clayton and Siddique, 1999)

Samplers 1 and 3 are likely to fail in the natural stiff clays during sampling due to the strains imposed on the specimen, with sampler 3 being by far the worst offender. The thin-wall push samplers outperform all geometries modelled and sampler 4 should be used as a minimum if stiffness parameters are required. For normally consolidated and lightly overconsolidated clays Hight (2000) notes that the strains at the periphery of the sample during sampling cause a zone of remoulded soil which combines with shear-induced pore water pressures which increase across the sample and are at their highest at the periphery. This leads to an overall reduction in mean effective stress caused by an increased water content in the centre of the sample as the sample re-equilibrates (due to the highly disturbed periphery). The same outcomes are true for overconsolidated clays along with damaged material structure and fabric.

Due to the high permeability of sands tube sampling will be 'drained' and both volumetric and shear strains will occur (Hight, 2000). Levels of sample disturbance will vary with the in situ density of the material and destructuring/density changes will result due to the yield strain at the particle contacts being very low.

For all tube sampling methods it is imperative that the base of the borehole is cleared of debris before the sample is taken. If this does not occur, the tube will be filled with varying amounts of highly disturbed material which may have very high moisture contents, leading to incorrect borehole logs and unrepresentative soil parameters.

On extraction from the borehole the sample tubes should be cleaned and any excess water immediately removed. The intact samples recovered should have their ends painted with low-melting-point wax and an identification label (indelible ink) placed in the top of the tube and marked as the 'top'. In cases where full recovery has not occurred, an inert non-compressible material should be used to fill the void left in the tube before the sealing caps are put in place. This is carried out to prevent the intact sample sliding around in the tube. The tubes should be stored and transported upright with the top of the sample uppermost. This is most important where soft samples have been taken which may try to 'flow' down the tube if knocked or vibrated during handling or transportation. Even the more competent samples may suffer from fissures opening up if stored on their side when moved or transported. The end sealing caps should be clearly marked with all the samples' details and the correct way up, again with indelible ink. The tubes should be preferably marked on both ends just in case the identification from one end is unreadable or removed (markings on the side of the tube are often rubbed off during handling and transport). A waterproof identification label should also be placed within the top of each tube as a failsafe.

49.15. Rotary core samples

This sampling method is probably the most common high-quality method used extensively in the UK, and in many circumstances is superior to thin-wall push samples. The advantage here is that the material is removed from the cutting face of the drill bit and so the sampler does not densify the soil as it is inserted but simply 'reams' a stick of intact material from the ground. The drawback is that some lubrication in the form of drill flush is required, which means the possible addition of moisture to the sample; however, with sufficient expertise and thought this can largely be overcome.

Rotary drill rigs come in various sizes, depending upon access and depth of hole required. A lorry-mounted rotary rig can be seen in **Figure 49.18**, while a smaller tracked rotary rig (**Figure 49.19**) can be used for poorly accessible terrain, slopes or where there is limited access.

Figure 49.18 Lorry-mounted rotary drill rig (Courtesy of Fugro Ltd)

Figure 49.19 Track-mounted rotary drill rig (Courtesy of Fugro Ltd)

These rigs can use a variety of drill bits and drill flush agents depending upon the types of material encountered. They are far too numerous to list here, but if in doubt drilling trials can be written into the ground investigation plan in order to verify sample quality and recovery before the main exploration phase begins. Typically triple barrel core tubes are used in which the innermost barrel is a semi-rigid plastic liner which enables the sample to be removed without undue stress. Using such equipment correctly and safely requires specialist contractors with knowledge and experience, and care should be taken when choosing such contractors. With this method it is possible to sample not only rock, but firm through stiff to hard clays and compact sands. With the correct drill equipment, cutting heads and flush recovery should be very good.

Following is an overview of the field sampling procedure used to identify the main processes. Variations are allowable, depending upon particular requirements and conditions, but these must be agreed with or scheduled by the client.

(*a*) Gain a suitable sample from the ground.

(*b*) Clean the sample in such a way that will preserve its strength and fabric, retaining the properties of the same material in situ.

(*c*) Subsample and/or trim the specimen to a size (normally H:D = 3:1) that is suitable for a laboratory test (triaxial or other).

(*d*) Preserve the specimen so that it may be stored until testing is required.

(*e*) Protect the sample from the effects of time and any changes to environmental conditions which may alter the properties or integrity of the sample, such as storage temperatures, cycling ambient temperatures, transport shocks and vibrations, ultraviolet light.

The full specification for the preparation and storage of samples from drilling a rotary core ready to be tested in the laboratory may take the form of (but not be limited to) the following.

- Remove the core from its liner immediately on extraction from the ground in order to remove drill fluids used and protect the natural moisture content and physical integrity of the core. The core liner should be split diametrically in two halves by using some form of counter-rotating opposing blades set so that they cut the plastic tube (liner) without cutting into or marking the sample within. The use of sharp knives for this operation should not only be prohibited for health and safety reasons but also due to the force required to penetrate and then pull along the tube. This may not only damage the sample within but there is a high risk of operator/spectator injury when slips occur.
- Clean off by wiping with an absorbent cloth any drill fluid or flush/water from the outside of the core.
- Log the core and subsample by carefully cutting for laboratory testing. Samples intended for triaxial testing should have an approximate length to diameter ratio of 3:1 (which will allow for later trimming in the laboratory).
- The subsamples should then have their extremities prepared for sealing. Rotary cored samples with very high permeabilities (e.g. compact sands) should have any drill flush removed from their extremity and immediately be sealed. Clays should have the outer 5 mm of the subsample carefully trimmed off in a soil lathe in order to expose 'fresh' material which has not been contaminated with drill fluids. This process should be carried out very quickly for both material types, but for slightly different reasons. For sandy materials it is important to prevent moisture loss (due to the relatively high void ratio and permeability) and for clay-type samples to prevent moisture ingress softening the sample and reducing its effective stress. Care should be taken that this process is carried out swiftly and in an environment suitable to reduce evaporation and localised heating of the sample.
- The subsample should then be sealed in such a way as to trap minimal air, not only to maintain its natural moisture content but also its physical integrity for storage and transport. Often samples are wrapped in a layer of aluminium foil for the first layer, which is fine for most sample types as it can be lightly moulded to expel air and maintain contact with the sample surface as an impermeable barrier. Some thought should be given, however, as salt water and some alkali pore water along with alkali minerals (gypsum) can react with the aluminium, causing loss of sealing and reaction with the sample itself. In these instances non-permeable plastic film should be used as the first layer. Traditionally (and because it is commonly available from the local shop) many core sealing operations are carried out using plastic food wrap. The only problem with this is that this stretchy plastic 'food wrap' is by its nature osmotic or semi-permeable, and therefore not ideal. It has been found that the plastic film used to wrap pallets is not only stretchy, strong and seals against itself, but is non-osmotic and impermeable. The sample should then have a layer of such material wrapped two to three times around to completely encase it. A label should be enclosed providing the sample identity and orientation. The complete sample should then be coated with low-melting-point wax (which is usually a mixture of 50% petrolatum and 50% paraffin wax), which is quite soft and 'sticky'. *Not pure candle wax* as its melting point is too high. This wax should be heated only to the temperature required for melting and not to the point of boiling. This is not only a health and safety issue but the idea is that as soon as the wax contacts the cooler specimen it will set while transferring minimal heat to the sample. The sample should not be dipped in the hot wax pot but should be painted with a brush dipped in the warm wax. These alternating coatings/layers can be repeated in order to protect the sample further, but the sample identity should always be clearly visible. Strong tape such as carpet tape should then be wrapped around the ends of the sample to protect the wax and layered coatings from damage. Finally, the sample may be placed in a split core liner which is then taped in order to give support to the sample. This core liner (or suitably rigid material) should be cut to the length of the sample so that the sample is retained rigidly and is unable to move about within its support. Caps should be placed over the ends in order to complete the encapsulation. The sample should be relabelled in an indelible fashion with full identification and orientation (top/bottom).
- In essence the sample is to be preserved as near as possible to the condition as it was in situ, but isotropically destressed. This preservation must be able to withstand handling transportation and storage, which in some instances may be for some appreciable time.

49.16. Transport

Transport is a simple process, but very often overlooked. The samples should be transported in such a way that they are not shocked, dropped or vibrated. If sample integrity is an issue then the samples should be protected accordingly. Personal delivery and transport within the chain is often the only way to maintain high integrity. External transport contractors often do not appreciate the care required when handling a piece of 'soil', so this should be avoided. Many couriers see such parcels as 'heavy' rather than highly fragile scientific material. The samples should be transported 'upright' and in a way that they are supported laterally (padded boxes) to prevent toppling, rattling and vibration. Even when transported personally, the samples should not be placed in footwells where they are near heating/cooling vents which may locally heat/cool the samples. This completes the chain of custody to the testing laboratory, which will then test the samples for the parameters required.

49.17. The testing laboratory

There are many testing houses available throughout the world, and they offer testing to many levels and standards. Care should be taken in choosing the testing laboratory as it is relatively simple for them to buy the equipment to carry out tests but this does not mean they are proficient in that particular test. Make sure you are clear about the parameters you require from your tests and the identity of the samples to be tested. Bear in mind that you will need sufficient soil samples of the correct quality in order for the tests to be representative. It is often a good idea to visit prospective laboratories, if possible, and again this promotes the sharing of information and keeping up to date with the latest testing and contractual developments. You will also be able to assess the level of expertise, the equipment and processes involved with your proposed testing. Some tests may be beyond the scope of external accreditation bodies (and the accreditors!) and in this instance the reputation and experience of the staff carrying out the tests will prevail.

Over the years the author has seen some 'very interesting' practices carried out in a range of laboratories, some with high levels of external accreditation. These accreditations are not a guarantee of high quality but proof of a level of proficiency and management system on the day of the audit. Procedures, systems and tests are audited (often by external auditors) so that a certificate can be issued providing evidence of compliance and repeatability. However, this does not necessarily mean that the result issued will be correct. It is very possible to follow an incorrect method repeatedly and gain the incorrect results repeatedly and still be accredited for this! There are many good laboratories out there. Find them and use them.

The Eurocodes are addressing the standardisation of modern drilling and sampling practices and their relative merits with regard to sample quality for laboratory testing. Hopefully the demise of the U100 plastic liner sampling system has arrived as samples from such tubes are 'highly disturbed' and are therefore of little value for laboratory testing apart from index properties. Strength, compressibility and permeability determinations will be in varying degrees of error to the material in situ. Where the intention is to study the variation of these parameters with depth, U100 tube samples are not good and often contribute to the 'scatter' in plots with depth. Unfortunately the same seems to be showing in the general quality seen from UT100 tubes, especially when they are fabricated from steel. This, along with other 'mishaps' and lapses in attention or detail in the chain from sampling to testing all adds to the error band or 'scatter'. Much of this can be avoided very simply through care and attention to detail. Unfortunately this is sometimes lost in the pressure to 'get the job done'. The parameters required from these samples for our design should be representative of the material in situ and not the environments which the material has been exposed to on its journey from the ground to the laboratory (and subsequent test methods).

This completes the physical sampling/testing loop of ground investigation. You should have spare material if possible for alternative or repeat tests, should you find the need for additional testing. It is very expensive to re-drill or have further boreholes sampled at a later date. All data from testing should be available if required for further analysis, even down to the weights for moisture contents or the cone values and moisture contents for Atterberg determinations.

It is understood that in some instances and in some areas around the world some deviations from these guidelines will be required; however, the same principles and ideals should be followed. Notes should be kept as to the methods used and the environment in which the samples were taken and sealed, including dates, times and personnel involved. Everything that we do as professionals should be able to withstand scrutiny, but also provide sufficient data to repeat or improve our activities in the interests of a forward-looking science. Attention to detail in the complete chain of custody will pay dividends in the quality of the subsequent tests and parameters derived.

References

ASTM (American Society for Testing and Materials) (2013) ASTM D5311/D5311M-13: Standard methods for load controlled cyclic triaxial strength of soil. ASTM, West Conshohocken, PA. (Note, this is withdrawn at present (2022) but will be reissued under a new date.)

ASTM (2017) ASTM D6528-17: Standard test method for consolidated undrained direct simple shear testing of fine grain soils. ASTM, West Conshohocken, PA.

ASTM (2020) ASTM D3999-91/D3999M-11e1: Standard test methods for the determination of the modulus and damping properties of soils using the cyclic triaxial apparatus. ASTM, West Conshohocken, PA. (Note, this is withdrawn at present (2022) but will be reissued under a new date.)

ASTM (2021) ASTM D4015-21: Standard test methods for modulus and damping of soils by fixed-base resonant column devices. ASTM, West Conshohocken, PA.

Atkinson JH (2007) *An Introduction to the Mechanics of Soils and Foundations*. Routledge, London, UK.

BSI (British Standards Institution) (2018) BS EN ISO 17892:2018: Geotechnical investigation and testing – Laboratory testing of soil. BSI, London, UK.

BSI (2021) BS EN ISO 22475-1:2021: Geotechnical investigation and testing – Sampling methods and groundwater measurements. BSI, London, UK.

BSI (2022) BS1377: Methods for soil testing. BSI, London, UK. (Parts 1–4 are being re-released as this chapter is going to print.)

Clayton CRI and Siddique A (1999) Tube sampling disturbance: forgotten truths and new perspectives. *Proceedings of the Institution of Civil Engineers, Geotechnical Engineering* **137**: 127–135.

Harrison IR (1991) A pushed thinwall sampling system for stiff clays. *Ground Engineering*, April: pp. 30–34.

Head KH (2014) *Manual of Soil Laboratory Testing, 3 vols*. Pentech Press, London, UK.

Hight DW (2000) Sampling methods: evaluation of disturbance and new practical techniques for high quality sampling in soils. Keynote lecture. *Proceedings of the 7th National Congress of the Portuguese Geotechnical Society, Porto, Portugal*, vol. 3, pp. 1275–1309.

Jardine R, Chow F, Overy R and Standing J (2005). *ICP Methods for Driven Piles in Sands and Clays*. Thomas Telford, London, UK.

Further reading

ASTM (2019) ASTM D8295-19: Standard test method for determination of shear wave velocity and initial shear modulus in soil specimens using bender elements. ASTM, West Conshohocken, PA.

BSI (2007) BS EN 1997-2:2007: Eurocode 7: Geotechnical design – Part 2: Ground investigation and testing. BSI, London, UK.

Clayton CRI, Simons NE and Matthews MC (1982) *Site Investigation*. Granada, London, UK.

Simons NE, Menzies B and Matthews MC (2002) *A Short Course in Geotechnical Site Investigation*. Thomas Telford, London, UK.

Useful websites

ASTM International (formally known as the American Society for Testing and Materials), contains many internationally recognised references for soil and rock testing: http://www.astm.org

British Geotechnical Association, contains information for updates of many relevant technical standards and links to many other sites of interest: http://www.britishgeotech.org

British Standards Institution, references for UK and European Standards, including training and accreditation: http://www.bsigroup.com

Engineering Group of the Geological Society (EGGS), many useful references for rock behaviour and categorisation: http://www.geolsoc.org.uk/engineering

International Society for Rock Mechanics, contains the European suggested methods for various rock tests (the 'Blue Book'): http://www.isrm.net

ISO Standards: http://www.iso.org

It is recommended this chapter is read in conjunction with

- Chapter 17 *Strength and deformation behaviour of soils*
- Chapter 18 *Rock behaviour*

All chapters in this book rely on the guidance in Sections 1 *Context* and 2 *Fundamental principles*. A sound knowledge of ground investigation is required for all geotechnical works, as set out in Section 4 *Site investigation*.

Appendix A
Standard soil and rock tests

The standards named here are generally applicable to the UK and should be used in association with the present guidelines required to apply to 'BS EN1997-2: 2007: Eurocode 7: Geotechnical design – Part 2: Ground investigation and testing'. Eurocode 7 is applicable to European construction and may be accepted in other parts of the world. It should be noted that different standards may apply in different locations around the world. The other major standards applied worldwide are the ASTM standards, and these should be applied or used where applicable.

The list here is not exhaustive and is only an indication of some of the more common soil and rock tests available. It should also be noted here that the European Standards for the identification and classification of soils and rocks (BS EN ISO 14688-1 (2017), BS EN ISO 14688-2 (2017) and BS EN ISO 14689-1 (2017)) implemented into UK practice in 2007 have all been incorporated into BS 5930:2015 Amendment 1, which incorporates a revised section 6 (published in 2007). Earlier versions of BS 5930:1990 do not meet the requirements of the new Eurocodes and would not comply with the recent code changes.

In the UK, BS 1377 has been the standard for soil testing for many years; however, at the time of writing it is under heavy revision and so careful checks should always be made in order to use the latest revision. BS 1377:1990 comprised nine parts, which at present are being compressed into four parts only. These are

- Part 1: General requirements and sample preparation (BS 1377-1:2016)
- Part 2: Classification tests and determination of geotechnical properties (BS 1377-2:2022)
- Part 3: Chemical and electrochemical tests (BS 1377-3:2018 + A1:2021)
- Part 4: In situ tests (BS 1377-9:1990 – still current at the time of writing).

From the above it can be seen that all laboratory testing to BS 1377 will now be dealt with in the first two parts now in circulation. The other main alternatives used in other parts of the world are ISO 17892:2015 (also known in the UK as BS EN ISO 17892:2014–2019), which is split into 12 sections covering a very similar range of testing as covered in BS 1377. The most commonly used world standards are covered by ASTM International (American Society for Testing and Materials). The number of published standards are vast, and so the particular ASTM 'D' number should be researched on the ASTM website under the test type/name for the most up-to-date reference.

Common laboratory rock tests showing the standards generally used worldwide are listed in **Table 49.A1**.

Rock test	Reference
Preparation of rock core specimens and determination of dimensional and shape tolerances	ASTM D4543-19 or ISRM suggested methods (2007)
Determination of water content	ASTM D2216-10 or ISRM suggested methods (2007)
Determination of porosity/density using buoyancy technique (for both regular and irregular shapes)	ISRM suggested method (2007)
Determination of slake durability index	ASTM D4644-16 or ISRM suggested methods (2007)
Determination of point load strength for diametral and axial tests	ASTM D5731-16
Determination of splitting (Brazilian) tensile strength of intact rock core specimens	ASTM D2936-20
Determination of compressive strength and elastic moduli of intact rock core specimens under varying states of stress and temperatures	ASTM D7012-17

Table 49.A1 Common laboratory rock tests

doi: 10.1680/icemge.66816.0747

Chapter 50
Geotechnical reporting

Phil Smith Geotechnical Consulting Group, London, UK

CONTENTS

Geotechnical reports exist in a number of different forms, each with its own distinct purpose. However, all geotechnical reports should be written with an understanding of the project that they have been prepared for and the use to which the report will be put. Geotechnical reports of all forms must be clear and concise, such that the information contained within them is accessible and unambiguous. If the project programme is such that some information to be contained within the report is required urgently, it may be appropriate to produce multiple reports, or addenda to a main report, rather than delay issuing the time-critical information. Since a report can provide useful information long after it was first commissioned it is important to ensure that all reporting is properly archived, and thought is given to how the information contained within the report will remain accessible.

50.1. Introduction

The purpose of a geotechnical report is to convey some form of information, data or similar. Such information may be purely factual in nature, but may also comprise interpretation or opinion. It may be of a purely technical nature or may be intended to address commercial aspects of a construction project. The information can be presented in the form of the written word, in hard copy or electronically, or consist of numerical data, tabulated or reported using digital methods, or comprise graphical output.

50.2. Factual reporting

50.2.1 Introduction

Traditionally, the term ‘the factual report’ has been used for the output of most ground investigations undertaken in the UK. When used in this way, the term is not compliant with the current iteration of Eurocode 7. However, it is important to remember that a factual report is ultimately a report containing factual information and is not necessarily limited to reporting of ground investigation activities. Moreover, it remains common for site investigation contractors to be instructed to produce only the factual data component of a ground investigation report (GIR, discussed later), as is permissible under Eurocode 7. Thus, the term ‘the factual report’ remains in common parlance.

The primary purpose of a geotechnical factual report is to accurately and precisely describe the ground ‘as found’: as it occurs in situ. While this description should be made with the knowledge that the end-user of the report is likely to make use of the report to design and build some form of construction, a factual report is not interpretative: the report should be factually accurate and as complete as possible, and not written on the assumption that construction will be of a certain form or use certain techniques. Site investigation works are often undertaken at an early stage in a project, and the information from the site and ground investigations may result in changes to the layout, nature or methodology of construction of the proposed works. Hence a factual report should always be a complete record of all factual information available. If interpretation and recommended parameters are needed, this requires completion of an interpretative report or full GIR (see below).

Thus, a factual report should be a clear, concise, complete and accurate description of the ground ‘as found’. The nature of this description is dependent on the methods and techniques used to obtain the information being reported.

It is important to realise that the true behaviour of the ground in situ, and its response to any construction work, can rarely if ever be precisely determined through investigative techniques. All methods of inspection, sampling and testing of the soil that are available, by their nature, are limited to testing a limited volume of soil; this may amount to many tens of cubic metres of soil for in situ testing, or as little as a few grams of soil in a laboratory test. However, no test truly determines the response of the entire in situ soil mass for the full range of conditions. The difference in volume of soil tested can result in apparently factual data that are inconsistent; for example, it is well known that the factual result of laboratory permeability testing gives a different indication for a soil’s permeability than a variable head test undertaken in a borehole. This is a reflection of the scale effects that apply: the structure of a small intact laboratory sample is much less likely to contain large fissures than the volume of soil affected by an in situ variable head test, and as a result often gives permeability results two orders of magnitude lower than the field test. Both tests are ‘correct’ and give factual information; neither necessarily gives an entirely true and accurate value for the soil permeability that may be applicable in relation to the planned construction works. Clearly, any report giving the measured permeability of the soil would need to give complete information on the nature of the test used to determine permeability, so that the end-user would have some indication of whether the test was obtained from a small sample or larger soil mass, and could understand and apply the test data appropriately.

Factual reporting also needs to provide complete information about the nature of the techniques used to recover and test samples, since it is necessary to allow for how such techniques may have affected the information obtained. For example, soil and rock core recovered from rotary drilled boreholes may have drilling-induced fractures which are not indicative of the state of the material in situ; methods of sampling can induce stresses and/or deformations in a sample, affect moisture content, soil suction and more. Thus, reporting a laboratory test without reporting the method by which the soil sample was obtained may render the entire test worthless if the result may have been significantly affected by sampling disturbance.

In undertaking factual reporting, there is already much guidance. The majority of ground investigation techniques used on site, and laboratory test methods applied to determine soil parameters, are defined through British Standards, many of which are now identified as BS EN ISO standards, being the British implementation of European Standards produced under the auspices of the European Committee for Standardization (CEN). Beyond this, there are national accreditation schemes, applicable to laboratory testing in particular (both geotechnical and environmental), which provide further guidance on the format and content of test reports. Other publications, for example the *UK Specification for Ground Investigation* (AGS, 2022), also include guidance as to what data should be reported as part of a factual report. However, there are still many techniques or tests where the available guidance is incomplete or absent, and particular project/client requirements or new research leading to changes in the understanding of how a test or technique works may result in the requirement to report factual information in a non-standard manner.

50.2.2 Methods of obtaining details

The information provided in a factual report obviously depends on the nature of the works being reported.

50.2.2.1 Non-intrusive

There are a variety of non-intrusive works that may be undertaken for engineering or geotechnical purposes that require reporting. Most typical would be a simple site walkover. This may be of a relatively small site, but could equally extend to very large worksites, particularly for larger infrastructure projects, where the site may be very linear in nature. Where sites are extensive the simple walkover may be supported or even replaced by aerial photographic survey. The use of drones for survey purposes is becoming increasingly common.

The site walkover, the aerial photographic surveys and analysis of LiDAR (LIght Detection And Ranging) data are fundamentally visual surveys, aiming to identify and report the same types of features. The intention is to identify and report those features of the ground that constitute potential hazards to the proposed project, such as areas of unstable land. Reporting such surveys therefore requires that all indications of current or previous ground movement be identified: distorted fences, tilted walls or trees and slip-scars are all obvious features. Aspects of site drainage are also important to identify, and this can be done through reporting of obvious drains, streams or larger watercourses, including those dry at the time of the survey. However, general geomorphology and land use (particularly the nature of the vegetation growing in a location) may also provide useful indications of the presence or absence of groundwater.

A specific form of visual survey is geological or geomorphological mapping. In many cases, the nature of the underlying geology can be adequately determined by a desk study, but there may be instances where accurate identification of geological boundaries or other features are required.

There is no convenient 'universal' checklist in completing and reporting a visual site inspection: on one site the nature of the vegetation may be irrelevant, on another it may give critical guidance to the presence of groundwater. A visual site inspection report needs to be as complete and accurate as possible, and to be undertaken with knowledge and understanding of the proposed project; each site inspection report must by its nature be as individual as the site to which it applies, always bearing in mind that development proposals may change, and that therefore the inspection should be as comprehensive as possible. A revision to EC7, due to be published in 2024, is likely to contain much more prescriptive guidance on what constitutes acceptable site inspection than typical current practice.

It should be noted that non-intrusive surveys of this form typically form part of a desk study. Desk study reports may include factual and interpretative aspects; for a more comprehensive discussion on the contents of a desk study, refer to Chapter 43 *Preliminary studies*.

There is also the potential for geophysical surveys to be undertaken, which may be by surface (using hand-held or vehicle-mounted equipment) or airborne. Geophysical surveying may also be undertaken in a marine/over-water environment from boats (see also Chapter 45 *Geophysical exploration and remote sensing*).

Where geophysics methods have been undertaken on a site, the information to be reported clearly depends on the particular method or methods employed. However, geophysical investigations are frequently undertaken and reported by specialist subcontractors, and the reports normally include some degree of data processing or interpretation, since the raw data is often not a direct measure of a soil engineering parameter. The geophysical report should contain all relevant details of calibration applicable to the test, and the full test data should be included within the report (possibly as an appendix; due to the volume of data an electronic format will likely be most suitable), such that further analysis/re-analysis of the test data can be undertaken.

50.2.2.2 Large-face excavations

'Large-face excavations' refers to any situation where it is possible to view a section through the ground. Most commonly in ground investigations this is obtained through the excavation of trial pits and trial trenches, for which there are codes and standards defining what should be reported. However, good information on the soil stratigraphy can also be obtained from natural exposures (e.g. sea cliffs), and from large-scale man-made excavations. Such excavations may take the form of large box excavations or vertical shafts, giving a section through the ground from the surface. They might be linear features at ground surface (cuttings, canals) or at depth (tunnels). Quarries and sand pits (whether operational or closed) also frequently provide potential for inspection of sections through the ground.

Whatever form the excavation takes, certain basic information should always be reported. The codes for reporting of trial pits should always be reviewed for any such excavation, since regardless of how the excavation was formed, it is, in effect, a trial pit. Precise location (in plan and elevation) must be given in relation to a commonly used grid and datum (in the UK, Ordnance Survey National Grid and Newlyn Datum), along with details of date and weather. The excavation faces should be sketched (with scale/dimensions shown), and it is good practice to include photographs as well. Where samples are taken, it is important to accurately locate where these were taken, showing face of excavation, depth and orientation; again, photography can be useful here, with 'before' and 'after' images of where the sample was taken. Similarly, any in situ testing needs to be accurately located. Reporting of large-face excavations should also include details of how the face was created, when and by whom (as applicable). The absence of something within a large-face excavation can often be as important as its presence; for example, it is important to note where seepage occurs, but where there is no water present the fact that the face is dry should be recorded.

50.2.2.3 Intrusive investigations

Intrusive investigation generally means boreholes or trial pits of some form. There are a variety of techniques for forming boreholes, but any borehole formed for construction purposes should be undertaken in accordance with the British/European Standard, and will therefore be reported accordingly. If reporting is to deviate from the appropriate standard, it needs to be specifically instructed. It is important therefore to know what standard the work is being undertaken to and to be familiar with the reporting requirements of that standard before the work begins, so that if variation from the standard is required it can be instructed in good time.

In situ testing methods vary greatly in the scale of the sample they test: pocket penetrometers and hand vane tests assess a small volume of soil at shallow depth (typically within the wall of a trial pit); cone penetration testing investigates a relatively small area of soil, but can test a continuous column of soil tens of metres deep; a plate loading test may be carried out at ground surface or in a shallow excavation, but affects a volume of soil to some depth below the level of the test plate. Hence, the reporting of such testing is specific to the method being applied. Most such field testing is fully defined by British/European Standards, which accordingly define the reporting requirements. (See also Chapters 46 *Ground exploration*, 47 *Field geotechnical testing* and 48 *Geo-environmental testing*).

While in some cases the data from this type of test are relatively straightforward to report (for example, a pocket penetrometer), in other cases considerable volumes of data may be created by the test – for example, pressuremeter tests. These tests require consideration of how all the data can be reported, and often require some degree of electronic/digital reporting as well as any printed report (see Section 50.5).

Calibration records, where applicable, should always be included in any field test report. Where the report output is generated by processing of field readings, the actual instrument data should always also be made available as part of the factual reporting, to allow for re-analysis of the data if some discrepancy is detected at a later date. Additionally, any problems encountered during the testing or unexpected results from the tests should be highlighted when reporting in situ testing.

It is common for the more complex in situ tests (e.g. pressuremeter testing) to be reported in a sub-report appended to the main factual report. The specialist test report is often a combined factual and interpretative report even when submitted as part of the factual report, since the direct output of the tests is rarely in a format that is directly useful to an engineer: the tests may give pressures within the instrument or displacements of parts of the instrument, which are interpreted through appropriate calibration factors to give soil material properties.

Where in situ testing is undertaken in completed boreholes, reporting requirements are typically given by the relevant standards. For geophysical testing undertaken in completed boreholes, tests typically give data that vary with depth. The data reported should always include sufficient information on the equipment used and details of installation so that the test could be repeated from the reported information. As with non-intrusive geophysics, the specialist nature of intrusive geophysical investigations means it is common for the report of the fieldwork to combine the factual information with an interpretation of this information, with this combined factual/interpretative geophysics report included as a discrete subsection within the factual report. Again, calibration information and raw data from the tests should be included within the report.

50.2.3 Reporting of field techniques

As detailed above, practically all field investigation techniques that may be used and will need reporting are covered by a British and/or European Standard, and thus the reporting requirements are fully defined. However, while such standards provide guidance to best practice, variation from them is possible. Regardless of whether a standard is being strictly followed or not, certain data must always be reported.

Any factual report of a field investigation technique needs to clearly state what method was used to obtain the data. Moreover, the actual type of equipment used should also be reported. Different sizes or makes of plant may perform differently, providing different results; for example, it has long been recognised that the type of standard penetration test (SPT) rod used during a test can affect the result. Further, accurate recording of the type of plant – and where appropriate the individual rig – provides useful quality control on the works, and can be vital in interpreting the data recovered if anomalous readings are obtained.

Similarly, the actual operatives of any plant should be recorded. It is possible to see apparent operator-dependent results in some cases; for example, using the same type of plant across the same site, a case has been observed by the author where SPT results from one driller were consistently and significantly higher than those obtained by a second driller, despite boring in the same ground conditions. The project concerned was undertaken before the routine measurement of SPT hammer energy ratio, and had this form of calibration been available at the time of the works this may have explained the variation in the recorded data. However, by knowing which driller worked on which borehole, it was possible to account for the apparently inconsistent test data from across the site.

All investigation locations need to be accurately located and orientated (if applicable). Someone picking up the report of the work at any point in the future, be it 1 year's time or 100 years' time, should be able to work out exactly where the work was undertaken. For this reason, simple sketches of the worksite on their own are not adequate: road layouts, buildings and trees shown on such sketches may completely change if the site is redeveloped, making locating the works impossible, while cut or fill operations can change ground elevations by many metres.

It is therefore good practice to ensure that all works are surveyed to a recognised datum. Within the UK, the best and most obvious choice is the Ordnance Survey (OS) National Grid, since this is well established, widely accessible and unlikely to become redundant within the foreseeable future. Specific project grid/data may be used, but care should be exercised in selecting this option, since details of the project grid may become unavailable in the future (through disposal or other loss of records), leading to the reported information becoming unusable. Where project-specific grids/data are used, this should be clearly stated with full details provided.

For larger-scale projects, or those extending beyond the boundaries of the OS (or other) national grid, latitude and longitude information may be provided, along with height above mean sea level, typically determined through use of a global positioning system (GPS). However, even here, there are different systems of mapping and determination of coordinates, so the equipment and methodology used in determining the site position should be explicitly stated.

While sketch plans alone are not adequate, such a sketch, or even better a scalable drawing or site plan, should be included in any report. Such drawings can be particularly valuable where works have some degree of orientation: trial trenches or inclined boreholes, for example, will have a definite orientation, while geological mapping of strata is likely to require reporting of stratigraphic dip direction.

The date of all works should always be given, with the text of any factual report stating the date of the full work period, and individual field test or borehole logs showing the date for that particular operation. Where applicable, times should also be given; an example of where this would be appropriate is where groundwater levels are to be monitored close to a tidal waterway and so may show tidal variation.

Weather conditions are also an external factor that should be reported: results of field testing or monitoring may be influenced by temperature, moisture or atmospheric pressure.

Any test or method of investigation reported should always include full records of any appropriate calibration factor that has been applied to the test, and also full records of cleaning, maintenance or calibration of any instrumentation used, if applicable: some instrumentation is capable of showing significant drift from true results if not regularly and correctly maintained, so without the calibration records the data from such instrumentation cannot be treated as being reliable.

The absence of some material or behaviour should be reported where appropriate. By way of example, any intrusive investigation into the ground capable of identifying groundwater should always state where it was encountered, or that it was not encountered. When field monitoring is undertaken, the methodology detection limit should be stated; values below this limit should be reported as 'not detected' rather than as zero.

Individual features of the report must always be tailored to the information being reported. While data such as borehole logs are typically reported to standard scales (often either 5 m per page or 10 m per page), the important thing is that the information is reported clearly and unambiguously. If, for example, the concentration of data obtained from a borehole to be reported makes a 4 m per page scale most suitable, this is the best scale to use, and the scale used should then be clearly stated.

Where field tests are undertaken after completion of the main fieldworks, such as permeability tests in a groundwater monitoring instrument or ground gas monitoring, the above reporting requirements still apply.

Reporting of field tests undertaken after the main fieldwork period does lead to potential issues over timing of reporting, as discussed in Sections 50.4 and 50.6.

50.2.4 Reporting of laboratory tests

Since the vast majority of laboratory tests are undertaken in accordance with an established code or standard, the reporting requirements are generally also prescribed. However, variations from the standard reporting format may be undertaken where specific project requirements demand it. In such cases, care needs to be taken to ensure that such non-standard reporting is undertaken consistently throughout. It should also be noted that any variation from the established codes/standards reporting requirements will prevent the test being reported as being to the standard, even though the actual testing phase was compliant.

Non-standard reporting may be required because of a perceived weakness in the standard for reporting, perhaps because the precision of the standard reporting fails to provide sufficient discretion for the project underway. Alternatively, it may represent improvements in the theoretical understanding of a particular test or soil behaviour in general, which has yet to be incorporated into the standards.

As with field monitoring techniques, laboratory tests may have applicable calibration factors that apply to the output data, calibration requirements for the equipment, or method detection limits indicating the smallest quantity that can be detected. Such information should be reported. Where a quantity is less than the detection limit, it is not correct to report zero concentration. For further discussion on laboratory testing, see Chapter 49 *Sampling and laboratory testing*.

50.2.5 Multi-stage investigations and gap analysis

It is good practice to undertake ground investigations in a phased, multi-stage manner. When this is done, each phase of the investigation will need to be reported upon completion. A decision will need to be taken as to whether each phase of the investigation is reported separately or whether there is a single factual report that is to be updated and reissued after each phase of works; generally, the latter option is likely to be preferred, since it avoids the risk of individual volumes of the report becoming unavailable.

Where multi-stage investigations are undertaken, some form of assessment of what data/information is still needed after each phase of works is required. The nature of this assessment should be appropriate to the individual case, but it may be appropriate to undertake and formally report it in the form of a gap analysis, particularly if this information needs to be communicated to third parties.

50.3. Interpretative reporting

Interpretative reporting follows on from the factual report, and provides the manner in which the strictly factual data can be related to the specific project for which they have been obtained.

The exact content of an interpretative report is, however, still open to some variation: the report may be being produced for a fully scoped and planned development, or the proposals may still be quite vague. In the first instance, the proposals may already have determined that the foundations are to be piled, and the interpretative report is thus required to give guidance on the details of the piles likely to be required; in the latter case, part of the function of the interpretative report may be to provide recommendations as to the basic nature of the foundations (raft, piles, etc.) best suited for the proposed development. It is thus important that the requirements of the client in respect of the content and use of the interpretative report are well established and understood.

Interpretative reports are commonly also used to give guidance on design parameters for the soil. Again, where this is to be done, the client's requirements need to be fully understood. Design parameters can be presented as recommended values, design lines (possibly showing variation of a parameter with depth), upper bound and lower bound lines for maximum/minimum credible values, upper bound and lower bound lines for maximum/minimum possible values and so on. The nature of the proposed development and the expected method of design will affect which of these formats for reporting data is required. It is also possible that specific design parameters are not required, only combined plots showing the actual field and laboratory test data, enabling a design consultant to review the data and select their own design values. Alternatively, particular design codes, standards or methodologies may specify how geotechnical design parameters should be selected. The data available may also affect the method of selecting design parameters: many field tests tend to give results exhibiting significant scatter in value, where boundary lines may be more appropriate than a single design line. Conversely, if the available data are very limited in quantity, a conservatively selected single value may be more appropriate.

Interpretative reports also enable details of the factual report to be reviewed and possibly explained. A factual report may contain borehole logs, and such logs may be present on a cross-section, but it is incorrect to interpret the geology between boreholes in a factual report. In an interpretative report, interpreting the geological stratigraphy between boreholes is often a fundamental part of producing the report, since it provides the understanding of the form and nature of the soil mass, contributing towards the formation of the 'Conceptual Site Model'. Through such a process, various geohazards may be identified, such as faults identified from vertically displaced stratigraphic boundaries, or buried river channels with possible high-volume groundwater flow from unexpected soil types encountered in a borehole.

Completing the interpretative report to a useful standard is not an exercise in repeating information from the factual report, but requires an intelligent and informed assessment of that factual information relative to the proposed development for which the information has been gathered. Moreover, the briefing to produce the report may require that it allows for a variety of design methods, construction techniques, building layouts and so on, since all aspects of site investigation and reporting tend to occur early in a project's life, when quite fundamental changes to the project may still occur.

Interpretative reports often include recommendations for further investigations, where the assessment of the available factual information reveals deficiencies in the quality or quantity of the available data, thus incorporating elements of gap analysis (as per Section 50.2.5).

It should be noted that an interpretative report may be combined with a factual report, where the requirement of the report is both to provide full and complete factual information and to include an assessment of the data meeting the standards of an interpretative report, and the nature and scope of the investigation makes a single combined report appropriate.

50.4. Other geotechnical reports

The most common form of geotechnical reports are the factual and interpretative reports that result from some form of site investigation. However, there are a number of other geotechnical reports which may be encountered.

Where geotechnical design is to be in accordance with Eurocode 7, ground investigation reporting requires the production of GIRs and geotechnical design reports (GDRs). The GIR broadly combines the factual report with interpretation to generate design parameters which are then fed into the GDR; the GIR in fact forms part of the GDR. Eurocode 7 defines the format for presenting information, and requires known limitations of test results to be stated, so that users of the data have an indication of their reliability. The GDR provides the foundation design and recommendations that may formerly have been (partly) encountered in an interpretative report. It must include the assumptions and data that feed into the design, the methods used in the design and verification of safety and serviceability. It also requires that supervision, monitoring and maintenance requirements of the completed structure be reported and provided to the owner/client. The specific requirements of the GIR and GDR are stated more fully in EC7.

It should be noted that EC7 is at the time of writing being revised, and is expected to be published in 2024. It is anticipated that the revised Eurocode will include expanded requirements in relation to the content of various geotechnical reports.

It is often most efficient to combine all investigation and reporting of a site into one work package. Thus both factual and interpretative reporting may be required to consider factors such as geology and geohazards, environmental contaminants (in soil, water and air), broader environmental issues (e.g. flood risk, mine reports, naturally occurring radon gas) and heritage and archaeological issues (including possible issues of unexploded ordnance). Such issues should typically have been identified in the desk study (see Chapter 43 *Preliminary studies*), and may have been adequately addressed at that stage, but there is the potential for specific site works related to these aspects which will then need to be reported, and the implications of what was found will need to be discussed.

A geotechnical baseline report (GBR) is a specialised form of geotechnical report produced for commercial rather than technical purposes. Its aim is to contractually define which potentially adverse construction encounters are 'owned' by the employer and which by the contractor. They draw on and interpret the available data to define baseline conditions relevant to the proposed construction. These baseline conditions are applicable to a specific contract and to specific construction techniques. They establish what conditions a contractor should expect to encounter in the ground when undertaking the works. If conditions are worse than these, and the contractor can demonstrate a resulting loss or delay, then a compensation event or claim may be made. A GBR is not a traditional interpretative report, nor is it a basis for design; it is only a means by which the allocation of ground risk is assigned between contractor and employer.

Another specialist form of report is a risk register. A risk register is not specifically a geotechnical report. The risk register is, as its name indicates, a means of identifying and tracking all risks to the project. Typically, the register would detail the nature of the risk, the likelihood of it being encountered and its potential impact on the project, to give an overall risk status, using standard risk assessment procedures. Recommendations for how to mitigate the risk may be given, with the party responsible for undertaking the mitigation identified and the residual risk after mitigation being stated, along with where that residual risk lies. While a risk register is not specifically a geotechnical report, geotechnical-related risks are often some of the more significant to a project due to uncertainty about the ground; hence risk registers routinely require geotechnical input.

Where field monitoring or instrumentation is installed, there will generally be a requirement for ongoing monitoring. Such monitoring needs to be reported. The frequency and manner in which these reports are issued will depend on the frequency and duration of the monitoring, and the requirement of the project. The results of each site monitoring visit should generally be provided to the engineer (or other specified party) within a day or two of the visit, but the contractor would normally only be expected to produce a factual report on the monitoring at the end of the monitoring period. However, monitoring of instrumentation can continue into the construction and post-construction period, in which circumstances reporting may need to be more formal and frequent. This may require, for example, formal issue of daily groundwater data or real-time remote monitoring of displacements of a retaining wall. In such cases, reporting is most likely to be electronic, perhaps with a summary printed report issued periodically. Where this type of data is being generated and issued, it is important to ensure that the critical data are prominent, and that the ability to generate and issue huge quantities of data is not allowed to swamp the recipient of the report, potentially resulting in significant information not being identified and acted upon. The requirements for frequency and format of monitoring reporting will generally be identified in the specification for the broader works.

50.5. Electronic data

The traditional method of delivering a report, whether factual or interpretative, has been in the form of a bound hard copy. Smaller reports may constitute only a few pages, and be issued as 'letter reports'; large investigations may require reports consisting of many volumes. However, hard copy reports alone are not convenient, since it is not easy to extract and manipulate data. Additionally, producing several copies of multi-volume reports typically requires a significant quantity of paper to be used, which may be at odds with modern standards of good environmental practice. For these reasons, there is an increasing acceptance of the need for electronic data transfer and reporting using electronic media.

The increased use of electronic reporting methods opens up the possibility of solely electronic reporting with no physical (hard copy) report being produced. The implications of this approach on what then forms the 'master copy' of the report have yet to be established.

The most common form of electronic data transfer in the UK is AGS: that is, data transfer according to the Association of Geotechnical and Geoenvironmental Specialists' Electronic Transfer of Geotechnical and Geoenvironmental Data format (AGS, 2020). This provides a common framework for transfer of geotechnical and geo-environmental data, enabling the data to be readily transferred and manipulated, or input to a database. While in theory the use of this format means that data can be transferred seamlessly, the reality is that considerable work may still be required in manipulating data before they can be used.

While there is an increasing tendency for laboratories and ground investigation contractors to utilise systems that automatically generate AGS data during the reporting process, this is not universal. Some organisations generate AGS by manually inputting data. As a result, there is the potential for data to be mis-entered, leading to factual inaccuracies and inconsistencies to develop within the data file. There is software available that can check for the latter, but it is very difficult to confirm that factual inaccuracies are not present in the data. Knowledge of how the data are generated is therefore vital, as is information regarding the standard of the quality management system employed by the company or individual generating the data. It is good practice to spot-check any electronic data received against the master copy, but if there is any doubt as to the quality control of the data-generating process, extensive checking is vital. Such checking is laborious and time-consuming, and unfortunately often not carried out to a sufficient standard. However, no electronic data should be used if there is any doubt as to their accuracy.

While the AGS format is very comprehensive, there are still tests or aspects of tests that cannot be adequately described using it. Geophysics results in particular are not readily reported through AGS. In such situations, it is normally appropriate to use a spreadsheet format for the data. The need to thoroughly check all such data remains.

While electronic data formats allow for rapid and easy data transfer, the real value is in the ability to create databases of project information that can be easily interrogated to enable specific data to be accessed in their entirety and with minimal delay. However, the ease with which data can be extracted means that where a database exists, it may be relied upon to such an extent that it becomes the sole source of data, and the data may not be referenced against the master copy. It is imperative that where a database exists, it is strictly controlled. While the ability to extract data needs to be limited only to the normal operational procedures of the company involved, subject to any appropriate commercial confidentiality, the ability to input or change data within a database needs to be controlled and logged, and any data to be entered into the database need to have been checked fully beforehand. Back-up copies of the database should be kept to enable lost or corrupted data to be recovered.

A form of digital reporting that is now routinely encountered is that of the portable document format (PDF) file, for which there are a number of applications to create and read files. It is increasingly common practice for the master hard copy report to be provided in PDF format also, or more typically instead of a hard copy. PDF reports should be produced directly from the original word-processing software used to create the report, not as scans. Such electronic reports are very valuable, since they can generally be transferred easily by CD, flash drive (USB memory stick), online file transfer site or email, and they enable multiple copies of a report to be created and issued without the expense or environmental impact of multiple printed copies. However, a good-quality electronic copy of a report requires some degree of processing; for example, this file format allows pages to be bookmarked. Large reports where this has not been done are considerably less useful than where the report is fully and sensibly bookmarked. While a PDF file provides a convenient form for storage and transfer of a report, the information cannot generally be readily extracted and manipulated, so the provision of a PDF report can supplement, but does not replace, AGS format data transfer.

Another form of digital data is digital photography and digital imaging. Provision of core photographs has long been a standard requirement of any ground investigation involving rotary drilling for core; the widespread introduction of good-quality digital cameras has resulted in the majority of such core photographs being provided not just as printed copies, but also or even solely as digital image files, which, as with electronic format reports, offers considerable advantages in copying or transferring the images. Care needs to be taken in the use of digital photography that any printed image is a true likeness of the actual soil, since the print is often made using a standard office printer, where the quality of colour reproduction may fluctuate. However, the same issue can occur in developing and printing of conventional film, so is not an issue that prohibits use of digital imaging. Digital photographs also offer the opportunity to manipulate colour or contrast of the image, which can be of great assistance in examining and identifying details of soil structure. Similarly, sample photography is now most likely undertaken with digital cameras, offering the same advantages and limitations as for core photography.

Where digital images are provided, the image file should be named in a sensible and consistent manner, such that it is obvious from the file name what the image shows.

A relatively new variant of core photography is the use of high-resolution core scanning. This generates a high-quality image of the entire core, which with the appropriate software can be readily manipulated. The quality and resolution of the image is such that small details of the soil structure can be identified, and the images are far superior to traditional core photography. However, the availability of the scanners is currently limited, the scanning process is more involved than for photographs and requires more resources on site, and the digital size of the images is very large, which may affect storage or transfer of the images.

The above points refer primarily to the use of electronic data transfer/storage of traditional geotechnical factual data. The development of digital technology enables a range of new information to be obtained. For example, it is now possible to obtain digital data from monitoring instruments on rotary borehole rigs which show factors such as drilling advance rates, rotational speeds and various pressures. Such information appears so far to be principally of value to the drilling contractor while doing the work on site, but may prove to have value in interpreting the ground conditions. Since the use of this form of instrumented plant has not yet become widespread, such data are not, at the time of writing, routinely available. However, it illustrates how the range of data that may be available for reporting is not a constant, but will change as technology and methods of working change.

In any form of electronic or digital data storage or transfer there are a number of issues that must be addressed. AGS is a standard format across the industry, and the widespread use of spreadsheet software makes it likely that any spreadsheet-based data will also be readily accessible. However, other electronic data formats may require specialist software to enable the data to be viewed/extracted. If the software is hard and/or expensive to acquire, difficult to use or consumes a lot of digital storage space on the computer, it may not be possible for all would-be users of the data to run the software. Even where the software is common, care must be taken to allow for different versions of the software. While newer versions of software are commonly written to be compatible with earlier versions, if the data are produced using the latest version of a piece of common software, a user operating an older version may be unable to fully access it.

This leads to two related issues that need to be considered when considering electronic reporting. If a project is expected to be running over a prolonged period, it is possible that commercially available software used to report data at the start of the project will be upgraded during the course of the works. A decision will then need to be taken as to whether to upgrade the software and accept that there will be some degree of inconsistency in the project dataset, or continue to operate using the older software, which may lead to problems if companies generating the data have updated their systems or if the software version ceases to be supported by the manufacturer.

Thought must also be given to the long-term availability of any software. Reports from ground investigations are typically used relatively soon after they are completed, but can continue to be of use for many years after. If software is not common, then the problems of having the appropriate software available to access the data are likely to become more pronounced with time. It is possible that if the software operation requires a licence to function and the supplying company has ceased trading, the software will be completely unusable and hence the digital data will be lost. Even if this does not occur, if data are supplied in an obscure format it is important to record what the format and appropriate software to access the data is, such that at any future time it is possible to identify this and access the data.

Having focused on the potential difficulties of long-term data access due to software, it is appropriate to mention long-term data storage as a further issue affecting reporting. The lifespan of the printed page is well proven; the lifespan of a CD-ROM, or a flash drive or magnetic storage tape is less well demonstrated. If digital data are to be archived, thought needs to be given to the environmental conditions in which they will be stored, such that the lifespan of the storage medium is maximised. Consideration also needs to be given to security and back-up copies. Archived data need to be accessible to only those who have authorisation to access the data, but depending on the value of the data concerned thought should be given to having back-up copies stored separately from the main archive.

One issue applicable – but not unique – to electronic reporting is the need for personnel using software to be adequately trained. Knowing how a piece of software functions makes it less likely that errors will be made in generating/inputting data, and gives the operator a better idea of where problems are most likely to occur. It is important therefore that personnel involved in reporting are familiar with any software being used.

If the software is being used for any sort of analytical or interpretative function, then there is also a requirement to ensure that its operation is adequately validated. Commercial software does not always provide sufficient details or validation of its operation that its use would meet a reasonable quality system without further proof of its reliability.

50.5.1 Public archiving

The British Geological Survey (BGS) have long been a repository for boreholes completed across the UK, and these data have for many years been available online in the form of PDF files of borehole logs. More recently, the BGS have held and made available AGS data from ground investigations undertaken in the UK. All ground practitioners undertaking ground investigations in the UK should routinely consider supplying their investigation data to BGS, in AGS format and/or as PDFs of the logs.

50.6. Reporting production and timescale

Having considered the nature of what may be reported in a geotechnical report, and how that information may be reported, consideration needs to be given to the practicalities of the report.

A geotechnical report from a ground investigation is often required early in a project's life, and may be a requirement before the proposed design can be progressed. Therefore, the timing of the report is important. Where ground investigation works are very large, the time required to complete and issue a report may be considerable, as potentially will be the time needed to fully check and correct the report. Allowance needs to be made in any project programme for this time period. In some circumstances it may be appropriate to issue reports in stages: long-term monitoring by definition is completed many months after the fieldwork stage of a ground investigation is completed, so it would be inappropriate to delay issuing the factual report from the fieldwork until the monitoring is complete. Similarly but less obviously, some laboratory testing (particularly drained tests on clay soil specimens) can take prolonged periods to complete. Where the available laboratory resource to complete these tests is limited and multiple tests are required, it may take several months before this part of the laboratory test programme is completed. In such cases, the demand for basic stratigraphic information may require that these laboratory tests are reported as a later addendum to the factual report.

The programming of production and issue of the interpretative report can also be a significant factor. It is not uncommon for the interpretative report to draw on data from more than one report: there may be a desk study, several phases of ground investigations for the project concerned and a variety of historical data. As previously noted, the interpretative report is in some ways harder to produce than the factual report, since it is not a simple statement of fact, and therefore may take considerably longer to complete than the factual reports upon which it is based.

When generating a report, it is necessary to consider how many copies of the report are required. It will rarely be sufficient to generate only a single copy for the client: additional copies may be needed by one or more design engineers, and other interested parties (architects, insurers, etc.). However, efforts should be made to ensure that the number of hard copies produced is the minimum actually required, both for commercial and environmental reasons. Production of electronic copies of the report is valuable in this respect, since it allows the report to be readily issued as required, in whole or in part. Many recipients find electronic copies more convenient than hard copies, and there should be a presumption that data and reports will generally be issued in electronic format (AGS + PDF). Subject to the client's requirements, solely electronic reporting may be acceptable and is encouraged.

In some circumstances, the issue of payment for the report also needs to be carefully considered. The typical factual report from a small ground investigation will be relatively straightforward, and a simple lump sum for its production may be appropriate. However, if the works are larger in scope or undertaken under a term contract and may vary considerably in nature and extent, such a mechanism may be inequitable. Allowance for reporting costs to be based on the value of the fieldwork undertaken is in some circumstances appropriate.

References

AGS (Association of Geotechnical and Geoenvironmental Specialists) (2020) *Electronic Transfer of Geotechnical and Geoenvironmental Data*, edn. 4.1. AGS, Beckenham, UK.

AGS (2022) *UK Specification for Ground Investigation*, 3rd edn. ICE Publishing, London, UK.

Further reading

Association of British Insurers and British Tunnelling Society (2003) *Joint Code of Practice for Risk Management of Tunnel Works in the UK*. British Tunnelling Society, London, UK.

AGS (2017) *A Client's Guide: The Purpose and Use of a Geotechnical Report*. AGS, Beckenham, UK.

AGS (2017) *AGS Guide to Ground Investigation Reports*. AGS, Beckenham, Kent.

BRE (Building Research Establishment) (1987) *Site Investigation for Low-Rise Building: Desk Studies*. BRE Digest 318. IHS BRE Press, London, UK.

BRE (1989) *Site Investigation for Low-Rise Building: The Walk-Over Survey*. BRE Digest 348. IHS BRE Press, London, UK.

BRE (1993) *Site Investigation for Low-Rise Building: Trial Pits*. BRE Digest 381. IHS BRE Press, London, UK.

BRE (1993) *Site Investigation for Low-Rise Building: Soil Description*. BRE Digest 383. IHS BRE Press, London, UK.

BRE (1995) *Site Investigation for Low-Rise Building: Direct Investigations*. BRE Digest 411. IHS BRE Press, London, UK.

BRE (2002) *Optimising Ground Investigation*. BRE Digest 472. IHS BRE Press, London, UK.

BSI (British Standards Institution) (various years) BS 1377: Methods of test for soils for civil engineering purposes (various parts). BSI, London, UK.

BSI (various years) BS EN ISO 17892: Geotechnical investigation and testing – Laboratory testing of soil (various parts). BSI, London, UK.

BSI (various years) BS EN ISO 22476: Geotechnical investigation and testing – Field testing (various parts). BSI, London, UK.

BSI (2007) BS EN 1997-2:2007: Eurocode 7: Geotechnical design – Part 2: Ground investigation and testing. BSI, London, UK.
BSI (2009) NA to BS EN 1997-2:2007. UK National Annex to Eurocode 7: Geotechnical design – Part 2: Ground investigation and testing. BSI, London, UK.
BSI (2013) BS EN 1997-1+A1:2013: Eurocode 7: Geotechnical design – Part 1: General rules. BSI, London, UK.
BSI (2018) BS EN ISO 14688-1:2018: Geotechnical investigation and testing: Identification and classification of soil – Part 1: Identification and description. BSI, London, UK.
BSI (2018) BS EN ISO 14688-2:2018: Geotechnical investigation and testing: Identification and classification of soil – Part 2: Principles for a classification. BSI, London, UK.
BSI (2018) BS EN ISO 14689:2018: Geotechnical investigation and testing: Identification, description and classification of rock. BSI, London, UK.
BSI (2020) BS 5930+A1:2020: Code of practice – Ground investigations. BSI, London, UK.
BSI (2021) BS EN ISO 22475-1:2021: Geotechnical investigation and testing: Sampling methods and groundwater measurements – Technical principles for the sampling of soil, rock and groundwater. BSI, London, UK.
BSI (2022) NA+A2 to BS EN 1997-1+A1:2013. UK National Annex to Eurocode 7: Geotechnical design – Part 1: General rules. BSI, London, UK.
Clayton CRI, Matthews MC and Simons NE (1987) *Site Investigation*, 2nd edn. Blackwell Science, Oxford, UK.
Davis J, Essex R, Farooq I and Drake A (2023) *Geotechnical Baseline Reporting: A Good Practice Guide*. CIRIA Report C807. CIRIA, London, UK.
Driscoll R, Scott P and Powell J (2008) *EC7: Implications for UK Practice – Eurocode 7 Geotechnical Design*. CIRIA Report C641. CIRIA, London, UK.
McDowell PW, Barker RD, Butcher AP, *et al.* (2002) *Geophysics in Engineering Investigations*. CIRIA Report C562. CIRIA, London, UK.
Site Investigation Steering Group (2013) *Site Investigation in Construction: Effective Site Investigation*. Thomas Telford, London, UK.

Useful websites

Association of Geotechnical and Geoenvironmental Specialists (AGS): http://www.ags.org.uk
British Geological Survey (BGS): http://www.bgs.ac.uk

It is recommended this chapter is read in conjunction with

- Chapter 9 *Foundation design decisions*
- Chapter 44 *Planning, procurement and management*
- Chapter 52 *Foundation types and conceptual design principles*

All chapters in this book rely on the guidance in Sections 1 *Context* and 2 *Fundamental principles*. A sound knowledge of ground investigation is required for all geotechnical works, as set out in Section 4 *Site investigation*.

Index

Page numbers in **bold** refer to figures and *italics* refer to tables.